Some Physical Constants

Quantity	Symbol	Value[a]
Atomic mass unit	u	$1.660\ 538\ 782\ (83) \times 10^{-27}$ kg
		$931.494\ 028\ (23)$ MeV/c^2
Avogadro's number	N_A	$6.022\ 141\ 79\ (30) \times 10^{23}$ particles/mol
Bohr magneton	$\mu_B = \dfrac{e\hbar}{2m_e}$	$9.274\ 009\ 15\ (23) \times 10^{-24}$ J/T
Bohr radius	$a_0 = \dfrac{\hbar^2}{m_e e^2 k_e}$	$5.291\ 772\ 085\ 9\ (36) \times 10^{-11}$ m
Boltzmann's constant	$k_B = \dfrac{R}{N_A}$	$1.380\ 650\ 4\ (24) \times 10^{-23}$ J/K
Compton wavelength	$\lambda_C = \dfrac{h}{m_e c}$	$2.426\ 310\ 217\ 5\ (33) \times 10^{-12}$ m
Coulomb constant	$k_e = \dfrac{1}{4\pi\epsilon_0}$	$8.987\ 551\ 788 \ldots \times 10^9$ N·m^2/C^2 (exact)
Deuteron mass	m_d	$3.343\ 583\ 20\ (17) \times 10^{-27}$ kg
		$2.013\ 553\ 212\ 724\ (78)$ u
Electron mass	m_e	$9.109\ 382\ 15\ (45) \times 10^{-31}$ kg
		$5.485\ 799\ 094\ 3\ (23) \times 10^{-4}$ u
		$0.510\ 998\ 910\ (13)$ MeV/c^2
Electron volt	eV	$1.602\ 176\ 487\ (40) \times 10^{-19}$ J
Elementary charge	e	$1.602\ 176\ 487\ (40) \times 10^{-19}$ C
Gas constant	R	$8.314\ 472\ (15)$ J/mol·K
Gravitational constant	G	$6.674\ 28\ (67) \times 10^{-11}$ N·m^2/kg^2
Neutron mass	m_n	$1.674\ 927\ 211\ (84) \times 10^{-27}$ kg
		$1.008\ 664\ 915\ 97\ (43)$ u
		$939.565\ 346\ (23)$ MeV/c^2
Nuclear magneton	$\mu_n = \dfrac{e\hbar}{2m_p}$	$5.050\ 783\ 24\ (13) \times 10^{-27}$ J/T
Permeability of free space	μ_0	$4\pi \times 10^{-7}$ T·m/A (exact)
Permittivity of free space	$\epsilon_0 = \dfrac{1}{\mu_0 c^2}$	$8.854\ 187\ 817 \ldots \times 10^{-12}$ C^2/N·m^2 (exact)
Planck's constant	h	$6.626\ 068\ 96\ (33) \times 10^{-34}$ J·s
	$\hbar = \dfrac{h}{2\pi}$	$1.054\ 571\ 628\ (53) \times 10^{-34}$ J·s
Proton mass	m_p	$1.672\ 621\ 637\ (83) \times 10^{-27}$ kg
		$1.007\ 276\ 466\ 77\ (10)$ u
		$938.272\ 013\ (23)$ MeV/c^2
Rydberg constant	R_H	$1.097\ 373\ 156\ 852\ 7\ (73) \times 10^7$ m^{-1}
Speed of light in vacuum	c	$2.997\ 924\ 58 \times 10^8$ m/s (exact)

Note: These constants are the values recommended in 2006 by CODATA, based on a least-squares adjustment of data from different measurements. For a more complete list, see P. J. Mohr, B. N. Taylor, and D. B. Newell, "CODATA Recommended Values of the Fundamental Physical Constants: 2006." *Rev. Mod. Phys.* **80:**2, 633–730, 2008.

[a]The numbers in parentheses for the values represent the uncertainties of the last two digits.

Solar System Data

Body	Mass (kg)	Mean Radius (m)	Period (s)	Mean Distance from the Sun (m)
Mercury	3.30×10^{23}	2.44×10^6	7.60×10^6	5.79×10^{10}
Venus	4.87×10^{24}	6.05×10^6	1.94×10^7	1.08×10^{11}
Earth	5.97×10^{24}	6.37×10^6	3.156×10^7	1.496×10^{11}
Mars	6.42×10^{23}	3.39×10^6	5.94×10^7	2.28×10^{11}
Jupiter	1.90×10^{27}	6.99×10^7	3.74×10^8	7.78×10^{11}
Saturn	5.68×10^{26}	5.82×10^7	9.29×10^8	1.43×10^{12}
Uranus	8.68×10^{25}	2.54×10^7	2.65×10^9	2.87×10^{12}
Neptune	1.02×10^{26}	2.46×10^7	5.18×10^9	4.50×10^{12}
Pluto[a]	1.25×10^{22}	1.20×10^6	7.82×10^9	5.91×10^{12}
Moon	7.35×10^{22}	1.74×10^6	—	—
Sun	1.989×10^{30}	6.96×10^8	—	—

[a]In August 2006, the International Astronomical Union adopted a definition of a planet that separates Pluto from the other eight planets. Pluto is now defined as a "dwarf planet" (like the asteroid Ceres).

Physical Data Often Used

Average Earth–Moon distance	3.84×10^8 m
Average Earth–Sun distance	1.496×10^{11} m
Average radius of the Earth	6.37×10^6 m
Density of air (20°C and 1 atm)	1.20 kg/m^3
Density of air (0°C and 1 atm)	1.29 kg/m^3
Density of water (20°C and 1 atm)	1.00×10^3 kg/m^3
Free-fall acceleration	9.80 m/s^2
Mass of the Earth	5.97×10^{24} kg
Mass of the Moon	7.35×10^{22} kg
Mass of the Sun	1.99×10^{30} kg
Standard atmospheric pressure	1.013×10^5 Pa

Note: These values are the ones used in the text.

Some Prefixes for Powers of Ten

Power	Prefix	Abbreviation	Power	Prefix	Abbreviation
10^{-24}	yocto	y	10^1	deka	da
10^{-21}	zepto	z	10^2	hecto	h
10^{-18}	atto	a	10^3	kilo	k
10^{-15}	femto	f	10^6	mega	M
10^{-12}	pico	p	10^9	giga	G
10^{-9}	nano	n	10^{12}	tera	T
10^{-6}	micro	μ	10^{15}	peta	P
10^{-3}	milli	m	10^{18}	exa	E
10^{-2}	centi	c	10^{21}	zetta	Z
10^{-1}	deci	d	10^{24}	yotta	Y

Physics

for Scientists and Engineers with Modern Physics
Volume 2

NINTH EDITION

Technology Update

Raymond A. Serway

Emeritus, James Madison University

John W. Jewett, Jr.

Emeritus, California State Polytechnic University, Pomona

With contributions from Vahé Peroomian,
University of California at Los Angeles

About the Cover

The cover shows a view inside the new railway departures concourse opened in March 2012 at the Kings Cross Station in London. The wall of the older structure (completed in 1852) is visible at the left. The sweeping shell-like roof is claimed by the architect to be the largest single-span station structure in Europe. Many principles of physics are required to design and construct such an open semicircular roof with a radius of 74 meters and containing over 2 000 triangular panels. Other principles of physics are necessary to develop the lighting design, optimize the acoustics, and integrate the new structure with existing infrastructure, historic buildings, and railway platforms.

© Ashley Cooper/Corbis

CENGAGE
Learning·

Australia • Brazil • Japan • Korea • Mexico • Singapore • Spain • United Kingdom • United States

CENGAGE
Learning®

Physics for Scientists and Engineers with Modern Physics, Volume 2, Ninth Edition Technology Update
Raymond A. Serway and John W. Jewett, Jr.

Product Director: Mary Finch

Senior Product Manager: Charles Hartford

Managing Developer: Peter McGahey

Content Developer: Ed Dodd

Product Development Specialist: Nicole Hurst

Media Developer: John Wimer

Senior Marketing Manager: Janet del Mundo

Content Project Manager: Alison Eigel Zade

Senior Art Director: Cate Barr

Manufacturing Planner: Sandee Milewski

IP Analyst: Farah J. Fard

IP Project Manager: Christine Myaskovsky

Production Service: Lachina Publishing Services

Text and Cover Designer: Roy Neuhaus

Cover Image: Kings Cross railway station, London, UK

Cover Image Credit: © Ashley Cooper/Corbis

Compositor: Lachina Publishing Services

For product information and technology assistance, contact us at
Cengage Learning Customer & Sales Support, 1-800-354-9706

For permission to use material from this text or product,
submit all requests online at **www.cengage.com/permissions**
Further permissions questions can be emailed to
permissionrequest@cengage.com

Library of Congress Control Number: 2014947990

ISBN: 978-1-305-11641-2

Cengage Learning
20 Channel Center Street
Boston, MA 02210
USA

Cengage Learning is a leading provider of customized learning solutions with office locations around the globe, including Singapore, the United Kingdom, Australia, Mexico, Brazil, and Japan. Locate your local office at **international.cengage.com/region**.

Cengage Learning products are represented in Canada by Nelson Education, Ltd.

For your course and learning solutions, visit **www.cengage.com**.

Purchase any of our products at your local college store or at our preferred online store **www.CengageBrain.com**.

Instructors: Please visit **login.cengage.com** and log in to access instructor-specific resources.

Printed in the United States of America
Print Number: 01   Print Year: 2014

We dedicate this book to our wives,
Elizabeth and Lisa, and all our children and
grandchildren for their loving understanding
when we spent time on writing
instead of being with them.

Brief Contents

Contents

PART 5
Light and Optics 1057

PART 6
Modern Physics 1191

Appendices

About the Authors

Raymond A. Serway received his doctorate at Illinois Institute of Technology and is Professor Emeritus at James Madison University. In 2011, he was awarded with an honorary doctorate degree from his alma mater, Utica College. He received the 1990 Madison Scholar Award at James Madison University, where he taught for 17 years. Dr. Serway began his teaching career at Clarkson University, where he conducted research and taught from 1967 to 1980. He was the recipient of the Distinguished Teaching Award at Clarkson University in 1977 and the Alumni Achievement Award from Utica College in 1985. As Guest Scientist at the IBM Research Laboratory in Zurich, Switzerland, he worked with K. Alex Müller, 1987 Nobel Prize recipient. Dr. Serway also was a visiting scientist at Argonne National Laboratory, where he collaborated with his mentor and friend, the late Dr. Sam Marshall. Dr. Serway is the coauthor of *College Physics,* Ninth Edition; *Principles of Physics,* Fifth Edition; *Essentials of College Physics; Modern Physics,* Third Edition; and the high school textbook *Physics,* published by Holt McDougal. In addition, Dr. Serway has published more than 40 research papers in the field of condensed matter physics and has given more than 60 presentations at professional meetings. Dr. Serway and his wife, Elizabeth, enjoy traveling, playing golf, fishing, gardening, singing in the church choir, and especially spending quality time with their four children, ten grandchildren, and a recent great grandson.

John W. Jewett, Jr. earned his undergraduate degree in physics at Drexel University and his doctorate at Ohio State University, specializing in optical and magnetic properties of condensed matter. Dr. Jewett began his academic career at Richard Stockton College of New Jersey, where he taught from 1974 to 1984. He is currently Emeritus Professor of Physics at California State Polytechnic University, Pomona. Through his teaching career, Dr. Jewett has been active in promoting effective physics education. In addition to receiving four National Science Foundation grants in physics education, he helped found and direct the Southern California Area Modern Physics Institute (SCAMPI) and Science IMPACT (Institute for Modern Pedagogy and Creative Teaching). Dr. Jewett's honors include the Stockton Merit Award at Richard Stockton College in 1980, selection as Outstanding Professor at California State Polytechnic University for 1991–1992, and the Excellence in Undergraduate Physics Teaching Award from the American Association of Physics Teachers (AAPT) in 1998. In 2010, he received an Alumni Lifetime Achievement Award from Drexel University in recognition of his contributions in physics education. He has given more than 100 presentations both domestically and abroad, including multiple presentations at national meetings of the AAPT. He has also published 25 research papers in condensed matter physics and physics education research. Dr. Jewett is the author of *The World of Physics: Mysteries, Magic, and Myth,* which provides many connections between physics and everyday experiences. In addition to his work as the coauthor for *Physics for Scientists and Engineers,* he is also the coauthor on *Principles of Physics,* Fifth Edition, as well as *Global Issues,* a four-volume set of instruction manuals in integrated science for high school. Dr. Jewett enjoys playing keyboard with his all-physicist band, traveling, underwater photography, learning foreign languages, and collecting antique quack medical devices that can be used as demonstration apparatus in physics lectures. Most importantly, he relishes spending time with his wife, Lisa, and their children and grandchildren.

Preface

In writing this Technology Update for the Ninth Edition of *Physics for Scientists and Engineers,* we continue our ongoing efforts to improve the clarity of presentation and include new pedagogical features that help support the learning and teaching processes. Drawing on positive feedback from users of the eighth edition, data gathered from both professors and students who use Enhanced WebAssign, as well as reviewers' suggestions, we have refined the text to better meet the needs of students and teachers.

This textbook is intended for a course in introductory physics for students majoring in science or engineering. The entire contents of the book in its extended version could be covered in a three-semester course, but it is possible to use the material in shorter sequences with the omission of selected chapters and sections. The mathematical background of the student taking this course should ideally include one semester of calculus. If that is not possible, the student should be enrolled in a concurrent course in introductory calculus.

 ## Content

The material in this book covers fundamental topics in classical physics and provides an introduction to modern physics. The book is divided into six parts. Part 1 (Chapters 1 to 14) deals with the fundamentals of Newtonian mechanics and the physics of fluids; Part 2 (Chapters 15 to 18) covers oscillations, mechanical waves, and sound; Part 3 (Chapters 19 to 22) addresses heat and thermodynamics; Part 4 (Chapters 23 to 34) treats electricity and magnetism; Part 5 (Chapters 35 to 38) covers light and optics; and Part 6 (Chapters 39 to 46) deals with relativity and modern physics.

 ## Objectives

This introductory physics textbook has three main objectives: to provide the student with a clear and logical presentation of the basic concepts and principles of physics, to strengthen an understanding of the concepts and principles through a broad range of interesting real-world applications, and to develop strong problem-solving skills through an effectively organized approach. To meet these objectives, we emphasize well-organized physical arguments and a focused problem-solving strategy. At the same time, we attempt to motivate the student through practical examples that demonstrate the role of physics in other disciplines, including engineering, chemistry, and medicine.

 ## Changes for the Ninth Edition Technology Update

After surveying Ninth Edition users and other instructors teaching the calculus-based introductory physics course, we've built on our existing suite of online assets by adding all-new Integrated Tutorials and PreLecture Explorations. The Integrated Tutorials emphasize the conceptual thinking necessary to good problem-solving, while the PreLecture Explorations provide additional opportunities to engage students in important concepts prior to lecture. The following offerings are available exclusively via Enhanced WebAssign:

- *An Expanded Offering of All-New Integrated Tutorials.* Many users told us that, while their students appreciated the fact that our tutorials are book-specific, additional unique tutorials *not* keyed to existing end-of-chapter problems would also facilitate student learning. For this Technology Update, we have added numerous new Integrated Tutorials, while maintaining our unique philosophy of utilizing the text's specific problem-solving steps and notation. These Integrated Tutorials strengthen students' problem-solving skills by guiding them through the steps in the book's problem-solving process, and include meaningful feedback at each step so students can practice the problem-solving process and improve their skills. The feedback also addresses student preconceptions and helps them to catch algebraic and other mathematical errors. Solutions are carried out symbolically as long as possible, with numerical values substituted at the end. This feature promotes conceptual understanding above memorization, helps students understand the effects of changing the values of each variable in the problem, avoids unnecessary repetitive substitution of the same numbers, and eliminates round-off errors.
- *New PreLecture Explorations.* This new online asset has been very favorably received by our users, so we've expanded the number of PreLecture Explorations to provide instructors with more assignments covering a wider variety of topics. Additionally, all of the PreLecture Explorations have been updated and converted to HTML so that they can be accessed across a variety of platforms and devices, with additional parameters added to enhance investigation of a physical phenomenon. Students can make predictions, change the parameters, and then observe the results. Every PreLecture Exploration comes with conceptual and analytical questions that guide students to a deeper understanding and help promote a robust physical intuition.
- *Personal Study Plan.* The Personal Study Plan in Enhanced WebAssign provides chapter and section assessments that show students what material they know and what areas require more work. For items that they answer incorrectly, students can click on links to related study resources such as videos, tutorials, or reading materials. Color-coded progress indicators let them see how well they are doing on different topics. Instructors can decide what chapters and sections to include—and whether to include the plan as part of the final grade or as a study guide with no scoring involved.

Changes in the Ninth Edition

A large number of changes and improvements were made for the Ninth Edition of this text. Some of the new features are based on our experiences and on current trends in science education. Other changes were incorporated in response to comments and suggestions offered by users of the Eighth Edition and by reviewers of the manuscript. The features listed here represent the major changes in the Ninth Edition.

Enhanced Integration of the Analysis Model Approach to Problem Solving. Students are faced with hundreds of problems during their physics courses. A relatively small number of fundamental principles form the basis of these problems. When faced with a new problem, a physicist forms a *model* of the problem that can be solved in a simple way by identifying the fundamental principle that is applicable in the problem. For example, many problems involve conservation of energy, Newton's second law, or kinematic equations. Because the physicist has studied these principles and their applications extensively, he or she can apply this knowledge as a model for solving a new problem. Although it would be ideal for students to follow this same process, most students have difficulty becoming familiar with the entire palette of fundamental principles that are available. It is easier for students to identify a *situation* rather than a fundamental principle.

The *Analysis Model approach* we focus on in this revision lays out a standard set of situations that appear in most physics problems. These situations are based on an entity in one of four simplification models: particle, system, rigid object, and wave. Once the simplification model is identified, the student thinks about what the entity is doing or how it interacts with its environment. This leads the student to identify a particular Analysis Model for the problem. For example, if an object is falling, the object is recognized as a particle experiencing an acceleration due to gravity that is constant. The student has learned that the Analysis Model of a *particle under constant acceleration* describes this situation. Furthermore, this model has a small number of equations associated with it for use in starting problems, the kinematic equations presented in Chapter 2. Therefore, an understanding of the situation has led to an Analysis Model, which then identifies a very small number of equations to start the problem, rather than the myriad equations that students see in the text. In this way, the use of Analysis Models leads the student to identify the fundamental principle. As the student gains more experience, he or she will lean less on the Analysis Model approach and begin to identify fundamental principles directly.

To better integrate the Analysis Model approach for this edition, **Analysis Model descriptive boxes** have been added at the end of any section that introduces a new Analysis Model. This feature recaps the Analysis Model introduced in the section and provides examples of the types of problems that a student could solve using the Analysis Model. These boxes function as a "refresher" before students see the Analysis Models in use in the worked examples for a given section.

Worked examples in the text that utilize Analysis Models are now designated with an AM icon for ease of reference. The solutions of these examples integrate the Analysis Model approach to problem solving. The approach is further reinforced in the end-of-chapter summary under the heading *Analysis Models for Problem Solving,* and through the new **Analysis Model Tutorials** that are based on selected end-of-chapter problems and appear in Enhanced WebAssign.

Analysis Model Tutorials. John Jewett developed 165 tutorials from selected end-of-chapter problems (indicated in each chapter's problem set with an AMT icon) that strengthen students' problem-solving skills by guiding them through the steps in the problem-solving process. Important first steps include making predictions and focusing on physics concepts before solving the problem quantitatively. A critical component of these tutorials is the selection of an appropriate Analysis Model to describe what is going on in the problem. This step allows students to make the important link between the situation in the problem and the mathematical representation of the situation. Analysis Model tutorials include meaningful feedback at each step to help students practice the problem-solving process and improve their skills. In addition, the feedback addresses student misconceptions and helps them to catch algebraic and other mathematical errors. Solutions are carried out symbolically as long as possible, with numerical values substituted at the end. This feature helps students understand the effects of changing the values of each variable in the problem, avoids unnecessary repetitive substitution of the same numbers, and eliminates round-off errors. Feedback at the end of the tutorial encourages students to compare the final answer with their original predictions.

Annotated Instructor's Edition. New for this edition, the Annotated Instructor's Edition provides instructors with teaching tips and other notes on how to utilize the textbook in the classroom, via cyan annotations. Additionally, the full complement of icons describing the various types of problems will be included in the questions/problems sets (the Student Edition contains only those icons needed by students).

New Master Its Added in Enhanced WebAssign. Approximately 50 new Master Its in Enhanced WebAssign have been added for this edition to the end-of-chapter problem sets.

 # Chapter-by-Chapter Changes

The list below highlights some of the major changes for the Ninth Edition.

Chapter 23

- A *new* analysis model has been introduced: *Particle in a Field (Electrical)*. This model follows on the introduction of the Particle in a Field (Gravitational) model introduced in Chapter 13. An Analysis Model descriptive box has been added, in Section 23.4. In addition, a new summary flash card has been added at the end of the chapter, and textual material has been revised to make reference to the new model.
- A new What If? has been added to Example 23.9 in order to make a connection to infinite planes of charge, to be further studied in later chapters.
- Several textual sections and worked examples have been revised to make more explicit references to analysis models.
- One new Master It was added to the end-of-chapter problems set.
- Four new Analysis Model Tutorials were added for this chapter in Enhanced WebAssign.

Chapter 24

- Section 24.1 has been significantly revised to clarify the geometry of area elements through which electric field lines pass to generate an electric flux.
- Two new figures have been added to Example 24.5 to further explore the electric fields due to single and paired infinite planes of charge.
- Two new Master Its were added to the end-of-chapter problems set.
- Four new Analysis Model Tutorials were added for this chapter in Enhanced WebAssign.

Chapter 25

- Sections 25.1 and 25.2 have been significantly revised to make connections to the new particle in a field analysis models introduced in Chapters 13 and 23.
- Example 25.4 has been moved so as to appear after the Problem-Solving Strategy in Section 25.5, allowing students to compare electric fields due to a small number of charges and a continuous charge distribution.
- Two new Master Its were added to the end-of-chapter problems set.
- Four new Analysis Model Tutorials were added for this chapter in Enhanced WebAssign.

Chapter 26

- The discussion of series and parallel capacitors in Section 26.3 has been revised for clarity.
- The discussion of potential energy associated with an electric dipole in an electric field in Section 26.6 has been revised for clarity.
- Four new Analysis Model Tutorials were added for this chapter in Enhanced WebAssign.

Chapter 27

- The discussion of the Drude model for electrical conduction in Section 27.3 has been revised to follow the outline of structural models introduced in Chapter 21.

- Several textual sections have been revised to make more explicit references to analysis models.
- Five new Master Its were added to the end-of-chapter problems set.
- Four new Analysis Model Tutorials were added for this chapter in Enhanced WebAssign.

Chapter 28

- The discussion of series and parallel resistors in Section 28.2 has been revised for clarity.
- Time-varying charge, current, and voltage have been represented with lower-case letters for clarity in distinguishing them from constant values.
- Five new Master Its were added to the end-of-chapter problems set.
- Two new Analysis Model Tutorials were added for this chapter in Enhanced WebAssign.

Chapter 29

- A *new* analysis model has been introduced: *Particle in a Field (Magnetic)*. This model follows on the introduction of the Particle in a Field (Gravitational) model introduced in Chapter 13 and the Particle in a Field (Electrical) model in Chapter 23. An Analysis Model descriptive box has been added, in Section 29.1. In addition, a new summary flash card has been added at the end of the chapter, and textual material has been revised to make reference to the new model.
- One new Master It was added to the end-of-chapter problems set.
- Six new Analysis Model Tutorials were added for this chapter in Enhanced WebAssign.

Chapter 30

- Several textual sections have been revised to make more explicit references to analysis models.
- One new Master It was added to the end-of-chapter problems set.
- Four new Analysis Model Tutorials were added for this chapter in Enhanced WebAssign.

Chapter 31

- Several textual sections have been revised to make more explicit references to analysis models.
- One new Master It was added to the end-of-chapter problems set.
- Four new Analysis Model Tutorials were added for this chapter in Enhanced WebAssign.

Chapter 32

- Several textual sections have been revised to make more explicit references to analysis models.
- Time-varying charge, current, and voltage have been represented with lower-case letters for clarity in distinguishing them from constant values.
- Two new Master Its were added to the end-of-chapter problems set.
- Three new Analysis Model Tutorials were added for this chapter in Enhanced WebAssign.

Chapter 33

- Phasor colors have been revised in many figures to improve clarity of presentation.
- Three new Analysis Model Tutorials were added for this chapter in Enhanced WebAssign.

Chapter 34

- Several textual sections have been revised to make more explicit references to analysis models.
- The status of spacecraft related to solar sailing has been updated in Section 34.5.
- Six new Analysis Model Tutorials were added for this chapter in Enhanced WebAssign.

Chapter 35

- Two new Analysis Model descriptive boxes have been added, in Sections 35.4 and 35.5.
- Several textual sections and worked examples have been revised to make more explicit references to analysis models.
- Five new Master Its were added to the end-of-chapter problems set.
- Four new Analysis Model Tutorials were added for this chapter in Enhanced WebAssign.

Chapter 36

- The discussion of the Keck Telescope in Section 36.10 has been updated, and a new figure from the Keck has been included, representing the first-ever direct optical image of a solar system beyond ours.
- Five new Master Its were added to the end-of-chapter problems set.
- Three new Analysis Model Tutorials were added for this chapter in Enhanced WebAssign.

Chapter 37

- An Analysis Model descriptive box has been added, in Section 37.2.
- The discussion of the Laser Interferometer Gravitational-Wave Observatory (LIGO) in Section 37.6 has been updated.
- Three new Master Its were added to the end-of-chapter problems set.
- Four new Analysis Model Tutorials were added for this chapter in Enhanced WebAssign.

Chapter 38

- Four new Master Its were added to the end-of-chapter problems set.
- Three new Analysis Model Tutorials were added for this chapter in Enhanced WebAssign.

Chapter 39

- Several textual sections have been revised to make more explicit references to analysis models.
- Sections 39.8 and 39.9 from the Eighth Edition have been combined into one section.
- Five new Master Its were added to the end-of-chapter problems set.
- Four new Analysis Model Tutorials were added for this chapter in Enhanced WebAssign.

Chapter 40

- The discussion of the Planck model for blackbody radiation in Section 40.1 has been revised to follow the outline of structural models introduced in Chapter 21.
- The discussion of the Einstein model for the photoelectric effect in Section 40.2 has been revised to follow the outline of structural models introduced in Chapter 21.

- Several textual sections have been revised to make more explicit references to analysis models.
- Two new Master Its were added to the end-of-chapter problems set.
- Two new Analysis Model Tutorials were added for this chapter in Enhanced WebAssign.

Chapter 41

- An Analysis Model descriptive box has been added, in Section 41.2.
- One new Analysis Model Tutorial was added for this chapter in Enhanced WebAssign.

Chapter 42

- The discussion of the Bohr model for the hydrogen atom in Section 42.3 has been revised to follow the outline of structural models introduced in Chapter 21.
- In Section 42.7, the tendency for atomic systems to drop to their lowest energy levels is related to the new discussion of the second law of thermodynamics appearing in Chapter 22.
- The discussion of the applications of lasers in Section 42.10 has been updated to include laser diodes, carbon dioxide lasers, and excimer lasers.
- Several textual sections have been revised to make more explicit references to analysis models.
- Five new Master Its were added to the end-of-chapter problems set.
- Three new Analysis Model Tutorials were added for this chapter in Enhanced WebAssign.

Chapter 43

- A new discussion of the contribution of carbon dioxide molecules in the atmosphere to global warming has been added to Section 43.2. A new figure has been added, showing the increasing concentration of carbon dioxide in the past decades.
- A new discussion of graphene (Nobel Prize in Physics, 2010) and its properties has been added to Section 43.4.
- The discussion of worldwide photovoltaic power plants in Section 43.7 has been updated.
- The discussion of transistor density on microchips in Section 43.7 has been updated.
- Several textual sections and worked examples have been revised to make more explicit references to analysis models.
- One new Analysis Model Tutorial was added for this chapter in Enhanced WebAssign.

Chapter 44

- Data for the helium-4 atom were added to Table 44.1.
- Several textual sections have been revised to make more explicit references to analysis models.
- Three new Master Its were added to the end-of-chapter problems set.
- Two new Analysis Model Tutorials were added for this chapter in Enhanced WebAssign.

Chapter 45

- Discussion of the March 2011 nuclear disaster after the earthquake and tsunami in Japan was added to Section 45.3.
- The discussion of the International Thermonuclear Experimental Reactor (ITER) in Section 45.4 has been updated.

- The discussion of the National Ignition Facility (NIF) in Section 45.4 has been updated.
- The discussion of radiation dosage in Section 45.5 has been cast in terms of SI units grays and sieverts.
- Section 45.6 from the Eighth Edition has been deleted.
- Four new Master Its were added to the end-of-chapter problems set.
- One new Analysis Model Tutorial was added for this chapter in Enhanced WebAssign.

Chapter 46

- A discussion of the ALICE (A Large Ion Collider Experiment) project searching for a quark–gluon plasma at the Large Hadron Collider (LHC) has been added to Section 46.9.
- A discussion of the July 2012 announcement of the discovery of a Higgs-like particle from the ATLAS (A Toroidal LHC Apparatus) and CMS (Compact Muon Solenoid) projects at the Large Hadron Collider (LHC) has been added to Section 46.10.
- A discussion of closures of colliders due to the beginning of operations at the Large Hadron Collider (LHC) has been added to Section 46.10.
- A discussion of recent missions and the new Planck mission to study the cosmic background radiation has been added to Section 46.11.
- Several textual sections have been revised to make more explicit references to analysis models.
- One new Master It was added to the end-of-chapter problems set.
- One new Analysis Model Tutorial was added for this chapter in Enhanced WebAssign.

▮ Text Features

Most instructors believe that the textbook selected for a course should be the student's primary guide for understanding and learning the subject matter. Furthermore, the textbook should be easily accessible and should be styled and written to facilitate instruction and learning. With these points in mind, we have included many pedagogical features, listed below, that are intended to enhance its usefulness to both students and instructors.

Problem Solving and Conceptual Understanding

General Problem-Solving Strategy. A general strategy outlined at the end of Chapter 2 (pages 45–47) provides students with a structured process for solving problems. In all remaining chapters, the strategy is employed explicitly in every example so that students learn how it is applied. Students are encouraged to follow this strategy when working end-of-chapter problems.

Worked Examples. All in-text worked examples are presented in a two-column format to better reinforce physical concepts. The left column shows textual information that describes the steps for solving the problem. The right column shows the mathematical manipulations and results of taking these steps. This layout facilitates matching the concept with its mathematical execution and helps students organize their work. The examples closely follow the General Problem-Solving Strategy introduced in Chapter 2 to reinforce effective problem-solving habits. All worked examples in the text may be assigned for homework in Enhanced WebAssign. A sample of a worked example can be found on the next page.

　　Examples consist of two types. The first (and most common) example type presents a problem and numerical answer. The second type of example is conceptual in nature. To accommodate increased emphasis on understanding physical concepts, the many conceptual examples are labeled as such and are designed to help

Example 3.2   A Vacation Trip

A car travels 20.0 km due north and then 35.0 km in a direction 60.0° west of north as shown in Figure 3.11a. Find the magnitude and direction of the car's resultant displacement.

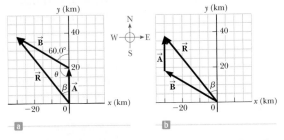

SOLUTION

Conceptualize The vectors $\vec{A}$ and $\vec{B}$ drawn in Figure 3.11a help us conceptualize the problem. The resultant vector $\vec{R}$ has also been drawn. We expect its magnitude to be a few tens of kilometers. The angle β that the resultant vector makes with the y axis is expected to be less than 60°, the angle that vector $\vec{B}$ makes with the y axis.

Figure 3.11 (Example 3.2) (a) Graphical method for finding the resultant displacement vector $\vec{R} = \vec{A} + \vec{B}$. (b) Adding the vectors in reverse order $(\vec{B} + \vec{A})$ gives the same result for $\vec{R}$.

Categorize We can categorize this example as a simple analysis problem in vector addition. The displacement $\vec{R}$ is the resultant when the two individual displacements $\vec{A}$ and $\vec{B}$ are added. We can further categorize it as a problem about the analysis of triangles, so we appeal to our expertise in geometry and trigonometry.

Analyze In this example, we show two ways to analyze the problem of finding the resultant of two vectors. The first way is to solve the problem geometrically, using graph paper and a protractor to measure the magnitude of $\vec{R}$ and its direction in Figure 3.11a. (In fact, even when you know you are going to be carrying out a calculation, you should sketch the vectors to check your results.) With an ordinary ruler and protractor, a large diagram typically gives answers to two-digit but not to three-digit precision. Try using these tools on $\vec{R}$ in Figure 3.11a and compare to the trigonometric analysis below!

The second way to solve the problem is to analyze it using algebra and trigonometry. The magnitude of $\vec{R}$ can be obtained from the law of cosines as applied to the triangle in Figure 3.11a (see Appendix B.4).

Use $R^2 = A^2 + B^2 - 2AB\cos\theta$ from the law of cosines to find R:

$$R = \sqrt{A^2 + B^2 - 2AB\cos\theta}$$

Substitute numerical values, noting that $\theta = 180° - 60° = 120°$:

$$R = \sqrt{(20.0 \text{ km})^2 + (35.0 \text{ km})^2 - 2(20.0 \text{ km})(35.0 \text{ km})\cos 120°}$$
$$= 48.2 \text{ km}$$

Use the law of sines (Appendix B.4) to find the direction of $\vec{R}$ measured from the northerly direction:

$$\frac{\sin\beta}{B} = \frac{\sin\theta}{R}$$

$$\sin\beta = \frac{B}{R}\sin\theta = \frac{35.0 \text{ km}}{48.2 \text{ km}}\sin 120° = 0.629$$

$$\beta = 38.9°$$

The resultant displacement of the car is 48.2 km in a direction 38.9° west of north.

Finalize Does the angle β that we calculated agree with an estimate made by looking at Figure 3.11a or with an actual angle measured from the diagram using the graphical method? Is it reasonable that the magnitude of $\vec{R}$ is larger than that of both $\vec{A}$ and $\vec{B}$? Are the units of $\vec{R}$ correct?

Although the head to tail method of adding vectors works well, it suffers from two disadvantages. First, some

people find using the laws of cosines and sines to be awkward. Second, a triangle only results if you are adding two vectors. If you are adding three or more vectors, the resulting geometric shape is usually not a triangle. In Section 3.4, we explore a new method of adding vectors that will address both of these disadvantages.

WHAT IF? Suppose the trip were taken with the two vectors in reverse order: 35.0 km at 60.0° west of north first and then 20.0 km due north. How would the magnitude and the direction of the resultant vector change?

Answer They would not change. The commutative law for vector addition tells us that the order of vectors in an addition is irrelevant. Graphically, Figure 3.11b shows that the vectors added in the reverse order give us the same resultant vector.

students focus on the physical situation in the problem. Worked examples in the text that utilize Analysis Models are now designated with an **AM** icon for ease of reference, and the solutions of these examples now more thoroughly integrate the Analysis Model approach to problem solving.

Based on reviewer feedback from the Eighth Edition, we have made careful revisions to the worked examples so that the solutions are presented symbolically as far as possible, with numerical values substituted at the end. This approach will help students think symbolically when they solve problems instead of unnecessarily inserting numbers into intermediate equations.

What If? Approximately one-third of the worked examples in the text contain a What If? feature. At the completion of the example solution, a What If? question offers a variation on the situation posed in the text of the example. This feature encourages students to think about the results of the example, and it also assists in conceptual understanding of the principles. What If? questions also prepare students to encounter novel problems that may be included on exams. Some of the end-of-chapter problems also include this feature.

Quick Quizzes. Students are provided an opportunity to test their understanding of the physical concepts presented through Quick Quizzes. The questions require students to make decisions on the basis of sound reasoning, and some of the questions have been written to help students overcome common misconceptions. Quick Quizzes have been cast in an objective format, including multiple-choice, true–false, and ranking. Answers to all Quick Quiz questions are found at the end of the text. Many instructors choose to use such questions in a "peer instruction" teaching style or with the use of personal response system "clickers," but they can be used in standard quiz format as well. An example of a Quick Quiz follows below.

Quick Quiz 7.5 A dart is inserted into a spring-loaded dart gun by pushing the spring in by a distance x. For the next loading, the spring is compressed a distance $2x$. How much faster does the second dart leave the gun compared with the first? **(a)** four times as fast **(b)** two times as fast **(c)** the same **(d)** half as fast **(e)** one-fourth as fast

---

Pitfall Prevention 16.2

Two Kinds of Speed/Velocity
Do not confuse v, the speed of the wave as it propagates along the string, with v_y, the transverse velocity of a point on the string. The speed v is constant for a uniform medium, whereas v_y varies sinusoidally.

Pitfall Preventions. More than two hundred Pitfall Preventions (such as the one to the left) are provided to help students avoid common mistakes and misunderstandings. These features, which are placed in the margins of the text, address both common student misconceptions and situations in which students often follow unproductive paths.

Summaries. Each chapter contains a summary that reviews the important concepts and equations discussed in that chapter. The summary is divided into three sections: Definitions, Concepts and Principles, and Analysis Models for Problem Solving. In each section, flash card–type boxes focus on each separate definition, concept, principle, or analysis model.

Questions and Problems Sets. For the Ninth Edition, the authors reviewed each question and problem and incorporated revisions designed to improve both readability and assignability. More than 10% of the problems are new to this edition.

Questions. The Questions section is divided into two sections: *Objective Questions* and *Conceptual Questions*. The instructor may select items to assign as homework or use in the classroom, possibly with "peer instruction" methods and possibly with personal response systems. More than 900 Objective and Conceptual Questions are included in this edition. Answers for selected questions are included in the *Student Solutions Manual/Study Guide*, and answers for all questions are found in the *Instructor's Solutions Manual*.

Objective Questions are multiple-choice, true–false, ranking, or other multiple guess–type questions. Some require calculations designed to facilitate students' familiarity with the equations, the variables used, the concepts the variables represent, and the relationships between the concepts. Others are more conceptual in nature and are designed to encourage conceptual thinking. Objective Questions are also written with the personal response system user in mind, and most of the questions could easily be used in these systems.

Conceptual Questions are more traditional short-answer and essay-type questions that require students to think conceptually about a physical situation.

Problems. An extensive set of problems is included at the end of each chapter; in all, this edition contains more than 3 700 problems. Answers for odd-numbered problems are provided at the end of the book. Full solutions for approximately 20% of the problems are included in the *Student Solutions Manual/Study Guide,* and solutions for all problems are found in the *Instructor's Solutions Manual.*

The end-of-chapter problems are organized by the sections in each chapter (about two-thirds of the problems are keyed to specific sections of the chapter). Within each section, the problems now "platform" students to higher-order thinking by presenting all the straightforward problems in the section first, followed by the intermediate problems. (The problem numbers for straightforward problems are printed in **black;** intermediate-level problems are in **blue.**) The *Additional Problems* section contains problems that are not keyed to specific sections. At the end of each chapter is the *Challenge Problems* section, which gathers the most difficult problems for a given chapter in one place. (Challenge Problems have problem numbers marked in **red.**)

There are several kinds of problems featured in this text:

Q/C *Quantitative/Conceptual problems* (indicated in the Annotated Instructor's Edition) contain parts that ask students to think both quantitatively and conceptually. An example of a Quantitative/Conceptual problem appears here:

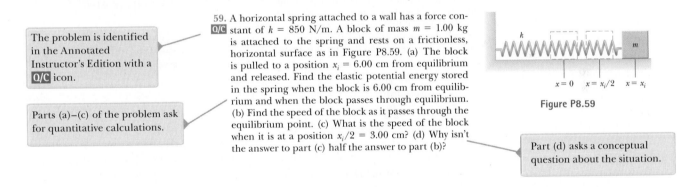

The problem is identified in the Annotated Instructor's Edition with a **Q/C** icon.

Parts (a)–(c) of the problem ask for quantitative calculations.

59. A horizontal spring attached to a wall has a force constant of $k = 850$ N/m. A block of mass $m = 1.00$ kg is attached to the spring and rests on a frictionless, horizontal surface as in Figure P8.59. (a) The block is pulled to a position $x_i = 6.00$ cm from equilibrium and released. Find the elastic potential energy stored in the spring when the block is 6.00 cm from equilibrium and when the block passes through equilibrium. (b) Find the speed of the block as it passes through the equilibrium point. (c) What is the speed of the block when it is at a position $x_i/2 = 3.00$ cm? (d) Why isn't the answer to part (c) half the answer to part (b)?

Figure P8.59

Part (d) asks a conceptual question about the situation.

S *Symbolic problems* (indicated in the Annotated Instructor's Edition) ask students to solve a problem using only symbolic manipulation. Reviewers of the Eighth Edition (as well as the majority of respondents to a large survey) asked specifically for an increase in the number of symbolic problems found in the text because it better reflects the way instructors want their students to think when solving physics problems. An example of a Symbolic problem appears here:

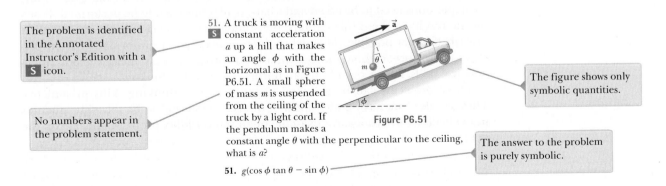

The problem is identified in the Annotated Instructor's Edition with a **S** icon.

No numbers appear in the problem statement.

51. A truck is moving with constant acceleration a up a hill that makes an angle ϕ with the horizontal as in Figure P6.51. A small sphere of mass m is suspended from the ceiling of the truck by a light cord. If the pendulum makes a constant angle θ with the perpendicular to the ceiling, what is a?

51. $g(\cos \phi \tan \theta - \sin \phi)$

Figure P6.51

The figure shows only symbolic quantities.

The answer to the problem is purely symbolic.

GP *Guided Problems* help students break problems into steps. A physics problem typically asks for one physical quantity in a given context. Often, however, several concepts must be used and a number of calculations are required to obtain that final answer. Many students are not accustomed to this level of complexity and often don't know where to start. A Guided Problem breaks a standard problem into smaller steps, enabling students to grasp all the concepts and strategies required to arrive at a correct solution. Unlike standard physics problems, guidance is often built into the problem statement. Guided Problems are reminiscent of how a student might interact with a professor in an office visit. These problems (there is one in every chapter of the text) help train students to break down complex problems into a series of simpler problems, an essential problem-solving skill. An example of a Guided Problem appears here:

The problem is identified with a GP icon.

38. A uniform beam resting on two pivots has a length $L =$
GP 6.00 m and mass $M = 90.0$ kg. The pivot under the left end exerts a normal force n_1 on the beam, and the second pivot located a distance $\ell = 4.00$ m from the left end exerts a normal force n_2. A woman of mass $m = 55.0$ kg steps onto the left end of the beam and begins walking to the right as in Figure P12.38. The goal is to find the woman's position when the beam begins to tip. (a) What is the appropriate analysis model for the beam before it begins to tip? (b) Sketch a force diagram for the beam, labeling the gravitational and normal forces acting on the beam and placing the woman a distance x to the right of the first pivot, which is the origin. (c) Where is the woman when the normal force n_1 is the greatest? (d) What is n_1 when the beam is about to tip? (e) Use Equation 12.1 to find the value of n_2 when the beam is about to tip. (f) Using the result of part (d) and Equation 12.2, with torques computed around the second pivot, find the woman's position x when the beam is about to tip. (g) Check the answer to part (e) by computing torques around the first pivot point.

The goal of the problem is identified.

Analysis begins by identifying the appropriate analysis model.

Students are provided with suggestions for steps to solve the problem.

The calculation associated with the goal is requested.

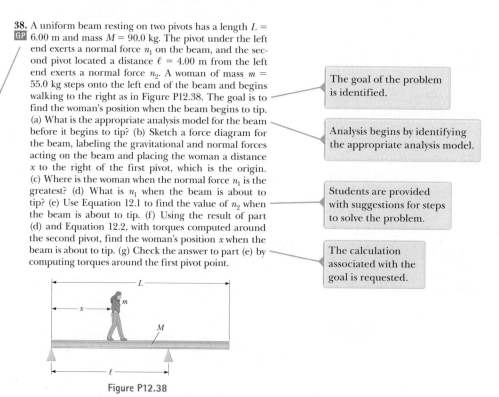

Figure P12.38

Impossibility problems. Physics education research has focused heavily on the problem-solving skills of students. Although most problems in this text are structured in the form of providing data and asking for a result of computation, two problems in each chapter, on average, are structured as impossibility problems. They begin with the phrase *Why is the following situation impossible?* That is followed by the description of a situation. The striking aspect of these problems is that no question is asked of the students, other than that in the initial italics. The student must determine what questions need to be asked and what calculations need to be performed. Based on the results of these calculations, the student must determine why the situation described is not possible. This determination may require information from personal experience, common sense, Internet or print research, measurement, mathematical skills, knowledge of human norms, or scientific thinking.

These problems can be assigned to build critical thinking skills in students. They are also fun, having the aspect of physics "mysteries" to be solved by students individually or in groups. An example of an impossibility problem appears on the next page.

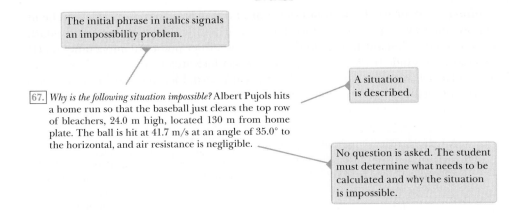

Paired problems. These problems are otherwise identical, one asking for a numerical solution and one asking for a symbolic derivation. There are now three pairs of these problems in most chapters, indicated in the Annotated Instructor's Edition by cyan shading in the end-of-chapter problems set.

Biomedical problems. These problems (indicated in the Annotated Instructor's Edition with a **BIO** icon) highlight the relevance of physics principles to those students taking this course who are majoring in one of the life sciences.

Review problems. Many chapters include review problems requiring the student to combine concepts covered in the chapter with those discussed in previous chapters. These problems (marked **Review**) reflect the cohesive nature of the principles in the text and verify that physics is not a scattered set of ideas. When facing a real-world issue such as global warming or nuclear weapons, it may be necessary to call on ideas in physics from several parts of a textbook such as this one.

"Fermi problems." One or more problems in most chapters ask the student to reason in order-of-magnitude terms.

Design problems. Several chapters contain problems that ask the student to determine design parameters for a practical device so that it can function as required.

Calculus-based problems. Every chapter contains at least one problem applying ideas and methods from differential calculus and one problem using integral calculus.

Integration with Enhanced WebAssign. The textbook's tight integration with Enhanced WebAssign content facilitates an online learning environment that helps students improve their problem-solving skills and gives them a variety of tools to meet their individual learning styles. Extensive user data gathered by WebAssign were used to ensure that the problems most often assigned were retained for this new edition. In each chapter's problems set, the top quartile of problems assigned in Enhanced WebAssign have cyan-shaded problem numbers in the Annotated Instructor's Edition for easy identification, allowing professors to quickly and easily find the most popular problems assigned in Enhanced WebAssign. The new, unique Integrated Tutorials in Enhanced WebAssign and the Analysis Model tutorials added for this edition have already been discussed (see pages ix–x). Master It tutorials help students solve problems by having them work through a stepped-out solution. Problems with Master It tutorials are indicated in each chapter's problem set with a **M** icon. In addition, Watch It solution videos are indicated in each chapter's problem set with a **W** icon and explain fundamental problem-solving strategies to help students step through the problem.

Artwork. Every piece of artwork in the Ninth Edition is in a modern style that helps express the physics principles at work in a clear and precise fashion. *Focus pointers* are included with many figures in the text; these either point out important aspects of a figure or guide students through a process illustrated by the artwork or photo. This format helps those students who are more visual learners. An example of a figure with a focus pointer appears below.

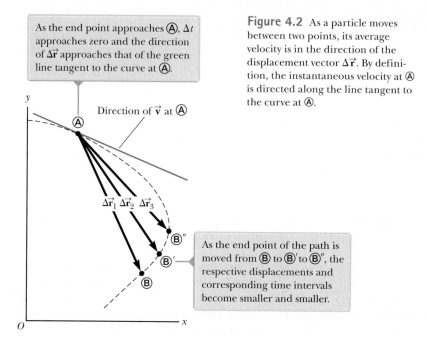

As the end point approaches Ⓐ, Δt approaches zero and the direction of $\Delta \vec{r}$ approaches that of the green line tangent to the curve at Ⓐ.

Figure 4.2 As a particle moves between two points, its average velocity is in the direction of the displacement vector $\Delta \vec{r}$. By definition, the instantaneous velocity at Ⓐ is directed along the line tangent to the curve at Ⓐ.

Direction of $\vec{v}$ at Ⓐ

$\Delta \vec{r}_1 \ \Delta \vec{r}_2 \ \Delta \vec{r}_3$

As the end point of the path is moved from Ⓑ to Ⓑ′ to Ⓑ″, the respective displacements and corresponding time intervals become smaller and smaller.

Math Appendix. The math appendix (Appendix B), a valuable tool for students, shows the math tools in a physics context. This resource is ideal for students who need a quick review on topics such as algebra, trigonometry, and calculus.

Helpful Features

Style. To facilitate rapid comprehension, we have written the book in a clear, logical, and engaging style. We have chosen a writing style that is somewhat informal and relaxed so that students will find the text appealing and enjoyable to read. New terms are carefully defined, and we have avoided the use of jargon.

Important Definitions and Equations. Most important definitions are set in **bold-face** or are highlighted with a background screen for added emphasis and ease of review. Similarly, important equations are also highlighted with a background screen to facilitate location.

Marginal Notes. Comments and notes appearing in the margin with a ▶ icon can be used to locate important statements, equations, and concepts in the text.

Pedagogical Use of Color. Readers should consult the **pedagogical color chart** (inside the front cover) for a listing of the color-coded symbols used in the text diagrams. This system is followed consistently throughout the text.

Mathematical Level. We have introduced calculus gradually, keeping in mind that students often take introductory courses in calculus and physics concurrently. Most steps are shown when basic equations are developed, and reference is often made to mathematical appendices near the end of the textbook. Although vectors are discussed in detail in Chapter 3, vector products are introduced later in the text, where they are needed in physical applications. The dot product is introduced in Chapter 7, which addresses energy of a system; the cross product is introduced in Chapter 11, which deals with angular momentum.

Significant Figures. In both worked examples and end-of-chapter problems, significant figures have been handled with care. Most numerical examples are worked to either two or three significant figures, depending on the precision of the data provided. End-of-chapter problems regularly state data and answers to three-digit precision. When carrying out estimation calculations, we shall typically work with a single significant figure. (More discussion of significant figures can be found in Chapter 1, pages 11–13.)

Units. The international system of units (SI) is used throughout the text. The U.S. customary system of units is used only to a limited extent in the chapters on mechanics and thermodynamics.

Appendices and Endpapers. Several appendices are provided near the end of the textbook. Most of the appendix material represents a review of mathematical concepts and techniques used in the text, including scientific notation, algebra, geometry, trigonometry, differential calculus, and integral calculus. Reference to these appendices is made throughout the text. Most mathematical review sections in the appendices include worked examples and exercises with answers. In addition to the mathematical reviews, the appendices contain tables of physical data, conversion factors, and the SI units of physical quantities as well as a periodic table of the elements. Other useful information—fundamental constants and physical data, planetary data, a list of standard prefixes, mathematical symbols, the Greek alphabet, and standard abbreviations of units of measure—appears on the endpapers.

 ## CengageCompose Options for *Physics for Scientists and Engineers*

Would you like to easily create your own personalized text, selecting the elements that meet your specific learning objectives?

CengageCompose puts the power of the vast Cengage Learning library of learning content at your fingertips to create exactly the text you need. The all-new, Web-based CengageCompose site lets you quickly scan content and review materials to pick what you need for your text. Site tools let you easily assemble the modular learning units into the order you want and immediately provide you with an online copy for review. Add enrichment content like case studies, exercises, and lab materials to further build your ideal learning materials. Even choose from hundreds of vivid, art-rich, customizable, full-color covers.

Cengage Learning offers the fastest and easiest way to create unique customized learning materials delivered the way you want. For more information about custom publishing options, visit **www.cengage.com/custom** or contact your local Cengage Learning representative.

 ## Course Solutions That Fit Your Teaching Goals and Your Students' Learning Needs

Recent advances in educational technology have made homework management systems and audience response systems powerful and affordable tools to enhance the way you teach your course. Whether you offer a more traditional text-based course, are interested in using or are currently using an online homework management system such as Enhanced WebAssign, or are ready to turn your lecture into an interactive learning environment, you can be confident that the text's proven content provides the foundation for each and every component of our technology and ancillary package.

Homework Management Systems

Enhanced WebAssign for *Physics for Scientists and Engineers*, Ninth Edition. Exclusively from Cengage Learning, Enhanced WebAssign offers an extensive online program for physics to encourage the practice that's so critical for concept mastery. The meticulously crafted pedagogy and exercises in our proven texts become even more effective in Enhanced WebAssign. Enhanced WebAssign includes the Cengage YouBook, a highly customizable, interactive eBook. WebAssign includes:

- **All of the quantitative end-of-chapter problems**
- **Selected problems enhanced with targeted feedback.** An example of targeted feedback appears below:

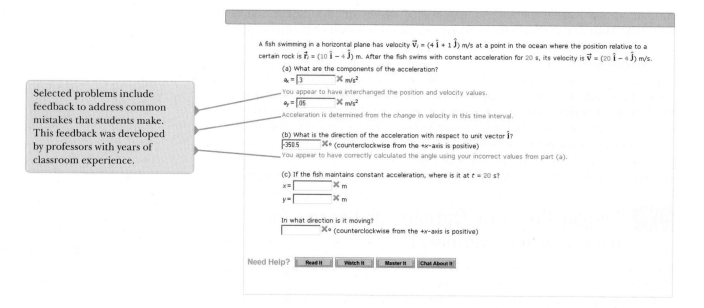

- **Master It tutorials** (indicated in the text by a M icon), to help students work through the problem one step at a time. An example of a Master It tutorial appears below:

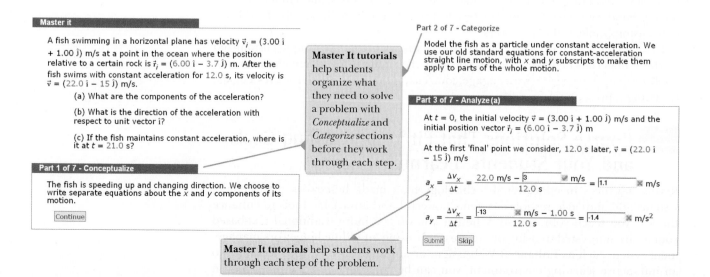

- **Watch It solution videos** (indicated in the text by a [W] icon) that explain fundamental problem-solving strategies, to help students step through the problem. In addition, instructors can choose to include video hints of problem-solving strategies. A screen shot from a Watch It solution video appears below:

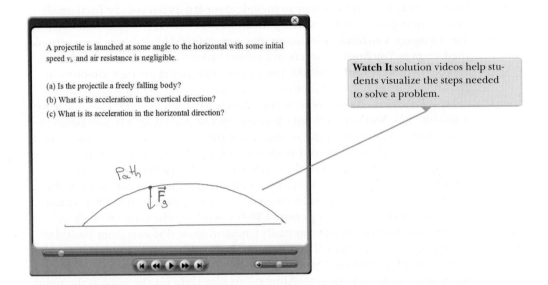

A projectile is launched at some angle to the horizontal with some initial speed v_i and air resistance is negligible.

(a) Is the projectile a freely falling body?
(b) What is its acceleration in the vertical direction?
(c) What is its acceleration in the horizontal direction?

Watch It solution videos help students visualize the steps needed to solve a problem.

- **All-New Integrated Tutorials.** These tutorials provide opportunities for students to strengthen their problem-solving skills by guiding them through the steps in the book's problem-solving process, and include meaningful feedback at each step so students can practice the problem-solving process and improve their skills. Solutions are carried out symbolically as long as possible, with numerical values substituted at the end.
- **Analysis Model Tutorials.** These tutorials (indicated in each chapter's problem set with an [AMT] icon) strengthen students' problem-solving skills by focusing on physics concepts and making predictions before solving a problem quantitatively. A critical component of these tutorials is the selection of an appropriate Analysis Model to describe what is going on in the problem. This step allows students to make the important link between the situation in the problem and the mathematical representation of the situation. Analysis Model tutorials include meaningful feedback at each step to help students practice the problem-solving process and improve their skills. Feedback at the end of the tutorial encourages students to compare the final answer with their original predictions.
- **Concept Checks**
- **Most worked examples,** enhanced with hints and feedback, to help strengthen students' problem-solving skills
- **Every Quick Quiz,** giving your students ample opportunity to test their conceptual understanding
- **PreLecture Explorations.** As explained on pages ix and x, we've expanded the number of PreLecture Explanations offered, allowing for greater topic coverage. Additionally, all of the PreLecture Explorations have been updated and converted to HTML, making them more stable and available across a variety of platforms and devices, with additional parameters added to enhance investigation of a physical phenomenon. Students can make predictions, change the parameters, and then observe the results. Every PreLecture Exploration comes with conceptual and analytical questions that guide students to a deeper understanding and help promote a robust physical intuition.

- **Personal Study Plan.** The Personal Study Plan in Enhanced WebAssign provides chapter and section assessments that show students what material they know and what areas require more work. For items that they answer incorrectly, students can click on links to related study resources such as videos, tutorials, or reading materials. Color-coded progress indicators let them see how well they are doing on different topics. You decide what chapters and sections to include—and whether to include the plan as part of the final grade or as a study guide with no scoring involved.

- **The Cengage YouBook.** WebAssign has a customizable and interactive eBook, the **Cengage YouBook,** that lets you tailor the textbook to fit your course and connect with your students. You can remove and rearrange chapters in the table of contents and tailor assigned readings that match your syllabus exactly. Powerful editing tools let you change as much as you'd like—or leave it just like it is. You can highlight key passages or add sticky notes to pages to comment on a concept in the reading, and then share any of these individual notes and highlights with your students, or keep them personal. You can also edit narrative content in the textbook by adding a text box or striking out text. With a handy link tool, you can drop in an icon at any point in the eBook that lets you link to your own lecture notes, audio summaries, video lectures, or other files on a personal Web site or anywhere on the Web. A simple YouTube widget lets you easily find and embed videos from YouTube directly into eBook pages. The Cengage YouBook helps students go beyond just reading the textbook. Students can also highlight the text, add their own notes, and bookmark the text. Animations play right on the page at the point of learning so that they're not speed bumps to reading but true enhancements. Please visit **www.webassign.net/brookscole** to view an interactive demonstration of Enhanced WebAssign.

- Offered exclusively in WebAssign, **Quick Prep** for physics is algebra and trigonometry math remediation within the context of physics applications and principles. Quick Prep helps students succeed by using narratives illustrated throughout with video examples. The Master It tutorial problems allow students to assess and retune their understanding of the material. The Practice Problems that go along with each tutorial allow both the student and the instructor to test the student's understanding of the material.

    Quick Prep includes the following features:

    - 67 interactive tutorials
    - 67 additional practice problems
    - A thorough overview of each topic, including video examples
    - Can be taken before the semester begins or during the first few weeks of the course
    - Can also be assigned alongside each chapter for "just in time" remediation

Topics include units, scientific notation, and significant figures; the motion of objects along a line; functions; approximation and graphing; probability and error; vectors, displacement, and velocity; spheres; force and vector projections.

CengageBrain.com

CENGAGE brain.com

On **CengageBrain.com** students will be able to save up to 60% on their course materials through our full spectrum of options. Students will have the option to rent their textbooks, purchase print textbooks, e-textbooks, or individual e-chapters and audio books all for substantial savings over average retail prices. **CengageBrain.com** also includes access to Cengage Learning's broad range of homework and study tools and features a selection of free content.

Lecture Presentation Resources

Instructor's Companion Site for Physics for Scientists and Engineers, Ninth Edition. Bringing physics principles and concepts to life in your lectures has never been easier! The full-featured Instructor's Companion Site provides everything you need for *Physics for Scientists and Engineers*, Ninth Edition. Key content includes the *Instructor's Solutions Manual*, art and images from the text, a physics movie library, and more.

Cengage Learning Testing Powered by Cognero is a flexible, online system that allows you to author, edit, and manage test bank content, create multiple test versions in an instant, and deliver tests from your LMS, your classroom, or wherever you want. No special installs or downloads needed, you can create tests from anywhere with Internet access. Cognero brings simplicity at every step, with pre-loaded test questions from the test bank from *Physics for Scientists and Engineers*, a desktop-inspired interface, a full-featured test generator, and cross-platform compatibility.

Supporting Materials for the Instructor

Please visit **http://www.cengage.com/physics/serway/PSEtechupdate9e** for information about instructor resources for this text, including custom versions.

Student Resources

Please visit **http://www.cengage.com/physics/serway/PSEtechupdate9e** for information about student resources for this text.

Laboratory Resources

Physics Laboratory Manual, Third Edition by David Loyd (Angelo State University) supplements the learning of basic physical principles while introducing laboratory procedures and equipment. Each chapter includes a prelaboratory assignment, objectives, an equipment list, the theory behind the experiment, experimental procedures, graphing exercises, and questions. A laboratory report form is included with each experiment so that the student can record data, calculations, and experimental results. Students are encouraged to apply statistical analysis to their data. A complete *Instructor's Manual* is also available to facilitate use of this lab manual.

Physics Laboratory Experiments, Eighth Edition by Jerry D. Wilson (Lander College) and Cecilia A. Hernández (American River College). This market-leading manual for the first-year physics laboratory course offers a wide range of class-tested experiments designed specifically for use in small to mid-size lab programs. A series of integrated experiments emphasizes the use of computerized instrumentation and includes a set of "computer-assisted experiments" that allow students and instructors to gain experience with modern equipment. It also lets instructors determine the appropriate balance of traditional versus computer-based experiments for their courses. By analyzing data through two different methods, students gain a greater understanding of the concepts behind the experiments. The Eighth Edition is updated with four new economical labs to accommodate shrinking department budgets and thirty new Pre-Lab Demonstrations, designed to capture students' interest prior to the lab and requiring only widely available materials and items.

Teaching Options

The topics in this textbook are presented in the following sequence: classical mechanics, oscillations and mechanical waves, and heat and thermodynamics, followed by electricity and magnetism, electromagnetic waves, optics, relativity, and modern physics. This presentation represents a traditional sequence, with the subject of mechanical waves being presented before electricity and magnetism. Some instructors may prefer to discuss both mechanical and electromagnetic waves together after completing electricity and magnetism. In this case, Chapters 16 through 18 could be covered along with Chapter 34. The chapter on relativity is placed near the end of the text because this topic often is treated as an introduction to the era of "modern physics." If time permits, instructors may choose to cover Chapter 39 after completing Chapter 13 as a conclusion to the material on Newtonian mechanics. For those instructors teaching a two-semester sequence, some sections and chapters in Volume 2 could be deleted without any loss of continuity. The following sections can be considered optional for this purpose:

25.7 The Millikan Oil-Drop Experiment
25.8 Applications of Electrostatics
26.7 An Atomic Description of Dielectrics
27.5 Superconductors
28.5 Household Wiring and Electrical Safety
29.3 Applications Involving Charged Particles Moving in a Magnetic Field
29.6 The Hall Effect
30.6 Magnetism in Matter
31.6 Eddy Currents
33.9 Rectifiers and Filters
34.6 Production of Electromagnetic Waves by an Antenna
36.5 Lens Aberrations
36.6 The Camera

36.7 The Eye
36.8 The Simple Magnifier
36.9 The Compound Microscope
36.10 The Telescope
38.5 Diffraction of X-Rays by Crystals
39.9 The General Theory of Relativity
41.6 Applications of Tunneling
42.9 Spontaneous and Stimulated Transitions
42.10 Lasers
43.7 Semiconductor Devices
43.8 Superconductivity
44.8 Nuclear Magnetic Resonance and Magnetic Resonance Imaging
45.5 Radiation Damage
45.6 Uses of Radiation

 # Acknowledgments

This Ninth Edition of *Physics for Scientists and Engineers* was prepared with the guidance and assistance of many professors who reviewed selections of the manuscript, the prerevision text, or both. We wish to acknowledge the following scholars and express our sincere appreciation for their suggestions, criticisms, and encouragement:

Benjamin C. Bromley, *University of Utah;* Elena Flitsiyan, *University of Central Florida;* Yuankun Lin, *University of North Texas;* Allen Mincer, *New York University;* Yibin Pan, *University of Wisconsin–Madison;* N. M. Ravindra, *New Jersey Institute of Technology;* Masao Sako, *University of Pennsylvania;* Charles Stone, *Colorado School of Mines;* Robert Weidman, *Michigan Technological University;* Michael Winokur, *University of Wisconsin–Madison*

Prior to our work on this revision, we conducted a survey of professors; their feedback and suggestions helped shape this revision, and so we would like to thank the survey participants:

Elise Adamson, *Wayland Baptist University;* Saul Adelman, *The Citadel;* Yiyan Bai, *Houston Community College;* Philip Blanco, *Grossmont College;* Ken Bolland, *Ohio State University;* Michael Butros, *Victor Valley College;* Brian Carter, *Grossmont College;* Jennifer Cash, *South Carolina State University;* Soumitra Chattopadhyay, *Georgia Highlands College;* John Cooper, *Brazosport College;* Gregory Dolise, *Harrisburg Area Community College;* Mike Durren, *Lake Michigan College;* Tim Farris, *Volunteer State Community College;* Mirela Fetea, *University of Richmond;* Susan Foreman, *Danville Area Community College;* Richard Gottfried, *Frederick Community College;* Christopher Gould, *University of Southern California;* Benjamin Grinstein, *University of California, San Diego;* Wayne Guinn, *Lon Morris College;* Joshua Guttman, *Bergen Community College;* Carlos Handy, *Texas Southern University;* David Heskett, *University of Rhode Island;* Ed Hungerford, *University of Houston;* Matthew Hyre, *Northwestern College;* Charles Johnson, *South Georgia College;* Lynne Lawson, *Providence College;* Byron Leles, *Northeast Alabama Community College;* Rizwan Mahmood, *Slippery Rock University;* Virginia Makepeace, *Kankakee Community College;* David Marasco, *Foothill College;* Richard McCorkle, *University of Rhode Island;* Brian Moudry, *Davis & Elkins College;* Charles Nickles, *University of Massachusetts Dartmouth;* Terrence O'Neill, *Riverside Community College;* Grant O'Rielly, *University of Massachusetts Dartmouth;* Michael Ottinger, *Missouri Western State University;* Michael Panunto, *Butte College;* Eugenia Peterson, *Richard J. Daley College;* Robert Pompi, *Binghamton University, State University of New York;* Ralph Popp, *Mercer County Community College;* Craig Rabatin, *West Virginia University at Parkersburg;* Marilyn Rands, *Lawrence Technological University;* Christina Reeves-Shull, *Cedar Valley College;* John Rollino, *Rutgers University, Newark;* Rich Schelp, *Erskine College;* Mark Semon, *Bates College;* Walther Spjeldvik, *Weber State University;* Mark Spraker, *North Georgia College and State University;* Julie Talbot, *University of West Georgia;* James Tressel, *Massasoit Community College;* Bruce Unger, *Wenatchee Valley College;* Joan Vogtman, *Potomac State College*

This title was carefully checked for accuracy by Grant Hart, *Brigham Young University;* James E. Rutledge, *University of California at Irvine;* and Som Tyagi, *Drexel University.* We thank them for their diligent efforts under schedule pressure.

Belal Abas, Zinoviy Akkerman, Eric Boyd, Hal Falk, Melanie Martin, Steve McCauley, and Glenn Stracher made corrections to problems taken from previous editions. Harvey Leff provided invaluable guidance on the restructuring of the discussion of entropy in Chapter 22. We are grateful to authors John R. Gordon and Vahé Peroomian for preparing the *Student Solutions Manual/Study Guide* and to Vahé Peroomian for preparing an excellent *Instructor's Solutions Manual.* Susan English carefully edited and improved the test bank. Linnea Cookson provided an excellent accuracy check of the Analysis Model tutorials.

Special thanks and recognition go to the professional staff at Cengage Learning—in particular, Charles Hartford, Ed Dodd, Rebecca Berardy Schwartz, Alison Eigel Zade, Cate Barr, and Brendan Killion (who managed the ancillary program)—for their fine work during the development, production, and promotion of this textbook. We recognize the skilled production service and excellent artwork provided by the staff at Lachina Publishing Services and the dedicated photo research efforts of Christopher Arena at the Bill Smith Group.

Finally, we are deeply indebted to our wives, children, and grandchildren for their love, support, and long-term sacrifices.

Raymond A. Serway
St. Petersburg, Florida

John W. Jewett, Jr.
Anaheim, California

To the Student

It is appropriate to offer some words of advice that should be of benefit to you, the student. Before doing so, we assume you have read the Preface, which describes the various features of the text and support materials that will help you through the course.

 ## How to Study

Instructors are often asked, "How should I study physics and prepare for examinations?" There is no simple answer to this question, but we can offer some suggestions based on our own experiences in learning and teaching over the years.

First and foremost, maintain a positive attitude toward the subject matter, keeping in mind that physics is the most fundamental of all natural sciences. Other science courses that follow will use the same physical principles, so it is important that you understand and are able to apply the various concepts and theories discussed in the text.

 ## Concepts and Principles

It is essential that you understand the basic concepts and principles before attempting to solve assigned problems. You can best accomplish this goal by carefully reading the textbook before you attend your lecture on the covered material. When reading the text, you should jot down those points that are not clear to you. Also be sure to make a diligent attempt at answering the questions in the Quick Quizzes as you come to them in your reading. We have worked hard to prepare questions that help you judge for yourself how well you understand the material. Study the **What If?** features that appear in many of the worked examples carefully. They will help you extend your understanding beyond the simple act of arriving at a numerical result. The Pitfall Preventions will also help guide you away from common misunderstandings about physics. During class, take careful notes and ask questions about those ideas that are unclear to you. Keep in mind that few people are able to absorb the full meaning of scientific material after only one reading; several readings of the text and your notes may be necessary. Your lectures and laboratory work supplement the textbook and should clarify some of the more difficult material. You should minimize your memorization of material. Successful memorization of passages from the text, equations, and derivations does not necessarily indicate that you understand the material. Your understanding of the material will be enhanced through a combination of efficient study habits, discussions with other students and with instructors, and your ability to solve the problems presented in the textbook. Ask questions whenever you believe that clarification of a concept is necessary.

 ## Study Schedule

It is important that you set up a regular study schedule, preferably a daily one. Make sure that you read the syllabus for the course and adhere to the schedule set by your instructor. The lectures will make much more sense if you read the corresponding text material *before* attending them. As a general rule, you should devote about two hours of study time for each hour you are in class. If you are having trouble with the course, seek the advice of the instructor or other students who have taken the course. You may find it necessary to seek further instruction from experienced students. Very often, instructors offer review sessions in addition to regular class periods. Avoid the practice of delaying study until a day or two before an exam. More often than not, this approach has disastrous results. Rather than undertake an all-night study session before a test, briefly review the basic concepts and equations, and then get a good night's rest. If you believe that you need additional help in understanding the concepts, in preparing for exams, or in problem solving, we suggest that you acquire a copy of the *Student Solutions Manual/Study Guide* that accompanies this textbook.

Please visit **http://www.cengage.com/physics/serway/PSEtechupdate9e** for information about student resources for this text.

Use the Features

You should make full use of the various features of the text discussed in the Preface. For example, marginal notes are useful for locating and describing important equations and concepts, and **boldface** indicates important definitions. Many useful tables are contained in the appendices, but most are incorporated in the text where they are most often referenced. Appendix B is a convenient review of mathematical tools used in the text.

Answers to Quick Quizzes and odd-numbered problems are given at the end of the textbook, and solutions to selected end-of-chapter questions and problems are provided in the *Student Solutions Manual/Study Guide*. The table of contents provides an overview of the entire text, and the index enables you to locate specific material quickly. Footnotes are sometimes used to supplement the text or to cite other references on the subject discussed.

After reading a chapter, you should be able to define any new quantities introduced in that chapter and discuss the principles and assumptions that were used to arrive at certain key relations. The chapter summaries and the review sections of the *Student Solutions Manual/Study Guide* should help you in this regard. In some cases, you may find it necessary to refer to the textbook's index to locate certain topics. You should be able to associate with each physical quantity the correct symbol used to represent that quantity and the unit in which the quantity is specified. Furthermore, you should be able to express each important equation in concise and accurate prose.

Problem Solving

R. P. Feynman, Nobel laureate in physics, once said, "You do not know anything until you have practiced." In keeping with this statement, we strongly advise you to develop the skills necessary to solve a wide range of problems. Your ability to solve problems will be one of the main tests of your knowledge of physics; therefore, you should try to solve as many problems as possible. It is essential that you understand basic concepts and principles before attempting to solve problems. It is good practice to try to find alternate solutions to the same problem. For example, you can solve problems in mechanics using Newton's laws, but very often an alternative method that draws on energy considerations is more direct. You should not deceive yourself into thinking that you understand a problem merely because you have seen it solved in class. You must be able to solve the problem and similar problems on your own.

The approach to solving problems should be carefully planned. A systematic plan is especially important when a problem involves several concepts. First, read the problem several times until you are confident you understand what is being asked. Look for any key words that will help you interpret the problem and perhaps allow you to make certain assumptions. Your ability to interpret a question properly is an integral part of problem solving. Second, you should acquire the habit of writing down the information given in a problem and those quantities that need to be found; for example, you might construct a table listing both the quantities given and the quantities to be found. This procedure is sometimes used in the worked examples of the textbook. Finally, after you have decided on the method you believe is appropriate for a given problem, proceed with your solution. The General Problem-Solving Strategy will guide you through complex problems. If you follow the steps of this procedure *(Conceptualize, Categorize, Analyze, Finalize)*, you will find it easier to come up with a solution and gain more from your efforts. This strategy, located at the end of Chapter 2 (pages 45–47), is used in all worked examples in the remaining chapters so that you can learn how to apply it. Specific problem-solving strategies for certain types of situations are included in the text and appear with a special heading. These specific strategies follow the outline of the General Problem-Solving Strategy.

Often, students fail to recognize the limitations of certain equations or physical laws in a particular situation. It is very important that you understand and remember the assumptions that underlie a particular theory or formalism. For example, certain equations in kinematics apply only to a particle moving with constant acceleration. These equations are not valid for describing motion whose acceleration is not constant, such as the motion of an object connected to a spring or the motion of an object through a fluid. Study the Analysis Models for Problem Solving in the chapter summaries carefully so that you know how each model can be applied to a specific situation. The analysis models provide you with a logical structure for solving problems and help you develop your thinking skills to become more like those of a physicist. Use the analysis model approach to save you hours of looking for the correct equation and to make you a faster and more efficient problem solver.

 ## Experiments

Physics is a science based on experimental observations. Therefore, we recommend that you try to supplement the text by performing various types of "hands-on" experiments either at home or in the laboratory. These experiments can be used to test ideas and models discussed in class or in the textbook. For example, the common Slinky toy is excellent for studying traveling waves, a ball swinging on the end of a long string can be used to investigate pendulum motion, various masses attached to the end of a vertical spring or rubber band can be used to determine its elastic nature, an old pair of polarized sunglasses and some discarded lenses and a magnifying glass are the components of various experiments in optics, and an approximate measure of the free-fall acceleration can be determined simply by measuring with a stopwatch the time interval required for a ball to drop from a known height. The list of such experiments is endless. When physical models are not available, be imaginative and try to develop models of your own.

 ## New Media

If available, we strongly encourage you to use the **Enhanced WebAssign** product that is available with this textbook. It is far easier to understand physics if you see it in action, and the materials available in Enhanced WebAssign will enable you to become a part of that action.

It is our sincere hope that you will find physics an exciting and enjoyable experience and that you will benefit from this experience, regardless of your chosen profession. Welcome to the exciting world of physics!

The scientist does not study nature because it is useful; he studies it because he delights in it, and he delights in it because it is beautiful. If nature were not beautiful, it would not be worth knowing, and if nature were not worth knowing, life would not be worth living.

—Henri Poincaré

Electricity and Magnetism

A Transrapid maglev train pulls into a station in Shanghai, China. The word *maglev* is an abbreviated form of *magnetic levitation*. This train makes no physical contact with its rails; its weight is totally supported by electromagnetic forces. In this part of the book, we will study these forces. *(OTHK/Asia Images/Jupiterimages)*

We now study the branch of physics concerned with electric and magnetic phe- nomena. The laws of electricity and magnetism play a central role in the operation of such devices as smartphones, televisions, electric motors, computers, high-energy accelerators, and other electronic devices. More fundamentally, the interatomic and intermolecular forces responsible for the formation of solids and liquids are electric in origin.

Evidence in Chinese documents suggests magnetism was observed as early as 2000 BC. The ancient Greeks observed electric and magnetic phenomena possibly as early as 700 BC. The Greeks knew about magnetic forces from observations that the naturally occurring stone *magnetite* (Fe_3O_4) is attracted to iron. (The word *electric* comes from *elecktron*, the Greek word for "amber." The word *magnetic* comes from *Magnesia*, the name of the district of Greece where magnetite was first found.)

Not until the early part of the nineteenth century did scientists establish that electricity and magnetism are related phenomena. In 1819, Hans Oersted discovered that a compass needle is deflected when placed near a circuit carrying an electric current. In 1831, Michael Faraday and, almost simultaneously, Joseph Henry showed that when a wire is moved near a magnet (or, equivalently, when a magnet is moved near a wire), an electric current is established in the wire. In 1873, James Clerk Maxwell used these observations and other experimental facts as a basis for formulating the laws of electromagnetism as we know them today. (*Electromagnetism* is a name given to the combined study of electricity and magnetism.)

Maxwell's contributions to the field of electromagnetism were especially significant because the laws he formulated are basic to *all* forms of electromagnetic phenomena. His work is as important as Newton's work on the laws of motion and the theory of gravitation. ■

Electric Fields

This young woman is enjoying the effects of electrically charging her body. Each individual hair on her head becomes charged and exerts a repulsive force on the other hairs, resulting in the "stand-up" hairdo seen here. *(Ted Kinsman / Photo Researchers, Inc.)*

In this chapter, we begin the study of electromagnetism. The first link that we will make to our previous study is through the concept of *force*. The electromagnetic force between charged particles is one of the fundamental forces of nature. We begin by describing some basic properties of one manifestation of the electromagnetic force, the electric force. We then discuss Coulomb's law, which is the fundamental law governing the electric force between any two charged particles. Next, we introduce the concept of an electric field associated with a charge distribution and describe its effect on other charged particles. We then show how to use Coulomb's law to calculate the electric field for a given charge distribution. The chapter concludes with a discussion of the motion of a charged particle in a uniform electric field.

The second link between electromagnetism and our previous study is through the concept of *energy*. We will discuss that connection in Chapter 25.

23.1 Properties of Electric Charges

A number of simple experiments demonstrate the existence of electric forces. For example, after rubbing a balloon on your hair on a dry day, you will find that the balloon attracts bits of paper. The attractive force is often strong enough to suspend the paper from the balloon.

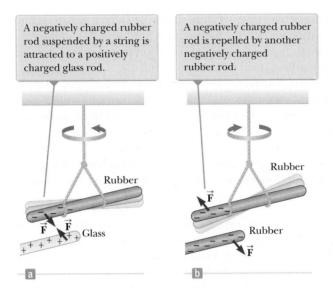

A negatively charged rubber rod suspended by a string is attracted to a positively charged glass rod.

A negatively charged rubber rod is repelled by another negatively charged rubber rod.

Rubber

Rubber

$\vec{F}$

Rubber

$\vec{F}$ $\vec{F}$ Glass

$\vec{F}$

a

b

Figure 23.1 The electric force between (a) oppositely charged objects and (b) like-charged objects.

When materials behave in this way, they are said to be *electrified* or to have become **electrically charged.** You can easily electrify your body by vigorously rubbing your shoes on a wool rug. Evidence of the electric charge on your body can be detected by lightly touching (and startling) a friend. Under the right conditions, you will see a spark when you touch and both of you will feel a slight tingle. (Experiments such as these work best on a dry day because an excessive amount of moisture in the air can cause any charge you build up to "leak" from your body to the Earth.)

In a series of simple experiments, it was found that there are two kinds of electric charges, which were given the names **positive** and **negative** by Benjamin Franklin (1706–1790). Electrons are identified as having negative charge, and protons are positively charged. To verify that there are two types of charge, suppose a hard rubber rod that has been rubbed on fur is suspended by a string as shown in Figure 23.1. When a glass rod that has been rubbed on silk is brought near the rubber rod, the two attract each other (Fig. 23.1a). On the other hand, if two charged rubber rods (or two charged glass rods) are brought near each other as shown in Figure 23.1b, the two repel each other. This observation shows that the rubber and glass have two different types of charge on them. On the basis of these observations, we conclude that **charges of the same sign repel one another and charges with opposite signs attract one another.**

Using the convention suggested by Franklin, the electric charge on the glass rod is called positive and that on the rubber rod is called negative. Therefore, any charged object attracted to a charged rubber rod (or repelled by a charged glass rod) must have a positive charge, and any charged object repelled by a charged rubber rod (or attracted to a charged glass rod) must have a negative charge.

Another important aspect of electricity that arises from experimental observations is that **electric charge is always conserved** in an isolated system. That is, when one object is rubbed against another, charge is not created in the process. The electrified state is due to a *transfer* of charge from one object to the other. One object gains some amount of negative charge while the other gains an equal amount of positive charge. For example, when a glass rod is rubbed on silk as in Figure 23.2, the silk obtains a negative charge equal in magnitude to the positive charge on the glass rod. We now know from our understanding of atomic structure that electrons are transferred in the rubbing process from the glass to the silk. Similarly, when rubber is rubbed on fur, electrons are transferred from the fur to the rubber, giving the rubber a net negative charge and the fur a net positive charge. This process works because neutral, uncharged matter contains as many positive charges (protons within atomic nuclei)

Because of conservation of charge, each electron adds negative charge to the silk and an equal positive charge is left on the glass rod.

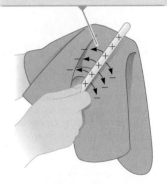

Figure 23.2 When a glass rod is rubbed with silk, electrons are transferred from the glass to the silk.

◀ Electric charge is conserved

The neutral sphere has equal numbers of positive and negative charges.

a

Electrons redistribute when a charged rod is brought close.

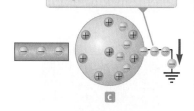

b

Some electrons leave the grounded sphere through the ground wire.

c

The excess positive charge is nonuniformly distributed.

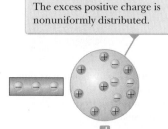

d

The remaining electrons redistribute uniformly, and there is a net uniform distribution of positive charge on the sphere.

e

Figure 23.3 Charging a metallic object by *induction*. (a) A neutral metallic sphere. (b) A charged rubber rod is placed near the sphere. (c) The sphere is grounded. (d) The ground connection is removed. (e) The rod is removed.

as negative charges (electrons). Conservation of electric charge for an isolated system is like conservation of energy, momentum, and angular momentum, but we don't identify an analysis model for this conservation principle because it is not used often enough in the mathematical solution to problems.

In 1909, Robert Millikan (1868–1953) discovered that electric charge always occurs as integral multiples of a fundamental amount of charge e (see Section 25.7). In modern terms, the electric charge q is said to be **quantized,** where q is the standard symbol used for charge as a variable. That is, electric charge exists as discrete "packets," and we can write $q = \pm Ne$, where N is some integer. Other experiments in the same period showed that the electron has a charge $-e$ and the proton has a charge of equal magnitude but opposite sign $+e$. Some particles, such as the neutron, have no charge.

Quick **Quiz 23.1** Three objects are brought close to each other, two at a time. When objects A and B are brought together, they repel. When objects B and C are brought together, they also repel. Which of the following are true? **(a)** Objects A and C possess charges of the same sign. **(b)** Objects A and C possess charges of opposite sign. **(c)** All three objects possess charges of the same sign. **(d)** One object is neutral. **(e)** Additional experiments must be performed to determine the signs of the charges.

23.2 Charging Objects by Induction

It is convenient to classify materials in terms of the ability of electrons to move through the material:

> Electrical **conductors** are materials in which some of the electrons are free electrons[1] that are not bound to atoms and can move relatively freely through the material; electrical **insulators** are materials in which all electrons are bound to atoms and cannot move freely through the material.

Materials such as glass, rubber, and dry wood fall into the category of electrical insulators. When such materials are charged by rubbing, only the area rubbed becomes charged and the charged particles are unable to move to other regions of the material.

In contrast, materials such as copper, aluminum, and silver are good electrical conductors. When such materials are charged in some small region, the charge readily distributes itself over the entire surface of the material.

Semiconductors are a third class of materials, and their electrical properties are somewhere between those of insulators and those of conductors. Silicon and germanium are well-known examples of semiconductors commonly used in the fabrication of a variety of electronic chips used in computers, cellular telephones, and home theater systems. The electrical properties of semiconductors can be changed over many orders of magnitude by the addition of controlled amounts of certain atoms to the materials.

To understand how to charge a conductor by a process known as **induction,** consider a neutral (uncharged) conducting sphere insulated from the ground as shown in Figure 23.3a. There are an equal number of electrons and protons in the sphere if the charge on the sphere is exactly zero. When a negatively charged rubber rod is brought near the sphere, electrons in the region nearest the rod experience a repulsive force and migrate to the opposite side of the sphere. This migration leaves

---

[1]A metal atom contains one or more outer electrons, which are weakly bound to the nucleus. When many atoms combine to form a metal, the *free electrons* are these outer electrons, which are not bound to any one atom. These electrons move about the metal in a manner similar to that of gas molecules moving in a container.

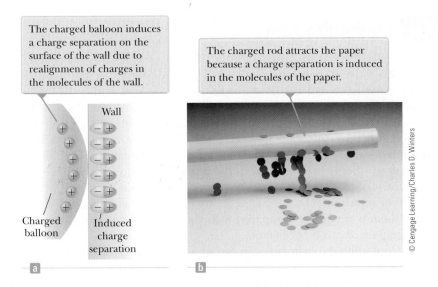

The charged balloon induces a charge separation on the surface of the wall due to realignment of charges in the molecules of the wall.

Wall

Charged balloon

Induced charge separation

The charged rod attracts the paper because a charge separation is induced in the molecules of the paper.

© Cengage Learning/Charles D. Winters

a

b

Figure 23.4 (a) A charged balloon is brought near an insulating wall. (b) A charged rod is brought close to bits of paper.

the side of the sphere near the rod with an effective positive charge because of the diminished number of electrons as in Figure 23.3b. (The left side of the sphere in Figure 23.3b is positively charged *as if* positive charges moved into this region, but remember that only electrons are free to move.) This process occurs even if the rod never actually touches the sphere. If the same experiment is performed with a conducting wire connected from the sphere to the Earth (Fig. 23.3c), some of the electrons in the conductor are so strongly repelled by the presence of the negative charge in the rod that they move out of the sphere through the wire and into the Earth. The symbol ⏚ at the end of the wire in Figure 23.3c indicates that the wire is connected to **ground,** which means a reservoir, such as the Earth, that can accept or provide electrons freely with negligible effect on its electrical characteristics. If the wire to ground is then removed (Fig. 23.3d), the conducting sphere contains an excess of *induced* positive charge because it has fewer electrons than it needs to cancel out the positive charge of the protons. When the rubber rod is removed from the vicinity of the sphere (Fig. 23.3e), this induced positive charge remains on the ungrounded sphere. Notice that the rubber rod loses none of its negative charge during this process.

Charging an object by induction requires no contact with the object inducing the charge. That is in contrast to charging an object by rubbing (that is, by *conduction*), which does require contact between the two objects.

A process similar to induction in conductors takes place in insulators. In most neutral molecules, the center of positive charge coincides with the center of negative charge. In the presence of a charged object, however, these centers inside each molecule in an insulator may shift slightly, resulting in more positive charge on one side of the molecule than on the other. This realignment of charge within individual molecules produces a layer of charge on the surface of the insulator as shown in Figure 23.4a. The proximity of the positive charges on the surface of the object and the negative charges on the surface of the insulator results in an attractive force between the object and the insulator. Your knowledge of induction in insulators should help you explain why a charged rod attracts bits of electrically neutral paper as shown in Figure 23.4b.

Quick Quiz 23.2 Three objects are brought close to one another, two at a time. When objects A and B are brought together, they attract. When objects B and C are brought together, they repel. Which of the following are necessarily true? **(a)** Objects A and C possess charges of the same sign. **(b)** Objects A and C possess charges of opposite sign. **(c)** All three objects possess charges of the same sign. **(d)** One object is neutral. **(e)** Additional experiments must be performed to determine information about the charges on the objects.

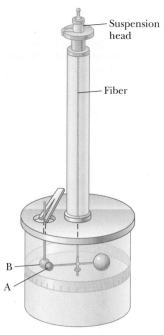

Figure 23.5 Coulomb's balance, used to establish the inverse-square law for the electric force.

23.3 Coulomb's Law

Charles Coulomb measured the magnitudes of the electric forces between charged objects using the torsion balance, which he invented (Fig. 23.5). The operating principle of the torsion balance is the same as that of the apparatus used by Cavendish to measure the density of the Earth (see Section 13.1), with the electrically neutral spheres replaced by charged ones. The electric force between charged spheres A and B in Figure 23.5 causes the spheres to either attract or repel each other, and the resulting motion causes the suspended fiber to twist. Because the restoring torque of the twisted fiber is proportional to the angle through which the fiber rotates, a measurement of this angle provides a quantitative measure of the electric force of attraction or repulsion. Once the spheres are charged by rubbing, the electric force between them is very large compared with the gravitational attraction, and so the gravitational force can be neglected.

From Coulomb's experiments, we can generalize the properties of the **electric force** (sometimes called the *electrostatic force*) between two stationary charged particles. We use the term **point charge** to refer to a charged particle of zero size. The electrical behavior of electrons and protons is very well described by modeling them as point charges. From experimental observations, we find that the magnitude of the electric force (sometimes called the *Coulomb force*) between two point charges is given by **Coulomb's law.**

Coulomb's law ▶

$$F_e = k_e \frac{|q_1||q_2|}{r^2} \tag{23.1}$$

where k_e is a constant called the **Coulomb constant.** In his experiments, Coulomb was able to show that the value of the exponent of r was 2 to within an uncertainty of a few percent. Modern experiments have shown that the exponent is 2 to within an uncertainty of a few parts in 10^{16}. Experiments also show that the electric force, like the gravitational force, is conservative.

The value of the Coulomb constant depends on the choice of units. The SI unit of charge is the **coulomb** (C). The Coulomb constant k_e in SI units has the value

Coulomb constant ▶

$$k_e = 8.987\,6 \times 10^9 \, \text{N} \cdot \text{m}^2/\text{C}^2 \tag{23.2}$$

This constant is also written in the form

$$k_e = \frac{1}{4\pi\epsilon_0} \tag{23.3}$$

where the constant ϵ_0 (Greek letter epsilon) is known as the **permittivity of free space** and has the value

$$\epsilon_0 = 8.854\,2 \times 10^{-12} \, \text{C}^2/\text{N} \cdot \text{m}^2 \tag{23.4}$$

The smallest unit of free charge e known in nature,[2] the charge on an electron ($-e$) or a proton ($+e$), has a magnitude

$$e = 1.602\,18 \times 10^{-19} \, \text{C} \tag{23.5}$$

Therefore, 1 C of charge is approximately equal to the charge of 6.24×10^{18} electrons or protons. This number is very small when compared with the number of free electrons in 1 cm³ of copper, which is on the order of 10^{23}. Nevertheless, 1 C is a substantial amount of charge. In typical experiments in which a rubber or glass rod is charged by friction, a net charge on the order of 10^{-6} C is obtained. In other

© INTERFOTO/Alamy

Charles Coulomb
French physicist (1736–1806)
Coulomb's major contributions to science were in the areas of electrostatics and magnetism. During his lifetime, he also investigated the strengths of materials, thereby contributing to the field of structural mechanics. In ergonomics, his research provided an understanding of the ways in which people and animals can best do work.

---

[2]No unit of charge smaller than e has been detected on a free particle; current theories, however, propose the existence of particles called *quarks* having charges $-e/3$ and $2e/3$. Although there is considerable experimental evidence for such particles inside nuclear matter, *free* quarks have never been detected. We discuss other properties of quarks in Chapter 46.

Table 23.1	Charge and Mass of the Electron, Proton, and Neutron	
Particle	Charge (C)	Mass (kg)
Electron (e)	$-1.602\ 176\ 5 \times 10^{-19}$	$9.109\ 4 \times 10^{-31}$
Proton (p)	$+1.602\ 176\ 5 \times 10^{-19}$	$1.672\ 62 \times 10^{-27}$
Neutron (n)	0	$1.674\ 93 \times 10^{-27}$

words, only a very small fraction of the total available charge is transferred between the rod and the rubbing material.

The charges and masses of the electron, proton, and neutron are given in Table 23.1. Notice that the electron and proton are identical in the magnitude of their charge but vastly different in mass. On the other hand, the proton and neutron are similar in mass but vastly different in charge. Chapter 46 will help us understand these interesting properties.

Example 23.1   The Hydrogen Atom

The electron and proton of a hydrogen atom are separated (on the average) by a distance of approximately 5.3×10^{-11} m. Find the magnitudes of the electric force and the gravitational force between the two particles.

SOLUTION

Conceptualize Think about the two particles separated by the very small distance given in the problem statement. In Chapter 13, we mentioned that the gravitational force between an electron and a proton is very small compared to the electric force between them, so we expect this to be the case with the results of this example.

Categorize The electric and gravitational forces will be evaluated from universal force laws, so we categorize this example as a substitution problem.

Use Coulomb's law to find the magnitude of the electric force:

$$F_e = k_e \frac{|e||-e|}{r^2} = (8.988 \times 10^9\ \text{N} \cdot \text{m}^2/\text{C}^2) \frac{(1.60 \times 10^{-19}\ \text{C})^2}{(5.3 \times 10^{-11}\ \text{m})^2}$$

$$= \boxed{8.2 \times 10^{-8}\ \text{N}}$$

Use Newton's law of universal gravitation and Table 23.1 (for the particle masses) to find the magnitude of the gravitational force:

$$F_g = G \frac{m_e m_p}{r^2}$$

$$= (6.674 \times 10^{-11}\ \text{N} \cdot \text{m}^2/\text{kg}^2) \frac{(9.11 \times 10^{-31}\ \text{kg})(1.67 \times 10^{-27}\ \text{kg})}{(5.3 \times 10^{-11}\ \text{m})^2}$$

$$= \boxed{3.6 \times 10^{-47}\ \text{N}}$$

The ratio $F_e/F_g \approx 2 \times 10^{39}$. Therefore, the gravitational force between charged atomic particles is negligible when compared with the electric force. Notice the similar forms of Newton's law of universal gravitation and Coulomb's law of electric forces. Other than the magnitude of the forces between elementary particles, what is a fundamental difference between the two forces?

When dealing with Coulomb's law, remember that force is a vector quantity and must be treated accordingly. Coulomb's law expressed in vector form for the electric force exerted by a charge q_1 on a second charge q_2, written $\vec{\mathbf{F}}_{12}$, is

$$\vec{\mathbf{F}}_{12} = k_e \frac{q_1 q_2}{r^2} \hat{\mathbf{r}}_{12} \qquad \text{(23.6)}$$

◀ **Vector form of Coulomb's law**

where $\hat{\mathbf{r}}_{12}$ is a unit vector directed from q_1 toward q_2 as shown in Figure 23.6a (page 696). Because the electric force obeys Newton's third law, the electric force exerted by q_2 on q_1 is equal in magnitude to the force exerted by q_1 on q_2 and in the opposite direction; that is, $\vec{\mathbf{F}}_{21} = -\vec{\mathbf{F}}_{12}$. Finally, Equation 23.6 shows that if q_1 and q_2 have the

Figure 23.6 Two point charges separated by a distance r exert a force on each other that is given by Coulomb's law. The force $\vec{\mathbf{F}}_{21}$ exerted by q_2 on q_1 is equal in magnitude and opposite in direction to the force $\vec{\mathbf{F}}_{12}$ exerted by q_1 on q_2.

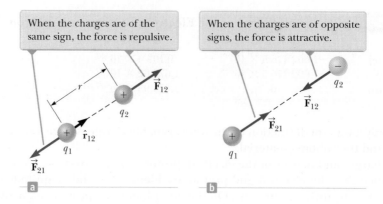

same sign as in Figure 23.6a, the product $q_1 q_2$ is positive and the electric force on one particle is directed away from the other particle. If q_1 and q_2 are of opposite sign as shown in Figure 23.6b, the product $q_1 q_2$ is negative and the electric force on one particle is directed toward the other particle. These signs describe the *relative* direction of the force but not the *absolute* direction. A negative product indicates an attractive force, and a positive product indicates a repulsive force. The *absolute* direction of the force on a charge depends on the location of the other charge. For example, if an x axis lies along the two charges in Figure 23.6a, the product $q_1 q_2$ is positive, but $\vec{\mathbf{F}}_{12}$ points in the positive x direction and $\vec{\mathbf{F}}_{21}$ points in the negative x direction.

When more than two charges are present, the force between any pair of them is given by Equation 23.6. Therefore, the resultant force on any one of them equals the vector sum of the forces exerted by the other individual charges. For example, if four charges are present, the resultant force exerted by particles 2, 3, and 4 on particle 1 is

$$\vec{\mathbf{F}}_1 = \vec{\mathbf{F}}_{21} + \vec{\mathbf{F}}_{31} + \vec{\mathbf{F}}_{41}$$

Quick Quiz 23.3 Object A has a charge of $+2\ \mu\text{C}$, and object B has a charge of $+6\ \mu\text{C}$. Which statement is true about the electric forces on the objects?
(a) $\vec{\mathbf{F}}_{AB} = -3\vec{\mathbf{F}}_{BA}$ **(b)** $\vec{\mathbf{F}}_{AB} = -\vec{\mathbf{F}}_{BA}$ **(c)** $3\vec{\mathbf{F}}_{AB} = -\vec{\mathbf{F}}_{BA}$ **(d)** $\vec{\mathbf{F}}_{AB} = 3\vec{\mathbf{F}}_{BA}$
(e) $\vec{\mathbf{F}}_{AB} = \vec{\mathbf{F}}_{BA}$ **(f)** $3\vec{\mathbf{F}}_{AB} = \vec{\mathbf{F}}_{BA}$

Example 23.2   **Find the Resultant Force**

Consider three point charges located at the corners of a right triangle as shown in Figure 23.7, where $q_1 = q_3 = 5.00\ \mu\text{C}$, $q_2 = -2.00\ \mu\text{C}$, and $a = 0.100$ m. Find the resultant force exerted on q_3.

SOLUTION

Conceptualize Think about the net force on q_3. Because charge q_3 is near two other charges, it will experience two electric forces. These forces are exerted in different directions as shown in Figure 23.7. Based on the forces shown in the figure, estimate the direction of the net force vector.

Categorize Because two forces are exerted on charge q_3, we categorize this example as a vector addition problem.

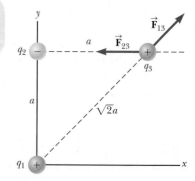

Figure 23.7 (Example 23.2) The force exerted by q_1 on q_3 is $\vec{\mathbf{F}}_{13}$. The force exerted by q_2 on q_3 is $\vec{\mathbf{F}}_{23}$. The resultant force $\vec{\mathbf{F}}_3$ exerted on q_3 is the vector sum $\vec{\mathbf{F}}_{13} + \vec{\mathbf{F}}_{23}$.

Analyze The directions of the individual forces exerted by q_1 and q_2 on q_3 are shown in Figure 23.7. The force $\vec{\mathbf{F}}_{23}$ exerted by q_2 on q_3 is attractive because q_2 and q_3 have opposite signs. In the coordinate system shown in Figure 23.7, the attractive force $\vec{\mathbf{F}}_{23}$ is to the left (in the negative x direction).

The force $\vec{\mathbf{F}}_{13}$ exerted by q_1 on q_3 is repulsive because both charges are positive. The repulsive force $\vec{\mathbf{F}}_{13}$ makes an angle of $45.0°$ with the x axis.

▶ **23.2** continued

Use Equation 23.1 to find the magnitude of $\vec{F}_{23}$:

$$F_{23} = k_e \frac{|q_2||q_3|}{a^2}$$

$$= (8.988 \times 10^9 \text{ N} \cdot \text{m}^2/\text{C}^2) \frac{(2.00 \times 10^{-6} \text{ C})(5.00 \times 10^{-6} \text{ C})}{(0.100 \text{ m})^2} = 8.99 \text{ N}$$

Find the magnitude of the force $\vec{F}_{13}$:

$$F_{13} = k_e \frac{|q_1||q_3|}{(\sqrt{2}\, a)^2}$$

$$= (8.988 \times 10^9 \text{ N} \cdot \text{m}^2/\text{C}^2) \frac{(5.00 \times 10^{-6} \text{ C})(5.00 \times 10^{-6} \text{ C})}{2(0.100 \text{ m})^2} = 11.2 \text{ N}$$

Find the x and y components of the force $\vec{F}_{13}$:

$$F_{13x} = (11.2 \text{ N}) \cos 45.0° = 7.94 \text{ N}$$
$$F_{13y} = (11.2 \text{ N}) \sin 45.0° = 7.94 \text{ N}$$

Find the components of the resultant force acting on q_3:

$$F_{3x} = F_{13x} + F_{23x} = 7.94 \text{ N} + (-8.99 \text{ N}) = -1.04 \text{ N}$$
$$F_{3y} = F_{13y} + F_{23y} = 7.94 \text{ N} + 0 = 7.94 \text{ N}$$

Express the resultant force acting on q_3 in unit-vector form:

$$\vec{F}_3 = (-1.04\hat{\mathbf{i}} + 7.94\hat{\mathbf{j}}) \text{ N}$$

Finalize The net force on q_3 is upward and toward the left in Figure 23.7. If q_3 moves in response to the net force, the distances between q_3 and the other charges change, so the net force changes. Therefore, if q_3 is free to move, it can be modeled as a particle under a net force as long as it is recognized that the force exerted on q_3 is *not* constant. As a reminder, we display most numerical values to three significant figures, which leads to operations such as 7.94 N + (−8.99 N) = −1.04 N above. If you carry all intermediate results to more significant figures, you will see that this operation is correct.

WHAT IF? What if the signs of all three charges were changed to the opposite signs? How would that affect the result for $\vec{F}_3$?

Answer The charge q_3 would still be attracted toward q_2 and repelled from q_1 with forces of the same magnitude. Therefore, the final result for $\vec{F}_3$ would be the same.

Example 23.3 **Where Is the Net Force Zero?** AM

Three point charges lie along the x axis as shown in Figure 23.8. The positive charge $q_1 = 15.0 \, \mu\text{C}$ is at $x = 2.00$ m, the positive charge $q_2 = 6.00 \, \mu\text{C}$ is at the origin, and the net force acting on q_3 is zero. What is the x coordinate of q_3?

SOLUTION

Conceptualize Because q_3 is near two other charges, it experiences two electric forces. Unlike the preceding example, however, the forces lie along the same line in this problem as indicated in Figure 23.8. Because q_3 is negative and q_1 and q_2 are positive, the forces $\vec{F}_{13}$ and $\vec{F}_{23}$ are both attractive. Because q_2 is the smaller charge, the position of q_3 at which the force is zero should be closer to q_2 than to q_1.

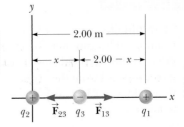

Figure 23.8 (Example 23.3) Three point charges are placed along the x axis. If the resultant force acting on q_3 is zero, the force $\vec{F}_{13}$ exerted by q_1 on q_3 must be equal in magnitude and opposite in direction to the force $\vec{F}_{23}$ exerted by q_2 on q_3.

Categorize Because the net force on q_3 is zero, we model the point charge as a *particle in equilibrium*.

Analyze Write an expression for the net force on charge q_3 when it is in equilibrium:

$$\vec{F}_3 = \vec{F}_{23} + \vec{F}_{13} = -k_e \frac{|q_2||q_3|}{x^2} \hat{\mathbf{i}} + k_e \frac{|q_1||q_3|}{(2.00 - x)^2} \hat{\mathbf{i}} = 0$$

Move the second term to the right side of the equation and set the coefficients of the unit vector $\hat{\mathbf{i}}$ equal:

$$k_e \frac{|q_2||q_3|}{x^2} = k_e \frac{|q_1||q_3|}{(2.00 - x)^2}$$

continued

▶ **23.3** continued

Eliminate k_e and $|q_3|$ and rearrange the equation:

$$(2.00 - x)^2 |q_2| = x^2 |q_1|$$

Take the square root of both sides of the equation:

$$(2.00 - x)\sqrt{|q_2|} = \pm x\sqrt{|q_1|}$$

Solve for x:

$$x = \frac{2.00\sqrt{|q_2|}}{\sqrt{|q_2|} \pm \sqrt{|q_1|}}$$

Substitute numerical values, choosing the plus sign:

$$x = \frac{2.00\sqrt{6.00 \times 10^{-6}\,\text{C}}}{\sqrt{6.00 \times 10^{-6}\,\text{C}} + \sqrt{15.0 \times 10^{-6}\,\text{C}}} = 0.775\,\text{m}$$

Finalize Notice that the movable charge is indeed closer to q_2 as we predicted in the Conceptualize step. The second solution to the equation (if we choose the negative sign) is $x = -3.44$ m. That is another location where the *magnitudes* of the forces on q_3 are equal, but both forces are in the same direction, so they do not cancel.

WHAT IF?   Suppose q_3 is constrained to move only along the x axis. From its initial position at $x = 0.775$ m, it is pulled a small distance along the x axis. When released, does it return to equilibrium, or is it pulled farther from equilibrium? That is, is the equilibrium stable or unstable?

Answer If q_3 is moved to the right, $\vec{F}_{13}$ becomes larger and $\vec{F}_{23}$ becomes smaller. The result is a net force to the right, in the same direction as the displacement. Therefore, the charge q_3 would continue to move to the right and the equilibrium is *unstable*. (See Section 7.9 for a review of stable and unstable equilibria.)

If q_3 is constrained to stay at a *fixed x* coordinate but allowed to move up and down in Figure 23.8, the equilibrium is stable. In this case, if the charge is pulled upward (or downward) and released, it moves back toward the equilibrium position and oscillates about this point.

Example 23.4   **Find the Charge on the Spheres**   AM

Two identical small charged spheres, each having a mass of 3.00×10^{-2} kg, hang in equilibrium as shown in Figure 23.9a. The length L of each string is 0.150 m, and the angle θ is 5.00°. Find the magnitude of the charge on each sphere.

SOLUTION

Conceptualize Figure 23.9a helps us conceptualize this example. The two spheres exert repulsive forces on each other. If they are held close to each other and released, they move outward from the center and settle into the configuration in Figure 23.9a after the oscillations have vanished due to air resistance.

Categorize The key phrase "in equilibrium" helps us model each sphere as a *particle in equilibrium*. This example is similar to the particle in equilibrium problems in Chapter 5 with the added feature that one of the forces on a sphere is an electric force.

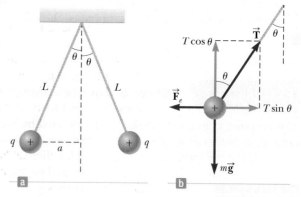

Figure 23.9 (Example 23.4) (a) Two identical spheres, each carrying the same charge q, suspended in equilibrium. (b) Diagram of the forces acting on the sphere on the left part of (a).

Analyze The force diagram for the left-hand sphere is shown in Figure 23.9b. The sphere is in equilibrium under the application of the force $\vec{T}$ from the string, the electric force $\vec{F}_e$ from the other sphere, and the gravitational force $m\vec{g}$.

From the particle in equilibrium model, set the net force on the left-hand sphere equal to zero for each component:

(1) $\sum F_x = T \sin\theta - F_e = 0 \;\rightarrow\; T \sin\theta = F_e$

(2) $\sum F_y = T \cos\theta - mg = 0 \;\rightarrow\; T \cos\theta = mg$

Divide Equation (1) by Equation (2) to find F_e:

(3) $\tan\theta = \dfrac{F_e}{mg} \;\rightarrow\; F_e = mg \tan\theta$

▶ **23.4** continued

Use the geometry of the right triangle in Figure 23.9a to find a relationship between a, L, and θ:

$$(4)\quad \sin\theta = \frac{a}{L} \rightarrow a = L\sin\theta$$

Solve Coulomb's law (Eq. 23.1) for the charge $|q|$ on each sphere and substitute from Equations (3) and (4):

$$|q| = \sqrt{\frac{F_e r^2}{k_e}} = \sqrt{\frac{F_e(2a)^2}{k_e}} = \sqrt{\frac{mg\tan\theta(2L\sin\theta)^2}{k_e}}$$

Substitute numerical values:

$$|q| = \sqrt{\frac{(3.00\times10^{-2}\,\text{kg})(9.80\,\text{m/s}^2)\tan(5.00°)[2(0.150\,\text{m})\sin(5.00°)]^2}{8.988\times10^9\,\text{N}\cdot\text{m}^2/\text{C}^2}}$$

$$= 4.42\times10^{-8}\,\text{C}$$

Finalize If the sign of the charges were not given in Figure 23.9, we could not determine them. In fact, the sign of the charge is not important. The situation is the same whether both spheres are positively charged or negatively charged.

WHAT IF? Suppose your roommate proposes solving this problem without the assumption that the charges are of equal magnitude. She claims the symmetry of the problem is destroyed if the charges are not equal, so the strings would make two different angles with the vertical and the problem would be much more complicated. How would you respond?

Answer The symmetry is not destroyed and the angles are not different. Newton's third law requires the magnitudes of the electric forces on the two spheres to be the same, regardless of the equality or nonequality of the charges. The solution to the example remains the same with one change: the value of $|q|$ in the solution is replaced by $\sqrt{|q_1 q_2|}$ in the new situation, where q_1 and q_2 are the values of the charges on the two spheres. The symmetry of the problem would be destroyed if the *masses* of the spheres were not the same. In this case, the strings would make different angles with the vertical and the problem would be more complicated.

23.4 Analysis Model: Particle in a Field (Electric)

In Section 5.1, we discussed the differences between contact forces and field forces. Two field forces—the gravitational force in Chapter 13 and the electric force here—have been introduced into our discussions so far. As pointed out earlier, field forces can act through space, producing an effect even when no physical contact occurs between interacting objects. Such an interaction can be modeled as a two-step process: a source particle establishes a field, and then a charged particle interacts with the field and experiences a force. The gravitational field $\vec{g}$ at a point in space due to a source particle was defined in Section 13.4 to be equal to the gravitational force $\vec{F}_g$ acting on a test particle of mass m divided by that mass: $\vec{g} \equiv \vec{F}_g/m$. Then the force exerted by the field is $\vec{F} = m\vec{g}$ (Eq. 5.5).

The concept of a field was developed by Michael Faraday (1791–1867) in the context of electric forces and is of such practical value that we shall devote much attention to it in the next several chapters. In this approach, an **electric field** is said to exist in the region of space around a charged object, the **source charge.** The presence of the electric field can be detected by placing a **test charge** in the field and noting the electric force on it. As an example, consider Figure 23.10, which shows a small positive test charge q_0 placed near a second object carrying a much greater positive charge Q. We define the electric field due to the source charge at the location of the test charge to be the electric force on the test charge *per unit charge*, or, to be more specific, the **electric field vector** $\vec{E}$ at a point in space is defined as the electric force $\vec{F}_e$ acting on a positive test charge q_0 placed at that point divided by the test charge:[3]

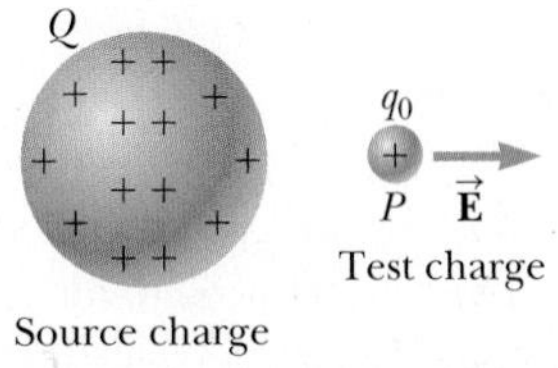

Figure 23.10 A small positive test charge q_0 placed at point P near an object carrying a much larger positive charge Q experiences an electric field $\vec{E}$ at point P established by the source charge Q. We will *always* assume that the test charge is so small that the field of the source charge is unaffected by its presence.

$$\vec{E} \equiv \frac{\vec{F}_e}{q_0} \tag{23.7}$$

◀ **Definition of electric field**

[3]When using Equation 23.7, we must assume the test charge q_0 is small enough that it does not disturb the charge distribution responsible for the electric field. If the test charge is great enough, the charge on the metallic sphere is redistributed and the electric field it sets up is different from the field it sets up in the presence of the much smaller test charge.

This dramatic photograph captures a lightning bolt striking a tree near some rural homes. Lightning is associated with very strong electric fields in the atmosphere.

The vector $\vec{\mathbf{E}}$ has the SI units of newtons per coulomb (N/C). The direction of $\vec{\mathbf{E}}$ as shown in Figure 23.10 is the direction of the force a positive test charge experiences when placed in the field. Note that $\vec{\mathbf{E}}$ is the field produced by some charge or charge distribution *separate from* the test charge; it is not the field produced by the test charge itself. Also note that the existence of an electric field is a property of its source; the presence of the test charge is not necessary for the field to exist. The test charge serves as a *detector* of the electric field: an electric field exists at a point if a test charge at that point experiences an electric force.

If an arbitrary charge q is placed in an electric field $\vec{\mathbf{E}}$, it experiences an electric force given by

$$\vec{\mathbf{F}}_e = q\vec{\mathbf{E}} \tag{23.8}$$

This equation is the mathematical representation of the electric version of the **particle in a field** analysis model. If q is positive, the force is in the same direction as the field. If q is negative, the force and the field are in opposite directions. Notice the similarity between Equation 23.8 and the corresponding equation from the gravitational version of the particle in a field model, $\vec{\mathbf{F}}_g = m\vec{\mathbf{g}}$ (Section 5.5). Once the magnitude and direction of the electric field are known at some point, the electric force exerted on *any* charged particle placed at that point can be calculated from Equation 23.8.

To determine the direction of an electric field, consider a point charge q as a source charge. This charge creates an electric field at all points in space surrounding it. A test charge q_0 is placed at point P, a distance r from the source charge, as in Figure 23.11a. We imagine using the test charge to determine the direction of the electric force and therefore that of the electric field. According to Coulomb's law, the force exerted by q on the test charge is

$$\vec{\mathbf{F}}_e = k_e \frac{qq_0}{r^2} \hat{\mathbf{r}}$$

where $\hat{\mathbf{r}}$ is a unit vector directed from q toward q_0. This force in Figure 23.11a is directed away from the source charge q. Because the electric field at P, the position of the test charge, is defined by $\vec{\mathbf{E}} = \vec{\mathbf{F}}_e / q_0$, the electric field at P created by q is

$$\vec{\mathbf{E}} = k_e \frac{q}{r^2} \hat{\mathbf{r}} \tag{23.9}$$

If the source charge q is positive, Figure 23.11b shows the situation with the test charge removed: the source charge sets up an electric field at P, directed away from q. If q is

Pitfall Prevention 23.1

Particles Only Equation 23.8 is valid only for a *particle* of charge q, that is, an object of zero size. For a charged *object* of finite size in an electric field, the field may vary in magnitude and direction over the size of the object, so the corresponding force equation may be more complicated.

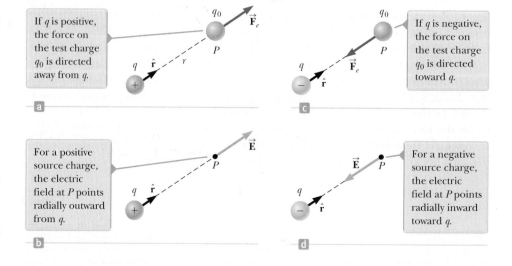

Figure 23.11 (a), (c) When a test charge q_0 is placed near a source charge q, the test charge experiences a force. (b), (d) At a point P near a source charge q, there exists an electric field.

negative as in Figure 23.11c, the force on the test charge is toward the source charge, so the electric field at P is directed toward the source charge as in Figure 23.11d.

To calculate the electric field at a point P due to a small number of point charges, we first calculate the electric field vectors at P individually using Equation 23.9 and then add them vectorially. In other words, at any point P, the total electric field due to a group of source charges equals the vector sum of the electric fields of all the charges. This superposition principle applied to fields follows directly from the vector addition of electric forces. Therefore, the electric field at point P due to a group of source charges can be expressed as the vector sum

$$\vec{\mathbf{E}} = k_e \sum_i \frac{q_i}{r_i^2} \hat{\mathbf{r}}_i \qquad (23.10)$$

◀ **Electric field due to a finite number of point charges**

where r_i is the distance from the ith source charge q_i to the point P and $\hat{\mathbf{r}}_i$ is a unit vector directed from q_i toward P.

In Example 23.6, we explore the electric field due to two charges using the superposition principle. Part (B) of the example focuses on an **electric dipole,** which is defined as a positive charge q and a negative charge $-q$ separated by a distance $2a$. The electric dipole is a good model of many molecules, such as hydrochloric acid (HCl). Neutral atoms and molecules behave as dipoles when placed in an external electric field. Furthermore, many molecules, such as HCl, are permanent dipoles. The effect of such dipoles on the behavior of materials subjected to electric fields is discussed in Chapter 26.

Quick Quiz 23.4 A test charge of $+3 \ \mu C$ is at a point P where an external electric field is directed to the right and has a magnitude of 4×10^6 N/C. If the test charge is replaced with another test charge of $-3 \ \mu C$, what happens to the external electric field at P? **(a)** It is unaffected. **(b)** It reverses direction. **(c)** It changes in a way that cannot be determined.

Analysis Model   **Particle in a Field (Electric)**

Imagine an object with charge that we call a *source charge*. The source charge establishes an **electric field $\vec{\mathbf{E}}$** throughout space. Now imagine a particle with charge q is placed in that field. The particle interacts with the electric field so that the particle experiences an electric force given by

$$\vec{\mathbf{F}}_e = q\vec{\mathbf{E}} \qquad (23.8)$$

Examples:

- an electron moves between the deflection plates of a cathode ray oscilloscope and is deflected from its original path
- charged ions experience an electric force from the electric field in a velocity selector before entering a mass spectrometer (Chapter 29)
- an electron moves around the nucleus in the electric field established by the proton in a hydrogen atom as modeled by the Bohr theory (Chapter 42)
- a hole in a semiconducting material moves in response to the electric field established by applying a voltage to the material (Chapter 43)

Example 23.5   **A Suspended Water Droplet**   AM

A water droplet of mass 3.00×10^{-12} kg is located in the air near the ground during a stormy day. An atmospheric electric field of magnitude 6.00×10^3 N/C points vertically downward in the vicinity of the water droplet. The droplet remains suspended at rest in the air. What is the electric charge on the droplet?

SOLUTION

Conceptualize Imagine the water droplet hovering at rest in the air. This situation is not what is normally observed, so something must be holding the water droplet up.

continued

▶ **23.5** continued

Categorize The droplet can be modeled as a particle and is described by two analysis models associated with fields: the *particle in a field (gravitational)* and the *particle in a field (electric)*. Furthermore, because the droplet is subject to forces but remains at rest, it is also described by the *particle in equilibrium* model.

Analyze

Write Newton's second law from the particle in equilibrium model in the vertical direction:

(1) $\sum F_y = 0 \rightarrow F_e - F_g = 0$

Using the two particle in a field models mentioned in the Categorize step, substitute for the forces in Equation (1), recognizing that the vertical component of the electric field is negative:

$q(-E) - mg = 0$

Solve for the charge on the water droplet:

$q = -\dfrac{mg}{E}$

Substitute numerical values:

$q = -\dfrac{(3.00 \times 10^{-12}\,\text{kg})(9.80\,\text{m/s}^2)}{6.00 \times 10^3\,\text{N/C}} = \boxed{-4.90 \times 10^{-15}\,\text{C}}$

Finalize Noting the smallest unit of free charge in Equation 23.5, the charge on the water droplet is a large number of these units. Notice that the electric *force* is upward to balance the downward gravitational force. The problem statement claims that the electric *field* is in the downward direction. Therefore, the charge found above is negative so that the electric force is in the direction opposite to the electric field.

Example 23.6 **Electric Field Due to Two Charges**

Charges q_1 and q_2 are located on the x axis, at distances a and b, respectively, from the origin as shown in Figure 23.12.

(A) Find the components of the net electric field at the point P, which is at position $(0, y)$.

SOLUTION

Conceptualize Compare this example with Example 23.2. There, we add vector forces to find the net force on a charged particle. Here, we add electric field vectors to find the net electric field at a point in space. If a charged particle were placed at P, we could use the particle in a field model to find the electric force on the particle.

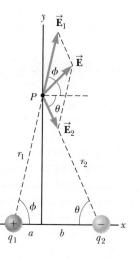

Figure 23.12 (Example 23.6) The total electric field $\vec{\mathbf{E}}$ at P equals the vector sum $\vec{\mathbf{E}}_1 + \vec{\mathbf{E}}_2$, where $\vec{\mathbf{E}}_1$ is the field due to the positive charge q_1 and $\vec{\mathbf{E}}_2$ is the field due to the negative charge q_2.

Categorize We have two source charges and wish to find the resultant electric field, so we categorize this example as one in which we can use the superposition principle represented by Equation 23.10.

Analyze Find the magnitude of the electric field at P due to charge q_1:

$E_1 = k_e\dfrac{|q_1|}{r_1^2} = k_e\dfrac{|q_1|}{a^2 + y^2}$

Find the magnitude of the electric field at P due to charge q_2:

$E_2 = k_e\dfrac{|q_2|}{r_2^2} = k_e\dfrac{|q_2|}{b^2 + y^2}$

Write the electric field vectors for each charge in unit-vector form:

$\vec{\mathbf{E}}_1 = k_e\dfrac{|q_1|}{a^2 + y^2}\cos\phi\,\hat{\mathbf{i}} + k_e\dfrac{|q_1|}{a^2 + y^2}\sin\phi\,\hat{\mathbf{j}}$

$\vec{\mathbf{E}}_2 = k_e\dfrac{|q_2|}{b^2 + y^2}\cos\theta\,\hat{\mathbf{i}} - k_e\dfrac{|q_2|}{b^2 + y^2}\sin\theta\,\hat{\mathbf{j}}$

▶ **23.6** continued

Write the components of the net electric field vector:

(1) $\quad E_x = E_{1x} + E_{2x} = k_e \dfrac{|q_1|}{a^2 + y^2} \cos \phi + k_e \dfrac{|q_2|}{b^2 + y^2} \cos \theta$

(2) $\quad E_y = E_{1y} + E_{2y} = k_e \dfrac{|q_1|}{a^2 + y^2} \sin \phi - k_e \dfrac{|q_2|}{b^2 + y^2} \sin \theta$

(B) Evaluate the electric field at point P in the special case that $|q_1| = |q_2|$ and $a = b$.

SOLUTION

Conceptualize Figure 23.13 shows the situation in this special case. Notice the symmetry in the situation and that the charge distribution is now an electric dipole.

Categorize Because Figure 23.13 is a special case of the general case shown in Figure 23.12, we can categorize this example as one in which we can take the result of part (A) and substitute the appropriate values of the variables.

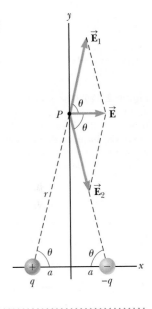

Figure 23.13 (Example 23.6) When the charges in Figure 23.12 are of equal magnitude and equidistant from the origin, the situation becomes symmetric as shown here.

Analyze Based on the symmetry in Figure 23.13, evaluate Equations (1) and (2) from part (A) with $a = b$, $|q_1| = |q_2| = q$, and $\phi = \theta$:

(3) $\quad E_x = k_e \dfrac{q}{a^2 + y^2} \cos \theta + k_e \dfrac{q}{a^2 + y^2} \cos \theta = 2k_e \dfrac{q}{a^2 + y^2} \cos \theta$

$E_y = k_e \dfrac{q}{a^2 + y^2} \sin \theta - k_e \dfrac{q}{a^2 + y^2} \sin \theta = 0$

From the geometry in Figure 23.13, evaluate $\cos \theta$:

(4) $\quad \cos \theta = \dfrac{a}{r} = \dfrac{a}{(a^2 + y^2)^{1/2}}$

Substitute Equation (4) into Equation (3):

$E_x = 2k_e \dfrac{q}{a^2 + y^2} \left[\dfrac{a}{(a^2 + y^2)^{1/2}} \right] = k_e \dfrac{2aq}{(a^2 + y^2)^{3/2}}$

(C) Find the electric field due to the electric dipole when point P is a distance $y \gg a$ from the origin.

SOLUTION

In the solution to part (B), because $y \gg a$, neglect a^2 compared with y^2 and write the expression for E in this case:

(5) $\quad E \approx k_e \dfrac{2aq}{y^3}$

Finalize From Equation (5), we see that at points far from a dipole but along the perpendicular bisector of the line joining the two charges, the magnitude of the electric field created by the dipole varies as $1/r^3$, whereas the more slowly varying field of a point charge varies as $1/r^2$ (see Eq. 23.9). That is because at distant points, the fields of the two charges of equal magnitude and opposite sign almost cancel each other. The $1/r^3$ variation in E for the dipole also is obtained for a distant point along the x axis and for any general distant point.

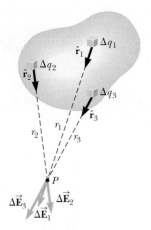

Figure 23.14 The electric field at P due to a continuous charge distribution is the vector sum of the fields $\Delta\vec{\mathbf{E}}_i$ due to all the elements Δq_i of the charge distribution. Three sample elements are shown.

23.5 Electric Field of a Continuous Charge Distribution

Equation 23.10 is useful for calculating the electric field due to a small number of charges. In many cases, we have a continuous distribution of charge rather than a collection of discrete charges. The charge in these situations can be described as continuously distributed along some line, over some surface, or throughout some volume.

To set up the process for evaluating the electric field created by a continuous charge distribution, let's use the following procedure. First, divide the charge distribution into small elements, each of which contains a small charge Δq as shown in Figure 23.14. Next, use Equation 23.9 to calculate the electric field due to one of these elements at a point P. Finally, evaluate the total electric field at P due to the charge distribution by summing the contributions of all the charge elements (that is, by applying the superposition principle).

The electric field at P due to one charge element carrying charge Δq is

$$\Delta \vec{\mathbf{E}} = k_e \frac{\Delta q}{r^2} \hat{\mathbf{r}}$$

where r is the distance from the charge element to point P and $\hat{\mathbf{r}}$ is a unit vector directed from the element toward P. The total electric field at P due to all elements in the charge distribution is approximately

$$\vec{\mathbf{E}} \approx k_e \sum_i \frac{\Delta q_i}{r_i^2} \hat{\mathbf{r}}_i$$

where the index i refers to the ith element in the distribution. Because the number of elements is very large and the charge distribution is modeled as continuous, the total field at P in the limit $\Delta q_i \rightarrow 0$ is

$$\vec{\mathbf{E}} = k_e \lim_{\Delta q_i \rightarrow 0} \sum_i \frac{\Delta q_i}{r_i^2} \hat{\mathbf{r}}_i = k_e \int \frac{dq}{r^2} \hat{\mathbf{r}} \qquad \text{(23.11)}$$

◀ **Electric field due to a continuous charge distribution**

where the integration is over the entire charge distribution. The integration in Equation 23.11 is a vector operation and must be treated appropriately.

Let's illustrate this type of calculation with several examples in which the charge is distributed on a line, on a surface, or throughout a volume. When performing such calculations, it is convenient to use the concept of a *charge density* along with the following notations:

- If a charge Q is uniformly distributed throughout a volume V, the **volume charge density** ρ is defined by

◀ **Volume charge density**

$$\rho \equiv \frac{Q}{V}$$

where ρ has units of coulombs per cubic meter (C/m³).

- If a charge Q is uniformly distributed on a surface of area A, the **surface charge density** σ (Greek letter sigma) is defined by

◀ **Surface charge density**

$$\sigma \equiv \frac{Q}{A}$$

where σ has units of coulombs per square meter (C/m²).

- If a charge Q is uniformly distributed along a line of length ℓ, the **linear charge density** λ is defined by

◀ **Linear charge density**

$$\lambda \equiv \frac{Q}{\ell}$$

where λ has units of coulombs per meter (C/m).

• If the charge is nonuniformly distributed over a volume, surface, or line, the amounts of charge dq in a small volume, surface, or length element are

$$dq = \rho \, dV \qquad dq = \sigma \, dA \qquad dq = \lambda \, d\ell$$

Problem-Solving Strategy   Calculating the Electric Field

The following procedure is recommended for solving problems that involve the determination of an electric field due to individual charges or a charge distribution.

1. Conceptualize. Establish a mental representation of the problem: think carefully about the individual charges or the charge distribution and imagine what type of electric field it would create. Appeal to any symmetry in the arrangement of charges to help you visualize the electric field.

2. Categorize. Are you analyzing a group of individual charges or a continuous charge distribution? The answer to this question tells you how to proceed in the Analyze step.

3. Analyze.

(a) If you are analyzing a group of individual charges, use the superposition principle: when several point charges are present, the resultant field at a point in space is the *vector sum* of the individual fields due to the individual charges (Eq. 23.10). Be very careful in the manipulation of vector quantities. It may be useful to review the material on vector addition in Chapter 3. Example 23.6 demonstrated this procedure.

(b) If you are analyzing a continuous charge distribution, the superposition principle is applied by replacing the vector sums for evaluating the total electric field from individual charges by vector integrals. The charge distribution is divided into infinitesimal pieces, and the vector sum is carried out by integrating over the entire charge distribution (Eq. 23.11). Examples 23.7 through 23.9 demonstrate such procedures.

Consider symmetry when dealing with either a distribution of point charges or a continuous charge distribution. Take advantage of any symmetry in the system you observed in the Conceptualize step to simplify your calculations. The cancellation of field components perpendicular to the axis in Example 23.8 is an example of the application of symmetry.

4. Finalize. Check to see if your electric field expression is consistent with the mental representation and if it reflects any symmetry that you noted previously. Imagine varying parameters such as the distance of the observation point from the charges or the radius of any circular objects to see if the mathematical result changes in a reasonable way.

Example 23.7   The Electric Field Due to a Charged Rod

A rod of length ℓ has a uniform positive charge per unit length λ and a total charge Q. Calculate the electric field at a point P that is located along the long axis of the rod and a distance a from one end (Fig. 23.15).

SOLUTION

Conceptualize The field $d\vec{\mathbf{E}}$ at P due to each segment of charge on the rod is in the negative x direction because every segment carries a positive charge. Figure 23.15 shows the appropriate geometry. In our result, we expect the electric field to become smaller as the distance a becomes larger because point P is farther from the charge distribution.

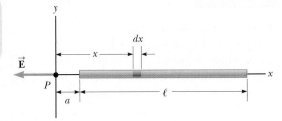

Figure 23.15 (Example 23.7) The electric field at P due to a uniformly charged rod lying along the x axis.

continued

▶ **23.7** continued

Categorize Because the rod is continuous, we are evaluating the field due to a continuous charge distribution rather than a group of individual charges. Because every segment of the rod produces an electric field in the negative x direction, the sum of their contributions can be handled without the need to add vectors.

Analyze Let's assume the rod is lying along the x axis, dx is the length of one small segment, and dq is the charge on that segment. Because the rod has a charge per unit length λ, the charge dq on the small segment is $dq = \lambda\, dx$.

Find the magnitude of the electric field at P due to one segment of the rod having a charge dq:

$$dE = k_e \frac{dq}{x^2} = k_e \frac{\lambda\, dx}{x^2}$$

Find the total field at P using[4] Equation 23.11:

$$E = \int_a^{\ell + a} k_e \lambda \frac{dx}{x^2}$$

Noting that k_e and $\lambda = Q/\ell$ are constants and can be removed from the integral, evaluate the integral:

$$E = k_e \lambda \int_a^{\ell + a} \frac{dx}{x^2} = k_e \lambda \left[-\frac{1}{x} \right]_a^{\ell + a}$$

$$(1) \quad E = k_e \frac{Q}{\ell} \left(\frac{1}{a} - \frac{1}{\ell + a} \right) = \boxed{\frac{k_e Q}{a(\ell + a)}}$$

Finalize We see that our prediction is correct; if a becomes larger, the denominator of the fraction grows larger, and E becomes smaller. On the other hand, if $a \to 0$, which corresponds to sliding the bar to the left until its left end is at the origin, then $E \to \infty$. That represents the condition in which the observation point P is at zero distance from the charge at the end of the rod, so the field becomes infinite. We explore large values of a below.

WHAT IF? Suppose point P is very far away from the rod. What is the nature of the electric field at such a point?

Answer If P is far from the rod ($a \gg \ell$), then ℓ in the denominator of Equation (1) can be neglected and $E \approx k_e Q/a^2$. That is exactly the form you would expect for a point charge. Therefore, at large values of a/ℓ, the charge distribution appears to be a point charge of magnitude Q; the point P is so far away from the rod we cannot distinguish that it has a size. The use of the limiting technique ($a/\ell \to \infty$) is often a good method for checking a mathematical expression.

Example 23.8 **The Electric Field of a Uniform Ring of Charge**

A ring of radius a carries a uniformly distributed positive total charge Q. Calculate the electric field due to the ring at a point P lying a distance x from its center along the central axis perpendicular to the plane of the ring (Fig. 23.16a).

SOLUTION

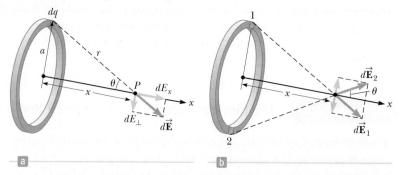

Figure 23.16 (Example 23.8) A uniformly charged ring of radius a. (a) The field at P on the x axis due to an element of charge dq. (b) The total electric field at P is along the x axis. The perpendicular component of the field at P due to segment 1 is canceled by the perpendicular component due to segment 2.

Conceptualize Figure 23.16a shows the electric field contribution $d\vec{\mathbf{E}}$ at P due to a single segment of charge at the top of the ring. This field vector can be resolved into components dE_x parallel to

---

[4]To carry out integrations such as this one, first express the charge element dq in terms of the other variables in the integral. (In this example, there is one variable, x, so we made the change $dq = \lambda\, dx$.) The integral must be over scalar quantities; therefore, express the electric field in terms of components, if necessary. (In this example, the field has only an x component, so this detail is of no concern.) Then, reduce your expression to an integral over a single variable (or to multiple integrals, each over a single variable). In examples that have spherical or cylindrical symmetry, the single variable is a radial coordinate.

▶ **23.8** continued

the axis of the ring and $dE_\perp$ perpendicular to the axis. Figure 23.16b shows the electric field contributions from two segments on opposite sides of the ring. Because of the symmetry of the situation, the perpendicular components of the field cancel. That is true for all pairs of segments around the ring, so we can ignore the perpendicular component of the field and focus solely on the parallel components, which simply add.

Categorize Because the ring is continuous, we are evaluating the field due to a continuous charge distribution rather than a group of individual charges.

Analyze Evaluate the parallel component of an electric field contribution from a segment of charge dq on the ring:

$$(1) \quad dE_x = k_e \frac{dq}{r^2} \cos\theta = k_e \frac{dq}{a^2 + x^2} \cos\theta$$

From the geometry in Figure 23.16a, evaluate $\cos\theta$:

$$(2) \quad \cos\theta = \frac{x}{r} = \frac{x}{(a^2 + x^2)^{1/2}}$$

Substitute Equation (2) into Equation (1):

$$dE_x = k_e \frac{dq}{a^2 + x^2} \left[\frac{x}{(a^2 + x^2)^{1/2}} \right] = \frac{k_e x}{(a^2 + x^2)^{3/2}} dq$$

All segments of the ring make the same contribution to the field at P because they are all equidistant from this point. Integrate over the circumference of the ring to obtain the total field at P:

$$E_x = \int \frac{k_e x}{(a^2 + x^2)^{3/2}} dq = \frac{k_e x}{(a^2 + x^2)^{3/2}} \int dq$$

$$(3) \quad E = \frac{k_e x}{(a^2 + x^2)^{3/2}} Q$$

Finalize This result shows that the field is zero at $x = 0$. Is that consistent with the symmetry in the problem? Furthermore, notice that Equation (3) reduces to $k_e Q / x^2$ if $x \gg a$, so the ring acts like a point charge for locations far away from the ring. From a faraway point, we cannot distinguish the ring shape of the charge.

WHAT IF? Suppose a negative charge is placed at the center of the ring in Figure 23.16 and displaced slightly by a distance $x \ll a$ along the x axis. When the charge is released, what type of motion does it exhibit?

Answer In the expression for the field due to a ring of charge, let $x \ll a$, which results in

$$E_x = \frac{k_e Q}{a^3} x$$

Therefore, from Equation 23.8, the force on a charge $-q$ placed near the center of the ring is

$$F_x = -\frac{k_e q Q}{a^3} x$$

Because this force has the form of Hooke's law (Eq. 15.1), the motion of the negative charge is described with the *particle in simple harmonic motion model!*

Example 23.9 The Electric Field of a Uniformly Charged Disk

A disk of radius R has a uniform surface charge density σ. Calculate the electric field at a point P that lies along the central perpendicular axis of the disk and a distance x from the center of the disk (Fig. 23.17).

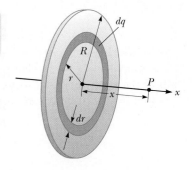

Figure 23.17 (Example 23.9) A uniformly charged disk of radius R. The electric field at an axial point P is directed along the central axis, perpendicular to the plane of the disk.

SOLUTION

Conceptualize If the disk is considered to be a set of concentric rings, we can use our result from Example 23.8—which gives the field created by a single ring of radius a—and sum the contributions of all rings making up the disk. By symmetry, the field at an axial point must be along the central axis.

continued

▶ **23.9** continued

Categorize Because the disk is continuous, we are evaluating the field due to a continuous charge distribution rather than a group of individual charges.

Analyze Find the amount of charge dq on the surface area of a ring of radius r and width dr as shown in Figure 23.17:

$$dq = \sigma\, dA = \sigma(2\pi r\, dr) = 2\pi\sigma r\, dr$$

Use this result in the equation given for E_x in Example 23.8 (with a replaced by r and Q replaced by dq) to find the field due to the ring:

$$dE_x = \frac{k_e x}{(r^2 + x^2)^{3/2}}(2\pi\sigma r\, dr)$$

To obtain the total field at P, integrate this expression over the limits $r = 0$ to $r = R$, noting that x is a constant in this situation:

$$E_x = k_e x\pi\sigma \int_0^R \frac{2r\, dr}{(r^2 + x^2)^{3/2}}$$

$$= k_e x\pi\sigma \int_0^R (r^2 + x^2)^{-3/2} d(r^2)$$

$$= k_e x\pi\sigma \left[\frac{(r^2 + x^2)^{-1/2}}{-1/2}\right]_0^R = 2\pi k_e \sigma \left[1 - \frac{x}{(R^2 + x^2)^{1/2}}\right]$$

Finalize This result is valid for all values of $x > 0$. For large values of x, the result above can be evaluated by a series expansion and shown to be equivalent to the electric field of a point charge Q. We can calculate the field close to the disk along the axis by assuming $x \ll R$; in this case, the expression in brackets reduces to unity to give us the near-field approximation

$$E = 2\pi k_e \sigma = \frac{\sigma}{2\epsilon_0}$$

where ϵ_0 is the permittivity of free space. In Chapter 24, we obtain the same result for the field created by an infinite plane of charge with uniform surface charge density.

WHAT IF? What if we let the radius of the disk grow so that the disk becomes an infinite plane of charge?

Answer The result of letting $R \to \infty$ in the final result of the example is that the magnitude of the electric field becomes

$$E = 2\pi k_e \sigma = \frac{\sigma}{2\epsilon_0}$$

This is the same expression that we obtained for $x \ll R$. If $R \to \infty$, *everywhere* is near-field—the result is independent of the position at which you measure the electric field. Therefore, the electric field due to an infinite plane of charge is uniform throughout space.

An infinite plane of charge is impossible in practice. If two planes of charge are placed close to each other, however, with one plane positively charged, and the other negatively, the electric field between the plates is very close to uniform at points far from the edges. Such a configuration will be investigated in Chapter 26.

23.6 Electric Field Lines

We have defined the electric field in the mathematical representation with Equation 23.7. Let's now explore a means of visualizing the electric field in a pictorial representation. A convenient way of visualizing electric field patterns is to draw lines, called **electric field lines** and first introduced by Faraday, that are related to the electric field in a region of space in the following manner:

- The electric field vector $\vec{\mathbf{E}}$ is tangent to the electric field line at each point. The line has a direction, indicated by an arrowhead, that is the same as that of the electric field vector. The direction of the line is that of the force on a positive charge placed in the field according to the particle in a field model.

- The number of lines per unit area through a surface perpendicular to the lines is proportional to the magnitude of the electric field in that region. Therefore, the field lines are close together where the electric field is strong and far apart where the field is weak.

These properties are illustrated in Figure 23.18. The density of field lines through surface A is greater than the density of lines through surface B. Therefore, the magnitude of the electric field is larger on surface A than on surface B. Furthermore, because the lines at different locations point in different directions, the field is nonuniform.

Is this relationship between strength of the electric field and the density of field lines consistent with Equation 23.9, the expression we obtained for E using Coulomb's law? To answer this question, consider an imaginary spherical surface of radius r concentric with a point charge. From symmetry, we see that the magnitude of the electric field is the same everywhere on the surface of the sphere. The number of lines N that emerge from the charge is equal to the number that penetrate the spherical surface. Hence, the number of lines per unit area on the sphere is $N/4\pi r^2$ (where the surface area of the sphere is $4\pi r^2$). Because E is proportional to the number of lines per unit area, we see that E varies as $1/r^2$; this finding is consistent with Equation 23.9.

Representative electric field lines for the field due to a single positive point charge are shown in Figure 23.19a. This two-dimensional drawing shows only the field lines that lie in the plane containing the point charge. The lines are actually directed radially outward from the charge in all directions; therefore, instead of the flat "wheel" of lines shown, you should picture an entire spherical distribution of lines. Because a positive charge placed in this field would be repelled by the positive source charge, the lines are directed radially away from the source charge. The electric field lines representing the field due to a single negative point charge are directed toward the charge (Fig. 23.19b). In either case, the lines are along the radial direction and extend all the way to infinity. Notice that the lines become closer together as they approach the charge, indicating that the strength of the field increases as we move toward the source charge.

The rules for drawing electric field lines are as follows:

- The lines must begin on a positive charge and terminate on a negative charge. In the case of an excess of one type of charge, some lines will begin or end infinitely far away.

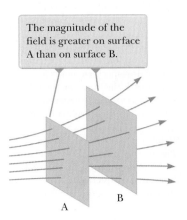

The magnitude of the field is greater on surface A than on surface B.

Figure 23.18 Electric field lines penetrating two surfaces.

Pitfall Prevention 23.2
Electric Field Lines Are Not Paths of Particles! Electric field lines represent the field at various locations. Except in very special cases, they *do not* represent the path of a charged particle moving in an electric field.

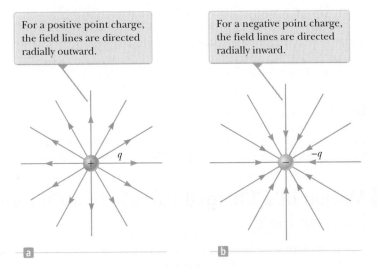

For a positive point charge, the field lines are directed radially outward.

For a negative point charge, the field lines are directed radially inward.

q

$-q$

a

b

Figure 23.19 The electric field lines for a point charge. Notice that the figures show only those field lines that lie in the plane of the page.

Pitfall Prevention 23.3
Electric Field Lines Are Not Real
Electric field lines are not material objects. They are used only as a pictorial representation to provide a qualitative description of the electric field. Only a finite number of lines from each charge can be drawn, which makes it appear as if the field were quantized and exists only in certain parts of space. The field, in fact, is continuous, existing at every point. You should avoid obtaining the wrong impression from a two-dimensional drawing of field lines used to describe a three-dimensional situation.

- The number of lines drawn leaving a positive charge or approaching a negative charge is proportional to the magnitude of the charge.
- No two field lines can cross.

We choose the number of field lines starting from any object with a positive charge q_+ to be Cq_+ and the number of lines ending on any object with a negative charge q_- to be $C|q_-|$, where C is an arbitrary proportionality constant. Once C is chosen, the number of lines is fixed. For example, in a two-charge system, if object 1 has charge Q_1 and object 2 has charge Q_2, the ratio of number of lines in contact with the charges is $N_2/N_1 = |Q_2/Q_1|$. The electric field lines for two point charges of equal magnitude but opposite signs (an electric dipole) are shown in Figure 23.20. Because the charges are of equal magnitude, the number of lines that begin at the positive charge must equal the number that terminate at the negative charge. At points very near the charges, the lines are nearly radial, as for a single isolated charge. The high density of lines between the charges indicates a region of strong electric field.

Figure 23.21 shows the electric field lines in the vicinity of two equal positive point charges. Again, the lines are nearly radial at points close to either charge, and the same number of lines emerges from each charge because the charges are equal in magnitude. Because there are no negative charges available, the electric field lines end infinitely far away. At great distances from the charges, the field is approximately equal to that of a single point charge of magnitude $2q$.

Finally, in Figure 23.22, we sketch the electric field lines associated with a positive charge $+2q$ and a negative charge $-q$. In this case, the number of lines leaving $+2q$ is twice the number terminating at $-q$. Hence, only half the lines that leave the positive charge reach the negative charge. The remaining half terminate on a negative charge we assume to be at infinity. At distances much greater than the charge separation, the electric field lines are equivalent to those of a single charge $+q$.

Quick Quiz 23.5 Rank the magnitudes of the electric field at points A, B, and C shown in Figure 23.21 (greatest magnitude first).

The number of field lines leaving the positive charge equals the number terminating at the negative charge.

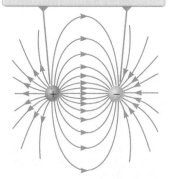

Figure 23.20 The electric field lines for two point charges of equal magnitude and opposite sign (an electric dipole).

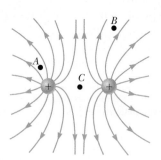

Figure 23.21 The electric field lines for two positive point charges. (The locations A, B, and C are discussed in Quick Quiz 23.5.)

Two field lines leave $+2q$ for every one that terminates on $-q$.

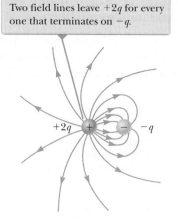

Figure 23.22 The electric field lines for a point charge $+2q$ and a second point charge $-q$.

23.7 Motion of a Charged Particle in a Uniform Electric Field

When a particle of charge q and mass m is placed in an electric field $\vec{\mathbf{E}}$, the electric force exerted on the charge is $q\vec{\mathbf{E}}$ according to Equation 23.8 in the particle in a field model. If that is the only force exerted on the particle, it must be the net force, and it causes the particle to accelerate according to the particle under a net force model. Therefore,

$$\vec{F}_e = q\vec{E} = m\vec{a}$$

and the acceleration of the particle is

$$\vec{a} = \frac{q\vec{E}}{m} \tag{23.12}$$

If $\vec{E}$ is uniform (that is, constant in magnitude and direction), and the particle is free to move, the electric force on the particle is constant and we can apply the particle under constant acceleration model to the motion of the particle. Therefore, the particle in this situation is described by *three* analysis models: particle in a field, particle under a net force, and particle under constant acceleration! If the particle has a positive charge, its acceleration is in the direction of the electric field. If the particle has a negative charge, its acceleration is in the direction opposite the electric field.

> **Pitfall Prevention 23.4**
> **Just Another Force** Electric forces and fields may seem abstract to you. Once $\vec{F}_e$ is evaluated, however, it causes a particle to move according to our well-established models of forces and motion from Chapters 2 through 6. Keeping this link with the past in mind should help you solve problems in this chapter.

Example 23.10   An Accelerating Positive Charge: Two Models   AM

A uniform electric field $\vec{E}$ is directed along the x axis between parallel plates of charge separated by a distance d as shown in Figure 23.23. A positive point charge q of mass m is released from rest at a point Ⓐ next to the positive plate and accelerates to a point Ⓑ next to the negative plate.

(A) Find the speed of the particle at Ⓑ by modeling it as a particle under constant acceleration.

SOLUTION

Conceptualize When the positive charge is placed at Ⓐ, it experiences an electric force toward the right in Figure 23.23 due to the electric field directed toward the right. As a result, it will accelerate to the right and arrive at Ⓑ with some speed.

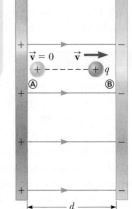

Figure 23.23 (Example 23.10) A positive point charge q in a uniform electric field $\vec{E}$ undergoes constant acceleration in the direction of the field.

Categorize Because the electric field is uniform, a constant electric force acts on the charge. Therefore, as suggested in the discussion preceding the example and in the problem statement, the point charge can be modeled as a charged *particle under constant acceleration.*

Analyze Use Equation 2.17 to express the velocity of the particle as a function of position:

$$v_f^2 = v_i^2 + 2a(x_f - x_i) = 0 + 2a(d - 0) = 2ad$$

Solve for v_f and substitute for the magnitude of the acceleration from Equation 23.12:

$$v_f = \sqrt{2ad} = \sqrt{2\left(\frac{qE}{m}\right)d} = \boxed{\sqrt{\frac{2qEd}{m}}}$$

(B) Find the speed of the particle at Ⓑ by modeling it as a nonisolated system in terms of energy.

SOLUTION

Categorize The problem statement tells us that the charge is a *nonisolated system* for *energy*. The electric force, like any force, can do work on a system. Energy is transferred to the system of the charge by work done by the electric force exerted on the charge. The initial configuration of the system is when the particle is at rest at Ⓐ, and the final configuration is when it is moving with some speed at Ⓑ.

continued

▶ **23.10** continued

Analyze Write the appropriate reduction of the conservation of energy equation, Equation 8.2, for the system of the charged particle:

$$W = \Delta K$$

Replace the work and kinetic energies with values appropriate for this situation:

$$F_e \Delta x = K_{\circledB} - K_{\circledA} = \tfrac{1}{2}mv_f^2 - 0 \;\rightarrow\; v_f = \sqrt{\frac{2F_e \Delta x}{m}}$$

Substitute for the magnitude of the electric force F_e from the particle in a field model and the displacement Δx:

$$v_f = \sqrt{\frac{2(qE)(d)}{m}} = \boxed{\sqrt{\frac{2qEd}{m}}}$$

Finalize The answer to part (B) is the same as that for part (A), as we expect. This problem can be solved with different approaches. We saw the same possibilities with mechanical problems.

Example 23.11 An Accelerated Electron

An electron enters the region of a uniform electric field as shown in Figure 23.24, with $v_i = 3.00 \times 10^6$ m/s and $E = 200$ N/C. The horizontal length of the plates is $\ell = 0.100$ m.

(A) Find the acceleration of the electron while it is in the electric field.

SOLUTION

Conceptualize This example differs from the preceding one because the velocity of the charged particle is initially perpendicular to the electric field lines. (In Example 23.10, the velocity of the charged particle is always parallel to the electric field lines.) As a result, the electron in this example follows a curved path as shown in Figure 23.24. The motion of the electron is the same as that of a massive particle projected horizontally in a gravitational field near the surface of the Earth.

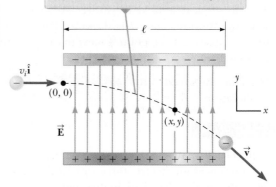

The electron undergoes a downward acceleration (opposite $\vec{\mathbf{E}}$), and its motion is parabolic while it is between the plates.

Figure 23.24 (Example 23.11) An electron is projected horizontally into a uniform electric field produced by two charged plates.

Categorize The electron is a *particle in a field (electric)*. Because the electric field is uniform, a constant electric force is exerted on the electron. To find the acceleration of the electron, we can model it as a *particle under a net force*.

Analyze From the particle in a field model, we know that the direction of the electric force on the electron is downward in Figure 23.24, opposite the direction of the electric field lines. From the particle under a net force model, therefore, the acceleration of the electron is downward.

The particle under a net force model was used to develop Equation 23.12 in the case in which the electric force on a particle is the only force. Use this equation to evaluate the y component of the acceleration of the electron:

$$a_y = -\frac{eE}{m_e}$$

Substitute numerical values:

$$a_y = -\frac{(1.60 \times 10^{-19}\,\text{C})(200\,\text{N/C})}{9.11 \times 10^{-31}\,\text{kg}} = \boxed{-3.51 \times 10^{13}\,\text{m/s}^2}$$

(B) Assuming the electron enters the field at time $t = 0$, find the time at which it leaves the field.

SOLUTION

Categorize Because the electric force acts only in the vertical direction in Figure 23.24, the motion of the particle in the horizontal direction can be analyzed by modeling it as a *particle under constant velocity*.

▶ **23.11** continued

Analyze Solve Equation 2.7 for the time at which the electron arrives at the right edges of the plates:

$$x_f = x_i + v_x t \rightarrow t = \frac{x_f - x_i}{v_x}$$

Substitute numerical values:

$$t = \frac{\ell - 0}{v_x} = \frac{0.100 \text{ m}}{3.00 \times 10^6 \text{ m/s}} = \boxed{3.33 \times 10^{-8} \text{ s}}$$

(C) Assuming the vertical position of the electron as it enters the field is $y_i = 0$, what is its vertical position when it leaves the field?

SOLUTION

Categorize Because the electric force is constant in Figure 23.24, the motion of the particle in the vertical direction can be analyzed by modeling it as a *particle under constant acceleration.*

Analyze Use Equation 2.16 to describe the position of the particle at any time t:

$$y_f = y_i + v_{yi} t + \tfrac{1}{2} a_y t^2$$

Substitute numerical values:

$$y_f = 0 + 0 + \tfrac{1}{2}(-3.51 \times 10^{13} \text{ m/s}^2)(3.33 \times 10^{-8} \text{ s})^2$$

$$= -0.019\,5 \text{ m} = \boxed{-1.95 \text{ cm}}$$

Finalize If the electron enters just below the negative plate in Figure 23.24 and the separation between the plates is less than the value just calculated, the electron will strike the positive plate.

Notice that we have used *four* analysis models to describe the electron in the various parts of this problem. We have neglected the gravitational force acting on the electron, which represents a good approximation when dealing with atomic particles. For an electric field of 200 N/C, the ratio of the magnitude of the electric force eE to the magnitude of the gravitational force mg is on the order of 10^{12} for an electron and on the order of 10^9 for a proton.

Summary

Definitions

The **electric field** $\vec{\mathbf{E}}$ at some point in space is defined as the electric force $\vec{\mathbf{F}}_e$ that acts on a small positive test charge placed at that point divided by the magnitude q_0 of the test charge:

$$\vec{\mathbf{E}} \equiv \frac{\vec{\mathbf{F}}_e}{q_0} \qquad (23.7)$$

Concepts and Principles

Electric charges have the following important properties:

- Charges of opposite sign attract one another, and charges of the same sign repel one another.
- The total charge in an isolated system is conserved.
- Charge is quantized.

Conductors are materials in which electrons move freely. **Insulators** are materials in which electrons do not move freely.

continued

Coulomb's law states that the electric force exerted by a point charge q_1 on a second point charge q_2 is

$$\vec{\mathbf{F}}_{12} = k_e \frac{q_1 q_2}{r^2} \hat{\mathbf{r}}_{12} \qquad \textbf{(23.6)}$$

where r is the distance between the two charges and $\hat{\mathbf{r}}_{12}$ is a unit vector directed from q_1 toward q_2. The constant k_e, which is called the **Coulomb constant,** has the value $k_e = 8.988 \times 10^9$ N · m²/C².

The electric field due to a group of point charges can be obtained by using the superposition principle. That is, the total electric field at some point equals the vector sum of the electric fields of all the charges:

$$\vec{\mathbf{E}} = k_e \sum_i \frac{q_i}{r_i^2} \hat{\mathbf{r}}_i \qquad \textbf{(23.10)}$$

At a distance r from a point charge q, the electric field due to the charge is

$$\vec{\mathbf{E}} = k_e \frac{q}{r^2} \hat{\mathbf{r}} \qquad \textbf{(23.9)}$$

where $\hat{\mathbf{r}}$ is a unit vector directed from the charge toward the point in question. The electric field is directed radially outward from a positive charge and radially inward toward a negative charge.

The electric field at some point due to a continuous charge distribution is

$$\vec{\mathbf{E}} = k_e \int \frac{dq}{r^2} \hat{\mathbf{r}} \qquad \textbf{(23.11)}$$

where dq is the charge on one element of the charge distribution and r is the distance from the element to the point in question.

Analysis Models for Problem Solving

Particle in a Field (Electric) A source particle with some electric charge establishes an **electric field $\vec{\mathbf{E}}$** throughout space. When a particle with charge q is placed in that field, it experiences an electric force given by

$$\vec{\mathbf{F}}_e = q\vec{\mathbf{E}} \qquad \textbf{(23.8)}$$

Objective Questions

1. denotes answer available in *Student Solutions Manual/Study Guide*

1. A free electron and a free proton are released in identical electric fields. **(i)** How do the magnitudes of the electric force exerted on the two particles compare? (a) It is millions of times greater for the electron. (b) It is thousands of times greater for the electron. (c) They are equal. (d) It is thousands of times smaller for the electron. (e) It is millions of times smaller for the electron. **(ii)** Compare the magnitudes of their accelerations. Choose from the same possibilities as in part (i).

2. What prevents gravity from pulling you through the ground to the center of the Earth? Choose the best answer. (a) The density of matter is too great. (b) The positive nuclei of your body's atoms repel the positive nuclei of the atoms of the ground. (c) The density of the ground is greater than the density of your body. (d) Atoms are bound together by chemical bonds. (e) Electrons on the ground's surface and the surface of your feet repel one another.

3. A very small ball has a mass of 5.00×10^{-3} kg and a charge of 4.00 μC. What magnitude electric field directed upward will balance the weight of the ball so that the ball is suspended motionless above the ground? (a) 8.21×10^2 N/C (b) 1.22×10^4 N/C (c) 2.00×10^{-2} N/C (d) 5.11×10^6 N/C (e) 3.72×10^3 N/C

4. An electron with a speed of 3.00×10^6 m/s moves into a uniform electric field of magnitude 1.00×10^3 N/C. The field lines are parallel to the electron's velocity and pointing in the same direction as the velocity. How far does the electron travel before it is brought to rest? (a) 2.56 cm (b) 5.12 cm (c) 11.2 cm (d) 3.34 m (e) 4.24 m

5. A point charge of −4.00 nC is located at (0, 1.00) m. What is the x component of the electric field due to the point charge at (4.00, −2.00) m? (a) 1.15 N/C (b) −0.864 N/C (c) 1.44 N/C (d) −1.15 N/C (e) 0.864 N/C

6. A circular ring of charge with radius b has total charge q uniformly distributed around it. What is the magnitude of the electric field at the center of the ring? (a) 0 (b) $k_e q/b^2$ (c) $k_e q^2/b^2$ (d) $k_e q^2/b$ (e) none of those answers

7. What happens when a charged insulator is placed near an uncharged metallic object? (a) They repel each other. (b) They attract each other. (c) They may attract or repel each other, depending on whether the charge on the insulator is positive or negative. (d) They exert no electrostatic force on each other. (e) The charged insulator always spontaneously discharges.

8. Estimate the magnitude of the electric field due to the proton in a hydrogen atom at a distance of 5.29×10^{-11} m, the expected position of the electron in the atom. (a) 10^{-11} N/C (b) 10^8 N/C (c) 10^{14} N/C (d) 10^6 N/C (e) 10^{12} N/C

9. **(i)** A metallic coin is given a positive electric charge. Does its mass (a) increase measurably, (b) increase by an amount too small to measure directly, (c) remain unchanged, (d) decrease by an amount too small to measure directly, or (e) decrease measurably? **(ii)** Now the coin is given a negative electric charge. What happens to its mass? Choose from the same possibilities as in part (i).

10. Assume the charged objects in Figure OQ23.10 are fixed. Notice that there is no sight line from the location of q_2 to the location of q_1. If you were at q_1, you would be unable to see q_2 because it is behind q_3. How would you calculate the electric force exerted on the object with charge q_1? (a) Find only the force exerted by q_2 on charge q_1. (b) Find only the force exerted by q_3 on charge q_1. (c) Add the force that q_2 would exert by itself on charge q_1 to the force that q_3 would exert by itself on charge q_1. (d) Add the force that q_3 would exert by itself to a certain fraction of the force that q_2 would exert by itself. (e) There is no definite way to find the force on charge q_1.

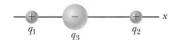

Figure OQ23.10

11. Three charged particles are arranged on corners of a square as shown in Figure OQ23.11, with charge $-Q$ on both the particle at the upper left corner and the particle at the lower right corner and with charge $+2Q$ on the particle at the lower left corner. **(i)** What is the direction of the electric field at the upper right corner, which is a point in empty space? (a) It is upward and to the right. (b) It is straight to the right. (c) It is straight downward. (d) It is downward and to the left. (e) It is perpendicular to the plane of the picture and outward. **(ii)** Suppose the $+2Q$ charge at the lower left corner is removed. Then does the magnitude of the field at the upper right corner (a) become larger, (b) become smaller, (c) stay the same, or (d) change unpredictably?

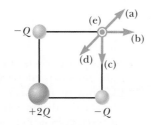

Figure OQ23.11

12. Two point charges attract each other with an electric force of magnitude F. If the charge on one of the particles is reduced to one-third its original value and the distance between the particles is doubled, what is the resulting magnitude of the electric force between them? (a) $\frac{1}{12}F$ (b) $\frac{1}{3}F$ (c) $\frac{1}{6}F$ (d) $\frac{3}{4}F$ (e) $\frac{3}{2}F$

13. Assume a uniformly charged ring of radius R and charge Q produces an electric field E_{ring} at a point P on its axis, at distance x away from the center of the ring as in Figure OQ23.13a. Now the same charge Q is spread uniformly over the circular area the ring encloses, forming a flat disk of charge with the same radius as in Figure OQ23.13b. How does the field E_{disk} produced by the disk at P compare with the field produced by the ring at the same point? (a) $E_{disk} < E_{ring}$ (b) $E_{disk} = E_{ring}$ (c) $E_{disk} > E_{ring}$ (d) impossible to determine

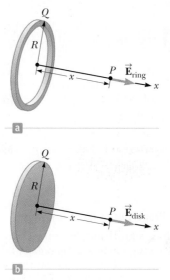

Figure OQ23.13

14. An object with negative charge is placed in a region of space where the electric field is directed vertically upward. What is the direction of the electric force exerted on this charge? (a) It is up. (b) It is down. (c) There is no force. (d) The force can be in any direction.

15. The magnitude of the electric force between two protons is 2.30×10^{-26} N. How far apart are they? (a) 0.100 m (b) 0.022 0 m (c) 3.10 m (d) 0.005 70 m (e) 0.480 m

---

Conceptual Questions   **1.** denotes answer available in *Student Solutions Manual/Study Guide*

1. (a) Would life be different if the electron were positively charged and the proton were negatively charged? (b) Does the choice of signs have any bearing on physical and chemical interactions? Explain your answers.

2. A charged comb often attracts small bits of dry paper that then fly away when they touch the comb. Explain why that occurs.

3. A person is placed in a large, hollow, metallic sphere that is insulated from ground. If a large charge is placed on the sphere, will the person be harmed upon touching the inside of the sphere?

4. A student who grew up in a tropical country and is studying in the United States may have no experience with static electricity sparks and shocks until his or her first American winter. Explain.

5. If a suspended object A is attracted to a charged object B, can we conclude that A is charged? Explain.

6. Consider point A in Figure CQ23.6 located an arbitrary distance from two positive point charges in otherwise empty space. (a) Is it possible for an electric field to exist at point A in empty space? Explain. (b) Does charge exist at this point? Explain. (c) Does a force exist at this point? Explain.

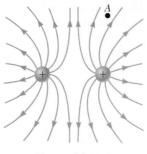

Figure CQ23.6

7. In fair weather, there is an electric field at the surface of the Earth, pointing down into the ground. What is the sign of the electric charge on the ground in this situation?

8. Why must hospital personnel wear special conducting shoes while working around oxygen in an operating room? What might happen if the personnel wore shoes with rubber soles?

9. A balloon clings to a wall after it is negatively charged by rubbing. (a) Does that occur because the wall is positively charged? (b) Why does the balloon eventually fall?

10. Consider two electric dipoles in empty space. Each dipole has zero net charge. (a) Does an electric force exist between the dipoles; that is, can two objects with zero net charge exert electric forces on each other? (b) If so, is the force one of attraction or of repulsion?

11. A glass object receives a positive charge by rubbing it with a silk cloth. In the rubbing process, have protons been added to the object or have electrons been removed from it?

Problems

ENHANCED **WebAssign** The problems found in this chapter may be assigned online in Enhanced WebAssign

AMT Analysis Model tutorial available in Enhanced WebAssign

GP Guided Problem

1. straightforward; 2. intermediate; 3. challenging

M Master It tutorial available in Enhanced WebAssign

1. full solution available in the *Student Solutions Manual/Study Guide*

W Watch It video solution available in Enhanced WebAssign

Section 23.1 Properties of Electric Charges

1. Find to three significant digits the charge and the mass of the following particles. *Suggestion:* Begin by looking up the mass of a neutral atom on the periodic table of the elements in Appendix C. (a) an ionized hydrogen atom, represented as H^+ (b) a singly ionized sodium atom, Na^+ (c) a chloride ion Cl^- (d) a doubly ionized calcium atom, $Ca^{++} = Ca^{2+}$ (e) the center of an ammonia molecule, modeled as an N^{3-} ion (f) quadruply ionized nitrogen atoms, N^{4+}, found in plasma in a hot star (g) the nucleus of a nitrogen atom (h) the molecular ion H_2O^-

2. (a) Calculate the number of electrons in a small, electrically neutral silver pin that has a mass of 10.0 g. Silver has 47 electrons per atom, and its molar mass is 107.87 g/mol. (b) Imagine adding electrons to the pin until the negative charge has the very large value 1.00 mC. How many electrons are added for every 10^9 electrons already present?

Section 23.2 Charging Objects by Induction
Section 23.3 Coulomb's Law

3. Two protons in an atomic nucleus are typically separated by a distance of 2×10^{-15} m. The electric repulsive force between the protons is huge, but the attractive nuclear force is even stronger and keeps the nucleus from bursting apart. What is the magnitude of the electric force between two protons separated by 2.00×10^{-15} m?

4. A charged particle A exerts a force of 2.62 μN to the right on charged particle B when the particles are 13.7 mm apart. Particle B moves straight away from A to make the distance between them 17.7 mm. What vector force does it then exert on A?

5. In a thundercloud, there may be electric charges of $+40.0$ C near the top of the cloud and -40.0 C near the bottom of the cloud. These charges are separated by 2.00 km. What is the electric force on the top charge?

6. (a) Find the magnitude of the electric force between a Na^+ ion and a Cl^- ion separated by 0.50 nm. (b) Would the answer change if the sodium ion were replaced by Li^+ and the chloride ion by Br^-? Explain.

7. **Review.** A molecule of DNA (deoxyribonucleic acid) is 2.17 μm long. The ends of the molecule become singly ionized: negative on one end, positive on the other. The helical molecule acts like a spring and compresses 1.00% upon becoming charged. Determine the effective spring constant of the molecule.

8. Nobel laureate Richard Feynman (1918–1988) once said that if two persons stood at arm's length from each other and each person had 1% more electrons than protons, the force of repulsion between them would be enough to lift a "weight" equal to that of the entire Earth. Carry out an order-of-magnitude calculation to substantiate this assertion.

9. A 7.50-nC point charge is located 1.80 m from a 4.20-nC point charge. (a) Find the magnitude of the

electric force that one particle exerts on the other. (b) Is the force attractive or repulsive?

10. (a) Two protons in a molecule are 3.80×10^{-10} m [W] apart. Find the magnitude of the electric force exerted by one proton on the other. (b) State how the magnitude of this force compares with the magnitude of the gravitational force exerted by one proton on the other. (c) **What If?** What must be a particle's charge-to-mass ratio if the magnitude of the gravitational force between two of these particles is equal to the magnitude of electric force between them?

11. Three point charges are arranged as shown in Figure [M] P23.11. Find (a) the magnitude and (b) the direction of the electric force on the particle at the origin.

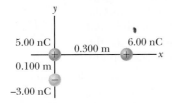

Figure P23.11 Problems 11 and 35.

12. Three point charges lie along a straight line as shown in Figure P23.12, where $q_1 = 6.00$ μC, $q_2 = 1.50$ μC, and $q_3 = -2.00$ μC. The separation distances are $d_1 = 3.00$ cm and $d_2 = 2.00$ cm. Calculate the magnitude and direction of the net electric force on (a) q_1, (b) q_2, and (c) q_3.

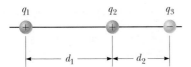

Figure P23.12

13. Two small beads having positive charges $q_1 = 3q$ and [W] $q_2 = q$ are fixed at the opposite ends of a horizontal insulating rod of length $d = 1.50$ m. The bead with charge q_1 is at the origin. As shown in Figure P23.13, a third small, charged bead is free to slide on the rod. (a) At what position x is the third bead in equilibrium? (b) Can the equilibrium be stable?

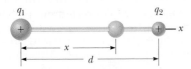

Figure P23.13 Problems 13 and 14.

14. Two small beads having charges q_1 and q_2 of the same sign are fixed at the opposite ends of a horizontal insulating rod of length d. The bead with charge q_1 is at the origin. As shown in Figure P23.13, a third small, charged bead is free to slide on the rod. (a) At what position x is the third bead in equilibrium? (b) Can the equilibrium be stable?

15. Three charged particles are located at the corners of [M] an equilateral triangle as shown in Figure P23.15. Calculate the total electric force on the 7.00-μC charge.

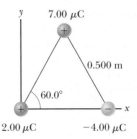

Figure P23.15 Problems 15 and 30.

16. Two small metallic spheres, each of mass $m = 0.200$ g, are suspended as pendulums by light strings of length L as shown in Figure P23.16. The spheres are given the same electric charge of 7.2 nC, and they come to equilibrium when each string is at an angle of $\theta = 5.00°$ with the vertical. How long are the strings?

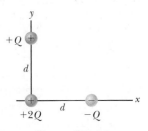

Figure P23.16

17. Review. In the Bohr theory of the hydrogen atom, an electron moves in a circular orbit about a proton, where the radius of the orbit is 5.29×10^{-11} m. (a) Find the magnitude of the electric force exerted on each particle. (b) If this force causes the centripetal acceleration of the electron, what is the speed of the electron?

18. Particle A of charge 3.00×10^{-4} C is at the origin, par-[GP] ticle B of charge -6.00×10^{-4} C is at (4.00 m, 0), and particle C of charge 1.00×10^{-4} C is at (0, 3.00 m). We wish to find the net electric force on C. (a) What is the x component of the electric force exerted by A on C? (b) What is the y component of the force exerted by A on C? (c) Find the magnitude of the force exerted by B on C. (d) Calculate the x component of the force exerted by B on C. (e) Calculate the y component of the force exerted by B on C. (f) Sum the two x components from parts (a) and (d) to obtain the resultant x component of the electric force acting on C. (g) Similarly, find the y component of the resultant force vector acting on C. (h) Find the magnitude and direction of the resultant electric force acting on C.

19. A point charge $+2Q$ is at the origin and a point charge $-Q$ is located along the x axis at $x = d$ as in Figure P23.19. Find a symbolic expression for the net force on a third point charge $+Q$ located along the y axis at $y = d$.

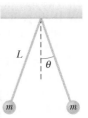

Figure P23.19

20. Review. Two identical particles, each having charge $+q$, are fixed in space and separated by a distance d. A third particle with charge $-Q$ is free to move and lies initially at rest on the perpendicular bisector of the two fixed charges a distance x from the midpoint between those charges (Fig. P23.20). (a) Show that if x is small compared with d, the motion of $-Q$ is simple harmonic along the perpendicular bisector. (b) Determine the period of that motion. (c) How fast will the charge $-Q$ be moving when it is at the midpoint between the two fixed charges if initially it is released at a distance $a \ll d$ from the midpoint?

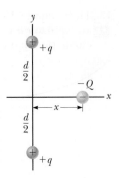

Figure P23.20

21. Two identical conducting small spheres are placed with their centers 0.300 m apart. One is given a charge of 12.0 nC and the other a charge of -18.0 nC. (a) Find the electric force exerted by one sphere on the other. (b) **What If?** The spheres are connected by a conducting wire. Find the electric force each exerts on the other after they have come to equilibrium.

22. *Why is the following situation impossible?* Two identical dust particles of mass 1.00 μg are floating in empty space, far from any external sources of large gravitational or electric fields, and at rest with respect to each other. Both particles carry electric charges that are identical in magnitude and sign. The gravitational and electric forces between the particles happen to have the same magnitude, so each particle experiences zero net force and the distance between the particles remains constant.

Section 23.4 Analysis Model: Particle in a Field (Electric)

23. What are the magnitude and direction of the electric field that will balance the weight of (a) an electron and (b) a proton? (You may use the data in Table 23.1.)

24. A small object of mass 3.80 g and charge -18.0 μC is suspended motionless above the ground when immersed in a uniform electric field perpendicular to the ground. What are the magnitude and direction of the electric field?

25. Four charged particles are at the corners of a square of side a as shown in Figure P23.25. Determine (a) the electric field at the location of charge q and (b) the total electric force exerted on q.

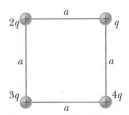

Figure P23.25

26. Three point charges lie along a circle of radius r at angles of 30°, 150°, and 270° as shown in Figure P23.26. Find a symbolic expression for the resultant electric field at the center of the circle.

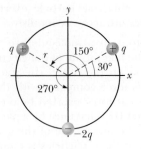

Figure P23.26

27. Two equal positively charged particles are at opposite corners of a trapezoid as shown in Figure P23.27. Find symbolic expressions for the total electric field at (a) the point P and (b) the point P'.

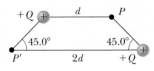

Figure P23.27

28. Consider n equal positively charged particles each of magnitude Q/n placed symmetrically around a circle of radius a. (a) Calculate the magnitude of the electric field at a point a distance x from the center of the circle and on the line passing through the center and perpendicular to the plane of the circle. (b) Explain why this result is identical to the result of the calculation done in Example 23.8.

29. In Figure P23.29, determine the point (other than infinity) at which the electric field is zero.

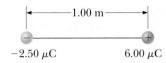

Figure P23.29

30. Three charged particles are at the corners of an equilateral triangle as shown in Figure P23.15. (a) Calculate the electric field at the position of the 2.00-μC charge due to the 7.00-μC and -4.00-μC charges. (b) Use your answer to part (a) to determine the force on the 2.00-μC charge.

31. Three point charges are located on a circular arc as shown in Figure P23.31. (a) What is the total electric field at P, the center of the arc? (b) Find the electric force that would be exerted on a -5.00-nC point charge placed at P.

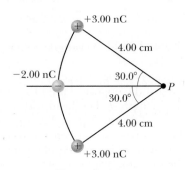

Figure P23.31

32. Two charged particles are located on the x axis. The first is a charge $+Q$ at $x = -a$. The second is an unknown charge located at $x = +3a$. The net electric field these charges produce at the origin has a magnitude of $2k_eQ/a^2$. Explain how many values are possible for the unknown charge and find the possible values.

33. A small, 2.00-g plastic ball is suspended by a 20.0-cm-long string in a uniform electric field as shown in Figure P23.33. If the ball is in equilibrium when the string makes a 15.0° angle with the vertical, what is the net charge on the ball?

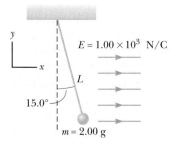

$E = 1.00 \times 10^3$ N/C

L

15.0°

$m = 2.00$ g

Figure P23.33

34. Two 2.00-μC point charges are located on the x axis. One is at $x = 1.00$ m, and the other is at $x = -1.00$ m. (a) Determine the electric field on the y axis at $y = 0.500$ m. (b) Calculate the electric force on a -3.00-μC charge placed on the y axis at $y = 0.500$ m.

35. Three point charges are arranged as shown in Figure P23.11. (a) Find the vector electric field that the 6.00-nC and -3.00-nC charges together create at the origin. (b) Find the vector force on the 5.00-nC charge.

36. Consider the electric dipole shown in Figure P23.36. Show that the electric field at a *distant* point on the $+x$ axis is $E_x \approx 4k_eqa/x^3$.

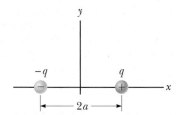

Figure P23.36

Section 23.5 **Electric Field of a Continuous Charge Distribution**

37. A rod 14.0 cm long is uniformly charged and has a total charge of -22.0 μC. Determine (a) the magnitude and (b) the direction of the electric field along the axis of the rod at a point 36.0 cm from its center.

38. A uniformly charged disk of radius 35.0 cm carries charge with a density of 7.90×10^{-3} C/m². Calculate the electric field on the axis of the disk at (a) 5.00 cm, (b) 10.0 cm, (c) 50.0 cm, and (d) 200 cm from the center of the disk.

39. A uniformly charged ring of radius 10.0 cm has a total charge of 75.0 μC. Find the electric field on the axis of

the ring at (a) 1.00 cm, (b) 5.00 cm, (c) 30.0 cm, and (d) 100 cm from the center of the ring.

40. The electric field along the axis of a uniformly charged disk of radius R and total charge Q was calculated in Example 23.9. Show that the electric field at distances x that are large compared with R approaches that of a particle with charge $Q = \sigma\pi R^2$. *Suggestion:* First show that $x/(x^2 + R^2)^{1/2} = (1 + R^2/x^2)^{-1/2}$ and use the binomial expansion $(1 + \delta)^n \approx 1 + n\delta$, when $\delta \ll 1$.

41. Example 23.9 derives the exact expression for the electric field at a point on the axis of a uniformly charged disk. Consider a disk of radius $R = 3.00$ cm having a uniformly distributed charge of $+5.20$ μC. (a) Using the result of Example 23.9, compute the electric field at a point on the axis and 3.00 mm from the center. (b) **What If?** Explain how the answer to part (a) compares with the field computed from the near-field approximation $E = \sigma/2\epsilon_0$. (We derived this expression in Example 23.9.) (c) Using the result of Example 23.9, compute the electric field at a point on the axis and 30.0 cm from the center of the disk. (d) **What If?** Explain how the answer to part (c) compares with the electric field obtained by treating the disk as a $+5.20$-μC charged particle at a distance of 30.0 cm.

42. A uniformly charged rod of length L and total charge Q lies along the x axis as shown in Figure P23.42. (a) Find the components of the electric field at the point P on the y axis a distance d from the origin. (b) What are the approximate values of the field components when $d \gg L$? Explain why you would expect these results.

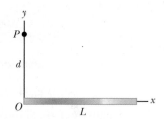

Figure P23.42

43. A continuous line of charge lies along the x axis, extending from $x = +x_0$ to positive infinity. The line carries positive charge with a uniform linear charge density λ_0. What are (a) the magnitude and (b) the direction of the electric field at the origin?

44. A thin rod of length ℓ and uniform charge per unit length λ lies along the x axis as shown in Figure P23.44. (a) Show that the electric field at P, a distance d from the rod along its perpendicular bisector, has no x

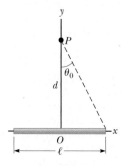

Figure P23.44

component and is given by $E = 2k_e\lambda \sin\theta_0/d$. (b) **What If?** Using your result to part (a), show that the field of a rod of infinite length is $E = 2k_e\lambda/d$.

45. A uniformly charged insulating rod of length 14.0 cm is bent into the shape of a semicircle as shown in Figure P23.45. The rod has a total charge of $-7.50\ \mu C$. Find (a) the magnitude and (b) the direction of the electric field at O, the center of the semicircle.

• O

Figure P23.45

46. (a) Consider a uniformly charged, thin-walled, right circular cylindrical shell having total charge Q, radius R, and length ℓ. Determine the electric field at a point a distance d from the right side of the cylinder as shown in Figure P23.46. *Suggestion:* Use the result of Example 23.8 and treat the cylinder as a collection of ring charges. (b) **What If?** Consider now a solid cylinder with the same dimensions and carrying the same charge, uniformly distributed through its volume. Use the result of Example 23.9 to find the field it creates at the same point.

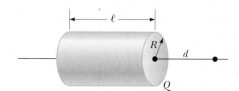

Figure P23.46

Section 23.6 Electric Field Lines

47. A negatively charged rod of finite length carries charge with a uniform charge per unit length. Sketch the electric field lines in a plane containing the rod.

48. A positively charged disk has a uniform charge per unit area σ as described in Example 23.9. Sketch the electric field lines in a plane perpendicular to the plane of the disk passing through its center.

49. Figure P23.49 shows the electric field lines for two charged particles separated by a small distance. (a) Determine the ratio q_1/q_2. (b) What are the signs of q_1 and q_2?

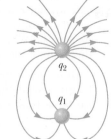

Figure P23.49

50. Three equal positive charges q are at the corners of an equilateral triangle of side a as shown in Figure P23.50. Assume the three charges together create an electric field. (a) Sketch the field lines in the plane of the charges. (b) Find the location of one point (other than ∞) where the electric field is zero. What are (c) the magnitude and (d) the direction of the electric field at P due to the two charges at the base?

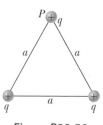

Figure P23.50

Section 23.7 Motion of a Charged Particle in a Uniform Electric Field

51. A proton accelerates from rest in a uniform electric field of 640 N/C. At one later moment, its speed is 1.20 Mm/s (nonrelativistic because v is much less than the speed of light). (a) Find the acceleration of the proton. (b) Over what time interval does the proton reach this speed? (c) How far does it move in this time interval? (d) What is its kinetic energy at the end of this interval?

52. A proton is projected in the positive x direction into a region of a uniform electric field $\vec{E} = (-6.00 \times 10^5)\,\hat{i}\,N/C$ at $t = 0$. The proton travels 7.00 cm as it comes to rest. Determine (a) the acceleration of the proton, (b) its initial speed, and (c) the time interval over which the proton comes to rest.

53. An electron and a proton are each placed at rest in a uniform electric field of magnitude 520 N/C. Calculate the speed of each particle 48.0 ns after being released.

54. Protons are projected with an initial speed $v_i = 9.55$ km/s from a field-free region through a plane and into a region where a uniform electric field $\vec{E} = -720\hat{j}\,N/C$ is present above the plane as shown in Figure P23.54. The initial velocity vector of the protons makes an angle θ with the plane. The protons are to hit a target that lies at a horizontal distance of $R = 1.27$ mm from the point where the protons cross the plane and enter the electric field. We wish to find the angle θ at which the protons must pass through the plane to strike the target. (a) What analysis model describes the horizontal motion of the protons above the plane? (b) What analysis model describes the vertical motion of the protons above the plane? (c) Argue that Equation 4.13 would be applicable to the protons in this situation. (d) Use Equation 4.13 to write an expression for R in terms of v_i, E, the charge and mass of the proton, and the angle θ. (e) Find the two possible values of the angle θ. (f) Find the time interval during which the proton is above the plane in Figure P23.54 for each of the two possible values of θ.

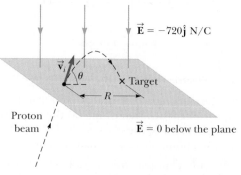

Figure P23.54

55. The electrons in a particle beam each have a kinetic energy K. What are (a) the magnitude and (b) the direction of the electric field that will stop these electrons in a distance d?

56. Two horizontal metal plates, each 10.0 cm square, are aligned 1.00 cm apart with one above the other. They are given equal-magnitude charges of opposite sign so that a uniform downward electric field of 2.00×10^3 N/C exists in the region between them. A particle of mass 2.00×10^{-16} kg and with a positive charge of 1.00×10^{-6} C leaves the center of the bottom negative plate with an initial speed of 1.00×10^5 m/s at an angle of 37.0° above the horizontal. (a) Describe the trajectory of the particle. (b) Which plate does it strike? (c) Where does it strike, relative to its starting point?

57. A proton moves at 4.50×10^5 m/s in the horizontal direction. It enters a uniform vertical electric field with a magnitude of 9.60×10^3 N/C. Ignoring any gravitational effects, find (a) the time interval required for the proton to travel 5.00 cm horizontally, (b) its vertical displacement during the time interval in which it travels 5.00 cm horizontally, and (c) the horizontal and vertical components of its velocity after it has traveled 5.00 cm horizontally.

Additional Problems

58. Three solid plastic cylinders all have radius 2.50 cm and length 6.00 cm. Find the charge of each cylinder given the following additional information about each one. Cylinder (a) carries charge with uniform density 15.0 nC/m² everywhere on its surface. Cylinder (b) carries charge with uniform density 15.0 nC/m² on its curved lateral surface only. Cylinder (c) carries charge with uniform density 500 nC/m³ throughout the plastic.

59. Consider an infinite number of identical particles, each with charge q, placed along the x axis at distances a, $2a$, $3a$, $4a$, ... from the origin. What is the electric field at the origin due to this distribution? *Suggestion:* Use

$$1 + \frac{1}{2^2} + \frac{1}{3^2} + \frac{1}{4^2} + \cdots = \frac{\pi^2}{6}$$

60. A particle with charge -3.00 nC is at the origin, and a particle with negative charge of magnitude Q is at $x = 50.0$ cm. A third particle with a positive charge is in equilibrium at $x = 20.9$ cm. What is Q?

61. A small block of mass m and charge Q is placed on an insulated, frictionless, inclined plane of angle θ as in Figure P23.61. An electric field is applied parallel to the incline. (a) Find an expression for the magnitude of the electric field that enables the block to remain at rest. (b) If $m = 5.40$ g, $Q = -7.00$ μC, and $\theta = 25.0°$, determine the magnitude and the direction of the electric field that enables the block to remain at rest on the incline.

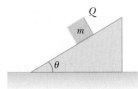

Figure P23.61

62. A small sphere of charge $q_1 = 0.800$ μC hangs from the end of a spring as in Figure P23.62a. When another small sphere of charge $q_2 = -0.600$ μC is held beneath the first sphere as in Figure P23.62b, the spring stretches by $d = 3.50$ cm from its original length and reaches a new equilibrium position with a separation between the charges of $r = 5.00$ cm. What is the force constant of the spring?

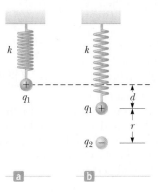

Figure P23.62

63. A line of charge starts at $x = +x_0$ and extends to positive infinity. The linear charge density is $\lambda = \lambda_0 x_0 / x$, where λ_0 is a constant. Determine the electric field at the origin.

64. A small sphere of mass $m = 7.50$ g and charge $q_1 = 32.0$ nC is attached to the end of a string and hangs vertically as in Figure P23.64. A second charge of equal mass and charge $q_2 = -58.0$ nC is located below the first charge a distance $d = 2.00$ cm below the first charge as in Figure P23.64. (a) Find the tension in the string. (b) If the string can withstand a maximum tension of 0.180 N, what is the smallest value d can have before the string breaks?

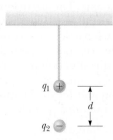

Figure P23.64

65. A uniform electric field of magnitude 640 N/C exists between two parallel plates that are 4.00 cm apart. A proton is released from rest at the positive plate at the same instant an electron is released from rest at the negative plate. (a) Determine the distance from the positive plate at which the two pass each other. Ignore the electrical attraction between the proton and electron. (b) **What If?** Repeat part (a) for a sodium ion (Na⁺) and a chloride ion (Cl⁻).

66. Two small silver spheres, each with a mass of 10.0 g, are separated by 1.00 m. Calculate the fraction of the electrons in one sphere that must be transferred to the other to produce an attractive force of 1.00×10^4 N (about 1 ton) between the spheres. The number of electrons per atom of silver is 47.

67. A charged cork ball of mass 1.00 g is suspended on a light string in the presence of a uniform electric field as shown in Figure P23.67. When $\vec{E} = (3.00\hat{i} + 5.00\hat{j}) \times 10^5$ N/C, the ball is in equilibrium at $\theta = 37.0°$. Find (a) the charge on the ball and (b) the tension in the string.

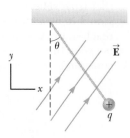

Figure P23.67
Problems 67 and 68.

68. A charged cork ball of mass m is suspended on a light string in the presence of a uniform electric field as shown in Figure P23.67. When $\vec{E} = A\hat{i} + B\hat{j}$, where A and B are positive quantities, the ball is in equilibrium at the angle θ. Find (a) the charge on the ball and (b) the tension in the string.

69. Three charged particles are aligned along the x axis as shown in Figure P23.69. Find the electric field at (a) the position (2.00 m, 0) and (b) the position (0, 2.00 m).

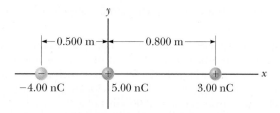

Figure P23.69

70. Two point charges $q_A = -12.0$ μC and $q_B = 45.0$ μC and a third particle with unknown charge q_C are located on the x axis. The particle q_A is at the origin, and q_B is at $x = 15.0$ cm. The third particle is to be placed so that each particle is in equilibrium under the action of the electric forces exerted by the other two particles. (a) Is this situation possible? If so, is it possible in more than one way? Explain. Find (b) the required location and (c) the magnitude and the sign of the charge of the third particle.

71. A line of positive charge is formed into a semicircle of radius $R = 60.0$ cm as shown in Figure P23.71. The charge per unit length along the semicircle is described by the expression $\lambda = \lambda_0 \cos \theta$. The total charge on the semicircle is 12.0 μC. Calculate the total force on a charge of 3.00 μC placed at the center of curvature P.

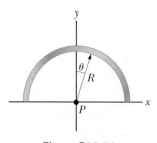

Figure P23.71

72. Four identical charged particles ($q = +10.0$ μC) are located on the corners of a rectangle as shown in Figure P23.72. The dimensions of the rectangle are $L = 60.0$ cm and $W = 15.0$ cm. Calculate (a) the magnitude and (b) the direction of the total electric force exerted on the charge at the lower left corner by the other three charges.

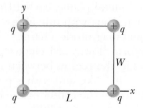

Figure P23.72

73. Two small spheres hang in equilibrium at the bottom ends of threads, 40.0 cm long, that have their top ends tied to the same fixed point. One sphere has mass 2.40 g and charge +300 nC. The other sphere has the same mass and charge +200 nC. Find the distance between the centers of the spheres.

74. *Why is the following situation impossible?* An electron enters a region of uniform electric field between two parallel plates. The plates are used in a cathode-ray tube to adjust the position of an electron beam on a distant fluorescent screen. The magnitude of the electric field between the plates is 200 N/C. The plates are 0.200 m in length and are separated by 1.50 cm. The electron enters the region at a speed of 3.00×10^6 m/s, traveling parallel to the plane of the plates in the direction of their length. It leaves the plates heading toward its correct location on the fluorescent screen.

75. **Review.** Two identical blocks resting on a frictionless, horizontal surface are connected by a light spring having a spring constant $k = 100$ N/m and an unstretched length $L_i = 0.400$ m as shown in Figure P23.75a. A charge Q is slowly placed on each block, causing the spring to stretch to an equilibrium length $L = 0.500$ m as shown in Figure P23.75b. Determine the value of Q, modeling the blocks as charged particles.

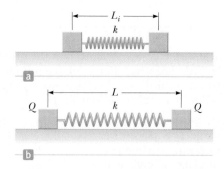

Figure P23.75 Problems 75 and 76.

76. **Review.** Two identical blocks resting on a frictionless, horizontal surface are connected by a light spring having a spring constant k and an unstretched length L_i as shown in Figure P23.75a. A charge Q is slowly placed on each block, causing the spring to stretch to an equilibrium length L as shown in Figure P23.75b. Determine the value of Q, modeling the blocks as charged particles.

77. Three identical point charges, each of mass $m = 0.100$ kg, hang from three strings as shown in Figure

P23.77. If the lengths of the left and right strings are each $L = 30.0$ cm and the angle θ is 45.0°, determine the value of q.

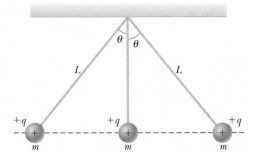

Figure P23.77

78. Show that the maximum magnitude E_{max} of the electric field along the axis of a uniformly charged ring occurs at $x = a/\sqrt{2}$ (see Fig. 23.16) and has the value $Q/(6\sqrt{3}\pi\epsilon_0 a^2)$.

79. Two hard rubber spheres, each of mass $m = 15.0$ g, are rubbed with fur on a dry day and are then suspended with two insulating strings of length $L = 5.00$ cm whose support points are a distance $d = 3.00$ cm from each other as shown in Figure P23.79. During the rubbing process, one sphere receives exactly twice the charge of the other. They are observed to hang at equilibrium, each at an angle of $\theta = 10.0°$ with the vertical. Find the amount of charge on each sphere.

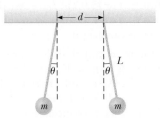

Figure P23.79

80. Two identical beads each have a mass m and charge q. When placed in a hemispherical bowl of radius R with frictionless, nonconducting walls, the beads move, and at equilibrium, they are a distance d apart (Fig. P23.80). (a) Determine the charge q on each bead. (b) Determine the charge required for d to become equal to $2R$.

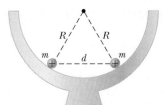

Figure P23.80

81. Two small spheres of mass m are suspended from strings of length ℓ that are connected at a common point. One sphere has charge Q and the other charge $2Q$. The strings make angles θ_1 and θ_2 with the vertical.

(a) Explain how θ_1 and θ_2 are related. (b) Assume θ_1 and θ_2 are small. Show that the distance r between the spheres is approximately

$$r \approx \left(\frac{4k_eQ^2\ell}{mg}\right)^{1/3}$$

82. **Review.** A negatively charged particle $-q$ is placed at the center of a uniformly charged ring, where the ring has a total positive charge Q as shown in Figure P23.82. The particle, confined to move along the x axis, is moved a small distance x along the axis (where $x \ll a$) and released. Show that the particle oscillates in simple harmonic motion with a frequency given by

$$f = \frac{1}{2\pi}\left(\frac{k_eqQ}{ma^3}\right)^{1/2}$$

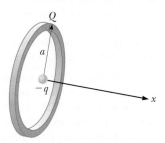

Figure P23.82

83. **Review.** A 1.00-g cork ball with charge 2.00 μC is suspended vertically on a 0.500-m-long light string in the presence of a uniform, downward-directed electric field of magnitude $E = 1.00 \times 10^5$ N/C. If the ball is displaced slightly from the vertical, it oscillates like a simple pendulum. (a) Determine the period of this oscillation. (b) Should the effect of gravitation be included in the calculation for part (a)? Explain.

Challenge Problems

84. Identical thin rods of length $2a$ carry equal charges $+Q$ uniformly distributed along their lengths. The rods lie along the x axis with their centers separated by a distance $b > 2a$ (Fig. P23.84). Show that the magnitude of the force exerted by the left rod on the right one is

$$F = \left(\frac{k_eQ^2}{4a^2}\right)\ln\left(\frac{b^2}{b^2 - 4a^2}\right)$$

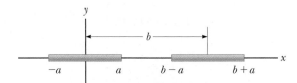

Figure P23.84

85. Eight charged particles, each of magnitude q, are located on the corners of a cube of edge s as shown in Figure P23.85 (page 724). (a) Determine the x, y, and z components of the total force exerted by the other charges on the charge located at point A. What are

(b) the magnitude and (c) the direction of this total force?

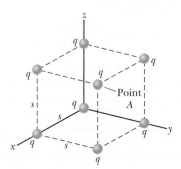

Figure P23.85 Problems 85 and 86.

86. Consider the charge distribution shown in Figure P23.85. (a) Show that the magnitude of the electric field at the center of any face of the cube has a value of $2.18k_eq/s^2$. (b) What is the direction of the electric field at the center of the top face of the cube?

87. **Review.** An electric dipole in a uniform horizontal electric field is displaced slightly from its equilibrium position as shown in Figure P23.87, where θ is small. The separation of the charges is $2a$, and each of the two particles has mass m. (a) Assuming the dipole is released from this position, show that its angular orientation exhibits simple harmonic motion with a frequency

$$f = \frac{1}{2\pi}\sqrt{\frac{qE}{ma}}$$

What If? (b) Suppose the masses of the two charged particles in the dipole are not the same even though each particle continues to have charge q. Let the masses of the particles be m_1 and m_2. Show that the frequency of the oscillation in this case is

$$f = \frac{1}{2\pi}\sqrt{\frac{qE(m_1 + m_2)}{2am_1m_2}}$$

Figure P23.87

88. Inez is putting up decorations for her sister's quinceañera (fifteenth birthday party). She ties three light silk ribbons together to the top of a gateway and hangs a rubber balloon from each ribbon (Fig. P23.88). To include the effects of the gravitational and buoyant forces on it, each balloon can be modeled as a particle of mass 2.00 g, with its center 50.0 cm from the point of support. Inez rubs the whole surface of each balloon with her woolen scarf, making the balloons hang separately with gaps between them. Looking directly upward from below the balloons, Inez notices that the centers of the hanging balloons form a horizontal equilateral triangle with sides 30.0 cm long. What is the common charge each balloon carries?

Figure P23.88

89. A line of charge with uniform density 35.0 nC/m lies along the line $y = -15.0$ cm between the points with coordinates $x = 0$ and $x = 40.0$ cm. Find the electric field it creates at the origin.

90. A particle of mass m and charge q moves at high speed along the x axis. It is initially near $x = -\infty$, and it ends up near $x = +\infty$. A second charge Q is fixed at the point $x = 0$, $y = -d$. As the moving charge passes the stationary charge, its x component of velocity does not change appreciably, but it acquires a small velocity in the y direction. Determine the angle through which the moving charge is deflected from the direction of its initial velocity.

91. Two particles, each with charge 52.0 nC, are located on the y axis at $y = 25.0$ cm and $y = -25.0$ cm. (a) Find the vector electric field at a point on the x axis as a function of x. (b) Find the field at $x = 36.0$ cm. (c) At what location is the field $1.00\hat{i}$ kN/C? You may need a computer to solve this equation. (d) At what location is the field $16.0\hat{i}$ kN/C?

Gauss's Law

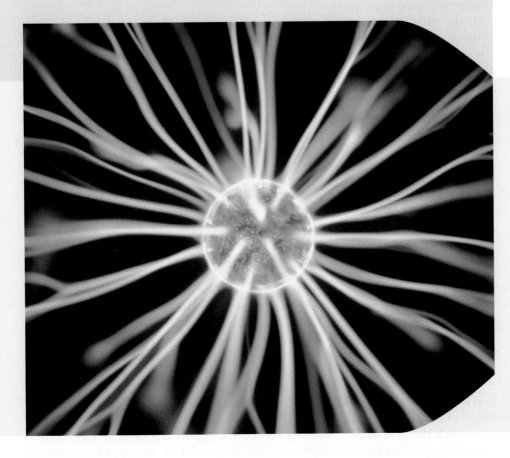

In Chapter 23, we showed how to calculate the electric field due to a given charge distribution by integrating over the distribution. In this chapter, we describe *Gauss's law* and an alternative procedure for calculating electric fields. Gauss's law is based on the inverse-square behavior of the electric force between point charges. Although Gauss's law is a direct consequence of Coulomb's law, it is more convenient for calculating the electric fields of highly symmetric charge distributions and makes it possible to deal with complicated problems using qualitative reasoning. As we show in this chapter, Gauss's law is important in understanding and verifying the properties of conductors in electrostatic equilibrium.

In a tabletop plasma ball, the colorful lines emanating from the sphere give evidence of strong electric fields. Using Gauss's law, we show in this chapter that the electric field surrounding a uniformly charged sphere is identical to that of a point charge. *(Steve Cole/Getty Images)*

24.1 Electric Flux

The concept of electric field lines was described qualitatively in Chapter 23. We now treat electric field lines in a more quantitative way.

Consider an electric field that is uniform in both magnitude and direction as shown in Figure 24.1. The field lines penetrate a rectangular surface of area A, whose plane is oriented perpendicular to the field. Recall from Section 23.6 that the number of lines per unit area (in other words, the *line density*) is proportional to the magnitude of the electric field. Therefore, the total number of lines penetrating the surface is proportional to the product EA. This product of the magnitude of the electric field E and surface area A perpendicular to the field is called the **electric flux** Φ_E (uppercase Greek letter phi):

$$\Phi_E = EA \tag{24.1}$$

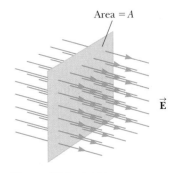

Figure 24.1 Field lines representing a uniform electric field penetrating a plane of area A perpendicular to the field.

725

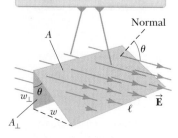

The number of field lines that go through the area $A_\perp$ is the same as the number that go through area A.

Figure 24.2 Field lines representing a uniform electric field penetrating an area A whose normal is at an angle θ to the field.

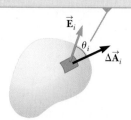

The electric field makes an angle θ_i with the vector $\Delta\vec{\mathbf{A}}_i$, defined as being normal to the surface element.

Figure 24.3 A small element of surface area ΔA_i in an electric field.

Definition of electric flux ▶

From the SI units of E and A, we see that Φ_E has units of newton meters squared per coulomb (N · m²/C). Electric flux is proportional to the number of electric field lines penetrating some surface.

If the surface under consideration is not perpendicular to the field, the flux through it must be less than that given by Equation 24.1. Consider Figure 24.2, where the normal to the surface of area A is at an angle θ to the uniform electric field. Notice that the number of lines that cross this area A is equal to the number of lines that cross the area $A_\perp$, which is a projection of area A onto a plane oriented perpendicular to the field. The area A is the product of the length and the width of the surface: $A = \ell w$. At the left edge of the figure, we see that the widths of the surfaces are related by $w_\perp = w \cos\theta$. The area $A_\perp$ is given by $A_\perp = \ell w_\perp = \ell w \cos\theta$ and we see that the two areas are related by $A_\perp = A \cos\theta$. Because the flux through A equals the flux through $A_\perp$, the flux through A is

$$\Phi_E = EA_\perp = EA \cos\theta \tag{24.2}$$

From this result, we see that the flux through a surface of fixed area A has a maximum value EA when the surface is perpendicular to the field (when the normal to the surface is parallel to the field, that is, when $\theta = 0°$ in Fig. 24.2); the flux is zero when the surface is parallel to the field (when the normal to the surface is perpendicular to the field, that is, when $\theta = 90°$).

In this discussion, the angle θ is used to describe the orientation of the surface of area A. We can also interpret the angle as that between the electric field vector and the normal to the surface. In this case, the product $E \cos\theta$ in Equation 24.2 is the component of the electric field perpendicular to the surface. The flux through the surface can then be written $\Phi_E = (E \cos\theta)A = E_n A$, where we use E_n as the component of the electric field normal to the surface.

We assumed a uniform electric field in the preceding discussion. In more general situations, the electric field may vary over a large surface. Therefore, the definition of flux given by Equation 24.2 has meaning only for a small element of area over which the field is approximately constant. Consider a general surface divided into a large number of small elements, each of area ΔA_i. It is convenient to define a vector $\Delta\vec{\mathbf{A}}_i$ whose magnitude represents the area of the ith element of the large surface and whose direction is defined to be *perpendicular* to the surface element as shown in Figure 24.3. The electric field $\vec{\mathbf{E}}_i$ at the location of this element makes an angle θ_i with the vector $\Delta\vec{\mathbf{A}}_i$. The electric flux $\Phi_{E,i}$ through this element is

$$\Phi_{E,i} = E_i \Delta A_i \cos\theta_i = \vec{\mathbf{E}}_i \cdot \Delta\vec{\mathbf{A}}_i$$

where we have used the definition of the scalar product of two vectors ($\vec{\mathbf{A}} \cdot \vec{\mathbf{B}} \equiv AB \cos\theta$; see Chapter 7). Summing the contributions of all elements gives an approximation to the total flux through the surface:

$$\Phi_E \approx \sum \vec{\mathbf{E}}_i \cdot \Delta\vec{\mathbf{A}}_i$$

If the area of each element approaches zero, the number of elements approaches infinity and the sum is replaced by an integral. Therefore, the general definition of electric flux is

$$\Phi_E \equiv \int_{surface} \vec{\mathbf{E}} \cdot d\vec{\mathbf{A}} \tag{24.3}$$

Equation 24.3 is a *surface integral*, which means it must be evaluated over the surface in question. In general, the value of Φ_E depends both on the field pattern and on the surface.

We are often interested in evaluating the flux through a *closed surface*, defined as a surface that divides space into an inside and an outside region so that one cannot move from one region to the other without crossing the surface. The surface of a sphere, for example, is a closed surface. By convention, if the area element in Equa-

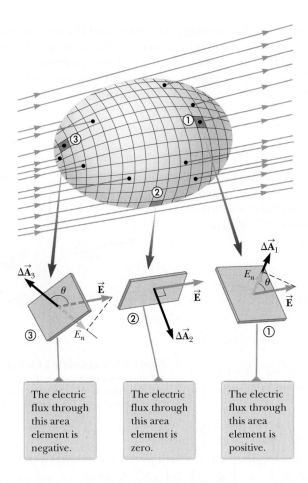

Figure 24.4 A closed surface in an electric field. The area vectors are, by convention, normal to the surface and point outward.

The electric flux through this area element is negative.

The electric flux through this area element is zero.

The electric flux through this area element is positive.

tion 24.3 is part of a closed surface, the direction of the area vector is chosen so that the vector points outward from the surface. If the area element is not part of a closed surface, the direction of the area vector is chosen so that the angle between the area vector and the electric field vector is less than or equal to 90°.

Consider the closed surface in Figure 24.4. The vectors $\Delta \vec{A}_i$ point in different directions for the various surface elements, but for each element they are normal to the surface and point outward. At the element labeled ①, the field lines are crossing the surface from the inside to the outside and $\theta < 90°$; hence, the flux $\Phi_{E,1} = \vec{E} \cdot \Delta \vec{A}_1$ through this element is positive. For element ②, the field lines graze the surface (perpendicular to $\Delta \vec{A}_2$); therefore, $\theta = 90°$ and the flux is zero. For elements such as ③, where the field lines are crossing the surface from outside to inside, $180° > \theta > 90°$ and the flux is negative because $\cos \theta$ is negative. The *net* flux through the surface is proportional to the net number of lines leaving the surface, where the net number means *the number of lines leaving the surface minus the number of lines entering the surface*. If more lines are leaving than entering, the net flux is positive. If more lines are entering than leaving, the net flux is negative. Using the symbol $\oint$ to represent an integral over a closed surface, we can write the net flux Φ_E through a closed surface as

$$\Phi_E = \oint \vec{E} \cdot d\vec{A} = \oint E_n \, dA \tag{24.4}$$

where E_n represents the component of the electric field normal to the surface.

Quick Quiz 24.1 Suppose a point charge is located at the center of a spherical surface. The electric field at the surface of the sphere and the total flux through the sphere are determined. Now the radius of the sphere is halved.

What happens to the flux through the sphere and the magnitude of the electric field at the surface of the sphere? **(a)** The flux and field both increase. **(b)** The flux and field both decrease. **(c)** The flux increases, and the field decreases. **(d)** The flux decreases, and the field increases. **(e)** The flux remains the same, and the field increases. **(f)** The flux decreases, and the field remains the same.

Example 24.1   Flux Through a Cube

Consider a uniform electric field $\vec{\mathbf{E}}$ oriented in the x direction in empty space. A cube of edge length ℓ is placed in the field, oriented as shown in Figure 24.5. Find the net electric flux through the surface of the cube.

SOLUTION

Conceptualize Examine Figure 24.5 carefully. Notice that the electric field lines pass through two faces perpendicularly and are parallel to four other faces of the cube.

Categorize We evaluate the flux from its definition, so we categorize this example as a substitution problem.

The flux through four of the faces (③, ④, and the unnumbered faces) is zero because $\vec{\mathbf{E}}$ is parallel to the four faces and therefore perpendicular to $d\vec{\mathbf{A}}$ on these faces.

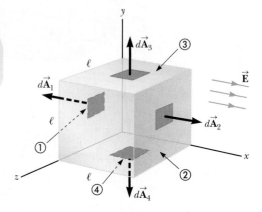

Figure 24.5 (Example 24.1) A closed surface in the shape of a cube in a uniform electric field oriented parallel to the x axis. Side ④ is the bottom of the cube, and side ① is opposite side ②.

Write the integrals for the net flux through faces ① and ②:

$$\Phi_E = \int_1 \vec{\mathbf{E}} \cdot d\vec{\mathbf{A}} + \int_2 \vec{\mathbf{E}} \cdot d\vec{\mathbf{A}}$$

For face ①, $\vec{\mathbf{E}}$ is constant and directed inward but $d\vec{\mathbf{A}}_1$ is directed outward ($\theta = 180°$). Find the flux through this face:

$$\int_1 \vec{\mathbf{E}} \cdot d\vec{\mathbf{A}} = \int_1 E(\cos 180°)\, dA = -E \int_1 dA = -EA = -E\ell^2$$

For face ②, $\vec{\mathbf{E}}$ is constant and outward and in the same direction as $d\vec{\mathbf{A}}_2$ ($\theta = 0°$). Find the flux through this face:

$$\int_2 \vec{\mathbf{E}} \cdot d\vec{\mathbf{A}} = \int_2 E(\cos 0°)\, dA = E \int_2 dA = +EA = E\ell^2$$

Find the net flux by adding the flux over all six faces:

$$\Phi_E = -E\ell^2 + E\ell^2 + 0 + 0 + 0 + 0 = \boxed{0}$$

24.2 Gauss's Law

In this section, we describe a general relationship between the net electric flux through a closed surface (often called a *gaussian surface*) and the charge enclosed by the surface. This relationship, known as *Gauss's law*, is of fundamental importance in the study of electric fields.

Consider a positive point charge q located at the center of a sphere of radius r as shown in Figure 24.6. From Equation 23.9, we know that the magnitude of the electric field everywhere on the surface of the sphere is $E = k_e q/r^2$. The field lines are directed radially outward and hence are perpendicular to the surface at every point on the surface. That is, at each surface point, $\vec{\mathbf{E}}$ is parallel to the vector $\Delta\vec{\mathbf{A}}_i$ representing a local element of area ΔA_i surrounding the surface point. Therefore,

$$\vec{\mathbf{E}} \cdot \Delta\vec{\mathbf{A}}_i = E\,\Delta A_i$$

and, from Equation 24.4, we find that the net flux through the gaussian surface is

$$\Phi_E = \oint \vec{\mathbf{E}} \cdot d\vec{\mathbf{A}} = \oint E\, dA = E \oint dA$$

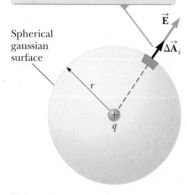

When the charge is at the center of the sphere, the electric field is everywhere normal to the surface and constant in magnitude.

Spherical gaussian surface

Figure 24.6 A spherical gaussian surface of radius r surrounding a positive point charge q.

where we have moved E outside of the integral because, by symmetry, E is constant over the surface. The value of E is given by $E = k_e q/r^2$. Furthermore, because the surface is spherical, $\oint dA = A = 4\pi r^2$. Hence, the net flux through the gaussian surface is

$$\Phi_E = k_e \frac{q}{r^2}(4\pi r^2) = 4\pi k_e q$$

Recalling from Equation 23.3 that $k_e = 1/4\pi\epsilon_0$, we can write this equation in the form

$$\Phi_E = \frac{q}{\epsilon_0} \qquad \textbf{(24.5)}$$

Equation 24.5 shows that the net flux through the spherical surface is proportional to the charge inside the surface. The flux is independent of the radius r because the area of the spherical surface is proportional to r^2, whereas the electric field is proportional to $1/r^2$. Therefore, in the product of area and electric field, the dependence on r cancels.

Now consider several closed surfaces surrounding a charge q as shown in Figure 24.7. Surface S_1 is spherical, but surfaces S_2 and S_3 are not. From Equation 24.5, the flux that passes through S_1 has the value q/ϵ_0. As discussed in the preceding section, flux is proportional to the number of electric field lines passing through a surface. The construction shown in Figure 24.7 shows that the number of lines through S_1 is equal to the number of lines through the nonspherical surfaces S_2 and S_3. Therefore,

> the net flux through *any* closed surface surrounding a point charge q is given by q/ϵ_0 and is independent of the shape of that surface.

Now consider a point charge located *outside* a closed surface of arbitrary shape as shown in Figure 24.8. As can be seen from this construction, any electric field line entering the surface leaves the surface at another point. The number of electric field lines entering the surface equals the number leaving the surface. Therefore, the net electric flux through a closed surface that surrounds no charge is zero. Applying this result to Example 24.1, we see that the net flux through the cube is zero because there is no charge inside the cube.

Let's extend these arguments to two generalized cases: (1) that of many point charges and (2) that of a continuous distribution of charge. We once again use the superposition principle, which states that the electric field due to many charges is

Karl Friedrich Gauss
German mathematician and astronomer (1777–1855)
Gauss received a doctoral degree in mathematics from the University of Helmstedt in 1799. In addition to his work in electromagnetism, he made contributions to mathematics and science in number theory, statistics, non-Euclidean geometry, and cometary orbital mechanics. He was a founder of the German Magnetic Union, which studies the Earth's magnetic field on a continual basis.

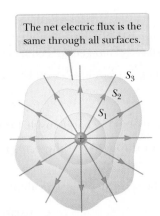

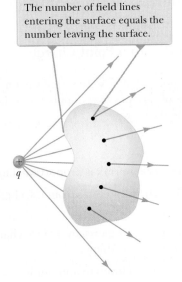

The net electric flux is the same through all surfaces.

The number of field lines entering the surface equals the number leaving the surface.

Figure 24.7 Closed surfaces of various shapes surrounding a positive charge.

Figure 24.8 A point charge located *outside* a closed surface.

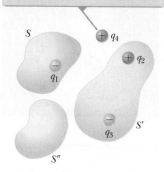

Charge q_4 does not contribute to the flux through any surface because it is outside all surfaces.

Figure 24.9 The net electric flux through any closed surface depends only on the charge *inside* that surface. The net flux through surface S is q_1/ϵ_0, the net flux through surface S' is $(q_2 + q_3)/\epsilon_0$, and the net flux through surface S'' is zero.

the vector sum of the electric fields produced by the individual charges. Therefore, the flux through any closed surface can be expressed as

$$\oint \vec{E} \cdot d\vec{A} = \oint (\vec{E}_1 + \vec{E}_2 + \cdots) \cdot d\vec{A}$$

where $\vec{E}$ is the total electric field at any point on the surface produced by the vector addition of the electric fields at that point due to the individual charges. Consider the system of charges shown in Figure 24.9. The surface S surrounds only one charge, q_1; hence, the net flux through S is q_1/ϵ_0. The flux through S due to charges q_2, q_3, and q_4 outside it is zero because each electric field line from these charges that enters S at one point leaves it at another. The surface S' surrounds charges q_2 and q_3; hence, the net flux through it is $(q_2 + q_3)/\epsilon_0$. Finally, the net flux through surface S'' is zero because there is no charge inside this surface. That is, *all* the electric field lines that enter S'' at one point leave at another. Charge q_4 does not contribute to the net flux through any of the surfaces.

The mathematical form of **Gauss's law** is a generalization of what we have just described and states that the net flux through *any* closed surface is

$$\Phi_E = \oint \vec{E} \cdot d\vec{A} = \frac{q_{in}}{\epsilon_0} \qquad \textbf{(24.6)}$$

where $\vec{E}$ represents the electric field at any point on the surface and q_{in} represents the net charge inside the surface.

When using Equation 24.6, you should note that although the charge q_{in} is the net charge inside the gaussian surface, $\vec{E}$ represents the *total electric field*, which includes contributions from charges both inside and outside the surface.

In principle, Gauss's law can be solved for $\vec{E}$ to determine the electric field due to a system of charges or a continuous distribution of charge. In practice, however, this type of solution is applicable only in a limited number of highly symmetric situations. In the next section, we use Gauss's law to evaluate the electric field for charge distributions that have spherical, cylindrical, or planar symmetry. If one chooses the gaussian surface surrounding the charge distribution carefully, the integral in Equation 24.6 can be simplified and the electric field determined.

Quick Quiz 24.2 If the net flux through a gaussian surface is *zero*, the following four statements *could be true*. Which of the statements *must be true*? **(a)** There are no charges inside the surface. **(b)** The net charge inside the surface is zero. **(c)** The electric field is zero everywhere on the surface. **(d)** The number of electric field lines entering the surface equals the number leaving the surface.

Conceptual Example 24.2 **Flux Due to a Point Charge**

A spherical gaussian surface surrounds a point charge q. Describe what happens to the total flux through the surface if **(A)** the charge is tripled, **(B)** the radius of the sphere is doubled, **(C)** the surface is changed to a cube, and **(D)** the charge is moved to another location inside the surface.

SOLUTION

(A) The flux through the surface is tripled because flux is proportional to the amount of charge inside the surface.

(B) The flux does not change because all electric field lines from the charge pass through the sphere, regardless of its radius.

(C) The flux does not change when the shape of the gaussian surface changes because all electric field lines from the charge pass through the surface, regardless of its shape.

(D) The flux does not change when the charge is moved to another location inside that surface because Gauss's law refers to the total charge enclosed, regardless of where the charge is located inside the surface.

24.3 Application of Gauss's Law to Various Charge Distributions

As mentioned earlier, Gauss's law is useful for determining electric fields when the charge distribution is highly symmetric. The following examples demonstrate ways of choosing the gaussian surface over which the surface integral given by Equation 24.6 can be simplified and the electric field determined. In choosing the surface, always take advantage of the symmetry of the charge distribution so that E can be removed from the integral. The goal in this type of calculation is to determine a surface for which each portion of the surface satisfies one or more of the following conditions:

1. The value of the electric field can be argued by symmetry to be constant over the portion of the surface.
2. The dot product in Equation 24.6 can be expressed as a simple algebraic product $E \, dA$ because $\vec{\mathbf{E}}$ and $d\vec{\mathbf{A}}$ are parallel.
3. The dot product in Equation 24.6 is zero because $\vec{\mathbf{E}}$ and $d\vec{\mathbf{A}}$ are perpendicular.
4. The electric field is zero over the portion of the surface.

Different portions of the gaussian surface can satisfy different conditions as long as every portion satisfies at least one condition. All four conditions are used in examples throughout the remainder of this chapter and will be identified by number. If the charge distribution does not have sufficient symmetry such that a gaussian surface that satisfies these conditions can be found, Gauss's law is still true, but is not useful for determining the electric field for that charge distribution.

> **Pitfall Prevention 24.2**
> **Gaussian Surfaces Are Not Real**
> A gaussian surface is an imaginary surface you construct to satisfy the conditions listed here. It does not have to coincide with a physical surface in the situation.

Example 24.3   A Spherically Symmetric Charge Distribution

An insulating solid sphere of radius a has a uniform volume charge density ρ and carries a total positive charge Q (Fig. 24.10).

(A) Calculate the magnitude of the electric field at a point outside the sphere.

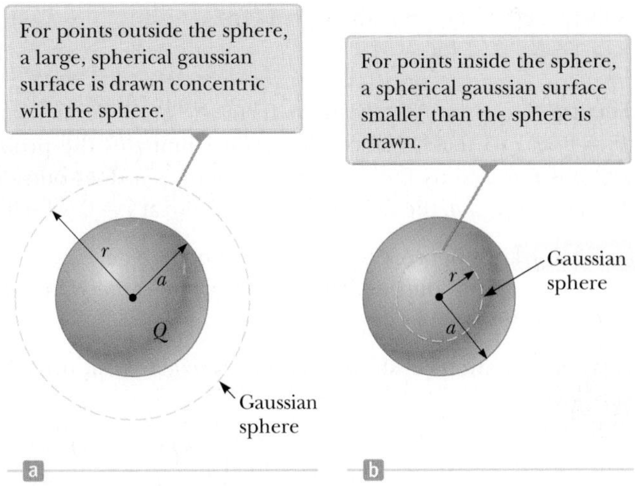

SOLUTION

Conceptualize Notice how this problem differs from our previous discussion of Gauss's law. The electric field due to point charges was discussed in Section 24.2. Now we are considering the electric field due to a distribution of charge. We found the field for various distributions of charge in Chapter 23 by integrating over the distribution. This example demonstrates a difference from our discussions in Chapter 23. In this chapter, we find the electric field using Gauss's law.

Categorize Because the charge is distributed uniformly throughout the sphere, the charge distribution has spherical symmetry and we can apply Gauss's law to find the electric field.

For points outside the sphere, a large, spherical gaussian surface is drawn concentric with the sphere.

For points inside the sphere, a spherical gaussian surface smaller than the sphere is drawn.

Figure 24.10 (Example 24.3) A uniformly charged insulating sphere of radius a and total charge Q. In diagrams such as this one, the dotted line represents the intersection of the gaussian surface with the plane of the page.

Analyze To reflect the spherical symmetry, let's choose a spherical gaussian surface of radius r, concentric with the sphere, as shown in Figure 24.10a. For this choice, condition (2) is satisfied everywhere on the surface and $\vec{\mathbf{E}} \cdot d\vec{\mathbf{A}} = E \, dA$.

continued

▶ **24.3** continued

Replace $\vec{\mathbf{E}} \cdot d\vec{\mathbf{A}}$ in Gauss's law with $E\, dA$:

$$\Phi_E = \oint \vec{\mathbf{E}} \cdot d\vec{\mathbf{A}} = \oint E\, dA = \frac{Q}{\epsilon_0}$$

By symmetry, E has the same value everywhere on the surface, which satisfies condition (1), so we can remove E from the integral:

$$\oint E\, dA = E \oint dA = E(4\pi r^2) = \frac{Q}{\epsilon_0}$$

Solve for E:

$$(1)\quad E = \frac{Q}{4\pi\epsilon_0 r^2} = k_e \frac{Q}{r^2} \quad (\text{for } r > a)$$

Finalize This field is identical to that for a point charge. Therefore, **the electric field due to a uniformly charged sphere in the region external to the sphere is** *equivalent* **to that of a point charge located at the center of the sphere.**

(B) Find the magnitude of the electric field at a point inside the sphere.

SOLUTION

Analyze In this case, let's choose a spherical gaussian surface having radius $r < a$, concentric with the insulating sphere (Fig. 24.10b). Let V' be the volume of this smaller sphere. To apply Gauss's law in this situation, recognize that the charge q_{in} within the gaussian surface of volume V' is less than Q.

Calculate q_{in} by using $q_{in} = \rho V'$:

$$q_{in} = \rho V' = \rho\left(\tfrac{4}{3}\pi r^3\right)$$

Notice that conditions (1) and (2) are satisfied everywhere on the gaussian surface in Figure 24.10b. Apply Gauss's law in the region $r < a$:

$$\oint E\, dA = E \oint dA = E(4\pi r^2) = \frac{q_{in}}{\epsilon_0}$$

Solve for E and substitute for q_{in}:

$$E = \frac{q_{in}}{4\pi\epsilon_0 r^2} = \frac{\rho\left(\tfrac{4}{3}\pi r^3\right)}{4\pi\epsilon_0 r^2} = \frac{\rho}{3\epsilon_0} r$$

Substitute $\rho = Q/\tfrac{4}{3}\pi a^3$ and $\epsilon_0 = 1/4\pi k_e$:

$$(2)\quad E = \frac{Q/\tfrac{4}{3}\pi a^3}{3(1/4\pi k_e)} r = k_e \frac{Q}{a^3} r \quad (\text{for } r < a)$$

Finalize This result for E differs from the one obtained in part (A). It shows that $E \rightarrow 0$ as $r \rightarrow 0$. Therefore, the result eliminates the problem that would exist at $r = 0$ if E varied as $1/r^2$ inside the sphere as it does outside the sphere. That is, if $E \propto 1/r^2$ for $r < a$, the field would be infinite at $r = 0$, which is physically impossible.

WHAT IF? Suppose the radial position $r = a$ is approached from inside the sphere and from outside. Do we obtain the same value of the electric field from both directions?

Answer Equation (1) shows that the electric field approaches a value from the outside given by

$$E = \lim_{r \to a}\left(k_e \frac{Q}{r^2}\right) = k_e \frac{Q}{a^2}$$

From the inside, Equation (2) gives

$$E = \lim_{r \to a}\left(k_e \frac{Q}{a^3} r\right) = k_e \frac{Q}{a^3} a = k_e \frac{Q}{a^2}$$

Therefore, the value of the field is the same as the surface is approached from both directions. A plot of E versus r is shown in Figure 24.11. Notice that the magnitude of the field is continuous.

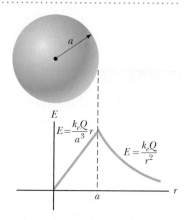

Figure 24.11 (Example 24.3) A plot of E versus r for a uniformly charged insulating sphere. The electric field inside the sphere ($r < a$) varies linearly with r. The field outside the sphere ($r > a$) is the same as that of a point charge Q located at $r = 0$.

Example 24.4 | A Cylindrically Symmetric Charge Distribution

Find the electric field a distance r from a line of positive charge of infinite length and constant charge per unit length λ (Fig. 24.12a).

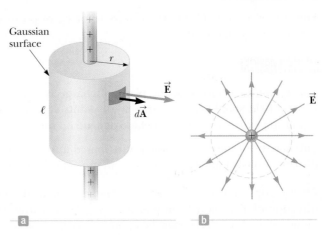

Figure 24.12 (Example 24.4) (a) An infinite line of charge surrounded by a cylindrical gaussian surface concentric with the line. (b) An end view shows that the electric field at the cylindrical surface is constant in magnitude and perpendicular to the surface.

SOLUTION

Conceptualize The line of charge is *infinitely* long. Therefore, the field is the same at all points equidistant from the line, regardless of the vertical position of the point in Figure 24.12a. We expect the field to become weaker as we move farther away from the line of charge.

Categorize Because the charge is distributed uniformly along the line, the charge distribution has cylindrical symmetry and we can apply Gauss's law to find the electric field.

Analyze The symmetry of the charge distribution requires that $\vec{E}$ be perpendicular to the line charge and directed outward as shown in Figure 24.12b. To reflect the symmetry of the charge distribution, let's choose a cylindrical gaussian surface of radius r and length ℓ that is coaxial with the line charge. For the curved part of this surface, $\vec{E}$ is constant in magnitude and perpendicular to the surface at each point, satisfying conditions (1) and (2). Furthermore, the flux through the ends of the gaussian cylinder is zero because $\vec{E}$ is parallel to these surfaces. That is the first application we have seen of condition (3).

We must take the surface integral in Gauss's law over the entire gaussian surface. Because $\vec{E} \cdot d\vec{A}$ is zero for the flat ends of the cylinder, however, we restrict our attention to only the curved surface of the cylinder.

Apply Gauss's law and conditions (1) and (2) for the curved surface, noting that the total charge inside our gaussian surface is $\lambda\ell$:

$$\Phi_E = \oint \vec{E} \cdot d\vec{A} = E \oint dA = EA = \frac{q_{in}}{\epsilon_0} = \frac{\lambda\ell}{\epsilon_0}$$

Substitute the area $A = 2\pi r\ell$ of the curved surface:

$$E(2\pi r\ell) = \frac{\lambda\ell}{\epsilon_0}$$

Solve for the magnitude of the electric field:

$$E = \frac{\lambda}{2\pi\epsilon_0 r} = 2k_e \frac{\lambda}{r} \qquad \text{(24.7)}$$

Finalize This result shows that the electric field due to a cylindrically symmetric charge distribution varies as $1/r$, whereas the field external to a spherically symmetric charge distribution varies as $1/r^2$. Equation 24.7 can also be derived by direct integration over the charge distribution. (See Problem 44 in Chapter 23.)

WHAT IF? What if the line segment in this example were not infinitely long?

Answer If the line charge in this example were of finite length, the electric field would not be given by Equation 24.7. A finite line charge does not possess sufficient symmetry to make use of Gauss's law because the magnitude of the electric field is no longer constant over the surface of the gaussian cylinder: the field near the ends of the line would be different from that far from the ends. Therefore, condition (1) would not be satisfied in this situation. Furthermore, $\vec{E}$ is not perpendicular to the cylindrical surface at all points: the field vectors near the ends would have a component parallel to the line. Therefore, condition (2) would not be satisfied. For points close to a finite line charge and far from the ends, Equation 24.7 gives a good approximation of the value of the field.

It is left for you to show (see Problem 33) that the electric field inside a uniformly charged rod of finite radius and infinite length is proportional to r.

Example 24.5 A Plane of Charge

Find the electric field due to an infinite plane of positive charge with uniform surface charge density σ.

SOLUTION

Conceptualize Notice that the plane of charge is *infinitely* large. Therefore, the electric field should be the same at all points equidistant from the plane. How would you expect the electric field to depend on the distance from the plane?

Categorize Because the charge is distributed uniformly on the plane, the charge distribution is symmetric; hence, we can use Gauss's law to find the electric field.

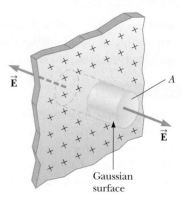

Figure 24.13 (Example 24.5) A cylindrical gaussian surface penetrating an infinite plane of charge. The flux is EA through each end of the gaussian surface and zero through its curved surface.

Analyze By symmetry, $\vec{E}$ must be perpendicular to the plane at all points. The direction of $\vec{E}$ is away from positive charges, indicating that the direction of $\vec{E}$ on one side of the plane must be opposite its direction on the other side as shown in Figure 24.13. A gaussian surface that reflects the symmetry is a small cylinder whose axis is perpendicular to the plane and whose ends each have an area A and are equidistant from the plane. Because $\vec{E}$ is parallel to the curved surface of the cylinder—and therefore perpendicular to $d\vec{A}$ at all points on this surface—condition (3) is satisfied and there is no contribution to the surface integral from this surface. For the flat ends of the cylinder, conditions (1) and (2) are satisfied. The flux through each end of the cylinder is EA; hence, the total flux through the entire gaussian surface is just that through the ends, $\Phi_E = 2EA$.

Write Gauss's law for this surface, noting that the enclosed charge is $q_{in} = \sigma A$:

$$\Phi_E = 2EA = \frac{q_{in}}{\epsilon_0} = \frac{\sigma A}{\epsilon_0}$$

Solve for E:

$$E = \frac{\sigma}{2\epsilon_0} \qquad \textbf{(24.8)}$$

Finalize Because the distance from each flat end of the cylinder to the plane does not appear in Equation 24.8, we conclude that $E = \sigma/2\epsilon_0$ at *any* distance from the plane. That is, the field is uniform everywhere. Figure 24.14 shows this uniform field due to an infinite plane of charge, seen edge-on.

WHAT IF? Suppose two infinite planes of charge are parallel to each other, one positively charged and the other negatively charged. The surface charge densities of both planes are of the same magnitude. What does the electric field look like in this situation?

Answer We first addressed this configuration in the **What If?** section of Example 23.9. The electric fields due to the two planes add in the region between the planes, resulting in a uniform field of magnitude σ/ϵ_0, and cancel elsewhere to give a field of zero. Figure 24.15 shows the field lines for such a configuration. This method is a practical way to achieve uniform electric fields with finite-sized planes placed close to each other.

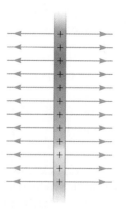

Figure 24.14 (Example 24.5) The electric field lines due to an infinite plane of positive charge.

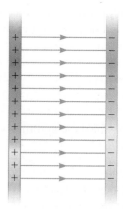

Figure 24.15 (Example 24.5) The electric field lines between two infinite planes of charge, one positive and one negative. In practice, the field lines near the edges of finite-sized sheets of charge will curve outward.

Conceptual Example 24.6 Don't Use Gauss's Law Here!

Explain why Gauss's law cannot be used to calculate the electric field near an electric dipole, a charged disk, or a triangle with a point charge at each corner.

▶ **24.6** continued

The charge distributions of all these configurations do not have sufficient symmetry to make the use of Gauss's law practical. We cannot find a closed surface surrounding any of these distributions for which all portions of the surface satisfy one or more of conditions (1) through (4) listed at the beginning of this section.

24.4 Conductors in Electrostatic Equilibrium

As we learned in Section 23.2, a good electrical conductor contains charges (electrons) that are not bound to any atom and therefore are free to move about within the material. When there is no net motion of charge within a conductor, the conductor is in **electrostatic equilibrium.** A conductor in electrostatic equilibrium has the following properties:

◀ **Properties of a conductor in electrostatic equilibrium**

1. The electric field is zero everywhere inside the conductor, whether the conductor is solid or hollow.
2. If the conductor is isolated and carries a charge, the charge resides on its surface.
3. The electric field at a point just outside a charged conductor is perpendicular to the surface of the conductor and has a magnitude σ/ϵ_0, where σ is the surface charge density at that point.
4. On an irregularly shaped conductor, the surface charge density is greatest at locations where the radius of curvature of the surface is smallest.

We verify the first three properties in the discussion that follows. The fourth property is presented here (but not verified until we have studied the appropriate material in Chapter 25) to provide a complete list of properties for conductors in electrostatic equilibrium.

We can understand the first property by considering a conducting slab placed in an external field $\vec{\mathbf{E}}$ (Fig. 24.16). The electric field inside the conductor *must* be zero, assuming electrostatic equilibrium exists. If the field were not zero, free electrons in the conductor would experience an electric force ($\vec{\mathbf{F}} = q\vec{\mathbf{E}}$) and would accelerate due to this force. This motion of electrons, however, would mean that the conductor is not in electrostatic equilibrium. Therefore, the existence of electrostatic equilibrium is consistent only with a zero field in the conductor.

Let's investigate how this zero field is accomplished. Before the external field is applied, free electrons are uniformly distributed throughout the conductor. When the external field is applied, the free electrons accelerate to the left in Figure 24.16, causing a plane of negative charge to accumulate on the left surface. The movement of electrons to the left results in a plane of positive charge on the right surface. These planes of charge create an additional electric field inside the conductor that opposes the external field. As the electrons move, the surface charge densities on the left and right surfaces increase until the magnitude of the internal field equals that of the external field, resulting in a net field of zero inside the conductor. The time it takes a good conductor to reach equilibrium is on the order of 10^{-16} s, which for most purposes can be considered instantaneous.

If the conductor is hollow, the electric field inside the conductor is also zero, whether we consider points in the conductor or in the cavity within the conductor. The zero value of the electric field in the cavity is easiest to argue with the concept of electric potential, so we will address this issue in Section 25.6.

Gauss's law can be used to verify the second property of a conductor in electrostatic equilibrium. Figure 24.17 shows an arbitrarily shaped conductor. A gaussian

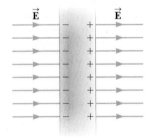

Figure 24.16 A conducting slab in an external electric field $\vec{\mathbf{E}}$. The charges induced on the two surfaces of the slab produce an electric field that opposes the external field, giving a resultant field of zero inside the slab.

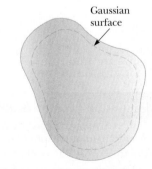

Figure 24.17 A conductor of arbitrary shape. The broken line represents a gaussian surface that can be just inside the conductor's surface.

surface is drawn inside the conductor and can be very close to the conductor's surface. As we have just shown, the electric field everywhere inside the conductor is zero when it is in electrostatic equilibrium. Therefore, the electric field must be zero at every point on the gaussian surface, in accordance with condition (4) in Section 24.3, and the net flux through this gaussian surface is zero. From this result and Gauss's law, we conclude that the net charge inside the gaussian surface is zero. Because there can be no net charge inside the gaussian surface (which is arbitrarily close to the conductor's surface), any net charge on the conductor must reside on its surface. Gauss's law does not indicate how this excess charge is distributed on the conductor's surface, only that it resides exclusively on the surface.

To verify the third property, let's begin with the perpendicularity of the field to the surface. If the field vector $\vec{E}$ had a component parallel to the conductor's surface, free electrons would experience an electric force and move along the surface; in such a case, the conductor would not be in equilibrium. Therefore, the field vector must be perpendicular to the surface.

To determine the magnitude of the electric field, we use Gauss's law and draw a gaussian surface in the shape of a small cylinder whose end faces are parallel to the conductor's surface (Fig. 24.18). Part of the cylinder is just outside the conductor, and part is inside. The field is perpendicular to the conductor's surface from the condition of electrostatic equilibrium. Therefore, condition (3) in Section 24.3 is satisfied for the curved part of the cylindrical gaussian surface: there is no flux through this part of the gaussian surface because $\vec{E}$ is parallel to the surface. There is no flux through the flat face of the cylinder inside the conductor because here $\vec{E} = 0$, which satisfies condition (4). Hence, the net flux through the gaussian surface is equal to that through only the flat face outside the conductor, where the field is perpendicular to the gaussian surface. Using conditions (1) and (2) for this face, the flux is EA, where E is the electric field just outside the conductor and A is the area of the cylinder's face. Applying Gauss's law to this surface gives

$$\Phi_E = \oint E\, dA = EA = \frac{q_{\text{in}}}{\epsilon_0} = \frac{\sigma A}{\epsilon_0}$$

where we have used $q_{\text{in}} = \sigma A$. Solving for E gives for the electric field immediately outside a charged conductor:

$$E = \frac{\sigma}{\epsilon_0} \qquad \textbf{(24.9)}$$

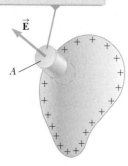

The flux through the gaussian surface is EA.

$\vec{E}$

A

Figure 24.18 A gaussian surface in the shape of a small cylinder is used to calculate the electric field immediately outside a charged conductor.

Quick Quiz 24.3 Your younger brother likes to rub his feet on the carpet and then touch you to give you a shock. While you are trying to escape the shock treatment, you discover a hollow metal cylinder in your basement, large enough to climb inside. In which of the following cases will you *not* be shocked? **(a)** You climb inside the cylinder, making contact with the inner surface, and your charged brother touches the outer metal surface. **(b)** Your charged brother is inside touching the inner metal surface and you are outside, touching the outer metal surface. **(c)** Both of you are outside the cylinder, touching its outer metal surface but not touching each other directly.

Example 24.7   A Sphere Inside a Spherical Shell

A solid insulating sphere of radius a carries a net positive charge Q uniformly distributed throughout its volume. A conducting spherical shell of inner radius b and outer radius c is concentric with the solid sphere and carries a net charge $-2Q$. Using Gauss's law, find the electric field in the regions labeled ①, ②, ③, and ④ in Figure 24.19 and the charge distribution on the shell when the entire system is in electrostatic equilibrium.

▶ **24.7** continued

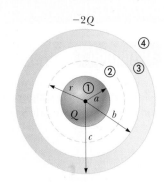

SOLUTION

Conceptualize Notice how this problem differs from Example 24.3. The charged sphere in Figure 24.10 appears in Figure 24.19, but it is now surrounded by a shell carrying a charge $-2Q$. Think about how the presence of the shell will affect the electric field of the sphere.

Categorize The charge is distributed uniformly throughout the sphere, and we know that the charge on the conducting shell distributes itself uniformly on the surfaces. Therefore, the system has spherical symmetry and we can apply Gauss's law to find the electric field in the various regions.

Figure 24.19 (Example 24.7) An insulating sphere of radius a and carrying a charge Q surrounded by a conducting spherical shell carrying a charge $-2Q$.

Analyze In region ②—between the surface of the solid sphere and the inner surface of the shell—we construct a spherical gaussian surface of radius r, where $a < r < b$, noting that the charge inside this surface is $+Q$ (the charge on the solid sphere). Because of the spherical symmetry, the electric field lines must be directed radially outward and be constant in magnitude on the gaussian surface.

The charge on the conducting shell creates zero electric field in the region $r < b$, so the shell has no effect on the field in region ② due to the sphere. Therefore, write an expression for the field in region ② as that due to the sphere from part (A) of Example 24.3:

$$E_2 = k_e \frac{Q}{r^2} \quad (\text{for } a < r < b)$$

Because the conducting shell creates zero field inside itself, it also has no effect on the field inside the sphere. Therefore, write an expression for the field in region ① as that due to the sphere from part (B) of Example 24.3:

$$E_1 = k_e \frac{Q}{a^3} r \quad (\text{for } r < a)$$

In region ④, where $r > c$, construct a spherical gaussian surface; this surface surrounds a total charge $q_{in} = Q + (-2Q) = -Q$. Therefore, model the charge distribution as a sphere with charge $-Q$ and write an expression for the field in region ④ from part (A) of Example 24.3:

$$E_4 = -k_e \frac{Q}{r^2} \quad (\text{for } r > c)$$

In region ③, the electric field must be zero because the spherical shell is a conductor in equilibrium:

$$E_3 = 0 \quad (\text{for } b < r < c)$$

Construct a gaussian surface of radius r in region ③, where $b < r < c$, and note that q_{in} must be zero because $E_3 = 0$. Find the amount of charge q_{inner} on the inner surface of the shell:

$$q_{in} = q_{sphere} + q_{inner}$$
$$q_{inner} = q_{in} - q_{sphere} = 0 - Q = -Q$$

Finalize The charge on the inner surface of the spherical shell must be $-Q$ to cancel the charge $+Q$ on the solid sphere and give zero electric field in the material of the shell. Because the net charge on the shell is $-2Q$, its outer surface must carry a charge $-Q$.

WHAT IF? How would the results of this problem differ if the sphere were conducting instead of insulating?

Answer The only change would be in region ①, where $r < a$. Because there can be no charge inside a conductor in electrostatic equilibrium, $q_{in} = 0$ for a gaussian surface of radius $r < a$; therefore, on the basis of Gauss's law and symmetry, $E_1 = 0$. In regions ②, ③, and ④, there would be no way to determine from observations of the electric field whether the sphere is conducting or insulating.

Summary

Definition

Electric flux is proportional to the number of electric field lines that penetrate a surface. If the electric field is uniform and makes an angle θ with the normal to a surface of area A, the electric flux through the surface is

$$\Phi_E = EA \cos \theta \tag{24.2}$$

In general, the electric flux through a surface is

$$\Phi_E \equiv \int_{\text{surface}} \vec{\mathbf{E}} \cdot d\vec{\mathbf{A}} \tag{24.3}$$

Concepts and Principles

Gauss's law says that the net electric flux Φ_E through any closed gaussian surface is equal to the *net* charge q_{in} inside the surface divided by ϵ_0:

$$\Phi_E = \oint \vec{\mathbf{E}} \cdot d\vec{\mathbf{A}} = \frac{q_{\text{in}}}{\epsilon_0} \tag{24.6}$$

Using Gauss's law, you can calculate the electric field due to various symmetric charge distributions.

A conductor in **electrostatic equilibrium** has the following properties:

1. The electric field is zero everywhere inside the conductor, whether the conductor is solid or hollow.
2. If the conductor is isolated and carries a charge, the charge resides on its surface.
3. The electric field at a point just outside a charged conductor is perpendicular to the surface of the conductor and has a magnitude σ/ϵ_0, where σ is the surface charge density at that point.
4. On an irregularly shaped conductor, the surface charge density is greatest at locations where the radius of curvature of the surface is smallest.

Objective Questions

1. denotes answer available in *Student Solutions Manual/Study Guide*

1. A cubical gaussian surface surrounds a long, straight, charged filament that passes perpendicularly through two opposite faces. No other charges are nearby. **(i)** Over how many of the cube's faces is the electric field zero? (a) 0 (b) 2 (c) 4 (d) 6 **(ii)** Through how many of the cube's faces is the electric flux zero? Choose from the same possibilities as in part (i).

2. A coaxial cable consists of a long, straight filament surrounded by a long, coaxial, cylindrical conducting shell. Assume charge Q is on the filament, zero net charge is on the shell, and the electric field is $E_1\hat{\mathbf{i}}$ at a particular point P midway between the filament and the inner surface of the shell. Next, you place the cable into a uniform external field $-E\hat{\mathbf{i}}$. What is the x component of the electric field at P then? (a) 0 (b) between 0 and E_1 (c) E_1 (d) between 0 and $-E_1$ (e) $-E_1$

3. In which of the following contexts can Gauss's law *not* be readily applied to find the electric field? (a) near a long, uniformly charged wire (b) above a large, uniformly charged plane (c) inside a uniformly charged ball (d) outside a uniformly charged sphere (e) Gauss's law can be readily applied to find the electric field in all these contexts.

4. A particle with charge q is located inside a cubical gaussian surface. No other charges are nearby. **(i)** If the particle is at the center of the cube, what is the flux through each one of the faces of the cube? (a) 0 (b) $q/2\epsilon_0$ (c) $q/6\epsilon_0$ (d) $q/8\epsilon_0$ (e) depends on the size of the cube **(ii)** If the particle can be moved to any point within the cube, what maximum value can the flux through one face approach? Choose from the same possibilities as in part (i).

5. Charges of 3.00 nC, -2.00 nC, -7.00 nC, and 1.00 nC are contained inside a rectangular box with length 1.00 m, width 2.00 m, and height 2.50 m. Outside the box are charges of 1.00 nC and 4.00 nC. What is the electric flux through the surface of the box? (a) 0 (b) -5.64×10^2 N $\cdot$ m^2/C (c) -1.47×10^3 N $\cdot$ m^2/C (d) 1.47×10^3 N $\cdot$ m^2/C (e) 5.64×10^2 N $\cdot$ m^2/C

6. A large, metallic, spherical shell has no net charge. It is supported on an insulating stand and has a small hole at the top. A small tack with charge Q is lowered on a silk thread through the hole into the interior of the shell. **(i)** What is the charge on the inner surface of the shell, (a) Q (b) $Q/2$ (c) 0 (d) $-Q/2$ or (e) $-Q$? Choose your answers to the following questions from

the same possibilities. **(ii)** What is the charge on the outer surface of the shell? **(iii)** The tack is now allowed to touch the interior surface of the shell. After this contact, what is the charge on the tack? **(iv)** What is the charge on the inner surface of the shell now? **(v)** What is the charge on the outer surface of the shell now?

7. Two solid spheres, both of radius 5 cm, carry identical total charges of 2 μC. Sphere A is a good conductor. Sphere B is an insulator, and its charge is distributed uniformly throughout its volume. **(i)** How do the magnitudes of the electric fields they separately create at a radial distance of 6 cm compare? (a) $E_A > E_B = 0$ (b) $E_A > E_B > 0$ (c) $E_A = E_B > 0$ (d) $0 < E_A < E_B$ (e) $0 = E_A < E_B$ **(ii)** How do the magnitudes of the electric fields they separately create at radius 4 cm compare? Choose from the same possibilities as in part (i).

8. A uniform electric field of 1.00 N/C is set up by a uniform distribution of charge in the xy plane. What is the electric field inside a metal ball placed 0.500 m above the xy plane? (a) 1.00 N/C (b) −1.00 N/C (c) 0 (d) 0.250 N/C (e) varies depending on the position inside the ball

9. A solid insulating sphere of radius 5 cm carries electric charge uniformly distributed throughout its volume. Concentric with the sphere is a conducting spherical shell with no net charge as shown in Figure OQ24.9. The inner radius of the shell is 10 cm, and the outer radius is 15 cm. No other charges are nearby. (a) Rank the magnitude of the electric field at points A (at radius 4 cm), B (radius 8 cm), C (radius 12 cm), and D (radius 16 cm) from largest to smallest. Display any cases of equality in your ranking. (b) Similarly rank the electric flux through concentric spherical surfaces through points A, B, C, and D.

Figure OQ24.9

10. A cubical gaussian surface is bisected by a large sheet of charge, parallel to its top and bottom faces. No other charges are nearby. **(i)** Over how many of the cube's faces is the electric field zero? (a) 0 (b) 2 (c) 4 (d) 6 **(ii)** Through how many of the cube's faces is the electric flux zero? Choose from the same possibilities as in part (i).

11. Rank the electric fluxes through each gaussian surface shown in Figure OQ24.11 from largest to smallest. Display any cases of equality in your ranking.

Figure OQ24.11

Conceptual Questions   **1.** denotes answer available in *Student Solutions Manual/Study Guide*

1. Consider an electric field that is uniform in direction throughout a certain volume. Can it be uniform in magnitude? Must it be uniform in magnitude? Answer these questions (a) assuming the volume is filled with an insulating material carrying charge described by a volume charge density and (b) assuming the volume is empty space. State reasoning to prove your answers.

2. A cubical surface surrounds a point charge q. Describe what happens to the total flux through the surface if (a) the charge is doubled, (b) the volume of the cube is doubled, (c) the surface is changed to a sphere, (d) the charge is moved to another location inside the surface, and (e) the charge is moved outside the surface.

3. A uniform electric field exists in a region of space containing no charges. What can you conclude about the net electric flux through a gaussian surface placed in this region of space?

4. If the total charge inside a closed surface is known but the distribution of the charge is unspecified, can you use Gauss's law to find the electric field? Explain.

5. Explain why the electric flux through a closed surface with a given enclosed charge is independent of the size or shape of the surface.

6. If more electric field lines leave a gaussian surface than enter it, what can you conclude about the net charge enclosed by that surface?

7. A person is placed in a large, hollow, metallic sphere that is insulated from ground. (a) If a large charge is placed on the sphere, will the person be harmed upon touching the inside of the sphere? (b) Explain what will happen if the person also has an initial charge whose sign is opposite that of the charge on the sphere.

8. Consider two identical conducting spheres whose surfaces are separated by a small distance. One sphere is given a large net positive charge, and the other is given a small net positive charge. It is found that the force between the spheres is attractive even though they both have net charges of the same sign. Explain how this attraction is possible.

9. A common demonstration involves charging a rubber balloon, which is an insulator, by rubbing it on your hair and then touching the balloon to a ceiling or wall, which is also an insulator. Because of the electrical attraction between the charged balloon and the neutral wall, the balloon sticks to the wall. Imagine now that we have two infinitely large, flat sheets of insulating

material. One is charged, and the other is neutral. If these sheets are brought into contact, does an attractive force exist between them as there was for the balloon and the wall?

10. On the basis of the repulsive nature of the force between like charges and the freedom of motion of charge within a conductor, explain why excess charge on an isolated conductor must reside on its surface.

11. The Sun is lower in the sky during the winter than it is during the summer. (a) How does this change affect the flux of sunlight hitting a given area on the surface of the Earth? (b) How does this change affect the weather?

Problems

ENHANCED **WebAssign**   The problems found in this chapter may be assigned online in Enhanced WebAssign

1. straightforward; **2.** intermediate;
3. challenging

[1.] full solution available in the *Student Solutions Manual/Study Guide*

AMT Analysis Model tutorial available in Enhanced WebAssign

GP Guided Problem

M Master It tutorial available in Enhanced WebAssign

W Watch It video solution available in Enhanced WebAssign

Section 24.1 Electric Flux

1. A flat surface of area 3.20 m² is rotated in a uniform electric field of magnitude $E = 6.20 \times 10^5$ N/C. Determine the electric flux through this area (a) when the electric field is perpendicular to the surface and (b) when the electric field is parallel to the surface.

2. A vertical electric field of magnitude 2.00×10^4 N/C **W** exists above the Earth's surface on a day when a thunderstorm is brewing. A car with a rectangular size of 6.00 m by 3.00 m is traveling along a dry gravel roadway sloping downward at 10.0°. Determine the electric flux through the bottom of the car.

3. A 40.0-cm-diameter circular loop is rotated in a uni- **M** form electric field until the position of maximum electric flux is found. The flux in this position is measured to be 5.20×10^5 N · m²/C. What is the magnitude of the electric field?

4. Consider a closed triangular box resting within a hori- **W** zontal electric field of magnitude $E = 7.80 \times 10^4$ N/C as shown in Figure P24.4. Calculate the electric flux through (a) the vertical rectangular surface, (b) the slanted surface, and (c) the entire surface of the box.

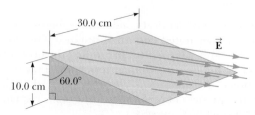

Figure P24.4

5. An electric field of magnitude 3.50 kN/C is applied **M** along the *x* axis. Calculate the electric flux through a rectangular plane 0.350 m wide and 0.700 m long (a) if the plane is parallel to the *yz* plane, (b) if the plane is parallel to the *xy* plane, and (c) if the plane contains the *y* axis and its normal makes an angle of 40.0° with the *x* axis.

6. A nonuniform electric field is given by the expression

$$\vec{E} = ay\,\hat{i} + bz\,\hat{j} + cx\,\hat{k}$$

where *a*, *b*, and *c* are constants. Determine the electric flux through a rectangular surface in the *xy* plane, extending from $x = 0$ to $x = w$ and from $y = 0$ to $y = h$.

Section 24.2 Gauss's Law

7. An uncharged, nonconducting, hollow sphere of radius 10.0 cm surrounds a 10.0-μC charge located at the origin of a Cartesian coordinate system. A drill with a radius of 1.00 mm is aligned along the *z* axis, and a hole is drilled in the sphere. Calculate the electric flux through the hole.

8. Find the net electric flux through the spherical closed surface shown in Figure P24.8. The two charges on the right are inside the spherical surface.

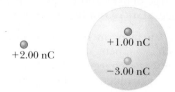

Figure P24.8

9. The following charges are located inside a submarine: **M** 5.00 μC, −9.00 μC, 27.0 μC, and −84.0 μC. (a) Calculate the net electric flux through the hull of the submarine. (b) Is the number of electric field lines leaving the submarine greater than, equal to, or less than the number entering it?

10. The electric field everywhere on the surface of a **W** thin, spherical shell of radius 0.750 m is of magnitude 890 N/C and points radially toward the center of the sphere. (a) What is the net charge within the sphere's surface? (b) What is the distribution of the charge inside the spherical shell?

11. Four closed surfaces, S_1
Ⓦ through S_4, together with
the charges $-2Q$, Q, and
$-Q$ are sketched in Figure
P24.11. (The colored lines
are the intersections of the
surfaces with the page.)
Find the electric flux
through each surface.

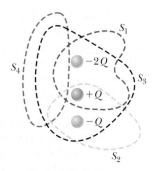

Figure P24.11

12. A charge of 170 μC is at the
center of a cube of edge
80.0 cm. No other charges
are nearby. (a) Find the
flux through each face of the cube. (b) Find the flux
through the whole surface of the cube. (c) **What If?**
Would your answers to either part (a) or part (b) change
if the charge were not at the center? Explain.

13. In the air over a particular region at an altitude of
500 m above the ground, the electric field is 120 N/C
directed downward. At 600 m above the ground, the
electric field is 100 N/C downward. What is the average
volume charge density in the layer of air between these
two elevations? Is it positive or negative?

14. A particle with charge of 12.0 μC is placed at the center of a spherical shell of radius 22.0 cm. What is the
total electric flux through (a) the surface of the shell
and (b) any hemispherical surface of the shell? (c) Do
the results depend on the radius? Explain.

15. (a) Find the net electric
flux through the cube
shown in Figure P24.15.
(b) Can you use Gauss's
law to find the electric
field on the surface of
this cube? Explain.

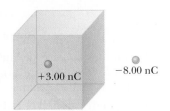

Figure P24.15

16. (a) A particle with charge
q is located a distance
d from an infinite plane. Determine the electric flux
through the plane due to the charged particle. (b) **What
If?** A particle with charge q is located a *very small* distance from the center of a *very large* square on the line
perpendicular to the square and going through its center. Determine the approximate electric flux through
the square due to the charged particle. (c) How do the
answers to parts (a) and (b) compare? Explain.

17. An infinitely long line charge having a uniform charge
per unit length λ lies a distance d from point O as
shown in Figure P24.17. Determine the total electric
flux through the surface of a sphere of radius R cen-

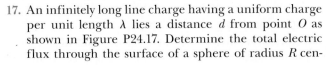

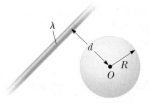

Figure P24.17

tered at O resulting from this line charge. Consider
both cases, where (a) $R < d$ and (b) $R > d$.

18. Find the net electric flux through (a) the closed spherical surface in a uniform electric field shown in Figure
P24.18a and (b) the closed cylindrical surface shown in
Figure P24.18b. (c) What can you conclude about the
charges, if any, inside the cylindrical surface?

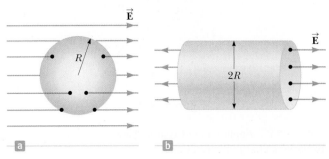

Figure P24.18

19. A particle with charge
$Q = 5.00$ μC is located
at the center of a cube
of edge $L = 0.100$ m. In
addition, six other iden-
tical charged particles
having $q = -1.00$ μC
are positioned sym-
metrically around Q as
shown in Figure P24.19.
Determine the electric
flux through one face
of the cube.

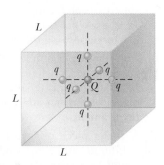

Figure P24.19
Problems 19 and 20.

20. A particle with charge
Q is located at the center of a cube of edge L. In addition, six other identical charged particles q are positioned symmetrically around Q as shown in Figure
P24.19. For each of these particles, q is a negative number. Determine the electric flux through one face of
the cube.

21. A particle with charge
Q is located a small distance δ immediately
above the center of
the flat face of a hemi-
sphere of radius R as
shown in Figure P24.21.
What is the electric flux
(a) through the curved
surface and (b) through
the flat face as $\delta \to 0$?

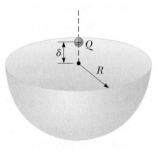

Figure P24.21

22. Figure P24.22 (page 742) represents the top view of a
cubic gaussian surface in a uniform electric field $\vec{E}$ oriented parallel to the top and bottom faces of the cube.
The field makes an angle θ with side ①, and the area of
each face is A. In symbolic form, find the electric flux
through (a) face ①, (b) face ②, (c) face ③, (d) face ④,
and (e) the top and bottom faces of the cube. (f) What

is the net electric flux through the cube? (g) How much charge is enclosed within the gaussian surface?

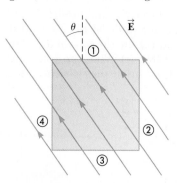

Figure P24.22

Section 24.3 Application of Gauss's Law to Various Charge Distributions

23. In nuclear fission, a nucleus of uranium-238, which contains 92 protons, can divide into two smaller spheres, each having 46 protons and a radius of 5.90×10^{-15} m. What is the magnitude of the repulsive electric force pushing the two spheres apart?

24. The charge per unit length on a long, straight filament is $-90.0\ \mu\text{C/m}$. Find the electric field (a) 10.0 cm, (b) 20.0 cm, and (c) 100 cm from the filament, where distances are measured perpendicular to the length of the filament.

25. A 10.0-g piece of Styrofoam carries a net charge of $-0.700\ \mu\text{C}$ and is suspended in equilibrium above the center of a large, horizontal sheet of plastic that has a uniform charge density on its surface. What is the charge per unit area on the plastic sheet?

26. Determine the magnitude of the electric field at the surface of a lead-208 nucleus, which contains 82 protons and 126 neutrons. Assume the lead nucleus has a volume 208 times that of one proton and consider a proton to be a sphere of radius 1.20×10^{-15} m.

27. A large, flat, horizontal sheet of charge has a charge per unit area of $9.00\ \mu\text{C/m}^2$. Find the electric field just above the middle of the sheet.

28. Suppose you fill two rubber balloons with air, suspend both of them from the same point, and let them hang down on strings of equal length. You then rub each with wool or on your hair so that the balloons hang apart with a noticeable separation between them. Make order-of-magnitude estimates of (a) the force on each, (b) the charge on each, (c) the field each creates at the center of the other, and (d) the total flux of electric field created by each balloon. In your solution, state the quantities you take as data and the values you measure or estimate for them.

29. Consider a thin, spherical shell of radius 14.0 cm with a total charge of $32.0\ \mu\text{C}$ distributed uniformly on its surface. Find the electric field (a) 10.0 cm and (b) 20.0 cm from the center of the charge distribution.

30. A nonconducting wall carries charge with a uniform density of $8.60\ \mu\text{C/cm}^2$. (a) What is the electric field 7.00 cm in front of the wall if 7.00 cm is small compared

with the dimensions of the wall? (b) Does your result change as the distance from the wall varies? Explain.

31. A uniformly charged, straight filament 7.00 m in length has a total positive charge of $2.00\ \mu\text{C}$. An uncharged cardboard cylinder 2.00 cm in length and 10.0 cm in radius surrounds the filament at its center, with the filament as the axis of the cylinder. Using reasonable approximations, find (a) the electric field at the surface of the cylinder and (b) the total electric flux through the cylinder.

32. Assume the magnitude of the electric field on each face of the cube of edge $L = 1.00$ m in Figure P24.32 is uniform and the directions of the fields on each face are as indicated. Find (a) the net electric flux through the cube and (b) the net charge inside the cube. (c) Could the net charge be a single point charge?

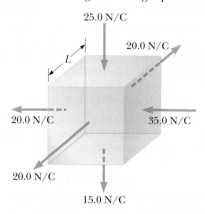

Figure P24.32

33. Consider a long, cylindrical charge distribution of radius R with a uniform charge density ρ. Find the electric field at distance r from the axis, where $r < R$.

34. A cylindrical shell of radius 7.00 cm and length 2.40 m has its charge uniformly distributed on its curved surface. The magnitude of the electric field at a point 19.0 cm radially outward from its axis (measured from the midpoint of the shell) is 36.0 kN/C. Find (a) the net charge on the shell and (b) the electric field at a point 4.00 cm from the axis, measured radially outward from the midpoint of the shell.

35. A solid sphere of radius 40.0 cm has a total positive charge of $26.0\ \mu\text{C}$ uniformly distributed throughout its volume. Calculate the magnitude of the electric field at (a) 0 cm, (b) 10.0 cm, (c) 40.0 cm, and (d) 60.0 cm from the center of the sphere.

36. Review. A particle with a charge of -60.0 nC is placed at the center of a nonconducting spherical shell of inner radius 20.0 cm and outer radius 25.0 cm. The spherical shell carries charge with a uniform density of $-1.33\ \mu\text{C/m}^3$. A proton moves in a circular orbit just outside the spherical shell. Calculate the speed of the proton.

Section 24.4 Conductors in Electrostatic Equilibrium

37. A long, straight metal rod has a radius of 5.00 cm and a charge per unit length of 30.0 nC/m. Find the electric field (a) 3.00 cm, (b) 10.0 cm, and (c) 100 cm from the axis of the rod, where distances are measured perpendicular to the rod's axis.

38. *Why is the following situation impossible?* A solid copper sphere of radius 15.0 cm is in electrostatic equilibrium and carries a charge of 40.0 nC. Figure P24.38 shows the magnitude of the electric field as a function of radial position r measured from the center of the sphere.

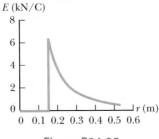

Figure P24.38

39. A solid metallic sphere of radius a carries total charge Q. No other charges are nearby. The electric field just outside its surface is $k_e Q / a^2$ radially outward. At this close point, the uniformly charged surface of the sphere looks exactly like a uniform flat sheet of charge. Is the electric field here given by σ / ϵ_0 or by $\sigma / 2\epsilon_0$?

40. A positively charged particle is at a distance $R/2$ from the center of an uncharged thin, conducting, spherical shell of radius R. Sketch the electric field lines set up by this arrangement both inside and outside the shell.

41. A very large, thin, flat plate of aluminum of area A has a total charge Q uniformly distributed over its surfaces. Assuming the same charge is spread uniformly over the *upper* surface of an otherwise identical glass plate, compare the electric fields just above the center of the upper surface of each plate.

42. In a certain region of space, the electric field is $\vec{E} = 6.00 \times 10^3 \, x^2 \hat{i}$, where $\vec{E}$ is in newtons per coulomb and x is in meters. Electric charges in this region are at rest and remain at rest. (a) Find the volume density of electric charge at $x = 0.300$ m. *Suggestion:* Apply Gauss's law to a box between $x = 0.300$ m and $x = 0.300$ m $+ \, dx$. (b) Could this region of space be inside a conductor?

43. Two identical conducting spheres each having a radius of 0.500 cm are connected by a light, 2.00-m-long conducting wire. A charge of 60.0 μC is placed on one of the conductors. Assume the surface distribution of charge on each sphere is uniform. Determine the tension in the wire.

44. A square plate of copper with 50.0-cm sides has no net charge and is placed in a region of uniform electric field of 80.0 kN/C directed perpendicularly to the plate. Find (a) the charge density of each face of the plate and (b) the total charge on each face.

45. A long, straight wire is surrounded by a hollow metal cylinder whose axis coincides with that of the wire. The wire has a charge per unit length of λ, and the cylinder has a net charge per unit length of 2λ. From this information, use Gauss's law to find (a) the charge per unit length on the inner surface of the cylinder, (b) the charge per unit length on the outer surface of the cylinder, and (c) the electric field outside the cylinder a distance r from the axis.

46. A thin, square, conducting plate 50.0 cm on a side lies in the xy plane. A total charge of 4.00×10^{-8} C is placed on the plate. Find (a) the charge density on each face of the plate, (b) the electric field just above the plate, and (c) the electric field just below the plate. You may assume the charge density is uniform.

47. A solid conducting sphere of radius 2.00 cm has a charge of 8.00 μC. A conducting spherical shell of inner radius 4.00 cm and outer radius 5.00 cm is concentric with the solid sphere and has a charge of -4.00 μC. Find the electric field at (a) $r = 1.00$ cm, (b) $r = 3.00$ cm, (c) $r = 4.50$ cm, and (d) $r = 7.00$ cm from the center of this charge configuration.

Additional Problems

48. Consider a plane surface in a uniform electric field as in Figure P24.48, where $d = 15.0$ cm and $\theta = 70.0°$. If the net flux through the surface is 6.00 N · m²/C, find the magnitude of the electric field.

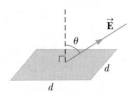

Figure P24.48
Problems 48 and 49.

49. Find the electric flux through the plane surface shown in Figure P24.48 if $\theta = 60.0°$, $E = 350$ N/C, and $d = 5.00$ cm. The electric field is uniform over the entire area of the surface.

50. A hollow, metallic, spherical shell has exterior radius 0.750 m, carries no net charge, and is supported on an insulating stand. The electric field everywhere just outside its surface is 890 N/C radially toward the center of the sphere. Explain what you can conclude about (a) the amount of charge on the exterior surface of the sphere and the distribution of this charge, (b) the amount of charge on the interior surface of the sphere and its distribution, and (c) the amount of charge inside the shell and its distribution.

51. A sphere of radius $R = 1.00$ m surrounds a particle with charge $Q = 50.0$ μC located at its center as shown in Figure P24.51. Find the electric flux through a circular cap of half-angle $\theta = 45.0°$.

52. A sphere of radius R surrounds a particle with charge Q located at its center as shown in Figure P24.51. Find the electric flux through a circular cap of half-angle θ.

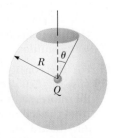

Figure P24.51
Problems 51 and 52.

53. A very large conducting plate lying in the xy plane carries a charge per unit area of σ. A second such plate located above the first plate at $z = z_0$ and oriented parallel to the xy plane carries a charge per unit area of -2σ. Find the electric field for (a) $z < 0$, (b) $0 < z < z_0$, and (c) $z > z_0$.

54. A solid, insulating sphere of radius a has a uniform charge density throughout its volume and a total charge Q. Concentric with this sphere is an uncharged, conducting, hollow sphere whose inner and outer radii are b and c as shown in Figure P24.54 (page 744). We wish to

understand completely the charges and electric fields at all locations. (a) Find the charge contained within a sphere of radius $r < a$. (b) From this value, find the magnitude of the electric field for $r < a$. (c) What charge is contained within a sphere of radius r when $a < r < b$? (d) From this value, find the magnitude of the electric field for r when $a < r < b$. (e) Now consider r when $b < r < c$. What is the magnitude of the electric field for this range of values of r? (f) From this value, what must be the charge on the inner surface of the hollow sphere? (g) From part (f), what must be the charge on the outer surface of the hollow sphere? (h) Consider the three spherical surfaces of radii a, b, and c. Which of these surfaces has the largest magnitude of surface charge density?

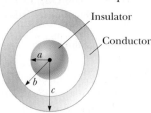

Figure P24.54
Problems 54, 55, and 57.

55. A solid insulating sphere of radius $a = 5.00$ cm carries a net positive charge of $Q = 3.00$ μC uniformly distributed throughout its volume. Concentric with this sphere is a conducting spherical shell with inner radius $b = 10.0$ cm and outer radius $c = 15.0$ cm as shown in Figure P24.54, having net charge $q = -1.00$ μC. Prepare a graph of the magnitude of the electric field due to this configuration versus r for $0 < r < 25.0$ cm.

56. Two infinite, nonconducting sheets of charge are parallel to each other as shown in Figure P24.56. The sheet on the left has a uniform surface charge density σ, and the one on the right has a uniform charge density $-\sigma$. Calculate the electric field at points (a) to the left of, (b) in between, and (c) to the right of the two sheets. (d) **What If?** Find the electric fields in all three regions if both sheets have *positive* uniform surface charge densities of value σ.

Figure P24.56

57. For the configuration shown in Figure P24.54, suppose $a = 5.00$ cm, $b = 20.0$ cm, and $c = 25.0$ cm. Furthermore, suppose the electric field at a point 10.0 cm from the center is measured to be 3.60×10^3 N/C radially inward and the electric field at a point 50.0 cm from the center is of magnitude 200 N/C and points radially outward. From this information, find (a) the charge on the insulating sphere, (b) the net charge on the hollow conducting sphere, (c) the charge on the inner surface of the hollow conducting sphere, and (d) the charge on the outer surface of the hollow conducting sphere.

58. An insulating solid sphere of radius a has a uniform volume charge density and carries a total positive charge Q. A spherical gaussian surface of radius r, which shares a common center with the insulating sphere, is inflated starting from $r = 0$. (a) Find an expression for the electric flux passing through the surface of the gaussian sphere as a function of r for $r < a$. (b) Find an expression for the electric flux for $r > a$. (c) Plot the flux versus r.

59. A uniformly charged spherical shell with positive surface charge density σ contains a circular hole in its surface. The radius r of the hole is small compared with the radius R of the sphere. What is the electric field at the center of the hole? *Suggestion:* This problem can be solved by using the principle of superposition.

60. An infinitely long, cylindrical, insulating shell of inner radius a and outer radius b has a uniform volume charge density ρ. A line of uniform linear charge density λ is placed along the axis of the shell. Determine the electric field for (a) $r < a$, (b) $a < r < b$, and (c) $r > b$.

Challenge Problems

61. A slab of insulating material has a nonuniform positive charge density $\rho = Cx^2$, where x is measured from the center of the slab as shown in Figure P24.61 and C is a constant. The slab is infinite in the y and z directions. Derive expressions for the electric field in (a) the exterior regions ($|x| > d/2$) and (b) the interior region of the slab ($-d/2 < x < d/2$).

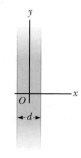

Figure P24.61
Problems 61 and 69.

62. **Review.** An early (incorrect) model of the hydrogen atom, suggested by J. J. Thomson, proposed that a positive cloud of charge $+e$ was uniformly distributed throughout the volume of a sphere of radius R, with the electron (an equal-magnitude negatively charged particle $-e$) at the center. (a) Using Gauss's law, show that the electron would be in equilibrium at the center and, if displaced from the center a distance $r < R$, would experience a restoring force of the form $F = -Kr$, where K is a constant. (b) Show that $K = k_e e^2/R^3$. (c) Find an expression for the frequency f of simple harmonic oscillations that an electron of mass m_e would undergo if displaced a small distance ($< R$) from the center and released. (d) Calculate a numerical value for R that would result in a frequency of 2.47×10^{15} Hz, the frequency of the light radiated in the most intense line in the hydrogen spectrum.

63. A closed surface with dimensions $a = b = 0.400$ m and $c = 0.600$ m is located as shown in Figure P24.63. The left edge of the closed surface is located at position $x = a$. The electric field throughout the region is nonuniform and is given by $\vec{E} = (3.00 + 2.00x^2)\hat{i}$ N/C, where x is in meters. (a) Calculate the net electric flux

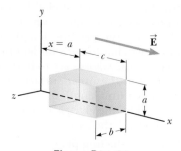

Figure P24.63

leaving the closed surface. (b) What net charge is enclosed by the surface?

64. A sphere of radius $2a$ is made of a nonconducting material that has a uniform volume charge density ρ. Assume the material does not affect the electric field. A spherical cavity of radius a is now removed from the sphere as shown in Figure P24.64. Show that the electric field within the cavity is uniform and is given by $E_x = 0$ and $E_y = \rho a/3\epsilon_0$.

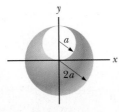

Figure P24.64

65. A spherically symmetric charge distribution has a charge density given by $\rho = a/r$, where a is constant. Find the electric field within the charge distribution as a function of r. *Note:* The volume element dV for a spherical shell of radius r and thickness dr is equal to $4\pi r^2 dr$.

66. A solid insulating sphere of radius R has a nonuniform charge density that varies with r according to the expression $\rho = Ar^2$, where A is a constant and $r < R$ is measured from the center of the sphere. (a) Show that the magnitude of the electric field outside $(r > R)$ the sphere is $E = AR^5/5\epsilon_0 r^2$. (b) Show that the magnitude of the electric field inside $(r < R)$ the sphere is $E = Ar^3/5\epsilon_0$. *Note:* The volume element dV for a spherical shell of radius r and thickness dr is equal to $4\pi r^2 dr$.

67. An infinitely long insulating cylinder of radius R has a volume charge density that varies with the radius as

$$\rho = \rho_0\left(a - \frac{r}{b}\right)$$

where ρ_0, a, and b are positive constants and r is the distance from the axis of the cylinder. Use Gauss's law to determine the magnitude of the electric field at radial distances (a) $r < R$ and (b) $r > R$.

68. A particle with charge Q is located on the axis of a circle of radius R at a distance b from the plane of the circle (Fig. P24.68). Show that if one-fourth of the electric flux from the charge passes through the circle, then $R = \sqrt{3}b$.

Figure P24.68

69. **Review.** A slab of insulating material (infinite in the y and z directions) has a thickness d and a uniform positive charge density ρ. An edge view of the slab is shown in Figure P24.61. (a) Show that the magnitude of the electric field a distance x from its center and inside the slab is $E = \rho x/\epsilon_0$. (b) **What If?** Suppose an electron of charge $-e$ and mass m_e can move freely within the slab. It is released from rest at a distance x from the center. Show that the electron exhibits simple harmonic motion with a frequency

$$f = \frac{1}{2\pi}\sqrt{\frac{\rho e}{m_e \epsilon_0}}$$

Electric Potential

Processes occurring during thunderstorms cause large differences in electric potential between a thundercloud and the ground. The result of this potential difference is an electrical discharge that we call lightning, such as this display. Notice at the left that a downward channel of lightning (a *stepped leader*) is about to make contact with a channel coming up from the ground (a *return stroke*). (Costazzurra/Shutterstock.com)

In Chapter 23, we linked our new study of electromagnetism to our earlier studies of *force*. Now we make a new link to our earlier investigations into *energy*. The concept of potential energy was introduced in Chapter 7 in connection with such conservative forces as the gravitational force and the elastic force exerted by a spring. By using the law of conservation of energy, we could solve various problems in mechanics that were not solvable with an approach using forces. The concept of potential energy is also of great value in the study of electricity. Because the electrostatic force is conservative, electrostatic phenomena can be conveniently described in terms of an electric potential energy. This idea enables us to define a quantity known as *electric potential*. Because the electric potential at any point in an electric field is a scalar quantity, we can use it to describe electrostatic phenomena more simply than if we were to rely only on the electric field and electric forces. The concept of electric potential is of great practical value in the operation of electric circuits and devices that we will study in later chapters.

25.1 Electric Potential and Potential Difference

When a charge q is placed in an electric field $\vec{\mathbf{E}}$ created by some source charge distribution, the particle in a field model tells us that there is an electric force $q\vec{\mathbf{E}}$

acting on the charge. This force is conservative because the force between charges described by Coulomb's law is conservative. Let us identify the charge and the field as a system. If the charge is free to move, it will do so in response to the electric force. Therefore, the electric field will be doing work on the charge. This work is *internal* to the system. This situation is similar to that in a gravitational system: When an object is released near the surface of the Earth, the gravitational force does work on the object. This work is internal to the object–Earth system as discussed in Sections 7.7 and 7.8.

When analyzing electric and magnetic fields, it is common practice to use the notation $d\vec{s}$ to represent an infinitesimal displacement vector that is oriented tangent to a path through space. This path may be straight or curved, and an integral performed along this path is called either a *path integral* or a *line integral* (the two terms are synonymous).

For an infinitesimal displacement $d\vec{s}$ of a point charge q immersed in an electric field, the work done within the charge–field system by the electric field on the charge is $W_{int} = \vec{F}_e \cdot d\vec{s} = q\vec{E} \cdot d\vec{s}$. Recall from Equation 7.26 that internal work done in a system is equal to the negative of the change in the potential energy of the system: $W_{int} = -\Delta U$. Therefore, as the charge q is displaced, the electric potential energy of the charge–field system is changed by an amount $dU = -W_{int} = -q\vec{E} \cdot d\vec{s}$. For a finite displacement of the charge from some point Ⓐ in space to some other point Ⓑ, the change in electric potential energy of the system is

$$\Delta U = -q \int_{Ⓐ}^{Ⓑ} \vec{E} \cdot d\vec{s} \tag{25.1}$$

◀ Change in electric potential energy of a system

The integration is performed along the path that q follows as it moves from Ⓐ to Ⓑ. Because the force $q\vec{E}$ is conservative, this line integral does not depend on the path taken from Ⓐ to Ⓑ.

For a given position of the charge in the field, the charge–field system has a potential energy U relative to the configuration of the system that is defined as $U = 0$. Dividing the potential energy by the charge gives a physical quantity that depends only on the source charge distribution and has a value at every point in an electric field. This quantity is called the **electric potential** (or simply the **potential**) V:

$$V = \frac{U}{q} \tag{25.2}$$

Because potential energy is a scalar quantity, electric potential also is a scalar quantity.

The **potential difference** $\Delta V = V_{Ⓑ} - V_{Ⓐ}$ between two points Ⓐ and Ⓑ in an electric field is defined as the change in electric potential energy of the system when a charge q is moved between the points (Eq. 25.1) divided by the charge:

$$\Delta V \equiv \frac{\Delta U}{q} = -\int_{Ⓐ}^{Ⓑ} \vec{E} \cdot d\vec{s} \tag{25.3}$$

◀ Potential difference between two points

In this definition, the infinitesimal displacement $d\vec{s}$ is interpreted as the displacement between two points in space rather than the displacement of a point charge as in Equation 25.1.

Just as with potential energy, only *differences* in electric potential are meaningful. We often take the value of the electric potential to be zero at some convenient point in an electric field.

Potential difference should not be confused with difference in potential energy. The potential *difference* between Ⓐ and Ⓑ exists solely because of a source charge and depends on the source charge distribution (consider points Ⓐ and Ⓑ in the discussion above *without* the presence of the charge q). For a potential *energy* to exist, we must have a system of two or more charges. The potential

Pitfall Prevention 25.1
Potential and Potential Energy
The *potential is characteristic of the field only*, independent of a charged particle that may be placed in the field. *Potential energy is characteristic of the charge-field system* due to an interaction between the field and a charged particle placed in the field.

energy belongs to the system and changes only if a charge is moved relative to the rest of the system. This situation is similar to that for the electric field. An electric *field* exists solely because of a source charge. An electric *force* requires two charges: the source charge to set up the field and another charge placed within that field.

Let's now consider the situation in which an external agent moves the charge in the field. If the agent moves the charge from Ⓐ to Ⓑ without changing the kinetic energy of the charge, the agent performs work that changes the potential energy of the system: $W = \Delta U$. From Equation 25.3, the work done by an external agent in moving a charge q through an electric field at constant velocity is

$$W = q\,\Delta V \tag{25.4}$$

Because electric potential is a measure of potential energy per unit charge, the SI unit of both electric potential and potential difference is joules per coulomb, which is defined as a **volt** (V):

$$1\,\text{V} \equiv 1\,\text{J/C}$$

That is, as we can see from Equation 25.4, 1 J of work must be done to move a 1-C charge through a potential difference of 1 V.

Equation 25.3 shows that potential difference also has units of electric field times distance. It follows that the SI unit of electric field (N/C) can also be expressed in volts per meter:

$$1\,\text{N/C} = 1\,\text{V/m}$$

Therefore, we can state a new interpretation of the electric field:

> The electric field is a measure of the rate of change of the electric potential with respect to position.

A unit of energy commonly used in atomic and nuclear physics is the **electron volt** (eV), which is defined as the energy a charge–field system gains or loses when a charge of magnitude e (that is, an electron or a proton) is moved through a potential difference of 1 V. Because 1 V = 1 J/C and the fundamental charge is equal to 1.60×10^{-19} C, the electron volt is related to the joule as follows:

$$1\,\text{eV} = 1.60 \times 10^{-19}\,\text{C} \cdot \text{V} = 1.60 \times 10^{-19}\,\text{J} \tag{25.5}$$

For instance, an electron in the beam of a typical dental x-ray machine may have a speed of 1.4×10^8 m/s. This speed corresponds to a kinetic energy 1.1×10^{-14} J (using relativistic calculations as discussed in Chapter 39), which is equivalent to 6.7×10^4 eV. Such an electron has to be accelerated from rest through a potential difference of 67 kV to reach this speed.

Ⓠuick Quiz 25.1 In Figure 25.1, two points Ⓐ and Ⓑ are located within a region in which there is an electric field. **(i)** How would you describe the potential difference $\Delta V = V_Ⓑ - V_Ⓐ$? (a) It is positive. (b) It is negative. (c) It is zero. **(ii)** A negative charge is placed at Ⓐ and then moved to Ⓑ. How would you describe the change in potential energy of the charge–field system for this process? Choose from the same possibilities.

25.2 Potential Difference in a Uniform Electric Field

Equations 25.1 and 25.3 hold in all electric fields, whether uniform or varying, but they can be simplified for the special case of a uniform field. First, consider a uniform electric field directed along the negative y axis as shown in Figure 25.2a. Let's calculate the potential difference between two points Ⓐ and Ⓑ separated by a dis-

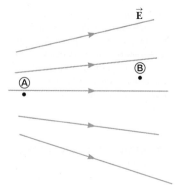

Figure 25.1 (Quick Quiz 25.1) Two points in an electric field.

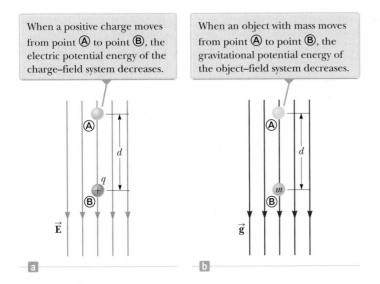

When a positive charge moves from point Ⓐ to point Ⓑ, the electric potential energy of the charge–field system decreases.

When an object with mass moves from point Ⓐ to point Ⓑ, the gravitational potential energy of the object–field system decreases.

Figure 25.2 (a) When the electric field $\vec{\mathbf{E}}$ is directed downward, point Ⓑ is at a lower electric potential than point Ⓐ. (b) A gravitational analog to the situation in (a).

tance d, where the displacement $\vec{\mathbf{s}}$ points from Ⓐ toward Ⓑ and is parallel to the field lines. Equation 25.3 gives

$$V_{Ⓑ} - V_{Ⓐ} = \Delta V = -\int_{Ⓐ}^{Ⓑ} \vec{\mathbf{E}} \cdot d\vec{\mathbf{s}} = -\int_{Ⓐ}^{Ⓑ} E\, ds\, (\cos 0°) = -\int_{Ⓐ}^{Ⓑ} E\, ds$$

Because E is constant, it can be removed from the integral sign, which gives

$$\Delta V = -E \int_{Ⓐ}^{Ⓑ} ds$$

$$\Delta V = -Ed \qquad (25.6)$$

◀ **Potential difference between two points in a uniform electric field**

The negative sign indicates that the electric potential at point Ⓑ is lower than at point Ⓐ; that is, $V_{Ⓑ} < V_{Ⓐ}$. Electric field lines *always* point in the direction of decreasing electric potential as shown in Figure 25.2a.

Now suppose a charge q moves from Ⓐ to Ⓑ. We can calculate the change in the potential energy of the charge–field system from Equations 25.3 and 25.6:

$$\Delta U = q\, \Delta V = -qEd \qquad (25.7)$$

This result shows that if q is positive, then ΔU is negative. Therefore, in a system consisting of a positive charge and an electric field, the electric potential energy of the system decreases when the charge moves in the direction of the field. If a positive charge is released from rest in this electric field, it experiences an electric force $q\vec{\mathbf{E}}$ in the direction of $\vec{\mathbf{E}}$ (downward in Fig. 25.2a). Therefore, it accelerates downward, gaining kinetic energy. As the charged particle gains kinetic energy, the electric potential energy of the charge–field system decreases by an equal amount. This equivalence should not be surprising; it is simply conservation of mechanical energy in an isolated system as introduced in Chapter 8.

Figure 25.2b shows an analogous situation with a gravitational field. When a particle with mass m is released in a gravitational field, it accelerates downward, gaining kinetic energy. At the same time, the gravitational potential energy of the object–field system decreases.

The comparison between a system of a positive charge residing in an electrical field and an object with mass residing in a gravitational field in Figure 25.2 is useful for conceptualizing electrical behavior. The electrical situation, however, has one feature that the gravitational situation does not: the charge can be negative. If q is negative, then ΔU in Equation 25.7 is positive and the situation is reversed.

Pitfall Prevention 25.4

The Sign of ΔV The negative sign in Equation 25.6 is due to the fact that we started at point Ⓐ and moved to a new point in the *same* direction as the electric field lines. If we started from Ⓑ and moved to Ⓐ, the potential difference would be $+Ed$. In a uniform electric field, the magnitude of the potential difference is Ed and the sign can be determined by the direction of travel.

Figure 25.3 A uniform electric field directed along the positive x axis. Three points in the electric field are labeled.

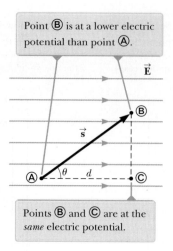

Point Ⓑ is at a lower electric potential than point Ⓐ.

$\vec{E}$

Points Ⓑ and Ⓒ are at the *same* electric potential.

A system consisting of a negative charge and an electric field *gains* electric potential energy when the charge moves in the direction of the field. If a negative charge is released from rest in an electric field, it accelerates in a direction *opposite* the direction of the field. For the negative charge to move in the direction of the field, an external agent must apply a force and do positive work on the charge.

Now consider the more general case of a charged particle that moves between Ⓐ and Ⓑ in a uniform electric field such that the vector $\vec{s}$ is *not* parallel to the field lines as shown in Figure 25.3. In this case, Equation 25.3 gives

◀ **Change in potential between two points in a uniform electric field**

$$\Delta V = -\int_{Ⓐ}^{Ⓑ} \vec{E} \cdot d\vec{s} = -\vec{E} \cdot \int_{Ⓐ}^{Ⓑ} d\vec{s} = -\vec{E} \cdot \vec{s} \qquad (25.8)$$

where again $\vec{E}$ was removed from the integral because it is constant. The change in potential energy of the charge–field system is

$$\Delta U = q\Delta V = -q\vec{E} \cdot \vec{s} \qquad (25.9)$$

Finally, we conclude from Equation 25.8 that all points in a plane perpendicular to a uniform electric field are at the same electric potential. We can see that in Figure 25.3, where the potential difference $V_Ⓑ - V_Ⓐ$ is equal to the potential difference $V_Ⓒ - V_Ⓐ$. (Prove this fact to yourself by working out two dot products for $\vec{E} \cdot \vec{s}$: one for $\vec{s}_{Ⓐ \to Ⓑ}$, where the angle θ between $\vec{E}$ and $\vec{s}$ is arbitrary as shown in Figure 25.3, and one for $\vec{s}_{Ⓐ \to Ⓒ}$, where $\theta = 0$.) Therefore, $V_Ⓑ = V_Ⓒ$. The name **equipotential surface** is given to any surface consisting of a continuous distribution of points having the same electric potential.

The equipotential surfaces associated with a uniform electric field consist of a family of parallel planes that are all perpendicular to the field. Equipotential surfaces associated with fields having other symmetries are described in later sections.

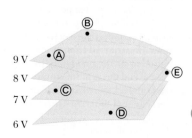

Figure 25.4 (Quick Quiz 25.2) Four equipotential surfaces.

Quick Quiz 25.2 The labeled points in Figure 25.4 are on a series of equipotential surfaces associated with an electric field. Rank (from greatest to least) the work done by the electric field on a positively charged particle that moves from Ⓐ to Ⓑ, from Ⓑ to Ⓒ, from Ⓒ to Ⓓ, and from Ⓓ to Ⓔ.

| Example 25.1 | The Electric Field Between Two Parallel Plates of Opposite Charge |

A battery has a specified potential difference ΔV between its terminals and establishes that potential difference between conductors attached to the terminals. A 12-V battery is connected between two parallel plates as shown in Figure 25.5. The separation between the plates is $d = 0.30$ cm, and we assume the electric field between the plates to be uniform. (This assumption is reasonable if the plate separation is small relative to the plate dimensions and we do not consider locations near the plate edges.) Find the magnitude of the electric field between the plates.

▶ **25.1** continued

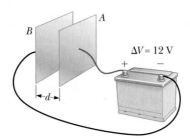

Figure 25.5 (Example 25.1) A 12-V battery connected to two parallel plates. The electric field between the plates has a magnitude given by the potential difference ΔV divided by the plate separation d.

SOLUTION

Conceptualize In Example 24.5, we illustrated the uniform electric field between parallel plates. The new feature to this problem is that the electric field is related to the new concept of electric potential.

Categorize The electric field is evaluated from a relationship between field and potential given in this section, so we categorize this example as a substitution problem.

Use Equation 25.6 to evaluate the magnitude of the electric field between the plates:

$$E = \frac{|V_B - V_A|}{d} = \frac{12\ \text{V}}{0.30 \times 10^{-2}\ \text{m}} = \boxed{4.0 \times 10^3\ \text{V/m}}$$

The configuration of plates in Figure 25.5 is called a *parallel-plate capacitor* and is examined in greater detail in Chapter 26.

Example 25.2   Motion of a Proton in a Uniform Electric Field   AM

A proton is released from rest at point Ⓐ in a uniform electric field that has a magnitude of 8.0×10^4 V/m (Fig. 25.6). The proton undergoes a displacement of magnitude $d = 0.50$ m to point Ⓑ in the direction of $\vec{\mathbf{E}}$. Find the speed of the proton after completing the displacement.

SOLUTION

Conceptualize Visualize the proton in Figure 25.6 moving downward through the potential difference. The situation is analogous to an object falling through a gravitational field. Also compare this example to Example 23.10 where a positive charge was moving in a uniform electric field. In that example, we applied the particle under constant acceleration and nonisolated system models. Now that we have investigated electric potential energy, what model can we use here?

Categorize The system of the proton and the two plates in Figure 25.6 does not interact with the environment, so we model it as an *isolated system* for *energy*.

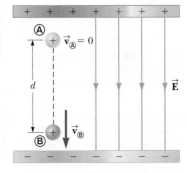

Figure 25.6 (Example 25.2) A proton accelerates from Ⓐ to Ⓑ in the direction of the electric field.

....

Analyze

Write the appropriate reduction of Equation 8.2, the conservation of energy equation, for the isolated system of the charge and the electric field:

$$\Delta K + \Delta U = 0$$

Substitute the changes in energy for both terms:

$$(\tfrac{1}{2}mv^2 - 0) + e\,\Delta V = 0$$

Solve for the final speed of the proton and substitute for ΔV from Equation 25.6:

$$v = \sqrt{\frac{-2e\,\Delta V}{m}} = \sqrt{\frac{-2e(-Ed)}{m}} = \sqrt{\frac{2eEd}{m}}$$

Substitute numerical values:

$$v = \sqrt{\frac{2(1.6 \times 10^{-19}\ \text{C})(8.0 \times 10^4\ \text{V})(0.50\ \text{m})}{1.67 \times 10^{-27}\ \text{kg}}}$$

$$= \boxed{2.8 \times 10^6\ \text{m/s}}$$

continued

▶ **25.2** continued

Finalize Because ΔV is negative for the field, ΔU is also negative for the proton–field system. The negative value of ΔU means the potential energy of the system decreases as the proton moves in the direction of the electric field. As the proton accelerates in the direction of the field, it gains kinetic energy while the electric potential energy of the system decreases at the same time.

Figure 25.6 is oriented so that the proton moves downward. The proton's motion is analogous to that of an object falling in a gravitational field. Although the gravitational field is always downward at the surface of the Earth, an electric field can be in any direction, depending on the orientation of the plates creating the field. Therefore, Figure 25.6 could be rotated 90° or 180° and the proton could move horizontally or upward in the electric field!

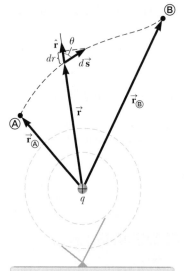

The two dashed circles represent intersections of spherical equipotential surfaces with the page.

Figure 25.7 The potential difference between points Ⓐ and Ⓑ due to a point charge q depends *only* on the initial and final radial coordinates $r_Ⓐ$ and $r_Ⓑ$.

Pitfall Prevention 25.5

Similar Equation Warning Do not confuse Equation 25.11 for the electric potential of a point charge with Equation 23.9 for the electric field of a point charge. Potential is proportional to $1/r$, whereas the magnitude of the field is proportional to $1/r^2$. The effect of a charge on the space surrounding it can be described in two ways. The charge sets up a vector electric field $\vec{E}$, which is related to the force experienced by a charge placed in the field. It also sets up a scalar potential V, which is related to the potential energy of the two-charge system when a charge is placed in the field.

25.3 Electric Potential and Potential Energy Due to Point Charges

As discussed in Section 23.4, an isolated positive point charge q produces an electric field directed radially outward from the charge. To find the electric potential at a point located a distance r from the charge, let's begin with the general expression for potential difference, Equation 25.3,

$$V_Ⓑ - V_Ⓐ = -\int_Ⓐ^Ⓑ \vec{E} \cdot d\vec{s}$$

where Ⓐ and Ⓑ are the two arbitrary points shown in Figure 25.7. At any point in space, the electric field due to the point charge is $\vec{E} = (k_e q/r^2)\hat{r}$ (Eq. 23.9), where $\hat{r}$ is a unit vector directed radially outward from the charge. Therefore, the quantity $\vec{E} \cdot d\vec{s}$ can be expressed as

$$\vec{E} \cdot d\vec{s} = k_e \frac{q}{r^2} \hat{r} \cdot d\vec{s}$$

Because the magnitude of $\hat{r}$ is 1, the dot product $\hat{r} \cdot d\vec{s} = ds \cos \theta$, where θ is the angle between $\hat{r}$ and $d\vec{s}$. Furthermore, $ds \cos \theta$ is the projection of $d\vec{s}$ onto $\hat{r}$; therefore, $ds \cos \theta = dr$. That is, any displacement $d\vec{s}$ along the path from point Ⓐ to point Ⓑ produces a change dr in the magnitude of $\vec{r}$, the position vector of the point relative to the charge creating the field. Making these substitutions, we find that $\vec{E} \cdot d\vec{s} = (k_e q/r^2)dr$; hence, the expression for the potential difference becomes

$$V_Ⓑ - V_Ⓐ = -k_e q \int_{r_Ⓐ}^{r_Ⓑ} \frac{dr}{r^2} = k_e \frac{q}{r} \Big|_{r_Ⓐ}^{r_Ⓑ}$$

$$V_Ⓑ - V_Ⓐ = k_e q \left[\frac{1}{r_Ⓑ} - \frac{1}{r_Ⓐ} \right] \qquad (25.10)$$

Equation 25.10 shows us that the integral of $\vec{E} \cdot d\vec{s}$ is *independent* of the path between points Ⓐ and Ⓑ. Multiplying by a charge q_0 that moves between points Ⓐ and Ⓑ, we see that the integral of $q_0 \vec{E} \cdot d\vec{s}$ is also independent of path. This latter integral, which is the work done by the electric force on the charge q_0, shows that the electric force is conservative (see Section 7.7). We define a field that is related to a conservative force as a **conservative field**. Therefore, Equation 25.10 tells us that the electric field of a fixed point charge q is conservative. Furthermore, Equation 25.10 expresses the important result that the potential difference between any two points Ⓐ and Ⓑ in a field created by a point charge depends only on the radial coordinates $r_Ⓐ$ and $r_Ⓑ$. It is customary to choose the reference of electric potential for a point charge to be $V = 0$ at $r_Ⓐ = \infty$. With this reference choice, the electric potential due to a point charge at any distance r from the charge is

$$V = k_e \frac{q}{r} \qquad (25.11)$$

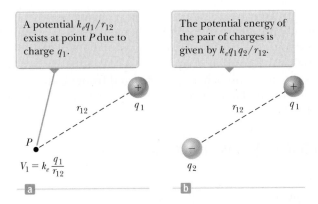

Figure 25.8 (a) Charge q_1 establishes an electric potential V_1 at point P. (b) Charge q_2 is brought from infinity to point P.

We obtain the electric potential resulting from two or more point charges by applying the superposition principle. That is, the total electric potential at some point P due to several point charges is the sum of the potentials due to the individual charges. For a group of point charges, we can write the total electric potential at P as

$$V = k_e \sum_i \frac{q_i}{r_i} \qquad (25.12)$$

◀ **Electric potential due to several point charges**

Figure 25.8a shows a charge q_1, which sets up an electric field throughout space. The charge also establishes an electric potential at all points, including point P, where the electric potential is V_1. Now imagine that an external agent brings a charge q_2 from infinity to point P. The work that must be done to do this is given by Equation 25.4, $W = q_2 \Delta V$. This work represents a transfer of energy across the boundary of the two-charge system, and the energy appears in the system as potential energy U when the particles are separated by a distance r_{12} as in Figure 25.8b. From Equation 8.2, we have $W = \Delta U$. Therefore, the **electric potential energy** of a pair of point charges[1] can be found as follows:

$$\Delta U = W = q_2 \Delta V \quad \rightarrow \quad U - 0 = q_2 \left(k_e \frac{q_1}{r_{12}} - 0 \right)$$

$$U = k_e \frac{q_1 q_2}{r_{12}} \qquad (25.13)$$

If the charges are of the same sign, then U is positive. Positive work must be done by an external agent on the system to bring the two charges near each other (because charges of the same sign repel). If the charges are of opposite sign, as in Figure 25.8b, then U is negative. Negative work is done by an external agent against the attractive force between the charges of opposite sign as they are brought near each other; a force must be applied opposite the displacement to prevent q_2 from accelerating toward q_1.

If the system consists of more than two charged particles, we can obtain the total potential energy of the system by calculating U for every *pair* of charges and summing the terms algebraically. For example, the total potential energy of the system of three charges shown in Figure 25.9 is

$$U = k_e \left(\frac{q_1 q_2}{r_{12}} + \frac{q_1 q_3}{r_{13}} + \frac{q_2 q_3}{r_{23}} \right) \qquad (25.14)$$

The potential energy of this system of charges is given by Equation 25.14.

Physically, this result can be interpreted as follows. Imagine q_1 is fixed at the position shown in Figure 25.9 but q_2 and q_3 are at infinity. The work an external agent must do to bring q_2 from infinity to its position near q_1 is $k_e q_1 q_2 / r_{12}$, which is the first term in Equation 25.14. The last two terms represent the work required to bring q_3 from infinity to its position near q_1 and q_2. (The result is independent of the order in which the charges are transported.)

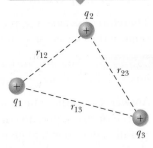

Figure 25.9 Three point charges are fixed at the positions shown.

[1]The expression for the electric potential energy of a system made up of two point charges, Equation 25.13, is of the *same* form as the equation for the gravitational potential energy of a system made up of two point masses, $-Gm_1 m_2/r$ (see Chapter 13). The similarity is not surprising considering that both expressions are derived from an inverse-square force law.

In Figure 25.8b, take q_2 to be a negative source charge and q_1 to be a second charge whose sign can be changed. **(i)** If q_1 is initially positive and is changed to a charge of the same magnitude but negative, what happens to the potential at the position of q_1 due to q_2? (a) It increases. (b) It decreases. (c) It remains the same. **(ii)** When q_1 is changed from positive to negative, what happens to the potential energy of the two-charge system? Choose from the same possibilities.

Example 25.3   The Electric Potential Due to Two Point Charges

As shown in Figure 25.10a, a charge $q_1 = 2.00~\mu C$ is located at the origin and a charge $q_2 = -6.00~\mu C$ is located at (0, 3.00) m.

(A) Find the total electric potential due to these charges at the point P, whose coordinates are (4.00, 0) m.

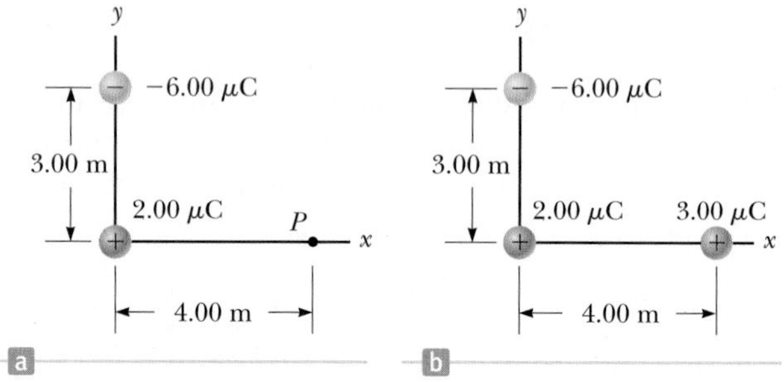

SOLUTION

Conceptualize Recognize first that the 2.00-μC and −6.00-μC charges are source charges and set up an electric field as well as a potential at all points in space, including point P.

Categorize The potential is evaluated using an equation developed in this chapter, so we categorize this example as a substitution problem.

Figure 25.10 (Example 25.3) (a) The electric potential at P due to the two charges q_1 and q_2 is the algebraic sum of the potentials due to the individual charges. (b) A third charge $q_3 = 3.00~\mu C$ is brought from infinity to point P.

Use Equation 25.12 for the system of two source charges:

$$V_P = k_e\left(\frac{q_1}{r_1} + \frac{q_2}{r_2}\right)$$

Substitute numerical values:

$$V_P = (8.988 \times 10^9~\text{N} \cdot \text{m}^2/\text{C}^2)\left(\frac{2.00 \times 10^{-6}~\text{C}}{4.00~\text{m}} + \frac{-6.00 \times 10^{-6}~\text{C}}{5.00~\text{m}}\right)$$

$$= \boxed{-6.29 \times 10^3~\text{V}}$$

(B) Find the change in potential energy of the system of two charges plus a third charge $q_3 = 3.00~\mu C$ as the latter charge moves from infinity to point P (Fig. 25.10b).

SOLUTION

Assign $U_i = 0$ for the system to the initial configuration in which the charge q_3 is at infinity. Use Equation 25.2 to evaluate the potential energy for the configuration in which the charge is at P:

$$U_f = q_3 V_P$$

Substitute numerical values to evaluate ΔU:

$$\Delta U = U_f - U_i = q_3 V_P - 0 = (3.00 \times 10^{-6}~\text{C})(-6.29 \times 10^3~\text{V})$$

$$= \boxed{-1.89 \times 10^{-2}~\text{J}}$$

Therefore, because the potential energy of the system has decreased, an external agent has to do positive work to remove the charge q_3 from point P back to infinity.

WHAT IF? You are working through this example with a classmate and she says, "Wait a minute! In part (B), we ignored the potential energy associated with the pair of charges q_1 and q_2!" How would you respond?

Answer Given the statement of the problem, it is not necessary to include this potential energy because part (B) asks for the *change* in potential energy of the system as q_3 is brought in from infinity. Because the configuration of charges q_1 and q_2 does not change in the process, there is no ΔU associated with these charges. Had part (B) asked to find the change in potential energy when *all three* charges start out infinitely far apart and are then brought to the positions in Figure 25.10b, however, you would have to calculate the change using Equation 25.14.

25.4 Obtaining the Value of the Electric Field from the Electric Potential

The electric field $\vec{E}$ and the electric potential V are related as shown in Equation 25.3, which tells us how to find ΔV if the electric field $\vec{E}$ is known. What if the situation is reversed? How do we calculate the value of the electric field if the electric potential is known in a certain region?

From Equation 25.3, the potential difference dV between two points a distance ds apart can be expressed as

$$dV = -\vec{E} \cdot d\vec{s} \tag{25.15}$$

If the electric field has only one component E_x, then $\vec{E} \cdot d\vec{s} = E_x\, dx$. Therefore, Equation 25.15 becomes $dV = -E_x\, dx$, or

$$E_x = -\frac{dV}{dx} \tag{25.16}$$

That is, the x component of the electric field is equal to the negative of the derivative of the electric potential with respect to x. Similar statements can be made about the y and z components. Equation 25.16 is the mathematical statement of the electric field being a measure of the rate of change with position of the electric potential as mentioned in Section 25.1.

Experimentally, electric potential and position can be measured easily with a voltmeter (a device for measuring potential difference) and a meterstick. Consequently, an electric field can be determined by measuring the electric potential at several positions in the field and making a graph of the results. According to Equation 25.16, the slope of a graph of V versus x at a given point provides the magnitude of the electric field at that point.

Imagine starting at a point and then moving through a displacement $d\vec{s}$ along an equipotential surface. For this motion, $dV = 0$ because the potential is constant along an equipotential surface. From Equation 25.15, we see that $dV = -\vec{E} \cdot d\vec{s} = 0$; therefore, because the dot product is zero, $\vec{E}$ must be perpendicular to the displacement along the equipotential surface. This result shows that the equipotential surfaces must always be perpendicular to the electric field lines passing through them.

As mentioned at the end of Section 25.2, the equipotential surfaces associated with a uniform electric field consist of a family of planes perpendicular to the field lines. Figure 25.11a shows some representative equipotential surfaces for this situation.

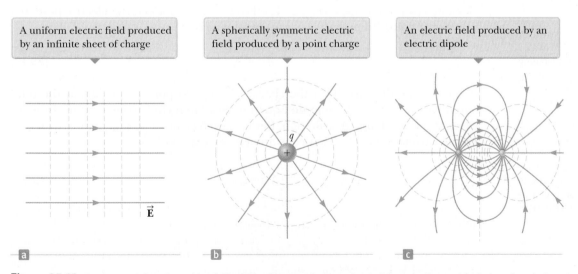

| A uniform electric field produced by an infinite sheet of charge | A spherically symmetric electric field produced by a point charge | An electric field produced by an electric dipole |

a   b   c

Figure 25.11 Equipotential surfaces (the dashed blue lines are intersections of these surfaces with the page) and electric field lines. In all cases, the equipotential surfaces are *perpendicular* to the electric field lines at every point.

If the charge distribution creating an electric field has spherical symmetry such that the volume charge density depends only on the radial distance r, the electric field is radial. In this case, $\vec{\mathbf{E}} \cdot d\vec{\mathbf{s}} = E_r \, dr$, and we can express dV as $dV = -E_r \, dr$. Therefore,

$$E_r = -\frac{dV}{dr} \tag{25.17}$$

For example, the electric potential of a point charge is $V = k_e q/r$. Because V is a function of r only, the potential function has spherical symmetry. Applying Equation 25.17, we find that the magnitude of the electric field due to the point charge is $E_r = k_e q/r^2$, a familiar result. Notice that the potential changes only in the radial direction, not in any direction perpendicular to r. Therefore, V (like E_r) is a function only of r, which is again consistent with the idea that equipotential surfaces are perpendicular to field lines. In this case, the equipotential surfaces are a family of spheres concentric with the spherically symmetric charge distribution (Fig. 25.11b). The equipotential surfaces for an electric dipole are sketched in Figure 25.11c.

In general, the electric potential is a function of all three spatial coordinates. If $V(r)$ is given in terms of the Cartesian coordinates, the electric field components E_x, E_y, and E_z can readily be found from $V(x, y, z)$ as the partial derivatives[2]

▶ **Finding the electric field from the potential**

$$E_x = -\frac{\partial V}{\partial x} \qquad E_y = -\frac{\partial V}{\partial y} \qquad E_z = -\frac{\partial V}{\partial z} \tag{25.18}$$

Quick Quiz 25.4 In a certain region of space, the electric potential is zero everywhere along the x axis. **(i)** From this information, you can conclude that the x component of the electric field in this region is (a) zero, (b) in the positive x direction, or (c) in the negative x direction. **(ii)** Suppose the electric potential is $+2$ V everywhere along the x axis. From the same choices, what can you conclude about the x component of the electric field now?

25.5 Electric Potential Due to Continuous Charge Distributions

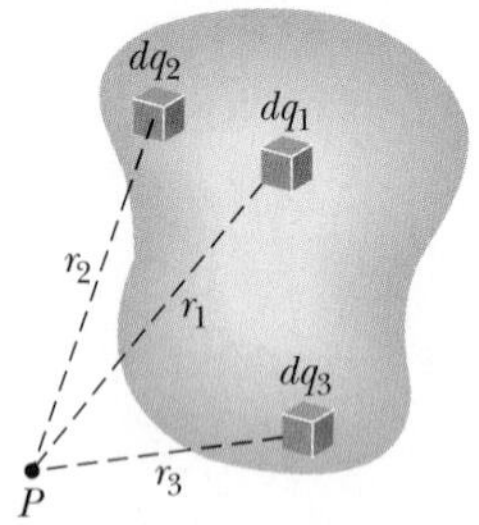

In Section 25.3, we found how to determine the electric potential due to a small number of charges. What if we wish to find the potential due to a continuous distribution of charge? The electric potential in this situation can be calculated using two different methods. The first method is as follows. If the charge distribution is known, we consider the potential due to a small charge element dq, treating this element as a point charge (Fig. 25.12). From Equation 25.11, the electric potential dV at some point P due to the charge element dq is

$$dV = k_e \frac{dq}{r} \tag{25.19}$$

Figure 25.12 The electric potential at point P due to a continuous charge distribution can be calculated by dividing the charge distribution into elements of charge dq and summing the electric potential contributions over all elements. Three sample elements of charge are shown.

where r is the distance from the charge element to point P. To obtain the total potential at point P, we integrate Equation 25.19 to include contributions from all elements of the charge distribution. Because each element is, in general, a different distance from point P and k_e is constant, we can express V as

▶ **Electric potential due to a continuous charge distribution**

$$V = k_e \int \frac{dq}{r} \tag{25.20}$$

---

[2]In vector notation, $\vec{\mathbf{E}}$ is often written in Cartesian coordinate systems as

$$\vec{\mathbf{E}} = -\nabla V = -\left(\hat{\mathbf{i}} \frac{\partial}{\partial x} + \hat{\mathbf{j}} \frac{\partial}{\partial y} + \hat{\mathbf{k}} \frac{\partial}{\partial z} \right) V$$

where ∇ is called the *gradient operator*.

In effect, we have replaced the sum in Equation 25.12 with an integral. In this expression for V, the electric potential is taken to be zero when point P is infinitely far from the charge distribution.

The second method for calculating the electric potential is used if the electric field is already known from other considerations such as Gauss's law. If the charge distribution has sufficient symmetry, we first evaluate $\vec{\mathbf{E}}$ using Gauss's law and then substitute the value obtained into Equation 25.3 to determine the potential difference ΔV between any two points. We then choose the electric potential V to be zero at some convenient point.

Problem-Solving Strategy | Calculating Electric Potential

The following procedure is recommended for solving problems that involve the determination of an electric potential due to a charge distribution.

1. Conceptualize. Think carefully about the individual charges or the charge distribution you have in the problem and imagine what type of potential would be created. Appeal to any symmetry in the arrangement of charges to help you visualize the potential.

2. Categorize. Are you analyzing a group of individual charges or a continuous charge distribution? The answer to this question will tell you how to proceed in the *Analyze* step.

3. Analyze. When working problems involving electric potential, remember that it is a *scalar quantity*, so there are no components to consider. Therefore, when using the superposition principle to evaluate the electric potential at a point, simply take the algebraic sum of the potentials due to each charge. You must keep track of signs, however.

As with potential energy in mechanics, only *changes* in electric potential are significant; hence, the point where the potential is set at zero is arbitrary. When dealing with point charges or a finite-sized charge distribution, we usually define $V = 0$ to be at a point infinitely far from the charges. If the charge distribution itself extends to infinity, however, some other nearby point must be selected as the reference point.

(a) *If you are analyzing a group of individual charges:* Use the superposition principle, which states that when several point charges are present, the resultant potential at a point P in space is the *algebraic sum* of the individual potentials at P due to the individual charges (Eq. 25.12). Example 25.4 below demonstrates this procedure.

(b) *If you are analyzing a continuous charge distribution:* Replace the sums for evaluating the total potential at some point P from individual charges by integrals (Eq. 25.20). The total potential at P is obtained by integrating over the entire charge distribution. For many problems, it is possible in performing the integration to express dq and r in terms of a single variable. To simplify the integration, give careful consideration to the geometry involved in the problem. Examples 25.5 through 25.7 demonstrate such a procedure.

To obtain the potential from the electric field: Another method used to obtain the potential is to start with the definition of the potential difference given by Equation 25.3. If $\vec{\mathbf{E}}$ is known or can be obtained easily (such as from Gauss's law), the line integral of $\vec{\mathbf{E}} \cdot d\vec{\mathbf{s}}$ can be evaluated.

4. Finalize. Check to see if your expression for the potential is consistent with your mental representation and reflects any symmetry you noted previously. Imagine varying parameters such as the distance of the observation point from the charges or the radius of any circular objects to see if the mathematical result changes in a reasonable way.

Example 25.4   The Electric Potential Due to a Dipole

An electric dipole consists of two charges of equal magnitude and opposite sign separated by a distance $2a$ as shown in Figure 25.13. The dipole is along the x axis and is centered at the origin.

(A) Calculate the electric potential at point P on the y axis.

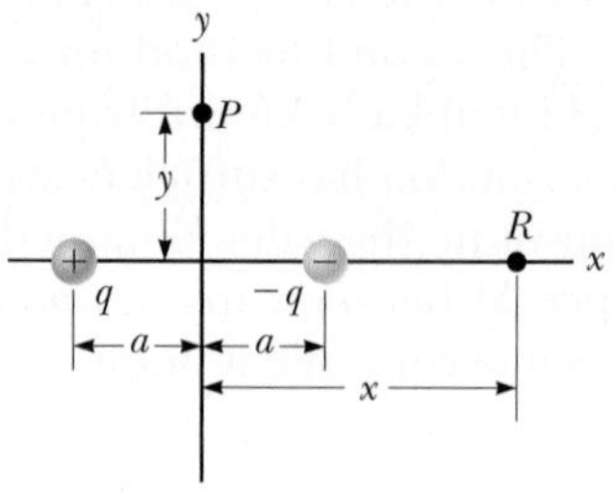

SOLUTION

Conceptualize Compare this situation to that in part (B) of Example 23.6. It is the same situation, but here we are seeking the electric potential rather than the electric field.

Categorize We categorize the problem as one in which we have a small number of particles rather than a continuous distribution of charge. The electric potential can be evaluated by summing the potentials due to the individual charges.

Figure 25.13 (Example 25.4) An electric dipole located on the x axis.

Analyze Use Equation 25.12 to find the electric potential at P due to the two charges:

$$V_P = k_e \sum_i \frac{q_i}{r_i} = k_e \left(\frac{q}{\sqrt{a^2 + y^2}} + \frac{-q}{\sqrt{a^2 + y^2}} \right) = \boxed{0}$$

(B) Calculate the electric potential at point R on the positive x axis.

SOLUTION

Use Equation 25.12 to find the electric potential at R due to the two charges:

$$V_R = k_e \sum_i \frac{q_i}{r_i} = k_e \left(\frac{-q}{x - a} + \frac{q}{x + a} \right) = \boxed{- \frac{2k_e qa}{x^2 - a^2}}$$

(C) Calculate V and E_x at a point on the x axis far from the dipole.

SOLUTION

For point R far from the dipole such that $x \gg a$, neglect a^2 in the denominator of the answer to part (B) and write V in this limit:

$$V_R = \lim_{x \gg a} \left(- \frac{2k_e qa}{x^2 - a^2} \right) \approx \boxed{- \frac{2k_e qa}{x^2}} \quad (x \gg a)$$

Use Equation 25.16 and this result to calculate the x component of the electric field at a point on the x axis far from the dipole:

$$E_x = - \frac{dV}{dx} = - \frac{d}{dx} \left(- \frac{2k_e qa}{x^2} \right)$$

$$= 2k_e qa \frac{d}{dx} \left(\frac{1}{x^2} \right) = \boxed{- \frac{4k_e qa}{x^3}} \quad (x \gg a)$$

Finalize The potentials in parts (B) and (C) are negative because points on the positive x axis are closer to the negative charge than to the positive charge. For the same reason, the x component of the electric field is negative. Notice that we have a $1/r^3$ falloff of the electric field with distance far from the dipole, similar to the behavior of the electric field on the y axis in Example 23.6.

WHAT IF? Suppose you want to find the electric field at a point P on the y axis. In part (A), the electric potential was found to be zero for all values of y. Is the electric field zero at all points on the y axis?

Answer No. That there is no change in the potential along the y axis tells us only that the y component of the electric field is zero. Look back at Figure 23.13 in Example 23.6. We showed there that the electric field of a dipole on the y axis has only an x component. We could not find the x component in the current example because we do not have an expression for the potential near the y axis as a function of x.

Example 25.5	Electric Potential Due to a Uniformly Charged Ring

(A) Find an expression for the electric potential at a point P located on the perpendicular central axis of a uniformly charged ring of radius a and total charge Q.

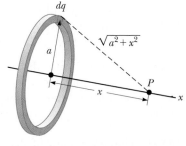

SOLUTION

Conceptualize Study Figure 25.14, in which the ring is oriented so that its plane is perpendicular to the x axis and its center is at the origin. Notice that the symmetry of the situation means that all the charges on the ring are the same distance from point P. Compare this example to Example 23.8. Notice that no vector considerations are necessary here because electric potential is a scalar.

Figure 25.14 (Example 25.5) A uniformly charged ring of radius a lies in a plane perpendicular to the x axis. All elements dq of the ring are the same distance from a point P lying on the x axis.

Categorize Because the ring consists of a continuous distribution of charge rather than a set of discrete charges, we must use the integration technique represented by Equation 25.20 in this example.

Analyze We take point P to be at a distance x from the center of the ring as shown in Figure 25.14.

Use Equation 25.20 to express V in terms of the geometry:

$$V = k_e \int \frac{dq}{r} = k_e \int \frac{dq}{\sqrt{a^2 + x^2}}$$

Noting that a and x do not vary for an integration over the ring, bring $\sqrt{a^2 + x^2}$ in front of the integral sign and integrate over the ring:

$$V = \frac{k_e}{\sqrt{a^2 + x^2}} \int dq = \frac{k_e Q}{\sqrt{a^2 + x^2}} \qquad \textbf{(25.21)}$$

(B) Find an expression for the magnitude of the electric field at point P.

SOLUTION

From symmetry, notice that along the x axis $\vec{E}$ can have only an x component. Therefore, apply Equation 25.16 to Equation 25.21:

$$E_x = -\frac{dV}{dx} = -k_e Q \frac{d}{dx}(a^2 + x^2)^{-1/2}$$

$$= -k_e Q(-\tfrac{1}{2})(a^2 + x^2)^{-3/2}(2x)$$

$$E_x = \frac{k_e x}{(a^2 + x^2)^{3/2}} Q \qquad \textbf{(25.22)}$$

Finalize The only variable in the expressions for V and E_x is x. That is not surprising because our calculation is valid only for points along the x axis, where y and z are both zero. This result for the electric field agrees with that obtained by direct integration (see Example 23.8). For practice, use the result of part (B) in Equation 25.3 to verify that the potential is given by the expression in part (A).

Example 25.6	Electric Potential Due to a Uniformly Charged Disk

A uniformly charged disk has radius R and surface charge density σ.

(A) Find the electric potential at a point P along the perpendicular central axis of the disk.

SOLUTION

Conceptualize If we consider the disk to be a set of concentric rings, we can use our result from Example 25.5—which gives the potential due to a ring of radius a—and sum the contributions of all rings making up the disk. Figure
continued

▶ **25.6** c o n t i n u e d

25.15 shows one such ring. Because point P is on the central axis of the disk, symmetry again tells us that all points in a given ring are the same distance from P.

Categorize Because the disk is continuous, we evaluate the potential due to a continuous charge distribution rather than a group of individual charges.

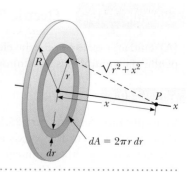

Figure 25.15 (Example 25.6) A uniformly charged disk of radius R lies in a plane perpendicular to the x axis. The calculation of the electric potential at any point P on the x axis is simplified by dividing the disk into many rings of radius r and width dr, with area $2\pi r\, dr$.

Analyze Find the amount of charge dq on a ring of radius r and width dr as shown in Figure 25.15:

$$dq = \sigma\, dA = \sigma(2\pi r\, dr) = 2\pi\sigma r\, dr$$

Use this result in Equation 25.21 in Example 25.5 (with a replaced by the variable r and Q replaced by the differential dq) to find the potential due to the ring:

$$dV = \frac{k_e\, dq}{\sqrt{r^2 + x^2}} = \frac{k_e 2\pi\sigma r\, dr}{\sqrt{r^2 + x^2}}$$

To obtain the total potential at P, integrate this expression over the limits $r = 0$ to $r = R$, noting that x is a constant:

$$V = \pi k_e \sigma \int_0^R \frac{2r\, dr}{\sqrt{r^2 + x^2}} = \pi k_e \sigma \int_0^R (r^2 + x^2)^{-1/2}\, 2r\, dr$$

This integral is of the common form $\int u^n\, du$, where $n = -\frac{1}{2}$ and $u = r^2 + x^2$, and has the value $u^{n+1}/(n + 1)$. Use this result to evaluate the integral:

$$V = \boxed{2\pi k_e \sigma [(R^2 + x^2)^{1/2} - x]} \tag{25.23}$$

(B) Find the x component of the electric field at a point P along the perpendicular central axis of the disk.

S O L U T I O N

As in Example 25.5, use Equation 25.16 to find the electric field at any axial point:

$$E_x = -\frac{dV}{dx} = \boxed{2\pi k_e \sigma \left[1 - \frac{x}{(R^2 + x^2)^{1/2}}\right]} \tag{25.24}$$

Finalize Compare Equation 25.24 with the result of Example 23.9. They are the same. The calculation of V and $\vec{E}$ for an arbitrary point off the x axis is more difficult to perform because of the absence of symmetry and we do not treat that situation in this book.

Example 25.7 **Electric Potential Due to a Finite Line of Charge**

A rod of length ℓ located along the x axis has a total charge Q and a uniform linear charge density λ. Find the electric potential at a point P located on the y axis a distance a from the origin (Fig. 25.16).

S O L U T I O N

Conceptualize The potential at P due to every segment of charge on the rod is positive because every segment carries a positive charge. Notice that we have no symmetry to appeal to here, but the simple geometry should make the problem solvable.

Categorize Because the rod is continuous, we evaluate the potential due to a continuous charge distribution rather than a group of individual charges.

Analyze In Figure 25.16, the rod lies along the x axis, dx is the length of one small segment, and dq is the charge on that segment. Because the rod has a charge per unit length λ, the charge dq on the small segment is $dq = \lambda\, dx$.

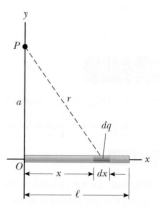

Figure 25.16 (Example 25.7) A uniform line charge of length ℓ located along the x axis. To calculate the electric potential at P, the line charge is divided into segments each of length dx and each carrying a charge $dq = \lambda\, dx$.

▶ **25.7** continued

Find the potential at P due to one segment of the rod at an arbitrary position x:

$$dV = k_e \frac{dq}{r} = k_e \frac{\lambda \, dx}{\sqrt{a^2 + x^2}}$$

Find the total potential at P by integrating this expression over the limits $x = 0$ to $x = \ell$:

$$V = \int_0^\ell k_e \frac{\lambda \, dx}{\sqrt{a^2 + x^2}}$$

Noting that k_e and $\lambda = Q/\ell$ are constants and can be removed from the integral, evaluate the integral with the help of Appendix B:

$$V = k_e \lambda \int_0^\ell \frac{dx}{\sqrt{a^2 + x^2}} = k_e \frac{Q}{\ell} \ln (x + \sqrt{a^2 + x^2}) \Big|_0^\ell$$

Evaluate the result between the limits:
$$V = k_e \frac{Q}{\ell} [\ln (\ell + \sqrt{a^2 + \ell^2}) - \ln a] = k_e \frac{Q}{\ell} \ln \left(\frac{\ell + \sqrt{a^2 + \ell^2}}{a} \right) \quad \text{(25.25)}$$

Finalize If $\ell \ll a$, the potential at P should approach that of a point charge because the rod is very short compared to the distance from the rod to P. By using a series expansion for the natural logarithm from Appendix B.5, it is easy to show that Equation 25.25 becomes $V = k_e Q/a$.

WHAT IF? What if you were asked to find the electric field at point P? Would that be a simple calculation?

Answer Calculating the electric field by means of Equation 23.11 would be a little messy. There is no symmetry to appeal to, and the integration over the line of charge would represent a vector addition of electric fields at point P. Using Equation 25.18, you could find E_y by replacing a with y in Equation 25.25 and performing the differentiation with respect to y. Because the charged rod in Figure

25.16 lies entirely to the right of $x = 0$, the electric field at point P would have an x component to the left if the rod is charged positively. You cannot use Equation 25.18 to find the x component of the field, however, because the potential due to the rod was evaluated at a specific value of x ($x = 0$) rather than a general value of x. You would have to find the potential as a function of both x and y to be able to find the x and y components of the electric field using Equation 25.18.

25.6 Electric Potential Due to a Charged Conductor

In Section 24.4, we found that when a solid conductor in equilibrium carries a net charge, the charge resides on the conductor's outer surface. Furthermore, the electric field just outside the conductor is perpendicular to the surface and the field inside is zero.

We now generate another property of a charged conductor, related to electric potential. Consider two points Ⓐ and Ⓑ on the surface of a charged conductor as shown in Figure 25.17. Along a surface path connecting these points, $\vec{E}$ is always

> **Pitfall Prevention 25.6**
> **Potential May Not Be Zero** The electric potential inside the conductor is not necessarily zero in Figure 25.17, even though the electric field is zero. Equation 25.15 shows that a zero value of the field results in no *change* in the potential from one point to another inside the conductor. Therefore, the potential everywhere inside the conductor, including the surface, has the same value, which may or may not be zero, depending on where the zero of potential is defined.

> Notice from the spacing of the positive signs that the surface charge density is nonuniform.

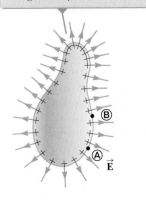

Figure 25.17 An arbitrarily shaped conductor carrying a positive charge. When the conductor is in electrostatic equilibrium, all the charge resides at the surface, $\vec{E} = 0$ inside the conductor, and the direction of $\vec{E}$ immediately outside the conductor is perpendicular to the surface. The electric potential is constant inside the conductor and is equal to the potential at the surface.

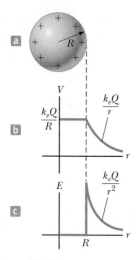

Figure 25.18 (a) The excess charge on a conducting sphere of radius R is uniformly distributed on its surface. (b) Electric potential versus distance r from the center of the charged conducting sphere. (c) Electric field magnitude versus distance r from the center of the charged conducting sphere.

perpendicular to the displacement $d\vec{s}$; therefore, $\vec{E} \cdot d\vec{s} = 0$. Using this result and Equation 25.3, we conclude that the potential difference between Ⓐ and Ⓑ is necessarily zero:

$$V_{Ⓑ} - V_{Ⓐ} = -\int_{Ⓐ}^{Ⓑ} \vec{E} \cdot d\vec{s} = 0$$

This result applies to any two points on the surface. Therefore, V is constant everywhere on the surface of a charged conductor in equilibrium. That is,

> the surface of any charged conductor in electrostatic equilibrium is an equipotential surface: every point on the surface of a charged conductor in equilibrium is at the same electric potential. Furthermore, because the electric field is zero inside the conductor, the electric potential is constant everywhere inside the conductor and equal to its value at the surface.

Because of the constant value of the potential, no work is required to move a charge from the interior of a charged conductor to its surface.

Consider a solid metal conducting sphere of radius R and total positive charge Q as shown in Figure 25.18a. As determined in part (A) of Example 24.3, the electric field outside the sphere is $k_e Q / r^2$ and points radially outward. Because the field outside a spherically symmetric charge distribution is identical to that of a point charge, we expect the potential to also be that of a point charge, $k_e Q / r$. At the surface of the conducting sphere in Figure 25.18a, the potential must be $k_e Q / R$. Because the entire sphere must be at the same potential, the potential at any point within the sphere must also be $k_e Q / R$. Figure 25.18b is a plot of the electric potential as a function of r, and Figure 25.18c shows how the electric field varies with r.

When a net charge is placed on a spherical conductor, the surface charge density is uniform as indicated in Figure 25.18a. If the conductor is nonspherical as in Figure 25.17, however, the surface charge density is high where the radius of curvature is small (as noted in Section 24.4) and low where the radius of curvature is large. Because the electric field immediately outside the conductor is proportional to the surface charge density, the electric field is large near convex points having small radii of curvature and reaches very high values at sharp points. In Example 25.8, the relationship between electric field and radius of curvature is explored mathematically.

Example 25.8   Two Connected Charged Spheres

Two spherical conductors of radii r_1 and r_2 are separated by a distance much greater than the radius of either sphere. The spheres are connected by a conducting wire as shown in Figure 25.19. The charges on the spheres in equilibrium are q_1 and q_2, respectively, and they are uniformly charged. Find the ratio of the magnitudes of the electric fields at the surfaces of the spheres.

SOLUTION

Conceptualize Imagine the spheres are much farther apart than shown in Figure 25.19. Because they are so far apart, the field of one does not affect the charge distribution on the other. The conducting wire between them ensures that both spheres have the same electric potential.

Categorize Because the spheres are so far apart, we model the charge distribution on them as spherically symmetric, and we can model the field and potential outside the spheres to be that due to point charges.

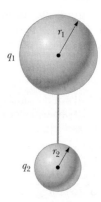

Figure 25.19 (Example 25.8) Two charged spherical conductors connected by a conducting wire. The spheres are at the *same* electric potential V.

Analyze Set the electric potentials at the surfaces of the spheres equal to each other:

$$V = k_e \frac{q_1}{r_1} = k_e \frac{q_2}{r_2}$$

▶ **25.8** continued

Solve for the ratio of charges on the spheres:

(1) $\dfrac{q_1}{q_2} = \dfrac{r_1}{r_2}$

Write expressions for the magnitudes of the electric fields at the surfaces of the spheres:

$E_1 = k_e \dfrac{q_1}{r_1^2}$ and $E_2 = k_e \dfrac{q_2}{r_2^2}$

Evaluate the ratio of these two fields:

$\dfrac{E_1}{E_2} = \dfrac{q_1}{q_2} \dfrac{r_2^2}{r_1^2}$

Substitute for the ratio of charges from Equation (1):

(2) $\dfrac{E_1}{E_2} = \dfrac{r_1}{r_2} \dfrac{r_2^2}{r_1^2} = \boxed{\dfrac{r_2}{r_1}}$

Finalize The field is stronger in the vicinity of the smaller sphere even though the electric potentials at the surfaces of both spheres are the same. If $r_2 \rightarrow 0$, then $E_2 \rightarrow \infty$, verifying the statement above that the electric field is very large at sharp points.

A Cavity Within a Conductor

Suppose a conductor of arbitrary shape contains a cavity as shown in Figure 25.20. Let's assume no charges are inside the cavity. In this case, the electric field inside the cavity must be *zero* regardless of the charge distribution on the outside surface of the conductor as we mentioned in Section 24.4. Furthermore, the field in the cavity is zero even if an electric field exists outside the conductor.

To prove this point, remember that every point on the conductor is at the same electric potential; therefore, any two points Ⓐ and Ⓑ on the cavity's surface must be at the same potential. Now imagine a field $\vec{E}$ exists in the cavity and evaluate the potential difference $V_{\text{Ⓑ}} - V_{\text{Ⓐ}}$ defined by Equation 25.3:

$$V_{\text{Ⓑ}} - V_{\text{Ⓐ}} = -\int_{\text{Ⓐ}}^{\text{Ⓑ}} \vec{E} \cdot d\vec{s}$$

Because $V_{\text{Ⓑ}} - V_{\text{Ⓐ}} = 0$, the integral of $\vec{E} \cdot d\vec{s}$ must be zero for all paths between any two points Ⓐ and Ⓑ on the conductor. The only way that can be true for *all* paths is if $\vec{E}$ is zero *everywhere* in the cavity. Therefore, a cavity surrounded by conducting walls is a field-free region as long as no charges are inside the cavity.

> The electric field in the cavity is zero regardless of the charge on the conductor.

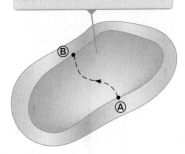

Figure 25.20 A conductor in electrostatic equilibrium containing a cavity.

Corona Discharge

A phenomenon known as **corona discharge** is often observed near a conductor such as a high-voltage power line. When the electric field in the vicinity of the conductor is sufficiently strong, electrons resulting from random ionizations of air molecules near the conductor accelerate away from their parent molecules. These rapidly moving electrons can ionize additional molecules near the conductor, creating more free electrons. The observed glow (or corona discharge) results from the recombination of these free electrons with the ionized air molecules. If a conductor has an irregular shape, the electric field can be very high near sharp points or edges of the conductor; consequently, the ionization process and corona discharge are most likely to occur around such points.

Corona discharge is used in the electrical transmission industry to locate broken or faulty components. For example, a broken insulator on a transmission tower has sharp edges where corona discharge is likely to occur. Similarly, corona discharge will occur at the sharp end of a broken conductor strand. Observation of these discharges is difficult because the visible radiation emitted is weak and most of the radiation is in the ultraviolet. (We will discuss ultraviolet radiation and other portions of the electromagnetic spectrum in Section 34.7.) Even use of traditional ultraviolet cameras is of little help because the radiation from the corona

Figure 25.21 Schematic drawing of the Millikan oil-drop apparatus.

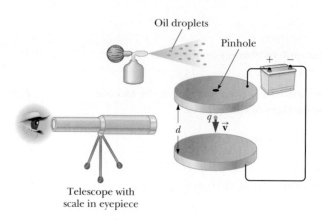

Oil droplets

Pinhole

Telescope with scale in eyepiece

discharge is overwhelmed by ultraviolet radiation from the Sun. Newly developed dual-spectrum devices combine a narrow-band ultraviolet camera with a visible-light camera to show a daylight view of the corona discharge in the actual location on the transmission tower or cable. The ultraviolet part of the camera is designed to operate in a wavelength range in which radiation from the Sun is very weak.

25.7 The Millikan Oil-Drop Experiment

Robert Millikan performed a brilliant set of experiments from 1909 to 1913 in which he measured e, the magnitude of the elementary charge on an electron, and demonstrated the quantized nature of this charge. His apparatus, diagrammed in Figure 25.21, contains two parallel metallic plates. Oil droplets from an atomizer are allowed to pass through a small hole in the upper plate. Millikan used x-rays to ionize the air in the chamber so that freed electrons would adhere to the oil drops, giving them a negative charge. A horizontally directed light beam is used to illuminate the oil droplets, which are viewed through a telescope whose long axis is perpendicular to the light beam. When viewed in this manner, the droplets appear as shining stars against a dark background and the rate at which individual drops fall can be determined.

Let's assume a single drop having a mass m and carrying a charge q is being viewed and its charge is negative. If no electric field is present between the plates, the two forces acting on the charge are the gravitational force $m\vec{g}$ acting downward[3] and a viscous drag force $\vec{F}_D$ acting upward as indicated in Figure 25.22a. The drag force is proportional to the drop's speed as discussed in Section 6.4. When the drop reaches its terminal speed v_T the two forces balance each other ($mg = F_D$).

Now suppose a battery connected to the plates sets up an electric field between the plates such that the upper plate is at the higher electric potential. In this case, a third force $q\vec{E}$ acts on the charged drop. The particle in a field model applies twice to the particle: it is in a gravitational field and an electric field. Because q is negative and $\vec{E}$ is directed downward, this electric force is directed upward as shown in Figure 25.22b. If this upward force is strong enough, the drop moves upward and the drag force $\vec{F}'_D$ acts downward. When the upward electric force $q\vec{E}$ balances the sum of the gravitational force and the downward drag force $\vec{F}'_D$, the drop reaches a new terminal speed v'_T in the upward direction.

With the field turned on, a drop moves slowly upward, typically at rates of hundredths of a centimeter per second. The rate of fall in the absence of a field is comparable. Hence, one can follow a single droplet for hours, alternately rising and falling, by simply turning the electric field on and off.

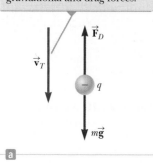

With the electric field off, the droplet falls at terminal velocity $\vec{v}_T$ under the influence of the gravitational and drag forces.

a

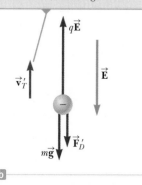

When the electric field is turned on, the droplet moves upward at terminal velocity $\vec{v}'_T$ under the influence of the electric, gravitational, and drag forces.

b

Figure 25.22 The forces acting on a negatively charged oil droplet in the Millikan experiment.

---

[3]There is also a buoyant force on the oil drop due to the surrounding air. This force can be incorporated as a correction in the gravitational force $m\vec{g}$ on the drop, so we will not consider it in our analysis.

After recording measurements on thousands of droplets, Millikan and his coworkers found that all droplets, to within about 1% precision, had a charge equal to some integer multiple of the elementary charge e:

$$q = ne \quad n = 0, -1, -2, -3, \ldots$$

where $e = 1.60 \times 10^{-19}$ C. Millikan's experiment yields conclusive evidence that charge is quantized. For this work, he was awarded the Nobel Prize in Physics in 1923.

25.8 Applications of Electrostatics

The practical application of electrostatics is represented by such devices as lightning rods and electrostatic precipitators and by such processes as xerography and the painting of automobiles. Scientific devices based on the principles of electrostatics include electrostatic generators, the field-ion microscope, and ion-drive rocket engines. Details of two devices are given below.

The Van de Graaff Generator

Experimental results show that when a charged conductor is placed in contact with the inside of a hollow conductor, all the charge on the charged conductor is transferred to the hollow conductor. In principle, the charge on the hollow conductor and its electric potential can be increased without limit by repetition of the process.

In 1929, Robert J. Van de Graaff (1901–1967) used this principle to design and build an electrostatic generator, and a schematic representation of it is given in Figure 25.23. This type of generator was once used extensively in nuclear physics research. Charge is delivered continuously to a high-potential electrode by means of a moving belt of insulating material. The high-voltage electrode is a hollow metal dome mounted on an insulating column. The belt is charged at point Ⓐ by means of a corona discharge between comb-like metallic needles and a grounded grid. The needles are maintained at a positive electric potential of typically 10^4 V. The positive charge on the moving belt is transferred to the dome by a second comb of needles at point Ⓑ. Because the electric field inside the dome is negligible, the positive charge on the belt is easily transferred to the conductor regardless of its potential. In practice, it is possible to increase the electric potential of the dome until electrical discharge occurs through the air. Because the "breakdown" electric field in air is about 3×10^6 V/m, a sphere 1.00 m in radius can be raised to a maximum potential of 3×10^6 V. The potential can be increased further by increasing the dome's radius and placing the entire system in a container filled with high-pressure gas.

Van de Graaff generators can produce potential differences as large as 20 million volts. Protons accelerated through such large potential differences receive enough energy to initiate nuclear reactions between themselves and various target nuclei. Smaller generators are often seen in science classrooms and museums. If a person insulated from the ground touches the sphere of a Van de Graaff generator, his or her body can be brought to a high electric potential. The person's hair acquires a net positive charge, and each strand is repelled by all the others as in the opening photograph of Chapter 23.

The Electrostatic Precipitator

One important application of electrical discharge in gases is the *electrostatic precipitator*. This device removes particulate matter from combustion gases, thereby reducing air pollution. Precipitators are especially useful in coal-burning power plants and industrial operations that generate large quantities of smoke. Current systems are able to eliminate more than 99% of the ash from smoke.

Figure 25.24a (page 766) shows a schematic diagram of an electrostatic precipitator. A high potential difference (typically 40 to 100 kV) is maintained between

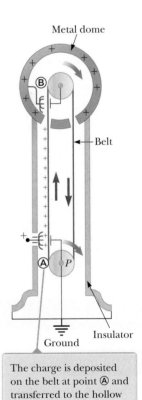

The charge is deposited on the belt at point Ⓐ and transferred to the hollow conductor at point Ⓑ.

Figure 25.23 Schematic diagram of a Van de Graaff generator. Charge is transferred to the metal dome at the top by means of a moving belt.

The high negative electric potential maintained on the central wire creates a corona discharge in the vicinity of the wire.

Insulator

Battery

Clean air out

Dirty air in

Weight

Dirt out

a

b

c

Figure 25.24 (a) Schematic diagram of an electrostatic precipitator. Compare the air pollution when the electrostatic precipitator is (b) operating and (c) turned off.

a wire running down the center of a duct and the walls of the duct, which are grounded. The wire is maintained at a negative electric potential with respect to the walls, so the electric field is directed toward the wire. The values of the field near the wire become high enough to cause a corona discharge around the wire; the air near the wire contains positive ions, electrons, and such negative ions as O_2^-. The air to be cleaned enters the duct and moves near the wire. As the electrons and negative ions created by the discharge are accelerated toward the outer wall by the electric field, the dirt particles in the air become charged by collisions and ion capture. Because most of the charged dirt particles are negative, they too are drawn to the duct walls by the electric field. When the duct is periodically shaken, the particles break loose and are collected at the bottom.

In addition to reducing the level of particulate matter in the atmosphere (compare Figs. 25.24b and c), the electrostatic precipitator recovers valuable materials in the form of metal oxides.

Summary

Definitions

The **potential difference** ΔV between points Ⓐ and Ⓑ in an electric field $\vec{\mathbf{E}}$ is defined as

$$\Delta V \equiv \frac{\Delta U}{q} = -\int_{Ⓐ}^{Ⓑ} \vec{\mathbf{E}} \cdot d\vec{\mathbf{s}} \qquad (25.3)$$

where ΔU is given by Equation 25.1 on page 767. The **electric potential** $V = U/q$ is a scalar quantity and has the units of joules per coulomb, where $1\ \text{J/C} \equiv 1\ \text{V}$.

An **equipotential surface** is one on which all points are at the same electric potential. Equipotential surfaces are perpendicular to electric field lines.

Concepts and Principles

When a positive charge q is moved between points Ⓐ and Ⓑ in an electric field $\vec{E}$, the change in the potential energy of the charge–field system is

$$\Delta U = -q \int_{Ⓐ}^{Ⓑ} \vec{E} \cdot d\vec{s} \qquad \textbf{(25.1)}$$

The potential difference between two points separated by a distance d in a uniform electric field $\vec{E}$ is

$$\Delta V = -Ed \qquad \textbf{(25.6)}$$

if the direction of travel between the points is in the same direction as the electric field.

If we define $V = 0$ at $r = \infty$, the electric potential due to a point charge at any distance r from the charge is

$$V = k_e \frac{q}{r} \qquad \textbf{(25.11)}$$

The electric potential associated with a group of point charges is obtained by summing the potentials due to the individual charges.

The **electric potential energy** associated with a pair of point charges separated by a distance r_{12} is

$$U = k_e \frac{q_1 q_2}{r_{12}} \qquad \textbf{(25.13)}$$

We obtain the potential energy of a distribution of point charges by summing terms like Equation 25.13 over all pairs of particles.

If the electric potential is known as a function of coordinates x, y, and z, we can obtain the components of the electric field by taking the negative derivative of the electric potential with respect to the coordinates. For example, the x component of the electric field is

$$E_x = -\frac{dV}{dx} \qquad \textbf{(25.16)}$$

The electric potential due to a continuous charge distribution is

$$V = k_e \int \frac{dq}{r} \qquad \textbf{(25.20)}$$

Every point on the surface of a charged conductor in electrostatic equilibrium is at the same electric potential. The potential is constant everywhere inside the conductor and equal to its value at the surface.

Objective Questions

1. denotes answer available in *Student Solutions Manual/Study Guide*

1. In a certain region of space, the electric field is zero. From this fact, what can you conclude about the electric potential in this region? (a) It is zero. (b) It does not vary with position. (c) It is positive. (d) It is negative. (e) None of those answers is necessarily true.

2. Consider the equipotential surfaces shown in Figure 25.4. In this region of space, what is the approximate direction of the electric field? (a) It is out of the page. (b) It is into the page. (c) It is toward the top of the page. (d) It is toward the bottom of the page. (e) The field is zero.

3. **(i)** A metallic sphere A of radius 1.00 cm is several centimeters away from a metallic spherical shell B of radius 2.00 cm. Charge 450 nC is placed on A, with no charge on B or anywhere nearby. Next, the two objects are joined by a long, thin, metallic wire (as shown in Fig. 25.19), and finally the wire is removed. How is the charge shared between A and B? (a) 0 on A, 450 nC on B (b) 90.0 nC on A and 360 nC on B, with equal surface charge densities (c) 150 nC on A and 300 nC on B (d) 225 nC on A and 225 nC on B (e) 450 nC on A and 0 on B **(ii)** A metallic sphere A of radius 1 cm with charge 450 nC hangs on an insulating thread inside an uncharged thin metallic spherical shell B of radius 2 cm. Next, A is made temporarily to touch the inner surface of B. How is the charge then shared between

them? Choose from the same possibilities. Arnold Arons, the only physics teacher yet to have his picture on the cover of *Time* magazine, suggested the idea for this question.

4. The electric potential at $x = 3.00$ m is 120 V, and the electric potential at $x = 5.00$ m is 190 V. What is the x component of the electric field in this region, assuming the field is uniform? (a) 140 N/C (b) −140 N/C (c) 35.0 N/C (d) −35.0 N/C (e) 75.0 N/C

5. Rank the potential energies of the four systems of particles shown in Figure OQ25.5 from largest to smallest. Include equalities if appropriate.

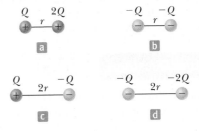

Figure OQ25.5

6. In a certain region of space, a uniform electric field is in the x direction. A particle with negative charge is carried from $x = 20.0$ cm to $x = 60.0$ cm. **(i)** Does

the electric potential energy of the charge–field system (a) increase, (b) remain constant, (c) decrease, or (d) change unpredictably? **(ii)** Has the particle moved to a position where the electric potential is (a) higher than before, (b) unchanged, (c) lower than before, or (d) unpredictable?

7. Rank the electric potentials at the four points shown in Figure OQ25.7 from largest to smallest.

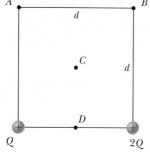

Figure OQ25.7

8. An electron in an x-ray machine is accelerated through a potential difference of 1.00×10^4 V before it hits the target. What is the kinetic energy of the electron in electron volts? (a) 1.00×10^4 eV (b) 1.60×10^{-15} eV (c) 1.60×10^{-22} eV (d) 6.25×10^{22} eV (e) 1.60×10^{-19} eV

9. Rank the electric potential energies of the systems of charges shown in Figure OQ25.9 from largest to smallest. Indicate equalities if appropriate.

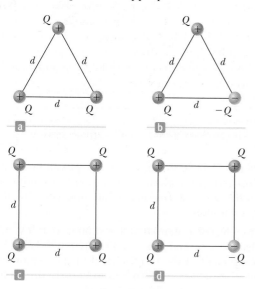

Figure OQ25.9

10. Four particles are positioned on the rim of a circle. The charges on the particles are $+0.500\ \mu C$, $+1.50\ \mu C$, $-1.00\ \mu C$, and $-0.500\ \mu C$. If the electric potential at the center of the circle due to the $+0.500\ \mu C$ charge alone is 4.50×10^4 V, what is the total electric potential at the center due to the four charges? (a) 18.0×10^4 V (b) 4.50×10^4 V (c) 0 (d) -4.50×10^4 V (e) 9.00×10^4 V

11. A proton is released from rest at the origin in a uniform electric field in the positive x direction with magnitude 850 N/C. What is the change in the electric potential energy of the proton–field system when the proton travels to $x = 2.50$ m? (a) 3.40×10^{-16} J (b) -3.40×10^{-16} J (c) 2.50×10^{-16} J (d) -2.50×10^{-16} J (e) -1.60×10^{-19} J

12. A particle with charge -40.0 nC is on the x axis at the point with coordinate $x = 0$. A second particle, with charge -20.0 nC, is on the x axis at $x = 0.500$ m. **(i)** Is the point at a finite distance where the electric field is zero (a) to the left of $x = 0$, (b) between $x = 0$ and $x = 0.500$ m, or (c) to the right of $x = 0.500$ m? **(ii)** Is the electric potential zero at this point? (a) No; it is positive. (b) Yes. (c) No; it is negative. **(iii)** Is there a point at a finite distance where the electric potential is zero? (a) Yes; it is to the left of $x = 0$. (b) Yes; it is between $x = 0$ and $x = 0.500$ m. (c) Yes; it is to the right of $x = 0.500$ m. (d) No.

13. A filament running along the x axis from the origin to $x = 80.0$ cm carries electric charge with uniform density. At the point P with coordinates ($x = 80.0$ cm, $y = 80.0$ cm), this filament creates electric potential 100 V. Now we add another filament along the y axis, running from the origin to $y = 80.0$ cm, carrying the same amount of charge with the same uniform density. At the same point P, is the electric potential created by the pair of filaments (a) greater than 200 V, (b) 200 V, (c) 100 V, (d) between 0 and 200 V, or (e) 0?

14. In different experimental trials, an electron, a proton, or a doubly charged oxygen atom (O^{--}), is fired within a vacuum tube. The particle's trajectory carries it through a point where the electric potential is 40.0 V and then through a point at a different potential. Rank each of the following cases according to the change in kinetic energy of the particle over this part of its flight from the largest increase to the largest decrease in kinetic energy. In your ranking, display any cases of equality. (a) An electron moves from 40.0 V to 60.0 V. (b) An electron moves from 40.0 V to 20.0 V. (c) A proton moves from 40.0 V to 20.0 V. (d) A proton moves from 40.0 V to 10.0 V. (e) An O^{--} ion moves from 40.0 V to 60.0 V.

15. A helium nucleus (charge $= 2e$, mass $= 6.63 \times 10^{-27}$ kg) traveling at 6.20×10^5 m/s enters an electric field, traveling from point Ⓐ, at a potential of 1.50×10^3 V, to point Ⓑ, at 4.00×10^3 V. What is its speed at point Ⓑ? (a) 7.91×10^5 m/s (b) 3.78×10^5 m/s (c) 2.13×10^5 m/s (d) 2.52×10^6 m/s (e) 3.01×10^8 m/s

Conceptual Questions

1. denotes answer available in *Student Solutions Manual/Study Guide*

1. What determines the maximum electric potential to which the dome of a Van de Graaff generator can be raised?

2. Describe the motion of a proton (a) after it is released from rest in a uniform electric field. Describe the changes (if any) in (b) its kinetic energy and (c) the electric potential energy of the proton–field system.

3. When charged particles are separated by an infinite distance, the electric potential energy of the pair is zero. When the particles are brought close, the elec-

tric potential energy of a pair with the same sign is positive, whereas the electric potential energy of a pair with opposite signs is negative. Give a physical explanation of this statement.

4. Study Figure 23.3 and the accompanying text discussion of charging by induction. When the grounding wire is touched to the rightmost point on the sphere in Figure 23.3c, electrons are drained away from the sphere to leave the sphere positively charged. Suppose the grounding wire is touched to the leftmost point on the sphere instead. (a) Will electrons still drain away, moving closer to the negatively charged rod as they do so? (b) What kind of charge, if any, remains on the sphere?

5. Distinguish between electric potential and electric potential energy.

6. Describe the equipotential surfaces for (a) an infinite line of charge and (b) a uniformly charged sphere.

Problems

Section 25.1 Electric Potential and Potential Difference

Section 25.2 Potential Difference in a Uniform Electric Field

1. Oppositely charged parallel plates are separated **M** by 5.33 mm. A potential difference of 600 V exists between the plates. (a) What is the magnitude of the electric field between the plates? (b) What is the magnitude of the force on an electron between the plates? (c) How much work must be done on the electron to move it to the negative plate if it is initially positioned 2.90 mm from the positive plate?

2. A uniform electric field of magnitude 250 V/m is directed in the positive x direction. A $+12.0$-μC charge moves from the origin to the point $(x, y) = (20.0$ cm, 50.0 cm). (a) What is the change in the potential energy of the charge–field system? (b) Through what potential difference does the charge move?

3. (a) Calculate the speed of a proton that is accelerated **M** from rest through an electric potential difference of 120 V. (b) Calculate the speed of an electron that is accelerated through the same electric potential difference.

4. How much work is done (by a battery, generator, or **W** some other source of potential difference) in moving Avogadro's number of electrons from an initial point where the electric potential is 9.00 V to a point where the electric potential is -5.00 V? (The potential in each case is measured relative to a common reference point.)

5. A uniform electric field **W** of magnitude 325 V/m is directed in the negative y direction in Figure P25.5. The coordinates of point

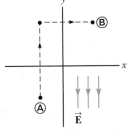

Figure P25.5

Ⓐ are $(-0.200, -0.300)$ m, and those of point Ⓑ are $(0.400, 0.500)$ m. Calculate the electric potential difference $V_Ⓑ - V_Ⓐ$ using the dashed-line path.

6. Starting with the definition of work, prove that at every point on an equipotential surface, the surface must be perpendicular to the electric field there.

7. An electron moving parallel to the x axis has an ini- **AMT** tial speed of 3.70×10^6 m/s at the origin. Its speed is **M** reduced to 1.40×10^5 m/s at the point $x = 2.00$ cm. (a) Calculate the electric potential difference between the origin and that point. (b) Which point is at the higher potential?

8. (a) Find the electric potential difference ΔV_e required to stop an electron (called a "stopping potential") moving with an initial speed of 2.85×10^7 m/s. (b) Would a proton traveling at the same speed require a greater or lesser magnitude of electric potential difference? Explain. (c) Find a symbolic expression for the ratio of the proton stopping potential and the electron stopping potential, $\Delta V_p/\Delta V_e$.

9. A particle having charge $q = +2.00$ μC and mass $m =$ **AMT** 0.010 0 kg is connected to a string that is $L = 1.50$ m long and tied to the pivot point P in Figure P25.9. The particle, string, and pivot point all lie on a frictionless,

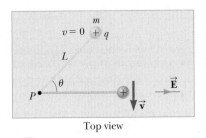

Top view

Figure P25.9

horizontal table. The particle is released from rest when the string makes an angle $\theta = 60.0°$ with a uniform electric field of magnitude $E = 300$ V/m. Determine the speed of the particle when the string is parallel to the electric field.

10. **Review.** A block having
GP mass m and charge $+Q$ is connected to an insulating spring having a force constant k. The block lies on a frictionless, insulating, horizontal track, and the system is immersed in a

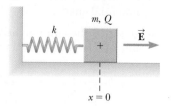

Figure P25.10

uniform electric field of magnitude E directed as shown in Figure P25.10. The block is released from rest when the spring is unstretched (at $x = 0$). We wish to show that the ensuing motion of the block is simple harmonic. (a) Consider the system of the block, the spring, and the electric field. Is this system isolated or nonisolated? (b) What kinds of potential energy exist within this system? (c) Call the initial configuration of the system that existing just as the block is released from rest. The final configuration is when the block momentarily comes to rest again. What is the value of x when the block comes to rest momentarily? (d) At some value of x we will call $x = x_0$, the block has zero net force on it. What analysis model describes the particle in this situation? (e) What is the value of x_0? (f) Define a new coordinate system x' such that $x' = x - x_0$. Show that x' satisfies a differential equation for simple harmonic motion. (g) Find the period of the simple harmonic motion. (h) How does the period depend on the electric field magnitude?

11. An insulating rod having linear charge density $\lambda = 40.0$ μC/m and linear mass density $\mu = 0.100$ kg/m is released from rest in a uniform electric field $E = 100$ V/m directed perpendicular to the rod (Fig. P25.11). (a) Determine the speed of the rod after it has traveled 2.00 m. (b) **What If?** How does your answer to part (a) change if the electric field is not perpendicular to the rod? Explain.

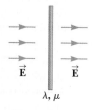

Figure P25.11

Section 25.3 Electric Potential and Potential Energy Due to Point Charges

Note: Unless stated otherwise, assume the reference level of potential is $V = 0$ at $r = \infty$.

12. (a) Calculate the electric potential 0.250 cm from an electron. (b) What is the electric potential difference between two points that are 0.250 cm and 0.750 cm from an electron? (c) How would the answers change if the electron were replaced with a proton?

13. Two point charges are on the y axis. A 4.50-μC charge is located at $y = 1.25$ cm, and a -2.24-μC charge is located at $y = -1.80$ cm. Find the total electric potential at (a) the origin and (b) the point whose coordinates are (1.50 cm, 0).

14. The two charges in Figure P25.14 are separated by $d = 2.00$ cm. Find the electric potential at (a) point A and (b) point B, which is halfway between the charges.

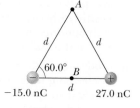

Figure P25.14

15. Three positive charges are located at the corners of an equilateral triangle as in Figure P25.15. Find an expression for the electric potential at the center of the triangle.

16. Two point charges $Q_1 = +5.00$ nC
M and $Q_2 = -3.00$ nC are separated by 35.0 cm. (a) What is the electric potential at a point midway between the charges? (b) What is the potential energy of the pair of charges? What is the significance of the algebraic sign of your answer?

Figure P25.15

17. Two particles, with charges of 20.0 nC and -20.0 nC, are placed at the points with coordinates (0, 4.00 cm) and (0, -4.00 cm) as shown in Figure P25.17. A particle with charge 10.0 nC is located at the origin. (a) Find the electric potential energy of the configuration of the three fixed charges. (b) A fourth particle, with a mass of 2.00×10^{-13} kg and a charge of 40.0 nC, is released from rest at the point (3.00 cm, 0). Find its speed after it has moved freely to a very large distance away.

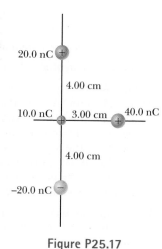

Figure P25.17

18. The two charges in Figure P25.18 are separated by a distance $d = 2.00$ cm, and $Q = +5.00$ nC. Find (a) the electric potential at A, (b) the electric potential at B, and (c) the electric potential difference between B and A.

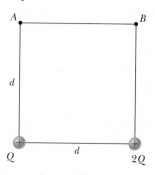

Figure P25.18

19. Given two particles with 2.00-μC charges as shown in
W Figure P25.19 and a particle with charge $q = 1.28 \times 10^{-18}$ C at the origin, (a) what is the net force exerted

by the two 2.00-μC charges on the charge q? (b) What is the electric field at the origin due to the two 2.00-μC particles? (c) What is the electric potential at the origin due to the two 2.00-μC particles?

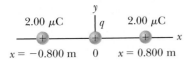

$2.00\ \mu C$ q $2.00\ \mu C$

$x = -0.800$ m 0 $x = 0.800$ m

Figure P25.19

20. At a certain distance from a charged particle, the magnitude of the electric field is 500 V/m and the electric potential is −3.00 kV. (a) What is the distance to the particle? (b) What is the magnitude of the charge?

21. Four point charges each having charge Q are located at the corners of a square having sides of length a. Find expressions for (a) the total electric potential at the center of the square due to the four charges and (b) the work required to bring a fifth charge q from infinity to the center of the square.

22. The three charged particles in Figure P25.22 are at the vertices of an isosceles triangle (where d = 2.00 cm). Taking q = 7.00 μC, calculate the electric potential at point A, the midpoint of the base.

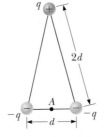

$2d$

$-q$ A $-q$

$\leftarrow d \rightarrow$

Figure P25.22

23. A particle with charge $+q$ is at the origin. A particle with charge $-2q$ is at x = 2.00 m on the x axis. (a) For what finite value(s) of x is the electric field zero? (b) For what finite value(s) of x is the electric potential zero?

24. Show that the amount of work required to assemble four identical charged particles of magnitude Q at the corners of a square of side s is $5.41k_eQ^2/s$.

25. Two particles each with charge $+2.00\ \mu$C are located on the x axis. One is at x = 1.00 m, and the other is at x = −1.00 m. (a) Determine the electric potential on the y axis at y = 0.500 m. (b) Calculate the change in electric potential energy of the system as a third charged particle of −3.00 μC is brought from infinitely far away to a position on the y axis at y = 0.500 m.

26. Two charged particles of equal magnitude are located along the y axis equal distances above and below the x axis as shown in Figure P25.26. (a) Plot a graph of the electric potential at points along the x axis over the interval $-3a < x < 3a$. You should plot the potential in units of k_eQ/a. (b) Let the charge of the particle located at $y = -a$ be negative. Plot the potential along the y axis over the interval $-4a < y < 4a$.

$+Q$
a
x
a
$+Q$

Figure P25.26

27. Four identical charged particles (q = +10.0 μC) are located on the corners of a rectangle as shown in Figure P25.27. The dimensions of the rectangle are L = 60.0 cm and W = 15.0 cm. Calculate the change in

electric potential energy of the system as the particle at the lower left corner in Figure P25.27 is brought to this position from infinitely far away. Assume the other three particles in Figure P25.27 remain fixed in position.

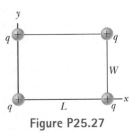

Figure P25.27

28. Three particles with equal positive charges q are at the corners of an equilateral triangle of side a as shown in Figure P25.28. (a) At what point, if any, in the plane of the particles is the electric potential zero? (b) What is the electric potential at the position of one of the particles due to the other two particles in the triangle?

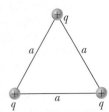

Figure P25.28

29. Five particles with equal negative charges $-q$ are placed symmetrically around a circle of radius R. Calculate the electric potential at the center of the circle.

30. Review. A light, unstressed spring has length d. Two identical particles, each with charge q, are connected to the opposite ends of the spring. The particles are held stationary a distance d apart and then released at the same moment. The system then oscillates on a frictionless, horizontal table. The spring has a bit of internal kinetic friction, so the oscillation is damped. The particles eventually stop vibrating when the distance between them is $3d$. Assume the system of the spring and two charged particles is isolated. Find the increase in internal energy that appears in the spring during the oscillations.

31. Review. Two insulating spheres have radii 0.300 cm and 0.500 cm, masses 0.100 kg and 0.700 kg, and uniformly distributed charges −2.00 μC and 3.00 μC. They are released from rest when their centers are separated by 1.00 m. (a) How fast will each be moving when they collide? (b) **What If?** If the spheres were conductors, would the speeds be greater or less than those calculated in part (a)? Explain.

32. Review. Two insulating spheres have radii r_1 and r_2, masses m_1 and m_2, and uniformly distributed charges $-q_1$ and q_2. They are released from rest when their centers are separated by a distance d. (a) How fast is each moving when they collide? (b) **What If?** If the spheres were conductors, would their speeds be greater or less than those calculated in part (a)? Explain.

33. How much work is required to assemble eight identical charged particles, each of magnitude q, at the corners of a cube of side s?

34. Four identical particles, each having charge q and mass m, are released from rest at the vertices of a square of side L. How fast is each particle moving when their distance from the center of the square doubles?

35. In 1911, Ernest Rutherford and his assistants Geiger and Marsden conducted an experiment in which they scattered alpha particles (nuclei of helium atoms) from thin sheets of gold. An alpha particle, having charge $+2e$ and mass 6.64×10^{-27} kg, is a product of certain radioactive decays. The results of the experiment led Rutherford to the idea that most of an atom's mass is in a very small nucleus, with electrons in orbit around it. (This is the planetary model of the atom, which we'll study in Chapter 42.) Assume an alpha particle, initially very far from a stationary gold nucleus, is fired with a velocity of 2.00×10^7 m/s directly toward the nucleus (charge $+79e$). What is the smallest distance between the alpha particle and the nucleus before the alpha particle reverses direction? Assume the gold nucleus remains stationary.

Section 25.4 Obtaining the Value of the Electric Field from the Electric Potential

36. Figure P25.36 represents a graph of the electric potential in a region of space versus position x, where the electric field is parallel to the x axis. Draw a graph of the x component of the electric field versus x in this region.

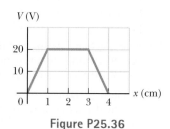

Figure P25.36

37. The potential in a region between $x = 0$ and $x = 6.00$ m is $V = a + bx$, where $a = 10.0$ V and $b = -7.00$ V/m. Determine (a) the potential at $x = 0$, 3.00 m, and 6.00 m and (b) the magnitude and direction of the electric field at $x = 0$, 3.00 m, and 6.00 m.

38. An electric field in a region of space is parallel to the x axis. The electric potential varies with position as shown in Figure P25.38. Graph the x component of the electric field versus position in this region of space.

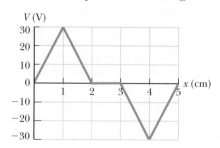

Figure P25.38

39. Over a certain region of space, the electric potential is $V = 5x - 3x^2y + 2yz^2$. (a) Find the expressions for the x, y, and z components of the electric field over this region. (b) What is the magnitude of the field at the point P that has coordinates $(1.00, 0, -2.00)$ m?

40. Figure P25.40 shows several equipotential lines, each labeled by its potential in volts. The distance between the lines of the square grid represents 1.00 cm. (a) Is the magnitude of the field larger at A or at B? Explain how you can tell. (b) Explain what you can determine

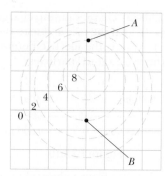

Numerical values are in volts.

Figure P25.40

about $\vec{E}$ at B. (c) Represent what the electric field looks like by drawing at least eight field lines.

41. The electric potential inside a charged spherical conductor of radius R is given by $V = k_eQ/R$, and the potential outside is given by $V = k_eQ/r$. Using $E_r = -dV/dr$, derive the electric field (a) inside and (b) outside this charge distribution.

42. It is shown in Example 25.7 that the potential at a point P a distance a above one end of a uniformly charged rod of length ℓ lying along the x axis is

$$V = k_e \frac{Q}{\ell} \ln\left(\frac{\ell + \sqrt{a^2 + \ell^2}}{a}\right)$$

Use this result to derive an expression for the y component of the electric field at P.

Section 25.5 Electric Potential Due to Continuous Charge Distributions

43. Consider a ring of radius R with the total charge Q spread uniformly over its perimeter. What is the potential difference between the point at the center of the ring and a point on its axis a distance $2R$ from the center?

44. A uniformly charged insulating rod of length 14.0 cm is bent into the shape of a semicircle as shown in Figure P25.44. The rod has a total charge of -7.50 μC. Find the electric potential at O, the center of the semicircle.

Figure P25.44

45. A rod of length L (Fig. P25.45) lies along the x axis with its left end at the origin. It has a nonuniform charge

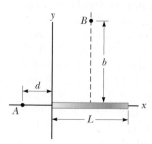

Figure P25.45 Problems 45 and 46.

density $\lambda = \alpha x$, where α is a positive constant. (a) What are the units of α? (b) Calculate the electric potential at A.

46. For the arrangement described in Problem 45, calculate the electric potential at point B, which lies on the perpendicular bisector of the rod a distance b above the x axis.

47. A wire having a uniform linear charge density λ is bent **W** into the shape shown in Figure P25.47. Find the electric potential at point O.

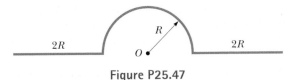

Figure P25.47

Section 25.6 Electric Potential Due to a Charged Conductor

48. The electric field magnitude on the surface of an irregularly shaped conductor varies from 56.0 kN/C to 28.0 kN/C. Can you evaluate the electric potential on the conductor? If so, find its value. If not, explain why not.

49. How many electrons should be removed from an initially uncharged spherical conductor of radius 0.300 m to produce a potential of 7.50 kV at the surface?

50. A spherical conductor has a radius of 14.0 cm and a **M** charge of 26.0 μC. Calculate the electric field and the electric potential at (a) $r = 10.0$ cm, (b) $r = 20.0$ cm, and (c) $r = 14.0$ cm from the center.

51. Electric charge can accumulate on an airplane in flight. You may have observed needle-shaped metal extensions on the wing tips and tail of an airplane. Their purpose is to allow charge to leak off before much of it accumulates. The electric field around the needle is much larger than the field around the body of the airplane and can become large enough to produce dielectric breakdown of the air, discharging the airplane. To model this process, assume two charged spherical conductors are connected by a long conducting wire and a 1.20-μC charge is placed on the combination. One sphere, representing the body of the airplane, has a radius of 6.00 cm; the other, representing the tip of the needle, has a radius of 2.00 cm. (a) What is the electric potential of each sphere? (b) What is the electric field at the surface of each sphere?

Section 25.8 Applications of Electrostatics

52. Lightning can be studied **M** with a Van de Graaff generator, which consists of a spherical dome on which charge is continuously deposited by a moving belt. Charge can be added until the electric field at the surface of the dome becomes equal to the

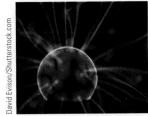

Figure P25.52

dielectric strength of air. Any more charge leaks off in sparks as shown in Figure P25.52. Assume the dome has a diameter of 30.0 cm and is surrounded by dry air with a "breakdown" electric field of 3.00×10^6 V/m. (a) What is the maximum potential of the dome? (b) What is the maximum charge on the dome?

Additional Problems

53. *Why is the following situation impossible?* In the Bohr model of the hydrogen atom, an electron moves in a circular orbit about a proton. The model states that the electron can exist only in certain allowed orbits around the proton: those whose radius r satisfies $r = n^2(0.052\ 9$ nm), where $n = 1, 2, 3, \ldots$. For one of the possible allowed states of the atom, the electric potential energy of the system is -13.6 eV.

54. **Review.** In fair weather, the electric field in the air at a particular location immediately above the Earth's surface is 120 N/C directed downward. (a) What is the surface charge density on the ground? Is it positive or negative? (b) Imagine the surface charge density is uniform over the planet. What then is the charge of the whole surface of the Earth? (c) What is the Earth's electric potential due to this charge? (d) What is the difference in potential between the head and the feet of a person 1.75 m tall? (Ignore any charges in the atmosphere.) (e) Imagine the Moon, with 27.3% of the radius of the Earth, had a charge 27.3% as large, with the same sign. Find the electric force the Earth would then exert on the Moon. (f) State how the answer to part (e) compares with the gravitational force the Earth exerts on the Moon.

55. **Review.** From a large distance away, a particle of mass 2.00 g and charge 15.0 μC is fired at $21.0\hat{\mathbf{i}}$ m/s straight toward a second particle, originally stationary but free to move, with mass 5.00 g and charge 8.50 μC. Both particles are constrained to move only along the x axis. (a) At the instant of closest approach, both particles will be moving at the same velocity. Find this velocity. (b) Find the distance of closest approach. After the interaction, the particles will move far apart again. At this time, find the velocity of (c) the 2.00-g particle and (d) the 5.00-g particle.

56. **Review.** From a large distance away, a particle of mass m_1 and positive charge q_1 is fired at speed v in the positive x direction straight toward a second particle, originally stationary but free to move, with mass m_2 and positive charge q_2. Both particles are constrained to move only along the x axis. (a) At the instant of closest approach, both particles will be moving at the same velocity. Find this velocity. (b) Find the distance of closest approach. After the interaction, the particles will move far apart again. At this time, find the velocity of (c) the particle of mass m_1 and (d) the particle of mass m_2.

57. The liquid-drop model of the atomic nucleus suggests **M** high-energy oscillations of certain nuclei can split the nucleus into two unequal fragments plus a few

neutrons. The fission products acquire kinetic energy from their mutual Coulomb repulsion. Assume the charge is distributed uniformly throughout the volume of each spherical fragment and, immediately before separating, each fragment is at rest and their surfaces are in contact. The electrons surrounding the nucleus can be ignored. Calculate the electric potential energy (in electron volts) of two spherical fragments from a uranium nucleus having the following charges and radii: $38e$ and 5.50×10^{-15} m, and $54e$ and 6.20×10^{-15} m.

58. On a dry winter day, you scuff your leather-soled shoes across a carpet and get a shock when you extend the tip of one finger toward a metal doorknob. In a dark room, you see a spark perhaps 5 mm long. Make order-of-magnitude estimates of (a) your electric potential and (b) the charge on your body before you touch the doorknob. Explain your reasoning.

59. The electric potential immediately outside a charged conducting sphere is 200 V, and 10.0 cm farther from the center of the sphere the potential is 150 V. Determine (a) the radius of the sphere and (b) the charge on it. The electric potential immediately outside another charged conducting sphere is 210 V, and 10.0 cm farther from the center the magnitude of the electric field is 400 V/m. Determine (c) the radius of the sphere and (d) its charge on it. (e) Are the answers to parts (c) and (d) unique?

60. (a) Use the exact result from Example 25.4 to find the electric potential created by the dipole described in the example at the point $(3a, 0)$. (b) Explain how this answer compares with the result of the approximate expression that is valid when x is much greater than a.

61. Calculate the work that must be done on charges brought from infinity to charge a spherical shell of radius $R = 0.100$ m to a total charge $Q = 125\ \mu$C.

62. Calculate the work that must be done on charges brought from infinity to charge a spherical shell of radius R to a total charge Q.

63. The electric potential everywhere on the xy plane is

$$V = \frac{36}{\sqrt{(x+1)^2 + y^2}} - \frac{45}{\sqrt{x^2 + (y-2)^2}}$$

where V is in volts and x and y are in meters. Determine the position and charge on each of the particles that create this potential.

64. *Why is the following situation impossible?* You set up an apparatus in your laboratory as follows. The x axis is the symmetry axis of a stationary, uniformly charged ring of radius $R = 0.500$ m and charge $Q = 50.0\ \mu$C (Fig. P25.64). You place a particle with charge

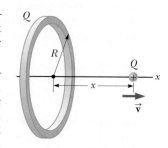

Figure P25.64

$Q = 50.0\ \mu$C and mass $m = 0.100$ kg at the center of the ring and arrange for it to be constrained to move only along the x axis. When it is displaced slightly, the particle is repelled by the ring and accelerates along the x axis. The particle moves faster than you expected and strikes the opposite wall of your laboratory at 40.0 m/s.

65. From Gauss's law, the electric field set up by a uniform line of charge is

$$\vec{\mathbf{E}} = \left(\frac{\lambda}{2\pi\epsilon_0 r}\right)\hat{\mathbf{r}}$$

where $\hat{\mathbf{r}}$ is a unit vector pointing radially away from the line and λ is the linear charge density along the line. Derive an expression for the potential difference between $r = r_1$ and $r = r_2$.

66. A uniformly charged filament lies along the x axis between $x = a = 1.00$ m and $x = a + \ell = 3.00$ m as shown in Figure P25.66. The total charge on the filament is 1.60 nC. Calculate successive approximations for the electric potential at the origin by modeling the filament as (a) a single charged particle at $x = 2.00$ m, (b) two 0.800-nC charged particles at $x = 1.5$ m and $x = 2.5$ m, and (c) four 0.400-nC charged particles at $x = 1.25$ m, $x = 1.75$ m, $x = 2.25$ m, and $x = 2.75$ m. (d) Explain how the results compare with the potential given by the exact expression

$$V = \frac{k_e Q}{\ell} \ln\left(\frac{\ell + a}{a}\right)$$

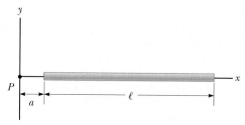

Figure P25.66

67. The thin, uniformly charged rod shown in Figure P25.67 has a linear charge density λ. Find an expression for the electric potential at P.

Figure P25.67

68. A Geiger–Mueller tube is a radiation detector that consists of a closed, hollow, metal cylinder (the cathode) of inner radius r_a and a coaxial cylindrical wire (the anode) of radius r_b (Fig. P25.68a). The charge per unit length on the anode is λ, and the charge per unit length on the cathode is $-\lambda$. A gas fills the space between the electrodes. When the tube is in use (Fig. P25.68b) and a high-energy elementary particle passes through this space, it can ionize an atom of the gas. The strong electric field makes the resulting ion and electron accelerate in opposite directions. They strike other molecules of the gas to ionize them, producing an avalanche of electrical discharge. The

pulse of electric current between the wire and the cylinder is counted by an external circuit. (a) Show that the magnitude of the electric potential difference between the wire and the cylinder is

$$\Delta V = 2k_e \lambda \ln\left(\frac{r_a}{r_b}\right)$$

(b) Show that the magnitude of the electric field in the space between cathode and anode is

$$E = \frac{\Delta V}{\ln\left(r_a/r_b\right)}\left(\frac{1}{r}\right)$$

where r is the distance from the axis of the anode to the point where the field is to be calculated.

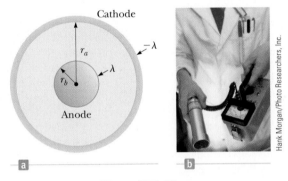

Figure P25.68

69. **Review.** Two parallel plates having charges of equal magnitude but opposite sign are separated by 12.0 cm. Each plate has a surface charge density of 36.0 nC/m². A proton is released from rest at the positive plate. Determine (a) the magnitude of the electric field between the plates from the charge density, (b) the potential difference between the plates, (c) the kinetic energy of the proton when it reaches the negative plate, (d) the speed of the proton just before it strikes the negative plate, (e) the acceleration of the proton, and (f) the force on the proton. (g) From the force, find the magnitude of the electric field. (h) How does your value of the electric field compare with that found in part (a)?

70. When an uncharged conducting sphere of radius a is placed at the origin of an xyz coordinate system that lies in an initially uniform electric field $\vec{\mathbf{E}} = E_0\hat{\mathbf{k}}$, the resulting electric potential is $V(x, y, z) = V_0$ for points inside the sphere and

$$V(x, y, z) = V_0 - E_0 z + \frac{E_0 a^3 z}{(x^2 + y^2 + z^2)^{3/2}}$$

for points outside the sphere, where V_0 is the (constant) electric potential on the conductor. Use this equation to determine the x, y, and z components of the resulting electric field (a) inside the sphere and (b) outside the sphere.

Challenge Problems

71. An electric dipole is located along the y axis as shown in Figure P25.71. The magnitude of its electric dipole moment is defined as $p = 2aq$. (a) At a point P, which

is far from the dipole ($r \gg a$), show that the electric potential is

$$V = \frac{k_e p \cos\theta}{r^2}$$

(b) Calculate the radial component E_r and the perpendicular component E_θ of the associated electric field. Note that $E_\theta = -(1/r)(\partial V/\partial \theta)$. Do these results seem reasonable for (c) $\theta = 90°$ and $0°$? (d) For $r = 0$? (e) For the dipole arrangement shown in Figure P25.71, express V in terms of Cartesian coordinates using $r = (x^2 + y^2)^{1/2}$ and

$$\cos\theta = \frac{y}{(x^2 + y^2)^{1/2}}$$

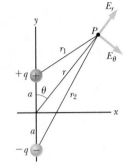

Figure P25.71

(f) Using these results and again taking $r \gg a$, calculate the field components E_x and E_y.

72. A solid sphere of radius R has a uniform charge density ρ and total charge Q. Derive an expression for its total electric potential energy. *Suggestion:* Imagine the sphere is constructed by adding successive layers of concentric shells of charge $dq = (4\pi r^2\, dr)\rho$ and use $dU = V\, dq$.

73. A disk of radius R (Fig. P25.73) has a nonuniform surface charge density $\sigma = Cr$, where C is a constant and r is measured from the center of the disk to a point on the surface of the disk. Find (by direct integration) the electric potential at P.

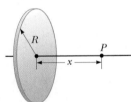

Figure P25.73

74. Four balls, each with mass m, are connected by four nonconducting strings to form a square with side a as shown in Figure P25.74. The assembly is placed on a nonconducting, frictionless, horizontal surface. Balls 1 and 2 each have charge q, and balls 3 and 4 are uncharged. After the string connecting balls 1 and 2 is cut, what is the maximum speed of balls 3 and 4?

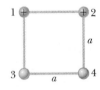

Figure P25.74

75. (a) A uniformly charged cylindrical shell with no end caps has total charge Q, radius R, and length h. Determine the electric potential at a point a distance d from the right end of the cylinder as shown in Figure P25.75.

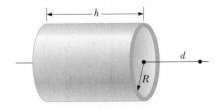

Figure P25.75

Suggestion: Use the result of Example 25.5 by treating the cylinder as a collection of ring charges. (b) **What If?** Use the result of Example 25.6 to solve the same problem for a solid cylinder.

76. As shown in Figure P25.76, two large, parallel, vertical conducting plates separated by distance d are charged so that their potentials are $+V_0$ and $-V_0$. A small conducting ball of mass m and radius R (where $R << d$) hangs midway between the plates. The thread of length L supporting the ball is a conducting wire connected to ground, so the potential of the ball is fixed at $V = 0$. The ball hangs straight down in stable equilibrium when V_0 is sufficiently small. Show that the equilibrium of the ball is unstable if V_0 exceeds the critical value $[k_e d^2 mg/(4RL)]^{1/2}$. *Suggestion:* Consider the forces on the ball when it is displaced a distance $x << L$.

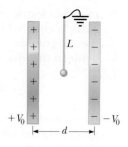

Figure P25.76

77. A particle with charge q is located at $x = -R$, and a particle with charge $-2q$ is located at the origin. Prove that the equipotential surface that has zero potential is a sphere centered at $(-4R/3, 0, 0)$ and having a radius $r = \frac{2}{3}R$.

Capacitance and Dielectrics

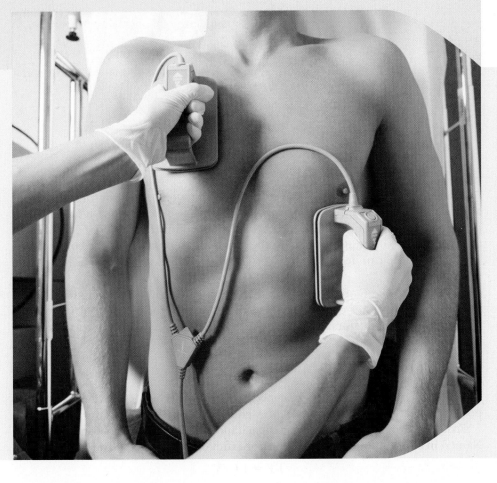

In this chapter, we introduce the first of three simple circuit elements that can be connected with wires to form an electric circuit. Electric circuits are the basis for the vast majority of the devices used in our society. Here we shall discuss *capacitors*, devices that store electric charge. This discussion is followed by the study of *resistors* in Chapter 27 and *inductors* in Chapter 32. In later chapters, we will study more sophisticated circuit elements such as *diodes* and *transistors*.

Capacitors are commonly used in a variety of electric circuits. For instance, they are used to tune the frequency of radio receivers, as filters in power supplies, to eliminate sparking in automobile ignition systems, and as energy-storing devices in electronic flash units.

When a patient receives a shock from a defibrillator, the energy delivered to the patient is initially stored in a *capacitor.* We will study capacitors and capacitance in this chapter. *(Andrew Olney/Getty Images)*

26.1 Definition of Capacitance

Consider two conductors as shown in Figure 26.1 (page 778). Such a combination of two conductors is called a **capacitor.** The conductors are called *plates*. If the conductors carry charges of equal magnitude and opposite sign, a potential difference ΔV exists between them.

Figure 26.1 A capacitor consists of two conductors.

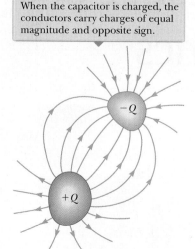

When the capacitor is charged, the conductors carry charges of equal magnitude and opposite sign.

What determines how much charge is on the plates of a capacitor for a given voltage? Experiments show that the quantity of charge Q on a capacitor[1] is linearly proportional to the potential difference between the conductors; that is, $Q \propto \Delta V$. The proportionality constant depends on the shape and separation of the conductors.[2] This relationship can be written as $Q = C\,\Delta V$ if we define capacitance as follows:

> The **capacitance** C of a capacitor is defined as the ratio of the magnitude of the charge on either conductor to the magnitude of the potential difference between the conductors:
>
> $$C \equiv \frac{Q}{\Delta V}$$
>
> **(26.1)**

◀ Definition of capacitance

By definition *capacitance is always a positive quantity*. Furthermore, the charge Q and the potential difference ΔV are always expressed in Equation 26.1 as positive quantities.

From Equation 26.1, we see that capacitance has SI units of coulombs per volt. Named in honor of Michael Faraday, the SI unit of capacitance is the **farad** (F):

$$1\ \text{F} = 1\ \text{C/V}$$

The farad is a very large unit of capacitance. In practice, typical devices have capacitances ranging from microfarads (10^{-6} F) to picofarads (10^{-12} F). We shall use the symbol μF to represent microfarads. In practice, to avoid the use of Greek letters, physical capacitors are often labeled "mF" for microfarads and "mmF" for micromicrofarads or, equivalently, "pF" for picofarads.

Let's consider a capacitor formed from a pair of parallel plates as shown in Figure 26.2. Each plate is connected to one terminal of a battery, which acts as a source of potential difference. If the capacitor is initially uncharged, the battery establishes an electric field in the connecting wires when the connections are made. Let's focus on the plate connected to the negative terminal of the battery. The electric field in the wire applies a force on electrons in the wire immediately outside this plate; this force causes the electrons to move onto the plate. The movement continues until the plate, the wire, and the terminal are all at the same electric potential. Once this equilibrium situation is attained, a potential difference no longer exists between the terminal and the plate; as a result, no electric field is present in the wire and

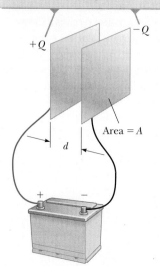

When the capacitor is connected to the terminals of a battery, electrons transfer between the plates and the wires so that the plates become charged.

Figure 26.2 A parallel-plate capacitor consists of two parallel conducting plates, each of area A, separated by a distance d.

---

[1] Although the total charge on the capacitor is zero (because there is as much excess positive charge on one conductor as there is excess negative charge on the other), it is common practice to refer to the magnitude of the charge on either conductor as "the charge on the capacitor."

[2] The proportionality between Q and ΔV can be proven from Coulomb's law or by experiment.

the electrons stop moving. The plate now carries a negative charge. A similar process occurs at the other capacitor plate, where electrons move from the plate to the wire, leaving the plate positively charged. In this final configuration, the potential difference across the capacitor plates is the same as that between the terminals of the battery.

Quick Quiz 26.1 A capacitor stores charge Q at a potential difference ΔV. What happens if the voltage applied to the capacitor by a battery is doubled to $2\,\Delta V$? **(a)** The capacitance falls to half its initial value, and the charge remains the same. **(b)** The capacitance and the charge both fall to half their initial values. **(c)** The capacitance and the charge both double. **(d)** The capacitance remains the same, and the charge doubles.

26.2 Calculating Capacitance

We can derive an expression for the capacitance of a pair of oppositely charged conductors having a charge of magnitude Q in the following manner. First we calculate the potential difference using the techniques described in Chapter 25. We then use the expression $C = Q/\Delta V$ to evaluate the capacitance. The calculation is relatively easy if the geometry of the capacitor is simple.

Although the most common situation is that of two conductors, a single conductor also has a capacitance. For example, imagine a single spherical, charged conductor. The electric field lines around this conductor are exactly the same as if there were a conducting, spherical shell of infinite radius, concentric with the sphere and carrying a charge of the same magnitude but opposite sign. Therefore, we can identify the imaginary shell as the second conductor of a two-conductor capacitor. The electric potential of the sphere of radius a is simply $k_e Q/a$ (see Section 25.6), and setting $V = 0$ for the infinitely large shell gives

> **Pitfall Prevention 26.3**
> **Too Many Cs** Do not confuse an italic C for capacitance with a non-italic C for the unit coulomb.

$$C = \frac{Q}{\Delta V} = \frac{Q}{k_e Q/a} = \frac{a}{k_e} = 4\pi\epsilon_0 a \qquad (26.2)$$

◀ **Capacitance of an isolated charged sphere**

This expression shows that the capacitance of an isolated, charged sphere is proportional to its radius and is independent of both the charge on the sphere and its potential, as is the case with all capacitors. Equation 26.1 is the general definition of capacitance in terms of electrical parameters, but the capacitance of a given capacitor will depend only on the geometry of the plates.

The capacitance of a pair of conductors is illustrated below with three familiar geometries, namely, parallel plates, concentric cylinders, and concentric spheres. In these calculations, we assume the charged conductors are separated by a vacuum.

Parallel–Plate Capacitors

Two parallel, metallic plates of equal area A are separated by a distance d as shown in Figure 26.2. One plate carries a charge $+Q$, and the other carries a charge $-Q$. The surface charge density on each plate is $\sigma = Q/A$. If the plates are very close together (in comparison with their length and width), we can assume the electric field is uniform between the plates and zero elsewhere. According to the What If? feature of Example 24.5, the value of the electric field between the plates is

$$E = \frac{\sigma}{\epsilon_0} = \frac{Q}{\epsilon_0 A}$$

Because the field between the plates is uniform, the magnitude of the potential difference between the plates equals Ed (see Eq. 25.6); therefore,

$$\Delta V = Ed = \frac{Qd}{\epsilon_0 A}$$

Substituting this result into Equation 26.1, we find that the capacitance is

$$C = \frac{Q}{\Delta V} = \frac{Q}{Qd/\epsilon_0 A}$$

◄ Capacitance of parallel plates

$$C = \frac{\epsilon_0 A}{d} \qquad (26.3)$$

That is, the capacitance of a parallel-plate capacitor is proportional to the area of its plates and inversely proportional to the plate separation.

Let's consider how the geometry of these conductors influences the capacity of the pair of plates to store charge. As a capacitor is being charged by a battery, electrons flow into the negative plate and out of the positive plate. If the capacitor plates are large, the accumulated charges are able to distribute themselves over a substantial area and the amount of charge that can be stored on a plate for a given potential difference increases as the plate area is increased. Therefore, it is reasonable that the capacitance is proportional to the plate area A as in Equation 26.3.

Now consider the region that separates the plates. Imagine moving the plates closer together. Consider the situation before any charges have had a chance to move in response to this change. Because no charges have moved, the electric field between the plates has the same value but extends over a shorter distance. Therefore, the magnitude of the potential difference between the plates $\Delta V = Ed$ (Eq. 25.6) is smaller. The difference between this new capacitor voltage and the terminal voltage of the battery appears as a potential difference across the wires connecting the battery to the capacitor, resulting in an electric field in the wires that drives more charge onto the plates and increases the potential difference between the plates. When the potential difference between the plates again matches that of the battery, the flow of charge stops. Therefore, moving the plates closer together causes the charge on the capacitor to increase. If d is increased, the charge decreases. As a result, the inverse relationship between C and d in Equation 26.3 is reasonable.

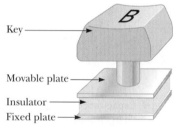

Key

Movable plate

Insulator

Fixed plate

Figure 26.3 (Quick Quiz 26.2) One type of computer keyboard button.

Quick Quiz 26.2 Many computer keyboard buttons are constructed of capacitors as shown in Figure 26.3. When a key is pushed down, the soft insulator between the movable plate and the fixed plate is compressed. When the key is pressed, what happens to the capacitance? **(a)** It increases. **(b)** It decreases. **(c)** It changes in a way you cannot determine because the electric circuit connected to the keyboard button may cause a change in ΔV.

Example 26.1 **The Cylindrical Capacitor**

A solid cylindrical conductor of radius a and charge Q is coaxial with a cylindrical shell of negligible thickness, radius $b > a$, and charge $-Q$ (Fig. 26.4a). Find the capacitance of this cylindrical capacitor if its length is ℓ.

SOLUTION

Conceptualize Recall that any pair of conductors qualifies as a capacitor, so the system described in this example therefore qualifies. Figure 26.4b helps visualize the electric field between the conductors. We expect the capacitance to depend only on geometric factors, which, in this case, are a, b, and ℓ.

Categorize Because of the cylindrical symmetry of the system, we can use results from previous studies of cylindrical systems to find the capacitance.

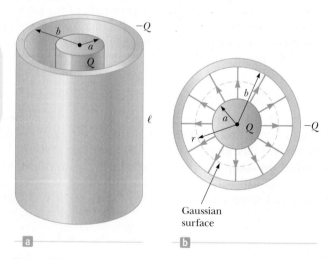

Figure 26.4 (Example 26.1) (a) A cylindrical capacitor consists of a solid cylindrical conductor of radius a and length ℓ surrounded by a coaxial cylindrical shell of radius b. (b) End view. The electric field lines are radial. The dashed line represents the end of a cylindrical gaussian surface of radius r and length ℓ.

▶ **26.1** continued

Analyze Assuming ℓ is much greater than a and b, we can neglect end effects. In this case, the electric field is perpendicular to the long axis of the cylinders and is confined to the region between them (Fig. 26.4b).

Write an expression for the potential difference between the two cylinders from Equation 25.3:

$$V_b - V_a = -\int_a^b \vec{\mathbf{E}} \cdot d\vec{\mathbf{s}}$$

Apply Equation 24.7 for the electric field outside a cylindrically symmetric charge distribution and notice from Figure 26.4b that $\vec{\mathbf{E}}$ is parallel to $d\vec{\mathbf{s}}$ along a radial line:

$$V_b - V_a = -\int_a^b E_r\,dr = -2k_e\lambda\int_a^b \frac{dr}{r} = -2k_e\lambda\ln\left(\frac{b}{a}\right)$$

Substitute the absolute value of ΔV into Equation 26.1 and use $\lambda = Q/\ell$:

$$C = \frac{Q}{\Delta V} = \frac{Q}{(2k_e Q/\ell)\ln(b/a)} = \frac{\ell}{2k_e\ln(b/a)} \quad \text{(26.4)}$$

Finalize The capacitance depends on the radii a and b and is proportional to the length of the cylinders. Equation 26.4 shows that the capacitance per unit length of a combination of concentric cylindrical conductors is

$$\frac{C}{\ell} = \frac{1}{2k_e\ln(b/a)} \quad \text{(26.5)}$$

An example of this type of geometric arrangement is a *coaxial cable,* which consists of two concentric cylindrical conductors separated by an insulator. You probably have a coaxial cable attached to your television set if you are a subscriber to cable television. The coaxial cable is especially useful for shielding electrical signals from any possible external influences.

WHAT IF? Suppose $b = 2.00a$ for the cylindrical capacitor. You would like to increase the capacitance, and you can do so by choosing to increase either ℓ by 10% or a by 10%. Which choice is more effective at increasing the capacitance?

Answer According to Equation 26.4, C is proportional to ℓ, so increasing ℓ by 10% results in a 10% increase in C. For the result of the change in a, let's use Equation 26.4 to set up a ratio of the capacitance C' for the enlarged cylinder radius a' to the original capacitance:

$$\frac{C'}{C} = \frac{\ell/2k_e\ln(b/a')}{\ell/2k_e\ln(b/a)} = \frac{\ln(b/a)}{\ln(b/a')}$$

We now substitute $b = 2.00a$ and $a' = 1.10a$, representing a 10% increase in a:

$$\frac{C'}{C} = \frac{\ln(2.00a/a)}{\ln(2.00a/1.10a)} = \frac{\ln 2.00}{\ln 1.82} = 1.16$$

which corresponds to a 16% increase in capacitance. Therefore, it is more effective to increase a than to increase ℓ.

Note two more extensions of this problem. First, it is advantageous to increase a only for a range of relationships between a and b. If $b > 2.85a$, increasing ℓ by 10% is more effective than increasing a (see Problem 70). Second, if b decreases, the capacitance increases. Increasing a or decreasing b has the effect of bringing the plates closer together, which increases the capacitance.

Example 26.2   **The Spherical Capacitor**

A spherical capacitor consists of a spherical conducting shell of radius b and charge $-Q$ concentric with a smaller conducting sphere of radius a and charge Q (Fig. 26.5, page 782). Find the capacitance of this device.

SOLUTION

Conceptualize As with Example 26.1, this system involves a pair of conductors and qualifies as a capacitor. We expect the capacitance to depend on the spherical radii a and b.

continued

▶ **26.2** c o n t i n u e d

Categorize Because of the spherical symmetry of the system, we can use results from previous studies of spherical systems to find the capacitance.

Analyze As shown in Chapter 24, the direction of the electric field outside a spherically symmetric charge distribution is radial and its magnitude is given by the expression $E = k_e Q / r^2$. In this case, this result applies to the field *between* the spheres ($a < r < b$).

Figure 26.5 (Example 26.2)
A spherical capacitor consists of an inner sphere of radius a surrounded by a concentric spherical shell of radius b. The electric field between the spheres is directed radially outward when the inner sphere is positively charged.

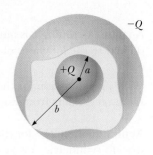

Write an expression for the potential difference between the two conductors from Equation 25.3:

$$V_b - V_a = -\int_a^b \vec{\mathbf{E}} \cdot d\vec{\mathbf{s}}$$

Apply the result of Example 24.3 for the electric field outside a spherically symmetric charge distribution and note that $\vec{\mathbf{E}}$ is parallel to $d\vec{\mathbf{s}}$ along a radial line:

$$V_b - V_a = -\int_a^b E_r\, dr = -k_e Q \int_a^b \frac{dr}{r^2} = k_e Q \left[\frac{1}{r} \right]_a^b$$

$$(1) \quad V_b - V_a = k_e Q \left(\frac{1}{b} - \frac{1}{a} \right) = k_e Q \frac{a - b}{ab}$$

Substitute the absolute value of ΔV into Equation 26.1:

$$C = \frac{Q}{\Delta V} = \frac{Q}{|V_b - V_a|} = \boxed{\frac{ab}{k_e(b - a)}} \qquad \textbf{(26.6)}$$

Finalize The capacitance depends on a and b as expected. The potential difference between the spheres in Equation (1) is negative because Q is positive and $b > a$. Therefore, in Equation 26.6, when we take the absolute value, we change $a - b$ to $b - a$. The result is a positive number.

WHAT IF?   If the radius b of the outer sphere approaches infinity, what does the capacitance become?

Answer In Equation 26.6, we let $b \rightarrow \infty$:

$$C = \lim_{b \to \infty} \frac{ab}{k_e(b - a)} = \frac{ab}{k_e(b)} = \frac{a}{k_e} = 4\pi\epsilon_0 a$$

Notice that this expression is the same as Equation 26.2, the capacitance of an isolated spherical conductor.

26.3 Combinations of Capacitors

Two or more capacitors often are combined in electric circuits. We can calculate the equivalent capacitance of certain combinations using methods described in this section. Throughout this section, we assume the capacitors to be combined are initially uncharged.

In studying electric circuits, we use a simplified pictorial representation called a **circuit diagram.** Such a diagram uses **circuit symbols** to represent various circuit elements. The circuit symbols are connected by straight lines that represent the wires between the circuit elements. The circuit symbols for capacitors, batteries, and switches as well as the color codes used for them in this text are given in Figure 26.6. The symbol for the capacitor reflects the geometry of the most common model for a capacitor, a pair of parallel plates. The positive terminal of the battery is at the higher potential and is represented in the circuit symbol by the longer line.

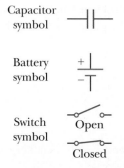

Figure 26.6 Circuit symbols for capacitors, batteries, and switches. Notice that capacitors are in blue, batteries are in green, and switches are in red. The closed switch can carry current, whereas the open one cannot.

Parallel Combination

Two capacitors connected as shown in Figure 26.7a are known as a **parallel combination** of capacitors. Figure 26.7b shows a circuit diagram for this combination of capacitors. The left plates of the capacitors are connected to the positive terminal of the battery by a conducting wire and are therefore both at the same electric potential

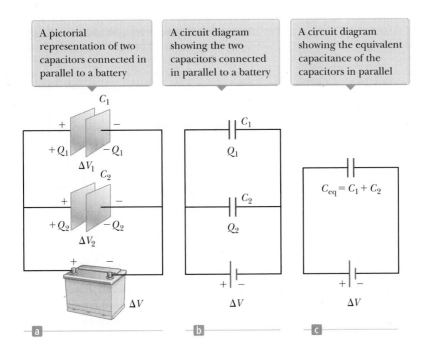

A pictorial representation of two capacitors connected in parallel to a battery

A circuit diagram showing the two capacitors connected in parallel to a battery

A circuit diagram showing the equivalent capacitance of the capacitors in parallel

Figure 26.7 Two capacitors connected in parallel. All three diagrams are equivalent.

as the positive terminal. Likewise, the right plates are connected to the negative terminal and so are both at the same potential as the negative terminal. Therefore, the individual potential differences across capacitors connected in parallel are the same and are equal to the potential difference applied across the combination. That is,

$$\Delta V_1 = \Delta V_2 = \Delta V$$

where ΔV is the battery terminal voltage.

After the battery is attached to the circuit, the capacitors quickly reach their maximum charge. Let's call the maximum charges on the two capacitors Q_1 and Q_2, where $Q_1 = C_1 \Delta V_1$ and $Q_2 = C_2 \Delta V_2$. The *total charge* Q_{tot} stored by the two capacitors is the sum of the charges on the individual capacitors:

$$Q_{tot} = Q_1 + Q_2 = C_1 \Delta V_1 + C_2 \Delta V_2 \qquad \textbf{(26.7)}$$

Suppose you wish to replace these two capacitors by one *equivalent capacitor* having a capacitance C_{eq} as in Figure 26.7c. The effect this equivalent capacitor has on the circuit must be exactly the same as the effect of the combination of the two individual capacitors. That is, the equivalent capacitor must store charge Q_{tot} when connected to the battery. Figure 26.7c shows that the voltage across the equivalent capacitor is ΔV because the equivalent capacitor is connected directly across the battery terminals. Therefore, for the equivalent capacitor,

$$Q_{tot} = C_{eq} \Delta V$$

Substituting this result into Equation 26.7 gives

$$C_{eq} \Delta V = C_1 \Delta V_1 + C_2 \Delta V_2$$

$$C_{eq} = C_1 + C_2 \quad (\text{parallel combination})$$

where we have canceled the voltages because they are all the same. If this treatment is extended to three or more capacitors connected in parallel, the **equivalent capacitance** is found to be

$$C_{eq} = C_1 + C_2 + C_3 + \cdots \quad (\text{parallel combination}) \qquad \textbf{(26.8)}$$

◀ **Equivalent capacitance for capacitors in parallel**

Therefore, the equivalent capacitance of a parallel combination of capacitors is (1) the algebraic sum of the individual capacitances and (2) greater than any of

Figure 26.8 Two capacitors connected in series. All three diagrams are equivalent.

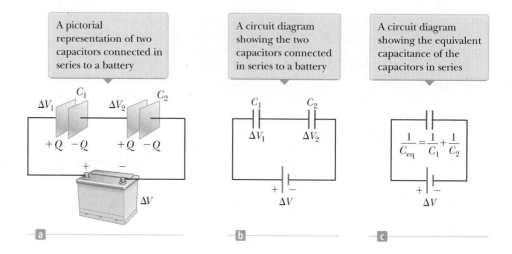

the individual capacitances. Statement (2) makes sense because we are essentially combining the areas of all the capacitor plates when they are connected with conducting wire, and capacitance of parallel plates is proportional to area (Eq. 26.3).

Series Combination

Two capacitors connected as shown in Figure 26.8a and the equivalent circuit diagram in Figure 26.8b are known as a **series combination** of capacitors. The left plate of capacitor 1 and the right plate of capacitor 2 are connected to the terminals of a battery. The other two plates are connected to each other and to nothing else; hence, they form an isolated system that is initially uncharged and must continue to have zero net charge. To analyze this combination, let's first consider the uncharged capacitors and then follow what happens immediately after a battery is connected to the circuit. When the battery is connected, electrons are transferred out of the left plate of C_1 and into the right plate of C_2. As this negative charge accumulates on the right plate of C_2, an equivalent amount of negative charge is forced off the left plate of C_2, and this left plate therefore has an excess positive charge. The negative charge leaving the left plate of C_2 causes negative charges to accumulate on the right plate of C_1. As a result, both right plates end up with a charge $-Q$ and both left plates end up with a charge $+Q$. Therefore, the charges on capacitors connected in series are the same:

$$Q_1 = Q_2 = Q$$

where Q is the charge that moved between a wire and the connected outside plate of one of the capacitors.

Figure 26.8a shows the individual voltages ΔV_1 and ΔV_2 across the capacitors. These voltages add to give the total voltage ΔV_{tot} across the combination:

$$\Delta V_{tot} = \Delta V_1 + \Delta V_2 = \frac{Q_1}{C_1} + \frac{Q_2}{C_2} \tag{26.9}$$

In general, the total potential difference across any number of capacitors connected in series is the sum of the potential differences across the individual capacitors.

Suppose the equivalent single capacitor in Figure 26.8c has the same effect on the circuit as the series combination when it is connected to the battery. After it is fully charged, the equivalent capacitor must have a charge of $-Q$ on its right plate and a charge of $+Q$ on its left plate. Applying the definition of capacitance to the circuit in Figure 26.8c gives

$$\Delta V_{tot} = \frac{Q}{C_{eq}}$$

Substituting this result into Equation 26.9, we have

$$\frac{Q}{C_{eq}} = \frac{Q_1}{C_1} + \frac{Q_2}{C_2}$$

Canceling the charges because they are all the same gives

$$\frac{1}{C_{eq}} = \frac{1}{C_1} + \frac{1}{C_2} \quad \text{(series combination)}$$

When this analysis is applied to three or more capacitors connected in series, the relationship for the **equivalent capacitance** is

$$\frac{1}{C_{eq}} = \frac{1}{C_1} + \frac{1}{C_2} + \frac{1}{C_3} + \cdots \quad \text{(series combination)} \qquad (26.10)$$

◀ **Equivalent capacitance for capacitors in series**

This expression shows that (1) the inverse of the equivalent capacitance is the algebraic sum of the inverses of the individual capacitances and (2) the equivalent capacitance of a series combination is always less than any individual capacitance in the combination.

Quick Quiz 26.3 Two capacitors are identical. They can be connected in series or in parallel. If you want the *smallest* equivalent capacitance for the combination, how should you connect them? **(a)** in series **(b)** in parallel **(c)** either way because both combinations have the same capacitance

---

Example 26.3 **Equivalent Capacitance**

Find the equivalent capacitance between a and b for the combination of capacitors shown in Figure 26.9a. All capacitances are in microfarads.

SOLUTION

Conceptualize Study Figure 26.9a carefully and make sure you understand how the capacitors are connected. Verify that there are only series and parallel connections between capacitors.

Categorize Figure 26.9a shows that the circuit contains both series and parallel connections, so we use the rules for series and parallel combinations discussed in this section.

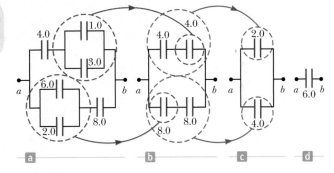

Figure 26.9 (Example 26.3) To find the equivalent capacitance of the capacitors in (a), we reduce the various combinations in steps as indicated in (b), (c), and (d), using the series and parallel rules described in the text. All capacitances are in microfarads.

Analyze Using Equations 26.8 and 26.10, we reduce the combination step by step as indicated in the figure. As you follow along below, notice that in each step we replace the combination of two capacitors in the circuit diagram with a single capacitor having the equivalent capacitance.

The 1.0-μF and 3.0-μF capacitors (upper red-brown circle in Fig. 26.9a) are in parallel. Find the equivalent capacitance from Equation 26.8:

$$C_{eq} = C_1 + C_2 = 4.0 \ \mu\text{F}$$

The 2.0-μF and 6.0-μF capacitors (lower red-brown circle in Fig. 26.9a) are also in parallel:

$$C_{eq} = C_1 + C_2 = 8.0 \ \mu\text{F}$$

The circuit now looks like Figure 26.9b. The two 4.0-μF capacitors (upper green circle in Fig. 26.9b) are in series. Find the equivalent capacitance from Equation 26.10:

$$\frac{1}{C_{eq}} = \frac{1}{C_1} + \frac{1}{C_2} = \frac{1}{4.0 \ \mu\text{F}} + \frac{1}{4.0 \ \mu\text{F}} = \frac{1}{2.0 \ \mu\text{F}}$$

$$C_{eq} = 2.0 \ \mu\text{F}$$

continued

▶ **26.3** continued

The two 8.0-μF capacitors (lower green circle in Fig. 26.9b) are also in series. Find the equivalent capacitance from Equation 26.10:

$$\frac{1}{C_{eq}} = \frac{1}{C_1} + \frac{1}{C_2} = \frac{1}{8.0 \ \mu F} + \frac{1}{8.0 \ \mu F} = \frac{1}{4.0 \ \mu F}$$

$$C_{eq} = 4.0 \ \mu F$$

The circuit now looks like Figure 26.9c. The 2.0-μF and 4.0-μF capacitors are in parallel:

$$C_{eq} = C_1 + C_2 = \boxed{6.0 \ \mu F}$$

Finalize This final value is that of the single equivalent capacitor shown in Figure 26.9d. For further practice in treating circuits with combinations of capacitors, imagine a battery is connected between points a and b in Figure 26.9a so that a potential difference ΔV is established across the combination. Can you find the voltage across and the charge on each capacitor?

26.4 Energy Stored in a Charged Capacitor

Because positive and negative charges are separated in the system of two conductors in a capacitor, electric potential energy is stored in the system. Many of those who work with electronic equipment have at some time verified that a capacitor can store energy. If the plates of a charged capacitor are connected by a conductor such as a wire, charge moves between each plate and its connecting wire until the capacitor is uncharged. The discharge can often be observed as a visible spark. If you accidentally touch the opposite plates of a charged capacitor, your fingers act as a pathway for discharge and the result is an electric shock. The degree of shock you receive depends on the capacitance and the voltage applied to the capacitor. Such a shock could be dangerous if high voltages are present as in the power supply of a home theater system. Because the charges can be stored in a capacitor even when the system is turned off, unplugging the system does not make it safe to open the case and touch the components inside.

Figure 26.10a shows a battery connected to a single parallel-plate capacitor with a switch in the circuit. Let us identify the circuit as a system. When the switch is closed (Fig. 26.10b), the battery establishes an electric field in the wires and charges

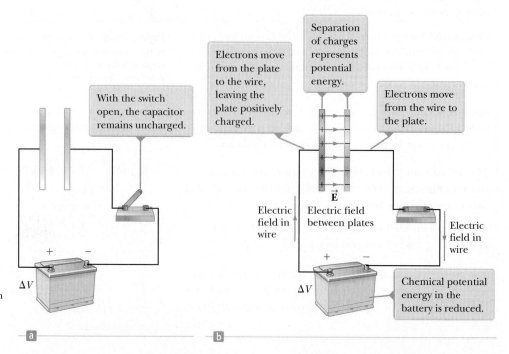

Figure 26.10 (a) A circuit consisting of a capacitor, a battery, and a switch. (b) When the switch is closed, the battery establishes an electric field in the wire and the capacitor becomes charged.

With the switch open, the capacitor remains uncharged.

Electrons move from the plate to the wire, leaving the plate positively charged.

Separation of charges represents potential energy.

Electrons move from the wire to the plate.

Electric field in wire

Electric field between plates

$\vec{E}$

Electric field in wire

Chemical potential energy in the battery is reduced.

ΔV

ΔV

a

b

flow between the wires and the capacitor. As that occurs, there is a transformation of energy within the system. Before the switch is closed, energy is stored as chemical potential energy in the battery. This energy is transformed during the chemical reaction that occurs within the battery when it is operating in an electric circuit. When the switch is closed, some of the chemical potential energy in the battery is transformed to electric potential energy associated with the separation of positive and negative charges on the plates.

To calculate the energy stored in the capacitor, we shall assume a charging process that is different from the actual process described in Section 26.1 but that gives the same final result. This assumption is justified because the energy in the final configuration does not depend on the actual charge-transfer process.[3] Imagine the plates are disconnected from the battery and you transfer the charge mechanically through the space between the plates as follows. You grab a small amount of positive charge on one plate and apply a force that causes this positive charge to move over to the other plate. Therefore, you do work on the charge as it is transferred from one plate to the other. At first, no work is required to transfer a small amount of charge dq from one plate to the other,[4] but once this charge has been transferred, a small potential difference exists between the plates. Therefore, work must be done to move additional charge through this potential difference. As more and more charge is transferred from one plate to the other, the potential difference increases in proportion and more work is required. The overall process is described by the nonisolated system model for energy. Equation 8.2 reduces to $W = \Delta U_E$; the work done on the system by the external agent appears as an increase in electric potential energy in the system.

Suppose q is the charge on the capacitor at some instant during the charging process. At the same instant, the potential difference across the capacitor is $\Delta V = q/C$. This relationship is graphed in Figure 26.11. From Section 25.1, we know that the work necessary to transfer an increment of charge dq from the plate carrying charge $-q$ to the plate carrying charge q (which is at the higher electric potential) is

$$dW = \Delta V\, dq = \frac{q}{C}\, dq$$

The work required to transfer the charge dq is the area of the tan rectangle in Figure 26.11. Because 1 V = 1 J/C, the unit for the area is the joule. The total work required to charge the capacitor from $q = 0$ to some final charge $q = Q$ is

$$W = \int_0^Q \frac{q}{C}\, dq = \frac{1}{C}\int_0^Q q\, dq = \frac{Q^2}{2C}$$

The work done in charging the capacitor appears as electric potential energy U_E stored in the capacitor. Using Equation 26.1, we can express the potential energy stored in a charged capacitor as

$$U_E = \frac{Q^2}{2C} = \tfrac{1}{2}Q\,\Delta V = \tfrac{1}{2}C(\Delta V)^2 \qquad \textbf{(26.11)}$$

◀ **Energy stored in a charged capacitor**

Because the curve in Figure 26.11 is a straight line, the total area under the curve is that of a triangle of base Q and height ΔV.

Equation 26.11 applies to any capacitor, regardless of its geometry. For a given capacitance, the stored energy increases as the charge and the potential difference increase. In practice, there is a limit to the maximum energy (or charge) that can be stored because, at a sufficiently large value of ΔV, discharge ultimately occurs

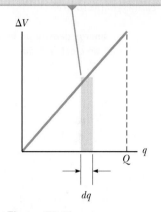

The work required to move charge dq through the potential difference ΔV across the capacitor plates is given approximately by the area of the shaded rectangle.

Figure 26.11 A plot of potential difference versus charge for a capacitor is a straight line having slope $1/C$.

---

[3]This discussion is similar to that of state variables in thermodynamics. The change in a state variable such as temperature is independent of the path followed between the initial and final states. The potential energy of a capacitor (or any system) is also a state variable, so its change does not depend on the process followed to charge the capacitor.

[4]We shall use lowercase q for the time-varying charge on the capacitor while it is charging to distinguish it from uppercase Q, which is the total charge on the capacitor after it is completely charged.

between the plates. For this reason, capacitors are usually labeled with a maximum operating voltage.

We can consider the energy in a capacitor to be stored in the electric field created between the plates as the capacitor is charged. This description is reasonable because the electric field is proportional to the charge on the capacitor. For a parallel-plate capacitor, the potential difference is related to the electric field through the relationship $\Delta V = Ed$. Furthermore, its capacitance is $C = \epsilon_0 A/d$ (Eq. 26.3). Substituting these expressions into Equation 26.11 gives

$$U_E = \tfrac{1}{2}\left(\frac{\epsilon_0 A}{d}\right)(Ed)^2 = \tfrac{1}{2}(\epsilon_0 Ad)E^2 \tag{26.12}$$

Because the volume occupied by the electric field is Ad, the *energy per unit volume* $u_E = U_E/Ad$, known as the *energy density*, is

▶ Energy density in an electric field

$$u_E = \tfrac{1}{2}\epsilon_0 E^2 \tag{26.13}$$

Although Equation 26.13 was derived for a parallel-plate capacitor, the expression is generally valid regardless of the source of the electric field. That is, the energy density in any electric field is proportional to the square of the magnitude of the electric field at a given point.

Quick Quiz 26.4 You have three capacitors and a battery. In which of the following combinations of the three capacitors is the maximum possible energy stored when the combination is attached to the battery? **(a)** series **(b)** parallel **(c)** no difference because both combinations store the same amount of energy

---

Example 26.4   **Rewiring Two Charged Capacitors**

Two capacitors C_1 and C_2 (where $C_1 > C_2$) are charged to the same initial potential difference ΔV_i. The charged capacitors are removed from the battery, and their plates are connected with opposite polarity as in Figure 26.12a. The switches S_1 and S_2 are then closed as in Figure 26.12b.

(A) Find the final potential difference ΔV_f between a and b after the switches are closed.

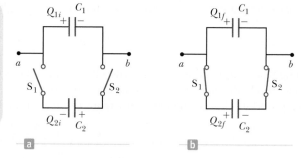

Figure 26.12 (Example 26.4) (a) Two capacitors are charged to the same initial potential difference and connected together with plates of opposite sign to be in contact when the switches are closed. (b) When the switches are closed, the charges redistribute.

SOLUTION

Conceptualize Figure 26.12 helps us understand the initial and final configurations of the system. When the switches are closed, the charge on the system will redistribute between the capacitors until both capacitors have the same potential difference. Because $C_1 > C_2$, more charge exists on C_1 than on C_2, so the final configuration will have positive charge on the left plates as shown in Figure 26.12b.

Categorize In Figure 26.12b, it might appear as if the capacitors are connected in parallel, but there is no battery in this circuit to apply a voltage across the combination. Therefore, we *cannot* categorize this problem as one in which capacitors are connected in parallel. We *can* categorize it as a problem involving an isolated system for electric charge. The left-hand plates of the capacitors form an isolated system because they are not connected to the right-hand plates by conductors.

Analyze Write an expression for the total charge on the left-hand plates of the system before the switches are closed, noting that a negative sign for Q_{2i} is necessary because the charge on the left plate of capacitor C_2 is negative:

$$(1) \quad Q_i = Q_{1i} + Q_{2i} = C_1\,\Delta V_i - C_2\,\Delta V_i = (C_1 - C_2)\Delta V_i$$

▶ **26.4** continued

After the switches are closed, the charges on the individual capacitors change to new values Q_{1f} and Q_{2f} such that the potential difference is again the same across both capacitors, with a value of ΔV_f. Write an expression for the total charge on the left-hand plates of the system after the switches are closed:

(2) $Q_f = Q_{1f} + Q_{2f} = C_1 \Delta V_f + C_2 \Delta V_f = (C_1 + C_2)\Delta V_f$

Because the system is isolated, the initial and final charges on the system must be the same. Use this condition and Equations (1) and (2) to solve for ΔV_f:

$Q_f = Q_i \rightarrow (C_1 + C_2)\, \Delta V_f = (C_1 - C_2)\, \Delta V_i$

(3) $\Delta V_f = \left(\dfrac{C_1 - C_2}{C_1 + C_2} \right) \Delta V_i$

(B) Find the total energy stored in the capacitors before and after the switches are closed and determine the ratio of the final energy to the initial energy.

SOLUTION

Use Equation 26.11 to find an expression for the total energy stored in the capacitors before the switches are closed:

(4) $U_i = \tfrac{1}{2}C_1(\Delta V_i)^2 + \tfrac{1}{2}C_2(\Delta V_i)^2 = \tfrac{1}{2}(C_1 + C_2)(\Delta V_i)^2$

Write an expression for the total energy stored in the capacitors after the switches are closed:

$U_f = \tfrac{1}{2}C_1(\Delta V_f)^2 + \tfrac{1}{2}C_2(\Delta V_f)^2 = \tfrac{1}{2}(C_1 + C_2)(\Delta V_f)^2$

Use the results of part (A) to rewrite this expression in terms of ΔV_i:

(5) $U_f = \tfrac{1}{2}(C_1 + C_2)\left[\left(\dfrac{C_1 - C_2}{C_1 + C_2}\right)\Delta V_i\right]^2 = \tfrac{1}{2}\dfrac{(C_1 - C_2)^2(\Delta V_i)^2}{C_1 + C_2}$

Divide Equation (5) by Equation (4) to obtain the ratio of the energies stored in the system:

$\dfrac{U_f}{U_i} = \dfrac{\tfrac{1}{2}(C_1 - C_2)^2(\Delta V_i)^2/(C_1 + C_2)}{\tfrac{1}{2}(C_1 + C_2)(\Delta V_i)^2}$

(6) $\dfrac{U_f}{U_i} = \left(\dfrac{C_1 - C_2}{C_1 + C_2}\right)^2$

Finalize The ratio of energies is *less* than unity, indicating that the final energy is *less* than the initial energy. At first, you might think the law of energy conservation has been violated, but that is not the case. The "missing" energy is transferred out of the system by the mechanism of electromagnetic waves (T_{ER} in Eq. 8.2), as we shall see in Chapter 34. Therefore, this system is isolated for electric charge, but nonisolated for energy.

WHAT IF? What if the two capacitors have the same capacitance? What would you expect to happen when the switches are closed?

Answer Because both capacitors have the same initial potential difference applied to them, the charges on the identical capacitors have the same magnitude. When the capacitors with opposite polarities are connected together, the equal-magnitude charges should cancel each other, leaving the capacitors uncharged.

Let's test our results to see if that is the case mathematically. In Equation (1), because the capacitances are equal, the initial charge Q_i on the system of left-hand plates is zero. Equation (3) shows that $\Delta V_f = 0$, which is consistent with uncharged capacitors. Finally, Equation (5) shows that $U_f = 0$, which is also consistent with uncharged capacitors.

One device in which capacitors have an important role is the portable *defibrillator* (see the chapter-opening photo on page 777). When cardiac fibrillation (random contractions) occurs, the heart produces a rapid, irregular pattern of beats. A fast discharge of energy through the heart can return the organ to its normal beat pattern. Emergency medical teams use portable defibrillators that contain batteries capable of charging a capacitor to a high voltage. (The circuitry actually permits the capacitor to be charged to a much higher voltage than that of the battery.) Up to 360 J is stored in the electric field of a large capacitor in a defibrillator when it is fully charged. The stored energy is released through the heart by conducting electrodes, called paddles, which are placed on both sides of the victim's chest. The defibrillator can deliver the energy to a patient in about 2 ms (roughly equivalent to 3 000 times the power delivered to a 60-W lightbulb!). The paramedics must wait between applications of the energy because of the time interval necessary for the capacitors to become fully charged. In this application and others (e.g., camera flash units and lasers used for fusion experiments), capacitors serve as energy reservoirs that can be slowly charged and then quickly discharged to provide large amounts of energy in a short pulse.

26.5 Capacitors with Dielectrics

Pitfall Prevention 26.5

Is the Capacitor Connected to a Battery? For problems in which a capacitor is modified (by insertion of a dielectric, for example), you must note whether modifications to the capacitor are being made while the capacitor is connected to a battery or after it is disconnected. If the capacitor remains connected to the battery, the voltage across the capacitor necessarily remains the same. If you disconnect the capacitor from the battery before making any modifications to the capacitor, the capacitor is an isolated system for electric charge and its charge remains the same.

A **dielectric** is a nonconducting material such as rubber, glass, or waxed paper. We can perform the following experiment to illustrate the effect of a dielectric in a capacitor. Consider a parallel-plate capacitor that without a dielectric has a charge Q_0 and a capacitance C_0. The potential difference across the capacitor is $\Delta V_0 = Q_0/C_0$. Figure 26.13a illustrates this situation. The potential difference is measured by a device called a *voltmeter*. Notice that no battery is shown in the figure; also, we must assume no charge can flow through an ideal voltmeter. Hence, there is no path by which charge can flow and alter the charge on the capacitor. If a dielectric is now inserted between the plates as in Figure 26.13b, the voltmeter indicates that the voltage between the plates decreases to a value ΔV. The voltages with and without the dielectric are related by a factor κ as follows:

$$\Delta V = \frac{\Delta V_0}{\kappa}$$

Because $\Delta V < \Delta V_0$, we see that $\kappa > 1$. The dimensionless factor κ is called the **dielectric constant** of the material. The dielectric constant varies from one material to another. In this section, we analyze this change in capacitance in terms of electrical parameters such as electric charge, electric field, and potential difference; Section 26.7 describes the microscopic origin of these changes.

Because the charge Q_0 on the capacitor does not change, the capacitance must change to the value

$$C = \frac{Q_0}{\Delta V} = \frac{Q_0}{\Delta V_0/\kappa} = \kappa \frac{Q_0}{\Delta V_0}$$

◀ Capacitance of a capacitor filled with a material of dielectric constant κ

$$C = \kappa C_0 \tag{26.14}$$

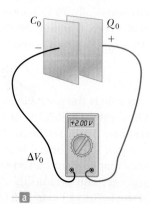
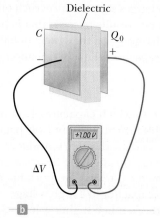

The potential difference across the charged capacitor is initially ΔV_0.

After the dielectric is inserted between the plates, the charge remains the same, but the potential difference decreases and the capacitance increases.

Dielectric

C_0 Q_0 C Q_0

+2.00 V   +1.00 V

ΔV_0 ΔV

Figure 26.13 A charged capacitor (a) before and (b) after insertion of a dielectric between the plates.

a   b

That is, the capacitance *increases* by the factor κ when the dielectric completely fills the region between the plates.[5] Because $C_0 = \epsilon_0 A/d$ (Eq. 26.3) for a parallel-plate capacitor, we can express the capacitance of a parallel-plate capacitor filled with a dielectric as

$$C = \kappa \frac{\epsilon_0 A}{d} \qquad (26.15)$$

From Equation 26.15, it would appear that the capacitance could be made very large by inserting a dielectric between the plates and decreasing d. In practice, the lowest value of d is limited by the electric discharge that could occur through the dielectric medium separating the plates. For any given separation d, the maximum voltage that can be applied to a capacitor without causing a discharge depends on the **dielectric strength** (maximum electric field) of the dielectric. If the magnitude of the electric field in the dielectric exceeds the dielectric strength, the insulating properties break down and the dielectric begins to conduct.

Physical capacitors have a specification called by a variety of names, including *working voltage, breakdown voltage,* and *rated voltage*. This parameter represents the largest voltage that can be applied to the capacitor without exceeding the dielectric strength of the dielectric material in the capacitor. Consequently, when selecting a capacitor for a given application, you must consider its capacitance as well as the expected voltage across the capacitor in the circuit, making sure the expected voltage is smaller than the rated voltage of the capacitor.

Insulating materials have values of κ greater than unity and dielectric strengths greater than that of air as Table 26.1 indicates. Therefore, a dielectric provides the following advantages:

- An increase in capacitance
- An increase in maximum operating voltage
- Possible mechanical support between the plates, which allows the plates to be close together without touching, thereby decreasing d and increasing C

Table 26.1   **Approximate Dielectric Constants and Dielectric Strengths of Various Materials at Room Temperature**

Material	Dielectric Constant κ	Dielectric Strength[a] (10^6 V/m)
Air (dry)	1.000 59	3
Bakelite	4.9	24
Fused quartz	3.78	8
Mylar	3.2	7
Neoprene rubber	6.7	12
Nylon	3.4	14
Paper	3.7	16
Paraffin-impregnated paper	3.5	11
Polystyrene	2.56	24
Polyvinyl chloride	3.4	40
Porcelain	6	12
Pyrex glass	5.6	14
Silicone oil	2.5	15
Strontium titanate	233	8
Teflon	2.1	60
Vacuum	1.000 00	—
Water	80	—

[a]The dielectric strength equals the maximum electric field that can exist in a dielectric without electrical breakdown. These values depend strongly on the presence of impurities and flaws in the materials.

---

[5] If the dielectric is introduced while the potential difference is held constant by a battery, the charge increases to a value $Q = \kappa Q_0$. The additional charge comes from the wires attached to the capacitor, and the capacitance again increases by the factor κ.

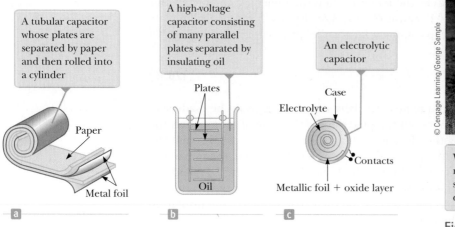

A tubular capacitor whose plates are separated by paper and then rolled into a cylinder

Paper

Metal foil

A high-voltage capacitor consisting of many parallel plates separated by insulating oil

Plates

Oil

An electrolytic capacitor

Case

Electrolyte

Contacts

Metallic foil + oxide layer

a   b   c

Figure 26.14 Three commercial capacitor designs.

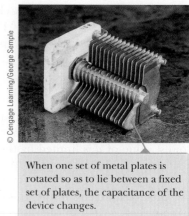

When one set of metal plates is rotated so as to lie between a fixed set of plates, the capacitance of the device changes.

Figure 26.15 A variable capacitor.

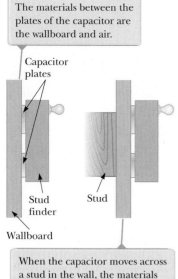

The materials between the plates of the capacitor are the wallboard and air.

Capacitor plates

Stud finder

Stud

Wallboard

When the capacitor moves across a stud in the wall, the materials between the plates are the wallboard and the wood stud. The change in the dielectric constant causes a signal light to illuminate.

a   b

Figure 26.16 (Quick Quiz 26.5) A stud finder.

Types of Capacitors

Many capacitors are built into integrated circuit chips, but some electrical devices still use stand-alone capacitors. Commercial capacitors are often made from metallic foil interlaced with thin sheets of either paraffin-impregnated paper or Mylar as the dielectric material. These alternate layers of metallic foil and dielectric are rolled into a cylinder to form a small package (Fig. 26.14a). High-voltage capacitors commonly consist of a number of interwoven metallic plates immersed in silicone oil (Fig. 26.14b). Small capacitors are often constructed from ceramic materials.

Often, an *electrolytic capacitor* is used to store large amounts of charge at relatively low voltages. This device, shown in Figure 26.14c, consists of a metallic foil in contact with an *electrolyte*, a solution that conducts electricity by virtue of the motion of ions contained in the solution. When a voltage is applied between the foil and the electrolyte, a thin layer of metal oxide (an insulator) is formed on the foil, and this layer serves as the dielectric. Very large values of capacitance can be obtained in an electrolytic capacitor because the dielectric layer is very thin and therefore the plate separation is very small.

Electrolytic capacitors are not reversible as are many other capacitors. They have a polarity, which is indicated by positive and negative signs marked on the device. When electrolytic capacitors are used in circuits, the polarity must be correct. If the polarity of the applied voltage is the opposite of what is intended, the oxide layer is removed and the capacitor conducts electricity instead of storing charge.

Variable capacitors (typically 10 to 500 pF) usually consist of two interwoven sets of metallic plates, one fixed and the other movable, and contain air as the dielectric (Fig. 26.15). These types of capacitors are often used in radio tuning circuits.

Quick Quiz 26.5 If you have ever tried to hang a picture or a mirror, you know it can be difficult to locate a wooden stud in which to anchor your nail or screw. A carpenter's stud finder is a capacitor with its plates arranged side by side instead of facing each other as shown in Figure 26.16. When the device is moved over a stud, does the capacitance **(a)** increase or **(b)** decrease?

Example 26.5 **Energy Stored Before and After**

A parallel-plate capacitor is charged with a battery to a charge Q_0. The battery is then removed, and a slab of material that has a dielectric constant κ is inserted between the plates. Identify the system as the capacitor and the dielectric. Find the energy stored in the system before and after the dielectric is inserted.

▶ **26.5** continued

SOLUTION

Conceptualize Think about what happens when the dielectric is inserted between the plates. Because the battery has been removed, the charge on the capacitor must remain the same. We know from our earlier discussion, however, that the capacitance must change. Therefore, we expect a change in the energy of the system.

Categorize Because we expect the energy of the system to change, we model it as a *nonisolated system* for *energy* involving a capacitor and a dielectric.

Analyze From Equation 26.11, find the energy stored in the absence of the dielectric:

$$U_0 = \frac{Q_0^2}{2C_0}$$

Find the energy stored in the capacitor after the dielectric is inserted between the plates:

$$U = \frac{Q_0^2}{2C}$$

Use Equation 26.14 to replace the capacitance C:

$$U = \frac{Q_0^2}{2\kappa C_0} = \frac{U_0}{\kappa}$$

Finalize Because $\kappa > 1$, the final energy is less than the initial energy. We can account for the decrease in energy of the system by performing an experiment and noting that the dielectric, when inserted, is pulled into the device. To keep the dielectric from accelerating, an external agent must do negative work on the dielectric. Equation 8.2 becomes $\Delta U = W$, where both sides of the equation are negative.

26.6 Electric Dipole in an Electric Field

We have discussed the effect on the capacitance of placing a dielectric between the plates of a capacitor. In Section 26.7, we shall describe the microscopic origin of this effect. Before we can do so, however, let's expand the discussion of the electric dipole introduced in Section 23.4 (see Example 23.6). The electric dipole consists of two charges of equal magnitude and opposite sign separated by a distance $2a$ as shown in Figure 26.17. The **electric dipole moment** of this configuration is defined as the vector $\vec{\mathbf{p}}$ directed from $-q$ toward $+q$ along the line joining the charges and having magnitude

$$p \equiv 2aq \tag{26.16}$$

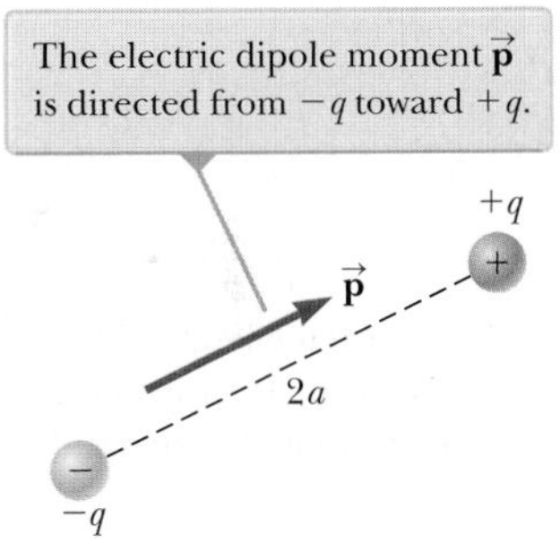

Now suppose an electric dipole is placed in a uniform electric field $\vec{\mathbf{E}}$ and makes an angle θ with the field as shown in Figure 26.18. We identify $\vec{\mathbf{E}}$ as the field *external* to the dipole, established by some other charge distribution, to distinguish it from the field *due to* the dipole, which we discussed in Section 23.4.

Each of the charges is modeled as a particle in an electric field. The electric forces acting on the two charges are equal in magnitude ($F = qE$) and opposite in direction as shown in Figure 26.18. Therefore, the net force on the dipole is zero. The two forces produce a net torque on the dipole, however; the dipole is therefore described by the rigid object under a net torque model. As a result, the dipole rotates in the direction that brings the dipole moment vector into greater alignment with the field. The torque due to the force on the positive charge about an axis through O in Figure 26.18 has magnitude $Fa \sin \theta$, where $a \sin \theta$ is the moment arm of F about O. This force tends to produce a clockwise rotation. The torque about O on the negative charge is also of magnitude $Fa \sin \theta$; here again, the force tends to produce a clockwise rotation. Therefore, the magnitude of the net torque about O is

$$\tau = 2Fa \sin \theta$$

Because $F = qE$ and $p = 2aq$, we can express τ as

$$\tau = 2aqE \sin \theta = pE \sin \theta \tag{26.17}$$

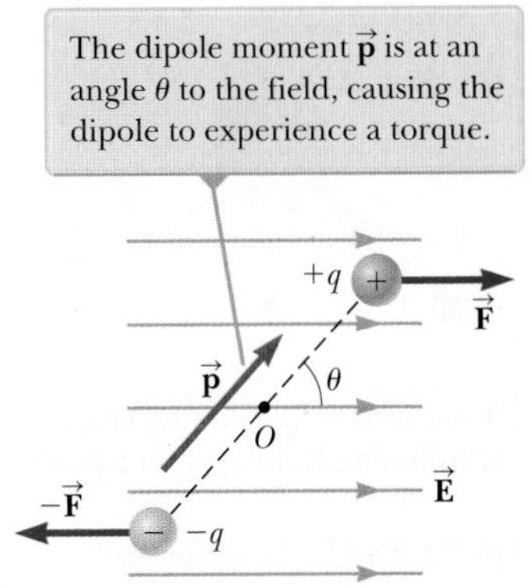

The electric dipole moment $\vec{\mathbf{p}}$ is directed from $-q$ toward $+q$.

Figure 26.17 An electric dipole consists of two charges of equal magnitude and opposite sign separated by a distance of $2a$.

The dipole moment $\vec{\mathbf{p}}$ is at an angle θ to the field, causing the dipole to experience a torque.

Figure 26.18 An electric dipole in a uniform external electric field.

Based on this expression, it is convenient to express the torque in vector form as the cross product of the vectors $\vec{\mathbf{p}}$ and $\vec{\mathbf{E}}$:

Torque on an electric dipole ▶
in an external electric field

$$\vec{\boldsymbol{\tau}} = \vec{\mathbf{p}} \times \vec{\mathbf{E}} \tag{26.18}$$

We can also model the system of the dipole and the external electric field as an isolated system for energy. Let's determine the potential energy of the system as a function of the dipole's orientation with respect to the field. To do so, recognize that work must be done by an external agent to rotate the dipole through an angle so as to cause the dipole moment vector to become less aligned with the field. The work done is then stored as electric potential energy in the system. Notice that this potential energy is associated with a *rotational* configuration of the system. Previously, we have seen potential energies associated with *translational* configurations: an object with mass was moved in a gravitational field, a charge was moved in an electric field, or a spring was extended. The work dW required to rotate the dipole through an angle $d\theta$ is $dW = \tau\, d\theta$ (see Eq. 10.25). Because $\tau = pE \sin\theta$ and the work results in an increase in the electric potential energy U, we find that for a rotation from θ_i to θ_f, the change in potential energy of the system is

$$U_f - U_i = \int_{\theta_i}^{\theta_f} \tau\, d\theta = \int_{\theta_i}^{\theta_f} pE \sin\theta\, d\theta = pE \int_{\theta_i}^{\theta_f} \sin\theta\, d\theta$$

$$= pE[-\cos\theta]_{\theta_i}^{\theta_f} = pE(\cos\theta_i - \cos\theta_f)$$

The term that contains $\cos\theta_i$ is a constant that depends on the initial orientation of the dipole. It is convenient to choose a reference angle of $\theta_i = 90°$ so that $\cos\theta_i = \cos 90° = 0$. Furthermore, let's choose $U_i = 0$ at $\theta_i = 90°$ as our reference value of potential energy. Hence, we can express a general value of $U_E = U_f$ as

$$U_E = -pE \cos\theta \tag{26.19}$$

We can write this expression for the potential energy of a dipole in an electric field as the dot product of the vectors $\vec{\mathbf{p}}$ and $\vec{\mathbf{E}}$:

Potential energy of the ▶
system of an electric dipole
in an external electric field

$$U_E = -\vec{\mathbf{p}} \cdot \vec{\mathbf{E}} \tag{26.20}$$

To develop a conceptual understanding of Equation 26.19, compare it with the expression for the potential energy of the system of an object in the Earth's gravitational field, $U_g = mgy$ (Eq. 7.19). First, both expressions contain a parameter of the entity placed in the field: mass for the object, dipole moment for the dipole. Second, both expressions contain the field, g for the object, E for the dipole. Finally, both expressions contain a configuration description: translational position y for the object, rotational position θ for the dipole. In both cases, once the configuration is changed, the system tends to return to the original configuration when the object is released: the object of mass m falls toward the ground, and the dipole begins to rotate back toward the configuration in which it is aligned with the field.

Molecules are said to be *polarized* when a separation exists between the average position of the negative charges and the average position of the positive charges in the molecule. In some molecules such as water, this condition is always present; such molecules are called **polar molecules.** Molecules that do not possess a permanent polarization are called **nonpolar molecules.**

We can understand the permanent polarization of water by inspecting the geometry of the water molecule. The oxygen atom in the water molecule is bonded to the hydrogen atoms such that an angle of 105° is formed between the two bonds (Fig. 26.19). The center of the negative charge distribution is near the oxygen atom, and the center of the positive charge distribution lies at a point midway along the line joining the hydrogen atoms (the point labeled ✕ in Fig. 26.19). We can model the water molecule and other polar molecules as dipoles because the average positions of the positive and negative charges act as point charges. As a result, we can apply our discussion of dipoles to the behavior of polar molecules.

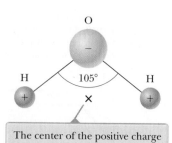

The center of the positive charge distribution is at the point ✕.

Figure 26.19 The water molecule, H_2O, has a permanent polarization resulting from its nonlinear geometry.

Washing with soap and water is a household scenario in which the dipole structure of water is exploited. Grease and oil are made up of nonpolar molecules, which are generally not attracted to water. Plain water is not very useful for removing this type of grime. Soap contains long molecules called *surfactants*. In a long molecule, the polarity characteristics of one end of the molecule can be different from those at the other end. In a surfactant molecule, one end acts like a nonpolar molecule and the other acts like a polar molecule. The nonpolar end can attach to a grease or oil molecule, and the polar end can attach to a water molecule. Therefore, the soap serves as a chain, linking the dirt and water molecules together. When the water is rinsed away, the grease and oil go with it.

A symmetric molecule (Fig. 26.20a) has no permanent polarization, but polarization can be induced by placing the molecule in an electric field. A field directed to the left as in Figure 26.20b causes the center of the negative charge distribution to shift to the right relative to the positive charges. This *induced polarization* is the effect that predominates in most materials used as dielectrics in capacitors.

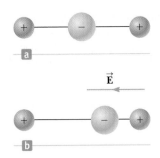

Figure 26.20 (a) A linear symmetric molecule has no permanent polarization. (b) An external electric field induces a polarization in the molecule.

Example 26.6 The H_2O Molecule

The water (H_2O) molecule has an electric dipole moment of 6.3×10^{-30} C · m. A sample contains 10^{21} water molecules, with the dipole moments all oriented in the direction of an electric field of magnitude 2.5×10^5 N/C. How much work is required to rotate the dipoles from this orientation ($\theta = 0°$) to one in which all the moments are perpendicular to the field ($\theta = 90°$)?

SOLUTION

Conceptualize When all the dipoles are aligned with the electric field, the dipoles–electric field system has the minimum potential energy. This energy has a negative value given by the product of the right side of Equation 26.19, evaluated at 0°, and the number N of dipoles.

Categorize The combination of the dipoles and the electric field is identified as a system. We use the *nonisolated system* model because an external agent performs work on the system to change its potential energy.

Analyze Write the appropriate reduction of the conservation of energy equation, Equation 8.2, for this situation:

(1) $\Delta U_E = W$

Use Equation 26.19 to evaluate the initial and final potential energies of the system and Equation (1) to calculate the work required to rotate the dipoles:

$W = U_{90°} - U_{0°} = (-NpE \cos 90°) - (-NpE \cos 0°)$

$= NpE = (10^{21})(6.3 \times 10^{-30} \text{ C · m})(2.5 \times 10^5 \text{ N/C})$

$= \boxed{1.6 \times 10^{-3} \text{ J}}$

Finalize Notice that the work done on the system is positive because the potential energy of the system has been raised from a negative value to a value of zero.

26.7 An Atomic Description of Dielectrics

In Section 26.5, we found that the potential difference ΔV_0 between the plates of a capacitor is reduced to $\Delta V_0 / \kappa$ when a dielectric is introduced. The potential difference is reduced because the magnitude of the electric field decreases between the plates. In particular, if $\vec{E}_0$ is the electric field without the dielectric, the field in the presence of a dielectric is

$$\vec{E} = \frac{\vec{E}_0}{\kappa} \tag{26.21}$$

First consider a dielectric made up of polar molecules placed in the electric field between the plates of a capacitor. The dipoles (that is, the polar molecules making

Figure 26.21 (a) Polar molecules in a dielectric. (b) An electric field is applied to the dielectric. (c) Details of the electric field inside the dielectric.

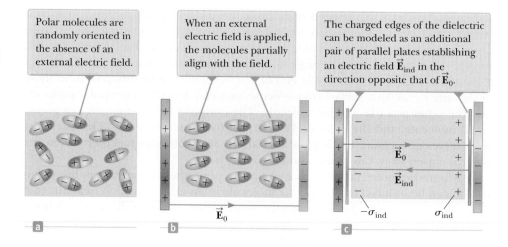

Polar molecules are randomly oriented in the absence of an external electric field.

When an external electric field is applied, the molecules partially align with the field.

The charged edges of the dielectric can be modeled as an additional pair of parallel plates establishing an electric field $\vec{E}_{ind}$ in the direction opposite that of $\vec{E}_0$.

up the dielectric) are randomly oriented in the absence of an electric field as shown in Figure 26.21a. When an external field $\vec{E}_0$ due to charges on the capacitor plates is applied, a torque is exerted on the dipoles, causing them to partially align with the field as shown in Figure 26.21b. The dielectric is now polarized. The degree of alignment of the molecules with the electric field depends on temperature and the magnitude of the field. In general, the alignment increases with decreasing temperature and with increasing electric field.

If the molecules of the dielectric are nonpolar, the electric field due to the plates produces an induced polarization in the molecule. These induced dipole moments tend to align with the external field, and the dielectric is polarized. Therefore, a dielectric can be polarized by an external field regardless of whether the molecules in the dielectric are polar or nonpolar.

With these ideas in mind, consider a slab of dielectric material placed between the plates of a capacitor so that it is in a uniform electric field $\vec{E}_0$ as shown in Figure 26.21b. The electric field due to the plates is directed to the right and polarizes the dielectric. The net effect on the dielectric is the formation of an *induced* positive surface charge density σ_{ind} on the right face and an equal-magnitude negative surface charge density $-\sigma_{ind}$ on the left face as shown in Figure 26.21c. Because we can model these surface charge distributions as being due to charged parallel plates, the induced surface charges on the dielectric give rise to an induced electric field $\vec{E}_{ind}$ in the direction opposite the external field $\vec{E}_0$. Therefore, the net electric field $\vec{E}$ in the dielectric has a magnitude

$$E = E_0 - E_{ind} \tag{26.22}$$

The induced charge density σ_{ind} on the dielectric is *less* than the charge density σ on the plates.

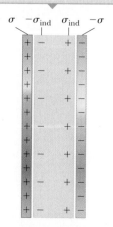

Figure 26.22 Induced charge on a dielectric placed between the plates of a charged capacitor.

In the parallel-plate capacitor shown in Figure 26.22, the external field E_0 is related to the charge density σ on the plates through the relationship $E_0 = \sigma/\epsilon_0$. The induced electric field in the dielectric is related to the induced charge density σ_{ind} through the relationship $E_{ind} = \sigma_{ind}/\epsilon_0$. Because $E = E_0/\kappa = \sigma/\kappa\epsilon_0$, substitution into Equation 26.22 gives

$$\frac{\sigma}{\kappa\epsilon_0} = \frac{\sigma}{\epsilon_0} - \frac{\sigma_{ind}}{\epsilon_0}$$

$$\sigma_{ind} = \left(\frac{\kappa - 1}{\kappa}\right)\sigma \tag{26.23}$$

Because $\kappa > 1$, this expression shows that the charge density σ_{ind} induced on the dielectric is less than the charge density σ on the plates. For instance, if $\kappa = 3$, the induced charge density is two-thirds the charge density on the plates. If no dielectric is present, then $\kappa = 1$ and $\sigma_{ind} = 0$ as expected. If the dielectric is replaced by an electrical conductor for which $E = 0$, however, Equation 26.22 indicates that $E_0 = E_{ind}$, which corresponds to $\sigma_{ind} = \sigma$. That is, the surface charge induced on the conductor is equal in magnitude but opposite in sign to that on the plates, resulting in a net electric field of zero in the conductor (see Fig. 24.16).

Example 26.7 Effect of a Metallic Slab

A parallel-plate capacitor has a plate separation d and plate area A. An uncharged metallic slab of thickness a is inserted midway between the plates.

(A) Find the capacitance of the device.

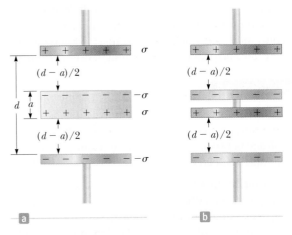

Figure 26.23 (Example 26.7) (a) A parallel-plate capacitor of plate separation d partially filled with a metallic slab of thickness a. (b) The equivalent circuit of the device in (a) consists of two capacitors in series, each having a plate separation $(d - a)/2$.

SOLUTION

Conceptualize Figure 26.23a shows the metallic slab between the plates of the capacitor. Any charge that appears on one plate of the capacitor must induce a charge of equal magnitude and opposite sign on the near side of the slab as shown in Figure 26.23a. Consequently, the net charge on the slab remains zero and the electric field inside the slab is zero.

Categorize The planes of charge on the metallic slab's upper and lower edges are identical to the distribution of charges on the plates of a capacitor. The metal between the slab's edges serves only to make an electrical connection between the edges. Therefore, we can model the edges of the slab as conducting planes and the bulk of the slab as a wire. As a result, the capacitor in Figure 26.23a is equivalent to two capacitors in series, each having a plate separation $(d - a)/2$ as shown in Figure 26.23b.

Analyze Use Equation 26.3 and the rule for adding two capacitors in series (Eq. 26.10) to find the equivalent capacitance in Figure 26.23b:

$$\frac{1}{C} = \frac{1}{C_1} + \frac{1}{C_2} = \frac{1}{\dfrac{\epsilon_0 A}{(d - a)/2}} + \frac{1}{\dfrac{\epsilon_0 A}{(d - a)/2}}$$

$$C = \frac{\epsilon_0 A}{d - a}$$

(B) Show that the capacitance of the original capacitor is unaffected by the insertion of the metallic slab if the slab is infinitesimally thin.

SOLUTION

In the result for part (A), let $a \to 0$:

$$C = \lim_{a \to 0} \left(\frac{\epsilon_0 A}{d - a} \right) = \frac{\epsilon_0 A}{d}$$

Finalize The result of part (B) is the original capacitance before the slab is inserted, which tells us that we can insert an infinitesimally thin metallic sheet between the plates of a capacitor without affecting the capacitance. We use this fact in the next example.

WHAT IF? What if the metallic slab in part (A) is not midway between the plates? How would that affect the capacitance?

Answer Let's imagine moving the slab in Figure 26.23a upward so that the distance between the upper edge of the slab and the upper plate is b. Then, the distance between the lower edge of the slab and the lower plate is $d - b - a$. As in part (A), we find the total capacitance of the series combination:

$$\frac{1}{C} = \frac{1}{C_1} + \frac{1}{C_2} = \frac{1}{\epsilon_0 A / b} + \frac{1}{\epsilon_0 A / (d - b - a)}$$

$$= \frac{b}{\epsilon_0 A} + \frac{d - b - a}{\epsilon_0 A} = \frac{d - a}{\epsilon_0 A} \rightarrow C = \frac{\epsilon_0 A}{d - a}$$

which is the same result as found in part (A). The capacitance is independent of the value of b, so it does not matter where the slab is located. In Figure 26.23b, when the central structure is moved up or down, the decrease in plate separation of one capacitor is compensated by the increase in plate separation for the other.

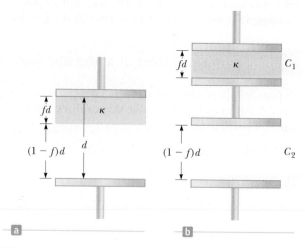

| Example 26.8 | A Partially Filled Capacitor |

A parallel-plate capacitor with a plate separation d has a capacitance C_0 in the absence of a dielectric. What is the capacitance when a slab of dielectric material of dielectric constant κ and thickness fd is inserted between the plates (Fig. 26.24a), where f is a fraction between 0 and 1?

SOLUTION

Conceptualize In our previous discussions of dielectrics between the plates of a capacitor, the dielectric filled the volume between the plates. In this example, only part of the volume between the plates contains the dielectric material.

Categorize In Example 26.7, we found that an infinitesimally thin metallic sheet inserted between the plates of a capacitor does not affect the capacitance. Imagine sliding an infinitesimally thin metallic slab along the bottom face of the dielectric shown in Figure 26.24a. We can model this system as a series combination of two capacitors as shown in Figure 26.24b. One capacitor has a plate separation fd and is filled with a dielectric; the other has a plate separation $(1 - f)d$ and has air between its plates.

Figure 26.24 (Example 26.8) (a) A parallel-plate capacitor of plate separation d partially filled with a dielectric of thickness fd. (b) The equivalent circuit of the capacitor consists of two capacitors connected in series.

Analyze Evaluate the two capacitances in Figure 26.24b from Equation 26.15:

$$C_1 = \frac{\kappa \epsilon_0 A}{fd} \quad \text{and} \quad C_2 = \frac{\epsilon_0 A}{(1 - f)d}$$

Find the equivalent capacitance C from Equation 26.10 for two capacitors combined in series:

$$\frac{1}{C} = \frac{1}{C_1} + \frac{1}{C_2} = \frac{fd}{\kappa \epsilon_0 A} + \frac{(1 - f)d}{\epsilon_0 A}$$

$$\frac{1}{C} = \frac{fd}{\kappa \epsilon_0 A} + \frac{\kappa(1 - f)d}{\kappa \epsilon_0 A} = \frac{f + \kappa(1 - f)}{\kappa} \frac{d}{\epsilon_0 A}$$

Invert and substitute for the capacitance without the dielectric, $C_0 = \epsilon_0 A/d$:

$$C = \frac{\kappa}{f + \kappa(1 - f)} \frac{\epsilon_0 A}{d} = \boxed{\frac{\kappa}{f + \kappa(1 - f)} C_0}$$

Finalize Let's test this result for some known limits. If $f \to 0$, the dielectric should disappear. In this limit, $C \to C_0$, which is consistent with a capacitor with air between the plates. If $f \to 1$, the dielectric fills the volume between the plates. In this limit, $C \to \kappa C_0$, which is consistent with Equation 26.14.

Summary

Definitions

A **capacitor** consists of two conductors carrying charges of equal magnitude and opposite sign. The **capacitance** C of any capacitor is the ratio of the charge Q on either conductor to the potential difference ΔV between them:

$$C \equiv \frac{Q}{\Delta V} \qquad \text{(26.1)}$$

The capacitance depends only on the geometry of the conductors and not on an external source of charge or potential difference. The SI unit of capacitance is coulombs per volt, or the **farad** (F): 1 F = 1 C/V.

The **electric dipole moment** $\vec{\mathbf{p}}$ of an electric dipole has a magnitude

$$p \equiv 2aq \qquad \text{(26.16)}$$

where $2a$ is the distance between the charges q and $-q$. The direction of the electric dipole moment vector is from the negative charge toward the positive charge.

Concepts and Principles

If two or more capacitors are connected in parallel, the potential difference is the same across all capacitors. The equivalent capacitance of a **parallel combination** of capacitors is

$$C_{eq} = C_1 + C_2 + C_3 + \cdots \qquad \textbf{(26.8)}$$

If two or more capacitors are connected in series, the charge is the same on all capacitors, and the equivalent capacitance of the **series combination** is given by

$$\frac{1}{C_{eq}} = \frac{1}{C_1} + \frac{1}{C_2} + \frac{1}{C_3} + \cdots \qquad \textbf{(26.10)}$$

These two equations enable you to simplify many electric circuits by replacing multiple capacitors with a single equivalent capacitance.

Energy is stored in a charged capacitor because the charging process is equivalent to the transfer of charges from one conductor at a lower electric potential to another conductor at a higher potential. The energy stored in a capacitor of capacitance C with charge Q and potential difference ΔV is

$$U_E = \frac{Q^2}{2C} = \tfrac{1}{2}Q\,\Delta V = \tfrac{1}{2}C\,(\Delta V)^2 \qquad \textbf{(26.11)}$$

When a dielectric material is inserted between the plates of a capacitor, the capacitance increases by a dimensionless factor κ, called the **dielectric constant:**

$$C = \kappa C_0 \qquad \textbf{(26.14)}$$

where C_0 is the capacitance in the absence of the dielectric.

The torque acting on an electric dipole in a uniform electric field $\vec{\mathbf{E}}$ is

$$\vec{\boldsymbol{\tau}} = \vec{\mathbf{p}} \times \vec{\mathbf{E}} \qquad \textbf{(26.18)}$$

The potential energy of the system of an electric dipole in a uniform external electric field $\vec{\mathbf{E}}$ is

$$U_E = -\vec{\mathbf{p}} \cdot \vec{\mathbf{E}} \qquad \textbf{(26.20)}$$

Objective Questions

1. denotes answer available in *Student Solutions Manual/Study Guide*

1. A fully charged parallel-plate capacitor remains connected to a battery while you slide a dielectric between the plates. Do the following quantities (a) increase, (b) decrease, or (c) stay the same? **(i)** C **(ii)** Q **(iii)** ΔV **(iv)** the energy stored in the capacitor

2. By what factor is the capacitance of a metal sphere multiplied if its volume is tripled? (a) 3 (b) $3^{1/3}$ (c) 1 (d) $3^{-1/3}$ (e) $\frac{1}{3}$

3. An electronics technician wishes to construct a parallel-plate capacitor using rutile ($\kappa = 100$) as the dielectric. The area of the plates is $1.00\ \text{cm}^2$. What is the capacitance if the rutile thickness is $1.00\ \text{mm}$? (a) 88.5 pF (b) 177 pF (c) $8.85\ \mu\text{F}$ (d) $100\ \mu\text{F}$ (e) $35.4\ \mu\text{F}$

4. A parallel-plate capacitor is connected to a battery. What happens to the stored energy if the plate separation is doubled while the capacitor remains connected to the battery? (a) It remains the same. (b) It is doubled. (c) It decreases by a factor of 2. (d) It decreases by a factor of 4. (e) It increases by a factor of 4.

5. If three unequal capacitors, initially uncharged, are connected in series across a battery, which of the following statements is true? (a) The equivalent capacitance is greater than any of the individual capacitances. (b) The largest voltage appears across the smallest capacitance. (c) The largest voltage appears across the largest capacitance. (d) The capacitor with the largest capacitance has the greatest charge. (e) The capacitor with the smallest capacitance has the smallest charge.

6. Assume a device is designed to obtain a large potential difference by first charging a bank of capacitors connected in parallel and then activating a switch arrangement that in effect disconnects the capacitors from the charging source and from each other and reconnects them all in a series arrangement. The group of charged capacitors is then discharged in series. What is the maximum potential difference that can be obtained in this manner by using ten $500\text{-}\mu\text{F}$ capacitors and an 800-V charging source? (a) 500 V (b) 8.00 kV (c) 400 kV (d) 800 V (e) 0

7. **(i)** What happens to the magnitude of the charge on each plate of a capacitor if the potential difference between the conductors is doubled? (a) It becomes four times larger. (b) It becomes two times larger. (c) It is unchanged. (d) It becomes one-half as large. (e) It becomes one-fourth as large. **(ii)** If the potential difference across a capacitor is doubled, what happens to the energy stored? Choose from the same possibilities as in part (i).

8. A capacitor with very large capacitance is in series with another capacitor with very small capacitance. What is the equivalent capacitance of the combination? (a) slightly greater than the capacitance of the large capacitor (b) slightly less than the capacitance of the large capacitor (c) slightly greater than the capacitance of the small capacitor (d) slightly less than the capacitance of the small capacitor

9. A parallel-plate capacitor filled with air carries a charge Q. The battery is disconnected, and a slab of material with dielectric constant $\kappa = 2$ is inserted between the plates. Which of the following statements is true? (a) The voltage across the capacitor decreases by a factor of 2. (b) The voltage across the capacitor is doubled. (c) The charge on the plates is doubled. (d) The charge on the plates decreases by a factor of 2. (e) The electric field is doubled.

10. (i) A battery is attached to several different capacitors connected in parallel. Which of the following statements is true? (a) All capacitors have the same charge, and the equivalent capacitance is greater than the capacitance of any of the capacitors in the group. (b) The capacitor with the largest capacitance carries the smallest charge. (c) The potential difference across each capacitor is the same, and the equivalent capacitance is greater than any of the capacitors in the group. (d) The capacitor with the smallest capacitance carries the largest charge. (e) The potential differences across the capacitors are the same only if the capacitances are the same. (ii) The capacitors are reconnected in series, and the combination is again connected to the battery. From the same choices, choose the one that is true.

11. A parallel-plate capacitor is charged and then is disconnected from the battery. By what factor does the stored energy change when the plate separation is then doubled? (a) It becomes four times larger. (b) It becomes two times larger. (c) It stays the same. (d) It becomes one-half as large. (e) It becomes one-fourth as large.

12. (i) Rank the following five capacitors from greatest to smallest capacitance, noting any cases of equality. (a) a 20-μF capacitor with a 4-V potential difference between its plates (b) a 30-μF capacitor with charges of magnitude 90 μC on each plate (c) a capacitor with charges of magnitude 80 μC on its plates, differing by 2 V in potential, (d) a 10-μF capacitor storing energy 125 μJ (e) a capacitor storing energy 250 μJ with a 10-V potential difference (ii) Rank the same capacitors in part (i) from largest to smallest according to the potential difference between the plates. (iii) Rank the capacitors in part (i) in the order of the magnitudes of the charges on their plates. (iv) Rank the capacitors in part (i) in the order of the energy they store.

13. True or False? (a) From the definition of capacitance $C = Q/\Delta V$, it follows that an uncharged capacitor has a capacitance of zero. (b) As described by the definition of capacitance, the potential difference across an uncharged capacitor is zero.

14. You charge a parallel-plate capacitor, remove it from the battery, and prevent the wires connected to the plates from touching each other. When you increase the plate separation, do the following quantities (a) increase, (b) decrease, or (c) stay the same? (i) C (ii) Q (iii) E between the plates (iv) ΔV

Conceptual Questions

1. denotes answer available in *Student Solutions Manual/Study Guide*

1. (a) Why is it dangerous to touch the terminals of a high-voltage capacitor even after the voltage source that charged the capacitor is disconnected from the capacitor? (b) What can be done to make the capacitor safe to handle after the voltage source has been removed?

2. Assume you want to increase the maximum operating voltage of a parallel-plate capacitor. Describe how you can do that with a fixed plate separation.

3. If you were asked to design a capacitor in which small size and large capacitance were required, what would be the two most important factors in your design?

4. Explain why a dielectric increases the maximum operating voltage of a capacitor even though the physical size of the capacitor doesn't change.

5. Explain why the work needed to move a particle with charge Q through a potential difference ΔV is $W = Q \Delta V$, whereas the energy stored in a charged capacitor is $U_E = \frac{1}{2} Q \Delta V$. Where does the factor $\frac{1}{2}$ come from?

6. An air-filled capacitor is charged, then disconnected from the power supply, and finally connected to a voltmeter. Explain how and why the potential difference changes when a dielectric is inserted between the plates of the capacitor.

7. The sum of the charges on both plates of a capacitor is zero. What does a capacitor store?

8. Because the charges on the plates of a parallel-plate capacitor are opposite in sign, they attract each other. Hence, it would take positive work to increase the plate separation. What type of energy in the system changes due to the external work done in this process?

Problems

 WebAssign The problems found in this chapter may be assigned online in Enhanced WebAssign

1. straightforward; 2. intermediate;
3. challenging

1. full solution available in the *Student Solutions Manual/Study Guide*

AMT Analysis Model tutorial available in Enhanced WebAssign

GP Guided Problem

M Master It tutorial available in Enhanced WebAssign

W Watch It video solution available in Enhanced WebAssign

Section 26.1 Definition of Capacitance

1. (a) When a battery is connected to the plates of a 3.00-μF capacitor, it stores a charge of 27.0 μC. What is the voltage of the battery? (b) If the same capacitor is connected to another battery and 36.0 μC of charge is stored on the capacitor, what is the voltage of the battery?

2. Two conductors having net charges of $+10.0$ μC and -10.0 μC have a potential difference of 10.0 V between them. (a) Determine the capacitance of the system. (b) What is the potential difference between the two conductors if the charges on each are increased to $+100$ μC and -100 μC?

3. (a) How much charge is on each plate of a 4.00-μF capacitor when it is connected to a 12.0-V battery? (b) If this same capacitor is connected to a 1.50-V battery, what charge is stored?

Section 26.2 Calculating Capacitance

4. An air-filled spherical capacitor is constructed with inner- and outer-shell radii of 7.00 cm and 14.0 cm, respectively. (a) Calculate the capacitance of the device. (b) What potential difference between the spheres results in a 4.00-μC charge on the capacitor?

5. A 50.0-m length of coaxial cable has an inner conductor that has a diameter of 2.58 mm and carries a charge of 8.10 μC. The surrounding conductor has an inner diameter of 7.27 mm and a charge of -8.10 μC. Assume the region between the conductors is air. (a) What is the capacitance of this cable? (b) What is the potential difference between the two conductors?

6. (a) Regarding the Earth and a cloud layer 800 m above the Earth as the "plates" of a capacitor, calculate the capacitance of the Earth–cloud layer system. Assume the cloud layer has an area of 1.00 km^2 and the air between the cloud and the ground is pure and dry. Assume charge builds up on the cloud and on the ground until a uniform electric field of 3.00×10^6 N/C throughout the space between them makes the air break down and conduct electricity as a lightning bolt. (b) What is the maximum charge the cloud can hold?

7. When a potential difference of 150 V is applied to the plates of a parallel-plate capacitor, the plates carry a surface charge density of 30.0 nC/cm^2. What is the spacing between the plates?

8. An air-filled parallel-plate capacitor has plates of area 2.30 cm^2 separated by 1.50 mm. (a) Find the value of its capacitance. The capacitor is connected to a 12.0-V battery. (b) What is the charge on the capacitor? (c) What is the magnitude of the uniform electric field between the plates?

9. An air-filled capacitor consists of two parallel plates, each with an area of 7.60 cm^2, separated by a distance of 1.80 mm. A 20.0-V potential difference is applied to these plates. Calculate (a) the electric field between the plates, (b) the surface charge density, (c) the capacitance, and (d) the charge on each plate.

10. A variable air capacitor used in a radio tuning circuit is made of N semicircular plates, each of radius R and positioned a distance d from its neighbors, to which it is electrically connected. As shown in Figure P26.10, a second identical set of plates is enmeshed with the first set. Each plate in the second set is halfway between two plates of the first set. The second set can rotate as a unit. Determine the capacitance as a function of the angle of rotation θ, where $\theta = 0$ corresponds to the maximum capacitance.

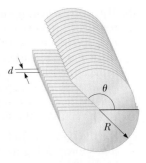

Figure P26.10

11. An isolated, charged conducting sphere of radius 12.0 cm creates an electric field of 4.90×10^4 N/C at a distance 21.0 cm from its center. (a) What is its surface charge density? (b) What is its capacitance?

12. **Review.** A small object of mass m carries a charge q and is suspended by a thread between the vertical plates of a parallel-plate capacitor. The plate separation is d. If the thread makes an angle θ with the vertical, what is the potential difference between the plates?

Section 26.3 Combinations of Capacitors

13. Two capacitors, $C_1 = 5.00$ μF and $C_2 = 12.0$ μF, are connected in parallel, and the resulting combination is connected to a 9.00-V battery. Find (a) the equivalent capacitance of the combination, (b) the potential difference across each capacitor, and (c) the charge stored on each capacitor.

14. **What If?** The two capacitors of Problem 13 ($C_1 = 5.00$ μF and $C_2 = 12.0$ μF) are now connected in series and to a 9.00-V battery. Find (a) the equivalent capacitance of the combination, (b) the potential difference across each capacitor, and (c) the charge on each capacitor.

15. Find the equivalent capacitance of a 4.20-μF capacitor and an 8.50-μF capacitor when they are connected (a) in series and (b) in parallel.

16. Given a 2.50-μF capacitor, a 6.25-μF capacitor, and a 6.00-V battery, find the charge on each capacitor if you connect them (a) in series across the battery and (b) in parallel across the battery.

17. According to its design specification, the timer circuit delaying the closing of an elevator door is to have a capacitance of 32.0 μF between two points A and B. When one circuit is being constructed, the inexpensive but durable capacitor installed between these two points is found to have capacitance 34.8 μF. To meet the specification, one additional capacitor can be placed between the two points. (a) Should it be in series or in parallel with the 34.8-μF capacitor? (b) What should be its capacitance? (c) **What If?** The next circuit comes down the assembly line with capacitance 29.8 μF between A and B. To meet the specification, what additional capacitor should be installed in series or in parallel in that circuit?

18. *Why is the following situation impossible?* A technician is testing a circuit that contains a capacitance C. He realizes that a better design for the circuit would include a capacitance $\frac{7}{3}C$ rather than C. He has three additional capacitors, each with capacitance C. By combining these additional capacitors in a certain combination that is then placed in parallel with the original capacitor, he achieves the desired capacitance.

19. For the system of four capacitors shown in Figure P26.19, find (a) the equivalent capacitance of the system, (b) the charge on each capacitor, and (c) the potential difference across each capacitor.

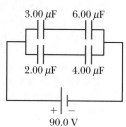

Figure P26.19
Problems 19 and 56.

20. Three capacitors are connected to a battery as shown in Figure P26.20. Their capacitances are $C_1 = 3C$, $C_2 = C$, and $C_3 = 5C$. (a) What is the equivalent capacitance of this set of capacitors? (b) State the ranking of the capacitors according to the charge they store from largest to smallest. (c) Rank the capacitors according to the potential differences across them from largest to smallest. (d) **What If?** Assume C_3 is increased. Explain what happens to the charge stored by each capacitor.

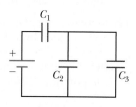

Figure P26.20

21. A group of identical capacitors is connected first in series and then in parallel. The combined capacitance in parallel is 100 times larger than for the series connection. How many capacitors are in the group?

22. (a) Find the equivalent capacitance between points a and b for the group of capacitors connected as shown in Figure P26.22. Take $C_1 = 5.00~\mu F$, $C_2 = 10.0~\mu F$, and $C_3 = 2.00~\mu F$. (b) What charge is stored on C_3 if the potential difference between points a and b is 60.0 V?

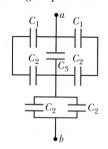

Figure P26.22

23. Four capacitors are connected as shown in Figure P26.23. (a) Find the equivalent capacitance between points a and b. (b) Calculate the charge on each capacitor, taking $\Delta V_{ab} = 15.0$ V.

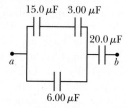

Figure P26.23

24. Consider the circuit shown in Figure P26.24, where $C_1 = 6.00~\mu F$, $C_2 = 3.00~\mu F$, and $\Delta V = 20.0$ V. Capacitor C_1 is first charged by closing switch S_1. Switch S_1 is then opened, and the charged capacitor is connected to the uncharged capacitor by closing S_2. Calculate (a) the initial charge acquired by C_1 and (b) the final charge on each capacitor.

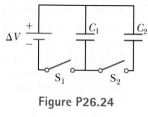

Figure P26.24

25. Find the equivalent capacitance between points a and b in the combination of capacitors shown in Figure P26.25.

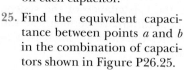

Figure P26.25

26. Find (a) the equivalent capacitance of the capacitors in Figure P26.26, (b) the charge on each capacitor, and (c) the potential difference across each capacitor.

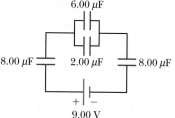

Figure P26.26

27. Two capacitors give an equivalent capacitance of 9.00 pF when connected in parallel and an equivalent capacitance of 2.00 pF when connected in series. What is the capacitance of each capacitor?

28. Two capacitors give an equivalent capacitance of C_p when connected in parallel and an equivalent capacitance of C_s when connected in series. What is the capacitance of each capacitor?

29. Consider three capacitors C_1, C_2, and C_3 and a battery. If only C_1 is connected to the battery, the charge on C_1 is 30.8 μC. Now C_1 is disconnected, discharged, and connected in series with C_2. When the series combination of C_2 and C_1 is connected across the battery, the charge on C_1 is 23.1 μC. The circuit is disconnected, and both capacitors are discharged. Next, C_3, C_1, and the battery are connected in series, resulting in a charge on C_1 of 25.2 μC. If, after being disconnected and discharged, C_1, C_2, and C_3 are connected in series with one another and with the battery, what is the charge on C_1?

Section 26.4 Energy Stored in a Charged Capacitor

30. The immediate cause of many deaths is ventricular fibrillation, which is an uncoordinated quivering of the heart. An electric shock to the chest can cause momentary paralysis of the heart muscle, after which the heart sometimes resumes its proper beating. One type of *defibrillator* (chapter-opening photo, page 777) applies a strong electric shock to the chest over a time interval of a few milliseconds. This device contains a

capacitor of several microfarads, charged to several thousand volts. Electrodes called paddles are held against the chest on both sides of the heart, and the capacitor is discharged through the patient's chest. Assume an energy of 300 J is to be delivered from a 30.0-μF capacitor. To what potential difference must it be charged?

31. A 12.0-V battery is connected to a capacitor, resulting in 54.0 μC of charge stored on the capacitor. How much energy is stored in the capacitor?

32. (a) A 3.00-μF capacitor is connected to a 12.0-V battery.
[W] How much energy is stored in the capacitor? (b) Had the capacitor been connected to a 6.00-V battery, how much energy would have been stored?

33. As a person moves about in a dry environment, electric charge accumulates on the person's body. Once it is at high voltage, either positive or negative, the body can discharge via sparks and shocks. Consider a human body isolated from ground, with the typical capacitance 150 pF. (a) What charge on the body will produce a potential of 10.0 kV? (b) Sensitive electronic devices can be destroyed by electrostatic discharge from a person. A particular device can be destroyed by a discharge releasing an energy of 250 μJ. To what voltage on the body does this situation correspond?

34. Two capacitors, $C_1 = 18.0$ μF and $C_2 = 36.0$ μF, are connected in series, and a 12.0-V battery is connected across the two capacitors. Find (a) the equivalent capacitance and (b) the energy stored in this equivalent capacitance. (c) Find the energy stored in each individual capacitor. (d) Show that the sum of these two energies is the same as the energy found in part (b). (e) Will this equality always be true, or does it depend on the number of capacitors and their capacitances? (f) If the same capacitors were connected in parallel, what potential difference would be required across them so that the combination stores the same energy as in part (a)? (g) Which capacitor stores more energy in this situation, C_1 or C_2?

35. Two identical parallel-plate capacitors, each with capacitance 10.0 μF, are charged to potential difference 50.0 V and then disconnected from the battery. They are then connected to each other in parallel with plates of like sign connected. Finally, the plate separation in one of the capacitors is doubled. (a) Find the total energy of the system of two capacitors *before* the plate separation is doubled. (b) Find the potential difference across each capacitor *after* the plate separation is doubled. (c) Find the total energy of the system *after* the plate separation is doubled. (d) Reconcile the difference in the answers to parts (a) and (c) with the law of conservation of energy.

36. Two identical parallel-plate capacitors, each with capacitance C, are charged to potential difference ΔV and then disconnected from the battery. They are then connected to each other in parallel with plates of like sign connected. Finally, the plate separation in one of the capacitors is doubled. (a) Find the total energy of the system of two capacitors *before* the plate separation is

doubled. (b) Find the potential difference across each capacitor *after* the plate separation is doubled. (c) Find the total energy of the system *after* the plate separation is doubled. (d) Reconcile the difference in the answers to parts (a) and (c) with the law of conservation of energy.

37. Two capacitors, $C_1 = 25.0$ μF and $C_2 = 5.00$ μF, are connected in parallel and charged with a 100-V power supply. (a) Draw a circuit diagram and (b) calculate the total energy stored in the two capacitors. (c) **What If?** What potential difference would be required across the same two capacitors connected in series for the combination to store the same amount of energy as in part (b)? (d) Draw a circuit diagram of the circuit described in part (c).

38. A parallel-plate capacitor has a charge Q and plates of area A. What force acts on one plate to attract it toward the other plate? Because the electric field between the plates is $E = Q/A\epsilon_0$, you might think the force is $F = QE = Q^2/A\epsilon_0$. This conclusion is wrong because the field E includes contributions from both plates, and the field created by the positive plate cannot exert any force on the positive plate. Show that the force exerted on each plate is actually $F = Q^2/2A\epsilon_0$. *Suggestion:* Let $C = \epsilon_0 A/x$ for an arbitrary plate separation x and note that the work done in separating the two charged plates is $W = \int F\, dx$.

39. **Review.** A storm cloud and the ground represent the
[AMT] plates of a capacitor. During a storm, the capacitor has a potential difference of 1.00×10^8 V between its plates and a charge of 50.0 C. A lightning strike delivers 1.00% of the energy of the capacitor to a tree on the ground. How much sap in the tree can be boiled away? Model the sap as water initially at 30.0°C. Water has a specific heat of 4 186 J/kg · °C, a boiling point of 100°C, and a latent heat of vaporization of 2.26×10^6 J/kg.

40. Consider two conducting spheres with radii R_1 and
[GP] R_2 separated by a distance much greater than either radius. A total charge Q is shared between the spheres. We wish to show that when the electric potential energy of the system has a minimum value, the potential difference between the spheres is zero. The total charge Q is equal to $q_1 + q_2$, where q_1 represents the charge on the first sphere and q_2 the charge on the second. Because the spheres are very far apart, you can assume the charge of each is uniformly distributed over its surface. (a) Show that the energy associated with a single conducting sphere of radius R and charge q surrounded by a vacuum is $U = k_e q^2/2R$. (b) Find the total energy of the system of two spheres in terms of q_1, the total charge Q, and the radii R_1 and R_2. (c) To minimize the energy, differentiate the result to part (b) with respect to q_1 and set the derivative equal to zero. Solve for q_1 in terms of Q and the radii. (d) From the result to part (c), find the charge q_2. (e) Find the potential of each sphere. (f) What is the potential difference between the spheres?

41. **Review.** The circuit in Figure P26.41 (page 804) con-
[AMT] sists of two identical, parallel metal plates connected to identical metal springs, a switch, and a 100-V battery.

With the switch open, the plates are uncharged, are separated by a distance $d = 8.00$ mm, and have a capacitance $C = 2.00$ μF. When the switch is closed, the distance between the plates decreases by a factor of 0.500. (a) How much charge collects on each plate? (b) What is the spring constant for each spring?

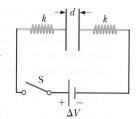

Figure P26.41

Section 26.5 Capacitors with Dielectrics

42. A supermarket sells rolls of aluminum foil, plastic wrap, and waxed paper. (a) Describe a capacitor made from such materials. Compute order-of-magnitude estimates for (b) its capacitance and (c) its breakdown voltage.

43. (a) How much charge can be placed on a capacitor with W air between the plates before it breaks down if the area of each plate is 5.00 cm^2? (b) **What If?** Find the maximum charge if polystyrene is used between the plates instead of air.

44. The voltage across an air-filled parallel-plate capacitor is measured to be 85.0 V. When a dielectric is inserted and completely fills the space between the plates as in Figure P26.44, the voltage drops to 25.0 V. (a) What is the dielectric constant of the inserted material? (b) Can you identify the dielectric? If so, what is it? (c) If the dielectric does not completely fill the space between the plates, what could you conclude about the voltage across the plates?

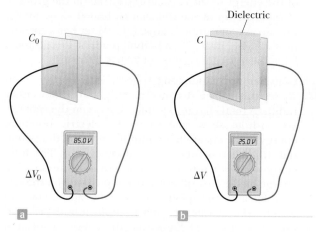

Figure P26.44

45. Determine (a) the capacitance and (b) the maximum W potential difference that can be applied to a Teflon-filled parallel-plate capacitor having a plate area of 1.75 cm^2 and a plate separation of 0.040 0 mm.

46. A commercial capacitor is to be constructed as shown in Figure P26.46. This particular capacitor is made from two strips of aluminum foil separated by a strip of paraffin-coated paper. Each strip of foil and paper is 7.00 cm wide. The foil is 0.004 00 mm thick, and the paper is 0.025 0 mm thick and has a dielectric constant of 3.70. What length should the strips have if a capaci-

tance of 9.50×10^{-8} F is desired before the capacitor is rolled up? (Adding a second strip of paper and rolling the capacitor would effectively double its capacitance by allowing charge storage on both sides of each strip of foil.)

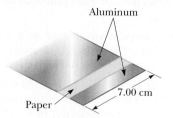

Figure P26.46

47. A parallel-plate capacitor in air has a plate separation of 1.50 cm and a plate area of 25.0 cm^2. The plates are charged to a potential difference of 250 V and disconnected from the source. The capacitor is then immersed in distilled water. Assume the liquid is an insulator. Determine (a) the charge on the plates before and after immersion, (b) the capacitance and potential difference after immersion, and (c) the change in energy of the capacitor.

48. Each capacitor in the combination shown in Figure P26.48 has a breakdown voltage of 15.0 V. What is the breakdown voltage of the combination?

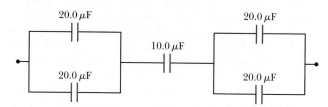

Figure P26.48

49. A 2.00-nF parallel-plate capacitor is charged to an ini-AMT tial potential difference $\Delta V_i = 100$ V and is then iso-lated. The dielectric material between the plates is mica, with a dielectric constant of 5.00. (a) How much work is required to withdraw the mica sheet? (b) What is the potential difference across the capacitor after the mica is withdrawn?

Section 26.6 Electric Dipole in an Electric Field

50. A small, rigid object carries positive and negative M 3.50-nC charges. It is oriented so that the positive charge has coordinates $(-1.20$ mm, 1.10 mm$)$ and the negative charge is at the point $(1.40$ mm, -1.30 mm$)$. (a) Find the electric dipole moment of the object. The object is placed in an electric field $\vec{\mathbf{E}} = (7.80 \times 10^3\,\hat{\mathbf{i}} - 4.90 \times 10^3\,\hat{\mathbf{j}})$ N/C. (b) Find the torque acting on the object. (c) Find the potential energy of the object–field system when the object is in this orientation. (d) Assuming the orientation of the object can change, find the difference between the maximum and minimum potential energies of the system.

51. An infinite line of positive charge lies along the y axis, AMT with charge density $\lambda = 2.00$ μC/m. A dipole is placed with its center along the x axis at x = 25.0 cm. The dipole consists of two charges ±10.0 μC separated by 2.00 cm. The axis of the dipole makes an angle of 35.0° with the x axis, and the positive charge is farther from the line of charge than the negative charge. Find the net force exerted on the dipole.

52. A small object with electric dipole moment $\vec{p}$ is placed in a nonuniform electric field $\vec{E} = E(x)\hat{i}$. That is, the field is in the x direction, and its magnitude depends only on the coordinate x. Let θ represent the angle between the dipole moment and the x direction. Prove that the net force on the dipole is

$$F = p\left(\frac{dE}{dx}\right)\cos\theta$$

acting in the direction of increasing field.

Section 26.7 An Atomic Description of Dielectrics

53. The general form of Gauss's law describes how a charge creates an electric field in a material, as well as in vacuum:

$$\int \vec{E} \cdot d\vec{A} = \frac{q_{in}}{\epsilon}$$

where $\epsilon = \kappa\epsilon_0$ is the permittivity of the material. (a) A sheet with charge Q uniformly distributed over its area A is surrounded by a dielectric. Show that the sheet creates a uniform electric field at nearby points with magnitude $E = Q/2A\epsilon$. (b) Two large sheets of area A, carrying opposite charges of equal magnitude Q, are a small distance d apart. Show that they create uniform electric field in the space between them with magnitude $E = Q/A\epsilon$. (c) Assume the negative plate is at zero potential. Show that the positive plate is at potential $Qd/A\epsilon$. (d) Show that the capacitance of the pair of plates is given by $C = A\epsilon/d = \kappa A\epsilon_0/d$.

Additional Problems

54. Find the equivalent capacitance of the group of capacitors shown in Figure P26.54.

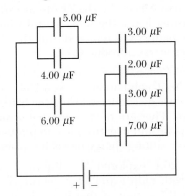

5.00 μF

3.00 μF

4.00 μF

2.00 μF

3.00 μF

6.00 μF

7.00 μF

+ −

Figure P26.54

55. Four parallel metal plates P_1, P_2, P_3, and P_4, each of area 7.50 cm², are separated successively by a distance $d = 1.19$ mm as shown in Figure P26.55. Plate P_1 is connected to the negative terminal of a battery, and P_2 is connected to the positive terminal. The

battery maintains a potential difference of 12.0 V. (a) If P_3 is connected to the negative terminal, what is the capacitance of the three-plate system $P_1P_2P_3$? (b) What is the charge on P_2? (c) If P_4 is now connected to the positive terminal, what is the capacitance of the four-plate system $P_1P_2P_3P_4$? (d) What is the charge on P_4?

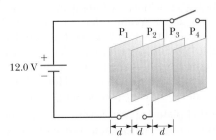

P_1 P_2 P_3 P_4

12.0 V

d d d

Figure P26.55

56. For the system of four capacitors shown in Figure P26.19, find (a) the total energy stored in the system and (b) the energy stored by each capacitor. (c) Compare the sum of the answers in part (b) with your result to part (a) and explain your observation.

57. A uniform electric field $E = 3\,000$ V/m exists within a certain region. What volume of space contains an energy equal to 1.00×10^{-7} J? Express your answer in cubic meters and in liters.

58. Two large, parallel metal plates, each of area A, are oriented horizontally and separated by a distance $3d$. A grounded conducting wire joins them, and initially each plate carries no charge. Now a third identical plate carrying charge Q is inserted between the two plates, parallel to them and located a distance d from the upper plate as shown in Figure P26.58. (a) What induced charge appears on each of the two original plates? (b) What potential difference appears between the middle plate and each of the other plates?

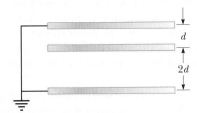

d

2d

Figure P26.58

59. A parallel-plate capacitor is constructed using a dielectric material whose dielectric constant is 3.00 and whose dielectric strength is 2.00×10^8 V/m. The desired capacitance is 0.250 μF, and the capacitor must withstand a maximum potential difference of 4.00 kV. Find the minimum area of the capacitor plates.

60. *Why is the following situation impossible?* A 10.0-μF capacitor has plates with vacuum between them. The capacitor is charged so that it stores 0.050 0 J of energy. A particle with charge −3.00 μC is fired from the positive plate toward the negative plate with an initial kinetic energy equal to 1.00×10^{-4} J. The particle arrives at the negative plate with a reduced kinetic energy.

61. A model of a red blood cell portrays the cell as a capacitor with two spherical plates. It is a positively charged conducting liquid sphere of area A, separated by an insulating membrane of thickness t from the surrounding negatively charged conducting fluid. Tiny electrodes introduced into the cell show a potential difference of 100 mV across the membrane. Take the membrane's thickness as 100 nm and its dielectric constant as 5.00. (a) Assume that a typical red blood cell has a mass of 1.00×10^{-12} kg and density 1 100 kg/m³. Calculate its volume and its surface area. (b) Find the capacitance of the cell. (c) Calculate the charge on the surfaces of the membrane. How many electronic charges does this charge represent?

62. A parallel-plate capacitor with vacuum between its horizontal plates has a capacitance of 25.0 μF. A nonconducting liquid with dielectric constant 6.50 is poured into the space between the plates, filling up a fraction f of its volume. (a) Find the new capacitance as a function of f. (b) What should you expect the capacitance to be when $f = 0$? Does your expression from part (a) agree with your answer? (c) What capacitance should you expect when $f = 1$? Does the expression from part (a) agree with your answer?

63. A 10.0-μF capacitor is charged to 15.0 V. It is next connected in series with an uncharged 5.00-μF capacitor. The series combination is finally connected across a 50.0-V battery as diagrammed in Figure P26.63. Find the new potential differences across the 5.00-μF and 10.0-μF capacitors after the switch is thrown closed.

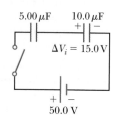

Figure P26.63

64. Assume that the internal diameter of the Geiger–Mueller tube described in Problem 68 in Chapter 25 is 2.50 cm and that the wire along the axis has a diameter of 0.200 mm. The dielectric strength of the gas between the central wire and the cylinder is 1.20×10^6 V/m. Use the result of that problem to calculate the maximum potential difference that can be applied between the wire and the cylinder before breakdown occurs in the gas.

65. Two square plates of sides ℓ are placed parallel to each other with separation d as suggested in Figure P26.65. You may assume d is much less than ℓ. The plates carry uniformly distributed static charges $+Q_0$ and $-Q_0$. A block of metal has width ℓ, length ℓ, and thickness slightly less than d. It is inserted a distance x into the space between the plates. The charges on the plates remain uniformly distributed as the block slides in. In a static situation, a metal prevents an electric field from penetrating inside it. The metal can be thought of as a perfect dielectric, with $\kappa \to \infty$. (a) Calculate the stored energy in the system as a function of x. (b) Find the direction and magnitude of the force that acts on the metallic block. (c) The area of the advancing front face of the block is essentially equal to ℓd. Considering the force on the block as acting on this face, find the stress (force per area)

on it. (d) Express the energy density in the electric field between the charged plates in terms of Q_0, ℓ, d, and ϵ_0. (e) Explain how the answers to parts (c) and (d) compare with each other.

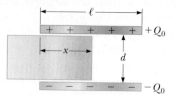

Figure P26.65

66. (a) Two spheres have radii a and b, and their centers are a distance d apart. Show that the capacitance of this system is

$$C = \frac{4\pi\epsilon_0}{\dfrac{1}{a} + \dfrac{1}{b} - \dfrac{2}{d}}$$

provided d is large compared with a and b. *Suggestion:* Because the spheres are far apart, assume the potential of each equals the sum of the potentials due to each sphere. (b) Show that as d approaches infinity, the above result reduces to that of two spherical capacitors in series.

67. A capacitor of unknown capacitance has been charged to a potential difference of 100 V and then disconnected from the battery. When the charged capacitor is then connected in parallel to an uncharged 10.0-μF capacitor, the potential difference across the combination is 30.0 V. Calculate the unknown capacitance.

68. A parallel-plate capacitor of plate separation d is charged to a potential difference ΔV_0. A dielectric slab of thickness d and dielectric constant κ is introduced between the plates while the battery remains connected to the plates. (a) Show that the ratio of energy stored after the dielectric is introduced to the energy stored in the empty capacitor is $U/U_0 = \kappa$. (b) Give a physical explanation for this increase in stored energy. (c) What happens to the charge on the capacitor? *Note:* This situation is not the same as in Example 26.5, in which the battery was removed from the circuit before the dielectric was introduced.

69. Capacitors $C_1 = 6.00$ μF and $C_2 = 2.00$ μF are charged as a parallel combination across a 250-V battery. The capacitors are disconnected from the battery and from each other. They are then connected positive plate to negative plate and negative plate to positive plate. Calculate the resulting charge on each capacitor.

70. Example 26.1 explored a cylindrical capacitor of length ℓ with radii a and b for the two conductors. In the What If? section of that example, it was claimed that increasing ℓ by 10% is more effective in terms of increasing the capacitance than increasing a by 10% if $b > 2.85a$. Verify this claim mathematically.

71. To repair a power supply for a stereo amplifier, an electronics technician needs a 100-μF capacitor capable of withstanding a potential difference of 90 V between the

plates. The immediately available supply is a box of five 100-μF capacitors, each having a maximum voltage capability of 50 V. (a) What combination of these capacitors has the proper electrical characteristics? Will the technician use all the capacitors in the box? Explain your answers. (b) In the combination of capacitors obtained in part (a), what will be the maximum voltage across each of the capacitors used?

Challenge Problems

72. The inner conductor of a coaxial cable has a radius of 0.800 mm, and the outer conductor's inside radius is 3.00 mm. The space between the conductors is filled with polyethylene, which has a dielectric constant of 2.30 and a dielectric strength of 18.0×10^6 V/m. What is the maximum potential difference this cable can withstand?

73. Some physical systems possessing capacitance continuously distributed over space can be modeled as an infinite array of discrete circuit elements. Examples are a microwave waveguide and the axon of a nerve cell. To practice analysis of an infinite array, determine the equivalent capacitance C between terminals X and Y of the infinite set of capacitors represented in Figure P26.73. Each capacitor has capacitance C_0. *Suggestions:* Imagine that the ladder is cut at the line AB and note that the equivalent capacitance of the infinite section to the right of AB is also C.

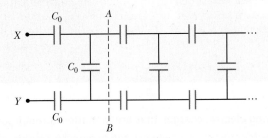

Figure P26.73

74. Consider two long, parallel, and oppositely charged wires of radius r with their centers separated by a distance D that is much larger than r. Assuming the charge is distributed uniformly on the surface of each wire, show that the capacitance per unit length of this pair of wires is

$$\frac{C}{\ell} = \frac{\pi \epsilon_0}{\ln (D/r)}$$

75. Determine the equivalent capacitance of the combination shown in Figure P26.75. *Suggestion:* Consider the symmetry involved.

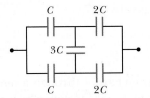

Figure P26.75

76. A parallel-plate capacitor with plates of area LW and plate separation t has the region between its plates filled with wedges of two dielectric materials as shown in Figure P26.76. Assume t is much less than both L and W. (a) Determine its capacitance. (b) Should the capacitance be the same if the labels κ_1 and κ_2 are interchanged? Demonstrate that your expression does or does not have this property. (c) Show that if κ_1 and κ_2 approach equality to a common value κ, your result becomes the same as the capacitance of a capacitor containing a single dielectric: $C = \kappa \epsilon_0 LW/t$.

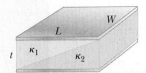

Figure P26.76

77. Calculate the equivalent capacitance between points a and b in Figure P26.77. Notice that this system is not a simple series or parallel combination. *Suggestion:* Assume a potential difference ΔV between points a and b. Write expressions for ΔV_{ab} in terms of the charges and capacitances for the various possible pathways from a to b and require conservation of charge for those capacitor plates that are connected to each other.

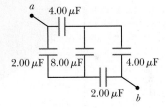

Figure P26.77

78. A capacitor is constructed from two square, metallic plates of sides ℓ and separation d. Charges $+Q$ and $-Q$ are placed on the plates, and the power supply is then removed. A material of dielectric constant κ is inserted a distance x into the capacitor as shown in Figure P26.78. Assume d is much smaller than x. (a) Find the equivalent capacitance of the device. (b) Calculate the energy stored in the capacitor. (c) Find the direction and magnitude of the force exerted by the plates on the dielectric. (d) Obtain a numerical value for the force when $x = \ell/2$, assuming $\ell = 5.00$ cm, $d = 2.00$ mm, the dielectric is glass ($\kappa = 4.50$), and the capacitor was charged to 2.00×10^3 V before the dielectric was inserted. *Suggestion:* The system can be considered as two capacitors connected in parallel.

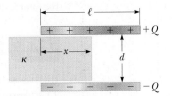

Figure P26.78

Current and Resistance

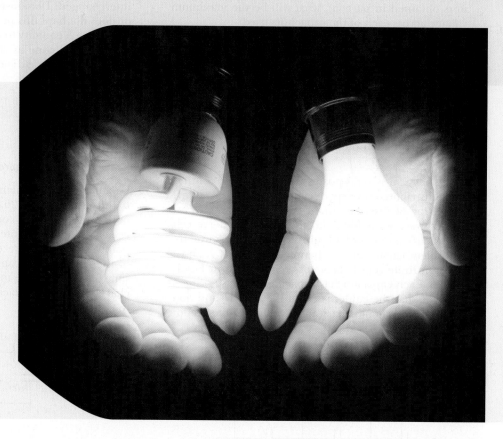

These two lightbulbs provide similar power output by visible light (electromagnetic radiation). The compact fluorescent bulb on the left, however, produces this light output with far less input by electrical transmission than the incandescent bulb on the right. The fluorescent bulb, therefore, is less costly to operate and saves valuable resources needed to generate electricity. *(Christina Richards/ Shutterstock.com)*

We now consider situations involving electric charges that are in motion through some region of space. We use the term *electric current*, or simply *current*, to describe the rate of flow of charge. Most practical applications of electricity deal with electric currents, including a variety of home appliances. For example, the voltage from a wall plug produces a current in the coils of a toaster when it is turned on. In these common situations, current exists in a conductor such as a copper wire. Currents can also exist outside a conductor. For instance, a beam of electrons in a particle accelerator constitutes a current.

This chapter begins with the definition of current. A microscopic description of current is given, and some factors that contribute to the opposition to the flow of charge in conductors are discussed. A classical model is used to describe electrical conduction in metals, and some limitations of this model are cited. We also define electrical resistance and introduce a new circuit element, the resistor. We conclude by discussing the rate at which energy is transferred to a device in an electric circuit. The energy transfer mechanism in Equation 8.2 that corresponds to this process is electrical transmission T_{ET}.

27.1 Electric Current

In this section, we study the flow of electric charges through a piece of material. The amount of flow depends on both the material through which the charges are passing and the potential difference across the material. Whenever there is a net flow of charge through some region, an electric *current* is said to exist.

It is instructive to draw an analogy between water flow and current. The flow of water in a plumbing pipe can be quantified by specifying the amount of water that emerges from a faucet during a given time interval, often measured in liters per minute. A river current can be characterized by describing the rate at which the water flows past a particular location. For example, the flow over the brink at Niagara Falls is maintained at rates between 1 400 m³/s and 2 800 m³/s.

There is also an analogy between thermal conduction and current. In Section 20.7, we discussed the flow of energy by heat through a sample of material. The rate of energy flow is determined by the material as well as the temperature difference across the material as described by Equation 20.15.

To define current quantitatively, suppose charges are moving perpendicular to a surface of area A as shown in Figure 27.1. (This area could be the cross-sectional area of a wire, for example.) The **current** is defined as the rate at which charge flows through this surface. If ΔQ is the amount of charge that passes through this surface in a time interval Δt, the **average current** I_{avg} is equal to the charge that passes through A per unit time:

$$I_{avg} = \frac{\Delta Q}{\Delta t} \qquad (27.1)$$

If the rate at which charge flows varies in time, the current varies in time; we define the **instantaneous current** I as the limit of the average current as $\Delta t \to 0$:

$$I \equiv \frac{dQ}{dt} \qquad (27.2)$$

◀ Electric current

The SI unit of current is the **ampere** (A):

$$1\ A = 1\ C/s \qquad (27.3)$$

That is, 1 A of current is equivalent to 1 C of charge passing through a surface in 1 s.

The charged particles passing through the surface in Figure 27.1 can be positive, negative, or both. It is conventional to assign to the current the same direction as the flow of positive charge. In electrical conductors such as copper or aluminum, the current results from the motion of negatively charged electrons. Therefore, in an ordinary conductor, the direction of the current is opposite the direction of flow of electrons. For a beam of positively charged protons in an accelerator, however, the current is in the direction of motion of the protons. In some cases—such as those involving gases and electrolytes, for instance—the current is the result of the flow of both positive and negative charges. It is common to refer to a moving charge (positive or negative) as a mobile **charge carrier.**

If the ends of a conducting wire are connected to form a loop, all points on the loop are at the same electric potential; hence, the electric field is zero within and at the surface of the conductor. Because the electric field is zero, there is no net transport of charge through the wire; therefore, there is no current. If the ends of the conducting wire are connected to a battery, however, all points on the loop are not at the same potential. The battery sets up a potential difference between the ends of the loop, creating an electric field within the wire. The electric field exerts forces on the electrons in the wire, causing them to move in the wire and therefore creating a current.

Microscopic Model of Current

We can relate current to the motion of the charge carriers by describing a microscopic model of conduction in a metal. Consider the current in a cylindrical

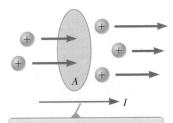

The direction of the current is the direction in which positive charges flow when free to do so.

Figure 27.1 Charges in motion through an area A. The time rate at which charge flows through the area is defined as the current I.

Pitfall Prevention 27.1

"Current Flow" Is Redundant
The phrase *current flow* is commonly used, although it is technically incorrect because current *is* a flow (of charge). This wording is similar to the phrase *heat transfer*, which is also redundant because heat *is* a transfer (of energy). We will avoid this phrase and speak of *flow of charge* or *charge flow.*

Pitfall Prevention 27.2

Batteries Do Not Supply Electrons
A battery does not supply electrons to the circuit. It establishes the electric field that exerts a force on electrons already in the wires and elements of the circuit.

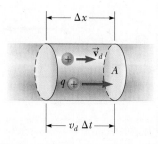

Figure 27.2 A segment of a uniform conductor of cross-sectional area A.

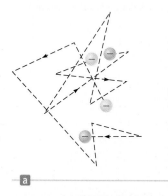

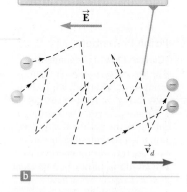

The random motion of the charge carriers is modified by the field, and they have a drift velocity opposite the direction of the electric field.

Figure 27.3 (a) A schematic diagram of the random motion of two charge carriers in a conductor in the absence of an electric field. The drift velocity is zero. (b) The motion of the charge carriers in a conductor in the presence of an electric field. Because of the acceleration of the charge carriers due to the electric force, the paths are actually parabolic. The drift speed, however, is much smaller than the average speed, so the parabolic shape is not visible on this scale.

conductor of cross-sectional area A (Fig. 27.2). The volume of a segment of the conductor of length Δx (between the two circular cross sections shown in Fig. 27.2) is $A \, \Delta x$. If n represents the number of mobile charge carriers per unit volume (in other words, the charge carrier density), the number of carriers in the segment is $nA \, \Delta x$. Therefore, the total charge ΔQ in this segment is

$$\Delta Q = (nA \, \Delta x)q$$

where q is the charge on each carrier. If the carriers move with a velocity $\vec{v}_d$ parallel to the axis of the cylinder, the magnitude of the displacement they experience in the x direction in a time interval Δt is $\Delta x = v_d \, \Delta t$. Let Δt be the time interval required for the charge carriers in the segment to move through a displacement whose magnitude is equal to the length of the segment. This time interval is also the same as that required for all the charge carriers in the segment to pass through the circular area at one end. With this choice, we can write ΔQ as

$$\Delta Q = (nAv_d \, \Delta t)q$$

Dividing both sides of this equation by Δt, we find that the average current in the conductor is

$$I_{\text{avg}} = \frac{\Delta Q}{\Delta t} = nqv_d A \tag{27.4}$$

In reality, the speed of the charge carriers v_d is an average speed called the **drift speed.** To understand the meaning of drift speed, consider a conductor in which the charge carriers are free electrons. If the conductor is isolated—that is, the potential difference across it is zero—these electrons undergo random motion that is analogous to the motion of gas molecules. The electrons collide repeatedly with the metal atoms, and their resultant motion is complicated and zigzagged as in Figure 27.3a. As discussed earlier, when a potential difference is applied across the conductor (for example, by means of a battery), an electric field is set up in the conductor; this field exerts an electric force on the electrons, producing a current. In addition to the zigzag motion due to the collisions with the metal atoms, the electrons move slowly along the conductor (in a direction opposite that of $\vec{E}$) at the **drift velocity** $\vec{v}_d$ as shown in Figure 27.3b.

You can think of the atom–electron collisions in a conductor as an effective internal friction (or drag force) similar to that experienced by a liquid's molecules flowing through a pipe stuffed with steel wool. The energy transferred from the electrons to the metal atoms during collisions causes an increase in the atom's vibrational energy and a corresponding increase in the conductor's temperature.

Quick Quiz 27.1 Consider positive and negative charges moving horizontally through the four regions shown in Figure 27.4. Rank the current in these four regions from highest to lowest.

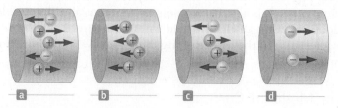

Figure 27.4 (Quick Quiz 27.1) Charges move through four regions.

Example 27.1   **Drift Speed in a Copper Wire**

The 12-gauge copper wire in a typical residential building has a cross-sectional area of 3.31×10^{-6} m^2. It carries a constant current of 10.0 A. What is the drift speed of the electrons in the wire? Assume each copper atom contributes one free electron to the current. The density of copper is 8.92 g/cm^3.

SOLUTION

Conceptualize Imagine electrons following a zigzag motion such as that in Figure 27.3a, with a drift velocity parallel to the wire superimposed on the motion as in Figure 27.3b. As mentioned earlier, the drift speed is small, and this example helps us quantify the speed.

Categorize We evaluate the drift speed using Equation 27.4. Because the current is constant, the average current during any time interval is the same as the constant current: $I_{\text{avg}} = I$.

Analyze The periodic table of the elements in Appendix C shows that the molar mass of copper is $M = 63.5$ g/mol. Recall that 1 mol of any substance contains Avogadro's number of atoms ($N_A = 6.02 \times 10^{23}$ mol^{-1}).

Use the molar mass and the density of copper to find the volume of 1 mole of copper:

$$V = \frac{M}{\rho}$$

From the assumption that each copper atom contributes one free electron to the current, find the electron density in copper:

$$n = \frac{N_A}{V} = \frac{N_A \rho}{M}$$

Solve Equation 27.4 for the drift speed and substitute for the electron density:

$$v_d = \frac{I_{\text{avg}}}{nqA} = \frac{I}{nqA} = \frac{IM}{qAN_A \rho}$$

Substitute numerical values:

$$v_d = \frac{(10.0 \text{ A})(0.063\,5 \text{ kg/mol})}{(1.60 \times 10^{-19} \text{ C})(3.31 \times 10^{-6} \text{ m}^2)(6.02 \times 10^{23} \text{ mol}^{-1})(8\,920 \text{ kg/m}^3)}$$

$$= 2.23 \times 10^{-4} \text{ m/s}$$

Finalize This result shows that typical drift speeds are very small. For instance, electrons traveling with a speed of 2.23×10^{-4} m/s would take about 75 min to travel 1 m! You might therefore wonder why a light turns on almost instantaneously when its switch is thrown. In a conductor, changes in the electric field that drives the free electrons according to the particle in a field model travel through the conductor with a speed close to that of light. So, when you flip on a light switch, electrons already in the filament of the lightbulb experience electric forces and begin moving after a time interval on the order of nanoseconds.

27.2 Resistance

In Section 24.4, we argued that the electric field inside a conductor is zero. This statement is true, however, *only* if the conductor is in static equilibrium as stated in that discussion. The purpose of this section is to describe what happens when there is a nonzero electric field in the conductor. As we saw in Section 27.1, a current exists in the wire in this case.

Consider a conductor of cross-sectional area A carrying a current I. The **current density** J in the conductor is defined as the current per unit area. Because the current $I = nqv_d A$, the current density is

$$J \equiv \frac{I}{A} = nqv_d \qquad \text{(27.5)} \qquad \blacktriangleleft \text{ Current density}$$

Georg Simon Ohm
German physicist (1789–1854)
Ohm, a high school teacher and later a professor at the University of Munich, formulated the concept of resistance and discovered the proportionalities expressed in Equations 27.6 and 27.7.

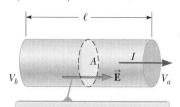

A potential difference $\Delta V = V_b - V_a$ maintained across the conductor sets up an electric field $\vec{\mathbf{E}}$, and this field produces a current I that is proportional to the potential difference.

Figure 27.5 A uniform conductor of length ℓ and cross-sectional area A.

where J has SI units of amperes per meter squared. This expression is valid only if the current density is uniform and only if the surface of cross-sectional area A is perpendicular to the direction of the current.

A current density and an electric field are established in a conductor whenever a potential difference is maintained across the conductor. In some materials, the current density is proportional to the electric field:

$$J = \sigma E \qquad (27.6)$$

where the constant of proportionality σ is called the **conductivity** of the conductor.[1] Materials that obey Equation 27.6 are said to follow **Ohm's law,** named after Georg Simon Ohm. More specifically, Ohm's law states the following:

> For many materials (including most metals), the ratio of the current density to the electric field is a constant σ that is independent of the electric field producing the current.

Materials and devices that obey Ohm's law and hence demonstrate this simple relationship between E and J are said to be *ohmic*. Experimentally, however, it is found that not all materials and devices have this property. Those that do not obey Ohm's law are said to be *nonohmic*. Ohm's law is not a fundamental law of nature; rather, it is an empirical relationship valid only for certain situations.

We can obtain an equation useful in practical applications by considering a segment of straight wire of uniform cross-sectional area A and length ℓ as shown in Figure 27.5. A potential difference $\Delta V = V_b - V_a$ is maintained across the wire, creating in the wire an electric field and a current. If the field is assumed to be uniform, the magnitude of the potential difference across the wire is related to the field within the wire through Equation 25.6,

$$\Delta V = E\ell$$

Therefore, we can express the current density (Eq. 27.6) in the wire as

$$J = \sigma \frac{\Delta V}{\ell}$$

Because $J = I/A$, the potential difference across the wire is

$$\Delta V = \frac{\ell}{\sigma} J = \left(\frac{\ell}{\sigma A} \right) I = R I$$

The quantity $R = \ell/\sigma A$ is called the **resistance** of the conductor. We define the resistance as the ratio of the potential difference across a conductor to the current in the conductor:

$$R \equiv \frac{\Delta V}{I} \qquad (27.7)$$

We will use this equation again and again when studying electric circuits. This result shows that resistance has SI units of volts per ampere. One volt per ampere is defined to be one **ohm** (Ω):

$$1\ \Omega \equiv 1\ \text{V/A} \qquad (27.8)$$

Equation 27.7 shows that if a potential difference of 1 V across a conductor causes a current of 1 A, the resistance of the conductor is 1 Ω. For example, if an electrical appliance connected to a 120-V source of potential difference carries a current of 6 A, its resistance is 20 Ω.

Most electric circuits use circuit elements called **resistors** to control the current in the various parts of the circuit. As with capacitors in Chapter 26, many resistors are built into integrated circuit chips, but stand-alone resistors are still available and

Pitfall Prevention 27.3
Equation 27.7 Is Not Ohm's Law
Many individuals call Equation 27.7 Ohm's law, but that is incorrect. This equation is simply the definition of resistance, and it provides an important relationship between voltage, current, and resistance. Ohm's law is related to a proportionality of J to E (Eq. 27.6) or, equivalently, of I to ΔV, which, from Equation 27.7, indicates that the resistance is constant, independent of the applied voltage. We will see some devices for which Equation 27.7 correctly describes their resistance, but that do *not* obey Ohm's law.

[1]Do not confuse conductivity σ with surface charge density, for which the same symbol is used.

Table 27.1	**Color Coding for Resistors**		
Color	**Number**	**Multiplier**	**Tolerance**
Black	0	1	
Brown	1	10^1	
Red	2	10^2	
Orange	3	10^3	
Yellow	4	10^4	
Green	5	10^5	
Blue	6	10^6	
Violet	7	10^7	
Gray	8	10^8	
White	9	10^9	
Gold		10^{-1}	5%
Silver		10^{-2}	10%
Colorless			20%

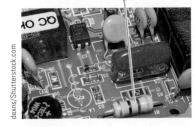

The colored bands on this resistor are yellow, violet, black, and gold.

Figure 27.6 A close-up view of a circuit board shows the color coding on a resistor. The gold band on the left tells us that the resistor is oriented "backward" in this view and we need to read the colors from right to left.

widely used. Two common types are the *composition resistor*, which contains carbon, and the *wire-wound resistor*, which consists of a coil of wire. Values of resistors in ohms are normally indicated by color coding as shown in Figure 27.6 and Table 27.1. The first two colors on a resistor give the first two digits in the resistance value, with the decimal place to the right of the second digit. The third color represents the power of 10 for the multiplier of the resistance value. The last color is the tolerance of the resistance value. As an example, the four colors on the resistor at the bottom of Figure 27.6 are yellow (= 4), violet (= 7), black (= 10^0), and gold (= 5%), and so the resistance value is $47 \times 10^0 = 47\ \Omega$ with a tolerance value of 5% = 2 Ω.

The inverse of conductivity is **resistivity**[2] ρ:

$$\rho = \frac{1}{\sigma} \qquad (27.9)$$

◀ Resistivity is the inverse of conductivity

where ρ has the units ohm $\cdot$ meters ($\Omega \cdot$ m). Because $R = \ell/\sigma A$, we can express the resistance of a uniform block of material along the length ℓ as

$$R = \rho\frac{\ell}{A} \qquad (27.10)$$

◀ Resistance of a uniform material along the length ℓ

Every ohmic material has a characteristic resistivity that depends on the properties of the material and on temperature. In addition, as you can see from Equation 27.10, the resistance of a sample of the material depends on the geometry of the sample as well as on the resistivity of the material. Table 27.2 (page 814) gives the resistivities of a variety of materials at 20°C. Notice the enormous range, from very low values for good conductors such as copper and silver to very high values for good insulators such as glass and rubber. An ideal conductor would have zero resistivity, and an ideal insulator would have infinite resistivity.

Equation 27.10 shows that the resistance of a given cylindrical conductor such as a wire is proportional to its length and inversely proportional to its cross-sectional area. If the length of a wire is doubled, its resistance doubles. If its cross-sectional area is doubled, its resistance decreases by one half. The situation is analogous to the flow of a liquid through a pipe. As the pipe's length is increased, the resistance to flow increases. As the pipe's cross-sectional area is increased, more liquid crosses a given cross section of the pipe per unit time interval. Therefore, more liquid flows for the same pressure differential applied to the pipe, and the resistance to flow decreases.

Ohmic materials and devices have a linear current–potential difference relationship over a broad range of applied potential differences (Fig. 27.7a, page 814). The slope of the I-versus-ΔV curve in the linear region yields a value for $1/R$. Nonohmic

Pitfall Prevention 27.4
Resistance and Resistivity Resistivity is a property of a *substance*, whereas resistance is a property of an *object*. We have seen similar pairs of variables before. For example, density is a property of a substance, whereas mass is a property of an object. Equation 27.10 relates resistance to resistivity, and Equation 1.1 relates mass to density.

---

[2]Do not confuse resistivity ρ with mass density or charge density, for which the same symbol is used.

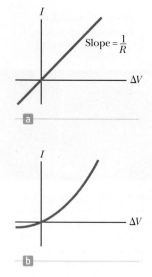

Figure 27.7 (a) The current–potential difference curve for an ohmic material. The curve is linear, and the slope is equal to the inverse of the resistance of the conductor. (b) A nonlinear current–potential difference curve for a junction diode. This device does not obey Ohm's law.

Table 27.2	Resistivities and Temperature Coefficients of Resistivity for Various Materials	
Material	Resistivity[a] $(\Omega \cdot m)$	Temperature Coefficient[b] $\alpha \, [(°C)^{-1}]$
Silver	1.59×10^{-8}	3.8×10^{-3}
Copper	1.7×10^{-8}	3.9×10^{-3}
Gold	2.44×10^{-8}	3.4×10^{-3}
Aluminum	2.82×10^{-8}	3.9×10^{-3}
Tungsten	5.6×10^{-8}	4.5×10^{-3}
Iron	10×10^{-8}	5.0×10^{-3}
Platinum	11×10^{-8}	3.92×10^{-3}
Lead	22×10^{-8}	3.9×10^{-3}
Nichrome[c]	1.00×10^{-6}	0.4×10^{-3}
Carbon	3.5×10^{-5}	-0.5×10^{-3}
Germanium	0.46	-48×10^{-3}
Silicon[d]	2.3×10^{3}	-75×10^{-3}
Glass	10^{10} to 10^{14}	
Hard rubber	$\sim 10^{13}$	
Sulfur	10^{15}	
Quartz (fused)	75×10^{16}	

[a] All values at 20°C. All elements in this table are assumed to be free of impurities.
[b] See Section 27.4.
[c] A nickel–chromium alloy commonly used in heating elements. The resistivity of Nichrome varies with composition and ranges between 1.00×10^{-6} and $1.50 \times 10^{-6} \, \Omega \cdot m$.
[d] The resistivity of silicon is very sensitive to purity. The value can be changed by several orders of magnitude when it is doped with other atoms.

materials have a nonlinear current–potential difference relationship. One common semiconducting device with nonlinear I-versus-ΔV characteristics is the *junction diode* (Fig. 27.7b). The resistance of this device is low for currents in one direction (positive ΔV) and high for currents in the reverse direction (negative ΔV). In fact, most modern electronic devices, such as transistors, have nonlinear current–potential difference relationships; their proper operation depends on the particular way they violate Ohm's law.

Quick Quiz 27.2 A cylindrical wire has a radius r and length ℓ. If both r and ℓ are doubled, does the resistance of the wire (a) increase, (b) decrease, or (c) remain the same?

Quick Quiz 27.3 In Figure 27.7b, as the applied voltage increases, does the resistance of the diode (a) increase, (b) decrease, or (c) remain the same?

Example 27.2 The Resistance of Nichrome Wire

The radius of 22-gauge Nichrome wire is 0.32 mm.

(A) Calculate the resistance per unit length of this wire.

SOLUTION

Conceptualize Table 27.2 shows that Nichrome has a resistivity two orders of magnitude larger than the best conductors in the table. Therefore, we expect it to have some special practical applications that the best conductors may not have.

Categorize We model the wire as a cylinder so that a simple geometric analysis can be applied to find the resistance.

Analyze Use Equation 27.10 and the resistivity of Nichrome from Table 27.2 to find the resistance per unit length:

$$\frac{R}{\ell} = \frac{\rho}{A} = \frac{\rho}{\pi r^2} = \frac{1.0 \times 10^{-6} \, \Omega \cdot m}{\pi (0.32 \times 10^{-3} \, m)^2} = 3.1 \, \Omega/m$$

▶ **27.2** continued

(B) If a potential difference of 10 V is maintained across a 1.0-m length of the Nichrome wire, what is the current in the wire?

SOLUTION

Analyze Use Equation 27.7 to find the current:

$$I = \frac{\Delta V}{R} = \frac{\Delta V}{(R/\ell)\ell} = \frac{10 \text{ V}}{(3.1 \text{ }\Omega/\text{m})(1.0 \text{ m})} = \boxed{3.2 \text{ A}}$$

Finalize Because of its high resistivity and resistance to oxidation, Nichrome is often used for heating elements in toasters, irons, and electric heaters.

WHAT IF? What if the wire were composed of copper instead of Nichrome? How would the values of the resistance per unit length and the current change?

Answer Table 27.2 shows us that copper has a resistivity two orders of magnitude smaller than that for Nichrome. Therefore, we expect the answer to part (A) to be smaller and the answer to part (B) to be larger. Calculations show that a copper wire of the same radius would have a resistance per unit length of only 0.053 Ω/m. A 1.0-m length of copper wire of the same radius would carry a current of 190 A with an applied potential difference of 10 V.

Example 27.3 **The Radial Resistance of a Coaxial Cable**

Coaxial cables are used extensively for cable television and other electronic applications. A coaxial cable consists of two concentric cylindrical conductors. The region between the conductors is completely filled with polyethylene plastic as shown in Figure 27.8a. Current leakage through the plastic, in the *radial* direction, is unwanted. (The cable is designed to conduct current along its length, but that is *not* the current being considered here.) The radius of the inner conductor is $a = 0.500$ cm, the radius of the outer conductor is $b = 1.75$ cm, and the length is $L = 15.0$ cm. The resistivity of the plastic is 1.0×10^{13} Ω · m. Calculate the resistance of the plastic between the two conductors.

SOLUTION

Conceptualize Imagine two currents as suggested in the text of the problem. The desired current is along the cable, carried within the conductors. The undesired current corresponds to leakage through the plastic, and its direction is radial.

Categorize Because the resistivity and the geometry of the plastic are known, we categorize this problem as one in which we find the resistance of the plastic from these parameters. Equation 27.10, however, represents the resistance of a block of material. We have a more complicated geometry in this situation. Because the area through which the charges pass depends on the radial position, we must use integral calculus to determine the answer.

Analyze We divide the plastic into concentric cylindrical shells of infinitesimal thickness dr (Fig. 27.8b). Any charge passing from the inner to the outer conductor must move radially through this shell. Use a differential form of Equation 27.10, replacing ℓ with dr for the length variable: $dR = \rho \, dr/A$, where dR is the resistance of a shell of plastic of thickness dr and surface area A.

Write an expression for the resistance of our hollow cylindrical shell of plastic representing the area as the surface area of the shell:

$$dR = \frac{\rho \, dr}{A} = \frac{\rho}{2\pi r L} \, dr$$

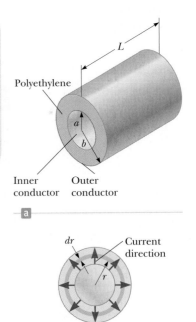

Figure 27.8 (Example 27.3) A coaxial cable. (a) Polyethylene plastic fills the gap between the two conductors. (b) End view, showing current leakage.

continued

▶ **27.3** continued

Integrate this expression from $r = a$ to $r = b$:

$$(1) \quad R = \int dR = \frac{\rho}{2\pi L} \int_a^b \frac{dr}{r} = \frac{\rho}{2\pi L} \ln\left(\frac{b}{a}\right)$$

Substitute the values given:

$$R = \frac{1.0 \times 10^{13}\ \Omega \cdot \text{m}}{2\pi(0.150\ \text{m})} \ln\left(\frac{1.75\ \text{cm}}{0.500\ \text{cm}}\right) = \boxed{1.33 \times 10^{13}\ \Omega}$$

Finalize Let's compare this resistance to that of the inner copper conductor of the cable along the 15.0-cm length.

Use Equation 27.10 to find the resistance of the copper cylinder:

$$R_{\text{Cu}} = \rho \frac{\ell}{A} = (1.7 \times 10^{-8}\ \Omega \cdot \text{m})\left[\frac{0.150\ \text{m}}{\pi(5.00 \times 10^{-3}\ \text{m})^2}\right]$$

$$= 3.2 \times 10^{-5}\ \Omega$$

This resistance is 18 orders of magnitude smaller than the radial resistance. Therefore, almost all the current corresponds to charge moving along the length of the cable, with a very small fraction leaking in the radial direction.

WHAT IF? Suppose the coaxial cable is enlarged to twice the overall diameter with two possible choices: (1) the ratio b/a is held fixed, or (2) the difference $b - a$ is held fixed. For which choice does the leakage current between the inner and outer conductors increase when the voltage is applied between them?

Answer For the current to increase, the resistance must decrease. For choice (1), in which b/a is held fixed, Equa-

tion (1) shows that the resistance is unaffected. For choice (2), we do not have an equation involving the difference $b - a$ to inspect. Looking at Figure 27.8b, however, we see that increasing b and a while holding the difference constant results in charge flowing through the same thickness of plastic but through a larger area perpendicular to the flow. This larger area results in lower resistance and a higher current.

27.3 A Model for Electrical Conduction

In this section, we describe a structural model of electrical conduction in metals that was first proposed by Paul Drude (1863–1906) in 1900. (See Section 21.1 for a review of structural models.) This model leads to Ohm's law and shows that resistivity can be related to the motion of electrons in metals. Although the Drude model described here has limitations, it introduces concepts that are applied in more elaborate treatments.

Following the outline of structural models from Section 21.1, the Drude model for electrical conduction has the following properties:

1. *Physical components:*
   Consider a conductor as a regular array of atoms plus a collection of free electrons, which are sometimes called *conduction* electrons. We identify the system as the combination of the atoms and the conduction electrons. The conduction electrons, although bound to their respective atoms when the atoms are not part of a solid, become free when the atoms condense into a solid.
2. *Behavior of the components:*
   (a) In the absence of an electric field, the conduction electrons move in random directions through the conductor (Fig. 27.3a). The situation is similar to the motion of gas molecules confined in a vessel. In fact, some scientists refer to conduction electrons in a metal as an *electron gas*.
   (b) When an electric field is applied to the system, the free electrons drift slowly in a direction opposite that of the electric field (Fig. 27.3b), with an average drift speed v_d that is much smaller (typically 10^{-4} m/s) than their average speed v_{avg} between collisions (typically 10^6 m/s).
   (c) The electron's motion after a collision is independent of its motion before the collision. The excess energy acquired by the electrons due to

the work done on them by the electric field is transferred to the atoms of the conductor when the electrons and atoms collide.

With regard to property 2(c) above, the energy transferred to the atoms causes the internal energy of the system and, therefore, the temperature of the conductor to increase.

We are now in a position to derive an expression for the drift velocity, using several of our analysis models. When a free electron of mass m_e and charge q $(= -e)$ is subjected to an electric field $\vec{\mathbf{E}}$, it is described by the particle in a field model and experiences a force $\vec{\mathbf{F}} = q\vec{\mathbf{E}}$. The electron is a particle under a net force, and its acceleration can be found from Newton's second law, $\sum \vec{\mathbf{F}} = m\vec{\mathbf{a}}$:

$$\vec{\mathbf{a}} = \frac{\sum \vec{\mathbf{F}}}{m} = \frac{q\vec{\mathbf{E}}}{m_e} \tag{27.11}$$

Because the electric field is uniform, the electron's acceleration is constant, so the electron can be modeled as a particle under constant acceleration. If $\vec{\mathbf{v}}_i$ is the electron's initial velocity the instant after a collision (which occurs at a time defined as $t = 0$), the velocity of the electron at a very short time t later (immediately before the next collision occurs) is, from Equation 4.8,

$$\vec{\mathbf{v}}_f = \vec{\mathbf{v}}_i + \vec{\mathbf{a}}t = \vec{\mathbf{v}}_i + \frac{q\vec{\mathbf{E}}}{m_e} t \tag{27.12}$$

Let's now take the average value of $\vec{\mathbf{v}}_f$ for all the electrons in the wire over all possible collision times t and all possible values of $\vec{\mathbf{v}}_i$. Assuming the initial velocities are randomly distributed over all possible directions (property 2(a) above), the average value of $\vec{\mathbf{v}}_i$ is zero. The average value of the second term of Equation 27.12 is $(q\vec{\mathbf{E}}/m_e)\tau$, where τ is the *average time interval between successive collisions*. Because the average value of $\vec{\mathbf{v}}_f$ is equal to the drift velocity,

$$\vec{\mathbf{v}}_{f,\text{avg}} = \vec{\mathbf{v}}_d = \frac{q\vec{\mathbf{E}}}{m_e} \tau \tag{27.13}$$

◀ Drift velocity in terms of microscopic quantities

The value of τ depends on the size of the metal atoms and the number of electrons per unit volume. We can relate this expression for drift velocity in Equation 27.13 to the current in the conductor. Substituting the magnitude of the velocity from Equation 27.13 into Equation 27.4, the average current in the conductor is given by

$$I_{\text{avg}} = nq\left(\frac{qE}{m_e}\tau\right)A = \frac{nq^2E}{m_e}\tau A \tag{27.14}$$

Because the current density J is the current divided by the area A,

$$J = \frac{nq^2E}{m_e}\tau \tag{27.15}$$

◀ Current density in terms of microscopic quantities

where n is the number of electrons per unit volume. Comparing this expression with Ohm's law, $J = \sigma E$, we obtain the following relationships for conductivity and resistivity of a conductor:

$$\sigma = \frac{nq^2\tau}{m_e} \tag{27.15}$$

◀ Conductivity in terms of microscopic quantities

$$\rho = \frac{1}{\sigma} = \frac{m_e}{nq^2\tau} \tag{27.16}$$

◀ Resistivity in terms of microscopic quantities

According to this classical model, conductivity and resistivity do not depend on the strength of the electric field. This feature is characteristic of a conductor obeying Ohm's law.

The model shows that the resistivity can be calculated from a knowledge of the density of the electrons, their charge and mass, and the average time interval τ between collisions. This time interval is related to the average distance between collisions ℓ_{avg} (the *mean free path*) and the average speed v_{avg} through the expression[3]

$$\tau = \frac{\ell_{avg}}{v_{avg}} \qquad \text{(27.17)}$$

Although this structural model of conduction is consistent with Ohm's law, it does not correctly predict the values of resistivity or the behavior of the resistivity with temperature. For example, the results of classical calculations for v_{avg} using the ideal gas model for the electrons are about a factor of ten smaller than the actual values, which results in incorrect predictions of values of resistivity from Equation 27.16. Furthermore, according to Equations 27.16 and 27.17, the resistivity is predicted to vary with temperature as does v_{avg}, which, according to an ideal-gas model (Chapter 21, Eq. 21.43), is proportional to $\sqrt{T}$. This behavior is in disagreement with the experimentally observed linear dependence of resistivity with temperature for pure metals. (See Section 27.4.) Because of these incorrect predictions, we must modify our structural model. We shall call the model that we have developed so far the *classical* model for electrical conduction. To account for the incorrect predictions of the classical model, we develop it further into a *quantum mechanical* model, which we shall describe briefly.

We discussed two important simplification models in earlier chapters, the particle model and the wave model. Although we discussed these two simplification models separately, quantum physics tells us that this separation is not so clear-cut. As we shall discuss in detail in Chapter 40, particles have wave-like properties. The predictions of some models can only be matched to experimental results if the model includes the wave-like behavior of particles. The structural model for electrical conduction in metals is one of these cases.

Let us imagine that the electrons moving through the metal have wave-like properties. If the array of atoms in a conductor is regularly spaced (that is, periodic), the wave-like character of the electrons makes it possible for them to move freely through the conductor and a collision with an atom is unlikely. For an idealized conductor, no collisions would occur, the mean free path would be infinite, and the resistivity would be zero. Electrons are scattered only if the atomic arrangement is irregular (not periodic), as a result of structural defects or impurities, for example. At low temperatures, the resistivity of metals is dominated by scattering caused by collisions between the electrons and impurities. At high temperatures, the resistivity is dominated by scattering caused by collisions between the electrons and the atoms of the conductor, which are continuously displaced as a result of thermal agitation, destroying the perfect periodicity. The thermal motion of the atoms makes the structure irregular (compared with an atomic array at rest), thereby reducing the electron's mean free path.

Although it is beyond the scope of this text to show this modification in detail, the classical model modified with the wave-like character of the electrons results in predictions of resistivity values that are in agreement with measured values and predicts a linear temperature dependence. Quantum notions had to be introduced in Chapter 21 to understand the temperature behavior of molar specific heats of gases. Here we have another case in which quantum physics is necessary for the model to agree with experiment. Although classical physics can explain a tremendous range of phenomena, we continue to see hints that quantum physics must be incorporated into our models. We shall study quantum physics in detail in Chapters 40 through 46.

---

[3]Recall that the average speed of a group of particles depends on the temperature of the group (Chapter 21) and is not the same as the drift speed v_d.

27.4 Resistance and Temperature

Over a limited temperature range, the resistivity of a conductor varies approximately linearly with temperature according to the expression

$$\rho = \rho_0[1 + \alpha(T - T_0)] \tag{27.18}$$

◀ **Variation of ρ with temperature**

where ρ is the resistivity at some temperature T (in degrees Celsius), ρ_0 is the resistivity at some reference temperature T_0 (usually taken to be 20°C), and α is the **temperature coefficient of resistivity.** From Equation 27.18, the temperature coefficient of resistivity can be expressed as

$$\alpha = \frac{1}{\rho_0}\frac{\Delta\rho}{\Delta T} \tag{27.19}$$

◀ **Temperature coefficient of resistivity**

where $\Delta\rho = \rho - \rho_0$ is the change in resistivity in the temperature interval $\Delta T = T - T_0$.

The temperature coefficients of resistivity for various materials are given in Table 27.2. Notice that the unit for α is degrees Celsius^{-1} $[(°C)^{-1}]$. Because resistance is proportional to resistivity (Eq. 27.10), the variation of resistance of a sample is

$$R = R_0[1 + \alpha(T - T_0)] \tag{27.20}$$

where R_0 is the resistance at temperature T_0. Use of this property enables precise temperature measurements through careful monitoring of the resistance of a probe made from a particular material.

For some metals such as copper, resistivity is nearly proportional to temperature as shown in Figure 27.9. A nonlinear region always exists at very low temperatures, however, and the resistivity usually reaches some finite value as the temperature approaches absolute zero. This residual resistivity near absolute zero is caused primarily by the collision of electrons with impurities and imperfections in the metal. In contrast, high-temperature resistivity (the linear region) is predominantly characterized by collisions between electrons and metal atoms.

Notice that three of the α values in Table 27.2 are negative, indicating that the resistivity of these materials decreases with increasing temperature. This behavior is indicative of a class of materials called *semiconductors*, first introduced in Section 23.2, and is due to an increase in the density of charge carriers at higher temperatures.

Because the charge carriers in a semiconductor are often associated with impurity atoms (as we discuss in more detail in Chapter 43), the resistivity of these materials is very sensitive to the type and concentration of such impurities.

Quick Quiz 27.4 When does an incandescent lightbulb carry more current, **(a)** immediately after it is turned on and the glow of the metal filament is increasing or **(b)** after it has been on for a few milliseconds and the glow is steady?

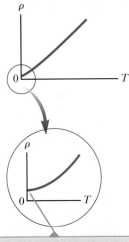

As T approaches absolute zero, the resistivity approaches a nonzero value.

Figure 27.9 Resistivity versus temperature for a metal such as copper. The curve is linear over a wide range of temperatures, and ρ increases with increasing temperature.

27.5 Superconductors

There is a class of metals and compounds whose resistance decreases to zero when they are below a certain temperature T_c, known as the **critical temperature.** These materials are known as **superconductors.** The resistance–temperature graph for a superconductor follows that of a normal metal at temperatures above T_c (Fig. 27.10). When the temperature is at or below T_c, the resistivity drops suddenly to zero. This phenomenon was discovered in 1911 by Dutch physicist Heike Kamerlingh-Onnes (1853–1926) as he worked with mercury, which is a superconductor below 4.2 K. Measurements have shown that the resistivities of superconductors below their T_c values are less than 4×10^{-25} $\Omega \cdot$ m, or approximately 10^{17} times smaller than the resistivity of copper. In practice, these resistivities are considered to be zero.

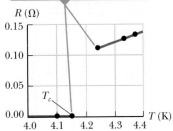

The resistance drops discontinuously to zero at T_c, which is 4.15 K for mercury.

Figure 27.10 Resistance versus temperature for a sample of mercury (Hg). The graph follows that of a normal metal above the critical temperature T_c.

A small permanent magnet levitated above a disk of the superconductor $YBa_2Cu_3O_7$, which is in liquid nitrogen at 77 K.

Table 27.3	**Critical Temperatures for Various Superconductors**
Material	T_c **(K)**
$HgBa_2Ca_2Cu_3O_8$	134
Tl—Ba—Ca—Cu—O	125
Bi—Sr—Ca—Cu—O	105
$YBa_2Cu_3O_7$	92
Nb_3Ge	23.2
Nb_3Sn	18.05
Nb	9.46
Pb	7.18
Hg	4.15
Sn	3.72
Al	1.19
Zn	0.88

Today, thousands of superconductors are known, and as Table 27.3 illustrates, the critical temperatures of recently discovered superconductors are substantially higher than initially thought possible. Two kinds of superconductors are recognized. The more recently identified ones are essentially ceramics with high critical temperatures, whereas superconducting materials such as those observed by Kamerlingh-Onnes are metals. If a room-temperature superconductor is ever identified, its effect on technology could be tremendous.

The value of T_c is sensitive to chemical composition, pressure, and molecular structure. Copper, silver, and gold, which are excellent conductors, do not exhibit superconductivity.

One truly remarkable feature of superconductors is that once a current is set up in them, it persists *without any applied potential difference* (because $R = 0$). Steady currents have been observed to persist in superconducting loops for several years with no apparent decay!

An important and useful application of superconductivity is in the development of superconducting magnets, in which the magnitudes of the magnetic field are approximately ten times greater than those produced by the best normal electromagnets. Such superconducting magnets are being considered as a means of storing energy. Superconducting magnets are currently used in medical magnetic resonance imaging, or MRI, units, which produce high-quality images of internal organs without the need for excessive exposure of patients to x-rays or other harmful radiation.

27.6 Electrical Power

In typical electric circuits, energy T_{ET} is transferred by electrical transmission from a source such as a battery to some device such as a lightbulb or a radio receiver. Let's determine an expression that will allow us to calculate the rate of this energy transfer. First, consider the simple circuit in Figure 27.11, where energy is delivered to a resistor. (Resistors are designated by the circuit symbol —WW—.) Because the connecting wires also have resistance, some energy is delivered to the wires and some to the resistor. Unless noted otherwise, we shall assume the resistance of the wires is small compared with the resistance of the circuit element so that the energy delivered to the wires is negligible.

Imagine following a positive quantity of charge Q moving clockwise around the circuit in Figure 27.11 from point a through the battery and resistor back to point a. We identify the entire circuit as our system. As the charge moves from a to b through the battery, the electric potential energy of the system *increases* by an amount $Q\,\Delta V$

The direction of the effective flow of positive charge is clockwise.

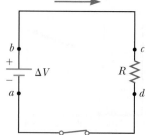

Figure 27.11 A circuit consisting of a resistor of resistance R and a battery having a potential difference ΔV across its terminals.

while the chemical potential energy in the battery *decreases* by the same amount. (Recall from Eq. 25.3 that $\Delta U = q\,\Delta V$.) As the charge moves from *c* to *d* through the resistor, however, the electric potential energy of the system decreases due to collisions of electrons with atoms in the resistor. In this process, the electric potential energy is transformed to internal energy corresponding to increased vibrational motion of the atoms in the resistor. Because the resistance of the interconnecting wires is neglected, no energy transformation occurs for paths *bc* and *da*. When the charge returns to point *a*, the net result is that some of the chemical potential energy in the battery has been delivered to the resistor and resides in the resistor as internal energy E_{int} associated with molecular vibration.

The resistor is normally in contact with air, so its increased temperature results in a transfer of energy by heat *Q* into the air. In addition, the resistor emits thermal radiation T_{ER}, representing another means of escape for the energy. After some time interval has passed, the resistor reaches a constant temperature. At this time, the input of energy from the battery is balanced by the output of energy from the resistor by heat and radiation, and the resistor is a nonisolated system in steady state. Some electrical devices include *heat sinks*[4] connected to parts of the circuit to prevent these parts from reaching dangerously high temperatures. Heat sinks are pieces of metal with many fins. Because the metal's high thermal conductivity provides a rapid transfer of energy by heat away from the hot component and the large number of fins provides a large surface area in contact with the air, energy can transfer by radiation and into the air by heat at a high rate.

Let's now investigate the rate at which the electric potential energy of the system decreases as the charge *Q* passes through the resistor:

$$\frac{dU}{dt} = \frac{d}{dt}(Q\,\Delta V) = \frac{dQ}{dt}\Delta V = I\,\Delta V$$

where *I* is the current in the circuit. The system regains this potential energy when the charge passes through the battery, at the expense of chemical energy in the battery. The rate at which the potential energy of the system decreases as the charge passes through the resistor is equal to the rate at which the system gains internal energy in the resistor. Therefore, the power *P*, representing the rate at which energy is delivered to the resistor, is

$$P = I\,\Delta V \qquad (27.21)$$

We derived this result by considering a battery delivering energy to a resistor. Equation 27.21, however, can be used to calculate the power delivered by a voltage source to *any* device carrying a current *I* and having a potential difference ΔV between its terminals.

Using Equation 27.21 and $\Delta V = IR$ for a resistor, we can express the power delivered to the resistor in the alternative forms

$$P = I^2 R = \frac{(\Delta V)^2}{R} \qquad (27.22)$$

When *I* is expressed in amperes, ΔV in volts, and *R* in ohms, the SI unit of power is the watt, as it was in Chapter 8 in our discussion of mechanical power. The process by which energy is transformed to internal energy in a conductor of resistance *R* is often called *joule heating*;[5] this transformation is also often referred to as an I^2R loss.

---

[4]This usage is another misuse of the word *heat* that is ingrained in our common language.

[5]It is commonly called *joule heating* even though the process of heat does not occur when energy delivered to a resistor appears as internal energy. It is another example of incorrect usage of the word *heat* that has become entrenched in our language.

Figure 27.12 These power lines transfer energy from the electric company to homes and businesses. The energy is transferred at a very high voltage, possibly hundreds of thousands of volts in some cases. Even though it makes power lines very dangerous, the high voltage results in less loss of energy due to resistance in the wires.

When transporting energy by electricity through power lines (Fig. 27.12), you should not assume the lines have zero resistance. Real power lines do indeed have resistance, and power is delivered to the resistance of these wires. Utility companies seek to minimize the energy transformed to internal energy in the lines and maximize the energy delivered to the consumer. Because $P = I\,\Delta V$, the same amount of energy can be transported either at high currents and low potential differences or at low currents and high potential differences. Utility companies choose to transport energy at low currents and high potential differences primarily for economic reasons. Copper wire is very expensive, so it is cheaper to use high-resistance wire (that is, wire having a small cross-sectional area; see Eq. 27.10). Therefore, in the expression for the power delivered to a resistor, $P = I^2R$, the resistance of the wire is fixed at a relatively high value for economic considerations. The I^2R loss can be reduced by keeping the current I as low as possible, which means transferring the energy at a high voltage. In some instances, power is transported at potential differences as great as 765 kV. At the destination of the energy, the potential difference is usually reduced to 4 kV by a device called a *transformer*. Another transformer drops the potential difference to 240 V for use in your home. Of course, each time the potential difference decreases, the current increases by the same factor and the power remains the same. We shall discuss transformers in greater detail in Chapter 33.

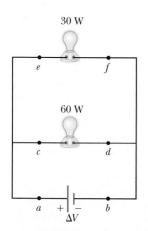

Figure 27.13 (Quick Quiz 27.5) Two lightbulbs connected across the same potential difference.

Quick Quiz 27.5 For the two lightbulbs shown in Figure 27.13, rank the current values at points *a* through *f* from greatest to least.

Example 27.4 **Power in an Electric Heater**

An electric heater is constructed by applying a potential difference of 120 V across a Nichrome wire that has a total resistance of 8.00 Ω. Find the current carried by the wire and the power rating of the heater.

SOLUTION

Conceptualize As discussed in Example 27.2, Nichrome wire has high resistivity and is often used for heating elements in toasters, irons, and electric heaters. Therefore, we expect the power delivered to the wire to be relatively high.

Categorize We evaluate the power from Equation 27.22, so we categorize this example as a substitution problem.

Use Equation 27.7 to find the current in the wire:

$$I = \frac{\Delta V}{R} = \frac{120\text{ V}}{8.00\ \Omega} = \boxed{15.0\text{ A}}$$

Find the power rating using the expression $P = I^2R$ from Equation 27.22:

$$P = I^2R = (15.0\text{ A})^2(8.00\ \Omega) = 1.80 \times 10^3\text{ W} = \boxed{1.80\text{ kW}}$$

WHAT IF? What if the heater were accidentally connected to a 240-V supply? (That is difficult to do because the shape and orientation of the metal contacts in 240-V plugs are different from those in 120-V plugs.) How would that affect the current carried by the heater and the power rating of the heater, assuming the resistance remains constant?

Answer If the applied potential difference were doubled, Equation 27.7 shows that the current would double. According to Equation 27.22, $P = (\Delta V)^2/R$, the power would be four times larger.

Example 27.5	**Linking Electricity and Thermodynamics** AM

An immersion heater must increase the temperature of 1.50 kg of water from 10.0°C to 50.0°C in 10.0 min while operating at 110 V.

(A) What is the required resistance of the heater?

SOLUTION

Conceptualize An immersion heater is a resistor that is inserted into a container of water. As energy is delivered to the immersion heater, raising its temperature, energy leaves the surface of the resistor by heat, going into the water. When the immersion heater reaches a constant temperature, the rate of energy delivered to the resistance by electrical transmission (T_{ET}) is equal to the rate of energy delivered by heat (Q) to the water.

Categorize This example allows us to link our new understanding of power in electricity with our experience with specific heat in thermodynamics (Chapter 20). The water is a *nonisolated system*. Its internal energy is rising because of energy transferred into the water by heat from the resistor, so Equation 8.2 reduces to $\Delta E_{int} = Q$. In our model, we assume the energy that enters the water from the heater remains in the water.

· ·

Analyze To simplify the analysis, let's ignore the initial period during which the temperature of the resistor increases and also ignore any variation of resistance with temperature. Therefore, we imagine a constant rate of energy transfer for the entire 10.0 min.

Set the rate of energy delivered to the resistor equal to the rate of energy Q entering the water by heat:

$$P = \frac{(\Delta V)^2}{R} = \frac{Q}{\Delta t}$$

Use Equation 20.4, $Q = mc\,\Delta T$, to relate the energy input by heat to the resulting temperature change of the water and solve for the resistance:

$$\frac{(\Delta V)^2}{R} = \frac{mc\,\Delta T}{\Delta t} \rightarrow R = \frac{(\Delta V)^2\,\Delta t}{mc\,\Delta T}$$

Substitute the values given in the statement of the problem:

$$R = \frac{(110\text{ V})^2(600\text{ s})}{(1.50\text{ kg})(4\,186\text{ J/kg}\cdot{}^\circ\text{C})(50.0^\circ\text{C} - 10.0^\circ\text{C})} = \boxed{28.9\ \Omega}$$

(B) Estimate the cost of heating the water.

SOLUTION

Multiply the power by the time interval to find the amount of energy transferred to the resistor:

$$T_{ET} = P\,\Delta t = \frac{(\Delta V)^2}{R}\,\Delta t = \frac{(110\text{ V})^2}{28.9\ \Omega}(10.0\text{ min})\left(\frac{1\text{ h}}{60.0\text{ min}}\right)$$

$$= 69.8\text{ Wh} = 0.069\,8\text{ kWh}$$

Find the cost knowing that energy is purchased at an estimated price of 11¢ per kilowatt-hour:

$$\text{Cost} = (0.069\,8\text{ kWh})(\$0.11/\text{kWh}) = \$0.008 = \boxed{0.8¢}$$

Finalize The cost to heat the water is very low, less than one cent. In reality, the cost is higher because some energy is transferred from the water into the surroundings by heat and electromagnetic radiation while its temperature is increasing. If you have electrical devices in your home with power ratings on them, use this power rating and an approximate time interval of use to estimate the cost for one use of the device.

Summary

Definitions

The electric **current** I in a conductor is defined as

$$I \equiv \frac{dQ}{dt} \tag{27.2}$$

where dQ is the charge that passes through a cross section of the conductor in a time interval dt. The SI unit of current is the **ampere** (A), where 1 A = 1 C/s.

continued

The **current density** J in a conductor is the current per unit area:

$$J \equiv \frac{I}{A} \quad \text{(27.5)}$$

The **resistance** R of a conductor is defined as

$$R \equiv \frac{\Delta V}{I} \quad \text{(27.7)}$$

where ΔV is the potential difference across the conductor and I is the current it carries. The SI unit of resistance is volts per ampere, which is defined to be 1 **ohm** (Ω); that is, $1\ \Omega = 1$ V/A.

Concepts and Principles

The average current in a conductor is related to the motion of the charge carriers through the relationship

$$I_{avg} = nqv_d A \quad \text{(27.4)}$$

where n is the density of charge carriers, q is the charge on each carrier, v_d is the drift speed, and A is the cross-sectional area of the conductor.

The current density in an ohmic conductor is proportional to the electric field according to the expression

$$J = \sigma E \quad \text{(27.6)}$$

The proportionality constant σ is called the **conductivity** of the material of which the conductor is made. The inverse of σ is known as **resistivity** ρ (that is, $\rho = 1/\sigma$). Equation 27.6 is known as **Ohm's law,** and a material is said to obey this law if the ratio of its current density to its applied electric field is a constant that is independent of the applied field.

For a uniform block of material of cross-sectional area A and length ℓ, the resistance over the length ℓ is

$$R = \rho\frac{\ell}{A} \quad \text{(27.10)}$$

where ρ is the resistivity of the material.

In a classical model of electrical conduction in metals, the electrons are treated as molecules of a gas. In the absence of an electric field, the average velocity of the electrons is zero. When an electric field is applied, the electrons move (on average) with a **drift velocity** $\vec{v}_d$ that is opposite the electric field. The drift velocity is given by

$$\vec{v}_d = \frac{q\vec{E}}{m_e}\tau \quad \text{(27.13)}$$

where q is the electron's charge, m_e is the mass of the electron, and τ is the average time interval between electron–atom collisions. According to this model, the resistivity of the metal is

$$\rho = \frac{m_e}{nq^2\tau} \quad \text{(27.16)}$$

where n is the number of free electrons per unit volume.

The resistivity of a conductor varies approximately linearly with temperature according to the expression

$$\rho = \rho_0[1 + \alpha(T - T_0)] \quad \text{(27.18)}$$

where ρ_0 is the resistivity at some reference temperature T_0 and α is the **temperature coefficient of resistivity.**

If a potential difference ΔV is maintained across a circuit element, the **power,** or rate at which energy is supplied to the element, is

$$P = I\,\Delta V \quad \text{(27.21)}$$

Because the potential difference across a resistor is given by $\Delta V = IR$, we can express the power delivered to a resistor as

$$P = I^2 R = \frac{(\Delta V)^2}{R} \quad \text{(27.22)}$$

The energy delivered to a resistor by electrical transmission T_{ET} appears in the form of internal energy E_{int} in the resistor.

Objective Questions

1. denotes answer available in *Student Solutions Manual/Study Guide*

1. Car batteries are often rated in ampere-hours. Does this information designate the amount of (a) current, (b) power, (c) energy, (d) charge, or (e) potential the battery can supply?

2. Two wires A and B with circular cross sections are made of the same metal and have equal lengths, but the resistance of wire A is three times greater than that of wire B. **(i)** What is the ratio of the cross-sectional area of A to that of B? (a) 3 (b) $\sqrt{3}$ (c) 1 (d) $1/\sqrt{3}$ (e) $\frac{1}{3}$ **(ii)** What is the ratio of the radius of A to that of B? Choose from the same possibilities as in part (i).

3. A cylindrical metal wire at room temperature is carrying electric current between its ends. One end is at potential $V_A = 50$ V, and the other end is at potential $V_B = 0$ V. Rank the following actions in terms of the change that each one separately would produce in the current from the greatest increase to the greatest decrease. In your ranking, note any cases of equality. (a) Make $V_A = 150$ V with $V_B = 0$ V. (b) Adjust V_A to triple the power with which the wire converts electrically transmitted energy into internal energy. (c) Double the radius of the wire. (d) Double the length of the wire. (e) Double the Celsius temperature of the wire.

4. A current-carrying ohmic metal wire has a cross-sectional area that gradually becomes smaller from one end of the wire to the other. The current has the same value for each section of the wire, so charge does not accumulate at any one point. **(i)** How does the drift speed vary along the wire as the area becomes smaller? (a) It increases. (b) It decreases. (c) It remains constant. **(ii)** How does the resistance per unit length vary along the wire as the area becomes smaller? Choose from the same possibilities as in part (i).

5. A potential difference of 1.00 V is maintained across a 10.0-Ω resistor for a period of 20.0 s. What total charge passes by a point in one of the wires connected to the resistor in this time interval? (a) 200 C (b) 20.0 C (c) 2.00 C (d) 0.005 00 C (e) 0.050 0 C

6. Three wires are made of copper having circular cross sections. Wire 1 has a length L and radius r. Wire 2 has a length L and radius $2r$. Wire 3 has a length $2L$ and radius $3r$. Which wire has the smallest resistance? (a) wire 1 (b) wire 2 (c) wire 3 (d) All have the same resistance. (e) Not enough information is given to answer the question.

7. A metal wire of resistance R is cut into three equal pieces that are then placed together side by side to form a new cable with a length equal to one-third the original length. What is the resistance of this new cable? (a) $\frac{1}{9}R$ (b) $\frac{1}{3}R$ (c) R (d) $3R$ (e) $9R$

8. A metal wire has a resistance of 10.0 Ω at a temperature of 20.0°C. If the same wire has a resistance of 10.6 Ω at 90.0°C, what is the resistance of this wire when its temperature is −20.0°C? (a) 0.700 Ω (b) 9.66 Ω (c) 10.3 Ω (d) 13.8 Ω (e) 6.59 Ω

9. The current-versus-voltage behavior of a certain electrical device is shown in Figure OQ27.9. When the potential difference across the device is 2 V, what is its resistance? (a) 1 Ω (b) $\frac{3}{4}$ Ω (c) $\frac{4}{3}$ Ω (d) undefined (e) none of those answers

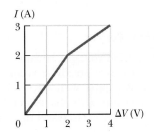

Figure OQ27.9

10. Two conductors made of the same material are connected across the same potential difference. Conductor A has twice the diameter and twice the length of conductor B. What is the ratio of the power delivered to A to the power delivered to B? (a) 8 (b) 4 (c) 2 (d) 1 (e) $\frac{1}{2}$

11. Two conducting wires A and B of the same length and radius are connected across the same potential difference. Conductor A has twice the resistivity of conductor B. What is the ratio of the power delivered to A to the power delivered to B? (a) 2 (b) $\sqrt{2}$ (c) 1 (d) $1/\sqrt{2}$ (e) $\frac{1}{2}$

12. Two lightbulbs both operate on 120 V. One has a power of 25 W and the other 100 W. **(i)** Which lightbulb has higher resistance? (a) The dim 25-W lightbulb does. (b) The bright 100-W lightbulb does. (c) Both are the same. **(ii)** Which lightbulb carries more current? Choose from the same possibilities as in part (i).

13. Wire B has twice the length and twice the radius of wire A. Both wires are made from the same material. If wire A has a resistance R, what is the resistance of wire B? (a) $4R$ (b) $2R$ (c) R (d) $\frac{1}{2}R$ (e) $\frac{1}{4}R$

Conceptual Questions
[1.] denotes answer available in *Student Solutions Manual/Study Guide*

1. If you were to design an electric heater using Nichrome wire as the heating element, what parameters of the wire could you vary to meet a specific power output such as 1 000 W?

2. What factors affect the resistance of a conductor?

3. When the potential difference across a certain conductor is doubled, the current is observed to increase by a factor of 3. What can you conclude about the conductor?

4. Over the time interval after a difference in potential is applied between the ends of a wire, what would happen to the drift velocity of the electrons in a wire and to the current in the wire if the electrons could move freely without resistance through the wire?

5. How does the resistance for copper and for silicon change with temperature? Why are the behaviors of these two materials different?

6. Use the atomic theory of matter to explain why the resistance of a material should increase as its temperature increases.

7. If charges flow very slowly through a metal, why does it not require several hours for a light to come on when you throw a switch?

8. Newspaper articles often contain statements such as "10 000 volts of electricity surged through the victim's body." What is wrong with this statement?

Problems

WebAssign The problems found in this chapter may be assigned online in Enhanced WebAssign

1. straightforward; **2.** intermediate;
3. challenging

1. full solution available in the *Student Solutions Manual/Study Guide*

AMT Analysis Model tutorial available in Enhanced WebAssign

GP Guided Problem

M Master It tutorial available in Enhanced WebAssign

W Watch It video solution available in Enhanced WebAssign

Section 27.1 Electric Current

1. A 200-km-long high-voltage transmission line 2.00 cm in diameter carries a steady current of 1 000 A. If the conductor is copper with a free charge density of 8.50×10^{28} electrons per cubic meter, how many years does it take one electron to travel the full length of the cable?

2. A small sphere that carries a charge q is whirled in a circle at the end of an insulating string. The angular frequency of revolution is ω. What average current does this revolving charge represent?

3. An aluminum wire having a cross-sectional area equal to 4.00×10^{-6} m^2 carries a current of 5.00 A. The density of aluminum is 2.70 g/cm^3. Assume each aluminum atom supplies one conduction electron per atom. Find the drift speed of the electrons in the wire.

4. In the Bohr model of the hydrogen atom (which will be covered in detail in Chapter 42), an electron in the lowest energy state moves at a speed of 2.19×10^6 m/s in a circular path of radius 5.29×10^{-11} m. What is the effective current associated with this orbiting electron?

5. A proton beam in an accelerator carries a current of 125 μA. If the beam is incident on a target, how many protons strike the target in a period of 23.0 s?

6. A copper wire has a circular cross section with a radius of 1.25 mm. (a) If the wire carries a current of 3.70 A, find the drift speed of the electrons in this wire. (b) All other things being equal, what happens to the drift speed in wires made of metal having a larger number of conduction electrons per atom than copper? Explain.

7. Suppose the current in a conductor decreases exponentially with time according to the equation $I(t) = I_0 e^{-t/\tau}$, where I_0 is the initial current (at $t = 0$) and τ is a constant having dimensions of time. Consider a fixed observation point within the conductor. (a) How much charge passes this point between $t = 0$ and $t = \tau$? (b) How much charge passes this point between $t = 0$ and $t = 10\tau$? (c) **What If?** How much charge passes this point between $t = 0$ and $t = \infty$?

8. Figure P27.8 represents a section of a conductor of nonuniform diameter carrying a current of $I = 5.00$ A. The radius of cross-section A_1 is $r_1 = 0.400$ cm. (a) What is the magnitude of the current density across A_1? The radius r_2 at A_2 is larger than the radius r_1 at A_1.

(b) Is the current at A_2 larger, smaller, or the same? (c) Is the current density at A_2 larger, smaller, or the same? Assume $A_2 = 4A_1$. Specify the (d) radius, (e) current, and (f) current density at A_2.

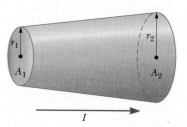

Figure P27.8

9. The quantity of charge q (in coulombs) that has passed through a surface of area 2.00 cm^2 varies with time according to the equation $q = 4t^3 + 5t + 6$, where t is in seconds. (a) What is the instantaneous current through the surface at $t = 1.00$ s? (b) What is the value of the current density?

10. A Van de Graaff generator produces a beam of 2.00-MeV deuterons, which are heavy hydrogen nuclei containing a proton and a neutron. (a) If the beam current is 10.0 μA, what is the average separation of the deuterons? (b) Is the electrical force of repulsion among them a significant factor in beam stability? Explain.

11. The electron beam emerging from a certain high-energy electron accelerator has a circular cross section of radius 1.00 mm. (a) The beam current is 8.00 μA. Find the current density in the beam assuming it is uniform throughout. (b) The speed of the electrons is so close to the speed of light that their speed can be taken as 300 Mm/s with negligible error. Find the electron density in the beam. (c) Over what time interval does Avogadro's number of electrons emerge from the accelerator?

12. An electric current in a conductor varies with time according to the expression $I(t) = 100 \sin (120\pi t)$, where I is in amperes and t is in seconds. What is the total charge passing a given point in the conductor from $t = 0$ to $t = \frac{1}{240}$ s?

13. A teapot with a surface area of 700 cm^2 is to be plated with silver. It is attached to the negative electrode of an electrolytic cell containing silver nitrate (Ag$^+$NO$_3^-$). The cell is powered by a 12.0-V battery and has a resistance of 1.80 Ω. If the density of silver is 10.5 × 10^3 kg/m^3, over what time interval does a 0.133-mm layer of silver build up on the teapot?

Section 27.2 Resistance

14. A lightbulb has a resistance of 240 Ω when operating **W** with a potential difference of 120 V across it. What is the current in the lightbulb?

15. A wire 50.0 m long and 2.00 mm in diameter is con- **M** nected to a source with a potential difference of 9.11 V, and the current is found to be 36.0 A. Assume a temperature of 20.0°C and, using Table 27.2, identify the metal out of which the wire is made.

16. A 0.900-V potential difference is maintained across a 1.50-m length of tungsten wire that has a cross-sectional area of 0.600 mm^2. What is the current in the wire?

17. An electric heater carries a current of 13.5 A when operating at a voltage of 120 V. What is the resistance of the heater?

18. Aluminum and copper wires of equal length are found to have the same resistance. What is the ratio of their radii?

19. Suppose you wish to fabricate a uniform wire from **M** 1.00 g of copper. If the wire is to have a resistance of $R = 0.500$ Ω and all the copper is to be used, what must be (a) the length and (b) the diameter of this wire?

20. Suppose you wish to fabricate a uniform wire from a mass m of a metal with density ρ_m and resistivity ρ. If the wire is to have a resistance of R and all the metal is to be used, what must be (a) the length and (b) the diameter of this wire?

21. A portion of Nichrome wire of radius 2.50 mm is to be used in winding a heating coil. If the coil must draw a current of 9.25 A when a voltage of 120 V is applied across its ends, find (a) the required resistance of the coil and (b) the length of wire you must use to wind the coil.

Section 27.3 A Model for Electrical Conduction

22. If the current carried by a conductor is doubled, what happens to (a) the charge carrier density, (b) the current density, (c) the electron drift velocity, and (d) the average time interval between collisions?

23. A current density of 6.00 × 10^{-13} A/m^2 exists in the atmosphere at a location where the electric field is 100 V/m. Calculate the electrical conductivity of the Earth's atmosphere in this region.

24. An iron wire has a cross-sectional area equal to 5.00 × **GP** 10^{-6} m^2. Carry out the following steps to determine the drift speed of the conduction electrons in the wire if it carries a current of 30.0 A. (a) How many kilograms are there in 1.00 mole of iron? (b) Starting with the density of iron and the result of part (a), compute the molar density of iron (the number of moles of iron per cubic meter). (c) Calculate the number density of

iron atoms using Avogadro's number. (d) Obtain the number density of conduction electrons given that there are two conduction electrons per iron atom. (e) Calculate the drift speed of conduction electrons in this wire.

25. If the magnitude of the drift velocity of free electrons **M** in a copper wire is 7.84 × 10^{-4} m/s, what is the electric field in the conductor?

Section 27.4 Resistance and Temperature

26. A certain lightbulb has a tungsten filament with a resistance of 19.0 Ω when at 20.0°C and 140 Ω when hot. Assume the resistivity of tungsten varies linearly with temperature even over the large temperature range involved here. Find the temperature of the hot filament.

27. What is the fractional change in the resistance of an iron filament when its temperature changes from 25.0°C to 50.0°C?

28. While taking photographs in Death Valley on a day when the temperature is 58.0°C, Bill Hiker finds that a certain voltage applied to a copper wire produces a current of 1.00 A. Bill then travels to Antarctica and applies the same voltage to the same wire. What current does he register there if the temperature is −88.0°C? Assume that no change occurs in the wire's shape and size.

29. If a certain silver wire has a resistance of 6.00 Ω at 20.0°C, what resistance will it have at 34.0°C?

30. Plethysmographs are devices used for measuring changes in the volume of internal organs or limbs. In one form of this device, a rubber capillary tube with an inside diameter of 1.00 mm is filled with mercury at 20.0°C. The resistance of the mercury is measured with the aid of electrodes sealed into the ends of the tube. If 100 cm of the tube is wound in a helix around a patient's upper arm, the blood flow during a heartbeat causes the arm to expand, stretching the length of the tube by 0.040 0 cm. From this observation and assuming cylindrical symmetry, you can find the change in volume of the arm, which gives an indication of blood flow. Taking the resistivity of mercury to be 9.58 × 10^{-7} Ω · m, calculate (a) the resistance of the mercury and (b) the fractional change in resistance during the heartbeat. *Hint:* The fraction by which the cross-sectional area of the mercury column decreases is the fraction by which the length increases because the volume of mercury is constant.

31. (a) A 34.5-m length of copper wire at 20.0°C has a **M** radius of 0.25 mm. If a potential difference of 9.00 V is applied across the length of the wire, determine the current in the wire. (b) If the wire is heated to 30.0°C while the 9.00-V potential difference is maintained, what is the resulting current in the wire?

32. An engineer needs a resistor with a zero overall temperature coefficient of resistance at 20.0°C. She designs a pair of circular cylinders, one of carbon and one of Nichrome as shown in Figure P27.32 (page 828). The device must have an overall resistance of $R_1 + R_2 = 10.0 \, \Omega$ independent of temperature and a uniform radius of $r = 1.50$ mm. Ignore thermal expansion of the cylinders and assume both are always at the same temperature. (a) Can she meet the design goal with this method? (b) If so, state what you can determine about the lengths ℓ_1 and ℓ_2 of each segment. If not, explain.

Figure P27.32

33. An aluminum wire with a diameter of 0.100 mm has a uniform electric field of 0.200 V/m imposed along its entire length. The temperature of the wire is 50.0°C. Assume one free electron per atom. (a) Use the information in Table 27.2 to determine the resistivity of aluminum at this temperature. (b) What is the current density in the wire? (c) What is the total current in the wire? (d) What is the drift speed of the conduction electrons? (e) What potential difference must exist between the ends of a 2.00-m length of the wire to produce the stated electric field?

34. **Review.** An aluminum rod has a resistance of 1.23 Ω at 20.0°C. Calculate the resistance of the rod at 120°C by accounting for the changes in both the resistivity and the dimensions of the rod. The coefficient of linear expansion for aluminum is $2.40 \times 10^{-6} \, (°C)^{-1}$.

35. At what temperature will aluminum have a resistivity that is three times the resistivity copper has at room temperature?

Section 27.6 Electrical Power

36. Assume that global lightning on the Earth constitutes a constant current of 1.00 kA between the ground and an atmospheric layer at potential 300 kV. (a) Find the power of terrestrial lightning. (b) For comparison, find the power of sunlight falling on the Earth. Sunlight has an intensity of 1 370 W/m² above the atmosphere. Sunlight falls perpendicularly on the circular projected area that the Earth presents to the Sun.

37. In a hydroelectric installation, a turbine delivers 1 500 hp to a generator, which in turn transfers 80.0% of the mechanical energy out by electrical transmission. Under these conditions, what current does the generator deliver at a terminal potential difference of 2 000 V?

38. A Van de Graaff generator (see Fig. 25.23) is operating so that the potential difference between the high-potential electrode Ⓑ and the charging needles at Ⓐ is 15.0 kV. Calculate the power required to drive the belt against electrical forces at an instant when the effective current delivered to the high-potential electrode is 500 μA.

39. A certain waffle iron is rated at 1.00 kW when connected to a 120-V source. (a) What current does the waffle iron carry? (b) What is its resistance?

40. The potential difference across a resting neuron in the human body is about 75.0 mV and carries a current of about 0.200 mA. How much power does the neuron release?

41. Suppose your portable DVD player draws a current of 350 mA at 6.00 V. How much power does the player require?

42. **Review.** A well-insulated electric water heater warms 109 kg of water from 20.0°C to 49.0°C in 25.0 min. Find the resistance of its heating element, which is connected across a 240-V potential difference.

43. A 100-W lightbulb connected to a 120-V source experiences a voltage surge that produces 140 V for a moment. By what percentage does its power output increase? Assume its resistance does not change.

44. The cost of energy delivered to residences by electrical transmission varies from \$0.070/kWh to \$0.258/kWh throughout the United States; \$0.110/kWh is the average value. At this average price, calculate the cost of (a) leaving a 40.0-W porch light on for two weeks while you are on vacation, (b) making a piece of dark toast in 3.00 min with a 970-W toaster, and (c) drying a load of clothes in 40.0 min in a 5.20×10^3-W dryer.

45. Batteries are rated in terms of ampere-hours (A · h). For example, a battery that can produce a current of 2.00 A for 3.00 h is rated at 6.00 A · h. (a) What is the total energy, in kilowatt-hours, stored in a 12.0-V battery rated at 55.0 A · h? (b) At \$0.110 per kilowatt-hour, what is the value of the electricity produced by this battery?

46. Residential building codes typically require the use of 12-gauge copper wire (diameter 0.205 cm) for wiring receptacles. Such circuits carry currents as large as 20.0 A. If a wire of smaller diameter (with a higher gauge number) carried that much current, the wire could rise to a high temperature and cause a fire. (a) Calculate the rate at which internal energy is produced in 1.00 m of 12-gauge copper wire carrying 20.0 A. (b) **What If?** Repeat the calculation for a 12-gauge aluminum wire. (c) Explain whether a 12-gauge aluminum wire would be as safe as a copper wire.

47. Assuming the cost of energy from the electric company is \$0.110/kWh, compute the cost per day of operating a lamp that draws a current of 1.70 A from a 110-V line.

48. An 11.0-W energy-efficient fluorescent lightbulb is designed to produce the same illumination as a conventional 40.0-W incandescent lightbulb. Assuming a cost of \$0.110/kWh for energy from the electric company, how much money does the user of the energy-efficient bulb save during 100 h of use?

49. A coil of Nichrome wire is 25.0 m long. The wire has a diameter of 0.400 mm and is at 20.0°C. If it carries a current of 0.500 A, what are (a) the magnitude of the electric field in the wire and (b) the power delivered to it? (c) **What If?** If the temperature is increased to 340°C and the potential difference across the wire remains constant, what is the power delivered?

50. **Review.** A rechargeable battery of mass 15.0 g delivers an average current of 18.0 mA to a portable DVD player at 1.60 V for 2.40 h before the battery must be

recharged. The recharger maintains a potential difference of 2.30 V across the battery and delivers a charging current of 13.5 mA for 4.20 h. (a) What is the efficiency of the battery as an energy storage device? (b) How much internal energy is produced in the battery during one charge–discharge cycle? (c) If the battery is surrounded by ideal thermal insulation and has an effective specific heat of 975 J/kg · °C, by how much will its temperature increase during the cycle?

51. A 500-W heating coil designed to operate from 110 V is made of Nichrome wire 0.500 mm in diameter. (a) Assuming the resistivity of the Nichrome remains constant at its 20.0°C value, find the length of wire used. (b) **What If?** Now consider the variation of resistivity with temperature. What power is delivered to the coil of part (a) when it is warmed to 1 200°C?

52. *Why is the following situation impossible?* A politician is decrying wasteful uses of energy and decides to focus on energy used to operate plug-in electric clocks in the United States. He estimates there are 270 million of these clocks, approximately one clock for each person in the population. The clocks transform energy taken in by electrical transmission at the average rate 2.50 W. The politician gives a speech in which he complains that, at today's electrical rates, the nation is losing $100 million every year to operate these clocks.

53. A certain toaster has a heating element made of Nichrome wire. When the toaster is first connected to a 120-V source (and the wire is at a temperature of 20.0°C), the initial current is 1.80 A. The current decreases as the heating element warms up. When the toaster reaches its final operating temperature, the current is 1.53 A. (a) Find the power delivered to the toaster when it is at its operating temperature. (b) What is the final temperature of the heating element?

54. Make an order-of-magnitude estimate of the cost of one person's routine use of a handheld hair dryer for 1 year. If you do not use a hair dryer yourself, observe or interview someone who does. State the quantities you estimate and their values.

55. **Review.** The heating element of an electric coffee maker operates at 120 V and carries a current of 2.00 A. Assuming the water absorbs all the energy delivered to the resistor, calculate the time interval during which the temperature of 0.500 kg of water rises from room temperature (23.0°C) to the boiling point.

56. A 120-V motor has mechanical power output of 2.50 hp. It is 90.0% efficient in converting power that it takes in by electrical transmission into mechanical power. (a) Find the current in the motor. (b) Find the energy delivered to the motor by electrical transmission in 3.00 h of operation. (c) If the electric company charges $0.110/kWh, what does it cost to run the motor for 3.00 h?

Additional Problems

57. A particular wire has a resistivity of 3.0×10^{-8} Ω · m and a cross-sectional area of 4.0×10^{-6} m^2. A length of this wire is to be used as a resistor that will receive 48 W of power when connected across a 20-V battery. What length of wire is required?

58. Determine the temperature at which the resistance of an aluminum wire will be twice its value at 20.0°C. Assume its coefficient of resistivity remains constant.

59. A car owner forgets to turn off the headlights of his car while it is parked in his garage. If the 12.0-V battery in his car is rated at 90.0 A · h and each headlight requires 36.0 W of power, how long will it take the battery to completely discharge?

60. Lightbulb A is marked "25 W 120 V," and lightbulb B is marked "100 W 120 V." These labels mean that each lightbulb has its respective power delivered to it when it is connected to a constant 120-V source. (a) Find the resistance of each lightbulb. (b) During what time interval does 1.00 C pass into lightbulb A? (c) Is this charge different upon its exit versus its entry into the lightbulb? Explain. (d) In what time interval does 1.00 J pass into lightbulb A? (e) By what mechanisms does this energy enter and exit the lightbulb? Explain. (f) Find the cost of running lightbulb A continuously for 30.0 days, assuming the electric company sells its product at $0.110 per kWh.

61. One wire in a high-voltage transmission line carries 1 000 A starting at 700 kV for a distance of 100 mi. If the resistance in the wire is 0.500 Ω/mi, what is the power loss due to the resistance of the wire?

62. An experiment is conducted to measure the electrical resistivity of Nichrome in the form of wires with different lengths and cross-sectional areas. For one set of measurements, a student uses 30-gauge wire, which has a cross-sectional area of 7.30×10^{-8} m^2. The student measures the potential difference across the wire and the current in the wire with a voltmeter and an ammeter, respectively. (a) For each set of measurements given in the table taken on wires of three different lengths, calculate the resistance of the wires and the corresponding values of the resistivity. (b) What is the average value of the resistivity? (c) Explain how this value compares with the value given in Table 27.2.

L (m)	ΔV (V)	I (A)	R (Ω)	ρ (Ω · m)
0.540	5.22	0.72		
1.028	5.82	0.414		
1.543	5.94	0.281		

63. A charge Q is placed on a capacitor of capacitance C. The capacitor is connected into the circuit shown in Figure P27.63, with an open switch, a resistor, and an initially uncharged capacitor of capacitance $3C$. The

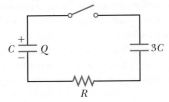

Figure P27.63

switch is then closed, and the circuit comes to equilibrium. In terms of Q and C, find (a) the final potential difference between the plates of each capacitor, (b) the charge on each capacitor, and (c) the final energy stored in each capacitor. (d) Find the internal energy appearing in the resistor.

64. **Review.** An office worker uses an immersion heater to warm 250 g of water in a light, covered, insulated cup from 20.0°C to 100°C in 4.00 min. The heater is a Nichrome resistance wire connected to a 120-V power supply. Assume the wire is at 100°C throughout the 4.00-min time interval. (a) Specify a relationship between a diameter and a length that the wire can have. (b) Can it be made from less than 0.500 cm³ of Nichrome?

65. An x-ray tube used for cancer therapy operates at 4.00 MV with electrons constituting a beam current of 25.0 mA striking a metal target. Nearly all the power in the beam is transferred to a stream of water flowing through holes drilled in the target. What rate of flow, in kilograms per second, is needed if the rise in temperature of the water is not to exceed 50.0°C?

66. An all-electric car (not a hybrid) is designed to run **AMT** from a bank of 12.0-V batteries with total energy storage of 2.00×10^7 J. If the electric motor draws 8.00 kW **M** as the car moves at a steady speed of 20.0 m/s, (a) what is the current delivered to the motor? (b) How far can the car travel before it is "out of juice"?

67. A straight, cylindrical wire lying along the x axis has a length of 0.500 m and a diameter of 0.200 mm. It is made of a material described by Ohm's law with a resistivity of $\rho = 4.00 \times 10^{-8}$ Ω · m. Assume a potential of 4.00 V is maintained at the left end of the wire at $x = 0$. Also assume $V = 0$ at $x = 0.500$ m. Find (a) the magnitude and direction of the electric field in the wire, (b) the resistance of the wire, (c) the magnitude and direction of the electric current in the wire, and (d) the current density in the wire. (e) Show that $E = \rho J$.

68. A straight, cylindrical wire lying along the x axis has a length L and a diameter d. It is made of a material described by Ohm's law with a resistivity ρ. Assume potential V is maintained at the left end of the wire at $x = 0$. Also assume the potential is zero at $x = L$. In terms of L, d, V, ρ, and physical constants, derive expressions for (a) the magnitude and direction of the electric field in the wire, (b) the resistance of the wire, (c) the magnitude and direction of the electric current in the wire, and (d) the current density in the wire. (e) Show that $E = \rho J$.

69. An electric utility company supplies a customer's house **W** from the main power lines (120 V) with two copper wires, each of which is 50.0 m long and has a resistance of 0.108 Ω per 300 m. (a) Find the potential difference at the customer's house for a load current of 110 A. For this load current, find (b) the power delivered to the customer and (c) the rate at which internal energy is produced in the copper wires.

70. The strain in a wire can be monitored and computed by measuring the resistance of the wire. Let L_i represent the original length of the wire, A_i its original cross-sectional area, $R_i = \rho L_i / A_i$ the original resistance between its ends, and $\delta = \Delta L / L_i = (L - L_i)/L_i$ the strain resulting from the application of tension. Assume the resistivity and the volume of the wire do not change as the wire stretches. (a) Show that the resistance between the ends of the wire under strain is given by $R = R_i(1 + 2\delta + \delta^2)$. (b) If the assumptions are precisely true, is this result exact or approximate? Explain your answer.

71. An oceanographer is studying how the ion concentration in seawater depends on depth. She makes a measurement by lowering into the water a pair of concentric metallic cylinders (Fig. P27.71) at the end of a cable and taking data to determine the resistance between these electrodes as a function of depth. The water between the two cylinders forms a cylindrical shell of inner radius r_a, outer radius r_b, and length L much larger than r_b. The scientist applies a potential difference ΔV between the inner and outer surfaces, producing an outward radial current I. Let ρ represent the resistivity of the water. (a) Find the resistance of the water between the cylinders in terms of L, ρ, r_a, and r_b. (b) Express the resistivity of the water in terms of the measured quantities L, r_a, r_b, ΔV, and I.

Figure P27.71

72. *Why is the following situation impossible?* An inquisitive physics student takes a 100-W incandescent lightbulb out of its socket and measures its resistance with an ohmmeter. He measures a value of 10.5 Ω. He is able to connect an ammeter to the lightbulb socket to correctly measure the current drawn by the bulb while operating. Inserting the bulb back into the socket and operating the bulb from a 120-V source, he measures the current to be 11.4 A.

73. The temperature coefficients of resistivity α in Table 27.2 are based on a reference temperature T_0 of 20.0°C. Suppose the coefficients were given the symbol α' and were based on a T_0 of 0°C. What would the coefficient α' for silver be? *Note:* The coefficient α satisfies $\rho = \rho_0[1 + \alpha(T - T_0)]$, where ρ_0 is the resistivity of the material at $T_0 = 20.0$°C. The coefficient α' must satisfy the expression $\rho = \rho_0'[1 + \alpha'T]$, where ρ_0' is the resistivity of the material at 0°C.

74. A close analogy exists between the flow of energy by heat because of a temperature difference (see Section 20.7) and the flow of electric charge because of a

potential difference. In a metal, energy dQ and electrical charge dq are both transported by free electrons. Consequently, a good electrical conductor is usually a good thermal conductor as well. Consider a thin conducting slab of thickness dx, area A, and electrical conductivity σ, with a potential difference dV between opposite faces. (a) Show that the current $I = dq/dt$ is given by the equation on the left:

Charge conduction   Thermal conduction

$$\frac{dq}{dt} = \sigma A \left| \frac{dV}{dx} \right| \qquad \frac{dQ}{dt} = kA \left| \frac{dT}{dx} \right|$$

In the analogous thermal conduction equation on the right (Eq. 20.15), the rate dQ/dt of energy flow by heat (in SI units of joules per second) is due to a temperature gradient dT/dx in a material of thermal conductivity k. (b) State analogous rules relating the direction of the electric current to the change in potential and relating the direction of energy flow to the change in temperature.

75. **Review.** When a straight wire is warmed, its resistance is given by $R = R_0[1 + \alpha(T - T_0)]$ according to Equation 27.20, where α is the temperature coefficient of resistivity. This expression needs to be modified if we include the change in dimensions of the wire due to thermal expansion. For a copper wire of radius 0.100 0 mm and length 2.000 m, find its resistance at 100.0°C, including the effects of both thermal expansion and temperature variation of resistivity. Assume the coefficients are known to four significant figures.

76. **Review.** When a straight wire is warmed, its resistance is given by $R = R_0[1 + \alpha(T - T_0)]$ according to Equation 27.20, where α is the temperature coefficient of resistivity. This expression needs to be modified if we include the change in dimensions of the wire due to thermal expansion. Find a more precise expression for the resistance, one that includes the effects of changes in the dimensions of the wire when it is warmed. Your final expression should be in terms of R_0, T, T_0, the temperature coefficient of resistivity α, and the coefficient of linear expansion α'.

77. **Review.** A parallel-plate capacitor consists of square plates of edge length ℓ that are separated by a distance d, where $d \ll \ell$. A potential difference ΔV is maintained between the plates. A material of dielectric constant κ fills half the space between the plates. The dielectric slab is withdrawn from the capacitor as shown in Figure P27.77. (a) Find the capacitance when

the left edge of the dielectric is at a distance x from the center of the capacitor. (b) If the dielectric is removed at a constant speed v, what is the current in the circuit as the dielectric is being withdrawn?

78. The dielectric material between the plates of a parallel-plate capacitor always has some nonzero conductivity σ. Let A represent the area of each plate and d the distance between them. Let κ represent the dielectric constant of the material. (a) Show that the resistance R and the capacitance C of the capacitor are related by

$$RC = \frac{\kappa \epsilon_0}{\sigma}$$

(b) Find the resistance between the plates of a 14.0-nF capacitor with a fused quartz dielectric.

79. Gold is the most ductile of all metals. For example, one gram of gold can be drawn into a wire 2.40 km long. The density of gold is 19.3×10^3 kg/m³, and its resistivity is 2.44×10^{-8} Ω · m. What is the resistance of such a wire at 20.0°C?

80. The current–voltage characteristic curve for a semiconductor diode as a function of temperature T is given by

$$I = I_0(e^{e\,\Delta V/k_B T} - 1)$$

Here the first symbol e represents Euler's number, the base of natural logarithms. The second e is the magnitude of the electron charge, the k_B stands for Boltzmann's constant, and T is the absolute temperature. (a) Set up a spreadsheet to calculate I and $R = \Delta V/I$ for $\Delta V = 0.400$ V to 0.600 V in increments of 0.005 V. Assume $I_0 = 1.00$ nA. (b) Plot R versus ΔV for $T = 280$ K, 300 K, and 320 K.

81. The potential difference across the filament of a lightbulb is maintained at a constant value while equilibrium temperature is being reached. The steady-state current in the bulb is only one-tenth of the current drawn by the bulb when it is first turned on. If the temperature coefficient of resistivity for the bulb at 20.0°C is 0.004 50 (°C)$^{-1}$ and the resistance increases linearly with increasing temperature, what is the final operating temperature of the filament?

Challenge Problems

82. A more general definition of the temperature coefficient of resistivity is

$$\alpha = \frac{1}{\rho}\frac{d\rho}{dT}$$

where ρ is the resistivity at temperature T. (a) Assuming α is constant, show that

$$\rho = \rho_0 e^{\alpha(T - T_0)}$$

where ρ_0 is the resistivity at temperature T_0. (b) Using the series expansion $e^x \approx 1 + x$ for $x \ll 1$, show that the resistivity is given approximately by the expression

$$\rho = \rho_0[1 + \alpha(T - T_0)] \quad \text{for } \alpha(T - T_0) \ll 1$$

83. A spherical shell with inner radius r_a and outer radius r_b is formed from a material of resistivity ρ. It carries

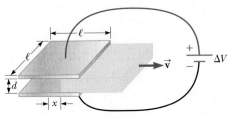

Figure P27.77

current radially, with uniform density in all directions. Show that its resistance is

$$R = \frac{\rho}{4\pi}\left(\frac{1}{r_a} - \frac{1}{r_b}\right)$$

84. Material with uniform resistivity ρ is formed into a wedge as shown in Figure P27.84. Show that the resistance between face A and face B of this wedge is

$$R = \rho \frac{L}{w(y_2 - y_1)} \ln \frac{y_2}{y_1}$$

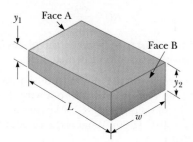

Figure P27.84

85. A material of resistivity ρ is formed into the shape of a truncated cone of height h as shown in Figure P27.85. The bottom end has radius b, and the top end has radius a. Assume the current is distributed uniformly over any circular cross section of the cone so that the current density does not depend on radial position. (The current density does vary with position along the axis of the cone.) Show that the resistance between the two ends is

$$R = \frac{\rho}{\pi}\left(\frac{h}{ab}\right)$$

Figure P27.85

Direct-Current Circuits

In this chapter, we analyze simple electric circuits that contain batteries, resistors, and capacitors in various combinations. Some circuits contain resistors that can be combined using simple rules. The analysis of more complicated circuits is simplified using *Kirchhoff's rules,* which follow from the laws of conservation of energy and conservation of electric charge for isolated systems. Most of the circuits analyzed are assumed to be in *steady state,* which means that currents in the circuit are constant in magnitude and direction. A current that is constant in direction is called a *direct current* (DC). We will study *alternating current* (AC), in which the current changes direction periodically, in Chapter 33. Finally, we discuss electrical circuits in the home.

A technician repairs a connection on a circuit board from a computer. In our lives today, we use various items containing electric circuits, including many with circuit boards much smaller than the board shown in the photograph. These include handheld game players, cell phones, and digital cameras. In this chapter, we study simple types of circuits and learn how to analyze them. *(Trombax/Shutterstock.com)*

28.1 Electromotive Force

In Section 27.6, we discussed a circuit in which a battery produces a current. We will generally use a battery as a source of energy for circuits in our discussion. Because the potential difference at the battery terminals is constant in a particular circuit, the current in the circuit is constant in magnitude and direction and is called **direct current.** A battery is called either a *source of electromotive force* or, more commonly, a *source of emf.* (The phrase *electromotive force* is an unfortunate historical term, describing not a force, but rather a potential difference in volts.) The **emf $\mathcal{E}$** of a battery is **the maximum possible voltage the battery can provide between its terminals.** You can think of a source of emf as a "charge pump." When an electric potential difference exists between two points, the source moves charges "uphill" from the lower potential to the higher.

We shall generally assume the connecting wires in a circuit have no resistance. The positive terminal of a battery is at a higher potential than the negative terminal.

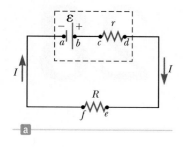

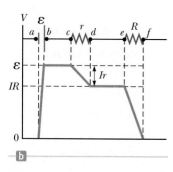

Figure 28.1 (a) Circuit diagram of a source of emf $\mathcal{E}$ (in this case, a battery), of internal resistance r, connected to an external resistor of resistance R. (b) Graphical representation showing how the electric potential changes as the circuit in (a) is traversed clockwise.

Because a real battery is made of matter, there is resistance to the flow of charge within the battery. This resistance is called **internal resistance** r. For an idealized battery with zero internal resistance, the potential difference across the battery (called its *terminal voltage*) equals its emf. For a real battery, however, the terminal voltage is *not* equal to the emf for a battery in a circuit in which there is a current. To understand why, consider the circuit diagram in Figure 28.1a. We model the battery as shown in the diagram; it is represented by the dashed rectangle containing an ideal, resistance-free emf $\mathcal{E}$ in series with an internal resistance r. A resistor of resistance R is connected across the terminals of the battery. Now imagine moving through the battery from a to d and measuring the electric potential at various locations. Passing from the negative terminal to the positive terminal, the potential increases by an amount $\mathcal{E}$. As we move through the resistance r, however, the potential *decreases* by an amount Ir, where I is the current in the circuit. Therefore, the terminal voltage of the battery $\Delta V = V_d - V_a$ is

$$\Delta V = \mathcal{E} - Ir \tag{28.1}$$

From this expression, notice that $\mathcal{E}$ is equivalent to the **open-circuit voltage,** that is, the terminal voltage when the current is zero. The emf is the voltage labeled on a battery; for example, the emf of a D cell is 1.5 V. The actual potential difference between a battery's terminals depends on the current in the battery as described by Equation 28.1. Figure 28.1b is a graphical representation of the changes in electric potential as the circuit is traversed in the clockwise direction.

Figure 28.1a shows that the terminal voltage ΔV must equal the potential difference across the external resistance R, often called the **load resistance.** The load resistor might be a simple resistive circuit element as in Figure 28.1a, or it could be the resistance of some electrical device (such as a toaster, electric heater, or lightbulb) connected to the battery (or, in the case of household devices, to the wall outlet). The resistor represents a *load* on the battery because the battery must supply energy to operate the device containing the resistance. The potential difference across the load resistance is $\Delta V = IR$. Combining this expression with Equation 28.1, we see that

$$\mathcal{E} = IR + Ir \tag{28.2}$$

Figure 28.1a shows a graphical representation of this equation. Solving for the current gives

$$I = \frac{\mathcal{E}}{R + r} \tag{28.3}$$

Equation 28.3 shows that the current in this simple circuit depends on both the load resistance R external to the battery and the internal resistance r. If R is much greater than r, as it is in many real-world circuits, we can neglect r.

Multiplying Equation 28.2 by the current I in the circuit gives

$$I\mathcal{E} = I^2R + I^2r \tag{28.4}$$

Equation 28.4 indicates that because power $P = I\,\Delta V$ (see Eq. 27.21), the total power output $I\mathcal{E}$ associated with the emf of the battery is delivered to the external load resistance in the amount I^2R and to the internal resistance in the amount I^2r.

Quick Quiz 28.1 To maximize the percentage of the power from the emf of a battery that is delivered to a device external to the battery, what should the internal resistance of the battery be? **(a)** It should be as low as possible. **(b)** It should be as high as possible. **(c)** The percentage does not depend on the internal resistance.

Pitfall Prevention 28.1

What Is Constant in a Battery?
It is a common misconception that a battery is a source of constant current. Equation 28.3 shows that is not true. The current in the circuit depends on the resistance R connected to the battery. It is also not true that a battery is a source of constant terminal voltage as shown by Equation 28.1. **A battery is a source of constant emf.**

Example 28.1   Terminal Voltage of a Battery

A battery has an emf of 12.0 V and an internal resistance of 0.050 0 Ω. Its terminals are connected to a load resistance of 3.00 Ω.

▶ **28.1** continued

(A) Find the current in the circuit and the terminal voltage of the battery.

SOLUTION

Conceptualize Study Figure 28.1a, which shows a circuit consistent with the problem statement. The battery delivers energy to the load resistor.

Categorize This example involves simple calculations from this section, so we categorize it as a substitution problem.

Use Equation 28.3 to find the current in the circuit:

$$I = \frac{\mathcal{E}}{R + r} = \frac{12.0 \text{ V}}{3.00 \ \Omega + 0.050 \ 0 \ \Omega} = \boxed{3.93 \text{ A}}$$

Use Equation 28.1 to find the terminal voltage:

$$\Delta V = \mathcal{E} - Ir = 12.0 \text{ V} - (3.93 \text{ A})(0.050 \ 0 \ \Omega) = \boxed{11.8 \text{ V}}$$

To check this result, calculate the voltage across the load resistance R:

$$\Delta V = IR = (3.93 \text{ A})(3.00 \ \Omega) = 11.8 \text{ V}$$

(B) Calculate the power delivered to the load resistor, the power delivered to the internal resistance of the battery, and the power delivered by the battery.

SOLUTION

Use Equation 27.22 to find the power delivered to the load resistor:

$$P_R = I^2 R = (3.93 \text{ A})^2 (3.00 \ \Omega) = \boxed{46.3 \text{ W}}$$

Find the power delivered to the internal resistance:

$$P_r = I^2 r = (3.93 \text{ A})^2 (0.050 \ 0 \ \Omega) = \boxed{0.772 \text{ W}}$$

Find the power delivered by the battery by adding these quantities:

$$P = P_R + P_r = 46.3 \text{ W} + 0.772 \text{ W} = \boxed{47.1 \text{ W}}$$

WHAT IF? As a battery ages, its internal resistance increases. Suppose the internal resistance of this battery rises to $2.00 \ \Omega$ toward the end of its useful life. How does that alter the battery's ability to deliver energy?

Answer Let's connect the same 3.00-Ω load resistor to the battery.

Find the new current in the battery:

$$I = \frac{\mathcal{E}}{R + r} = \frac{12.0 \text{ V}}{3.00 \ \Omega + 2.00 \ \Omega} = 2.40 \text{ A}$$

Find the new terminal voltage:

$$\Delta V = \mathcal{E} - Ir = 12.0 \text{ V} - (2.40 \text{ A})(2.00 \ \Omega) = 7.2 \text{ V}$$

Find the new powers delivered to the load resistor and internal resistance:

$$P_R = I^2 R = (2.40 \text{ A})^2 (3.00 \ \Omega) = 17.3 \text{ W}$$
$$P_r = I^2 r = (2.40 \text{ A})^2 (2.00 \ \Omega) = 11.5 \text{ W}$$

In this situation, the terminal voltage is only 60% of the emf. Notice that 40% of the power from the battery is delivered to the internal resistance when r is $2.00 \ \Omega$. When r is $0.050 \ 0 \ \Omega$ as in part (B), this percentage is only 1.6%. Consequently, even though the emf remains fixed, the increasing internal resistance of the battery significantly reduces the battery's ability to deliver energy to an external load.

Example 28.2 **Matching the Load**

Find the load resistance R for which the maximum power is delivered to the load resistance in Figure 28.1a.

SOLUTION

Conceptualize Think about varying the load resistance in Figure 28.1a and the effect on the power delivered to the load resistance. When R is large, there is very little current, so the power $I^2 R$ delivered to the load resistor is small.

continued

▶ **28.2** continued

When R is small, let's say $R \ll r$, the current is large and the power delivered to the internal resistance is $I^2 r \gg I^2 R$. Therefore, the power delivered to the load resistor is small compared to that delivered to the internal resistance. For some intermediate value of the resistance R, the power must maximize.

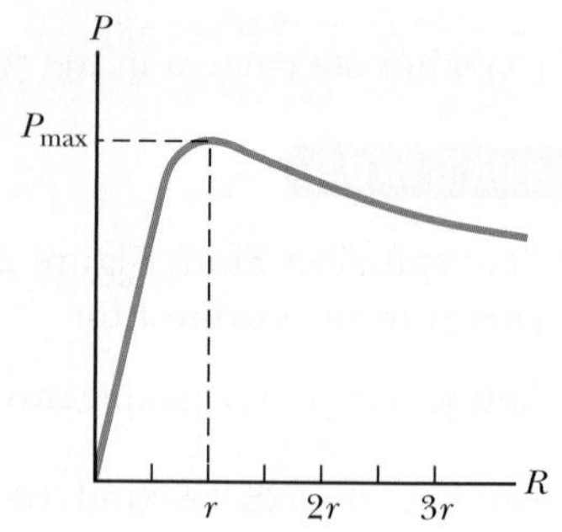

Categorize We categorize this example as an analysis problem because we must undertake a procedure to maximize the power. The circuit is the same as that in Example 28.1. The load resistance R in this case, however, is a variable.

Figure 28.2 (Example 28.2) Graph of the power P delivered by a battery to a load resistor of resistance R as a function of R.

Analyze Find the power delivered to the load resistance using Equation 27.22, with I given by Equation 28.3:

$$(1) \quad P = I^2 R = \frac{\mathbf{\mathcal{E}}^2 R}{(R + r)^2}$$

Differentiate the power with respect to the load resistance R and set the derivative equal to zero to maximize the power:

$$\frac{dP}{dR} = \frac{d}{dR}\left[\frac{\mathbf{\mathcal{E}}^2 R}{(R + r)^2}\right] = \frac{d}{dR}\left[\mathbf{\mathcal{E}}^2 R (R + r)^{-2}\right] = 0$$

$$[\mathbf{\mathcal{E}}^2 (R + r)^{-2}] + [\mathbf{\mathcal{E}}^2 R(-2)(R + r)^{-3}] = 0$$

$$\frac{\mathbf{\mathcal{E}}^2 (R + r)}{(R + r)^3} - \frac{2\mathbf{\mathcal{E}}^2 R}{(R + r)^3} = \frac{\mathbf{\mathcal{E}}^2 (r - R)}{(R + r)^3} = 0$$

Solve for R:

$$R = r$$

Finalize To check this result, let's plot P versus R as in Figure 28.2. The graph shows that P reaches a maximum value at $R = r$. Equation (1) shows that this maximum value is $P_{max} = \mathbf{\mathcal{E}}^2/4r$.

28.2 Resistors in Series and Parallel

When two or more resistors are connected together as are the incandescent lightbulbs in Figure 28.3a, they are said to be in a **series combination.** Figure 28.3b is the circuit diagram for the lightbulbs, shown as resistors, and the battery. What if you wanted to replace the series combination with a single resistor that would draw the same current from the battery? What would be its value? In a series connection, if an amount of charge Q exits resistor R_1, charge Q must also enter the second resistor R_2. Otherwise, charge would accumulate on the wire between the resistors. Therefore, the same amount of charge passes through both resistors in a given time interval and the currents are the same in both resistors:

$$I = I_1 = I_2$$

where I is the current leaving the battery, I_1 is the current in resistor R_1, and I_2 is the current in resistor R_2.

The potential difference applied across the series combination of resistors divides between the resistors. In Figure 28.3b, because the voltage drop[1] from a to b equals $I_1 R_1$ and the voltage drop from b to c equals $I_2 R_2$, the voltage drop from a to c is

$$\Delta V = \Delta V_1 + \Delta V_2 = I_1 R_1 + I_2 R_2$$

The potential difference across the battery is also applied to the **equivalent resistance** R_{eq} in Figure 28.3c:

$$\Delta V = I R_{eq}$$

---

[1]The term *voltage drop* is synonymous with a decrease in electric potential across a resistor. It is often used by individuals working with electric circuits.

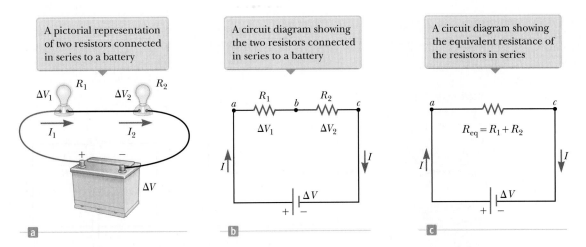

Figure 28.3 Two lightbulbs with resistances R_1 and R_2 connected in series. All three diagrams are equivalent.

where the equivalent resistance has the same effect on the circuit as the series combination because it results in the same current I in the battery. Combining these equations for ΔV gives

$$IR_{eq} = I_1 R_1 + I_2 R_2 \quad \rightarrow \quad R_{eq} = R_1 + R_2 \qquad \textbf{(28.5)}$$

where we have canceled the currents I, I_1, and I_2 because they are all the same. We see that we can replace the two resistors in series with a single equivalent resistance whose value is the *sum* of the individual resistances.

The equivalent resistance of three or more resistors connected in series is

$$R_{eq} = R_1 + R_2 + R_3 + \cdots \qquad \textbf{(28.6)}$$

This relationship indicates that the equivalent resistance of a series combination of resistors is the numerical sum of the individual resistances and is always greater than any individual resistance.

Looking back at Equation 28.3, we see that the denominator of the right-hand side is the simple algebraic sum of the external and internal resistances. That is consistent with the internal and external resistances being in series in Figure 28.1a.

If the filament of one lightbulb in Figure 28.3 were to fail, the circuit would no longer be complete (resulting in an open-circuit condition) and the second lightbulb would also go out. This fact is a general feature of a series circuit: if one device in the series creates an open circuit, all devices are inoperative.

◀ **The equivalent resistance of a series combination of resistors**

Quick **Quiz** 28.2 With the switch in the circuit of Figure 28.4a closed, there is no current in R_2 because the current has an alternate zero-resistance path through the switch. There is current in R_1, and this current is measured with the ammeter (a device for measuring current) at the bottom of the circuit. If the switch is opened (Fig. 28.4b), there is current in R_2. What happens to the reading on the ammeter when the switch is opened? **(a)** The reading goes up. **(b)** The reading goes down. **(c)** The reading does not change.

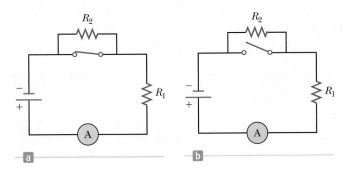

Figure 28.4 (Quick Quiz 28.2) What happens when the switch is opened?

Pitfall Prevention 28.2

Lightbulbs Don't Burn We will describe the end of the life of an incandescent lightbulb by saying *the filament fails* rather than by saying the lightbulb "burns out." The word *burn* suggests a combustion process, which is not what occurs in a lightbulb. The failure of a lightbulb results from the slow sublimation of tungsten from the very hot filament over the life of the lightbulb. The filament eventually becomes very thin because of this process. The mechanical stress from a sudden temperature increase when the lightbulb is turned on causes the thin filament to break.

Pitfall Prevention 28.3

Local and Global Changes A local change in one part of a circuit may result in a global change throughout the circuit. For example, if a single resistor is changed in a circuit containing several resistors and batteries, the currents in all resistors and batteries, the terminal voltages of all batteries, and the voltages across all resistors may change as a result.

Figure 28.5 Two lightbulbs with resistances R_1 and R_2 connected in parallel. All three diagrams are equivalent.

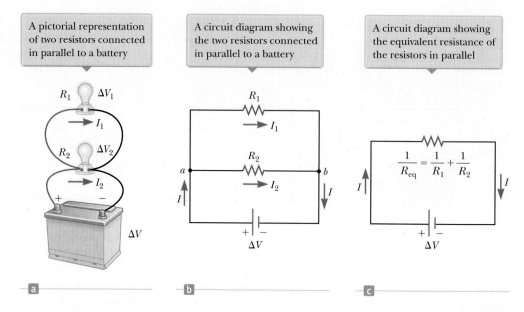

A pictorial representation of two resistors connected in parallel to a battery

A circuit diagram showing the two resistors connected in parallel to a battery

A circuit diagram showing the equivalent resistance of the resistors in parallel

Pitfall Prevention 28.4

Current Does Not Take the Path of Least Resistance You may have heard the phrase "current takes the path of least resistance" (or similar wording) in reference to a parallel combination of current paths such that there are two or more paths for the current to take. Such wording is incorrect. The current takes *all* paths. Those paths with lower resistance have larger currents, but even very high resistance paths carry *some* of the current. In theory, if current has a choice between a zero-resistance path and a finite resistance path, all the current takes the path of zero resistance; a path with zero resistance, however, is an idealization.

Now consider two resistors in a **parallel combination** as shown in Figure 28.5. As with the series combination, what is the value of the single resistor that could replace the combination and draw the same current from the battery? Notice that both resistors are connected directly across the terminals of the battery. Therefore, the potential differences across the resistors are the same:

$$\Delta V = \Delta V_1 = \Delta V_2$$

where ΔV is the terminal voltage of the battery.

When charges reach point a in Figure 28.5b, they split into two parts, with some going toward R_1 and the rest going toward R_2. A **junction** is any such point in a circuit where a current can split. This split results in less current in each individual resistor than the current leaving the battery. Because electric charge is conserved, the current I that enters point a must equal the total current leaving that point:

$$I = I_1 + I_2 = \frac{\Delta V_1}{R_1} + \frac{\Delta V_2}{R_2}$$

where I_1 is the current in R_1 and I_2 is the current in R_2.

The current in the **equivalent resistance** R_{eq} in Figure 28.5c is

$$I = \frac{\Delta V}{R_{eq}}$$

where the equivalent resistance has the same effect on the circuit as the two resistors in parallel; that is, the equivalent resistance draws the same current I from the battery. Combining these equations for I, we see that the equivalent resistance of two resistors in parallel is given by

$$\frac{\Delta V}{R_{eq}} = \frac{\Delta V_1}{R_1} + \frac{\Delta V_2}{R_2} \quad \rightarrow \quad \frac{1}{R_{eq}} = \frac{1}{R_1} + \frac{1}{R_2} \tag{28.7}$$

where we have canceled ΔV, ΔV_1, and ΔV_2 because they are all the same.

An extension of this analysis to three or more resistors in parallel gives

$$\frac{1}{R_{eq}} = \frac{1}{R_1} + \frac{1}{R_2} + \frac{1}{R_3} + \cdots \tag{28.8}$$

▶ The equivalent resistance of a parallel combination of resistors

This expression shows that the inverse of the equivalent resistance of two or more resistors in a parallel combination is equal to the sum of the inverses of the individual resistances. Furthermore, the equivalent resistance is always less than the smallest resistance in the group.

Household circuits are always wired such that the appliances are connected in parallel. Each device operates independently of the others so that if one is switched off, the others remain on. In addition, in this type of connection, all the devices operate on the same voltage.

Let's consider two examples of practical applications of series and parallel circuits. Figure 28.6 illustrates how a three-way incandescent lightbulb is constructed to provide three levels of light intensity.[2] The socket of the lamp is equipped with a three-way switch for selecting different light intensities. The lightbulb contains two filaments. When the lamp is connected to a 120-V source, one filament receives 100 W of power and the other receives 75 W. The three light intensities are made possible by applying the 120 V to one filament alone, to the other filament alone, or to the two filaments in parallel. When switch S_1 is closed and switch S_2 is opened, current exists only in the 75-W filament. When switch S_1 is open and switch S_2 is closed, current exists only in the 100-W filament. When both switches are closed, current exists in both filaments and the total power is 175 W.

If the filaments were connected in series and one of them were to break, no charges could pass through the lightbulb and it would not glow, regardless of the switch position. If, however, the filaments were connected in parallel and one of them (for example, the 75-W filament) were to break, the lightbulb would continue to glow in two of the switch positions because current exists in the other (100-W) filament.

As a second example, consider strings of incandescent lights that are used for many ornamental purposes such as decorating Christmas trees. Over the years, both parallel and series connections have been used for strings of lights. Because series-wired lightbulbs operate with less energy per bulb and at a lower temperature, they are safer than parallel-wired lightbulbs for indoor Christmas-tree use. If, however, the filament of a single lightbulb in a series-wired string were to fail (or if the lightbulb were removed from its socket), all the lights on the string would go out. The popularity of series-wired light strings diminished because troubleshooting a failed lightbulb is a tedious, time-consuming chore that involves trial-and-error substitution of a good lightbulb in each socket along the string until the defective one is found.

In a parallel-wired string, each lightbulb operates at 120 V. By design, the lightbulbs are brighter and hotter than those on a series-wired string. As a result, they are inherently more dangerous (more likely to start a fire, for instance), but if one lightbulb in a parallel-wired string fails or is removed, the rest of the lightbulbs continue to glow.

To prevent the failure of one lightbulb from causing the entire string to go out, a new design was developed for so-called miniature lights wired in series. When the filament breaks in one of these miniature lightbulbs, the break in the filament represents the largest resistance in the series, much larger than that of the intact filaments. As a result, most of the applied 120 V appears across the lightbulb with the broken filament. Inside the lightbulb, a small jumper loop covered by an insulating material is wrapped around the filament leads. When the filament fails and 120 V appears across the lightbulb, an arc burns the insulation on the jumper and connects the filament leads. This connection now completes the circuit through the lightbulb even though its filament is no longer active (Fig. 28.7, page 840).

When a lightbulb fails, the resistance across its terminals is reduced to almost zero because of the alternate jumper connection mentioned in the preceding paragraph. All the other lightbulbs not only stay on, but they glow more brightly because

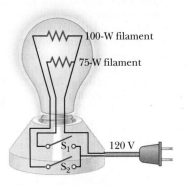

Figure 28.6 A three-way incandescent lightbulb.

---

[2]The three-way lightbulb and other household devices actually operate on alternating current (AC), to be introduced in Chapter 33.

Figure 28.7 (a) Schematic diagram of a modern "miniature" incandescent holiday lightbulb, with a jumper connection to provide a current path if the filament breaks. (b) A holiday lightbulb with a broken filament. (c) A Christmas-tree lightbulb.

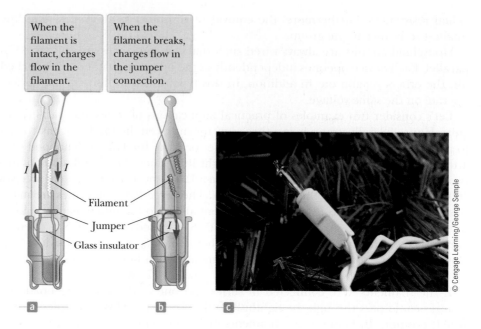

When the filament is intact, charges flow in the filament.

When the filament breaks, charges flow in the jumper connection.

Filament

Jumper

Glass insulator

a   b   c

© Cengage Learning/George Semple

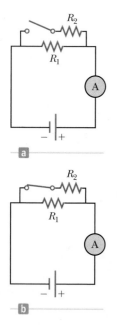

a

b

Figure 28.8 (Quick Quiz 28.3) What happens when the switch is closed?

the total resistance of the string is reduced and consequently the current in each remaining lightbulb increases. Each lightbulb operates at a slightly higher temperature than before. As more lightbulbs fail, the current keeps rising, the filament of each remaining lightbulb operates at a higher temperature, and the lifetime of the lightbulb is reduced. For this reason, you should check for failed (nonglowing) lightbulbs in such a series-wired string and replace them as soon as possible, thereby maximizing the lifetimes of all the lightbulbs.

Q uick Quiz 28.3 With the switch in the circuit of Figure 28.8a open, there is no current in R_2. There is current in R_1, however, and it is measured with the ammeter at the right side of the circuit. If the switch is closed (Fig. 28.8b), there is current in R_2. What happens to the reading on the ammeter when the switch is closed? **(a)** The reading increases. **(b)** The reading decreases. **(c)** The reading does not change.

Q uick Quiz 28.4 Consider the following choices: (a) increases, (b) decreases, (c) remains the same. From these choices, choose the best answer for the following situations. **(i)** In Figure 28.3, a third resistor is added in series with the first two. What happens to the current in the battery? **(ii)** What happens to the terminal voltage of the battery? **(iii)** In Figure 28.5, a third resistor is added in parallel with the first two. What happens to the current in the battery? **(iv)** What happens to the terminal voltage of the battery?

Conceptual Example 28.3   Landscape Lights

A homeowner wishes to install low-voltage landscape lighting in his back yard. To save money, he purchases inexpensive 18-gauge cable, which has a relatively high resistance per unit length. This cable consists of two side-by-side wires separated by insulation, like the cord on an appliance. He runs a 200-foot length of this cable from the power supply to the farthest point at which he plans to position a light fixture. He attaches light fixtures across the two wires on the cable at 10-foot intervals so that the light fixtures are in parallel. Because of the cable's resistance, the brightness of the lightbulbs in the fixtures is not as desired. Which of the following problems does the homeowner have? (a) All the lightbulbs glow equally less brightly than they would if lower-resistance cable had been used. (b) The brightness of the lightbulbs decreases as you move farther from the power supply.

▶ **28.3** continued

SOLUTION

A circuit diagram for the system appears in Figure 28.9. The horizontal resistors with letter subscripts (such as R_A) represent the resistance of the wires in the cable between the light fixtures, and the vertical resistors with number subscripts (such as R_1) represent the resistance of the light fixtures themselves. Part of the terminal voltage of the power supply is dropped across resistors R_A and R_B. Therefore, the voltage across light fixture R_1 is less than the terminal voltage. There is a further voltage drop across resistors R_C and R_D. Consequently, the voltage across light fixture R_2 is smaller than that across R_1. This pattern continues down the line of light fixtures, so the correct choice is (b). Each successive light fixture has a smaller voltage across it and glows less brightly than the one before.

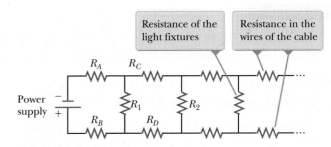

Figure 28.9 (Conceptual Example 28.3) The circuit diagram for a set of landscape light fixtures connected in parallel across the two wires of a two-wire cable.

Example 28.4 Find the Equivalent Resistance

Four resistors are connected as shown in Figure 28.10a.

(A) Find the equivalent resistance between points a and c.

SOLUTION

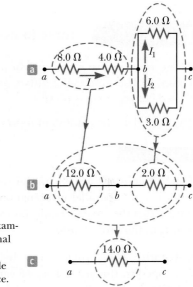

Conceptualize Imagine charges flowing into and through this combination from the left. All charges must pass from a to b through the first two resistors, but the charges split at b into two different paths when encountering the combination of the 6.0-Ω and the 3.0-Ω resistors.

Categorize Because of the simple nature of the combination of resistors in Figure 28.10, we categorize this example as one for which we can use the rules for series and parallel combinations of resistors.

Figure 28.10 (Example 28.4) The original network of resistors is reduced to a single equivalent resistance.

Analyze The combination of resistors can be reduced in steps as shown in Figure 28.10.

Find the equivalent resistance between a and b of the 8.0-Ω and 4.0-Ω resistors, which are in series (left-hand red-brown circles):

$$R_{eq} = 8.0 \ \Omega + 4.0 \ \Omega = 12.0 \ \Omega$$

Find the equivalent resistance between b and c of the 6.0-Ω and 3.0-Ω resistors, which are in parallel (right-hand red-brown circles):

$$\frac{1}{R_{eq}} = \frac{1}{6.0 \ \Omega} + \frac{1}{3.0 \ \Omega} = \frac{3}{6.0 \ \Omega}$$

$$R_{eq} = \frac{6.0 \ \Omega}{3} = 2.0 \ \Omega$$

The circuit of equivalent resistances now looks like Figure 28.10b. The 12.0-Ω and 2.0-Ω resistors are in series (green circles). Find the equivalent resistance from a to c:

$$R_{eq} = 12.0 \ \Omega + 2.0 \ \Omega = \boxed{14.0 \ \Omega}$$

This resistance is that of the single equivalent resistor in Figure 28.10c.

(B) What is the current in each resistor if a potential difference of 42 V is maintained between a and c?

continued

▶ **28.4** continued

SOLUTION

The currents in the 8.0-Ω and 4.0-Ω resistors are the same because they are in series. In addition, they carry the same current that would exist in the 14.0-Ω equivalent resistor subject to the 42-V potential difference.

Use Equation 27.7 ($R = \Delta V/I$) and the result from part (A) to find the current in the 8.0-Ω and 4.0-Ω resistors:

$$I = \frac{\Delta V_{ac}}{R_{eq}} = \frac{42 \text{ V}}{14.0 \text{ } \Omega} = \boxed{3.0 \text{ A}}$$

Set the voltages across the resistors in parallel in Figure 28.10a equal to find a relationship between the currents:

$$\Delta V_1 = \Delta V_2 \rightarrow (6.0 \text{ } \Omega)I_1 = (3.0 \text{ } \Omega)I_2 \rightarrow I_2 = 2I_1$$

Use $I_1 + I_2 = 3.0$ A to find I_1:

$$I_1 + I_2 = 3.0 \text{ A} \rightarrow I_1 + 2I_1 = 3.0 \text{ A} \rightarrow I_1 = \boxed{1.0 \text{ A}}$$

Find I_2:

$$I_2 = 2I_1 = 2(1.0 \text{ A}) = \boxed{2.0 \text{ A}}$$

Finalize As a final check of our results, note that $\Delta V_{bc} = (6.0 \text{ } \Omega)I_1 = (3.0 \text{ } \Omega)I_2 = 6.0$ V and $\Delta V_{ab} = (12.0 \text{ } \Omega)I = 36$ V; therefore, $\Delta V_{ac} = \Delta V_{ab} + \Delta V_{bc} = 42$ V, as it must.

Example 28.5 Three Resistors in Parallel

Three resistors are connected in parallel as shown in Figure 28.11a. A potential difference of 18.0 V is maintained between points a and b.

(A) Calculate the equivalent resistance of the circuit.

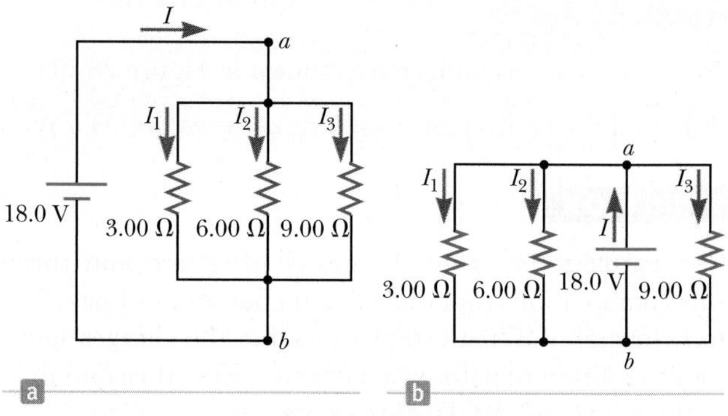

SOLUTION

Conceptualize Figure 28.11a shows that we are dealing with a simple parallel combination of three resistors. Notice that the current I splits into three currents I_1, I_2, and I_3 in the three resistors.

Categorize This problem can be solved with rules developed in this section, so we categorize it as a substitution problem. Because the three resistors are connected in parallel, we can use the rule for resistors in parallel, Equation 28.8, to evaluate the equivalent resistance.

Figure 28.11 (Example 28.5) (a) Three resistors connected in parallel. The voltage across each resistor is 18.0 V. (b) Another circuit with three resistors and a battery. Is it equivalent to the circuit in (a)?

Use Equation 28.8 to find R_{eq}:

$$\frac{1}{R_{eq}} = \frac{1}{3.00 \text{ } \Omega} + \frac{1}{6.00 \text{ } \Omega} + \frac{1}{9.00 \text{ } \Omega} = \frac{11}{18.0 \text{ } \Omega}$$

$$R_{eq} = \frac{18.0 \text{ } \Omega}{11} = \boxed{1.64 \text{ } \Omega}$$

(B) Find the current in each resistor.

SOLUTION

The potential difference across each resistor is 18.0 V. Apply the relationship $\Delta V = IR$ to find the currents:

$$I_1 = \frac{\Delta V}{R_1} = \frac{18.0 \text{ V}}{3.00 \text{ } \Omega} = \boxed{6.00 \text{ A}}$$

$$I_2 = \frac{\Delta V}{R_2} = \frac{18.0 \text{ V}}{6.00 \text{ } \Omega} = \boxed{3.00 \text{ A}}$$

$$I_3 = \frac{\Delta V}{R_3} = \frac{18.0 \text{ V}}{9.00 \text{ } \Omega} = \boxed{2.00 \text{ A}}$$

(C) Calculate the power delivered to each resistor and the total power delivered to the combination of resistors.

▶ **28.5** continued

SOLUTION

Apply the relationship $P = I^2R$ to each resistor using the currents calculated in part (B):

3.00-Ω: $P_1 = I_1^2R_1 = (6.00 \text{ A})^2(3.00 \text{ Ω}) =$ 108 W

6.00-Ω: $P_2 = I_2^2R_2 = (3.00 \text{ A})^2(6.00 \text{ Ω}) =$ 54 W

9.00-Ω: $P_3 = I_3^2R_3 = (2.00 \text{ A})^2(9.00 \text{ Ω}) =$ 36 W

These results show that the smallest resistor receives the most power. Summing the three quantities gives a total power of 198 W. We could have calculated this final result from part (A) by considering the equivalent resistance as follows: $P = (\Delta V)^2/R_{eq} = (18.0 \text{ V})^2/1.64 \text{ Ω} = 198 \text{ W}$.

WHAT IF? What if the circuit were as shown in Figure 28.11b instead of as in Figure 28.11a? How would that affect the calculation?

Answer There would be no effect on the calculation. The physical placement of the battery is not important. Only the electrical arrangement is important. In Figure 28.11b, the battery still maintains a potential difference of 18.0 V between points a and b, so the two circuits in the figure are electrically identical.

28.3 Kirchhoff's Rules

As we saw in the preceding section, combinations of resistors can be simplified and analyzed using the expression $\Delta V = IR$ and the rules for series and parallel combinations of resistors. Very often, however, it is not possible to reduce a circuit to a single loop using these rules. The procedure for analyzing more complex circuits is made possible by using the following two principles, called **Kirchhoff's rules.**

> **1. Junction rule.** At any junction, the sum of the currents must equal zero:
>
> $$\sum_{\text{junction}} I = 0 \qquad \text{(28.9)}$$
>
> **2. Loop rule.** The sum of the potential differences across all elements around any closed circuit loop must be zero:
>
> $$\sum_{\text{closed loop}} \Delta V = 0 \qquad \text{(28.10)}$$

Kirchhoff's first rule is a statement of conservation of electric charge. All charges that enter a given point in a circuit must leave that point because charge cannot build up or disappear at a point. Currents directed into the junction are entered into the sum in the junction rule as $+I$, whereas currents directed out of a junction are entered as $-I$. Applying this rule to the junction in Figure 28.12a gives

$$I_1 - I_2 - I_3 = 0$$

Figure 28.12b represents a mechanical analog of this situation, in which water flows through a branched pipe having no leaks. Because water does not build up anywhere in the pipe, the flow rate into the pipe on the left equals the total flow rate out of the two branches on the right.

Kirchhoff's second rule follows from the law of conservation of energy for an isolated system. Let's imagine moving a charge around a closed loop of a circuit. When the charge returns to the starting point, the charge–circuit system must have the same total energy as it had before the charge was moved. The sum of the increases in energy as the charge passes through some circuit elements must equal the sum of the decreases in energy as it passes through other elements. The potential energy of the system decreases whenever the charge moves through a potential drop $-IR$ across a resistor or whenever it moves in the reverse direction through a

The amount of charge flowing out of the branches on the right must equal the amount flowing into the single branch on the left.

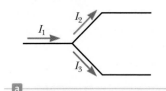

a

The amount of water flowing out of the branches on the right must equal the amount flowing into the single branch on the left.

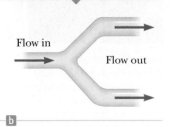

Flow in

Flow out

b

Figure 28.12 (a) Kirchhoff's junction rule. (b) A mechanical analog of the junction rule.

In each diagram, $\Delta V = V_b - V_a$ and the circuit element is traversed from a to b, left to right.

a
$$\Delta V = -IR$$

b
$$\Delta V = +IR$$

c
$$\Delta V = +\mathcal{E}$$

d
$$\Delta V = -\mathcal{E}$$

Figure 28.13 Rules for determining the signs of the potential differences across a resistor and a battery. (The battery is assumed to have no internal resistance.)

source of emf. The potential energy increases whenever the charge passes through a battery from the negative terminal to the positive terminal.

When applying Kirchhoff's second rule, imagine *traveling* around the loop and consider changes in *electric potential* rather than the changes in *potential energy* described in the preceding paragraph. Imagine traveling through the circuit elements in Figure 28.13 toward the right. The following sign conventions apply when using the second rule:

- Charges move from the high-potential end of a resistor toward the low-potential end, so if a resistor is traversed in the direction of the current, the potential difference ΔV across the resistor is $-IR$ (Fig. 28.13a).
- If a resistor is traversed in the direction *opposite* the current, the potential difference ΔV across the resistor is $+IR$ (Fig. 28.13b).
- If a source of emf (assumed to have zero internal resistance) is traversed in the direction of the emf (from negative to positive), the potential difference ΔV is $+\mathcal{E}$ (Fig. 28.13c).
- If a source of emf (assumed to have zero internal resistance) is traversed in the direction opposite the emf (from positive to negative), the potential difference ΔV is $-\mathcal{E}$ (Fig. 28.13d).

There are limits on the number of times you can usefully apply Kirchhoff's rules in analyzing a circuit. You can use the junction rule as often as you need as long as you include in it a current that has not been used in a preceding junction-rule equation. In general, the number of times you can use the junction rule is one fewer than the number of junction points in the circuit. You can apply the loop rule as often as needed as long as a new circuit element (resistor or battery) or a new current appears in each new equation. In general, to solve a particular circuit problem, the number of independent equations you need to obtain from the two rules equals the number of unknown currents.

Complex networks containing many loops and junctions generate a great number of independent linear equations and a correspondingly great number of unknowns. Such situations can be handled formally through the use of matrix algebra. Computer software can also be used to solve for the unknowns.

The following examples illustrate how to use Kirchhoff's rules. In all cases, it is assumed the circuits have reached steady-state conditions; in other words, the currents in the various branches are constant. Any capacitor acts as an open branch in a circuit; that is, the current in the branch containing the capacitor is zero under steady-state conditions.

Gustav Kirchhoff
German Physicist (1824–1887)
Kirchhoff, a professor at Heidelberg, and Robert Bunsen invented the spectroscope and founded the science of spectroscopy, which we shall study in Chapter 42. They discovered the elements cesium and rubidium and invented astronomical spectroscopy.

Problem-Solving Strategy Kirchhoff's Rules

The following procedure is recommended for solving problems that involve circuits that cannot be reduced by the rules for combining resistors in series or parallel.

1. Conceptualize. Study the circuit diagram and make sure you recognize all elements in the circuit. Identify the polarity of each battery and try to imagine the directions in which the current would exist in the batteries.

2. Categorize. Determine whether the circuit can be reduced by means of combining series and parallel resistors. If so, use the techniques of Section 28.2. If not, apply Kirchhoff's rules according to the *Analyze* step below.

3. Analyze. Assign labels to all known quantities and symbols to all unknown quantities. You must assign *directions* to the currents in each part of the circuit. Although the assignment of current directions is arbitrary, you must adhere *rigorously* to the directions you assign when you apply Kirchhoff's rules.

Apply the junction rule (Kirchhoff's first rule) to all junctions in the circuit except one. Now apply the loop rule (Kirchhoff's second rule) to as many loops in

▶ Problem-Solving Strategy continued

the circuit as are needed to obtain, in combination with the equations from the junction rule, as many equations as there are unknowns. To apply this rule, you must choose a direction in which to travel around the loop (either clockwise or counterclockwise) and correctly identify the change in potential as you cross each element. Be careful with signs!

Solve the equations simultaneously for the unknown quantities.

4. Finalize. Check your numerical answers for consistency. Do not be alarmed if any of the resulting currents have a negative value. That only means you have guessed the direction of that current incorrectly, but *its magnitude will be correct*.

Example 28.6 A Single-Loop Circuit

A single-loop circuit contains two resistors and two batteries as shown in Figure 28.14. (Neglect the internal resistances of the batteries.) Find the current in the circuit.

SOLUTION

Conceptualize Figure 28.14 shows the polarities of the batteries and a guess at the direction of the current. The 12-V battery is the stronger of the two, so the current should be counterclockwise. Therefore, we expect our guess for the direction of the current to be wrong, but we will continue and see how this incorrect guess is represented by our final answer.

Categorize We do not need Kirchhoff's rules to analyze this simple circuit, but let's use them anyway simply to see how they are applied. There are no junctions in this single-loop circuit; therefore, the current is the same in all elements.

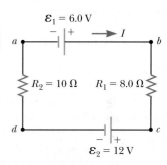

Figure 28.14 (Example 28.6) A series circuit containing two batteries and two resistors, where the polarities of the batteries are in opposition.

Analyze Let's assume the current is clockwise as shown in Figure 28.14. Traversing the circuit in the clockwise direction, starting at a, we see that $a \rightarrow b$ represents a potential difference of $+\mathcal{E}_1$, $b \rightarrow c$ represents a potential difference of $-IR_1$, $c \rightarrow d$ represents a potential difference of $-\mathcal{E}_2$, and $d \rightarrow a$ represents a potential difference of $-IR_2$.

Apply Kirchhoff's loop rule to the single loop in the circuit:

$$\sum \Delta V = 0 \quad \rightarrow \quad \mathcal{E}_1 - IR_1 - \mathcal{E}_2 - IR_2 = 0$$

Solve for I and use the values given in Figure 28.14:

$$(1) \quad I = \frac{\mathcal{E}_1 - \mathcal{E}_2}{R_1 + R_2} = \frac{6.0 \text{ V} - 12 \text{ V}}{8.0 \ \Omega + 10 \ \Omega} = \boxed{-0.33 \text{ A}}$$

Finalize The negative sign for I indicates that the direction of the current is opposite the assumed direction. The emfs in the numerator subtract because the batteries in Figure 28.14 have opposite polarities. The resistances in the denominator add because the two resistors are in series.

WHAT IF? What if the polarity of the 12.0-V battery were reversed? How would that affect the circuit?

Answer Although we could repeat the Kirchhoff's rules calculation, let's instead examine Equation (1) and modify it accordingly. Because the polarities of the two batteries are now in the same direction, the signs of $\mathcal{E}_1$ and $\mathcal{E}_2$ are the same and Equation (1) becomes

$$I = \frac{\mathcal{E}_1 + \mathcal{E}_2}{R_1 + R_2} = \frac{6.0 \text{ V} + 12 \text{ V}}{8.0 \ \Omega + 10 \ \Omega} = 1.0 \text{ A}$$

Example 28.7 A Multiloop Circuit

Find the currents I_1, I_2, and I_3 in the circuit shown in Figure 28.15 on page 846.

continued

▶ **28.7** continued

SOLUTION

Conceptualize Imagine physically rearranging the circuit while keeping it electrically the same. Can you rearrange it so that it consists of simple series or parallel combinations of resistors? You should find that you cannot. (If the 10.0-V battery were removed and replaced by a wire from *b* to the 6.0-Ω resistor, the circuit would consist of only series and parallel combinations.)

Categorize We cannot simplify the circuit by the rules associated with combining resistances in series and in parallel. Therefore, this problem is one in which we must use Kirchhoff's rules.

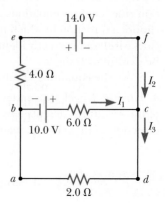

Figure 28.15 (Example 28.7) A circuit containing different branches.

Analyze We arbitrarily choose the directions of the currents as labeled in Figure 28.15.

Apply Kirchhoff's junction rule to junction *c*:

(1) $I_1 + I_2 - I_3 = 0$

We now have one equation with three unknowns: I_1, I_2, and I_3. There are three loops in the circuit: *abcda*, *befcb*, and *aefda*. We need only two loop equations to determine the unknown currents. (The third equation would give no new information.) Let's choose to traverse these loops in the clockwise direction. Apply Kirchhoff's loop rule to loops *abcda* and *befcb*:

abcda: (2) $10.0\text{ V} - (6.0\ \Omega)I_1 - (2.0\ \Omega)I_3 = 0$

befcb: $-(4.0\ \Omega)I_2 - 14.0\text{ V} + (6.0\ \Omega)I_1 - 10.0\text{ V} = 0$

(3) $-24.0\text{ V} + (6.0\ \Omega)I_1 - (4.0\ \Omega)I_2 = 0$

Solve Equation (1) for I_3 and substitute into Equation (2):

$10.0\text{ V} - (6.0\ \Omega)I_1 - (2.0\ \Omega)(I_1 + I_2) = 0$

(4) $10.0\text{ V} - (8.0\ \Omega)I_1 - (2.0\ \Omega)I_2 = 0$

Multiply each term in Equation (3) by 4 and each term in Equation (4) by 3:

(5) $-96.0\text{ V} + (24.0\ \Omega)I_1 - (16.0\ \Omega)I_2 = 0$

(6) $30.0\text{ V} - (24.0\ \Omega)I_1 - (6.0\ \Omega)I_2 = 0$

Add Equation (6) to Equation (5) to eliminate I_1 and find I_2:

$-66.0\text{ V} - (22.0\ \Omega)I_2 = 0$

$I_2 = \boxed{-3.0\text{ A}}$

Use this value of I_2 in Equation (3) to find I_1:

$-24.0\text{ V} + (6.0\ \Omega)I_1 - (4.0\ \Omega)(-3.0\text{ A}) = 0$

$-24.0\text{ V} + (6.0\ \Omega)I_1 + 12.0\text{ V} = 0$

$I_1 = \boxed{2.0\text{ A}}$

Use Equation (1) to find I_3:

$I_3 = I_1 + I_2 = 2.0\text{ A} - 3.0\text{ A} = \boxed{-1.0\text{ A}}$

Finalize Because our values for I_2 and I_3 are negative, the directions of these currents are opposite those indicated in Figure 28.15. The numerical values for the currents are correct. Despite the incorrect direction, we *must* continue to use these negative values in subsequent calculations because our equations were established with our original choice of direction. What would have happened had we left the current directions as labeled in Figure 28.15 but traversed the loops in the opposite direction?

28.4 RC Circuits

So far, we have analyzed direct-current circuits in which the current is constant. In DC circuits containing capacitors, the current is always in the same direction but may vary in magnitude at different times. A circuit containing a series combination of a resistor and a capacitor is called an **RC circuit.**

Charging a Capacitor

Figure 28.16 shows a simple series *RC* circuit. Let's assume the capacitor in this circuit is initially uncharged. There is no current while the switch is open (Fig. 28.16a). If the switch is thrown to position *a* at $t = 0$ (Fig. 28.16b), however, charge begins to flow, setting up a current in the circuit, and the capacitor begins to charge.[3] Notice that during charging, charges do not jump across the capacitor plates because the gap between the plates represents an open circuit. Instead, charge is transferred between each plate and its connecting wires due to the electric field established in the wires by the battery until the capacitor is fully charged. As the plates are being charged, the potential difference across the capacitor increases. The value of the maximum charge on the plates depends on the voltage of the battery. Once the maximum charge is reached, the current in the circuit is zero because the potential difference across the capacitor matches that supplied by the battery.

To analyze this circuit quantitatively, let's apply Kirchhoff's loop rule to the circuit after the switch is thrown to position *a*. Traversing the loop in Figure 28.16b clockwise gives

$$\mathcal{E} - \frac{q}{C} - iR = 0 \qquad \textbf{(28.11)}$$

where q/C is the potential difference across the capacitor and iR is the potential difference across the resistor. We have used the sign conventions discussed earlier for the signs on $\mathcal{E}$ and iR. The capacitor is traversed in the direction from the positive plate to the negative plate, which represents a decrease in potential. Therefore, we use a negative sign for this potential difference in Equation 28.11. Note that lowercase q and i are *instantaneous* values that depend on time (as opposed to steady-state values) as the capacitor is being charged.

We can use Equation 28.11 to find the initial current I_i in the circuit and the maximum charge Q_{max} on the capacitor. At the instant the switch is thrown to position *a* ($t = 0$), the charge on the capacitor is zero. Equation 28.11 shows that the initial current I_i in the circuit is a maximum and is given by

$$I_i = \frac{\mathcal{E}}{R} \qquad \text{(current at } t = 0\text{)} \qquad \textbf{(28.12)}$$

At this time, the potential difference from the battery terminals appears entirely across the resistor. Later, when the capacitor is charged to its maximum value Q_{max}, charges cease to flow, the current in the circuit is zero, and the potential difference from the battery terminals appears entirely across the capacitor. Substituting $i = 0$ into Equation 28.11 gives the maximum charge on the capacitor:

$$Q_{max} = C\mathcal{E} \qquad \text{(maximum charge)} \qquad \textbf{(28.13)}$$

To determine analytical expressions for the time dependence of the charge and current, we must solve Equation 28.11, a single equation containing two variables q and i. The current in all parts of the series circuit must be the same. Therefore, the current in the resistance R must be the same as the current between each capacitor plate and the wire connected to it. This current is equal to the time rate of change of the charge on the capacitor plates. Therefore, we substitute $i = dq/dt$ into Equation 28.11 and rearrange the equation:

$$\frac{dq}{dt} = \frac{\mathcal{E}}{R} - \frac{q}{RC}$$

To find an expression for q, we solve this separable differential equation as follows. First combine the terms on the right-hand side:

$$\frac{dq}{dt} = \frac{C\mathcal{E}}{RC} - \frac{q}{RC} = -\frac{q - C\mathcal{E}}{RC}$$

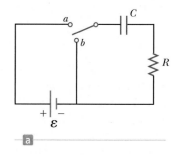

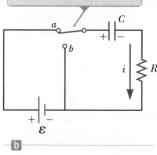

When the switch is thrown to position *a*, the capacitor begins to charge up.

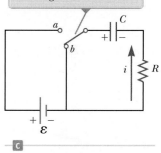

When the switch is thrown to position *b*, the capacitor discharges.

Figure 28.16 A capacitor in series with a resistor, switch, and battery.

---

[3]In previous discussions of capacitors, we assumed a steady-state situation, in which no current was present in any branch of the circuit containing a capacitor. Now we are considering the case *before* the steady-state condition is realized; in this situation, charges are moving and a current exists in the wires connected to the capacitor.

Multiply this equation by dt and divide by $q - C\mathcal{E}$:

$$\frac{dq}{q - C\mathcal{E}} = -\frac{1}{RC}\,dt$$

Integrate this expression, using $q = 0$ at $t = 0$:

$$\int_0^q \frac{dq}{q - C\mathcal{E}} = -\frac{1}{RC}\int_0^t dt$$

$$\ln\left(\frac{q - C\mathcal{E}}{-C\mathcal{E}}\right) = -\frac{t}{RC}$$

From the definition of the natural logarithm, we can write this expression as

▶ **Charge as a function of time for a capacitor being charged**

$$q(t) = C\mathcal{E}(1 - e^{-t/RC}) = Q_{max}(1 - e^{-t/RC}) \tag{28.14}$$

where e is the base of the natural logarithm and we have made the substitution from Equation 28.13.

We can find an expression for the charging current by differentiating Equation 28.14 with respect to time. Using $i = dq/dt$, we find that

▶ **Current as a function of time for a capacitor being charged**

$$i(t) = \frac{\mathcal{E}}{R}\,e^{-t/RC} \tag{28.15}$$

Plots of capacitor charge and circuit current versus time are shown in Figure 28.17. Notice that the charge is zero at $t = 0$ and approaches the maximum value $C\mathcal{E}$ as $t \to \infty$. The current has its maximum value $I_i = \mathcal{E}/R$ at $t = 0$ and decays exponentially to zero as $t \to \infty$. The quantity RC, which appears in the exponents of Equations 28.14 and 28.15, is called the **time constant** τ of the circuit:

$$\tau = RC \tag{28.16}$$

The time constant represents the time interval during which the current decreases to $1/e$ of its initial value; that is, after a time interval τ, the current decreases to $i = e^{-1}I_i = 0.368I_i$. After a time interval 2τ, the current decreases to $i = e^{-2}I_i = 0.135I_i$, and so forth. Likewise, in a time interval τ, the charge increases from zero to $C\mathcal{E}[1 - e^{-1}] = 0.632C\mathcal{E}$.

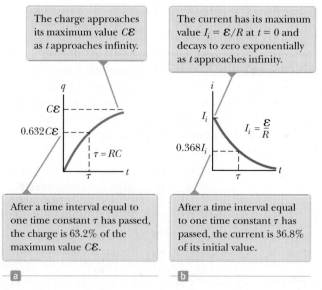

The charge approaches its maximum value $C\mathcal{E}$ as t approaches infinity.

The current has its maximum value $I_i = \mathcal{E}/R$ at $t = 0$ and decays to zero exponentially as t approaches infinity.

After a time interval equal to one time constant τ has passed, the charge is 63.2% of the maximum value $C\mathcal{E}$.

After a time interval equal to one time constant τ has passed, the current is 36.8% of its initial value.

Figure 28.17 (a) Plot of capacitor charge versus time for the circuit shown in Figure 28.16b. (b) Plot of current versus time for the circuit shown in Figure 28.16b.

The following dimensional analysis shows that τ has units of time:

$$[\tau] = [RC] = \left[\left(\frac{\Delta V}{I}\right)\left(\frac{Q}{\Delta V}\right)\right] = \left[\frac{Q}{Q/\Delta t}\right] = [\Delta t] = T$$

Because $\tau = RC$ has units of time, the combination t/RC is dimensionless, as it must be to be an exponent of e in Equations 28.14 and 28.15.

The energy supplied by the battery during the time interval required to fully charge the capacitor is $Q_{max}\mathcal{E} = C\mathcal{E}^2$. After the capacitor is fully charged, the energy stored in the capacitor is $\frac{1}{2}Q_{max}\mathcal{E} = \frac{1}{2}C\mathcal{E}^2$, which is only half the energy output of the battery. It is left as a problem (Problem 68) to show that the remaining half of the energy supplied by the battery appears as internal energy in the resistor.

Discharging a Capacitor

Imagine that the capacitor in Figure 28.16b is completely charged. An initial potential difference Q_i/C exists across the capacitor, and there is zero potential difference across the resistor because $i = 0$. If the switch is now thrown to position b at $t = 0$ (Fig. 28.16c), the capacitor begins to discharge through the resistor. At some time t during the discharge, the current in the circuit is i and the charge on the capacitor is q. The circuit in Figure 28.16c is the same as the circuit in Figure 28.16b except for the absence of the battery. Therefore, we eliminate the emf $\mathcal{E}$ from Equation 28.11 to obtain the appropriate loop equation for the circuit in Figure 28.16c:

$$-\frac{q}{C} - iR = 0 \tag{28.17}$$

When we substitute $i = dq/dt$ into this expression, it becomes

$$-R\frac{dq}{dt} = \frac{q}{C}$$

$$\frac{dq}{q} = -\frac{1}{RC}\,dt$$

Integrating this expression using $q = Q_i$ at $t = 0$ gives

$$\int_{Q_i}^{q} \frac{dq}{q} = -\frac{1}{RC}\int_0^t dt$$

$$\ln\left(\frac{q}{Q_i}\right) = -\frac{t}{RC}$$

$$q(t) = Q_i e^{-t/RC} \tag{28.18}$$

◀ **Charge as a function of time for a discharging capacitor**

Differentiating Equation 28.18 with respect to time gives the instantaneous current as a function of time:

$$i(t) = -\frac{Q_i}{RC}\,e^{-t/RC} \tag{28.19}$$

◀ **Current as a function of time for a discharging capacitor**

where $Q_i/RC = I_i$ is the initial current. The negative sign indicates that as the capacitor discharges, the current direction is opposite its direction when the capacitor was being charged. (Compare the current directions in Figs. 28.16b and 28.16c.) Both the charge on the capacitor and the current decay exponentially at a rate characterized by the time constant $\tau = RC$.

Quick Quiz 28.5 Consider the circuit in Figure 28.18 and assume the battery has no internal resistance. **(i)** Just after the switch is closed, what is the current in the battery? (a) 0 (b) $\mathcal{E}/2R$ (c) $2\mathcal{E}/R$ (d) $\mathcal{E}/R$ (e) impossible to determine **(ii)** After a very long time, what is the current in the battery? Choose from the same choices.

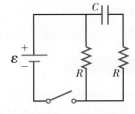

Figure 28.18 (Quick Quiz 28.5) How does the current vary after the switch is closed?

Conceptual Example 28.8   Intermittent Windshield Wipers

Many automobiles are equipped with windshield wipers that can operate intermittently during a light rainfall. How does the operation of such wipers depend on the charging and discharging of a capacitor?

SOLUTION

The wipers are part of an *RC* circuit whose time constant can be varied by selecting different values of *R* through a multiposition switch. As the voltage across the capacitor increases, the capacitor reaches a point at which it discharges and triggers the wipers. The circuit then begins another charging cycle. The time interval between the individual sweeps of the wipers is determined by the value of the time constant.

Example 28.9   Charging a Capacitor in an *RC* Circuit

An uncharged capacitor and a resistor are connected in series to a battery as shown in Figure 28.16, where $\mathcal{E} = 12.0$ V, $C = 5.00\ \mu$F, and $R = 8.00 \times 10^5\ \Omega$. The switch is thrown to position *a*. Find the time constant of the circuit, the maximum charge on the capacitor, the maximum current in the circuit, and the charge and current as functions of time.

SOLUTION

Conceptualize Study Figure 28.16 and imagine throwing the switch to position *a* as shown in Figure 28.16b. Upon doing so, the capacitor begins to charge.

Categorize We evaluate our results using equations developed in this section, so we categorize this example as a substitution problem.

Evaluate the time constant of the circuit from Equation 28.16:

$$\tau = RC = (8.00 \times 10^5\ \Omega)(5.00 \times 10^{-6}\ \text{F}) = \boxed{4.00\ \text{s}}$$

Evaluate the maximum charge on the capacitor from Equation 28.13:

$$Q_{max} = C\mathcal{E} = (5.00\ \mu\text{F})(12.0\ \text{V}) = \boxed{60.0\ \mu\text{C}}$$

Evaluate the maximum current in the circuit from Equation 28.12:

$$I_i = \frac{\mathcal{E}}{R} = \frac{12.0\ \text{V}}{8.00 \times 10^5\ \Omega} = \boxed{15.0\ \mu\text{A}}$$

Use these values in Equations 28.14 and 28.15 to find the charge and current as functions of time:

(1) $q(t) = \boxed{60.0(1 - e^{-t/4.00})}$

(2) $i(t) = \boxed{15.0e^{-t/4.00}}$

In Equations (1) and (2), *q* is in microcoulombs, *i* is in microamperes, and *t* is in seconds.

Example 28.10   Discharging a Capacitor in an *RC* Circuit

Consider a capacitor of capacitance *C* that is being discharged through a resistor of resistance *R* as shown in Figure 28.16c.

(A) After how many time constants is the charge on the capacitor one-fourth its initial value?

SOLUTION

Conceptualize Study Figure 28.16 and imagine throwing the switch to position *b* as shown in Figure 28.16c. Upon doing so, the capacitor begins to discharge.

Categorize We categorize the example as one involving a discharging capacitor and use the appropriate equations.

▶ **28.10** continued

Analyze Substitute $q(t) = Q_i/4$ into Equation 28.18:

$$\frac{Q_i}{4} = Q_i e^{-t/RC}$$

$$\tfrac{1}{4} = e^{-t/RC}$$

Take the logarithm of both sides of the equation and solve for t:

$$-\ln 4 = -\frac{t}{RC}$$

$$t = RC\ln 4 = 1.39RC = \boxed{1.39\tau}$$

(B) The energy stored in the capacitor decreases with time as the capacitor discharges. After how many time constants is this stored energy one-fourth its initial value?

SOLUTION

Use Equations 26.11 and 28.18 to express the energy stored in the capacitor at any time t:

$$(1) \quad U(t) = \frac{q^2}{2C} = \frac{Q_i^2}{2C} e^{-2t/RC}$$

Substitute $U(t) = \tfrac{1}{4}(Q_i^2/2C)$ into Equation (1):

$$\tfrac{1}{4}\frac{Q_i^2}{2C} = \frac{Q_i^2}{2C} e^{-2t/RC}$$

$$\tfrac{1}{4} = e^{-2t/RC}$$

Take the logarithm of both sides of the equation and solve for t:

$$-\ln 4 = -\frac{2t}{RC}$$

$$t = \tfrac{1}{2}RC\ln 4 = 0.693RC = \boxed{0.693\tau}$$

Finalize Notice that because the energy depends on the square of the charge, the energy in the capacitor drops more rapidly than the charge on the capacitor.

WHAT IF? What if you want to describe the circuit in terms of the time interval required for the charge to fall to one-half its original value rather than by the time constant τ? That would give a parameter for the circuit called its *half-life $t_{1/2}$*. How is the half-life related to the time constant?

Answer In one half-life, the charge falls from Q_i to $Q_i/2$. Therefore, from Equation 28.18,

$$\frac{Q_i}{2} = Q_i e^{-t_{1/2}/RC} \quad \rightarrow \quad \tfrac{1}{2} = e^{-t_{1/2}/RC}$$

which leads to

$$t_{1/2} = 0.693\tau$$

The concept of half-life will be important to us when we study nuclear decay in Chapter 44. The radioactive decay of an unstable sample behaves in a mathematically similar manner to a discharging capacitor in an *RC* circuit.

Example 28.11   **Energy Delivered to a Resistor**   AM

A 5.00-μF capacitor is charged to a potential difference of 800 V and then discharged through a resistor. How much energy is delivered to the resistor in the time interval required to fully discharge the capacitor?

SOLUTION

Conceptualize In Example 28.10, we considered the energy decrease in a discharging capacitor to a value of one-fourth the initial energy. In this example, the capacitor fully discharges.

Categorize We solve this example using two approaches. The first approach is to model the circuit as an *isolated system* for *energy*. Because energy in an isolated system is conserved, the initial electric potential energy U_E stored in the

continued

▶ **28.11** continued

capacitor is transformed into internal energy $E_{int} = E_R$ in the resistor. The second approach is to model the resistor as a *nonisolated system* for *energy*. Energy enters the resistor by electrical transmission from the capacitor, causing an increase in the resistor's internal energy.

Analyze We begin with the isolated system approach.

Write the appropriate reduction of the conservation of energy equation, Equation 8.2:

$$\Delta U + \Delta E_{int} = 0$$

Substitute the initial and final values of the energies:

$$(0 - U_E) + (E_{int} - 0) = 0 \;\; \rightarrow \;\; E_R = U_E$$

Use Equation 26.11 for the electric potential energy in the capacitor:

$$E_R = \tfrac{1}{2}C\mathcal{E}^2$$

Substitute numerical values:

$$E_R = \tfrac{1}{2}(5.00 \times 10^{-6}\,\text{F})(800\,\text{V})^2 = \boxed{1.60\,\text{J}}$$

The second approach, which is more difficult but perhaps more instructive, is to note that as the capacitor discharges through the resistor, the rate at which energy is delivered to the resistor by electrical transmission is i^2R, where i is the instantaneous current given by Equation 28.19.

Evaluate the energy delivered to the resistor by integrating the power over all time because it takes an infinite time interval for the capacitor to completely discharge:

$$P = \frac{dE}{dt} \;\; \rightarrow \;\; E_R = \int_0^\infty P\,dt$$

Substitute for the power delivered to the resistor:

$$E_R = \int_0^\infty i^2R\,dt$$

Substitute for the current from Equation 28.19:

$$E_R = \int_0^\infty \left(-\frac{Q_i}{RC}\,e^{-t/RC}\right)^2 R\,dt = \frac{Q_i^2}{RC^2}\int_0^\infty e^{-2t/RC}\,dt = \frac{\mathcal{E}^2}{R}\int_0^\infty e^{-2t/RC}\,dt$$

Substitute the value of the integral, which is $RC/2$ (see Problem 44):

$$E_R = \frac{\mathcal{E}^2}{R}\left(\frac{RC}{2}\right) = \tfrac{1}{2}C\mathcal{E}^2$$

Finalize This result agrees with that obtained using the isolated system approach, as it must. We can use this second approach to find the total energy delivered to the resistor at *any* time after the switch is closed by simply replacing the upper limit in the integral with that specific value of t.

28.5 Household Wiring and Electrical Safety

Many considerations are important in the design of an electrical system of a home that will provide adequate electrical service for the occupants while maximizing their safety. We discuss some aspects of a home electrical system in this section.

Household Wiring

Household circuits represent a practical application of some of the ideas presented in this chapter. In our world of electrical appliances, it is useful to understand the power requirements and limitations of conventional electrical systems and the safety measures that prevent accidents.

In a conventional installation, the utility company distributes electric power to individual homes by means of a pair of wires, with each home connected in parallel to these wires. One wire is called the *live wire*[4] as illustrated in Figure 28.19, and the other is called the *neutral wire*. The neutral wire is grounded; that is, its electric potential is taken to be zero. The potential difference between the live and neutral wires is approximately 120 V. This voltage alternates in time, and the potential of the live wire oscillates relative to ground. Much of what we have learned so far for the constant-emf situation (direct current) can also be applied to the alternating current that power companies supply to businesses and households. (Alternating voltage and current are discussed in Chapter 33.)

To record a household's energy consumption, a meter is connected in series with the live wire entering the house. After the meter, the wire splits so that there are several separate circuits in parallel distributed throughout the house. Each circuit contains a circuit breaker (or, in older installations, a fuse). A circuit breaker is a special switch that opens if the current exceeds the rated value for the circuit breaker. The wire and circuit breaker for each circuit are carefully selected to meet the current requirements for that circuit. If a circuit is to carry currents as large as 30 A, a heavy wire and an appropriate circuit breaker must be selected to handle this current. A circuit used to power only lamps and small appliances often requires only 20 A. Each circuit has its own circuit breaker to provide protection for that part of the entire electrical system of the house.

As an example, consider a circuit in which a toaster oven, a microwave oven, and a coffee maker are connected (corresponding to R_1, R_2, and R_3 in Fig. 28.19). We can calculate the current in each appliance by using the expression $P = I\,\Delta V$. The toaster oven, rated at 1 000 W, draws a current of 1 000 W/120 V = 8.33 A. The microwave oven, rated at 1 300 W, draws 10.8 A, and the coffee maker, rated at 800 W, draws 6.67 A. When the three appliances are operated simultaneously, they draw a total current of 25.8 A. Therefore, the circuit must be wired to handle at least this much current. If the rating of the circuit breaker protecting the circuit is too small—say, 20 A—the breaker will be tripped when the third appliance is turned on, preventing all three appliances from operating. To avoid this situation, the toaster oven and coffee maker can be operated on one 20-A circuit and the microwave oven on a separate 20-A circuit.

Many heavy-duty appliances such as electric ranges and clothes dryers require 240 V for their operation. The power company supplies this voltage by providing a third wire that is 120 V below ground potential (Fig. 28.20). The potential difference between this live wire and the other live wire (which is 120 V above ground potential) is 240 V. An appliance that operates from a 240-V line requires half as much current compared with operating it at 120 V; therefore, smaller wires can be used in the higher-voltage circuit without overheating.

Electrical Safety

When the live wire of an electrical outlet is connected directly to ground, the circuit is completed and a *short-circuit condition* exists. A short circuit occurs when almost zero resistance exists between two points at different potentials, and the result is a very large current. When that happens accidentally, a properly operating circuit breaker opens the circuit and no damage is done. A person in contact with ground, however, can be electrocuted by touching the live wire of a frayed cord or other exposed conductor. An exceptionally effective (and dangerous!) ground contact is made when the person either touches a water pipe (normally at ground potential) or stands on the ground with wet feet. The latter situation represents effective ground contact because normal, nondistilled water is a conductor due to the large number of ions associated with impurities. This situation should be avoided at all cost.

Figure 28.19 Wiring diagram for a household circuit. The resistances represent appliances or other electrical devices that operate with an applied voltage of 120 V.

© Cengage Learning/George Semple

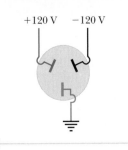

Figure 28.20 (a) An outlet for connection to a 240-V supply. (b) The connections for each of the openings in a 240-V outlet.

---

[4]*Live wire* is a common expression for a conductor whose electric potential is above or below ground potential.

Figure 28.21 (a) A diagram of the circuit for an electric drill with only two connecting wires. The normal current path is from the live wire through the motor connections and back to ground through the neutral wire. (b) This shock can be avoided by connecting the drill case to ground through a third ground wire. The wire colors represent electrical standards in the United States: the "hot" wire is black, the ground wire is green, and the neutral wire is white (shown as gray in the figure).

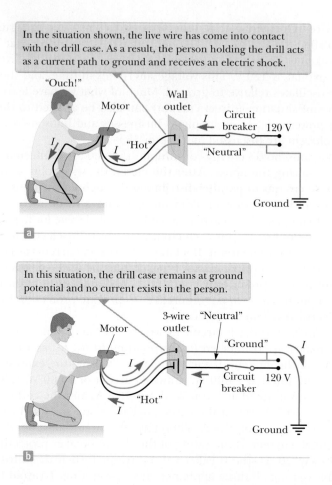

In the situation shown, the live wire has come into contact with the drill case. As a result, the person holding the drill acts as a current path to ground and receives an electric shock.

In this situation, the drill case remains at ground potential and no current exists in the person.

Electric shock can result in fatal burns or can cause the muscles of vital organs such as the heart to malfunction. The degree of damage to the body depends on the magnitude of the current, the length of time it acts, the part of the body touched by the live wire, and the part of the body in which the current exists. Currents of 5 mA or less cause a sensation of shock, but ordinarily do little or no damage. If the current is larger than about 10 mA, the muscles contract and the person may be unable to release the live wire. If the body carries a current of about 100 mA for only a few seconds, the result can be fatal. Such a large current paralyzes the respiratory muscles and prevents breathing. In some cases, currents of approximately 1 A can produce serious (and sometimes fatal) burns. In practice, no contact with live wires is regarded as safe whenever the voltage is greater than 24 V.

Many 120-V outlets are designed to accept a three-pronged power cord. (This feature is required in all new electrical installations.) One of these prongs is the live wire at a nominal potential of 120 V. The second is the neutral wire, nominally at 0 V, which carries current to ground. Figure 28.21a shows a connection to an electric drill with only these two wires. If the live wire accidentally makes contact with the casing of the electric drill (which can occur if the wire insulation wears off), current can be carried to ground by way of the person, resulting in an electric shock. The third wire in a three-pronged power cord, the round prong, is a safety ground wire that normally carries no current. It is both grounded and connected directly to the casing of the appliance. If the live wire is accidentally shorted to the casing in this situation, most of the current takes the low-resistance path through the appliance to ground as shown in Figure 28.21b.

Special power outlets called *ground-fault circuit interrupters*, or GFCIs, are used in kitchens, bathrooms, basements, exterior outlets, and other hazardous areas of homes. These devices are designed to protect persons from electric shock by sensing small currents (< 5 mA) leaking to ground. (The principle of their operation is described in Chapter 31.) When an excessive leakage current is detected, the current is shut off in less than 1 ms.

Summary

Definition

The **emf** of a battery is equal to the voltage across its terminals when the current is zero. That is, the emf is equivalent to the **open-circuit voltage** of the battery.

Concepts and Principles

The **equivalent resistance** of a set of resistors connected in a **series combination** is

$$R_{eq} = R_1 + R_2 + R_3 + \cdots \quad \textbf{(28.6)}$$

The **equivalent resistance** of a set of resistors connected in a **parallel combination** is found from the relationship

$$\frac{1}{R_{eq}} = \frac{1}{R_1} + \frac{1}{R_2} + \frac{1}{R_3} + \cdots \quad \textbf{(28.8)}$$

Circuits involving more than one loop are conveniently analyzed with the use of **Kirchhoff's rules:**

1. **Junction rule.** At any junction, the sum of the currents must equal zero:

$$\sum_{\text{junction}} I = 0 \quad \textbf{(28.9)}$$

2. **Loop rule.** The sum of the potential differences across all elements around any circuit loop must be zero:

$$\sum_{\text{closed loop}} \Delta V = 0 \quad \textbf{(28.10)}$$

When a resistor is traversed in the direction of the current, the potential difference ΔV across the resistor is $-IR$. When a resistor is traversed in the direction opposite the current, $\Delta V = +IR$. When a source of emf is traversed in the direction of the emf (negative terminal to positive terminal), the potential difference is $+\mathcal{E}$. When a source of emf is traversed opposite the emf (positive to negative), the potential difference is $-\mathcal{E}$.

If a capacitor is charged with a battery through a resistor of resistance R, the charge on the capacitor and the current in the circuit vary in time according to the expressions

$$q(t) = Q_{max}(1 - e^{-t/RC}) \quad \textbf{(28.14)}$$

$$i(t) = \frac{\mathcal{E}}{R} e^{-t/RC} \quad \textbf{(28.15)}$$

where $Q_{max} = C\mathcal{E}$ is the maximum charge on the capacitor. The product RC is called the **time constant** τ of the circuit.

If a charged capacitor of capacitance C is discharged through a resistor of resistance R, the charge and current decrease exponentially in time according to the expressions

$$q(t) = Q_i e^{-t/RC} \quad \textbf{(28.18)}$$

$$i(t) = -\frac{Q_i}{RC} e^{-t/RC} \quad \textbf{(28.19)}$$

where Q_i is the initial charge on the capacitor and Q_i/RC is the initial current in the circuit.

Objective Questions   1. denotes answer available in *Student Solutions Manual/Study Guide*

1. Is a circuit breaker wired (a) in series with the device it is protecting, (b) in parallel, or (c) neither in series or in parallel, or (d) is it impossible to tell?

2. A battery has some internal resistance. **(i)** Can the potential difference across the terminals of the battery be equal to its emf? (a) no (b) yes, if the battery is absorbing energy by electrical transmission (c) yes, if more than one wire is connected to each terminal (d) yes, if the current in the battery is zero (e) yes, with no special condition required. **(ii)** Can the terminal voltage exceed the emf? Choose your answer from the same possibilities as in part (i).

3. The terminals of a battery are connected across two resistors in series. The resistances of the resistors are not the same. Which of the following statements are correct? Choose all that are correct. (a) The resistor with the smaller resistance carries more current than the other resistor. (b) The resistor with the larger resistance carries less current than the other resistor. (c) The current in each resistor is the same. (d) The potential difference across each resistor is the same. (e) The potential difference is greatest across the resistor closest to the positive terminal.

4. When operating on a 120-V circuit, an electric heater receives 1.30×10^3 W of power, a toaster receives 1.00×10^3 W, and an electric oven receives 1.54×10^3 W. If all three appliances are connected in parallel on a 120-V circuit and turned on, what is the total current drawn from an external source? (a) 24.0 A (b) 32.0 A (c) 40.0 A (d) 48.0 A (e) none of those answers

5. If the terminals of a battery with zero internal resistance are connected across two identical resistors in series, the total power delivered by the battery is 8.00 W. If the same battery is connected across the same resistors in parallel, what is the total power delivered by the battery? (a) 16.0 W (b) 32.0 W (c) 2.00 W (d) 4.00 W (e) none of those answers

6. Several resistors are connected in series. Which of the following statements is correct? Choose all that are correct. (a) The equivalent resistance is greater than any of the resistances in the group. (b) The equivalent resistance is less than any of the resistances in the group. (c) The equivalent resistance depends on the voltage applied across the group. (d) The equivalent resistance is equal to the sum of the resistances in the group. (e) None of those statements is correct.

7. What is the time constant of the circuit shown in Figure OQ28.7? Each of the five resistors has resistance R, and each of the five capacitors has capacitance C. The internal resistance of the battery is negligible. (a) RC (b) $5RC$ (c) $10RC$ (d) $25RC$ (e) none of those answers

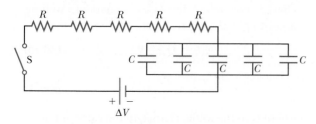

Figure OQ28.7

8. When resistors with different resistances are connected in series, which of the following must be the same for each resistor? Choose all correct answers. (a) potential difference (b) current (c) power delivered (d) charge entering each resistor in a given time interval (e) none of those answers

9. When resistors with different resistances are connected in parallel, which of the following must be the same for each resistor? Choose all correct answers. (a) potential difference (b) current (c) power delivered (d) charge entering each resistor in a given time interval (e) none of those answers

10. The terminals of a battery are connected across two resistors in parallel. The resistances of the resistors are not the same. Which of the following statements is correct? Choose all that are correct. (a) The resistor with the larger resistance carries more current than the other resistor. (b) The resistor with the larger resistance carries less current than the other resistor. (c) The potential difference across each resistor is the same. (d) The potential difference across the larger resistor is greater than the potential difference across the smaller resistor. (e) The potential difference is greater across the resistor closer to the battery.

11. Are the two headlights of a car wired (a) in series with each other, (b) in parallel, or (c) neither in series nor in parallel, or (d) is it impossible to tell?

12. In the circuit shown in Figure OQ28.12, each battery is delivering energy to the circuit by electrical transmission. All the resistors have equal resistance. **(i)** Rank the electric potentials at points a, b, c, d, and e from highest to lowest, noting any cases of equality in the ranking. **(ii)** Rank the magnitudes of the currents at the same points from greatest to least, noting any cases of equality.

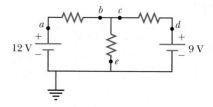

Figure OQ28.12

13. Several resistors are connected in parallel. Which of the following statements are correct? Choose all that are correct. (a) The equivalent resistance is greater than any of the resistances in the group. (b) The equivalent resistance is less than any of the resistances in the group. (c) The equivalent resistance depends on the voltage applied across the group. (d) The equivalent resistance is equal to the sum of the resistances in the group. (e) None of those statements is correct.

14. A circuit consists of three identical lamps connected to a battery as in Figure OQ28.14. The battery has some internal resistance. The switch S, originally open, is closed. **(i)** What then happens to the brightness of lamp B? (a) It increases. (b) It decreases somewhat. (c) It does not change. (d) It drops to zero. For parts (ii) to (vi), choose from the same possibilities (a) through (d). **(ii)** What happens to the brightness of lamp C? **(iii)** What happens to the current in the battery? **(iv)** What happens to the potential difference across lamp A? **(v)** What happens to the potential difference

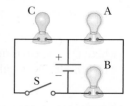

Figure OQ28.14

across lamp C? **(vi)** What happens to the total power delivered to the lamps by the battery?

15. A series circuit consists of three identical lamps connected to a battery as shown in Figure OQ28.15. The switch S, originally open, is closed. **(i)** What then happens to the brightness of lamp B? (a) It increases. (b) It decreases somewhat. (c) It does not change. (d) It drops to zero. For parts (ii) to (vi), choose from the same possibilities (a) through (d). **(ii)** What happens to the brightness of lamp C? **(iii)** What happens to the current in the battery? **(iv)** What happens to the potential difference across

lamp A? **(v)** What happens to the potential difference across lamp C? **(vi)** What happens to the total power delivered to the lamps by the battery?

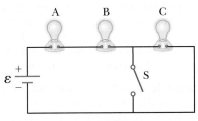

Figure OQ28.15

Conceptual Questions

1. denotes answer available in *Student Solutions Manual/Study Guide*

1. Suppose a parachutist lands on a high-voltage wire and grabs the wire as she prepares to be rescued. (a) Will she be electrocuted? (b) If the wire then breaks, should she continue to hold onto the wire as she falls to the ground? Explain.

2. A student claims that the second of two lightbulbs in series is less bright than the first because the first lightbulb uses up some of the current. How would you respond to this statement?

3. Why is it possible for a bird to sit on a high-voltage wire without being electrocuted?

4. Given three lightbulbs and a battery, sketch as many different electric circuits as you can.

5. A ski resort consists of a few chairlifts and several interconnected downhill runs on the side of a mountain, with a lodge at the bottom. The chairlifts are analogous to batteries, and the runs are analogous to resistors. Describe how two runs can be in series. Describe how three runs can be in parallel. Sketch a junction between one chairlift and two runs. State Kirchhoff's junction rule for ski resorts. One of the skiers happens to be carrying a skydiver's altimeter. She never takes the same set of chairlifts and runs twice, but keeps passing you at the fixed location where you are working. State Kirchhoff's loop rule for ski resorts.

6. Referring to Figure CQ28.6, describe what happens to the lightbulb after the switch is closed. Assume the capacitor has a large capacitance and is initially uncharged. Also assume the light illuminates when connected directly across the battery terminals.

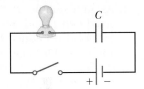

Figure CQ28.6

7. So that your grandmother can listen to *A Prairie Home Companion*, you take her bedside radio to the hospital where she is staying. You are required to have a maintenance worker test the radio for electrical safety. Finding that it develops 120 V on one of its knobs, he does not let you take it to your grandmother's room. Your grandmother complains that she has had the radio for many years and nobody has ever gotten a shock from it. You end up having to buy a new plastic radio. (a) Why is your grandmother's old radio dangerous in a hospital room? (b) Will the old radio be safe back in her bedroom?

8. (a) What advantage does 120-V operation offer over 240 V? (b) What disadvantages does it have?

9. Is the direction of current in a battery always from the negative terminal to the positive terminal? Explain.

10. Compare series and parallel resistors to the series and parallel rods in Figure 20.13 on page 610. How are the situations similar?

Problems

 WebAssign The problems found in this chapter may be assigned online in Enhanced WebAssign

1. straightforward; **2.** intermediate;
3. challenging

1. full solution available in the *Student Solutions Manual/Study Guide*

AMT Analysis Model tutorial available in Enhanced WebAssign

GP Guided Problem

M Master It tutorial available in Enhanced WebAssign

W Watch It video solution available in Enhanced WebAssign

Section 28.1 Electromotive Force

1. A battery has an emf of 15.0 V. The terminal voltage **M** of the battery is 11.6 V when it is delivering 20.0 W of

power to an external load resistor R. (a) What is the value of R? (b) What is the internal resistance of the battery?

2. Two 1.50-V batteries—with their positive terminals **AMT** in the same direction—are inserted in series into a flashlight. One battery has an internal resistance of 0.255 Ω, and the other has an internal resistance of 0.153 Ω. When the switch is closed, the bulb carries a current of 600 mA. (a) What is the bulb's resistance? (b) What fraction of the chemical energy transformed appears as internal energy in the batteries?

3. An automobile battery has an emf of 12.6 V and **W** an internal resistance of 0.080 0 Ω. The headlights together have an equivalent resistance of 5.00 Ω (assumed constant). What is the potential difference across the headlight bulbs (a) when they are the only load on the battery and (b) when the starter motor is operated, with 35.0 A of current in the motor?

4. As in Example 28.2, consider a power supply with fixed emf ε and internal resistance r causing current in a load resistance R. In this problem, R is fixed and r is a variable. The efficiency is defined as the energy delivered to the load divided by the energy delivered by the emf. (a) When the internal resistance is adjusted for maximum power transfer, what is the efficiency? (b) What should be the internal resistance for maximum possible efficiency? (c) When the electric company sells energy to a customer, does it have a goal of high efficiency or of maximum power transfer? Explain. (d) When a student connects a loudspeaker to an amplifier, does she most want high efficiency or high power transfer? Explain.

Section 28.2 Resistors in Series and Parallel

5. Three 100-Ω resistors are connected as shown in Fig- **W** ure P28.5. The maximum power that can safely be delivered to any one resistor is 25.0 W. (a) What is the maximum potential difference that can be applied to the terminals a and b? (b) For the voltage determined in part (a), what is the power delivered to each resistor? (c) What is the total power delivered to the combination of resistors?

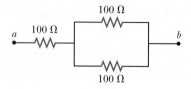

Figure P28.5

6. A lightbulb marked "75 W [at] 120 V" is screwed into a socket at one end of a long extension cord, in which each of the two conductors has resistance 0.800 Ω. The other end of the extension cord is plugged into a 120-V outlet. (a) Explain why the actual power delivered to the lightbulb cannot be 75 W in this situation. (b) Draw a circuit diagram. (c) Find the actual power delivered to the lightbulb in this circuit.

7. What is the equivalent resistance of the combination of identical resistors between points a and b in Figure P28.7?

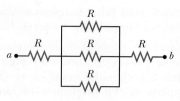

Figure P28.7

8. Consider the two circuits shown in Figure P28.8 in which the batteries are identical. The resistance of each lightbulb is R. Neglect the internal resistances of the batteries. (a) Find expressions for the currents in each lightbulb. (b) How does the brightness of B compare with that of C? Explain. (c) How does the brightness of A compare with that of B and of C? Explain.

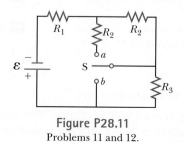

Figure P28.8

9. Consider the circuit shown in Figure P28.9. Find **M** (a) the current in the 20.0-Ω resistor and (b) the potential difference between points a and b.

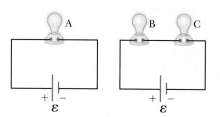

Figure P28.9

10. (a) You need a 45-Ω resistor, but the stockroom has only 20-Ω and 50-Ω resistors. How can the desired resistance be achieved under these circumstances? (b) What can you do if you need a 35-Ω resistor?

11. A battery with $\varepsilon = 6.00$ V and no internal resistance supplies current to the circuit shown in Figure P28.11. When the double-throw switch S is open as shown in the figure, the current in the battery is 1.00 mA. When the switch is closed in position a, the current in the

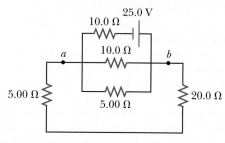

Figure P28.11
Problems 11 and 12.

battery is 1.20 mA. When the switch is closed in position b, the current in the battery is 2.00 mA. Find the resistances (a) R_1, (b) R_2, and (c) R_3.

12. A battery with emf $\mathcal{E}$ and no internal resistance supplies current to the circuit shown in Figure P28.11. When the double-throw switch S is open as shown in the figure, the current in the battery is I_0. When the switch is closed in position a, the current in the battery is I_a. When the switch is closed in position b, the current in the battery is I_b. Find the resistances (a) R_1, (b) R_2, and (c) R_3.

13. (a) Find the equivalent resistance between points a and b in Figure P28.13. (b) Calculate the current in each resistor if a potential difference of 34.0 V is applied between points a and b.

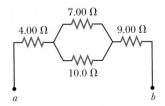

Figure P28.13

14. (a) When the switch S in the circuit of Figure P28.14 is closed, will the equivalent resistance between points a and b increase or decrease? State your reasoning. (b) Assume the equivalent resistance drops by 50.0% when the switch is closed. Determine the value of R.

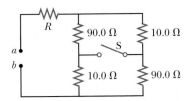

Figure P28.14

15. Two resistors connected in series have an equivalent resistance of 690 Ω. When they are connected in parallel, their equivalent resistance is 150 Ω. Find the resistance of each resistor.

16. Four resistors are connected to a battery as shown in Figure P28.16. (a) Determine the potential difference across each resistor in terms of $\mathcal{E}$. (b) Determine the current in each resistor in terms of I. (c) **What If?** If R_3 is increased, explain what happens to the current in each of the resistors. (d) In the limit that $R_3 \rightarrow \infty$, what are the new values of the current in each resistor in terms of I, the original current in the battery?

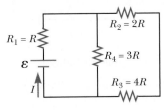

Figure P28.16

17. Consider the combination of resistors shown in Figure P28.17. (a) Find the equivalent resistance between points a and b. (b) If a voltage of 35.0 V is applied between points a and b, find the current in each resistor.

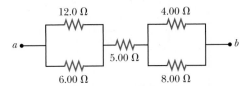

Figure P28.17

18. For the purpose of measuring the electric resistance of shoes through the body of the wearer standing on a metal ground plate, the American National Standards Institute (ANSI) specifies the circuit shown in Figure P28.18. The potential difference ΔV across the 1.00-MΩ resistor is measured with an ideal voltmeter. (a) Show that the resistance of the footwear is

$$R_{shoes} = \frac{50.0 \text{ V} - \Delta V}{\Delta V}$$

(b) In a medical test, a current through the human body should not exceed 150 μA. Can the current delivered by the ANSI-specified circuit exceed 150 μA? To decide, consider a person standing barefoot on the ground plate.

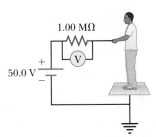

Figure P28.18

19. Calculate the power delivered to each resistor in the circuit shown in Figure P28.19.

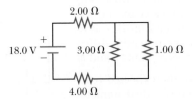

Figure P28.19

20. *Why is the following situation impossible?* A technician is testing a circuit that contains a resistance R. He realizes that a better design for the circuit would include a resistance $\frac{7}{3}R$ rather than R. He has three additional resistors, each with resistance R. By combining these additional resistors in a certain combination that is then placed in series with the original resistor, he achieves the desired resistance.

21. Consider the circuit shown in Figure P28.21 on page 860. (a) Find the voltage across the 3.00-Ω resistor. (b) Find the current in the 3.00-Ω resistor.

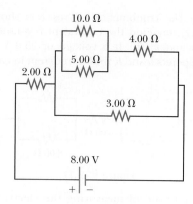

Figure P28.21

Section 28.3 Kirchhoff's Rules

22. In Figure P28.22, show how to add just enough ammeters to measure every different current. Show how
to add just enough voltmeters to measure the potential difference across each resistor and across each
battery.

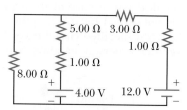

Figure P28.22 Problems 22 and 23.

23. The circuit shown in Figure P28.22 is connected for
[M] 2.00 min. (a) Determine the current in each branch of
the circuit. (b) Find the energy delivered by each battery. (c) Find the energy delivered to each resistor.
(d) Identify the type of energy storage transformation
that occurs in the operation of the circuit. (e) Find the
total amount of energy transformed into internal
energy in the resistors.

24. For the circuit shown in Figure P28.24, calculate (a) the
current in the 2.00-Ω resistor
and (b) the potential difference between points a and b.

25. What are the expected read-
[M] ings of (a) the ideal ammeter
and (b) the ideal voltmeter
in Figure P28.25?

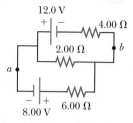

Figure P28.24

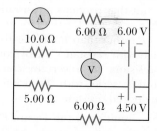

Figure P28.25

26. The following equations describe an electric circuit:

$$-I_1 (220\ \Omega) + 5.80\ \text{V} - I_2 (370\ \Omega) = 0$$
$$+I_2 (370\ \Omega) + I_3 (150\ \Omega) - 3.10\ \text{V} = 0$$
$$I_1 + I_3 - I_2 = 0$$

(a) Draw a diagram of the circuit. (b) Calculate the
unknowns and identify the physical meaning of each
unknown.

27. Taking $R = 1.00$ kΩ and $\varepsilon = 250$ V in Figure P28.27,
determine the direction and magnitude of the current
in the horizontal wire between a and e.

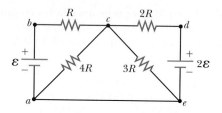

Figure P28.27

28. Jumper cables are connected from a fresh battery in
one car to charge a dead battery in another car. Figure P28.28 shows the circuit diagram for this situation.
While the cables are connected, the ignition switch of
the car with the dead battery is closed and the starter is
activated to start the engine. Determine the current in
(a) the starter and (b) the dead battery. (c) Is the dead
battery being charged while the starter is operating?

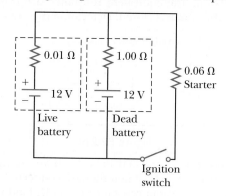

Figure P28.28

29. The ammeter shown in Figure P28.29 reads 2.00 A.
[W] Find (a) I_1, (b) I_2, and (c) ε.

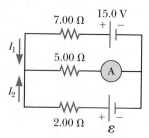

Figure P28.29

30. In the circuit of Figure P28.30, determine (a) the cur-
[W] rent in each resistor and (b) the potential difference
across the 200-Ω resistor.

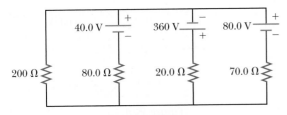

Figure P28.30

31. Using Kirchhoff's rules, (a) find the current in each
    [M] resistor shown in Figure P28.31 and (b) find the poten-
    tial difference between points c and f.

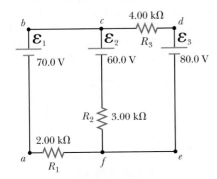

Figure P28.31

32. In the circuit of Figure P28.32, the current $I_1 = 3.00$ A
    and the values of $\mathcal{E}$ for the ideal battery and R are
    unknown. What are the currents (a) I_2 and (b) I_3?
    (c) Can you find the values of $\mathcal{E}$ and R? If so, find their
    values. If not, explain.

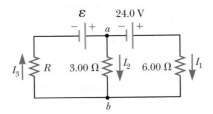

Figure P28.32

33. In Figure P28.33, find (a) the current in each resistor
    and (b) the power delivered to each resistor.

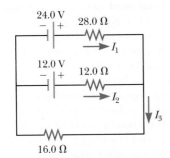

Figure P28.33

34. For the circuit shown in Figure P28.34, we wish to
    [GP] find the currents I_1, I_2, and I_3. Use Kirchhoff's rules to
    obtain equations for (a) the upper loop, (b) the lower

loop, and (c) the junction on the left side. In each case,
suppress units for clarity and simplify, combining the
terms. (d) Solve the junction equation for I_3. (e) Using
the equation found in part (d), eliminate I_3 from the
equation found in part (b). (f) Solve the equations
found in parts (a) and (e) simultaneously for the two
unknowns I_1 and I_2. (g) Substitute the answers found
in part (f) into the junction equation found in part (d),
solving for I_3. (h) What is the significance of the negative answer for I_2?

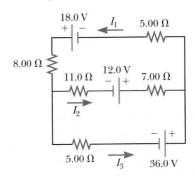

Figure P28.34

35. Find the potential difference across each resistor in
    [M] Figure P28.35.

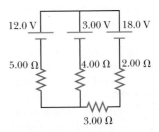

Figure P28.35

36. (a) Can the circuit shown in Figure P28.36 be reduced
    to a single resistor connected to a battery? Explain.
    Calculate the currents (b) I_1, (c) I_2, and (d) I_3.

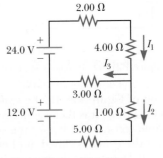

Figure P28.36

Section 28.4 *RC* Circuits

37. An uncharged capacitor and a resistor are connected
    in series to a source of emf. If $\mathcal{E} = 9.00$ V, $C = 20.0$ μF,
    and $R = 100$ Ω, find (a) the time constant of the cir-
    cuit, (b) the maximum charge on the capacitor, and
    (c) the charge on the capacitor at a time equal to one
    time constant after the battery is connected.

38. Consider a series RC circuit as in Figure P28.38 for [W] which $R = 1.00$ MΩ, $C = 5.00$ μF, and $\varepsilon = 30.0$ V. Find (a) the time constant of the circuit and (b) the maximum charge on the capacitor after the switch is thrown closed. (c) Find the current in the resistor 10.0 s after the switch is closed.

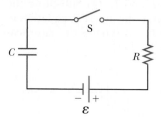

Figure P28.38
Problems 38, 67, and 68.

39. A 2.00-nF capacitor with an initial charge of 5.10 μC [W] is discharged through a 1.30-kΩ resistor. (a) Calculate the current in the resistor 9.00 μs after the resistor is connected across the terminals of the capacitor. (b) What charge remains on the capacitor after 8.00 μs? (c) What is the maximum current in the resistor?

40. A 10.0-μF capacitor is charged by a 10.0-V battery through a resistance R. The capacitor reaches a potential difference of 4.00 V in a time interval of 3.00 s after charging begins. Find R.

41. In the circuit of Figure P28.41, the switch S has been [W] open for a long time. It is then suddenly closed. Take $\varepsilon = 10.0$ V, $R_1 = 50.0$ kΩ, $R_2 = 100$ kΩ, and $C = 10.0$ μF. Determine the time constant (a) before the switch is closed and (b) after the switch is closed. (c) Let the switch be closed at $t = 0$. Determine the current in the switch as a function of time.

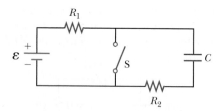

Figure P28.41 Problems 41 and 42.

42. In the circuit of Figure P28.41, the switch S has been open for a long time. It is then suddenly closed. Determine the time constant (a) before the switch is closed and (b) after the switch is closed. (c) Let the switch be closed at $t = 0$. Determine the current in the switch as a function of time.

43. The circuit in Figure P28.43 has been connected for a [M] long time. (a) What is the potential difference across the capacitor? (b) If the battery is disconnected from the circuit, over what time interval does the capacitor discharge to one-tenth its initial voltage?

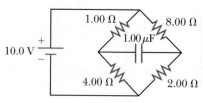

Figure P28.43

44. Show that the integral $\int_0^\infty e^{-2t/RC}\, dt$ in Example 28.11 has the value $\frac{1}{2}RC$.

45. A charged capacitor is connected to a resistor and switch as in Figure P28.45. The circuit has a time constant of 1.50 s. Soon after the switch is closed, the charge on the capacitor is 75.0% of its initial charge. (a) Find the time interval required for the capacitor to reach this charge. (b) If $R = 250$ kΩ, what is the value of C?

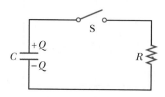

Figure P28.45

Section 28.5 Household Wiring and Electrical Safety

46. An electric heater is rated at 1.50×10^3 W, a toaster [M] at 750 W, and an electric grill at 1.00×10^3 W. The three appliances are connected to a common 120-V household circuit. (a) How much current does each draw? (b) If the circuit is protected with a 25.0-A circuit breaker, will the circuit breaker be tripped in this situation? Explain your answer.

47. A heating element in a stove is designed to receive [M] 3 000 W when connected to 240 V. (a) Assuming the resistance is constant, calculate the current in the heating element if it is connected to 120 V. (b) Calculate the power it receives at that voltage.

48. Turn on your desk lamp. Pick up the cord, with your thumb and index finger spanning the width of the cord. (a) Compute an order-of-magnitude estimate for the current in your hand. Assume the conductor inside the lamp cord next to your thumb is at potential $\sim 10^2$ V at a typical instant and the conductor next to your index finger is at ground potential (0 V). The resistance of your hand depends strongly on the thickness and the moisture content of the outer layers of your skin. Assume the resistance of your hand between fingertip and thumb tip is $\sim 10^4$ Ω. You may model the cord as having rubber insulation. State the other quantities you measure or estimate and their values. Explain your reasoning. (b) Suppose your body is isolated from any other charges or currents. In order-of-magnitude terms, estimate the potential difference between your thumb where it contacts the cord and your finger where it touches the cord.

Additional Problems

49. Assume you have a battery of emf ε and three identical lightbulbs, each having constant resistance R. What is the total power delivered by the battery if the lightbulbs are connected (a) in series and (b) in parallel? (c) For which connection will the lightbulbs shine the brightest?

50. Find the equivalent resistance between points a and b in Figure P28.50.

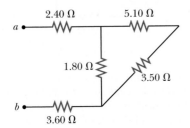

Figure P28.50

51. Four 1.50-V AA batteries in series are used to power a small radio. If the batteries can move a charge of 240 C, how long will they last if the radio has a resistance of 200 Ω?

52. Four resistors are connected in parallel across a 9.20-V battery. They carry currents of 150 mA, 45.0 mA, 14.0 mA, and 4.00 mA. If the resistor with the largest resistance is replaced with one having twice the resistance, (a) what is the ratio of the new current in the battery to the original current? (b) **What If?** If instead the resistor with the smallest resistance is replaced with one having twice the resistance, what is the ratio of the new total current to the original current? (c) On a February night, energy leaves a house by several energy leaks, including 1.50×10^3 W by conduction through the ceiling, 450 W by infiltration (airflow) around the windows, 140 W by conduction through the basement wall above the foundation sill, and 40.0 W by conduction through the plywood door to the attic. To produce the biggest saving in heating bills, which one of these energy transfers should be reduced first? Explain how you decide. Clifford Swartz suggested the idea for this problem.

53. The circuit in Figure P28.53 has been connected for several seconds. Find the current (a) in the 4.00-V bat-

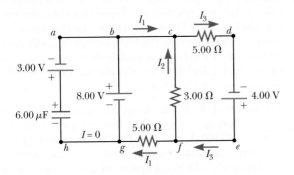

Figure P28.53

tery, (b) in the 3.00-Ω resistor, (c) in the 8.00-V battery, and (d) in the 3.00-V battery. (e) Find the charge on the capacitor.

54. The circuit in Figure P28.54a consists of three resistors and one battery with no internal resistance. (a) Find the current in the 5.00-Ω resistor. (b) Find the power delivered to the 5.00-Ω resistor. (c) In each of the circuits in Figures P28.54b, P28.54c, and P28.54d, an additional 15.0-V battery has been inserted into the circuit. Which diagram or diagrams represent a circuit that requires the use of Kirchhoff's rules to find the currents? Explain why. (d) In which of these three new circuits is the smallest amount of power delivered to the 10.0-Ω resistor? (You need not calculate the power in each circuit if you explain your answer.)

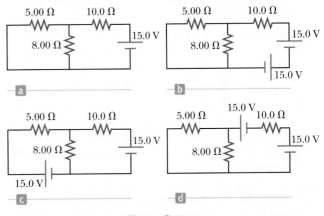

Figure P28.54

55. For the circuit shown in Figure P28.55, the ideal voltmeter reads 6.00 V and the ideal ammeter reads 3.00 mA. Find (a) the value of R, (b) the emf of the battery, and (c) the voltage across the 3.00-kΩ resistor.

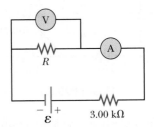

Figure P28.55

56. The resistance between terminals a and b in Figure P28.56 is 75.0 Ω. If the resistors labeled R have the same value, determine R.

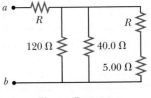

Figure P28.56

57. (a) Calculate the potential difference between points a and b in Figure P28.57 and (b) identify which point is at the higher potential.

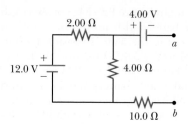

Figure P28.57

58. *Why is the following situation impossible?* A battery has an emf of $\varepsilon = 9.20$ V and an internal resistance of $r = 1.20\ \Omega$. A resistance R is connected across the battery and extracts from it a power of $P = 21.2$ W.

59. A rechargeable battery has an emf of 13.2 V and an internal resistance of $0.850\ \Omega$. It is charged by a 14.7-V power supply for a time interval of 1.80 h. After charging, the battery returns to its original state as it delivers a constant current to a load resistor over 7.30 h. Find the efficiency of the battery as an energy storage device. (The efficiency here is defined as the energy delivered to the load during discharge divided by the energy delivered by the 14.7-V power supply during the charging process.)

60. Find (a) the equivalent resistance of the circuit in Figure P28.60, (b) the potential difference across each resistor, (c) each current indicated in Figure P28.60, and (d) the power delivered to each resistor.

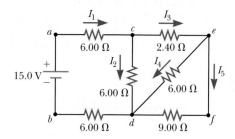

Figure P28.60

61. When two unknown resistors are connected in series with a battery, the battery delivers 225 W and carries a total current of 5.00 A. For the same total current, 50.0 W is delivered when the resistors are connected in parallel. Determine the value of each resistor.

62. When two unknown resistors are connected in series with a battery, the battery delivers total power P_s and carries a total current of I. For the same total current, a total power P_p is delivered when the resistors are connected in parallel. Determine the value of each resistor.

63. The pair of capacitors in Figure P28.63 are fully charged by a 12.0-V battery. The battery is disconnected, and the switch is then closed. After 1.00 ms has elapsed, (a) how much charge remains on the 3.00-μF

capacitor? (b) How much charge remains on the 2.00-μF capacitor? (c) What is the current in the resistor at this time?

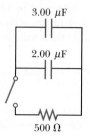

Figure P28.63

64. A power supply has an open-circuit voltage of 40.0 V and an internal resistance of $2.00\ \Omega$. It is used to charge two storage batteries connected in series, each having an emf of 6.00 V and internal resistance of $0.300\ \Omega$. If the charging current is to be 4.00 A, (a) what additional resistance should be added in series? At what rate does the internal energy increase in (b) the supply, (c) in the batteries, and (d) in the added series resistance? (e) At what rate does the chemical energy increase in the batteries?

65. The circuit in Figure P28.65 contains two resistors, $R_1 = 2.00$ kΩ and $R_2 = 3.00$ kΩ, and two capacitors, $C_1 = 2.00\ \mu$F and $C_2 = 3.00\ \mu$F, connected to a battery with emf $\varepsilon = 120$ V. If there are no charges on the capacitors before switch S is closed, determine the charges on capacitors (a) C_1 and (b) C_2 as functions of time, after the switch is closed.

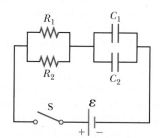

Figure P28.65

66. Two resistors R_1 and R_2 are in parallel with each other. Together they carry total current I. (a) Determine the current in each resistor. (b) Prove that this division of the total current I between the two resistors results in less power delivered to the combination than any other division. It is a general principle that *current in a direct current circuit distributes itself so that the total power delivered to the circuit is a minimum.*

67. The values of the components in a simple series RC circuit containing a switch (Fig. P28.38) are $C = 1.00\ \mu$F, $R = 2.00 \times 10^6\ \Omega$, and $\varepsilon = 10.0$ V. At the instant 10.0 s after the switch is closed, calculate (a) the charge on the capacitor, (b) the current in the resistor, (c) the rate at which energy is being stored in the capacitor, and (d) the rate at which energy is being delivered by the battery.

68. A battery is used to charge a capacitor through a resistor as shown in Figure P28.38. Show that half the energy supplied by the battery appears as internal energy in the resistor and half is stored in the capacitor.

69. A young man owns a canister vacuum cleaner marked "535 W [at] 120 V" and a Volkswagen Beetle, which he wishes to clean. He parks the car in his apartment parking lot and uses an inexpensive extension cord 15.0 m long to plug in the vacuum cleaner. You may assume the cleaner has constant resistance. (a) If the resistance of each of the two conductors in the extension cord is 0.900 Ω, what is the actual power delivered to the cleaner? (b) If instead the power is to be at least 525 W, what must be the diameter of each of two identical copper conductors in the cord he buys? (c) Repeat part (b) assuming the power is to be at least 532 W.

70. (a) Determine the equilibrium charge on the capacitor in the circuit of Figure P28.70 as a function of R. (b) Evaluate the charge when $R = 10.0\ \Omega$. (c) Can the charge on the capacitor be zero? If so, for what value of R? (d) What is the maximum possible magnitude of the charge on the capacitor? For what value of R is it achieved? (e) Is it experimentally meaningful to take $R = \infty$? Explain your answer. If so, what charge magnitude does it imply?

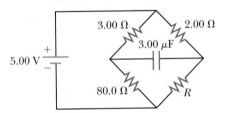

Figure P28.70

71. Switch S shown in Figure P28.71 has been closed for a long time, and the electric circuit carries a constant current. Take $C_1 = 3.00\ \mu F$, $C_2 = 6.00\ \mu F$, $R_1 = 4.00\ k\Omega$, and $R_2 = 7.00\ k\Omega$. The power delivered to R_2 is 2.40 W. (a) Find the charge on C_1. (b) Now the switch is opened. After many milliseconds, by how much has the charge on C_2 changed?

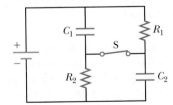

Figure P28.71

72. Three identical 60.0-W, 120-V lightbulbs are connected
M across a 120-V power source as shown in Figure P28.72. Assuming the resistance of each lightbulb is constant (even though in reality the resistance might increase markedly with current), find (a) the total power supplied by the power source and (b) the potential difference across each lightbulb.

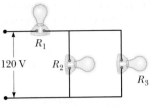

Figure P28.72

73. A regular tetrahedron is a pyramid with a triangular base and triangular sides as shown in Figure P28.73. Imagine the six straight lines in Figure P28.73 are each 10.0-Ω resistors, with junctions at the four vertices. A 12.0-V battery is connected to any two of the vertices. Find (a) the equivalent resistance of the tetrahedron between these vertices and (b) the current in the battery.

Figure P28.73

74. An ideal voltmeter connected across a certain fresh 9-V battery reads 9.30 V, and an ideal ammeter briefly connected across the same battery reads 3.70 A. We say the battery has an open-circuit voltage of 9.30 V and a short-circuit current of 3.70 A. Model the battery as a source of emf $\mathcal{E}$ in series with an internal resistance r as in Figure 28.1a. Determine both (a) $\mathcal{E}$ and (b) r. An experimenter connects two of these identical batteries together as shown in Figure P28.74. Find (c) the open-circuit voltage and (d) the short-circuit current of the pair of connected batteries. (e) The experimenter connects a 12.0-Ω resistor between the exposed terminals of the connected batteries. Find the current in the resistor. (f) Find the power delivered to the resistor. (g) The experimenter connects a second identical resistor in parallel with the first. Find the power delivered to each resistor. (h) Because the same pair of batteries is connected across both resistors as was connected across the single resistor, why is the power in part (g) not the same as that in part (f)?

Figure P28.74

75. In Figure P28.75 on page 866, suppose the switch has been closed for a time interval sufficiently long for the capacitor to become fully charged. Find (a) the

steady-state current in each resistor and (b) the charge Q_{max} on the capacitor. (c) The switch is now opened at $t = 0$. Write an equation for the current in R_2 as a function of time and (d) find the time interval required for the charge on the capacitor to fall to one-fifth its initial value.

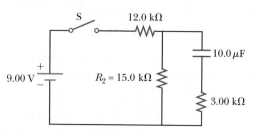

Figure P28.75

76. Figure P28.76 shows a circuit model for the transmission of an electrical signal such as cable TV to a large number of subscribers. Each subscriber connects a load resistance R_L between the transmission line and the ground. The ground is assumed to be at zero potential and able to carry any current between any ground connections with negligible resistance. The resistance of the transmission line between the connection points of different subscribers is modeled as the constant resistance R_T. Show that the equivalent resistance across the signal source is

$$R_{eq} = \tfrac{1}{2}\left[(4R_T R_L + R_T^2)^{1/2} + R_T\right]$$

Suggestion: Because the number of subscribers is large, the equivalent resistance would not change noticeably if the first subscriber canceled the service. Consequently, the equivalent resistance of the section of the circuit to the right of the first load resistor is nearly equal to R_{eq}.

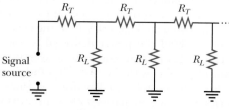

Figure P28.76

77. The student engineer of a campus radio station wishes to verify the effectiveness of the lightning rod on the antenna mast (Fig. P28.77). The unknown resistance R_x is between points C and E. Point E is a true ground, but it is inaccessible for direct measurement because this stratum is several meters below the Earth's surface. Two identical rods are driven into the ground at A and B, introducing an unknown resistance R_y. The procedure is as follows. Measure resistance R_1 between points A and B, then connect A and B with a heavy conducting wire and measure resistance R_2 between points A and C. (a) Derive an equation for R_x in terms of the

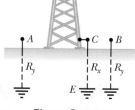

Figure P28.77

observable resistances, R_1 and R_2. (b) A satisfactory ground resistance would be $R_x < 2.00\ \Omega$. Is the grounding of the station adequate if measurements give $R_1 = 13.0\ \Omega$ and $R_2 = 6.00\ \Omega$? Explain.

78. The circuit shown in Figure P28.78 is set up in the laboratory to measure an unknown capacitance C in series with a resistance $R = 10.0$ MΩ powered by a battery whose emf is 6.19 V. The data given in the table are the measured voltages across the capacitor as a function of time, where $t = 0$ represents the instant at which the switch is thrown to position b. (a) Construct a graph of $\ln(\mathcal{E}/\Delta v)$ versus t and perform a linear least-squares fit to the data. (b) From the slope of your graph, obtain a value for the time constant of the circuit and a value for the capacitance.

Δv (V)	t (s)	$\ln(\mathcal{E}/\Delta v)$
6.19	0	
5.55	4.87	
4.93	11.1	
4.34	19.4	
3.72	30.8	
3.09	46.6	
2.47	67.3	
1.83	102.2	

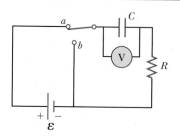

Figure P28.78

79. An electric teakettle has a multiposition switch and two heating coils. When only one coil is switched on, the well-insulated kettle brings a full pot of water to a boil over the time interval Δt. When only the other coil is switched on, it takes a time interval of $2\,\Delta t$ to boil the same amount of water. Find the time interval required to boil the same amount of water if both coils are switched on (a) in a parallel connection and (b) in a series connection.

80. A voltage ΔV is applied to a series configuration of n resistors, each of resistance R. The circuit components are reconnected in a parallel configuration, and voltage ΔV is again applied. Show that the power delivered to the series configuration is $1/n^2$ times the power delivered to the parallel configuration.

81. In places such as hospital operating rooms or factories for electronic circuit boards, electric sparks must be avoided. A person standing on a grounded floor and touching nothing else can typically have a body capacitance of 150 pF, in parallel with a foot capacitance of 80.0 pF produced by the dielectric soles of his or her shoes. The person acquires static electric charge from interactions with his or her surroundings. The static charge flows to ground through the equivalent resistance of the two

shoe soles in parallel with each other. A pair of rubber-soled street shoes can present an equivalent resistance of 5.00×10^3 MΩ. A pair of shoes with special static-dissipative soles can have an equivalent resistance of 1.00 MΩ. Consider the person's body and shoes as forming an *RC* circuit with the ground. (a) How long does it take the rubber-soled shoes to reduce a person's potential from 3.00×10^3 V to 100 V? (b) How long does it take the static-dissipative shoes to do the same thing?

Challenge Problems

82. The switch in Figure P28.82a closes when $\Delta V_c > \frac{2}{3}\Delta V$ and opens when $\Delta V_c < \frac{1}{3}\Delta V$. The ideal voltmeter reads a potential difference as plotted in Figure P28.82b. What is the period T of the waveform in terms of R_1, R_2, and C?

83. The resistor R in Figure P28.83 receives 20.0 W of power. Determine the value of R.

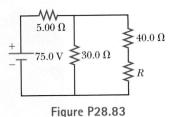

Figure P28.83

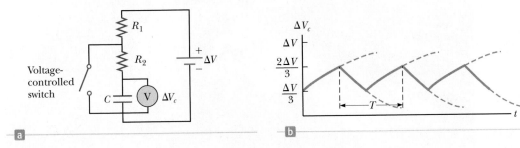

Figure P28.82

Magnetic Fields

An engineer performs a test on the electronics associated with one of the superconducting magnets in the Large Hadron Collider at the European Laboratory for Particle Physics, operated by the European Organization for Nuclear Research (CERN). The magnets are used to control the motion of charged particles in the accelerator. We will study the effects of magnetic fields on moving charged particles in this chapter. *(CERN)*

Many historians of science believe that the compass, which uses a magnetic needle, was used in China as early as the 13th century BC, its invention being of Arabic or Indian origin. The early Greeks knew about magnetism as early as 800 BC. They discovered that the stone magnetite (Fe_3O_4) attracts pieces of iron. Legend ascribes the name *magnetite* to the shepherd Magnes, the nails of whose shoes and the tip of whose staff stuck fast to chunks of magnetite while he pastured his flocks.

In 1269, Pierre de Maricourt of France found that the directions of a needle near a spherical natural magnet formed lines that encircled the sphere and passed through two points diametrically opposite each other, which he called the *poles* of the magnet. Subsequent experiments showed that every magnet, regardless of its shape, has two poles, called *north* (N) and *south* (S) poles, that exert forces on other magnetic poles similar to the way electric charges exert forces on one another. That is, like poles (N–N or S–S) repel each other, and opposite poles (N–S) attract each other.

The poles received their names because of the way a magnet, such as that in a compass, behaves in the presence of the Earth's magnetic field. If a bar magnet is suspended from its midpoint and can swing freely in a horizontal plane, it will rotate until its north pole points to the Earth's geographic North Pole and its south pole points to the Earth's geographic South Pole.[1]

In 1600, William Gilbert (1540–1603) extended de Maricourt's experiments to a variety of materials. He knew that a compass needle orients in preferred directions, so he suggested that the Earth itself is a large, permanent magnet. In 1750, experimenters used a torsion balance to show that magnetic poles exert attractive or repulsive forces on each other and that these forces vary as the inverse square of the distance between interacting poles. Although the force between two magnetic poles is otherwise similar to the force between two electric charges, electric charges can be isolated (witness the electron and proton), whereas a single magnetic pole has never been isolated. That is, magnetic poles are always found in pairs. All attempts thus far to detect an isolated magnetic pole have been unsuccessful. No matter how many times a permanent magnet is cut in two, each piece always has a north and a south pole.[2]

The relationship between magnetism and electricity was discovered in 1819 when, during a lecture demonstration, Hans Christian Oersted found that an electric current in a wire deflected a nearby compass needle.[3] In the 1820s, further connections between electricity and magnetism were demonstrated independently by Faraday and Joseph Henry (1797–1878). They showed that an electric current can be produced in a circuit either by moving a magnet near the circuit or by changing the current in a nearby circuit. These observations demonstrate that a changing magnetic field creates an electric field. Years later, theoretical work by Maxwell showed that the reverse is also true: a changing electric field creates a magnetic field.

This chapter examines the forces that act on moving charges and on current-carrying wires in the presence of a magnetic field. The source of the magnetic field is described in Chapter 30.

Hans Christian Oersted
Danish Physicist and Chemist (1777–1851)
Oersted is best known for observing that a compass needle deflects when placed near a wire carrying a current. This important discovery was the first evidence of the connection between electric and magnetic phenomena. Oersted was also the first to prepare pure aluminum.

29.1 Analysis Model: Particle in a Field (Magnetic)

In our study of electricity, we described the interactions between charged objects in terms of electric fields. Recall that an electric field surrounds any electric charge. In addition to containing an electric field, the region of space surrounding any *moving* electric charge also contains a **magnetic field.** A magnetic field also surrounds a magnetic substance making up a permanent magnet.

Historically, the symbol $\vec{B}$ has been used to represent a magnetic field, and we use this notation in this book. The direction of the magnetic field $\vec{B}$ at any location is the direction in which a compass needle points at that location. As with the electric field, we can represent the magnetic field by means of drawings with *magnetic field lines*.

Figure 29.1 shows how the magnetic field lines of a bar magnet can be traced with the aid of a compass. Notice that the magnetic field lines outside the magnet

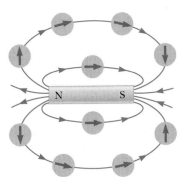

Figure 29.1 Compass needles can be used to trace the magnetic field lines in the region outside a bar magnet.

---

[1]The Earth's geographic North Pole is magnetically a south pole, whereas the Earth's geographic South Pole is magnetically a north pole. Because *opposite* magnetic poles attract each other, the pole on a magnet that is attracted to the Earth's geographic North Pole is the magnet's *north* pole and the pole attracted to the Earth's geographic South Pole is the magnet's *south* pole.

[2]There is some theoretical basis for speculating that magnetic *monopoles*—isolated north or south poles—may exist in nature, and attempts to detect them are an active experimental field of investigation.

[3]The same discovery was reported in 1802 by an Italian jurist, Gian Domenico Romagnosi, but was overlooked, probably because it was published in an obscure journal.

Figure 29.2 Magnetic field patterns can be displayed with iron filings sprinkled on paper near magnets.

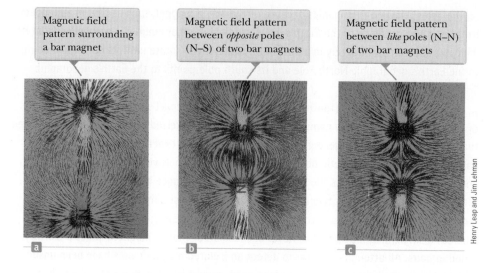

Magnetic field pattern surrounding a bar magnet

Magnetic field pattern between *opposite* poles (N–S) of two bar magnets

Magnetic field pattern between *like* poles (N–N) of two bar magnets

a   b   c

Henry Leap and Jim Lehman

point away from the north pole and toward the south pole. One can display magnetic field patterns of a bar magnet using small iron filings as shown in Figure 29.2.

When we speak of a compass magnet having a north pole and a south pole, it is more proper to say that it has a "north-seeking" pole and a "south-seeking" pole. This wording means that the north-seeking pole points to the north geographic pole of the Earth, whereas the south-seeking pole points to the south geographic pole. Because the north pole of a magnet is attracted toward the north geographic pole of the Earth, the Earth's south magnetic pole is located near the north geographic pole and the Earth's north magnetic pole is located near the south geographic pole. In fact, the configuration of the Earth's magnetic field, pictured in Figure 29.3, is very much like the one that would be achieved by burying a gigantic bar magnet deep in the Earth's interior. If a compass needle is supported by bearings that allow it to rotate in the vertical plane as well as in the horizontal plane, the needle is horizontal with respect to the Earth's surface only near the equator. As the compass is moved northward, the needle rotates so that it points more and more toward the Earth's surface. Finally, at a point near Hudson Bay in Canada, the north pole of the needle points directly downward. This site, first found in 1832, is considered to be the location of the south magnetic pole of the Earth. It is approximately 1 300 mi from the Earth's geographic

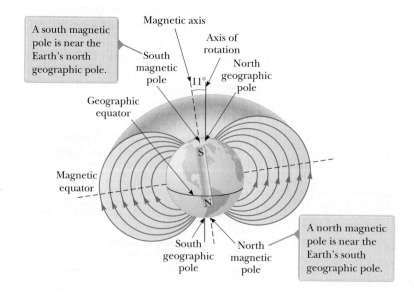

A south magnetic pole is near the Earth's north geographic pole.

Magnetic axis

Axis of rotation

South magnetic pole

North geographic pole

11°

Geographic equator

Magnetic equator

South geographic pole

North magnetic pole

A north magnetic pole is near the Earth's south geographic pole.

Figure 29.3 The Earth's magnetic field lines.

North Pole, and its exact position varies slowly with time. Similarly, the north magnetic pole of the Earth is about 1 200 mi away from the Earth's geographic South Pole.

Although the Earth's magnetic field pattern is similar to the one that would be set up by a bar magnet deep within the Earth, it is easy to understand why the source of this magnetic field cannot be large masses of permanently magnetized material. The Earth does have large deposits of iron ore deep beneath its surface, but the high temperatures in the Earth's core prevent the iron from retaining any permanent magnetization. Scientists consider it more likely that the source of the Earth's magnetic field is convection currents in the Earth's core. Charged ions or electrons circulating in the liquid interior could produce a magnetic field just like a current loop does, as we shall see in Chapter 30. There is also strong evidence that the magnitude of a planet's magnetic field is related to the planet's rate of rotation. For example, Jupiter rotates faster than the Earth, and space probes indicate that Jupiter's magnetic field is stronger than the Earth's. Venus, on the other hand, rotates more slowly than the Earth, and its magnetic field is found to be weaker. Investigation into the cause of the Earth's magnetism is ongoing.

The direction of the Earth's magnetic field has reversed several times during the last million years. Evidence for this reversal is provided by basalt, a type of rock that contains iron. Basalt forms from material spewed forth by volcanic activity on the ocean floor. As the lava cools, it solidifies and retains a picture of the Earth's magnetic field direction. The rocks are dated by other means to provide a time line for these periodic reversals of the magnetic field.

We can quantify the magnetic field $\vec{B}$ by using our model of a particle in a field, like the model discussed for gravity in Chapter 13 and for electricity in Chapter 23. The existence of a magnetic field at some point in space can be determined by measuring the **magnetic force** $\vec{F}_B$ exerted on an appropriate test particle placed at that point. This process is the same one we followed in defining the electric field in Chapter 23. If we perform such an experiment by placing a particle with charge q in the magnetic field, we find the following results that are similar to those for experiments on electric forces:

- The magnetic force is proportional to the charge q of the particle.
- The magnetic force on a negative charge is directed opposite to the force on a positive charge moving in the same direction.
- The magnetic force is proportional to the magnitude of the magnetic field vector $\vec{B}$.

We also find the following results, which are *totally different* from those for experiments on electric forces:

- The magnetic force is proportional to the speed v of the particle.
- If the velocity vector makes an angle θ with the magnetic field, the magnitude of the magnetic force is proportional to $\sin \theta$.
- When a charged particle moves *parallel* to the magnetic field vector, the magnetic force on the charge is zero.
- When a charged particle moves in a direction *not* parallel to the magnetic field vector, the magnetic force acts in a direction perpendicular to both $\vec{v}$ and $\vec{B}$; that is, the magnetic force is perpendicular to the plane formed by $\vec{v}$ and $\vec{B}$.

These results show that the magnetic force on a particle is more complicated than the electric force. The magnetic force is distinctive because it depends on the velocity of the particle and because its direction is perpendicular to both $\vec{v}$ and $\vec{B}$. Figure 29.4 (page 872) shows the details of the direction of the magnetic force on a charged

Figure 29.4 (a) The direction of the magnetic force $\vec{F}_B$ acting on a charged particle moving with a velocity $\vec{v}$ in the presence of a magnetic field $\vec{B}$. (b) Magnetic forces on positive and negative charges. The dashed lines show the paths of the particles, which are investigated in Section 29.2.

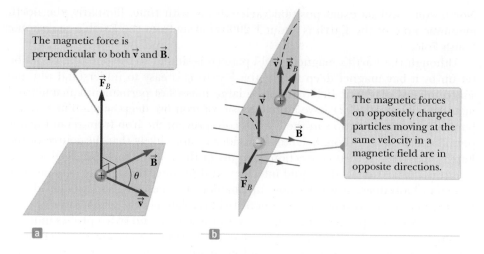

The magnetic force is perpendicular to both $\vec{v}$ and $\vec{B}$.

The magnetic forces on oppositely charged particles moving at the same velocity in a magnetic field are in opposite directions.

particle. Despite this complicated behavior, these observations can be summarized in a compact way by writing the magnetic force in the form

▶ Vector expression for the magnetic force on a charged particle moving in a magnetic field

$$\vec{F}_B = q\vec{v} \times \vec{B} \tag{29.1}$$

which by definition of the cross product (see Section 11.1) is perpendicular to both $\vec{v}$ and $\vec{B}$. We can regard this equation as an operational definition of the magnetic field at some point in space. That is, the magnetic field is defined in terms of the force acting on a moving charged particle. Equation 29.1 is the mathematical representation of the magnetic version of the **particle in a field** analysis model.

Figure 29.5 reviews two right-hand rules for determining the direction of the cross product $\vec{v} \times \vec{B}$ and determining the direction of $\vec{F}_B$. The rule in Figure 29.5a depends on our right-hand rule for the cross product in Figure 11.2. Point the four fingers of your right hand along the direction of $\vec{v}$ with the palm facing $\vec{B}$ and curl them toward $\vec{B}$. Your extended thumb, which is at a right angle to your fingers, points in the direction of $\vec{v} \times \vec{B}$. Because $\vec{F}_B = q\vec{v} \times \vec{B}$, $\vec{F}_B$ is in the direction of your thumb if q is positive and is opposite the direction of your thumb if q is negative. (If you need more help understanding the cross product, you should review Section 11.1, including Fig. 11.2.)

An alternative rule is shown in Figure 29.5b. Here the thumb points in the direction of $\vec{v}$ and the extended fingers in the direction of $\vec{B}$. Now, the force $\vec{F}_B$ on a positive charge extends outward from the palm. The advantage of this rule is that the force on the charge is in the direction you would push on something with your

Figure 29.5 Two right-hand rules for determining the direction of the magnetic force $\vec{F}_B = q\vec{v} \times \vec{B}$ acting on a particle with charge q moving with a velocity $\vec{v}$ in a magnetic field $\vec{B}$. (a) In this rule, the magnetic force is in the direction in which your thumb points. (b) In this rule, the magnetic force is in the direction of your palm, as if you are pushing the particle with your hand.

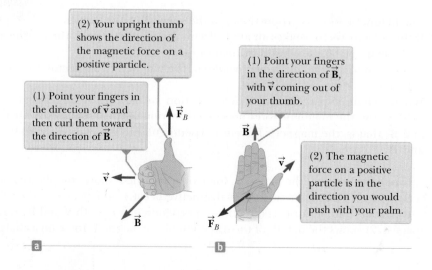

(2) Your upright thumb shows the direction of the magnetic force on a positive particle.

(1) Point your fingers in the direction of $\vec{v}$ and then curl them toward the direction of $\vec{B}$.

(1) Point your fingers in the direction of $\vec{B}$, with $\vec{v}$ coming out of your thumb.

(2) The magnetic force on a positive particle is in the direction you would push with your palm.

hand: outward from your palm. The force on a negative charge is in the opposite direction. You can use either of these two right-hand rules.

The magnitude of the magnetic force on a charged particle is

$$F_B = |q|vB \sin \theta \qquad \text{(29.2)}$$

◀ **Magnitude of the magnetic force on a charged particle moving in a magnetic field**

where θ is the smaller angle between $\vec{v}$ and $\vec{B}$. From this expression, we see that F_B is zero when $\vec{v}$ is parallel or antiparallel to $\vec{B}$ ($\theta = 0$ or $180°$) and maximum when $\vec{v}$ is perpendicular to $\vec{B}$ ($\theta = 90°$).

Let's compare the important differences between the electric and magnetic versions of the particle in a field model:

- The electric force vector is along the direction of the electric field, whereas the magnetic force vector is perpendicular to the magnetic field.
- The electric force acts on a charged particle regardless of whether the particle is moving, whereas the magnetic force acts on a charged particle only when the particle is in motion.
- The electric force does work in displacing a charged particle, whereas the magnetic force associated with a steady magnetic field does no work when a particle is displaced because the force is perpendicular to the displacement of its point of application.

From the last statement and on the basis of the work–kinetic energy theorem, we conclude that the kinetic energy of a charged particle moving through a magnetic field cannot be altered by the magnetic field alone. The field can alter the direction of the velocity vector, but it cannot change the speed or kinetic energy of the particle.

From Equation 29.2, we see that the SI unit of magnetic field is the newton per coulomb-meter per second, which is called the **tesla** (T):

$$1 \text{ T} = 1 \frac{\text{N}}{\text{C} \cdot \text{m/s}}$$

◀ **The tesla**

Because a coulomb per second is defined to be an ampere,

$$1 \text{ T} = 1 \frac{\text{N}}{\text{A} \cdot \text{m}}$$

A non-SI magnetic-field unit in common use, called the *gauss* (G), is related to the tesla through the conversion $1 \text{ T} = 10^4 \text{ G}$. Table 29.1 shows some typical values of magnetic fields.

Q uick Quiz 29.1 An electron moves in the plane of this paper toward the top of the page. A magnetic field is also in the plane of the page and directed toward the right. What is the direction of the magnetic force on the electron? **(a)** toward the top of the page **(b)** toward the bottom of the page **(c)** toward the left edge of the page **(d)** toward the right edge of the page **(e)** upward out of the page **(f)** downward into the page

Table 29.1 **Some Approximate Magnetic Field Magnitudes**

Source of Field	Field Magnitude (T)
Strong superconducting laboratory magnet	30
Strong conventional laboratory magnet	2
Medical MRI unit	1.5
Bar magnet	10^{-2}
Surface of the Sun	10^{-2}
Surface of the Earth	0.5×10^{-4}
Inside human brain (due to nerve impulses)	10^{-13}

Analysis Model	**Particle in a Field (Magnetic)**

Imagine some source (which we will investigate later) establishes a **magnetic field** $\vec{\mathbf{B}}$ throughout space. Now imagine a particle with charge q is placed in that field. The particle interacts with the magnetic field so that the particle experiences a magnetic force given by

$$\vec{\mathbf{F}}_B = q\vec{\mathbf{v}} \times \vec{\mathbf{B}} \qquad \textbf{(29.1)}$$

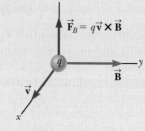

Examples:

- an ion moves in a circular path in the magnetic field of a mass spectrometer (Section 29.3)
- a coil in a motor rotates in response to the magnetic field in the motor (Chapter 31)
- a magnetic field is used to separate particles emitted by radioactive sources (Chapter 44)
- in a bubble chamber, particles created in collisions follow curved paths in a magnetic field, allowing the particles to be identified (Chapter 46)

Example 29.1	**An Electron Moving in a Magnetic Field** **AM**

An electron in an old-style television picture tube moves toward the front of the tube with a speed of 8.0×10^6 m/s along the x axis (Fig. 29.6). Surrounding the neck of the tube are coils of wire that create a magnetic field of magnitude 0.025 T, directed at an angle of 60° to the x axis and lying in the xy plane. Calculate the magnetic force on the electron.

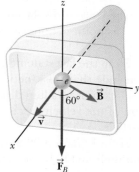

Figure 29.6 (Example 29.1) The magnetic force $\vec{\mathbf{F}}_B$ acting on the electron is in the negative z direction when $\vec{\mathbf{v}}$ and $\vec{\mathbf{B}}$ lie in the xy plane.

SOLUTION

Conceptualize Recall that the magnetic force on a charged particle is perpendicular to the plane formed by the velocity and magnetic field vectors. Use one of the right-hand rules in Figure 29.5 to convince yourself that the direction of the force on the electron is downward in Figure 29.6.

Categorize We evaluate the magnetic force using the *magnetic* version of the *particle in a field* model.

Analyze Use Equation 29.2 to find the magnitude of the magnetic force:

$$F_B = |q|vB \sin\theta$$
$$= (1.6 \times 10^{-19}\ \text{C})(8.0 \times 10^6\ \text{m/s})(0.025\ \text{T})(\sin 60°)$$
$$= \boxed{2.8 \times 10^{-14}\ \text{N}}$$

Finalize For practice using the vector product, evaluate this force in vector notation using Equation 29.1. The magnitude of the magnetic force may seem small to you, but remember that it is acting on a very small particle, the electron. To convince yourself that this is a substantial force for an electron, calculate the initial acceleration of the electron due to this force.

29.2 Motion of a Charged Particle in a Uniform Magnetic Field

Before we continue our discussion, some explanation of the notation used in this book is in order. To indicate the direction of $\vec{\mathbf{B}}$ in illustrations, we sometimes present perspective views such as those in Figure 29.6. If $\vec{\mathbf{B}}$ lies in the plane of the page or is present in a perspective drawing, we use green vectors or green field lines with arrowheads. In nonperspective illustrations, we depict a magnetic field perpendicular to and directed out of the page with a series of green dots, which represent the tips of arrows coming toward you (see Fig. 29.7a). In this case, the field is labeled

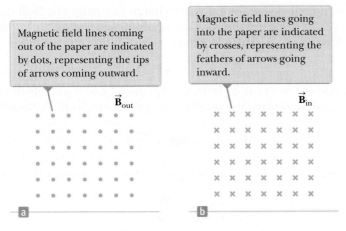

Figure 29.7 Representations of magnetic field lines perpendicular to the page.

Magnetic field lines coming out of the paper are indicated by dots, representing the tips of arrows coming outward.

$\vec{B}_{out}$

Magnetic field lines going into the paper are indicated by crosses, representing the feathers of arrows going inward.

$\vec{B}_{in}$

$\vec{B}_{out}$. If $\vec{B}$ is directed perpendicularly into the page, we use green crosses, which represent the feathered tails of arrows fired away from you, as in Figure 29.7b. In this case, the field is labeled $\vec{B}_{in}$, where the subscript "in" indicates "into the page." The same notation with crosses and dots is also used for other quantities that might be perpendicular to the page such as forces and current directions.

In Section 29.1, we found that the magnetic force acting on a charged particle moving in a magnetic field is perpendicular to the particle's velocity and consequently the work done by the magnetic force on the particle is zero. Now consider the special case of a positively charged particle moving in a uniform magnetic field with the initial velocity vector of the particle perpendicular to the field. Let's assume the direction of the magnetic field is into the page as in Figure 29.8. The particle in a field model tells us that the magnetic force on the particle is perpendicular to both the magnetic field lines and the velocity of the particle. The fact that there is a force on the particle tells us to apply the particle under a net force model to the particle. As the particle changes the direction of its velocity in response to the magnetic force, the magnetic force remains perpendicular to the velocity. As we found in Section 6.1, if the force is always perpendicular to the velocity, the path of the particle is a circle! Figure 29.8 shows the particle moving in a circle in a plane perpendicular to the magnetic field. Although magnetism and magnetic forces may be new and unfamiliar to you now, we see a magnetic effect that results in something with which we are familiar: the particle in uniform circular motion model!

The particle moves in a circle because the magnetic force $\vec{F}_B$ is perpendicular to $\vec{v}$ and $\vec{B}$ and has a constant magnitude qvB. As Figure 29.8 illustrates, the

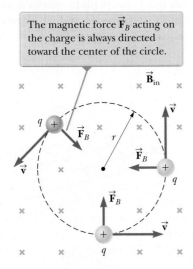

The magnetic force $\vec{F}_B$ acting on the charge is always directed toward the center of the circle.

$\vec{B}_{in}$

Figure 29.8 When the velocity of a charged particle is perpendicular to a uniform magnetic field, the particle moves in a circular path in a plane perpendicular to $\vec{B}$.

rotation is counterclockwise for a positive charge in a magnetic field directed into the page. If q were negative, the rotation would be clockwise. We use the particle under a net force model to write Newton's second law for the particle:

$$\sum F = F_B = ma$$

Because the particle moves in a circle, we also model it as a particle in uniform circular motion and we replace the acceleration with centripetal acceleration:

$$F_B = qvB = \frac{mv^2}{r}$$

This expression leads to the following equation for the radius of the circular path:

$$r = \frac{mv}{qB} \qquad \textbf{(29.3)}$$

That is, the radius of the path is proportional to the linear momentum mv of the particle and inversely proportional to the magnitude of the charge on the particle and to the magnitude of the magnetic field. The angular speed of the particle (from Eq. 10.10) is

$$\omega = \frac{v}{r} = \frac{qB}{m} \qquad \textbf{(29.4)}$$

The period of the motion (the time interval the particle requires to complete one revolution) is equal to the circumference of the circle divided by the speed of the particle:

$$T = \frac{2\pi r}{v} = \frac{2\pi}{\omega} = \frac{2\pi m}{qB} \qquad \textbf{(29.5)}$$

These results show that the angular speed of the particle and the period of the circular motion do not depend on the speed of the particle or on the radius of the orbit. The angular speed ω is often referred to as the **cyclotron frequency** because charged particles circulate at this angular frequency in the type of accelerator called a *cyclotron*, which is discussed in Section 29.3.

   If a charged particle moves in a uniform magnetic field with its velocity at some arbitrary angle with respect to $\vec{\mathbf{B}}$, its path is a helix. For example, if the field is directed in the x direction as shown in Figure 29.9, there is no component of force in the x direction. As a result, $a_x = 0$, and the x component of velocity remains constant. The charged particle is a particle in equilibrium in this direction. The magnetic force $q\vec{\mathbf{v}} \times \vec{\mathbf{B}}$ causes the components v_y and v_z to change in time, however, and the resulting motion is a helix whose axis is parallel to the magnetic field. The projection of the path onto the yz plane (viewed along the x axis) is a circle. (The projections of the path onto the xy and xz planes are sinusoids!) Equations 29.3 to 29.5 still apply provided v is replaced by $v_\perp = \sqrt{v_y^2 + v_z^2}$.

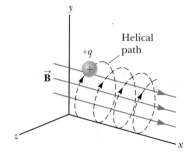

Figure 29.9 A charged particle having a velocity vector that has a component parallel to a uniform magnetic field moves in a helical path.

$\bigcirc$ uick Quiz 29.2 A charged particle is moving perpendicular to a magnetic field in a circle with a radius *r*. **(i)** An identical particle enters the field, with $\vec{\mathbf{v}}$ perpendicular to $\vec{\mathbf{B}}$, but with a higher speed than the first particle. Compared with the radius of the circle for the first particle, is the radius of the circular path for the second particle (a) smaller, (b) larger, or (c) equal in size? **(ii)** The magnitude of the magnetic field is increased. From the same choices, compare the radius of the new circular path of the first particle with the radius of its initial path.

Example 29.2   **A Proton Moving Perpendicular to a Uniform Magnetic Field**   AM

A proton is moving in a circular orbit of radius 14 cm in a uniform 0.35-T magnetic field perpendicular to the velocity of the proton. Find the speed of the proton.

SOLUTION

Conceptualize From our discussion in this section, we know the proton follows a circular path when moving perpendicular to a uniform magnetic field. In Chapter 39, we will learn that the highest possible speed for a particle is the speed of light, 3.00×10^8 m/s, so the speed of the particle in this problem must come out to be smaller than that value.

Categorize The proton is described by both the *particle in a field* model and the *particle in uniform circular motion* model. These models led to Equation 29.3.

Analyze

Solve Equation 29.3 for the speed of the particle:

$$v = \frac{qBr}{m_p}$$

Substitute numerical values:

$$v = \frac{(1.60 \times 10^{-19}\,\text{C})(0.35\,\text{T})(0.14\,\text{m})}{1.67 \times 10^{-27}\,\text{kg}}$$

$$= 4.7 \times 10^6\,\text{m/s}$$

Finalize The speed is indeed smaller than the speed of light, as required.

WHAT IF?   What if an electron, rather than a proton, moves in a direction perpendicular to the same magnetic field with this same speed? Will the radius of its orbit be different?

Answer An electron has a much smaller mass than a proton, so the magnetic force should be able to change its velocity much more easily than that for the proton. Therefore, we expect the radius to be smaller. Equation 29.3 shows that *r* is proportional to *m* with *q*, *B*, and *v* the same for the electron as for the proton. Consequently, the radius will be smaller by the same factor as the ratio of masses m_e/m_p.

Example 29.3   **Bending an Electron Beam**   AM

In an experiment designed to measure the magnitude of a uniform magnetic field, electrons are accelerated from rest through a potential difference of 350 V and then enter a uniform magnetic field that is perpendicular to the velocity vector of the electrons. The electrons travel along a curved path because of the magnetic force exerted on them, and the radius of the path is measured to be 7.5 cm. (Such a curved beam of electrons is shown in Fig. 29.10.)

(A) What is the magnitude of the magnetic field?

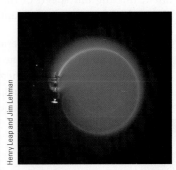

Henry Leap and Jim Lehman

Figure 29.10 (Example 29.3) The bending of an electron beam in a magnetic field.

continued

▶ **29.3** continued

SOLUTION

Conceptualize This example involves electrons accelerating from rest due to an electric force and then moving in a circular path due to a magnetic force. With the help of Figures 29.8 and 29.10, visualize the circular motion of the electrons.

Categorize Equation 29.3 shows that we need the speed v of the electron to find the magnetic field magnitude, and v is not given. Consequently, we must find the speed of the electron based on the potential difference through which it is accelerated. To do so, we categorize the first part of the problem by modeling an electron and the electric field as an *isolated system* in terms of *energy*. Once the electron enters the magnetic field, we categorize the second part of the problem as one involving a *particle in a field* and a *particle in uniform circular motion*, as we have done in this section.

Analyze Write the appropriate reduction of the conservation of energy equation, Equation 8.2, for the electron–electric field system:

$$\Delta K + \Delta U = 0$$

Substitute the appropriate initial and final energies:

$$\left(\tfrac{1}{2}m_e v^2 - 0\right) + (q\,\Delta V) = 0$$

Solve for the speed of the electron:

$$v = \sqrt{\frac{-2q\,\Delta V}{m_e}}$$

Substitute numerical values:

$$v = \sqrt{\frac{-2(-1.60 \times 10^{-19}\ \text{C})(350\ \text{V})}{9.11 \times 10^{-31}\ \text{kg}}} = 1.11 \times 10^7\ \text{m/s}$$

Now imagine the electron entering the magnetic field with this speed. Solve Equation 29.3 for the magnitude of the magnetic field:

$$B = \frac{m_e v}{er}$$

Substitute numerical values:

$$B = \frac{(9.11 \times 10^{-31}\ \text{kg})(1.11 \times 10^7\ \text{m/s})}{(1.60 \times 10^{-19}\ \text{C})(0.075\ \text{m})} = \boxed{8.4 \times 10^{-4}\ \text{T}}$$

(B) What is the angular speed of the electrons?

SOLUTION

Use Equation 10.10:

$$\omega = \frac{v}{r} = \frac{1.11 \times 10^7\ \text{m/s}}{0.075\ \text{m}} = \boxed{1.5 \times 10^8\ \text{rad/s}}$$

Finalize The angular speed can be represented as $\omega = (1.5 \times 10^8\ \text{rad/s})(1\ \text{rev}/2\pi\ \text{rad}) = 2.4 \times 10^7\ \text{rev/s}$. The electrons travel around the circle 24 million times per second! This answer is consistent with the very high speed found in part (A).

WHAT IF? What if a sudden voltage surge causes the accelerating voltage to increase to 400 V? How does that affect the angular speed of the electrons, assuming the magnetic field remains constant?

Answer The increase in accelerating voltage ΔV causes the electrons to enter the magnetic field with a higher speed v. This higher speed causes them to travel in a circle with a larger radius r. The angular speed is the ratio of v to r. Both v and r increase by the same factor, so the effects cancel and the angular speed remains the same. Equation 29.4 is an expression for the cyclotron frequency, which is the same as the angular speed of the electrons. The cyclotron frequency depends only on the charge q, the magnetic field B, and the mass m_e, none of which have changed. Therefore, the voltage surge has no effect on the angular speed. (In reality, however, the voltage surge may also increase the magnetic field if the magnetic field is powered by the same source as the accelerating voltage. In that case, the angular speed increases according to Eq. 29.4.)

When charged particles move in a nonuniform magnetic field, the motion is complex. For example, in a magnetic field that is strong at the ends and weak in the middle such as that shown in Figure 29.11, the particles can oscillate between two positions. A charged particle starting at one end spirals along the field lines until it reaches the other end, where it reverses its path and spirals back. This configura-

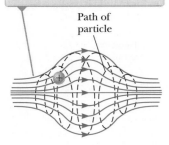

The magnetic force exerted on the particle near either end of the bottle has a component that causes the particle to spiral back toward the center.

Path of particle

Figure 29.11 A charged particle moving in a nonuniform magnetic field (a magnetic bottle) spirals about the field and oscillates between the endpoints.

Figure 29.12 The Van Allen belts are made up of charged particles trapped by the Earth's nonuniform magnetic field. The magnetic field lines are in green, and the particle paths are dashed black lines.

tion is known as a *magnetic bottle* because charged particles can be trapped within it. The magnetic bottle has been used to confine a *plasma,* a gas consisting of ions and electrons. Such a plasma-confinement scheme could fulfill a crucial role in the control of nuclear fusion, a process that could supply us in the future with an almost endless source of energy. Unfortunately, the magnetic bottle has its problems. If a large number of particles are trapped, collisions between them cause the particles to eventually leak from the system.

The Van Allen radiation belts consist of charged particles (mostly electrons and protons) surrounding the Earth in doughnut-shaped regions (Fig. 29.12). The particles, trapped by the Earth's nonuniform magnetic field, spiral around the field lines from pole to pole, covering the distance in only a few seconds. These particles originate mainly from the Sun, but some come from stars and other heavenly objects. For this reason, the particles are called *cosmic rays.* Most cosmic rays are deflected by the Earth's magnetic field and never reach the atmosphere. Some of the particles become trapped, however, and it is these particles that make up the Van Allen belts. When the particles are located over the poles, they sometimes collide with atoms in the atmosphere, causing the atoms to emit visible light. Such collisions are the origin of the beautiful aurora borealis, or northern lights, in the northern hemisphere and the aurora australis in the southern hemisphere. Auroras are usually confined to the polar regions because the Van Allen belts are nearest the Earth's surface there. Occasionally, though, solar activity causes larger numbers of charged particles to enter the belts and significantly distort the normal magnetic field lines associated with the Earth. In these situations, an aurora can sometimes be seen at lower latitudes.

29.3 Applications Involving Charged Particles Moving in a Magnetic Field

A charge moving with a velocity $\vec{\mathbf{v}}$ in the presence of both an electric field $\vec{\mathbf{E}}$ and a magnetic field $\vec{\mathbf{B}}$ is described by two particle in a field models. It experiences both an electric force $q\vec{\mathbf{E}}$ and a magnetic force $q\vec{\mathbf{v}} \times \vec{\mathbf{B}}$. The total force (called the Lorentz force) acting on the charge is

$$\vec{\mathbf{F}} = q\vec{\mathbf{E}} + q\vec{\mathbf{v}} \times \vec{\mathbf{B}} \qquad (29.6)$$

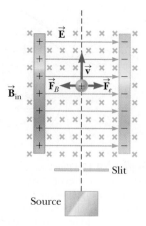

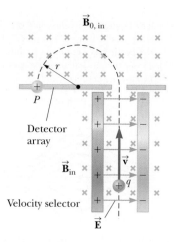

Figure 29.13 A velocity selector. When a positively charged particle is moving with velocity $\vec{v}$ in the presence of a magnetic field directed into the page and an electric field directed to the right, it experiences an electric force $q\vec{E}$ to the right and a magnetic force $q\vec{v} \times \vec{B}$ to the left.

Figure 29.14 A mass spectrometer. Positively charged particles are sent first through a velocity selector and then into a region where the magnetic field $\vec{B}_0$ causes the particles to move in a semicircular path and strike a detector array at P.

Velocity Selector

In many experiments involving moving charged particles, it is important that all particles move with essentially the same velocity, which can be achieved by applying a combination of an electric field and a magnetic field oriented as shown in Figure 29.13. A uniform electric field is directed to the right (in the plane of the page in Fig. 29.13), and a uniform magnetic field is applied in the direction perpendicular to the electric field (into the page in Fig. 29.13). If q is positive and the velocity $\vec{v}$ is upward, the magnetic force $q\vec{v} \times \vec{B}$ is to the left and the electric force $q\vec{E}$ is to the right. When the magnitudes of the two fields are chosen so that $qE = qvB$, the forces cancel. The charged particle is modeled as a particle in equilibrium and moves in a straight vertical line through the region of the fields. From the expression $qE = qvB$, we find that

$$v = \frac{E}{B}$$

(29.7)

Only those particles having this speed pass undeflected through the mutually perpendicular electric and magnetic fields. The magnetic force exerted on particles moving at speeds greater than that is stronger than the electric force, and the particles are deflected to the left. Those moving at slower speeds are deflected to the right.

The Mass Spectrometer

A **mass spectrometer** separates ions according to their mass-to-charge ratio. In one version of this device, known as the *Bainbridge mass spectrometer,* a beam of ions first passes through a velocity selector and then enters a second uniform magnetic field $\vec{B}_0$ that has the same direction as the magnetic field in the selector (Fig. 29.14). Upon entering the second magnetic field, the ions are described by the particle in uniform circular motion model. They move in a semicircle of radius r before striking a detector array at P. If the ions are positively charged, the beam deflects to the left as Figure 29.14 shows. If the ions are negatively charged, the beam deflects to the right. From Equation 29.3, we can express the ratio m/q as

$$\frac{m}{q} = \frac{rB_0}{v}$$

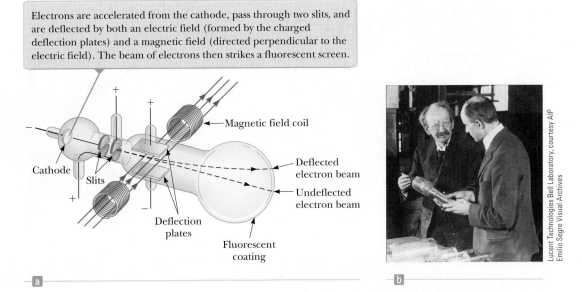

Electrons are accelerated from the cathode, pass through two slits, and are deflected by both an electric field (formed by the charged deflection plates) and a magnetic field (directed perpendicular to the electric field). The beam of electrons then strikes a fluorescent screen.

Magnetic field coil

Cathode

Slits

Deflection plates

Deflected electron beam

Undeflected electron beam

Fluorescent coating

Lucent Technologies Bell Laboratory, courtesy AIP Emilio Segre Visual Archives

Figure 29.15 (a) Thomson's apparatus for measuring e/m_e. (b) J. J. Thomson (*left*) in the Cavendish Laboratory, University of Cambridge. The man on the right, Frank Baldwin Jewett, is a distant relative of John W. Jewett, Jr., coauthor of this text.

Using Equation 29.7 gives

$$\frac{m}{q} = \frac{rB_0 B}{E} \tag{29.8}$$

Therefore, we can determine m/q by measuring the radius of curvature and knowing the field magnitudes B, B_0, and E. In practice, one usually measures the masses of various isotopes of a given ion, with the ions all carrying the same charge q. In this way, the mass ratios can be determined even if q is unknown.

A variation of this technique was used by J. J. Thomson (1856–1940) in 1897 to measure the ratio e/m_e for electrons. Figure 29.15a shows the basic apparatus he used. Electrons are accelerated from the cathode and pass through two slits. They then drift into a region of perpendicular electric and magnetic fields. The magnitudes of the two fields are first adjusted to produce an undeflected beam. When the magnetic field is turned off, the electric field produces a measurable beam deflection that is recorded on the fluorescent screen. From the size of the deflection and the measured values of E and B, the charge-to-mass ratio can be determined. The results of this crucial experiment represent the discovery of the electron as a fundamental particle of nature.

The Cyclotron

A **cyclotron** is a device that can accelerate charged particles to very high speeds. The energetic particles produced are used to bombard atomic nuclei and thereby produce nuclear reactions of interest to researchers. A number of hospitals use cyclotron facilities to produce radioactive substances for diagnosis and treatment.

Both electric and magnetic forces play key roles in the operation of a cyclotron, a schematic drawing of which is shown in Figure 29.16a (page 882). The charges move inside two semicircular containers D_1 and D_2, referred to as *dees* because of their shape like the letter D. A high-frequency alternating potential difference is applied to the dees, and a uniform magnetic field is directed perpendicular to them. A positive ion released at P near the center of the magnet in one dee moves in a semicircular path (indicated by the dashed black line in the drawing) and arrives back at the gap in a time interval $T/2$, where T is the time interval needed to make one complete trip around the two dees, given by Equation 29.5. The frequency

Pitfall Prevention 29.1

The Cyclotron Is Not the Only Type of Particle Accelerator The cyclotron is important historically because it was the first particle accelerator to produce particles with very high speeds. Cyclotrons still play important roles in medical applications and some research activities. Many other research activities make use of a different type of accelerator called a *synchrotron*.

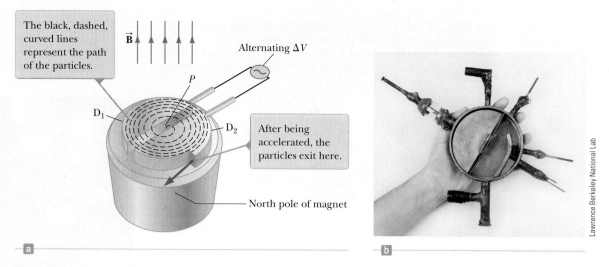

The black, dashed, curved lines represent the path of the particles.

$\vec{B}$

Alternating ΔV

P

D_1

D_2

After being accelerated, the particles exit here.

North pole of magnet

a

b

Figure 29.16 (a) A cyclotron consists of an ion source at P, two dees D_1 and D_2 across which an alternating potential difference is applied, and a uniform magnetic field. (The south pole of the magnet is not shown.) (b) The first cyclotron, invented by E. O. Lawrence and M. S. Livingston in 1934.

of the applied potential difference is adjusted so that the polarity of the dees is reversed in the same time interval during which the ion travels around one dee. If the applied potential difference is adjusted such that D_1 is at a lower electric potential than D_2 by an amount ΔV, the ion accelerates across the gap to D_1 and its kinetic energy increases by an amount $q \Delta V$. It then moves around D_1 in a semicircular path of greater radius (because its speed has increased). After a time interval $T/2$, it again arrives at the gap between the dees. By this time, the polarity across the dees has again been reversed and the ion is given another "kick" across the gap. The motion continues so that for each half-circle trip around one dee, the ion gains additional kinetic energy equal to $q \Delta V$. When the radius of its path is nearly that of the dees, the energetic ion leaves the system through the exit slit. The cyclotron's operation depends on T being independent of the speed of the ion and of the radius of the circular path (Eq. 29.5).

We can obtain an expression for the kinetic energy of the ion when it exits the cyclotron in terms of the radius R of the dees. From Equation 29.3, we know that $v = qBR/m$. Hence, the kinetic energy is

$$K = \tfrac{1}{2}mv^2 = \frac{q^2 B^2 R^2}{2m} \tag{29.9}$$

When the energy of the ions in a cyclotron exceeds about 20 MeV, relativistic effects come into play. (Such effects are discussed in Chapter 39.) Observations show that T increases and the moving ions do not remain in phase with the applied potential difference. Some accelerators overcome this problem by modifying the period of the applied potential difference so that it remains in phase with the moving ions.

29.4 Magnetic Force Acting on a Current-Carrying Conductor

If a magnetic force is exerted on a single charged particle when the particle moves through a magnetic field, it should not surprise you that a current-carrying wire also experiences a force when placed in a magnetic field. The current is a collection of many charged particles in motion; hence, the resultant force exerted by the field on the wire is the vector sum of the individual forces exerted on all the charged particles making up the current. The force exerted on the particles is transmitted to the wire when the particles collide with the atoms making up the wire.

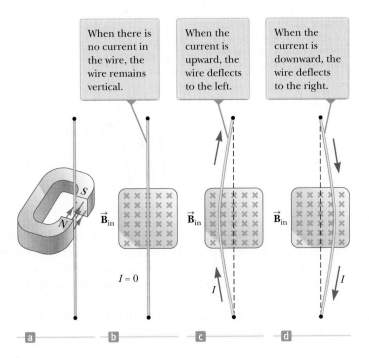

When there is no current in the wire, the wire remains vertical.

When the current is upward, the wire deflects to the left.

When the current is downward, the wire deflects to the right.

$I = 0$

$\vec{\mathbf{B}}_{in}$ $\vec{\mathbf{B}}_{in}$ $\vec{\mathbf{B}}_{in}$

I I

a   b   c   d

Figure 29.17 (a) A wire suspended vertically between the poles of a magnet. (b)–(d) The setup shown in (a) as seen looking at the south pole of the magnet so that the magnetic field (green crosses) is directed into the page.

One can demonstrate the magnetic force acting on a current-carrying conductor by hanging a wire between the poles of a magnet as shown in Figure 29.17a. For ease in visualization, part of the horseshoe magnet in part (a) is removed to show the end face of the south pole in parts (b) through (d) of Figure 29.17. The magnetic field is directed into the page and covers the region within the shaded squares. When the current in the wire is zero, the wire remains vertical as in Figure 29.17b. When the wire carries a current directed upward as in Figure 29.17c, however, the wire deflects to the left. If the current is reversed as in Figure 29.17d, the wire deflects to the right.

Let's quantify this discussion by considering a straight segment of wire of length L and cross-sectional area A carrying a current I in a uniform magnetic field $\vec{\mathbf{B}}$ as in Figure 29.18. According to the magnetic version of the particle in a field model, the magnetic force exerted on a charge q moving with a drift velocity $\vec{\mathbf{v}}_d$ is $q\vec{\mathbf{v}}_d \times \vec{\mathbf{B}}$. To find the total force acting on the wire, we multiply the force $q\vec{\mathbf{v}}_d \times \vec{\mathbf{B}}$ exerted on one charge by the number of charges in the segment. Because the volume of the segment is AL, the number of charges in the segment is nAL, where n is the number of mobile charge carriers per unit volume. Hence, the total magnetic force on the segment of wire of length L is

$$\vec{\mathbf{F}}_B = (q\vec{\mathbf{v}}_d \times \vec{\mathbf{B}})nAL$$

We can write this expression in a more convenient form by noting that, from Equation 27.4, the current in the wire is $I = nqv_dA$. Therefore,

$$\vec{\mathbf{F}}_B = I\vec{\mathbf{L}} \times \vec{\mathbf{B}} \tag{29.10}$$

where $\vec{\mathbf{L}}$ is a vector that points in the direction of the current I and has a magnitude equal to the length L of the segment. This expression applies only to a straight segment of wire in a uniform magnetic field.

Now consider an arbitrarily shaped wire segment of uniform cross section in a magnetic field as shown in Figure 29.19 (page 884). It follows from Equation 29.10 that the magnetic force exerted on a small segment of vector length $d\vec{\mathbf{s}}$ in the presence of a field $\vec{\mathbf{B}}$ is

$$d\vec{\mathbf{F}}_B = I\,d\vec{\mathbf{s}} \times \vec{\mathbf{B}} \tag{29.11}$$

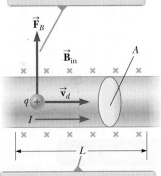

The average magnetic force exerted on a charge moving in the wire is $q\vec{\mathbf{v}}_d \times \vec{\mathbf{B}}$.

$\vec{\mathbf{F}}_B$ $\vec{\mathbf{B}}_{in}$ A

$q\,+$ $\vec{\mathbf{v}}_d$

I

L

The magnetic force on the wire segment of length L is $I\vec{\mathbf{L}} \times \vec{\mathbf{B}}$.

Figure 29.18 A segment of a current-carrying wire in a magnetic field $\vec{\mathbf{B}}$.

◀ Force on a segment of current-carrying wire in a uniform magnetic field

Figure 29.19 A wire segment of arbitrary shape carrying a current I in a magnetic field $\vec{\mathbf{B}}$ experiences a magnetic force.

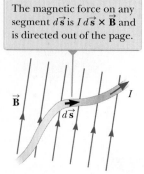

The magnetic force on any segment $d\vec{\mathbf{s}}$ is $I\,d\vec{\mathbf{s}} \times \vec{\mathbf{B}}$ and is directed out of the page.

where $d\vec{\mathbf{F}}_B$ is directed out of the page for the directions of $\vec{\mathbf{B}}$ and $d\vec{\mathbf{s}}$ in Figure 29.19. Equation 29.11 can be considered as an alternative definition of $\vec{\mathbf{B}}$. That is, we can define the magnetic field $\vec{\mathbf{B}}$ in terms of a measurable force exerted on a current element, where the force is a maximum when $\vec{\mathbf{B}}$ is perpendicular to the element and zero when $\vec{\mathbf{B}}$ is parallel to the element.

To calculate the total force $\vec{\mathbf{F}}_B$ acting on the wire shown in Figure 29.19, we integrate Equation 29.11 over the length of the wire:

$$\vec{\mathbf{F}}_B = I \int_a^b d\vec{\mathbf{s}} \times \vec{\mathbf{B}} \tag{29.12}$$

where a and b represent the endpoints of the wire. When this integration is carried out, the magnitude of the magnetic field and the direction the field makes with the vector $d\vec{\mathbf{s}}$ may differ at different points.

Quick Quiz 29.3 A wire carries current in the plane of this paper toward the top of the page. The wire experiences a magnetic force toward the right edge of the page. Is the direction of the magnetic field causing this force **(a)** in the plane of the page and toward the left edge, **(b)** in the plane of the page and toward the bottom edge, **(c)** upward out of the page, or **(d)** downward into the page?

Example 29.4   Force on a Semicircular Conductor

A wire bent into a semicircle of radius R forms a closed circuit and carries a current I. The wire lies in the xy plane, and a uniform magnetic field is directed along the positive y axis as in Figure 29.20. Find the magnitude and direction of the magnetic force acting on the straight portion of the wire and on the curved portion.

SOLUTION

Conceptualize Using the right-hand rule for cross products, we see that the force $\vec{\mathbf{F}}_1$ on the straight portion of the wire is out of the page and the force $\vec{\mathbf{F}}_2$ on the curved portion is into the page. Is $\vec{\mathbf{F}}_2$ larger in magnitude than $\vec{\mathbf{F}}_1$ because the length of the curved portion is longer than that of the straight portion?

Categorize Because we are dealing with a current-carrying wire in a magnetic field rather than a single charged particle, we must use Equation 29.12 to find the total force on each portion of the wire.

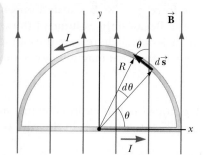

Figure 29.20 (Example 29.4) The magnetic force on the straight portion of the loop is directed out of the page, and the magnetic force on the curved portion is directed into the page.

Analyze Notice that $d\vec{\mathbf{s}}$ is perpendicular to $\vec{\mathbf{B}}$ everywhere on the straight portion of the wire. Use Equation 29.12 to find the force on this portion:

$$\vec{\mathbf{F}}_1 = I \int_a^b d\vec{\mathbf{s}} \times \vec{\mathbf{B}} = I \int_{-R}^R B\,dx\,\hat{\mathbf{k}} = \boxed{2IRB\,\hat{\mathbf{k}}}$$

▶ **29.4** continued

To find the magnetic force on the curved part, first write an expression for the magnetic force $d\vec{F}_2$ on the element $d\vec{s}$ in Figure 29.20:

(1) $d\vec{F}_2 = I\,d\vec{s} \times \vec{B} = -IB\sin\theta\,ds\,\hat{k}$

From the geometry in Figure 29.20, write an expression for ds:

(2) $ds = R\,d\theta$

Substitute Equation (2) into Equation (1) and integrate over the angle θ from 0 to π:

$$\vec{F}_2 = -\int_0^\pi IRB\sin\theta\,d\theta\,\hat{k} = -IRB\int_0^\pi \sin\theta\,d\theta\,\hat{k} = -IRB[-\cos\theta]_0^\pi\,\hat{k}$$
$$= IRB(\cos\pi - \cos 0)\hat{k} = IRB(-1-1)\hat{k} = \boxed{-2IRB\,\hat{k}}$$

....

Finalize Two very important general statements follow from this example. First, the force on the curved portion is the same in magnitude as the force on a straight wire between the same two points. In general, the magnetic force on a curved current-carrying wire in a uniform magnetic field is equal to that on a straight wire connecting the endpoints and carrying the same current. Furthermore, $\vec{F}_1 + \vec{F}_2 = 0$ is also a general result: the net magnetic force acting on any closed current loop in a uniform magnetic field is zero.

29.5 Torque on a Current Loop in a Uniform Magnetic Field

In Section 29.4, we showed how a magnetic force is exerted on a current-carrying conductor placed in a magnetic field. With that as a starting point, we now show that a torque is exerted on a current loop placed in a magnetic field.

Consider a rectangular loop carrying a current I in the presence of a uniform magnetic field directed parallel to the plane of the loop as shown in Figure 29.21a. No magnetic forces act on sides ① and ③ because these wires are parallel to the field; hence, $\vec{L} \times \vec{B} = 0$ for these sides. Magnetic forces do, however, act on sides ② and ④ because these sides are oriented perpendicular to the field. The magnitude of these forces is, from Equation 29.10,

$$F_2 = F_4 = IaB$$

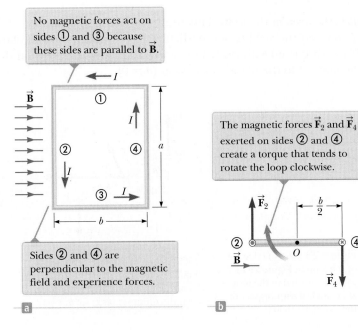

No magnetic forces act on sides ① and ③ because these sides are parallel to $\vec{B}$.

Sides ② and ④ are perpendicular to the magnetic field and experience forces.

The magnetic forces $\vec{F}_2$ and $\vec{F}_4$ exerted on sides ② and ④ create a torque that tends to rotate the loop clockwise.

Figure 29.21 (a) Overhead view of a rectangular current loop in a uniform magnetic field. (b) Edge view of the loop sighting down sides ② and ④. The purple dot in the left circle represents current in wire ② coming toward you; the purple cross in the right circle represents current in wire ④ moving away from you.

The direction of $\vec{\mathbf{F}}_2$, the magnetic force exerted on wire ②, is out of the page in the view shown in Figure 29.20a and that of $\vec{\mathbf{F}}_4$, the magnetic force exerted on wire ④, is into the page in the same view. If we view the loop from side ③ and sight along sides ② and ④, we see the view shown in Figure 29.21b, and the two magnetic forces $\vec{\mathbf{F}}_2$ and $\vec{\mathbf{F}}_4$ are directed as shown. Notice that the two forces point in opposite directions but are *not* directed along the same line of action. If the loop is pivoted so that it can rotate about point O, these two forces produce about O a torque that rotates the loop clockwise. The magnitude of this torque τ_{max} is

$$\tau_{max} = F_2 \frac{b}{2} + F_4 \frac{b}{2} = (IaB)\frac{b}{2} + (IaB)\frac{b}{2} = IabB$$

where the moment arm about O is $b/2$ for each force. Because the area enclosed by the loop is $A = ab$, we can express the maximum torque as

$$\tau_{max} = IAB \qquad\qquad \textbf{(29.13)}$$

This maximum-torque result is valid only when the magnetic field is parallel to the plane of the loop. The sense of the rotation is clockwise when viewed from side ③ as indicated in Figure 29.21b. If the current direction were reversed, the force directions would also reverse and the rotational tendency would be counterclockwise.

Now suppose the uniform magnetic field makes an angle $\theta < 90°$ with a line perpendicular to the plane of the loop as in Figure 29.22. For convenience, let's assume $\vec{\mathbf{B}}$ is perpendicular to sides ② and ④. In this case, the magnetic forces $\vec{\mathbf{F}}_1$ and $\vec{\mathbf{F}}_3$ exerted on sides ① and ③ cancel each other and produce no torque because they act along the same line. The magnetic forces $\vec{\mathbf{F}}_2$ and $\vec{\mathbf{F}}_4$ acting on sides ② and ④, however, produce a torque about *any point*. Referring to the edge view shown in Figure 29.22, we see that the moment arm of $\vec{\mathbf{F}}_2$ about the point O is equal to $(b/2) \sin \theta$. Likewise, the moment arm of $\vec{\mathbf{F}}_4$ about O is also equal to $(b/2) \sin \theta$. Because $F_2 = F_4 = IaB$, the magnitude of the net torque about O is

$$\tau = F_2 \frac{b}{2} \sin \theta + F_4 \frac{b}{2} \sin \theta$$

$$= IaB\left(\frac{b}{2} \sin \theta\right) + IaB\left(\frac{b}{2} \sin \theta\right) = IabB \sin \theta$$

$$= IAB \sin \theta$$

where $A = ab$ is the area of the loop. This result shows that the torque has its maximum value IAB when the field is perpendicular to the normal to the plane of the loop ($\theta = 90°$) as discussed with regard to Figure 29.21 and is zero when the field is parallel to the normal to the plane of the loop ($\theta = 0$).

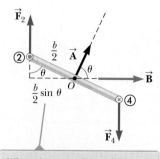

Figure 29.22 An edge view of the loop in Figure 29.21 with the normal to the loop at an angle θ with respect to the magnetic field.

When the normal to the loop makes an angle θ with the magnetic field, the moment arm for the torque is $(b/2) \sin \theta$.

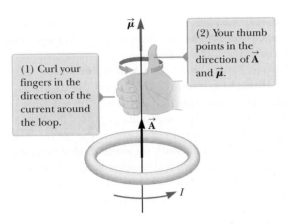

Figure 29.23 Right-hand rule for determining the direction of the vector $\vec{\mathbf{A}}$ for a current loop. The direction of the magnetic moment $\vec{\boldsymbol{\mu}}$ is the same as the direction of $\vec{\mathbf{A}}$.

A convenient vector expression for the torque exerted on a loop placed in a uniform magnetic field $\vec{\mathbf{B}}$ is

$$\vec{\boldsymbol{\tau}} = I\vec{\mathbf{A}} \times \vec{\mathbf{B}} \qquad (29.14)$$

◀ **Torque on a current loop in a magnetic field**

where $\vec{\mathbf{A}}$, the vector shown in Figure 29.22, is perpendicular to the plane of the loop and has a magnitude equal to the area of the loop. To determine the direction of $\vec{\mathbf{A}}$, use the right-hand rule described in Figure 29.23. When you curl the fingers of your right hand in the direction of the current in the loop, your thumb points in the direction of $\vec{\mathbf{A}}$. Figure 29.22 shows that the loop tends to rotate in the direction of decreasing values of θ (that is, such that the area vector $\vec{\mathbf{A}}$ rotates toward the direction of the magnetic field).

The product $I\vec{\mathbf{A}}$ is defined to be the **magnetic dipole moment** $\vec{\boldsymbol{\mu}}$ (often simply called the "magnetic moment") of the loop:

$$\vec{\boldsymbol{\mu}} \equiv I\vec{\mathbf{A}} \qquad (29.15)$$

◀ **Magnetic dipole moment of a current loop**

The SI unit of magnetic dipole moment is the ampere-meter^2 (A · m^2). If a coil of wire contains N loops of the same area, the magnetic moment of the coil is

$$\vec{\boldsymbol{\mu}}_{\text{coil}} = NI\vec{\mathbf{A}} \qquad (29.16)$$

Using Equation 29.15, we can express the torque exerted on a current-carrying loop in a magnetic field $\vec{\mathbf{B}}$ as

$$\vec{\boldsymbol{\tau}} = \vec{\boldsymbol{\mu}} \times \vec{\mathbf{B}} \qquad (29.17)$$

◀ **Torque on a magnetic moment in a magnetic field**

This result is analogous to Equation 26.18, $\vec{\boldsymbol{\tau}} = \vec{\mathbf{p}} \times \vec{\mathbf{E}}$, for the torque exerted on an electric dipole in the presence of an electric field $\vec{\mathbf{E}}$, where $\vec{\mathbf{p}}$ is the electric dipole moment.

Although we obtained the torque for a particular orientation of $\vec{\mathbf{B}}$ with respect to the loop, the equation $\vec{\boldsymbol{\tau}} = \vec{\boldsymbol{\mu}} \times \vec{\mathbf{B}}$ is valid for any orientation. Furthermore, although we derived the torque expression for a rectangular loop, the result is valid for a loop of any shape. The torque on an N-turn coil is given by Equation 29.17 by using Equation 29.16 for the magnetic moment.

In Section 26.6, we found that the potential energy of a system of an electric dipole in an electric field is given by $U_E = -\vec{\mathbf{p}} \cdot \vec{\mathbf{E}}$. This energy depends on the orientation of the dipole in the electric field. Likewise, the potential energy of a system of a magnetic dipole in a magnetic field depends on the orientation of the dipole in the magnetic field and is given by

Potential energy of a system of a magnetic moment in

$$U_B = -\vec{\boldsymbol{\mu}} \cdot \vec{\mathbf{B}} \qquad (29.18)$$

◀ **a magnetic field**

This expression shows that the system has its lowest energy $U_{min} = -\mu B$ when $\vec{\mu}$ points in the same direction as $\vec{B}$. The system has its highest energy $U_{max} = +\mu B$ when $\vec{\mu}$ points in the direction opposite $\vec{B}$.

Imagine the loop in Figure 29.22 is pivoted at point O on sides ① and ③, so that it is free to rotate. If the loop carries current and the magnetic field is turned on, the loop is modeled as a rigid object under a net torque, with the torque given by Equation 29.17. The torque on the current loop causes the loop to rotate; this effect is exploited practically in a **motor.** Energy enters the motor by electrical transmission, and the rotating coil can do work on some device external to the motor. For example, the motor in a car's electrical window system does work on the windows, applying a force on them and moving them up or down through some displacement. We will discuss motors in more detail in Section 31.5.

Quick Quiz 29.4 **(i)** Rank the magnitudes of the torques acting on the rectangular loops (a), (b), and (c) shown edge-on in Figure 29.24 from highest to lowest. All loops are identical and carry the same current. **(ii)** Rank the magnitudes of the net forces acting on the rectangular loops shown in Figure 29.24 from highest to lowest.

Figure 29.24 (Quick Quiz 29.4) Which current loop (seen edge-on) experiences the greatest torque, (a), (b), or (c)? Which experiences the greatest net force?

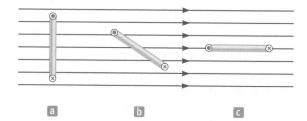

Example 29.5   The Magnetic Dipole Moment of a Coil

A rectangular coil of dimensions 5.40 cm × 8.50 cm consists of 25 turns of wire and carries a current of 15.0 mA. A 0.350-T magnetic field is applied parallel to the plane of the coil.

(A) Calculate the magnitude of the magnetic dipole moment of the coil.

SOLUTION

Conceptualize The magnetic moment of the coil is independent of any magnetic field in which the loop resides, so it depends only on the geometry of the loop and the current it carries.

Categorize We evaluate quantities based on equations developed in this section, so we categorize this example as a substitution problem.

Use Equation 29.16 to calculate the magnetic moment associated with a coil consisting of N turns:

$$\mu_{coil} = NIA = (25)(15.0 \times 10^{-3}\,\text{A})(0.054\,0\,\text{m})(0.085\,0\,\text{m})$$
$$= \boxed{1.72 \times 10^{-3}\,\text{A}\cdot\text{m}^2}$$

(B) What is the magnitude of the torque acting on the loop?

SOLUTION

Use Equation 29.17, noting that $\vec{B}$ is perpendicular to $\vec{\mu}_{coil}$:

$$\tau = \mu_{coil}B = (1.72 \times 10^{-3}\,\text{A}\cdot\text{m}^2)(0.350\,\text{T})$$
$$= \boxed{6.02 \times 10^{-4}\,\text{N}\cdot\text{m}}$$

Example 29.6 Rotating a Coil

Consider the loop of wire in Figure 29.25a. Imagine it is pivoted along side ④, which is parallel to the z axis and fastened so that side ④ remains fixed and the rest of the loop hangs vertically in the gravitational field of the Earth but can rotate around side ④ (Fig. 29.25b). The mass of the loop is 50.0 g, and the sides are of lengths $a = 0.200$ m and $b = 0.100$ m. The loop carries a current of 3.50 A and is immersed in a vertical uniform magnetic field of magnitude 0.010 0 T in the positive y direction (Fig. 29.25c). What angle does the plane of the loop make with the vertical?

SOLUTION

Conceptualize In the edge view of Figure 29.25b, notice that the magnetic moment of the loop is to the left. Therefore, when the loop is in the magnetic field, the magnetic torque on the loop causes it to rotate in a clockwise direction around side ④, which we choose as the rotation axis. Imagine the loop making this clockwise rotation so that the plane of the loop is at some angle θ to the vertical as in Figure 29.25c. The gravitational force on the loop exerts a torque that would cause a rotation in the counterclockwise direction if the magnetic field were turned off.

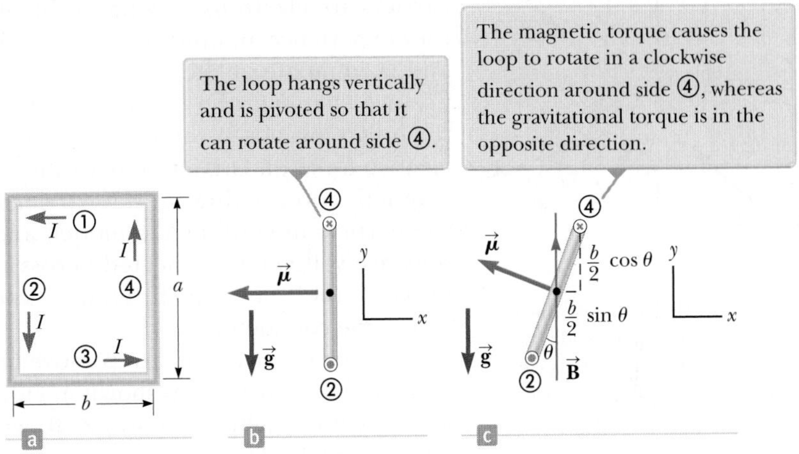

The loop hangs vertically and is pivoted so that it can rotate around side ④.

The magnetic torque causes the loop to rotate in a clockwise direction around side ④, whereas the gravitational torque is in the opposite direction.

Figure 29.25 (Example 29.6) (a) The dimensions of a rectangular current loop. (b) Edge view of the loop sighting down sides ② and ④. (c) An edge view of the loop in (b) rotated through an angle with respect to the horizontal when it is placed in a magnetic field.

Categorize At some angle of the loop, the two torques described in the Conceptualize step are equal in magnitude and the loop is at rest. We therefore model the loop as a *rigid object in equilibrium*.

Analyze Evaluate the magnetic torque on the loop about side ④ from Equation 29.17:

$$\vec{\boldsymbol{\tau}}_B = \vec{\boldsymbol{\mu}} \times \vec{\mathbf{B}} = -\mu B \sin(90° - \theta)\hat{\mathbf{k}} = -IAB \cos\theta\,\hat{\mathbf{k}} = -IabB \cos\theta\,\hat{\mathbf{k}}$$

Evaluate the gravitational torque on the loop, noting that the gravitational force can be modeled to act at the center of the loop:

$$\vec{\boldsymbol{\tau}}_g = \vec{\mathbf{r}} \times m\vec{\mathbf{g}} = mg\frac{b}{2}\sin\theta\,\hat{\mathbf{k}}$$

From the rigid body in equilibrium model, add the torques and set the net torque equal to zero:

$$\sum \vec{\boldsymbol{\tau}} = -IabB \cos\theta\,\hat{\mathbf{k}} + mg\frac{b}{2}\sin\theta\,\hat{\mathbf{k}} = 0$$

Solve for θ:

$$IabB \cos\theta = mg\frac{b}{2}\sin\theta \;\;\rightarrow\;\; \tan\theta = \frac{2IaB}{mg}$$

$$\theta = \tan^{-1}\left(\frac{2IaB}{mg}\right)$$

Substitute numerical values:

$$\theta = \tan^{-1}\left[\frac{2(3.50\text{ A})(0.200\text{ m})(0.010\ 0\text{ T})}{(0.050\ 0\text{ kg})(9.80\text{ m/s}^2)}\right] = \boxed{1.64°}$$

Finalize The angle is relatively small, so the loop still hangs almost vertically. If the current I or the magnetic field B is increased, however, the angle increases as the magnetic torque becomes stronger.

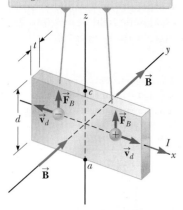

When I is in the x direction and $\vec{\mathbf{B}}$ in the y direction, both positive and negative charge carriers are deflected upward in the magnetic field.

Figure 29.26 To observe the Hall effect, a magnetic field is applied to a current-carrying conductor. The Hall voltage is measured between points a and c.

29.6 The Hall Effect

When a current-carrying conductor is placed in a magnetic field, a potential difference is generated in a direction perpendicular to both the current and the magnetic field. This phenomenon, first observed by Edwin Hall (1855–1938) in 1879, is known as the *Hall effect*. The arrangement for observing the Hall effect consists of a flat conductor carrying a current I in the x direction as shown in Figure 29.26. A uniform magnetic field $\vec{\mathbf{B}}$ is applied in the y direction. If the charge carriers are electrons moving in the negative x direction with a drift velocity $\vec{\mathbf{v}}_d$, they experience an upward magnetic force $\vec{\mathbf{F}}_B = q\vec{\mathbf{v}}_d \times \vec{\mathbf{B}}$, are deflected upward, and accumulate at the upper edge of the flat conductor, leaving an excess of positive charge at the lower edge (Fig. 29.27a). This accumulation of charge at the edges establishes an electric field in the conductor and increases until the electric force on carriers remaining in the bulk of the conductor balances the magnetic force acting on the carriers. The electrons can now be described by the particle in equilibrium model, and they are no longer deflected upward. A sensitive voltmeter connected across the sample as shown in Figure 29.27 can measure the potential difference, known as the **Hall voltage** ΔV_H, generated across the conductor.

If the charge carriers are positive and hence move in the positive x direction (for rightward current) as shown in Figures 29.26 and 29.27b, they also experience an upward magnetic force $q\vec{\mathbf{v}}_d \times \vec{\mathbf{B}}$, which produces a buildup of positive charge on the upper edge and leaves an excess of negative charge on the lower edge. Hence, the sign of the Hall voltage generated in the sample is opposite the sign of the Hall voltage resulting from the deflection of electrons. The sign of the charge carriers can therefore be determined from measuring the polarity of the Hall voltage.

In deriving an expression for the Hall voltage, first note that the magnetic force exerted on the carriers has magnitude qv_dB. In equilibrium, this force is balanced by the electric force qE_H, where E_H is the magnitude of the electric field due to the charge separation (sometimes referred to as the *Hall field*). Therefore,

$$qv_dB = qE_H$$

$$E_H = v_dB$$

If d is the width of the conductor, the Hall voltage is

$$\Delta V_H = E_H d = v_d B d \tag{29.19}$$

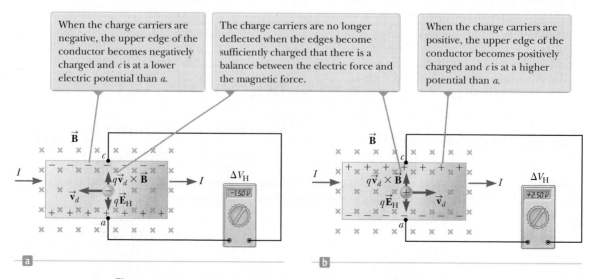

When the charge carriers are negative, the upper edge of the conductor becomes negatively charged and c is at a lower electric potential than a.

The charge carriers are no longer deflected when the edges become sufficiently charged that there is a balance between the electric force and the magnetic force.

When the charge carriers are positive, the upper edge of the conductor becomes positively charged and c is at a higher potential than a.

Figure 29.27 The sign of the Hall voltage depends on the sign of the charge carriers.

Therefore, the measured Hall voltage gives a value for the drift speed of the charge carriers if d and B are known.

We can obtain the charge-carrier density n by measuring the current in the sample. From Equation 27.4, we can express the drift speed as

$$v_d = \frac{I}{nqA} \qquad \text{(29.20)}$$

where A is the cross-sectional area of the conductor. Substituting Equation 29.20 into Equation 29.19 gives

$$\Delta V_H = \frac{IBd}{nqA} \qquad \text{(29.21)}$$

Because $A = td$, where t is the thickness of the conductor, we can also express Equation 29.21 as

$$\Delta V_H = \frac{IB}{nqt} = \frac{R_H IB}{t} \qquad \text{(29.22)} \qquad \blacktriangleleft \text{ The Hall voltage}$$

where $R_H = 1/nq$ is called the **Hall coefficient.** This relationship shows that a properly calibrated conductor can be used to measure the magnitude of an unknown magnetic field.

Because all quantities in Equation 29.22 other than nq can be measured, a value for the Hall coefficient is readily obtainable. The sign and magnitude of R_H give the sign of the charge carriers and their number density. In most metals, the charge carriers are electrons and the charge-carrier density determined from Hall-effect measurements is in good agreement with calculated values for such metals as lithium (Li), sodium (Na), copper (Cu), and silver (Ag), whose atoms each give up one electron to act as a current carrier. In this case, n is approximately equal to the number of conducting electrons per unit volume. This classical model, however, is not valid for metals such as iron (Fe), bismuth (Bi), and cadmium (Cd) or for semiconductors. These discrepancies can be explained only by using a model based on the quantum nature of solids.

Example 29.7   The Hall Effect for Copper

A rectangular copper strip 1.5 cm wide and 0.10 cm thick carries a current of 5.0 A. Find the Hall voltage for a 1.2-T magnetic field applied in a direction perpendicular to the strip.

SOLUTION

Conceptualize Study Figures 29.26 and 29.27 carefully and make sure you understand that a Hall voltage is developed between the top and bottom edges of the strip.

Categorize We evaluate the Hall voltage using an equation developed in this section, so we categorize this example as a substitution problem.

Assuming one electron per atom is available for conduction, find the charge-carrier density in terms of the molar mass M and density ρ of copper:

$$n = \frac{N_A}{V} = \frac{N_A \rho}{M}$$

Substitute this result into Equation 29.22:

$$\Delta V_H = \frac{IB}{nqt} = \frac{MIB}{N_A \rho qt}$$

Substitute numerical values:

$$\Delta V_H = \frac{(0.063\ 5\ \text{kg/mol})(5.0\ \text{A})(1.2\ \text{T})}{(6.02 \times 10^{23}\ \text{mol}^{-1})(8\ 920\ \text{kg/m}^3)(1.60 \times 10^{-19}\ \text{C})(0.001\ 0\ \text{m})}$$

$$= \boxed{0.44\ \mu\text{V}}$$

continued

▶ **29.7** continued

Such an extremely small Hall voltage is expected in good conductors. (Notice that the width of the conductor is not needed in this calculation.)

WHAT IF?   What if the strip has the same dimensions but is made of a semiconductor? Will the Hall voltage be smaller or larger?

Answer In semiconductors, n is much smaller than it is in metals that contribute one electron per atom to the current; hence, the Hall voltage is usually larger because it varies as the inverse of n. Currents on the order of 0.1 mA are generally used for such materials. Consider a piece of silicon that has the same dimensions as the copper strip in this example and whose value for n is 1.0×10^{20} electrons/m^3. Taking $B = 1.2$ T and $I = 0.10$ mA, we find that $\Delta V_H = 7.5$ mV. A potential difference of this magnitude is readily measured.

Summary

Definition

■ The **magnetic dipole moment** $\vec{\mu}$ of a loop carrying a current I is

$$\vec{\mu} \equiv I\vec{A} \tag{29.15}$$

where the area vector $\vec{A}$ is perpendicular to the plane of the loop and $|\vec{A}|$ is equal to the area of the loop. The SI unit of $\vec{\mu}$ is A · m^2.

Concepts and Principles

■ If a charged particle moves in a uniform magnetic field so that its initial velocity is perpendicular to the field, the particle moves in a circle, the plane of which is perpendicular to the magnetic field. The radius of the circular path is

$$r = \frac{mv}{qB} \tag{29.3}$$

where m is the mass of the particle and q is its charge. The angular speed of the charged particle is

$$\omega = \frac{qB}{m} \tag{29.4}$$

■ If a straight conductor of length L carries a current I, the force exerted on that conductor when it is placed in a uniform magnetic field $\vec{B}$ is

$$\vec{F}_B = I\vec{L} \times \vec{B} \tag{29.10}$$

where the direction of $\vec{L}$ is in the direction of the current and $|\vec{L}| = L$.

■ If an arbitrarily shaped wire carrying a current I is placed in a magnetic field, the magnetic force exerted on a very small segment $d\vec{s}$ is

$$d\vec{F}_B = I\,d\vec{s} \times \vec{B} \tag{29.11}$$

To determine the total magnetic force on the wire, one must integrate Equation 29.11 over the wire, keeping in mind that both $\vec{B}$ and $d\vec{s}$ may vary at each point.

■ The torque $\vec{\tau}$ on a current loop placed in a uniform magnetic field $\vec{B}$ is

$$\vec{\tau} = \vec{\mu} \times \vec{B} \tag{29.17}$$

■ The potential energy of the system of a magnetic dipole in a magnetic field is

$$U_B = -\vec{\mu} \cdot \vec{B} \tag{29.18}$$

Analysis Models for Problem Solving

Particle in a Field (Magnetic) A source (to be discussed in Chapter 30) establishes a **magnetic field** $\vec{\mathbf{B}}$ throughout space. When a particle with charge q and moving with velocity $\vec{\mathbf{v}}$ is placed in that field, it experiences a magnetic force given by

$$\vec{\mathbf{F}}_B = q\vec{\mathbf{v}} \times \vec{\mathbf{B}} \qquad\qquad \textbf{(29.1)}$$

The direction of this magnetic force is perpendicular both to the velocity of the particle and to the magnetic field. The magnitude of this force is

$$F_B = |q|vB\sin\theta \qquad\qquad \textbf{(29.2)}$$

where θ is the smaller angle between $\vec{\mathbf{v}}$ and $\vec{\mathbf{B}}$. The SI unit of $\vec{\mathbf{B}}$ is the **tesla** (T), where $1\text{ T} = 1\text{ N/A}\cdot\text{m}$.

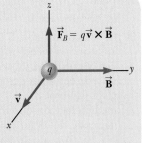

Objective Questions   1. denotes answer available in *Student Solutions Manual/Study Guide*

Objective Questions 3, 4, and 6 in Chapter 11 can be assigned with this chapter as review for the vector product.

1. A spatially uniform magnetic field cannot exert a magnetic force on a particle in which of the following circumstances? There may be more than one correct statement. (a) The particle is charged. (b) The particle moves perpendicular to the magnetic field. (c) The particle moves parallel to the magnetic field. (d) The magnitude of the magnetic field changes with time. (e) The particle is at rest.

2. Rank the magnitudes of the forces exerted on the following particles from largest to smallest. In your ranking, display any cases of equality. (a) an electron moving at 1 Mm/s perpendicular to a 1-mT magnetic field (b) an electron moving at 1 Mm/s parallel to a 1-mT magnetic field (c) an electron moving at 2 Mm/s perpendicular to a 1-mT magnetic field (d) a proton moving at 1 Mm/s perpendicular to a 1-mT magnetic field (e) a proton moving at 1 Mm/s at a 45° angle to a 1-mT magnetic field

3. A particle with electric charge is fired into a region of space where the electric field is zero. It moves in a straight line. Can you conclude that the magnetic field in that region is zero? (a) Yes, you can. (b) No; the field might be perpendicular to the particle's velocity. (c) No; the field might be parallel to the particle's velocity. (d) No; the particle might need to have charge of the opposite sign to have a force exerted on it. (e) No; an observation of an object with *electric* charge gives no information about a *magnetic* field.

4. A proton moving horizontally enters a region where a uniform magnetic field is directed perpendicular to the proton's velocity as shown in Figure OQ29.4. After the proton enters the field, does it (a) deflect downward, with its speed remaining constant; (b) deflect upward, moving in a semicircular path with constant speed, and exit the field moving to the left; (c) continue to move in the horizontal direction with constant velocity; (d) move in a circular orbit and become trapped by the field; or (e) deflect out of the plane of the paper?

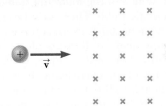

Figure OQ29.4

5. At a certain instant, a proton is moving in the positive x direction through a magnetic field in the negative z direction. What is the direction of the magnetic force exerted on the proton? (a) positive z direction (b) negative z direction (c) positive y direction (d) negative y direction (e) The force is zero.

6. A thin copper rod 1.00 m long has a mass of 50.0 g. What is the minimum current in the rod that would allow it to levitate above the ground in a magnetic field of magnitude 0.100 T? (a) 1.20 A (b) 2.40 A (c) 4.90 A (d) 9.80 A (e) none of those answers

7. Electron A is fired horizontally with speed 1.00 Mm/s into a region where a vertical magnetic field exists. Electron B is fired along the same path with speed 2.00 Mm/s. **(i)** Which electron has a larger magnetic force exerted on it? (a) A does. (b) B does. (c) The forces have the same nonzero magnitude. (d) The forces are both zero. **(ii)** Which electron has a path that curves more sharply? (a) A does. (b) B does. (c) The particles follow the same curved path. (d) The particles continue to go straight.

8. Classify each of the following statements as a characteristic (a) of electric forces only, (b) of magnetic forces only, (c) of both electric and magnetic forces, or (d) of neither electric nor magnetic forces. **(i)** The force is proportional to the magnitude of the field exerting it. **(ii)** The force is proportional to the magnitude of the charge of the object on which the force is exerted. **(iii)** The force exerted on a negatively charged object is opposite in direction to the force on a positive charge. **(iv)** The force exerted on a stationary charged object is nonzero. **(v)** The force exerted on a moving charged

object is zero. **(vi)** The force exerted on a charged object is proportional to its speed. **(vii)** The force exerted on a charged object cannot alter the object's speed. **(viii)** The magnitude of the force depends on the charged object's direction of motion.

9. An electron moves horizontally across the Earth's equator at a speed of 2.50×10^6 m/s and in a direction 35.0° N of E. At this point, the Earth's magnetic field has a direction due north, is parallel to the surface, and has a value of 3.00×10^{-5} T. What is the force acting on the electron due to its interaction with the Earth's magnetic field? (a) 6.88×10^{-18} N due west (b) 6.88×10^{-18} N toward the Earth's surface (c) 9.83×10^{-18} N toward the Earth's surface (d) 9.83×10^{-18} N away from the Earth's surface (e) 4.00×10^{-18} N away from the Earth's surface

10. A charged particle is traveling through a uniform magnetic field. Which of the following statements are true of the magnetic field? There may be more than one correct statement. (a) It exerts a force on the particle parallel to the field. (b) It exerts a force on the particle along the direction of its motion. (c) It increases the kinetic energy of the particle. (d) It exerts a force that is perpendicular to the direction of motion. (e) It does not change the magnitude of the momentum of the particle.

11. In the velocity selector shown in Figure 29.13, electrons with speed $v = E/B$ follow a straight path. Electrons moving significantly faster than this speed through the same selector will move along what kind of path? (a) a circle (b) a parabola (c) a straight line (d) a more complicated trajectory

12. Answer each question yes or no. Assume the motions and currents mentioned are along the x axis and fields are in the y direction. (a) Does an electric field exert a force on a stationary charged object? (b) Does a magnetic field do so? (c) Does an electric field exert a force on a moving charged object? (d) Does a magnetic field do so? (e) Does an electric field exert a force on a straight current-carrying wire? (f) Does a magnetic field do so? (g) Does an electric field exert a force on a beam of moving electrons? (h) Does a magnetic field do so?

13. A magnetic field exerts a torque on each of the current-carrying single loops of wire shown in Figure OQ29.13. The loops lie in the xy plane, each carrying the same magnitude current, and the uniform magnetic field points in the positive x direction. Rank the loops by the magnitude of the torque exerted on them by the field from largest to smallest.

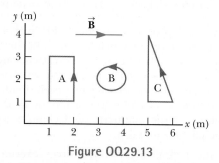

Figure OQ29.13

Conceptual Questions

1. denotes answer available in *Student Solutions Manual/Study Guide*

1. Can a constant magnetic field set into motion an electron initially at rest? Explain your answer.

2. Explain why it is not possible to determine the charge and the mass of a charged particle separately by measuring accelerations produced by electric and magnetic forces on the particle.

3. Is it possible to orient a current loop in a uniform magnetic field such that the loop does not tend to rotate? Explain.

4. How can the motion of a moving charged particle be used to distinguish between a magnetic field and an electric field? Give a specific example to justify your argument.

5. How can a current loop be used to determine the presence of a magnetic field in a given region of space?

6. Charged particles from outer space, called cosmic rays, strike the Earth more frequently near the poles than near the equator. Why?

7. Two charged particles are projected in the same direction into a magnetic field perpendicular to their velocities. If the particles are deflected in opposite directions, what can you say about them?

Problems

WebAssign The problems found in this chapter may be assigned online in Enhanced WebAssign

1. straightforward; **2.** intermediate;
3. challenging

1. full solution available in the *Student Solutions Manual/Study Guide*

AMT Analysis Model tutorial available in Enhanced WebAssign

GP Guided Problem

M Master It tutorial available in Enhanced WebAssign

W Watch It video solution available in Enhanced WebAssign

Section 29.1 Analysis Model: Particle in a Field (Magnetic)

> Problems 1–4, 6–7, and 10 in Chapter 11 can be assigned with this section as review for the vector product.

1. At the equator, near the surface of the Earth, the magnetic field is approximately 50.0 μT northward, and the electric field is about 100 N/C downward in fair weather. Find the gravitational, electric, and magnetic forces on an electron in this environment, assuming that the electron has an instantaneous velocity of 6.00×10^6 m/s directed to the east.

2. Determine the initial direction of the deflection of **W** charged particles as they enter the magnetic fields shown in Figure P29.2.

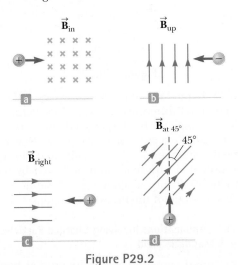

Figure P29.2

3. Find the direction of the magnetic field acting on a positively charged particle moving in the various situations shown in Figure P29.3 if the direction of the magnetic force acting on it is as indicated.

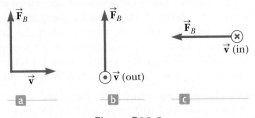

Figure P29.3

4. Consider an electron near the Earth's equator. In which direction does it tend to deflect if its velocity is (a) directed downward? (b) Directed northward? (c) Directed westward? (d) Directed southeastward?

5. A proton is projected into a magnetic field that is directed along the positive x axis. Find the direction of the magnetic force exerted on the proton for each of the following directions of the proton's velocity: (a) the positive y direction, (b) the negative y direction, (c) the positive x direction.

6. A proton moving at 4.00×10^6 m/s through a magnetic field of magnitude 1.70 T experiences a magnetic force of magnitude 8.20×10^{-13} N. What is the angle between the proton's velocity and the field?

7. An electron is accelerated through 2.40×10^3 V from **W** rest and then enters a uniform 1.70-T magnetic field. What are (a) the maximum and (b) the minimum values of the magnetic force this particle experiences?

8. A proton moves with a velocity of $\vec{\mathbf{v}} = $ **W** $(2\hat{\mathbf{i}} - 4\hat{\mathbf{j}} + \hat{\mathbf{k}})$ m/s in a region in which the magnetic field is $\vec{\mathbf{B}} = (\hat{\mathbf{i}} + 2\hat{\mathbf{j}} - \hat{\mathbf{k}})$ T. What is the magnitude of the magnetic force this particle experiences?

9. A proton travels with a speed of 5.02×10^6 m/s in a **AMT** direction that makes an angle of 60.0° with the direction of a magnetic field of magnitude 0.180 T in the positive x direction. What are the magnitudes of (a) the magnetic force on the proton and (b) the proton's acceleration?

10. A laboratory electromagnet produces a magnetic field of magnitude 1.50 T. A proton moves through this field with a speed of 6.00×10^6 m/s. (a) Find the magnitude of the maximum magnetic force that could be exerted on the proton. (b) What is the magnitude of the maximum acceleration of the proton? (c) Would the field exert the same magnetic force on an electron moving through the field with the same speed? (d) Would the electron experience the same acceleration? Explain.

11. A proton moves perpendicular to a uniform magnetic **M** field $\vec{\mathbf{B}}$ at a speed of 1.00×10^7 m/s and experiences an acceleration of 2.00×10^{13} m/s² in the positive x direction when its velocity is in the positive z direction. Determine the magnitude and direction of the field.

12. **Review.** A charged particle of mass 1.50 g is moving at a speed of 1.50×10^4 m/s. Suddenly, a uniform magnetic field of magnitude 0.150 mT in a direction perpendicular to the particle's velocity is turned on and then turned off in a time interval of 1.00 s. During this time interval, the magnitude and direction of the velocity of the particle undergo a negligible change, but the particle moves by a distance of 0.150 m in a direction perpendicular to the velocity. Find the charge on the particle.

Section 29.2 Motion of a Charged Particle in a Uniform Magnetic Field

13. An electron moves in a circular path perpendicular to a uniform magnetic field with a magnitude of 2.00 mT. If the speed of the electron is 1.50×10^7 m/s, determine (a) the radius of the circular path and (b) the time interval required to complete one revolution.

14. An accelerating voltage of 2.50×10^3 V is applied to an electron gun, producing a beam of electrons originally traveling horizontally north in vacuum toward the center of a viewing screen 35.0 cm away. What are (a) the magnitude and (b) the direction of the deflection on

the screen caused by the Earth's gravitational field? What are (c) the magnitude and (d) the direction of the deflection on the screen caused by the vertical component of the Earth's magnetic field, taken as 20.0 μT down? (e) Does an electron in this vertical magnetic field move as a projectile, with constant vector acceleration perpendicular to a constant northward component of velocity? (f) Is it a good approximation to assume it has this projectile motion? Explain.

15. A proton (charge $+e$, mass m_p), a deuteron (charge $+e$, mass $2m_p$), and an alpha particle (charge $+2e$, mass $4m_p$) are accelerated from rest through a common potential difference ΔV. Each of the particles enters a uniform magnetic field $\vec{B}$, with its velocity in a direction perpendicular to $\vec{B}$. The proton moves in a circular path of radius r_p. In terms of r_p, determine (a) the radius r_d of the circular orbit for the deuteron and (b) the radius r_α for the alpha particle.

16. A particle with charge q and kinetic energy K travels in a uniform magnetic field of magnitude B. If the particle moves in a circular path of radius R, find expressions for (a) its speed and (b) its mass.

17. **Review.** One electron collides elastically with a second **AMT** electron initially at rest. After the collision, the radii of their trajectories are 1.00 cm and 2.40 cm. The trajectories are perpendicular to a uniform magnetic field of magnitude 0.044 0 T. Determine the energy (in keV) of the incident electron.

18. **Review.** One electron collides elastically with a second electron initially at rest. After the collision, the radii of their trajectories are r_1 and r_2. The trajectories are perpendicular to a uniform magnetic field of magnitude B. Determine the energy of the incident electron.

19. **Review.** An electron moves in a circular path perpendicular to a constant magnetic field of magnitude 1.00 mT. The angular momentum of the electron about the center of the circle is 4.00×10^{-25} kg · m²/s. Determine (a) the radius of the circular path and (b) the speed of the electron.

20. **Review.** A 30.0-g metal ball having net charge $Q = 5.00\ \mu$C is thrown out of a window horizontally north at a speed $v = 20.0$ m/s. The window is at a height $h = 20.0$ m above the ground. A uniform, horizontal magnetic field of magnitude $B = 0.010\ 0$ T is perpendicular to the plane of the ball's trajectory and directed toward the west. (a) Assuming the ball follows the same trajectory as it would in the absence of the magnetic field, find the magnetic force acting on the ball just before it hits the ground. (b) Based on the result of part (a), is it justified for three-significant-digit precision to assume the trajectory is unaffected by the magnetic field? Explain.

21. A cosmic-ray proton in interstellar space has an energy **M** of 10.0 MeV and executes a circular orbit having a radius equal to that of Mercury's orbit around the Sun (5.80×10^{10} m). What is the magnetic field in that region of space?

22. Assume the region to the right of a certain plane contains a uniform magnetic field of magnitude 1.00 mT and the field is zero in the region to the left of the plane as shown in Figure P29.22. An electron, originally traveling perpendicular to the boundary plane, passes into the region of the field. (a) Determine the time interval required for the electron to leave the "field-filled" region, noting that the electron's path is a semicircle. (b) Assuming the maximum depth of penetration into the field is 2.00 cm, find the kinetic energy of the electron.

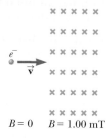

$B = 0$ $\qquad B = 1.00$ mT

Figure P29.22

23. A singly charged ion of mass m is accelerated from rest by a potential difference ΔV. It is then deflected by a uniform magnetic field (perpendicular to the ion's velocity) into a semicircle of radius R. Now a doubly charged ion of mass m' is accelerated through the same potential difference and deflected by the same magnetic field into a semicircle of radius $R' = 2R$. What is the ratio of the masses of the ions?

Section 29.3 Applications Involving Charged Particles Moving in a Magnetic Field

24. A cyclotron designed to accelerate protons has a mag- **M** netic field of magnitude 0.450 T over a region of radius 1.20 m. What are (a) the cyclotron frequency and (b) the maximum speed acquired by the protons?

25. Consider the mass spectrometer shown schematically **W** in Figure 29.14. The magnitude of the electric field between the plates of the velocity selector is 2.50×10^3 V/m, and the magnetic field in both the velocity selector and the deflection chamber has a magnitude of 0.035 0 T. Calculate the radius of the path for a singly charged ion having a mass $m = 2.18 \times 10^{-26}$ kg.

26. Singly charged uranium-238 ions are accelerated through a potential difference of 2.00 kV and enter a uniform magnetic field of magnitude 1.20 T directed perpendicular to their velocities. (a) Determine the radius of their circular path. (b) Repeat this calculation for uranium-235 ions. (c) **What If?** How does the ratio of these path radii depend on the accelerating voltage? (d) On the magnitude of the magnetic field?

27. A cyclotron (Fig. 29.16) designed to accelerate protons has an outer radius of 0.350 m. The protons are emitted nearly at rest from a source at the center and are accelerated through 600 V each time they cross the gap between the dees. The dees are between the poles of an electromagnet where the field is 0.800 T. (a) Find the cyclotron frequency for the protons in

this cyclotron. Find (b) the speed at which protons exit the cyclotron and (c) their maximum kinetic energy. (d) How many revolutions does a proton make in the cyclotron? (e) For what time interval does the proton accelerate?

28. A particle in the cyclotron shown in Figure 29.16a gains energy $q \, \Delta V$ from the alternating power supply each time it passes from one dee to the other. The time interval for each full orbit is

$$T = \frac{2\pi}{\omega} = \frac{2\pi m}{qB}$$

so the particle's average rate of increase in energy is

$$\frac{2q \, \Delta V}{T} = \frac{q^2 B \, \Delta V}{\pi m}$$

Notice that this power input is constant in time. On the other hand, the rate of increase in the radius r of its path is *not* constant. (a) Show that the rate of increase in the radius r of the particle's path is given by

$$\frac{dr}{dt} = \frac{1}{r} \frac{\Delta V}{\pi B}$$

(b) Describe how the path of the particles in Figure 29.16a is consistent with the result of part (a). (c) At what rate is the radial position of the protons in a cyclotron increasing immediately before the protons leave the cyclotron? Assume the cyclotron has an outer radius of 0.350 m, an accelerating voltage of $\Delta V = 600$ V, and a magnetic field of magnitude 0.800 T. (d) By how much does the radius of the protons' path increase during their last full revolution?

29. A velocity selector consists of electric and magnetic fields described by the expressions $\vec{\mathbf{E}} = E\hat{\mathbf{k}}$ and $\vec{\mathbf{B}} = B\hat{\mathbf{j}}$, with $B = 15.0$ mT. Find the value of E such that a 750-eV electron moving in the negative x direction is undeflected.

30. In his experiments on "cathode rays" during which he discovered the electron, J. J. Thomson showed that the same beam deflections resulted with tubes having cathodes made of *different* materials and containing *various* gases before evacuation. (a) Are these observations important? Explain your answer. (b) When he applied various potential differences to the deflection plates and turned on the magnetic coils, alone or in combination with the deflection plates, Thomson observed that the fluorescent screen continued to show a *single small* glowing patch. Argue whether his observation is important. (c) Do calculations to show that the charge-to-mass ratio Thomson obtained was huge compared with that of any macroscopic object or of any ionized atom or molecule. How can one make sense of this comparison? (d) Could Thomson observe any deflection of the beam due to gravitation? Do a calculation to argue for your answer. *Note:* To obtain a visibly glowing patch on the fluorescent screen, the potential difference between the slits and the cathode must be 100 V or more.

31. The picture tube in an old black-and-white television uses magnetic deflection coils rather than electric

deflection plates. Suppose an electron beam is accelerated through a 50.0-kV potential difference and then through a region of uniform magnetic field 1.00 cm wide. The screen is located 10.0 cm from the center of the coils and is 50.0 cm wide. When the field is turned off, the electron beam hits the center of the screen. Ignoring relativistic corrections, what field magnitude is necessary to deflect the beam to the side of the screen?

Section 29.4 Magnetic Force Acting on a Current-Carrying Conductor

32. A straight wire carrying a 3.00-A current is placed in a uniform magnetic field of magnitude 0.280 T directed perpendicular to the wire. (a) Find the magnitude of the magnetic force on a section of the wire having a length of 14.0 cm. (b) Explain why you can't determine the direction of the magnetic force from the information given in the problem.

33. A conductor carrying a current $I = 15.0$ A is directed along the positive x axis and perpendicular to a uniform magnetic field. A magnetic force per unit length of 0.120 N/m acts on the conductor in the negative y direction. Determine (a) the magnitude and (b) the direction of the magnetic field in the region through which the current passes.

34. A wire 2.80 m in length carries a current of 5.00 A in a region where a uniform magnetic field has a magnitude of 0.390 T. Calculate the magnitude of the magnetic force on the wire assuming the angle between the magnetic field and the current is (a) 60.0°, (b) 90.0°, and (c) 120°.

35. A wire carries a steady current of 2.40 A. A straight section of the wire is 0.750 m long and lies along the x axis within a uniform magnetic field, $\vec{\mathbf{B}} = 1.60\hat{\mathbf{k}}$ T. If the current is in the positive x direction, what is the magnetic force on the section of wire?

36. *Why is the following situation impossible?* Imagine a copper wire with radius 1.00 mm encircling the Earth at its magnetic equator, where the field direction is horizontal. A power supply delivers 100 MW to the wire to maintain a current in it, in a direction such that the magnetic force from the Earth's magnetic field is upward. Due to this force, the wire is levitated immediately above the ground.

37. **Review.** A rod of mass 0.720 kg and radius 6.00 cm rests on two parallel rails (Fig. P29.37) that are $d = 12.0$ cm apart and $L = 45.0$ cm long. The rod carries a

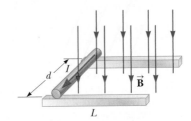

Figure P29.37 Problems 37 and 38.

current of $I = 48.0$ A in the direction shown and rolls along the rails without slipping. A uniform magnetic field of magnitude 0.240 T is directed perpendicular to the rod and the rails. If it starts from rest, what is the speed of the rod as it leaves the rails?

38. **Review.** A rod of mass m and radius R rests on two parallel rails (Fig. P29.37) that are a distance d apart and have a length L. The rod carries a current I in the direction shown and rolls along the rails without slipping. A uniform magnetic field B is directed perpendicular to the rod and the rails. If it starts from rest, what is the speed of the rod as it leaves the rails?

39. A wire having a mass per unit length of 0.500 g/cm **M** carries a 2.00-A current horizontally to the south. What are (a) the direction and (b) the magnitude of the minimum magnetic field needed to lift this wire vertically upward?

40. Consider the system pictured in Figure P29.40. A 15.0-cm horizontal wire of mass 15.0 g is placed between two thin, vertical conductors, and a uniform magnetic field acts perpendicular to the page. The wire is free to move vertically without friction on the two vertical conductors. When a 5.00-A current is directed as shown in the figure, the horizontal wire moves upward at constant velocity in the presence of gravity. (a) What forces act on the horizontal wire, and (b) under what condition is the wire able to move upward at constant velocity? (c) Find the magnitude and direction of the minimum magnetic field required to move the wire at constant speed. (d) What happens if the magnetic field exceeds this minimum value?

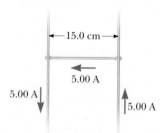

Figure P29.40

41. A horizontal power line of length 58.0 m carries a current of 2.20 kA northward as shown in Figure P29.41. The Earth's magnetic field at this location has a magnitude of 5.00×10^{-5} T. The field at this location is directed toward the north at an angle 65.0° below the

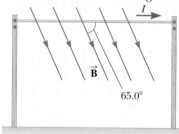

Figure P29.41

power line. Find (a) the magnitude and (b) the direction of the magnetic force on the power line.

42. A strong magnet is placed under a horizontal conducting ring of radius r that carries current I as shown in Figure P29.42. If the magnetic field $\vec{B}$ makes an angle θ with the vertical at the ring's location, what are (a) the magnitude and (b) the direction of the resultant magnetic force on the ring?

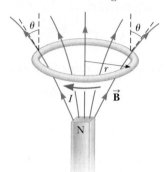

Figure P29.42

43. Assume the Earth's magnetic field is 52.0 μT northward at 60.0° below the horizontal in Atlanta, Georgia. A tube in a neon sign stretches between two diagonally opposite corners of a shop window—which lies in a north–south vertical plane—and carries current 35.0 mA. The current enters the tube at the bottom south corner of the shop's window. It exits at the opposite corner, which is 1.40 m farther north and 0.850 m higher up. Between these two points, the glowing tube spells out DONUTS. Determine the total vector magnetic force on the tube. *Hint:* You may use the first "important general statement" presented in the Finalize section of Example 29.4.

44. In Figure P29.44, the cube is 40.0 cm on each edge. Four straight segments of wire—*ab*, *bc*, *cd*, and *da*—form a closed loop that carries a current $I = 5.00$ A in the direction shown. A uniform magnetic field of magnitude $B = 0.020\ 0$ T is in the positive y direction. Determine the magnetic force vector on (a) *ab*, (b) *bc*, (c) *cd*, and (d) *da*. (e) Explain how you could find the force exerted on the fourth of these segments from the forces on the other three, without further calculation involving the magnetic field.

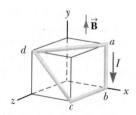

Figure P29.44

Section 29.5 Torque on a Current Loop in a Uniform Magnetic Field

45. A typical magnitude of the external magnetic field in a cardiac catheter ablation procedure using remote

magnetic navigation is $B = 0.080$ T. Suppose that the permanent magnet in the catheter used in the procedure is inside the left atrium of the heart and subject to this external magnetic field. The permanent magnet has a magnetic moment of 0.10 A · m². The orientation of the permanent magnet is $30°$ from the direction of the external magnetic field lines. (a) What is the magnitude of the torque on the tip of the catheter containing this permanent magnet? (b) What is the potential energy of the system consisting of the permanent magnet in the catheter and the magnetic field provided by the external magnets?

46. A 50.0-turn circular coil of radius 5.00 cm can be oriented in any direction in a uniform magnetic field having a magnitude of 0.500 T. If the coil carries a current of 25.0 mA, find the magnitude of the maximum possible torque exerted on the coil.

47. A magnetized sewing needle has a magnetic moment of 9.70 mA · m². At its location, the Earth's magnetic field is 55.0 μT northward at $48.0°$ below the horizontal. Identify the orientations of the needle that represent (a) the minimum potential energy and (b) the maximum potential energy of the needle–field system. (c) How much work must be done on the system to move the needle from the minimum to the maximum potential energy orientation?

48. A current of 17.0 mA is maintained in a single circular loop of 2.00 m circumference. A magnetic field of 0.800 T is directed parallel to the plane of the loop. (a) Calculate the magnetic moment of the loop. (b) What is the magnitude of the torque exerted by the magnetic field on the loop?

49. An eight-turn coil encloses an elliptical area having a major axis of 40.0 cm and a minor axis of 30.0 cm (Fig. P29.49). The coil lies in the plane of the page and has a 6.00-A current flowing clockwise around it. If the coil is in a uniform magnetic field of 2.00×10^{-4} T directed toward the left of the page, what is the magnitude of the torque on the coil? *Hint:* The area of an ellipse is $A = \pi ab$, where a and b are, respectively, the semimajor and semiminor axes of the ellipse.

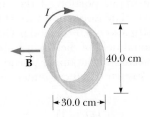

Figure P29.49

50. The rotor in a certain electric motor is a flat, rectangular coil with 80 turns of wire and dimensions 2.50 cm by 4.00 cm. The rotor rotates in a uniform magnetic field of 0.800 T. When the plane of the rotor is perpendicular to the direction of the magnetic field, the rotor carries a current of 10.0 mA. In this orientation, the magnetic moment of the rotor is directed opposite the magnetic field. The rotor then turns through one-

half revolution. This process is repeated to cause the rotor to turn steadily at an angular speed of 3.60×10^3 rev/min. (a) Find the maximum torque acting on the rotor. (b) Find the peak power output of the motor. (c) Determine the amount of work performed by the magnetic field on the rotor in every full revolution. (d) What is the average power of the motor?

51. A rectangular coil consists of $N = 100$ closely wrapped turns and has dimensions $a = 0.400$ m and $b = 0.300$ m. The coil is hinged along the y axis, and its plane makes an angle $\theta = 30.0°$ with the x axis (Fig. P29.51). (a) What is the magnitude of the torque exerted on the coil by a uniform magnetic field $B = 0.800$ T directed in the positive x direction when the current is $I = 1.20$ A in the direction shown? (b) What is the expected direction of rotation of the coil?

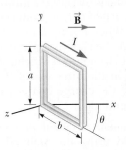

Figure P29.51

52. A rectangular loop of wire has dimensions 0.500 m by 0.300 m. The loop is pivoted at the x axis and lies in the xy plane as shown in Figure P29.52. A uniform magnetic field of magnitude 1.50 T is directed at an angle of $40.0°$ with respect to the y axis with field lines parallel to the yz plane. The loop carries a current of 0.900 A in the direction shown. (Ignore gravitation.) We wish to evaluate the torque on the current loop. (a) What is the direction of the magnetic force exerted on wire segment ab? (b) What is the direction of the torque associated with this force about an axis through the origin? (c) What is the direction of the magnetic force exerted on segment cd? (d) What is the direction of the torque associated with this force about an axis through the origin? (e) Can the forces examined in parts (a) and (c) combine to cause the loop to rotate around the x axis? (f) Can they affect the motion of the loop in any way? Explain. (g) What is the direction of the magnetic force exerted on segment bc? (h) What is the direction of the torque associated with this force about an axis through the origin? (i) What is the torque on segment ad about an axis through the origin? (j) From the point of view of Figure P29.52, once the loop is released from rest at

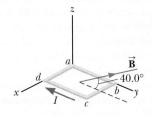

Figure P29.52

the position shown, will it rotate clockwise or counter-clockwise around the *x* axis? (k) Compute the magnitude of the magnetic moment of the loop. (l) What is the angle between the magnetic moment vector and the magnetic field? (m) Compute the torque on the loop using the results to parts (k) and (l).

53. A wire is formed into a circle having a diameter of **[W]** 10.0 cm and is placed in a uniform magnetic field of 3.00 mT. The wire carries a current of 5.00 A. Find (a) the maximum torque on the wire and (b) the range of potential energies of the wire–field system for different orientations of the circle.

Section 29.6 The Hall Effect

54. A Hall-effect probe operates with a 120-mA current. When the probe is placed in a uniform magnetic field of magnitude 0.080 0 T, it produces a Hall voltage of 0.700 μV. (a) When it is used to measure an unknown magnetic field, the Hall voltage is 0.330 μV. What is the magnitude of the unknown field? (b) The thickness of the probe in the direction of $\vec{B}$ is 2.00 mm. Find the density of the charge carriers, each of which has charge of magnitude *e*.

55. In an experiment designed to measure the Earth's **[M]** magnetic field using the Hall effect, a copper bar 0.500 cm thick is positioned along an east–west direction. Assume $n = 8.46 \times 10^{28}$ electrons/m^3 and the plane of the bar is rotated to be perpendicular to the direction of $\vec{B}$. If a current of 8.00 A in the conductor results in a Hall voltage of 5.10×10^{-12} V, what is the magnitude of the Earth's magnetic field at this location?

Additional Problems

56. Carbon-14 and carbon-12 ions (each with charge of magnitude *e*) are accelerated in a cyclotron. If the cyclotron has a magnetic field of magnitude 2.40 T, what is the difference in cyclotron frequencies for the two ions?

57. In Niels Bohr's 1913 model of the hydrogen atom, the single electron is in a circular orbit of radius 5.29×10^{-11} m and its speed is 2.19×10^6 m/s. (a) What is the magnitude of the magnetic moment due to the electron's motion? (b) If the electron moves in a horizontal circle, counterclockwise as seen from above, what is the direction of this magnetic moment vector?

58. Heart–lung machines and artificial kidney machines employ electromagnetic blood pumps. The blood is confined to an electrically insulating tube, cylindrical in practice but represented here for simplicity as a rectangle of interior width *w* and height *h*. Figure P29.58 shows a rectangular section of blood within the tube. Two electrodes fit into the top and the bottom of the tube. The potential difference between them establishes an electric current through the blood, with current density *J* over the section of length *L* shown in Figure P29.58. A perpendicular magnetic field exists in the same region. (a) Explain why this arrangement produces on the liquid a force that is directed along the length of the

pipe. (b) Show that the section of liquid in the magnetic field experiences a pressure increase *JLB*. (c) After the blood leaves the pump, is it charged? (d) Is it carrying current? (e) Is it magnetized? (The same electromagnetic pump can be used for any fluid that conducts electricity, such as liquid sodium in a nuclear reactor.)

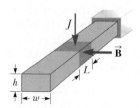

Figure P29.58

59. A particle with positive charge $q = 3.20 \times 10^{-19}$ C **[M]** moves with a velocity $\vec{v} = (2\hat{i} + 3\hat{j} - \hat{k})$ m/s through a region where both a uniform magnetic field and a uniform electric field exist. (a) Calculate the total force on the moving particle (in unit-vector notation), taking $\vec{B} = (2\hat{i} + 4\hat{j} + \hat{k})$ T and $\vec{E} = (4\hat{i} - \hat{j} - 2\hat{k})$ V/m. (b) What angle does the force vector make with the positive *x* axis?

60. Figure 29.11 shows a charged particle traveling in a nonuniform magnetic field forming a magnetic bottle. (a) Explain why the positively charged particle in the figure must be moving clockwise when viewed from the right of the figure. The particle travels along a helix whose radius decreases and whose pitch decreases as the particle moves into a stronger magnetic field. If the particle is moving to the right along the *x* axis, its velocity in this direction will be reduced to zero and it will be reflected from the right-hand side of the bottle, acting as a "magnetic mirror." The particle ends up bouncing back and forth between the ends of the bottle. (b) Explain qualitatively why the axial velocity is reduced to zero as the particle moves into the region of strong magnetic field at the end of the bottle. (c) Explain why the tangential velocity increases as the particle approaches the end of the bottle. (d) Explain why the orbiting particle has a magnetic dipole moment.

61. Review. The upper portion of the circuit in Figure **[AMT]** P29.61 is fixed. The horizontal wire at the bottom has a mass of 10.0 g and is 5.00 cm long. This wire hangs in the gravitational field of the Earth from identical light springs connected to the upper portion of the circuit. The springs stretch 0.500 cm under the weight of the

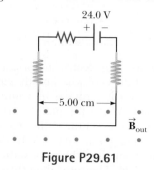

Figure P29.61

wire, and the circuit has a total resistance of 12.0 Ω. When a magnetic field is turned on, directed out of the page, the springs stretch an additional 0.300 cm. Only the horizontal wire at the bottom of the circuit is in the magnetic field. What is the magnitude of the magnetic field?

62. Within a cylindrical region of space of radius 100 Mm, a magnetic field is uniform with a magnitude 25.0 μT and oriented parallel to the axis of the cylinder. The magnetic field is zero outside this cylinder. A cosmic-ray proton traveling at one-tenth the speed of light is heading directly toward the center of the cylinder, moving perpendicular to the cylinder's axis. (a) Find the radius of curvature of the path the proton follows when it enters the region of the field. (b) Explain whether the proton will arrive at the center of the cylinder.

63. **Review.** A proton is at rest at the plane boundary of a region containing a uniform magnetic field B (Fig. P29.63). An alpha particle moving horizontally makes a head-on elastic collision with the proton. Immediately after the collision, both particles enter the magnetic field, moving perpendicular to the direction of the field. The radius of the proton's trajectory is R. The mass of the alpha particle is four times that of the proton, and its charge is twice that of the proton. Find the radius of the alpha particle's trajectory.

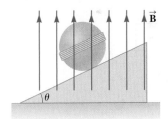

Figure P29.63

64. (a) A proton moving with velocity $\vec{\mathbf{v}} = v_i\hat{\mathbf{i}}$ experiences a magnetic force $\vec{\mathbf{F}} = F_i\hat{\mathbf{j}}$. Explain what you can and cannot infer about $\vec{\mathbf{B}}$ from this information. (b) **What If?** In terms of F_i, what would be the force on a proton in the same field moving with velocity $\vec{\mathbf{v}} = -v_i\hat{\mathbf{i}}$? (c) What would be the force on an electron in the same field moving with velocity $\vec{\mathbf{v}} = -v_i\hat{\mathbf{i}}$?

65. **Review.** A 0.200-kg metal rod carrying a current of 10.0 A glides on two horizontal rails 0.500 m apart. If the coefficient of kinetic friction between the rod and rails is 0.100, what vertical magnetic field is required to keep the rod moving at a constant speed?

66. **Review.** A metal rod of mass m carrying a current I glides on two horizontal rails a distance d apart. If the coefficient of kinetic friction between the rod and rails is μ, what vertical magnetic field is required to keep the rod moving at a constant speed?

67. A proton having an initial velocity of $20.0\hat{\mathbf{i}}$ Mm/s enters a uniform magnetic field of magnitude 0.300 T

with a direction perpendicular to the proton's velocity. It leaves the field-filled region with velocity $-20.0\hat{\mathbf{j}}$ Mm/s. Determine (a) the direction of the magnetic field, (b) the radius of curvature of the proton's path while in the field, (c) the distance the proton traveled in the field, and (d) the time interval during which the proton is in the field.

68. Model the electric motor in a handheld electric mixer as a single flat, compact, circular coil carrying electric current in a region where a magnetic field is produced by an external permanent magnet. You need consider only one instant in the operation of the motor. (We will consider motors again in Chapter 31.) Make order-of-magnitude estimates of (a) the magnetic field, (b) the torque on the coil, (c) the current in the coil, (d) the coil's area, and (e) the number of turns in the coil. The input power to the motor is electric, given by $P = I\Delta V$, and the useful output power is mechanical, $P = \tau\omega$.

69. A nonconducting sphere has mass 80.0 g and radius 20.0 cm. A flat, compact coil of wire with five turns is wrapped tightly around it, with each turn concentric with the sphere. The sphere is placed on an inclined plane that slopes downward to the left (Fig. P29.69), making an angle θ with the horizontal so that the coil is parallel to the inclined plane. A uniform magnetic field of 0.350 T vertically upward exists in the region of the sphere. (a) What current in the coil will enable the sphere to rest in equilibrium on the inclined plane? (b) Show that the result does not depend on the value of θ.

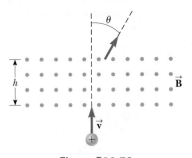

Figure P29.69

70. *Why is the following situation impossible?* Figure P29.70 shows an experimental technique for altering the direction of travel for a charged particle. A particle of charge $q = 1.00$ μC and mass $m = 2.00 \times 10^{-13}$ kg enters the bottom of the region of uniform magnetic field at speed $v = 2.00 \times 10^5$ m/s, with a velocity vector

Figure P29.70

perpendicular to the field lines. The magnetic force on the particle causes its direction of travel to change so that it leaves the region of the magnetic field at the top traveling at an angle from its original direction. The magnetic field has magnitude $B = 0.400$ T and is directed out of the page. The length h of the magnetic field region is 0.110 m. An experimenter performs the technique and measures the angle θ at which the particles exit the top of the field. She finds that the angles of deviation are exactly as predicted.

71. Figure P29.71 shows a schematic representation of an apparatus that can be used to measure magnetic fields. A rectangular coil of wire contains N turns and has a width w. The coil is attached to one arm of a balance and is suspended between the poles of a magnet. The magnetic field is uniform and perpendicular to the plane of the coil. The system is first balanced when the current in the coil is zero. When the switch is closed and the coil carries a current I, a mass m must be added to the right side to balance the system. (a) Find an expression for the magnitude of the magnetic field. (b) Why is the result independent of the vertical dimensions of the coil? (c) Suppose the coil has 50 turns and a width of 5.00 cm. When the switch is closed, the coil carries a current of 0.300 A, and a mass of 20.0 g must be added to the right side to balance the system. What is the magnitude of the magnetic field?

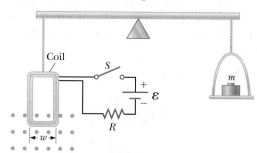

Figure P29.71

72. A heart surgeon monitors the flow rate of blood through an artery using an electromagnetic flowmeter (Fig. P29.72). Electrodes A and B make contact with the outer surface of the blood vessel, which has a diameter of 3.00 mm. (a) For a magnetic field magnitude of 0.040 0 T, an emf of 160 μV appears between the electrodes. Calculate the speed of the blood. (b) Explain why electrode A has to be positive as shown. (c) Does the sign of the emf depend on whether the mobile ions in the blood are predominantly positively or negatively charged? Explain.

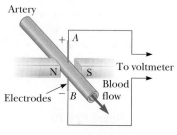

Figure P29.72

73. A uniform magnetic field of magnitude 0.150 T is directed along the positive x axis. A positron moving at a speed of 5.00×10^6 m/s enters the field along a direction that makes an angle of $\theta = 85.0°$ with the x axis (Fig. P29.73). The motion of the particle is expected to be a helix as described in Section 29.2. Calculate (a) the pitch p and (b) the radius r of the trajectory as defined in Figure P29.73.

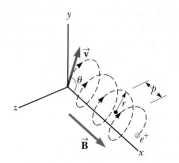

Figure P29.73

74. **Review.** (a) Show that a magnetic dipole in a uniform magnetic field, displaced from its equilibrium orientation and released, can oscillate as a torsional pendulum (Section 15.5) in simple harmonic motion. (b) Is this statement true for all angular displacements, for all displacements less than 180°, or only for small angular displacements? Explain. (c) Assume the dipole is a compass needle—a light bar magnet—with a magnetic moment of magnitude μ. It has moment of inertia I about its center, where it is mounted on a frictionless, vertical axle, and it is placed in a horizontal magnetic field of magnitude B. Determine its frequency of oscillation. (d) Explain how the compass needle can be conveniently used as an indicator of the magnitude of the external magnetic field. (e) If its frequency is 0.680 Hz in the Earth's local field, with a horizontal component of 39.2 μT, what is the magnitude of a field parallel to the needle in which its frequency of oscillation is 4.90 Hz?

75. The accompanying table shows measurements of the Hall voltage and corresponding magnetic field for a probe used to measure magnetic fields. (a) Plot these data and deduce a relationship between the two variables. (b) If the measurements were taken with a current of 0.200 A and the sample is made from a material having a charge-carrier density of 1.00×10^{26} carriers/m^3, what is the thickness of the sample?

ΔV_H (μV)	B (T)
0	0.00
11	0.10
19	0.20
28	0.30
42	0.40
50	0.50
61	0.60
68	0.70
79	0.80
90	0.90
102	1.00

76. A metal rod having a mass per unit length λ carries a current I. The rod hangs from two wires in a uniform vertical magnetic field as shown in Figure P29.76. The wires make an angle θ with the vertical when in equilibrium. Determine the magnitude of the magnetic field.

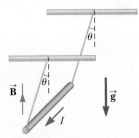

Figure P29.76

Challenge Problems

77. Consider an electron orbiting a proton and maintained in a fixed circular path of radius $R = 5.29 \times 10^{-11}$ m by the Coulomb force. Treat the orbiting particle as a current loop. Calculate the resulting torque when the electron–proton system is placed in a magnetic field of 0.400 T directed perpendicular to the magnetic moment of the loop.

78. Protons having a kinetic energy of 5.00 MeV (1 eV = 1.60×10^{-19} J) are moving in the positive x direction and enter a magnetic field $\vec{\mathbf{B}} = 0.050\ 0\hat{\mathbf{k}}$ T directed out of the plane of the page and extending from $x = 0$ to $x = 1.00$ m as shown in Figure P29.78. (a) Ignoring relativistic effects, find the angle α between the initial velocity vector of the proton beam and the velocity vector after the beam emerges from the field. (b) Calculate the y component of the protons' momenta as they leave the magnetic field.

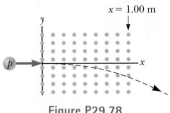

Figure P29.78

79. **Review.** A wire having a linear mass density of 1.00 g/cm is placed on a horizontal surface that has a coefficient of kinetic friction of 0.200. The wire carries a current of 1.50 A toward the east and slides horizontally to the north at constant velocity. What are (a) the magnitude and (b) the direction of the smallest magnetic field that enables the wire to move in this fashion?

80. A proton moving in the plane of the page has a kinetic energy of 6.00 MeV. A magnetic field of magnitude $B = 1.00$ T is directed into the page. The proton enters the magnetic field with its velocity vector at an angle $\theta = 45.0°$ to the linear boundary of the field as shown in Figure P29.80. (a) Find x, the distance from the point of entry to where the proton will leave the field. (b) Determine θ, the angle between the boundary and the proton's velocity vector as it leaves the field.

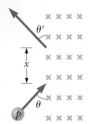

Figure P29.80

Sources of the Magnetic Field

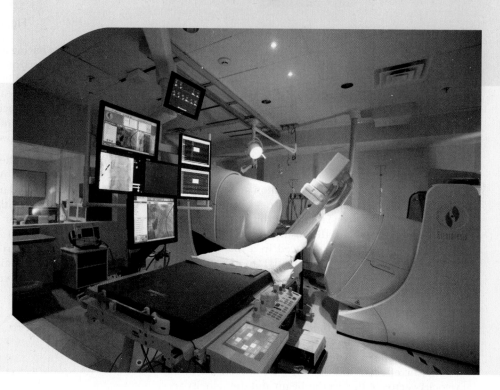

A cardiac catheterization laboratory stands ready to receive a patient suffering from atrial fibrillation. The large white objects on either side of the operating table are strong magnets that place the patient in a magnetic field. The electrophysiologist performing a catheter ablation procedure sits at a computer in the room to the left. With guidance from the magnetic field, he or she uses a joystick and other controls to thread the magnetically sensitive tip of a cardiac catheter through blood vessels and into the chambers of the heart. *(© Courtesy of Stereotaxis, Inc.)*

In Chapter 29, we discussed the magnetic force exerted on a charged particle moving in a magnetic field. To complete the description of the magnetic interaction, this chapter explores the origin of the magnetic field, moving charges. We begin by showing how to use the law of Biot and Savart to calculate the magnetic field produced at some point in space by a small current element. This formalism is then used to calculate the total magnetic field due to various current distributions. Next, we show how to determine the force between two current-carrying conductors, leading to the definition of the ampere. We also introduce Ampère's law, which is useful in calculating the magnetic field of a highly symmetric configuration carrying a steady current.

This chapter is also concerned with the complex processes that occur in magnetic materials. All magnetic effects in matter can be explained on the basis of atomic magnetic moments, which arise both from the orbital motion of electrons and from an intrinsic property of electrons known as spin.

30.1 The Biot–Savart Law

Shortly after Oersted's discovery in 1819 that a compass needle is deflected by a current-carrying conductor, Jean-Baptiste Biot (1774–1862) and Félix Savart (1791–1841) performed quantitative experiments on the force exerted by an electric current on a nearby magnet. From their experimental results, Biot and Savart arrived at a mathematical expression that gives the magnetic field at some point in space in terms of the current that produces the field. That expression is based on the following experimental observations for the magnetic field $d\vec{\mathbf{B}}$ at a point P associated with a length element $d\vec{\mathbf{s}}$ of a wire carrying a steady current I (Fig. 30.1):

- The vector $d\vec{\mathbf{B}}$ is perpendicular both to $d\vec{\mathbf{s}}$ (which points in the direction of the current) and to the unit vector $\hat{\mathbf{r}}$ directed from $d\vec{\mathbf{s}}$ toward P.
- The magnitude of $d\vec{\mathbf{B}}$ is inversely proportional to r^2, where r is the distance from $d\vec{\mathbf{s}}$ to P.
- The magnitude of $d\vec{\mathbf{B}}$ is proportional to the current I and to the magnitude ds of the length element $d\vec{\mathbf{s}}$.
- The magnitude of $d\vec{\mathbf{B}}$ is proportional to $\sin\theta$, where θ is the angle between the vectors $d\vec{\mathbf{s}}$ and $\hat{\mathbf{r}}$.

These observations are summarized in the mathematical expression known today as the **Biot–Savart law:**

$$d\vec{\mathbf{B}} = \frac{\mu_0}{4\pi}\frac{I\,d\vec{\mathbf{s}} \times \hat{\mathbf{r}}}{r^2} \tag{30.1}$$

◀ Biot–Savart law

where μ_0 is a constant called the **permeability of free space:**

$$\mu_0 = 4\pi \times 10^{-7}\ \text{T}\cdot\text{m/A} \tag{30.2}$$

◀ Permeability of free space

Notice that the field $d\vec{\mathbf{B}}$ in Equation 30.1 is the field created at a point by the current in only a small length element $d\vec{\mathbf{s}}$ of the conductor. To find the *total* magnetic field $\vec{\mathbf{B}}$ created at some point by a current of finite size, we must sum up contributions from all current elements $I\,d\vec{\mathbf{s}}$ that make up the current. That is, we must evaluate $\vec{\mathbf{B}}$ by integrating Equation 30.1:

$$\vec{\mathbf{B}} = \frac{\mu_0 I}{4\pi}\int\frac{d\vec{\mathbf{s}} \times \hat{\mathbf{r}}}{r^2} \tag{30.3}$$

where the integral is taken over the entire current distribution. This expression must be handled with special care because the integrand is a cross product and therefore a vector quantity. We shall see one case of such an integration in Example 30.1.

Although the Biot–Savart law was discussed for a current-carrying wire, it is also valid for a current consisting of charges flowing through space such as the particle beam in an accelerator. In that case, $d\vec{\mathbf{s}}$ represents the length of a small segment of space in which the charges flow.

Interesting similarities and differences exist between Equation 30.1 for the magnetic field due to a current element and Equation 23.9 for the electric field due to a point charge. The magnitude of the magnetic field varies as the inverse square of the distance from the source, as does the electric field due to a point charge. The directions of the two fields are quite different, however. The electric field created by a point charge is radial, but the magnetic field created by a current element is perpendicular to both the length element $d\vec{\mathbf{s}}$ and the unit vector $\hat{\mathbf{r}}$ as described by the cross product in Equation 30.1. Hence, if the conductor lies in the plane of the page as shown in Figure 30.1, $d\vec{\mathbf{B}}$ points out of the page at P and into the page at P'.

Another difference between electric and magnetic fields is related to the source of the field. An electric field is established by an isolated electric charge. The Biot–Savart law gives the magnetic field of an isolated current element at some point, but such an isolated current element cannot exist the way an isolated electric charge can. A current element *must* be part of an extended current distribution because a complete circuit is needed for charges to flow. Therefore,

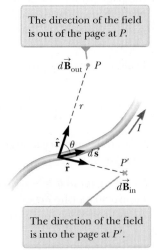

The direction of the field is out of the page at P.

The direction of the field is into the page at P'.

Figure 30.1 The magnetic field $d\vec{\mathbf{B}}$ at a point due to the current I through a length element* $d\vec{\mathbf{s}}$ is given by the Biot–Savart law.

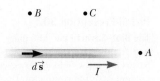

Figure 30.2 (Quick Quiz 30.1) Where is the magnetic field due to the current element the greatest?

the Biot–Savart law (Eq. 30.1) is only the first step in a calculation of a magnetic field; it must be followed by an integration over the current distribution as in Equation 30.3.

Quick Quiz 30.1 Consider the magnetic field due to the current in the wire shown in Figure 30.2. Rank the points A, B, and C in terms of magnitude of the magnetic field that is due to the current in just the length element $d\vec{s}$ shown from greatest to least.

Example 30.1 | **Magnetic Field Surrounding a Thin, Straight Conductor**

Consider a thin, straight wire of finite length carrying a constant current I and placed along the x axis as shown in Figure 30.3. Determine the magnitude and direction of the magnetic field at point P due to this current.

SOLUTION

Conceptualize From the Biot–Savart law, we expect that the magnitude of the field is proportional to the current in the wire and decreases as the distance a from the wire to point P increases. We also expect the field to depend on the angles θ_1 and θ_2 in Figure 30.3b. We place the origin at O and let point P be along the positive y axis, with $\hat{\mathbf{k}}$ being a unit vector pointing out of the page.

Categorize We are asked to find the magnetic field due to a simple current distribution, so this example is a typical problem for which the Biot–Savart law is appropriate. We must find the field contribution from a small element of current and then integrate over the current distribution.

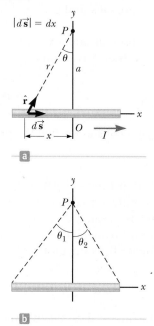

Figure 30.3 (Example 30.1) (a) A thin, straight wire carrying a current I. (b) The angles θ_1 and θ_2 used for determining the net field.

Analyze Let's start by considering a length element $d\vec{s}$ located a distance r from P. The direction of the magnetic field at point P due to the current in this element is out of the page because $d\vec{s} \times \hat{\mathbf{r}}$ is out of the page. In fact, because *all* the current elements $I\,d\vec{s}$ lie in the plane of the page, they all produce a magnetic field directed out of the page at point P. Therefore, the direction of the magnetic field at point P is out of the page and we need only find the magnitude of the field.

Evaluate the cross product in the Biot–Savart law:

$$d\vec{s} \times \hat{\mathbf{r}} = |d\vec{s} \times \hat{\mathbf{r}}|\hat{\mathbf{k}} = \left[dx \sin\left(\frac{\pi}{2} - \theta\right)\right]\hat{\mathbf{k}} = (dx \cos\theta)\hat{\mathbf{k}}$$

Substitute into Equation 30.1:

$$(1) \quad d\vec{\mathbf{B}} = (dB)\hat{\mathbf{k}} = \frac{\mu_0 I}{4\pi} \frac{dx \cos\theta}{r^2}\hat{\mathbf{k}}$$

From the geometry in Figure 30.3a, express r in terms of θ:

$$(2) \quad r = \frac{a}{\cos\theta}$$

Notice that $\tan\theta = -x/a$ from the right triangle in Figure 30.3a (the negative sign is necessary because $d\vec{s}$ is located at a negative value of x) and solve for x:

$$x = -a\tan\theta$$

Find the differential dx:

$$(3) \quad dx = -a\sec^2\theta\,d\theta = -\frac{a\,d\theta}{\cos^2\theta}$$

Substitute Equations (2) and (3) into the expression for the z component of the field from Equation (1):

$$(4) \quad dB = -\frac{\mu_0 I}{4\pi}\left(\frac{a\,d\theta}{\cos^2\theta}\right)\left(\frac{\cos^2\theta}{a^2}\right)\cos\theta = -\frac{\mu_0 I}{4\pi a}\cos\theta\,d\theta$$

▶ **30.1** continued

Integrate Equation (4) over all length elements on the wire, where the subtending angles range from θ_1 to θ_2 as defined in Figure 30.3b:

$$B = -\frac{\mu_0 I}{4\pi a} \int_{\theta_1}^{\theta_2} \cos\theta \, d\theta = \frac{\mu_0 I}{4\pi a}(\sin\theta_1 - \sin\theta_2) \qquad \textbf{(30.4)}$$

....

Finalize We can use this result to find the magnitude of the magnetic field of *any* straight current-carrying wire if we know the geometry and hence the angles θ_1 and θ_2. Consider the special case of an infinitely long, straight wire. If the wire in Figure 30.3b becomes infinitely long, we see that $\theta_1 = \pi/2$ and $\theta_2 = -\pi/2$ for length elements ranging between positions $x = -\infty$ and $x = +\infty$. Because $(\sin\theta_1 - \sin\theta_2) = [\sin\pi/2 - \sin(-\pi/2)] = 2$, Equation 30.4 becomes

$$B = \frac{\mu_0 I}{2\pi a} \qquad \textbf{(30.5)}$$

Equations 30.4 and 30.5 both show that the magnitude of the magnetic field is proportional to the current and decreases with increasing distance from the wire, as expected. Equation 30.5 has the same mathematical form as the expression for the magnitude of the electric field due to a long charged wire (see Eq. 24.7).

Example 30.2 Magnetic Field Due to a Curved Wire Segment

Calculate the magnetic field at point O for the current-carrying wire segment shown in Figure 30.4. The wire consists of two straight portions and a circular arc of radius a, which subtends an angle θ.

SOLUTION

Conceptualize The magnetic field at O due to the current in the straight segments AA' and CC' is zero because $d\vec{\mathbf{s}}$ is parallel to $\hat{\mathbf{r}}$ along these paths, which means that $d\vec{\mathbf{s}} \times \hat{\mathbf{r}} = 0$ for these paths. Therefore, we expect the magnetic field at O to be due only to the current in the curved portion of the wire.

Categorize Because we can ignore segments AA' and CC', this example is categorized as an application of the Biot–Savart law to the curved wire segment AC.

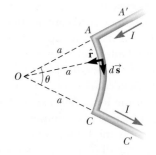

Figure 30.4 (Example 30.2) The length of the curved segment AC is s.

....

Analyze Each length element $d\vec{\mathbf{s}}$ along path AC is at the same distance a from O, and the current in each contributes a field element $d\vec{\mathbf{B}}$ directed into the page at O. Furthermore, at every point on AC, $d\vec{\mathbf{s}}$ is perpendicular to $\hat{\mathbf{r}}$; hence, $|d\vec{\mathbf{s}} \times \hat{\mathbf{r}}| = ds$.

From Equation 30.1, find the magnitude of the field at O due to the current in an element of length ds:

$$dB = \frac{\mu_0}{4\pi}\frac{I\,ds}{a^2}$$

Integrate this expression over the curved path AC, noting that I and a are constants:

$$B = \frac{\mu_0 I}{4\pi a^2}\int ds = \frac{\mu_0 I}{4\pi a^2}s$$

From the geometry, note that $s = a\theta$ and substitute:

$$B = \frac{\mu_0 I}{4\pi a^2}(a\theta) = \frac{\mu_0 I}{4\pi a}\theta \qquad \textbf{(30.6)}$$

....

Finalize Equation 30.6 gives the magnitude of the magnetic field at O. The direction of $\vec{\mathbf{B}}$ is into the page at O because $d\vec{\mathbf{s}} \times \hat{\mathbf{r}}$ is into the page for every length element.

WHAT IF? What if you were asked to find the magnetic field at the center of a circular wire loop of radius R that carries a current I? Can this question be answered at this point in our understanding of the source of magnetic fields?

continued

▶ **30.2** c o n t i n u e d

Answer Yes, it can. The straight wires in Figure 30.4 do not contribute to the magnetic field. The only contribution is from the curved segment. As the angle θ increases, the curved segment becomes a full circle when $\theta = 2\pi$. Therefore, you can find the magnetic field at the center of a wire loop by letting $\theta = 2\pi$ in Equation 30.6:

$$B = \frac{\mu_0 I}{4\pi a} 2\pi = \frac{\mu_0 I}{2a}$$

This result is a limiting case of a more general result discussed in Example 30.3.

Example 30.3   Magnetic Field on the Axis of a Circular Current Loop

Consider a circular wire loop of radius a located in the yz plane and carrying a steady current I as in Figure 30.5. Calculate the magnetic field at an axial point P a distance x from the center of the loop.

SOLUTION

Conceptualize Compare this problem to Example 23.8 for the electric field due to a ring of charge. Figure 30.5 shows the magnetic field contribution $d\vec{\mathbf{B}}$ at P due to a single current element at the top of the ring. This field vector can be resolved into components dB_x parallel to the axis of the ring and $dB_\perp$ perpendicular to the axis. Think about the magnetic field contributions from a current element at the bottom of the loop. Because of the symmetry of the situation, the perpendicular components of the field due to elements at the top and bottom of the ring cancel. This cancellation occurs for all pairs of segments around the ring, so we can ignore the perpendicular component of the field and focus solely on the parallel components, which simply add.

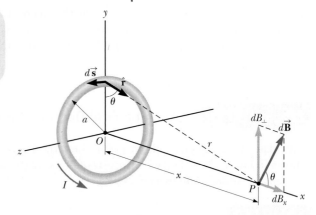

Figure 30.5 (Example 30.3) Geometry for calculating the magnetic field at a point P lying on the axis of a current loop. By symmetry, the total field $\vec{\mathbf{B}}$ is along this axis.

Categorize We are asked to find the magnetic field due to a simple current distribution, so this example is a typical problem for which the Biot–Savart law is appropriate.

....

Analyze In this situation, every length element $d\vec{\mathbf{s}}$ is perpendicular to the vector $\hat{\mathbf{r}}$ at the location of the element. Therefore, for any element, $|d\vec{\mathbf{s}} \times \hat{\mathbf{r}}| = (ds)(1)\sin 90° = ds$. Furthermore, all length elements around the loop are at the same distance r from P, where $r^2 = a^2 + x^2$.

Use Equation 30.1 to find the magnitude of $d\vec{\mathbf{B}}$ due to the current in any length element $d\vec{\mathbf{s}}$:

$$dB = \frac{\mu_0 I}{4\pi} \frac{|d\vec{\mathbf{s}} \times \hat{\mathbf{r}}|}{r^2} = \frac{\mu_0 I}{4\pi} \frac{ds}{(a^2 + x^2)}$$

Find the x component of the field element:

$$dB_x = \frac{\mu_0 I}{4\pi} \frac{ds}{(a^2 + x^2)} \cos\theta$$

Integrate over the entire loop:

$$B_x = \oint dB_x = \frac{\mu_0 I}{4\pi} \oint \frac{ds \cos\theta}{a^2 + x^2}$$

From the geometry, evaluate $\cos\theta$:

$$\cos\theta = \frac{a}{(a^2 + x^2)^{1/2}}$$

Substitute this expression for $\cos\theta$ into the integral and note that x and a are both constant:

$$B_x = \frac{\mu_0 I}{4\pi} \oint \frac{ds}{a^2 + x^2} \left[\frac{a}{(a^2 + x^2)^{1/2}}\right] = \frac{\mu_0 I}{4\pi} \frac{a}{(a^2 + x^2)^{3/2}} \oint ds$$

Integrate around the loop:

$$B_x = \frac{\mu_0 I}{4\pi} \frac{a}{(a^2 + x^2)^{3/2}} (2\pi a) = \frac{\mu_0 I a^2}{2(a^2 + x^2)^{3/2}} \qquad \textbf{(30.7)}$$

....

▶ **30.3** continued

Finalize To find the magnetic field at the center of the loop, set $x = 0$ in Equation 30.7. At this special point,

$$B = \frac{\mu_0 I}{2a} \quad (\text{at } x = 0) \tag{30.8}$$

which is consistent with the result of the **What If?** feature of Example 30.2.

The pattern of magnetic field lines for a circular current loop is shown in Figure 30.6a. For clarity, the lines are drawn for only the plane that contains the axis of the loop. The field-line pattern is axially symmetric and looks like the pattern around a bar magnet, which is shown in Figure 30.6b.

WHAT IF? What if we consider points on the x axis very far from the loop? How does the magnetic field behave at these distant points?

Answer In this case, in which $x \gg a$, we can neglect the term a^2 in the denominator of Equation 30.7 and obtain

$$B \approx \frac{\mu_0 I a^2}{2x^3} \quad (\text{for } x \gg a) \tag{30.9}$$

The magnitude of the magnetic moment μ of the loop is defined as the product of current and loop area (see Eq. 29.15): $\mu = I(\pi a^2)$ for our circular loop. We can express Equation 30.9 as

$$B \approx \frac{\mu_0}{2\pi} \frac{\mu}{x^3} \tag{30.10}$$

This result is similar in form to the expression for the electric field due to an electric dipole, $E = k_e(p/y^3)$ (see Example 23.6), where $p = 2aq$ is the electric dipole moment as defined in Equation 26.16.

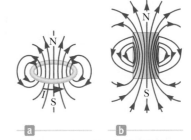

Figure 30.6 (Example 30.3) (a) Magnetic field lines surrounding a current loop. (b) Magnetic field lines surrounding a bar magnet. Notice the similarity between this line pattern and that of a current loop.

30.2 The Magnetic Force Between Two Parallel Conductors

In Chapter 29, we described the magnetic force that acts on a current-carrying conductor placed in an external magnetic field. Because a current in a conductor sets up its own magnetic field, it is easy to understand that two current-carrying conductors exert magnetic forces on each other. One wire establishes the magnetic field and the other wire is modeled as a collection of particles in a magnetic field. Such forces between wires can be used as the basis for defining the ampere and the coulomb.

Consider two long, straight, parallel wires separated by a distance a and carrying currents I_1 and I_2 in the same direction as in Figure 30.7. Let's determine the force exerted on one wire due to the magnetic field set up by the other wire. Wire 2, which carries a current I_2 and is identified arbitrarily as the source wire, creates a magnetic field $\vec{\mathbf{B}}_2$ at the location of wire 1, the test wire. The magnitude of this magnetic field is the same at all points on wire 1. The direction of $\vec{\mathbf{B}}_2$ is perpendicular to wire 1 as shown in Figure 30.7. According to Equation 29.10, the magnetic force on a length ℓ of wire 1 is $\vec{\mathbf{F}}_1 = I_1 \vec{\ell} \times \vec{\mathbf{B}}_2$. Because $\vec{\ell}$ is perpendicular to $\vec{\mathbf{B}}_2$ in this situation, the magnitude of $\vec{\mathbf{F}}_1$ is $F_1 = I_1 \ell B_2$. Because the magnitude of $\vec{\mathbf{B}}_2$ is given by Equation 30.5,

$$F_1 = I_1 \ell B_2 = I_1 \ell \left(\frac{\mu_0 I_2}{2\pi a} \right) = \frac{\mu_0 I_1 I_2}{2\pi a} \ell \tag{30.11}$$

The direction of $\vec{\mathbf{F}}_1$ is toward wire 2 because $\vec{\ell} \times \vec{\mathbf{B}}_2$ is in that direction. When the field set up at wire 2 by wire 1 is calculated, the force $\vec{\mathbf{F}}_2$ acting on wire 2 is found to be equal in magnitude and opposite in direction to $\vec{\mathbf{F}}_1$, which is what we expect because Newton's third law must be obeyed. When the currents are in opposite directions (that is, when one of the currents is reversed in Fig. 30.7), the forces

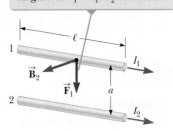

The field $\vec{\mathbf{B}}_2$ due to the current in wire 2 exerts a magnetic force of magnitude $F_1 = I_1 \ell B_2$ on wire 1.

Figure 30.7 Two parallel wires that each carry a steady current exert a magnetic force on each other. The force is attractive if the currents are parallel (as shown) and repulsive if the currents are antiparallel.

are reversed and the wires repel each other. Hence, parallel conductors carrying currents in the *same* direction *attract* each other, and parallel conductors carrying currents in *opposite* directions *repel* each other.

Because the magnitudes of the forces are the same on both wires, we denote the magnitude of the magnetic force between the wires as simply F_B. We can rewrite this magnitude in terms of the force per unit length:

$$\frac{F_B}{\ell} = \frac{\mu_0 I_1 I_2}{2\pi a} \tag{30.12}$$

The force between two parallel wires is used to define the **ampere** as follows:

◀ Definition of the ampere

> When the magnitude of the force per unit length between two long, parallel wires that carry identical currents and are separated by 1 m is 2×10^{-7} N/m, the current in each wire is defined to be 1 A.

The value 2×10^{-7} N/m is obtained from Equation 30.12 with $I_1 = I_2 = 1$ A and $a = 1$ m. Because this definition is based on a force, a mechanical measurement can be used to standardize the ampere. For instance, the National Institute of Standards and Technology uses an instrument called a *current balance* for primary current measurements. The results are then used to standardize other, more conventional instruments such as ammeters.

The SI unit of charge, the **coulomb,** is defined in terms of the ampere: When a conductor carries a steady current of 1 A, the quantity of charge that flows through a cross section of the conductor in 1 s is 1 C.

In deriving Equations 30.11 and 30.12, we assumed both wires are long compared with their separation distance. In fact, only one wire needs to be long. The equations accurately describe the forces exerted on each other by a long wire and a straight, parallel wire of limited length ℓ.

Quick Quiz 30.2 A loose spiral spring carrying no current is hung from a ceiling. When a switch is thrown so that a current exists in the spring, do the coils **(a)** move closer together, **(b)** move farther apart, or **(c)** not move at all?

Example 30.4   **Suspending a Wire**   `AM`

Two infinitely long, parallel wires are lying on the ground a distance $a = 1.00$ cm apart as shown in Figure 30.8a. A third wire, of length $L = 10.0$ m and mass 400 g, carries a current of $I_1 = 100$ A and is levitated above the first two wires, at a horizontal position midway between them. The infinitely long wires carry equal currents I_2 in the same direction, but in the direction opposite that in the levitated wire. What current must the infinitely long wires carry so that the three wires form an equilateral triangle?

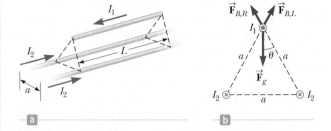

Figure 30.8 (Example 30.4) (a) Two current-carrying wires lie on the ground and suspend a third wire in the air by magnetic forces. (b) End view. In the situation described in the example, the three wires form an equilateral triangle. The two magnetic forces on the levitated wire are $\vec{\mathbf{F}}_{B,L}$, the force due to the left-hand wire on the ground, and $\vec{\mathbf{F}}_{B,R}$, the force due to the right-hand wire. The gravitational force $\vec{\mathbf{F}}_g$ on the levitated wire is also shown.

SOLUTION

Conceptualize Because the current in the short wire is opposite those in the long wires, the short wire is repelled from both of the others. Imagine the currents in the long wires in Figure 30.8a are increased. The repulsive force becomes stronger, and the levitated wire rises to the point at which the wire is once again levitated in equilibrium at a higher position. Figure 30.8b shows the desired situation with the three wires forming an equilateral triangle.

Categorize Because the levitated wire is subject to forces but does not accelerate, it is modeled as a *particle in equilibrium.*

▶ **30.4** continued

Analyze The horizontal components of the magnetic forces on the levitated wire cancel. The vertical components are both positive and add together. Choose the z axis to be upward through the top wire in Figure 30.8b and in the plane of the page.

Find the total magnetic force in the upward direction on the levitated wire:

$$\vec{\mathbf{F}}_B = 2\left(\frac{\mu_0 I_1 I_2}{2\pi a}\,\ell\right)\cos\theta\,\hat{\mathbf{k}} = \frac{\mu_0 I_1 I_2}{\pi a}\ell\cos\theta\,\hat{\mathbf{k}}$$

Find the gravitational force on the levitated wire:

$$\vec{\mathbf{F}}_g = -mg\hat{\mathbf{k}}$$

Apply the particle in equilibrium model by adding the forces and setting the net force equal to zero:

$$\sum\vec{\mathbf{F}} = \vec{\mathbf{F}}_B + \vec{\mathbf{F}}_g = \frac{\mu_0 I_1 I_2}{\pi a}\ell\cos\theta\,\hat{\mathbf{k}} - mg\hat{\mathbf{k}} = 0$$

Solve for the current in the wires on the ground:

$$I_2 = \frac{mg\pi a}{\mu_0 I_1 \ell\cos\theta}$$

Substitute numerical values:

$$I_2 = \frac{(0.400\ \text{kg})(9.80\ \text{m/s}^2)\pi(0.010\ 0\ \text{m})}{(4\pi\times10^{-7}\ \text{T}\cdot\text{m/A})(100\ \text{A})(10.0\ \text{m})\cos 30.0°}$$

$$= \boxed{113\ \text{A}}$$

Finalize The currents in all wires are on the order of 10^2 A. Such large currents would require specialized equipment. Therefore, this situation would be difficult to establish in practice. Is the equilibrium of wire 1 stable or unstable?

30.3 Ampère's Law

Looking back, we can see that the result of Example 30.1 is important because a current in the form of a long, straight wire occurs often. Figure 30.9 is a perspective view of the magnetic field surrounding a long, straight, current-carrying wire. Because of the wire's symmetry, the magnetic field lines are circles concentric with the wire and lie in planes perpendicular to the wire. The magnitude of $\vec{\mathbf{B}}$ is constant on any circle of radius a and is given by Equation 30.5. A convenient rule for determining the direction of $\vec{\mathbf{B}}$ is to grasp the wire with the right hand, positioning the thumb along the direction of the current. The four fingers wrap in the direction of the magnetic field.

Figure 30.9 also shows that the magnetic field line has no beginning and no end. Rather, it forms a closed loop. That is a major difference between magnetic field lines and electric field lines, which begin on positive charges and end on negative charges. We will explore this feature of magnetic field lines further in Section 30.5.

Oersted's 1819 discovery about deflected compass needles demonstrates that a current-carrying conductor produces a magnetic field. Figure 30.10a (page 912) shows how this effect can be demonstrated in the classroom. Several compass needles are placed in a horizontal plane near a long, vertical wire. When no current is present in the wire, all the needles point in the same direction (that of the horizontal component of the Earth's magnetic field) as expected. When the wire carries a strong, steady current, the needles all deflect in a direction tangent to the circle as in Figure 30.10b. These observations demonstrate that the direction of the magnetic field produced by the current in the wire is consistent with the right-hand rule described in Figure 30.9. When the current is reversed, the needles in Figure 30.10b also reverse.

Now let's evaluate the product $\vec{\mathbf{B}}\cdot d\vec{\mathbf{s}}$ for a small length element $d\vec{\mathbf{s}}$ on the circular path defined by the compass needles and sum the products for all elements

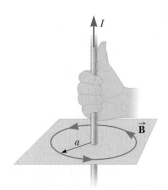

Figure 30.9 The right-hand rule for determining the direction of the magnetic field surrounding a long, straight wire carrying a current. Notice that the magnetic field lines form circles around the wire.

Andre–Marie Ampère
French Physicist (1775–1836)
Ampère is credited with the discovery of electromagnetism, which is the relationship between electric currents and magnetic fields. Ampère's genius, particularly in mathematics, became evident by the time he was 12 years old; his personal life, however, was filled with tragedy. His father, a wealthy city official, was guillotined during the French Revolution, and his wife died young, in 1803. Ampère died at the age of 61 of pneumonia.

Pitfall Prevention 30.2
Avoiding Problems with Signs When using Ampère's law, apply the following right-hand rule. Point your thumb in the direction of the current through the amperian loop. Your curled fingers then point in the direction that you should integrate when traversing the loop to avoid having to define the current as negative.

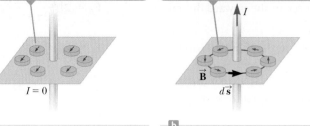

When no current is present in the wire, all compass needles point in the same direction (toward the Earth's north pole).

When the wire carries a strong current, the compass needles deflect in a direction tangent to the circle, which is the direction of the magnetic field created by the current.

$I = 0$

I

$\vec{\mathbf{B}}$

$d\vec{\mathbf{s}}$

a **b** **c**

Figure 30.10 (a) and (b) Compasses show the effects of the current in a nearby wire. (c) Circular magnetic field lines surrounding a current-carrying conductor, displayed with iron filings.

over the closed circular path.[1] Along this path, the vectors $d\vec{\mathbf{s}}$ and $\vec{\mathbf{B}}$ are parallel at each point (see Fig. 30.10b), so $\vec{\mathbf{B}} \cdot d\vec{\mathbf{s}} = B\,ds$. Furthermore, the magnitude of $\vec{\mathbf{B}}$ is constant on this circle and is given by Equation 30.5. Therefore, the sum of the products $B\,ds$ over the closed path, which is equivalent to the line integral of $\vec{\mathbf{B}} \cdot d\vec{\mathbf{s}}$, is

$$\oint \vec{\mathbf{B}} \cdot d\vec{\mathbf{s}} = B \oint ds = \frac{\mu_0 I}{2\pi r}(2\pi r) = \mu_0 I$$

where $\oint ds = 2\pi r$ is the circumference of the circular path of radius r. Although this result was calculated for the special case of a circular path surrounding a wire of infinite length, it holds for a closed path of *any* shape (an *amperian loop*) surrounding a current that exists in an unbroken circuit. The general case, known as **Ampère's law,** can be stated as follows:

> The line integral of $\vec{\mathbf{B}} \cdot d\vec{\mathbf{s}}$ around any closed path equals $\mu_0 I$, where I is the total steady current passing through any surface bounded by the closed path:
>
> $$\oint \vec{\mathbf{B}} \cdot d\vec{\mathbf{s}} = \mu_0 I \qquad (30.13)$$

◀ Ampère's law

Ampère's law describes the creation of magnetic fields by all continuous current configurations, but at our mathematical level it is useful only for calculating the magnetic field of current configurations having a high degree of symmetry. Its use is similar to that of Gauss's law in calculating electric fields for highly symmetric charge distributions.

Quick Quiz 30.3 Rank the magnitudes of $\oint \vec{\mathbf{B}} \cdot d\vec{\mathbf{s}}$ for the closed paths *a* through *d* in Figure 30.11 from greatest to least.

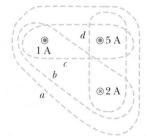

$1\,A$ d $5\,A$ c b a $2\,A$

Figure 30.11 (Quick Quiz 30.3) Four closed paths around three current-carrying wires.

---

[1]You may wonder why we would choose to evaluate this scalar product. The origin of Ampère's law is in 19th-century science, in which a "magnetic charge" (the supposed analog to an isolated electric charge) was imagined to be moved around a circular field line. The work done on the charge was related to $\vec{\mathbf{B}} \cdot d\vec{\mathbf{s}}$, just as the work done moving an electric charge in an electric field is related to $\vec{\mathbf{E}} \cdot d\vec{\mathbf{s}}$. Therefore, Ampère's law, a valid and useful principle, arose from an erroneous and abandoned work calculation!

Quick Quiz 30.4 Rank the magnitudes of $\oint \vec{\mathbf{B}} \cdot d\vec{\mathbf{s}}$ for the closed paths a through d in Figure 30.12 from greatest to least.

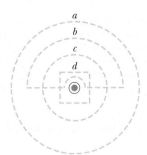

Figure 30.12 (Quick Quiz 30.4) Several closed paths near a single current-carrying wire.

Example 30.5   **The Magnetic Field Created by a Long Current-Carrying Wire**

A long, straight wire of radius R carries a steady current I that is uniformly distributed through the cross section of the wire (Fig. 30.13). Calculate the magnetic field a distance r from the center of the wire in the regions $r \geq R$ and $r < R$.

SOLUTION

Conceptualize Study Figure 30.13 to understand the structure of the wire and the current in the wire. The current creates magnetic fields everywhere, both inside and outside the wire. Based on our discussions about long, straight wires, we expect the magnetic field lines to be circles centered on the central axis of the wire.

Categorize Because the wire has a high degree of symmetry, we categorize this example as an Ampère's law problem. For the $r \geq R$ case, we should arrive at the same result as was obtained in Example 30.1, where we applied the Biot–Savart law to the same situation.

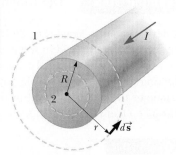

Figure 30.13 (Example 30.5) A long, straight wire of radius R carrying a steady current I uniformly distributed across the cross section of the wire. The magnetic field at any point can be calculated from Ampère's law using a circular path of radius r, concentric with the wire.

Analyze For the magnetic field exterior to the wire, let us choose for our path of integration circle 1 in Figure 30.13. From symmetry, $\vec{\mathbf{B}}$ must be constant in magnitude and parallel to $d\vec{\mathbf{s}}$ at every point on this circle.

Note that the total current passing through the plane of the circle is I and apply Ampère's law:

$$\oint \vec{\mathbf{B}} \cdot d\vec{\mathbf{s}} = B \oint ds = B(2\pi r) = \mu_0 I$$

Solve for B:

$$B = \frac{\mu_0 I}{2\pi r} \quad \text{(for } r \geq R\text{)} \tag{30.14}$$

Now consider the interior of the wire, where $r < R$. Here the current I' passing through the plane of circle 2 is less than the total current I.

Set the ratio of the current I' enclosed by circle 2 to the entire current I equal to the ratio of the area πr^2 enclosed by circle 2 to the cross-sectional area πR^2 of the wire:

$$\frac{I'}{I} = \frac{\pi r^2}{\pi R^2}$$

Solve for I':

$$I' = \frac{r^2}{R^2} I$$

Apply Ampère's law to circle 2:

$$\oint \vec{\mathbf{B}} \cdot d\vec{\mathbf{s}} = B(2\pi r) = \mu_0 I' = \mu_0 \left(\frac{r^2}{R^2} I \right)$$

Solve for B:

$$B = \left(\frac{\mu_0 I}{2\pi R^2} \right) r \quad \text{(for } r < R\text{)} \tag{30.15}$$

continued

▶ **30.5** continued

Finalize The magnetic field exterior to the wire is identical in form to Equation 30.5. As is often the case in highly symmetric situations, it is much easier to use Ampère's law than the Biot–Savart law (Example 30.1). The magnetic field interior to the wire is similar in form to the expression for the electric field inside a uniformly charged sphere (see Example 24.3). The magnitude of the magnetic field versus r for this configuration is plotted in Figure 30.14. Inside the

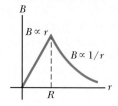

Figure 30.14 (Example 30.5) Magnitude of the magnetic field versus r for the wire shown in Figure 30.13. The field is proportional to r inside the wire and varies as $1/r$ outside the wire.

wire, $B \rightarrow 0$ as $r \rightarrow 0$. Furthermore, Equations 30.14 and 30.15 give the same value of the magnetic field at $r = R$, demonstrating that the magnetic field is continuous at the surface of the wire.

Example 30.6 The Magnetic Field Created by a Toroid

A device called a *toroid* (Fig. 30.15) is often used to create an almost uniform magnetic field in some enclosed area. The device consists of a conducting wire wrapped around a ring (a *torus*) made of a nonconducting material. For a toroid having N closely spaced turns of wire, calculate the magnetic field in the region occupied by the torus, a distance r from the center.

SOLUTION

Conceptualize Study Figure 30.15 carefully to understand how the wire is wrapped around the torus. The torus could be a solid material or it could be air, with a stiff wire wrapped into the shape shown in Figure 30.15 to form an empty toroid. Imagine each turn of the wire to be a circular loop as in Example 30.3. The magnetic field at the center of the loop is perpendicular to the plane of the loop. Therefore, the magnetic field lines of the collection of loops will form circles within the toroid such as suggested by loop 1 in Figure 30.15.

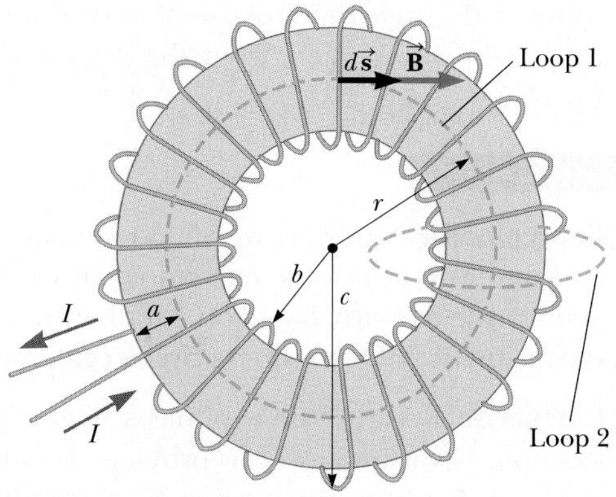

Categorize Because the toroid has a high degree of symmetry, we categorize this example as an Ampère's law problem.

Figure 30.15 (Example 30.6) A toroid consisting of many turns of wire. If the turns are closely spaced, the magnetic field in the interior of the toroid is tangent to the dashed circle (loop 1) and varies as $1/r$. The dimension a is the cross-sectional radius of the torus. The field outside the toroid is very small and can be described by using the amperian loop (loop 2) at the right side, perpendicular to the page.

Analyze Consider the circular amperian loop (loop 1) of radius r in the plane of Figure 30.15. By symmetry, the magnitude of the field is constant on this circle and tangent to it, so $\vec{\mathbf{B}} \cdot d\vec{\mathbf{s}} = B\,ds$. Furthermore, the wire passes through the loop N times, so the total current through the loop is NI.

Apply Ampère's law to loop 1:

$$\oint \vec{\mathbf{B}} \cdot d\vec{\mathbf{s}} = B \oint ds = B(2\pi r) = \mu_0 NI$$

Solve for B:

$$B = \frac{\mu_0 NI}{2\pi r} \tag{30.16}$$

Finalize This result shows that B varies as $1/r$ and hence is *nonuniform* in the region occupied by the torus. If, however, r is very large compared with the cross-sectional radius a of the torus, the field is approximately uniform inside the torus.

For an ideal toroid, in which the turns are closely spaced, the external magnetic field is close to zero, but it is not exactly zero. In Figure 30.15, imagine the radius r

of amperian loop 1 to be either smaller than b or larger than c. In either case, the loop encloses zero net current, so $\oint \vec{\mathbf{B}} \cdot d\vec{\mathbf{s}} = 0$. You might think this result proves that $\vec{\mathbf{B}} = 0$, but it does not. Consider the amperian loop (loop 2) on the right side of the toroid in Figure 30.15. The plane of this loop is perpendicular to the page, and the toroid passes through the loop. As charges enter the toroid as indicated by the current directions in Figure 30.15,

▶ **30.6** continued

they work their way counterclockwise around the toroid. Therefore, there is a counterclockwise current around the toroid, so that a current passes through amperian loop 2! This current is small, but not zero. As a result, the toroid acts as a current loop and produces a weak external field of the form shown in Figure 30.6. The reason $\oint \vec{\textbf{B}} \cdot d\vec{\textbf{s}} = 0$ for amperian loop 1 of radius $r < b$ or $r > c$ is that the field lines are perpendicular to $d\vec{\textbf{s}}$, *not* because $\vec{\textbf{B}} = 0$.

30.4 The Magnetic Field of a Solenoid

A **solenoid** is a long wire wound in the form of a helix. With this configuration, a reasonably uniform magnetic field can be produced in the space surrounded by the turns of wire—which we shall call the *interior* of the solenoid—when the solenoid carries a current. When the turns are closely spaced, each can be approximated as a circular loop; the net magnetic field is the vector sum of the fields resulting from all the turns.

Figure 30.16 shows the magnetic field lines surrounding a loosely wound solenoid. The field lines in the interior are nearly parallel to one another, are uniformly distributed, and are close together, indicating that the field in this space is strong and almost uniform.

If the turns are closely spaced and the solenoid is of finite length, the external magnetic field lines are as shown in Figure 30.17a. This field line distribution is similar to that surrounding a bar magnet (Fig. 30.17b). Hence, one end of the solenoid behaves like the north pole of a magnet and the opposite end behaves like the south pole. As the length of the solenoid increases, the interior field becomes more uniform and the exterior field becomes weaker. An *ideal solenoid* is approached when the turns are closely spaced and the length is much greater than the radius of the turns. Figure 30.18 (page 916) shows a longitudinal cross section of part of such a solenoid carrying a current *I*. In this case, the external field is close to zero and the interior field is uniform over a great volume.

Consider the amperian loop (loop 1) perpendicular to the page in Figure 30.18 (page 916), surrounding the ideal solenoid. This loop encloses a small

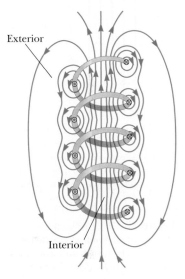

Figure 30.16 The magnetic field lines for a loosely wound solenoid.

The magnetic field lines resemble those of a bar magnet, meaning that the solenoid effectively has north and south poles.

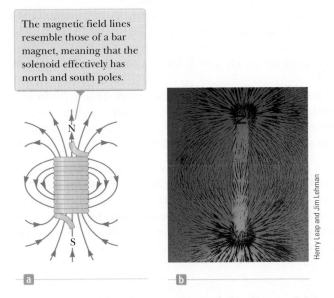

Figure 30.17 (a) Magnetic field lines for a tightly wound solenoid of finite length, carrying a steady current. The field in the interior space is strong and nearly uniform. (b) The magnetic field pattern of a bar magnet, displayed with small iron filings on a sheet of paper.

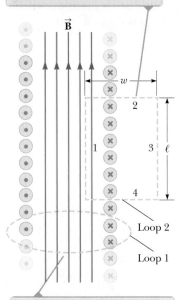

Ampère's law applied to the rectangular dashed path can be used to calculate the magnitude of the interior field.

Ampère's law applied to the circular path whose plane is perpendicular to the page can be used to show that there is a weak field outside the solenoid.

Figure 30.18 Cross-sectional view of an ideal solenoid, where the interior magnetic field is uniform and the exterior field is close to zero.

▶ **Magnetic field inside a solenoid**

current as the charges in the wire move coil by coil along the length of the solenoid. Therefore, there is a nonzero magnetic field outside the solenoid. It is a weak field, with circular field lines, like those due to a line of current as in Figure 30.9. For an ideal solenoid, this weak field is the only field external to the solenoid.

We can use Ampère's law to obtain a quantitative expression for the interior magnetic field in an ideal solenoid. Because the solenoid is ideal, $\vec{B}$ in the interior space is uniform and parallel to the axis and the magnetic field lines in the exterior space form circles around the solenoid. The planes of these circles are perpendicular to the page. Consider the rectangular path (loop 2) of length ℓ and width w shown in Figure 30.18. Let's apply Ampère's law to this path by evaluating the integral of $\vec{B} \cdot d\vec{s}$ over each side of the rectangle. The contribution along side 3 is zero because the external magnetic field lines are perpendicular to the path in this region. The contributions from sides 2 and 4 are both zero, again because $\vec{B}$ is perpendicular to $d\vec{s}$ along these paths, both inside and outside the solenoid. Side 1 gives a contribution to the integral because along this path $\vec{B}$ is uniform and parallel to $d\vec{s}$. The integral over the closed rectangular path is therefore

$$\oint \vec{B} \cdot d\vec{s} = \int_{\text{path 1}} \vec{B} \cdot d\vec{s} = B \int_{\text{path 1}} ds = B\ell$$

The right side of Ampère's law involves the total current I through the area bounded by the path of integration. In this case, the total current through the rectangular path equals the current through each turn multiplied by the number of turns. If N is the number of turns in the length ℓ, the total current through the rectangle is NI. Therefore, Ampère's law applied to this path gives

$$\oint \vec{B} \cdot d\vec{s} = B\ell = \mu_0 NI$$

$$B = \mu_0 \frac{N}{\ell} I = \mu_0 nI \tag{30.17}$$

where $n = N/\ell$ is the number of turns per unit length.

We also could obtain this result by reconsidering the magnetic field of a toroid (see Example 30.6). If the radius r of the torus in Figure 30.15 containing N turns is much greater than the toroid's cross-sectional radius a, a short section of the toroid approximates a solenoid for which $n = N/2\pi r$. In this limit, Equation 30.16 agrees with Equation 30.17.

Equation 30.17 is valid only for points near the center (that is, far from the ends) of a very long solenoid. As you might expect, the field near each end is smaller than the value given by Equation 30.17. As the length of a solenoid increases, the magnitude of the field at the end approaches half the magnitude at the center (see Problem 69).

Quick Quiz 30.5 Consider a solenoid that is very long compared with its radius. Of the following choices, what is the most effective way to increase the magnetic field in the interior of the solenoid? **(a)** double its length, keeping the number of turns per unit length constant **(b)** reduce its radius by half, keeping the number of turns per unit length constant **(c)** overwrap the entire solenoid with an additional layer of current-carrying wire

30.5 Gauss's Law in Magnetism

The flux associated with a magnetic field is defined in a manner similar to that used to define electric flux (see Eq. 24.3). Consider an element of area dA on an arbitrarily shaped surface as shown in Figure 30.19. If the magnetic field at this element is $\vec{B}$, the magnetic flux through the element is $\vec{B} \cdot d\vec{A}$, where $d\vec{A}$ is a vector that is perpendicular to the surface and has a magnitude equal to the area dA. Therefore, the total magnetic flux Φ_B through the surface is

$$\Phi_B \equiv \int \vec{B} \cdot d\vec{A} \qquad \text{(30.18)}$$

◀ **Definition of magnetic flux**

Consider the special case of a plane of area A in a uniform field $\vec{B}$ that makes an angle θ with $d\vec{A}$. The magnetic flux through the plane in this case is

$$\Phi_B = BA \cos \theta \qquad \text{(30.19)}$$

If the magnetic field is parallel to the plane as in Figure 30.20a, then $\theta = 90°$ and the flux through the plane is zero. If the field is perpendicular to the plane as in Figure 30.20b, then $\theta = 0$ and the flux through the plane is BA (the maximum value).

The unit of magnetic flux is $\text{T} \cdot \text{m}^2$, which is defined as a *weber* (Wb); 1 Wb = $1 \text{ T} \cdot \text{m}^2$.

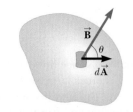

Figure 30.19 The magnetic flux through an area element dA is $\vec{B} \cdot d\vec{A} = B\, dA \cos \theta$, where $d\vec{A}$ is a vector perpendicular to the surface.

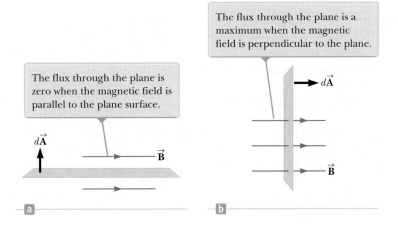

The flux through the plane is zero when the magnetic field is parallel to the plane surface.

The flux through the plane is a maximum when the magnetic field is perpendicular to the plane.

Figure 30.20 Magnetic flux through a plane lying in a magnetic field.

Example 30.7 **Magnetic Flux Through a Rectangular Loop**

A rectangular loop of width a and length b is located near a long wire carrying a current I (Fig. 30.21). The distance between the wire and the closest side of the loop is c. The wire is parallel to the long side of the loop. Find the total magnetic flux through the loop due to the current in the wire.

SOLUTION

Conceptualize As we saw in Section 30.3, the magnetic field lines due to the wire will be circles, many of which will pass through the rectangular loop. We know that the magnetic field is a function of distance r from a long wire. Therefore, the magnetic field varies over the area of the rectangular loop.

Categorize Because the magnetic field varies over the area of the loop, we must integrate over this area to find the total flux. That identifies this as an analysis problem.

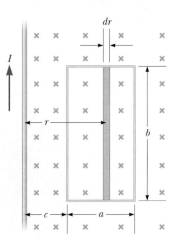

Figure 30.21 (Example 30.7) The magnetic field due to the wire carrying a current I is not uniform over the rectangular loop.

Analyze Noting that $\vec{B}$ is parallel to $d\vec{A}$ at any point within the loop, find the magnetic flux through the rectangular area using Equation 30.18 and incorporate Equation 30.14 for the magnetic field:

$$\Phi_B = \int \vec{B} \cdot d\vec{A} = \int B\, dA = \int \frac{\mu_0 I}{2\pi r}\, dA$$

continued

▶ **30.7** continued

Express the area element (the tan strip in Fig. 30.21) as $dA = b\,dr$ and substitute:

$$\Phi_B = \int \frac{\mu_0 I}{2\pi r} b\,dr = \frac{\mu_0 I b}{2\pi} \int \frac{dr}{r}$$

Integrate from $r = c$ to $r = a + c$:

$$\Phi_B = \frac{\mu_0 I b}{2\pi} \int_c^{a+c} \frac{dr}{r} = \frac{\mu_0 I b}{2\pi} \ln r \Big|_c^{a+c}$$

$$= \frac{\mu_0 I b}{2\pi} \ln \left(\frac{a+c}{c} \right) = \frac{\mu_0 I b}{2\pi} \ln \left(1 + \frac{a}{c} \right)$$

Finalize Notice how the flux depends on the size of the loop. Increasing either a or b increases the flux as expected. If c becomes large such that the loop is very far from the wire, the flux approaches zero, also as expected. If c goes to zero, the flux becomes infinite. In principle, this infinite value occurs because the field becomes infinite at $r = 0$ (assuming an infinitesimally thin wire). That will not happen in reality because the thickness of the wire prevents the left edge of the loop from reaching $r = 0$.

In Chapter 24, we found that the electric flux through a closed surface surrounding a net charge is proportional to that charge (Gauss's law). In other words, the number of electric field lines leaving the surface depends only on the net charge within it. This behavior exists because electric field lines originate and terminate on electric charges.

The situation is quite different for magnetic fields, which are continuous and form closed loops. In other words, as illustrated by the magnetic field lines of a current in Figure 30.9 and of a bar magnet in Figure 30.22, magnetic field lines do not begin or end at any point. For any closed surface such as the one outlined by the dashed line in Figure 30.22, the number of lines entering the surface equals the number leaving the surface; therefore, the net magnetic flux is zero. In contrast, for a closed surface surrounding one charge of an electric dipole (Fig. 30.23), the net electric flux is not zero.

Gauss's law in magnetism states that

Gauss's law in magnetism ▶

the net magnetic flux through any closed surface is always zero:

$$\oint \vec{\mathbf{B}} \cdot d\vec{\mathbf{A}} = 0 \tag{30.20}$$

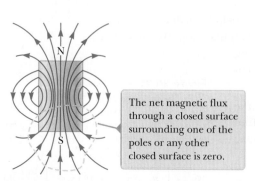

Figure 30.22 The magnetic field lines of a bar magnet form closed loops. (The dashed line represents the intersection of a closed surface with the page.)

The net magnetic flux through a closed surface surrounding one of the poles or any other closed surface is zero.

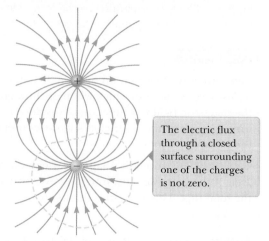

The electric flux through a closed surface surrounding one of the charges is not zero.

Figure 30.23 The electric field lines surrounding an electric dipole begin on the positive charge and terminate on the negative charge.

This statement represents that isolated magnetic poles (monopoles) have never been detected and perhaps do not exist. Nonetheless, scientists continue the search because certain theories that are otherwise successful in explaining fundamental physical behavior suggest the possible existence of magnetic monopoles.

30.6 Magnetism in Matter

The magnetic field produced by a current in a coil of wire gives us a hint as to what causes certain materials to exhibit strong magnetic properties. Earlier we found that a solenoid like the one shown in Figure 30.17a has a north pole and a south pole. In general, *any* current loop has a magnetic field and therefore has a magnetic dipole moment, including the atomic-level current loops described in some models of the atom.

The Magnetic Moments of Atoms

Let's begin our discussion with a classical model of the atom in which electrons move in circular orbits around the much more massive nucleus. In this model, an orbiting electron constitutes a tiny current loop (because it is a moving charge), and the magnetic moment of the electron is associated with this orbital motion. Although this model has many deficiencies, some of its predictions are in good agreement with the correct theory, which is expressed in terms of quantum physics.

In our classical model, we assume an electron is a particle in uniform circular motion: it moves with constant speed v in a circular orbit of radius r about the nucleus as in Figure 30.24. The current I associated with this orbiting electron is its charge e divided by its period T. Using Equation 4.15 from the particle in uniform circular motion model, $T = 2\pi r/v$, gives

$$I = \frac{e}{T} = \frac{ev}{2\pi r}$$

The magnitude of the magnetic moment associated with this current loop is given by $\mu = IA$, where $A = \pi r^2$ is the area enclosed by the orbit. Therefore,

$$\mu = IA = \left(\frac{ev}{2\pi r}\right)\pi r^2 = \tfrac{1}{2}evr \tag{30.21}$$

Because the magnitude of the orbital angular momentum of the electron is given by $L = m_e vr$ (Eq. 11.12 with $\phi = 90°$), the magnetic moment can be written as

$$\mu = \left(\frac{e}{2m_e}\right)L \tag{30.22}$$

◀ **Orbital magnetic moment**

This result demonstrates that the magnetic moment of the electron is proportional to its orbital angular momentum. Because the electron is negatively charged, the vectors $\vec{\mu}$ and $\vec{L}$ point in *opposite* directions. Both vectors are perpendicular to the plane of the orbit as indicated in Figure 30.24.

A fundamental outcome of quantum physics is that orbital angular momentum is quantized and is equal to multiples of $\hbar = h/2\pi = 1.05 \times 10^{-34}\,\text{J}\cdot\text{s}$, where h is Planck's constant (see Chapter 40). The smallest nonzero value of the electron's magnetic moment resulting from its orbital motion is

$$\mu = \sqrt{2}\,\frac{e}{2m_e}\hbar \tag{30.23}$$

We shall see in Chapter 42 how expressions such as Equation 30.23 arise.

Because all substances contain electrons, you may wonder why most substances are not magnetic. The main reason is that, in most substances, the magnetic

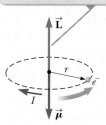

The electron has an angular momentum $\vec{L}$ in one direction and a magnetic moment $\vec{\mu}$ in the opposite direction.

Figure 30.24 An electron moving in the direction of the gray arrow in a circular orbit of radius r. Because the electron carries a negative charge, the direction of the current due to its motion about the nucleus is opposite the direction of that motion.

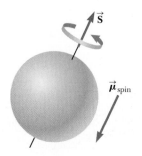

Figure 30.25 Classical model of a spinning electron. We can adopt this model to remind ourselves that electrons have an intrinsic angular momentum. The model should not be pushed too far, however; it gives an incorrect magnitude for the magnetic moment, incorrect quantum numbers, and too many degrees of freedom.

Table 30.1 **Magnetic Moments of Some Atoms and Ions**

Atom or Ion	Magnetic Moment $(10^{-24}\,\text{J/T})$
H	9.27
He	0
Ne	0
Ce^{3+}	19.8
Yb^{3+}	37.1

moment of one electron in an atom is canceled by that of another electron orbiting in the opposite direction. The net result is that, for most materials, the magnetic effect produced by the orbital motion of the electrons is either zero or very small.

In addition to its orbital magnetic moment, an electron (as well as protons, neutrons, and other particles) has an intrinsic property called **spin** that also contributes to its magnetic moment. Classically, the electron might be viewed as spinning about its axis as shown in Figure 30.25, but you should be very careful with the classical interpretation. The magnitude of the angular momentum $\vec{S}$ associated with spin is on the same order of magnitude as the magnitude of the angular momentum $\vec{L}$ due to the orbital motion. The magnitude of the spin angular momentum of an electron predicted by quantum theory is

$$S = \frac{\sqrt{3}}{2}\hbar$$

The magnetic moment characteristically associated with the spin of an electron has the value

$$\mu_{\text{spin}} = \frac{e\hbar}{2m_e} \tag{30.24}$$

This combination of constants is called the **Bohr magneton** μ_{B}:

$$\mu_{\text{B}} = \frac{e\hbar}{2m_e} = 9.27 \times 10^{-24}\,\text{J/T} \tag{30.25}$$

Therefore, atomic magnetic moments can be expressed as multiples of the Bohr magneton. (Note that $1\,\text{J/T} = 1\,\text{A}\cdot\text{m}^2$.)

In atoms containing many electrons, the electrons usually pair up with their spins opposite each other; therefore, the spin magnetic moments cancel. Atoms containing an odd number of electrons, however, must have at least one unpaired electron and therefore some spin magnetic moment. The total magnetic moment of an atom is the vector sum of the orbital and spin magnetic moments, and a few examples are given in Table 30.1. Notice that helium and neon have zero moments because their individual spin and orbital moments cancel.

The nucleus of an atom also has a magnetic moment associated with its constituent protons and neutrons. The magnetic moment of a proton or neutron, however, is much smaller than that of an electron and can usually be neglected. We can understand this smaller value by inspecting Equation 30.25 and replacing the mass of the electron with the mass of a proton or a neutron. Because the masses of the proton and neutron are much greater than that of the electron, their magnetic moments are on the order of 10^3 times smaller than that of the electron.

Ferromagnetism

A small number of crystalline substances exhibit strong magnetic effects called **ferromagnetism.** Some examples of ferromagnetic substances are iron, cobalt, nickel, gadolinium, and dysprosium. These substances contain permanent atomic magnetic moments that tend to align parallel to each other even in a weak external magnetic field. Once the moments are aligned, the substance remains magnetized after the external field is removed. This permanent alignment is due to a strong coupling between neighboring moments, a coupling that can be understood only in quantum-mechanical terms.

All ferromagnetic materials are made up of microscopic regions called **domains,** regions within which all magnetic moments are aligned. These domains have volumes of about 10^{-12} to 10^{-8} m^3 and contain 10^{17} to 10^{21} atoms. The boundaries between the various domains having different orientations are called **domain walls.** In an unmagnetized sample, the magnetic moments in the domains are randomly oriented so that the net magnetic moment is zero as in Figure 30.26a. When the sample is placed in an external magnetic field $\vec{B}$, the size of those domains with magnetic moments aligned with the field grows, which results in a magnetized sample as in Figure 30.26b. As the external field becomes very strong as in Figure 30.26c, the domains in which the magnetic moments are not aligned with the field become very small. When the external field is removed, the sample may retain a net magnetization in the direction of the original field. At ordinary temperatures, thermal agitation is not sufficient to disrupt this preferred orientation of magnetic moments.

When the temperature of a ferromagnetic substance reaches or exceeds a critical temperature called the **Curie temperature,** the substance loses its residual magnetization. Below the Curie temperature, the magnetic moments are aligned and the substance is ferromagnetic. Above the Curie temperature, the thermal agitation is great enough to cause a random orientation of the moments and the substance becomes paramagnetic. Curie temperatures for several ferromagnetic substances are given in Table 30.2.

Paramagnetism

Paramagnetic substances have a weak magnetism resulting from the presence of atoms (or ions) that have permanent magnetic moments. These moments interact only weakly with one another and are randomly oriented in the absence of an external magnetic field. When a paramagnetic substance is placed in an external magnetic field, its atomic moments tend to line up with the field. This alignment process, however, must compete with thermal motion, which tends to randomize the magnetic moment orientations.

Diamagnetism

When an external magnetic field is applied to a diamagnetic substance, a weak magnetic moment is induced in the direction opposite the applied field, causing diamagnetic substances to be weakly repelled by a magnet. Although diamagnetism is present in all matter, its effects are much smaller than those of paramagnetism or ferromagnetism and are evident only when those other effects do not exist.

We can attain some understanding of diamagnetism by considering a classical model of two atomic electrons orbiting the nucleus in opposite directions but with the same speed. The electrons remain in their circular orbits because of the attractive electrostatic force exerted by the positively charged nucleus. Because the magnetic moments of the two electrons are equal in magnitude and opposite in direction, they cancel each other and the magnetic moment of the atom is zero. When an external magnetic field is applied, the electrons experience an additional magnetic force $q\vec{v} \times \vec{B}$. This added magnetic force combines with the electrostatic force to increase the orbital speed of the electron whose magnetic moment is antiparallel to the field and to decrease the speed of the electron whose magnetic moment is parallel to the field. As a result, the two magnetic moments of the electrons no longer cancel and the substance acquires a net magnetic moment that is opposite the applied field.

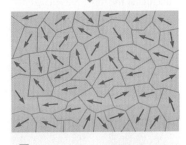

In an unmagnetized substance, the atomic magnetic dipoles are randomly oriented.

a

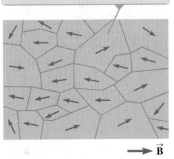

When an external field $\vec{B}$ is applied, the domains with components of magnetic moment in the same direction as $\vec{B}$ grow larger, giving the sample a net magnetization.

$\longrightarrow \vec{B}$

b

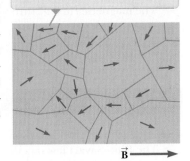

As the field is made even stronger, the domains with magnetic moment vectors not aligned with the external field become very small.

$\vec{B} \longrightarrow$

c

Figure 30.26 Orientation of magnetic dipoles before and after a magnetic field is applied to a ferromagnetic substance.

Table 30.2 **Curie Temperatures for Several Ferromagnetic Substances**

Substance	T_{Curie} (K)
Iron	1 043
Cobalt	1 394
Nickel	631
Gadolinium	317
Fe_2O_3	893

In the Meissner effect, the small magnet at the top induces currents in the superconducting disk below, which is cooled to −321°F (77 K). The currents create a repulsive magnetic force on the magnet causing it to levitate above the superconducting disk.

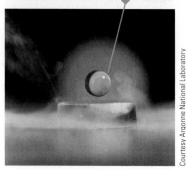

Liquid oxygen, a paramagnetic material, is attracted to the poles of a magnet.

The levitation force is exerted on the diamagnetic water molecules in the frog's body.

Courtesy Argonne National Laboratory

© Cengage Learning/Leon Lewandowski

Courtesy of Dr. Andre Geim, Manchester University

(*Left*) Paramagnetism. (*Right*) Diamagnetism: a frog is levitated in a 16-T magnetic field at the Nijmegen High Field Magnet Laboratory in the Netherlands.

Figure 30.27 An illustration of the Meissner effect, shown by this magnet suspended above a cooled ceramic superconductor disk, has become our most visual image of high-temperature superconductivity. Superconductivity is the loss of all resistance to electrical current and is a key to more-efficient energy use.

As you recall from Chapter 27, a superconductor is a substance in which the electrical resistance is zero below some critical temperature. Certain types of superconductors also exhibit perfect diamagnetism in the superconducting state. As a result, an applied magnetic field is expelled by the superconductor so that the field is zero in its interior. This phenomenon is known as the **Meissner effect.** If a permanent magnet is brought near a superconductor, the two objects repel each other. This repulsion is illustrated in Figure 30.27, which shows a small permanent magnet levitated above a superconductor maintained at 77 K.

Summary

Definition

The **magnetic flux** Φ_B through a surface is defined by the surface integral

$$\Phi_B \equiv \int \vec{\mathbf{B}} \cdot d\vec{\mathbf{A}} \qquad (30.18)$$

Concepts and Principles

The **Biot–Savart law** says that the magnetic field $d\vec{\mathbf{B}}$ at a point P due to a length element $d\vec{\mathbf{s}}$ that carries a steady current I is

$$d\vec{\mathbf{B}} = \frac{\mu_0}{4\pi} \frac{I\,d\vec{\mathbf{s}} \times \hat{\mathbf{r}}}{r^2} \qquad (30.1)$$

where μ_0 is the **permeability of free space,** r is the distance from the element to the point P, and $\hat{\mathbf{r}}$ is a unit vector pointing from $d\vec{\mathbf{s}}$ toward point P. We find the total field at P by integrating this expression over the entire current distribution.

The magnetic force per unit length between two parallel wires separated by a distance a and carrying currents I_1 and I_2 has a magnitude

$$\frac{F_B}{\ell} = \frac{\mu_0 I_1 I_2}{2\pi a} \qquad (30.12)$$

The force is attractive if the currents are in the same direction and repulsive if they are in opposite directions.

Ampère's law says that the line integral of $\vec{\mathbf{B}} \cdot d\vec{\mathbf{s}}$ around any closed path equals $\mu_0 I$, where I is the total steady current through any surface bounded by the closed path:

$$\oint \vec{\mathbf{B}} \cdot d\vec{\mathbf{s}} = \mu_0 I \qquad \textbf{(30.13)}$$

The magnitude of the magnetic field at a distance r from a long, straight wire carrying an electric current I is

$$B = \frac{\mu_0 I}{2\pi r} \qquad \textbf{(30.14)}$$

The field lines are circles concentric with the wire.

The magnitudes of the fields inside a toroid and solenoid are

$$B = \frac{\mu_0 N I}{2\pi r} \quad \text{(toroid)} \qquad \textbf{(30.16)}$$

$$B = \mu_0 \frac{N}{\ell} I = \mu_0 n I \quad \text{(solenoid)} \qquad \textbf{(30.17)}$$

where N is the total number of turns.

Gauss's law of magnetism states that the net magnetic flux through any closed surface is zero:

$$\oint \vec{\mathbf{B}} \cdot d\vec{\mathbf{A}} = 0 \qquad \textbf{(30.20)}$$

Substances can be classified into one of three categories that describe their magnetic behavior. **Diamagnetic** substances are those in which the magnetic moment is weak and opposite the applied magnetic field. **Paramagnetic** substances are those in which the magnetic moment is weak and in the same direction as the applied magnetic field. In **ferromagnetic** substances, interactions between atoms cause magnetic moments to align and create a strong magnetization that remains after the external field is removed.

Objective Questions

1. denotes answer available in *Student Solutions Manual/Study Guide*

1. **(i)** What happens to the magnitude of the magnetic field inside a long solenoid if the current is doubled? (a) It becomes four times larger. (b) It becomes twice as large. (c) It is unchanged. (d) It becomes one-half as large. (e) It becomes one-fourth as large. **(ii)** What happens to the field if instead the length of the solenoid is doubled, with the number of turns remaining the same? Choose from the same possibilities as in part (i). **(iii)** What happens to the field if the number of turns is doubled, with the length remaining the same? Choose from the same possibilities as in part (i). **(iv)** What happens to the field if the radius is doubled? Choose from the same possibilities as in part (i).

2. In Figure 30.7, assume $I_1 = 2.00$ A and $I_2 = 6.00$ A. What is the relationship between the magnitude F_1 of the force exerted on wire 1 and the magnitude F_2 of the force exerted on wire 2? (a) $F_1 = 6F_2$ (b) $F_1 = 3F_2$ (c) $F_1 = F_2$ (d) $F_1 = \frac{1}{3}F_2$ (e) $F_1 = \frac{1}{6}F_2$

3. Answer each question yes or no. (a) Is it possible for each of three stationary charged particles to exert a force of attraction on the other two? (b) Is it possible for each of three stationary charged particles to repel both of the other particles? (c) Is it possible for each of three current-carrying metal wires to attract the other two wires? (d) Is it possible for each of three current-carrying metal wires to repel the other two wires? André-Marie Ampère's experiments on electromagnetism are models of logical precision and included observation of the phenomena referred to in this question.

4. Two long, parallel wires each carry the same current I in the same direction (Fig. OQ30.4). Is the total magnetic field at the point P midway between the wires (a) zero, (b) directed into the page, (c) directed out of the page, (d) directed to the left, or (e) directed to the right?

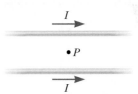

Figure OQ30.4

5. Two long, straight wires cross each other at a right angle, and each carries the same current I (Fig. OQ30.5). Which of the following statements is true regarding the total magnetic field due to the two wires at the various points in the figure? More than one statement may be correct. (a) The field is strongest at points B and D. (b) The field is strongest at points A and C. (c) The field is out of the page at point B and

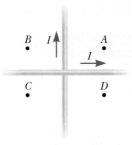

Figure OQ30.5

into the page at point *D*. (d) The field is out of the page at point *C* and out of the page at point *D*. (e) The field has the same magnitude at all four points.

6. A long, vertical, metallic wire carries downward electric current. **(i)** What is the direction of the magnetic field it creates at a point 2 cm horizontally east of the center of the wire? (a) north (b) south (c) east (d) west (e) up **(ii)** What would be the direction of the field if the current consisted of positive charges moving downward instead of electrons moving upward? Choose from the same possibilities as in part (i).

7. Suppose you are facing a tall makeup mirror on a vertical wall. Fluorescent tubes framing the mirror carry a clockwise electric current. **(i)** What is the direction of the magnetic field created by that current at the center of the mirror? (a) left (b) right (c) horizontally toward you (d) horizontally away from you (e) no direction because the field has zero magnitude **(ii)** What is the direction of the field the current creates at a point on the wall outside the frame to the right? Choose from the same possibilities as in part (i).

8. A long, straight wire carries a current *I* (Fig. OQ30.8). Which of the following statements is true regarding the magnetic field due to the wire? More than one statement may be correct. (a) The magnitude is proportional to I/r, and the direction is out of the page at *P*. (b) The magnitude is proportional to I/r^2, and the direction is out of the page at *P*. (c) The magnitude is proportional to I/r, and the direction is into the page at *P*. (d) The magnitude is proportional to I/r^2, and the direction is into the page at *P*. (e) The magnitude is proportional to *I*, but does not depend on *r*.

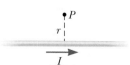

Figure OQ30.8

9. Two long, parallel wires carry currents of 20.0 A and 10.0 A in opposite directions (Fig. OQ30.9). Which of the following statements is true? More than one state-

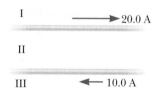

Figure OQ30.9 Objective Questions 9 and 10.

ment may be correct. (a) In region I, the magnetic field is into the page and is never zero. (b) In region II, the field is into the page and can be zero. (c) In region III, it is possible for the field to be zero. (d) In region I, the magnetic field is out of the page and is never zero. (e) There are no points where the field is zero.

10. Consider the two parallel wires carrying currents in opposite directions in Figure OQ30.9. Due to the magnetic interaction between the wires, does the lower wire experience a magnetic force that is (a) upward, (b) downward, (c) to the left, (d) to the right, or (e) into the paper?

11. What creates a magnetic field? More than one answer may be correct. (a) a stationary object with electric charge (b) a moving object with electric charge (c) a stationary conductor carrying electric current (d) a difference in electric potential (e) a charged capacitor disconnected from a battery and at rest *Note:* In Chapter 34, we will see that a changing electric field also creates a magnetic field.

12. A long solenoid with closely spaced turns carries electric current. Does each turn of wire exert (a) an attractive force on the next adjacent turn, (b) a repulsive force on the next adjacent turn, (c) zero force on the next adjacent turn, or (d) either an attractive or a repulsive force on the next turn, depending on the direction of current in the solenoid?

13. A uniform magnetic field is directed along the *x* axis. For what orientation of a flat, rectangular coil is the flux through the rectangle a maximum? (a) It is a maximum in the *xy* plane. (b) It is a maximum in the *xz* plane. (c) It is a maximum in the *yz* plane. (d) The flux has the same nonzero value for all these orientations. (e) The flux is zero in all cases.

14. Rank the magnitudes of the following magnetic fields from largest to smallest, noting any cases of equality. (a) the field 2 cm away from a long, straight wire carrying a current of 3 A (b) the field at the center of a flat, compact, circular coil, 2 cm in radius, with 10 turns, carrying a current of 0.3 A (c) the field at the center of a solenoid 2 cm in radius and 200 cm long, with 1 000 turns, carrying a current of 0.3 A (d) the field at the center of a long, straight, metal bar, 2 cm in radius, carrying a current of 300 A (e) a field of 1 mT

15. Solenoid A has length *L* and *N* turns, solenoid B has length 2*L* and *N* turns, and solenoid C has length *L*/2 and 2*N* turns. If each solenoid carries the same current, rank the magnitudes of the magnetic fields in the centers of the solenoids from largest to smallest.

Conceptual Questions

1. denotes answer available in *Student Solutions Manual/Study Guide*

1. Is the magnetic field created by a current loop uniform? Explain.

2. One pole of a magnet attracts a nail. Will the other pole of the magnet attract the nail? Explain. Also explain how a magnet sticks to a refrigerator door.

3. Compare Ampère's law with the Biot–Savart law. Which is more generally useful for calculating $\vec{B}$ for a current-carrying conductor?

4. A hollow copper tube carries a current along its length. Why is $B = 0$ inside the tube? Is *B* nonzero outside the tube?

5. Imagine you have a compass whose needle can rotate vertically as well as horizontally. Which way would the compass needle point if you were at the Earth's north magnetic pole?

6. Is Ampère's law valid for all closed paths surrounding a conductor? Why is it not useful for calculating $\vec{B}$ for all such paths?

7. A magnet attracts a piece of iron. The iron can then attract another piece of iron. On the basis of domain alignment, explain what happens in each piece of iron.

8. Why does hitting a magnet with a hammer cause the magnetism to be reduced?

9. The quantity $\int \vec{B} \cdot d\vec{s}$ in Ampère's law is called *magnetic circulation*. Figures 30.10 and 30.13 show paths around which the magnetic circulation is evaluated. Each of these paths encloses an area. What is the magnetic flux through each area? Explain your answer.

10. Figure CQ30.10 shows four permanent magnets, each having a hole through its center. Notice that the blue and yellow magnets are levitated above the red ones. (a) How does this levitation occur? (b) What purpose do the rods serve? (c) What can you say about the poles of the magnets from this observation? (d) If the blue magnet were inverted, what do you suppose would happen?

Figure CQ30.10

11. Explain why two parallel wires carrying currents in opposite directions repel each other.

12. Consider a magnetic field that is uniform in direction throughout a certain volume. (a) Can the field be uniform in magnitude? (b) Must it be uniform in magnitude? Give evidence for your answers.

Section 30.1 The Biot–Savart Law

1. Review. In studies of the possibility of migrating birds using the Earth's magnetic field for navigation, birds have been fitted with coils as "caps" and "collars" as shown in Figure P30.1. (a) If the identical coils have radii of 1.20 cm and are 2.20 cm apart, with 50 turns of wire apiece, what current should they both carry to produce a magnetic field of 4.50×10^{-5} T halfway between them? (b) If the resistance of each coil is 210 Ω, what voltage should the battery supplying each coil have? (c) What power is delivered to each coil?

Figure P30.1

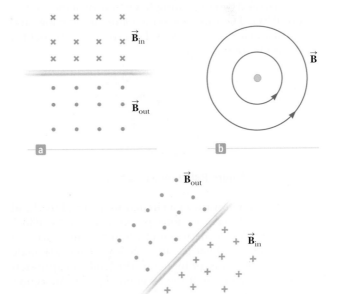
Figure P30.2

2. In each of parts (a) through (c) of Figure P30.2, find the direction of the current in the wire that would produce a magnetic field directed as shown.

3. Calculate the magnitude of the magnetic field at a point 25.0 cm from a long, thin conductor carrying a current of 2.00 A.

4. In 1962, measurements of the magnetic field of a large tornado were made at the Geophysical Observatory in Tulsa, Oklahoma. If the magnitude of the tornado's field was $B = 1.50 \times 10^{-8}$ T pointing north when the tornado was 9.00 km east of the observatory, what current was carried up or down the funnel of the tornado? Model the vortex as a long, straight wire carrying a current.

5. (a) A conducting loop in the shape of a square of edge length $\ell = 0.400$ m carries a current $I = 10.0$ A as shown in Figure P30.5. Calculate the magnitude and direction of the magnetic field at the center of the square. (b) **What If?** If this conductor is reshaped to form a circular loop and carries the same current, what is the value of the magnetic field at the center?

Figure P30.5

6. In Niels Bohr's 1913 model of the hydrogen atom, an electron circles the proton at a distance of 5.29×10^{-11} m with a speed of 2.19×10^6 m/s. Compute the magnitude of the magnetic field this motion produces at the location of the proton.

7. A conductor consists of a circular loop of radius $R = 15.0$ cm and two long, straight sections as shown in Figure P30.7. The wire lies in the plane of the paper and carries a current $I = 1.00$ A. Find the magnetic field at the center of the loop.

Figure P30.7 Problems 7 and 8.

8. A conductor consists of a circular loop of radius R and two long, straight sections as shown in Figure P30.7. The wire lies in the plane of the paper and carries a current I. (a) What is the direction of the magnetic field at the center of the loop? (b) Find an expression for the magnitude of the magnetic field at the center of the loop.

9. Two long, straight, parallel wires carry currents that are directed perpendicular to the page as shown in Figure P30.9. Wire 1 carries a current I_1 into the page (in the negative z direction) and passes through the x axis at $x = +a$. Wire 2 passes through the x axis at $x = -2a$ and carries an unknown cur-

rent I_2. The total magnetic field at the origin due to the current-carrying wires has the magnitude $2\mu_0 I_1/(2\pi a)$. The current I_2 can have either of two possible values. (a) Find the value of I_2 with the smaller magnitude, stating it in terms of I_1 and giving its direction. (b) Find the other possible value of I_2.

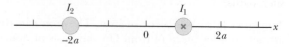

Figure P30.9

10. An infinitely long wire carrying a current I is bent at a right angle as shown in Figure P30.10. Determine the magnetic field at point P, located a distance x from the corner of the wire.

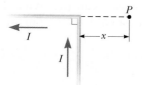

Figure P30.10

11. A long, straight wire carries a current I. A right-angle bend is made in the middle of the wire. The bend forms an arc of a circle of radius r as shown in Figure P30.11. Determine the magnetic field at point P, the center of the arc.

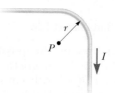

Figure P30.11

12. Consider a flat, circular current loop of radius R carrying a current I. Choose the x axis to be along the axis of the loop, with the origin at the loop's center. Plot a graph of the ratio of the magnitude of the magnetic field at coordinate x to that at the origin for $x = 0$ to $x = 5R$. It may be helpful to use a programmable calculator or a computer to solve this problem.

13. A current path shaped as shown in Figure P30.13 produces a magnetic field at P, the center of the arc. If the arc subtends an angle of $\theta = 30.0°$ and the radius of the arc is 0.600 m, what are the magnitude and

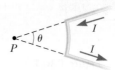

Figure P30.13

direction of the field produced at P if the current is 3.00 A?

14. One long wire carries current 30.0 A to the left along the x axis. A second long wire carries current 50.0 A to the right along the line ($y = 0.280$ m, $z = 0$). (a) Where in the plane of the two wires is the total magnetic field equal to zero? (b) A particle with a charge of -2.00 μC is moving with a velocity of $150\hat{\imath}$ Mm/s along the line ($y = 0.100$ m, $z = 0$). Calculate the vector magnetic force acting on the particle. (c) **What If?** A uniform electric field is applied to allow this particle to pass through this region undeflected. Calculate the required vector electric field.

15. Three long, parallel conductors each carry a current of $I = 2.00$ A. Figure P30.15 is an end view of the conductors, with each current coming out of the page. Taking $a = 1.00$ cm, determine the magnitude and direction of the magnetic field at (a) point A, (b) point B, and (c) point C.

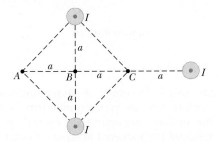

Figure P30.15

16. In a long, straight, vertical lightning stroke, electrons move downward and positive ions move upward and constitute a current of magnitude 20.0 kA. At a location 50.0 m east of the middle of the stroke, a free electron drifts through the air toward the west with a speed of 300 m/s. (a) Make a sketch showing the various vectors involved. Ignore the effect of the Earth's magnetic field. (b) Find the vector force the lightning stroke exerts on the electron. (c) Find the radius of the electron's path. (d) Is it a good approximation to model the electron as moving in a uniform field? Explain your answer. (e) If it does not collide with any obstacles, how many revolutions will the electron complete during the 60.0-μs duration of the lightning stroke?

17. Determine the magnetic field (in terms of I, a, and d) at the origin due to the current loop in Figure P30.17. The loop extends to infinity above the figure.

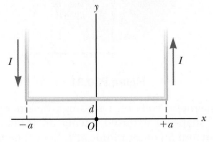

Figure P30.17

18. A wire carrying a current I is bent into the shape of an equilateral triangle of side L. (a) Find the magnitude of the magnetic field at the center of the triangle. (b) At a point halfway between the center and any vertex, is the field stronger or weaker than at the center? Give a qualitative argument for your answer.

19. The two wires shown in Figure P30.19 are separated by $d = 10.0$ cm and carry currents of $I = 5.00$ A in opposite directions. Find the magnitude and direction of the net magnetic field (a) at a point midway between the wires; (b) at point P_1, 10.0 cm to the right of the wire on the right; and (c) at point P_2, $2d = 20.0$ cm to the left of the wire on the left.

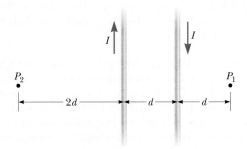

Figure P30.19

20. Two long, parallel wires carry currents of $I_1 = 3.00$ A and $I_2 = 5.00$ A in the directions indicated in Figure P30.20. (a) Find the magnitude and direction of the magnetic field at a point midway between the wires. (b) Find the magnitude and direction of the magnetic field at point P, located $d = 20.0$ cm above the wire carrying the 5.00-A current.

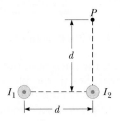

Figure P30.20

Section 30.2 The Magnetic Force Between Two Parallel Conductors

21. Two long, parallel conductors, separated by 10.0 cm, carry currents in the same direction. The first wire carries a current $I_1 = 5.00$ A, and the second carries $I_2 = 8.00$ A. (a) What is the magnitude of the magnetic field created by I_1 at the location of I_2? (b) What is the force per unit length exerted by I_1 on I_2? (c) What is the magnitude of the magnetic field created by I_2 at the location of I_1? (d) What is the force per length exerted by I_2 on I_1?

22. Two parallel wires separated by 4.00 cm repel each other with a force per unit length of 2.00×10^{-4} N/m. The current in one wire is 5.00 A. (a) Find the current in the other wire. (b) Are the currents in the same direction or in opposite directions? (c) What would happen if the direction of one current were reversed and doubled?

23. Two parallel wires are separated by 6.00 cm, each carrying 3.00 A of current in the same direction. (a) What is the magnitude of the force per unit length between the wires? (b) Is the force attractive or repulsive?

24. Two long wires hang vertically. Wire 1 carries an upward current of 1.50 A. Wire 2, 20.0 cm to the right of wire 1, carries a downward current of 4.00 A. A third wire, wire 3, is to be hung vertically and located such that when it carries a certain current, each wire experiences no net force. (a) Is this situation possible? Is it possible in more than one way? Describe (b) the position of wire 3 and (c) the magnitude and direction of the current in wire 3.

25. In Figure P30.25, the current in the long, straight wire is $I_1 = 5.00$ A and the wire lies in the plane of the rectangular loop, which carries a current $I_2 = 10.0$ A. The dimensions in the figure are $c = 0.100$ m, $a = 0.150$ m, and $\ell = 0.450$ m. Find the magnitude and direction of the net force exerted on the loop by the magnetic field created by the wire.

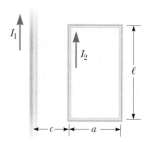

Figure P30.25 Problems 25 and 26.

26. In Figure P30.25, the current in the long, straight wire is I_1 and the wire lies in the plane of a rectangular loop, which carries a current I_2. The loop is of length ℓ and width a. Its left end is a distance c from the wire. Find the magnitude and direction of the net force exerted on the loop by the magnetic field created by the wire.

27. Two long, parallel wires are attracted to each other by a force per unit length of 320 μN/m. One wire carries a current of 20.0 A to the right and is located along the line $y = 0.500$ m. The second wire lies along the x axis. Determine the value of y for the line in the plane of the two wires along which the total magnetic field is zero.

28. *Why is the following situation impossible?* Two parallel copper conductors each have length $\ell = 0.500$ m and radius $r = 250$ μm. They carry currents $I = 10.0$ A in opposite directions and repel each other with a magnetic force $F_B = 1.00$ N.

29. The unit of magnetic flux is named for Wilhelm Weber. A practical-size unit of magnetic field is named for Johann Karl Friedrich Gauss. Along with their indi-

vidual accomplishments, Weber and Gauss built a telegraph in 1833 that consisted of a battery and switch, at one end of a transmission line 3 km long, operating an electromagnet at the other end. Suppose their transmission line was as diagrammed in Figure P30.29. Two long, parallel wires, each having a mass per unit length of 40.0 g/m, are supported in a horizontal plane by strings $\ell = 6.00$ cm long. When both wires carry the same current I, the wires repel each other so that the angle between the supporting strings is $\theta = 16.0°$. (a) Are the currents in the same direction or in opposite directions? (b) Find the magnitude of the current. (c) If this transmission line were taken to Mars, would the current required to separate the wires by the same angle be larger or smaller than that required on the Earth? Why?

Figure P30.29

Section 30.3 Ampère's Law

30. Niobium metal becomes a superconductor when cooled below 9 K. Its superconductivity is destroyed when the surface magnetic field exceeds 0.100 T. In the absence of any external magnetic field, determine the maximum current a 2.00-mm-diameter niobium wire can carry and remain superconducting.

31. Figure P30.31 is a cross-sectional view of a coaxial cable. The center conductor is surrounded by a rubber layer, an outer conductor, and another rubber layer. In a particular application, the current in the inner conductor is $I_1 = 1.00$ A out of the page and the current in the outer conductor is $I_2 = 3.00$ A into the page. Assuming the distance $d = 1.00$ mm, determine the magnitude and direction of the magnetic field at (a) point a and (b) point b.

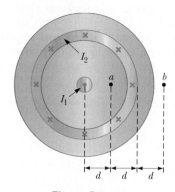

Figure P30.31

32. The magnetic coils of a tokamak fusion reactor are in the shape of a toroid having an inner radius of 0.700 m and an outer radius of 1.30 m. The toroid has 900 turns of large-diameter wire, each of which carries a current of 14.0 kA. Find the magnitude of the magnetic field inside the toroid along (a) the inner radius and (b) the outer radius.

33. A long, straight wire lies on a horizontal table and carries a current of 1.20 µA. In a vacuum, a proton moves parallel to the wire (opposite the current) with a constant speed of 2.30×10^4 m/s at a distance d above the wire. Ignoring the magnetic field due to the Earth, determine the value of d.

34. An infinite sheet of current lying in the yz plane carries a surface current of linear density J_s. The current is in the positive z direction, and J_s represents the current per unit length measured along the y axis. Figure P30.34 is an edge view of the sheet. Prove that the magnetic field near the sheet is parallel to the sheet and perpendicular to the current direction, with magnitude $\mu_0 J_s / 2$.

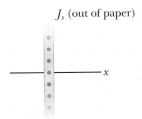

Figure P30.34

35. The magnetic field 40.0 cm away from a long, straight
W wire carrying current 2.00 A is 1.00 µT. (a) At what distance is it 0.100 µT? (b) **What If?** At one instant, the two conductors in a long household extension cord carry equal 2.00-A currents in opposite directions. The two wires are 3.00 mm apart. Find the magnetic field 40.0 cm away from the middle of the straight cord, in the plane of the two wires. (c) At what distance is it one-tenth as large? (d) The center wire in a coaxial cable carries current 2.00 A in one direction, and the sheath around it carries current 2.00 A in the opposite direction. What magnetic field does the cable create at points outside the cable?

36. A packed bundle of 100 long, straight, insulated wires forms a cylinder of radius $R = 0.500$ cm. If each wire carries 2.00 A, what are (a) the magnitude and (b) the direction of the magnetic force per unit length acting on a wire located 0.200 cm from the center of the bundle? (c) **What If?** Would a wire on the outer edge of the bundle experience a force greater or smaller than the value calculated in parts (a) and (b)? Give a qualitative argument for your answer.

37. The magnetic field created by a large current passing through plasma (ionized gas) can force current-carrying particles together. This *pinch effect* has been used in designing fusion reactors. It can be demonstrated by making an empty aluminum can carry a large current parallel to its axis. Let R represent the radius of the can and I the current, uniformly distributed over the can's curved wall. Determine the magnetic field (a) just inside the wall and (b) just outside. (c) Determine the pressure on the wall.

38. A long, cylindrical conductor of radius R carries a current I as shown in Figure P30.38. The current density J, however, is not uniform over the cross section of the conductor but rather is a function of the radius according to $J = br$, where b is a constant. Find an expression for the magnetic field magnitude B (a) at a distance $r_1 < R$ and (b) at a distance $r_2 > R$, measured from the center of the conductor.

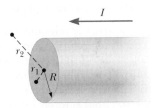

Figure P30.38

39. Four long, parallel conductors carry equal currents of
M $I = 5.00$ A. Figure P30.39 is an end view of the conductors. The current direction is into the page at points A and B and out of the page at points C and D. Calculate (a) the magnitude and (b) the direction of the magnetic field at point P, located at the center of the square of edge length $\ell = 0.200$ m.

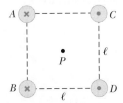

Figure P30.39

Section 30.4 The Magnetic Field of a Solenoid

40. A certain superconducting magnet in the form of a solenoid of length 0.500 m can generate a magnetic field of 9.00 T in its core when its coils carry a current of 75.0 A. Find the number of turns in the solenoid.

41. A long solenoid that has 1 000 turns uniformly dis-
M tributed over a length of 0.400 m produces a magnetic field of magnitude 1.00×10^{-4} T at its center. What current is required in the windings for that to occur?

42. You are given a certain volume of copper from which you can make copper wire. To insulate the wire, you can have as much enamel as you like. You will use the wire to make a tightly wound solenoid 20 cm long having the greatest possible magnetic field at the center and using a power supply that can deliver a current of 5 A. The solenoid can be wrapped with wire in one or more layers. (a) Should you make the wire long and thin or shorter and thick? Explain. (b) Should you make the radius of the solenoid small or large? Explain.

43. A single-turn square loop of wire, 2.00 cm on each edge,
W carries a clockwise current of 0.200 A. The loop is inside a solenoid, with the plane of the loop perpendicular to the magnetic field of the solenoid. The solenoid has

30.0 turns/cm and carries a clockwise current of 15.0 A. Find (a) the force on each side of the loop and (b) the torque acting on the loop.

44. A solenoid 10.0 cm in diameter and 75.0 cm long is made from copper wire of diameter 0.100 cm, with very thin insulation. The wire is wound onto a cardboard tube in a single layer, with adjacent turns touching each other. What power must be delivered to the solenoid if it is to produce a field of 8.00 mT at its center?

45. It is desired to construct a solenoid that will have a resistance of 5.00 Ω (at 20.0°C) and produce a magnetic field of 4.00×10^{-2} T at its center when it carries a current of 4.00 A. The solenoid is to be constructed from copper wire having a diameter of 0.500 mm. If the radius of the solenoid is to be 1.00 cm, determine (a) the number of turns of wire needed and (b) the required length of the solenoid.

Section 30.5 Gauss's Law in Magnetism

46. Consider the hemispherical closed surface in Figure P30.46. The hemisphere is in a uniform magnetic field that makes an angle θ with the vertical. Calculate the magnetic flux through (a) the flat surface S_1 and (b) the hemispherical surface S_2.

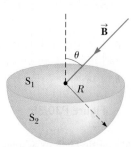

Figure P30.46

47. A cube of edge length $\ell = 2.50$ cm is positioned as shown in Figure P30.47. A uniform magnetic field given by $\vec{B} = (5\hat{i} + 4\hat{j} + 3\hat{k})$ T exists throughout the region. (a) Calculate the magnetic flux through the shaded face. (b) What is the total flux through the six faces?

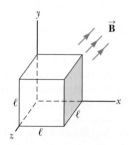

Figure P30.47

48. A solenoid of radius $r = 1.25$ cm and length $\ell = 30.0$ cm has 300 turns and carries 12.0 A. (a) Calculate the flux through the surface of a disk-shaped area of radius $R = 5.00$ cm that is positioned perpendicular to and centered on the axis of the solenoid as

shown in Figure P30.48a. (b) Figure P30.48b shows an enlarged end view of the same solenoid. Calculate the flux through the tan area, which is an annulus with an inner radius of $a = 0.400$ cm and an outer radius of $b = 0.800$ cm.

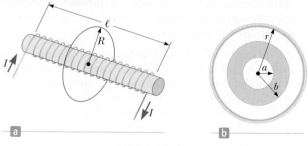

Figure P30.48

Section 30.6 Magnetism in Matter

49. The magnetic moment of the Earth is approximately 8.00×10^{22} A·m². Imagine that the planetary magnetic field were caused by the complete magnetization of a huge iron deposit with density 7 900 kg/m³ and approximately 8.50×10^{28} iron atoms/m³. (a) How many unpaired electrons, each with a magnetic moment of 9.27×10^{-24} A · m², would participate? (b) At two unpaired electrons per iron atom, how many kilograms of iron would be present in the deposit?

50. At *saturation,* when nearly all the atoms have their magnetic moments aligned, the magnetic field is equal to the permeability constant μ_0 multiplied by the magnetic moment per unit volume. In a sample of iron, where the number density of atoms is approximately 8.50×10^{28} atoms/m³, the magnetic field can reach 2.00 T. If each electron contributes a magnetic moment of 9.27×10^{-24} A·m² (1 Bohr magneton), how many electrons per atom contribute to the saturated field of iron?

Additional Problems

51. A 30.0-turn solenoid of length 6.00 cm produces a magnetic field of magnitude 2.00 mT at its center. Find the current in the solenoid.

52. A wire carries a 7.00-A current along the *x* axis, and another wire carries a 6.00-A current along the *y* axis, as shown in Figure P30.52. What is the magnetic field at point *P*, located at $x = 4.00$ m, $y = 3.00$ m?

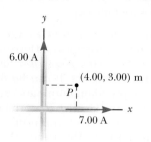

Figure P30.52

53. Suppose you install a compass on the center of a car's dashboard. (a) Assuming the dashboard is made mostly of plastic, compute an order-of-magnitude estimate for the magnetic field at this location produced by the current when you switch on the car's headlights. (b) How does this estimate compare with the Earth's magnetic field?

54. *Why is the following situation impossible?* The magnitude of the Earth's magnetic field at either pole is approximately 7.00×10^{-5} T. Suppose the field fades away to zero before its next reversal. Several scientists propose plans for artificially generating a replacement magnetic field to assist with devices that depend on the presence of the field. The plan that is selected is to lay a copper wire around the equator and supply it with a current that would generate a magnetic field of magnitude 7.00×10^{-5} T at the poles. (Ignore magnetization of any materials inside the Earth.) The plan is implemented and is highly successful.

55. [M] A nonconducting ring of radius 10.0 cm is uniformly charged with a total positive charge 10.0 μC. The ring rotates at a constant angular speed 20.0 rad/s about an axis through its center, perpendicular to the plane of the ring. What is the magnitude of the magnetic field on the axis of the ring 5.00 cm from its center?

56. A nonconducting ring of radius R is uniformly charged with a total positive charge q. The ring rotates at a constant angular speed ω about an axis through its center, perpendicular to the plane of the ring. What is the magnitude of the magnetic field on the axis of the ring a distance $\frac{1}{2}R$ from its center?

57. A very long, thin strip of metal of width w carries a current I along its length as shown in Figure P30.57. The current is distributed uniformly across the width of the strip. Find the magnetic field at point P in the diagram. Point P is in the plane of the strip at distance b away from its edge.

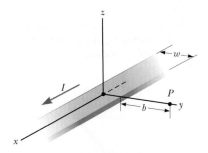

Figure P30.57

58. A circular coil of five turns and a diameter of 30.0 cm is oriented in a vertical plane with its axis perpendicular to the horizontal component of the Earth's magnetic field. A horizontal compass placed at the coil's center is made to deflect 45.0° from magnetic north by a current of 0.600 A in the coil. (a) What is the horizontal component of the Earth's magnetic field? (b) The current in the coil is switched off. A "dip

needle" is a magnetic compass mounted so that it can rotate in a vertical north–south plane. At this location, a dip needle makes an angle of 13.0° from the vertical. What is the total magnitude of the Earth's magnetic field at this location?

59. A very large parallel-plate capacitor has uniform charge per unit area $+\sigma$ on the upper plate and $-\sigma$ on the lower plate. The plates are horizontal, and both move horizontally with speed v to the right. (a) What is the magnetic field between the plates? (b) What is the magnetic field just above or just below the plates? (c) What are the magnitude and direction of the magnetic force per unit area on the upper plate? (d) At what extrapolated speed v will the magnetic force on a plate balance the electric force on the plate? *Suggestion:* Use Ampere's law and choose a path that closes between the plates of the capacitor.

60. Two circular coils of radius R, each with N turns, are perpendicular to a common axis. The coil centers are a distance R apart. Each coil carries a steady current I in the same direction as shown in Figure P30.60. (a) Show that the magnetic field on the axis at a distance x from the center of one coil is

$$B = \frac{N\mu_0 I R^2}{2}\left[\frac{1}{(R^2 + x^2)^{3/2}} + \frac{1}{(2R^2 + x^2 - 2Rx)^{3/2}}\right]$$

(b) Show that dB/dx and d^2B/dx^2 are both zero at the point midway between the coils. We may then conclude that the magnetic field in the region midway between the coils is uniform. Coils in this configuration are called *Helmholtz coils*.

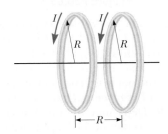

Figure P30.60 Problems 60 and 61.

61. Two identical, flat, circular coils of wire each have 100 turns and radius $R = 0.500$ m. The coils are arranged as a set of Helmholtz coils so that the separation distance between the coils is equal to the radius of the coils (see Fig. P30.60). Each coil carries current $I = 10.0$ A. Determine the magnitude of the magnetic field at a point on the common axis of the coils and halfway between them.

62. [AMT] Two circular loops are parallel, coaxial, and almost in contact, with their centers 1.00 mm apart (Fig. P30.62, page 932). Each loop is 10.0 cm in radius. The top loop carries a clockwise current of $I = 140$ A. The bottom loop carries a counterclockwise current of $I = 140$ A. (a) Calculate the magnetic force exerted by the bottom loop on the top loop. (b) Suppose a student thinks the first step in solving part (a) is to use Equation 30.7 to find the magnetic field created by one of the loops.

How would you argue for or against this idea? (c) The upper loop has a mass of 0.021 0 kg. Calculate its acceleration, assuming the only forces acting on it are the force in part (a) and the gravitational force.

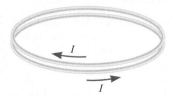

Figure P30.62

63. Two long, straight wires cross each other perpendicularly as shown in Figure P30.63. The wires are thin so that they are effectively in the same plane but do not touch. Find the magnetic field at a point 30.0 cm above the point of intersection of the wires along the z axis; that is, 30.0 cm out of the page, toward you.

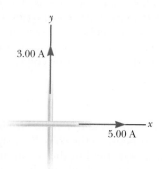

Figure P30.63

64. Two coplanar and concentric circular loops of wire carry currents of $I_1 = 5.00$ A and $I_2 = 3.00$ A in opposite directions as in Figure P30.64. If $r_1 = 12.0$ cm and $r_2 = 9.00$ cm, what are (a) the magnitude and (b) the direction of the net magnetic field at the center of the two loops? (c) Let r_1 remain fixed at 12.0 cm and let r_2 be a variable. Determine the value of r_2 such that the net field at the center of the loops is zero.

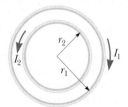

Figure P30.64

65. As seen in previous chapters, any object with electric charge, stationary or moving, other than the charged object that created the field, experiences a force in an electric field. Also, any object with electric charge, stationary or moving, can create an electric field (Chapter 23). Similarly, an electric current or a moving electric charge, other than the current or charge that created the field, experiences a force in a magnetic field (Chapter 29), and an electric current cre-

ates a magnetic field (Section 30.1). (a) To understand how a moving charge can also create a magnetic field, consider a particle with charge q moving with velocity $\vec{\mathbf{v}}$. Define the position vector $\vec{\mathbf{r}} = r\hat{\mathbf{r}}$ leading from the particle to some location. Show that the magnetic field at that location is

$$\vec{\mathbf{B}} = \frac{\mu_0}{4\pi} \frac{q\vec{\mathbf{v}} \times \hat{\mathbf{r}}}{r^2}$$

(b) Find the magnitude of the magnetic field 1.00 mm to the side of a proton moving at 2.00×10^7 m/s. (c) Find the magnetic force on a second proton at this point, moving with the same speed in the opposite direction. (d) Find the electric force on the second proton.

66. **Review.** Rail guns have been suggested for launching projectiles into space without chemical rockets. A tabletop model rail gun (Fig. P30.66) consists of two long, parallel, horizontal rails $\ell = 3.50$ cm apart, bridged by a bar of mass $m = 3.00$ g that is free to slide without friction. The rails and bar have low electric resistance, and the current is limited to a constant $I = 24.0$ A by a power supply that is far to the left of the figure, so it has no magnetic effect on the bar. Figure P30.66 shows the bar at rest at the midpoint of the rails at the moment the current is established. We wish to find the speed with which the bar leaves the rails after being released from the midpoint of the rails. (a) Find the magnitude of the magnetic field at a distance of 1.75 cm from a single long wire carrying a current of 2.40 A. (b) For purposes of evaluating the magnetic field, model the rails as infinitely long. Using the result of part (a), find the magnitude and direction of the magnetic field at the midpoint of the bar. (c) Argue that this value of the field will be the same at all positions of the bar to the right of the midpoint of the rails. At other points along the bar, the field is in the same direction as at the midpoint, but is larger in magnitude. Assume the average effective magnetic field along the bar is five times larger than the field at the midpoint. With this assumption, find (d) the magnitude and (e) the direction of the force on the bar. (f) Is the bar properly modeled as a particle under constant acceleration? (g) Find the velocity of the bar after it has traveled a distance $d = 130$ cm to the end of the rails.

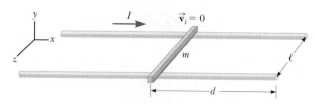

Figure P30.66

67. Fifty turns of insulated wire 0.100 cm in diameter are tightly wound to form a flat spiral. The spiral fills a disk surrounding a circle of radius 5.00 cm and extending to a radius 10.00 cm at the outer edge. Assume the wire carries a current I at the center of its cross section. Approximate each turn of wire as a circle. Then a loop

of current exists at radius 5.05 cm, another at 5.15 cm, and so on. Numerically calculate the magnetic field at the center of the coil.

68. An infinitely long, straight wire carrying a current I_1 is partially surrounded by a loop as shown in Figure P30.68. The loop has a length L and radius R, and it carries a current I_2. The axis of the loop coincides with the wire. Calculate the magnetic force exerted on the loop.

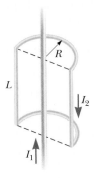

Figure P30.68

Challenge Problems

69. Consider a solenoid of length ℓ and radius a containing N closely spaced turns and carrying a steady current I. (a) In terms of these parameters, find the magnetic field at a point along the axis as a function of position x from the end of the solenoid. (b) Show that as ℓ becomes very long, B approaches $\mu_0 NI/2\ell$ at each end of the solenoid.

70. We have seen that a long solenoid produces a uniform magnetic field directed along the axis of a cylindrical region. To produce a uniform magnetic field directed parallel to a *diameter* of a cylindrical region, however, one can use the *saddle coils* illustrated in Figure P30.70. The loops are wrapped over a long, somewhat flattened tube. Figure P30.70a shows one wrapping of wire around the tube. This wrapping is continued in this manner until the visible side has many long sections of wire carrying current to the left in Figure P30.70a and the back side has many lengths carrying current to the right. The end view of the tube in Figure P30.70b shows these wires and the currents they carry. By wrapping the wires carefully, the distribution of wires can take the shape suggested in the end view such that the overall current distribution is approximately the superposition of two overlapping, circular cylinders of radius R (shown by the dashed lines) with uniformly distributed current, one toward you and one away from you. The current density J is the same for each cylinder. The center of one cylinder is described by a position vector $\vec{d}$ relative to the center of the other cylinder. Prove that the magnetic field inside the hollow tube is $\mu_0 Jd/2$ downward. *Suggestion:* The use of vector methods simplifies the calculation.

71. A thin copper bar of length $\ell = 10.0$ cm is supported horizontally by two (nonmagnetic) contacts at its ends. The bar carries a current of $I_1 = 100$ A in the negative x direction as shown in Figure P30.71. At a distance $h = 0.500$ cm below one end of the bar, a long, straight wire carries a current of $I_2 = 200$ A in the positive z direction. Determine the magnetic force exerted on the bar.

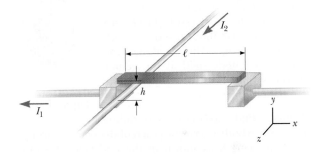

Figure P30.71

72. In Figure P30.72, both currents in the infinitely long wires are 8.00 A in the negative x direction. The wires are separated by the distance $2a = 6.00$ cm. (a) Sketch the magnetic field pattern in the yz plane. (b) What is the value of the magnetic field at the origin? (c) At $(y = 0, z \to \infty)$? (d) Find the magnetic field at points along the z axis as a function of z. (e) At what distance d along the positive z axis is the magnetic field a maximum? (f) What is this maximum value?

Figure P30.72

73. A wire carrying a current I is bent into the shape of an exponential spiral, $r = e^\theta$, from $\theta = 0$ to $\theta = 2\pi$ as suggested in Figure P30.73 (page 934). To complete a loop, the ends of the spiral are connected by a straight wire along the x axis. (a) The angle β between a radial

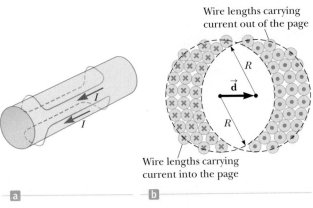

Wire lengths carrying current out of the page

Wire lengths carrying current into the page

Figure P30.70

line and its tangent line at any point on a curve $r = f(\theta)$ is related to the function by

$$\tan \beta = \frac{r}{dr/d\theta}$$

Use this fact to show that $\beta = \pi/4$. (b) Find the magnetic field at the origin.

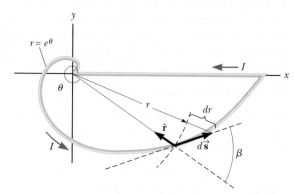

Figure P30.73

74. A sphere of radius R has a uniform volume charge density ρ. When the sphere rotates as a rigid object with angular speed ω about an axis through its center (Fig. P30.74), determine (a) the magnetic field at the center of the sphere and (b) the magnetic moment of the sphere.

Figure P30.74

75. A long, cylindrical conductor of radius a has two cylindrical cavities each of diameter a through its entire length as shown in the end view of Figure P30.75. A current I is directed out of the page and is uniform through a cross section of the conducting material. Find the magnitude and direction of the magnetic field in terms of μ_0, I, r, and a at (a) point P_1 and (b) point P_2.

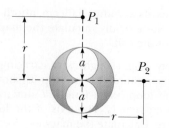

Figure P30.75

76. A wire is formed into the shape of a square of edge length L (Fig. P30.76). Show that when the current in the loop is I, the magnetic field at point P a distance x from the center of the square along its axis is

$$B = \frac{\mu_0 I L^2}{2\pi(x^2 + L^2/4)\sqrt{x^2 + L^2/2}}$$

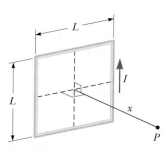

Figure P30.76

77. The magnitude of the force on a magnetic dipole $\vec{\mu}$ aligned with a nonuniform magnetic field in the positive x direction is $F_x = |\vec{\mu}| \, dB/dx$. Suppose two flat loops of wire each have radius R and carry a current I. (a) The loops are parallel to each other and share the same axis. They are separated by a variable distance $x \gg R$. Show that the magnetic force between them varies as $1/x^4$. (b) Find the magnitude of this force, taking $I = 10.0$ A, $R = 0.500$ cm, and $x = 5.00$ cm.

Faraday's Law

So far, our studies in electricity and magnetism have focused on the electric fields produced by stationary charges and the magnetic fields produced by moving charges. This chapter explores the effects produced by magnetic fields that vary in time.

Experiments conducted by Michael Faraday in England in 1831 and independently by Joseph Henry in the United States that same year showed that an emf can be induced in a circuit by a changing magnetic field. The results of these experiments led to a very basic and important law of electromagnetism known as *Faraday's law of induction.* An emf (and therefore a current as well) can be induced in various processes that involve a change in a magnetic flux.

31.1 Faraday's Law of Induction

To see how an emf can be induced by a changing magnetic field, consider the experimental results obtained when a loop of wire is connected to a sensitive ammeter as illustrated in Figure 31.1 (page 936). When a magnet is moved toward the loop, the reading on the ammeter changes from zero to a nonzero value, arbitrarily shown as negative in Figure 31.1a. When the magnet is brought to rest and held stationary relative to the loop (Fig. 31.1b), a reading of zero is observed. When the magnet is moved away from the loop, the reading on the ammeter changes to a positive value as shown in Figure 31.1c. Finally, when the magnet is held stationary and the loop

An artist's impression of the Skerries SeaGen Array, a tidal energy generator under development near the island of Anglesey, North Wales. When it is brought online, it will offer 10.5 MW of power from generators turned by tidal streams. The image shows the underwater blades that are driven by the tidal currents. The second blade system has been raised from the water for servicing. We will study generators in this chapter. *(Marine Current Turbines TM Ltd.)*

Michael Faraday
British Physicist and Chemist
(1791–1867)
Faraday is often regarded as the greatest experimental scientist of the 1800s.
His many contributions to the study of
electricity include the invention of the
electric motor, electric generator, and
transformer as well as the discovery
of electromagnetic induction and the
laws of electrolysis. Greatly influenced
by religion, he refused to work on the
development of poison gas for the British military.

is moved either toward or away from it, the reading changes from zero. From these
observations, we conclude that the loop detects that the magnet is moving relative to
it and we relate this detection to a change in magnetic field. Therefore, it seems that
a relationship exists between a current and a changing magnetic field.

These results are quite remarkable because a current is set up even though no
batteries are present in the circuit! We call such a current an *induced current* and say
that it is produced by an *induced emf*.

Now let's describe an experiment conducted by Faraday and illustrated in Figure
31.2. A primary coil is wrapped around an iron ring and connected to a switch and
a battery. A current in the coil produces a magnetic field when the switch is closed.
A secondary coil also is wrapped around the ring and is connected to a sensitive
ammeter. No battery is present in the secondary circuit, and the secondary coil is
not electrically connected to the primary coil. Any current detected in the secondary circuit must be induced by some external agent.

Initially, you might guess that no current is ever detected in the secondary circuit. Something quite amazing happens when the switch in the primary circuit is
either opened or thrown closed, however. At the instant the switch is closed, the
ammeter reading changes from zero momentarily and then returns to zero. At the
instant the switch is opened, the ammeter changes to a reading with the opposite
sign and again returns to zero. Finally, the ammeter reads zero when there is either
a steady current or no current in the primary circuit. To understand what happens
in this experiment, note that when the switch is closed, the current in the primary
circuit produces a magnetic field that penetrates the secondary circuit. Furthermore, when the switch is thrown closed, the magnetic field produced by the current in the primary circuit changes from zero to some value over some finite time,
and this changing field induces a current in the secondary circuit. Notice that no
current is induced in the secondary coil even when a steady current exists in the
primary coil. It is a *change* in the current in the primary coil that induces a current
in the secondary coil, not just the *existence* of a current.

As a result of these observations, Faraday concluded that an electric current can
be induced in a loop by a changing magnetic field. The induced current exists
only while the magnetic field through the loop is changing. Once the magnetic
field reaches a steady value, the current in the loop disappears. In effect, the loop
behaves as though a source of emf were connected to it for a short time. It is customary to say that an induced emf is produced in the loop by the changing magnetic field.

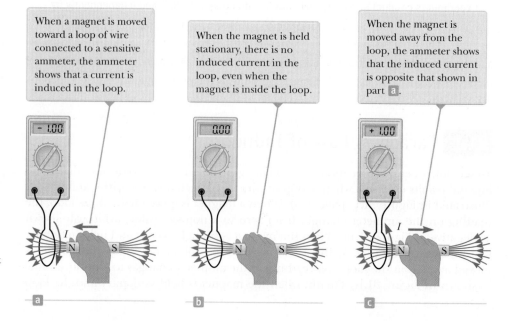

Figure 31.1 A simple experiment
showing that a current is induced
in a loop when a magnet is moved
toward or away from the loop.

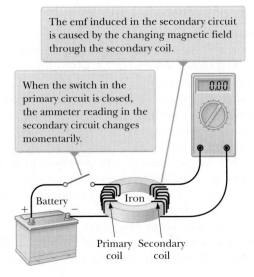

The emf induced in the secondary circuit is caused by the changing magnetic field through the secondary coil.

When the switch in the primary circuit is closed, the ammeter reading in the secondary circuit changes momentarily.

Battery

Iron

Primary coil Secondary coil

Figure 31.2 Faraday's experiment.

The experiments shown in Figures 31.1 and 31.2 have one thing in common: in each case, an emf is induced in a loop when the magnetic flux through the loop changes with time. In general, this emf is directly proportional to the time rate of change of the magnetic flux through the loop. This statement can be written mathematically as **Faraday's law of induction:**

$$\mathcal{E} = -\frac{d\Phi_B}{dt} \tag{31.1}$$

◀ Faraday's law of induction

where $\Phi_B = \int \vec{\mathbf{B}} \cdot d\vec{\mathbf{A}}$ is the magnetic flux through the loop. (See Section 30.5.)

If a coil consists of N loops with the same area and Φ_B is the magnetic flux through one loop, an emf is induced in every loop. The loops are in series, so their emfs add; therefore, the total induced emf in the coil is given by

$$\mathcal{E} = -N\frac{d\Phi_B}{dt} \tag{31.2}$$

The negative sign in Equations 31.1 and 31.2 is of important physical significance and will be discussed in Section 31.3.

Suppose a loop enclosing an area A lies in a uniform magnetic field $\vec{\mathbf{B}}$ as in Figure 31.3. The magnetic flux through the loop is equal to $BA \cos\theta$, where θ is the angle between the magnetic field and the normal to the loop; hence, the induced emf can be expressed as

$$\mathcal{E} = -\frac{d}{dt}(BA \cos\theta) \tag{31.3}$$

From this expression, we see that an emf can be induced in the circuit in several ways:

- The magnitude of $\vec{\mathbf{B}}$ can change with time.
- The area enclosed by the loop can change with time.
- The angle θ between $\vec{\mathbf{B}}$ and the normal to the loop can change with time.
- Any combination of the above can occur.

Quick Quiz 31.1 A circular loop of wire is held in a uniform magnetic field, with the plane of the loop perpendicular to the field lines. Which of the following will *not* cause a current to be induced in the loop? **(a)** crushing the loop **(b)** rotating the loop about an axis perpendicular to the field lines **(c)** keeping the orientation of the loop fixed and moving it along the field lines **(d)** pulling the loop out of the field

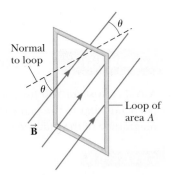

Normal to loop

θ

θ

$\vec{\mathbf{B}}$

Loop of area A

Figure 31.3 A conducting loop that encloses an area A in the presence of a uniform magnetic field $\vec{\mathbf{B}}$. The angle between $\vec{\mathbf{B}}$ and the normal to the loop is θ.

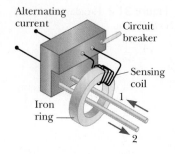

Figure 31.4 Essential components of a ground fault circuit interrupter.

Some Applications of Faraday's Law

The ground fault circuit interrupter (GFCI) is an interesting safety device that protects users of electrical appliances against electric shock. Its operation makes use of Faraday's law. In the GFCI shown in Figure 31.4, wire 1 leads from the wall outlet to the appliance to be protected and wire 2 leads from the appliance back to the wall outlet. An iron ring surrounds the two wires, and a sensing coil is wrapped around part of the ring. Because the currents in the wires are in opposite directions and of equal magnitude, there is zero net current flowing through the ring and the net magnetic flux through the sensing coil is zero. Now suppose the return current in wire 2 changes so that the two currents are not equal in magnitude. (That can happen if, for example, the appliance becomes wet, enabling current to leak to ground.) Then the net current through the ring is not zero and the magnetic flux through the sensing coil is no longer zero. Because household current is alternating (meaning that its direction keeps reversing), the magnetic flux through the sensing coil changes with time, inducing an emf in the coil. This induced emf is used to trigger a circuit breaker, which stops the current before it is able to reach a harmful level.

Another interesting application of Faraday's law is the production of sound in an electric guitar. The coil in this case, called the *pickup coil,* is placed near the vibrating guitar string, which is made of a metal that can be magnetized. A permanent magnet inside the coil magnetizes the portion of the string nearest the coil (Fig. 31.5a). When the string vibrates at some frequency, its magnetized segment produces a changing magnetic flux through the coil. The changing flux induces an emf in the coil that is fed to an amplifier. The output of the amplifier is sent to the loudspeakers, which produce the sound waves we hear.

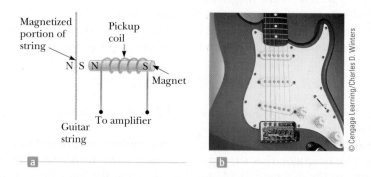

Figure 31.5 (a) In an electric guitar, a vibrating magnetized string induces an emf in a pickup coil. (b) The pickups (the circles beneath the metallic strings) of this electric guitar detect the vibrations of the strings and send this information through an amplifier and into speakers. (A switch on the guitar allows the musician to select which set of six pickups is used.)

Example 31.1   **Inducing an emf in a Coil**

A coil consists of 200 turns of wire. Each turn is a square of side $d = 18$ cm, and a uniform magnetic field directed perpendicular to the plane of the coil is turned on. If the field changes linearly from 0 to 0.50 T in 0.80 s, what is the magnitude of the induced emf in the coil while the field is changing?

SOLUTION

Conceptualize From the description in the problem, imagine magnetic field lines passing through the coil. Because the magnetic field is changing in magnitude, an emf is induced in the coil.

Categorize We will evaluate the emf using Faraday's law from this section, so we categorize this example as a substitution problem.

▶ **31.1** continued

Evaluate Equation 31.2 for the situation described here, noting that the magnetic field changes linearly with time:

$$|\boldsymbol{\varepsilon}| = N\frac{\Delta\Phi_B}{\Delta t} = N\frac{\Delta(BA)}{\Delta t} = NA\frac{\Delta B}{\Delta t} = Nd^2\frac{B_f - B_i}{\Delta t}$$

Substitute numerical values:

$$|\boldsymbol{\varepsilon}| = (200)(0.18 \text{ m})^2 \frac{(0.50 \text{ T} - 0)}{0.80 \text{ s}} = \boxed{4.0 \text{ V}}$$

WHAT IF?   What if you were asked to find the magnitude of the induced current in the coil while the field is changing? Can you answer that question?

Answer If the ends of the coil are not connected to a circuit, the answer to this question is easy: the current is zero! (Charges move within the wire of the coil, but they cannot move into or out of the ends of the coil.) For a steady current to exist, the ends of the coil must be connected to an external circuit. Let's assume the coil is connected to a circuit and the total resistance of the coil and the circuit is 2.0 Ω. Then, the magnitude of the induced current in the coil is

$$I = \frac{|\boldsymbol{\varepsilon}|}{R} = \frac{4.0 \text{ V}}{2.0 \text{ Ω}} = 2.0 \text{ A}$$

Example 31.2   **An Exponentially Decaying Magnetic Field**

A loop of wire enclosing an area A is placed in a region where the magnetic field is perpendicular to the plane of the loop. The magnitude of $\vec{B}$ varies in time according to the expression $B = B_{max}e^{-at}$, where a is some constant. That is, at $t = 0$, the field is B_{max}, and for $t > 0$, the field decreases exponentially (Fig. 31.6). Find the induced emf in the loop as a function of time.

Figure 31.6 (Example 31.2) Exponential decrease in the magnitude of the magnetic field through a loop with time. The induced emf and induced current in a conducting path attached to the loop vary with time in the same way.

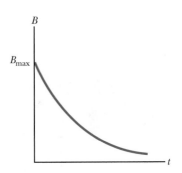

SOLUTION

Conceptualize The physical situation is similar to that in Example 31.1 except for two things: there is only one loop, and the field varies exponentially with time rather than linearly.

Categorize We will evaluate the emf using Faraday's law from this section, so we categorize this example as a substitution problem.

Evaluate Equation 31.1 for the situation described here:

$$\boldsymbol{\varepsilon} = -\frac{d\Phi_B}{dt} = -\frac{d}{dt}\left(AB_{max}\,e^{-at}\right) = -AB_{max}\frac{d}{dt}e^{-at} = \boxed{aAB_{max}\,e^{-at}}$$

This expression indicates that the induced emf decays exponentially in time. The maximum emf occurs at $t = 0$, where $\boldsymbol{\varepsilon}_{max} = aAB_{max}$. The plot of $\boldsymbol{\varepsilon}$ versus t is similar to the B-versus-t curve shown in Figure 31.6.

31.2 Motional emf

In Examples 31.1 and 31.2, we considered cases in which an emf is induced in a stationary circuit placed in a magnetic field when the field changes with time. In this section, we describe **motional emf,** the emf induced in a conductor moving through a constant magnetic field.

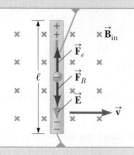

Due to the magnetic force on electrons, the ends of the conductor become oppositely charged, which establishes an electric field in the conductor.

Figure 31.7 A straight electrical conductor of length ℓ moving with a velocity $\vec{v}$ through a uniform magnetic field $\vec{B}$ directed perpendicular to $\vec{v}$.

The straight conductor of length ℓ shown in Figure 31.7 is moving through a uniform magnetic field directed into the page. For simplicity, let's assume the conductor is moving in a direction perpendicular to the field with constant velocity under the influence of some external agent. From the magnetic version of the particle in a field model, the electrons in the conductor experience a force $\vec{F}_B = q\vec{v} \times \vec{B}$ (Eq. 29.1) that is directed along the length ℓ, perpendicular to both $\vec{v}$ and $\vec{B}$. Under the influence of this force, the electrons move to the lower end of the conductor and accumulate there, leaving a net positive charge at the upper end. As a result of this charge separation, an electric field $\vec{E}$ is produced inside the conductor. Therefore, the electrons are also described by the electric version of the particle in a field model. The charges accumulate at both ends until the downward magnetic force qvB on charges remaining in the conductor is balanced by the upward electric force qE. The electrons are then described by the particle in equilibrium model. The condition for equilibrium requires that the forces on the electrons balance:

$$qE = qvB \quad \text{or} \quad E = vB$$

The magnitude of the electric field produced in the conductor is related to the potential difference across the ends of the conductor according to the relationship $\Delta V = E\ell$ (Eq. 25.6). Therefore, for the equilibrium condition,

$$\Delta V = E\ell = B\ell v \tag{31.4}$$

where the upper end of the conductor in Figure 31.7 is at a higher electric potential than the lower end. Therefore, a potential difference is maintained between the ends of the conductor as long as the conductor continues to move through the uniform magnetic field. If the direction of the motion is reversed, the polarity of the potential difference is also reversed.

A more interesting situation occurs when the moving conductor is part of a closed conducting path. This situation is particularly useful for illustrating how a changing magnetic flux causes an induced current in a closed circuit. Consider a circuit consisting of a conducting bar of length ℓ sliding along two fixed, parallel conducting rails as shown in Figure 31.8a. For simplicity, let's assume the bar has zero resistance and the stationary part of the circuit has a resistance R. A uniform and constant magnetic field $\vec{B}$ is applied perpendicular to the plane of the circuit. As the bar is pulled to the right with a velocity $\vec{v}$ under the influence of an applied force $\vec{F}_{app}$, free charges in the bar are moving particles in a magnetic field that experience a magnetic force directed along the length of the bar. This force sets up an induced current because the charges are free to move in the closed conducting path. In this case, the rate of change of magnetic flux through the circuit and the corresponding

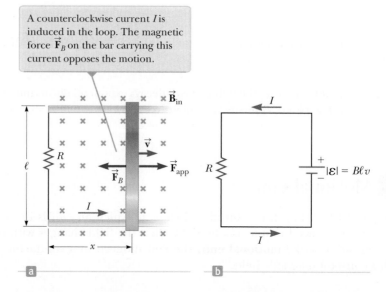

A counterclockwise current I is induced in the loop. The magnetic force $\vec{F}_B$ on the bar carrying this current opposes the motion.

Figure 31.8 (a) A conducting bar sliding with a velocity $\vec{v}$ along two conducting rails under the action of an applied force $\vec{F}_{app}$. (b) The equivalent circuit diagram for the setup shown in (a).

induced motional emf across the moving bar are proportional to the change in area of the circuit.

Because the area enclosed by the circuit at any instant is ℓx, where x is the position of the bar, the magnetic flux through that area is

$$\Phi_B = B\ell x$$

Using Faraday's law and noting that x changes with time at a rate $dx/dt = v$, we find that the induced motional emf is

$$\varepsilon = -\frac{d\Phi_B}{dt} = -\frac{d}{dt}(B\ell x) = -B\ell\frac{dx}{dt}$$

$$\varepsilon = -B\ell v \qquad\qquad (31.5) \qquad \blacktriangleleft \text{ Motional emf}$$

Because the resistance of the circuit is R, the magnitude of the induced current is

$$I = \frac{|\varepsilon|}{R} = \frac{B\ell v}{R} \qquad\qquad (31.6)$$

The equivalent circuit diagram for this example is shown in Figure 31.8b.

Let's examine the system using energy considerations. Because no battery is in the circuit, you might wonder about the origin of the induced current and the energy delivered to the resistor. We can understand the source of this current and energy by noting that the applied force does work on the conducting bar. Therefore, we model the circuit as a nonisolated system. The movement of the bar through the field causes charges to move along the bar with some average drift velocity; hence, a current is established. The change in energy in the system during some time interval must be equal to the transfer of energy into the system by work, consistent with the general principle of conservation of energy described by Equation 8.2. The appropriate reduction of Equation 8.2 is $W = \Delta E_{\text{int}}$, because the input energy appears as internal energy in the resistor.

Let's verify this equality mathematically. As the bar moves through the uniform magnetic field $\vec{\mathbf{B}}$, it experiences a magnetic force $\vec{\mathbf{F}}_B$ of magnitude $I\ell B$ (see Section 29.4). Because the bar moves with constant velocity, it is modeled as a particle in equilibrium and the magnetic force must be equal in magnitude and opposite in direction to the applied force, or to the left in Figure 31.8a. (If $\vec{\mathbf{F}}_B$ acted in the direction of motion, it would cause the bar to accelerate, violating the principle of conservation of energy.) Using Equation 31.6 and $F_{\text{app}} = F_B = I\ell B$, the power delivered by the applied force is

$$P = F_{\text{app}}v = (I\ell B)v = \frac{B^2\ell^2 v^2}{R} = \frac{\varepsilon^2}{R} \qquad\qquad (31.7)$$

From Equation 27.22, we see that this power input is equal to the rate at which energy is delivered to the resistor, consistent with the principle of conservation of energy.

> **Q**uick Quiz 31.2 In Figure 31.8a, a given applied force of magnitude F_{app} results in a constant speed v and a power input P. Imagine that the force is increased so that the constant speed of the bar is doubled to $2v$. Under these conditions, what are the new force and the new power input? **(a)** $2F$ and $2P$ **(b)** $4F$ and $2P$ **(c)** $2F$ and $4P$ **(d)** $4F$ and $4P$

Example 31.3	**Magnetic Force Acting on a Sliding Bar** AM

The conducting bar illustrated in Figure 31.9 (page 942) moves on two frictionless, parallel rails in the presence of a uniform magnetic field directed into the page. The bar has mass m, and its length is ℓ. The bar is given an initial velocity $\vec{\mathbf{v}}_i$ to the right and is released at $t = 0$.

continued

▶ **31.3** continued

(A) Using Newton's laws, find the velocity of the bar as a function of time.

SOLUTION

Conceptualize As the bar slides to the right in Figure 31.9, a counterclockwise current is established in the circuit consisting of the bar, the rails, and the resistor. The upward current in the bar results in a magnetic force to the left on the bar as shown in the figure. Therefore, the bar must slow down, so our mathematical solution should demonstrate that.

Categorize The text already categorizes this problem as one that uses Newton's laws. We model the bar as a *particle under a net force*.

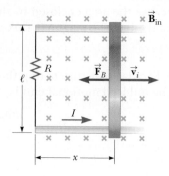

Figure 31.9 (Example 31.3) A conducting bar of length ℓ on two fixed conducting rails is given an initial velocity $\vec{v}_i$ to the right.

Analyze From Equation 29.10, the magnetic force is $F_B = -I\ell B$, where the negative sign indicates that the force is to the left. The magnetic force is the *only* horizontal force acting on the bar.

Using the particle under a net force model, apply Newton's second law to the bar in the horizontal direction:

$$F_x = ma \quad \rightarrow \quad -I\ell B = m\frac{dv}{dt}$$

Substitute $I = B\ell v/R$ from Equation 31.6:

$$m\frac{dv}{dt} = -\frac{B^2\ell^2}{R}v$$

Rearrange the equation so that all occurrences of the variable v are on the left and those of t are on the right:

$$\frac{dv}{v} = -\left(\frac{B^2\ell^2}{mR}\right)dt$$

Integrate this equation using the initial condition that $v = v_i$ at $t = 0$ and noting that $(B^2\ell^2/mR)$ is a constant:

$$\int_{v_i}^{v}\frac{dv}{v} = -\frac{B^2\ell^2}{mR}\int_0^t dt$$

$$\ln\left(\frac{v}{v_i}\right) = -\left(\frac{B^2\ell^2}{mR}\right)t$$

Define the constant $\tau = mR/B^2\ell^2$ and solve for the velocity:

$$(1) \quad v = v_i e^{-t/\tau}$$

Finalize This expression for v indicates that the velocity of the bar decreases with time under the action of the magnetic force as expected from our conceptualization of the problem.

(B) Show that the same result is found by using an energy approach.

SOLUTION

Categorize The text of this part of the problem tells us to use an energy approach for the same situation. We model the entire circuit in Figure 31.9 as an *isolated system*.

Analyze Consider the sliding bar as one system component possessing kinetic energy, which decreases because energy is transferring *out* of the bar by electrical transmission through the rails. The resistor is another system component possessing internal energy, which rises because energy is transferring *into* the resistor. Because energy is not leaving the system, the rate of energy transfer out of the bar equals the rate of energy transfer into the resistor.

Equate the power entering the resistor to that leaving the bar:

$$P_{\text{resistor}} = -P_{\text{bar}}$$

Substitute for the electrical power delivered to the resistor and the time rate of change of kinetic energy for the bar:

$$I^2 R = -\frac{d}{dt}\left(\tfrac{1}{2}mv^2\right)$$

Use Equation 31.6 for the current and carry out the derivative:

$$\frac{B^2\ell^2 v^2}{R} = -mv\frac{dv}{dt}$$

▶ **31.3** continued

Rearrange terms:

$$\frac{dv}{v} = -\left(\frac{B^2\ell^2}{mR}\right)dt$$

....

Finalize This result is the same expression to be integrated that we found in part (A).

WHAT IF? Suppose you wished to increase the distance through which the bar moves between the time it is initially projected and the time it essentially comes to rest. You can do so by changing one of three variables—v_i, R, or B—by a factor of 2 or $\frac{1}{2}$. Which variable should you change to maximize the distance, and would you double it or halve it?

Answer Increasing v_i would make the bar move farther. Increasing R would decrease the current and therefore the magnetic force, making the bar move farther. Decreasing B would decrease the magnetic force and make the bar move farther. Which method is most effective, though?

Use Equation (1) to find the distance the bar moves by integration:

$$v = \frac{dx}{dt} = v_i e^{-t/\tau}$$

$$x = \int_0^\infty v_i e^{-t/\tau}\, dt = -v_i\tau e^{-t/\tau}\Big|_0^\infty$$

$$= -v_i\tau(0-1) = v_i\tau = v_i\left(\frac{mR}{B^2\ell^2}\right)$$

This expression shows that doubling v_i or R will double the distance. Changing B by a factor of $\frac{1}{2}$, however, causes the distance to be four times as great!

Example 31.4   Motional emf Induced in a Rotating Bar

A conducting bar of length ℓ rotates with a constant angular speed ω about a pivot at one end. A uniform magnetic field $\vec{B}$ is directed perpendicular to the plane of rotation as shown in Figure 31.10. Find the motional emf induced between the ends of the bar.

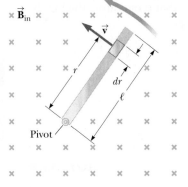

Figure 31.10 (Example 31.4) A conducting bar rotating around a pivot at one end in a uniform magnetic field that is perpendicular to the plane of rotation. A motional emf is induced between the ends of the bar.

SOLUTION

Conceptualize The rotating bar is different in nature from the sliding bar in Figure 31.8. Consider a small segment of the bar, however. It is a short length of conductor moving in a magnetic field and has an emf generated in it like the sliding bar. By thinking of each small segment as a source of emf, we see that all segments are in series and the emfs add.

Categorize Based on the conceptualization of the problem, we approach this example as we did in the discussion leading to Equation 31.5, with the added feature that the short segments of the bar are traveling in circular paths.

....

Analyze Evaluate the magnitude of the emf induced in a segment of the bar of length dr having a velocity $\vec{v}$ from Equation 31.5:

$$d\mathcal{E} = Bv\, dr$$

Find the total emf between the ends of the bar by adding the emfs induced across all segments:

$$\mathcal{E} = \int Bv\, dr$$

The tangential speed v of an element is related to the angular speed ω through the relationship $v = r\omega$ (Eq. 10.10); use that fact and integrate:

$$\mathcal{E} = B\int v\, dr = B\omega\int_0^\ell r\, dr = \tfrac{1}{2}B\omega\ell^2$$

continued

▶ **31.4** continued

Finalize In Equation 31.5 for a sliding bar, we can increase $\mathcal{E}$ by increasing B, ℓ, or v. Increasing any one of these variables by a given factor increases $\mathcal{E}$ by the same factor. Therefore, you would choose whichever of these three variables is most convenient to increase. For the rotating rod, however, there is an advantage to increasing the length of the rod to raise the emf because ℓ is squared. Doubling the length gives four times the emf, whereas doubling the angular speed only doubles the emf.

WHAT IF? Suppose, after reading through this example, you come up with a brilliant idea. A Ferris wheel has radial metallic spokes between the hub and the circular rim. These spokes move in the magnetic field of the Earth, so each spoke acts like the bar in Figure 31.10. You plan to use the emf generated by the rotation of the Ferris wheel to power the lightbulbs on the wheel. Will this idea work?

Answer Let's estimate the emf that is generated in this situation. We know the magnitude of the magnetic field of the Earth from Table 29.1: $B = 0.5 \times 10^{-4}$ T. A typical spoke on a Ferris wheel might have a length on the order of 10 m. Suppose the period of rotation is on the order of 10 s.

Determine the angular speed of the spoke:

$$\omega = \frac{2\pi}{T} = \frac{2\pi}{10\text{ s}} = 0.63\text{ s}^{-1} \sim 1\text{ s}^{-1}$$

Assume the magnetic field lines of the Earth are horizontal at the location of the Ferris wheel and perpendicular to the spokes. Find the emf generated:

$$\mathcal{E} = \tfrac{1}{2}B\omega\ell^2 = \tfrac{1}{2}(0.5 \times 10^{-4}\text{ T})(1\text{ s}^{-1})(10\text{ m})^2$$
$$= 2.5 \times 10^{-3}\text{ V} \sim 1\text{ mV}$$

This value is a tiny emf, far smaller than that required to operate lightbulbs.

An additional difficulty is related to energy. Even assuming you could find lightbulbs that operate using a potential difference on the order of millivolts, a spoke must be part of a circuit to provide a voltage to the lightbulbs. Consequently, the spoke must carry a current. Because this current-carrying spoke is in a magnetic field, a magnetic force is exerted on the spoke in the direction opposite its direction of motion. As a result, the motor of the Ferris wheel must supply more energy to perform work against this magnetic drag force. The motor must ultimately provide the energy that is operating the lightbulbs, and you have not gained anything for free!

31.3 Lenz's Law

Faraday's law (Eq. 31.1) indicates that the induced emf and the change in flux have opposite algebraic signs. This feature has a very real physical interpretation that has come to be known as **Lenz's law:**[1]

Lenz's law ▶

> The induced current in a loop is in the direction that creates a magnetic field that opposes the change in magnetic flux through the area enclosed by the loop.

That is, the induced current tends to keep the original magnetic flux through the loop from changing. We shall show that this law is a consequence of the law of conservation of energy.

To understand Lenz's law, let's return to the example of a bar moving to the right on two parallel rails in the presence of a uniform magnetic field (the *external* magnetic field, shown by the green crosses in Fig. 31.11a). As the bar moves to the right, the magnetic flux through the area enclosed by the circuit increases with time because the area increases. Lenz's law states that the induced current must be directed so that the magnetic field it produces opposes the change in the external magnetic flux. Because the magnetic flux due to an external field directed into the page is increasing, the induced current—if it is to oppose this change—must

---
[1]Developed by German physicist Heinrich Lenz (1804–1865).

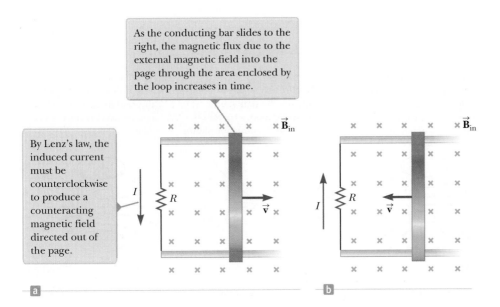

As the conducting bar slides to the right, the magnetic flux due to the external magnetic field into the page through the area enclosed by the loop increases in time.

By Lenz's law, the induced current must be counterclockwise to produce a counteracting magnetic field directed out of the page.

Figure 31.11 (a) Lenz's law can be used to determine the direction of the induced current. (b) When the bar moves to the left, the induced current must be clockwise. Why?

produce a field directed out of the page. Hence, the induced current must be directed counterclockwise when the bar moves to the right. (Use the right-hand rule to verify this direction.) If the bar is moving to the left as in Figure 31.11b, the external magnetic flux through the area enclosed by the loop decreases with time. Because the field is directed into the page, the direction of the induced current must be clockwise if it is to produce a field that also is directed into the page. In either case, the induced current attempts to maintain the original flux through the area enclosed by the current loop.

Let's examine this situation using energy considerations. Suppose the bar is given a slight push to the right. In the preceding analysis, we found that this motion sets up a counterclockwise current in the loop. What happens if we assume the current is clockwise such that the direction of the magnetic force exerted on the bar is to the right? This force would accelerate the rod and increase its velocity, which in turn would cause the area enclosed by the loop to increase more rapidly. The result would be an increase in the induced current, which would cause an increase in the force, which would produce an increase in the current, and so on. In effect, the system would acquire energy with no input of energy. This behavior is clearly inconsistent with all experience and violates the law of conservation of energy. Therefore, the current must be counterclockwise.

Quick Quiz 31.3 Figure 31.12 shows a circular loop of wire falling toward a wire carrying a current to the left. What is the direction of the induced current in the loop of wire? **(a)** clockwise **(b)** counterclockwise **(c)** zero **(d)** impossible to determine

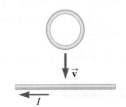

Figure 31.12 (Quick Quiz 31.3)

Conceptual Example 31.5   **Application of Lenz's Law**

A magnet is placed near a metal loop as shown in Figure 31.13a (page 946).

(A) Find the direction of the induced current in the loop when the magnet is pushed toward the loop.

SOLUTION

As the magnet moves to the right toward the loop, the external magnetic flux through the loop increases with time. To counteract this increase in flux due to a field toward the right, the induced current produces its own magnetic field to the left as illustrated in Figure 31.13b; hence, the induced current is in the direction shown. Knowing that like
continued

▶ **31.5** continued

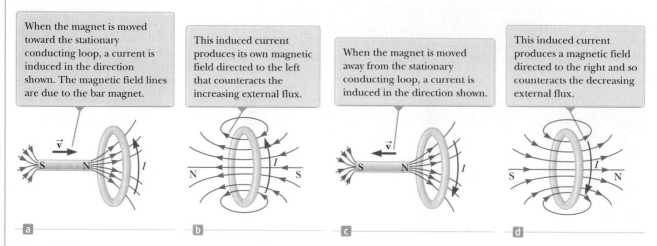

When the magnet is moved toward the stationary conducting loop, a current is induced in the direction shown. The magnetic field lines are due to the bar magnet.

This induced current produces its own magnetic field directed to the left that counteracts the increasing external flux.

When the magnet is moved away from the stationary conducting loop, a current is induced in the direction shown.

This induced current produces a magnetic field directed to the right and so counteracts the decreasing external flux.

Figure 31.13 (Conceptual Example 31.5) A moving bar magnet induces a current in a conducting loop.

magnetic poles repel each other, we conclude that the left face of the current loop acts like a north pole and the right face acts like a south pole.

(B) Find the direction of the induced current in the loop when the magnet is pulled away from the loop.

SOLUTION

If the magnet moves to the left as in Figure 31.13c, its flux through the area enclosed by the loop decreases in time. Now the induced current in the loop is in the direction shown in Figure 31.13d because this current direction produces a magnetic field in the same direction as the external field. In this case, the left face of the loop is a south pole and the right face is a north pole.

Conceptual Example 31.6   **A Loop Moving Through a Magnetic Field**

A rectangular metallic loop of dimensions ℓ and w and resistance R moves with constant speed v to the right as in Figure 31.14a. The loop passes through a uniform magnetic field $\vec{B}$ directed into the page and extending a distance $3w$ along the x axis. Define x as the position of the right side of the loop along the x axis.

(A) Plot the magnetic flux through the area enclosed by the loop as a function of x.

SOLUTION

Figure 31.14b shows the flux through the area enclosed by the loop as a function of x. Before the loop enters the field, the flux through the loop is zero. As the loop enters the field, the flux increases linearly with position until the left edge of the loop is just inside the field. Finally, the flux through the loop decreases linearly to zero as the loop leaves the field.

(B) Plot the induced motional emf in the loop as a function of x.

SOLUTION

Before the loop enters the field, no motional emf is induced in it because no field is present (Fig. 31.14c). As the right side of the loop enters the field, the magnetic flux directed into the page increases. Hence, according to Lenz's law, the induced current is counterclockwise because it must produce its own magnetic field directed out of the page. The motional emf $-B\ell v$ (from Eq. 31.5) arises from the magnetic force experienced by charges in the right side of the loop. When the loop is entirely in the field, the change in magnetic flux through the loop is zero; hence, the motional emf vanishes. That happens because once the left side of the loop enters the field, the motional emf induced in it

▶ **31.6** c o n t i n u e d

cancels the motional emf present in the right side of the loop. As the right side of the loop leaves the field, the flux through the loop begins to decrease, a clockwise current is induced, and the induced emf is $B\ell v$. As soon as the left side leaves the field, the emf decreases to zero.

(C) Plot the external applied force necessary to counter the magnetic force and keep v constant as a function of x.

SOLUTION

The external force that must be applied to the loop to maintain this motion is plotted in Figure 31.14d. Before the loop enters the field, no magnetic force acts on it; hence, the applied force must be zero if v is constant. When the right side of the loop enters the field, the applied force necessary to maintain constant speed must be equal in magnitude and opposite in direction to the magnetic force exerted on that side, so that the loop is a particle in equilibrium. When the loop is entirely in the field, the flux through the loop is not changing with time. Hence, the net emf induced in the loop is zero and the current also is zero. Therefore, no external force is needed to maintain the motion. Finally, as the right side leaves the field, the applied force must be equal in magnitude and opposite in direction to the magnetic force acting on the left side of the loop.

From this analysis, we conclude that power is supplied only when the loop is either entering or leaving the field. Furthermore, this example shows that the motional emf induced in the loop can be zero even when there is motion through the field! A motional emf is induced *only* when the magnetic flux through the loop *changes in time*.

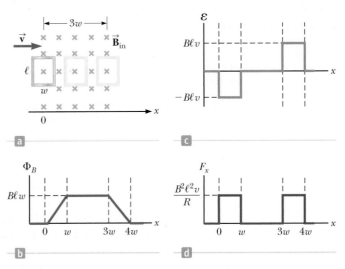

Figure 31.14 (Conceptual Example 31.6) (a) A conducting rectangular loop of width w and length ℓ moving with a velocity $\vec{v}$ through a uniform magnetic field extending a distance $3w$. (b) Magnetic flux through the area enclosed by the loop as a function of loop position. (c) Induced emf as a function of loop position. (d) Applied force required for constant velocity as a function of loop position.

31.4 Induced emf and Electric Fields

We have seen that a changing magnetic flux induces an emf and a current in a conducting loop. In our study of electricity, we related a current to an electric field that applies electric forces on charged particles. In the same way, we can relate an induced current in a conducting loop to an electric field by claiming that an electric field is created in the conductor as a result of the changing magnetic flux.

We also noted in our study of electricity that the existence of an electric field is independent of the presence of any test charges. This independence suggests that even in the absence of a conducting loop, a changing magnetic field generates an electric field in empty space.

This induced electric field is *nonconservative*, unlike the electrostatic field produced by stationary charges. To illustrate this point, consider a conducting loop of radius r situated in a uniform magnetic field that is perpendicular to the plane of the loop as in Figure 31.15. If the magnetic field changes with time, an emf $\mathcal{E} = -d\Phi_B/dt$ is, according to Faraday's law (Eq. 31.1), induced in the loop. The induction of a current in the loop implies the presence of an induced electric field $\vec{E}$, which must be tangent to the loop because that is the direction in which the charges in the wire move in response to the electric force. The work done by the electric field in moving a charge q once around the loop is equal to $q\mathcal{E}$. Because the electric force acting on the charge is $q\vec{E}$, the work done by the electric field in

If $\vec{B}$ changes in time, an electric field is induced in a direction tangent to the circumference of the loop.

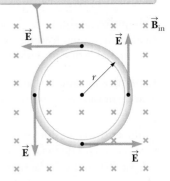

Figure 31.15 A conducting loop of radius r in a uniform magnetic field perpendicular to the plane of the loop.

Pitfall Prevention 31.1
Induced Electric Fields The changing magnetic field does *not* need to exist at the location of the induced electric field. In Figure 31.15, even a loop outside the region of magnetic field experiences an induced electric field.

moving the charge once around the loop is $qE(2\pi r)$, where $2\pi r$ is the circumference of the loop. These two expressions for the work done must be equal; therefore,

$$q\mathcal{E} = qE(2\pi r)$$

$$E = \frac{\mathcal{E}}{2\pi r}$$

Using this result along with Equation 31.1 and that $\Phi_B = BA = B\pi r^2$ for a circular loop, the induced electric field can be expressed as

$$E = -\frac{1}{2\pi r}\frac{d\Phi_B}{dt} = -\frac{r}{2}\frac{dB}{dt} \tag{31.8}$$

If the time variation of the magnetic field is specified, the induced electric field can be calculated from Equation 31.8.

The emf for any closed path can be expressed as the line integral of $\vec{E} \cdot d\vec{s}$ over that path: $\mathcal{E} = \oint \vec{E} \cdot d\vec{s}$. In more general cases, E may not be constant and the path may not be a circle. Hence, Faraday's law of induction, $\mathcal{E} = -d\Phi_B/dt$, can be written in the general form

Faraday's law in general form ▶

$$\oint \vec{E} \cdot d\vec{s} = -\frac{d\Phi_B}{dt} \tag{31.9}$$

The induced electric field $\vec{E}$ in Equation 31.9 is a nonconservative field that is generated by a changing magnetic field. The field $\vec{E}$ that satisfies Equation 31.9 cannot possibly be an electrostatic field because were the field electrostatic and hence conservative, the line integral of $\vec{E} \cdot d\vec{s}$ over a closed loop would be zero (Section 25.1), which would be in contradiction to Equation 31.9.

Example 31.7   **Electric Field Induced by a Changing Magnetic Field in a Solenoid**

A long solenoid of radius R has n turns of wire per unit length and carries a time-varying current that varies sinusoidally as $I = I_{max} \cos \omega t$, where I_{max} is the maximum current and ω is the angular frequency of the alternating current source (Fig. 31.16).

(A) Determine the magnitude of the induced electric field outside the solenoid at a distance $r > R$ from its long central axis.

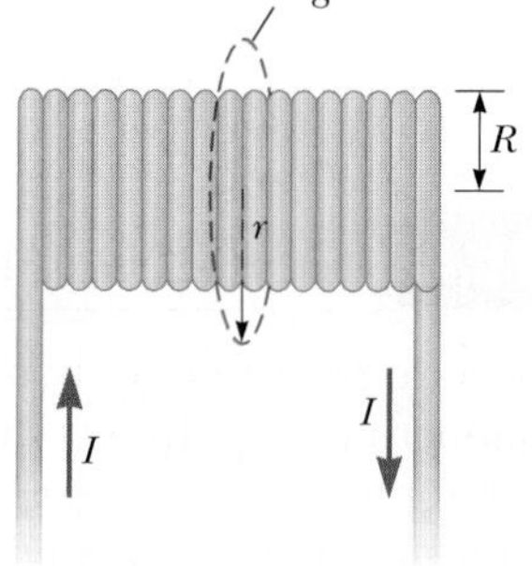

SOLUTION

Conceptualize Figure 31.16 shows the physical situation. As the current in the coil changes, imagine a changing magnetic field at all points in space as well as an induced electric field.

Categorize In this analysis problem, because the current varies in time, the magnetic field is changing, leading to an induced electric field as opposed to the electrostatic electric fields due to stationary electric charges.

Figure 31.16 (Example 31.7) A long solenoid carrying a time-varying current given by $I = I_{max} \cos \omega t$. An electric field is induced both inside and outside the solenoid.

Analyze First consider an external point and take the path for the line integral to be a circle of radius r centered on the solenoid as illustrated in Figure 31.16.

Evaluate the right side of Equation 31.9, noting that the magnetic field $\vec{B}$ inside the solenoid is perpendicular to the circle bounded by the path of integration:

$$(1) \quad -\frac{d\Phi_B}{dt} = -\frac{d}{dt}(B\pi R^2) = -\pi R^2 \frac{dB}{dt}$$

Evaluate the magnetic field inside the solenoid from Equation 30.17:

$$(2) \quad B = \mu_0 nI = \mu_0 nI_{max} \cos \omega t$$

▶ **31.7** continued

Substitute Equation (2) into Equation (1):

$$(3) \quad -\frac{d\Phi_B}{dt} = -\pi R^2 \mu_0 n I_{max} \frac{d}{dt}(\cos \omega t) = \pi R^2 \mu_0 n I_{max} \omega \sin \omega t$$

Evaluate the left side of Equation 31.9, noting that the magnitude of $\vec{E}$ is constant on the path of integration and $\vec{E}$ is tangent to it:

$$(4) \quad \oint \vec{E} \cdot d\vec{s} = E(2\pi r)$$

Substitute Equations (3) and (4) into Equation 31.9:

$$E(2\pi r) = \pi R^2 \mu_0 n I_{max} \omega \sin \omega t$$

Solve for the magnitude of the electric field:

$$E = \frac{\mu_0 n I_{max} \omega R^2}{2r} \sin \omega t \quad \text{(for } r > R)$$

Finalize This result shows that the amplitude of the electric field outside the solenoid falls off as $1/r$ and varies sinusoidally with time. It is proportional to the current I as well as to the frequency ω, consistent with the fact that a larger value of ω means more change in magnetic flux per unit time. As we will learn in Chapter 34, the time-varying electric field creates an additional contribution to the magnetic field. The magnetic field can be somewhat stronger than we first stated, both inside and outside the solenoid. The correction to the magnetic field is small if the angular frequency ω is small. At high frequencies, however, a new phenomenon can dominate: The electric and magnetic fields, each re-creating the other, constitute an electromagnetic wave radiated by the solenoid as we will study in Chapter 34.

(B) What is the magnitude of the induced electric field inside the solenoid, a distance r from its axis?

SOLUTION

Analyze For an interior point ($r < R$), the magnetic flux through an integration loop is given by $\Phi_B = B\pi r^2$.

Evaluate the right side of Equation 31.9:

$$(5) \quad -\frac{d\Phi_B}{dt} = -\frac{d}{dt}(B\pi r^2) = -\pi r^2 \frac{dB}{dt}$$

Substitute Equation (2) into Equation (5):

$$(6) \quad -\frac{d\Phi_B}{dt} = -\pi r^2 \mu_0 n I_{max} \frac{d}{dt}(\cos \omega t) = \pi r^2 \mu_0 n I_{max} \omega \sin \omega t$$

Substitute Equations (4) and (6) into Equation 31.9:

$$E(2\pi r) = \pi r^2 \mu_0 n I_{max} \omega \sin \omega t$$

Solve for the magnitude of the electric field:

$$E = \frac{\mu_0 n I_{max} \omega}{2} r \sin \omega t \quad \text{(for } r < R)$$

Finalize This result shows that the amplitude of the electric field induced inside the solenoid by the changing magnetic flux through the solenoid increases linearly with r and varies sinusoidally with time. As with the field outside the solenoid, the field inside is proportional to the current I and the frequency ω.

31.5 Generators and Motors

Electric generators are devices that take in energy by work and transfer it out by electrical transmission. To understand how they operate, let us consider the **alternating-current (AC) generator.** In its simplest form, it consists of a loop of wire rotated by some external means in a magnetic field (Fig. 31.17a, page 950).

In commercial power plants, the energy required to rotate the loop can be derived from a variety of sources. For example, in a hydroelectric plant, falling water directed against the blades of a turbine produces the rotary motion; in a coal-fired plant, the energy released by burning coal is used to convert water to steam, and this steam is directed against the turbine blades.

Figure 31.17 (a) Schematic diagram of an AC generator. (b) The alternating emf induced in the loop plotted as a function of time.

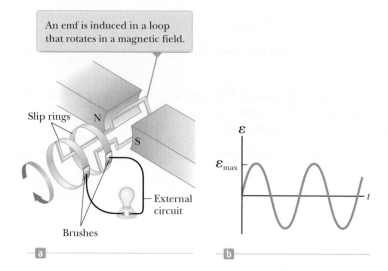

As a loop rotates in a magnetic field, the magnetic flux through the area enclosed by the loop changes with time, and this change induces an emf and a current in the loop according to Faraday's law. The ends of the loop are connected to slip rings that rotate with the loop. Connections from these slip rings, which act as output terminals of the generator, to the external circuit are made by stationary metallic brushes in contact with the slip rings.

Instead of a single turn, suppose a coil with N turns (a more practical situation), with the same area A, rotates in a magnetic field with a constant angular speed ω. If θ is the angle between the magnetic field and the normal to the plane of the coil as in Figure 31.18, the magnetic flux through the coil at any time t is

$$\Phi_B = BA \cos \theta = BA \cos \omega t$$

where we have used the relationship $\theta = \omega t$ between angular position and angular speed (see Eq. 10.3). (We have set the clock so that $t = 0$ when $\theta = 0$.) Hence, the induced emf in the coil is

$$\mathcal{E} = -N\frac{d\Phi_B}{dt} = -NBA \frac{d}{dt}(\cos \omega t) = NBA\omega \sin \omega t \qquad \text{(31.10)}$$

This result shows that the emf varies sinusoidally with time as plotted in Figure 31.17b. Equation 31.10 shows that the maximum emf has the value

$$\mathcal{E}_{max} = NBA\omega \qquad \text{(31.11)}$$

which occurs when $\omega t = 90°$ or $270°$. In other words, $\mathcal{E} = \mathcal{E}_{max}$ when the magnetic field is in the plane of the coil and the time rate of change of flux is a maximum. Furthermore, the emf is zero when $\omega t = 0$ or $180°$, that is, when $\vec{B}$ is perpendicular to the plane of the coil and the time rate of change of flux is zero.

The frequency for commercial generators in the United States and Canada is 60 Hz, whereas in some European countries it is 50 Hz. (Recall that $\omega = 2\pi f$, where f is the frequency in hertz.)

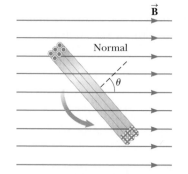

Figure 31.18 A cutaway view of a loop enclosing an area A and containing N turns, rotating with constant angular speed ω in a magnetic field. The emf induced in the loop varies sinusoidally in time.

Quick Quiz 31.4 In an AC generator, a coil with N turns of wire spins in a magnetic field. Of the following choices, which does *not* cause an increase in the emf generated in the coil? **(a)** replacing the coil wire with one of lower resistance **(b)** spinning the coil faster **(c)** increasing the magnetic field **(d)** increasing the number of turns of wire on the coil

Example 31.8 **emf Induced in a Generator**

The coil in an AC generator consists of 8 turns of wire, each of area $A = 0.090\ 0\ \text{m}^2$, and the total resistance of the wire is 12.0 Ω. The coil rotates in a 0.500-T magnetic field at a constant frequency of 60.0 Hz.

(A) Find the maximum induced emf in the coil.

SOLUTION

Conceptualize Study Figure 31.17 to make sure you understand the operation of an AC generator.

Categorize We evaluate parameters using equations developed in this section, so we categorize this example as a substitution problem.

Use Equation 31.11 to find the maximum induced emf:
$$\mathcal{E}_{max} = NBA\omega = NBA(2\pi f)$$

Substitute numerical values:
$$\mathcal{E}_{max} = 8(0.500\ \text{T})(0.090\ 0\ \text{m}^2)(2\pi)(60.0\ \text{Hz}) = \boxed{136\ \text{V}}$$

(B) What is the maximum induced current in the coil when the output terminals are connected to a low-resistance conductor?

SOLUTION

Use Equation 27.7 and the result to part (A):
$$I_{max} = \frac{\mathcal{E}_{max}}{R} = \frac{136\ \text{V}}{12.0\ \Omega} = \boxed{11.3\ \text{A}}$$

The **direct-current (DC) generator** is illustrated in Figure 31.19a. Such generators are used, for instance, in older cars to charge the storage batteries. The components are essentially the same as those of the AC generator except that the contacts to the rotating coil are made using a split ring called a *commutator*.

In this configuration, the output voltage always has the same polarity and pulsates with time as shown in Figure 31.19b. We can understand why by noting that the contacts to the split ring reverse their roles every half cycle. At the same time, the polarity of the induced emf reverses; hence, the polarity of the split ring (which is the same as the polarity of the output voltage) remains the same.

A pulsating DC current is not suitable for most applications. To obtain a steadier DC current, commercial DC generators use many coils and commutators distributed so that the sinusoidal pulses from the various coils are out of phase. When these pulses are superimposed, the DC output is almost free of fluctuations.

A **motor** is a device into which energy is transferred by electrical transmission while energy is transferred out by work. A motor is essentially a generator operating

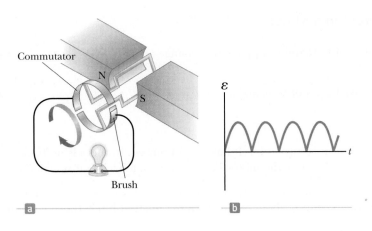

Figure 31.19 (a) Schematic diagram of a DC generator. (b) The magnitude of the emf varies in time, but the polarity never changes.

Figure 31.20 The engine compartment of a Toyota Prius, a hybrid vehicle.

in reverse. Instead of generating a current by rotating a coil, a current is supplied to the coil by a battery, and the torque acting on the current-carrying coil (Section 29.5) causes it to rotate.

Useful mechanical work can be done by attaching the rotating coil of a motor to some external device. As the coil rotates in a magnetic field, however, the changing magnetic flux induces an emf in the coil; consistent with Lenz's law, this induced emf always acts to reduce the current in the coil. The back emf increases in magnitude as the rotational speed of the coil increases. (The phrase *back emf* is used to indicate an emf that tends to reduce the supplied current.) Because the voltage available to supply current equals the difference between the supply voltage and the back emf, the current in the rotating coil is limited by the back emf.

When a motor is turned on, there is initially no back emf, and the current is very large because it is limited only by the resistance of the coil. As the coil begins to rotate, the induced back emf opposes the applied voltage and the current in the coil decreases. If the mechanical load increases, the motor slows down, which causes the back emf to decrease. This reduction in the back emf increases the current in the coil and therefore also increases the power needed from the external voltage source. For this reason, the power requirements for running a motor are greater for heavy loads than for light ones. If the motor is allowed to run under no mechanical load, the back emf reduces the current to a value just large enough to overcome energy losses due to internal energy and friction. If a very heavy load jams the motor so that it cannot rotate, the lack of a back emf can lead to dangerously high current in the motor's wire. This dangerous situation is explored in the What If? section of Example 31.9.

A modern application of motors in automobiles is seen in the development of *hybrid drive systems*. In these automobiles, a gasoline engine and an electric motor are combined to increase the fuel economy of the vehicle and reduce its emissions. Figure 31.20 shows the engine compartment of a Toyota Prius, one of the hybrids available in the United States. In this automobile, power to the wheels can come from either the gasoline engine or the electric motor. In normal driving, the electric motor accelerates the vehicle from rest until it is moving at a speed of about 15 mi/h (24 km/h). During this acceleration period, the engine is not running, so gasoline is not used and there is no emission. At higher speeds, the motor and engine work together so that the engine always operates at or near its most efficient speed. The result is a significantly higher gasoline mileage than that obtained by a traditional gasoline-powered automobile. When a hybrid vehicle brakes, the motor acts as a generator and returns some of the vehicle's kinetic energy back to the battery as stored energy. In a normal vehicle, this kinetic energy is not recovered because it is transformed to internal energy in the brakes and roadway.

Example 31.9   The Induced Current in a Motor

A motor contains a coil with a total resistance of 10 Ω and is supplied by a voltage of 120 V. When the motor is running at its maximum speed, the back emf is 70 V.

(A) Find the current in the coil at the instant the motor is turned on.

SOLUTION

Conceptualize Think about the motor just after it is turned on. It has not yet moved, so there is no back emf generated. As a result, the current in the motor is high. After the motor begins to turn, a back emf is generated and the current decreases.

Categorize We need to combine our new understanding about motors with the relationship between current, voltage, and resistance in this substitution problem.

▶ **31.9** continued

Evaluate the current in the coil from Equation 27.7 with no back emf generated:

$$I = \frac{\mathcal{E}}{R} = \frac{120 \text{ V}}{10 \text{ }\Omega} = \boxed{12 \text{ A}}$$

(B) Find the current in the coil when the motor has reached maximum speed.

SOLUTION

Evaluate the current in the coil with the maximum back emf generated:

$$I = \frac{\mathcal{E} - \mathcal{E}_{\text{back}}}{R} = \frac{120 \text{ V} - 70 \text{ V}}{10 \text{ }\Omega} = \frac{50 \text{ V}}{10 \text{ }\Omega} = \boxed{5.0 \text{ A}}$$

The current drawn by the motor when operating at its maximum speed is significantly less than that drawn before it begins to turn.

WHAT IF? Suppose this motor is in a circular saw. When you are operating the saw, the blade becomes jammed in a piece of wood and the motor cannot turn. By what percentage does the power input to the motor increase when it is jammed?

Answer You may have everyday experiences with motors becoming warm when they are prevented from turning. That is due to the increased power input to the motor. The higher rate of energy transfer results in an increase in the internal energy of the coil, an undesirable effect.

Set up the ratio of power input to the motor when jammed, using the current calculated in part (A), to that when it is not jammed, part (B):

$$\frac{P_{\text{jammed}}}{P_{\text{not jammed}}} = \frac{I_A^2 R}{I_B^2 R} = \frac{I_A^2}{I_B^2}$$

Substitute numerical values:

$$\frac{P_{\text{jammed}}}{P_{\text{not jammed}}} = \frac{(12 \text{ A})^2}{(5.0 \text{ A})^2} = 5.76$$

That represents a 476% increase in the input power! Such a high power input can cause the coil to become so hot that it is damaged.

31.6 Eddy Currents

As we have seen, an emf and a current are induced in a circuit by a changing magnetic flux. In the same manner, circulating currents called **eddy currents** are induced in bulk pieces of metal moving through a magnetic field. This phenomenon can be demonstrated by allowing a flat copper or aluminum plate attached at the end of a rigid bar to swing back and forth through a magnetic field (Fig. 31.21). As the plate enters the field, the changing magnetic flux induces an emf in the plate, which in turn causes the free electrons in the plate to move, producing the swirling eddy currents. According to Lenz's law, the direction of the eddy currents is such that they create magnetic fields that oppose the change that causes the currents. For this reason, the eddy currents must produce effective magnetic poles on the plate, which are repelled by the poles of the magnet; this situation gives rise to a repulsive force that opposes the motion of the plate. (If the opposite were true, the plate would accelerate and its energy would increase after each swing, in violation of the law of conservation of energy.)

As indicated in Figure 31.22a (page 954), with $\vec{\mathbf{B}}$ directed into the page, the induced eddy current is counterclockwise as the swinging plate enters the field at position 1 because the flux due to the external magnetic field into the page through the plate is increasing. Hence, by Lenz's law, the induced current must provide its own magnetic field out of the page. The opposite is true as the plate

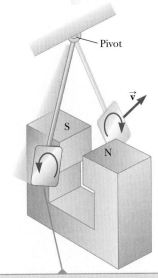

As the plate enters or leaves the field, the changing magnetic flux induces an emf, which causes eddy currents in the plate.

Figure 31.21 Formation of eddy currents in a conducting plate moving through a magnetic field.

Figure 31.22 When a conducting plate swings through a magnetic field, eddy currents are induced and the magnetic force $\vec{\mathbf{F}}_B$ on the plate opposes its velocity, causing it to eventually come to rest.

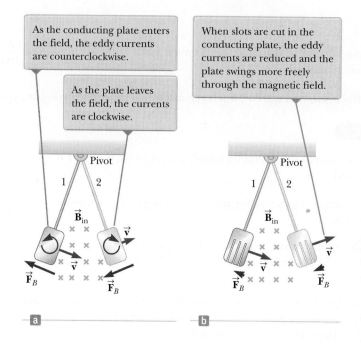

As the conducting plate enters the field, the eddy currents are counterclockwise.

As the plate leaves the field, the currents are clockwise.

When slots are cut in the conducting plate, the eddy currents are reduced and the plate swings more freely through the magnetic field.

leaves the field at position 2, where the current is clockwise. Because the induced eddy current always produces a magnetic retarding force $\vec{\mathbf{F}}_B$ when the plate enters or leaves the field, the swinging plate eventually comes to rest.

If slots are cut in the plate as shown in Figure 31.22b, the eddy currents and the corresponding retarding force are greatly reduced. We can understand this reduction in force by realizing that the cuts in the plate prevent the formation of any large current loops.

The braking systems on many subway and rapid-transit cars make use of electromagnetic induction and eddy currents. An electromagnet attached to the train is positioned near the steel rails. (An electromagnet is essentially a solenoid with an iron core.) The braking action occurs when a large current is passed through the electromagnet. The relative motion of the magnet and rails induces eddy currents in the rails, and the direction of these currents produces a drag force on the moving train. Because the eddy currents decrease steadily in magnitude as the train slows down, the braking effect is quite smooth. As a safety measure, some power tools use eddy currents to stop rapidly spinning blades once the device is turned off.

Eddy currents are often undesirable because they represent a transformation of mechanical energy to internal energy. To reduce this energy loss, conducting parts are often laminated; that is, they are built up in thin layers separated by a nonconducting material such as lacquer or a metal oxide. This layered structure prevents large current loops and effectively confines the currents to small loops in individual layers. Such a laminated structure is used in transformer cores (see Section 33.8) and motors to minimize eddy currents and thereby increase the efficiency of these devices.

Q uick Quiz 31.5 In an equal-arm balance from the early 20th century (Fig. 31.23), an aluminum sheet hangs from one of the arms and passes between the poles of a magnet, causing the oscillations of the balance to decay rapidly. In the absence of such magnetic braking, the oscillation might continue for a long time, and the experimenter would have to wait to take a reading. Why do the oscillations decay? **(a)** because the aluminum sheet is attracted to the magnet

(b) because currents in the aluminum sheet set up a magnetic field that opposes the oscillations **(c)** because aluminum is paramagnetic

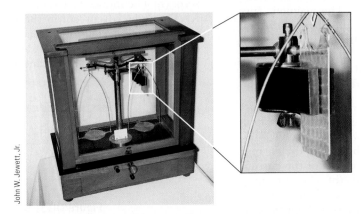

Figure 31.23 (Quick Quiz 31.5) In an old-fashioned equal-arm balance, an aluminum sheet hangs between the poles of a magnet.

Summary

Concepts and Principles

Faraday's law of induction states that the emf induced in a loop is directly proportional to the time rate of change of magnetic flux through the loop, or

$$\mathcal{E} = -\frac{d\Phi_B}{dt} \qquad (31.1)$$

where $\Phi_B = \int \vec{\mathbf{B}} \cdot d\vec{\mathbf{A}}$ is the magnetic flux through the loop.

When a conducting bar of length ℓ moves at a velocity $\vec{\mathbf{v}}$ through a magnetic field $\vec{\mathbf{B}}$, where $\vec{\mathbf{B}}$ is perpendicular to the bar and to $\vec{\mathbf{v}}$, the **motional emf** induced in the bar is

$$\mathcal{E} = -B\ell v \qquad (31.5)$$

Lenz's law states that the induced current and induced emf in a conductor are in such a direction as to set up a magnetic field that opposes the change that produced them.

A general form of **Faraday's law of induction** is

$$\oint \vec{\mathbf{E}} \cdot d\vec{\mathbf{s}} = -\frac{d\Phi_B}{dt} \qquad (31.9)$$

where $\vec{\mathbf{E}}$ is the nonconservative electric field that is produced by the changing magnetic flux.

Objective Questions   [1.] denotes answer available in *Student Solutions Manual/Study Guide*

1. Figure OQ31.1 is a graph of the magnetic flux through a certain coil of wire as a function of time during an interval while the radius of the coil is increased, the coil is rotated through 1.5 revolutions, and the external source of the magnetic field is turned off, in that order. Rank the emf induced in the coil at the instants marked A through E from the largest positive value to the largest-magnitude negative value. In your ranking,

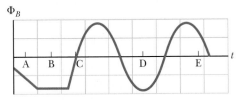

Figure OQ31.1

note any cases of equality and also any instants when the emf is zero.

2. A flat coil of wire is placed in a uniform magnetic field that is in the y direction. **(i)** The magnetic flux through the coil is a maximum if the plane of the coil is where? More than one answer may be correct. (a) in the xy plane (b) in the yz plane (c) in the xz plane (d) in any orientation, because it is a constant **(ii)** For what orientation is the flux zero? Choose from the same possibilities as in part (i).

3. A rectangular conducting loop is placed near a long wire carrying a current I as shown in Figure OQ31.3. If I decreases in time, what can be said of the current induced in the loop? (a) The direction of the current depends on the size of the loop. (b) The current is clockwise. (c) The current is counterclockwise. (d) The current is zero. (e) Nothing can be said about the current in the loop without more information.

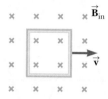

Figure OQ31.3

4. A circular loop of wire with a radius of 4.0 cm is in a uniform magnetic field of magnitude 0.060 T. The plane of the loop is perpendicular to the direction of the magnetic field. In a time interval of 0.50 s, the magnetic field changes to the opposite direction with a magnitude of 0.040 T. What is the magnitude of the average emf induced in the loop? (a) 0.20 V (b) 0.025 V (c) 5.0 mV (d) 1.0 mV (e) 0.20 mV

5. A square, flat loop of wire is pulled at constant velocity through a region of uniform magnetic field directed perpendicular to the plane of the loop as shown in Figure OQ31.5. Which of the following statements are correct? More than one statement may be correct. (a) Current is induced in the loop in the clockwise direction. (b) Current is induced in the loop in the counterclockwise direction. (c) No current is induced in the loop. (d) Charge separation occurs in the loop, with the top edge positive. (e) Charge separation occurs in the loop, with the top edge negative.

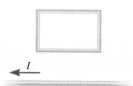

Figure OQ31.5

6. The bar in Figure OQ31.6 moves on rails to the right with a velocity $\vec{v}$, and a uniform, constant magnetic field is directed out of the page. Which of the following statements are correct? More than one statement may be correct. (a) The induced current in the loop is zero. (b) The induced current in the loop is clockwise. (c) The induced current in the loop is counterclockwise. (d) An external force is required to keep the bar moving at constant speed. (e) No force is required to keep the bar moving at constant speed.

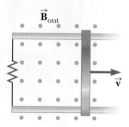

Figure OQ31.6

7. A bar magnet is held in a vertical orientation above a loop of wire that lies in the horizontal plane as shown in Figure OQ31.7. The south end of the magnet is toward the loop. After the magnet is dropped, what is true of the induced current in the loop as viewed from above? (a) It is clockwise as the magnet falls toward the loop. (b) It is counterclockwise as the magnet falls toward the loop. (c) It is clockwise after the magnet has moved through the loop and moves away from it. (d) It is always clockwise. (e) It is first counterclockwise as the magnet approaches the loop and then clockwise after it has passed through the loop.

Figure OQ31.7

8. What happens to the amplitude of the induced emf when the rate of rotation of a generator coil is doubled? (a) It becomes four times larger. (b) It becomes two times larger. (c) It is unchanged. (d) It becomes one-half as large. (e) It becomes one-fourth as large.

9. Two coils are placed near each other as shown in Figure OQ31.9. The coil on the left is connected to a battery and a switch, and the coil on the right is connected to a resistor. What is the direction of the cur-

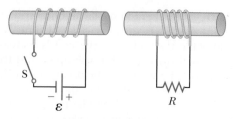

Figure OQ31.9

rent in the resistor **(i)** at an instant immediately after the switch is thrown closed, **(ii)** after the switch has been closed for several seconds, and **(iii)** at an instant after the switch has then been thrown open? Choose each answer from the possibilities (a) left, (b) right, or (c) the current is zero.

10. A circuit consists of a conducting movable bar and a lightbulb connected to two conducting rails as shown in Figure OQ31.10. An external magnetic field is directed perpendicular to the plane of the circuit. Which of the following actions will make the bulb light up? More than one statement may be correct. (a) The bar is

Figure OQ31.10

moved to the left. (b) The bar is moved to the right. (c) The magnitude of the magnetic field is increased. (d) The magnitude of the magnetic field is decreased. (e) The bar is lifted off the rails.

11. Two rectangular loops of wire lie in the same plane as shown in Figure OQ31.11. If the current *I* in the outer loop is counterclockwise and increases with time, what is true of the current induced in the inner loop? More than one statement may be correct. (a) It is zero. (b) It is clockwise. (c) It is counterclockwise. (d) Its magnitude depends on the dimensions of the loops. (e) Its direction depends on the dimensions of the loops.

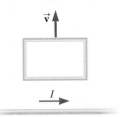

Figure OQ31.11

Conceptual Questions   **1.** denotes answer available in *Student Solutions Manual/Study Guide*

1. In Section 7.7, we defined conservative and nonconservative forces. In Chapter 23, we stated that an electric charge creates an electric field that produces a conservative force. Argue now that induction creates an electric field that produces a nonconservative force.

2. A spacecraft orbiting the Earth has a coil of wire in it. An astronaut measures a small current in the coil, although there is no battery connected to it and there are no magnets in the spacecraft. What is causing the current?

3. In a hydroelectric dam, how is energy produced that is then transferred out by electrical transmission? That is, how is the energy of motion of the water converted to energy that is transmitted by AC electricity?

4. A bar magnet is dropped toward a conducting ring lying on the floor. As the magnet falls toward the ring, does it move as a freely falling object? Explain.

5. A circular loop of wire is located in a uniform and constant magnetic field. Describe how an emf can be induced in the loop in this situation.

6. A piece of aluminum is dropped vertically downward between the poles of an electromagnet. Does the magnetic field affect the velocity of the aluminum?

7. What is the difference between magnetic flux and magnetic field?

8. When the switch in Figure CQ31.8a is closed, a current is set up in the coil and the metal ring springs upward (Fig. CQ31.8b). Explain this behavior.

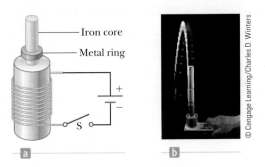

Figure CQ31.8 Conceptual Questions 8 and 9.

9. Assume the battery in Figure CQ31.8a is replaced by an AC source and the switch is held closed. If held down, the metal ring on top of the solenoid becomes hot. Why?

10. A loop of wire is moving near a long, straight wire carrying a constant current *I* as shown in Figure CQ31.10. (a) Determine the direction of the induced current in the loop as it moves away from the wire. (b) What would be the direction of the induced current in the loop if it were moving toward the wire?

Figure CQ31.10

Problems

WebAssign The problems found in this chapter may be assigned online in Enhanced WebAssign

1. straightforward; **2.** intermediate; **3.** challenging

1. full solution available in the *Student Solutions Manual/Study Guide*

AMT Analysis Model tutorial available in Enhanced WebAssign

GP Guided Problem

M Master It tutorial available in Enhanced WebAssign

W Watch It video solution available in Enhanced WebAssign

Section 31.1 **Faraday's Law of Induction**

1. A flat loop of wire consisting of a single turn of cross-sectional area 8.00 cm² is perpendicular to a magnetic field that increases uniformly in magnitude from 0.500 T to 2.50 T in 1.00 s. What is the resulting induced current if the loop has a resistance of 2.00 Ω?

2. An instrument based on induced emf has been used to measure projectile speeds up to 6 km/s. A small magnet is imbedded in the projectile as shown in Figure P31.2. The projectile passes through two coils separated by a distance *d*. As the projectile passes through each coil, a pulse of emf is induced in the coil. The time interval between pulses can be measured accurately with an oscilloscope, and thus the speed can be determined. (a) Sketch a graph of Δ*V* versus *t* for the arrangement shown. Consider a current that flows counterclockwise as viewed from the starting point of the projectile as positive. On your graph, indicate which pulse is from coil 1 and which is from coil 2. (b) If the pulse separation is 2.40 ms and *d* = 1.50 m, what is the projectile speed?

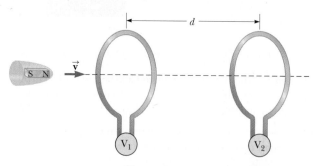

Figure P31.2

3. Transcranial magnetic stimulation (TMS) is a noninvasive technique used to stimulate regions of the human brain (Figure P31.3). In TMS, a small coil is placed on the scalp and a brief burst of current in the coil produces a rapidly changing magnetic field inside the brain. The induced emf can stimulate neuronal activity. (a) One such device generates an upward magnetic field within the brain that rises from zero to 1.50 T in 120 ms. Determine the induced emf around a horizontal circle of tissue of radius 1.60 mm. (b) **What If?** The field next changes to 0.500 T downward in 80.0 ms. How does the emf induced in this process compare with that in part (a)?

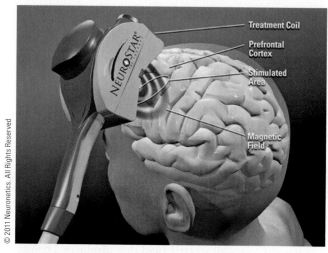

Figure P31.3 Problems 3 and 51. The magnetic coil of a Neurostar TMS apparatus is held near the head of a patient.

4. A 25-turn circular coil of wire has diameter 1.00 m. It **W** is placed with its axis along the direction of the Earth's magnetic field of 50.0 μT and then in 0.200 s is flipped 180°. An average emf of what magnitude is generated in the coil?

5. The flexible loop in Figure P31.5 has a radius of 12.0 cm and is in a magnetic field of magnitude 0.150 T. The loop is grasped at points *A* and *B* and stretched until its area is nearly zero. If it takes 0.200 s to close the loop, what is the magnitude of the average induced emf in it during this time interval?

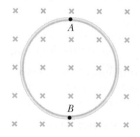

Figure P31.5 Problems 5 and 6.

6. A circular loop of wire of radius 12.0 cm is placed in a magnetic field directed perpendicular to the plane of the loop as in Figure P31.5. If the field decreases at the rate of 0.050 0 T/s in some time interval, find the magnitude of the emf induced in the loop during this interval.

7. To monitor the breathing of a hospital patient, a thin belt is girded around the patient's chest. The belt is a 200-turn coil. When the patient inhales, the area encircled by the coil increases by 39.0 cm². The magnitude of the Earth's magnetic field is 50.0 μT and makes an angle of 28.0° with the plane of the coil. Assuming a patient takes 1.80 s to inhale, find the average induced emf in the coil during this time interval.

8. A strong electromagnet produces a uniform magnetic
W field of 1.60 T over a cross-sectional area of 0.200 m². A coil having 200 turns and a total resistance of 20.0 Ω is placed around the electromagnet. The current in the electromagnet is then smoothly reduced until it reaches zero in 20.0 ms. What is the current induced in the coil?

9. A 30-turn circular coil of radius 4.00 cm and resistance
W 1.00 Ω is placed in a magnetic field directed perpendicular to the plane of the coil. The magnitude of the magnetic field varies in time according to the expression $B = 0.010\,0t + 0.040\,0t^2$, where B is in teslas and t is in seconds. Calculate the induced emf in the coil at $t = 5.00$ s.

10. Scientific work is currently under way to determine whether weak oscillating magnetic fields can affect human health. For example, one study found that drivers of trains had a higher incidence of blood cancer than other railway workers, possibly due to long exposure to mechanical devices in the train engine cab. Consider a magnetic field of magnitude 1.00×10^{-3} T, oscillating sinusoidally at 60.0 Hz. If the diameter of a red blood cell is 8.00 μm, determine the maximum emf that can be generated around the perimeter of a cell in this field.

11. An aluminum ring of radius $r_1 = 5.00$ cm and resistance
M 3.00×10^{-4} Ω is placed around one end of a long air-core solenoid with 1 000 turns per meter and radius $r_2 = 3.00$ cm as shown in Figure P31.11. Assume the axial component of the field produced by the solenoid is one-half as strong over the area of the end of the solenoid as at the center of the solenoid. Also assume the solenoid produces negligible field outside its cross-sectional area. The current in the solenoid is increasing at a rate of 270 A/s. (a) What is the induced current in the ring? At the center of the ring, what are (b) the magnitude and (c) the direction of the magnetic field produced by the induced current in the ring?

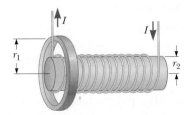

Figure P31.11 Problems 11 and 12.

12. An aluminum ring of radius r_1 and resistance R is placed around one end of a long air-core solenoid with n turns per meter and smaller radius r_2 as shown in Figure P31.11. Assume the axial component of the field produced by the solenoid over the area of the end of the solenoid is one-half as strong as at the center of the solenoid. Also assume the solenoid produces negligible field outside its cross-sectional area. The current in the solenoid is increasing at a rate of $\Delta I/\Delta t$. (a) What is the induced current in the ring? (b) At the center of the ring, what is the magnetic field produced by the induced current in the ring? (c) What is the direction of this field?

13. A loop of wire in the shape of a rectangle of width w
W and length L and a long, straight wire carrying a current I lie on a tabletop as shown in Figure P31.13. (a) Determine the magnetic flux through the loop due to the current I. (b) Suppose the current is changing with time according to $I = a + bt$, where a and b are constants. Determine the emf that is induced in the loop if $b = 10.0$ A/s, $h = 1.00$ cm, $w = 10.0$ cm, and $L = 1.00$ m. (c) What is the direction of the induced current in the rectangle?

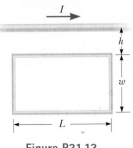

Figure P31.13

14. A coil of 15 turns and radius 10.0 cm surrounds a long
W solenoid of radius 2.00 cm and 1.00×10^3 turns/meter (Fig. P31.14). The current in the solenoid changes as $I = 5.00 \sin 120t$, where I is in amperes and t is in seconds. Find the induced emf in the 15-turn coil as a function of time.

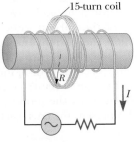

Figure P31.14

15. A square, single-turn wire loop $\ell = 1.00$ cm on a side is placed inside a solenoid that has a circular cross section of radius $r = 3.00$ cm as shown in the end view of Figure P31.15 (page 960). The solenoid is 20.0 cm long and wound with 100 turns of wire. (a) If the current in the solenoid is 3.00 A, what is the magnetic flux through the square loop? (b) If the current in the solenoid is reduced to zero in 3.00 s, what is the magnitude of the average induced emf in the square loop?

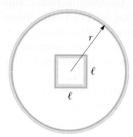

Figure P31.15

16. A long solenoid has $n = 400$ turns per meter and carries a current given by $I = 30.0(1 - e^{-1.60t})$, where I is in amperes and t is in seconds. Inside the solenoid and coaxial with it is a coil that has a radius of $R = 6.00$ cm and consists of a total of $N = 250$ turns of fine wire (Fig. P31.16). What emf is induced in the coil by the changing current?

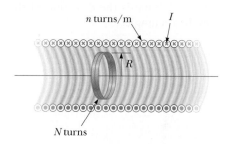

Figure P31.16

17. A coil formed by wrapping 50 turns of wire in the shape of a square is positioned in a magnetic field so that the normal to the plane of the coil makes an angle of 30.0° with the direction of the field. When the magnetic field is increased uniformly from 200 μT to 600 μT in 0.400 s, an emf of magnitude 80.0 mV is induced in the coil. What is the total length of the wire in the coil?

18. When a wire carries an AC current with a known frequency, you can use a *Rogowski coil* to determine the amplitude I_{max} of the current without disconnecting the wire to shunt the current through a meter. The Rogowski coil, shown in Figure P31.18, simply clips around the wire. It consists of a toroidal conductor wrapped around a circular return cord. Let n represent the number of turns in the toroid per unit distance along it. Let A represent the cross-sectional area of the

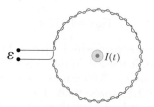

Figure P31.18

toroid. Let $I(t) = I_{max} \sin \omega t$ represent the current to be measured. (a) Show that the amplitude of the emf induced in the Rogowski coil is $\mathcal{E}_{max} = \mu_0 n A \omega I_{max}$. (b) Explain why the wire carrying the unknown current need not be at the center of the Rogowski coil and why the coil will not respond to nearby currents that it does not enclose.

19. A toroid having a rectangular cross section ($a = 2.00$ cm by $b = 3.00$ cm) and inner radius $R = 4.00$ cm consists of $N = 500$ turns of wire that carry a sinusoidal current $I = I_{max} \sin \omega t$, with $I_{max} = 50.0$ A and a frequency $f = \omega/2\pi = 60.0$ Hz. A coil that consists of $N' = 20$ turns of wire is wrapped around one section of the toroid as shown in Figure P31.19. Determine the emf induced in the coil as a function of time.

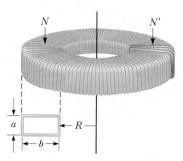

Figure P31.19

20. A piece of insulated wire is shaped into a figure eight as shown in Figure P31.20. For simplicity, model the two halves of the figure eight as circles. The radius of the upper circle is 5.00 cm and that of the lower circle is 9.00 cm. The wire has a uniform resistance per unit length of 3.00 Ω/m. A uniform magnetic field is applied perpendicular to the plane of the two circles, in the direction shown. The magnetic field is increasing at a constant rate of 2.00 T/s. Find (a) the magnitude and (b) the direction of the induced current in the wire.

Figure P31.20

Section 31.2 Motional emf

Section 31.3 Lenz's Law

> Problem 72 in Chapter 29 can be assigned with this section.

21. A helicopter (Fig. P31.21) has blades of length 3.00 m, extending out from a central hub and rotating at 2.00 rev/s. If the vertical component of the Earth's

magnetic field is 50.0 μT, what is the emf induced between the blade tip and the center hub?

Figure P31.21

22. Use Lenz's law to answer the following questions concerning the direction of induced currents. Express your answers in terms of the letter labels a and b in each part of Figure P31.22. (a) What is the direction of the induced current in the resistor R in Figure P31.22a when the bar magnet is moved to the left? (b) What is the direction of the current induced in the resistor R immediately after the switch S in Figure P31.22b is closed? (c) What is the direction of the induced current in the resistor R when the current I in Figure P31.22c decreases rapidly to zero?

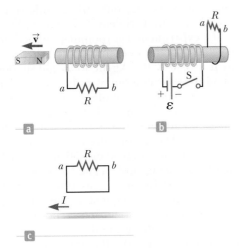

Figure P31.22

23. A truck is carrying a steel beam of length 15.0 m on a freeway. An accident causes the beam to be dumped off the truck and slide horizontally along the ground at a speed of 25.0 m/s. The velocity of the center of mass of the beam is northward while the length of the beam maintains an east–west orientation. The vertical component of the Earth's magnetic field at this location has a magnitude of 35.0 μT. What is the magnitude of the induced emf between the ends of the beam?

24. A small airplane with a wingspan of 14.0 m is flying due north at a speed of 70.0 m/s over a region where the vertical component of the Earth's magnetic field is 1.20 μT downward. (a) What potential difference is developed between the airplane's wingtips? (b) Which wingtip is at higher potential? (c) **What If?** How would the answers to parts (a) and (b) change if the plane turned to fly due east? (d) Can this emf be used to power a lightbulb in the passenger compartment? Explain your answer.

25. A 2.00-m length of wire is held in an east–west direction and moves horizontally to the north with a speed of 0.500 m/s. The Earth's magnetic field in this region is of magnitude 50.0 μT and is directed northward and 53.0° below the horizontal. (a) Calculate the magnitude of the induced emf between the ends of the wire and (b) determine which end is positive.

26. Consider the arrangement shown in Figure P31.26. Assume that $R = 6.00\ \Omega$, $\ell = 1.20$ m, and a uniform 2.50-T magnetic field is directed into the page. At what speed should the bar be moved to produce a current of 0.500 A in the resistor?

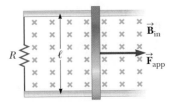

Figure P31.26 Problems 26 through 29.

27. Figure P31.26 shows a top view of a bar that can slide on two frictionless rails. The resistor is $R = 6.00\ \Omega$, and a 2.50-T magnetic field is directed perpendicularly downward, into the paper. Let $\ell = 1.20$ m. (a) Calculate the applied force required to move the bar to the right at a constant speed of 2.00 m/s. (b) At what rate is energy delivered to the resistor?

28. A metal rod of mass m slides without friction along two parallel horizontal rails, separated by a distance ℓ and connected by a resistor R, as shown in Figure P31.26. A uniform vertical magnetic field of magnitude B is applied perpendicular to the plane of the paper. The applied force shown in the figure acts only for a moment, to give the rod a speed v. In terms of m, ℓ, R, B, and v, find the distance the rod will then slide as it coasts to a stop.

29. A conducting rod of length ℓ moves on two horizontal, frictionless rails as shown in Figure P31.26. If a constant force of 1.00 N moves the bar at 2.00 m/s through a magnetic field $\vec{B}$ that is directed into the page, (a) what is the current through the 8.00-Ω resistor R? (b) What is the rate at which energy is delivered to the resistor? (c) What is the mechanical power delivered by the force $\vec{F}_{app}$?

30. *Why is the following situation impossible?* An automobile has a vertical radio antenna of length $\ell = 1.20$ m. The automobile travels on a curvy, horizontal road where the Earth's magnetic field has a magnitude of $B = 50.0\ \mu$T and is directed toward the north and downward at an angle of $\theta = 65.0°$ below the horizontal. The motional emf developed between the top and bottom of the antenna varies with the speed and direction of the automobile's travel and has a maximum value of 4.50 mV.

31. **Review.** Figure P31.31 shows a bar of mass $m = 0.200$ kg that can slide without friction on a pair of rails separated by a distance $\ell = 1.20$ m and located on an inclined plane that makes an angle $\theta = 25.0°$ with respect to the ground. The resistance of the resistor is $R = 1.00$ Ω and a uniform magnetic field of magnitude $B = 0.500$ T is directed downward, perpendicular to the ground, over the entire region through which the bar moves. With what constant speed v does the bar slide along the rails?

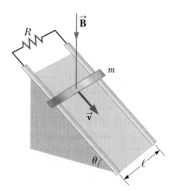

Figure P31.31 Problems 31 and 32.

32. **Review.** Figure P31.31 shows a bar of mass m that can slide without friction on a pair of rails separated by a distance ℓ and located on an inclined plane that makes an angle θ with respect to the ground. The resistance of the resistor is R, and a uniform magnetic field of magnitude B is directed downward, perpendicular to the ground, over the entire region through which the bar moves. With what constant speed v does the bar slide along the rails?

33. The *homopolar generator,* also called the *Faraday disk,* is a low-voltage, high-current electric generator. It consists of a rotating conducting disk with one stationary brush (a sliding electrical contact) at its axle and another at a point on its circumference as shown in Figure P31.33. A uniform magnetic field is applied perpendicular to the plane of the disk. Assume the field is 0.900 T, the angular speed is 3.20×10^3 rev/min, and the radius of

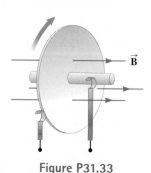

Figure P31.33

the disk is 0.400 m. Find the emf generated between the brushes. When superconducting coils are used to produce a large magnetic field, a homopolar generator can have a power output of several megawatts. Such a generator is useful, for example, in purifying metals by electrolysis. If a voltage is applied to the output terminals of the generator, it runs in reverse as a *homopolar motor* capable of providing great torque, useful in ship propulsion.

34. A conducting bar of length ℓ moves to the right on two frictionless rails as shown in Figure P31.34. A uniform magnetic field directed into the page has a magnitude of 0.300 T. Assume $R = 9.00$ Ω and $\ell = 0.350$ m. (a) At what constant speed should the bar move to produce an 8.50-mA current in the resistor? (b) What is the direction of the induced current? (c) At what rate is energy delivered to the resistor? (d) Explain the origin of the energy being delivered to the resistor.

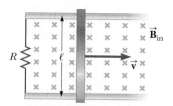

Figure P31.34

35. **Review.** After removing one string while restringing his acoustic guitar, a student is distracted by a video game. His experimentalist roommate notices his inattention and attaches one end of the string, of linear density $\mu = 3.00 \times 10^{-3}$ kg/m, to a rigid support. The other end passes over a pulley, a distance $\ell = 64.0$ cm from the fixed end, and an object of mass $m = 27.2$ kg is attached to the hanging end of the string. The roommate places a magnet across the string as shown in Figure P31.35. The magnet does not touch the string, but produces a uniform field of 4.50 mT over a 2.00-cm length of the string and negligible field elsewhere. Strumming the string sets it vibrating vertically at its fundamental (lowest) frequency. The section of the string in the magnetic field moves perpendicular to the field with a uniform amplitude of 1.50 cm. Find (a) the frequency and (b) the amplitude of the emf induced between the ends of the string.

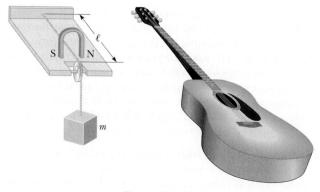

Figure P31.35

36. A rectangular coil with resistance R has N turns, each of length ℓ and width w as shown in Figure P31.36. The coil moves into a uniform magnetic field $\vec{\mathbf{B}}$ with constant velocity $\vec{\mathbf{v}}$. What are the magnitude and direction of the total magnetic force on the coil (a) as it enters the magnetic field, (b) as it moves within the field, and (c) as it leaves the field?

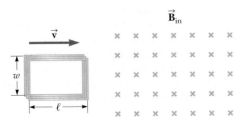

Figure P31.36

37. Two parallel rails with negligible resistance are 10.0 cm apart and are connected by a resistor of resistance $R_3 = 5.00\ \Omega$. The circuit also contains two metal rods having resistances of $R_1 = 10.0\ \Omega$ and $R_2 = 15.0\ \Omega$ sliding along the rails (Fig. P31.37). The rods are pulled away from the resistor at constant speeds of $v_1 = 4.00$ m/s and $v_2 = 2.00$ m/s, respectively. A uniform magnetic field of magnitude $B = 0.010\ 0$ T is applied perpendicular to the plane of the rails. Determine the current in R_3.

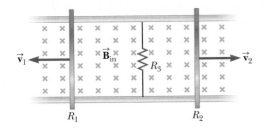

Figure P31.37

38. An astronaut is connected to her spacecraft by a 25.0-m-long tether cord as she and the spacecraft orbit the Earth in a circular path at a speed of 7.80×10^3 m/s. At one instant, the emf between the ends of a wire embedded in the cord is measured to be 1.17 V. Assume the long dimension of the cord is perpendicular to the Earth's magnetic field at that instant. Assume also the tether's center of mass moves with a velocity perpendicular to the Earth's magnetic field. (a) What is the magnitude of the Earth's field at this location? (b) Does the emf change as the system moves from one location to another? Explain. (c) Provide two conditions under which the emf would be zero even though the magnetic field is not zero.

Section 31.4 Induced emf and Electric Fields

39. Within the green dashed circle shown in Figure P31.39, the magnetic field changes with time according to the expression $B = 2.00t^3 - 4.00t^2 + 0.800$, where B is in teslas, t is in seconds, and $R = 2.50$ cm. When $t = 2.00$ s, calculate (a) the magnitude and (b) the direc-

tion of the force exerted on an electron located at point P_1, which is at a distance $r_1 = 5.00$ cm from the center of the circular field region. (c) At what instant is this force equal to zero?

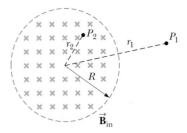

Figure P31.39 Problems 39 and 40.

40. A magnetic field directed into the page changes with time according to $B = 0.030\ 0t^2 + 1.40$, where B is in teslas and t is in seconds. The field has a circular cross section of radius $R = 2.50$ cm (see Fig. P31.39). When $t = 3.00$ s and $r_2 = 0.020\ 0$ m, what are (a) the magnitude and (b) the direction of the electric field at point P_2?

41. A long solenoid with 1.00×10^3 turns per meter and radius 2.00 cm carries an oscillating current $I = 5.00 \sin 100\pi t$, where I is in amperes and t is in seconds. (a) What is the electric field induced at a radius $r = 1.00$ cm from the axis of the solenoid? (b) What is the direction of this electric field when the current is increasing counterclockwise in the solenoid?

Section 31.5 Generators and Motors

Problems 50 and 68 in Chapter 29 can be assigned with this section.

42. A 100-turn square coil of side 20.0 cm rotates about a vertical axis at 1.50×10^3 rev/min as indicated in Figure P31.42. The horizontal component of the Earth's magnetic field at the coil's location is equal to 2.00×10^{-5} T. (a) Calculate the maximum emf induced in the coil by this field. (b) What is the orientation of the coil with respect to the magnetic field when the maximum emf occurs?

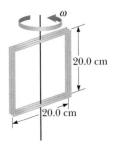

Figure P31.42

43. A generator produces 24.0 V when turning at 900 rev/min. What emf does it produce when turning at 500 rev/min?

44. Figure P31.44 (page 964) is a graph of the induced emf versus time for a coil of N turns rotating with angular speed ω in a uniform magnetic field directed perpendicular to the coil's axis of rotation. **What If?** Copy this sketch (on a larger scale) and on the same set of axes show the graph of emf versus t (a) if the number of turns in the coil is doubled, (b) if instead the angular speed is doubled, and (c) if the angular speed is doubled while the number of turns in the coil is halved.

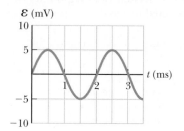

Figure P31.44

45. In a 250-turn automobile alternator, the magnetic flux
W in each turn is $\Phi_B = 2.50 \times 10^{-4} \cos \omega t$, where Φ_B is in webers, ω is the angular speed of the alternator, and t is in seconds. The alternator is geared to rotate three times for each engine revolution. When the engine is running at an angular speed of 1.00×10^3 rev/min, determine (a) the induced emf in the alternator as a function of time and (b) the maximum emf in the alternator.

46. In Figure P31.46, a semicircular conductor of radius $R = 0.250$ m is rotated about the axis AC at a constant rate of 120 rev/min. A uniform magnetic field of magnitude 1.30 T fills the entire region below the axis and is directed out of the page. (a) Calculate the maximum value of the emf induced between the ends of the conductor. (b) What is the value of the average induced emf for each complete rotation? (c) **What If?** How would your answers to parts (a) and (b) change if the magnetic field were allowed to extend a distance R above the axis of rotation? Sketch the emf versus time (d) when the field is as drawn in Figure P31.46 and (e) when the field is extended as described in part (c).

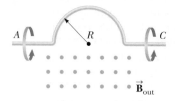

Figure P31.46

47. A long solenoid, with its axis along the x axis, consists of 200 turns per meter of wire that carries a steady current of 15.0 A. A coil is formed by wrapping 30 turns of thin wire around a circular frame that has a radius of 8.00 cm. The coil is placed inside the solenoid and mounted on an axis that is a diameter of the coil and coincides with the y axis. The coil is then rotated with an angular speed of 4.00π rad/s. The plane of the coil is in the yz plane at $t = 0$. Determine the emf generated in the coil as a function of time.

48. A motor in normal operation carries a direct current of 0.850 A when connected to a 120-V power supply. The resistance of the motor windings is 11.8 Ω. While in normal operation, (a) what is the back emf gener-

ated by the motor? (b) At what rate is internal energy produced in the windings? (c) **What If?** Suppose a malfunction stops the motor shaft from rotating. At what rate will internal energy be produced in the windings in this case? (Most motors have a thermal switch that will turn off the motor to prevent overheating when this stalling occurs.)

49. The rotating loop in an AC generator is a square 10.0 cm on each side. It is rotated at 60.0 Hz in a uniform field of 0.800 T. Calculate (a) the flux through the loop as a function of time, (b) the emf induced in the loop, (c) the current induced in the loop for a loop resistance of 1.00 Ω, (d) the power delivered to the loop, and (e) the torque that must be exerted to rotate the loop.

Section 31.6 Eddy Currents

50. Figure P31.50 represents an electromagnetic brake that uses eddy currents. An electromagnet hangs from a railroad car near one rail. To stop the car, a large current is sent through the coils of the electromagnet. The moving electromagnet induces eddy currents in the rails, whose fields oppose the change in the electromagnet's field. The magnetic fields of the eddy currents exert force on the current in the electromagnet, thereby slowing the car. The direction of the car's motion and the direction of the current in the electromagnet are shown correctly in the picture. Determine which of the eddy currents shown on the rails is correct. Explain your answer.

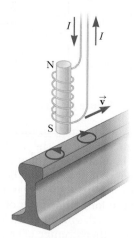

Figure P31.50

Additional Problems

51. Consider a transcranial magnetic stimulation (TMS) device (Figure P31.3) containing a coil with several turns of wire, each of radius 6.00 cm. In a circular area of the brain of radius 6.00 cm directly below and coaxial with the coil, the magnetic field changes at the rate of 1.00×10^4 T/s. Assume that this rate of change is the same everywhere inside the circular area. (a) What is the emf induced around the circumference of this circular area in the brain? (b) What electric field is induced on the circumference of this circular area?

52. Suppose you wrap wire onto the core from a roll of cellophane tape to make a coil. Describe how you can use a bar magnet to produce an induced voltage in the coil. What is the order of magnitude of the emf you generate? State the quantities you take as data and their values.

53. A circular coil enclosing an area of 100 cm² is made of 200 turns of copper wire (Figure P31.53). The wire making up the coil has no resistance; the ends of the wire are connected across a 5.00-Ω resistor to form a closed circuit. Initially, a 1.10-T uniform magnetic field points perpendicularly upward through the plane of the coil. The direction of the field then reverses so that the final magnetic field has a magnitude of 1.10 T and points downward through the coil. If the time interval required for the field to reverse directions is 0.100 s, what is the average current in the coil during that time?

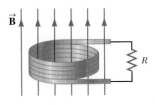

Figure P31.53

54. A circular loop of wire of resistance $R = 0.500 \ \Omega$ and radius $r = 8.00$ cm is in a uniform magnetic field directed out of the page as in Figure P31.54. If a clockwise current of $I = 2.50$ mA is induced in the loop, (a) is the magnetic field increasing or decreasing in time? (b) Find the rate at which the field is changing with time.

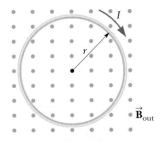

Figure P31.54

55. A rectangular loop of area $A = 0.160$ m² is placed in a region where the magnetic field is perpendicular to the plane of the loop. The magnitude of the field is allowed to vary in time according to $B = 0.350 \ e^{-t/2.00}$, where B is in teslas and t is in seconds. The field has the constant value 0.350 T for $t < 0$. What is the value for $\mathcal{E}$ at $t = 4.00$ s?

56. A rectangular loop of area A is placed in a region where the magnetic field is perpendicular to the plane of the loop. The magnitude of the field is allowed to vary in time according to $B = B_{max}e^{-t/\tau}$, where B_{max} and τ are constants. The field has the constant value B_{max} for $t < 0$. Find the emf induced in the loop as a function of time.

57. Strong magnetic fields are used in such medical procedures as magnetic resonance imaging, or MRI. A technician wearing a brass bracelet enclosing area 0.005 00 m²

places her hand in a solenoid whose magnetic field is 5.00 T directed perpendicular to the plane of the bracelet. The electrical resistance around the bracelet's circumference is 0.020 0 Ω. An unexpected power failure causes the field to drop to 1.50 T in a time interval of 20.0 ms. Find (a) the current induced in the bracelet and (b) the power delivered to the bracelet. *Note:* As this problem implies, you should not wear any metal objects when working in regions of strong magnetic fields.

58. Consider the apparatus shown in Figure P31.58 in which a conducting bar can be moved along two rails connected to a lightbulb. The whole system is immersed in a magnetic field of magnitude $B = 0.400$ T perpendicular and into the page. The distance between the horizontal rails is $\ell = 0.800$ m. The resistance of the lightbulb is $R = 48.0 \ \Omega$, assumed to be constant. The bar and rails have negligible resistance. The bar is moved toward the right by a constant force of magnitude $F = 0.600$ N. We wish to find the maximum power delivered to the lightbulb. (a) Find an expression for the current in the lightbulb as a function of B, ℓ, R, and v, the speed of the bar. (b) When the maximum power is delivered to the lightbulb, what analysis model properly describes the moving bar? (c) Use the analysis model in part (b) to find a numerical value for the speed v of the bar when the maximum power is being delivered to the lightbulb. (d) Find the current in the lightbulb when maximum power is being delivered to it. (e) Using $P = I^2R$, what is the maximum power delivered to the lightbulb? (f) What is the maximum mechanical input power delivered to the bar by the force F? (g) We have assumed the resistance of the lightbulb is constant. In reality, as the power delivered to the lightbulb increases, the filament temperature increases and the resistance increases. Does the speed found in part (c) change if the resistance increases and all other quantities are held constant? (h) If so, does the speed found in part (c) increase or decrease? If not, explain. (i) With the assumption that the resistance of the lightbulb increases as the current increases, does the power found in part (f) change? (j) If so, is the power found in part (f) larger or smaller? If not, explain.

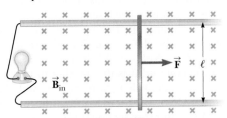

Figure P31.58

59. A guitar's steel string vibrates (see Fig. 31.5). The component of magnetic field perpendicular to the area of a pickup coil nearby is given by

$$B = 50.0 + 3.20 \sin 1 \ 046\pi t$$

where B is in milliteslas and t is in seconds. The circular pickup coil has 30 turns and radius 2.70 mm. Find the emf induced in the coil as a function of time.

60. *Why is the following situation impossible?* A conducting
AMT rectangular loop of mass $M = 0.100$ kg, resistance $R =$
$1.00\ \Omega$, and dimensions $w = 50.0$ cm by $\ell = 90.0$ cm is
held with its lower edge just above a region with a uniform magnetic field of magnitude $B = 1.00$ T as shown
in Figure P31.60. The loop is released from rest. Just as
the top edge of the loop reaches the region containing
the field, the loop moves with a speed 4.00 m/s.

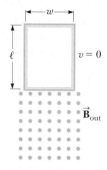

Figure P31.60

61. The circuit in Figure P31.61 is located in a magnetic
W field whose magnitude varies with time according to
the expression $B = 1.00 \times 10^{-3}\ t$, where B is in teslas
and t is in seconds. Assume the resistance per length of
the wire is $0.100\ \Omega/\text{m}$. Find the current in section PQ
of length $a = 65.0$ cm.

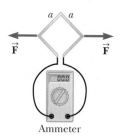

Figure P31.61

62. Magnetic field values are often determined by using a
device known as a *search coil.* This technique depends
on the measurement of the total charge passing
through a coil in a time interval during which the magnetic flux linking the windings changes either because
of the coil's motion or because of a change in the value
of B. (a) Show that as the flux through the coil changes
from Φ_1 to Φ_2, the charge transferred through the coil
is given by $Q = N(\Phi_2 - \Phi_1)/R$, where R is the resistance of the coil and N is the number of turns. (b) As
a specific example, calculate B when a total charge of
5.00×10^{-4} C passes through a 100-turn coil of resistance $200\ \Omega$ and cross-sectional area $40.0\ \text{cm}^2$ as it is
rotated in a uniform field from a position where the
plane of the coil is perpendicular to the field to a position where it is parallel to the field.

63. A conducting rod of length $\ell = 35.0$ cm is free to slide
on two parallel conducting bars as shown in Figure
P31.63. Two resistors $R_1 = 2.00\ \Omega$ and $R_2 = 5.00\ \Omega$ are
connected across the ends of the bars to form a loop.
A constant magnetic field $B = 2.50$ T is directed perpendicularly into the page. An external agent pulls the
rod to the left with a constant speed of $v = 8.00$ m/s.

Find (a) the currents in both resistors, (b) the total
power delivered to the resistance of the circuit, and
(c) the magnitude of the applied force that is needed
to move the rod with this constant velocity.

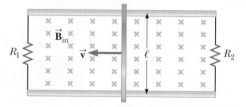

Figure P31.63

64. Review. A particle with a mass of 2.00×10^{-16} kg and
a charge of 30.0 nC starts from rest, is accelerated
through a potential difference ΔV, and is fired from
a small source in a region containing a uniform, constant magnetic field of magnitude 0.600 T. The particle's velocity is perpendicular to the magnetic field
lines. The circular orbit of the particle as it returns to
the location of the source encloses a magnetic flux of
$15.0\ \mu\text{Wb}$. (a) Calculate the particle's speed. (b) Calculate the potential difference through which the particle was accelerated inside the source.

65. The plane of a square loop of wire with edge length
M $a = 0.200$ m is oriented vertically and along an east–
west axis. The Earth's magnetic field at this point is of
magnitude $B = 35.0\ \mu\text{T}$ and is directed northward at
35.0° below the horizontal. The total resistance of the
loop and the wires connecting it to a sensitive ammeter
is $0.500\ \Omega$. If the loop is suddenly collapsed by horizontal forces as shown in Figure P31.65, what total charge
enters one terminal of the ammeter?

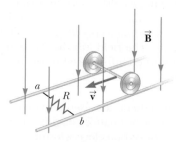

Figure P31.65

66. In Figure P31.66, the rolling axle, 1.50 m long, is
pushed along horizontal rails at a constant speed $v =$
3.00 m/s. A resistor $R = 0.400\ \Omega$ is connected to the
rails at points a and b, directly opposite each other.

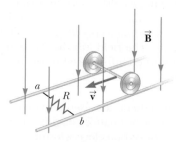

Figure P31.66

The wheels make good electrical contact with the rails, so the axle, rails, and R form a closed-loop circuit. The only significant resistance in the circuit is R. A uniform magnetic field $B = 0.080\ 0$ T is vertically downward. (a) Find the induced current I in the resistor. (b) What horizontal force F is required to keep the axle rolling at constant speed? (c) Which end of the resistor, a or b, is at the higher electric potential? (d) **What If?** After the axle rolls past the resistor, does the current in R reverse direction? Explain your answer.

67. Figure P31.67 shows a stationary conductor whose shape is similar to the letter e. The radius of its circular portion is $a = 50.0$ cm. It is placed in a constant magnetic field of 0.500 T directed out of the page. A straight conducting rod, 50.0 cm long, is pivoted about point O and rotates with a constant angular speed of 2.00 rad/s. (a) Determine the induced emf in the loop POQ. Note that the area of the loop is $\theta a^2/2$. (b) If all the conducting material has a resistance per length of 5.00 Ω/m, what is the induced current in the loop POQ at the instant 0.250 s after point P passes point Q?

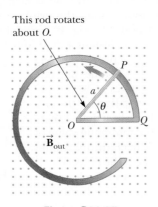

This rod rotates about O.

Figure P31.67

68. A conducting rod moves with a constant velocity in a direction perpendicular to a long, straight wire carrying a current I as shown in Figure P31.68. Show that the magnitude of the emf generated between the ends of the rod is

$$|\mathcal{E}| = \frac{\mu_0 v I \ell}{2\pi r}$$

In this case, note that the emf decreases with increasing r as you might expect.

Figure P31.68

69. A small, circular washer of radius $a = 0.500$ cm is held
AMT directly below a long, straight wire carrying a current of $I = 10.0$ A. The washer is located $h = 0.500$ m above the top of a table (Fig. P31.69). Assume the magnetic field is nearly constant over the area of the washer and equal to the magnetic field at the center of the washer. (a) If the washer is dropped from rest, what is the magnitude of the average induced emf in the washer over the time interval between its release and the moment it hits the tabletop? (b) What is the direction of the induced current in the washer?

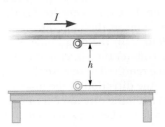

Figure P31.69

70. Figure P31.70 shows a compact, circular coil with 220 turns and radius 12.0 cm immersed in a uniform magnetic field parallel to the axis of the coil. The rate of change of the field has the constant magnitude 20.0 mT/s. (a) What additional information is necessary to determine whether the coil is carrying clockwise or counterclockwise current? (b) The coil overheats if more than 160 W of power is delivered to it. What resistance would the coil have at this critical point? (c) To run cooler, should it have lower resistance or higher resistance?

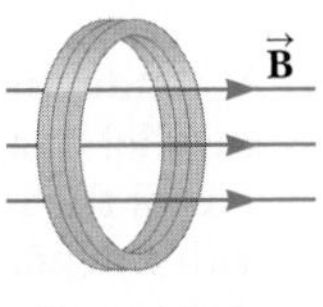

Figure P31.70

71. A rectangular coil of 60 turns, dimensions 0.100 m by 0.200 m, and total resistance 10.0 Ω rotates with angular speed 30.0 rad/s about the y axis in a region where a 1.00-T magnetic field is directed along the x axis. The time $t = 0$ is chosen to be at an instant when the plane of the coil is perpendicular to the direction of $\vec{B}$. Calculate (a) the maximum induced emf in the coil, (b) the maximum rate of change of magnetic flux through the coil, (c) the induced emf at $t = 0.050\ 0$ s, and (d) the torque exerted by the magnetic field on the coil at the instant when the emf is a maximum.

72. **Review.** In Figure P31.72, a uniform magnetic field decreases at a constant rate $dB/dt = -K$, where K is a positive constant. A circular loop of wire of radius a containing a resistance R and a capacitance C is placed

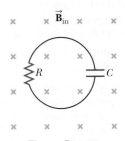

Figure P31.72

with its plane normal to the field. (a) Find the charge Q on the capacitor when it is fully charged. (b) Which plate, upper or lower, is at the higher potential? (c) Discuss the force that causes the separation of charges.

73. An N-turn square coil with side ℓ and resistance R is pulled to the right at constant speed v in the presence of a uniform magnetic field B acting perpendicular to the coil as shown in Figure P31.73. At $t = 0$, the right side of the coil has just departed the right edge of the field. At time t, the left side of the coil enters the region where $B = 0$. In terms of the quantities N, B, ℓ, v, and R, find symbolic expressions for (a) the magnitude of the induced emf in the loop during the time interval from $t = 0$ to t, (b) the magnitude of the induced current in the coil, (c) the power delivered to the coil, and (d) the force required to remove the coil from the field. (e) What is the direction of the induced current in the loop? (f) What is the direction of the magnetic force on the loop while it is being pulled out of the field?

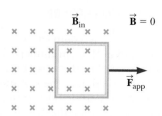

Figure P31.73

74. A conducting rod of length ℓ moves with velocity $\vec{v}$ parallel to a long wire carrying a steady current I. The axis of the rod is maintained perpendicular to the wire with the near end a distance r away (Fig. P31.74). Show that the magnitude of the emf induced in the rod is

$$|\mathcal{E}| = \frac{\mu_0 I v}{2\pi} \ln\left(1 + \frac{\ell}{r}\right)$$

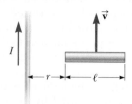

Figure P31.74

75. The magnetic flux through a metal ring varies with time t according to $\Phi_B = at^3 - bt^2$, where Φ_B is in webers, $a = 6.00$ s^{-3}, $b = 18.0$ s^{-2}, and t is in seconds. The resistance of the ring is 3.00 Ω. For the interval from $t = 0$ to $t = 2.00$ s, determine the maximum current induced in the ring.

76. A rectangular loop of dimensions ℓ and w moves with a constant velocity $\vec{v}$ away from a long wire that carries a cur-

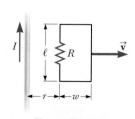

Figure P31.76

rent I in the plane of the loop (Fig. P31.76). The total resistance of the loop is R. Derive an expression that gives the current in the loop at the instant the near side is a distance r from the wire.

77. A long, straight wire carries a current given by $I = I_{max} \sin(\omega t + \phi)$. The wire lies in the plane of a rectangular coil of N turns of wire as shown in Figure P31.77. The quantities I_{max}, ω, and ϕ are all constants. Assume $I_{max} = 50.0$ A, $\omega = 200\pi$ s^{-1}, $N = 100$, $h = w = 5.00$ cm, and $L = 20.0$ cm. Determine the emf induced in the coil by the magnetic field created by the current in the straight wire.

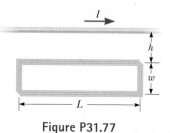

Figure P31.77

78. A thin wire $\ell = 30.0$ cm long is held parallel to and $d = 80.0$ cm above a long, thin wire carrying $I = 200$ A and fixed in position (Fig. P31.78). The 30.0-cm wire is released at the instant $t = 0$ and falls, remaining parallel to the current-carrying wire as it falls. Assume the falling wire accelerates at 9.80 m/s^2. (a) Derive an equation for the emf induced in it as a function of time. (b) What is the minimum value of the emf? (c) What is the maximum value? (d) What is the induced emf 0.300 s after the wire is released?

Figure P31.78

Challenge Problems

79. Two infinitely long solenoids (seen in cross section) pass through a circuit as shown in Figure P31.79. The magnitude of $\vec{B}$ inside each is the same and is increasing at the rate of 100 T/s. What is the current in each resistor?

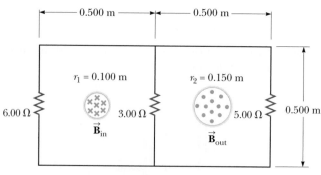

Figure P31.79

80. An *induction furnace* uses electromagnetic induction to produce eddy currents in a conductor, thereby raising the conductor's temperature. Commercial units

operate at frequencies ranging from 60 Hz to about 1 MHz and deliver powers from a few watts to several megawatts. Induction heating can be used for warming a metal pan on a kitchen stove. It can be used to avoid oxidation and contamination of the metal when welding in a vacuum enclosure. To explore induction heating, consider a flat conducting disk of radius R, thickness b, and resistivity ρ. A sinusoidal magnetic field $B_{max} \cos \omega t$ is applied perpendicular to the disk. Assume the eddy currents occur in circles concentric with the disk. (a) Calculate the average power delivered to the disk. (b) **What If?** By what factor does the power change when the amplitude of the field doubles? (c) When the frequency doubles? (d) When the radius of the disk doubles?

81. A bar of mass m and resistance R slides without friction in a horizontal plane, moving on parallel rails as shown in Figure P31.81. The rails are separated by a distance d. A battery that maintains a constant emf ε is connected between the rails, and a constant magnetic field $\vec{B}$ is directed perpendicularly out of the page. Assuming the bar starts from rest at time $t = 0$, show that at time t it moves with a speed

$$v = \frac{\varepsilon}{Bd}\left(1 - e^{-B^2 d^2 t / mR}\right)$$

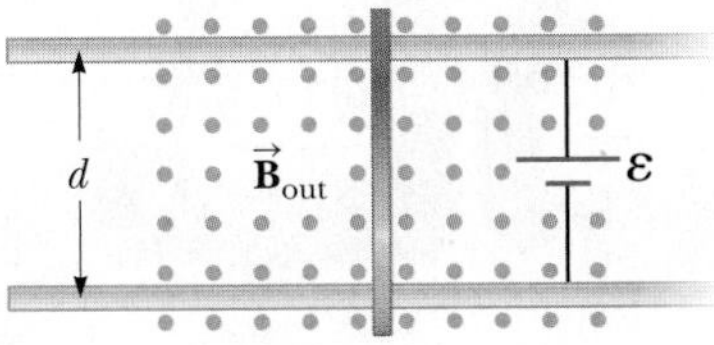

Figure P31.81

82. A *betatron* is a device that accelerates electrons to energies in the MeV range by means of electromagnetic induction. Electrons in a vacuum chamber are held in a circular orbit by a magnetic field perpendicular to the orbital plane. The magnetic field is gradually increased to induce an electric field around the orbit. (a) Show that the electric field is in the correct direction to make the electrons speed up. (b) Assume the radius of the orbit remains constant. Show that the average magnetic field over the area enclosed by the orbit must be twice as large as the magnetic field at the circle's circumference.

83. **Review.** The bar of mass m in Figure P31.83 is pulled horizontally across parallel, frictionless rails by a massless string that passes over a light, frictionless pulley and is attached to a suspended object of mass M. The uniform upward magnetic field has a magnitude B, and the distance between the rails is ℓ. The only significant electrical resistance is the load resistor R shown connecting the rails at one end. Assuming the suspended object is released with the bar at rest at $t = 0$, derive an expression that gives the bar's horizontal speed as a function of time.

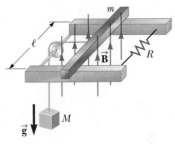

Figure P31.83

CHAPTER
32

Inductance

A treasure hunter uses a metal detector to search for buried objects at a beach. At the end of the metal detector is a coil of wire that is part of a circuit. When the coil comes near a metal object, the inductance of the coil is affected and the current in the circuit changes. This change triggers a signal in the earphones worn by the treasure hunter. We investigate inductance in this chapter. *(Andy Ryan/Stone/Getty Images)*

In Chapter 31, we saw that an emf and a current are induced in a loop of wire when the magnetic flux through the area enclosed by the loop changes with time. This phenomenon of electromagnetic induction has some practical consequences. In this chapter, we first describe an effect known as *self-induction*, in which a time-varying current in a circuit produces an induced emf opposing the emf that initially set up the time-varying current. Self-induction is the basis of the *inductor*, an electrical circuit element. We discuss the energy stored in the magnetic field of an inductor and the energy density associated with the magnetic field.

Next, we study how an emf is induced in a coil as a result of a changing magnetic flux produced by a second coil, which is the basic principle of *mutual induction*. Finally, we examine the characteristics of circuits that contain inductors, resistors, and capacitors in various combinations.

32.1 Self-Induction and Inductance

In this chapter, we need to distinguish carefully between emfs and currents that are caused by physical sources such as batteries and those that are induced by changing magnetic fields. When we use a term (such as *emf* or *current*) without an adjective, we are describing the parameters associated with a physical source. We use the adjective *induced* to describe those emfs and currents caused by a changing magnetic field.

Consider a circuit consisting of a switch, a resistor, and a source of emf as shown in Figure 32.1. The circuit diagram is represented in perspective to show the orientations of some of the magnetic field lines due to the current in the circuit. When the switch is thrown to its closed position, the current does not immediately jump from zero to its maximum value $\mathcal{E}/R$. Faraday's law of electromagnetic induction (Eq. 31.1) can be used to describe this effect as follows. As the current increases with time, the magnetic field lines surrounding the wires pass through the loop represented by the circuit itself. This magnetic field passing through the loop causes a magnetic flux through the loop. This increasing flux creates an induced emf in the circuit. The direction of the induced emf is such that it would cause an induced current in the loop (if the loop did not already carry a current), which would establish a magnetic field opposing the change in the original magnetic field. Therefore, the direction of the induced emf is opposite the direction of the emf of the battery, which results in a gradual rather than instantaneous increase in the current to its final equilibrium value. Because of the direction of the induced emf, it is also called a *back emf*, similar to that in a motor as discussed in Chapter 31. This effect is called **self-induction** because the changing flux through the circuit and the resultant induced emf arise from the circuit itself. The emf $\mathcal{E}_L$ set up in this case is called a **self-induced emf.**

To obtain a quantitative description of self-induction, recall from Faraday's law that the induced emf is equal to the negative of the time rate of change of the magnetic flux. The magnetic flux is proportional to the magnetic field, which in turn is proportional to the current in the circuit. Therefore, a self-induced emf is always proportional to the time rate of change of the current. For any loop of wire, we can write this proportionality as

$$\mathcal{E}_L = -L\frac{di}{dt} \qquad \text{(32.1)}$$

where L is a proportionality constant—called the **inductance** of the loop—that depends on the geometry of the loop and other physical characteristics. If we consider a closely spaced coil of N turns (a toroid or an ideal solenoid) carrying a current i and containing N turns, Faraday's law tells us that $\mathcal{E}_L = -N\,d\Phi_B/dt$. Combining this expression with Equation 32.1 gives

$$L = \frac{N\Phi_B}{i} \qquad \text{(32.2)}$$

◀ Inductance of an *N*-turn coil

where it is assumed the same magnetic flux passes through each turn and L is the inductance of the entire coil.

From Equation 32.1, we can also write the inductance as the ratio

$$L = -\frac{\mathcal{E}_L}{di/dt} \qquad \text{(32.3)}$$

Recall that resistance is a measure of the opposition to current as given by Equation 27.7, $R = \Delta V/I$; in comparison, Equation 32.3, being of the same mathematical form as Equation 27.7, shows us that inductance is a measure of the opposition to a *change* in current.

The SI unit of inductance is the **henry** (H), which as we can see from Equation 32.3 is 1 volt-second per ampere: $1\text{ H} = 1\text{ V} \cdot \text{s/A}$.

As shown in Example 32.1, the inductance of a coil depends on its geometry. This dependence is analogous to the capacitance of a capacitor depending on the geometry of its plates as we found in Equation 26.3 and the resistance of a resistor depending on the length and area of the conducting material in Equation 27.10. Inductance calculations can be quite difficult to perform for complicated geometries, but the examples below involve simple situations for which inductances are easily evaluated.

Joseph Henry
American Physicist (1797–1878)
Henry became the first director of the Smithsonian Institution and first president of the Academy of Natural Science. He improved the design of the electromagnet and constructed one of the first motors. He also discovered the phenomenon of self-induction, but he failed to publish his findings. The unit of inductance, the henry, is named in his honor.

After the switch is closed, the current produces a magnetic flux through the area enclosed by the loop. As the current increases toward its equilibrium value, this magnetic flux changes in time and induces an emf in the loop.

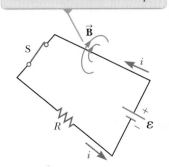

Figure 32.1 Self-induction in a simple circuit.

Quick Quiz 32.1 A coil with zero resistance has its ends labeled a and b. The potential at a is higher than at b. Which of the following could be consistent with this situation? **(a)** The current is constant and is directed from a to b. **(b)** The current is constant and is directed from b to a. **(c)** The current is increasing and is directed from a to b. **(d)** The current is decreasing and is directed from a to b. **(e)** The current is increasing and is directed from b to a. **(f)** The current is decreasing and is directed from b to a.

Example 32.1   Inductance of a Solenoid

Consider a uniformly wound solenoid having N turns and length ℓ. Assume ℓ is much longer than the radius of the windings and the core of the solenoid is air.

(A) Find the inductance of the solenoid.

SOLUTION

Conceptualize The magnetic field lines from each turn of the solenoid pass through all the turns, so an induced emf in each coil opposes changes in the current.

Categorize We categorize this example as a substitution problem. Because the solenoid is long, we can use the results for an ideal solenoid obtained in Chapter 30.

Find the magnetic flux through each turn of area A in the solenoid, using the expression for the magnetic field from Equation 30.17:

$$\Phi_B = BA = \mu_0 niA = \mu_0 \frac{N}{\ell} iA$$

Substitute this expression into Equation 32.2:

$$L = \frac{N\Phi_B}{i} = \mu_0 \frac{N^2}{\ell} A \qquad \textbf{(32.4)}$$

(B) Calculate the inductance of the solenoid if it contains 300 turns, its length is 25.0 cm, and its cross-sectional area is 4.00 cm².

SOLUTION

Substitute numerical values into Equation 32.4:

$$L = (4\pi \times 10^{-7}\,\text{T} \cdot \text{m/A}) \frac{300^2}{25.0 \times 10^{-2}\,\text{m}} (4.00 \times 10^{-4}\,\text{m}^2)$$

$$= 1.81 \times 10^{-4}\,\text{T} \cdot \text{m}^2/\text{A} = \boxed{0.181\,\text{mH}}$$

(C) Calculate the self-induced emf in the solenoid if the current it carries decreases at the rate of 50.0 A/s.

SOLUTION

Substitute $di/dt = -50.0$ A/s and the answer to part (B) into Equation 32.1:

$$\mathcal{E}_L = -L\frac{di}{dt} = -(1.81 \times 10^{-4}\,\text{H})(-50.0\,\text{A/s})$$

$$= \boxed{9.05\,\text{mV}}$$

The result for part (A) shows that L depends on geometry and is proportional to the square of the number of turns. Because $N = n\ell$, we can also express the result in the form

$$L = \mu_0 \frac{(n\ell)^2}{\ell} A = \mu_0 n^2 A\ell = \mu_0 n^2 V \qquad \textbf{(32.5)}$$

where $V = A\ell$ is the interior volume of the solenoid.

32.2 RL Circuits

If a circuit contains a coil such as a solenoid, the inductance of the coil prevents the current in the circuit from increasing or decreasing instantaneously. A circuit element that has a large inductance is called an **inductor** and has the circuit symbol —000—. We always assume the inductance of the remainder of a circuit is negligible compared with that of the inductor. Keep in mind, however, that even a circuit without a coil has some inductance that can affect the circuit's behavior.

Because the inductance of an inductor results in a back emf, an inductor in a circuit opposes changes in the current in that circuit. The inductor attempts to keep the current the same as it was before the change occurred. If the battery voltage in the circuit is increased so that the current rises, the inductor opposes this change and the rise is not instantaneous. If the battery voltage is decreased, the inductor causes a slow drop in the current rather than an immediate drop. Therefore, the inductor causes the circuit to be "sluggish" as it reacts to changes in the voltage.

Consider the circuit shown in Figure 32.2, which contains a battery of negligible internal resistance. This circuit is an ***RL* circuit** because the elements connected to the battery are a resistor and an inductor. The curved lines on switch S_2 suggest this switch can never be open; it is always set to either *a* or *b*. (If the switch is connected to neither *a* nor *b*, any current in the circuit suddenly stops.) Suppose S_2 is set to *a* and switch S_1 is open for $t < 0$ and then thrown closed at $t = 0$. The current in the circuit begins to increase, and a back emf (Eq. 32.1) that opposes the increasing current is induced in the inductor.

With this point in mind, let's apply Kirchhoff's loop rule to this circuit, traversing the circuit in the clockwise direction:

$$\mathcal{E} - iR - L\frac{di}{dt} = 0 \qquad (32.6)$$

where iR is the voltage drop across the resistor. (Kirchhoff's rules were developed for circuits with steady currents, but they can also be applied to a circuit in which the current is changing if we imagine them to represent the circuit at one *instant* of time.) Now let's find a solution to this differential equation, which is similar to that for the *RC* circuit (see Section 28.4).

A mathematical solution of Equation 32.6 represents the current in the circuit as a function of time. To find this solution, we change variables for convenience, letting $x = (\mathcal{E}/R) - i$, so $dx = -di$. With these substitutions, Equation 32.6 becomes

$$x + \frac{L}{R}\frac{dx}{dt} = 0$$

Rearranging and integrating this last expression gives

$$\int_{x_0}^{x}\frac{dx}{x} = -\frac{R}{L}\int_{0}^{t} dt$$

$$\ln\frac{x}{x_0} = -\frac{R}{L}t$$

where x_0 is the value of *x* at time $t = 0$. Taking the antilogarithm of this result gives

$$x = x_0 e^{-Rt/L}$$

Because $i = 0$ at $t = 0$, note from the definition of *x* that $x_0 = \mathcal{E}/R$. Hence, this last expression is equivalent to

$$\frac{\mathcal{E}}{R} - i = \frac{\mathcal{E}}{R}e^{-Rt/L}$$

$$i = \frac{\mathcal{E}}{R}(1 - e^{-Rt/L})$$

This expression shows how the inductor affects the current. The current does not increase instantly to its final equilibrium value when the switch is closed, but instead increases according to an exponential function. If the inductance is removed from the circuit, which corresponds to letting *L* approach zero, the exponential term

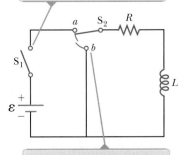

When switch S_1 is thrown closed, the current increases and an emf that opposes the increasing current is induced in the inductor.

When the switch S_2 is thrown to position *b*, the battery is no longer part of the circuit and the current decreases.

Figure 32.2 An *RL* circuit. When switch S_2 is in position *a*, the battery is in the circuit.

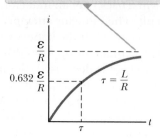

After switch S_1 is thrown closed at $t = 0$, the current increases toward its maximum value ε/R.

Figure 32.3 Plot of the current versus time for the *RL* circuit shown in Figure 32.2. The time constant τ is the time interval required for i to reach 63.2% of its maximum value.

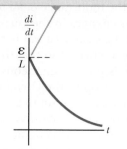

The time rate of change of current is a maximum at $t = 0$, which is the instant at which switch S_1 is thrown closed.

Figure 32.4 Plot of di/dt versus time for the *RL* circuit shown in Figure 32.2. The rate decreases exponentially with time as i increases toward its maximum value.

becomes zero and there is no time dependence of the current in this case; the current increases instantaneously to its final equilibrium value in the absence of the inductance.

We can also write this expression as

$$i = \frac{\varepsilon}{R}(1 - e^{-t/\tau})$$ (32.7)

where the constant τ is the **time constant** of the *RL* circuit:

$$\tau = \frac{L}{R}$$ (32.8)

Physically, τ is the time interval required for the current in the circuit to reach $(1 - e^{-1}) = 0.632 = 63.2\%$ of its final value ε/R. The time constant is a useful parameter for comparing the time responses of various circuits.

Figure 32.3 shows a graph of the current versus time in the *RL* circuit. Notice that the equilibrium value of the current, which occurs as t approaches infinity, is ε/R. That can be seen by setting di/dt equal to zero in Equation 32.6 and solving for the current i. (At equilibrium, the change in the current is zero.) Therefore, the current initially increases very rapidly and then gradually approaches the equilibrium value ε/R as t approaches infinity.

Let's also investigate the time rate of change of the current. Taking the first time derivative of Equation 32.7 gives

$$\frac{di}{dt} = \frac{\varepsilon}{L}e^{-t/\tau}$$ (32.9)

This result shows that the time rate of change of the current is a maximum (equal to ε/L) at $t = 0$ and falls off exponentially to zero as t approaches infinity (Fig. 32.4).

Now consider the *RL* circuit in Figure 32.2 again. Suppose switch S_2 has been set at position *a* long enough (and switch S_1 remains closed) to allow the current to reach its equilibrium value ε/R. In this situation, the circuit is described by the outer loop in Figure 32.2. If S_2 is thrown from *a* to *b*, the circuit is now described by only the right-hand loop in Figure 32.2. Therefore, the battery has been eliminated from the circuit. Setting $\varepsilon = 0$ in Equation 32.6 gives

$$iR + L\frac{di}{dt} = 0$$

It is left as a problem (Problem 22) to show that the solution of this differential equation is

$$i = \frac{\mathcal{E}}{R} e^{-t/\tau} = I_i e^{-t/\tau} \tag{32.10}$$

where $\mathcal{E}$ is the emf of the battery and $I_i = \mathcal{E}/R$ is the initial current at the instant the switch is thrown to b.

If the circuit did not contain an inductor, the current would immediately decrease to zero when the battery is removed. When the inductor is present, it opposes the decrease in the current and causes the current to decrease exponentially. A graph of the current in the circuit versus time (Fig. 32.5) shows that the current is continuously decreasing with time.

Quick Quiz 32.2 Consider the circuit in Figure 32.2 with S_1 open and S_2 at position a. Switch S_1 is now thrown closed. **(i)** At the instant it is closed, across which circuit element is the voltage equal to the emf of the battery? (a) the resistor (b) the inductor (c) both the inductor and resistor **(ii)** After a very long time, across which circuit element is the voltage equal to the emf of the battery? Choose from among the same answers.

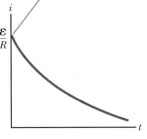

At $t = 0$, the switch is thrown to position b and the current has its maximum value $\mathcal{E}/R$.

Figure 32.5 Current versus time for the right-hand loop of the circuit shown in Figure 32.2. For $t < 0$, switch S_2 is at position a.

Example 32.2 **Time Constant of an *RL* Circuit**

Consider the circuit in Figure 32.2 again. Suppose the circuit elements have the following values: $\mathcal{E} = 12.0$ V, $R = 6.00\ \Omega$, and $L = 30.0$ mH.

(A) Find the time constant of the circuit.

SOLUTION

Conceptualize You should understand the operation and behavior of the circuit in Figure 32.2 from the discussion in this section.

Categorize We evaluate the results using equations developed in this section, so this example is a substitution problem.

Evaluate the time constant from Equation 32.8:

$$\tau = \frac{L}{R} = \frac{30.0 \times 10^{-3}\ \text{H}}{6.00\ \Omega} = 5.00\ \text{ms}$$

(B) Switch S_2 is at position a, and switch S_1 is thrown closed at $t = 0$. Calculate the current in the circuit at $t = 2.00$ ms.

SOLUTION

Evaluate the current at $t = 2.00$ ms from Equation 32.7:

$$i = \frac{\mathcal{E}}{R}\left(1 - e^{-t/\tau}\right) = \frac{12.0\ \text{V}}{6.00\ \Omega}\left(1 - e^{-2.00\ \text{ms}/5.00\ \text{ms}}\right) = 2.00\ \text{A}\left(1 - e^{-0.400}\right)$$
$$= 0.659\ \text{A}$$

(C) Compare the potential difference across the resistor with that across the inductor.

SOLUTION

At the instant the switch is closed, there is no current and therefore no potential difference across the resistor. At this instant, the battery voltage appears entirely across the inductor in the form of a back emf of 12.0 V as the inductor tries to maintain the zero-current condition. (The top end of the inductor in Fig. 32.2 is at a higher electric potential than the bottom end.) As time passes, the emf across the inductor decreases and the current in the resistor (and hence the voltage across it) increases as shown in Figure 32.6 (page 976). The sum of the two voltages at all times is 12.0 V.

WHAT IF? In Figure 32.6, the voltages across the resistor and inductor are equal at 3.4 ms. What if you wanted to delay the condition in which the voltages are equal to some later instant, such as $t = 10.0$ ms? Which parameter, L or R, would require the least adjustment, in terms of a percentage change, to achieve that?

continued

▶ **32.2** continued

Answer Figure 32.6 shows that the voltages are equal when the voltage across the inductor has fallen to half its original value. Therefore, the time interval required for the voltages to become equal is the *half-life* $t_{1/2}$ of the decay. We introduced the half-life in the What If? section of Example 28.10 to describe the exponential decay in *RC* circuits, where $t_{1/2} = 0.693\tau$.

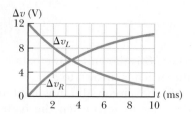

Figure 32.6 (Example 32.2) The time behavior of the voltages across the resistor and inductor in Figure 32.2 given the values provided in this example.

From the desired half-life of 10.0 ms, use the result from Example 28.10 to find the time constant of the circuit:

$$\tau = \frac{t_{1/2}}{0.693} = \frac{10.0 \text{ ms}}{0.693} = 14.4 \text{ ms}$$

Hold *L* fixed and find the value of *R* that gives this time constant:

$$\tau = \frac{L}{R} \rightarrow R = \frac{L}{\tau} = \frac{30.0 \times 10^{-3} \text{ H}}{14.4 \text{ ms}} = 2.08 \ \Omega$$

Now hold *R* fixed and find the appropriate value of *L*:

$$\tau = \frac{L}{R} \rightarrow L = \tau R = (14.4 \text{ ms})(6.00 \ \Omega) = 86.4 \times 10^{-3} \text{ H}$$

The change in *R* corresponds to a 65% decrease compared with the initial resistance. The change in *L* represents a 188% increase in inductance! Therefore, a much smaller percentage adjustment in *R* can achieve the desired effect than would an adjustment in *L*.

32.3 Energy in a Magnetic Field

A battery in a circuit containing an inductor must provide more energy than one in a circuit without the inductor. Consider Figure 32.2 with switch S_2 in position *a*. When switch S_1 is thrown closed, part of the energy supplied by the battery appears as internal energy in the resistance in the circuit, and the remaining energy is stored in the magnetic field of the inductor. Multiplying each term in Equation 32.6 by *i* and rearranging the expression gives

$$i\mathcal{E} = i^2 R + Li\frac{di}{dt} \tag{32.11}$$

Recognizing $i\mathcal{E}$ as the rate at which energy is supplied by the battery and $i^2 R$ as the rate at which energy is delivered to the resistor, we see that $Li(di/dt)$ must represent the rate at which energy is being stored in the inductor. If U_B is the energy stored in the inductor at any time, we can write the rate dU_B/dt at which energy is stored as

$$\frac{dU_B}{dt} = Li\frac{di}{dt}$$

To find the total energy stored in the inductor at any instant, let's rewrite this expression as $dU_B = Li \, di$ and integrate:

$$U_B = \int dU_B = \int_0^i Li \, di = L\int_0^i i \, di$$

◀ **Energy stored in an inductor**

$$U_B = \tfrac{1}{2}Li^2 \tag{32.12}$$

where *L* is constant and has been removed from the integral. Equation 32.12 represents the energy stored in the magnetic field of the inductor when the current is *i*. It is similar in form to Equation 26.11 for the energy stored in the electric field of a capacitor, $U_E = \tfrac{1}{2}C(\Delta V)^2$. In either case, energy is required to establish a field.

We can also determine the energy density of a magnetic field. For simplicity, consider a solenoid whose inductance is given by Equation 32.5:

$$L = \mu_0 n^2 V$$

The magnetic field of a solenoid is given by Equation 30.17:

$$B = \mu_0 n i$$

Substituting the expression for L and $i = B/\mu_0 n$ into Equation 32.12 gives

$$U_B = \tfrac{1}{2} L i^2 = \tfrac{1}{2} \mu_0 n^2 V \left(\frac{B}{\mu_0 n} \right)^2 = \frac{B^2}{2\mu_0} V \qquad \textbf{(32.13)}$$

The magnetic energy density, or the energy stored per unit volume in the magnetic field of the inductor, is $u_B = U_B / V$, or

$$u_B = \frac{B^2}{2\mu_0} \qquad \textbf{(32.14)} \qquad \blacktriangleleft \textbf{ Magnetic energy density}$$

Although this expression was derived for the special case of a solenoid, it is valid for any region of space in which a magnetic field exists. Equation 32.14 is similar in form to Equation 26.13 for the energy per unit volume stored in an electric field, $u_E = \tfrac{1}{2}\epsilon_0 E^2$. In both cases, the energy density is proportional to the square of the field magnitude.

Quick Quiz 32.3 You are performing an experiment that requires the highest-possible magnetic energy density in the interior of a very long current-carrying solenoid. Which of the following adjustments increases the energy density? (More than one choice may be correct.) **(a)** increasing the number of turns per unit length on the solenoid **(b)** increasing the cross-sectional area of the solenoid **(c)** increasing only the length of the solenoid while keeping the number of turns per unit length fixed **(d)** increasing the current in the solenoid

---

Example 32.3   **What Happens to the Energy in the Inductor?**   **AM**

Consider once again the RL circuit shown in Figure 32.2, with switch S_2 at position a and the current having reached its steady-state value. When S_2 is thrown to position b, the current in the right-hand loop decays exponentially with time according to the expression $i = I_i e^{-t/\tau}$, where $I_i = \mathcal{E}/R$ is the initial current in the circuit and $\tau = L/R$ is the time constant. Show that all the energy initially stored in the magnetic field of the inductor appears as internal energy in the resistor as the current decays to zero.

SOLUTION

Conceptualize Before S_2 is thrown to b, energy is being delivered at a constant rate to the resistor from the battery and energy is stored in the magnetic field of the inductor. After $t = 0$, when S_2 is thrown to b, the battery can no longer provide energy and energy is delivered to the resistor only from the inductor.

Categorize We model the right-hand loop of the circuit as an *isolated system* so that energy is transferred between components of the system but does not leave the system.

Analyze We begin by evaluating the energy delivered to the resistor, which appears as internal energy in the resistor.

Begin with Equation 27.22 and recognize that the rate of change of internal energy in the resistor is the power delivered to the resistor:

$$\frac{dE_{int}}{dt} = P = i^2 R$$

Substitute the current given by Equation 32.10 into this equation:

$$\frac{dE_{int}}{dt} = i^2 R = (I_i e^{-Rt/L})^2 R = I_i^2 R e^{-2Rt/L}$$

Solve for dE_{int} and integrate this expression over the limits $t = 0$ to $t \to \infty$:

$$E_{int} = \int_0^\infty I_i^2 R e^{-2Rt/L} \, dt = I_i^2 R \int_0^\infty e^{-2Rt/L} \, dt$$

The value of the definite integral can be shown to be $L/2R$ (see Problem 36). Use this result to evaluate E_{int}:

$$E_{int} = I_i^2 R \left(\frac{L}{2R} \right) = \tfrac{1}{2} L I_i^2$$

continued

▶ **32.3** continued

Finalize This result is equal to the initial energy stored in the magnetic field of the inductor, given by Equation 32.12, as we set out to prove.

Example 32.4 The Coaxial Cable

Coaxial cables are often used to connect electrical devices, such as your video system, and in receiving signals in television cable systems. Model a long coaxial cable as a thin, cylindrical conducting shell of radius b concentric with a solid cylinder of radius a as in Figure 32.7. The conductors carry the same current I in opposite directions. Calculate the inductance L of a length ℓ of this cable.

SOLUTION

Conceptualize Consider Figure 32.7. Although we do not have a visible coil in this geometry, imagine a thin, radial slice of the coaxial cable such as the light gold rectangle in Figure 32.7. If the inner and outer conductors are connected at the ends of the cable (above and below the figure), this slice represents one large conducting loop. The current in the loop sets up a magnetic field between the inner and outer conductors that passes through this loop. If the current changes, the magnetic field changes and the induced emf opposes the original change in the current in the conductors.

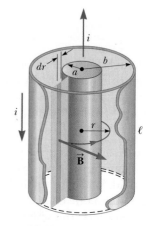

Figure 32.7 (Example 32.4) Section of a long coaxial cable. The inner and outer conductors carry equal currents in opposite directions.

Categorize We categorize this situation as one in which we must return to the fundamental definition of inductance, Equation 32.2.

Analyze We must find the magnetic flux through the light gold rectangle in Figure 32.7. Ampère's law (see Section 30.3) tells us that the magnetic field in the region between the conductors is due to the inner conductor alone and that its magnitude is $B = \mu_0 i/2\pi r$, where r is measured from the common center of the cylinders. A sample circular field line is shown in Figure 32.7, along with a field vector tangent to the field line. The magnetic field is zero outside the outer shell because the net current passing through the area enclosed by a circular path surrounding the cable is zero; hence, from Ampère's law, $\oint \vec{B} \cdot d\vec{s} = 0$.

The magnetic field is perpendicular to the light gold rectangle of length ℓ and width $b - a$, the cross section of interest. Because the magnetic field varies with radial position across this rectangle, we must use calculus to find the total magnetic flux.

Divide the light gold rectangle into strips of width dr such as the darker strip in Figure 32.7. Evaluate the magnetic flux through such a strip:

$$d\Phi_B = B\, dA = B\ell\, dr$$

Substitute for the magnetic field and integrate over the entire light gold rectangle:

$$\Phi_B = \int_a^b \frac{\mu_0 i}{2\pi r}\,\ell\, dr = \frac{\mu_0 i\ell}{2\pi} \int_a^b \frac{dr}{r} = \frac{\mu_0 i\ell}{2\pi} \ln\left(\frac{b}{a}\right)$$

Use Equation 32.2 to find the inductance of the cable:

$$L = \frac{\Phi_B}{i} = \frac{\mu_0 \ell}{2\pi} \ln\left(\frac{b}{a}\right)$$

Finalize The inductance depends only on geometric factors related to the cable. It increases if ℓ increases, if b increases, or if a decreases. This result is consistent with our conceptualization: any of these changes increases the size of the loop represented by our radial slice and through which the magnetic field passes, increasing the inductance.

32.4 Mutual Inductance

Very often, the magnetic flux through the area enclosed by a circuit varies with time because of time-varying currents in nearby circuits. This condition induces an emf through a process known as *mutual induction,* so named because it depends on the interaction of two circuits.

Consider the two closely wound coils of wire shown in cross-sectional view in Figure 32.8. The current i_1 in coil 1, which has N_1 turns, creates a magnetic field. Some of the magnetic field lines pass through coil 2, which has N_2 turns. The magnetic flux caused by the current in coil 1 and passing through coil 2 is represented by Φ_{12}. In analogy to Equation 32.2, we can identify the **mutual inductance** M_{12} of coil 2 with respect to coil 1:

$$M_{12} = \frac{N_2 \Phi_{12}}{i_1} \qquad \textbf{(32.15)}$$

Mutual inductance depends on the geometry of both circuits and on their orientation with respect to each other. As the circuit separation distance increases, the mutual inductance decreases because the flux linking the circuits decreases.

If the current i_1 varies with time, we see from Faraday's law and Equation 32.15 that the emf induced by coil 1 in coil 2 is

$$\mathcal{E}_2 = -N_2 \frac{d\Phi_{12}}{dt} = -N_2 \frac{d}{dt}\left(\frac{M_{12} i_1}{N_2}\right) = -M_{12} \frac{di_1}{dt} \qquad \textbf{(32.16)}$$

In the preceding discussion, it was assumed the current is in coil 1. Let's also imagine a current i_2 in coil 2. The preceding discussion can be repeated to show that there is a mutual inductance M_{21}. If the current i_2 varies with time, the emf induced by coil 2 in coil 1 is

$$\mathcal{E}_1 = -M_{21} \frac{di_2}{dt} \qquad \textbf{(32.17)}$$

In mutual induction, the emf induced in one coil is always proportional to the rate at which the current in the other coil is changing. Although the proportionality constants M_{12} and M_{21} have been treated separately, it can be shown that they are equal. Therefore, with $M_{12} = M_{21} = M$, Equations 32.16 and 32.17 become

$$\mathcal{E}_2 = -M \frac{di_1}{dt} \quad \text{and} \quad \mathcal{E}_1 = -M \frac{di_2}{dt}$$

These two equations are similar in form to Equation 32.1 for the self-induced emf $\mathcal{E} = -L\,(di/dt)$. The unit of mutual inductance is the henry.

Quick Quiz 32.4 In Figure 32.8, coil 1 is moved closer to coil 2, with the orientation of both coils remaining fixed. Because of this movement, the mutual induction of the two coils **(a)** increases, **(b)** decreases, or **(c)** is unaffected.

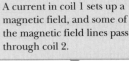

A current in coil 1 sets up a magnetic field, and some of the magnetic field lines pass through coil 2.

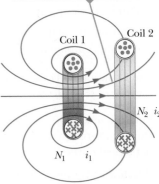

Figure 32.8 A cross-sectional view of two adjacent coils.

Example 32.5	**"Wireless" Battery Charger**

An electric toothbrush has a base designed to hold the toothbrush handle when not in use. As shown in Figure 32.9a, the handle has a cylindrical hole that fits loosely over a matching cylinder on the base. When the handle is placed on the base, a changing current in a solenoid inside the base cylinder induces a current in a coil inside the handle. This induced current charges the battery in the handle.

We can model the base as a solenoid of length ℓ with N_B turns (Fig. 32.9b), carrying a current i, and having a cross-sectional area A. The handle coil contains N_H turns and completely surrounds the base coil. Find the mutual inductance of the system.

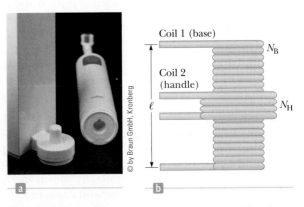

Figure 32.9 (Example 32.5) (a) This electric toothbrush uses the mutual induction of solenoids as part of its battery-charging system. (b) A coil of N_H turns wrapped around the center of a solenoid of N_B turns.

continued

▶ **32.5** continued

SOLUTION

Conceptualize Be sure you can identify the two coils in the situation and understand that a changing current in one coil induces a current in the second coil.

Categorize We will determine the result using concepts discussed in this section, so we categorize this example as a substitution problem.

Use Equation 30.17 to express the magnetic field in the interior of the base solenoid:

$$B = \mu_0 \frac{N_B}{\ell} i$$

Find the mutual inductance, noting that the magnetic flux Φ_{BH} through the handle's coil caused by the magnetic field of the base coil is BA:

$$M = \frac{N_H \Phi_{BH}}{i} = \frac{N_H \, BA}{i} = \boxed{\mu_0 \frac{N_B N_H}{\ell} A}$$

Wireless charging is used in a number of other "cordless" devices. One significant example is the inductive charging used by some manufacturers of electric cars that avoids direct metal-to-metal contact between the car and the charging apparatus.

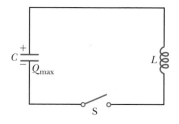

Figure 32.10 A simple LC circuit. The capacitor has an initial charge Q_{max}, and the switch is open for $t < 0$ and then closed at $t = 0$.

32.5 Oscillations in an *LC* Circuit

When a capacitor is connected to an inductor as illustrated in Figure 32.10, the combination is an ***LC* circuit.** If the capacitor is initially charged and the switch is then closed, both the current in the circuit and the charge on the capacitor oscillate between maximum positive and negative values. If the resistance of the circuit is zero, no energy is transformed to internal energy. In the following analysis, the resistance in the circuit is neglected. We also assume an idealized situation in which energy is not radiated away from the circuit. This radiation mechanism is discussed in Chapter 34.

Assume the capacitor has an initial charge Q_{max} (the maximum charge) and the switch is open for $t < 0$ and then closed at $t = 0$. Let's investigate what happens from an energy viewpoint.

When the capacitor is fully charged, the energy U in the circuit is stored in the capacitor's electric field and is equal to $Q_{max}^2/2C$ (Eq. 26.11). At this time, the current in the circuit is zero; therefore, no energy is stored in the inductor. After the switch is closed, the rate at which charges leave or enter the capacitor plates (which is also the rate at which the charge on the capacitor changes) is equal to the current in the circuit. After the switch is closed and the capacitor begins to discharge, the energy stored in its electric field decreases. The capacitor's discharge represents a current in the circuit, and some energy is now stored in the magnetic field of the inductor. Therefore, energy is transferred from the electric field of the capacitor to the magnetic field of the inductor. When the capacitor is fully discharged, it stores no energy. At this time, the current reaches its maximum value and all the energy in the circuit is stored in the inductor. The current continues in the same direction, decreasing in magnitude, with the capacitor eventually becoming fully charged again but with the polarity of its plates now opposite the initial polarity. This process is followed by another discharge until the circuit returns to its original state of maximum charge Q_{max} and the plate polarity shown in Figure 32.10. The energy continues to oscillate between inductor and capacitor.

The oscillations of the LC circuit are an electromagnetic analog to the mechanical oscillations of the particle in simple harmonic motion studied in Chapter 15. Much of what was discussed there is applicable to LC oscillations. For example, we investigated the effect of driving a mechanical oscillator with an external force,

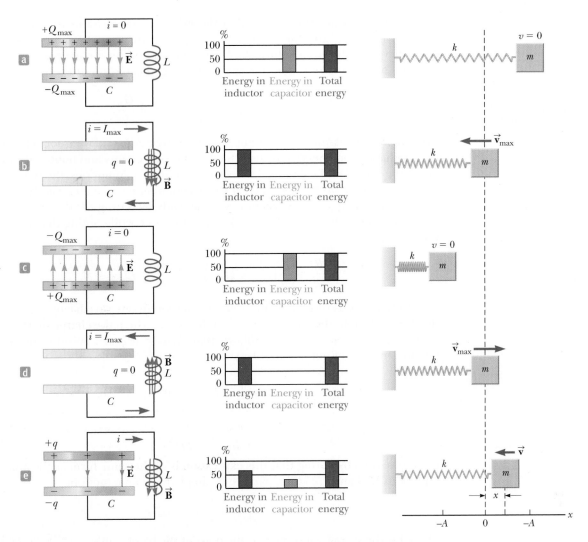

Figure 32.11 Energy transfer in a resistanceless, nonradiating *LC* circuit. The capacitor has a charge Q_{max} at $t = 0$, the instant at which the switch in Figure 32.10 is closed. The mechanical analog of this circuit is the particle in simple harmonic motion, represented by the block–spring system at the right of the figure. (a)–(d) At these special instants, all of the energy in the circuit resides in one of the circuit elements. (e) At an arbitrary instant, the energy is split between the capacitor and the inductor.

which leads to the phenomenon of *resonance*. The same phenomenon is observed in the *LC* circuit. (See Section 33.7.)

A representation of the energy transfer in an *LC* circuit is shown in Figure 32.11. As mentioned, the behavior of the circuit is analogous to that of the particle in simple harmonic motion studied in Chapter 15. For example, consider the block–spring system shown in Figure 15.10. The oscillations of this system are shown at the right of Figure 32.11. The potential energy $\frac{1}{2}kx^2$ stored in the stretched spring is analogous to the potential energy $Q_{max}^2/2C$ stored in the capacitor in Figure 32.11. The kinetic energy $\frac{1}{2}mv^2$ of the moving block is analogous to the magnetic energy $\frac{1}{2}Li^2$ stored in the inductor, which requires the presence of moving charges. In Figure 32.11a, all the energy is stored as electric potential energy in the capacitor at $t = 0$ (because $i = 0$), just as all the energy in a block–spring system is initially stored as potential energy in the spring if it is stretched and released at $t = 0$. In Figure 32.11b, all the energy is stored as magnetic energy $\frac{1}{2}LI_{max}^2$ in the inductor, where I_{max} is the maximum current. Figures 32.11c and 32.11d show subsequent quarter-cycle situations in which the energy is all electric or all magnetic. At intermediate points, part of the energy is electric and part is magnetic.

Consider some arbitrary time t after the switch is closed so that the capacitor has a charge $q < Q_{max}$ and the current is $i < I_{max}$. At this time, both circuit elements store energy, as shown in Figure 32.11e, but the sum of the two energies must equal the total initial energy U stored in the fully charged capacitor at $t = 0$:

Total energy stored in ▸
an *LC* circuit

$$U = U_E + U_B = \frac{q^2}{2C} + \tfrac{1}{2}Li^2 = \frac{Q_{max}^2}{2C} \tag{32.18}$$

Because we have assumed the circuit resistance to be zero and we ignore electromagnetic radiation, no energy is transformed to internal energy and none is transferred out of the system of the circuit. Therefore, with these assumptions, the system of the circuit is isolated: *the total energy of the system must remain constant in time.* We describe the constant energy of the system mathematically by setting $dU/dt = 0$. Therefore, by differentiating Equation 32.18 with respect to time while noting that q and i vary with time gives

$$\frac{dU}{dt} = \frac{d}{dt}\left(\frac{q^2}{2C} + \tfrac{1}{2}Li^2\right) = \frac{q}{C}\frac{dq}{dt} + Li\frac{di}{dt} = 0 \tag{32.19}$$

We can reduce this result to a differential equation in one variable by remembering that the current in the circuit is equal to the rate at which the charge on the capacitor changes: $i = dq/dt$. It then follows that $di/dt = d^2q/dt^2$. Substitution of these relationships into Equation 32.19 gives

$$\frac{q}{C} + L\frac{d^2q}{dt^2} = 0$$

$$\frac{d^2q}{dt^2} = -\frac{1}{LC}q \tag{32.20}$$

Let's solve for q by noting that this expression is of the same form as the analogous Equations 15.3 and 15.5 for a particle in simple harmonic motion:

$$\frac{d^2x}{dt^2} = -\frac{k}{m}x = -\omega^2 x$$

where k is the spring constant, m is the mass of the block, and $\omega = \sqrt{k/m}$. The solution of this mechanical equation has the general form (Eq. 15.6):

$$x = A\cos(\omega t + \phi)$$

where A is the amplitude of the simple harmonic motion (the maximum value of x), ω is the angular frequency of this motion, and ϕ is the phase constant; the values of A and ϕ depend on the initial conditions. Because Equation 32.20 is of the same mathematical form as the differential equation of the simple harmonic oscillator, it has the solution

Charge as a function of time ▸
for an ideal *LC* circuit

$$q = Q_{max}\cos(\omega t + \phi) \tag{32.21}$$

where Q_{max} is the maximum charge of the capacitor and the angular frequency ω is

Angular frequency of ▸
oscillation in an *LC* circuit

$$\omega = \frac{1}{\sqrt{LC}} \tag{32.22}$$

Note that the angular frequency of the oscillations depends solely on the inductance and capacitance of the circuit. Equation 32.22 gives the *natural frequency* of oscillation of the *LC* circuit.

Because q varies sinusoidally with time, the current in the circuit also varies sinusoidally with time. We can show that by differentiating Equation 32.21 with respect to time:

Current as a function of ▸
time for an ideal *LC* current

$$i = \frac{dq}{dt} = -\omega Q_{max}\sin(\omega t + \phi) \tag{32.23}$$

To determine the value of the phase angle ϕ, let's examine the initial conditions, which in our situation require that at $t = 0$, $i = 0$, and $q = Q_{max}$. Setting $i = 0$ at $t = 0$ in Equation 32.23 gives

$$0 = -\omega Q_{max} \sin \phi$$

which shows that $\phi = 0$. This value for ϕ also is consistent with Equation 32.21 and the condition that $q = Q_{max}$ at $t = 0$. Therefore, in our case, the expressions for q and i are

$$q = Q_{max} \cos \omega t \qquad \textbf{(32.24)}$$

$$i = -\omega Q_{max} \sin \omega t = -I_{max} \sin \omega t \qquad \textbf{(32.25)}$$

Graphs of q versus t and i versus t are shown in Figure 32.12. The charge on the capacitor oscillates between the extreme values Q_{max} and $-Q_{max}$, and the current oscillates between I_{max} and $-I_{max}$. Furthermore, the current is 90° out of phase with the charge. That is, when the charge is a maximum, the current is zero, and when the charge is zero, the current has its maximum value.

Let's return to the energy discussion of the *LC* circuit. Substituting Equations 32.24 and 32.25 in Equation 32.18, we find that the total energy is

$$U = U_E + U_B = \frac{Q_{max}^2}{2C} \cos^2 \omega t + \tfrac{1}{2} L I_{max}^2 \sin^2 \omega t \qquad \textbf{(32.26)}$$

This expression contains all the features described qualitatively at the beginning of this section. It shows that the energy of the *LC* circuit continuously oscillates between energy stored in the capacitor's electric field and energy stored in the inductor's magnetic field. When the energy stored in the capacitor has its maximum value $Q_{max}^2/2C$, the energy stored in the inductor is zero. When the energy stored in the inductor has its maximum value $\tfrac{1}{2} L I_{max}^2$, the energy stored in the capacitor is zero.

Plots of the time variations of U_E and U_B are shown in Figure 32.13. The sum $U_E + U_B$ is a constant and is equal to the total energy $Q_{max}^2/2C$, or $\tfrac{1}{2} L I_{max}^2$. Analytical verification is straightforward. The amplitudes of the two graphs in Figure 32.13 must be equal because the maximum energy stored in the capacitor (when $I = 0$) must equal the maximum energy stored in the inductor (when $q = 0$). This equality is expressed mathematically as

$$\frac{Q_{max}^2}{2C} = \frac{L I_{max}^2}{2}$$

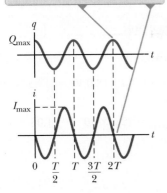

The charge q and the current i are 90° out of phase with each other.

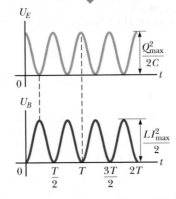

The sum of the two curves is a constant and is equal to the total energy stored in the circuit.

Figure 32.12 Graphs of charge versus time and current versus time for a resistanceless, nonradiating *LC* circuit.

Figure 32.13 Plots of U_E versus t and U_B versus t for a resistanceless, nonradiating *LC* circuit.

Using this expression in Equation 32.26 for the total energy gives

$$U = \frac{Q_{max}^2}{2C}(\cos^2 \omega t + \sin^2 \omega t) = \frac{Q_{max}^2}{2C} \tag{32.27}$$

because $\cos^2 \omega t + \sin^2 \omega t = 1$.

In our idealized situation, the oscillations in the circuit persist indefinitely; the total energy U of the circuit, however, remains constant only if energy transfers and transformations are neglected. In actual circuits, there is always some resistance and some energy is therefore transformed to internal energy. We mentioned at the beginning of this section that we are also ignoring radiation from the circuit. In reality, radiation is inevitable in this type of circuit, and the total energy in the circuit continuously decreases as a result of this process.

Quick Quiz 32.5 **(i)** At an instant of time during the oscillations of an LC circuit, the current is at its maximum value. At this instant, what happens to the voltage across the capacitor? (a) It is different from that across the inductor. (b) It is zero. (c) It has its maximum value. (d) It is impossible to determine. **(ii)** Now consider an instant when the current is momentarily zero. From the same choices, describe the magnitude of the voltage across the capacitor at this instant.

Example 32.6 **Oscillations in an *LC* Circuit**

In Figure 32.14, the battery has an emf of 12.0 V, the inductance is 2.81 mH, and the capacitance is 9.00 pF. The switch has been set to position a for a long time so that the capacitor is charged. The switch is then thrown to position b, removing the battery from the circuit and connecting the capacitor directly across the inductor.

(A) Find the frequency of oscillation of the circuit.

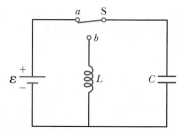

Figure 32.14 (Example 32.6) First the capacitor is fully charged with the switch set to position a. Then the switch is thrown to position b, and the battery is no longer in the circuit.

SOLUTION

Conceptualize When the switch is thrown to position b, the active part of the circuit is the right-hand loop, which is an LC circuit.

Categorize We use equations developed in this section, so we categorize this example as a substitution problem.

Use Equation 32.22 to find the frequency:

$$f = \frac{\omega}{2\pi} = \frac{1}{2\pi\sqrt{LC}}$$

Substitute numerical values:

$$f = \frac{1}{2\pi[(2.81 \times 10^{-3}\ \text{H})(9.00 \times 10^{-12}\ \text{F})]^{1/2}} = \boxed{1.00 \times 10^6\ \text{Hz}}$$

(B) What are the maximum values of charge on the capacitor and current in the circuit?

SOLUTION

Find the initial charge on the capacitor, which equals the maximum charge:

$$Q_{max} = C\,\Delta V = (9.00 \times 10^{-12}\ \text{F})(12.0\ \text{V}) = \boxed{1.08 \times 10^{-10}\ \text{C}}$$

Use Equation 32.25 to find the maximum current from the maximum charge:

$$I_{max} = \omega Q_{max} = 2\pi f Q_{max} = (2\pi \times 10^6\ \text{s}^{-1})(1.08 \times 10^{-10}\ \text{C})$$

$$= \boxed{6.79 \times 10^{-4}\ \text{A}}$$

32.6 The *RLC* Circuit

Let's now turn our attention to a more realistic circuit consisting of a resistor, an inductor, and a capacitor connected in series as shown in Figure 32.15. We assume the resistance of the resistor represents all the resistance in the circuit. Suppose the switch is at position *a* so that the capacitor has an initial charge Q_{max}. The switch is now thrown to position *b*. At this instant, the total energy stored in the capacitor and inductor is $Q_{max}^2/2C$. This total energy, however, is no longer constant as it was in the *LC* circuit because the resistor causes transformation to internal energy. (We continue to ignore electromagnetic radiation from the circuit in this discussion.) Because the rate of energy transformation to internal energy within a resistor is i^2R,

$$\frac{dU}{dt} = -i^2R$$

where the negative sign signifies that the energy *U* of the circuit is decreasing in time. Substituting $U = U_E + U_B$ gives

$$\frac{q}{C}\frac{dq}{dt} + Li\frac{di}{dt} = -i^2R \tag{32.28}$$

To convert this equation into a form that allows us to compare the electrical oscillations with their mechanical analog, we first use $i = dq/dt$ and move all terms to the left-hand side to obtain

$$Li\frac{d^2q}{dt^2} + i^2R + \frac{q}{C}i = 0$$

Now divide through by *i*:

$$L\frac{d^2q}{dt^2} + iR + \frac{q}{C} = 0$$

$$L\frac{d^2q}{dt^2} + R\frac{dq}{dt} + \frac{q}{C} = 0 \tag{32.29}$$

The *RLC* circuit is analogous to the damped harmonic oscillator discussed in Section 15.6 and illustrated in Figure 15.20. The equation of motion for a damped block–spring system is, from Equation 15.31,

$$m\frac{d^2x}{dt^2} + b\frac{dx}{dt} + kx = 0 \tag{32.30}$$

Comparing Equations 32.29 and 32.30, we see that *q* corresponds to the position *x* of the block at any instant, *L* to the mass *m* of the block, *R* to the damping coefficient *b*, and *C* to $1/k$, where *k* is the force constant of the spring. These and other relationships are listed in Table 32.1 on page 986.

Because the analytical solution of Equation 32.29 is cumbersome, we give only a qualitative description of the circuit behavior. In the simplest case, when $R = 0$, Equation 32.29 reduces to that of a simple *LC* circuit as expected, and the charge and the current oscillate sinusoidally in time. This situation is equivalent to removing all damping in the mechanical oscillator.

When *R* is small, a situation that is analogous to light damping in the mechanical oscillator, the solution of Equation 32.29 is

$$q = Q_{max}e^{-Rt/2L}\cos\omega_d t \tag{32.31}$$

where ω_d, the angular frequency at which the circuit oscillates, is given by

$$\omega_d = \left[\frac{1}{LC} - \left(\frac{R}{2L}\right)^2\right]^{1/2} \tag{32.32}$$

That is, the value of the charge on the capacitor undergoes a damped harmonic oscillation in analogy with a block–spring system moving in a viscous medium. Equation 32.32 shows that when $R \ll \sqrt{4L/C}$ (so that the second term in the

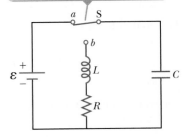

The switch is set first to position *a*, and the capacitor is charged. The switch is then thrown to position *b*.

Figure 32.15 A series *RLC* circuit.

Table 32.1   **Analogies Between the *RLC* Circuit and the Particle in Simple Harmonic Motion**

RLC Circuit		One-Dimensional Particle in Simple Harmonic Motion
Charge	$q \leftrightarrow x$	Position
Current	$i \leftrightarrow v_x$	Velocity
Potential difference	$\Delta V \leftrightarrow F_x$	Force
Resistance	$R \leftrightarrow b$	Viscous damping coefficient
Capacitance	$C \leftrightarrow 1/k$	(k = spring constant)
Inductance	$L \leftrightarrow m$	Mass
Current = time derivative of charge	$i = \dfrac{dq}{dt} \leftrightarrow v_x = \dfrac{dx}{dt}$	Velocity = time derivative of position
Rate of change of current = second time derivative of charge	$\dfrac{di}{dt} = \dfrac{d^2q}{dt^2} \leftrightarrow a_x = \dfrac{dv_x}{dt} = \dfrac{d^2x}{dt^2}$	Acceleration = second time derivative of position
Energy in inductor	$U_B = \frac{1}{2}Li^2 \leftrightarrow K = \frac{1}{2}mv^2$	Kinetic energy of moving object
Energy in capacitor	$U_E = \frac{1}{2}\dfrac{q^2}{C} \leftrightarrow U = \frac{1}{2}kx^2$	Potential energy stored in a spring
Rate of energy loss due to resistance	$i^2R \leftrightarrow bv^2$	Rate of energy loss due to friction
RLC circuit	$L\dfrac{d^2q}{dt^2} + R\dfrac{dq}{dt} + \dfrac{q}{C} = 0 \leftrightarrow m\dfrac{d^2x}{dt^2} + b\dfrac{dx}{dt} + kx = 0$	Damped object on a spring

brackets is much smaller than the first), the frequency ω_d of the damped oscillator is close to that of the undamped oscillator, $1/\sqrt{LC}$. Because $i = dq/dt$, it follows that the current also undergoes damped harmonic oscillation. A plot of the charge versus time for the damped oscillator is shown in Figure 32.16a, and an oscilloscope trace for a real *RLC* circuit is shown in Figure 32.16b. The maximum value of q decreases after each oscillation, just as the amplitude of a damped block–spring system decreases in time.

For larger values of R, the oscillations damp out more rapidly; in fact, there exists a critical resistance value $R_c = \sqrt{4L/C}$ above which no oscillations occur. A system with $R = R_c$ is said to be *critically damped*. When R exceeds R_c, the system is said to be *overdamped*.

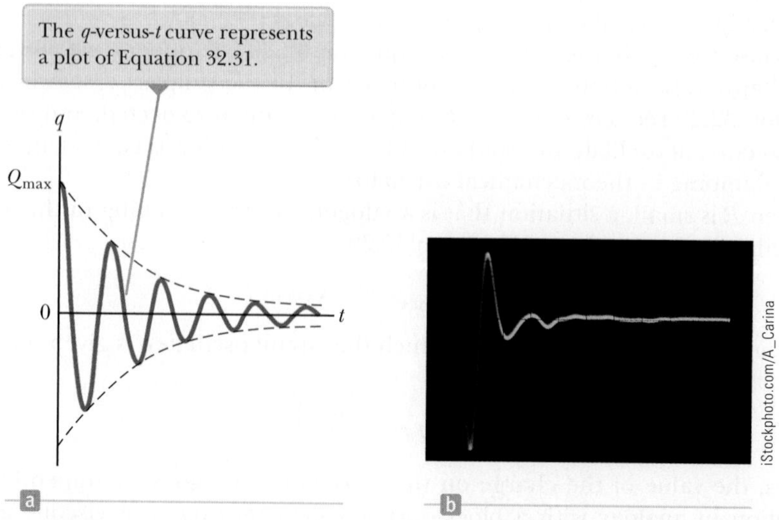

The q-versus-t curve represents a plot of Equation 32.31.

Figure 32.16 (a) Charge versus time for a damped *RLC* circuit. The charge decays in this way when $R < \sqrt{4L/C}$. (b) Oscilloscope pattern showing the decay in the oscillations of an *RLC* circuit.

Summary

Concepts and Principles

When the current in a loop of wire changes with time, an emf is induced in the loop according to Faraday's law. The **self-induced emf** is

$$\varepsilon_L = -L\frac{di}{dt} \qquad (32.1)$$

where L is the **inductance** of the loop. Inductance is a measure of how much opposition a loop offers to a change in the current in the loop. Inductance has the SI unit of **henry** (H), where $1\ \text{H} = 1\ \text{V} \cdot \text{s/A}$.

The inductance of any coil is

$$L = \frac{N\Phi_B}{i} \qquad (32.2)$$

where N is the total number of turns and Φ_B is the magnetic flux through the coil. The inductance of a device depends on its geometry. For example, the inductance of an air-core solenoid is

$$L = \mu_0 \frac{N^2}{\ell} A \qquad (32.4)$$

where ℓ is the length of the solenoid and A is the cross-sectional area.

If a resistor and inductor are connected in series to a battery of emf ε at time $t = 0$, the current in the circuit varies in time according to the expression

$$i = \frac{\varepsilon}{R}\left(1 - e^{-t/\tau}\right) \qquad (32.7)$$

where $\tau = L/R$ is the **time constant** of the RL circuit. If we replace the battery in the circuit by a resistanceless wire, the current decays exponentially with time according to the expression

$$i = \frac{\varepsilon}{R} e^{-t/\tau} \qquad (32.10)$$

where ε/R is the initial current in the circuit.

The energy stored in the magnetic field of an inductor carrying a current i is

$$U_B = \tfrac{1}{2}Li^2 \qquad (32.12)$$

This energy is the magnetic counterpart to the energy stored in the electric field of a charged capacitor.

The energy density at a point where the magnetic field is B is

$$u_B = \frac{B^2}{2\mu_0} \qquad (32.14)$$

The **mutual inductance** of a system of two coils is

$$M_{12} = \frac{N_2\Phi_{12}}{i_1} = M_{21} = \frac{N_1\Phi_{21}}{i_2} = M \qquad (32.15)$$

This mutual inductance allows us to relate the induced emf in a coil to the changing source current in a nearby coil using the relationships

$$\varepsilon_2 = -M_{12}\frac{di_1}{dt} \quad \text{and} \quad \varepsilon_1 = -M_{21}\frac{di_2}{dt} \qquad (32.16,\ 32.17)$$

In an LC circuit that has zero resistance and does not radiate electromagnetically (an idealization), the values of the charge on the capacitor and the current in the circuit vary sinusoidally in time at an angular frequency given by

$$\omega = \frac{1}{\sqrt{LC}} \qquad (32.22)$$

The energy in an LC circuit continuously transfers between energy stored in the capacitor and energy stored in the inductor.

In an RLC circuit with small resistance, the charge on the capacitor varies with time according to

$$q = Q_{max}e^{-Rt/2L}\cos\omega_d t \qquad (32.31)$$

where

$$\omega_d = \left[\frac{1}{LC} - \left(\frac{R}{2L}\right)^2\right]^{1/2} \qquad (32.32)$$

Objective Questions   [1.] denotes answer available in *Student Solutions Manual/Study Guide*

1. The centers of two circular loops are separated by a fixed distance. **(i)** For what relative orientation of the loops is their mutual inductance a maximum? (a) coaxial and lying in parallel planes (b) lying in the same plane (c) lying in perpendicular planes, with the center of one on the axis of the other (d) The orientation makes no difference. **(ii)** For what relative orientation is their mutual inductance a minimum? Choose from the same possibilities as in part (i).

2. A long, fine wire is wound into a coil with inductance 5 mH. The coil is connected across the terminals of a battery, and the current is measured a few seconds after the connection is made. The wire is unwound and wound again into a different coil with $L = 10$ mH. This second coil is connected across the same battery, and the current is measured in the same way. Compared with the current in the first coil, is the current in the second coil (a) four times as large, (b) twice as large, (c) unchanged, (d) half as large, or (e) one-fourth as large?

3. A solenoidal inductor for a printed circuit board is being redesigned. To save weight, the number of turns is reduced by one-half, with the geometric dimensions kept the same. By how much must the current change if the energy stored in the inductor is to remain the same? (a) It must be four times larger. (b) It must be two times larger. (c) It should be left the same. (d) It should be one-half as large. (e) No change in the current can compensate for the reduction in the number of turns.

4. In Figure OQ32.4, the switch is left in position *a* for a long time interval and is then quickly thrown to posi- tion *b*. Rank the magnitudes of the voltages across the four circuit elements a short time thereafter from the largest to the smallest.

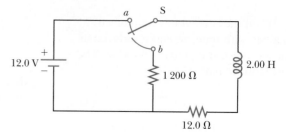

Figure OQ32.4

5. Two solenoids, A and B, are wound using equal lengths of the same kind of wire. The length of the axis of each solenoid is large compared with its diameter. The axial length of A is twice as large as that of B, and A has twice as many turns as B. What is the ratio of the inductance of solenoid A to that of solenoid B? (a) 4 (b) 2 (c) 1 (d) $\frac{1}{2}$ (e) $\frac{1}{4}$

6. If the current in an inductor is doubled, by what factor is the stored energy multiplied? (a) 4 (b) 2 (c) 1 (d) $\frac{1}{2}$ (e) $\frac{1}{4}$

7. Initially, an inductor with no resistance carries a constant current. Then the current is brought to a new constant value twice as large. *After* this change, when the current is constant at its higher value, what has happened to the emf in the inductor? (a) It is larger than before the change by a factor of 4. (b) It is larger by a factor of 2. (c) It has the same nonzero value. (d) It continues to be zero. (e) It has decreased.

Conceptual Questions   [1.] denotes answer available in *Student Solutions Manual/Study Guide*

1. Consider this thesis: "Joseph Henry, America's first professional physicist, caused a basic change in the human view of the Universe when he discovered self-induction during a school vacation at the Albany Academy about 1830. Before that time, one could think of the Universe as composed of only one thing: matter. The energy that temporarily maintains the current after a battery is removed from a coil, on the other hand, is not energy that belongs to any chunk of matter. It is energy in the massless magnetic field surrounding the coil. With Henry's discovery, Nature forced us to admit that the Universe consists of fields as well as matter." (a) Argue for or against the statement. (b) In your view, what makes up the Universe?

2. (a) What parameters affect the inductance of a coil? (b) Does the inductance of a coil depend on the current in the coil?

3. A switch controls the current in a circuit that has a large inductance. The electric arc at the switch (Fig.

CQ32.3) can melt and oxidize the contact surfaces, resulting in high resistivity of the contacts and eventual destruction of the switch. Is a spark more likely to be produced at the switch when the switch is being closed, when it is being opened, or does it not matter?

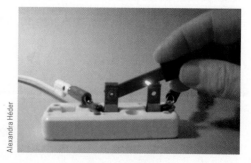

Figure CQ32.3

4. Consider the four circuits shown in Figure CQ32.4, each consisting of a battery, a switch, a lightbulb, a

resistor, and either a capacitor or an inductor. Assume the capacitor has a large capacitance and the inductor has a large inductance but no resistance. The lightbulb has high efficiency, glowing whenever it carries electric current. **(i)** Describe what the lightbulb does in each of circuits (a) through (d) after the switch is thrown closed. **(ii)** Describe what the lightbulb does in each of circuits (a) through (d) when, having been closed for a long time interval, the switch is opened.

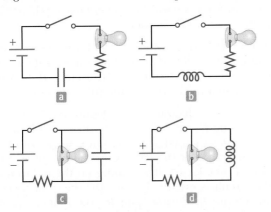

Figure CQ32.4

5. The current in a circuit containing a coil, a resistor, and a battery has reached a constant value. (a) Does the coil have an inductance? (b) Does the coil affect the value of the current?

6. (a) Can an object exert a force on itself? (b) When a coil induces an emf in itself, does it exert a force on itself?

7. The open switch in Figure CQ32.7 is thrown closed at $t = 0$. Before the switch is closed, the capacitor is uncharged and all currents are zero. Determine the currents in L, C, and R, the emf across L, and the potential differences across C and R (a) at the instant after the switch is closed and (b) long after it is closed.

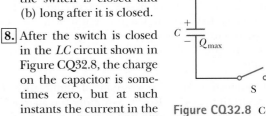

Figure CQ32.7

8. After the switch is closed in the LC circuit shown in Figure CQ32.8, the charge on the capacitor is sometimes zero, but at such instants the current in the circuit is not zero. How is this behavior possible?

Figure CQ32.8 Conceptual Question 8 and Problems 52, 54, and 55.

9. How can you tell whether an RLC circuit is overdamped or underdamped?

10. Discuss the similarities between the energy stored in the electric field of a charged capacitor and the energy stored in the magnetic field of a current-carrying coil.

Problems

WebAssign The problems found in this chapter may be assigned online in Enhanced WebAssign

1. straightforward; **2.** intermediate; **3.** challenging

1. full solution available in the *Student Solutions Manual/Study Guide*

AMT Analysis Model tutorial available in Enhanced WebAssign

GP Guided Problem

M Master It tutorial available in Enhanced WebAssign

W Watch It video solution available in Enhanced WebAssign

Section 32.1 Self-Induction and Inductance

1. A coil has an inductance of 3.00 mH, and the current in it changes from 0.200 A to 1.50 A in a time interval of 0.200 s. Find the magnitude of the average induced emf in the coil during this time interval.

2. A coiled telephone cord forms a spiral with 70.0 turns, a diameter of 1.30 cm, and an unstretched length of 60.0 cm. Determine the inductance of one conductor in the unstretched cord.

3. A 2.00-H inductor carries a steady current of 0.500 A. When the switch in the circuit is opened, the current is effectively zero after 10.0 ms. What is the average induced emf in the inductor during this time interval?

4. A solenoid of radius 2.50 cm has 400 turns and a length of 20.0 cm. Find (a) its inductance and (b) the rate at which current must change through it to produce an emf of 75.0 mV.

5. An emf of 24.0 mV is induced in a 500-turn coil when the current is changing at the rate of 10.0 A/s. What is the magnetic flux through each turn of the coil at an instant when the current is 4.00 A?

6. A 40.0-mA current is carried by a uniformly wound air-core solenoid with 450 turns, a 15.0-mm diameter, and 12.0-cm length. Compute (a) the magnetic field inside the solenoid, (b) the magnetic flux through each turn, and (c) the inductance of the solenoid. (d) **What If?** If the current were different, which of these quantities would change?

7. The current in a coil changes from 3.50 A to 2.00 A in the same direction in 0.500 s. If the average emf induced in the coil is 12.0 mV, what is the inductance of the coil?

8. A technician wraps wire around a tube of length 36.0 cm having a diameter of 8.00 cm. When the windings are evenly spread over the full length of the tube, the result is a solenoid containing 580 turns of wire. (a) Find the inductance of this solenoid. (b) If the current in this solenoid increases at the rate of 4.00 A/s, find the self-induced emf in the solenoid.

9. The current in a 90.0-mH inductor changes with time **W** as $i = 1.00t^2 - 6.00t$, where i is in amperes and t is in seconds. Find the magnitude of the induced emf at (a) $t = 1.00$ s and (b) $t = 4.00$ s. (c) At what time is the emf zero?

10. An inductor in the form of a solenoid contains 420 **M** turns and is 16.0 cm in length. A uniform rate of decrease of current through the inductor of 0.421 A/s induces an emf of 175 μV. What is the radius of the solenoid?

11. A self-induced emf in a solenoid of inductance L changes in time as $\mathcal{E} = \mathcal{E}_0 e^{-kt}$. Assuming the charge is finite, find the total charge that passes a point in the wire of the solenoid.

12. A toroid has a major radius R and a minor radius r and is tightly wound with N turns of wire on a hollow cardboard torus. Figure P32.12 shows half of this toroid, allowing us to see its cross section. If $R \gg r$, the magnetic field in the region enclosed by the wire is essentially the same as the magnetic field of a solenoid that has been bent into a large circle of radius R. Modeling the field as the uniform field of a long solenoid, show that the inductance of such a toroid is approximately

$$L \approx \tfrac{1}{2}\mu_0 N^2 \frac{r^2}{R}$$

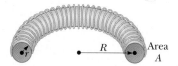

Figure P32.12

13. A 10.0-mH inductor carries a current $i = I_{max} \sin \omega t$, **M** with $I_{max} = 5.00$ A and $f = \omega/2\pi = 60.0$ Hz. What is the self-induced emf as a function of time?

14. The current in a 4.00 mH-inductor varies in time as shown in Figure P32.14. Construct a graph of the self-induced emf across the inductor over the time interval $t = 0$ to $t = 12.0$ ms.

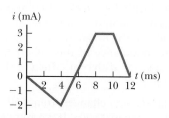

Figure P32.14

Section 32.2 **RL** Circuits

15. A 510-turn solenoid has a radius of 8.00 mm and an overall length of 14.0 cm. (a) What is its inductance? (b) If the solenoid is connected in series with a 2.50-Ω resistor and a battery, what is the time constant of the circuit?

16. A 12.0-V battery is connected into a series circuit con- **M** taining a 10.0-Ω resistor and a 2.00-H inductor. In what time interval will the current reach (a) 50.0% and (b) 90.0% of its final value?

17. A series RL circuit with $L = 3.00$ H and a series RC circuit with $C = 3.00$ μF have equal time constants. If the two circuits contain the same resistance R, (a) what is the value of R? (b) What is the time constant?

18. In the circuit diagrammed in Figure P32.18, take $\mathcal{E} = 12.0$ V and $R = 24.0$ Ω. Assume the switch is open for $t < 0$ and is closed at $t = 0$. On a single set of axes, sketch graphs of the current in the circuit as a function of time for $t \geq 0$, assuming (a) the inductance in the circuit is essentially zero, (b) the inductance has an intermediate value, and (c) the inductance has a very large value. Label the initial and final values of the current.

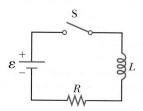

Figure P32.18
Problems 18, 20, 23, 24, and 27.

19. Consider the circuit shown in Figure P32.19. (a) When the switch is in position a, for what value of R will the circuit have a time constant of 15.0 μs? (b) What is the current in the inductor at the instant the switch is thrown to position b?

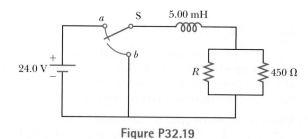

Figure P32.19

20. When the switch in Figure P32.18 is closed, the current takes 3.00 ms to reach 98.0% of its final value. If $R = 10.0$ Ω, what is the inductance?

21. A circuit consists of a coil, a switch, and a battery, all in series. The internal resistance of the battery is negligible compared with that of the coil. The switch is originally open. It is thrown closed, and after a time interval Δt, the current in the circuit reaches 80.0%

of its final value. The switch then remains closed for a time interval much longer than Δt. The wires connected to the terminals of the battery are then short-circuited with another wire and removed from the battery, so that the current is uninterrupted. (a) At an instant that is a time interval Δt after the short circuit, the current is what percentage of its maximum value? (b) At the moment $2\Delta t$ after the coil is short-circuited, the current in the coil is what percentage of its maximum value?

22. Show that $i = I_i e^{-t/\tau}$ is a solution of the differential equation

$$iR + L\frac{di}{dt} = 0$$

where I_i is the current at $t = 0$ and $\tau = L/R$.

23. In the circuit shown in Figure P32.18, let $L = 7.00$ H, **W** $R = 9.00\ \Omega$, and $\mathcal{E} = 120$ V. What is the self-induced emf 0.200 s after the switch is closed?

24. Consider the circuit in Figure P32.18, taking $\mathcal{E} = 6.00$ V, $L = 8.00$ mH, and $R = 4.00\ \Omega$. (a) What is the inductive time constant of the circuit? (b) Calculate the current in the circuit 250 μs after the switch is closed. (c) What is the value of the final steady-state current? (d) After what time interval does the current reach 80.0% of its maximum value?

25. The switch in Figure P32.25 is open for $t < 0$ and is then thrown closed at time $t = 0$. Assume $R = 4.00\ \Omega$, $L = 1.00$ H, and $\mathcal{E} = 10.0$ V. Find (a) the current in the inductor and (b) the current in the switch as functions of time thereafter.

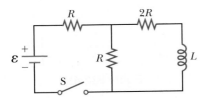

Figure P32.25 Problems 25, 26, and 64.

26. The switch in Figure P32.25 is open for $t < 0$ and is then thrown closed at time $t = 0$. Find (a) the current in the inductor and (b) the current in the switch as functions of time thereafter.

27. For the RL circuit shown in Figure P32.18, let the inductance be 3.00 H, the resistance 8.00 Ω, and the battery emf 36.0 V. (a) Calculate $\Delta V_R/\mathcal{E}_L$, that is, the ratio of the potential difference across the resistor to the emf across the inductor when the current is 2.00 A. (b) Calculate the emf across the inductor when the current is 4.50 A.

28. Consider the current pulse $i(t)$ shown in Figure P32.28a. The current begins at zero, becomes 10.0 A between $t = 0$ and $t = 200\ \mu$s, and then is zero once again. This pulse is applied to the input of the partial

circuit shown in Figure P32.28b. Determine the current in the inductor as a function of time.

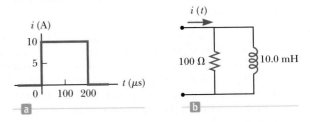

Figure P32.28

29. An inductor that has an inductance of 15.0 H and a resistance of 30.0 Ω is connected across a 100-V battery. What is the rate of increase of the current (a) at $t = 0$ and (b) at $t = 1.50$ s?

30. Two ideal inductors, L_1 and L_2, have *zero* internal resistance and are far apart, so their magnetic fields do not influence each other. (a) Assuming these inductors are connected in series, show that they are equivalent to a single ideal inductor having $L_{eq} = L_1 + L_2$. (b) Assuming these same two inductors are connected in parallel, show that they are equivalent to a single ideal inductor having $1/L_{eq} = 1/L_1 + 1/L_2$. (c) **What If?** Now consider two inductors L_1 and L_2 that have *nonzero* internal resistances R_1 and R_2, respectively. Assume they are still far apart, so their mutual inductance is zero, and assume they are connected in series. Show that they are equivalent to a single inductor having $L_{eq} = L_1 + L_2$ and $R_{eq} = R_1 + R_2$. (d) If these same inductors are now connected in parallel, is it necessarily true that they are equivalent to a single ideal inductor having $1/L_{eq} = 1/L_1 + 1/L_2$ and $1/R_{eq} = 1/R_1 + 1/R_2$? Explain your answer.

31. A 140-mH inductor and a 4.90-Ω resistor are con- **M** nected with a switch to a 6.00-V battery as shown in Figure P32.31. (a) After the switch is first thrown to a (connecting the battery), what time interval elapses before the current reaches 220 mA? (b) What is the current in the inductor 10.0 s after the switch is closed? (c) Now the switch is quickly thrown from a to b. What time interval elapses before the current in the inductor falls to 160 mA?

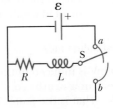

Figure P32.31

Section 32.3 Energy in a Magnetic Field

32. Calculate the energy associated with the magnetic field of a 200-turn solenoid in which a current of 1.75 A produces a magnetic flux of 3.70×10^{-4} T $\cdot$ m^2 in each turn.

33. An air-core solenoid with 68 turns is 8.00 cm long and has a diameter of 1.20 cm. When the solenoid carries a current of 0.770 A, how much energy is stored in its magnetic field?

34. A 10.0-V battery, a 5.00-Ω resistor, and a 10.0-H inductor are connected in series. After the current in the circuit has reached its maximum value, calculate (a) the power being supplied by the battery, (b) the power being delivered to the resistor, (c) the power being delivered to the inductor, and (d) the energy stored in the magnetic field of the inductor.

35. On a clear day at a certain location, a 100-V/m vertical electric field exists near the Earth's surface. At the same place, the Earth's magnetic field has a magnitude of 0.500×10^{-4} T. Compute the energy densities of (a) the electric field and (b) the magnetic field.

36. Complete the calculation in Example 32.3 by proving that

$$\int_0^\infty e^{-2Rt/L}\, dt = \frac{L}{2R}$$

37. A 24.0-V battery is connected in series with a resistor and an inductor, with $R = 8.00\ \Omega$ and $L = 4.00$ H, respectively. Find the energy stored in the inductor (a) when the current reaches its maximum value and (b) at an instant that is a time interval of one time constant after the switch is closed.

38. A flat coil of wire has an inductance of 40.0 mH and a resistance of 5.00 Ω. It is connected to a 22.0-V battery at the instant $t = 0$. Consider the moment when the current is 3.00 A. (a) At what rate is energy being delivered by the battery? (b) What is the power being delivered to the resistance of the coil? (c) At what rate is energy being stored in the magnetic field of the coil? (d) What is the relationship among these three power values? (e) Is the relationship described in part (d) true at other instants as well? (f) Explain the relationship at the moment immediately after $t = 0$ and at a moment several seconds later.

39. The magnetic field inside a superconducting solenoid is 4.50 T. The solenoid has an inner diameter of 6.20 cm and a length of 26.0 cm. Determine (a) the magnetic energy density in the field and (b) the energy stored in the magnetic field within the solenoid.

Section 32.4 Mutual Inductance

40. An emf of 96.0 mV is induced in the windings of a coil when the current in a nearby coil is increasing at the rate of 1.20 A/s. What is the mutual inductance of the two coils?

41. Two coils, held in fixed positions, have a mutual inductance of 100 μH. What is the peak emf in one coil when the current in the other coil is $i(t) = 10.0 \sin (1.00 \times 10^3 t)$, where i is in amperes and t is in seconds?

42. Two coils are close to each other. The first coil carries a current given by $i(t) = 5.00\, e^{-0.025\,0t} \sin 120\pi t$, where i

is in amperes and t is in seconds. At $t = 0.800$ s, the emf measured across the second coil is −3.20 V. What is the mutual inductance of the coils?

43. Two solenoids A and B, spaced close to each other and sharing the same cylindrical axis, have 400 and 700 turns, respectively. A current of 3.50 A in solenoid A produces an average flux of 300 μWb through each turn of A and a flux of 90.0 μWb through each turn of B. (a) Calculate the mutual inductance of the two solenoids. (b) What is the inductance of A? (c) What emf is induced in B when the current in A changes at the rate of 0.500 A/s?

44. Solenoid S_1 has N_1 turns, radius R_1, and length ℓ. It is so long that its magnetic field is uniform nearly everywhere inside it and is nearly zero outside. Solenoid S_2 has N_2 turns, radius $R_2 < R_1$, and the same length as S_1. It lies inside S_1, with their axes parallel. (a) Assume S_1 carries variable current i. Compute the mutual inductance characterizing the emf induced in S_2. (b) Now assume S_2 carries current i. Compute the mutual inductance to which the emf in S_1 is proportional. (c) State how the results of parts (a) and (b) compare with each other.

45. On a printed circuit board, a relatively long, straight conductor and a conducting rectangular loop lie in the same plane as shown in Figure P32.45. Taking $h = 0.400$ mm, $w = 1.30$ mm, and $\ell = 2.70$ mm, find their mutual inductance.

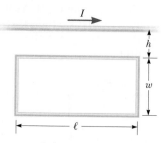

Figure P32.45

46. Two single-turn circular loops of wire have radii R and r, with $R \gg r$. The loops lie in the same plane and are concentric. (a) Show that the mutual inductance of the pair is approximately $M = \mu_0 \pi r^2 / 2R$. (b) Evaluate M for $r = 2.00$ cm and $R = 20.0$ cm.

Section 32.5 Oscillations in an *LC* Circuit

47. In the circuit of Figure P32.47, the battery emf is 50.0 V, the resistance is 250 Ω, and the capacitance is 0.500 μF. The switch S is closed for a long time

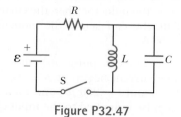

Figure P32.47

interval, and zero potential difference is measured across the capacitor. After the switch is opened, the potential difference across the capacitor reaches a maximum value of 150 V. What is the value of the inductance?

48. A 1.05-μH inductor is connected in series with a variable capacitor in the tuning section of a shortwave radio set. What capacitance tunes the circuit to the signal from a transmitter broadcasting at 6.30 MHz?

49. A 1.00-μF capacitor is charged by a 40.0-V power supply. The fully charged capacitor is then discharged through a 10.0-mH inductor. Find the maximum current in the resulting oscillations.

50. Calculate the inductance of an *LC* circuit that oscillates at 120 Hz when the capacitance is 8.00 μF.

51. An *LC* circuit consists of a 20.0-mH inductor and a 0.500-μF capacitor. If the maximum instantaneous current is 0.100 A, what is the greatest potential difference across the capacitor?

52. *Why is the following situation impossible?* The *LC* circuit shown in Figure CQ32.8 has $L = 30.0$ mH and $C = 50.0$ μF. The capacitor has an initial charge of 200 μC. The switch is closed, and the circuit undergoes undamped *LC* oscillations. At periodic instants, the energies stored by the capacitor and the inductor are equal, with each of the two components storing 250 μJ.

53. The switch in Figure P32.53 is connected to position *a* for a long time interval. At $t = 0$, the switch is thrown to position *b*. After this time, what are (a) the frequency of oscillation of the *LC* circuit, (b) the maximum charge that appears on the capacitor, (c) the maximum current in the inductor, and (d) the total energy the circuit possesses at $t = 3.00$ s?

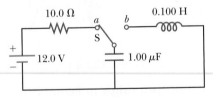

Figure P32.53

54. An *LC* circuit like that in Figure CQ32.8 consists of a 3.30-H inductor and an 840-pF capacitor that initially carries a 105-μC charge. The switch is open for $t < 0$ and is then thrown closed at $t = 0$. Compute the following quantities at $t = 2.00$ ms: (a) the energy stored in the capacitor, (b) the energy stored in the inductor, and (c) the total energy in the circuit.

55. An *LC* circuit like the one in Figure CQ32.8 contains an 82.0-mH inductor and a 17.0-μF capacitor that initially carries a 180-μC charge. The switch is open for $t < 0$ and is then thrown closed at $t = 0$. (a) Find the frequency (in hertz) of the resulting oscillations. At $t = 1.00$ ms, find (b) the charge on the capacitor and (c) the current in the circuit.

Section 32.6 The *RLC* Circuit

56. Show that Equation 32.28 in the text is Kirchhoff's loop rule as applied to the circuit in Figure P32.56 with the switch thrown to position *b*.

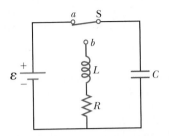

Figure P32.56 Problems 56 and 57.

57. In Figure P32.56, let $R = 7.60$ Ω, $L = 2.20$ mH, and $C = 1.80$ μF. (a) Calculate the frequency of the damped oscillation of the circuit when the switch is thrown to position *b*. (b) What is the critical resistance for damped oscillations?

58. Consider an *LC* circuit in which $L = 500$ mH and $C = 0.100$ μF. (a) What is the resonance frequency ω_0? (b) If a resistance of 1.00 kΩ is introduced into this circuit, what is the frequency of the damped oscillations? (c) By what percentage does the frequency of the damped oscillations differ from the resonance frequency?

59. Electrical oscillations are initiated in a series circuit containing a capacitance C, inductance L, and resistance R. (a) If $R \ll \sqrt{4L/C}$ (weak damping), what time interval elapses before the amplitude of the current oscillation falls to 50.0% of its initial value? (b) Over what time interval does the energy decrease to 50.0% of its initial value?

Additional Problems

60. **Review.** This problem extends the reasoning of Section 26.4, Problem 38 in Chapter 26, Problem 34 in Chapter 30, and Section 32.3. (a) Consider a capacitor with vacuum between its large, closely spaced, oppositely charged parallel plates. Show that the force on one plate can be accounted for by thinking of the electric field between the plates as exerting a "negative pressure" equal to the energy density of the electric field. (b) Consider two infinite plane sheets carrying electric currents in opposite directions with equal linear current densities J_s. Calculate the force per area acting on one sheet due to the magnetic field, of magnitude $\mu_0 J_s/2$, created by the other sheet. (c) Calculate the net magnetic field between the sheets and the field outside of the volume between them. (d) Calculate the energy density in the magnetic field between the sheets. (e) Show that the force on one sheet can be accounted for by thinking of the magnetic field between the sheets as exerting a positive pressure equal to its energy density. This result for magnetic pressure applies to all current configurations, not only to sheets of current.

61. A 1.00-mH inductor and a 1.00-μF capacitor are connected in series. The current in the circuit increases linearly in time as $i = 20.0t$, where i is in amperes and t is in seconds. The capacitor initially has no charge. Determine (a) the voltage across the inductor as a function of time, (b) the voltage across the capacitor as a function of time, and (c) the time when the energy stored in the capacitor first exceeds that in the inductor.

62. An inductor having inductance L and a capacitor having capacitance C are connected in series. The current in the circuit increases linearly in time as described by $i = Kt$, where K is a constant. The capacitor is initially uncharged. Determine (a) the voltage across the inductor as a function of time, (b) the voltage across the capacitor as a function of time, and (c) the time when the energy stored in the capacitor first exceeds that in the inductor.

63. A capacitor in a series LC circuit has an initial charge Q and is being discharged. When the charge on the capacitor is $Q/2$, find the flux through each of the N turns in the coil of the inductor in terms of Q, N, L, and C.

64. In the circuit diagrammed in Figure P32.25, assume the switch has been closed for a long time interval and is opened at $t = 0$. Also assume $R = 4.00\ \Omega$, $L = 1.00$ H, and $\mathcal{E} = 10.0$ V. (a) Before the switch is opened, does the inductor behave as an open circuit, a short circuit, a resistor of some particular resistance, or none of those choices? (b) What current does the inductor carry? (c) How much energy is stored in the inductor for $t < 0$? (d) After the switch is opened, what happens to the energy previously stored in the inductor? (e) Sketch a graph of the current in the inductor for $t \geq 0$. Label the initial and final values and the time constant.

65. When the current in the portion of the circuit shown in Figure P32.65 is 2.00 A and increases at a rate of 0.500 A/s, the measured voltage is $\Delta V_{ab} = 9.00$ V. When the current is 2.00 A and decreases at the rate of 0.500 A/s, the measured voltage is $\Delta V_{ab} = 5.00$ V. Calculate the values of (a) L and (b) R.

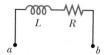

Figure P32.65

66. At the moment $t = 0$, a 24.0-V battery is connected to a 5.00-mH coil and a 6.00-Ω resistor. (a) Immediately thereafter, how does the potential difference across the resistor compare to the emf across the coil? (b) Answer the same question about the circuit several seconds later. (c) Is there an instant at which these two voltages are equal in magnitude? If so, when? Is there more than one such instant? (d) After a 4.00-A current is established in the resistor and coil, the battery is sud-

denly replaced by a short circuit. Answer parts (a), (b), and (c) again with reference to this new circuit.

67. (a) A flat, circular coil does not actually produce a uniform magnetic field in the area it encloses. Nevertheless, estimate the inductance of a flat, compact, circular coil with radius R and N turns by assuming the field at its center is uniform over its area. (b) A circuit on a laboratory table consists of a 1.50-volt battery, a 270-Ω resistor, a switch, and three 30.0-cm-long patch cords connecting them. Suppose the circuit is arranged to be circular. Think of it as a flat coil with one turn. Compute the order of magnitude of its inductance and (c) of the time constant describing how fast the current increases when you close the switch.

68. *Why is the following situation impossible?* You are working on an experiment involving a series circuit consisting of a charged 500-μF capacitor, a 32.0-mH inductor, and a resistor R. You discharge the capacitor through the inductor and resistor and observe the decaying oscillations of the current in the circuit. When the resistance R is 8.00 Ω, the decay in the oscillations is too slow for your experimental design. To make the decay faster, you double the resistance. As a result, you generate decaying oscillations of the current that are perfect for your needs.

69. A time-varying current i is sent through a 50.0-mH inductor from a source as shown in Figure P32.69a. The current is constant at $i = -1.00$ mA until $t = 0$ and then varies with time afterward as shown in Figure P32.69b. Make a graph of the emf across the inductor as a function of time.

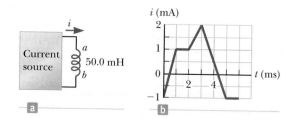

Figure P32.69

70. At $t = 0$, the open switch in Figure P32.70 is thrown closed. We wish to find a symbolic expression for the current in the inductor for time $t > 0$. Let this current be called i and choose it to be downward in the inductor in Figure P32.70. Identify i_1 as the current to the right through R_1 and i_2 as the current downward through R_2. (a) Use Kirchhoff's junction rule to find

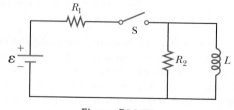

Figure P32.70

a relation among the three currents. (b) Use Kirchhoff's loop rule around the left loop to find another relationship. (c) Use Kirchhoff's loop rule around the outer loop to find a third relationship. (d) Eliminate i_1 and i_2 among the three equations to find an equation involving only the current i. (e) Compare the equation in part (d) with Equation 32.6 in the text. Use this comparison to rewrite Equation 32.7 in the text for the situation in this problem and show that

$$i(t) = \frac{\varepsilon}{R_1}\left[1 - e^{-(R'/L)t}\right]$$

where $R' = R_1R_2/(R_1 + R_2)$.

71. The toroid in Figure P32.71 consists of N turns and has a rectangular cross section. Its inner and outer radii are a and b, respectively. The figure shows half of the toroid to allow us to see its cross-section. Compute the inductance of a 500-turn toroid for which $a = 10.0$ cm, $b = 12.0$ cm, and $h = 1.00$ cm.

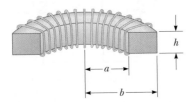

Figure P32.71 Problems 71 and 72.

72. The toroid in Figure P32.71 consists of N turns and has a rectangular cross section. Its inner and outer radii are a and b, respectively. Find the inductance of the toroid.

Problems 73 through 76 apply ideas from this and earlier chapters to some properties of superconductors, which were introduced in Section 27.5.

73. **Review.** A novel method of storing energy has been proposed. A huge underground superconducting coil, 1.00 km in diameter, would be fabricated. It would carry a maximum current of 50.0 kA through each winding of a 150-turn Nb₃Sn solenoid. (a) If the inductance of this huge coil were 50.0 H, what would be the total energy stored? (b) What would be the compressive force per unit length acting between two adjacent windings 0.250 m apart?

74. **Review.** In an experiment carried out by S. C. Collins between 1955 and 1958, a current was maintained in a superconducting lead ring for 2.50 yr with no observed loss, even though there was no energy input. If the inductance of the ring were 3.14×10^{-8} H and the sensitivity of the experiment were 1 part in 10^9, what was the maximum resistance of the ring? *Suggestion:* Treat the ring as an *RL* circuit carrying decaying current and recall that the approximation $e^{-x} \approx 1 - x$ is valid for small x.

75. **Review.** The use of superconductors has been proposed for power transmission lines. A single coaxial cable (Fig. P32.75) could carry a power of 1.00×10^3 MW (the output of a large power plant) at 200 kV, DC, over a distance of 1.00×10^3 km without loss. An inner wire of radius $a = 2.00$ cm, made from the superconductor Nb₃Sn, carries the current I in one direction. A surrounding superconducting cylinder of radius $b = 5.00$ cm would carry the return current I. In such a system, what is the magnetic field (a) at the surface of the inner conductor and (b) at the inner surface of the outer conductor? (c) How much energy would be stored in the magnetic field in the space between the conductors in a 1.00×10^3 km superconducting line? (d) What is the pressure exerted on the outer conductor due to the current in the inner conductor?

Figure P32.75

76. **Review.** A fundamental property of a type I superconducting material is *perfect diamagnetism*, or demonstration of the *Meissner effect*, illustrated in Figure 30.27 in Section 30.6 and described as follows. If a sample of superconducting material is placed into an externally produced magnetic field or is cooled to become superconducting while it is in a magnetic field, electric currents appear on the surface of the sample. The currents have precisely the strength and orientation required to make the total magnetic field be zero throughout the interior of the sample. This problem will help you understand the magnetic force that can then act on the sample. Compare this problem with Problem 65 in Chapter 26, pertaining to the force attracting a perfect dielectric into a strong electric field.

A vertical solenoid with a length of 120 cm and a diameter of 2.50 cm consists of 1 400 turns of copper wire carrying a counterclockwise current (when viewed from above) of 2.00 A as shown in Figure P32.76a (page 996). (a) Find the magnetic field in the vacuum inside the solenoid. (b) Find the energy density of the magnetic field. Now a superconducting bar 2.20 cm in diameter is inserted partway into the solenoid. Its upper end is far outside the solenoid, where the magnetic field is negligible. The lower end of the bar is deep inside the solenoid. (c) Explain how you identify the direction required for the current on the curved surface of the bar so that the total magnetic field is zero within the bar. The field created by the supercurrents is sketched in Figure P32.76b, and the total field is sketched in Figure P32.76c. (d) The field of the solenoid exerts a force on the current in the superconductor. Explain how you determine the direction of the force on the bar. (e) Noting that the units J/m³ of energy density are the same as the units N/m² of pressure, calculate the magnitude of the force by multiplying the energy density of the solenoid field times the area of the bottom end of the superconducting bar.

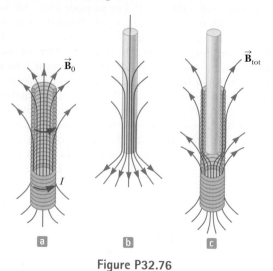

Figure P32.76

77. A wire of nonmagnetic material, with radius R, carries current uniformly distributed over its cross section. The total current carried by the wire is I. Show that the magnetic energy per unit length inside the wire is $\mu_0 I^2/16\pi$.

Challenge Problems

78. In earlier times when many households received non-digital television signals from an antenna, the lead-in wires from the antenna were often constructed in the form of two parallel wires (Fig. P32.78). The two wires carry currents of equal magnitude in opposite directions. The center-to-center separation of the wires is w, and a is their radius. Assume w is large enough compared with a that the wires carry the current uniformly distributed over their surfaces and negligible magnetic field exists inside the wires. (a) Why does this configuration of conductors have an inductance? (b) What constitutes the flux loop for this configuration? (c) Show that the inductance of a length x of this type of lead-in is

$$L = \frac{\mu_0 x}{\pi} \ln\left(\frac{w-a}{a}\right)$$

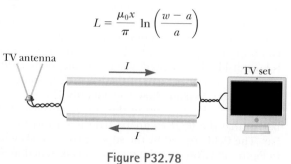

Figure P32.78

79. Assume the magnitude of the magnetic field outside a sphere of radius R is $B = B_0(R/r)^2$, where B_0 is a constant. (a) Determine the total energy stored in the

magnetic field outside the sphere. (b) Evaluate your result from part (a) for $B_0 = 5.00 \times 10^{-5}$ T and $R = 6.00 \times 10^6$ m, values appropriate for the Earth's magnetic field.

80. In Figure P32.80, the battery has emf $\mathcal{E} = 18.0$ V and the other circuit elements have values $L = 0.400$ H, $R_1 = 2.00$ kΩ, and $R_2 = 6.00$ kΩ. The switch is closed for $t < 0$, and steady-state conditions are established. The switch is then opened at $t = 0$. (a) Find the emf across L immediately after $t = 0$. (b) Which end of the coil, a or b, is at the higher potential? (c) Make graphs of the currents in R_1 and in R_2 as a function of time, treating the steady-state directions as positive. Show values before and after $t = 0$. (d) At what moment after $t = 0$ does the current in R_2 have the value 2.00 mA?

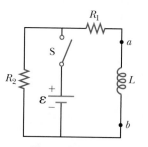

Figure P32.80

81. To prevent damage from arcing in an electric motor, a discharge resistor is sometimes placed in parallel with the armature. If the motor is suddenly unplugged while running, this resistor limits the voltage that appears across the armature coils. Consider a 12.0-V DC motor with an armature that has a resistance of 7.50 Ω and an inductance of 450 mH. Assume the magnitude of the self-induced emf in the armature coils is 10.0 V when the motor is running at normal speed. (The equivalent circuit for the armature is shown in Fig. P32.81.) Calculate the maximum resistance R that limits the voltage across the armature to 80.0 V when the motor is unplugged.

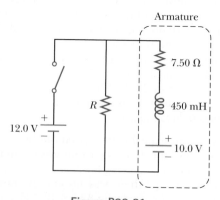

Figure P32.81

82. One application of an RL circuit is the generation of time-varying high voltage from a low-voltage source as shown in Figure P32.82. (a) What is the current in the circuit a long time after the switch has been in position *a*? (b) Now the switch is thrown quickly from *a* to *b*. Compute the initial voltage across each resistor and across the inductor. (c) How much time elapses before the voltage across the inductor drops to 12.0 V?

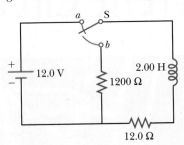

Figure P32.82

83. Two inductors having inductances L_1 and L_2 are connected in parallel as shown in Figure P32.83a. The mutual inductance between the two inductors is M. Determine the equivalent inductance L_{eq} for the system (Fig. P32.83b).

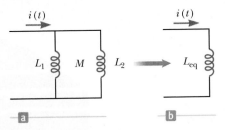

Figure P32.83

Alternating-Current Circuits

These large transformers are used to increase the voltage at a power plant for distribution of energy by electrical transmission to the power grid. Voltages can be changed relatively easily because power is distributed by alternating current rather than direct current. *(©Lester Lefkowitz/Getty Images)*

In this chapter, we describe alternating-current (AC) circuits. Every time you turn on a television set, a computer, or any of a multitude of other electrical appliances in a home, you are calling on alternating currents to provide the power to operate them. We begin our study by investigating the characteristics of simple series circuits that contain resistors, inductors, and capacitors and that are driven by a sinusoidal voltage. The primary aim of this chapter can be summarized as follows: if an AC source applies an alternating voltage to a series circuit containing resistors, inductors, and capacitors, we want to know the amplitude and time characteristics of the alternating current. We conclude this chapter with two sections concerning transformers, power transmission, and electrical filters.

33.1 AC Sources

An AC circuit consists of circuit elements and a power source that provides an alternating voltage Δv. This time-varying voltage from the source is described by

$$\Delta v = \Delta V_{max} \sin \omega t$$

where ΔV_{max} is the maximum output voltage of the source, or the **voltage amplitude.** There are various possibilities for AC sources, including generators as dis-

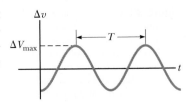

cussed in Section 31.5 and electrical oscillators. In a home, each electrical outlet serves as an AC source. Because the output voltage of an AC source varies sinusoidally with time, the voltage is positive during one half of the cycle and negative during the other half as in Figure 33.1. Likewise, the current in any circuit driven by an AC source is an alternating current that also varies sinusoidally with time.

From Equation 15.12, the angular frequency of the AC voltage is

$$\omega = 2\pi f = \frac{2\pi}{T}$$

where *f* is the frequency of the source and *T* is the period. The source determines the frequency of the current in any circuit connected to it. Commercial electric-power plants in the United States use a frequency of 60.0 Hz, which corresponds to an angular frequency of 377 rad/s.

33.2 Resistors in an AC Circuit

Consider a simple AC circuit consisting of a resistor and an AC source as shown in Figure 33.2. At any instant, the algebraic sum of the voltages around a closed loop in a circuit must be zero (Kirchhoff's loop rule). Therefore, $\Delta v + \Delta v_R = 0$ or, using Equation 27.7 for the voltage across the resistor,

$$\Delta v - i_R R = 0$$

If we rearrange this expression and substitute $\Delta V_{max} \sin \omega t$ for Δv, the instantaneous current in the resistor is

$$i_R = \frac{\Delta v}{R} = \frac{\Delta V_{max}}{R} \sin \omega t = I_{max} \sin \omega t \qquad (33.1)$$

where I_{max} is the maximum current:

$$I_{max} = \frac{\Delta V_{max}}{R} \qquad (33.2)$$

Equation 33.1 shows that the instantaneous voltage across the resistor is

$$\Delta v_R = i_R R = I_{max} R \sin \omega t \qquad (33.3)$$

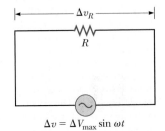

Figure 33.2 A circuit consisting of a resistor of resistance *R* connected to an AC source, designated by the symbol.

◀ **Maximum current in a resistor**

◀ **Voltage across a resistor**

A plot of voltage and current versus time for this circuit is shown in Figure 33.3a on page 1000. At point *a*, the current has a maximum value in one direction, arbitrarily called the positive direction. Between points *a* and *b*, the current is decreasing in magnitude but is still in the positive direction. At point *b*, the current is momentarily zero; it then begins to increase in the negative direction between points *b* and *c*. At point *c*, the current has reached its maximum value in the negative direction.

The current and voltage are in step with each other because they vary identically with time. Because i_R and Δv_R both vary as $\sin \omega t$ and reach their maximum values at the same time as shown in Figure 33.3a, they are said to be **in phase,** similar to the way two waves can be in phase as discussed in our study of wave motion in Chapter 18. Therefore, for a sinusoidal applied voltage, the current in a resistor is always in phase with the voltage across the resistor. For resistors in AC circuits, there are no new concepts to learn. Resistors behave essentially the same way in both DC and AC circuits. That, however, is not the case for capacitors and inductors.

To simplify our analysis of circuits containing two or more elements, we use a graphical representation called a *phasor diagram*. A **phasor** is a vector whose length is proportional to the maximum value of the variable it represents (ΔV_{max} for voltage and I_{max} for current in this discussion). The phasor rotates counterclockwise at an angular speed equal to the angular frequency associated with the variable. The

Figure 33.3 (a) Plots of the instantaneous current i_R and instantaneous voltage Δv_R across a resistor as functions of time. At time $t = T$, one cycle of the time-varying voltage and current has been completed. (b) Phasor diagram for the resistive circuit showing that the current is in phase with the voltage.

Pitfall Prevention 33.2
A Phasor Is Like a Graph An alternating voltage can be presented in different representations. One graphical representation is shown in Figure 33.1 in which the voltage is drawn in rectangular coordinates, with voltage on the vertical axis and time on the horizontal axis. Figure 33.3b shows another graphical representation. The phase space in which the phasor is drawn is similar to polar coordinate graph paper. The radial coordinate represents the amplitude of the voltage. The angular coordinate is the phase angle. The vertical-axis coordinate of the tip of the phasor represents the instantaneous value of the voltage. The horizontal coordinate represents nothing at all. As shown in Figure 33.3b, alternating currents can also be represented by phasors.

To help with this discussion of phasors, review Section 15.4, where we represented the simple harmonic motion of a real object by the projection of an imaginary object's uniform circular motion onto a coordinate axis. A phasor is a direct analog to this representation.

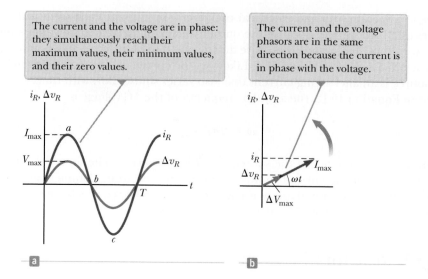

The current and the voltage are in phase: they simultaneously reach their maximum values, their minimum values, and their zero values.

The current and the voltage phasors are in the same direction because the current is in phase with the voltage.

projection of the phasor onto the vertical axis represents the instantaneous value of the quantity it represents.

Figure 33.3b shows voltage and current phasors for the circuit of Figure 33.2 at some instant of time. The projections of the phasor arrows onto the vertical axis are determined by a sine function of the angle of the phasor with respect to the horizontal axis. For example, the projection of the current phasor in Figure 33.3b is $I_{max} \sin \omega t$. Notice that this expression is the same as Equation 33.1. Therefore, the projections of phasors represent current values that vary sinusoidally in time. We can do the same with time-varying voltages. The advantage of this approach is that the phase relationships among currents and voltages can be represented as vector additions of phasors using the vector addition techniques discussed in Chapter 3.

In the case of the single-loop resistive circuit of Figure 33.2, the current and voltage phasors are in the same direction in Figure 33.3b because i_R and Δv_R are in phase. The current and voltage in circuits containing capacitors and inductors have different phase relationships.

Q uick Quiz 33.1 Consider the voltage phasor in Figure 33.4, shown at three instants of time. **(i)** Choose the part of the figure, (a), (b), or (c), that represents the instant of time at which the instantaneous value of the voltage has the largest magnitude. **(ii)** Choose the part of the figure that represents the instant of time at which the instantaneous value of the voltage has the smallest magnitude.

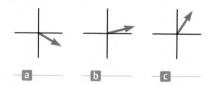

Figure 33.4 (Quick Quiz 33.1) A voltage phasor is shown at three instants of time, (a), (b), and (c).

For the simple resistive circuit in Figure 33.2, notice that the average value of the current over one cycle is zero. That is, the current is maintained in the positive direction for the same amount of time and at the same magnitude as it is maintained in the negative direction. The direction of the current, however, has no effect on the behavior of the resistor. We can understand this concept by realizing that collisions between electrons and the fixed atoms of the resistor result in an

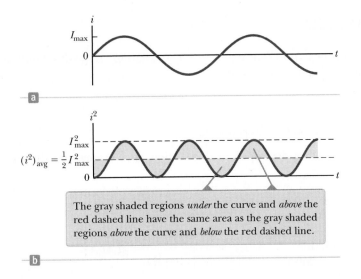

The gray shaded regions *under* the curve and *above* the red dashed line have the same area as the gray shaded regions *above* the curve and *below* the red dashed line.

increase in the resistor's temperature. Although this temperature increase depends on the magnitude of the current, it is independent of the current's direction.

We can make this discussion quantitative by recalling that the rate at which energy is delivered to a resistor is the power $P = i^2 R$, where i is the instantaneous current in the resistor. Because this rate is proportional to the square of the current, it makes no difference whether the current is direct or alternating, that is, whether the sign associated with the current is positive or negative. The temperature increase produced by an alternating current having a maximum value I_{max}, however, is not the same as that produced by a direct current equal to I_{max} because the alternating current has this maximum value for only an instant during each cycle (Fig. 33.5a). What is of importance in an AC circuit is an average value of current, referred to as the **rms current.** As we learned in Section 21.1, the notation *rms* stands for *root-mean-square,* which in this case means the square root of the mean (average) value of the square of the current: $I_{rms} = \sqrt{(i^2)_{avg}}$. Because i^2 varies as $\sin^2 \omega t$ and because the average value of i^2 is $\frac{1}{2} I_{max}^2$ (see Fig. 33.5b), the rms current is

$$I_{rms} = \frac{I_{max}}{\sqrt{2}} = 0.707 I_{max} \qquad (33.4)$$

◀ rms current

This equation states that an alternating current whose maximum value is 2.00 A delivers to a resistor the same power as a direct current that has a value of $(0.707)(2.00\ \text{A}) = 1.41$ A. The average power delivered to a resistor that carries an alternating current is

$$P_{avg} = I_{rms}^2 R$$

◀ Average power delivered to a resistor

Alternating voltage is also best discussed in terms of rms voltage, and the relationship is identical to that for current:

$$\Delta V_{rms} = \frac{\Delta V_{max}}{\sqrt{2}} = 0.707\ \Delta V_{max} \qquad (33.5)$$

◀ rms voltage

When we speak of measuring a 120-V alternating voltage from an electrical outlet, we are referring to an rms voltage of 120 V. A calculation using Equation 33.5 shows that such an alternating voltage has a maximum value of about 170 V. One reason rms values are often used when discussing alternating currents and voltages is that AC ammeters and voltmeters are designed to read rms values. Furthermore, with rms values, many of the equations we use have the same form as their direct-current counterparts.

Example 33.1 What Is the rms Current?

The voltage output of an AC source is given by the expression $\Delta v = 200 \sin \omega t$, where Δv is in volts. Find the rms current in the circuit when this source is connected to a 100-Ω resistor.

SOLUTION

Conceptualize Figure 33.2 shows the physical situation for this problem.

Categorize We evaluate the current with an equation developed in this section, so we categorize this example as a substitution problem.

Combine Equations 33.2 and 33.4 to find the rms current:

$$I_{rms} = \frac{I_{max}}{\sqrt{2}} = \frac{\Delta V_{max}}{\sqrt{2}\, R}$$

Comparing the expression for voltage output with the general form $\Delta v = \Delta V_{max} \sin \omega t$ shows that $\Delta V_{max} = 200$ V. Substitute numerical values:

$$I_{rms} = \frac{200\text{ V}}{\sqrt{2}\,(100\ \Omega)} = \boxed{1.41\text{ A}}$$

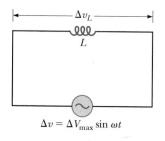

Figure 33.6 A circuit consisting of an inductor of inductance L connected to an AC source.

$\Delta v = \Delta V_{max} \sin \omega t$

33.3 Inductors in an AC Circuit

Now consider an AC circuit consisting only of an inductor connected to the terminals of an AC source as shown in Figure 33.6. Because $\Delta v_L = -L(di_L/dt)$ is the self-induced instantaneous voltage across the inductor (see Eq. 32.1), Kirchhoff's loop rule applied to this circuit gives $\Delta v + \Delta v_L = 0$, or

$$\Delta v - L\frac{di_L}{dt} = 0$$

Substituting $\Delta V_{max} \sin \omega t$ for Δv and rearranging gives

$$\Delta v = L\frac{di_L}{dt} = \Delta V_{max} \sin \omega t \tag{33.6}$$

Solving this equation for di_L gives

$$di_L = \frac{\Delta V_{max}}{L} \sin \omega t\, dt$$

Integrating this expression[1] gives the instantaneous current i_L in the inductor as a function of time:

$$i_L = \frac{\Delta V_{max}}{L} \int \sin \omega t\, dt = -\frac{\Delta V_{max}}{\omega L} \cos \omega t \tag{33.7}$$

Using the trigonometric identity $\cos \omega t = -\sin(\omega t - \pi/2)$, we can express Equation 33.7 as

Current in an inductor ▶

$$i_L = \frac{\Delta V_{max}}{\omega L} \sin\left(\omega t - \frac{\pi}{2}\right) \tag{33.8}$$

Comparing this result with Equation 33.6 shows that the instantaneous current i_L in the inductor and the instantaneous voltage Δv_L across the inductor are *out* of phase by $\pi/2$ rad $= 90°$.

A plot of voltage and current versus time is shown in Figure 33.7a. When the current i_L in the inductor is a maximum (point b in Fig. 33.7a), it is momentarily

---

[1]We neglect the constant of integration here because it depends on the initial conditions, which are not important for this situation.

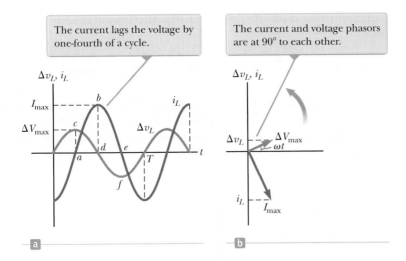

not changing, so the voltage across the inductor is zero (point d). At points such as a and e, the current is zero and the rate of change of current is at a maximum. Therefore, the voltage across the inductor is also at a maximum (points c and f). Notice that the voltage reaches its maximum value one-quarter of a period before the current reaches its maximum value. Therefore, for a sinusoidal applied voltage, the current in an inductor always *lags* behind the voltage across the inductor by 90° (one-quarter cycle in time).

As with the relationship between current and voltage for a resistor, we can represent this relationship for an inductor with a phasor diagram as in Figure 33.7b. The phasors are at 90° to each other, representing the 90° phase difference between current and voltage.

Equation 33.7 shows that the current in an inductive circuit reaches its maximum value when $\cos \omega t = \pm 1$:

$$I_{max} = \frac{\Delta V_{max}}{\omega L} \tag{33.9}$$

◀ **Maximum current in an inductor**

This expression is similar to the relationship between current, voltage, and resistance in a DC circuit, $I = \Delta V/R$ (Eq. 27.7). Because I_{max} has units of amperes and ΔV_{max} has units of volts, ωL must have units of ohms. Therefore, ωL has the same units as resistance and is related to current and voltage in the same way as resistance. It must behave in a manner similar to resistance in the sense that it represents opposition to the flow of charge. Because ωL depends on the applied frequency ω, the inductor *reacts* differently, in terms of offering opposition to current, for different frequencies. For this reason, we define ωL as the **inductive reactance** X_L:

$$X_L \equiv \omega L \tag{33.10}$$

◀ **Inductive reactance**

Therefore, we can write Equation 33.9 as

$$I_{max} = \frac{\Delta V_{max}}{X_L} \tag{33.11}$$

The expression for the rms current in an inductor is similar to Equation 33.11, with I_{max} replaced by I_{rms} and ΔV_{max} replaced by ΔV_{rms}.

Equation 33.10 indicates that, for a given applied voltage, the inductive reactance increases as the frequency increases. This conclusion is consistent with Faraday's law: the greater the rate of change of current in the inductor, the larger the back emf. The larger back emf translates to an increase in the reactance and a decrease in the current.

Using Equations 33.6 and 33.11, we find that the instantaneous voltage across the inductor is

Voltage across an inductor ▶

$$\Delta v_L = -L \frac{di_L}{dt} = -\Delta V_{max} \sin \omega t = -I_{max} X_L \sin \omega t \qquad \textbf{(33.12)}$$

Quick Quiz 33.2 Consider the AC circuit in Figure 33.8. The frequency of the AC source is adjusted while its voltage amplitude is held constant. When does the lightbulb glow the brightest? **(a)** It glows brightest at high frequencies. **(b)** It glows brightest at low frequencies. **(c)** The brightness is the same at all frequencies.

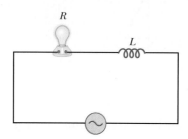

Figure 33.8 (Quick Quiz 33.2) At what frequencies does the lightbulb glow the brightest?

Example 33.2   **A Purely Inductive AC Circuit**

In a purely inductive AC circuit, $L = 25.0$ mH and the rms voltage is 150 V. Calculate the inductive reactance and rms current in the circuit if the frequency is 60.0 Hz.

SOLUTION

Conceptualize Figure 33.6 shows the physical situation for this problem. Keep in mind that inductive reactance increases with increasing frequency of the applied voltage.

Categorize We determine the reactance and the current from equations developed in this section, so we categorize this example as a substitution problem.

Use Equation 33.10 to find the inductive reactance:

$$X_L = \omega L = 2\pi f L = 2\pi (60.0 \text{ Hz})(25.0 \times 10^{-3} \text{ H})$$

$$= \boxed{9.42 \ \Omega}$$

Use an rms version of Equation 33.11 to find the rms current:

$$I_{rms} = \frac{\Delta V_{rms}}{X_L} = \frac{150 \text{ V}}{9.42 \ \Omega} = \boxed{15.9 \text{ A}}$$

WHAT IF?   If the frequency increases to 6.00 kHz, what happens to the rms current in the circuit?

Answer If the frequency increases, the inductive reactance also increases because the current is changing at a higher rate. The increase in inductive reactance results in a lower current.
   Let's calculate the new inductive reactance and the new rms current:

$$X_L = 2\pi (6.00 \times 10^3 \text{ Hz})(25.0 \times 10^{-3} \text{ H}) = 942 \ \Omega$$

$$I_{rms} = \frac{150 \text{ V}}{942 \ \Omega} = 0.159 \text{ A}$$

33.4 Capacitors in an AC Circuit

Figure 33.9 shows an AC circuit consisting of a capacitor connected across the terminals of an AC source. Kirchhoff's loop rule applied to this circuit gives $\Delta v + \Delta v_C = 0$, or

$$\Delta v - \frac{q}{C} = 0 \qquad \textbf{(33.13)}$$

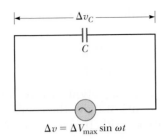

Figure 33.9 A circuit consisting of a capacitor of capacitance C connected to an AC source.

Substituting $\Delta V_{max} \sin \omega t$ for Δv and rearranging gives

$$q = C \Delta V_{max} \sin \omega t \qquad (33.14)$$

where q is the instantaneous charge on the capacitor. Differentiating Equation 33.14 with respect to time gives the instantaneous current in the circuit:

$$i_C = \frac{dq}{dt} = \omega C \Delta V_{max} \cos \omega t \qquad (33.15)$$

Using the trigonometric identity

$$\cos \omega t = \sin \left(\omega t + \frac{\pi}{2} \right)$$

we can express Equation 33.15 in the alternative form

$$i_C = \omega C \Delta V_{max} \sin \left(\omega t + \frac{\pi}{2} \right) \qquad (33.16)$$

◀ **Current in a capacitor**

Comparing this expression with $\Delta v = \Delta V_{max} \sin \omega t$ shows that the current is $\pi/2$ rad $= 90°$ out of phase with the voltage across the capacitor. A plot of current and voltage versus time (Fig. 33.10a) shows that the current reaches its maximum value one-quarter of a cycle sooner than the voltage reaches its maximum value.

Consider a point such as b in Figure 33.10a where the current is zero at this instant. That occurs when the capacitor reaches its maximum charge so that the voltage across the capacitor is a maximum (point d). At points such as a and e, the current is a maximum, which occurs at those instants when the charge on the capacitor reaches zero and the capacitor begins to recharge with the opposite polarity. When the charge is zero, the voltage across the capacitor is zero (points c and f).

As with inductors, we can represent the current and voltage for a capacitor on a phasor diagram. The phasor diagram in Figure 33.10b shows that for a sinusoidally applied voltage, the current always *leads* the voltage across a capacitor by 90°.

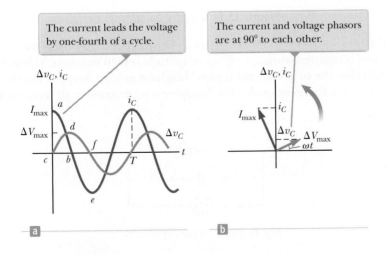

The current leads the voltage by one-fourth of a cycle.

The current and voltage phasors are at 90° to each other.

Figure 33.10 (a) Plots of the instantaneous current i_C and instantaneous voltage Δv_C across a capacitor as functions of time. (b) Phasor diagram for the capacitive circuit.

Equation 33.15 shows that the current in the circuit reaches its maximum value when $\cos \omega t = \pm 1$:

$$I_{max} = \omega C \, \Delta V_{max} = \frac{\Delta V_{max}}{(1/\omega C)} \tag{33.17}$$

As in the case with inductors, this looks like Equation 27.7, so the denominator plays the role of resistance, with units of ohms. We give the combination $1/\omega C$ the symbol X_C, and because this function varies with frequency, we define it as the **capacitive reactance:**

◀ Capacitive reactance

$$X_C \equiv \frac{1}{\omega C} \tag{33.18}$$

We can now write Equation 33.17 as

◀ Maximum current
in a capacitor

$$I_{max} = \frac{\Delta V_{max}}{X_C} \tag{33.19}$$

The rms current is given by an expression similar to Equation 33.19, with I_{max} replaced by I_{rms} and ΔV_{max} replaced by ΔV_{rms}.

Using Equation 33.19, we can express the instantaneous voltage across the capacitor as

◀ Voltage across a capacitor

$$\Delta v_C = \Delta V_{max} \sin \omega t = I_{max} X_C \sin \omega t \tag{33.20}$$

Equations 33.18 and 33.19 indicate that as the frequency of the voltage source increases, the capacitive reactance decreases and the maximum current therefore increases. The frequency of the current is determined by the frequency of the voltage source driving the circuit. As the frequency approaches zero, the capacitive reactance approaches infinity and the current therefore approaches zero. This conclusion makes sense because the circuit approaches direct current conditions as ω approaches zero and the capacitor represents an open circuit.

Quick **Quiz** 33.3 Consider the AC circuit in Figure 33.11. The frequency of the AC source is adjusted while its voltage amplitude is held constant. When does the lightbulb glow the brightest? **(a)** It glows brightest at high frequencies. **(b)** It glows brightest at low frequencies. **(c)** The brightness is the same at all frequencies.

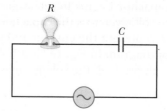

Figure 33.11 (Quick Quiz 33.3)

Quick **Quiz** 33.4 Consider the AC circuit in Figure 33.12. The frequency of the AC source is adjusted while its voltage amplitude is held constant. When does the lightbulb glow the brightest? **(a)** It glows brightest at high frequencies. **(b)** It glows brightest at low frequencies. **(c)** The brightness is the same at all frequencies.

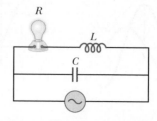

Figure 33.12 (Quick Quiz 33.4)

Example 33.3   A Purely Capacitive AC Circuit

An 8.00-μF capacitor is connected to the terminals of a 60.0-Hz AC source whose rms voltage is 150 V. Find the capacitive reactance and the rms current in the circuit.

SOLUTION

Conceptualize Figure 33.9 shows the physical situation for this problem. Keep in mind that capacitive reactance decreases with increasing frequency of the applied voltage.

Categorize We determine the reactance and the current from equations developed in this section, so we categorize this example as a substitution problem.

Use Equation 33.18 to find the capacitive reactance:

$$X_C = \frac{1}{\omega C} = \frac{1}{2\pi f C} = \frac{1}{2\pi(60.0 \text{ Hz})(8.00 \times 10^{-6} \text{ F})} = \boxed{332 \ \Omega}$$

Use an rms version of Equation 33.19 to find the rms current:

$$I_{\text{rms}} = \frac{\Delta V_{\text{rms}}}{X_C} = \frac{150 \text{ V}}{332 \ \Omega} = \boxed{0.452 \text{ A}}$$

WHAT IF? What if the frequency is doubled? What happens to the rms current in the circuit?

Answer If the frequency increases, the capacitive reactance decreases, which is just the opposite from the case of an inductor. The decrease in capacitive reactance results in an increase in the current.

Let's calculate the new capacitive reactance and the new rms current:

$$X_C = \frac{1}{\omega C} = \frac{1}{2\pi(120 \text{ Hz})(8.00 \times 10^{-6} \text{ F})} = 166 \ \Omega$$

$$I_{\text{rms}} = \frac{150 \text{ V}}{166 \ \Omega} = 0.904 \text{ A}$$

33.5 The *RLC* Series Circuit

In the previous sections, we considered individual circuit elements connected to an AC source. Figure 33.13a shows a circuit that contains a combination of circuit elements: a resistor, an inductor, and a capacitor connected in series across an alternating-voltage source. If the applied voltage varies sinusoidally with time, the instantaneous applied voltage is

$$\Delta v = \Delta V_{\text{max}} \sin \omega t$$

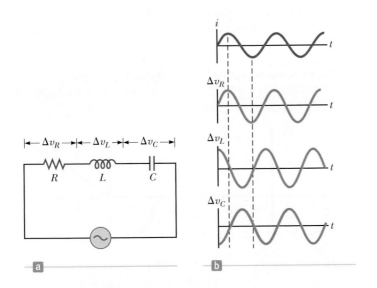

Figure 33.13 (a) A series circuit consisting of a resistor, an inductor, and a capacitor connected to an AC source. (b) Phase relationships between the current and the voltages in the individual circuit elements if they were connected alone to the AC source.

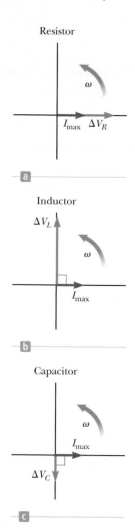

Figure 33.14 Phase relationships between the voltage and current phasors for (a) a resistor, (b) an inductor, and (c) a capacitor connected in series.

Figure 33.13b shows the voltage versus time across each element in the circuit and its phase relationships to the current if it were connected individually to the AC source, as discussed in Sections 33.2–33.4.

When the circuit elements are all connected together to the AC source, as in Figure 33.13a, the current in the circuit is given by

$$i = I_{max} \sin(\omega t - \phi)$$

where ϕ is some **phase angle** between the current and the applied voltage. Based on our discussions of phase in Sections 33.3 and 33.4, we expect that the current will generally not be in phase with the voltage in an *RLC* circuit.

Because the circuit elements in Figure 33.13a are in series, the current everywhere in the circuit must be the same at any instant. That is, the current at all points in a series AC circuit has the same amplitude and phase. Based on the preceding sections, we know that the voltage across each element has a different amplitude and phase. In particular, the voltage across the resistor is in phase with the current, the voltage across the inductor leads the current by 90°, and the voltage across the capacitor lags behind the current by 90°. Using these phase relationships, we can express the instantaneous voltages across the three circuit elements as

$$\Delta v_R = I_{max} R \sin \omega t = \Delta V_R \sin \omega t \qquad \textbf{(33.21)}$$

$$\Delta v_L = I_{max} X_L \sin\left(\omega t + \frac{\pi}{2}\right) = \Delta V_L \cos \omega t \qquad \textbf{(33.22)}$$

$$\Delta v_C = I_{max} X_C \sin\left(\omega t - \frac{\pi}{2}\right) = -\Delta V_C \cos \omega t \qquad \textbf{(33.23)}$$

The sum of these three voltages must equal the instantaneous voltage from the AC source, but it is important to recognize that because the three voltages have different phase relationships with the current, they cannot be added directly. Figure 33.14 represents the phasors at an instant at which the current in all three elements is momentarily zero. The zero current is represented by the current phasor along the horizontal axis in each part of the figure. Next the voltage phasor is drawn at the appropriate phase angle to the current for each element.

Because phasors are rotating vectors, the voltage phasors in Figure 33.14 can be combined using vector addition as in Figure 33.15. In Figure 33.15a, the voltage phasors in Figure 33.14 are combined on the same coordinate axes. Figure 33.15b shows the vector addition of the voltage phasors. The voltage phasors ΔV_L and ΔV_C are in *opposite* directions along the same line, so we can construct the difference phasor $\Delta V_L - \Delta V_C$, which is perpendicular to the phasor ΔV_R. This diagram shows that the vector sum of the voltage amplitudes ΔV_R, ΔV_L, and ΔV_C equals a phasor whose length is the maximum applied voltage ΔV_{max} and which makes an angle ϕ with the current phasor I_{max}. From the right triangle in Figure 33.15b, we see that

Figure 33.15 (a) Phasor diagram for the series *RLC* circuit shown in Figure 33.13a. (b) The inductance and capacitance phasors are added together and then added vectorially to the resistance phasor.

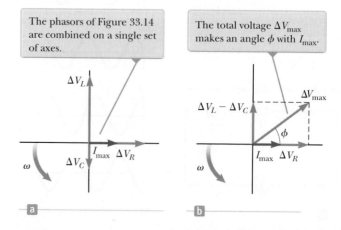

$$\Delta V_{\max} = \sqrt{\Delta V_R^2 + (\Delta V_L - \Delta V_C)^2} = \sqrt{(I_{\max}R)^2 + (I_{\max}X_L - I_{\max}X_C)^2}$$

$$\Delta V_{\max} = I_{\max}\sqrt{R^2 + (X_L - X_C)^2}$$

Therefore, we can express the maximum current as

$$I_{\max} = \frac{\Delta V_{\max}}{\sqrt{R^2 + (X_L - X_C)^2}} \qquad (33.24)$$

◀ Maximum current in an *RLC* circuit

Once again, this expression has the same mathematical form as Equation 27.7. The denominator of the fraction plays the role of resistance and is called the **impedance** *Z* of the circuit:

$$Z \equiv \sqrt{R^2 + (X_L - X_C)^2} \qquad (33.25)$$

◀ Impedance

where impedance also has units of ohms. Therefore, Equation 33.24 can be written in the form

$$I_{\max} = \frac{\Delta V_{\max}}{Z} \qquad (33.26)$$

Equation 33.26 is the AC equivalent of Equation 27.7. Note that the impedance and therefore the current in an AC circuit depend on the resistance, the inductance, the capacitance, and the frequency (because the reactances are frequency dependent).

From the right triangle in the phasor diagram in Figure 33.15b, the phase angle ϕ between the current and the voltage is found as follows:

$$\phi = \tan^{-1}\left(\frac{\Delta V_L - \Delta V_C}{\Delta V_R}\right) = \tan^{-1}\left(\frac{I_{\max}X_L - I_{\max}X_C}{I_{\max}R}\right)$$

$$\phi = \tan^{-1}\left(\frac{X_L - X_C}{R}\right) \qquad (33.27)$$

◀ Phase angle

When $X_L > X_C$ (which occurs at high frequencies), the phase angle is positive, signifying that the current lags the applied voltage as in Figure 33.15b. We describe this situation by saying that the circuit is *more inductive than capacitive*. When $X_L < X_C$, the phase angle is negative, signifying that the current leads the applied voltage, and the circuit is *more capacitive than inductive*. When $X_L = X_C$, the phase angle is zero and the circuit is *purely resistive*.

Quick Quiz 33.5 Label each part of Figure 33.16, (a), (b), and (c), as representing $X_L > X_C$, $X_L = X_C$, or $X_L < X_C$.

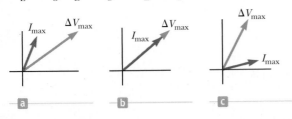

Figure 33.16 (Quick Quiz 33.5) Match the phasor diagrams to the relationships between the reactances.

Example 33.4 | **Analyzing a Series *RLC* Circuit**

A series *RLC* circuit has $R = 425\ \Omega$, $L = 1.25$ H, and $C = 3.50\ \mu$F. It is connected to an AC source with $f = 60.0$ Hz and $\Delta V_{\max} = 150$ V.

(A) Determine the inductive reactance, the capacitive reactance, and the impedance of the circuit. *continued*

▶ **33.4** continued

SOLUTION

Conceptualize The circuit of interest in this example is shown in Figure 33.13a. The current in the combination of the resistor, inductor, and capacitor oscillates at a particular phase angle with respect to the applied voltage.

Categorize The circuit is a simple series *RLC* circuit, so we can use the approach discussed in this section.

Analyze Find the angular frequency:

$$\omega = 2\pi f = 2\pi(60.0 \text{ Hz}) = 377 \text{ s}^{-1}$$

Use Equation 33.10 to find the inductive reactance:

$$X_L = \omega L = (377 \text{ s}^{-1})(1.25 \text{ H}) = \boxed{471 \ \Omega}$$

Use Equation 33.18 to find the capacitive reactance:

$$X_C = \frac{1}{\omega C} = \frac{1}{(377 \text{ s}^{-1})(3.50 \times 10^{-6} \text{ F})} = \boxed{758 \ \Omega}$$

Use Equation 33.25 to find the impedance:

$$Z = \sqrt{R^2 + (X_L - X_C)^2}$$

$$= \sqrt{(425 \ \Omega)^2 + (471 \ \Omega - 758 \ \Omega)^2} = \boxed{513 \ \Omega}$$

(B) Find the maximum current in the circuit.

SOLUTION

Use Equation 33.26 to find the maximum current:

$$I_{max} = \frac{\Delta V_{max}}{Z} = \frac{150 \text{ V}}{513 \ \Omega} = \boxed{0.293 \text{ A}}$$

(C) Find the phase angle between the current and voltage.

SOLUTION

Use Equation 33.27 to calculate the phase angle:

$$\phi = \tan^{-1}\left(\frac{X_L - X_C}{R}\right) = \tan^{-1}\left(\frac{471 \ \Omega - 758 \ \Omega}{425 \ \Omega}\right) = \boxed{-34.0°}$$

(D) Find the maximum voltage across each element.

SOLUTION

Use Equations 33.2, 33.11, and 33.19 to calculate the maximum voltages:

$$\Delta V_R = I_{max} R = (0.293 \text{ A})(425 \ \Omega) = \boxed{124 \text{ V}}$$

$$\Delta V_L = I_{max} X_L = (0.293 \text{ A})(471 \ \Omega) = \boxed{138 \text{ V}}$$

$$\Delta V_C = I_{max} X_C = (0.293 \text{ A})(758 \ \Omega) = \boxed{222 \text{ V}}$$

(E) What replacement value of *L* should an engineer analyzing the circuit choose such that the current leads the applied voltage by 30.0° rather than 34.0°? All other values in the circuit stay the same.

SOLUTION

Solve Equation 33.27 for the inductive reactance:

$$X_L = X_C + R \tan \phi$$

Substitute Equations 33.10 and 33.18 into this expression:

$$\omega L = \frac{1}{\omega C} + R \tan \phi$$

Solve for *L*:

$$L = \frac{1}{\omega}\left(\frac{1}{\omega C} + R \tan \phi\right)$$

Substitute the given values:

$$L = \frac{1}{(377 \text{ s}^{-1})}\left[\frac{1}{(377 \text{ s}^{-1})(3.50 \times 10^{-6} \text{ F})} + (425 \ \Omega) \tan (-30.0°)\right]$$

$$L = \boxed{1.36 \text{ H}}$$

Finalize Because the capacitive reactance is larger than the inductive reactance, the circuit is more capacitive than inductive. In this case, the phase angle ϕ is negative, so the current leads the applied voltage.

▶ **33.4** continued

Using Equations 33.21, 33.22, and 33.23, the instantaneous voltages across the three elements are

$$\Delta v_R = (124 \text{ V}) \sin 377t$$

$$\Delta v_L = (138 \text{ V}) \cos 377t$$

$$\Delta v_C = (-222 \text{ V}) \cos 377t$$

WHAT IF?   What if you added up the maximum voltages across the three circuit elements? Is that a physically meaningful quantity?

Answer The sum of the maximum voltages across the elements is $\Delta V_R + \Delta V_L + \Delta V_C = 484$ V. This sum is much greater than the maximum voltage of the source, 150 V. The sum of the maximum voltages is a meaningless quantity because when sinusoidally varying quantities are added, *both their amplitudes and their phases* must be taken into account. The maximum voltages across the various elements occur at different times. Therefore, the voltages must be added in a way that takes account of the different phases as shown in Figure 33.15.

33.6 Power in an AC Circuit

Now let's take an energy approach to analyzing AC circuits and consider the transfer of energy from the AC source to the circuit. The power delivered by a battery to an external DC circuit is equal to the product of the current and the terminal voltage of the battery. Likewise, the instantaneous power delivered by an AC source to a circuit is the product of the current and the applied voltage. For the *RLC* circuit shown in Figure 33.13a, we can express the instantaneous power P as

$$P = i\,\Delta v = I_{max} \sin(\omega t - \phi)\,\Delta V_{max} \sin \omega t$$

$$P = I_{max}\,\Delta V_{max} \sin \omega t \sin(\omega t - \phi) \tag{33.28}$$

This result is a complicated function of time and is therefore not very useful from a practical viewpoint. What is generally of interest is the average power over one or more cycles. Such an average can be computed by first using the trigonometric identity $\sin(\omega t - \phi) = \sin \omega t \cos \phi - \cos \omega t \sin \phi$. Substituting this identity into Equation 33.28 gives

$$P = I_{max}\,\Delta V_{max} \sin^2 \omega t \cos \phi - I_{max}\,\Delta V_{max} \sin \omega t \cos \omega t \sin \phi \tag{33.29}$$

Let's now take the time average of P over one or more cycles, noting that I_{max}, ΔV_{max}, ϕ, and ω are all constants. The time average of the first term on the right of the equal sign in Equation 33.29 involves the average value of $\sin^2 \omega t$, which is $\frac{1}{2}$. The time average of the second term on the right of the equal sign is identically zero because $\sin \omega t \cos \omega t = \frac{1}{2} \sin 2\omega t$, and the average value of $\sin 2\omega t$ is zero. Therefore, we can express the **average power** P_{avg} as

$$P_{avg} = \tfrac{1}{2} I_{max}\,\Delta V_{max} \cos \phi \tag{33.30}$$

It is convenient to express the average power in terms of the rms current and rms voltage defined by Equations 33.4 and 33.5:

$$P_{avg} = I_{rms}\,\Delta V_{rms} \cos \phi \tag{33.31}$$

◀ **Average power delivered to an *RLC* circuit**

where the quantity $\cos \phi$ is called the **power factor.** Figure 33.15b shows that the maximum voltage across the resistor is given by $\Delta V_R = \Delta V_{max} \cos \phi = I_{max} R$. Therefore, $\cos \phi = I_{max} R / \Delta V_{max} = R/Z$, and we can express P_{avg} as

$$P_{avg} = I_{rms}\,\Delta V_{rms} \cos \phi = I_{rms}\,\Delta V_{rms}\left(\frac{R}{Z}\right) = I_{rms}\left(\frac{\Delta V_{rms}}{Z}\right)R$$

Recognizing that $\Delta V_{rms}/Z = I_{rms}$ gives

$$P_{avg} = I_{rms}^2 R \qquad \text{(33.32)}$$

The average power delivered by the source is converted to internal energy in the resistor, just as in the case of a DC circuit. When the load is purely resistive, $\phi = 0$, $\cos \phi = 1$, and, from Equation 33.31, we see that

$$P_{avg} = I_{rms}\, \Delta V_{rms}$$

Note that no power losses are associated with pure capacitors and pure inductors in an AC circuit. To see why that is true, let's first analyze the power in an AC circuit containing only a source and a capacitor. When the current begins to increase in one direction in an AC circuit, charge begins to accumulate on the capacitor and a voltage appears across it. When this voltage reaches its maximum value, the energy stored in the capacitor as electric potential energy is $\frac{1}{2}C(\Delta V_{max})^2$. This energy storage, however, is only momentary. The capacitor is charged and discharged twice during each cycle: charge is delivered to the capacitor during two quarters of the cycle and is returned to the voltage source during the remaining two quarters. Therefore, the average power supplied by the source is zero. In other words, no power losses occur in a capacitor in an AC circuit.

Now consider the case of an inductor. When the current in an inductor reaches its maximum value, the energy stored in the inductor is a maximum and is given by $\frac{1}{2}LI_{max}^2$. When the current begins to decrease in the circuit, this stored energy in the inductor returns to the source as the inductor attempts to maintain the current in the circuit.

Equation 33.31 shows that the power delivered by an AC source to any circuit depends on the phase, a result that has many interesting applications. For example, a factory that uses large motors in machines, generators, or transformers has a large inductive load (because of all the windings). To deliver greater power to such devices in the factory without using excessively high voltages, technicians introduce capacitance in the circuits to shift the phase.

Quick Quiz 33.6 An AC source drives an RLC circuit with a fixed voltage amplitude. If the driving frequency is ω_1, the circuit is more capacitive than inductive and the phase angle is $-10°$. If the driving frequency is ω_2, the circuit is more inductive than capacitive and the phase angle is $+10°$. At what frequency is the largest amount of power delivered to the circuit? **(a)** It is largest at ω_1. **(b)** It is largest at ω_2. **(c)** The same amount of power is delivered at both frequencies.

Example 33.5   Average Power in an *RLC* Series Circuit

Calculate the average power delivered to the series *RLC* circuit described in Example 33.4.

SOLUTION

Conceptualize Consider the circuit in Figure 33.13a and imagine energy being delivered to the circuit by the AC source. Review Example 33.4 for other details about this circuit.

Categorize We find the result by using equations developed in this section, so we categorize this example as a substitution problem.

Use Equation 33.5 and the maximum voltage from Example 33.4 to find the rms voltage from the source:

$$\Delta V_{rms} = \frac{\Delta V_{max}}{\sqrt{2}} = \frac{150\ \text{V}}{\sqrt{2}} = 106\ \text{V}$$

Similarly, find the rms current in the circuit:

$$I_{rms} = \frac{I_{max}}{\sqrt{2}} = \frac{0.293\ \text{A}}{\sqrt{2}} = 0.207\ \text{A}$$

▶ **33.5** continued

Use Equation 33.31 to find the power delivered by the source:

$$P_{avg} = I_{rms} V_{rms} \cos \phi = (0.207\ \text{A})(106\ \text{V}) \cos (-34.0°)$$

$$= \boxed{18.2\ \text{W}}$$

33.7 Resonance in a Series *RLC* Circuit

We investigated resonance in mechanical oscillating systems in Chapter 15. As shown in Chapter 32, a series *RLC* circuit is an electrical oscillating system. Such a circuit is said to be **in resonance** when the driving frequency is such that the rms current has its maximum value. In general, the rms current can be written

$$I_{rms} = \frac{\Delta V_{rms}}{Z} \tag{33.33}$$

where *Z* is the impedance. Substituting the expression for *Z* from Equation 33.25 into Equation 33.33 gives

$$I_{rms} = \frac{\Delta V_{rms}}{\sqrt{R^2 + (X_L - X_C)^2}} \tag{33.34}$$

Because the impedance depends on the frequency of the source, the current in the *RLC* circuit also depends on the frequency. The angular frequency ω_0 at which $X_L - X_C = 0$ is called the **resonance frequency** of the circuit. To find ω_0, we set $X_L = X_C$, which gives $\omega_0 L = 1/\omega_0 C$, or

$$\omega_0 = \frac{1}{\sqrt{LC}} \tag{33.35}$$ ◀ **Resonance frequency**

This frequency also corresponds to the natural frequency of oscillation of an *LC* circuit (see Section 32.5). Therefore, the rms current in a series *RLC* circuit has its maximum value when the frequency of the applied voltage matches the natural oscillator frequency, which depends only on *L* and *C*. Furthermore, at the resonance frequency, the current is in phase with the applied voltage.

Quick Quiz 33.7 What is the impedance of a series *RLC* circuit at resonance?
(a) larger than *R* (b) less than *R* (c) equal to *R* (d) impossible to determine

A plot of rms current versus angular frequency for a series *RLC* circuit is shown in Figure 33.17a on page 1014. The data assume a constant $\Delta V_{rms} = 5.0$ mV, $L = 5.0\ \mu$H, and $C = 2.0$ nF. The three curves correspond to three values of *R*. In each case, the rms current has its maximum value at the resonance frequency ω_0. Furthermore, the curves become narrower and taller as the resistance decreases.

Equation 33.34 shows that when $R = 0$, the current becomes infinite at resonance. Real circuits, however, always have some resistance, which limits the value of the current to some finite value.

We can also calculate the average power as a function of frequency for a series *RLC* circuit. Using Equations 33.32, 33.33, and 33.25 gives

$$P_{avg} = I_{rms}^2 R = \frac{(\Delta V_{rms})^2}{Z^2} R = \frac{(\Delta V_{rms})^2 R}{R^2 + (X_L - X_C)^2} \tag{33.36}$$

Because $X_L = \omega L$, $X_C = 1/\omega C$, and $\omega_0^2 = 1/LC$, the term $(X_L - X_C)^2$ can be expressed as

$$(X_L - X_C)^2 = \left(\omega L - \frac{1}{\omega C} \right)^2 = \frac{L^2}{\omega^2} (\omega^2 - \omega_0^2)^2$$

Figure 33.17 (a) The rms current versus frequency for a series *RLC* circuit for three values of *R*. (b) Average power delivered to the circuit versus frequency for the series *RLC* circuit for three values of *R*.

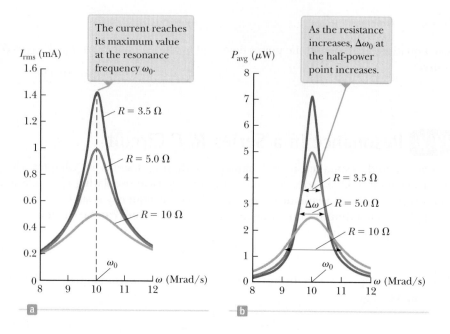

Using this result in Equation 33.36 gives

Average power as a function of frequency in an *RLC* circuit ▶

$$P_{\text{avg}} = \frac{(\Delta V_{\text{rms}})^2 R\omega^2}{R^2 \omega^2 + L^2(\omega^2 - \omega_0^2)^2}$$

(33.37)

Equation 33.37 shows that at resonance, when $\omega = \omega_0$, the average power is a maximum and has the value $(\Delta V_{\text{rms}})^2/R$. Figure 33.17b is a plot of average power versus frequency for three values of R in a series *RLC* circuit. As the resistance is made smaller, the curve becomes sharper in the vicinity of the resonance frequency. This curve sharpness is usually described by a dimensionless parameter known as the **quality factor**,[2] denoted by Q:

Quality factor ▶

$$Q = \frac{\omega_0}{\Delta\omega}$$

where $\Delta\omega$ is the width of the curve measured between the two values of ω for which P_{avg} has one-half its maximum value, called the *half-power points* (see Fig. 33.17b.) It is left as a problem (Problem 76) to show that the width at the half-power points has the value $\Delta\omega = R/L$ so that

$$Q = \frac{\omega_0 L}{R}$$

(33.38)

A radio's receiving circuit is an important application of a resonant circuit. The radio is tuned to a particular station (which transmits an electromagnetic wave or signal of a specific frequency) by varying a capacitor, which changes the receiving circuit's resonance frequency. When the circuit is driven by the electromagnetic oscillations a radio signal produces in an antenna, the tuner circuit responds with a large amplitude of electrical oscillation only for the station frequency that matches the resonance frequency. Therefore, only the signal from one radio station is passed on to the amplifier and loudspeakers even though signals from all stations are driving the circuit at the same time. Because many signals are often present over a range of frequencies, it is important to design a high-Q circuit to eliminate unwanted signals. In this manner, stations whose frequencies are near but not equal to the resonance frequency have a response at the receiver that is negligibly small relative to the signal that matches the resonance frequency.

---

[2]The quality factor is also defined as the ratio $2\pi E/\Delta E$, where E is the energy stored in the oscillating system and ΔE is the energy decrease per cycle of oscillation due to the resistance.

Example 33.6   **A Resonating Series *RLC* Circuit**

Consider a series *RLC* circuit for which $R = 150 \ \Omega$, $L = 20.0$ mH, $\Delta V_{rms} = 20.0$ V, and $\omega = 5\ 000$ s^{-1}. Determine the value of the capacitance for which the current is a maximum.

SOLUTION

Conceptualize Consider the circuit in Figure 33.13a and imagine varying the frequency of the AC source. The current in the circuit has its maximum value at the resonance frequency ω_0.

Categorize We find the result by using equations developed in this section, so we categorize this example as a substitution problem.

Use Equation 33.35 to solve for the required capacitance in terms of the resonance frequency:

$$\omega_0 = \frac{1}{\sqrt{LC}} \rightarrow C = \frac{1}{\omega_0^2 L}$$

Substitute numerical values:

$$C = \frac{1}{(5.00 \times 10^3 \ \text{s}^{-1})^2 (20.0 \times 10^{-3} \ \text{H})} = \boxed{2.00 \ \mu\text{F}}$$

33.8 The Transformer and Power Transmission

As discussed in Section 27.6, it is economical to use a high voltage and a low current to minimize the I^2R loss in transmission lines when electric power is transmitted over great distances. Consequently, 350-kV lines are common, and in many areas, even higher-voltage (765-kV) lines are used. At the receiving end of such lines, the consumer requires power at a low voltage (for safety and for efficiency in design). In practice, the voltage is decreased to approximately 20 000 V at a distribution substation, then to 4 000 V for delivery to residential areas, and finally to 120 V and 240 V at the customer's site. Therefore, a device is needed that can change the alternating voltage and current without causing appreciable changes in the power delivered. The AC transformer is that device.

In its simplest form, the **AC transformer** consists of two coils of wire wound around a core of iron as illustrated in Figure 33.18. (Compare this arrangement to Faraday's experiment in Figure 31.2.) The coil on the left, which is connected to the input alternating-voltage source and has N_1 turns, is called the *primary winding* (or the *primary*). The coil on the right, consisting of N_2 turns and connected to a load resistor R_L, is called the *secondary winding* (or the *secondary*). The purposes of the iron core are to increase the magnetic flux through the coil and to provide a medium in which nearly all the magnetic field lines through one coil pass through the other coil. Eddy-current losses are reduced by using a laminated core. Transformation of energy to internal energy in the finite resistance of the coil wires is usually quite small. Typical transformers have power efficiencies from 90% to

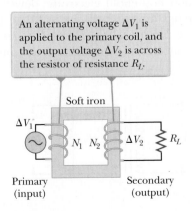

An alternating voltage ΔV_1 is applied to the primary coil, and the output voltage ΔV_2 is across the resistor of resistance R_L.

Soft iron

ΔV_1 N_1 N_2 ΔV_2 R_L

Primary
(input)

Secondary
(output)

Figure 33.18 An ideal transformer consists of two coils wound on the same iron core.

99%. In the discussion that follows, let's assume we are working with an *ideal transformer,* one in which the energy losses in the windings and core are zero.

Faraday's law states that the voltage Δv_1 across the primary is

$$\Delta v_1 = -N_1 \frac{d\Phi_B}{dt} \tag{33.39}$$

where Φ_B is the magnetic flux through each turn. If we assume all magnetic field lines remain within the iron core, the flux through each turn of the primary equals the flux through each turn of the secondary. Hence, the voltage across the secondary is

$$\Delta v_2 = -N_2 \frac{d\Phi_B}{dt} \tag{33.40}$$

Solving Equation 33.39 for $d\Phi_B/dt$ and substituting the result into Equation 33.40 gives

$$\Delta v_2 = \frac{N_2}{N_1} \Delta v_1 \tag{33.41}$$

When $N_2 > N_1$, the output voltage Δv_2 exceeds the input voltage Δv_1. This configuration is referred to as a *step-up transformer.* When $N_2 < N_1$, the output voltage is less than the input voltage, and we have a *step-down transformer.* A circuit diagram for a transformer connected to a load resistance is shown in Figure 33.19.

When a current I_1 exists in the primary circuit, a current I_2 is induced in the secondary. (In this discussion, uppercase I and ΔV refer to rms values.) If the load in the secondary circuit is a pure resistance, the induced current is in phase with the induced voltage. The power supplied to the secondary circuit must be provided by the AC source connected to the primary circuit. In an ideal transformer where there are no losses, the power $I_1 \Delta V_1$ supplied by the source is equal to the power $I_2 \Delta V_2$ in the secondary circuit. That is,

$$I_1 \Delta V_1 = I_2 \Delta V_2 \tag{33.42}$$

The value of the load resistance R_L determines the value of the secondary current because $I_2 = \Delta V_2/R_L$. Furthermore, the current in the primary is $I_1 = \Delta V_1/R_{eq}$, where

$$R_{eq} = \left(\frac{N_1}{N_2}\right)^2 R_L \tag{33.43}$$

is the equivalent resistance of the load resistance when viewed from the primary side. We see from this analysis that a transformer may be used to match resistances between the primary circuit and the load. In this manner, maximum power transfer can be achieved between a given power source and the load resistance. For example, a transformer connected between the 1-kΩ output of an audio amplifier and an 8-Ω speaker ensures that as much of the audio signal as possible is transferred into the speaker. In stereo terminology, this process is called *impedance matching.*

To operate properly, many common household electronic devices require low voltages. A small transformer that plugs directly into the wall like the one illustrated in Figure 33.20 can provide the proper voltage. The photograph shows the two windings wrapped around a common iron core that is found inside all these

Nikola Tesla
American Physicist (1856–1943)
Tesla was born in Croatia, but he spent most of his professional life as an inventor in the United States. He was a key figure in the development of alternating-current electricity, high-voltage transformers, and the transport of electrical power using AC transmission lines. Tesla's viewpoint was at odds with the ideas of Thomas Edison, who committed himself to the use of direct current in power transmission. Tesla's AC approach won out.

Figure 33.19 Circuit diagram for a transformer.

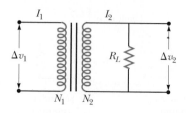

This transformer is smaller than the one in the opening photograph of this chapter. In addition, it is a step-down transformer. It drops the voltage from 4 000 V to 240 V for delivery to a group of residences.

The primary winding in this transformer is attached to the prongs of the plug, whereas the secondary winding is connected to the power cord on the right.

Figure 33.20 Electronic devices are often powered by AC adaptors containing transformers such as this one. These adaptors alter the AC voltage. In many applications, the adaptors also convert alternating current to direct current.

little "black boxes." This particular transformer converts the 120-V AC in the wall socket to 12.5-V AC. (Can you determine the ratio of the numbers of turns in the two coils?) Some black boxes also make use of diodes to convert the alternating current to direct current. (See Section 33.9.)

Example 33.7   **The Economics of AC Power**

An electricity-generating station needs to deliver energy at a rate of 20 MW to a city 1.0 km away. A common voltage for commercial power generators is 22 kV, but a step-up transformer is used to boost the voltage to 230 kV before transmission.

(A) If the resistance of the wires is 2.0 Ω and the energy costs are about 11¢/kWh, estimate the cost of the energy converted to internal energy in the wires during one day.

SOLUTION

Conceptualize The resistance of the wires is in series with the resistance representing the load (homes and businesses). Therefore, there is a voltage drop in the wires, which means that some of the transmitted energy is converted to internal energy in the wires and never reaches the load.

Categorize This problem involves finding the power delivered to a resistive load in an AC circuit. Let's ignore any capacitive or inductive characteristics of the load and set the power factor equal to 1.

Analyze Calculate I_{rms} in the wires from Equation 33.31:

$$I_{rms} = \frac{P_{avg}}{\Delta V_{rms}} = \frac{20 \times 10^6 \text{ W}}{230 \times 10^3 \text{ V}} = 87 \text{ A}$$

Determine the rate at which energy is delivered to the resistance in the wires from Equation 33.32:

$$P_{wires} = I_{rms}^2 R = (87 \text{ A})^2 (2.0 \text{ Ω}) = 15 \text{ kW}$$

Calculate the energy T_{ET} delivered to the wires over the course of a day:

$$T_{ET} = P_{wires} \Delta t = (15 \text{ kW})(24 \text{ h}) = \boxed{363 \text{ kWh}}$$

Find the cost of this energy at a rate of 11¢/kWh:

$$\text{Cost} = (363 \text{ kWh})(\$0.11/\text{kWh}) = \boxed{\$40}$$

(B) Repeat the calculation for the situation in which the power plant delivers the energy at its original voltage of 22 kV.

continued

▶ **33.7** continued

SOLUTION

Calculate I_{rms} in the wires from Equation 33.31:

$$I_{rms} = \frac{P_{avg}}{\Delta V_{rms}} = \frac{20 \times 10^6 \text{ W}}{22 \times 10^3 \text{ V}} = 909 \text{ A}$$

From Equation 33.32, determine the rate at which energy is delivered to the resistance in the wires:

$$P_{wires} = I_{rms}^2 R = (909 \text{ A})^2 (2.0 \text{ }\Omega) = 1.7 \times 10^3 \text{ kW}$$

Calculate the energy delivered to the wires over the course of a day:

$$T_{ET} = P_{wires} \Delta t = (1.7 \times 10^3 \text{ kW})(24 \text{ h}) = 4.0 \times 10^4 \text{ kWh}$$

Find the cost of this energy at a rate of 11¢/kWh:

$$\text{Cost} = (4.0 \times 10^4 \text{ kWh})(\$0.11/\text{kWh}) = \boxed{\$4.4 \times 10^3}$$

Finalize Notice the tremendous savings that are possible through the use of transformers and high-voltage transmission lines. Such savings in combination with the efficiency of using alternating current to operate motors led to the universal adoption of alternating current instead of direct current for commercial power grids.

33.9 Rectifiers and Filters

Portable electronic devices such as radios and laptop computers are often powered by direct current supplied by batteries. Many devices come with AC–DC converters such as that shown in Figure 33.20. Such a converter contains a transformer that steps the voltage down from 120 V to, typically, 6 V or 9 V and a circuit that converts alternating current to direct current. The AC–DC converting process is called **rectification,** and the converting device is called a **rectifier.**

The most important element in a rectifier circuit is a **diode,** a circuit element that conducts current in one direction but not the other. Most diodes used in modern electronics are semiconductor devices. The circuit symbol for a diode is ——▶|——, where the arrow indicates the direction of the current in the diode. A diode has low resistance to current in one direction (the direction of the arrow) and high resistance to current in the opposite direction. To understand how a diode rectifies a current, consider Figure 33.21a, which shows a diode and a resistor connected to the secondary of a transformer. The transformer reduces the voltage from 120-V AC to the lower voltage that is needed for the device having a resistance R (the load

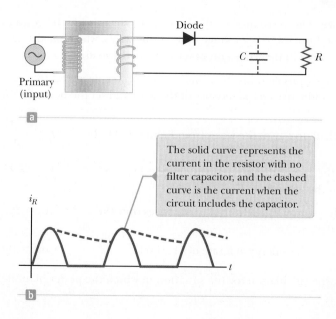

Figure 33.21 (a) A half-wave rectifier with an optional filter capacitor. (b) Current versus time in the resistor.

The solid curve represents the current in the resistor with no filter capacitor, and the dashed curve is the current when the circuit includes the capacitor.

resistance). Because the diode conducts current in only one direction, the alternating current in the load resistor is reduced to the form shown by the solid curve in Figure 33.21b. The diode conducts current only when the side of the symbol containing the arrowhead has a positive potential relative to the other side. In this situation, the diode acts as a *half-wave rectifier* because current is present in the circuit only during half of each cycle.

When a capacitor is added to the circuit as shown by the dashed lines and the capacitor symbol in Figure 33.21a, the circuit is a simple DC power supply. The time variation of the current in the load resistor (the dashed curve in Fig. 33.21b) is close to being zero, as determined by the RC time constant of the circuit. As the current in the circuit begins to rise at $t = 0$ in Figure 33.21b, the capacitor charges up. When the current begins to fall, however, the capacitor discharges through the resistor, so the current in the resistor does not fall as quickly as the current from the transformer.

The RC circuit in Figure 33.21a is one example of a **filter circuit,** which is used to smooth out or eliminate a time-varying signal. For example, radios are usually powered by a 60-Hz alternating voltage. After rectification, the voltage still contains a small AC component at 60 Hz (sometimes called *ripple*), which must be filtered. By "filtered," we mean that the 60-Hz ripple must be reduced to a value much less than that of the audio signal to be amplified because without filtering, the resulting audio signal includes an annoying hum at 60 Hz.

We can also design filters that respond differently to different frequencies. Consider the simple series RC circuit shown in Figure 33.22a. The input voltage is across the series combination of the two elements. The output is the voltage across the resistor. A plot of the ratio of the output voltage to the input voltage as a function of the logarithm of angular frequency (see Fig. 33.22b) shows that at low frequencies, ΔV_{out} is much smaller than ΔV_{in}, whereas at high frequencies, the two voltages are equal. Because the circuit preferentially passes signals of higher frequency while blocking low-frequency signals, the circuit is called an **RC high-pass filter.** (See Problem 54 for an analysis of this filter.)

Physically, a high-pass filter works because a capacitor "blocks out" direct current and AC current at low frequencies. At low frequencies, the capacitive reactance is large and much of the applied voltage appears across the capacitor rather than across the output resistor. As the frequency increases, the capacitive reactance drops and more of the applied voltage appears across the resistor.

Now consider the circuit shown in Figure 33.23a on page 1020, where we have interchanged the resistor and capacitor and where the output voltage is taken across the capacitor. At low frequencies, the reactance of the capacitor and the voltage across the capacitor is high. As the frequency increases, the voltage across the capacitor drops. Therefore, this filter is an **RC low-pass filter.** The ratio of output voltage to input voltage (see Problem 56), plotted as a function of the logarithm of ω in Figure 33.23b, shows this behavior.

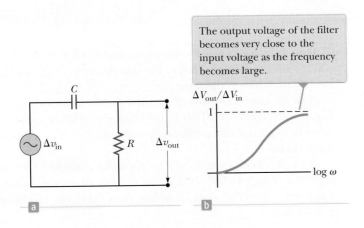

The output voltage of the filter becomes very close to the input voltage as the frequency becomes large.

Figure 33.22 (a) A simple RC high-pass filter. (b) Ratio of output voltage to input voltage for an RC high-pass filter as a function of the angular frequency of the AC source.

Figure 33.23 (a) A simple RC low-pass filter. (b) Ratio of output voltage to input voltage for an RC low-pass filter as a function of the angular frequency of the AC source.

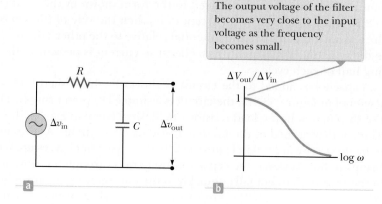

The output voltage of the filter becomes very close to the input voltage as the frequency becomes small.

You may be familiar with crossover networks, which are an important part of the speaker systems for high-quality audio systems. These networks use low-pass filters to direct low frequencies to a special type of speaker, the "woofer," which is designed to reproduce the low notes accurately. The high frequencies are sent by a high-pass filter to the "tweeter" speaker.

Summary

Definitions

In AC circuits that contain inductors and capacitors, it is useful to define the **inductive reactance** X_L and the **capacitive reactance** X_C as

$$X_L \equiv \omega L \qquad \text{(33.10)}$$

$$X_C \equiv \frac{1}{\omega C} \qquad \text{(33.18)}$$

where ω is the angular frequency of the AC source. The SI unit of reactance is the ohm.

The **impedance** Z of an RLC series AC circuit is

$$Z \equiv \sqrt{R^2 + (X_L - X_C)^2} \qquad \text{(33.25)}$$

This expression illustrates that we cannot simply add the resistance and reactances in a circuit. We must account for the applied voltage and current being out of phase, with the **phase angle** ϕ between the current and voltage being

$$\phi = \tan^{-1}\left(\frac{X_L - X_C}{R}\right) \qquad \text{(33.27)}$$

The sign of ϕ can be positive or negative, depending on whether X_L is greater or less than X_C. The phase angle is zero when $X_L = X_C$.

Concepts and Principles

The **rms current** and **rms voltage** in an AC circuit in which the voltages and current vary sinusoidally are given by

$$I_{rms} = \frac{I_{max}}{\sqrt{2}} = 0.707 I_{max} \qquad \text{(33.4)}$$

$$\Delta V_{rms} = \frac{\Delta V_{max}}{\sqrt{2}} = 0.707 \Delta V_{max} \qquad \text{(33.5)}$$

where I_{max} and ΔV_{max} are the maximum values.

If an AC circuit consists of a source and a resistor, the current is in phase with the voltage. That is, the current and voltage reach their maximum values at the same time.

If an AC circuit consists of a source and an inductor, the current lags the voltage by 90°. That is, the voltage reaches its maximum value one-quarter of a period before the current reaches its maximum value.

If an AC circuit consists of a source and a capacitor, the current leads the voltage by 90°. That is, the current reaches its maximum value one-quarter of a period before the voltage reaches its maximum value.

The **average power** delivered by the source in an RLC circuit is

$$P_{avg} = I_{rms} \, \Delta V_{rms} \cos \phi \qquad \text{(33.31)}$$

An equivalent expression for the average power is

$$P_{avg} = I_{rms}^2 \, R \qquad \text{(33.32)}$$

The average power delivered by the source results in increasing internal energy in the resistor. No power loss occurs in an ideal inductor or capacitor.

A series RLC circuit is in resonance when the inductive reactance equals the capacitive reactance. When this condition is met, the rms current given by Equation 33.34 has its maximum value. The **resonance frequency** ω_0 of the circuit is

$$\omega_0 = \frac{1}{\sqrt{LC}} \qquad \text{(33.35)}$$

The rms current in a series RLC circuit has its maximum value when the frequency of the source equals ω_0, that is, when the "driving" frequency matches the resonance frequency.

The rms current in a series RLC circuit is

$$I_{rms} = \frac{\Delta V_{rms}}{\sqrt{R^2 + (X_L - X_C)^2}} \qquad \text{(33.34)}$$

AC transformers allow for easy changes in alternating voltage according to

$$\Delta v_2 = \frac{N_2}{N_1} \Delta v_1 \qquad \text{(33.41)}$$

where N_1 and N_2 are the numbers of windings on the primary and secondary coils, respectively, and Δv_1 and Δv_2 are the voltages on these coils.

Objective Questions   **1.** denotes answer available in *Student Solutions Manual/Study Guide*

1. An inductor and a resistor are connected in series across an AC source as in Figure OQ33.1. Immediately after the switch is closed, which of the following statements is true? (a) The current in the circuit is $\Delta V/R$. (b) The voltage across the inductor is zero. (c) The current in the circuit is zero. (d) The voltage across the resistor is ΔV. (e) The voltage across the inductor is half its maximum value.

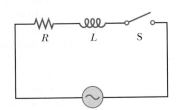

Figure OQ33.1

2. **(i)** When a particular inductor is connected to a source of sinusoidally varying emf with constant amplitude and a frequency of 60.0 Hz, the rms current is 3.00 A. What is the rms current if the source frequency is doubled? (a) 12.0 A (b) 6.00 A (c) 4.24 A (d) 3.00 A (e) 1.50 A **(ii)** Repeat part (i) assuming the load is a capacitor instead of an inductor. **(iii)** Repeat part (i) assuming the load is a resistor instead of an inductor.

3. A capacitor and a resistor are connected in series across an AC source as shown in Figure OQ33.3. After the switch is closed, which of the following statements is true? (a) The voltage across the capacitor lags the current by 90°. (b) The voltage across the resistor is out of phase with the current. (c) The voltage across the capacitor leads the current by 90°. (d) The current decreases as the frequency of the source is increased,

but its peak voltage remains the same. (e) None of those statements is correct.

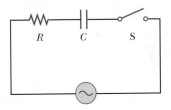

Figure OQ33.3

4. **(i)** What is the time average of the "square-wave" potential shown in Figure OQ33.4? (a) $\sqrt{2}\,\Delta V_{max}$ (b) ΔV_{max} (c) $\Delta V_{max}/\sqrt{2}$ (d) $\Delta V_{max}/2$ (e) $\Delta V_{max}/4$ **(ii)** What is the rms voltage? Choose from the same possibilities as in part (i).

Figure OQ33.4

5. If the voltage across a circuit element has its maximum value when the current in the circuit is zero, which of the following statements *must* be true? (a) The circuit element is a resistor. (b) The circuit element is a capacitor. (c) The circuit element is an inductor. (d) The current and voltage are 90° out of phase. (e) The current and voltage are 180° out of phase.

6. A sinusoidally varying potential difference has amplitude 170 V. **(i)** What is its minimum instantaneous value? (a) 170 V (b) 120 V (c) 0 (d) −120 V (e) −170 V (ii) What is its average value? (iii) What is its rms value? Choose from the same possibilities as in part (i) in each case.

7. A series *RLC* circuit contains a 20.0-Ω resistor, a 0.750-μF capacitor, and a 120-mH inductor. (i) If a sinusoidally varying rms voltage of 120 V at f = 500 Hz is applied across this combination of elements, what is the rms current in the circuit? (a) 2.33 A (b) 6.00 A (c) 10.0 A (d) 17.0 A (e) none of those answers (ii) **What If?** What is the rms current in the circuit when operating at its resonance frequency? Choose from the same possibilities as in part (i).

8. A resistor, a capacitor, and an inductor are connected in series across an AC source. Which of the following statements is *false*? (a) The instantaneous voltage across the capacitor lags the current by 90°. (b) The instantaneous voltage across the inductor leads the current by 90°. (c) The instantaneous voltage across the resistor is in phase with the current. (d) The voltages across the resistor, capacitor, and inductor are not in phase. (e) The rms voltage across the combination of the three elements equals the algebraic sum of the rms voltages across each element separately.

9. Under what conditions is the impedance of a series *RLC* circuit equal to the resistance in the circuit? (a) The driving frequency is lower than the resonance frequency. (b) The driving frequency is equal to the resonance frequency. (c) The driving frequency is higher than the resonance frequency. (d) always (e) never

10. What is the phase angle in a series *RLC* circuit at resonance? (a) 180° (b) 90° (c) 0 (d) −90° (e) None of those answers is necessarily correct.

11. A circuit containing an AC source, a capacitor, an inductor, and a resistor has a high-*Q* resonance at 1 000 Hz. From greatest to least, rank the following contributions to the impedance of the circuit at that frequency and at lower and higher frequencies. Note any cases of equality in your ranking. (a) X_C at 500 Hz (b) X_C at 1 500 Hz (c) X_L at 500 Hz (d) X_L at 1 500 Hz (e) R at 1 000 Hz

12. A 6.00-V battery is connected across the primary coil of a transformer having 50 turns. If the secondary coil of the transformer has 100 turns, what voltage appears across the secondary? (a) 24.0 V (b) 12.0 V (c) 6.00 V (d) 3.00 V (e) none of those answers

13. Do AC ammeters and voltmeters read (a) peak-to-valley, (b) maximum, (c) rms, or (d) average values?

Conceptual Questions

1. denotes answer available in *Student Solutions Manual/Study Guide*

1. (a) Explain how the quality factor is related to the response characteristics of a radio receiver. (b) Which variable most strongly influences the quality factor?

2. (a) Explain how the mnemonic "ELI the ICE man" can be used to recall whether current leads voltage or voltage leads current in *RLC* circuits. Note that E represents emf $\mathcal{E}$. (b) Explain how "CIVIL" works as another mnemonic device, where V represents voltage.

3. Why is the sum of the maximum voltages across each element in a series *RLC* circuit usually greater than the maximum applied voltage? Doesn't that inequality violate Kirchhoff's loop rule?

4. (a) Does the phase angle in an *RLC* series circuit depend on frequency? (b) What is the phase angle for the circuit when the inductive reactance equals the capacitive reactance?

5. Do some research to answer these questions: Who invented the metal detector? Why? What are its limitations?

6. As shown in Figure CQ33.6, a person pulls a vacuum cleaner at speed v across a horizontal floor, exerting on it a force of magnitude F directed upward at an angle θ with the horizontal. (a) At what rate is the person doing work on the cleaner? (b) State as completely as you can the analogy between power in this situation and in an electric circuit.

7. A certain power supply can be modeled as a source of emf in series with both a resistance of 10 Ω and an inductive reactance of 5 Ω. To obtain maximum power delivered to the load, it is found that the load should have a resistance of R_L = 10 Ω, an inductive reactance of zero, and a capacitive reactance of 5 Ω. (a) With this load, is the circuit in resonance? (b) With this load, what fraction of the average power put out by the source of emf is delivered to the load? (c) To increase the fraction of the power delivered to the load, how could the load be changed? You may wish to review Example 28.2 and Problem 4 in Chapter 28 on maximum power transfer in DC circuits.

8. Will a transformer operate if a battery is used for the input voltage across the primary? Explain.

9. (a) Why does a capacitor act as a short circuit at high frequencies? (b) Why does a capacitor act as an open circuit at low frequencies?

10. An ice storm breaks a transmission line and interrupts electric power to a town. A homeowner starts a gasoline-powered 120-V generator and clips its output terminals to "hot" and "ground" terminals of the electrical panel for his house. On a power pole down the block is a transformer designed to step down the voltage for household use. It has a ratio of turns N_1/N_2 of 100 to 1. A repairman climbs the pole. What voltage

Figure CQ33.6

will he encounter on the input side of the transformer? As this question implies, safety precautions must be taken in the use of home generators and during power failures in general.

Problems

Section 33.1 AC Sources

Section 33.2 Resistors in an AC Circuit

1. When an AC source is connected across a 12.0-Ω resistor, the rms current in the resistor is 8.00 A. Find (a) the rms voltage across the resistor, (b) the peak voltage of the source, (c) the maximum current in the resistor, and (d) the average power delivered to the resistor.

2. (a) What is the resistance of a lightbulb that uses an average power of 75.0 W when connected to a 60.0-Hz power source having a maximum voltage of 170 V? (b) **What If?** What is the resistance of a 100-W lightbulb?

3. An AC power supply produces a maximum voltage $\Delta V_{max} = 100$ V. This power supply is connected to a resistor $R = 24.0$ Ω, and the current and resistor voltage are measured with an ideal AC ammeter and voltmeter as shown in Figure P33.3. An ideal ammeter has zero resistance, and an ideal voltmeter has infinite resistance. What is the reading on (a) the ammeter and (b) the voltmeter?

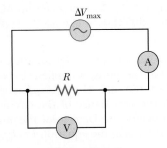

Figure P33.3

4. A certain lightbulb is rated at 60.0 W when operating at an rms voltage of 120 V. (a) What is the peak voltage applied across the bulb? (b) What is the resistance of the bulb? (c) Does a 100-W bulb have greater or less resistance than a 60.0-W bulb? Explain.

5. The current in the circuit shown in Figure P33.5 equals 60.0% of the peak current at $t = 7.00$ ms.

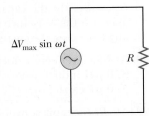

Figure P33.5
Problems 5 and 6.

What is the lowest source frequency that gives this current?

6. In the AC circuit shown in Figure P33.5, $R = 70.0$ Ω and the output voltage of the AC source is $\Delta V_{max} \sin \omega t$. (a) If $\Delta V_R = 0.250$ ΔV_{max} for the first time at $t = 0.010$ 0 s, what is the angular frequency of the source? (b) What is the next value of t for which $\Delta V_R = 0.250$ ΔV_{max}?

7. An audio amplifier, represented by the AC source and resistor in Figure P33.7, delivers to the speaker alternating voltage at audio frequencies. If the source voltage has an amplitude of 15.0 V, $R = 8.20$ Ω, and the speaker is equivalent to a resistance of 10.4 Ω, what is the time-averaged power transferred to it?

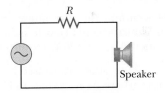

Figure P33.7

8. Figure P33.8 shows three lightbulbs connected to a 120-V AC (rms) household supply voltage. Bulbs 1 and 2 have a power rating of 150 W, and bulb 3 has a 100-W rating. Find (a) the rms current in each bulb and (b) the resistance of each bulb. (c) What is the total resistance of the combination of the three lightbulbs?

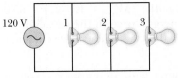

Figure P33.8

Section 33.3 Inductors in an AC Circuit

9. An inductor has a 54.0-Ω reactance when connected to a 60.0–Hz source. The inductor is removed and then connected to a 50.0-Hz source that produces a 100-V rms voltage. What is the maximum current in the inductor?

10. In a purely inductive AC circuit as shown in Figure P33.10 (page 1024), $\Delta V_{max} = 100$ V. (a) The maximum

current is 7.50 A at 50.0 Hz. Calculate the inductance L. (b) **What If?** At what angular frequency ω is the maximum current 2.50 A?

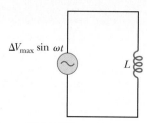

Figure P33.10   Problems 10 and 11.

11. For the circuit shown in Figure P33.10, $\Delta V_{max} = 80.0$ V, $\omega = 65.0\pi$ rad/s, and $L = 70.0$ mH. Calculate the current in the inductor at $t = 15.5$ ms.

12. An inductor is connected to an AC power supply having a maximum output voltage of 4.00 V at a frequency of 300 Hz. What inductance is needed to keep the rms current less than 2.00 mA?

13. An AC source has an output rms voltage of 78.0 V at a frequency of 80.0 Hz. If the source is connected across a 25.0-mH inductor, what are (a) the inductive reactance of the circuit, (b) the rms current in the circuit, and (c) the maximum current in the circuit?

14. A 20.0-mH inductor is connected to a North American electrical outlet ($\Delta V_{rms} = 120$ V, $f = 60.0$ Hz). Assuming the energy stored in the inductor is zero at $t = 0$, determine the energy stored at $t = \frac{1}{180}$ s.

15. **Review.** Determine the maximum magnetic flux through an inductor connected to a North American electrical outlet ($\Delta V_{rms} = 120$ V, $f = 60.0$ Hz).

16. The output voltage of an AC source is given by $\Delta v = 120 \sin 30.0\pi t$, where Δv is in volts and t is in seconds. The source is connected across a 0.500-H inductor. Find (a) the frequency of the source, (b) the rms voltage across the inductor, (c) the inductive reactance of the circuit, (d) the rms current in the inductor, and (e) the maximum current in the inductor.

Section 33.4 Capacitors in an AC Circuit

17. A 1.00-mF capacitor is connected to a North American electrical outlet ($\Delta V_{rms} = 120$ V, $f = 60.0$ Hz). Assuming the energy stored in the capacitor is zero at $t = 0$, determine the magnitude of the current in the wires at $t = \frac{1}{180}$ s.

18. An AC source with an output rms voltage of 36.0 V at a frequency of 60.0 Hz is connected across a 12.0-μF capacitor. Find (a) the capacitive reactance, (b) the rms current, and (c) the maximum current in the circuit. (d) Does the capacitor have its maximum charge when the current has its maximum value? Explain.

19. (a) For what frequencies does a 22.0-μF capacitor have a reactance below 175 Ω? (b) **What If?** What is the reactance of a 44.0-μF capacitor over this same frequency range?

20. A source delivers an AC voltage of the form $\Delta v = 98.0 \sin 80\pi t$, where Δv is in volts and t is in seconds, to a capacitor. The maximum current in the circuit is 0.500 A. Find (a) the rms voltage of the source, (b) the frequency of the source, and (c) the value of the capacitance.

21. What maximum current is delivered by an AC source with $\Delta V_{max} = 48.0$ V and $f = 90.0$ Hz when connected across a 3.70-μF capacitor?

22. A capacitor C is connected to a power supply that operates at a frequency f and produces an rms voltage ΔV. What is the maximum charge that appears on either capacitor plate?

23. What is the maximum current in a 2.20-μF capacitor when it is connected across (a) a North American electrical outlet having $\Delta V_{rms} = 120$ V and $f = 60.0$ Hz and (b) a European electrical outlet having $\Delta V_{rms} = 240$ V and $f = 50.0$ Hz?

Section 33.5 The *RLC* Series Circuit

24. An AC source with $\Delta V_{max} = 150$ V and $f = 50.0$ Hz is connected between points a and d in Figure P33.24. Calculate the maximum voltages between (a) points a and b, (b) points b and c, (c) points c and d, and (d) points b and d.

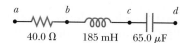

a b c d
40.0 Ω 185 mH   65.0 μF

Figure P33.24   Problems 24 and 81.

25. In addition to phasor diagrams showing voltages such as in Figure 33.15, we can draw phasor diagrams with resistance and reactances. The resultant of adding the phasors is the impedance. Draw to scale a phasor diagram showing Z, X_L, X_C, and ϕ for an AC series circuit for which $R = 300$ Ω, $C = 11.0$ μF, $L = 0.200$ H, and $f = 500/\pi$ Hz.

26. A sinusoidal voltage $\Delta v = 40.0 \sin 100t$, where Δv is in volts and t is in seconds, is applied to a series *RLC* circuit with $L = 160$ mH, $C = 99.0$ μF, and $R = 68.0$ Ω. (a) What is the impedance of the circuit? (b) What is the maximum current? Determine the numerical values for (c) ω and (d) ϕ in the equation $i = I_{max} \sin (\omega t - \phi)$.

27. A series AC circuit contains a resistor, an inductor of 150 mH, a capacitor of 5.00 μF , and a source with $\Delta V_{max} = 240$ V operating at 50.0 Hz. The maximum current in the circuit is 100 mA. Calculate (a) the inductive reactance, (b) the capacitive reactance, (c) the impedance, (d) the resistance in the circuit, and (e) the phase angle between the current and the source voltage.

28. At what frequency does the inductive reactance of a 57.0-μH inductor equal the capacitive reactance of a 57.0-μF capacitor?

29. An *RLC* circuit consists of a 150-Ω resistor, a 21.0-μF capacitor, and a 460-mH inductor connected in series with a 120-V, 60.0-Hz power supply. (a) What is the phase

angle between the current and the applied voltage? (b) Which reaches its maximum earlier, the current or the voltage?

30. Draw phasors to scale for the following voltages in SI units: (a) $25.0 \sin \omega t$ at $\omega t = 90.0°$, (b) $30.0 \sin \omega t$ at $\omega t = 60.0°$, and (c) $18.0 \sin \omega t$ at $\omega t = 300°$.

31. An inductor ($L = 400$ mH), a capacitor ($C = 4.43 \ \mu$F), [M] and a resistor ($R = 500 \ \Omega$) are connected in series. A 50.0-Hz AC source produces a peak current of 250 mA in the circuit. (a) Calculate the required peak voltage ΔV_{max}. (b) Determine the phase angle by which the current leads or lags the applied voltage.

32. A 60.0-Ω resistor is connected in series with a 30.0-μF capacitor and a source whose maximum voltage is 120 V, operating at 60.0 Hz. Find (a) the capacitive reactance of the circuit, (b) the impedance of the circuit, and (c) the maximum current in the circuit. (d) Does the voltage lead or lag the current? (e) How will adding an inductor in series with the existing resistor and capacitor affect the current? Explain.

33. Review. In an *RLC* series circuit that includes a source of alternating current operating at fixed frequency and voltage, the resistance R is equal to the inductive reactance. If the plate separation of the parallel-plate capacitor is reduced to one-half its original value, the current in the circuit doubles. Find the initial capacitive reactance in terms of R.

Section 33.6 Power in an AC Circuit

34. *Why is the following situation impossible?* A series circuit consists of an ideal AC source (no inductance or capacitance in the source itself) with an rms voltage of ΔV at a frequency f and a magnetic buzzer with a resistance R and an inductance L. By carefully adjusting the inductance L of the circuit, a power factor of exactly 1.00 is attained.

35. A series *RLC* circuit has a resistance of 45.0 Ω and an [W] impedance of 75.0 Ω. What average power is delivered to this circuit when $\Delta V_{rms} = 210$ V?

36. An AC voltage of the form $\Delta v = 100 \sin 1\ 000t$, where [W] Δv is in volts and t is in seconds, is applied to a series *RLC* circuit. Assume the resistance is 400 Ω, the capacitance is 5.00 μF, and the inductance is 0.500 H. Find the average power delivered to the circuit.

37. A series *RLC* circuit has a resistance of 22.0 Ω and an impedance of 80.0 Ω. If the rms voltage applied to the circuit is 160 V, what average power is delivered to the circuit?

38. An AC voltage of the form $\Delta v = 90.0 \sin 350t$, where Δv is in volts and t is in seconds, is applied to a series *RLC* circuit. If $R = 50.0 \ \Omega$, $C = 25.0 \ \mu$F, and $L = 0.200$ H, find (a) the impedance of the circuit, (b) the rms current in the circuit, and (c) the average power delivered to the circuit.

39. In a certain series *RLC* circuit, $I_{rms} = 9.00$ A, $\Delta V_{rms} =$ [W] 180 V, and the current leads the voltage by 37.0°.

(a) What is the total resistance of the circuit? (b) Calculate the reactance of the circuit ($X_L - X_C$).

40. Suppose you manage a factory that uses many electric [W] motors. The motors create a large inductive load to the electric power line as well as a resistive load. The electric company builds an extra-heavy distribution line to supply you with two components of current: one that is 90° out of phase with the voltage and another that is in phase with the voltage. The electric company charges you an extra fee for "reactive volt-amps" in addition to the amount you pay for the energy you use. You can avoid the extra fee by installing a capacitor between the power line and your factory. The following problem models this solution.

In an *RL* circuit, a 120-V (rms), 60.0-Hz source is in series with a 25.0-mH inductor and a 20.0-Ω resistor. What are (a) the rms current and (b) the power factor? (c) What capacitor must be added in series to make the power factor equal to 1? (d) To what value can the supply voltage be reduced if the power supplied is to be the same as before the capacitor was installed?

41. A diode is a device that allows current to be carried in only one direction (the direction indicated by the arrowhead in its circuit symbol). Find the average power delivered to the diode circuit of Figure P33.41 in terms of ΔV_{rms} and R.

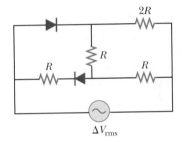

Figure P33.41

Section 33.7 Resonance in a Series *RLC* Circuit

42. A series *RLC* circuit has components with the follow- [W] ing values: $L = 20.0$ mH, $C = 100$ nF, $R = 20.0 \ \Omega$, and $\Delta V_{max} = 100$ V, with $\Delta v = \Delta V_{max} \sin \omega t$. Find (a) the resonant frequency of the circuit, (b) the amplitude of the current at the resonant frequency, (c) the Q of the circuit, and (d) the amplitude of the voltage across the inductor at resonance.

43. An *RLC* circuit is used in a radio to tune into an FM [AMT] station broadcasting at $f = 99.7$ MHz. The resistance [M] in the circuit is $R = 12.0 \ \Omega$, and the inductance is $L = 1.40 \ \mu$H. What capacitance should be used?

44. The *LC* circuit of a radar transmitter oscillates at 9.00 GHz. (a) What inductance is required for the circuit to resonate at this frequency if its capacitance is 2.00 pF? (b) What is the inductive reactance of the circuit at this frequency?

45. A 10.0-Ω resistor, 10.0-mH inductor, and 100-μF capacitor are connected in series to a 50.0-V (rms) source having variable frequency. If the operating frequency is twice the resonance frequency, find the energy delivered to the circuit during one period.

46. A resistor R, inductor L, and capacitor C are connected in series to an AC source of rms voltage ΔV and variable frequency. If the operating frequency is twice the resonance frequency, find the energy delivered to the circuit during one period.

47. **Review.** A radar transmitter contains an LC circuit oscillating at 1.00×10^{10} Hz. (a) For a one-turn loop having an inductance of 400 pH to resonate at this frequency, what capacitance is required in series with the loop? (b) The capacitor has square, parallel plates separated by 1.00 mm of air. What should the edge length of the plates be? (c) What is the common reactance of the loop and capacitor at resonance?

Section 33.8 The Transformer and Power Transmission

48. A step-down transformer is used for recharging the
W batteries of portable electronic devices. The turns ratio N_2/N_1 for a particular transformer used in a DVD player is 1:13. When used with 120-V (rms) household service, the transformer draws an rms current of 20.0 mA from the house outlet. Find (a) the rms output voltage of the transformer and (b) the power delivered to the DVD player.

49. The primary coil of a transformer has $N_1 = 350$ turns,
M and the secondary coil has $N_2 = 2\,000$ turns. If the input voltage across the primary coil is $\Delta v = 170 \cos \omega t$, where Δv is in volts and t is in seconds, what rms voltage is developed across the secondary coil?

50. A transmission line that has a resistance per unit
AMT length of 4.50×10^{-4} Ω/m is to be used to transmit 5.00 MW across 400 mi (6.44×10^5 m). The output voltage of the source is 4.50 kV. (a) What is the line loss if a transformer is used to step up the voltage to 500 kV? (b) What fraction of the input power is lost to the line under these circumstances? (c) **What If?** What difficulties would be encountered in attempting to transmit the 5.00 MW at the source voltage of 4.50 kV?

51. In the transformer shown in Figure P33.51, the load resistance R_L is 50.0 Ω. The turns ratio N_1/N_2 is 2.50, and the rms source voltage is $\Delta V_s = 80.0$ V. If a voltmeter across the load resistance measures an rms voltage of 25.0 V, what is the source resistance R_s?

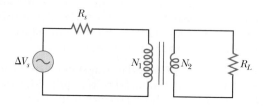

Figure P33.51

52. A person is working near the secondary of a transformer as shown in Figure P33.52. The primary voltage is 120 V at 60.0 Hz. The secondary voltage is

5 000 V. The capacitance C_s, which is the stray capacitance between the hand and the secondary winding, is 20.0 pF. Assuming the person has a body resistance to ground of $R_b = 50.0$ kΩ, determine the rms voltage across the body. *Suggestion:* Model the secondary of the transformer as an AC source.

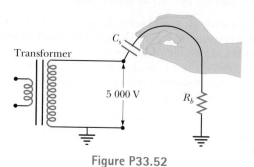

Figure P33.52

Section 33.9 Rectifiers and Filters

53. The RC high-pass filter shown in Figure P33.53
M has a resistance $R = 0.500$ Ω and a capacitance $C = 613$ μF. What is the ratio of the amplitude of the output voltage to that of the input voltage for this filter for a source frequency of 600 Hz?

54. Consider the RC high-pass filter circuit shown in Figure P33.53. (a) Find an expression for the ratio of the amplitude of the output voltage to that of the input voltage in terms of R, C, and the AC source frequency ω. (b) What value does this ratio approach as the frequency decreases toward zero? (c) What value does this ratio approach as the frequency increases without limit?

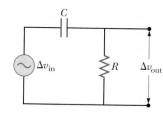

Figure P33.53
Problems 53 and 54.

55. One particular plug-in power supply for a radio looks similar to the one shown in Figure 33.20 and is marked with the following information: Input 120 V AC 8 W Output 9 V DC 300 mA. Assume these values are accurate to two digits. (a) Find the energy efficiency of the device when the radio is operating. (b) At what rate is energy wasted in the device when the radio is operating? (c) Suppose the input power to the transformer is 8.00 W when the radio is switched off and energy costs \$0.110/kWh from the electric company. Find the cost of having six such transformers around the house, each plugged in for 31 days.

56. Consider the filter circuit shown in Figure P33.56. (a) Show that the ratio of the amplitude of the output voltage to that of the input voltage is

$$\frac{\Delta V_{out}}{\Delta V_{in}} = \frac{1/\omega C}{\sqrt{R^2 + \left(\dfrac{1}{\omega C}\right)^2}}$$

(b) What value does this ratio approach as the frequency decreases toward zero? (c) What value does this ratio approach as the frequency increases without limit? (d) At what frequency is the ratio equal to one-half?

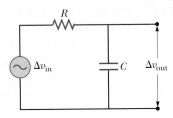

Figure P33.56

Additional Problems

57. A step-up transformer is designed to have an output voltage of 2 200 V (rms) when the primary is connected across a 110-V (rms) source. (a) If the primary winding has exactly 80 turns, how many turns are required on the secondary? (b) If a load resistor across the secondary draws a current of 1.50 A, what is the current in the primary, assuming ideal conditions? (c) **What If?** If the transformer actually has an efficiency of 95.0%, what is the current in the primary when the secondary current is 1.20 A?

58. *Why is the following situation impossible?* An *RLC* circuit is used in a radio to tune into a North American AM commercial radio station. The values of the circuit components are $R = 15.0$ Ω, $L = 2.80$ μH, and $C = 0.910$ pF.

59. Review. The voltage phasor diagram for a certain series *RLC* circuit is shown in Figure P33.59. The resistance of the circuit is 75.0 Ω, and the frequency is 60.0 Hz. Find (a) the maximum voltage ΔV_{max}, (b) the phase angle ϕ, (c) the maximum current, (d) the impedance, (e) the capacitance and (f) the inductance of the circuit, and (g) the average power delivered to the circuit.

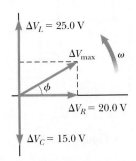

Figure P33.59

60. Consider a series *RLC* circuit having the parameters $R = 200$ Ω, $L = 663$ mH, and $C = 26.5$ μF. The applied voltage has an amplitude of 50.0 V and a frequency of 60.0 Hz. Find (a) the current I_{max} and its phase relative to the applied voltage Δv, (b) the maximum voltage ΔV_R across the resistor and its phase relative to the current, (c) the maximum voltage ΔV_C across the capacitor and its phase relative to the current, and (d) the maxi-

mum voltage ΔV_L across the inductor and its phase relative to the current.

61. Energy is to be transmitted over a pair of copper wires in a transmission line at the rate of 20.0 kW with only a 1.00% loss over a distance of 18.0 km at potential difference $\Delta V_{rms} = 1.50 \times 10^3$ V between the wires. Assuming the current density is uniform in the conductors, what is the diameter required for each of the two wires?

62. Energy is to be transmitted over a pair of copper wires in a transmission line at a rate P with only a fractional loss f over a distance ℓ at potential difference ΔV_{rms} between the wires. Assuming the current density is uniform in the conductors, what is the diameter required for each of the two wires?

63. A 400-Ω resistor, an inductor, and a capacitor are in series with an AC source. The reactance of the inductor is 700 Ω, and the circuit impedance is 760 Ω. (a) What are the possible values of the reactance of the capacitor? (b) If you find that the power delivered to the circuit decreases as you raise the frequency, what is the capacitive reactance in the original circuit? (c) Repeat part (a) assuming the resistance is 200 Ω instead of 400 Ω and the circuit impedance continues to be 760 Ω.

64. Show that the rms value for the sawtooth voltage shown in Figure P33.64 is $\Delta V_{max}/\sqrt{3}$.

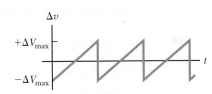

Figure P33.64

65. A transformer may be used to provide maximum power transfer between two AC circuits that have different impedances Z_1 and Z_2. This process is called *impedance matching*. (a) Show that the ratio of turns N_1/N_2 for this transformer is

$$\frac{N_1}{N_2} = \sqrt{\frac{Z_1}{Z_2}}$$

(b) Suppose you want to use a transformer as an impedance-matching device between an audio amplifier that has an output impedance of 8.00 kΩ and a speaker that has an input impedance of 8.00 Ω. What should your N_1/N_2 ratio be?

66. A capacitor, a coil, and two resistors of equal resistance are arranged in an AC circuit as shown in Figure P33.66 (page 1028). An AC source provides an emf of $\Delta V_{rms} = 20.0$ V at a frequency of 60.0 Hz. When the double-throw switch S is open as shown in the figure, the rms current is 183 mA. When the switch is closed in position *a*, the rms current is 298 mA. When the switch is closed in position *b*, the rms current is 137 mA. Determine the values of (a) R, (b) C, and (c) L. (d) Is more than one set of values possible? Explain.

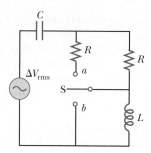

Figure P33.66

67. Marie Cornu, a physicist at the Polytechnic Institute in
    [W] Paris, invented phasors in about 1880. This problem helps you see their general utility in representing oscillations. Two mechanical vibrations are represented by the expressions

    $$y_1 = 12.0 \sin 4.50t$$

    and

    $$y_2 = 12.0 \sin (4.50t + 70.0°)$$

    where y_1 and y_2 are in centimeters and t is in seconds. Find the amplitude and phase constant of the sum of these functions (a) by using a trigonometric identity (as from Appendix B) and (b) by representing the oscillations as phasors. (c) State the result of comparing the answers to parts (a) and (b). (d) Phasors make it equally easy to add traveling waves. Find the amplitude and phase constant of the sum of the three waves represented by

    $$y_1 = 12.0 \sin (15.0x - 4.50t + 70.0°)$$

    $$y_2 = 15.5 \sin (15.0x - 4.50t - 80.0°)$$

    $$y_3 = 17.0 \sin (15.0x - 4.50t + 160°)$$

    where x, y_1, y_2, and y_3 are in centimeters and t is in seconds.

68. A series RLC circuit has resonance angular frequency 2.00×10^3 rad/s. When it is operating at some input frequency, $X_L = 12.0 \ \Omega$ and $X_C = 8.00 \ \Omega$. (a) Is this input frequency higher than, lower than, or the same as the resonance frequency? Explain how you can tell. (b) Explain whether it is possible to determine the values of both L and C. (c) If it is possible, find L and C. If it is not possible, give a compact expression for the condition that L and C must satisfy.

69. **Review.** One insulated conductor from a household
    [AMT] extension cord has a mass per length of 19.0 g/m. A section of this conductor is held under tension between two clamps. A subsection is located in a magnetic field of magnitude 15.3 mT directed perpendicular to the length of the cord. When the cord carries an AC current of 9.00 A at a frequency of 60.0 Hz, it vibrates in resonance in its simplest standing-wave vibration mode. (a) Determine the relationship that must be satisfied between the separation d of the clamps and

the tension T in the cord. (b) Determine one possible combination of values for these variables.

70. (a) Sketch a graph of the phase angle for an RLC series circuit as a function of angular frequency from zero to a frequency much higher than the resonance frequency. (b) Identify the value of ϕ at the resonance angular frequency ω_0. (c) Prove that the slope of the graph of ϕ versus ω at the resonance point is $2Q/\omega_0$.

71. In Figure P33.71, find the rms current delivered by the 45.0-V (rms) power supply when (a) the frequency is very large and (b) the frequency is very small.

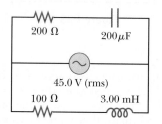

Figure P33.71

72. **Review.** In the circuit shown in Figure P33.72, assume all parameters except C are given. Find (a) the current in the circuit as a function of time and (b) the power delivered to the circuit. (c) Find the current as a function of time after *only* switch 1 is opened. (d) After switch 2 is *also* opened, the current and voltage are in phase. Find the capacitance C. Find (e) the impedance of the circuit when both switches are open, (f) the maximum energy stored in the capacitor during oscillations, and (g) the maximum energy stored in the inductor during oscillations. (h) Now the frequency of the voltage source is doubled. Find the phase difference between the current and the voltage. (i) Find the frequency that makes the inductive reactance one-half the capacitive reactance.

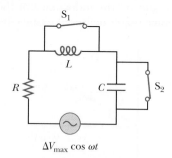

Figure P33.72

73. A series RLC circuit contains the following compo-
    [GP] nents: $R = 150 \ \Omega$, $L = 0.250$ H, $C = 2.00 \ \mu$F, and a source with $\Delta V_{max} = 210$ V operating at 50.0 Hz. Our goal is to find the phase angle, the power factor, and the power input for this circuit. (a) Find the inductive reactance in the circuit. (b) Find the capacitive reactance in the circuit. (c) Find the impedance in the circuit. (d) Calculate the maximum current in the circuit. (e) Determine the phase angle between the cur-

rent and source voltage. (f) Find the power factor for the circuit. (g) Find the power input to the circuit.

74. A series *RLC* circuit is operating at 2.00×10^3 Hz. At this frequency, $X_L = X_C = 1\ 884\ \Omega$. The resistance of the circuit is 40.0 Ω. (a) Prepare a table showing the values of X_L, X_C, and Z for $f = 300, 600, 800, 1.00 \times 10^3$, 1.50×10^3, 2.00×10^3, 3.00×10^3, 4.00×10^3, 6.00×10^3, and 1.00×10^4 Hz. (b) Plot on the same set of axes X_L, X_C, and Z as a function of ln f.

75. A series *RLC* circuit consists of an 8.00-Ω resistor, a 5.00-μF capacitor, and a 50.0-mH inductor. A variable-frequency source applies an emf of 400 V (rms) across the combination. Assuming the frequency is equal to one-half the resonance frequency, determine the power delivered to the circuit.

76. A series *RLC* circuit in which $R = 1.00\ \Omega$, $L = 1.00$ mH, and $C = 1.00$ nF is connected to an AC source delivering 1.00 V (rms). (a) Make a precise graph of the power delivered to the circuit as a function of the frequency and (b) verify that the full width of the resonance peak at half-maximum is $R/2\pi L$.

Challenge Problems

77. The resistor in Figure P33.77 represents the midrange speaker in a three-speaker system. Assume its resistance to be constant at 8.00 Ω. The source represents an audio amplifier producing signals of uniform amplitude $\Delta V_{\text{max}} = 10.0$ V at all audio frequencies. The inductor and capacitor are to function as a band-pass filter with $\Delta V_{\text{out}}/\Delta V_{\text{in}} = \frac{1}{2}$ at 200 Hz and at 4.00×10^3 Hz. Determine the required values of (a) L and (b) C. Find (c) the maximum value of the ratio $\Delta V_{\text{out}}/\Delta V_{\text{in}}$; (d) the frequency f_0 at which the ratio has its maximum value; (e) the phase shift between Δv_{in} and Δv_{out} at 200 Hz, at f_0, and at 4.00×10^3 Hz; and (f) the average power transferred to the speaker at 200 Hz, at f_0, and at

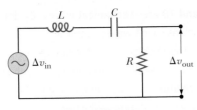

Figure P33.77

4.00×10^3 Hz. (g) Treating the filter as a resonant circuit, find its quality factor.

78. An 80.0-Ω resistor and a 200-mH inductor are connected in *parallel* across a 100-V (rms), 60.0-Hz source. (a) What is the rms current in the resistor? (b) By what angle does the total current lead or lag behind the voltage?

79. A voltage $\Delta v = 100 \sin \omega t$, where Δv is in volts and t is in seconds, is applied across a series combination of a 2.00-H inductor, a 10.0-μF capacitor, and a 10.0-Ω resistor. (a) Determine the angular frequency ω_0 at which the power delivered to the resistor is a maximum. (b) Calculate the average power delivered at that frequency. (c) Determine the two angular frequencies ω_1 and ω_2 at which the power is one-half the maximum value. *Note:* The Q of the circuit is $\omega_0/(\omega_2 - \omega_1)$.

80. Figure P33.80a shows a parallel *RLC* circuit. The instantaneous voltages (and rms voltages) across each of the three circuit elements are the same, and each is in phase with the current in the resistor. The currents in C and L lead or lag the current in the resistor as shown in the current phasor diagram, Figure P33.80b. (a) Show that the rms current delivered by the source is

$$I_{\text{rms}} = \Delta V_{\text{rms}} \left[\frac{1}{R^2} + \left(\omega C - \frac{1}{\omega L} \right)^2 \right]^{1/2}$$

(b) Show that the phase angle ϕ between ΔV_{rms} and I_{rms} is given by

$$\tan \phi = R \left(\frac{1}{X_C} - \frac{1}{X_L} \right)$$

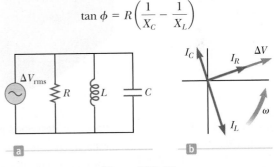

Figure P33.80

81. An AC source with $\Delta V_{\text{rms}} = 120$ V is connected between points a and d in Figure P33.24. At what frequency will it deliver a power of 250 W? Explain your answer.

Electromagnetic Waves

This image of the Crab Nebula taken with visible light shows a variety of colors, with each color representing a different wavelength of visible light. *(NASA, ESA, J. Hester, A. Loll (ASU))*

The waves described in Chapters 16, 17, and 18 are mechanical waves. By definition, the propagation of mechanical disturbances—such as sound waves, water waves, and waves on a string—requires the presence of a medium. This chapter is concerned with the properties of electromagnetic waves, which (unlike mechanical waves) can propagate through empty space.

We begin by considering Maxwell's contributions in modifying Ampère's law, which we studied in Chapter 30. We then discuss Maxwell's equations, which form the theoretical basis of all electromagnetic phenomena. These equations predict the existence of electromagnetic waves that propagate through space at the speed of light *c* according to the traveling wave analysis model. Heinrich Hertz confirmed Maxwell's prediction when he generated and detected electromagnetic waves in 1887. That discovery has led to many practical communication systems, including radio, television, cell phone systems, wireless Internet connectivity, and optoelectronics.

Next, we learn how electromagnetic waves are generated by oscillating electric charges. The waves radiated from the oscillating charges can be detected at great distances. Furthermore, because electromagnetic waves carry energy (T_{ER} in Eq. 8.2) and momentum, they can exert pressure on a surface. The chapter concludes with a description of the various frequency ranges in the electromagnetic spectrum.

34.1 Displacement Current and the General Form of Ampère's Law

In Chapter 30, we discussed using Ampère's law (Eq. 30.13) to analyze the magnetic fields created by currents:

$$\oint \vec{\mathbf{B}} \cdot d\vec{\mathbf{s}} = \mu_0 I$$

In this equation, the line integral is over any closed path through which conduction current passes, where conduction current is defined by the expression $I = dq/dt$. (In this section, we use the term *conduction current* to refer to the current carried by charge carriers in the wire to distinguish it from a new type of current we shall introduce shortly.) We now show that Ampère's law in this form is valid only if any electric fields present are constant in time. James Clerk Maxwell recognized this limitation and modified Ampère's law to include time-varying electric fields.

Consider a capacitor being charged as illustrated in Figure 34.1. When a conduction current is present, the charge on the positive plate changes, but no conduction current exists in the gap between the plates because there are no charge carriers in the gap. Now consider the two surfaces S_1 and S_2 in Figure 34.1, bounded by the same path P. Ampère's law states that $\oint \vec{\mathbf{B}} \cdot d\vec{\mathbf{s}}$ around this path must equal $\mu_0 I$, where I is the total current through *any* surface bounded by the path P.

When the path P is considered to be the boundary of S_1, $\oint \vec{\mathbf{B}} \cdot d\vec{\mathbf{s}} = \mu_0 I$ because the conduction current I passes through S_1. When the path is considered to be the boundary of S_2, however, $\oint \vec{\mathbf{B}} \cdot d\vec{\mathbf{s}} = 0$ because no conduction current passes through S_2. Therefore, we have a contradictory situation that arises from the discontinuity of the current! Maxwell solved this problem by postulating an additional term on the right side of Ampère's law, which includes a factor called the **displacement current** I_d defined as[1]

$$I_d \equiv \epsilon_0 \frac{d\Phi_E}{dt} \qquad\qquad \textbf{(34.1)} \qquad \blacktriangleleft \textbf{ Displacement current}$$

James Clerk Maxwell
Scottish Theoretical Physicist (1831–1879)
Maxwell developed the electromagnetic theory of light and the kinetic theory of gases, and explained the nature of Saturn's rings and color vision. Maxwell's successful interpretation of the electromagnetic field resulted in the field equations that bear his name. Formidable mathematical ability combined with great insight enabled him to lead the way in the study of electromagnetism and kinetic theory. He died of cancer before he was 50.

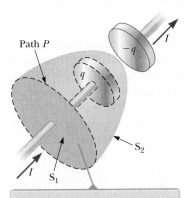

The conduction current I in the wire passes only through S_1, which leads to a contradiction in Ampère's law that is resolved only if one postulates a displacement current through S_2.

Figure 34.1 Two surfaces S_1 and S_2 near the plate of a capacitor are bounded by the same path P.

---

[1]*Displacement* in this context does not have the meaning it does in Chapter 2. Despite the inaccurate implications, the word is historically entrenched in the language of physics, so we continue to use it.

where ϵ_0 is the permittivity of free space (see Section 23.3) and $\Phi_E \equiv \int \vec{\mathbf{E}} \cdot d\vec{\mathbf{A}}$ is the electric flux (see Eq. 24.3) through the surface bounded by the path of integration.

As the capacitor is being charged (or discharged), the changing electric field between the plates may be considered equivalent to a current that acts as a continuation of the conduction current in the wire. When the expression for the displacement current given by Equation 34.1 is added to the conduction current on the right side of Ampère's law, the difficulty represented in Figure 34.1 is resolved. No matter which surface bounded by the path P is chosen, either a conduction current or a displacement current passes through it. With this new term I_d, we can express the general form of Ampère's law (sometimes called the **Ampère–Maxwell law**) as

Ampère–Maxwell law ▶

$$\oint \vec{\mathbf{B}} \cdot d\vec{\mathbf{s}} = \mu_0(I + I_d) = \mu_0 I + \mu_0 \epsilon_0 \frac{d\Phi_E}{dt} \qquad (34.2)$$

We can understand the meaning of this expression by referring to Figure 34.2. The electric flux through surface S is $\Phi_E = \int \vec{\mathbf{E}} \cdot d\vec{\mathbf{A}} = EA$, where A is the area of the capacitor plates and E is the magnitude of the uniform electric field between the plates. If q is the charge on the plates at any instant, then $E = q/(\epsilon_0 A)$ (see Section 26.2). Therefore, the electric flux through S is

$$\Phi_E = EA = \frac{q}{\epsilon_0}$$

Hence, the displacement current through S is

$$I_d = \epsilon_0 \frac{d\Phi_E}{dt} = \frac{dq}{dt} \qquad (34.3)$$

That is, the displacement current I_d through S is precisely equal to the conduction current I in the wires connected to the capacitor!

By considering surface S, we can identify the displacement current as the source of the magnetic field on the surface boundary. The displacement current has its physical origin in the time-varying electric field. The central point of this formalism is that magnetic fields are produced *both* by conduction currents *and* by time-varying electric fields. This result was a remarkable example of theoretical work by Maxwell, and it contributed to major advances in the understanding of electromagnetism.

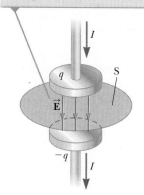

The electric field lines between the plates create an electric flux through surface S.

Figure 34.2 When a conduction current exists in the wires, a changing electric field $\vec{\mathbf{E}}$ exists between the plates of the capacitor.

Quick Quiz 34.1 In an *RC* circuit, the capacitor begins to discharge. **(i)** During the discharge, in the region of space between the plates of the capacitor, is there (a) conduction current but no displacement current, (b) displacement current but no conduction current, (c) both conduction and displacement current, or (d) no current of any type? **(ii)** In the same region of space, is there (a) an electric field but no magnetic field, (b) a magnetic field but no electric field, (c) both electric and magnetic fields, or (d) no fields of any type?

Example 34.1   **Displacement Current in a Capacitor**

A sinusoidally varying voltage is applied across a capacitor as shown in Figure 34.3. The capacitance is $C = 8.00~\mu\text{F}$, the frequency of the applied voltage is $f = 3.00$ kHz, and the voltage amplitude is $\Delta V_{max} = 30.0$ V. Find the displacement current in the capacitor.

SOLUTION

Conceptualize Figure 34.3 represents the circuit diagram for this situation. Figure 34.2 shows a close-up of the capacitor and the electric field between the plates.

Categorize We determine results using equations discussed in this section, so we categorize this example as a substitution problem.

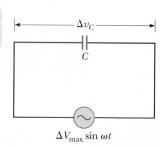

Figure 34.3 (Example 34.1)

▶ **34.1** continued

Evaluate the angular frequency of the source from Equation 15.12:

$$\omega = 2\pi f = 2\pi (3.00 \times 10^3 \text{ Hz}) = 1.88 \times 10^4 \text{ s}^{-1}$$

Use Equation 33.20 to express the potential difference in volts across the capacitor as a function of time in seconds:

$$\Delta v_C = \Delta V_{max} \sin \omega t = 30.0 \sin (1.88 \times 10^4\, t)$$

Use Equation 34.3 to find the displacement current in amperes as a function of time. Note that the charge on the capacitor is $q = C\Delta v_C$:

$$i_d = \frac{dq}{dt} = \frac{d}{dt}(C\,\Delta v_C) = C\frac{d}{dt}(\Delta V_{max} \sin \omega t)$$

$$= \omega C\,\Delta V_{max} \cos \omega t$$

Substitute numerical values:

$$i_d = (1.88 \times 10^4 \text{ s}^{-1})(8.00 \times 10^{-6} \text{ C})(30.0 \text{ V}) \cos (1.88 \times 10^4\, t)$$

$$= \boxed{4.51 \cos (1.88 \times 10^4\, t)}$$

34.2 Maxwell's Equations and Hertz's Discoveries

We now present four equations that are regarded as the basis of all electrical and magnetic phenomena. These equations, developed by Maxwell, are as fundamental to electromagnetic phenomena as Newton's laws are to mechanical phenomena. In fact, the theory that Maxwell developed was more far-reaching than even he imagined because it turned out to be in agreement with the special theory of relativity, as Einstein showed in 1905.

Maxwell's equations represent the laws of electricity and magnetism that we have already discussed, but they have additional important consequences. For simplicity, we present **Maxwell's equations** as applied to free space, that is, in the absence of any dielectric or magnetic material. The four equations are

$$\oint \vec{\mathbf{E}} \cdot d\vec{\mathbf{A}} = \frac{q}{\epsilon_0} \qquad \text{(34.4)} \qquad \blacktriangleleft \text{ Gauss's law}$$

$$\oint \vec{\mathbf{B}} \cdot d\vec{\mathbf{A}} = 0 \qquad \text{(34.5)} \qquad \blacktriangleleft \text{ Gauss's law in magnetism}$$

$$\oint \vec{\mathbf{E}} \cdot d\vec{\mathbf{s}} = -\frac{d\Phi_B}{dt} \qquad \text{(34.6)} \qquad \blacktriangleleft \text{ Faraday's law}$$

$$\oint \vec{\mathbf{B}} \cdot d\vec{\mathbf{s}} = \mu_0 I + \epsilon_0 \mu_0 \frac{d\Phi_E}{dt} \qquad \text{(34.7)} \qquad \blacktriangleleft \text{ Ampère–Maxwell law}$$

Equation 34.4 is Gauss's law: the total electric flux through any closed surface equals the net charge inside that surface divided by ϵ_0. This law relates an electric field to the charge distribution that creates it.

Equation 34.5 is Gauss's law in magnetism, and it states that the net magnetic flux through a closed surface is zero. That is, the number of magnetic field lines that enter a closed volume must equal the number that leave that volume, which implies that magnetic field lines cannot begin or end at any point. If they did, it would mean that isolated magnetic monopoles existed at those points. That isolated magnetic monopoles have not been observed in nature can be taken as a confirmation of Equation 34.5.

Equation 34.6 is Faraday's law of induction, which describes the creation of an electric field by a changing magnetic flux. This law states that the emf, which is the line integral of the electric field around any closed path, equals the rate of change of magnetic flux through any surface bounded by that path. One consequence of Faraday's law is the current induced in a conducting loop placed in a time-varying magnetic field.

Equation 34.7 is the Ampère–Maxwell law, discussed in Section 34.1, and it describes the creation of a magnetic field by a changing electric field and by electric current: the line integral of the magnetic field around any closed path is the sum of μ_0 multiplied by the net current through that path and $\epsilon_0\mu_0$ multiplied by the rate of change of electric flux through any surface bounded by that path.

Once the electric and magnetic fields are known at some point in space, the force acting on a particle of charge q can be calculated from the electric and magnetic versions of the particle in a field model:

Lorentz force law ▶

$$\vec{\mathbf{F}} = q\vec{\mathbf{E}} + q\vec{\mathbf{v}} \times \vec{\mathbf{B}} \tag{34.8}$$

This relationship is called the **Lorentz force law.** (We saw this relationship earlier as Eq. 29.6.) Maxwell's equations, together with this force law, completely describe all classical electromagnetic interactions in a vacuum.

Notice the symmetry of Maxwell's equations. Equations 34.4 and 34.5 are symmetric, apart from the absence of the term for magnetic monopoles in Equation 34.5. Furthermore, Equations 34.6 and 34.7 are symmetric in that the line integrals of $\vec{\mathbf{E}}$ and $\vec{\mathbf{B}}$ around a closed path are related to the rate of change of magnetic flux and electric flux, respectively. Maxwell's equations are of fundamental importance not only to electromagnetism, but to all science. Hertz once wrote, "One cannot escape the feeling that these mathematical formulas have an independent existence and an intelligence of their own, that they are wiser than we are, wiser even than their discoverers, that we get more out of them than we put into them."

In the next section, we show that Equations 34.6 and 34.7 can be combined to obtain a wave equation for both the electric field and the magnetic field. In empty space, where $q = 0$ and $I = 0$, the solution to these two equations shows that the speed at which electromagnetic waves travel equals the measured speed of light. This result led Maxwell to predict that light waves are a form of electromagnetic radiation.

Hertz performed experiments that verified Maxwell's prediction. The experimental apparatus Hertz used to generate and detect electromagnetic waves is shown schematically in Figure 34.4. An induction coil is connected to a transmitter made up of two spherical electrodes separated by a narrow gap. The coil provides short voltage surges to the electrodes, making one positive and the other negative. A spark is generated between the spheres when the electric field near either electrode surpasses the dielectric strength for air (3×10^6 V/m; see Table 26.1). Free electrons in a strong electric field are accelerated and gain enough energy to ionize any molecules they strike. This ionization provides more electrons, which can accelerate and cause further ionizations. As the air in the gap is ionized, it becomes a much better conductor and the discharge between the electrodes exhibits an oscillatory behavior at a very high frequency. From an electric-circuit viewpoint, this experimental apparatus is equivalent to an LC circuit in which the inductance is that of the coil and the capacitance is due to the spherical electrodes.

Because L and C are small in Hertz's apparatus, the frequency of oscillation is high, on the order of 100 MHz. (Recall from Eq. 32.22 that $\omega = 1/\sqrt{LC}$ for an LC circuit.) Electromagnetic waves are radiated at this frequency as a result of the oscillation of free charges in the transmitter circuit. Hertz was able to detect these waves using a single loop of wire with its own spark gap (the receiver). Such a receiver loop, placed several meters from the transmitter, has its own effective inductance, capacitance, and natural frequency of oscillation. In Hertz's experiment, sparks were induced across the gap of the receiving electrodes when the receiver's frequency was adjusted to match that of the transmitter. In this way,

The transmitter consists of two spherical electrodes connected to an induction coil, which provides short voltage surges to the spheres, setting up oscillations in the discharge between the electrodes.

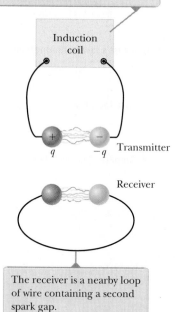

The receiver is a nearby loop of wire containing a second spark gap.

Figure 34.4 Schematic diagram of Hertz's apparatus for generating and detecting electromagnetic waves.

Hertz demonstrated that the oscillating current induced in the receiver was produced by electromagnetic waves radiated by the transmitter. His experiment is analogous to the mechanical phenomenon in which a tuning fork responds to acoustic vibrations from an identical tuning fork that is oscillating nearby.

In addition, Hertz showed in a series of experiments that the radiation generated by his spark-gap device exhibited the wave properties of interference, diffraction, reflection, refraction, and polarization, which are all properties exhibited by light as we shall see in Part 5. Therefore, it became evident that the radio-frequency waves Hertz was generating had properties similar to those of light waves and that they differed only in frequency and wavelength. Perhaps his most convincing experiment was the measurement of the speed of this radiation. Waves of known frequency were reflected from a metal sheet and created a standing-wave interference pattern whose nodal points could be detected. The measured distance between the nodal points enabled determination of the wavelength λ. Using the relationship $v = \lambda f$ (Eq. 16.12) from the traveling wave model, Hertz found that v was close to 3×10^8 m/s, the known speed c of visible light.

Heinrich Rudolf Hertz
German Physicist (1857–1894)
Hertz made his most important discovery of electromagnetic waves in 1887. After finding that the speed of an electromagnetic wave was the same as that of light, Hertz showed that electromagnetic waves, like light waves, could be reflected, refracted, and diffracted. The hertz, equal to one complete vibration or cycle per second, is named after him.

34.3 Plane Electromagnetic Waves

The properties of electromagnetic waves can be deduced from Maxwell's equations. One approach to deriving these properties is to solve the second-order differential equation obtained from Maxwell's third and fourth equations. A rigorous mathematical treatment of that sort is beyond the scope of this text. To circumvent this problem, let's assume the vectors for the electric field and magnetic field in an electromagnetic wave have a specific space–time behavior that is simple but consistent with Maxwell equations.

To understand the prediction of electromagnetic waves more fully, let's focus our attention on an electromagnetic wave that travels in the x direction (the *direction of propagation*). For this wave, the electric field $\vec{\mathbf{E}}$ is in the y direction and the magnetic field $\vec{\mathbf{B}}$ is in the z direction as shown in Figure 34.5. Such waves, in which the electric and magnetic fields are restricted to being parallel to a pair of perpendicular axes, are said to be **linearly polarized waves.** Furthermore, let's assume the field magnitudes E and B depend on x and t only, not on the y or z coordinate.

Let's also imagine that the source of the electromagnetic waves is such that a wave radiated from *any* position in the yz plane (not only from the origin as might be suggested by Fig. 34.5) propagates in the x direction and all such waves are emitted in phase. If we define a **ray** as the line along which the wave travels, all rays for these waves are parallel. This entire collection of waves is often called a **plane wave.** A surface connecting points of equal phase on all waves is a geometric plane called a **wave front,** introduced in Chapter 17. In comparison, a point source of radiation sends waves out radially in all directions. A surface connecting points of equal phase for this situation is a sphere, so this wave is called a **spherical wave.**

To generate the prediction of plane electromagnetic waves, we start with Faraday's law, Equation 34.6:

$$\oint \vec{\mathbf{E}} \cdot d\vec{\mathbf{s}} = -\frac{d\Phi_B}{dt}$$

To apply this equation to the wave in Figure 34.5, consider a rectangle of width dx and height ℓ lying in the xy plane as shown in Figure 34.6 (page 1036). Let's first evaluate the line integral of $\vec{\mathbf{E}} \cdot d\vec{\mathbf{s}}$ around this rectangle in the counterclockwise direction at an instant of time when the wave is passing through the rectangle. The contributions from the top and bottom of the rectangle are zero because $\vec{\mathbf{E}}$ is perpendicular to $d\vec{\mathbf{s}}$ for these paths. We can express the electric field on the right side of the rectangle as

$$E(x + dx) \approx E(x) + \frac{dE}{dx}\bigg|_{t \text{ constant}} dx = E(x) + \frac{\partial E}{\partial x} dx$$

> **Pitfall Prevention 34.1**
> **What Is "a" Wave?** What do we mean by a *single* wave? The word *wave* represents both the emission from a *single point* ("wave radiated from *any* position in the *yz* plane" in the text) and the collection of waves from *all points* on the source ("**plane wave**" in the text). You should be able to use this term in both ways and understand its meaning from the context.

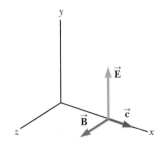

Figure 34.5 Electric and magnetic fields of an electromagnetic wave traveling at velocity $\vec{\mathbf{c}}$ in the positive x direction. The field vectors are shown at one instant of time and at one position in space. These fields depend on x and t.

According to Equation 34.11, this spatial variation in $\vec{\mathbf{E}}$ gives rise to a time-varying magnetic field along the z direction.

According to Equation 34.14, this spatial variation in $\vec{\mathbf{B}}$ gives rise to a time-varying electric field along the y direction.

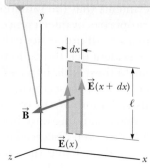

Figure 34.6 At an instant when a plane wave moving in the positive x direction passes through a rectangular path of width dx lying in the xy plane, the electric field in the y direction varies from $\vec{\mathbf{E}}(x)$ to $\vec{\mathbf{E}}(x + dx)$.

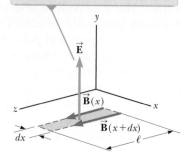

Figure 34.7 At an instant when a plane wave passes through a rectangular path of width dx lying in the xz plane, the magnetic field in the z direction varies from $\vec{\mathbf{B}}(x)$ to $\vec{\mathbf{B}}(x + dx)$.

where $E(x)$ is the field on the left side of the rectangle at this instant.[2] Therefore, the line integral over this rectangle is approximately

$$\oint \vec{\mathbf{E}} \cdot d\vec{\mathbf{s}} = [E(x + dx)]\ell - [E(x)]\ell \approx \ell\left(\frac{\partial E}{\partial x}\right) dx \qquad \textbf{(34.9)}$$

Because the magnetic field is in the z direction, the magnetic flux through the rectangle of area $\ell\,dx$ is approximately $\Phi_B = B\ell\,dx$ (assuming dx is very small compared with the wavelength of the wave). Taking the time derivative of the magnetic flux gives

$$\frac{d\Phi_B}{dt} = \ell\,dx\,\frac{dB}{dt}\bigg|_{x\,\text{constant}} = \ell\,dx\,\frac{\partial B}{\partial t} \qquad \textbf{(34.10)}$$

Substituting Equations 34.9 and 34.10 into Equation 34.6 gives

$$\ell\left(\frac{\partial E}{\partial x}\right) dx = -\ell\,dx\,\frac{\partial B}{\partial t}$$

$$\frac{\partial E}{\partial x} = -\frac{\partial B}{\partial t} \qquad \textbf{(34.11)}$$

In a similar manner, we can derive a second equation by starting with Maxwell's fourth equation in empty space (Eq. 34.7). In this case, the line integral of $\vec{\mathbf{B}} \cdot d\vec{\mathbf{s}}$ is evaluated around a rectangle lying in the xz plane and having width dx and length ℓ as in Figure 34.7. Noting that the magnitude of the magnetic field changes from $B(x)$ to $B(x + dx)$ over the width dx and that the direction for taking the line integral is counterclockwise when viewed from above in Figure 34.7, the line integral over this rectangle is found to be approximately

$$\oint \vec{\mathbf{B}} \cdot d\vec{\mathbf{s}} = [B(x)]\ell - [B(x + dx)]\ell \approx -\ell\left(\frac{\partial B}{\partial x}\right) dx \qquad \textbf{(34.12)}$$

---

[2]Because dE/dx in this equation is expressed as the change in E with x at a given instant t, dE/dx is equivalent to the partial derivative $\partial E/\partial x$. Likewise, dB/dt means the change in B with time at a particular position x; therefore, in Equation 34.10, we can replace dB/dt with $\partial B/\partial t$.

The electric flux through the rectangle is $\Phi_E = E\ell\,dx$, which, when differentiated with respect to time, gives

$$\frac{\partial \Phi_E}{\partial t} = \ell\,dx\,\frac{\partial E}{\partial t} \tag{34.13}$$

Substituting Equations 34.12 and 34.13 into Equation 34.7 gives

$$-\ell\left(\frac{\partial B}{\partial x}\right)dx = \mu_0\,\epsilon_0\,\ell\,dx\left(\frac{\partial E}{\partial t}\right)$$

$$\frac{\partial B}{\partial x} = -\mu_0\epsilon_0\,\frac{\partial E}{\partial t} \tag{34.14}$$

Taking the derivative of Equation 34.11 with respect to x and combining the result with Equation 34.14 gives

$$\frac{\partial^2 E}{\partial x^2} = -\frac{\partial}{\partial x}\left(\frac{\partial B}{\partial t}\right) = -\frac{\partial}{\partial t}\left(\frac{\partial B}{\partial x}\right) = -\frac{\partial}{\partial t}\left(-\mu_0\epsilon_0\,\frac{\partial E}{\partial t}\right)$$

$$\frac{\partial^2 E}{\partial x^2} = \mu_0\epsilon_0\,\frac{\partial^2 E}{\partial t^2} \tag{34.15}$$

In the same manner, taking the derivative of Equation 34.14 with respect to x and combining it with Equation 34.11 gives

$$\frac{\partial^2 B}{\partial x^2} = \mu_0\epsilon_0\,\frac{\partial^2 B}{\partial t^2} \tag{34.16}$$

Equations 34.15 and 34.16 both have the form of the linear wave equation[3] with the wave speed v replaced by c, where

$$c = \frac{1}{\sqrt{\mu_0\epsilon_0}} \tag{34.17}$$

◀ **Speed of electromagnetic waves**

Let's evaluate this speed numerically:

$$c = \frac{1}{\sqrt{(4\pi \times 10^{-7}\,\text{T}\cdot\text{m/A})(8.854\,19 \times 10^{-12}\,\text{C}^2/\text{N}\cdot\text{m}^2)}}$$

$$= 2.997\,92 \times 10^8\,\text{m/s}$$

Because this speed is precisely the same as the speed of light in empty space, we are led to believe (correctly) that light is an electromagnetic wave.

The simplest solution to Equations 34.15 and 34.16 is a sinusoidal wave for which the field magnitudes E and B vary with x and t according to the expressions

$$E = E_{max}\cos(kx - \omega t) \tag{34.18}$$

◀ **Sinusoidal electric and magnetic fields**

$$B = B_{max}\cos(kx - \omega t) \tag{34.19}$$

where E_{max} and B_{max} are the maximum values of the fields. The angular wave number is $k = 2\pi/\lambda$, where λ is the wavelength. The angular frequency is $\omega = 2\pi f$, where f is the wave frequency. According to the traveling wave model, the ratio ω/k equals the speed of an electromagnetic wave, c:

$$\frac{\omega}{k} = \frac{2\pi f}{2\pi/\lambda} = \lambda f = c$$

---

[3]The linear wave equation is of the form $(\partial^2 y/\partial x^2) = (1/v^2)(\partial^2 y/\partial t^2)$, where v is the speed of the wave and y is the wave function. The linear wave equation was introduced as Equation 16.27, and we suggest you review Section 16.6.

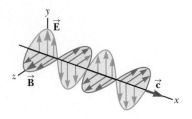

Figure 34.8 A sinusoidal electromagnetic wave moves in the positive x direction with a speed c.

where we have used Equation 16.12, $v = c = \lambda f$, which relates the speed, frequency, and wavelength of a sinusoidal wave. Therefore, for electromagnetic waves, the wavelength and frequency of these waves are related by

$$\lambda = \frac{c}{f} = \frac{3.00 \times 10^8 \text{ m/s}}{f} \qquad (34.20)$$

Figure 34.8 is a pictorial representation, at one instant, of a sinusoidal, linearly polarized electromagnetic wave moving in the positive x direction.

We can generate other mathematical representations of the traveling wave model for electromagnetic waves. Taking partial derivatives of Equations 34.18 (with respect to x) and 34.19 (with respect to t) gives

$$\frac{\partial E}{\partial x} = -kE_{max} \sin(kx - \omega t)$$

$$\frac{\partial B}{\partial t} = \omega B_{max} \sin(kx - \omega t)$$

Substituting these results into Equation 34.11 shows that, at any instant,

$$kE_{max} = \omega B_{max}$$

$$\frac{E_{max}}{B_{max}} = \frac{\omega}{k} = c$$

Using these results together with Equations 34.18 and 34.19 gives

$$\frac{E_{max}}{B_{max}} = \frac{E}{B} = c \qquad (34.21)$$

That is, at every instant, the ratio of the magnitude of the electric field to the magnitude of the magnetic field in an electromagnetic wave equals the speed of light.

Finally, note that electromagnetic waves obey the superposition principle as described in the waves in interference analysis model (which we discussed in Section 18.1 with respect to mechanical waves) because the differential equations involving E and B are linear equations. For example, we can add two waves with the same frequency and polarization simply by adding the magnitudes of the two electric fields algebraically.

> **Pitfall Prevention 34.2**
> $\vec{E}$ **Stronger Than** $\vec{B}$? Because the value of c is so large, some students incorrectly interpret Equation 34.21 as meaning that the electric field is much stronger than the magnetic field. Electric and magnetic fields are measured in different units, however, so they cannot be directly compared. In Section 34.4, we find that the electric and magnetic fields contribute equally to the wave's energy.

Quick Quiz 34.2 What is the phase difference between the sinusoidal oscillations of the electric and magnetic fields in Figure 34.8? **(a)** 180° **(b)** 90° **(c)** 0 **(d)** impossible to determine

Quick Quiz 34.3 An electromagnetic wave propagates in the negative y direction. The electric field at a point in space is momentarily oriented in the positive x direction. In which direction is the magnetic field at that point momentarily oriented? **(a)** the negative x direction **(b)** the positive y direction **(c)** the positive z direction **(d)** the negative z direction

---

Example 34.2   **An Electromagnetic Wave**

A sinusoidal electromagnetic wave of frequency 40.0 MHz travels in free space in the x direction as in Figure 34.9.

(A) Determine the wavelength and period of the wave.

▶ **34.2** continued

SOLUTION

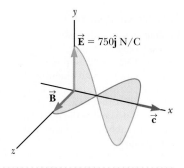

Conceptualize Imagine the wave in Figure 34.9 moving to the right along the x axis, with the electric and magnetic fields oscillating in phase.

Categorize We use the mathematical representation of the *traveling wave* model for electromagnetic waves.

Figure 34.9 (Example 34.2) At some instant, a plane electromagnetic wave moving in the x direction has a maximum electric field of 750 N/C in the positive y direction.

Analyze

Use Equation 34.20 to find the wavelength of the wave:

$$\lambda = \frac{c}{f} = \frac{3.00 \times 10^8 \text{ m/s}}{40.0 \times 10^6 \text{ Hz}} = \boxed{7.50 \text{ m}}$$

Find the period T of the wave as the inverse of the frequency:

$$T = \frac{1}{f} = \frac{1}{40.0 \times 10^6 \text{ Hz}} = \boxed{2.50 \times 10^{-8} \text{ s}}$$

(B) At some point and at some instant, the electric field has its maximum value of 750 N/C and is directed along the y axis. Calculate the magnitude and direction of the magnetic field at this position and time.

SOLUTION

Use Equation 34.21 to find the magnitude of the magnetic field:

$$B_{max} = \frac{E_{max}}{c} = \frac{750 \text{ N/C}}{3.00 \times 10^8 \text{ m/s}} = \boxed{2.50 \times 10^{-6} \text{ T}}$$

Because $\vec{\mathbf{E}}$ and $\vec{\mathbf{B}}$ must be perpendicular to each other and perpendicular to the direction of wave propagation (x in this case), we conclude that $\vec{\mathbf{B}}$ is in the z direction.

Finalize Notice that the wavelength is several meters. This is relatively long for an electromagnetic wave. As we will see in Section 34.7, this wave belongs to the radio range of frequencies.

34.4 Energy Carried by Electromagnetic Waves

In our discussion of the nonisolated system model for energy in Section 8.1, we identified electromagnetic radiation as one method of energy transfer across the boundary of a system. The amount of energy transferred by electromagnetic waves is symbolized as T_{ER} in Equation 8.2. The rate of transfer of energy by an electromagnetic wave is described by a vector $\vec{\mathbf{S}}$, called the **Poynting vector,** which is defined by the expression

$$\vec{\mathbf{S}} \equiv \frac{1}{\mu_0} \vec{\mathbf{E}} \times \vec{\mathbf{B}} \qquad \text{(34.22)}$$

◀ **Poynting vector**

The magnitude of the Poynting vector represents the rate at which energy passes through a unit surface area perpendicular to the direction of wave propagation. Therefore, the magnitude of $\vec{\mathbf{S}}$ represents *power per unit area.* The direction of the vector is along the direction of wave propagation (Fig. 34.10, page 1040). The SI units of $\vec{\mathbf{S}}$ are J/s · m² = W/m².

As an example, let's evaluate the magnitude of $\vec{\mathbf{S}}$ for a plane electromagnetic wave where $|\vec{\mathbf{E}} \times \vec{\mathbf{B}}| = EB$. In this case,

$$S = \frac{EB}{\mu_0} \qquad \text{(34.23)}$$

Pitfall Prevention 34.3
An Instantaneous Value The Poynting vector given by Equation 34.22 is time dependent. Its magnitude varies in time, reaching a maximum value at the same instant the magnitudes of $\vec{\mathbf{E}}$ and $\vec{\mathbf{B}}$ do. The *average* rate of energy transfer is given by Equation 34.24 on the next page.

Because $B = E/c$, we can also express this result as

$$S = \frac{E^2}{\mu_0 c} = \frac{cB^2}{\mu_0}$$

These equations for S apply at any instant of time and represent the *instantaneous* rate at which energy is passing through a unit area in terms of the instantaneous values of E and B.

What is of greater interest for a sinusoidal plane electromagnetic wave is the time average of S over one or more cycles, which is called the *wave intensity I.* (We discussed the intensity of sound waves in Chapter 17.) When this average is taken, we obtain an expression involving the time average of $\cos^2 (kx - \omega t)$, which equals $\frac{1}{2}$. Hence, the average value of S (in other words, the intensity of the wave) is

Wave intensity ▶

$$I = S_{\text{avg}} = \frac{E_{\text{max}} B_{\text{max}}}{2\mu_0} = \frac{E_{\text{max}}^2}{2\mu_0 c} = \frac{cB_{\text{max}}^2}{2\mu_0} \tag{34.24}$$

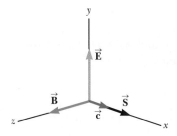

Figure 34.10 The Poynting vector $\vec{S}$ for a plane electromagnetic wave is along the direction of wave propagation.

Recall that the energy per unit volume associated with an electric field, which is the instantaneous energy density u_E, is given by Equation 26.13:

$$u_E = \tfrac{1}{2}\epsilon_0 E^2$$

Also recall that the instantaneous energy density u_B associated with a magnetic field is given by Equation 32.14:

$$u_B = \frac{B^2}{2\mu_0}$$

Because E and B vary with time for an electromagnetic wave, the energy densities also vary with time. Using the relationships $B = E/c$ and $c = 1/\sqrt{\mu_0\epsilon_0}$, the expression for u_B becomes

$$u_B = \frac{(E/c)^2}{2\mu_0} = \frac{\mu_0\epsilon_0}{2\mu_0}E^2 = \tfrac{1}{2}\epsilon_0 E^2$$

Comparing this result with the expression for u_E, we see that

$$u_B = u_E = \tfrac{1}{2}\epsilon_0 E^2 = \frac{B^2}{2\mu_0}$$

That is, the instantaneous energy density associated with the magnetic field of an electromagnetic wave equals the instantaneous energy density associated with the electric field. Hence, in a given volume, the energy is equally shared by the two fields.

The **total instantaneous energy density** u is equal to the sum of the energy densities associated with the electric and magnetic fields:

Total instantaneous ▶
energy density of an
electromagnetic wave

$$u = u_E + u_B = \epsilon_0 E^2 = \frac{B^2}{\mu_0}$$

When this total instantaneous energy density is averaged over one or more cycles of an electromagnetic wave, we again obtain a factor of $\frac{1}{2}$. Hence, for any electromagnetic wave, the total average energy per unit volume is

Average energy density of ▶
an electromagnetic wave

$$u_{\text{avg}} = \epsilon_0 (E^2)_{\text{avg}} = \tfrac{1}{2}\epsilon_0 E_{\text{max}}^2 = \frac{B_{\text{max}}^2}{2\mu_0} \tag{34.25}$$

Comparing this result with Equation 34.24 for the average value of S, we see that

$$I = S_{\text{avg}} = cu_{\text{avg}} \tag{34.26}$$

In other words, the intensity of an electromagnetic wave equals the average energy density multiplied by the speed of light.

The Sun delivers about 10^3 W/m^2 of energy to the Earth's surface via electromagnetic radiation. Let's calculate the total power that is incident on the roof of a home. The roof's dimensions are 8.00 m × 20.0 m. We assume the average magnitude of the Poynting vector for solar radiation at the surface of the Earth is $S_{avg} = 1\,000$ W/m^2. This average value represents the power per unit area, or the light intensity. Assuming the radiation is incident normal to the roof, we obtain

$$P_{avg} = S_{avg}A = (1\,000 \text{ W/m}^2)(8.00 \text{ m} \times 20.0 \text{ m}) = 1.60 \times 10^5 \text{ W}$$

This power is large compared with the power requirements of a typical home. If this power could be absorbed and made available to electrical devices, it would provide more than enough energy for the average home. Solar energy is not easily harnessed, however, and the prospects for large-scale conversion are not as bright as may appear from this calculation. For example, the efficiency of conversion from solar energy is typically 12–18% for photovoltaic cells, reducing the available power by an order of magnitude. Other considerations reduce the power even further. Depending on location, the radiation is most likely not incident normal to the roof and, even if it is, this situation exists for only a short time near the middle of the day. No energy is available for about half of each day during the nighttime hours, and cloudy days further reduce the available energy. Finally, while energy is arriving at a large rate during the middle of the day, some of it must be stored for later use, requiring batteries or other storage devices. All in all, complete solar operation of homes is not currently cost effective for most homes.

Example 34.3   Fields on the Page

Estimate the maximum magnitudes of the electric and magnetic fields of the light that is incident on this page because of the visible light coming from your desk lamp. Treat the lightbulb as a point source of electromagnetic radiation that is 5% efficient at transforming energy coming in by electrical transmission to energy leaving by visible light.

SOLUTION

Conceptualize The filament in your lightbulb emits electromagnetic radiation. The brighter the light, the larger the magnitudes of the electric and magnetic fields.

Categorize Because the lightbulb is to be treated as a point source, it emits equally in all directions, so the outgoing electromagnetic radiation can be modeled as a spherical wave.

Analyze Recall from Equation 17.13 that the wave intensity I a distance r from a point source is $I = P_{avg}/4\pi r^2$, where P_{avg} is the average power output of the source and $4\pi r^2$ is the area of a sphere of radius r centered on the source.

Set this expression for I equal to the intensity of an electromagnetic wave given by Equation 34.24:

$$I = \frac{P_{avg}}{4\pi r^2} = \frac{E_{max}^2}{2\mu_0 c}$$

Solve for the electric field magnitude:

$$E_{max} = \sqrt{\frac{\mu_0 c P_{avg}}{2\pi r^2}}$$

Let's make some assumptions about numbers to enter in this equation. The visible light output of a 60-W lightbulb operating at 5% efficiency is approximately 3.0 W by visible light. (The remaining energy transfers out of the lightbulb by thermal conduction and invisible radiation.) A reasonable distance from the lightbulb to the page might be 0.30 m.

Substitute these values:

$$E_{max} = \sqrt{\frac{(4\pi \times 10^{-7} \text{ T} \cdot \text{m/A})(3.00 \times 10^8 \text{ m/s})(3.0 \text{ W})}{2\pi(0.30 \text{ m})^2}}$$

$$= 45 \text{ V/m}$$

continued

▶ **34.3** continued

Use Equation 34.21 to find the magnetic field magnitude:

$$B_{max} = \frac{E_{max}}{c} = \frac{45 \text{ V/m}}{3.00 \times 10^8 \text{ m/s}} = \boxed{1.5 \times 10^{-7} \text{ T}}$$

Finalize This value of the magnetic field magnitude is two orders of magnitude smaller than the Earth's magnetic field.

34.5 Momentum and Radiation Pressure

Electromagnetic waves transport linear momentum as well as energy. As this momentum is absorbed by some surface, pressure is exerted on the surface. Therefore, the surface is a nonisolated system for momentum. In this discussion, let's assume the electromagnetic wave strikes the surface at normal incidence and transports a total energy T_{ER} to the surface in a time interval Δt. Maxwell showed that if the surface absorbs all the incident energy T_{ER} in this time interval (as does a black body, introduced in Section 20.7), the total momentum $\vec{\mathbf{p}}$ transported to the surface has a magnitude

◀ **Momentum transported to a perfectly absorbing surface**

$$p = \frac{T_{ER}}{c} \quad \text{(complete absorption)} \tag{34.27}$$

The pressure P exerted on the surface is defined as force per unit area F/A, which when combined with Newton's second law gives

$$P = \frac{F}{A} = \frac{1}{A}\frac{dp}{dt}$$

Substituting Equation 34.27 into this expression for pressure P gives

$$P = \frac{1}{A}\frac{dp}{dt} = \frac{1}{A}\frac{d}{dt}\left(\frac{T_{ER}}{c}\right) = \frac{1}{c}\frac{(dT_{ER}/dt)}{A}$$

We recognize $(dT_{ER}/dt)/A$ as the rate at which energy is arriving at the surface per unit area, which is the magnitude of the Poynting vector. Therefore, the radiation pressure P exerted on the perfectly absorbing surface is

◀ **Radiation pressure exerted on a perfectly absorbing surface**

$$P = \frac{S}{c} \quad \text{(complete absorption)} \tag{34.28}$$

Pitfall Prevention 34.5
So Many p's We have p for momentum and P for pressure, and they are both related to $\mathcal{P}$ for power! Be sure to keep all these symbols straight.

If the surface is a perfect reflector (such as a mirror) and incidence is normal, the momentum transported to the surface in a time interval Δt is twice that given by Equation 34.27. That is, the momentum transferred to the surface by the incoming light is $p = T_{ER}/c$ and that transferred by the reflected light is also $p = T_{ER}/c$. Therefore,

$$p = \frac{2T_{ER}}{c} \quad \text{(complete reflection)} \tag{34.29}$$

The radiation pressure exerted on a perfectly reflecting surface for normal incidence of the wave is

◀ **Radiation pressure exerted on a perfectly reflecting surface**

$$P = \frac{2S}{c} \quad \text{(complete reflection)} \tag{34.30}$$

The pressure on a surface having a reflectivity somewhere between these two extremes has a value between S/c and $2S/c$, depending on the properties of the surface.

Although radiation pressures are very small (about 5×10^{-6} N/m² for direct sunlight), *solar sailing* is a low-cost means of sending spacecraft to the planets. Large sheets experience radiation pressure from sunlight and are used in much the way canvas sheets are used on earthbound sailboats. In 2010, the Japan Aerospace Exploration Agency (JAXA) launched the first spacecraft to use solar sailing as its primary propulsion, *IKAROS* (Interplanetary Kite-craft Accelerated by Radiation of the Sun). Successful testing of this spacecraft would lead to a larger effort to send a spacecraft to Jupiter by radiation pressure later in the present decade.

Quick Quiz 34.4 To maximize the radiation pressure on the sails of a spacecraft using solar sailing, should the sheets be **(a)** very black to absorb as much sunlight as possible or **(b)** very shiny to reflect as much sunlight as possible?

Conceptual Example 34.4 Sweeping the Solar System

A great amount of dust exists in interplanetary space. Although in theory these dust particles can vary in size from molecular size to a much larger size, very little of the dust in our solar system is smaller than about 0.2 μm. Why?

SOLUTION

The dust particles are subject to two significant forces: the gravitational force that draws them toward the Sun and the radiation-pressure force that pushes them away from the Sun. The gravitational force is proportional to the cube of the radius of a spherical dust particle because it is proportional to the mass and therefore to the volume $4\pi r^3/3$ of the particle. The radiation pressure is proportional to the square of the radius because it depends on the planar cross section of the particle. For large particles, the gravitational force is greater than the force from radiation pressure. For particles having radii less than about 0.2 μm, the radiation-pressure force is greater than the gravitational force. As a result, these particles are swept out of our solar system by sunlight.

Example 34.5 Pressure from a Laser Pointer

When giving presentations, many people use a laser pointer to direct the attention of the audience to information on a screen. If a 3.0-mW pointer creates a spot on a screen that is 2.0 mm in diameter, determine the radiation pressure on a screen that reflects 70% of the light that strikes it. The power 3.0 mW is a time-averaged value.

SOLUTION

Conceptualize Imagine the waves striking the screen and exerting a radiation pressure on it. The pressure should not be very large.

Categorize This problem involves a calculation of radiation pressure using an approach like that leading to Equation 34.28 or Equation 34.30, but it is complicated by the 70% reflection.

Analyze We begin by determining the magnitude of the beam's Poynting vector.

Divide the time-averaged power delivered via the electromagnetic wave by the cross-sectional area of the beam:

$$S_{avg} = \frac{(Power)_{avg}}{A} = \frac{(Power)_{avg}}{\pi r^2} = \frac{3.0 \times 10^{-3}\,\text{W}}{\pi\left(\dfrac{2.0 \times 10^{-3}\,\text{m}}{2}\right)^2} = 955\,\text{W/m}^2$$

Now let's determine the radiation pressure from the laser beam. Equation 34.30 indicates that a completely reflected beam would apply an average pressure of $P_{avg} = 2S_{avg}/c$. We can model the actual reflection as follows. Imagine that the surface absorbs the beam, resulting in pressure $P_{avg} = S_{avg}/c$. Then the surface emits the beam, resulting in additional pressure $P_{avg} = S_{avg}/c$. If the surface emits only a fraction f of the beam (so that f is the amount of the incident beam reflected), the pressure due to the emitted beam is $P_{avg} = fS_{avg}/c$.

Use this model to find the total pressure on the surface due to absorption and re-emission (reflection):

$$P_{avg} = \frac{S_{avg}}{c} + f\frac{S_{avg}}{c} = (1 + f)\frac{S_{avg}}{c}$$

continued

▶ **34.5** continued

Evaluate this pressure for a beam that is 70% reflected:

$$P_{avg} = (1 + 0.70)\frac{955 \text{ W/m}^2}{3.0 \times 10^8 \text{ m/s}} = \boxed{5.4 \times 10^{-6} \text{ N/m}^2}$$

Finalize The pressure has an extremely small value, as expected. (Recall from Section 14.2 that atmospheric pressure is approximately 10^5 N/m^2.) Consider the magnitude of the Poynting vector, $S_{avg} = 955$ W/m^2. It is about the same as the intensity of sunlight at the Earth's surface. For this reason, it is not safe to shine the beam of a laser pointer into a person's eyes, which may be more dangerous than looking directly at the Sun.

WHAT IF?   What if the laser pointer is moved twice as far away from the screen? Does that affect the radiation pressure on the screen?

Answer Because a laser beam is popularly represented as a beam of light with constant cross section, you might think that the intensity of radiation, and therefore the radiation pressure, is independent of distance from the screen. A laser beam, however, does not have a constant cross section at all distances from the source; rather,

there is a small but measurable divergence of the beam. If the laser is moved farther away from the screen, the area of illumination on the screen increases, decreasing the intensity. In turn, the radiation pressure is reduced.

In addition, the doubled distance from the screen results in more loss of energy from the beam due to scattering from air molecules and dust particles as the light travels from the laser to the screen. This energy loss further reduces the radiation pressure on the screen.

34.6 Production of Electromagnetic Waves by an Antenna

Stationary charges and steady currents cannot produce electromagnetic waves. If the current in a wire changes with time, however, the wire emits electromagnetic waves. The fundamental mechanism responsible for this radiation is the acceleration of a charged particle. **Whenever a charged particle accelerates, energy is transferred away from the particle by electromagnetic radiation.**

Let's consider the production of electromagnetic waves by a *half-wave antenna*. In this arrangement, two conducting rods are connected to a source of alternating voltage (such as an *LC* oscillator) as shown in Figure 34.11. The length of each rod is equal to one-quarter the wavelength of the radiation emitted when the oscillator operates at frequency *f*. The oscillator forces charges to accelerate back and forth between the two rods. Figure 34.11 shows the configuration of the electric and magnetic fields at some instant when the current is upward. The separation of charges in the upper and lower portions of the antenna make the electric field lines resemble those of an electric dipole. (As a result, this type of antenna is sometimes called a *dipole antenna*.) Because these charges are continuously oscillating between the two rods, the antenna can be approximated by an oscillating electric dipole. The current representing the movement of charges between the ends of the antenna produces magnetic field lines forming concentric circles around the antenna that are perpendicular to the electric field lines at all points. The magnetic field is zero at all points along the axis of the antenna. Furthermore, $\vec{E}$ and $\vec{B}$ are 90° out of phase in time; for example, the current is zero when the charges at the outer ends of the rods are at a maximum.

At the two points where the magnetic field is shown in Figure 34.11, the Poynting vector $\vec{S}$ is directed radially outward, indicating that energy is flowing away from the antenna at this instant. At later times, the fields and the Poynting vector reverse direction as the current alternates. Because $\vec{E}$ and $\vec{B}$ are 90° out of phase at points near the dipole, the net energy flow is zero. From this fact, you might conclude (incorrectly) that no energy is radiated by the dipole.

Energy is indeed radiated, however. Because the dipole fields fall off as $1/r^3$ (as shown in Example 23.6 for the electric field of a static dipole), they are negligible at great distances from the antenna. At these great distances, something else causes

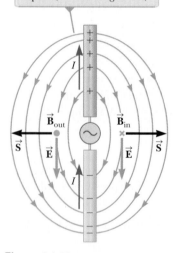

The electric field lines resemble those of an electric dipole (shown in Fig. 23.20).

Figure 34.11 A half-wave antenna consists of two metal rods connected to an alternating voltage source. This diagram shows $\vec{E}$ and $\vec{B}$ at an arbitrary instant when the current is upward.

a type of radiation different from that close to the antenna. The source of this radiation is the continuous induction of an electric field by the time-varying magnetic field and the induction of a magnetic field by the time-varying electric field, predicted by Equations 34.6 and 34.7. The electric and magnetic fields produced in this manner are in phase with each other and vary as $1/r$. The result is an outward flow of energy at all times.

The angular dependence of the radiation intensity produced by a dipole antenna is shown in Figure 34.12. Notice that the intensity and the power radiated are a maximum in a plane that is perpendicular to the antenna and passing through its midpoint. Furthermore, the power radiated is zero along the antenna's axis. A mathematical solution to Maxwell's equations for the dipole antenna shows that the intensity of the radiation varies as $(\sin^2 \theta)/r^2$, where θ is measured from the axis of the antenna.

Electromagnetic waves can also induce currents in a receiving antenna. The response of a dipole receiving antenna at a given position is a maximum when the antenna axis is parallel to the electric field at that point and zero when the axis is perpendicular to the electric field.

Quick Quiz 34.5 If the antenna in Figure 34.11 represents the source of a distant radio station, what would be the best orientation for your portable radio antenna located to the right of the figure? **(a)** up-down along the page **(b)** left-right along the page **(c)** perpendicular to the page

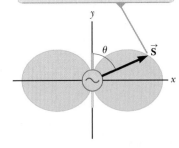

The distance from the origin to a point on the edge of the tan shape is proportional to the magnitude of the Poynting vector and the intensity of radiation in that direction.

Figure 34.12 Angular dependence of the intensity of radiation produced by an oscillating electric dipole.

34.7 The Spectrum of Electromagnetic Waves

The various types of electromagnetic waves are listed in Figure 34.13 (page 1046), which shows the **electromagnetic spectrum.** Notice the wide ranges of frequencies and wavelengths. No sharp dividing point exists between one type of wave and the next. Remember that all forms of the various types of radiation are produced by the same phenomenon: acceleration of electric charges. The names given to the types of waves are simply a convenient way to describe the region of the spectrum in which they lie.

Radio waves, whose wavelengths range from more than 10^4 m to about 0.1 m, are the result of charges accelerating through conducting wires. They are generated by such electronic devices as LC oscillators and are used in radio and television communication systems.

Microwaves have wavelengths ranging from approximately 0.3 m to 10^{-4} m and are also generated by electronic devices. Because of their short wavelengths, they are well suited for radar systems and for studying the atomic and molecular properties of matter. Microwave ovens are an interesting domestic application of these waves. It has been suggested that solar energy could be harnessed by beaming microwaves to the Earth from a solar collector in space.

Infrared waves have wavelengths ranging from approximately 10^{-3} m to the longest wavelength of visible light, 7×10^{-7} m. These waves, produced by molecules and room-temperature objects, are readily absorbed by most materials. The infrared (IR) energy absorbed by a substance appears as internal energy because the energy agitates the object's atoms, increasing their vibrational or translational motion, which results in a temperature increase. Infrared radiation has practical and scientific applications in many areas, including physical therapy, IR photography, and vibrational spectroscopy.

Visible light, the most familiar form of electromagnetic waves, is the part of the electromagnetic spectrum the human eye can detect. Light is produced by the rearrangement of electrons in atoms and molecules. The various wavelengths of visible light, which correspond to different colors, range from red ($\lambda \approx 7 \times 10^{-7}$ m) to violet ($\lambda \approx 4 \times 10^{-7}$ m). The sensitivity of the human eye is a function of wavelength, being a maximum at a wavelength of about 5.5×10^{-7} m. With that in mind, why do you suppose tennis balls often have a yellow-green color? Table 34.1 provides

Pitfall Prevention 34.6
"Heat Rays" Infrared rays are often called "heat rays," but this terminology is a misnomer. Although infrared radiation is used to raise or maintain temperature as in the case of keeping food warm with "heat lamps" at a fast-food restaurant, all wavelengths of electromagnetic radiation carry energy that can cause the temperature of a system to increase. As an example, consider a potato baking in your microwave oven.

Table 34.1 **Approximate Correspondence Between Wavelengths of Visible Light and Color**

Wavelength Range (nm)	Color Description
400–430	Violet
430–485	Blue
485–560	Green
560–590	Yellow
590–625	Orange
625–700	Red

Note: The wavelength ranges here are approximate. Different people will describe colors differently.

Figure 34.13 The electromagnetic spectrum.

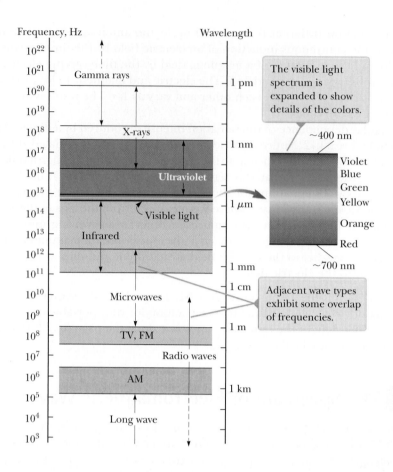

Raymond A. Serway

Wearing sunglasses that do not block ultraviolet (UV) light is worse for your eyes than wearing no sunglasses at all. The lenses of any sunglasses absorb some visible light, thereby causing the wearer's pupils to dilate. If the glasses do not also block UV light, more damage may be done to the lens of the eye because of the dilated pupils. If you wear no sunglasses at all, your pupils are contracted, you squint, and much less UV light enters your eyes. High-quality sunglasses block nearly all the eye-damaging UV light.

approximate correspondences between the wavelength of visible light and the color assigned to it by humans. Light is the basis of the science of optics and optical instruments, to be discussed in Chapters 35 through 38.

Ultraviolet waves cover wavelengths ranging from approximately 4×10^{-7} m to 6×10^{-10} m. The Sun is an important source of ultraviolet (UV) light, which is the main cause of sunburn. Sunscreen lotions are transparent to visible light but absorb most UV light. The higher a sunscreen's solar protection factor, or SPF, the greater the percentage of UV light absorbed. Ultraviolet rays have also been implicated in the formation of cataracts, a clouding of the lens inside the eye.

Most of the UV light from the Sun is absorbed by ozone (O_3) molecules in the Earth's upper atmosphere, in a layer called the stratosphere. This ozone shield converts lethal high-energy UV radiation to IR radiation, which in turn warms the stratosphere.

X-rays have wavelengths in the range from approximately 10^{-8} m to 10^{-12} m. The most common source of x-rays is the stopping of high-energy electrons upon bombarding a metal target. X-rays are used as a diagnostic tool in medicine and as a treatment for certain forms of cancer. Because x-rays can damage or destroy living tissues and organisms, care must be taken to avoid unnecessary exposure or overexposure. X-rays are also used in the study of crystal structure because x-ray wavelengths are comparable to the atomic separation distances in solids (about 0.1 nm).

Gamma rays are electromagnetic waves emitted by radioactive nuclei and during certain nuclear reactions. High-energy gamma rays are a component of cosmic rays that enter the Earth's atmosphere from space. They have wavelengths ranging from approximately 10^{-10} m to less than 10^{-14} m. Gamma rays are highly penetrating and produce serious damage when absorbed by living tissues. Consequently, those working near such dangerous radiation must be protected with heavily absorbing materials such as thick layers of lead.

Quick Quiz 34.6 In many kitchens, a microwave oven is used to cook food. The frequency of the microwaves is on the order of 10^{10} Hz. Are the wavelengths of these microwaves on the order of (a) kilometers, (b) meters, (c) centimeters, or (d) micrometers?

Quick Quiz 34.7 A radio wave of frequency on the order of 10^5 Hz is used to carry a sound wave with a frequency on the order of 10^3 Hz. Is the wavelength of this radio wave on the order of (a) kilometers, (b) meters, (c) centimeters, or (d) micrometers?

Summary

Definitions

In a region of space in which there is a changing electric field, there is a **displacement current** defined as

$$I_d \equiv \epsilon_0 \frac{d\Phi_E}{dt} \qquad (34.1)$$

where ϵ_0 is the permittivity of free space (see Section 23.3) and $\Phi_E = \int \vec{\mathbf{E}} \cdot d\vec{\mathbf{A}}$ is the electric flux.

The rate at which energy passes through a unit area by electromagnetic radiation is described by the **Poynting vector $\vec{\mathbf{S}}$**, where

$$\vec{\mathbf{S}} \equiv \frac{1}{\mu_0} \vec{\mathbf{E}} \times \vec{\mathbf{B}} \qquad (34.22)$$

Concepts and Principles

When used with the **Lorentz force law, $\vec{\mathbf{F}} = q\vec{\mathbf{E}} + q\vec{\mathbf{v}} \times \vec{\mathbf{B}}$, Maxwell's equations** describe all electromagnetic phenomena:

$$\oint \vec{\mathbf{E}} \cdot d\vec{\mathbf{A}} = \frac{q}{\epsilon_0} \qquad (34.4)$$

$$\oint \vec{\mathbf{E}} \cdot d\vec{\mathbf{s}} = -\frac{d\Phi_B}{dt} \qquad (34.6)$$

$$\oint \vec{\mathbf{B}} \cdot d\vec{\mathbf{A}} = 0 \qquad (34.5)$$

$$\oint \vec{\mathbf{B}} \cdot d\vec{\mathbf{s}} = \mu_0 I + \epsilon_0 \mu_0 \frac{d\Phi_E}{dt} \qquad (34.7)$$

Electromagnetic waves, which are predicted by Maxwell's equations, have the following properties and are described by the following mathematical representations of the traveling wave model for electromagnetic waves.

- The electric field and the magnetic field each satisfy a wave equation. These two wave equations, which can be obtained from Maxwell's third and fourth equations, are

$$\frac{\partial^2 E}{\partial x^2} = \mu_0 \epsilon_0 \frac{\partial^2 E}{\partial t^2} \qquad (34.15)$$

$$\frac{\partial^2 B}{\partial x^2} = \mu_0 \epsilon_0 \frac{\partial^2 B}{\partial t^2} \qquad (34.16)$$

- The waves travel through a vacuum with the speed of light c, where

$$c = \frac{1}{\sqrt{\mu_0 \epsilon_0}} \qquad (34.17)$$

- Numerically, the speed of electromagnetic waves in a vacuum is 3.00×10^8 m/s.
- The wavelength and frequency of electromagnetic waves are related by

$$\lambda = \frac{c}{f} = \frac{3.00 \times 10^8 \text{ m/s}}{f} \qquad (34.20)$$

- The electric and magnetic fields are perpendicular to each other and perpendicular to the direction of wave propagation.
- The instantaneous magnitudes of $\vec{\mathbf{E}}$ and $\vec{\mathbf{B}}$ in an electromagnetic wave are related by the expression

$$\frac{E}{B} = c \qquad (34.21)$$

- Electromagnetic waves carry energy.
- Electromagnetic waves carry momentum.

continued

Because electromagnetic waves carry momentum, they exert pressure on surfaces. If an electromagnetic wave whose Poynting vector is $\vec{S}$ is completely absorbed by a surface upon which it is normally incident, the radiation pressure on that surface is

$$P = \frac{S}{c} \quad \text{(complete absorption)} \quad \textbf{(34.28)}$$

If the surface totally reflects a normally incident wave, the pressure is doubled.

The electric and magnetic fields of a sinusoidal plane electromagnetic wave propagating in the positive x direction can be written as

$$E = E_{max} \cos (kx - \omega t) \quad \textbf{(34.18)}$$

$$B = B_{max} \cos (kx - \omega t) \quad \textbf{(34.19)}$$

where k is the angular wave number and ω is the angular frequency of the wave. These equations represent special solutions to the wave equations for E and B.

The average value of the Poynting vector for a plane electromagnetic wave has a magnitude

$$S_{avg} = \frac{E_{max} B_{max}}{2\mu_0} = \frac{E_{max}^2}{2\mu_0 c} = \frac{cB_{max}^2}{2\mu_0} \quad \textbf{(34.24)}$$

The intensity of a sinusoidal plane electromagnetic wave equals the average value of the Poynting vector taken over one or more cycles.

The electromagnetic spectrum includes waves covering a broad range of wavelengths, from long radio waves at more than 10^4 m to gamma rays at less than 10^{-14} m.

Objective Questions   1. denotes answer available in *Student Solutions Manual/Study Guide*

1. A spherical interplanetary grain of dust of radius 0.2 μm is at a distance r_1 from the Sun. The gravitational force exerted by the Sun on the grain just balances the force due to radiation pressure from the Sun's light. **(i)** Assume the grain is moved to a distance $2r_1$ from the Sun and released. At this location, what is the net force exerted on the grain? (a) toward the Sun (b) away from the Sun (c) zero (d) impossible to determine without knowing the mass of the grain **(ii)** Now assume the grain is moved back to its original location at r_1, compressed so that it crystallizes into a sphere with significantly higher density, and then released. In this situation, what is the net force exerted on the grain? Choose from the same possibilities as in part (i).

2. A small source radiates an electromagnetic wave with a single frequency into vacuum, equally in all directions. **(i)** As the wave moves, does its frequency (a) increase, (b) decrease, or (c) stay constant? Using the same choices, answer the same question about **(ii)** its wavelength, **(iii)** its speed, **(iv)** its intensity, and **(v)** the amplitude of its electric field.

3. A typical microwave oven operates at a frequency of 2.45 GHz. What is the wavelength associated with the electromagnetic waves in the oven? (a) 8.20 m (b) 12.2 cm (c) 1.20×10^8 m (d) 8.20×10^{-9} m (e) none of those answers

4. A student working with a transmitting apparatus like Heinrich Hertz's wishes to adjust the electrodes to generate electromagnetic waves with a frequency half as large as before. **(i)** How large should she make the effective capacitance of the pair of electrodes? (a) four times larger than before (b) two times larger than before (c) one-half as large as before (d) one-fourth as large as before (e) none of those answers **(ii)** After she makes the required adjustment, what will the wavelength of the transmitted wave be? Choose from the same possibilities as in part (i).

5. Assume you charge a comb by running it through your hair and then hold the comb next to a bar magnet. Do the electric and magnetic fields produced constitute an electromagnetic wave? (a) Yes they do, necessarily. (b) Yes they do because charged particles are moving inside the bar magnet. (c) They can, but only if the electric field of the comb and the magnetic field of the magnet are perpendicular. (d) They can, but only if both the comb and the magnet are moving. (e) They can, if either the comb or the magnet or both are accelerating.

6. Which of the following statements are true regarding electromagnetic waves traveling through a vacuum? More than one statement may be correct. (a) All waves have the same wavelength. (b) All waves have the same frequency. (c) All waves travel at 3.00×10^8 m/s. (d) The electric and magnetic fields associated with the waves are perpendicular to each other and to the direction of wave propagation. (e) The speed of the waves depends on their frequency.

7. A plane electromagnetic wave with a single frequency moves in vacuum in the positive x direction. Its amplitude is uniform over the yz plane. **(i)** As the wave moves, does its frequency (a) increase, (b) decrease, or (c) stay constant? Using the same choices, answer the same question about **(ii)** its wavelength, **(iii)** its speed, **(iv)** its intensity, and **(v)** the amplitude of its magnetic field.

8. Assume the amplitude of the electric field in a plane electromagnetic wave is E_1 and the amplitude of the magnetic field is B_1. The source of the wave is then adjusted so that the amplitude of the electric field doubles to become $2E_1$. **(i)** What happens to the amplitude of the magnetic field in this process? (a) It becomes four times larger. (b) It becomes two times larger. (c) It can stay constant. (d) It becomes one-half as large. (e) It becomes one-fourth as large. **(ii)** What happens to the intensity of the wave? Choose from the same possibilities as in part (i).

9. An electromagnetic wave with a peak magnetic field magnitude of 1.50×10^{-7} T has an associated peak electric field of what magnitude? (a) 0.500×10^{-15} N/C (b) 2.00×10^{-5} N/C (c) 2.20×10^4 N/C (d) 45.0 N/C (e) 22.0 N/C

10. **(i)** Rank the following kinds of waves according to their wavelength ranges from those with the largest typical or average wavelength to the smallest, noting any cases of equality: (a) gamma rays (b) microwaves (c) radio waves (d) visible light (e) x-rays **(ii)** Rank the kinds of waves according to their frequencies from highest to lowest. **(iii)** Rank the kinds of waves according to their speeds in vacuum from fastest to slowest.

11. Consider an electromagnetic wave traveling in the positive y direction. The magnetic field associated with the wave at some location at some instant points in the negative x direction as shown in Figure OQ34.11. What is the direction of the electric field at this position and at this instant? (a) the positive x direction (b) the positive y direction (c) the positive z direction (d) the negative z direction (e) the negative y direction

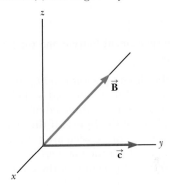

Figure OQ34.11

Conceptual Questions

1. denotes answer available in *Student Solutions Manual/Study Guide*

1. Suppose a creature from another planet has eyes that are sensitive to infrared radiation. Describe what the alien would see if it looked around your library. In particular, what would appear bright and what would appear dim?

2. For a given incident energy of an electromagnetic wave, why is the radiation pressure on a perfectly reflecting surface twice as great as that on a perfectly absorbing surface?

3. Radio stations often advertise "instant news." If that means you can hear the news the instant the radio announcer speaks it, is the claim true? What approximate time interval is required for a message to travel from Maine to California by radio waves? (Assume the waves can be detected at this range.)

4. List at least three differences between sound waves and light waves.

5. If a high-frequency current exists in a solenoid containing a metallic core, the core becomes warm due to induction. Explain why the material rises in temperature in this situation.

6. When light (or other electromagnetic radiation) travels across a given region, (a) what is it that oscillates? (b) What is it that is transported?

7. Why should an infrared photograph of a person look different from a photograph taken with visible light?

8. Do Maxwell's equations allow for the existence of magnetic monopoles? Explain.

9. Despite the advent of digital television, some viewers still use "rabbit ears" atop their sets (Fig. CQ34.9) instead of purchasing cable television service or satellite dishes. Certain orientations of the receiving antenna on a television set give better reception than others. Furthermore, the best orientation varies from station to station. Explain.

Figure CQ34.9 Conceptual Question 9 and Problem 78.

10. What does a radio wave do to the charges in the receiving antenna to provide a signal for your car radio?

11. Describe the physical significance of the Poynting vector.

12. An empty plastic or glass dish being removed from a microwave oven can be cool to the touch, even when food on an adjoining dish is hot. How is this phenomenon possible?

13. What new concept did Maxwell's generalized form of Ampère's law include?

Problems

The problems found in this chapter may be assigned online in Enhanced WebAssign

1. straightforward; **2.** intermediate; **3.** challenging

1. full solution available in the *Student Solutions Manual/Study Guide*

AMT Analysis Model tutorial available in Enhanced WebAssign

GP Guided Problem

M Master It tutorial available in Enhanced WebAssign

W Watch It video solution available in Enhanced WebAssign

Section 34.1 Displacement Current and the General Form of Ampère's Law

1. Consider the situation shown in Figure P34.1. An electric field of 300 V/m is confined to a circular area $d =$ 10.0 cm in diameter and directed outward perpendicular to the plane of the figure. If the field is increasing at a rate of 20.0 V/m · s, what are (a) the direction and (b) the magnitude of the magnetic field at the point P, $r =$ 15.0 cm from the center of the circle?

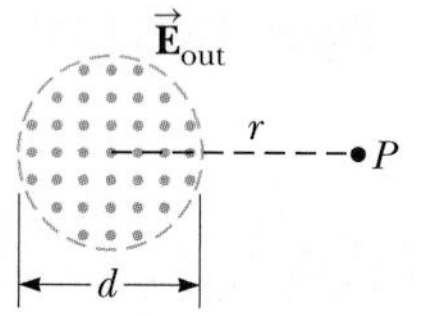

Figure P34.1

2. W A 0.200-A current is charging a capacitor that has circular plates 10.0 cm in radius. If the plate separation is 4.00 mm, (a) what is the time rate of increase of electric field between the plates? (b) What is the magnetic field between the plates 5.00 cm from the center?

3. M A 0.100-A current is charging a capacitor that has square plates 5.00 cm on each side. The plate separation is 4.00 mm. Find (a) the time rate of change of electric flux between the plates and (b) the displacement current between the plates.

Section 34.2 Maxwell's Equations and Hertz's Discoveries

4. AMT W An electron moves through a uniform electric field $\vec{E} = (2.50\hat{i} + 5.00\hat{j})$ V/m and a uniform magnetic field $\vec{B} = 0.400\hat{k}$ T. Determine the acceleration of the electron when it has a velocity $\vec{v} = 10.0\hat{i}$ m/s.

5. M A proton moves through a region containing a uniform electric field given by $\vec{E} = 50.0\hat{j}$ V/m and a uniform magnetic field $\vec{B} = (0.200\hat{i} + 0.300\hat{j} + 0.400\hat{k})$ T. Determine the acceleration of the proton when it has a velocity $\vec{v} = 200\hat{i}$ m/s.

6. A very long, thin rod carries electric charge with the linear density 35.0 nC/m. It lies along the x axis and moves in the x direction at a speed of 1.50×10^7 m/s. (a) Find the electric field the rod creates at the point ($x = 0$, $y = 20.0$ cm, $z = 0$). (b) Find the magnetic field it creates at the same point. (c) Find the force exerted on an electron at this point, moving with a velocity of $(2.40 \times 10^8)\hat{i}$ m/s.

Section 34.3 Plane Electromagnetic Waves

Note: Assume the medium is vacuum unless specified otherwise.

7. Suppose you are located 180 m from a radio transmitter. (a) How many wavelengths are you from the transmitter if the station calls itself 1150 AM? (The AM band frequencies are in kilohertz.) (b) What if this station is 98.1 FM? (The FM band frequencies are in megahertz.)

8. A diathermy machine, used in physiotherapy, generates electromagnetic radiation that gives the effect of "deep heat" when absorbed in tissue. One assigned frequency for diathermy is 27.33 MHz. What is the wavelength of this radiation?

9. The distance to the North Star, Polaris, is approximately 6.44×10^{18} m. (a) If Polaris were to burn out today, how many years from now would we see it disappear? (b) What time interval is required for sunlight to reach the Earth? (c) What time interval is required for a microwave signal to travel from the Earth to the Moon and back?

10. The red light emitted by a helium–neon laser has a wavelength of 632.8 nm. What is the frequency of the light waves?

11. **Review.** A standing-wave pattern is set up by radio waves between two metal sheets 2.00 m apart, which is the shortest distance between the plates that produces a standing-wave pattern. What is the frequency of the radio waves?

12. W An electromagnetic wave in vacuum has an electric field amplitude of 220 V/m. Calculate the amplitude of the corresponding magnetic field.

13. The speed of an electromagnetic wave traveling in a transparent nonmagnetic substance is $v = 1/\sqrt{\kappa\mu_0\epsilon_0}$, where κ is the dielectric constant of the substance. Determine the speed of light in water, which has a dielectric constant of 1.78 at optical frequencies.

14. A radar pulse returns to the transmitter–receiver after a total travel time of 4.00×10^{-4} s. How far away is the object that reflected the wave?

15. M Figure P34.15 shows a plane electromagnetic sinusoidal wave propagating in the x direction. Suppose the wavelength is 50.0 m and the electric field vibrates in the xy plane with an amplitude of 22.0 V/m. Calculate

(a) the frequency of the wave and (b) the magnetic field $\vec{B}$ when the electric field has its maximum value in the negative y direction. (c) Write an expression for $\vec{B}$ with the correct unit vector, with numerical values for B_{max}, k, and ω, and with its magnitude in the form

$$B = B_{max} \cos (kx - \omega t)$$

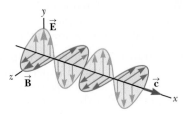

Figure P34.15 Problems 15 and 70.

16. Verify by substitution that the following equations are solutions to Equations 34.15 and 34.16, respectively:

$$E = E_{max} \cos (kx - \omega t)$$

$$B = B_{max} \cos (kx - \omega t)$$

17. **Review.** A microwave oven is powered by a magnetron, an electronic device that generates electromagnetic waves of frequency 2.45 GHz. The microwaves enter the oven and are reflected by the walls. The standing-wave pattern produced in the oven can cook food unevenly, with hot spots in the food at antinodes and cool spots at nodes, so a turntable is often used to rotate the food and distribute the energy. If a microwave oven intended for use with a turntable is instead used with a cooking dish in a fixed position, the antinodes can appear as burn marks on foods such as carrot strips or cheese. The separation distance between the burns is measured to be 6 cm ± 5%. From these data, calculate the speed of the microwaves.

18. *Why is the following situation impossible?* An electromagnetic wave travels through empty space with electric and magnetic fields described by

$$E = 9.00 \times 10^3 \cos [(9.00 \times 10^6)x - (3.00 \times 10^{15})t]$$

$$B = 3.00 \times 10^{-5} \cos [(9.00 \times 10^6)x - (3.00 \times 10^{15})t]$$

where all numerical values and variables are in SI units.

19. In SI units, the electric field in an electromagnetic wave is described by

$$E_y = 100 \sin (1.00 \times 10^7 \, x - \omega t)$$

Find (a) the amplitude of the corresponding magnetic field oscillations, (b) the wavelength λ, and (c) the frequency f.

Section 34.4 Energy Carried by Electromagnetic Waves

20. At what distance from the Sun is the intensity of sunlight three times the value at the Earth? (The average Earth–Sun separation is 1.496×10^{11} m.)

21. If the intensity of sunlight at the Earth's surface under a fairly clear sky is 1 000 W/m², how much electromagnetic energy per cubic meter is contained in sunlight?

22. The power of sunlight reaching each square meter of the Earth's surface on a clear day in the tropics is close to 1 000 W. On a winter day in Manitoba, the power concentration of sunlight can be 100 W/m². Many human activities are described by a power per unit area on the order of 10^2 W/m² or less. (a) Consider, for example, a family of four paying $66 to the electric company every 30 days for 600 kWh of energy carried by electrical transmission to their house, which has floor dimensions of 13.0 m by 9.50 m. Compute the power per unit area used by the family. (b) Consider a car 2.10 m wide and 4.90 m long traveling at 55.0 mi/h using gasoline having "heat of combustion" 44.0 MJ/kg with fuel economy 25.0 mi/gal. One gallon of gasoline has a mass of 2.54 kg. Find the power per unit area used by the car. (c) Explain why direct use of solar energy is not practical for running a conventional automobile. (d) What are some uses of solar energy that are more practical?

23. A community plans to build a facility to convert solar radiation to electrical power. The community requires 1.00 MW of power, and the system to be installed has an efficiency of 30.0% (that is, 30.0% of the solar energy incident on the surface is converted to useful energy that can power the community). Assuming sunlight has a constant intensity of 1 000 W/m², what must be the effective area of a perfectly absorbing surface used in such an installation?

24. In a region of free space, the electric field at an instant of time is $\vec{E} = (80.0\hat{i} + 32.0\hat{j} - 64.0\hat{k})$ N/C and the magnetic field is $\vec{B} = (0.200\hat{i} + 0.080 \, 0\hat{j} + 0.290\hat{k}) \, \mu$T. (a) Show that the two fields are perpendicular to each other. (b) Determine the Poynting vector for these fields.

25. When a high-power laser is used in the Earth's atmosphere, the electric field associated with the laser beam can ionize the air, turning it into a conducting plasma that reflects the laser light. In dry air at 0°C and 1 atm, electric breakdown occurs for fields with amplitudes above about 3.00 MV/m. (a) What laser beam intensity will produce such a field? (b) At this maximum intensity, what power can be delivered in a cylindrical beam of diameter 5.00 mm?

26. **Review.** Model the electromagnetic wave in a microwave oven as a plane traveling wave moving to the left, with an intensity of 25.0 kW/m². An oven contains two cubical containers of small mass, each full of water. One has an edge length of 6.00 cm, and the other, 12.0 cm. Energy falls perpendicularly on one face of each container. The water in the smaller container absorbs 70.0% of the energy that falls on it. The water in the larger container absorbs 91.0%. That is, the fraction 0.300 of the incoming microwave energy passes through a 6.00-cm thickness of water, and the fraction (0.300)(0.300) = 0.090 passes through a 12.0-cm thickness. Assume a negligible amount of energy leaves either container by heat. Find the temperature change of the water in each container over a time interval of 480 s.

27. High-power lasers in factories are used to cut through [W] cloth and metal (Fig. P34.27). One such laser has a beam diameter of 1.00 mm and generates an electric field having an amplitude of 0.700 MV/m at the target. Find (a) the amplitude of the magnetic field produced, (b) the intensity of the laser, and (c) the power delivered by the laser.

Philippe Plailly/SPL/Photo Researchers, Inc.

Figure P34.27

28. Consider a bright star in our night sky. Assume its distance from the Earth is 20.0 light-years (ly) and its power output is 4.00×10^{28} W, about 100 times that of the Sun. (a) Find the intensity of the starlight at the Earth. (b) Find the power of the starlight the Earth intercepts. One light-year is the distance traveled by light through a vacuum in one year.

29. What is the average magnitude of the Poynting vector [M] 5.00 mi from a radio transmitter broadcasting isotropically (equally in all directions) with an average power of 250 kW?

30. Assuming the antenna of a 10.0-kW radio station radiates spherical electromagnetic waves, (a) compute the maximum value of the magnetic field 5.00 km from the antenna and (b) state how this value compares with the surface magnetic field of the Earth.

31. Review. An AM radio station broadcasts isotropically [W] (equally in all directions) with an average power of 4.00 kW. A receiving antenna 65.0 cm long is at a location 4.00 mi from the transmitter. Compute the amplitude of the emf that is induced by this signal between the ends of the receiving antenna.

32. At what distance from a 100-W electromagnetic wave point source does $E_{max} = 15.0$ V/m?

33. The filament of an incandescent lamp has a 150-Ω resistance and carries a direct current of 1.00 A. The filament is 8.00 cm long and 0.900 mm in radius. (a) Calculate the Poynting vector at the surface of the filament, associated with the static electric field producing the current and the current's static magnetic field. (b) Find the magnitude of the static electric and magnetic fields at the surface of the filament.

34. At one location on the Earth, the rms value of the magnetic field caused by solar radiation is 1.80 μT. From this value, calculate (a) the rms electric field due to solar radiation, (b) the average energy density of the solar component of electromagnetic radiation at this location, and (c) the average magnitude of the Poynting vector for the Sun's radiation.

Section 34.5 Momentum and Radiation Pressure

35. A 25.0-mW laser beam of diameter 2.00 mm is reflected at normal incidence by a perfectly reflecting mirror. Calculate the radiation pressure on the mirror.

36. A radio wave transmits 25.0 W/m^2 of power per unit area. A flat surface of area A is perpendicular to the direction of propagation of the wave. Assuming the surface is a perfect absorber, calculate the radiation pressure on it.

37. A 15.0-mW helium–neon laser emits a beam of circular [M] cross section with a diameter of 2.00 mm. (a) Find the maximum electric field in the beam. (b) What total energy is contained in a 1.00-m length of the beam? (c) Find the momentum carried by a 1.00-m length of the beam.

38. A helium–neon laser emits a beam of circular cross section with a radius r and a power P. (a) Find the maximum electric field in the beam. (b) What total energy is contained in a length ℓ of the beam? (c) Find the momentum carried by a length ℓ of the beam.

39. A uniform circular disk of mass $m = 24.0$ g and radius [AMT] $r = 40.0$ cm hangs vertically from a fixed, frictionless, horizontal hinge at a point on its circumference as shown in Figure P34.39a. A beam of electromagnetic radiation with intensity 10.0 MW/m^2 is incident on the disk in a direction perpendicular to its surface. The disk is perfectly absorbing, and the resulting radiation pressure makes the disk rotate. Assuming the radiation is *always* perpendicular to the surface of the disk, find the angle θ through which the disk rotates from the vertical as it reaches its new equilibrium position shown in Figure 34.39b.

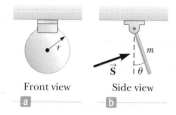

Front view   Side view

a   b

Figure P34.39

40. The intensity of sunlight at the Earth's distance from the Sun is 1 370 W/m^2. Assume the Earth absorbs all the sunlight incident upon it. (a) Find the total force the Sun exerts on the Earth due to radiation pressure. (b) Explain how this force compares with the Sun's gravitational attraction.

41. A plane electromagnetic wave of intensity 6.00 W/m^2, moving in the x direction, strikes a small perfectly reflecting pocket mirror, of area 40.0 cm^2, held in the yz plane. (a) What momentum does the wave transfer to the mirror each second? (b) Find the force the wave exerts on the mirror. (c) Explain the relationship between the answers to parts (a) and (b).

42. Assume the intensity of solar radiation incident on the upper atmosphere of the Earth is 1 370 W/m² and use data from Table 13.2 as necessary. Determine (a) the intensity of solar radiation incident on Mars, (b) the total power incident on Mars, and (c) the radiation force that acts on that planet if it absorbs nearly all the light. (d) State how this force compares with the gravitational attraction exerted by the Sun on Mars. (e) Compare the ratio of the gravitational force to the light-pressure force exerted on the Earth and the ratio of these forces exerted on Mars, found in part (d).

43. A possible means of space flight is to place a perfectly
AMT reflecting aluminized sheet into orbit around the Earth
W and then use the light from the Sun to push this "solar sail." Suppose a sail of area $A = 6.00 \times 10^5$ m² and mass $m = 6.00 \times 10^3$ kg is placed in orbit facing the Sun. Ignore all gravitational effects and assume a solar intensity of 1 370 W/m². (a) What force is exerted on the sail? (b) What is the sail's acceleration? (c) Assuming the acceleration calculated in part (b) remains constant, find the time interval required for the sail to reach the Moon, 3.84×10^8 m away, starting from rest at the Earth.

Section 34.6 Production of Electromagnetic Waves by an Antenna

44. Extremely low-frequency (ELF) waves that can penetrate the oceans are the only practical means of communicating with distant submarines. (a) Calculate the length of a quarter-wavelength antenna for a transmitter generating ELF waves of frequency 75.0 Hz into air. (b) How practical is this means of communication?

45. A Marconi antenna, used by most AM radio stations, consists of the top half of a Hertz antenna (also known as a half-wave antenna because its length is $\lambda/2$). The lower end of this Marconi (quarter-wave) antenna is connected to Earth ground, and the ground itself serves as the missing lower half. What are the heights of the Marconi antennas for radio stations broadcasting at (a) 560 kHz and (b) 1 600 kHz?

46. A large, flat sheet carries a uniformly distributed electric current with current per unit width J_s. This current creates a magnetic field on both sides of the sheet, parallel to the sheet and perpendicular to the current, with magnitude $B = \frac{1}{2}\mu_0 J_s$. If the current is in the y direction and oscillates in time according to

$$J_{max}(\cos \omega t)\hat{\mathbf{j}} = J_{max}[\cos(-\omega t)]\hat{\mathbf{j}}$$

the sheet radiates an electromagnetic wave. Figure P34.46 shows such a wave emitted from one point on the sheet chosen to be the origin. Such electromagnetic waves are emitted from all points on the sheet. The magnetic field of the wave to the right of the sheet is described by the wave function

$$\vec{\mathbf{B}} = \frac{1}{2}\mu_0 J_{max}[\cos(kx - \omega t)]\hat{\mathbf{k}}$$

(a) Find the wave function for the electric field of the wave to the right of the sheet. (b) Find the Poynting vector as a function of x and t. (c) Find the intensity of the wave. (d) **What If?** If the sheet is to emit radiation in each direction (normal to the plane of the sheet) with intensity 570 W/m², what maximum value of sinusoidal current density is required?

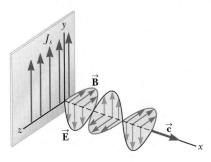

Figure P34.46

47. **Review.** Accelerating charges radiate electromagnetic waves. Calculate the wavelength of radiation produced by a proton in a cyclotron with a magnetic field of 0.350 T.

48. **Review.** Accelerating charges radiate electromagnetic waves. Calculate the wavelength of radiation produced by a proton of mass m_p moving in a circular path perpendicular to a magnetic field of magnitude B.

49. Two vertical radio-transmitting antennas are separated by half the broadcast wavelength and are driven in phase with each other. In what horizontal directions are (a) the strongest and (b) the weakest signals radiated?

Section 34.7 The Spectrum of Electromagnetic Waves

50. Compute an order-of-magnitude estimate for the fre-
W quency of an electromagnetic wave with wavelength equal to (a) your height and (b) the thickness of a sheet of paper. How is each wave classified on the electromagnetic spectrum?

51. What are the wavelengths of electromagnetic waves in free space that have frequencies of (a) 5.00×10^{19} Hz and (b) 4.00×10^9 Hz?

52. An important news announcement is transmitted by radio waves to people sitting next to their radios 100 km from the station and by sound waves to people sitting across the newsroom 3.00 m from the newscaster. Taking the speed of sound in air to be 343 m/s, who receives the news first? Explain.

53. In addition to cable and satellite broadcasts, television stations still use VHF and UHF bands for digitally broadcasting their signals. Twelve VHF television channels (channels 2 through 13) lie in the range of frequencies between 54.0 MHz and 216 MHz. Each channel is assigned a width of 6.00 MHz, with the two ranges 72.0–76.0 MHz and 88.0–174 MHz reserved for non-TV purposes. (Channel 2, for example, lies

between 54.0 and 60.0 MHz.) Calculate the broadcast wavelength range for (a) channel 4, (b) channel 6, and (c) channel 8.

Additional Problems

54. Classify waves with frequencies of 2 Hz, 2 kHz, 2 MHz, 2 GHz, 2 THz, 2 PHz, 2 EHz, 2 ZHz, and 2 YHz on the electromagnetic spectrum. Classify waves with wavelengths of 2 km, 2 m, 2 mm, 2 μm, 2 nm, 2 pm, 2 fm, and 2 am.

55. Assume the intensity of solar radiation incident on the cloud tops of the Earth is 1 370 W/m². (a) Taking the average Earth–Sun separation to be 1.496×10^{11} m, calculate the total power radiated by the Sun. Determine the maximum values of (b) the electric field and (c) the magnetic field in the sunlight at the Earth's location.

56. In 1965, Arno Penzias and Robert Wilson discovered the cosmic microwave radiation left over from the big bang expansion of the Universe. Suppose the energy density of this background radiation is 4.00×10^{-14} J/m³. Determine the corresponding electric field amplitude.

57. The eye is most sensitive to light having a frequency of 5.45×10^{14} Hz, which is in the green-yellow region of the visible electromagnetic spectrum. What is the wavelength of this light?

58. Write expressions for the electric and magnetic fields of a sinusoidal plane electromagnetic wave having an electric field amplitude of 300 V/m and a frequency of 3.00 GHz and traveling in the positive x direction.

59. One goal of the Russian space program is to illuminate dark northern cities with sunlight reflected to the Earth from a 200-m diameter mirrored surface in orbit. Several smaller prototypes have already been constructed and put into orbit. (a) Assume that sunlight with intensity 1 370 W/m² falls on the mirror nearly perpendicularly and that the atmosphere of the Earth allows 74.6% of the energy of sunlight to pass though it in clear weather. What is the power received by a city when the space mirror is reflecting light to it? (b) The plan is for the reflected sunlight to cover a circle of diameter 8.00 km. What is the intensity of light (the average magnitude of the Poynting vector) received by the city? (c) This intensity is what percentage of the vertical component of sunlight at St. Petersburg in January, when the sun reaches an angle of 7.00° above the horizon at noon?

60. A microwave source produces pulses of 20.0-GHz radiation, with each pulse lasting 1.00 ns. A parabolic reflector with a face area of radius 6.00 cm is used to focus the microwaves into a parallel beam of radiation as shown in Figure P34.60. The average power during each pulse is 25.0 kW. (a) What is the wavelength of these microwaves? (b) What is the total energy contained in each pulse? (c) Compute the average energy density inside each pulse. (d) Determine the amplitude

of the electric and magnetic fields in these microwaves. (e) Assuming that this pulsed beam strikes an absorbing surface, compute the force exerted on the surface during the 1.00-ns duration of each pulse.

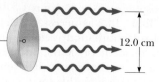

12.0 cm

Figure P34.60

61. The intensity of solar radiation at the top of the Earth's atmosphere is 1 370 W/m². Assuming 60% of the incoming solar energy reaches the Earth's surface and you absorb 50% of the incident energy, make an order-of-magnitude estimate of the amount of solar energy you absorb if you sunbathe for 60 minutes.

62. Two handheld radio transceivers with dipole antennas are separated by a large, fixed distance. If the transmitting antenna is vertical, what fraction of the maximum received power will appear in the receiving antenna when it is inclined from the vertical (a) by 15.0°? (b) By 45.0°? (c) By 90.0°?

63. Consider a small, spherical particle of radius r located in space a distance $R = 3.75 \times 10^{11}$ m from the Sun. Assume the particle has a perfectly absorbing surface and a mass density of $\rho = 1.50$ g/cm³. Use $S = 214$ W/m² as the value of the solar intensity at the location of the particle. Calculate the value of r for which the particle is in equilibrium between the gravitational force and the force exerted by solar radiation.

64. Consider a small, spherical particle of radius r located in space a distance R from the Sun, of mass M_S. Assume the particle has a perfectly absorbing surface and a mass density ρ. The value of the solar intensity at the particle's location is S. Calculate the value of r for which the particle is in equilibrium between the gravitational force and the force exerted by solar radiation. Your answer should be in terms of S, R, ρ, and other constants.

65. A dish antenna having a diameter of 20.0 m receives (at normal incidence) a radio signal from a distant source as shown in Figure P34.65. The radio signal is a continuous sinusoidal wave with amplitude $E_{max} = 0.200$ μV/m.

Figure P34.65

Assume the antenna absorbs all the radiation that falls on the dish. (a) What is the amplitude of the magnetic field in this wave? (b) What is the intensity of the radiation received by this antenna? (c) What is the power received by the antenna? (d) What force is exerted by the radio waves on the antenna?

66. The Earth reflects approximately 38.0% of the incident sunlight from its clouds and surface. (a) Given that the intensity of solar radiation at the top of the atmosphere is 1 370 W/m², find the radiation pressure on the Earth, in pascals, at the location where the Sun is straight overhead. (b) State how this quantity compares with normal atmospheric pressure at the Earth's surface, which is 101 kPa.

67. **Review.** A 1.00-m-diameter circular mirror focuses the Sun's rays onto a circular absorbing plate 2.00 cm in radius, which holds a can containing 1.00 L of water at 20.0°C. (a) If the solar intensity is 1.00 kW/m², what is the intensity on the absorbing plate? At the plate, what are the maximum magnitudes of the fields (b) $\vec{E}$ and (c) $\vec{B}$? (d) If 40.0% of the energy is absorbed, what time interval is required to bring the water to its boiling point?

68. (a) A stationary charged particle at the origin creates an electric flux of 487 N · m²/C through any closed surface surrounding the charge. Find the electric field it creates in the empty space around it as a function of radial distance r away from the particle. (b) A small source at the origin emits an electromagnetic wave with a single frequency into vacuum, equally in all directions, with power 25.0 W. Find the electric field amplitude as a function of radial distance away from the source. (c) At what distance is the amplitude of the electric field in the wave equal to 3.00 MV/m, representing the dielectric strength of air? (d) As the distance from the source doubles, what happens to the field amplitude? (e) State how the behavior shown in part (d) compares with the behavior of the field in part (a).

69. **Review.** (a) A homeowner has a solar water heater installed on the roof of his house (Fig. P34.69). The heater is a flat, closed box with excellent thermal insulation. Its interior is painted black, and its front face is made of insulating glass. Its emissivity for visible light

Figure P34.69

is 0.900, and its emissivity for infrared light is 0.700. Light from the noontime Sun is incident perpendicular to the glass with an intensity of 1 000 W/m², and no water enters or leaves the box. Find the steady-state temperature of the box's interior. (b) **What If?** The homeowner builds an identical box with no water tubes. It lies flat on the ground in front of the house. He uses it as a cold frame, where he plants seeds in early spring. Assuming the same noontime Sun is at an elevation angle of 50.0°, find the steady-state temperature of the interior of the box when its ventilation slots are tightly closed.

70. You may wish to review Sections 16.5 and 17.3 on the transport of energy by string waves and sound. Figure P34.15 is a graphical representation of an electromagnetic wave moving in the x direction. We wish to find an expression for the intensity of this wave by means of a different process from that by which Equation 34.24 was generated. (a) Sketch a graph of the electric field in this wave at the instant $t = 0$, letting your flat paper represent the xy plane. (b) Compute the energy density u_E in the electric field as a function of x at the instant $t = 0$. (c) Compute the energy density in the magnetic field u_B as a function of x at that instant. (d) Find the total energy density u as a function of x, expressed in terms of only the electric field amplitude. (e) The energy in a "shoebox" of length λ and frontal area A is $E_\lambda = \int_0^\lambda uA \, dx$. (The symbol E_λ for energy in a wavelength imitates the notation of Section 16.5.) Perform the integration to compute the amount of this energy in terms of A, λ, E_{max}, and universal constants. (f) We may think of the energy transport by the whole wave as a series of these shoeboxes going past as if carried on a conveyor belt. Each shoebox passes by a point in a time interval defined as the period $T = 1/f$ of the wave. Find the power the wave carries through area A. (g) The intensity of the wave is the power per unit area through which the wave passes. Compute this intensity in terms of E_{max} and universal constants. (h) Explain how your result compares with that given in Equation 34.24.

71. Lasers have been used to suspend spherical glass beads in the Earth's gravitational field. (a) A black bead has a radius of 0.500 mm and a density of 0.200 g/cm³. Determine the radiation intensity needed to support the bead. (b) What is the minimum power required for this laser?

72. Lasers have been used to suspend spherical glass beads in the Earth's gravitational field. (a) A black bead has radius r and density ρ. Determine the radiation intensity needed to support the bead. (b) What is the minimum power required for this laser?

73. **Review.** A 5.50-kg black cat and her four black kittens, each with mass 0.800 kg, sleep snuggled together on a mat on a cool night, with their bodies forming a hemisphere. Assume the hemisphere has a surface temperature of 31.0°C, an emissivity of 0.970, and a uniform density of 990 kg/m³. Find (a) the radius of the hemisphere, (b) the area of its curved surface, (c) the

radiated power emitted by the cats at their curved surface, and (d) the intensity of radiation at this surface. You may think of the emitted electromagnetic wave as having a single predominant frequency. Find (e) the amplitude of the electric field in the electromagnetic wave just outside the surface of the cozy pile and (f) the amplitude of the magnetic field. (g) **What If?** The next night, the kittens all sleep alone, curling up into separate hemispheres like their mother. Find the total radiated power of the family. (For simplicity, ignore the cats' absorption of radiation from the environment.)

74. The electromagnetic power radiated by a nonrelativistic particle with charge q moving with acceleration a is

$$P = \frac{q^2 a^2}{6 \pi \epsilon_0 c^3}$$

where ϵ_0 is the permittivity of free space (also called the permittivity of vacuum) and c is the speed of light in vacuum. (a) Show that the right side of this equation has units of watts. An electron is placed in a constant electric field of magnitude 100 N/C. Determine (b) the acceleration of the electron and (c) the electromagnetic power radiated by this electron. (d) **What If?** If a proton is placed in a cyclotron with a radius of 0.500 m and a magnetic field of magnitude 0.350 T, what electromagnetic power does this proton radiate just before leaving the cyclotron?

75. **Review.** Gliese 581c is the first Earth-like extrasolar terrestrial planet discovered. Its parent star, Gliese 581, is a red dwarf that radiates electromagnetic waves with power 5.00×10^{24} W, which is only 1.30% of the power of the Sun. Assume the emissivity of the planet is equal for infrared and for visible light and the planet has a uniform surface temperature. Identify (a) the projected area over which the planet absorbs light from Gliese 581 and (b) the radiating area of the planet. (c) If an average temperature of 287 K is necessary for life to exist on Gliese 581c, what should the radius of the planet's orbit be?

Challenge Problems

76. A plane electromagnetic wave varies sinusoidally at 90.0 MHz as it travels through vacuum along the positive x direction. The peak value of the electric field is 2.00 mV/m, and it is directed along the positive y direction. Find (a) the wavelength, (b) the period, and (c) the maximum value of the magnetic field. (d) Write expressions in SI units for the space and time variations of the electric field and of the magnetic field.

Include both numerical values and unit vectors to indicate directions. (e) Find the average power per unit area this wave carries through space. (f) Find the average energy density in the radiation (in joules per cubic meter). (g) What radiation pressure would this wave exert upon a perfectly reflecting surface at normal incidence?

77. A linearly polarized microwave of wavelength 1.50 cm is directed along the positive x axis. The electric field vector has a maximum value of 175 V/m and vibrates in the xy plane. Assuming the magnetic field component of the wave can be written in the form $B = B_{max} \sin(kx - \omega t)$, give values for (a) B_{max}, (b) k, and (c) ω. (d) Determine in which plane the magnetic field vector vibrates. (e) Calculate the average value of the Poynting vector for this wave. (f) If this wave were directed at normal incidence onto a perfectly reflecting sheet, what radiation pressure would it exert? (g) What acceleration would be imparted to a 500-g sheet (perfectly reflecting and at normal incidence) with dimensions of 1.00 m × 0.750 m?

78. **Review.** In the absence of cable input or a satellite dish, a television set can use a dipole-receiving antenna for VHF channels and a loop antenna for UHF channels. In Figure CQ34.9, the "rabbit ears" form the VHF antenna and the smaller loop of wire is the UHF antenna. The UHF antenna produces an emf from the changing magnetic flux through the loop. The television station broadcasts a signal with a frequency f, and the signal has an electric field amplitude E_{max} and a magnetic field amplitude B_{max} at the location of the receiving antenna. (a) Using Faraday's law, derive an expression for the amplitude of the emf that appears in a single-turn, circular loop antenna with a radius r that is small compared with the wavelength of the wave. (b) If the electric field in the signal points vertically, what orientation of the loop gives the best reception?

79. **Review.** An astronaut, stranded in space 10.0 m from her spacecraft and at rest relative to it, has a mass (including equipment) of 110 kg. Because she has a 100-W flashlight that forms a directed beam, she considers using the beam as a photon rocket to propel herself continuously toward the spacecraft. (a) Calculate the time interval required for her to reach the spacecraft by this method. (b) **What If?** Suppose she throws the 3.00-kg flashlight in the direction away from the spacecraft instead. After being thrown, the flashlight moves at 12.0 m/s relative to the recoiling astronaut. After what time interval will the astronaut reach the spacecraft?

Light and Optics

The Grand Tetons in western Wyoming are reflected in a smooth lake at sunset. The optical principles we study in this part of the book will explain the nature of the reflected image of the mountains and why the sky appears red. *(David Muench/ Terra/Corbis)*

Light is basic to almost all life on the Earth. For example, plants convert the energy transferred by sunlight to chemical energy through photosynthesis. In addition, light is the principal means by which we are able to transmit and receive information to and from objects around us and throughout the Universe. Light is a form of electromagnetic radiation and represents energy transfer from the source to the observer.

Many phenomena in our everyday life depend on the properties of light. When you watch a television or view photos on a computer monitor, you are seeing millions of colors formed from combinations of only three colors that are physically on the screen: red, blue, and green. The blue color of the daytime sky is a result of the optical phenomenon of *scattering* of light by air molecules, as are the red and orange colors of sunrises and sunsets. You see your image in your bathroom mirror in the morning or the images of other cars in your rearview mirror when you are driving. These images result from *reflection* of light. If you wear glasses or contact lenses, you are depending on *refraction* of light for clear vision. The colors of a rainbow result from *dispersion* of light as it passes through raindrops hovering in the sky after a rainstorm. If you have ever seen the colored circles of the glory surrounding the shadow of your airplane on clouds as you fly above them, you are seeing an effect that results from *interference* of light. The phenomena mentioned here have been studied by scientists and are well understood.

In the introduction to Chapter 35, we discuss the dual nature of light. In some cases, it is best to model light as a stream of particles; in others, a wave model works better. Chapters 35 through 38 concentrate on those aspects of light that are best understood through the wave model of light. In Part 6, we will investigate the particle nature of light. ■

The Nature of Light and the Principles of Ray Optics

This photograph of a rainbow shows the range of colors from red on the top to violet on the bottom. The appearance of the rainbow depends on three optical phenomena discussed in this chapter: reflection, refraction, and dispersion. The faint pastel-colored bows beneath the main rainbow are called supernumerary bows. They are formed by interference between rays of light leaving raindrops below those causing the main rainbow. *(John W. Jewett, Jr.)*

This first chapter on optics begins by introducing two historical models for light and discussing early methods for measuring the speed of light. Next we study the fundamental phenomena of geometric optics: reflection of light from a surface and refraction as the light crosses the boundary between two media. We also study the dispersion of light as it refracts into materials, resulting in visual displays such as the rainbow. Finally, we investigate the phenomenon of total internal reflection, which is the basis for the operation of optical fibers and the technology of fiber optics.

35.1 The Nature of Light

Before the beginning of the 19th century, light was considered to be a stream of particles that either was emitted by the object being viewed or emanated from the eyes of the viewer. Newton, the chief architect of the particle model of light, held that particles were emitted from a light source and that these particles stimulated the sense of sight upon entering the eye. Using this idea, he was able to explain reflection and refraction.

Most scientists accepted Newton's particle model. During Newton's lifetime, however, another model was proposed, one that argued that light might be some sort of wave motion. In 1678, Dutch physicist and astronomer Christian Huygens showed that a wave model of light could also explain reflection and refraction.

In 1801, Thomas Young (1773–1829) provided the first clear experimental demonstration of the wave nature of light. Young showed that under appropriate conditions light rays interfere with one another according to the waves in interference model, just like mechanical waves (Chapter 18). Such behavior could not be explained at that time by a particle model because there was no conceivable way in which two or more particles could come together and cancel one another. Additional developments during the 19th century led to the general acceptance of the wave model of light, the most important resulting from the work of Maxwell, who in 1873 asserted that light was a form of high-frequency electromagnetic wave. As discussed in Chapter 34, Hertz provided experimental confirmation of Maxwell's theory in 1887 by producing and detecting electromagnetic waves.

Although the wave model and the classical theory of electricity and magnetism were able to explain most known properties of light, they could not explain some subsequent experiments. The most striking phenomenon is the photoelectric effect, also discovered by Hertz: when light strikes a metal surface, electrons are sometimes ejected from the surface. As one example of the difficulties that arose, experiments showed that the kinetic energy of an ejected electron is independent of the light intensity. This finding contradicted the wave model, which held that a more intense beam of light should add more energy to the electron. Einstein proposed an explanation of the photoelectric effect in 1905 using a model based on the concept of quantization developed by Max Planck (1858–1947) in 1900. The quantization model assumes the energy of a light wave is present in particles called *photons;* hence, the energy is said to be quantized. According to Einstein's theory, the energy of a photon is proportional to the frequency of the electromagnetic wave:

$$E = hf \tag{35.1}$$

Photo Researchers, Inc.

Christian Huygens
Dutch Physicist and Astronomer
(1629–1695)
Huygens is best known for his contributions to the fields of optics and dynamics. To Huygens, light was a type of vibratory motion, spreading out and producing the sensation of light when impinging on the eye. On the basis of this theory, he deduced the laws of reflection and refraction and explained the phenomenon of double refraction.

◄ **Energy of a photon**

where the constant of proportionality $h = 6.63 \times 10^{-34}$ J · s is called *Planck's constant.* We study this theory in Chapter 40.

In view of these developments, light must be regarded as having a dual nature. Light exhibits the characteristics of a wave in some situations and the characteristics of a particle in other situations. Light is light, to be sure. The question "Is light a wave or a particle?" is inappropriate, however. Sometimes light acts like a wave, and other times it acts like a particle. In the next few chapters, we investigate the wave nature of light.

35.2 Measurements of the Speed of Light

Light travels at such a high speed (to three digits, $c = 3.00 \times 10^8$ m/s) that early attempts to measure its speed were unsuccessful. Galileo attempted to measure the speed of light by positioning two observers in towers separated by approximately 10 km. Each observer carried a shuttered lantern. One observer would open his lantern first, and then the other would open his lantern at the moment he saw the light from the first lantern. Galileo reasoned that by knowing the transit time of the light beams from one lantern to the other and the distance between the two lanterns, he could obtain the speed. His results were inconclusive. Today, we realize (as Galileo concluded) that it is impossible to measure the speed of light in this manner because the transit time for the light is so much less than the reaction time of the observers.

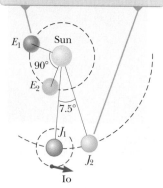

In the time interval during which the Earth travels 90° around the Sun (three months), Jupiter travels only about 7.5°.

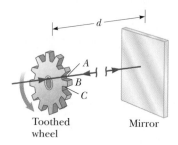

Figure 35.1 Roemer's method for measuring the speed of light (drawing not to scale).

Figure 35.2 Fizeau's method for measuring the speed of light using a rotating toothed wheel. The light source is considered to be at the location of the wheel; therefore, the distance d is known.

Roemer's Method

In 1675, Danish astronomer Ole Roemer (1644–1710) made the first successful estimate of the speed of light. Roemer's technique involved astronomical observations of Io, one of the moons of Jupiter. Io has a period of revolution around Jupiter of approximately 42.5 h. The period of revolution of Jupiter around the Sun is about 12 yr; therefore, as the Earth moves through 90° around the Sun, Jupiter revolves through only $(\frac{1}{12})90° = 7.5°$ (Fig. 35.1).

An observer using the orbital motion of Io as a clock would expect the orbit to have a constant period. After collecting data for more than a year, however, Roemer observed a systematic variation in Io's period. He found that the periods were longer than average when the Earth was receding from Jupiter and shorter than average when the Earth was approaching Jupiter. Roemer attributed this variation in period to the distance between the Earth and Jupiter changing from one observation to the next.

Using Roemer's data, Huygens estimated the lower limit for the speed of light to be approximately 2.3×10^8 m/s. This experiment is important historically because it demonstrated that light does have a finite speed and gave an estimate of this speed.

Fizeau's Method

The first successful method for measuring the speed of light by means of purely terrestrial techniques was developed in 1849 by French physicist Armand H. L. Fizeau (1819–1896). Figure 35.2 represents a simplified diagram of Fizeau's apparatus. The basic procedure is to measure the total time interval during which light travels from some point to a distant mirror and back. If d is the distance between the light source (considered to be at the location of the wheel) and the mirror and if the time interval for one round trip is Δt, the speed of light is $c = 2d/\Delta t$.

To measure the transit time, Fizeau used a rotating toothed wheel, which converts a continuous beam of light into a series of light pulses. The rotation of such a wheel controls what an observer at the light source sees. For example, if the pulse traveling toward the mirror and passing the opening at point A in Figure 35.2 should return to the wheel at the instant tooth B had rotated into position to cover the return path, the pulse would not reach the observer. At a greater rate of rotation, the opening at point C could move into position to allow the reflected pulse to reach the observer. Knowing the distance d, the number of teeth in the wheel, and the angular speed of the wheel, Fizeau arrived at a value of 3.1×10^8 m/s. Similar measurements made by subsequent investigators yielded more precise values for c, which led to the currently accepted value of $2.997\ 924\ 58 \times 10^8$ m/s.

Example 35.1   Measuring the Speed of Light with Fizeau's Wheel   AM

Assume Fizeau's wheel has 360 teeth and rotates at 27.5 rev/s when a pulse of light passing through opening A in Figure 35.2 is blocked by tooth B on its return. If the distance to the mirror is 7 500 m, what is the speed of light?

SOLUTION

Conceptualize Imagine a pulse of light passing through opening A in Figure 35.2 and reflecting from the mirror. By the time the pulse arrives back at the wheel, tooth B has rotated into the position previously occupied by opening A.

Categorize The wheel is a rigid object rotating at constant angular speed. We model the pulse of light as a *particle under constant speed.*

....

Analyze The wheel has 360 teeth, so it must have 360 openings. Therefore, because the light passes through opening A but is blocked by the tooth immediately adjacent to A, the wheel must rotate through an angular displacement of $\frac{1}{720}$ rev in the time interval during which the light pulse makes its round trip.

Use Equation 10.2, with the angular speed constant, to find the time interval for the pulse's round trip:

$$\Delta t = \frac{\Delta\theta}{\omega} = \frac{\frac{1}{720}\ \text{rev}}{27.5\ \text{rev/s}} = 5.05 \times 10^{-5}\ \text{s}$$

▶ **35.1** continued

From the particle under constant speed model, find the speed of the pulse of light:

$$c = \frac{2d}{\Delta t} = \frac{2(7\,500 \text{ m})}{5.05 \times 10^{-5} \text{ s}} = \boxed{2.97 \times 10^8 \text{ m/s}}$$

Finalize This result is very close to the actual value of the speed of light.

35.3 The Ray Approximation in Ray Optics

The field of **ray optics** (sometimes called *geometric optics*) involves the study of the propagation of light. Ray optics assumes light travels in a fixed direction in a straight line as it passes through a uniform medium and changes its direction when it meets the surface of a different medium or if the optical properties of the medium are nonuniform in either space or time. In our study of ray optics here and in Chapter 36, we use what is called the **ray approximation.** To understand this approximation, first notice that the rays of a given wave are straight lines perpendicular to the wave fronts as illustrated in Figure 35.3 for a plane wave. In the ray approximation, a wave moving through a medium travels in a straight line in the direction of its rays.

If the wave meets a barrier in which there is a circular opening whose diameter is much larger than the wavelength as in Figure 35.4a, the wave emerging from the opening continues to move in a straight line (apart from some small edge effects); hence, the ray approximation is valid. If the diameter of the opening is on the order of the wavelength as in Figure 35.4b, the waves spread out from the opening in all directions. This effect, called *diffraction,* will be studied in Chapter 37. Finally, if the opening is much smaller than the wavelength, the opening can be approximated as a point source of waves as shown in Fig. 35.4c.

Similar effects are seen when waves encounter an opaque object of dimension d. In that case, when $\lambda \ll d$, the object casts a sharp shadow.

The ray approximation and the assumption that $\lambda \ll d$ are used in this chapter and in Chapter 36, both of which deal with ray optics. This approximation is very good for the study of mirrors, lenses, prisms, and associated optical instruments such as telescopes, cameras, and eyeglasses.

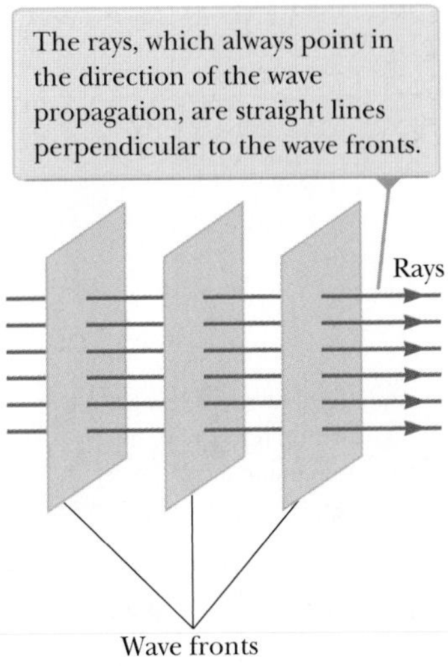

The rays, which always point in the direction of the wave propagation, are straight lines perpendicular to the wave fronts.

Rays

Wave fronts

Figure 35.3 A plane wave propagating to the right.

35.4 Analysis Model: Wave Under Reflection

We introduced the concept of reflection of waves in a discussion of waves on strings in Section 16.4. As with waves on strings, when a light ray traveling in one medium encounters a boundary with another medium, part of the incident light

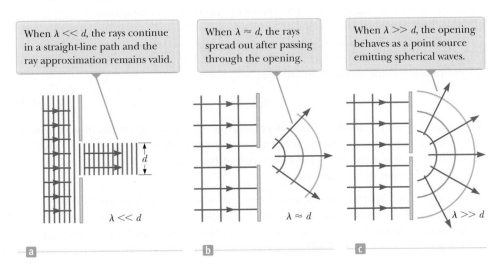

When $\lambda \ll d$, the rays continue in a straight-line path and the ray approximation remains valid.

When $\lambda \approx d$, the rays spread out after passing through the opening.

When $\lambda \gg d$, the opening behaves as a point source emitting spherical waves.

$\lambda \ll d$

$\lambda \approx d$

$\lambda \gg d$

a   b   c

Figure 35.4 A plane wave of wavelength λ is incident on a barrier in which there is an opening of diameter d.

Figure 35.5 Schematic representation of (a) specular reflection, where the reflected rays are all parallel to one another, and (b) diffuse reflection, where the reflected rays travel in random directions. (c) and (d) Photographs of specular and diffuse reflection using laser light.

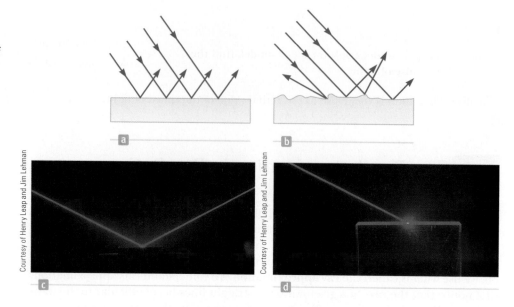

Courtesy of Henry Leap and Jim Lehman

Courtesy of Henry Leap and Jim Lehman

is reflected. For waves on a one-dimensional string, the reflected wave must necessarily be restricted to a direction along the string. For light waves traveling in three-dimensional space, no such restriction applies and the reflected light waves can be in directions different from the direction of the incident waves. Figure 35.5a shows several rays of a beam of light incident on a smooth, mirror-like, reflecting surface. The reflected rays are parallel to one another as indicated in the figure. The direction of a reflected ray is in the plane perpendicular to the reflecting surface that contains the incident ray. Reflection of light from such a smooth surface is called **specular reflection.** If the reflecting surface is rough as in Figure 35.5b, the surface reflects the rays not as a parallel set but in various directions. Reflection from any rough surface is known as **diffuse reflection.** A surface behaves as a smooth surface as long as the surface variations are much smaller than the wavelength of the incident light.

The difference between these two kinds of reflection explains why it is more difficult to see while driving on a rainy night than on a dry night. If the road is wet, the smooth surface of the water specularly reflects most of your headlight beams away from your car (and perhaps into the eyes of oncoming drivers). When the road is dry, its rough surface diffusely reflects part of your headlight beam back toward you, allowing you to see the road more clearly. Your bathroom mirror exhibits specular reflection, whereas light reflecting from this page experiences diffuse reflection. In this book, we restrict our study to specular reflection and use the term *reflection* to mean specular reflection.

Consider a light ray traveling in air and incident at an angle on a flat, smooth surface as shown in Figure 35.6. The incident and reflected rays make angles θ_1 and θ_1', respectively, where the angles are measured between the normal and the rays. (The normal is a line drawn perpendicular to the surface at the point where the incident ray strikes the surface.) Experiments and theory show that the angle of reflection equals the angle of incidence:

$$\theta_1' = \theta_1 \tag{35.2}$$

This relationship is called the **law of reflection.** Because reflection of waves from an interface between two media is a common phenomenon, we identify an analysis model for this situation: the **wave under reflection.** Equation 35.2 is the mathematical representation of this model.

> The incident ray, the reflected ray, and the normal all lie in the same plane, and $\theta_1' = \theta_1$.

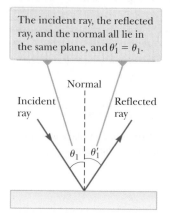

Normal

Incident ray   Reflected ray

θ_1 θ_1'

Figure 35.6 The wave under reflection model.

Pitfall Prevention 35.1

Subscript Notation The subscript 1 refers to parameters for the light in the initial medium. When light travels from one medium to another, we use the subscript 2 for the parameters associated with the light in the new medium. In this discussion, the light stays in the same medium, so we only have to use the subscript 1.

Law of reflection ▶

Quick Quiz 35.1 In the movies, you sometimes see an actor looking in a mirror and you can see his face in the mirror. It can be said with certainty that during the filming of such a scene, the actor sees in the mirror: **(a)** his face **(b)** your face **(c)** the director's face **(d)** the movie camera **(e)** impossible to determine

| **Example 35.2** | **The Double-Reflected Light Ray** | **AM** |

Two mirrors make an angle of 120° with each other as illustrated in Figure 35.7a. A ray is incident on mirror M_1 at an angle of 65° to the normal. Find the direction of the ray after it is reflected from mirror M_2.

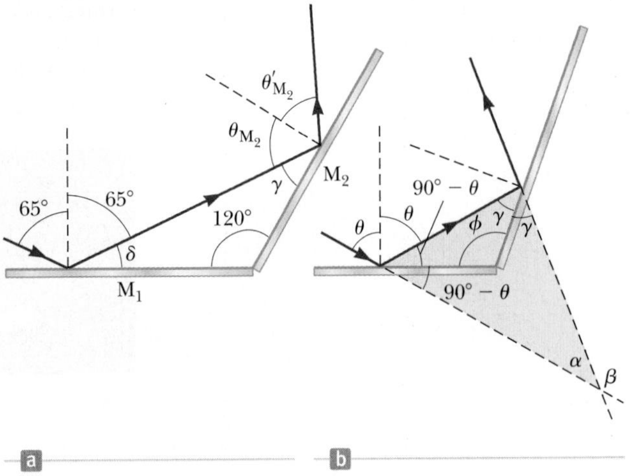

SOLUTION

Conceptualize Figure 35.7a helps conceptualize this situation. The incoming ray reflects from the first mirror, and the reflected ray is directed toward the second mirror. Therefore, there is a second reflection from the second mirror.

Categorize Because the interactions with both mirrors are simple reflections, we apply the *wave under reflection* model and some geometry.

Figure 35.7 (Example 35.2) (a) Mirrors M_1 and M_2 make an angle of 120° with each other. (b) The geometry for an arbitrary mirror angle.

Analyze From the law of reflection, the first reflected ray makes an angle of 65° with the normal.

Find the angle the first reflected ray makes with the horizontal:

$$\delta = 90° - 65° = 25°$$

From the triangle made by the first reflected ray and the two mirrors, find the angle the reflected ray makes with M_2:

$$\gamma = 180° - 25° - 120° = 35°$$

Find the angle the first reflected ray makes with the normal to M_2:

$$\theta_{M_2} = 90° - 35° = 55°$$

From the law of reflection, find the angle the second reflected ray makes with the normal to M_2:

$$\theta'_{M_2} = \theta_{M_2} = \boxed{55°}$$

Finalize Let's explore variations in the angle between the mirrors as follows.

WHAT IF?   If the incoming and outgoing rays in Figure 35.7a are extended behind the mirror, they cross at an angle of 60° and the overall change in direction of the light ray is 120°. This angle is the same as that between the mirrors. What if the angle between the mirrors is changed? Is the overall change in the direction of the light ray always equal to the angle between the mirrors?

Answer Making a general statement based on one data point or one observation is always a dangerous practice! Let's investigate the change in direction for a general situation. Figure 35.7b shows the mirrors at an arbitrary angle ϕ and the incoming light ray striking the mirror at an arbitrary angle θ with respect to the normal to the mirror surface. In accordance with the law of reflection and the sum of the interior angles of a triangle, the angle γ is given by $\gamma = 180° - (90° - \theta) - \phi = 90° + \theta - \phi$.

Consider the triangle highlighted in yellow in Figure 35.7b and determine α:

$$\alpha + 2\gamma + 2(90° - \theta) = 180° \quad \rightarrow \quad \alpha = 2(\theta - \gamma)$$

Notice from Figure 35.7b that the change in direction of the light ray is angle β. Use the geometry in the figure to solve for β:

$$\beta = 180° - \alpha = 180° - 2(\theta - \gamma)$$
$$= 180° - 2[\theta - (90° + \theta - \phi)] = 360° - 2\phi$$

Notice that β is not equal to ϕ. For $\phi = 120°$, we obtain $\beta = 120°$, which happens to be the same as the mirror angle; that is true only for this special angle between the mirrors, however. For example, if $\phi = 90°$, we obtain $\beta = 180°$. In that case, the light is reflected straight back to its origin.

   If the angle between two mirrors is 90°, the reflected beam returns to the source parallel to its original path as discussed in the What If? section of the preceding example. This phenomenon, called *retroreflection*, has many practical applications. If a third mirror is placed perpendicular to the first two so that the three form the

Figure 35.8 Applications of retroreflection.

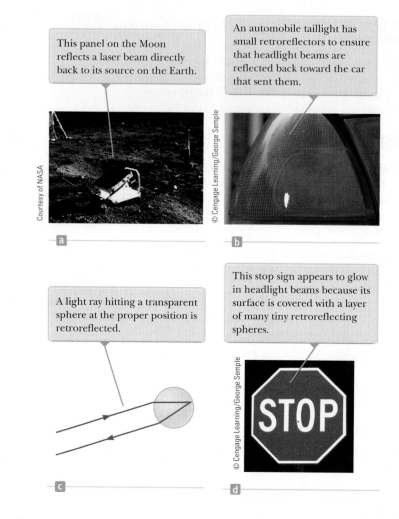

This panel on the Moon reflects a laser beam directly back to its source on the Earth.

An automobile taillight has small retroreflectors to ensure that headlight beams are reflected back toward the car that sent them.

A light ray hitting a transparent sphere at the proper position is retroreflected.

This stop sign appears to glow in headlight beams because its surface is covered with a layer of many tiny retroreflecting spheres.

This leg of an ant gives a scale for the size of the mirrors.

The mirror on the left is "on," and the one on the right is "off."

Figure 35.9 (a) An array of mirrors on the surface of a digital micromirror device. Each mirror has an area of approximately 16 μm². (b) A close-up view of two single micromirrors.

corner of a cube, retroreflection works in three dimensions. In 1969, a panel of many small reflectors was placed on the Moon by the *Apollo 11* astronauts (Fig. 35.8a). A laser beam from the Earth is reflected directly back on itself, and its transit time is measured. This information is used to determine the distance to the Moon with an uncertainty of 15 cm. (Imagine how difficult it would be to align a regular flat mirror on the Moon so that the reflected laser beam would hit a particular location on the Earth!) A more everyday application is found in automobile taillights. Part of the plastic making up the taillight is formed into many tiny cube corners (Fig. 35.8b) so that headlight beams from cars approaching from the rear are reflected back to the drivers. Instead of cube corners, small spherical bumps are sometimes used (Fig. 35.8c). Tiny clear spheres are used in a coating material found on many road signs. Due to retroreflection from these spheres, the stop sign in Figure 35.8d appears much brighter than it would if it were simply a flat, shiny surface. Retroreflectors are also used for reflective panels on running shoes and running clothing to allow joggers to be seen at night.

Another practical application of the law of reflection is the digital projection of movies, television shows, and computer presentations. A digital projector uses an optical semiconductor chip called a *digital micromirror device*. This device contains an array of tiny mirrors (Fig. 35.9a) that can be individually tilted by means of signals to an address electrode underneath the edge of the mirror. Each mirror corresponds to a pixel in the projected image. When the pixel corresponding to a given mirror is to be bright, the mirror is in the "on" position and is oriented so as to reflect light from a source illuminating the array to the screen (Fig. 35.9b). When the pixel for this mirror is to be dark, the mirror is "off" and is tilted so that the light is reflected away from the screen. The brightness of the pixel is determined by the total time interval during which the mirror is in the "on" position during the display of one image.

Digital movie projectors use three micromirror devices, one for each of the primary colors red, blue, and green, so that movies can be displayed with up to 35 trillion colors. Because information is stored as binary data, a digital movie does not degrade with time as does film. Furthermore, because the movie is entirely in the form of computer software, it can be delivered to theaters by means of satellites, optical discs, or optical fiber networks.

Analysis Model — Wave Under Reflection

Imagine a wave (electromagnetic or mechanical) traveling through space and striking a flat surface at an angle θ_1 with respect to the normal to the surface. The wave will reflect from the surface in a direction described by the **law of reflection**—the angle of reflection θ_1' equals the angle of incidence θ_1:

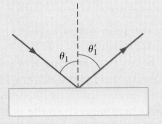

$$\theta_1' = \theta_1 \qquad \textbf{(35.2)}$$

Examples:

- sound waves from an orchestra reflect from a band-shell out to the audience
- a mirror is used to deflect a laser beam in a laser light show
- your bathroom mirror reflects light from your face back to you to form an image of your face (Chapter 36)
- x-rays reflected from a crystalline material create an optical pattern that can be used to understand the structure of the solid (Chapter 38)

35.5 Analysis Model: Wave Under Refraction

In addition to the phenomenon of reflection discussed for waves on strings in Section 16.4, we also found that some of the energy of the incident wave transmits into the new medium. For example, consider Figures 16.15 and 16.16, in which a pulse on a string approaching a junction with another string both reflects from and transmits past the junction and into the second string. Similarly, when a ray of light traveling through a transparent medium encounters a boundary leading into another transparent medium as shown in Figure 35.10, part of the energy is reflected and part enters the second medium. As with reflection, the direction of the transmitted wave exhibits an interesting behavior because of the three-dimensional nature of the light waves. The ray that enters the second medium changes its direction of propagation at the boundary and is said to be **refracted.** The incident ray, the reflected ray, and the refracted ray all lie in the same plane. The **angle of refraction,** θ_2 in Figure 35.10a, depends on the properties of the two media and on the angle of incidence θ_1 through the relationship

$$\frac{\sin \theta_2}{\sin \theta_1} = \frac{v_2}{v_1} \qquad \textbf{(35.3)}$$

where v_1 is the speed of light in the first medium and v_2 is the speed of light in the second medium.

The path of a light ray through a refracting surface is reversible. For example, the ray shown in Figure 35.10a travels from point A to point B. If the ray originated at B, it would travel along line BA to reach point A and the reflected ray would point downward and to the left in the glass.

Quick Quiz 35.2 If beam ① is the incoming beam in Figure 35.10b, which of the other four red lines are reflected beams and which are refracted beams?

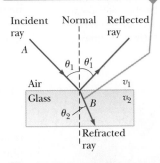

All rays and the normal lie in the same plane, and the refracted ray is bent toward the normal because $v_2 < v_1$.

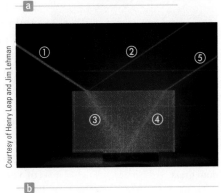

Figure 35.10 (a) The wave under refraction model. (b) Light incident on the Lucite block refracts both when it enters the block and when it leaves the block.

Courtesy of Henry Leap and Jim Lehman

Figure 35.11 The refraction of light as it (a) moves from air into glass and (b) moves from glass into air.

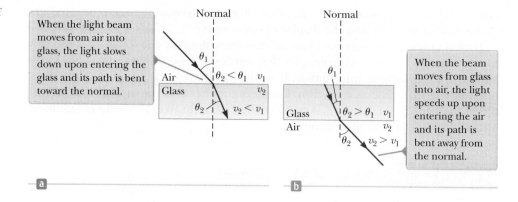

When the light beam moves from air into glass, the light slows down upon entering the glass and its path is bent toward the normal.

When the beam moves from glass into air, the light speeds up upon entering the air and its path is bent away from the normal.

From Equation 35.3, we can infer that when light moves from a material in which its speed is high to a material in which its speed is lower as shown in Figure 35.11a, the angle of refraction θ_2 is less than the angle of incidence θ_1 and the ray is bent *toward* the normal. If the ray moves from a material in which light moves slowly to a material in which it moves more rapidly as illustrated in Figure 35.11b, then θ_2 is greater than θ_1 and the ray is bent *away* from the normal.

The behavior of light as it passes from air into another substance and then re-emerges into air is often a source of confusion to students. When light travels in air, its speed is 3.00×10^8 m/s, but this speed is reduced to approximately 2×10^8 m/s when the light enters a block of glass. When the light re-emerges into air, its speed instantaneously increases to its original value of 3.00×10^8 m/s. This effect is far different from what happens, for example, when a bullet is fired through a block of wood. In that case, the speed of the bullet decreases as it moves through the wood because some of its original energy is used to tear apart the wood fibers. When the bullet enters the air once again, it emerges at a speed lower than it had when it entered the wood.

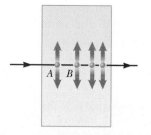

Figure 35.12 Light passing from one atom to another in a medium. The blue spheres are electrons, and the vertical arrows represent their oscillations.

To see why light behaves as it does, consider Figure 35.12, which represents a beam of light entering a piece of glass from the left. Once inside the glass, the light may encounter an electron bound to an atom, indicated as point A. Let's assume light is absorbed by the atom, which causes the electron to oscillate (a detail represented by the double-headed vertical arrows). The oscillating electron then acts as an antenna and radiates the beam of light toward an atom at B, where the light is again absorbed. The details of these absorptions and radiations are best explained in terms of quantum mechanics (Chapter 42). For now, it is sufficient to think of light passing from one atom to another through the glass. Although light travels from one atom to another at 3.00×10^8 m/s, the absorption and radiation that take place cause the *average* light speed through the material to fall to approximately 2×10^8 m/s. Once the light emerges into the air, absorption and radiation cease and the light travels at a constant speed of 3.00×10^8 m/s.

A mechanical analog of refraction is shown in Figure 35.13. When the left end of the rolling barrel reaches the grass, it slows down, whereas the right end remains on the concrete and moves at its original speed. This difference in speeds causes the barrel to pivot, which changes the direction of travel.

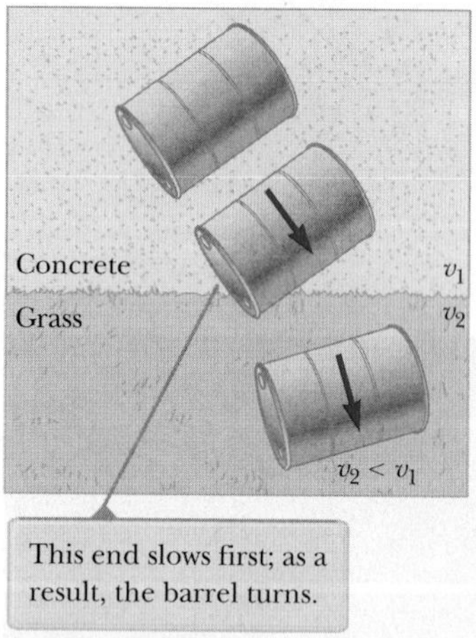

This end slows first; as a result, the barrel turns.

Figure 35.13 Overhead view of a barrel rolling from concrete onto grass.

Index of Refraction

In general, the speed of light in any material is *less* than its speed in vacuum. In fact, *light travels at its maximum speed c in vacuum*. It is convenient to define the **index of refraction** n of a medium to be the ratio

Index of refraction ▶

$$n \equiv \frac{\text{speed of light in vacuum}}{\text{speed of light in a medium}} \equiv \frac{c}{v} \qquad (35.4)$$

This definition shows that the index of refraction is a dimensionless number greater than unity because v is always less than c. Furthermore, n is equal to unity for vacuum. The indices of refraction for various substances are listed in Table 35.1.

As light travels from one medium to another, its frequency does not change but its wavelength does. To see why that is true, consider Figure 35.14. Waves pass an observer at point A in medium 1 with a certain frequency and are incident on the boundary between medium 1 and medium 2. The frequency with which the waves pass an observer at point B in medium 2 must equal the frequency at which they pass point A. If that were not the case, energy would be piling up or disappearing at the boundary. Because there is no mechanism for that to occur, the frequency must be a constant as a light ray passes from one medium into another. Therefore, because the relationship $v = \lambda f$ (Eq. 16.12) from the traveling wave model must be valid in both media and because $f_1 = f_2 = f$, we see that

$$v_1 = \lambda_1 f \quad \text{and} \quad v_2 = \lambda_2 f \tag{35.5}$$

Because $v_1 \neq v_2$, it follows that $\lambda_1 \neq \lambda_2$ as shown in Figure 35.14.

We can obtain a relationship between index of refraction and wavelength by dividing the first Equation 35.5 by the second and then using Equation 35.4:

$$\frac{\lambda_1}{\lambda_2} = \frac{v_1}{v_2} = \frac{c/n_1}{c/n_2} = \frac{n_2}{n_1} \tag{35.6}$$

This expression gives

$$\lambda_1 n_1 = \lambda_2 n_2$$

If medium 1 is vacuum or, for all practical purposes, air, then $n_1 = 1$. Hence, it follows from Equation 35.6 that the index of refraction of any medium can be expressed as the ratio

$$n = \frac{\lambda}{\lambda_n} \tag{35.7}$$

where λ is the wavelength of light in vacuum and λ_n is the wavelength of light in the medium whose index of refraction is n. From Equation 35.7, we see that because $n > 1$, $\lambda_n < \lambda$.

We are now in a position to express Equation 35.3 in an alternative form. Replacing the v_2/v_1 term in Equation 35.3 with n_1/n_2 from Equation 35.6 gives

$$n_1 \sin \theta_1 = n_2 \sin \theta_2 \tag{35.8}$$

The experimental discovery of this relationship is usually credited to Willebrord Snell (1591–1626) and it is therefore known as **Snell's law of refraction.** We shall

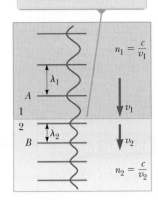

As a wave moves between the media, its wavelength changes but its frequency remains constant.

$$n_1 = \frac{c}{v_1}$$

$$n_2 = \frac{c}{v_2}$$

Figure 35.14 A wave travels from medium 1 to medium 2, in which it moves with lower speed.

Pitfall Prevention 35.2

An Inverse Relationship The index of refraction is *inversely* proportional to the wave speed. As the wave speed v decreases, the index of refraction n increases. Therefore, the higher the index of refraction of a material, the more it *slows down* light from its speed in vacuum. The more the light slows down, the more θ_2 differs from θ_1 in Equation 35.8.

◀ Snell's law of refraction

Pitfall Prevention 35.3

***n* Is Not an Integer Here** The symbol n has been used several times as an integer, such as in Chapter 18 to indicate the standing wave mode on a string or in an air column. The index of refraction n is *not* an integer.

Table 35.1   **Indices of Refraction**

Substance	Index of Refraction	Substance	Index of Refraction
Solids at 20°C		*Liquids at 20°C*	
Cubic zirconia	2.20	Benzene	1.501
Diamond (C)	2.419	Carbon disulfide	1.628
Fluorite (CaF$_2$)	1.434	Carbon tetrachloride	1.461
Fused quartz (SiO$_2$)	1.458	Ethyl alcohol	1.361
Gallium phosphide	3.50	Glycerin	1.473
Glass, crown	1.52	Water	1.333
Glass, flint	1.66		
Ice (H$_2$O)	1.309	*Gases at 0°C, 1 atm*	
Polystyrene	1.49	Air	1.000 293
Sodium chloride (NaCl)	1.544	Carbon dioxide	1.000 45

Note: All values are for light having a wavelength of 589 nm in vacuum.

examine this equation further in Section 35.6. Refraction of waves at an interface between two media is a common phenomenon, so we identify an analysis model for this situation: the **wave under refraction.** Equation 35.8 is the mathematical representation of this model for electromagnetic radiation. Other waves, such as seismic waves and sound waves, also exhibit refraction according to this model, and the mathematical representation of the model for these waves is Equation 35.3.

Quick Quiz 35.3 Light passes from a material with index of refraction 1.3 into one with index of refraction 1.2. Compared to the incident ray, what happens to the refracted ray? **(a)** It bends toward the normal. **(b)** It is undeflected. **(c)** It bends away from the normal.

Analysis Model **Wave Under Refraction**

Imagine a wave (electromagnetic or mechanical) traveling through space and striking a flat surface at an angle θ_1 with respect to the normal to the surface. Some of the energy of the wave refracts into the medium below the surface in a direction θ_2 described by the **law of refraction—**

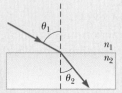

$$\frac{\sin \theta_2}{\sin \theta_1} = \frac{v_2}{v_1} \qquad (35.3)$$

where v_1 and v_2 are the speeds of the wave in medium 1 and medium 2, respectively.

For light waves, **Snell's law of refraction** states that

$$n_1 \sin \theta_1 = n_2 \sin \theta_2 \qquad (35.8)$$

where n_1 and n_2 are the indices of refraction in the two media.

Examples:

- sound waves moving upward from the shore of a lake refract in warmer layers of air higher above the lake and travel downward to a listener in a boat, making sounds from the shore louder than expected

- light from the sky approaches a hot roadway at a grazing angle and refracts upward so as to miss the roadway and enter a driver's eye, giving the illusion of a pool of water on the distant roadway

- light is sent over long distances in an optical fiber because of a difference in index of refraction between the fiber and the surrounding material (Section 35.8)

- a magnifying glass forms an enlarged image of a postage stamp due to refraction of light through the lens (Chapter 36)

Example 35.3 **Angle of Refraction for Glass** AM

A light ray of wavelength 589 nm traveling through air is incident on a smooth, flat slab of crown glass at an angle of 30.0° to the normal.

(A) Find the angle of refraction.

SOLUTION

Conceptualize Study Figure 35.11a, which illustrates the refraction process occurring in this problem. We expect that $\theta_2 < \theta_1$ because the speed of light is lower in the glass.

Categorize This is a typical problem in which we apply the *wave under refraction* model.

Analyze Rearrange Snell's law of refraction to find $\sin \theta_2$: $\qquad \sin \theta_2 = \dfrac{n_1}{n_2} \sin \theta_1$

Solve for θ_2: $\qquad \theta_2 = \sin^{-1}\left(\dfrac{n_1}{n_2} \sin \theta_1\right)$

Substitute indices of refraction from Table 35.1 and the incident angle: $\qquad \theta_2 = \sin^{-1}\left(\dfrac{1.00}{1.52} \sin 30.0°\right) = \boxed{19.2°}$

(B) Find the speed of this light once it enters the glass.

▶ **35.3** continued

SOLUTION

Solve Equation 35.4 for the speed of light in the glass:

$$v = \frac{c}{n}$$

Substitute numerical values:

$$v = \frac{3.00 \times 10^8 \text{ m/s}}{1.52} = \boxed{1.97 \times 10^8 \text{ m/s}}$$

(C) What is the wavelength of this light in the glass?

SOLUTION

Use Equation 35.7 to find the wavelength in the glass:

$$\lambda_n = \frac{\lambda}{n} = \frac{589 \text{ nm}}{1.52} = \boxed{388 \text{ nm}}$$

Finalize In part (A), note that $\theta_2 < \theta_1$, consistent with the slower speed of the light found in part (B). In part (C), we see that the wavelength of the light is shorter in the glass than in the air.

Example 35.4 **Light Passing Through a Slab** `AM`

A light beam passes from medium 1 to medium 2, with the latter medium being a thick slab of material whose index of refraction is n_2 (Fig. 35.15). Show that the beam emerging into medium 1 from the other side is parallel to the incident beam.

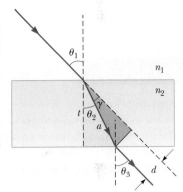

SOLUTION

Conceptualize Follow the path of the light beam as it enters and exits the slab of material in Figure 35.15, where we have assumed that $n_2 > n_1$. The ray bends toward the normal upon entering and away from the normal upon leaving.

Figure 35.15 (Example 35.4) The dashed line drawn parallel to the ray coming out the bottom of the slab represents the path the light would take were the slab not there.

Categorize Like Example 35.3, this is another typical problem in which we apply the *wave under refraction* model.

Analyze Apply Snell's law of refraction to the upper surface:

$$(1) \quad \sin \theta_2 = \frac{n_1}{n_2} \sin \theta_1$$

Apply Snell's law to the lower surface:

$$(2) \quad \sin \theta_3 = \frac{n_2}{n_1} \sin \theta_2$$

Substitute Equation (1) into Equation (2):

$$\sin \theta_3 = \frac{n_2}{n_1} \left(\frac{n_1}{n_2} \sin \theta_1 \right) = \sin \theta_1$$

Finalize Therefore, $\theta_3 = \theta_1$ and the slab does not alter the direction of the beam. It does, however, offset the beam parallel to itself by the distance d shown in Figure 35.15.

WHAT IF? What if the thickness t of the slab is doubled? Does the offset distance d also double?

Answer Consider the region of the light path within the slab in Figure 35.15. The distance a is the hypotenuse of two right triangles.

Find an expression for a from the yellow triangle:

$$a = \frac{t}{\cos \theta_2}$$

Find an expression for d from the red triangle:

$$d = a \sin \gamma = a \sin (\theta_1 - \theta_2)$$

Combine these equations:

$$d = \frac{t}{\cos \theta_2} \sin (\theta_1 - \theta_2)$$

For a given incident angle θ_1, the refracted angle θ_2 is determined solely by the index of refraction, so the offset distance d is proportional to t. If the thickness doubles, so does the offset distance.

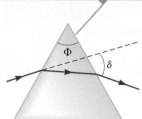

The apex angle Φ is the angle between the sides of the prism through which the light enters and leaves.

Figure 35.16 A prism refracts a single-wavelength light ray through an angle of deviation δ.

In Example 35.4, the light passes through a slab of material with parallel sides. What happens when light strikes a prism with nonparallel sides as shown in Figure 35.16? In this case, the outgoing ray does not propagate in the same direction as the incoming ray. A ray of single-wavelength light incident on the prism from the left emerges at angle δ from its original direction of travel. This angle δ is called the **angle of deviation.** The **apex angle** Φ of the prism, shown in the figure, is defined as the angle between the surface at which the light enters the prism and the second surface that the light encounters.

Example 35.5 Measuring *n* Using a Prism **AM**

Although we do not prove it here, the minimum angle of deviation δ_{min} for a prism occurs when the angle of incidence θ_1 is such that the refracted ray inside the prism makes the same angle with the normal to the two prism faces[1] as shown in Figure 35.17. Obtain an expression for the index of refraction of the prism material in terms of the minimum angle of deviation and the apex angle Φ.

SOLUTION

Conceptualize Study Figure 35.17 carefully and be sure you understand why the light ray comes out of the prism traveling in a different direction.

Categorize In this example, light enters a material through one surface and leaves the material at another surface. Let's apply the *wave under refraction* model to the light passing through the prism.

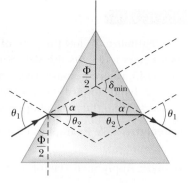

Figure 35.17 (Example 35.5) A light ray passing through a prism at the minimum angle of deviation δ_{min}.

Analyze Consider the geometry in Figure 35.17, where we have used symmetry to label several angles. The reproduction of the angle $\Phi/2$ at the location of the incoming light ray shows that $\theta_2 = \Phi/2$. The theorem that an exterior angle of any triangle equals the sum of the two opposite interior angles shows that $\delta_{min} = 2\alpha$. The geometry also shows that $\theta_1 = \theta_2 + \alpha$.

Combine these three geometric results:

$$\theta_1 = \theta_2 + \alpha = \frac{\Phi}{2} + \frac{\delta_{min}}{2} = \frac{\Phi + \delta_{min}}{2}$$

Apply the wave under refraction model at the left surface and solve for n:

$$(1.00) \sin \theta_1 = n \sin \theta_2 \rightarrow n = \frac{\sin \theta_1}{\sin \theta_2}$$

Substitute for the incident and refracted angles:

$$n = \frac{\sin \left(\dfrac{\Phi + \delta_{min}}{2} \right)}{\sin (\Phi/2)} \qquad \textbf{(35.9)}$$

[1]The details of this proof are available in texts on optics.

▶ **35.5** continued

Finalize Knowing the apex angle Φ of the prism and measuring δ_{min}, you can calculate the index of refraction of the prism material. Furthermore, a hollow prism can be used to determine the values of n for various liquids filling the prism.

35.6 Huygens's Principle

The laws of reflection and refraction were stated earlier in this chapter without proof. In this section, we develop these laws by using a geometric method proposed by Huygens in 1678. **Huygens's principle** is a geometric construction for using knowledge of an earlier wave front to determine the position of a new wave front at some instant:

> All points on a given wave front are taken as point sources for the production of spherical secondary waves, called wavelets, that propagate outward through a medium with speeds characteristic of waves in that medium. After some time interval has passed, the new position of the wave front is the surface tangent to the wavelets.

First, consider a plane wave moving through free space as shown in Figure 35.18a. At $t = 0$, the wave front is indicated by the plane labeled AA'. In Huygens's construction, each point on this wave front is considered a point source. For clarity, only three point sources on AA' are shown. With these sources for the wavelets, we draw circular arcs, each of radius $c\,\Delta t$, where c is the speed of light in vacuum and Δt is some time interval during which the wave propagates. The surface drawn tangent to these wavelets is the plane BB', which is the wave front at a later time, and is parallel to AA'. In a similar manner, Figure 35.18b shows Huygens's construction for a spherical wave.

Huygens's Principle Applied to Reflection and Refraction

We now derive the laws of reflection and refraction, using Huygens's principle.

For the law of reflection, refer to Figure 35.19. The line AB represents a plane wave front of the incident light just as ray 1 strikes the surface. At this instant, the wave at A sends out a Huygens wavelet (appearing at a later time as the light brown circular arc passing through D); the reflected light makes an angle γ' with the surface. At the

Pitfall Prevention 35.4
Of What Use Is Huygens's Principle? At this point, the importance of Huygens's principle may not be evident. Predicting the position of a future wave front may not seem to be very critical. We will use Huygens's principle here to generate the laws of reflection and refraction and in later chapters to explain additional wave phenomena for light.

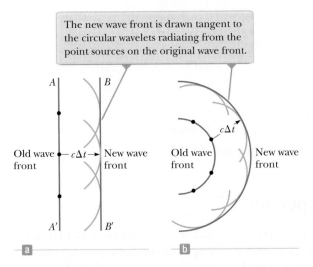

The new wave front is drawn tangent to the circular wavelets radiating from the point sources on the original wave front.

Figure 35.18 Huygens's construction for (a) a plane wave propagating to the right and (b) a spherical wave propagating to the right.

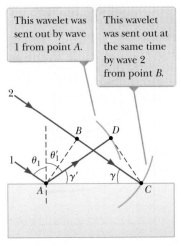

This wavelet was sent out by wave 1 from point A.

This wavelet was sent out at the same time by wave 2 from point B.

Figure 35.19 Huygens's construction for proving the law of reflection.

same time, the wave at B emits a Huygens wavelet (the light brown circular arc passing through C) with the incident light making an angle γ with the surface. Figure 35.19 shows these wavelets after a time interval Δt, after which ray 2 strikes the surface. Because both rays 1 and 2 move with the same speed, we must have $AD = BC = c\,\Delta t$.

The remainder of our analysis depends on geometry. Notice that the two triangles ABC and ADC are congruent because they have the same hypotenuse AC and because $AD = BC$. Figure 35.19 shows that

$$\cos\gamma = \frac{BC}{AC} \quad \text{and} \quad \cos\gamma' = \frac{AD}{AC}$$

where $\gamma = 90° - \theta_1$ and $\gamma' = 90° - \theta_1'$. Because $AD = BC$,

$$\cos\gamma = \cos\gamma'$$

Therefore,

$$\gamma = \gamma'$$
$$90° - \theta_1 = 90° - \theta_1'$$

and

$$\theta_1 = \theta_1'$$

which is the law of reflection.

Now let's use Huygens's principle to derive Snell's law of refraction. We focus our attention on the instant ray 1 strikes the surface and the subsequent time interval until ray 2 strikes the surface as in Figure 35.20. During this time interval, the wave at A sends out a Huygens wavelet (the light brown arc passing through D) and the light refracts into the material, making an angle θ_2 with the normal to the surface. In the same time interval, the wave at B sends out a Huygens wavelet (the light brown arc passing through C) and the light continues to propagate in the same direction. Because these two wavelets travel through different media, the radii of the wavelets are different. The radius of the wavelet from A is $AD = v_2\,\Delta t$, where v_2 is the wave speed in the second medium. The radius of the wavelet from B is $BC = v_1\,\Delta t$, where v_1 is the wave speed in the original medium.

From triangles ABC and ADC, we find that

$$\sin\theta_1 = \frac{BC}{AC} = \frac{v_1\,\Delta t}{AC} \quad \text{and} \quad \sin\theta_2 = \frac{AD}{AC} = \frac{v_2\,\Delta t}{AC}$$

Dividing the first equation by the second gives

$$\frac{\sin\theta_1}{\sin\theta_2} = \frac{v_1}{v_2}$$

From Equation 35.4, however, we know that $v_1 = c/n_1$ and $v_2 = c/n_2$. Therefore,

$$\frac{\sin\theta_1}{\sin\theta_2} = \frac{c/n_1}{c/n_2} = \frac{n_2}{n_1}$$

and

$$n_1 \sin\theta_1 = n_2 \sin\theta_2$$

which is Snell's law of refraction.

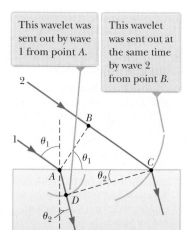

This wavelet was sent out by wave 1 from point A.

This wavelet was sent out at the same time by wave 2 from point B.

Figure 35.20 Huygens's construction for proving Snell's law of refraction.

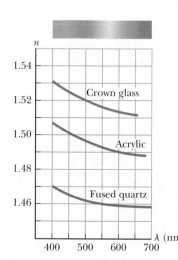

Figure 35.21 Variation of index of refraction with vacuum wavelength for three materials.

35.7 Dispersion

An important property of the index of refraction n is that, for a given material, the index varies with the wavelength of the light passing through the material as Figure 35.21 shows. This behavior is called **dispersion.** Because n is a function of wavelength, Snell's law of refraction indicates that light of different wavelengths is refracted at different angles when incident on a material.

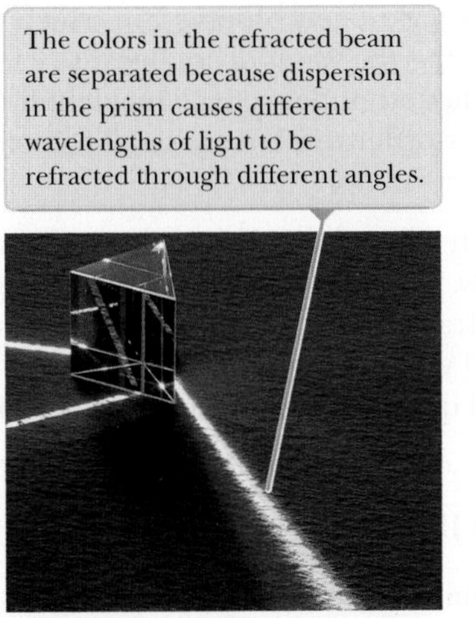

The colors in the refracted beam are separated because dispersion in the prism causes different wavelengths of light to be refracted through different angles.

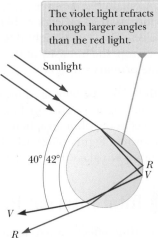

David Parker/Science Photo Library/Photo Researchers, Inc.

Figure 35.22 White light enters a glass prism at the upper left.

The violet light refracts through larger angles than the red light.

Sunlight

40° 42°

R
V

V

R

Figure 35.23 Path of sunlight through a spherical raindrop. Light following this path contributes to the visible rainbow.

Pitfall Prevention 35.5
A Rainbow of Many Light Rays
Pictorial representations such as Figure 35.23 are subject to misinterpretation. The figure shows one ray of light entering the raindrop and undergoing reflection and refraction, exiting the raindrop in a range of 40° to 42° from the entering ray. This illustration might be interpreted incorrectly as meaning that *all* light entering the raindrop exits in this small range of angles. In reality, light exits the raindrop over a much larger range of angles, from 0° to 42°. A careful analysis of the reflection and refraction from the spherical raindrop shows that the range of 40° to 42° is where the *highest-intensity light* exits the raindrop.

Figure 35.21 shows that the index of refraction generally decreases with increasing wavelength. For example, violet light refracts more than red light does when passing into a material.

Now suppose a beam of *white light* (a combination of all visible wavelengths) is incident on a prism as illustrated in Figure 35.22. Clearly, the angle of deviation δ depends on wavelength. The rays that emerge spread out in a series of colors known as the **visible spectrum.** These colors, in order of decreasing wavelength, are red, orange, yellow, green, blue, and violet. Newton showed that each color has a particular angle of deviation and that the colors can be recombined to form the original white light.

The dispersion of light into a spectrum is demonstrated most vividly in nature by the formation of a rainbow, which is often seen by an observer positioned between the Sun and a rain shower. To understand how a rainbow is formed, consider Figure 35.23. We will need to apply both the wave under reflection and wave under refraction models. A ray of sunlight (which is white light) passing overhead strikes a drop of water in the atmosphere and is refracted and reflected as follows. It is first refracted at the front surface of the drop, with the violet light deviating the most and the red light the least. At the back surface of the drop, the light is reflected and returns to the front surface, where it again undergoes refraction as it moves from water into air. The rays leave the drop such that the angle between the incident white light and the most intense returning violet ray is 40° and the angle between the incident white light and the most intense returning red ray is 42°. This small angular difference between the returning rays causes us to see a colored bow.

Now suppose an observer is viewing a rainbow as shown in Figure 35.24. If a raindrop high in the sky is being observed, the most intense red light returning from the drop reaches the observer because it is deviated the least; the most intense violet light, however, passes over the observer because it is deviated the most. Hence, the observer sees red light coming from this drop. Similarly, a drop lower in the sky directs the most intense violet light toward the observer and appears violet to the observer. (The most intense red light from this drop passes below the observer's eye and is not seen.) The most intense light from other colors of the spectrum reaches the observer from raindrops lying between these two extreme positions.

Figure 35.25 (page 1074) shows a *double rainbow.* The secondary rainbow is fainter than the primary rainbow, and the colors are reversed. The secondary rainbow arises from light that makes two reflections from the interior surface before exiting

The highest-intensity light traveling from higher raindrops toward the eyes of the observer is red, whereas the most intense light from lower drops is violet.

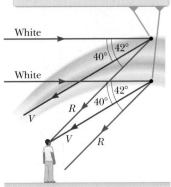

White

40° 42°

White

42°

R   40°

V   V   R

Figure 35.24 The formation of a rainbow seen by an observer standing with the Sun behind his back.

Figure 35.25 This photograph of a rainbow shows a distinct secondary rainbow with the colors reversed.

the raindrop. In the laboratory, rainbows have been observed in which the light makes more than 30 reflections before exiting the water drop. Because each reflection involves some loss of light due to refraction of part of the incident light out of the water drop, the intensity of these higher-order rainbows is small compared with that of the primary rainbow.

Quick Quiz 35.4 In photography, lenses in a camera use refraction to form an image on a light-sensitive surface. Ideally, you want all the colors in the light from the object being photographed to be refracted by the same amount. Of the materials shown in Figure 35.21, which would you choose for a single-element camera lens? **(a)** crown glass **(b)** acrylic **(c)** fused quartz **(d)** impossible to determine

35.8 Total Internal Reflection

An interesting effect called **total internal reflection** can occur when light is directed from a medium having a given index of refraction toward one having a lower index of refraction. Consider Figure 35.26a, in which a light ray travels in medium 1 and meets the boundary between medium 1 and medium 2, where n_1 is greater than n_2. In the figure, labels 1 through 5 indicate various possible directions of the ray consistent with the wave under refraction model. The refracted rays are bent away from the normal because n_1 is greater than n_2. At some particular angle of incidence θ_c, called the **critical angle,** the refracted light ray moves parallel to the boundary so that $\theta_2 = 90°$ (Fig. 35.26b). For angles of incidence greater than θ_c, the ray is entirely reflected at the boundary as shown by ray 5 in Figure 35.26a.

We can use Snell's law of refraction to find the critical angle. When $\theta_1 = \theta_c, \theta_2 = 90°$ and Equation 35.8 gives

$$n_1 \sin \theta_c = n_2 \sin 90° = n_2$$

Critical angle for total internal reflection ▶

$$\sin \theta_c = \frac{n_2}{n_1} \quad (\text{for } n_1 > n_2) \tag{35.10}$$

This equation can be used only when n_1 is greater than n_2. That is, total internal reflection occurs only when light is directed from a medium of a given index of refraction toward a medium of lower index of refraction. If n_1 were less than n_2,

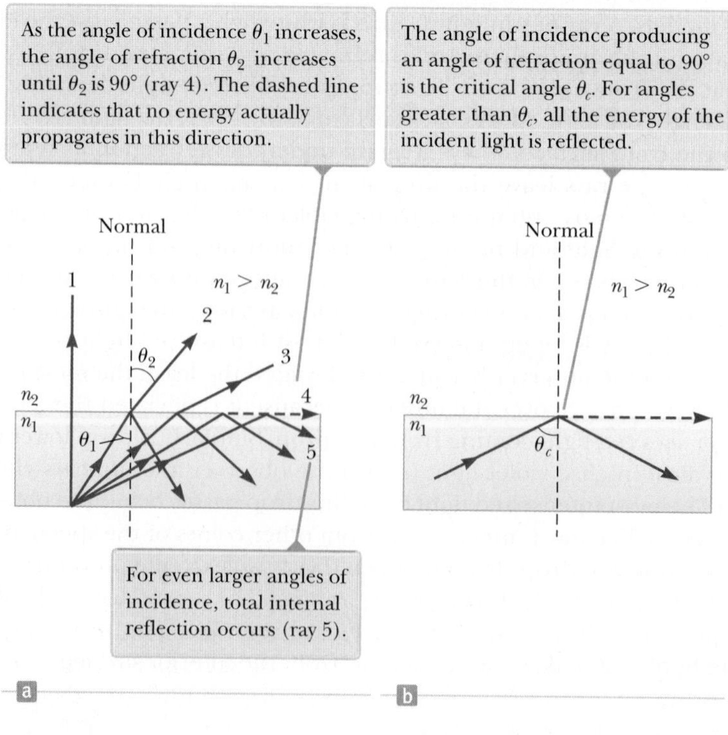

As the angle of incidence θ_1 increases, the angle of refraction θ_2 increases until θ_2 is 90° (ray 4). The dashed line indicates that no energy actually propagates in this direction.

The angle of incidence producing an angle of refraction equal to 90° is the critical angle θ_c. For angles greater than θ_c, all the energy of the incident light is reflected.

Normal

1

$n_1 > n_2$

2

θ_2

3

n_2

n_1

4

θ_1

5

Normal

$n_1 > n_2$

n_2

n_1

θ_c

For even larger angles of incidence, total internal reflection occurs (ray 5).

Figure 35.26 (a) Rays travel from a medium of index of refraction n_1 into a medium of index of refraction n_2, where $n_2 < n_1$. (b) Ray 4 is singled out.

a   b

Equation 35.10 would give sin $\theta_c > 1$, which is a meaningless result because the sine of an angle can never be greater than unity.

The critical angle for total internal reflection is small when n_1 is considerably greater than n_2. For example, the critical angle for a diamond in air is 24°. Any ray inside the diamond that approaches the surface at an angle greater than 24° is completely reflected back into the crystal. This property, combined with proper faceting, causes diamonds to sparkle. The angles of the facets are cut so that light is "caught" inside the crystal through multiple internal reflections. These multiple reflections give the light a long path through the medium, and substantial dispersion of colors occurs. By the time the light exits through the top surface of the crystal, the rays associated with different colors have been fairly widely separated from one another.

Cubic zirconia also has a high index of refraction and can be made to sparkle very much like a diamond. If a suspect jewel is immersed in corn syrup, the difference in n for the cubic zirconia and that for the corn syrup is small and the critical angle is therefore great. Hence, more rays escape sooner; as a result, the sparkle completely disappears. A real diamond does not lose all its sparkle when placed in corn syrup.

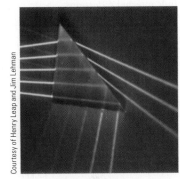

Courtesy of Henry Leap and Jim Lehman

Quick Quiz 35.5 In Figure 35.27, five light rays enter a glass prism from the left. **(i)** How many of these rays undergo total internal reflection at the slanted surface of the prism? (a) one (b) two (c) three (d) four (e) five **(ii)** Suppose the prism in Figure 35.27 can be rotated in the plane of the paper. For *all five* rays to experience total internal reflection from the slanted surface, should the prism be rotated (a) clockwise or (b) counterclockwise?

Figure 35.27 (Quick Quiz 35.5) Five nonparallel light rays enter a glass prism from the left.

Example 35.6 | **A View from the Fish's Eye**

Find the critical angle for an air–water boundary. (Assume the index of refraction of water is 1.33.)

SOLUTION

Conceptualize Study Figure 35.26 to understand the concept of total internal reflection and the significance of the critical angle.

Categorize We use concepts developed in this section, so we categorize this example as a substitution problem.

Apply Equation 35.10 to the air–water interface:

$$\sin \theta_c = \frac{n_2}{n_1} = \frac{1.00}{1.33} = 0.752$$

$$\theta_c = \boxed{48.8°}$$

WHAT IF? What if a fish in a still pond looks upward toward the water's surface at different angles relative to the surface as in Figure 35.28? What does it see?

Answer Because the path of a light ray is reversible, light traveling from medium 2 into medium 1 in Figure 35.26a follows the paths shown, but in the *opposite* direction. A fish looking upward toward the water surface as in Figure 35.28 can see out of the water if it looks toward the surface at an angle less than the critical angle. Therefore, when the fish's line of vision makes an angle of $\theta = 40°$ with the normal to the surface, for example, light from above the water reaches the fish's eye. At $\theta = 48.8°$, the critical angle for water, the light has to skim along the water's surface before being refracted to the fish's eye; at this angle, the fish can, in principle, see the entire shore of the pond. At angles greater than the critical angle, the light reaching the fish comes by means of total internal reflection at the surface. Therefore, at $\theta = 60°$, the fish sees a reflection of the bottom of the pond.

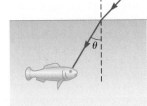

Figure 35.28 (Example 35.6) **What If?** A fish looks upward toward the water surface.

Optical Fibers

Another interesting application of total internal reflection is the use of glass or transparent plastic rods to "pipe" light from one place to another. As indicated in Figure 35.29 (page 1076), light is confined to traveling within a rod, even around curves, as the result of successive total internal reflections. Such a light pipe is flexible

Figure 35.29 Light travels in a curved transparent rod by multiple internal reflections.

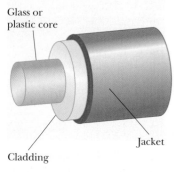

Glass or plastic core

Jacket

Cladding

Figure 35.30 The construction of an optical fiber. Light travels in the core, which is surrounded by a cladding and a protective jacket.

if thin fibers are used rather than thick rods. A flexible light pipe is called an **optical fiber.** If a bundle of parallel fibers is used to construct an optical transmission line, images can be transferred from one point to another. Part of the 2009 Nobel Prize in Physics was awarded to Charles K. Kao (b. 1933) for his discovery of how to transmit light signals over long distances through thin glass fibers. This discovery has led to the development of a sizable industry known as *fiber optics.*

A practical optical fiber consists of a transparent core surrounded by a *cladding,* a material that has a lower index of refraction than the core. The combination may be surrounded by a plastic *jacket* to prevent mechanical damage. Figure 35.30 shows a cutaway view of this construction. Because the index of refraction of the cladding is less than that of the core, light traveling in the core experiences total internal reflection if it arrives at the interface between the core and the cladding at an angle of incidence that exceeds the critical angle. In this case, light "bounces" along the core of the optical fiber, losing very little of its intensity as it travels.

Any loss in intensity in an optical fiber is essentially due to reflections from the two ends and absorption by the fiber material. Optical fiber devices are particularly useful for viewing an object at an inaccessible location. For example, physicians often use such devices to examine internal organs of the body or to perform surgery without making large incisions. Optical fiber cables are replacing copper wiring and coaxial cables for telecommunications because the fibers can carry a much greater volume of telephone calls or other forms of communication than electrical wires can.

Figure 35.31a shows a bundle of optical fibers gathered into an optical cable that can be used to carry communication signals. Figure 35.31b shows laser light following the curves of a coiled bundle by total internal reflection. Many computers and other electronic equipment now have optical ports as well as electrical ports for transferring information.

Figure 35.31 (a) Strands of glass optical fibers are used to carry voice, video, and data signals in telecommunication networks. (b) A bundle of optical fibers is illuminated by a laser.

Dennis O'Clair/Getty Images

a

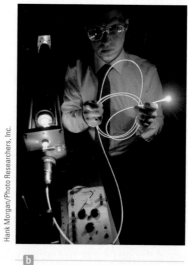

Hank Morgan/Photo Researchers, Inc.

b

Summary

Definition

The **index of refraction** n of a medium is defined by the ratio

$$n \equiv \frac{c}{v} \tag{35.4}$$

where c is the speed of light in vacuum and v is the speed of light in the medium.

Concepts and Principles

In geometric optics, we use the **ray approximation,** in which a wave travels through a uniform medium in straight lines in the direction of the rays.

Total internal reflection occurs when light travels from a medium of high index of refraction to one of lower index of refraction. The **critical angle** θ_c for which total internal reflection occurs at an interface is given by

$$\sin \theta_c = \frac{n_2}{n_1} \quad (\text{for } n_1 > n_2) \tag{35.10}$$

Analysis Models for Problem Solving

Wave Under Reflection. The **law of reflection** states that for a light ray (or other type of wave) incident on a smooth surface, the angle of reflection θ_1' equals the angle of incidence θ_1:

$$\theta_1' = \theta_1 \tag{35.2}$$

Wave Under Refraction. A wave crossing a boundary as it travels from medium 1 to medium 2 is **refracted.** The angle of refraction θ_2 is related to the incident angle θ_1 by the relationship

$$\frac{\sin \theta_2}{\sin \theta_1} = \frac{v_2}{v_1} \tag{35.3}$$

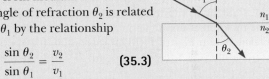

where v_1 and v_2 are the speeds of the wave in medium 1 and medium 2, respectively. The incident ray, the reflected ray, the refracted ray, and the normal to the surface all lie in the same plane.

For light waves, **Snell's law of refraction** states that

$$n_1 \sin \theta_1 = n_2 \sin \theta_2 \tag{35.8}$$

where n_1 and n_2 are the indices of refraction in the two media.

Objective Questions

1. denotes answer available in *Student Solutions Manual/Study Guide*

1. In each of the following situations, a wave passes through an opening in an absorbing wall. Rank the situations in order from the one in which the wave is best described by the ray approximation to the one in which the wave coming through the opening spreads out most nearly equally in all directions in the hemisphere beyond the wall. (a) The sound of a low whistle at 1 kHz passes through a doorway 1 m wide. (b) Red light passes through the pupil of your eye. (c) Blue light passes through the pupil of your eye. (d) The wave broadcast by an AM radio station passes through a doorway 1 m wide. (e) An x-ray passes through the space between bones in your elbow joint.

2. A source emits monochromatic light of wavelength 495 nm in air. When the light passes through a liquid, its wavelength reduces to 434 nm. What is the liquid's index of refraction? (a) 1.26 (b) 1.49 (c) 1.14 (d) 1.33 (e) 2.03

3. Carbon disulfide ($n = 1.63$) is poured into a container made of crown glass ($n = 1.52$). What is the critical angle for total internal reflection of a light ray in the liquid when it is incident on the liquid-to-glass surface? (a) 89.2° (b) 68.8° (c) 21.2° (d) 1.07° (e) 43.0°

4. A light wave moves between medium 1 and medium 2. Which of the following are correct statements relating its speed, frequency, and wavelength in the two media, the indices of refraction of the media, and the angles of incidence and refraction? More than one statement may be correct. (a) $v_1/\sin \theta_1 = v_2/\sin \theta_2$ (b) $\csc \theta_1/n_1 = \csc \theta_2/n_2$ (c) $\lambda_1/\sin \theta_1 = \lambda_2/\sin \theta_2$ (d) $f_1/\sin \theta_1 = f_2/\sin \theta_2$ (e) $n_1/\cos \theta_1 = n_2/\cos \theta_2$

5. What happens to a light wave when it travels from air into glass? (a) Its speed remains the same. (b) Its speed increases. (c) Its wavelength increases. (d) Its wavelength remains the same. (e) Its frequency remains the same.

6. The index of refraction for water is about $\frac{4}{3}$. What happens as a beam of light travels from air into water? (a) Its speed increases to $\frac{4}{3}c$, and its frequency decreases. (b) Its speed decreases to $\frac{3}{4}c$, and its wavelength decreases by a factor of $\frac{3}{4}$. (c) Its speed decreases to $\frac{3}{4}c$, and its wavelength increases by a factor of $\frac{4}{3}$. (d) Its speed and frequency remain the same. (e) Its speed decreases to $\frac{3}{4}c$, and its frequency increases.

7. Light can travel from air into water. Some possible paths for the light ray in the water are shown in Figure

OQ35.7. Which path will the light most likely follow? (a) *A* (b) *B* (c) *C* (d) *D* (e) *E*

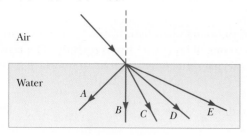

Figure OQ35.7

8. What is the order of magnitude of the time interval required for light to travel 10 km as in Galileo's attempt to measure the speed of light? (a) several seconds (b) several milliseconds (c) several microseconds (d) several nanoseconds

9. A light ray containing both blue and red wavelengths is incident at an angle on a slab of glass. Which of the sketches in Figure OQ35.9 represents the most likely outcome? (a) *A* (b) *B* (c) *C* (d) *D* (e) none of them

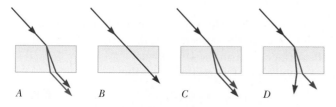

Figure OQ35.9

10. For the following questions, choose from the following possibilities: (a) yes; water (b) no; water (c) yes; air (d) no; air. **(i)** Can light undergo total internal reflection at a smooth interface between air and water? If so, in which medium must it be traveling originally? **(ii)** Can sound undergo total internal reflection at a smooth interface between air and water? If so, in which medium must it be traveling originally?

11. A light ray travels from vacuum into a slab of material with index of refraction n_1 at incident angle θ with respect to the surface. It subsequently passes into a second slab of material with index of refraction n_2 before passing back into vacuum again. The surfaces of the different materials are all parallel to one another. As the light exits the second slab, what can be said of the final angle ϕ that the outgoing light makes with the normal? (a) $\phi > \theta$ (b) $\phi < \theta$ (c) $\phi = \theta$ (d) The angle depends on the magnitudes of n_1 and n_2. (e) The angle depends on the wavelength of the light.

12. Suppose you find experimentally that two colors of light, A and B, originally traveling in the same direction in air, are sent through a glass prism, and A changes direction more than B. Which travels more slowly in the prism, A or B? Alternatively, is there insufficient information to determine which moves more slowly?

13. The core of an optical fiber transmits light with minimal loss if it is surrounded by what? (a) water (b) diamond (c) air (d) glass (e) fused quartz

14. Which color light refracts the most when entering crown glass from air at some incident angle θ with respect to the normal? (a) violet (b) blue (c) green (d) yellow (e) red

15. Light traveling in a medium of index of refraction n_1 is incident on another medium having an index of refraction n_2. Under which of the following conditions can total internal reflection occur at the interface of the two media? (a) The indices of refraction have the relation $n_2 > n_1$. (b) The indices of refraction have the relation $n_1 > n_2$. (c) Light travels slower in the second medium than in the first. (d) The angle of incidence is less than the critical angle. (e) The angle of incidence must equal the angle of refraction.

Conceptual Questions

1. denotes answer available in *Student Solutions Manual/Study Guide*

1. The level of water in a clear, colorless glass can easily be observed with the naked eye. The level of liquid helium in a clear glass vessel is extremely difficult to see with the naked eye. Explain.

2. A complete circle of a rainbow can sometimes be seen from an airplane. With a stepladder, a lawn sprinkler, and a sunny day, how can you show the complete circle to children?

3. You take a child for walks around the neighborhood. She loves to listen to echoes from houses when she shouts or when you clap loudly. A house with a large, flat front wall can produce an echo if you stand straight in front of it and reasonably far away. (a) Draw a bird's-eye view of the situation to explain the production of the echo. Shade the area where you can stand to hear the echo. For parts (b) through (e), explain your answers with diagrams. (b) **What If?** The

child helps you discover that a house with an L-shaped floor plan can produce echoes if you are standing in a wider range of locations. You can be standing at any reasonably distant location from which you can see the inside corner. Explain the echo in this case and compare with your diagram in part (a). (c) **What If?** What if the two wings of the house are not perpendicular? Will you and the child, standing close together, hear echoes? (d) **What If?** What if a rectangular house and its garage have perpendicular walls that would form an inside corner but have a breezeway between them so that the walls do not meet? Will the structure produce strong echoes for people in a wide range of locations?

4. The F-117A stealth fighter (Fig. CQ35.4) is specifically designed to be a *non*retroreflector of radar. What aspects of its design help accomplish this purpose?

Courtesy U.S. Air Force

Figure CQ35.4

5. Retroreflection by transparent spheres, mentioned in Section 35.4, can be observed with dewdrops. To do so, look at your head's shadow where it falls on dewy grass. The optical display around the shadow of your head is called *heiligenschein*, which is German for *holy light*. Renaissance artist Benvenuto Cellini described the phenomenon and his reaction in his *Autobiography*, at the end of Part One, and American philosopher Henry David Thoreau did the same in *Walden*, "Baker Farm," second paragraph. Do some Internet research to find out more about the heiligenschein.

6. Sound waves have much in common with light waves, including the properties of reflection and refraction. Give an example of each of these phenomena for sound waves.

7. Total internal reflection is applied in the periscope of a submerged submarine to let the user observe events above the water surface. In this device, two prisms are arranged as shown in Figure CQ35.7 so that an incident beam of light follows the path shown. Parallel tilted, silvered mirrors could be used, but glass prisms with no silvered surfaces give higher light throughput. Propose a reason for the higher efficiency.

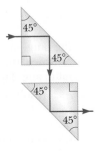

Figure CQ35.7

8. Explain why a diamond sparkles more than a glass crystal of the same shape and size.

9. A laser beam passing through a nonhomogeneous sugar solution follows a curved path. Explain.

10. The display windows of some department stores are slanted slightly inward at the bottom. This tilt is to decrease the glare from streetlights and the Sun, which would make it difficult for shoppers to see the display inside. Sketch a light ray reflecting from such a window to show how this design works.

11. At one restaurant, a worker uses colored chalk to write the daily specials on a blackboard illuminated with a spotlight. At another restaurant, a worker writes with colored grease pencils on a flat, smooth sheet of transparent acrylic plastic with an index of refraction 1.55. The panel hangs in front of a piece of black felt. Small, bright fluorescent tube lights are installed all along the edges of the sheet, inside an opaque channel. Figure CQ35.11 shows a cutaway view of the sign. (a) Explain why viewers at both restaurants see the letters shining against a black background. (b) Explain why the sign at the second restaurant may use less energy from the electric company than the illuminated blackboard at the first restaurant. (c) What would be a good choice for the index of refraction of the material in the grease pencils?

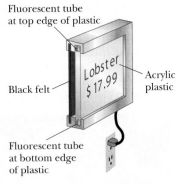

Fluorescent tube at top edge of plastic

Black felt

Acrylic plastic

Fluorescent tube at bottom edge of plastic

Figure CQ35.11

12. (a) Under what conditions is a mirage formed? While driving on a hot day, sometimes you see what appears to be water on the road far ahead. When you arrive at the location of the water, however, the road is perfectly dry. Explain this phenomenon. (b) The mirage called *fata morgana* often occurs over water or in cold regions covered with snow or ice. It can cause islands to sometimes become visible, even though they are not normally visible because they are below the horizon due to the curvature of the Earth. Explain this phenomenon.

13. Figure CQ35.13 shows a pencil partially immersed in a cup of water. Why does the pencil appear to be bent?

© Cengage Learning/Charles D. Winters

Figure CQ35.13

14. A scientific supply catalog advertises a material having an index of refraction of 0.85. Is that a good product to buy? Why or why not?

15. Why do astronomers looking at distant galaxies talk about looking backward in time?

16. Try this simple experiment on your own. Take two opaque cups, place a coin at the bottom of each cup near the edge, and fill one cup with water. Next, view the cups at some angle from the side so that the coin in water is just visible as shown on the left in Figure CQ35.16. Notice that the coin in air is not visible as shown on the right in Figure CQ35.16. Explain this observation.

© Cengage Learning/Ed Dodd

Figure CQ35.16

17. Figure CQ35.17a shows a desk ornament globe containing a photograph. The flat photograph is in air, inside a vertical slot located behind a water-filled compart-

ment having the shape of one half of a cylinder. Suppose you are looking at the center of the photograph and then rotate the globe about a vertical axis. You find that the center of the photograph disappears when you rotate the globe beyond a certain maximum angle (Fig. CQ35.17b). (a) Account for this phenomenon and (b) describe what you see when you turn the globe beyond this angle.

Courtesy of Edwin Lo

a   b

Figure CQ35.17

Problems

Section 35.1 The Nature of Light
Section 35.2 Measurements of the Speed of Light

1. Find the energy of (a) a photon having a frequency of 5.00×10^{17} Hz and (b) a photon having a wavelength of 3.00×10^2 nm. Express your answers in units of electron volts, noting that 1 eV = 1.60×10^{-19} J.

2. The *Apollo 11* astronauts set up a panel of efficient corner-cube retroreflectors on the Moon's surface (Fig. 35.8a). The speed of light can be found by measuring the time interval required for a laser beam to travel from the Earth, reflect from the panel, and return to the Earth. Assume this interval is measured to be 2.51 s at a station where the Moon is at the zenith and take the center-to-center distance from the Earth to the Moon to be equal to 3.84×10^8 m. (a) What is the measured speed of light? (b) Explain whether it is necessary to consider the sizes of the Earth and Moon in your calculation.

3. In an experiment to measure the speed of light using the apparatus of Armand H. L. Fizeau (see Fig. 35.2), the distance between light source and mirror was 11.45 km and the wheel had 720 notches. The experimentally determined value of c was 2.998×10^8 m/s when

the outgoing light passed through one notch and then returned through the next notch. Calculate the minimum angular speed of the wheel for this experiment.

4. As a result of his observations, Ole Roemer concluded that eclipses of Io by Jupiter were delayed by 22 min during a six-month period as the Earth moved from the point in its orbit where it is closest to Jupiter to the diametrically opposite point where it is farthest from Jupiter. Using the value 1.50×10^8 km as the average radius of the Earth's orbit around the Sun, calculate the speed of light from these data.

Section 35.3 The Ray Approximation in Ray Optics
Section 35.4 Analysis Model: Wave Under Reflection
Section 35.5 Analysis Model: Wave Under Refraction

Notes: You may look up indices of refraction in Table 35.1. Unless indicated otherwise, assume the medium surrounding a piece of material is air with $n = 1.000\ 293$.

5. The wavelength of red helium–neon laser light in air is 632.8 nm. (a) What is its frequency? (b) What is its wavelength in glass that has an index of refraction of 1.50? (c) What is its speed in the glass?

6. An underwater scuba diver sees the Sun at an apparent
W angle of 45.0° above the horizontal. What is the actual
elevation angle of the Sun above the horizontal?

7. A ray of light is incident on a flat surface of a block of
crown glass that is surrounded by water. The angle of
refraction is 19.6°. Find the angle of reflection.

8. Figure P35.8 shows a refracted light beam in linseed oil
W making an angle of $\phi = 20.0°$ with the normal line NN'.
The index of refraction of linseed oil is 1.48. Determine
the angles (a) θ and (b) θ'.

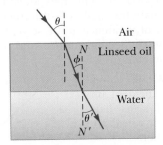

Figure P35.8

9. Find the speed of light in (a) flint glass, (b) water, and
(c) cubic zirconia.

10. A dance hall is built without pillars and with a horizontal ceiling 7.20 m above the floor. A mirror is fastened
flat against one section of the ceiling. Following an
earthquake, the mirror is in place and unbroken. An
engineer makes a quick check of whether the ceiling is
sagging by directing a vertical beam of laser light up at
the mirror and observing its reflection on the floor.
(a) Show that if the mirror has rotated to make an
angle ϕ with the horizontal, the normal to the mirror
makes an angle ϕ with the vertical. (b) Show that the
reflected laser light makes an angle 2ϕ with the vertical. (c) Assume the reflected laser light makes a spot
on the floor 1.40 cm away from the point vertically
below the laser. Find the angle ϕ.

11. A ray of light travels
from air into another
medium, making an
angle of $\theta_1 = 45.0°$ with
the normal as in Figure
P35.11. Find the angle of
refraction θ_2 if the second medium is (a) fused
quartz, (b) carbon disulfide, and (c) water.

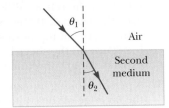

Figure P35.11

12. A ray of light strikes a flat block of glass ($n = 1.50$) of
thickness 2.00 cm at an angle of 30.0° with the normal.
Trace the light beam through the glass and find the
angles of incidence and refraction at each surface.

13. A prism that has an apex angle of 50.0° is made of cubic
M zirconia. What is its minimum angle of deviation?

14. A plane sound wave in air at 20°C, with wavelength
W 589 mm, is incident on a smooth surface of water at
25°C at an angle of incidence of 13.0°. Determine
(a) the angle of refraction for the sound wave and
(b) the wavelength of the sound in water. A narrow

beam of sodium yellow light, with wavelength 589 nm
in vacuum, is incident from air onto a smooth water
surface at an angle of incidence of 13.0°. Determine
(c) the angle of refraction and (d) the wavelength of
the light in water. (e) Compare and contrast the behavior of the sound and light waves in this problem.

15. A light ray initially in water enters a transparent substance at an angle of incidence of 37.0°, and the transmitted ray is refracted at an angle of 25.0°. Calculate
the speed of light in the transparent substance.

16. A laser beam is incident at an angle of 30.0° from the
vertical onto a solution of corn syrup in water. The
beam is refracted to 19.24° from the vertical. (a) What
is the index of refraction of the corn syrup solution?
Assume that the light is red, with vacuum wavelength
632.8 nm. Find its (b) wavelength, (c) frequency, and
(d) speed in the solution.

17. A ray of light strikes the midpoint of one face of an
M equiangular (60°–60°–60°) glass prism ($n = 1.5$) at an
angle of incidence of 30°. (a) Trace the path of the light
ray through the glass and find the angles of incidence
and refraction at each surface. (b) If a small fraction
of light is also reflected at each surface, what are the
angles of reflection at the surfaces?

18. The reflecting surfaces of two intersecting flat mirrors
are at an angle θ ($0° < \theta < 90°$) as shown in Figure
P35.18. For a light ray that strikes the horizontal mirror, show that the emerging ray will intersect the incident ray at an angle $\beta = 180° - 2\theta$.

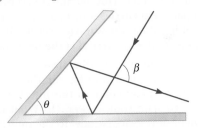

Figure P35.18

19. When you look through a window, by what time interval is the light you see delayed by having to go through
glass instead of air? Make an order-of-magnitude estimate on the basis of data you specify. By how many
wavelengths is it delayed?

20. Two flat, rectangular mirrors, both perpendicular to a
horizontal sheet of paper, are set edge to edge with
their reflecting surfaces perpendicular to each other.
(a) A light ray in the plane of the paper strikes one of
the mirrors at an arbitrary angle of incidence θ_1. Prove
that the final direction of the ray, after reflection from
both mirrors, is opposite its initial direction. (b) **What
If?** Now assume the paper is replaced with a third flat
mirror, touching edges with the other two and perpendicular to both, creating a corner-cube retroreflector
(Fig. 35.8a). A ray of light is incident from any direction within the octant of space bounded by the reflecting surfaces. Argue that the ray will reflect once from
each mirror and that its final direction will be opposite its original direction. The *Apollo 11* astronauts

placed a panel of corner-cube retroreflectors on the Moon. Analysis of timing data taken with it reveals that the radius of the Moon's orbit is increasing at the rate of 3.8 cm/yr as it loses kinetic energy because of tidal friction.

21. The two mirrors illustrated in Figure P35.21 meet at a right angle. The beam of light in the vertical plane indicated by the dashed lines strikes mirror 1 as shown. (a) Determine the distance the reflected light beam travels before striking mirror 2. (b) In what direction does the light beam travel after being reflected from mirror 2?

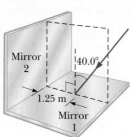

Figure P35.21

22. When the light ray illustrated in Figure P35.22 passes through the glass block of index of refraction $n = 1.50$, it is shifted laterally by the distance d. (a) Find the value of d. (b) Find the time interval required for the light to pass through the glass block.

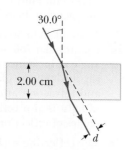

Figure P35.22

23. Two light pulses are emitted simultaneously from a source. Both pulses travel through the same total length of air to a detector, but mirrors shunt one pulse along a path that carries it through an extra length of 6.20 m of ice along the way. Determine the difference in the pulses' times of arrival at the detector.

24. Light passes from air into flint glass at a nonzero angle of incidence. (a) Is it possible for the component of its velocity perpendicular to the interface to remain constant? Explain your answer. (b) **What If?** Can the component of velocity parallel to the interface remain constant during refraction? Explain your answer.

25. A laser beam with vacuum wavelength 632.8 nm is incident from air onto a block of Lucite as shown in Figure 35.10b. The line of sight of the photograph is perpendicular to the plane in which the light moves. Find (a) the speed, (b) the frequency, and (c) the wavelength of the light in the Lucite. *Suggestion:* Use a protractor.

26. A narrow beam of ultrasonic waves reflects off the liver tumor illustrated in Figure P35.26. The speed of the wave is 10.0% less in the liver than in the surrounding medium. Determine the depth of the tumor.

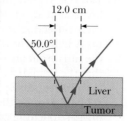

Figure P35.26

27. An opaque cylindrical tank with an open top has a diameter of 3.00 m and is completely filled with water. When the afternoon Sun reaches an angle of 28.0° above the horizon, sunlight ceases to illuminate any part of the bottom of the tank. How deep is the tank?

28. A triangular glass prism with apex angle 60.0° has an index of refraction of 1.50. (a) Show that if its angle of incidence on the first surface is $\theta_1 = 48.6°$, light will pass symmetrically through the prism as shown in Figure 35.17. (b) Find the angle of deviation δ_{min} for $\theta_1 = 48.6°$. (c) **What If?** Find the angle of deviation if the angle of incidence on the first surface is 45.6°. (d) Find the angle of deviation if $\theta_1 = 51.6°$.

29. Light of wavelength 700 nm is incident on the face of a fused quartz prism ($n = 1.458$ at 700 nm) at an incidence angle of 75.0°. The apex angle of the prism is 60.0°. Calculate the angle (a) of refraction at the first surface, (b) of incidence at the second surface, (c) of refraction at the second surface, and (d) between the incident and emerging rays.

30. Figure P35.30 shows a light ray incident on a series of slabs having different refractive indices, where $n_1 < n_2 < n_3 < n_4$. Notice that the path of the ray steadily bends toward the normal. If the variation in n were continuous, the path would form a smooth curve. Use this idea and a ray diagram to explain why you can see the Sun at sunset after it has fallen below the horizon.

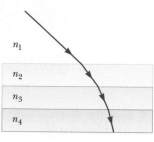

Figure P35.30

31. Three sheets of plastic have unknown indices of refraction. Sheet 1 is placed on top of sheet 2, and a laser beam is directed onto the sheets from above. The laser beam enters sheet 1 and then strikes the interface between sheet 1 and sheet 2 at an angle of 26.5° with the normal. The refracted beam in sheet 2 makes an angle of 31.7° with the normal. The experiment is repeated with sheet 3 on top of sheet 2, and, with the same angle of incidence on the sheet 3–sheet 2 interface, the refracted beam makes an angle of 36.7° with the normal. If the experiment is repeated again with sheet 1 on top of sheet 3, with that same angle of incidence on the sheet 1–sheet 3 interface, what is the expected angle of refraction in sheet 3?

32. A person looking into an empty container is able to see the far edge of the container's bottom as shown in Figure P35.32a. The height of the container is h, and its width is d. When the container is completely filled with a fluid of index of refraction n and viewed from the same angle, the person can see the center of a coin at

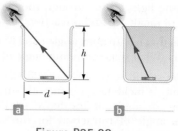

Figure P35.32

the middle of the container's bottom as shown in Figure P35.32b. (a) Show that the ratio h/d is given by

$$\frac{h}{d} = \sqrt{\frac{n^2 - 1}{4 - n^2}}$$

(b) Assuming the container has a width of 8.00 cm and is filled with water, use the expression above to find the height of the container. (c) For what range of values of n will the center of the coin not be visible for any values of h and d?

33. A laser beam is incident on a 45°–45°–90° prism perpendicular to one of its faces as shown in Figure P35.33. The transmitted beam that exits the hypotenuse of the prism makes an angle of $\theta = 15.0°$ with the direction of the incident beam. Find the index of refraction of the prism.

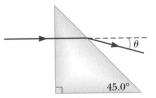

Figure P35.33

34. A submarine is 300 m horizontally from the shore of a freshwater lake and 100 m beneath the surface of the water. A laser beam is sent from the submarine so that the beam strikes the surface of the water 210 m from the shore. A building stands on the shore, and the laser beam hits a target at the top of the building. The goal is to find the height of the target above sea level. (a) Draw a diagram of the situation, identifying the two triangles that are important in finding the solution. (b) Find the angle of incidence of the beam striking the water–air interface. (c) Find the angle of refraction. (d) What angle does the refracted beam make with the horizontal? (e) Find the height of the target above sea level.

35. A beam of light both reflects and refracts at the surface between air and glass as shown in Figure P35.35. If the refractive index of the glass is n_g, find the angle of incidence θ_1 in the air that would result in the reflected ray and the refracted ray being perpendicular to each other.

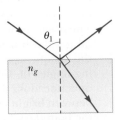

Figure P35.35

Section 35.6 Huygens's Principle

Section 35.7 Dispersion

36. The index of refraction for red light in water is 1.331 and that for blue light is 1.340. If a ray of white light enters the water at an angle of incidence of 83.0°, what are the underwater angles of refraction for the (a) red and (b) blue components of the light?

37. A light beam containing red and violet wavelengths is incident on a slab of quartz at an angle of incidence of 50.0°. The index of refraction of quartz is 1.455 at 600 nm (red light), and its index of refraction is 1.468 at 410 nm (violet light). Find the dispersion of the slab, which is defined as the difference in the angles of refraction for the two wavelengths.

38. The speed of a water wave is described by $v = \sqrt{gd}$, where d is the water depth, assumed to be small compared to the wavelength. Because their speed changes, water waves refract when moving into a region of different depth. (a) Sketch a map of an ocean beach on the eastern side of a landmass. Show contour lines of constant depth under water, assuming a reasonably uniform slope. (b) Suppose waves approach the coast from a storm far away to the north–northeast. Demonstrate that the waves move nearly perpendicular to the shoreline when they reach the beach. (c) Sketch a map of a coastline with alternating bays and headlands as suggested in Figure P35.38. Again make a reasonable guess about the shape of contour lines of constant depth. (d) Suppose waves approach the coast, carrying energy with uniform density along originally straight wave fronts. Show that the energy reaching the coast is concentrated at the headlands and has lower intensity in the bays.

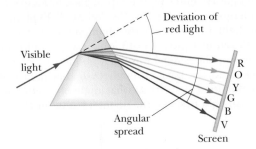

Andy Ryan/Stone/Getty Images

Figure P35.38

39. The index of refraction for violet light in silica flint glass is 1.66, and that for red light is 1.62. What is the angular spread of visible light passing through a prism of apex angle 60.0° if the angle of incidence is 50.0°? See Figure P35.39.

Figure P35.39 Problems 39 and 40.

40. The index of refraction for violet light in silica flint glass is n_V, and that for red light is n_R. What is the angular spread of visible light passing through a prism of apex angle Φ if the angle of incidence is θ? See Figure P35.39.

Section 35.8 Total Internal Reflection

41. A glass optical fiber ($n = 1.50$) is submerged in water ($n = 1.33$). What is the critical angle for light to stay inside the fiber?

42. For 589-nm light, calculate the critical angle for the
W   following materials surrounded by air: (a) cubic zirco-
     nia, (b) flint glass, and (c) ice.

43. A triangular glass prism with
M   apex angle $\Phi = 60.0°$ has an
     index of refraction $n = 1.50$
     (Fig. P35.43). What is the
     smallest angle of incidence
     θ_1 for which a light ray can
     emerge from the other side?

Figure P35.43
Problems 43 and 44.

44. A triangular glass prism with
     apex angle Φ has an index of
     refraction n (Fig. P35.43). What is the smallest angle of
     incidence θ_1 for which a light ray can emerge from the
     other side?

45. Assume a transparent rod of
     diameter $d = 2.00$ μm has an
     index of refraction of 1.36.
     Determine the maximum
     angle θ for which the light rays
     incident on the end of the rod
     in Figure P35.45 are subject

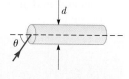

Figure P35.45

     to total internal reflection along the walls of the rod.
     Your answer defines the size of the *cone of acceptance* for
     the rod.

46. Consider a light ray traveling between air and a dia-
     mond cut in the shape shown in Figure P35.46. (a) Find
     the critical angle for total internal reflection for light
     in the diamond incident on the interface between the
     diamond and the outside air. (b) Consider the light ray
     incident normally on the top surface of the diamond as
     shown in Figure P35.46. Show that the light traveling
     toward point P in the diamond is totally reflected. **What
     If?** Suppose the diamond is immersed in water. (c) What
     is the critical angle at the diamond–water interface?
     (d) When the diamond is immersed in water, does
     the light ray entering the top surface in Figure P35.46
     undergo total internal reflection at P? Explain. (e) If
     the light ray entering the diamond remains vertical
     as shown in Figure P35.46, which way should the dia-
     mond in the water be rotated about an axis perpen-
     dicular to the page through O so that light will exit
     the diamond at P? (f) At what angle of rotation in part
     (e) will light first exit the diamond at point P?

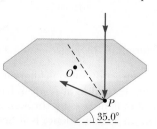

Figure P35.46

47. Consider a common mirage formed by superheated air
M   immediately above a roadway. A truck driver whose
     eyes are 2.00 m above the road, where $n = 1.000\ 293$,
     looks forward. She perceives the illusion of a patch of

water ahead on the road. The road appears wet only
beyond a point on the road at which her line of sight
makes an angle of 1.20° below the horizontal. Find the
index of refraction of the air immediately above the
road surface.

48. A room contains air in which the speed of sound is
     343 m/s. The walls of the room are made of concrete
     in which the speed of sound is 1 850 m/s. (a) Find the
     critical angle for total internal reflection of sound at
     the concrete–air boundary. (b) In which medium must
     the sound be initially traveling if it is to undergo total
     internal reflection? (c) "A bare concrete wall is a highly
     efficient mirror for sound." Give evidence for or
     against this statement.

49. An optical fiber has an index of
     refraction n and diameter d. It is
     surrounded by vacuum. Light is
     sent into the fiber along its axis as
     shown in Figure P35.49. (a) Find
     the smallest outside radius R_{min}
     permitted for a bend in the fiber if
     no light is to escape. (b) **What If?**
     What result does part (a) predict as
     d approaches zero? Is this behavior reasonable?
     Explain. (c) As n increases? (d) As n approaches 1?
     (e) Evaluate R_{min} assuming the fiber diameter is
     100 μm and its index of refraction is 1.40.

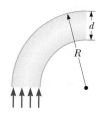

Figure P35.49

50. Around 1968, Richard A. Thorud,
     an engineer at The Toro Company,
     invented a gasoline gauge for small
     engines diagrammed in Figure
     P35.50. The gauge has no moving
     parts. It consists of a flat slab of trans-
     parent plastic fitting vertically into a
     slot in the cap on the gas tank. None
     of the plastic has a reflective coating.
     The plastic projects from the horizon-
     tal top down nearly to the bottom of

Figure P35.50

     the opaque tank. Its lower edge is cut with facets mak-
     ing angles of 45° with the horizontal. A lawn mower
     operator looks down from above and sees a boundary
     between bright and dark on the gauge. The location of
     the boundary, across the width of the plastic, indicates
     the quantity of gasoline in the tank. (a) Explain how the
     gauge works. (b) Explain the design requirements, if
     any, for the index of refraction of the plastic.

Additional Problems

51. A beam of light is incident from air on the surface of a
M   liquid. If the angle of incidence is 30.0° and the angle of
     refraction is 22.0°, find the critical angle for total inter-
     nal reflection for the liquid when surrounded by air.

52. Consider a horizontal interface between air above and
     glass of index of refraction 1.55 below. (a) Draw a light
     ray incident from the air at angle of incidence 30.0°.
     Determine the angles of the reflected and refracted
     rays and show them on the diagram. (b) **What If?** Now
     suppose the light ray is incident from the glass at an
     angle of 30.0°. Determine the angles of the reflected

and refracted rays and show all three rays on a new diagram. (c) For rays incident from the air onto the air–glass surface, determine and tabulate the angles of reflection and refraction for all the angles of incidence at 10.0° intervals from 0° to 90.0°. (d) Do the same for light rays coming up to the interface through the glass.

53. A small light fixture on the bottom of a swimming pool is 1.00 m below the surface. The light emerging from the still water forms a circle on the water surface. What is the diameter of this circle?

54. *Why is the following situation impossible?* While at the bottom of a calm freshwater lake, a scuba diver sees the Sun at an apparent angle of 38.0° above the horizontal.

55. A digital video disc (DVD) records information in a spiral track approximately 1 μm wide. The track consists of a series of pits in the information layer (Fig. P35.55a) that scatter light from a laser beam sharply focused on them. The laser shines in from below through transparent plastic of thickness $t = 1.20$ mm and index of refraction 1.55 (Fig. P35.55b). Assume the width of the laser beam at the information layer must be $a = 1.00$ μm to read from only one track and not from its neighbors. Assume the width of the beam as it enters the transparent plastic is $w = 0.700$ mm. A lens makes the beam converge into a cone with an apex angle $2\theta_1$ before it enters the DVD. Find the incidence angle θ_1 of the light at the edge of the conical beam. This design is relatively immune to small dust particles degrading the video quality.

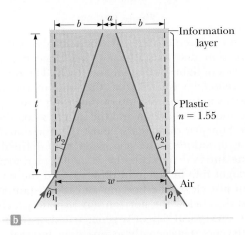

Andrew Syred/Photo Researchers, Inc.

Ⓐ

Ⓑ

Figure P35.55

56. How many times will the incident beam shown in Figure P35.56 be reflected by each of the parallel mirrors?

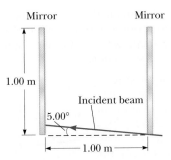

Figure P35.56

57. When light is incident normally on the interface between two transparent optical media, the intensity of the reflected light is given by the expression

$$S_1' = \left(\frac{n_2 - n_1}{n_2 + n_1}\right)^2 S_1$$

In this equation, S_1 represents the average magnitude of the Poynting vector in the incident light (the incident intensity), S_1' is the reflected intensity, and n_1 and n_2 are the refractive indices of the two media. (a) What fraction of the incident intensity is reflected for 589-nm light normally incident on an interface between air and crown glass? (b) Does it matter in part (a) whether the light is in the air or in the glass as it strikes the interface?

58. Refer to Problem 57 for its description of the reflected intensity of light normally incident on an interface between two transparent media. (a) For light normally incident on an interface between vacuum and a transparent medium of index n, show that the intensity S_2 of the transmitted light is given by $S_2/S_1 = 4n/(n + 1)^2$. (b) Light travels perpendicularly through a diamond slab, surrounded by air, with parallel surfaces of entry and exit. Apply the transmission fraction in part (a) to find the approximate overall transmission through the slab of diamond, as a percentage. Ignore light reflected back and forth within the slab.

59. A light ray enters the atmosphere of the Earth and descends vertically to the surface a distance $h = 100$ km below. The index of refraction where the light enters the atmosphere is 1.00, and it increases linearly with distance to have the value $n = 1.000\ 293$ at the Earth's surface. (a) Over what time interval does the light traverse this path? (b) By what percentage is the time interval larger than that required in the absence of the Earth's atmosphere?

60. A light ray enters the atmosphere of a planet and descends vertically to the surface a distance h below. The index of refraction where the light enters the atmosphere is 1.00, and it increases linearly with distance to have the value n at the planet surface. (a) Over what time interval does the light traverse this path? (b) By what fraction is the time interval larger than that required in the absence of an atmosphere?

61. A narrow beam of light is incident from air onto the surface of glass with index of refraction 1.56. Find the angle of incidence for which the corresponding angle of refraction is half the angle of incidence. *Suggestion:* You might want to use the trigonometric identity $\sin 2\theta = 2 \sin \theta \cos \theta$.

62. One technique for measuring the apex angle of a prism is shown in Figure P35.62. Two parallel rays of light are directed onto the apex of the prism so that the rays reflect from opposite faces of the prism. The angular separation γ of the two reflected rays can be measured. Show that $\phi = \frac{1}{2}\gamma$.

Figure P35.62

63. A thief hides a precious jewel by placing it on the bottom of a public swimming pool. He places a circular raft on the surface of the water directly above and centered over the jewel as shown in Figure P35.63. The surface of the water is calm. The raft, of diameter $d = 4.54$ m, prevents the jewel from being seen by any observer above the water, either on the raft or on the side of the pool. What is the maximum depth h of the pool for the jewel to remain unseen?

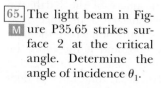

Figure P35.63

64. **Review.** A mirror is often "silvered" with aluminum. By adjusting the thickness of the metallic film, one can make a sheet of glass into a mirror that reflects anything between 3% and 98% of the incident light, transmitting the rest. Prove that it is impossible to construct a "one-way mirror" that would reflect 90% of the electromagnetic waves incident from one side and reflect 10% of those incident from the other side. *Suggestion:* Use Clausius's statement of the second law of thermodynamics.

65. The light beam in Figure P35.65 strikes surface 2 at the critical angle. Determine the angle of incidence θ_1.

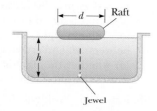

Figure P35.65

66. *Why is the following situation impossible?* A laser beam strikes one end of a slab of material of length $L = 42.0$ cm and thickness $t = 3.10$ mm as shown in Figure P35.66 (not to scale). It enters the material at the center of the left end, striking it at an angle of incidence of $\theta = 50.0°$. The index of refraction of the slab is $n = 1.48$. The light makes 85 internal reflections from the top and bottom of the slab before exiting at the other end.

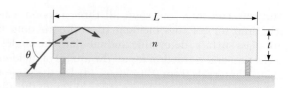

Figure P35.66

67. A 4.00-m-long pole stands vertically in a freshwater lake having a depth of 2.00 m. The Sun is 40.0° above the horizontal. Determine the length of the pole's shadow on the bottom of the lake.

68. A light ray of wavelength 589 nm is incident at an angle θ on the top surface of a block of polystyrene as shown in Figure P35.68. (a) Find the maximum value of θ for which the refracted ray undergoes total internal reflection at the point P located at the left vertical face of the block. **What If?** Repeat the calculation for the case in which the polystyrene block is immersed in (b) water and (c) carbon disulfide. Explain your answers.

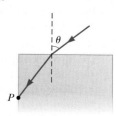

Figure P35.68

69. A light ray traveling in air is incident on one face of a right-angle prism with index of refraction $n = 1.50$ as shown in Figure P35.69, and the ray follows the path shown in the figure. Assuming $\theta = 60.0°$ and the base of the prism is mirrored, determine the angle ϕ made by the outgoing ray with the normal to the right face of the prism.

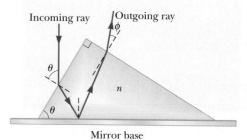

Figure P35.69

70. As sunlight enters the Earth's atmosphere, it changes direction due to the small difference between the speeds of light in vacuum and in air. The duration of an *optical* day is defined as the time interval between the instant when the top of the rising Sun is just visible above the horizon and the instant when the top of the Sun just disappears below the horizontal plane. The duration of the *geometric* day is defined as the time interval between the instant a mathematically straight line between an observer and the top of the Sun just clears the horizon and the instant this line just dips below the horizon. (a) Explain which is longer, an optical day or a geometric day. (b) Find the difference between these two time intervals. Model the Earth's atmosphere as uniform, with index of refraction 1.000 293, a sharply defined upper surface, and depth 8 614 m. Assume the observer is at the

Earth's equator so that the apparent path of the rising and setting Sun is perpendicular to the horizon.

71. A material having an index of refraction n is surrounded by vacuum and is in the shape of a quarter circle of radius R (Fig. P35.71). A light ray parallel to the base of the material is incident from the left at a distance L above the base and emerges from the material at the angle θ. Determine an expression for θ in terms of n, R, and L.

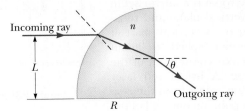

Figure P35.71

72. A ray of light passes from air into water. For its deviation angle $\delta = |\theta_1 - \theta_2|$ to be 10.0°, what must its angle of incidence be?

73. As shown in Figure P35.73, a light ray is incident normal to one face of a 30°–60°–90° block of flint glass (a prism) that is immersed in water. (a) Determine the exit angle θ_3 of the ray. (b) A substance is dissolved in the water to increase the index of refraction n_2. At what value of n_2 does total internal reflection cease at point P?

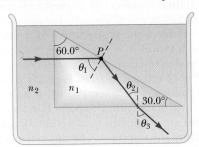

Figure P35.73

74. A transparent cylinder of radius $R = 2.00$ m has a mirrored surface on its right half as shown in Figure P35.74. A light ray traveling in air is incident on the left side of the cylinder. The incident light ray and exiting light ray are parallel, and $d = 2.00$ m. Determine the index of refraction of the material.

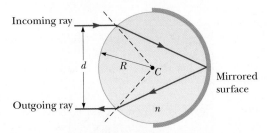

Figure P35.74

75. Figure P35.75 shows the path of a light beam through several slabs with different indices of refraction. (a) If $\theta_1 = 30.0°$, what is the angle θ_2 of the emerging beam?

(b) What must the incident angle θ_1 be to have total internal reflection at the surface between the medium with $n = 1.20$ and the medium with $n = 1.00$?

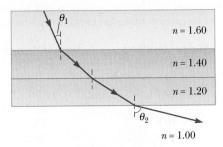

Figure P35.75

76. A. H. Pfund's method for measuring the index of refraction of glass is illustrated in Figure P35.76. One face of a slab of thickness t is painted white, and a small hole scraped clear at point P serves as a source of diverging rays when the slab is illuminated from below. Ray PBB' strikes the clear surface at the critical angle and is totally reflected, as are rays such as PCC'. Rays such as PAA' emerge from the clear surface. On the painted surface, there appears a dark circle of diameter d surrounded by an illuminated region, or halo. (a) Derive an equation for n in terms of the measured quantities d and t. (b) What is the diameter of the dark circle if $n = 1.52$ for a slab 0.600 cm thick? (c) If white light is used, dispersion causes the critical angle to depend on color. Is the inner edge of the white halo tinged with red light or with violet light? Explain.

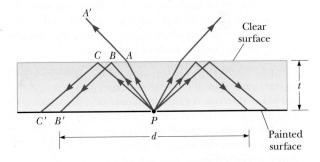

Figure P35.76

77. A light ray enters a rectangular block of plastic at an angle $\theta_1 = 45.0°$ and emerges at an angle $\theta_2 = 76.0°$ as shown in Figure P35.77. (a) Determine the index of refraction of the plastic. (b) If the light ray enters the plastic at a point $L = 50.0$ cm from the bottom edge, what time interval is required for the light ray to travel through the plastic?

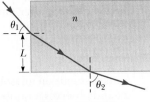

Figure P35.77

78. Students allow a narrow beam of laser light to strike a water surface. They measure the angle of refraction for selected angles of incidence and record the data shown in the accompanying table. (a) Use the data to

verify Snell's law of refraction by plotting the sine of the angle of incidence versus the sine of the angle of refraction. (b) Explain what the shape of the graph demonstrates. (c) Use the resulting plot to deduce the index of refraction of water, explaining how you do so.

Angle of Incidence (degrees)	Angle of Refraction (degrees)
10.0	7.5
20.0	15.1
30.0	22.3
40.0	28.7
50.0	35.2
60.0	40.3
70.0	45.3
80.0	47.7

79. The walls of an ancient shrine are perpendicular to the four cardinal compass directions. On the first day of spring, light from the rising Sun enters a rectangular window in the eastern wall. The light traverses 2.37 m horizontally to shine perpendicularly on the wall opposite the window. A tourist observes the patch of light moving across this western wall. (a) With what speed does the illuminated rectangle move? (b) The tourist holds a small, square mirror flat against the western wall at one corner of the rectangle of light. The mirror reflects light back to a spot on the eastern wall close beside the window. With what speed does the smaller square of light move across that wall? (c) Seen from a latitude of 40.0° north, the rising Sun moves through the sky along a line making a 50.0° angle with the southeastern horizon. In what direction does the rectangular patch of light on the western wall of the shrine move? (d) In what direction does the smaller square of light on the eastern wall move?

80. Figure P35.80 shows a top view of a square enclosure. The inner surfaces are plane mirrors. A ray of light enters a small hole in the center of one mirror. (a) At what angle θ must the ray enter if it exits through the hole after being reflected once by each of the other three mirrors? (b) **What If?** Are there other values of θ for which the ray can exit after multiple reflections? If so, sketch one of the ray's paths.

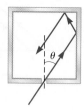

Figure P35.80

Challenge Problems

81. A hiker stands on an isolated mountain peak near sunset and observes a rainbow caused by water droplets in the air at a distance of 8.00 km along her line of sight to the most intense light from the rainbow. The valley is 2.00 km below the mountain peak and entirely flat. What fraction of the complete circular arc of the rainbow is visible to the hiker?

82. *Why is the following situation impossible?* The perpendicular distance of a lightbulb from a large plane mirror is twice the perpendicular distance of a person from the mirror. Light from the lightbulb reaches the person by two paths: (1) it travels to the mirror and reflects from the mirror to the person, and (2) it travels directly to the person without reflecting off the mirror. The total distance traveled by the light in the first case is 3.10 times the distance traveled by the light in the second case.

83. Figure P35.83 shows an overhead view of a room of square floor area and side L. At the center of the room is a mirror set in a vertical plane and rotating on a vertical shaft at angular speed ω about an axis coming out of the page. A bright red laser beam enters from the center point on one wall of the room and strikes the mirror. As the mirror rotates, the reflected laser beam creates a red spot sweeping across the walls of the room. (a) When the spot of light on the wall is at distance x from point O, what is its speed? (b) What value of x corresponds to the minimum value for the speed? (c) What is the minimum value for the speed? (d) What is the maximum speed of the spot on the wall? (e) In what time interval does the spot change from its minimum to its maximum speed?

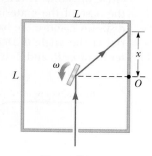

Figure P35.83

84. Pierre de Fermat (1601–1665) showed that whenever light travels from one point to another, its actual path is the path that requires the smallest time interval. This statement is known as *Fermat's principle*. The simplest example is for light propagating in a homogeneous medium. It moves in a straight line because a straight line is the shortest distance between two points. Derive Snell's law of refraction from Fermat's principle. Proceed as follows. In Figure P35.84, a light ray travels from point P in medium 1 to point Q in medium 2. The two points are, respectively, at perpendicular distances a and b from the interface. The displacement from P to Q has the component d parallel to the interface, and we let x represent the coordinate of the point where the ray enters the second medium. Let t = 0 be the instant the light starts from P. (a) Show that the time at which the light arrives at Q is

$$t = \frac{r_1}{v_1} + \frac{r_2}{v_2} = \frac{n_1\sqrt{a^2 + x^2}}{c} + \frac{n_2\sqrt{b^2 + (d - x)^2}}{c}$$

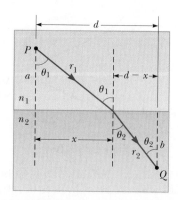

Figure P35.84 Problems 84 and 85.

(b) To obtain the value of x for which t has its minimum value, differentiate t with respect to x and set the derivative equal to zero. Show that the result implies

$$\frac{n_1 x}{\sqrt{a^2 + x^2}} = \frac{n_2(d - x)}{\sqrt{b^2 + (d - x)^2}}$$

(c) Show that this expression in turn gives Snell's law,

$$n_1 \sin \theta_1 = n_2 \sin \theta_2$$

85. Refer to Problem 84 for the statement of Fermat's principle of least time. Derive the law of reflection (Eq. 35.2) from Fermat's principle.

86. Suppose a luminous sphere of radius R_1 (such as the Sun) is surrounded by a uniform atmosphere of radius $R_2 > R_1$ and index of refraction n. When the sphere is viewed from a location far away in vacuum, what is its apparent radius (a) when $R_2 > nR_1$ and (b) when $R_2 < nR_1$?

87. This problem builds upon the results of Problems 57 and 58. Light travels perpendicularly through a diamond slab, surrounded by air, with parallel surfaces of entry and exit. The intensity of the transmitted light is what fraction of the incident intensity? Include the effects of light reflected back and forth inside the slab.

Image Formation

The light rays coming from the leaves in the background of this scene did not form a focused image in the camera that took this photograph. Consequently, the background appears very blurry. Light rays passing though the raindrop, however, have been altered so as to form a focused image of the background leaves for the camera. In this chapter, we investigate the formation of images as light rays reflect from mirrors and refract through lenses. *(Don Hammond Photography Ltd. RF)*

This chapter is concerned with the images that result when light rays encounter flat or curved surfaces between two media. Images can be formed by either reflection or refraction due to these surfaces. We can design mirrors and lenses to form images with desired characteristics. In this chapter, we continue to use the ray approximation and assume light travels in straight lines. We first study the formation of images by mirrors and lenses and techniques for locating an image and determining its size. Then we investigate how to combine these elements into several useful optical instruments such as microscopes and telescopes.

36.1 Images Formed by Flat Mirrors

Image formation by mirrors can be understood through the behavior of light rays as described by the wave under reflection analysis model. We begin by considering the simplest possible mirror, the flat mirror. Consider a point source of light placed at O in Figure 36.1, a distance p in front of a flat mirror. The distance p is called the **object distance.** Diverging light rays leave the source and are reflected from the mirror. Upon reflection, the rays continue to diverge. The dashed lines in Figure 36.1 are extensions of the diverging rays back to a point of intersection at *I*. The diverging rays appear to the viewer to originate at the point *I* behind the mirror. Point *I*, which is a distance *q* behind the mirror, is called the **image** of the object at *O*. The distance *q* is called the **image distance.** Regardless of the system under study, images can always be located by extending diverging rays back to a point at which they intersect. Images are located either at a point from which rays of light *actually* diverge or at a point from which they *appear* to diverge.

Images are classified as *real* or *virtual*. A **real image** is formed when all light rays pass through and diverge from the image point; a **virtual image** is formed when most if not all of the light rays do *not* pass through the image point but only appear to diverge from that point. The image formed by the mirror in Figure 36.1 is virtual. No light rays from the object exist behind the mirror, at the location of the image, so the light rays in front of the mirror only seem to be diverging from *I*. The image of an object seen in a flat mirror is *always* virtual. Real images can be displayed on a screen (as at a movie theater), but virtual images cannot be displayed on a screen. We shall see an example of a real image in Section 36.2.

We can use the simple geometry in Figure 36.2 to examine the properties of the images of extended objects formed by flat mirrors. Even though there are an infinite number of choices of direction in which light rays could leave each point on the object (represented by a gray arrow), we need to choose only two rays to determine where an image is formed. One of those rays starts at *P*, follows a path perpendicular to the mirror to *Q*, and reflects back on itself. The second ray follows the oblique path *PR* and reflects as shown in Figure 36.2 according to the law of reflection. An observer in front of the mirror would extend the two reflected rays back to the point at which they appear to have originated, which is point *P'* behind the mirror. A continuation of this process for points other than *P* on the object would result in a virtual image (represented by a pink arrow) of the entire object behind the mirror. Because triangles *PQR* and *P'QR* are congruent, *PQ* = *P'Q*, so |*p*| = |*q*|. Therefore, the image formed of an object placed in front of a flat mirror is as far behind the mirror as the object is in front of the mirror.

The geometry in Figure 36.2 also reveals that the object height *h* equals the image height *h'*. Let us define **lateral magnification** *M* of an image as follows:

$$M = \frac{\text{image height}}{\text{object height}} = \frac{h'}{h} \qquad (36.1)$$

This general definition of the lateral magnification for an image from any type of mirror is also valid for images formed by lenses, which we study in Section 36.4. For a flat mirror, *M* = +1 for any image because *h'* = *h*. The positive value of the magnification signifies that the image is upright. (By upright we mean that if the object arrow points upward as in Figure 36.2, so does the image arrow.)

A flat mirror produces an image that has an *apparent* left–right reversal. You can see this reversal by standing in front of a mirror and raising your right hand as shown in Figure 36.3. The image you see raises its left hand. Likewise, your hair appears to be parted on the side opposite your real part, and a mole on your right cheek appears to be on your left cheek.

This reversal is not *actually* a left–right reversal. Imagine, for example, lying on your left side on the floor with your body parallel to the mirror surface. Now your head is on the left and your feet are on the right. If you shake your feet, the image does not shake its head! If you raise your right hand, however, the image again raises its left hand. Therefore, the mirror again appears to produce a left–right reversal but in the up–down direction!

The reversal is actually a *front–back reversal*, caused by the light rays going forward toward the mirror and then reflecting back from it. An interesting

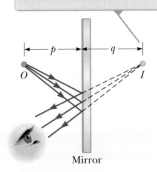

The image point *I* is located behind the mirror a distance *q* from the mirror. The image is virtual.

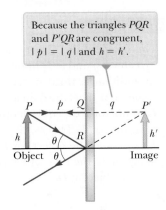

Figure 36.1 An image formed by reflection from a flat mirror.

Because the triangles *PQR* and *P'QR* are congruent, |*p*| = |*q*| and *h* = *h'*.

Figure 36.2 A geometric construction that is used to locate the image of an object placed in front of a flat mirror.

The thumb is on the left side of both real hands and on the left side of the image. That the thumb is not on the right side of the image indicates that there is no left-to-right reversal.

Figure 36.3 The image in the mirror of a person's right hand is reversed front to back, which makes the right hand appear to be a left hand.

exercise is to stand in front of a mirror while holding an overhead transparency in front of you so that you can read the writing on the transparency. You will also be able to read the writing on the image of the transparency. You may have had a similar experience if you have attached a transparent decal with words on it to the rear window of your car. If the decal can be read from outside the car, you can also read it when looking into your rearview mirror from inside the car.

Quick Quiz 36.1 You are standing approximately 2 m away from a mirror. The mirror has water spots on its surface. True or False: It is possible for you to see the water spots and your image both in focus at the same time.

Conceptual Example 36.1   **Multiple Images Formed by Two Mirrors**

Two flat mirrors are perpendicular to each other as in Figure 36.4, and an object is placed at point O. In this situation, multiple images are formed. Locate the positions of these images.

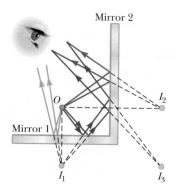

SOLUTION

The image of the object is at I_1 in mirror 1 (green rays) and at I_2 in mirror 2 (red rays). In addition, a third image is formed at I_3 (blue rays). This third image is the image of I_1 in mirror 2 or, equivalently, the image of I_2 in mirror 1. That is, the image at I_1 (or I_2) serves as the object for I_3. To form this image at I_3, the rays reflect twice after leaving the object at O.

Figure 36.4 (Conceptual Example 36.1) When an object is placed in front of two mutually perpendicular mirrors as shown, three images are formed. Follow the different-colored light rays to understand the formation of each image.

Conceptual Example 36.2   **The Tilting Rearview Mirror**

Most rearview mirrors in cars have a day setting and a night setting. The night setting greatly diminishes the intensity of the image so that lights from trailing vehicles do not temporarily blind the driver. How does such a mirror work?

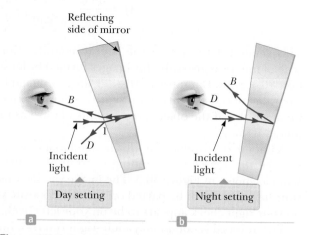

Figure 36.5 (Conceptual Example 36.2) Cross-sectional views of a rearview mirror.

SOLUTION

Figure 36.5 shows a cross-sectional view of a rearview mirror for each setting. The unit consists of a reflective coating on the back of a wedge of glass. In the day setting (Fig. 36.5a), the light from an object behind the car strikes the glass wedge at point 1. Most of the light enters the wedge, refracting as it crosses the front surface, and reflects from the back surface to return to the front surface, where it is refracted again as it re-enters the air as ray B (for *bright*). In addition, a small portion of the light is reflected at the front surface of the glass as indicated by ray D (for *dim*).

This dim reflected light is responsible for the image observed when the mirror is in the night setting (Fig. 36.5b). In that case, the wedge is rotated so that the path followed by the bright light (ray B) does not lead to the eye. Instead, the dim light reflected from the front surface of the wedge travels to the eye, and the brightness of trailing headlights does not become a hazard.

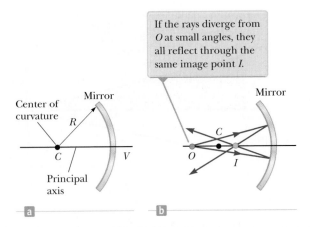

If the rays diverge from *O* at small angles, they all reflect through the same image point *I.*

Figure 36.6 (a) A concave mirror of radius *R*. The center of curvature *C* is located on the principal axis. (b) A point object placed at *O* in front of a concave spherical mirror of radius *R*, where *O* is any point on the principal axis farther than *R* from the mirror surface, forms a real image at *I*.

36.2 Images Formed by Spherical Mirrors

In the preceding section, we considered images formed by flat mirrors. Now we study images formed by curved mirrors. Although a variety of curvatures are possible, we will restrict our investigation to spherical mirrors. As its name implies, a **spherical mirror** has the shape of a section of a sphere.

Concave Mirrors

We first consider reflection of light from the inner, concave surface of a spherical mirror as shown in Figure 36.6. This type of reflecting surface is called a **concave mirror.** Figure 36.6a shows that the mirror has a radius of curvature *R*, and its center of curvature is point *C*. Point *V* is the center of the spherical section, and a line through *C* and *V* is called the **principal axis** of the mirror. Figure 36.6a shows a cross section of a spherical mirror, with its surface represented by the solid, curved dark blue line. (The lighter blue band represents the structural support for the mirrored surface, such as a curved piece of glass on which a silvered reflecting surface is deposited.) This type of mirror focuses incoming parallel rays to a point as demonstrated by the yellow light rays in Figure 36.7.

Now consider a point source of light placed at point *O* in Figure 36.6b, where *O* is any point on the principal axis to the left of *C*. Two diverging light rays that originate at *O* are shown. After reflecting from the mirror, these rays converge and cross at the image point *I*. They then continue to diverge from *I* as if an object were there. As a result, the image at point *I* is real.

In this section, we shall consider only rays that diverge from the object and make a small angle with the principal axis. Such rays are called **paraxial rays.** All paraxial rays reflect through the image point as shown in Figure 36.6b. Rays that are far from the principal axis such as those shown in Figure 36.8 converge to other points on the principal axis, producing a blurred image. This effect, called *spherical aberration*, is present to some extent for any spherical mirror and is discussed in Section 36.5.

If the object distance *p* and radius of curvature *R* are known, we can use Figure 36.9 (page 1094) to calculate the image distance *q*. By convention, these distances are measured from point *V*. Figure 36.9 shows two rays leaving the tip of the object. The red ray passes through the center of curvature *C* of the mirror, hitting the mirror perpendicular to the mirror surface and reflecting back on itself. The blue ray strikes the mirror at its center (point *V*) and reflects as shown, obeying the law of reflection. The image of the tip of the arrow is located at the point where these two rays intersect. From the large, red right triangle in Figure 36.9, we see that $\tan \theta = h/p$, and from the yellow right triangle, we see that $\tan \theta = -h'/q$. The

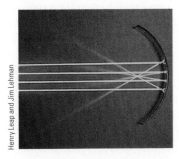

Figure 36.7 Reflection of parallel rays from a concave mirror.

The reflected rays intersect at different points on the principal axis.

Figure 36.8 A spherical concave mirror exhibits spherical aberration when light rays make large angles with the principal axis.

Henry Leap and Jim Lehman

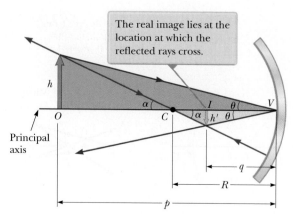

The real image lies at the location at which the reflected rays cross.

Figure 36.9 The image formed by a spherical concave mirror when the object *O* lies outside the center of curvature *C*. This geometric construction is used to derive Equation 36.4.

A satellite-dish antenna is a concave reflector for television signals from a satellite in orbit around the Earth. Because the satellite is so far away, the signals are carried by microwaves that are parallel when they arrive at the dish. These waves reflect from the dish and are focused on the receiver.

Mirror equation in terms of radius of curvature ▶

Focal length ▶

negative sign is introduced because the image is inverted, so h' is taken to be negative. Therefore, from Equation 36.1 and these results, we find that the magnification of the image is

$$M = \frac{h'}{h} = -\frac{q}{p}$$ (36.2)

Also notice from the green right triangle in Figure 36.9 and the smaller red right triangle that

$$\tan \alpha = \frac{-h'}{R - q} \quad \text{and} \quad \tan \alpha = \frac{h}{p - R}$$

from which it follows that

$$\frac{h'}{h} = -\frac{R - q}{p - R}$$ (36.3)

Comparing Equations 36.2 and 36.3 gives

$$\frac{R - q}{p - R} = \frac{q}{p}$$

Simple algebra reduces this expression to

$$\frac{1}{p} + \frac{1}{q} = \frac{2}{R}$$ (36.4)

which is called the *mirror equation*. We present a modified version of this equation shortly.

If the object is very far from the mirror—that is, if p is so much greater than R that p can be said to approach infinity—then $1/p \approx 0$, and Equation 36.4 shows that $q \approx R/2$. That is, when the object is very far from the mirror, the image point is halfway between the center of curvature and the center point on the mirror as shown in Figure 36.10. The incoming rays from the object are essentially parallel in this figure because the source is assumed to be very far from the mirror. The image point in this special case is called the **focal point** *F*, and the image distance is called the **focal length** *f*, where

$$f = \frac{R}{2}$$ (36.5)

The focal point is a distance *f* from the mirror, as noted in Figure 36.10. In Figure 36.7, the beams are traveling parallel to the principal axis and the mirror reflects all beams to the focal point.

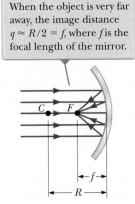

When the object is very far away, the image distance $q \approx R/2 = f$, where f is the focal length of the mirror.

Figure 36.10 Light rays from a distant object ($p \rightarrow \infty$) reflect from a concave mirror through the focal point F.

Pitfall Prevention 36.2

The *Focal* Point Is Not the *Focus* Point The focal point *is usually not* the point at which the light rays focus to form an image. The focal point is determined solely by the curvature of the mirror; it does not depend on the location of the object. In general, an image forms at a point different from the focal point of a mirror (or a lens), as in Figure 36.9. The *only* exception is when the object is located infinitely far away from the mirror.

Because the focal length is a parameter particular to a given mirror, it can be used to compare one mirror with another. Combining Equations 36.4 and 36.5, the **mirror equation** can be expressed in terms of the focal length:

$$\frac{1}{p} + \frac{1}{q} = \frac{1}{f}$$

(36.6)

◀ Mirror equation in terms of focal length

Notice that the focal length of a mirror depends only on the curvature of the mirror and not on the material from which the mirror is made because the formation of the image results from rays reflected from the surface of the material. The situation is different for lenses; in that case, the light actually passes through the material and the focal length depends on the type of material from which the lens is made. (See Section 36.4.)

Convex Mirrors

Figure 36.11 shows the formation of an image by a **convex mirror,** that is, one silvered so that light is reflected from the outer, convex surface. It is sometimes called a **diverging mirror** because the rays from any point on an object diverge after reflection as though they were coming from some point behind the mirror. The image in Figure 36.11 is virtual because the reflected rays only appear to originate at the image point as indicated by the dashed lines. Furthermore, the image is always upright and smaller than the object. This type of mirror is often used in stores to foil shoplifters. A single mirror can be used to survey a large field of view because it forms a smaller image of the interior of the store.

We do not derive any equations for convex spherical mirrors because Equations 36.2, 36.4, and 36.6 can be used for either concave or convex mirrors if we adhere to a strict sign convention. We will refer to the region in which light rays originate and move toward the mirror as the *front side* of the mirror and the other side as the

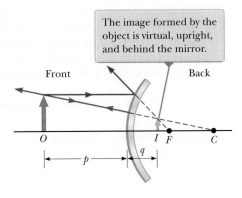

The image formed by the object is virtual, upright, and behind the mirror.

Front   Back

Figure 36.11 Formation of an image by a spherical convex mirror.

Pitfall Prevention 36.3

Watch Your Signs Success in working mirror problems (as well as problems involving refracting surfaces and thin lenses) is largely determined by proper sign choices when substituting into the equations. The best way to success is to work a multitude of problems on your own.

Table 36.1	Sign Conventions for Mirrors		
Quantity	Positive When . . .	Negative When . . .
Object location (p)	object is in front of mirror (real object).	object is in back of mirror (virtual object).
Image location (q)	image is in front of mirror (real image).	image is in back of mirror (virtual image).
Image height (h')	image is upright.	image is inverted.
Focal length (f) and radius (R)	mirror is concave.	mirror is convex.
Magnification (M)	image is upright.	image is inverted.

back side. For example, in Figures 36.9 and 36.11, the side to the left of the mirrors is the front side and the side to the right of the mirrors is the back side. Figure 36.12 states the sign conventions for object and image distances for any type of mirror, and Table 36.1 summarizes the sign conventions for all quantities. One entry in the table, a *virtual object,* is formally introduced in Section 36.4.

Front, or real, side — p and q positive — Incident light — Reflected light

Back, or virtual, side — p and q negative — No light

Flat, convex, or concave mirrored surface

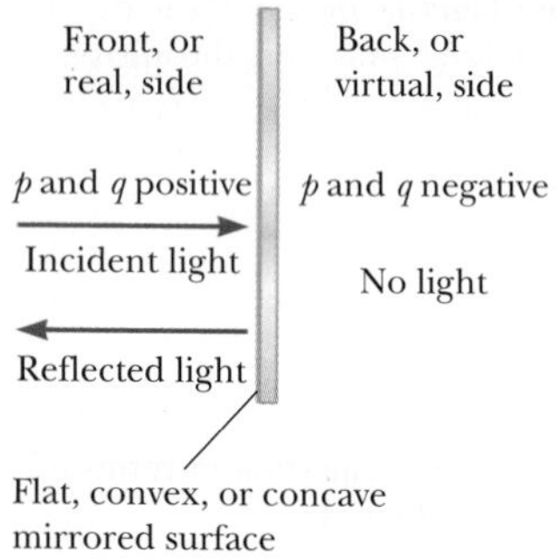

Figure 36.12 Signs of p and q for all types of mirrors.

Ray Diagrams for Mirrors

The positions and sizes of images formed by mirrors can be conveniently determined with *ray diagrams.* These pictorial representations reveal the nature of the image and can be used to check results calculated from the mathematical representation using the mirror and magnification equations. To draw a ray diagram, you must know the position of the object and the locations of the mirror's focal point and center of curvature. You then draw three rays to locate the image as shown by the examples in Figure 36.13. These rays all start from the same object point and are drawn as follows. You may choose any point on the object; here, let's choose the top of the object for simplicity. For concave mirrors (see Figs. 36.13a and 36.13b), draw the following three rays:

Pitfall Prevention 36.4

Choose a Small Number of Rays A *huge* number of light rays leave each point on an object (and pass through each point on an image). In a ray diagram, which displays the characteristics of the image, we choose only a few rays that follow simply stated rules. Locating the image by calculation complements the diagram.

- Ray 1 is drawn from the top of the object parallel to the principal axis and is reflected through the focal point F.
- Ray 2 is drawn from the top of the object through the focal point (or as if coming from the focal point if $p < f$) and is reflected parallel to the principal axis.
- Ray 3 is drawn from the top of the object through the center of curvature C (or as if coming from the center C if $p < 2f$) and is reflected back on itself.

The intersection of any two of these rays locates the image. The third ray serves as a check of the construction. The image point obtained in this fashion must always agree with the value of q calculated from the mirror equation. With concave mirrors, notice what happens as the object is moved closer to the mirror. The real, inverted image in Figure 36.13a moves to the left and becomes larger as the object approaches the focal point. When the object is at the focal point, the image is infinitely far to the left. When the object lies between the focal point and the mirror surface as shown in Figure 36.13b, however, the image is to the right, behind the object, and virtual, upright, and enlarged. This latter situation applies when you use a shaving mirror or a makeup mirror, both of which are concave. Your face is closer to the mirror than the focal point, and you see an upright, enlarged image of your face.

For convex mirrors (see Fig. 36.13c), draw the following three rays:

- Ray 1 is drawn from the top of the object parallel to the principal axis and is reflected *away from* the focal point F.

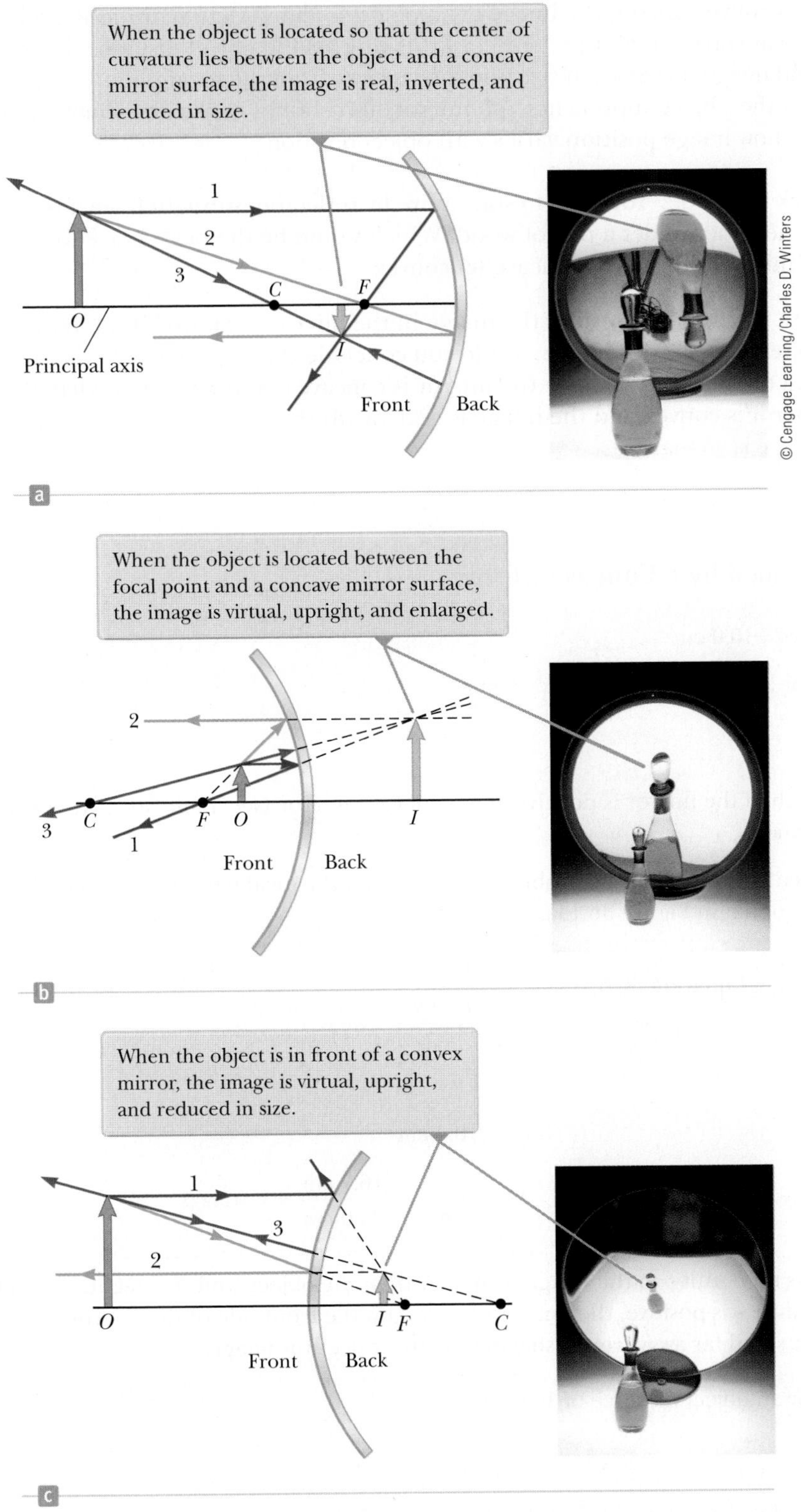

Figure 36.13 Ray diagrams for spherical mirrors along with corresponding photographs of the images of bottles.

When the object is located so that the center of curvature lies between the object and a concave mirror surface, the image is real, inverted, and reduced in size.

1

2

3

C F

O

Principal axis

I

Front   Back

© Cengage Learning/Charles D. Winters

a

When the object is located between the focal point and a concave mirror surface, the image is virtual, upright, and enlarged.

2

3   C F O I

1

Front   Back

b

When the object is in front of a convex mirror, the image is virtual, upright, and reduced in size.

1

3

2

O I F C

Front   Back

c

- Ray 2 is drawn from the top of the object toward the focal point on the back side of the mirror and is reflected parallel to the principal axis.
- Ray 3 is drawn from the top of the object toward the center of curvature C on the back side of the mirror and is reflected back on itself.

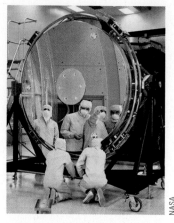

Figure 36.14 (Quick Quiz 36.3) What type of mirror is shown here?

In a convex mirror, the image of an object is always virtual, upright, and reduced in size as shown in Figure 36.13c. In this case, as the object distance decreases, the virtual image increases in size and moves away from the focal point toward the mirror as the object approaches the mirror. You should construct other diagrams to verify how image position varies with object position.

Quick Quiz 36.2 You wish to start a fire by reflecting sunlight from a mirror onto some paper under a pile of wood. Which would be the best choice for the type of mirror? **(a)** flat **(b)** concave **(c)** convex

Quick Quiz 36.3 Consider the image in the mirror in Figure 36.14. Based on the appearance of this image, would you conclude that **(a)** the mirror is concave and the image is real, **(b)** the mirror is concave and the image is virtual, **(c)** the mirror is convex and the image is real, or **(d)** the mirror is convex and the image is virtual?

Example 36.3 **The Image Formed by a Concave Mirror**

A spherical mirror has a focal length of +10.0 cm.

(A) Locate and describe the image for an object distance of 25.0 cm.

SOLUTION

Conceptualize Because the focal length of the mirror is positive, it is a concave mirror (see Table 36.1). We expect the possibilities of both real and virtual images.

Categorize Because the object distance in this part of the problem is larger than the focal length, we expect the image to be real. This situation is analogous to that in Figure 36.13a.

Analyze Find the image distance by using Equation 36.6:

$$\frac{1}{q} = \frac{1}{f} - \frac{1}{p}$$

$$\frac{1}{q} = \frac{1}{10.0 \text{ cm}} - \frac{1}{25.0 \text{ cm}}$$

$$q = \boxed{16.7 \text{ cm}}$$

Find the magnification of the image from Equation 36.2:

$$M = -\frac{q}{p} = -\frac{16.7 \text{ cm}}{25.0 \text{ cm}} = \boxed{-0.667}$$

Finalize The absolute value of M is less than unity, so the image is smaller than the object, and the negative sign for M tells us that the image is inverted. Because q is positive, the image is located on the front side of the mirror and is real. Look into the bowl of a shiny spoon or stand far away from a shaving mirror to see this image.

(B) Locate and describe the image for an object distance of 10.0 cm.

SOLUTION

Categorize Because the object is at the focal point, we expect the image to be infinitely far away.

Analyze Find the image distance by using Equation 36.6:

$$\frac{1}{q} = \frac{1}{f} - \frac{1}{p}$$

$$\frac{1}{q} = \frac{1}{10.0 \text{ cm}} - \frac{1}{10.0 \text{ cm}}$$

$$q = \boxed{\infty}$$

▶ **36.3** continued

Finalize This result means that rays originating from an object positioned at the focal point of a mirror are reflected so that the image is formed at an infinite distance from the mirror; that is, the rays travel parallel to one another after reflection. Such is the situation in a flashlight or an automobile headlight, where the bulb filament is placed at the focal point of a reflector, producing a parallel beam of light.

(C) Locate and describe the image for an object distance of 5.00 cm.

SOLUTION

Categorize Because the object distance is smaller than the focal length, we expect the image to be virtual. This situation is analogous to that in Figure 36.13b.

Analyze Find the image distance by using Equation 36.6:

$$\frac{1}{q} = \frac{1}{f} - \frac{1}{p}$$

$$\frac{1}{q} = \frac{1}{10.0 \text{ cm}} - \frac{1}{5.00 \text{ cm}}$$

$$q = -10.0 \text{ cm}$$

Find the magnification of the image from Equation 36.2:

$$M = -\frac{q}{p} = -\left(\frac{-10.0 \text{ cm}}{5.00 \text{ cm}}\right) = +2.00$$

Finalize The image is twice as large as the object, and the positive sign for M indicates that the image is upright (see Fig. 36.13b). The negative value of the image distance tells us that the image is virtual, as expected. Put your face close to a shaving mirror to see this type of image.

WHAT IF? Suppose you set up the bottle and mirror apparatus illustrated in Figure 36.13a and described here in part (A). While adjusting the apparatus, you accidentally bump the bottle and it begins to slide toward the mirror at speed v_p. How fast does the image of the bottle move?

Answer Solve the mirror equation, Equation 36.6, for q:

$$q = \frac{fp}{p - f}$$

Differentiate this equation with respect to time to find the velocity of the image:

$$(1) \quad v_q = \frac{dq}{dt} = \frac{d}{dt}\left(\frac{fp}{p - f}\right) = -\frac{f^2}{(p - f)^2}\frac{dp}{dt} = -\frac{f^2 v_p}{(p - f)^2}$$

Substitute numerical values from part (A):

$$v_q = -\frac{(10.0 \text{ cm})^2 \, v_p}{(25.0 \text{ cm} - 10.0 \text{ cm})^2} = -0.444 v_p$$

Therefore, the speed of the image is less than that of the object in this case.

We can see two interesting behaviors of the function for v_q in Equation (1). First, the velocity is negative regardless of the value of p or f. Therefore, if the object moves toward the mirror, the image moves toward the left in Figure 36.13 without regard for the side of the focal point at which the object is located or whether the mirror is concave or convex. Second, in the limit of $p \rightarrow 0$, the velocity v_q approaches $-v_p$. As the object moves very close to the mirror, the mirror looks like a plane mirror, the image is as far behind the mirror as the object is in front, and both the object and the image move with the same speed.

Example 36.4 **The Image Formed by a Convex Mirror**

An automobile rearview mirror as shown in Figure 36.15 (page 1100) shows an image of a truck located 10.0 m from the mirror. The focal length of the mirror is −0.60 m.

(A) Find the position of the image of the truck.

continued

▶ **36.4** continued

SOLUTION

Conceptualize This situation is depicted in Figure 36.13c.

Categorize Because the mirror is convex, we expect it to form an upright, reduced, virtual image for any object position.

Figure 36.15 (Example 36.4) An approaching truck is seen in a convex mirror on the right side of an automobile. Notice that the image of the truck is in focus, but the frame of the mirror is not, which demonstrates that the image is not at the same location as the mirror surface.

© Bo Zaunders/Corbis

Analyze Find the image distance by using Equation 36.6:

$$\frac{1}{q} = \frac{1}{f} - \frac{1}{p}$$

$$\frac{1}{q} = \frac{1}{-0.60 \text{ m}} - \frac{1}{10.0 \text{ m}}$$

$$q = \boxed{-0.57 \text{ m}}$$

(B) Find the magnification of the image.

SOLUTION

Analyze Use Equation 36.2:

$$M = -\frac{q}{p} = -\left(\frac{-0.57 \text{ m}}{10.0 \text{ m}}\right) = \boxed{+0.057}$$

Finalize The negative value of q in part (A) indicates that the image is virtual, or behind the mirror, as shown in Figure 36.13c. The magnification in part (B) indicates that the image is much smaller than the truck and is upright because M is positive. The image is reduced in size, so the truck appears to be farther away than it actually is. Because of the image's small size, these mirrors carry the inscription, "Objects in this mirror are closer than they appear." Look into your rearview mirror or the back side of a shiny spoon to see an image of this type.

36.3 Images Formed by Refraction

In this section, we describe how images are formed when light rays follow the wave under refraction model at the boundary between two transparent materials. Consider two transparent media having indices of refraction n_1 and n_2, where the boundary between the two media is a spherical surface of radius R (Fig. 36.16). We assume the object at O is in the medium for which the index of refraction is n_1. Let's consider the paraxial rays leaving O. As we shall see, all such rays are refracted at the spherical surface and focus at a single point I, the image point.

Figure 36.17 shows a single ray leaving point O and refracting to point I. Snell's law of refraction applied to this ray gives

$$n_1 \sin \theta_1 = n_2 \sin \theta_2$$

Because θ_1 and θ_2 are assumed to be small, we can use the small-angle approximation $\sin \theta \approx \theta$ (with angles in radians) and write Snell's law as

$$n_1 \theta_1 = n_2 \theta_2$$

We know that an exterior angle of any triangle equals the sum of the two opposite interior angles, so applying this rule to triangles OPC and PIC in Figure 36.17 gives

$$\theta_1 = \alpha + \beta$$

$$\beta = \theta_2 + \gamma$$

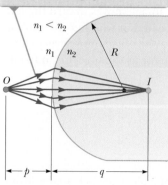

Rays making small angles with the principal axis diverge from a point object at O and are refracted through the image point I.

Figure 36.16 An image formed by refraction at a spherical surface.

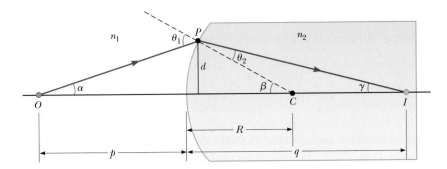

Figure 36.17 Geometry used to derive Equation 36.8, assuming $n_1 < n_2$.

Combining all three expressions and eliminating θ_1 and θ_2 gives

$$n_1\alpha + n_2\gamma = (n_2 - n_1)\beta \qquad \textbf{(36.7)}$$

Figure 36.17 shows three right triangles that have a common vertical leg of length d. For paraxial rays (unlike the relatively large-angle ray shown in Fig. 36.17), the horizontal legs of these triangles are approximately p for the triangle containing angle α, R for the triangle containing angle β, and q for the triangle containing angle γ. In the small-angle approximation, $\tan \theta \approx \theta$, so we can write the approximate relationships from these triangles as follows:

$$\tan \alpha \approx \alpha \approx \frac{d}{p} \qquad \tan \beta \approx \beta \approx \frac{d}{R} \qquad \tan \gamma \approx \gamma \approx \frac{d}{q}$$

Substituting these expressions into Equation 36.7 and dividing through by d gives

$$\frac{n_1}{p} + \frac{n_2}{q} = \frac{n_2 - n_1}{R} \qquad \textbf{(36.8)}$$

◄ Relation between object and image distance for a refracting surface

For a fixed object distance p, the image distance q is independent of the angle the ray makes with the axis. This result tells us that all paraxial rays focus at the same point I.

As with mirrors, we must use a sign convention to apply Equation 36.8 to a variety of cases. We define the side of the surface in which light rays originate as the front side. The other side is called the back side. In contrast with mirrors, where real images are formed in front of the reflecting surface, real images are formed by refraction of light rays to the back of the surface. Because of the difference in location of real images, the refraction sign conventions for q and R are opposite the reflection sign conventions. For example, q and R are both positive in Figure 36.17. The sign conventions for spherical refracting surfaces are summarized in Table 36.2.

We derived Equation 36.8 from an assumption that $n_1 < n_2$ in Figure 36.17. This assumption is not necessary, however. Equation 36.8 is valid regardless of which index of refraction is greater.

Table 36.2	**Sign Conventions for Refracting Surfaces**	
Quantity	**Positive When . . .**	**Negative When . . .**
Object location (p)	object is in front of surface (real object).	object is in back of surface (virtual object).
Image location (q)	image is in back of surface (real image).	image is in front of surface (virtual image).
Image height (h')	image is upright.	image is inverted.
Radius (R)	center of curvature is in back of surface.	center of curvature is in front of surface.

Flat Refracting Surfaces

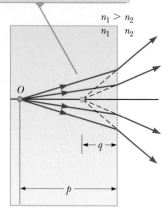

The image is virtual and on the same side of the surface as the object.

Figure 36.18 The image formed by a flat refracting surface. All rays are assumed to be paraxial.

If a refracting surface is flat, then R is infinite and Equation 36.8 reduces to

$$\frac{n_1}{p} = -\frac{n_2}{q}$$

$$q = -\frac{n_2}{n_1} p \tag{36.9}$$

From this expression, we see that the sign of q is opposite that of p. Therefore, according to Table 36.2, the image formed by a flat refracting surface is on the same side of the surface as the object as illustrated in Figure 36.18 for the situation in which the object is in the medium of index n_1 and n_1 is greater than n_2. In this case, a virtual image is formed between the object and the surface. If n_1 is less than n_2, the rays on the back side diverge from one another at smaller angles than those in Figure 36.18. As a result, the virtual image is formed to the left of the object.

Quick Quiz 36.4 In Figure 36.16, what happens to the image point I as the object point O is moved to the right from very far away to very close to the refracting surface? **(a)** It is always to the right of the surface. **(b)** It is always to the left of the surface. **(c)** It starts off to the left, and at some position of O, I moves to the right of the surface. **(d)** It starts off to the right, and at some position of O, I moves to the left of the surface.

Quick Quiz 36.5 In Figure 36.18, what happens to the image point I as the object point O moves toward the right-hand surface of the material of index of refraction n_1? **(a)** It always remains between O and the surface, arriving at the surface just as O does. **(b)** It moves toward the surface more slowly than O so that eventually O passes I. **(c)** It approaches the surface and then moves to the right of the surface.

Conceptual Example 36.5 Let's Go Scuba Diving!

Objects viewed under water with the naked eye appear blurred and out of focus. A scuba diver using a mask, however, has a clear view of underwater objects. Explain how that works, using the information that the indices of refraction of the cornea, water, and air are 1.376, 1.333, and 1.000 29, respectively.

SOLUTION

Because the cornea and water have almost identical indices of refraction, very little refraction occurs when a person under water views objects with the naked eye. In this case, light rays from an object focus behind the retina, resulting in a blurred image. When a mask is used, however, the air space between the eye and the mask surface provides the normal amount of refraction at the eye–air interface; consequently, the light from the object focuses on the retina.

Example 36.6 Gaze into the Crystal Ball

A set of coins is embedded in a spherical plastic paperweight having a radius of 3.0 cm. The index of refraction of the plastic is $n_1 = 1.50$. One coin is located 2.0 cm from the edge of the sphere (Fig. 36.19). Find the position of the image of the coin.

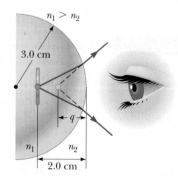

Figure 36.19 (Example 36.6) Light rays from a coin embedded in a plastic sphere form a virtual image between the surface of the object and the sphere surface. Because the object is inside the sphere, the front of the refracting surface is the *interior* of the sphere.

SOLUTION

Conceptualize Because $n_1 > n_2$, where $n_2 = 1.00$ is the index of refraction for air, the rays originating from the coin in Figure 36.19 are refracted away from the normal at the surface and diverge outward. Extending the outgoing rays backward shows an image point within the sphere.

▶ **36.6** continued

Categorize Because the light rays originate in one material and then pass through a curved surface into another material, this example involves an image formed by refraction.

Analyze Apply Equation 36.8, noting from Table 36.2 that R is negative:

$$\frac{n_2}{q} = \frac{n_2 - n_1}{R} - \frac{n_1}{p}$$

Substitute numerical values and solve for q:

$$\frac{1}{q} = \frac{1.00 - 1.50}{-3.0 \text{ cm}} - \frac{1.50}{2.0 \text{ cm}}$$

$$q = \boxed{-1.7 \text{ cm}}$$

Finalize The negative sign for q indicates that the image is in front of the surface; in other words, it is in the same medium as the object as shown in Figure 36.19. Therefore, the image must be virtual. (See Table 36.2.) The coin appears to be closer to the paperweight surface than it actually is.

Example 36.7   The One That Got Away

A small fish is swimming at a depth d below the surface of a pond (Fig. 36.20).

(A) What is the apparent depth of the fish as viewed from directly overhead?

SOLUTION

Conceptualize Because $n_1 > n_2$, where $n_2 = 1.00$ is the index of refraction for air, the rays originating from the fish in Figure 36.20a are refracted away from the normal at the surface and diverge outward. Extending the outgoing rays backward shows an image point under the water.

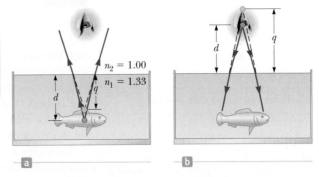

Figure 36.20 (Example 36.7) (a) The apparent depth q of the fish is less than the true depth d. All rays are assumed to be paraxial. (b) Your face appears to the fish to be higher above the surface than it is.

Categorize Because the refracting surface is flat, R is infinite. Hence, we can use Equation 36.9 to determine the location of the image with $p = d$.

Analyze Use the indices of refraction given in Figure 36.20a in Equation 36.9:

$$q = -\frac{n_2}{n_1} p = -\frac{1.00}{1.33} d = \boxed{-0.752d}$$

Finalize Because q is negative, the image is virtual as indicated by the dashed lines in Figure 36.20a. The apparent depth is approximately three-fourths the actual depth.

(B) If your face is a distance d above the water surface, at what apparent distance above the surface does the fish see your face?

SOLUTION

Conceptualize The light rays from your face are shown in Figure 36.20b. Because the rays refract toward the normal, your face appears higher above the surface than it actually is.

Categorize Because the refracting surface is flat, R is infinite. Hence, we can use Equation 36.9 to determine the location of the image with $p = d$.

Analyze Use Equation 36.9 to find the image distance:

$$q = -\frac{n_2}{n_1} p = -\frac{1.33}{1.00} d = \boxed{-1.33d}$$

Finalize The negative sign for q indicates that the image is in the medium from which the light originated, which is the air above the water.

continued

▶ **36.7** continued

WHAT IF?   What if you look more carefully at the fish and measure its apparent *height* from its upper fin to its lower fin? Is the apparent height h' of the fish different from the actual height h?

Answer   Because all points on the fish appear to be fractionally closer to the observer, we expect the height to be smaller. Let the distance d in Figure 36.20a be measured to the top fin and let the distance to the bottom fin be $d + h$. Then the images of the top and bottom of the fish are located at

$$q_{\text{top}} = -0.752d$$

$$q_{\text{bottom}} = -0.752(d + h)$$

The apparent height h' of the fish is

$$h' = q_{\text{top}} - q_{\text{bottom}} = -0.752d - [-0.752(d + h)] = 0.752h$$

Hence, the fish appears to be approximately three-fourths its actual height.

36.4 Images Formed by Thin Lenses

Lenses are commonly used to form images by refraction in optical instruments such as cameras, telescopes, and microscopes. Let's use what we just learned about images formed by refracting surfaces to help locate the image formed by a lens. Light passing through a lens experiences refraction at two surfaces. The development we shall follow is based on the notion that the image formed by one refracting surface serves as the object for the second surface. We shall analyze a thick lens first and then let the thickness of the lens be approximately zero.

Consider a lens having an index of refraction n and two spherical surfaces with radii of curvature R_1 and R_2 as in Figure 36.21. (Notice that R_1 is the radius of curvature of the lens surface the light from the object reaches first and R_2 is the radius of curvature of the other surface of the lens.) An object is placed at point O at a distance p_1 in front of surface 1.

Let's begin with the image formed by surface 1. Using Equation 36.8 and assuming $n_1 = 1$ because the lens is surrounded by air, we find that the image I_1 formed by surface 1 satisfies the equation

$$\frac{1}{p_1} + \frac{n}{q_1} = \frac{n - 1}{R_1} \tag{36.10}$$

where q_1 is the position of the image formed by surface 1. If the image formed by surface 1 is virtual (Fig. 36.21a), then q_1 is negative; it is positive if the image is real (Fig. 36.21b).

Now let's apply Equation 36.8 to surface 2, taking $n_1 = n$ and $n_2 = 1$. (We make this switch in index because the light rays approaching surface 2 are *in the material of the lens,* and this material has index n.) Taking p_2 as the object distance for surface 2 and q_2 as the image distance gives

$$\frac{n}{p_2} + \frac{1}{q_2} = \frac{1 - n}{R_2} \tag{36.11}$$

We now introduce mathematically that the image formed by the first surface acts as the object for the second surface. If the image from surface 1 is virtual as in Figure 36.21a, we see that p_2, measured from surface 2, is related to q_1 as $p_2 = -q_1 + t$, where t is the thickness of the lens. Because q_1 is negative, p_2 is a positive number. Figure 36.21b shows the case of the image from surface 1 being real. In this situation, q_1 is positive and $p_2 = -q_1 + t$, where the image from surface 1 acts as a **virtual object,** so p_2 is negative. Regardless of the type of image from surface 1, the same equation describes the location of the object for surface 2 based on our sign

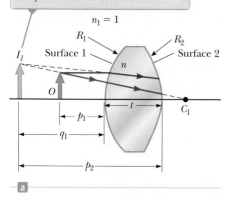

The image due to surface 1 is virtual, so I_1 is to the left of the surface.

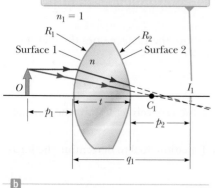

The image due to surface 1 is real, so I_1 is to the right of the surface.

Figure 36.21 To locate the image formed by a lens, we use the virtual image at I_1 formed by surface 1 as the object for the image formed by surface 2. The point C_1 is the center of curvature of surface 1.

convention. For a *thin* lens (one whose thickness is small compared with the radii of curvature), we can neglect t. In this approximation, $p_2 = -q_1$ for either type of image from surface 1. Hence, Equation 36.11 becomes

$$-\frac{n}{q_1} + \frac{1}{q_2} = \frac{1 - n}{R_2}$$ (36.12)

Adding Equations 36.10 and 36.12 gives

$$\frac{1}{p_1} + \frac{1}{q_2} = (n - 1)\left(\frac{1}{R_1} - \frac{1}{R_2}\right)$$ (36.13)

For a thin lens, we can omit the subscripts on p_1 and q_2 in Equation 36.13 and call the object distance p and the image distance q as in Figure 36.22. Hence, we can write Equation 36.13 as

$$\frac{1}{p} + \frac{1}{q} = (n - 1)\left(\frac{1}{R_1} - \frac{1}{R_2}\right)$$ (36.14)

This expression relates the image distance q of the image formed by a thin lens to the object distance p and to the lens properties (index of refraction and radii of curvature). It is valid only for paraxial rays and only when the lens thickness is much less than R_1 and R_2.

The **focal length** f of a thin lens is the image distance that corresponds to an infinite object distance, just as with mirrors. Letting p approach ∞ and q approach f in Equation 36.14, we see that the inverse of the focal length for a thin lens is

$$\boxed{\frac{1}{f} = (n - 1)\left(\frac{1}{R_1} - \frac{1}{R_2}\right)}$$ (36.15)

◀ Lens-makers' equation

This relationship is called the **lens-makers' equation** because it can be used to determine the values of R_1 and R_2 needed for a given index of refraction and a desired focal length f. Conversely, if the index of refraction and the radii of curvature of a lens are given, this equation can be used to find the focal length. If the lens is immersed in something other than air, this same equation can be used, with n interpreted as the *ratio* of the index of refraction of the lens material to that of the surrounding fluid.

Using Equation 36.15, we can write Equation 36.14 in a form identical to Equation 36.6 for mirrors:

$$\boxed{\frac{1}{p} + \frac{1}{q} = \frac{1}{f}}$$ (36.16)

This equation, called the **thin lens equation,** can be used to relate the image distance and object distance for a thin lens.

Because light can travel in either direction through a lens, each lens has two focal points, one for light rays passing through in one direction and one for rays passing through in the other direction. These two focal points are illustrated in Figure 36.23 for a plano-convex lens (a converging lens) and a plano-concave lens (a diverging lens).

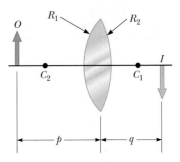

Figure 36.22 Simplified geometry for a thin lens.

Pitfall Prevention 36.5
A Lens Has Two Focal Points but Only One Focal Length A lens has a focal point on each side, front and back. There is only one focal length, however; each of the two focal points is located the same distance from the lens (Fig. 36.23). As a result, the lens forms an image of an object at the same point if it is turned around. In practice, that might not happen because real lenses are not infinitesimally thin.

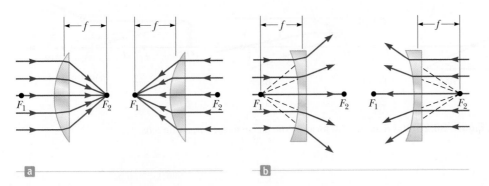

Figure 36.23 Parallel light rays pass through (a) a converging lens and (b) a diverging lens. The focal length is the same for light rays passing through a given lens in either direction. Both focal points F_1 and F_2 are the same distance from the lens.

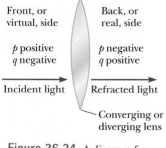

Front, or virtual, side   Back, or real, side

p positive
q negative

p negative
q positive

Incident light   Refracted light

Converging or diverging lens

Figure 36.24 A diagram for obtaining the signs of p and q for a thin lens. (This diagram also applies to a refracting surface.)

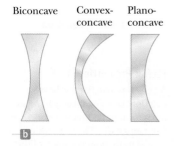

Biconvex   Convex-concave   Plano-convex

Ⓐ

Biconcave   Convex-concave   Plano-concave

Ⓑ

Figure 36.25 Various lens shapes. (a) Converging lenses have a positive focal length and are thickest at the middle. (b) Diverging lenses have a negative focal length and are thickest at the edges.

Table 36.3	**Sign Conventions for Thin Lenses**	
Quantity	**Positive When . . .**	**Negative When . . .**
Object location (p)	object is in front of lens (real object).	object is in back of lens (virtual object).
Image location (q)	image is in back of lens (real image).	image is in front of lens (virtual image).
Image height (h')	image is upright.	image is inverted.
R_1 and R_2	center of curvature is in back of lens.	center of curvature is in front of lens.
Focal length (f)	a converging lens.	a diverging lens.

Figure 36.24 is useful for obtaining the signs of p and q, and Table 36.3 gives the sign conventions for thin lenses. These sign conventions are the *same* as those for refracting surfaces (see Table 36.2).

Various lens shapes are shown in Figure 36.25. Notice that a converging lens is thicker at the center than at the edge, whereas a diverging lens is thinner at the center than at the edge.

Magnification of Images

Consider a thin lens through which light rays from an object pass. As with mirrors (Eq. 36.2), a geometric construction shows that the lateral magnification of the image is

$$M = \frac{h'}{h} = -\frac{q}{p} \tag{36.17}$$

From this expression, it follows that when M is positive, the image is upright and on the same side of the lens as the object. When M is negative, the image is inverted and on the side of the lens opposite the object.

Ray Diagrams for Thin Lenses

Ray diagrams are convenient for locating the images formed by thin lenses or systems of lenses. They also help clarify our sign conventions. Figure 36.26 shows such diagrams for three single-lens situations.

To locate the image of a *converging* lens (Figs. 36.26a and 36.26b), the following three rays are drawn from the top of the object:

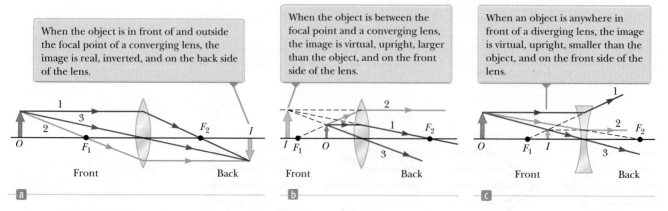

When the object is in front of and outside the focal point of a converging lens, the image is real, inverted, and on the back side of the lens.

When the object is between the focal point and a converging lens, the image is virtual, upright, larger than the object, and on the front side of the lens.

When an object is anywhere in front of a diverging lens, the image is virtual, upright, smaller than the object, and on the front side of the lens.

Figure 36.26 Ray diagrams for locating the image formed by a thin lens.

- Ray 1 is drawn parallel to the principal axis. After being refracted by the lens, this ray passes through the focal point on the back side of the lens.
- Ray 2 is drawn through the focal point on the front side of the lens (or as if coming from the focal point if $p < f$) and emerges from the lens parallel to the principal axis.
- Ray 3 is drawn through the center of the lens and continues in a straight line.

To locate the image of a *diverging* lens (Fig. 36.26c), the following three rays are drawn from the top of the object:

- Ray 1 is drawn parallel to the principal axis. After being refracted by the lens, this ray emerges directed away from the focal point on the front side of the lens.
- Ray 2 is drawn in the direction toward the focal point on the back side of the lens and emerges from the lens parallel to the principal axis.
- Ray 3 is drawn through the center of the lens and continues in a straight line.

For the converging lens in Figure 36.26a, where the object is to the left of the focal point ($p > f$), the image is real and inverted. When the object is between the focal point and the lens ($p < f$) as in Figure 36.26b, the image is virtual and upright. In that case, the lens acts as a magnifying glass, which we study in more detail in Section 36.8. For a diverging lens (Fig. 36.26c), the image is always virtual and upright, regardless of where the object is placed. These geometric constructions are reasonably accurate only if the distance between the rays and the principal axis is much less than the radii of the lens surfaces.

Refraction occurs only at the surfaces of the lens. A certain lens design takes advantage of this behavior to produce the *Fresnel lens*, a powerful lens without great thickness. Because only the surface curvature is important in the refracting qualities of the lens, material in the middle of a Fresnel lens is removed as shown in the cross sections of lenses in Figure 36.27. Because the edges of the curved segments cause some distortion, Fresnel lenses are generally used only in situations in which image quality is less important than reduction of weight. A classroom overhead projector often uses a Fresnel lens; the circular edges between segments of the lens can be seen by looking closely at the light projected onto a screen.

Quick Quiz 36.6 What is the focal length of a pane of window glass? **(a)** zero **(b)** infinity **(c)** the thickness of the glass **(d)** impossible to determine

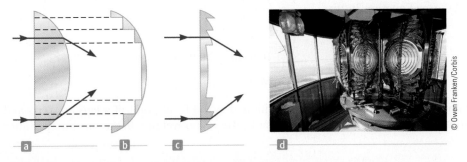

Figure 36.27 A side view of the construction of a Fresnel lens. (a) The thick lens refracts a light ray as shown. (b) Lens material in the bulk of the lens is cut away, leaving only the material close to the curved surface. (c) The small pieces of remaining material are moved to the left to form a flat surface on the left of the Fresnel lens with ridges on the right surface. From a front view, these ridges would be circular in shape. This new lens refracts light in the same way as the lens in (a). (d) A Fresnel lens used in a lighthouse shows several segments with the ridges discussed in (c).

Example 36.8   Images Formed by a Converging Lens

A converging lens has a focal length of 10.0 cm.

(A) An object is placed 30.0 cm from the lens. Construct a ray diagram, find the image distance, and describe the image.

SOLUTION

Conceptualize Because the lens is converging, the focal length is positive (see Table 36.3). We expect the possibilities of both real and virtual images.

Categorize Because the object distance is larger than the focal length, we expect the image to be real. The ray diagram for this situation is shown in Figure 36.28a.

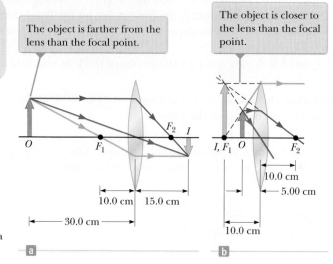

The object is farther from the lens than the focal point.

The object is closer to the lens than the focal point.

Figure 36.28
(Example 36.8) An image is formed by a converging lens.

Analyze Find the image distance by using Equation 36.16:

$$\frac{1}{q} = \frac{1}{f} - \frac{1}{p}$$

$$\frac{1}{q} = \frac{1}{10.0\ \text{cm}} - \frac{1}{30.0\ \text{cm}}$$

$$q = \boxed{+15.0\ \text{cm}}$$

Find the magnification of the image from Equation 36.17:

$$M = -\frac{q}{p} = -\frac{15.0\ \text{cm}}{30.0\ \text{cm}} = \boxed{-0.500}$$

Finalize The positive sign for the image distance tells us that the image is indeed real and on the back side of the lens. The magnification of the image tells us that the image is reduced in height by one half, and the negative sign for M tells us that the image is inverted.

(B) An object is placed 10.0 cm from the lens. Find the image distance and describe the image.

SOLUTION

Categorize Because the object is at the focal point, we expect the image to be infinitely far away.

Analyze Find the image distance by using Equation 36.16:

$$\frac{1}{q} = \frac{1}{f} - \frac{1}{p}$$

$$\frac{1}{q} = \frac{1}{10.0\ \text{cm}} - \frac{1}{10.0\ \text{cm}}$$

$$q = \boxed{\infty}$$

Finalize This result means that rays originating from an object positioned at the focal point of a lens are refracted so that the image is formed at an infinite distance from the lens; that is, the rays travel parallel to one another after refraction.

(C) An object is placed 5.00 cm from the lens. Construct a ray diagram, find the image distance, and describe the image.

SOLUTION

Categorize Because the object distance is smaller than the focal length, we expect the image to be virtual. The ray diagram for this situation is shown in Figure 36.28b.

▶ **36.8** continued

Analyze Find the image distance by using Equation 36.16:

$$\frac{1}{q} = \frac{1}{f} - \frac{1}{p}$$

$$\frac{1}{q} = \frac{1}{10.0 \text{ cm}} - \frac{1}{5.00 \text{ cm}}$$

$$q = \boxed{-10.0 \text{ cm}}$$

Find the magnification of the image from Equation 36.17:

$$M = -\frac{q}{p} = -\left(\frac{-10.0 \text{ cm}}{5.00 \text{ cm}}\right) = \boxed{+2.00}$$

Finalize The negative image distance tells us that the image is virtual and formed on the side of the lens from which the light is incident, the front side. The image is enlarged, and the positive sign for M tells us that the image is upright.

WHAT IF? What if the object moves right up to the lens surface so that $p \rightarrow 0$? Where is the image?

Answer In this case, because $p \ll R$, where R is either of the radii of the surfaces of the lens, the curvature of the lens can be ignored. The lens should appear to have the same effect as a flat piece of material, which suggests that the image is just on the front side of the lens, at $q = 0$. This conclusion can be verified mathematically by rearranging the thin lens equation:

$$\frac{1}{q} = \frac{1}{f} - \frac{1}{p}$$

If we let $p \rightarrow 0$, the second term on the right becomes very large compared with the first and we can neglect $1/f$. The equation becomes

$$\frac{1}{q} = -\frac{1}{p} \rightarrow q = -p = 0$$

Therefore, q is on the front side of the lens (because it has the opposite sign as p) and right at the lens surface.

Example 36.9 **Images Formed by a Diverging Lens**

A diverging lens has a focal length of 10.0 cm.

(A) An object is placed 30.0 cm from the lens. Construct a ray diagram, find the image distance, and describe the image.

SOLUTION

Conceptualize Because the lens is diverging, the focal length is negative (see Table 36.3). The ray diagram for this situation is shown in Figure 36.29a.

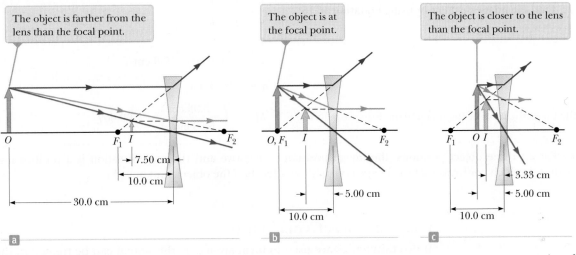

Figure 36.29 (Example 36.9) An image is formed by a diverging lens.

continued

▶ **36.9** continued

Categorize Because the lens is diverging, we expect it to form an upright, reduced, virtual image for any object position.

Analyze Find the image distance by using Equation 36.16:

$$\frac{1}{q} = \frac{1}{f} - \frac{1}{p}$$

$$\frac{1}{q} = \frac{1}{-10.0 \text{ cm}} - \frac{1}{30.0 \text{ cm}}$$

$$q = \boxed{-7.50 \text{ cm}}$$

Find the magnification of the image from Equation 36.17:

$$M = -\frac{q}{p} = -\left(\frac{-7.50 \text{ cm}}{30.0 \text{ cm}}\right) = \boxed{+0.250}$$

Finalize This result confirms that the image is virtual, smaller than the object, and upright. Look through the diverging lens in a door peephole to see this type of image.

(B) An object is placed 10.0 cm from the lens. Construct a ray diagram, find the image distance, and describe the image.

SOLUTION

The ray diagram for this situation is shown in Figure 36.29b.

Analyze Find the image distance by using Equation 36.16:

$$\frac{1}{q} = \frac{1}{f} - \frac{1}{p}$$

$$\frac{1}{q} = \frac{1}{-10.0 \text{ cm}} - \frac{1}{10.0 \text{ cm}}$$

$$q = \boxed{-5.00 \text{ cm}}$$

Find the magnification of the image from Equation 36.17:

$$M = -\frac{q}{p} = -\left(\frac{-5.00 \text{ cm}}{10.0 \text{ cm}}\right) = \boxed{+0.500}$$

Finalize Notice the difference between this situation and that for a converging lens. For a diverging lens, an object at the focal point does not produce an image infinitely far away.

(C) An object is placed 5.00 cm from the lens. Construct a ray diagram, find the image distance, and describe the image.

SOLUTION

The ray diagram for this situation is shown in Figure 36.29c.

Analyze Find the image distance by using Equation 36.16:

$$\frac{1}{q} = \frac{1}{f} - \frac{1}{p}$$

$$\frac{1}{q} = \frac{1}{-10.0 \text{ cm}} - \frac{1}{5.0 \text{ cm}}$$

$$q = \boxed{-3.33 \text{ cm}}$$

Find the magnification of the image from Equation 36.17:

$$M = -\left(\frac{-3.33 \text{ cm}}{5.00 \text{ cm}}\right) = \boxed{+0.667}$$

Finalize For all three object positions, the image position is negative and the magnification is a positive number smaller than 1, which confirms that the image is virtual, smaller than the object, and upright.

Combinations of Thin Lenses

If two thin lenses are used to form an image, the system can be treated in the following manner. First, the image formed by the first lens is located as if the second lens were not present. Then a ray diagram is drawn for the second lens, with the image formed by the first lens now serving as the object for the second lens. The second image formed is the final image of the system. If the image formed by the first lens lies on the back side of the second lens, that image is treated as a virtual object for the second lens (that is, in the thin lens equation, p is negative). The same procedure can be extended to a system of three or more lenses. Because the magnification due to the second lens is performed on the magnified image due to the first lens, the overall magnification of the image due to the combination of lenses is the product of the individual magnifications:

$$M = M_1 M_2 \tag{36.18}$$

This equation can be used for combinations of any optical elements such as a lens and a mirror. For more than two optical elements, the magnifications due to all elements are multiplied together.

Let's consider the special case of a system of two lenses of focal lengths f_1 and f_2 in contact with each other. If $p_1 = p$ is the object distance for the combination, application of the thin lens equation (Eq. 36.16) to the first lens gives

$$\frac{1}{p} + \frac{1}{q_1} = \frac{1}{f_1}$$

where q_1 is the image distance for the first lens. Treating this image as the object for the second lens, we see that the object distance for the second lens must be $p_2 = -q_1$. (The distances are the same because the lenses are in contact and assumed to be infinitesimally thin. The object distance is negative because the object is virtual if the image from the first lens is real.) Therefore, for the second lens,

$$\frac{1}{p_2} + \frac{1}{q_2} = \frac{1}{f_2} \quad \rightarrow \quad -\frac{1}{q_1} + \frac{1}{q} = \frac{1}{f_2}$$

where $q = q_2$ is the final image distance from the second lens, which is the image distance for the combination. Adding the equations for the two lenses eliminates q_1 and gives

$$\frac{1}{p} + \frac{1}{q} = \frac{1}{f_1} + \frac{1}{f_2}$$

If the combination is replaced with a single lens that forms an image at the same location, its focal length must be related to the individual focal lengths by the expression

$$\frac{1}{f} = \frac{1}{f_1} + \frac{1}{f_2} \tag{36.19}$$

◀ **Focal length for a combination of two thin lenses in contact**

Therefore, two thin lenses in contact with each other are equivalent to a single thin lens having a focal length given by Equation 36.19.

Example 36.10 Where Is the Final Image?

Two thin converging lenses of focal lengths $f_1 = 10.0$ cm and $f_2 = 20.0$ cm are separated by 20.0 cm as illustrated in Figure 36.30. An object is placed 30.0 cm to the left of lens 1. Find the position and the magnification of the final image.

SOLUTION

Conceptualize Imagine light rays passing through the first lens and forming a real image (because $p > f$) in the absence of a second lens. Figure 36.30 shows these light rays forming the inverted image I_1. Once the light rays converge to the image point, they do not stop. They continue through the image point and interact with the

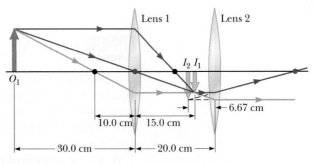

Figure 36.30 (Example 36.10) A combination of two converging lenses. The ray diagram shows the location of the final image (I_2) due to the combination of lenses. The black dots are the focal points of lens 1, and the red dots are the focal points of lens 2.

continued

▶ **36.10** continued

second lens. The rays leaving the image point behave in the same way as the rays leaving an object. Therefore, the image of the first lens serves as the object of the second lens.

Categorize We categorize this problem as one in which the thin lens equation is applied in a stepwise fashion to the two lenses.

Analyze Find the location of the image formed by lens 1 from the thin lens equation:

$$\frac{1}{q_1} = \frac{1}{f} - \frac{1}{p_1}$$

$$\frac{1}{q_1} = \frac{1}{10.0 \text{ cm}} - \frac{1}{30.0 \text{ cm}}$$

$$q_1 = +15.0 \text{ cm}$$

Find the magnification of the image from Equation 36.17:

$$M_1 = -\frac{q_1}{p_1} = -\frac{15.0 \text{ cm}}{30.0 \text{ cm}} = -0.500$$

The image formed by this lens acts as the object for the second lens. Therefore, the object distance for the second lens is $20.0 \text{ cm} - 15.0 \text{ cm} = 5.00 \text{ cm}$.

Find the location of the image formed by lens 2 from the thin lens equation:

$$\frac{1}{q_2} = \frac{1}{20.0 \text{ cm}} - \frac{1}{5.00 \text{ cm}}$$

$$q_2 = \boxed{-6.67 \text{ cm}}$$

Find the magnification of the image from Equation 36.17:

$$M_2 = -\frac{q_2}{p_2} = -\frac{(-6.67 \text{ cm})}{5.00 \text{ cm}} = +1.33$$

Find the overall magnification of the system from Equation 36.18:

$$M = M_1 M_2 = (-0.500)(1.33) = \boxed{-0.667}$$

Finalize The negative sign on the overall magnification indicates that the final image is inverted with respect to the initial object. Because the absolute value of the magnification is less than 1, the final image is smaller than the object.

Because q_2 is negative, the final image is on the front, or left, side of lens 2. These conclusions are consistent with the ray diagram in Figure 36.30.

WHAT IF? Suppose you want to create an upright image with this system of two lenses. How must the second lens be moved?

Answer Because the object is farther from the first lens than the focal length of that lens, the first image is inverted. Consequently, the second lens must invert the image once again so that the final image is upright. An inverted image is only formed by a converging lens if the object is outside the focal point. Therefore, the image formed by the first lens must be to the left of the focal point of the second lens in Figure 36.30. To make that happen, you must move the second lens at least as far away from the first lens as the sum $q_1 + f_2 = 15.0 \text{ cm} + 20.0 \text{ cm} = 35.0 \text{ cm}$.

36.5 Lens Aberrations

Our analysis of mirrors and lenses assumes rays make small angles with the principal axis and the lenses are thin. In this simple model, all rays leaving a point source focus at a single point, producing a sharp image. Clearly, that is not always true. When the approximations used in this analysis do not hold, imperfect images are formed.

A precise analysis of image formation requires tracing each ray, using Snell's law at each refracting surface and the law of reflection at each reflecting surface. This procedure shows that the rays from a point object do not focus at a single point, with the result that the image is blurred. The departures of actual images from the ideal predicted by our simplified model are called **aberrations.**

Spherical Aberration

Spherical aberration occurs because the focal points of rays far from the principal axis of a spherical lens (or mirror) are different from the focal points of rays of the same wavelength passing near the axis. Figure 36.31 illustrates spherical aberration for parallel rays passing through a converging lens. Rays passing through points near the center of the lens are imaged farther from the lens than rays passing through points near the edges. Figure 36.8 earlier in the chapter shows spherical aberration for light rays leaving a point object and striking a spherical mirror.

Many cameras have an adjustable aperture to control light intensity and reduce spherical aberration. (An aperture is an opening that controls the amount of light passing through the lens.) Sharper images are produced as the aperture size is reduced; with a small aperture, only the central portion of the lens is exposed to the light and therefore a greater percentage of the rays are paraxial. At the same time, however, less light passes through the lens. To compensate for this lower light intensity, a longer exposure time is used.

In the case of mirrors, spherical aberration can be minimized through the use of a parabolic reflecting surface rather than a spherical surface. Parabolic surfaces are not used often, however, because those with high-quality optics are very expensive to make. Parallel light rays incident on a parabolic surface focus at a common point, regardless of their distance from the principal axis. Parabolic reflecting surfaces are used in many astronomical telescopes to enhance image quality.

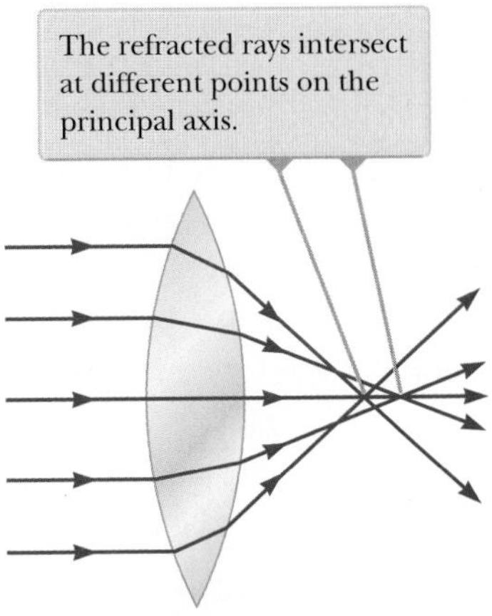

Chromatic Aberration

In Chapter 35, we described dispersion, whereby a material's index of refraction varies with wavelength. Because of this phenomenon, violet rays are refracted more than red rays when white light passes through a lens (Fig. 36.32). The figure shows that the focal length of a lens is greater for red light than for violet light. Other wavelengths (not shown in Fig. 36.32) have focal points intermediate between those of red and violet, which causes a blurred image and is called **chromatic aberration.**

Chromatic aberration for a diverging lens also results in a shorter focal length for violet light than for red light, but on the front side of the lens. Chromatic aberration can be greatly reduced by combining a converging lens made of one type of glass and a diverging lens made of another type of glass.

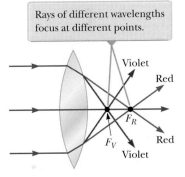

The refracted rays intersect at different points on the principal axis.

Figure 36.31 Spherical aberration caused by a converging lens. Does a diverging lens cause spherical aberration?

Rays of different wavelengths focus at different points.

Violet
Red
F_R
F_V
Red
Violet

Figure 36.32 Chromatic aberration caused by a converging lens.

36.6 The Camera

The photographic **camera** is a simple optical instrument whose essential features are shown in Figure 36.33. It consists of a light-tight chamber, a converging lens that produces a real image, and a light-sensitive component behind the lens on which the image is formed.

The image in a digital camera is formed on a *charge-coupled device* (CCD), which digitizes the image, turning it into binary code. (A CCD is described in Section 40.2.) The digital information is then stored on a memory chip for playback on the camera's display screen, or it can be downloaded to a computer. Film cameras are similar to digital cameras except that the light forms an image on light-sensitive film rather than on a CCD. The film must then be chemically processed to produce the image on paper. In the discussion that follows, we assume the camera is digital.

A camera is focused by varying the distance between the lens and the CCD. For proper focusing—which is necessary for the formation of sharp images—the lens-to-CCD distance depends on the object distance as well as the focal length of the lens.

The shutter, positioned behind the lens, is a mechanical device that is opened for selected time intervals, called *exposure times*. You can photograph moving objects by using short exposure times or photograph dark scenes (with low light levels) by using long exposure times. If this adjustment were not available, it would be impossible

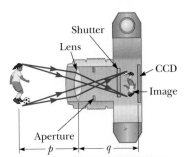

Shutter
Lens
CCD
Image
Aperture
p
q

Figure 36.33 Cross-sectional view of a simple digital camera. The CCD is the light-sensitive component of the camera. In a nondigital camera, the light from the lens falls onto photographic film. In reality, $p \gg q$.

to take stop-action photographs. For example, a rapidly moving vehicle could move enough in the time interval during which the shutter is open to produce a blurred image. Another major cause of blurred images is the movement of the camera while the shutter is open. To prevent such movement, either short exposure times or a tripod should be used, even for stationary objects. Typical shutter speeds (that is, exposure times) are $\frac{1}{30}$ s, $\frac{1}{60}$ s, $\frac{1}{125}$ s, and $\frac{1}{250}$ s. In practice, stationary objects are normally shot with an intermediate shutter speed of $\frac{1}{60}$ s.

The intensity I of the light reaching the CCD is proportional to the area of the lens. Because this area is proportional to the square of the diameter D, it follows that I is also proportional to D^2. Light intensity is a measure of the rate at which energy is received by the CCD per unit area of the image. Because the area of the image is proportional to q^2 and $q \approx f$ (when $p \gg f$, so p can be approximated as infinite), we conclude that the intensity is also proportional to $1/f^2$ and therefore that $I \propto D^2/f^2$.

The ratio f/D is called the **f-number** of a lens:

$$f\text{-number} \equiv \frac{f}{D} \tag{36.20}$$

Hence, the intensity of light incident on the CCD varies according to the following proportionality:

$$I \propto \frac{1}{(f/D)^2} \propto \frac{1}{(f\text{-number})^2} \tag{36.21}$$

The f-number is often given as a description of the lens's "speed." The lower the f-number, the wider the aperture and the higher the rate at which energy from the light exposes the CCD; therefore, a lens with a low f-number is a "fast" lens. The conventional notation for an f-number is "$f/$" followed by the actual number. For example, "$f/4$" means an f-number of 4; it *does not* mean to divide f by 4! Extremely fast lenses, which have f-numbers as low as approximately $f/1.2$, are expensive because it is very difficult to keep aberrations acceptably small with light rays passing through a large area of the lens. Camera lens systems (that is, combinations of lenses with adjustable apertures) are often marked with multiple f-numbers, usually $f/2.8$, $f/4$, $f/5.6$, $f/8$, $f/11$, and $f/16$. Any one of these settings can be selected by adjusting the aperture, which changes the value of D. Increasing the setting from one f-number to the next higher value (for example, from $f/2.8$ to $f/4$) decreases the area of the aperture by a factor of 2. The lowest f-number setting on a camera lens corresponds to a wide-open aperture and the use of the maximum possible lens area.

Simple cameras usually have a fixed focal length and a fixed aperture size, with an f-number of about $f/11$. This high value for the f-number allows for a large **depth of field,** meaning that objects at a wide range of distances from the lens form reasonably sharp images on the CCD. In other words, the camera does not have to be focused.

Quick Quiz 36.7 A camera can be modeled as a simple converging lens that focuses an image on the CCD, acting as the screen. A camera is initially focused on a distant object. To focus the image of an object close to the camera, must the lens be **(a)** moved away from the CCD, **(b)** left where it is, or **(c)** moved toward the CCD?

Example 36.11   **Finding the Correct Exposure Time**

The lens of a digital camera has a focal length of 55 mm and a speed (an f-number) of $f/1.8$. The correct exposure time for this speed under certain conditions is known to be $\frac{1}{500}$ s.

(A) Determine the diameter of the lens.

SOLUTION

Conceptualize Remember that the f-number for a lens relates its focal length to its diameter.

▶ **36.11** continued

Categorize We determine results using equations developed in this section, so we categorize this example as a substitution problem.

Solve Equation 36.20 for D and substitute numerical values:

$$D = \frac{f}{f\text{-number}} = \frac{55 \text{ mm}}{1.8} = \boxed{31 \text{ mm}}$$

(B) Calculate the correct exposure time if the f-number is changed to $f/4$ under the same lighting conditions.

SOLUTION

The total light energy hitting the CCD is proportional to the product of the intensity and the exposure time. If I is the light intensity reaching the CCD, the energy per unit area received by the CCD in a time interval Δt is proportional to $I \Delta t$. Comparing the two situations, we require that $I_1 \Delta t_1 = I_2 \Delta t_2$, where Δt_1 is the correct exposure time for $f/1.8$ and Δt_2 is the correct exposure time for $f/4$.

Use this result and substitute for I from Equation 36.21:

$$I_1 \Delta t_1 = I_2 \Delta t_2 \;\; \rightarrow \;\; \frac{\Delta t_1}{(f_1\text{-number})^2} = \frac{\Delta t_2}{(f_2\text{-number})^2}$$

Solve for Δt_2 and substitute numerical values:

$$\Delta t_2 = \left(\frac{f_2\text{-number}}{f_1\text{-number}}\right)^2 \Delta t_1 = \left(\frac{4}{1.8}\right)^2 \left(\tfrac{1}{500} \text{ s}\right) \approx \boxed{\tfrac{1}{100} \text{ s}}$$

As the aperture size is reduced, the exposure time must increase.

36.7 The Eye

Like a camera, a normal eye focuses light and produces a sharp image. The mechanisms by which the eye controls the amount of light admitted and adjusts to produce correctly focused images, however, are far more complex, intricate, and effective than those in even the most sophisticated camera. In all respects, the eye is a physiological wonder.

Figure 36.34 shows the basic parts of the human eye. Light entering the eye passes through a transparent structure called the *cornea* (Fig. 36.35), behind which are a clear liquid (the *aqueous humor*), a variable aperture (the *pupil*, which is an opening in the *iris*), and the *crystalline lens*. Most of the refraction occurs at the outer surface of the eye, where the cornea is covered with a film of tears. Relatively little refraction occurs in the crystalline lens because the aqueous humor in contact with the lens has an average index of refraction close to that of the lens. The iris, which is the colored portion of the eye, is a muscular diaphragm that controls pupil size. The iris regulates the amount of light entering the eye by dilating, or

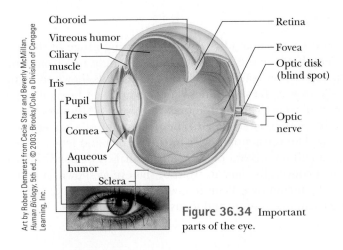

Figure **36.34** Important parts of the eye.

Art by Robert Demarest from Cecie Starr and Beverly McMillan, *Human Biology*, 5th ed., © 2003, Brooks/Cole, a Division of Cengage Learning, Inc.

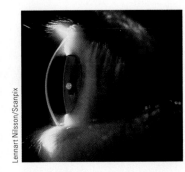

Lennart Nilsson/Scanpix

Figure **36.35** Close-up photograph of the cornea of the human eye.

opening, the pupil in low-light conditions and contracting, or closing, the pupil in high-light conditions. The *f*-number range of the human eye is approximately *f*/2.8 to *f*/16.

The cornea–lens system focuses light onto the back surface of the eye, the *retina*, which consists of millions of sensitive receptors called *rods* and *cones*. When stimulated by light, these receptors send impulses via the optic nerve to the brain, where an image is perceived. By this process, a distinct image of an object is observed when the image falls on the retina.

The eye focuses on an object by varying the shape of the pliable crystalline lens through a process called **accommodation.** The lens adjustments take place so swiftly that we are not even aware of the change. Accommodation is limited in that objects very close to the eye produce blurred images. The **near point** is the closest distance for which the lens can accommodate to focus light on the retina. This distance usually increases with age and has an average value of 25 cm. At age 10, the near point of the eye is typically approximately 18 cm. It increases to approximately 25 cm at age 20, to 50 cm at age 40, and to 500 cm or greater at age 60. The **far point** of the eye represents the greatest distance for which the lens of the relaxed eye can focus light on the retina. A person with normal vision can see very distant objects and therefore has a far point that can be approximated as infinity.

The retina is covered with two types of light-sensitive cells, called **rods** and **cones.** The rods are not sensitive to color but are more light sensitive than the cones. The rods are responsible for *scotopic vision,* or dark-adapted vision. Rods are spread throughout the retina and allow good peripheral vision for all light levels and motion detection in the dark. The cones are concentrated in the fovea. These cells are sensitive to different wavelengths of light. The three categories of these cells are called red, green, and blue cones because of the peaks of the color ranges to which they respond (Fig. 36.36). If the red and green cones are stimulated simultaneously (as would be the case if yellow light were shining on them), the brain interprets what is seen as yellow. If all three types of cones are stimulated by the separate colors red, blue, and green, white light is seen. If all three types of cones are stimulated by light that contains *all* colors, such as sunlight, again white light is seen.

Televisions and computer monitors take advantage of this visual illusion by having only red, green, and blue dots on the screen. With specific combinations of brightness in these three primary colors, our eyes can be made to see any color in the rainbow. Therefore, the yellow lemon you see in a television commercial is not actually yellow, it is red and green! The paper on which this page is printed is made of tiny, matted, translucent fibers that scatter light in all directions, and the resultant mixture of colors appears white to the eye. Snow, clouds, and white hair are not actually white. In fact, there is no such thing as a white pigment. The appearance of these things is a consequence of the scattering of light containing all colors, which we interpret as white.

Conditions of the Eye

When the eye suffers a mismatch between the focusing range of the lens–cornea system and the length of the eye, with the result that light rays from a near object reach the retina before they converge to form an image as shown in Figure 36.37a, the condition is known as **farsightedness** (or *hyperopia*). A farsighted person can usually see faraway objects clearly but not nearby objects. Although the near point of a normal eye is approximately 25 cm, the near point of a farsighted person is much farther away. The refracting power in the cornea and lens is insufficient to focus the light from all but distant objects satisfactorily. The condition can be corrected by placing a converging lens in front of the eye as shown in Figure 36.37b. The lens refracts the incoming rays more toward the principal axis before entering the eye, allowing them to converge and focus on the retina.

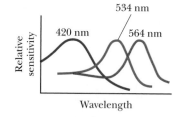

Figure 36.36 Approximate color sensitivity of the three types of cones in the retina.

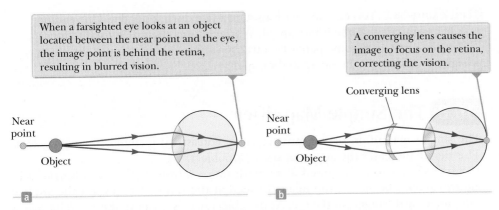

Figure 36.37 (a) An uncorrected farsighted eye. (b) A farsighted eye corrected with a converging lens.

A person with **nearsightedness** (or *myopia*), another mismatch condition, can focus on nearby objects but not on faraway objects. The far point of the nearsighted eye is not infinity and may be less than 1 m. The maximum focal length of the nearsighted eye is insufficient to produce a sharp image on the retina, and rays from a distant object converge to a focus in front of the retina. They then continue past that point, diverging before they finally reach the retina and causing blurred vision (Fig. 36.38a). Nearsightedness can be corrected with a diverging lens as shown in Figure 36.38b. The lens refracts the rays away from the principal axis before they enter the eye, allowing them to focus on the retina.

Beginning in middle age, most people lose some of their accommodation ability as their visual muscles weaken and the lens hardens. Unlike farsightedness, which is a mismatch between focusing power and eye length, **presbyopia** (literally, "old-age vision") is due to a reduction in accommodation ability. The cornea and lens do not have sufficient focusing power to bring nearby objects into focus on the retina. The symptoms are the same as those of farsightedness, and the condition can be corrected with converging lenses.

In eyes having a defect known as **astigmatism**, light from a point source produces a line image on the retina. This condition arises when the cornea, the lens, or both are not perfectly symmetric. Astigmatism can be corrected with lenses that have different curvatures in two mutually perpendicular directions.

Optometrists and ophthalmologists usually prescribe lenses[1] measured in **diopters:** the **power** P of a lens in diopters equals the inverse of the focal length in meters: $P = 1/f$. For example, a converging lens of focal length +20 cm has a power of +5.0 diopters, and a diverging lens of focal length −40 cm has a power of −2.5 diopters.

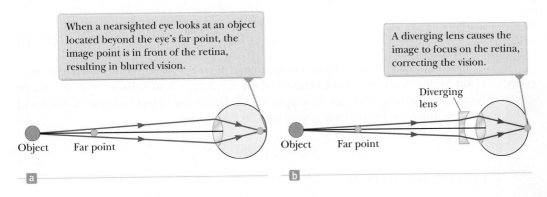

Figure 36.38 (a) An uncorrected nearsighted eye. (b) A nearsighted eye corrected with a diverging lens.

---

[1]The word *lens* comes from *lentil*, the name of an Italian legume. (You may have eaten lentil soup.) Early eyeglasses were called "glass lentils" because the biconvex shape of their lenses resembled the shape of a lentil. The first lenses for farsightedness and presbyopia appeared around 1280; concave eyeglasses for correcting nearsightedness did not appear until more than 100 years later.

Quick Quiz 36.8 Two campers wish to start a fire during the day. One camper is nearsighted, and one is farsighted. Whose glasses should be used to focus the Sun's rays onto some paper to start the fire? **(a)** either camper **(b)** the near-sighted camper **(c)** the farsighted camper

36.8 The Simple Magnifier

The simple magnifier, or magnifying glass, consists of a single converging lens. This device increases the apparent size of an object.

Suppose an object is viewed at some distance p from the eye as illustrated in Figure 36.39. The size of the image formed at the retina depends on the angle θ subtended by the object at the eye. As the object moves closer to the eye, θ increases and a larger image is observed. An average normal human eye, however, cannot focus on an object closer than about 25 cm, the near point (Fig. 36.40a). Therefore, θ is maximum at the near point.

To further increase the apparent angular size of an object, a converging lens can be placed in front of the eye as in Figure 36.40b, with the object located at point O, immediately inside the focal point of the lens. At this location, the lens forms a virtual, upright, enlarged image. We define **angular magnification** m as the ratio of the angle subtended by an object with a lens in use (angle θ in Fig. 36.40b) to the angle subtended by the object placed at the near point with no lens in use (angle θ_0 in Fig. 36.40a):

$$m \equiv \frac{\theta}{\theta_0} \tag{36.22}$$

The angular magnification is a maximum when the image is at the near point of the eye, that is, when $q = -25$ cm. The object distance corresponding to this image distance can be calculated from the thin lens equation:

$$\frac{1}{p} + \frac{1}{-25 \text{ cm}} = \frac{1}{f} \quad \rightarrow \quad p = \frac{25f}{25 + f}$$

where f is the focal length of the magnifier in centimeters. If we make the small-angle approximations

$$\tan \theta_0 \approx \theta_0 \approx \frac{h}{25} \quad \text{and} \quad \tan \theta \approx \theta \approx \frac{h}{p} \tag{36.23}$$

Equation 36.22 becomes

$$m_{\text{max}} = \frac{\theta}{\theta_0} = \frac{h/p}{h/25} = \frac{25}{p} = \frac{25}{25f/(25 + f)}$$

$$m_{\text{max}} = 1 + \frac{25 \text{ cm}}{f} \tag{36.24}$$

Although the eye can focus on an image formed anywhere between the near point and infinity, it is most relaxed when the image is at infinity. For the image formed by the magnifying lens to appear at infinity, the object has to be at the focal point of the lens. In this case, Equations 36.23 become

$$\theta_0 \approx \frac{h}{25} \quad \text{and} \quad \theta \approx \frac{h}{f}$$

and the magnification is

$$m_{\text{min}} = \frac{\theta}{\theta_0} = \frac{25 \text{ cm}}{f} \tag{36.25}$$

With a single lens, it is possible to obtain angular magnifications up to about 4 without serious aberrations. Magnifications up to about 20 can be achieved by using one or two additional lenses to correct for aberrations.

The size of the image formed on the retina depends on the angle θ subtended at the eye.

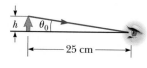

Figure 36.39 An observer looks at an object at distance p.

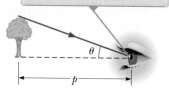

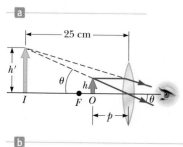

Figure 36.40 (a) An object placed at the near point of the eye ($p = 25$ cm) subtends an angle $\theta_0 \approx h/25$ cm at the eye. (b) An object placed near the focal point of a converging lens produces a magnified image that subtends an angle $\theta \approx h'/25$ cm at the eye.

© Cengage Learning/George Semple

A simple magnifier, also called a magnifying glass, is used to view an enlarged image of a portion of a map.

Magnification of a Lens

What is the maximum magnification that is possible with a lens having a focal length of 10 cm, and what is the magnification of this lens when the eye is relaxed?

Conceptualize Study Figure 36.40b for the situation in which a magnifying glass forms an enlarged image of an object placed inside the focal point. The maximum magnification occurs when the image is located at the near point of the eye. When the eye is relaxed, the image is at infinity.

Categorize We determine results using equations developed in this section, so we categorize this example as a substitution problem.

Evaluate the maximum magnification from Equation 36.24:

$$m_{max} = 1 + \frac{25 \text{ cm}}{f} = 1 + \frac{25 \text{ cm}}{10 \text{ cm}} = \boxed{3.5}$$

Evaluate the minimum magnification, when the eye is relaxed, from Equation 36.25:

$$m_{min} = \frac{25 \text{ cm}}{f} = \frac{25 \text{ cm}}{10 \text{ cm}} = \boxed{2.5}$$

36.9 The Compound Microscope

A simple magnifier provides only limited assistance in inspecting minute details of an object. Greater magnification can be achieved by combining two lenses in a device called a **compound microscope** shown in Figure 36.41a. It consists of one lens, the *objective*, that has a very short focal length $f_o < 1$ cm and a second lens, the *eyepiece*, that has a focal length f_e of a few centimeters. The two lenses are separated by a distance L that is much greater than either f_o or f_e. The object, which is placed just outside the focal point of the objective, forms a real, inverted image at I_1, and this image is located at or close to the focal point of the eyepiece. The eyepiece, which serves as a simple magnifier, produces at I_2 a virtual, enlarged image of I_1. The lateral magnification M_1 of the first image is $-q_1/p_1$. Notice from Figure 36.41a that q_1 is approximately equal to L and that the object is very close

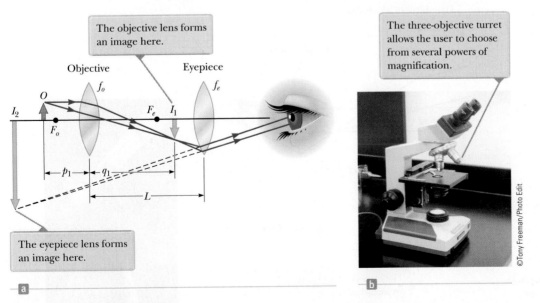

The objective lens forms an image here.

The three-objective turret allows the user to choose from several powers of magnification.

Objective   Eyepiece

f_o f_e

O

I_2

F_o F_e I_1

p_1 q_1

L

The eyepiece lens forms an image here.

©Tony Freeman/Photo Edit

a   b

Figure 36.41 (a) Diagram of a compound microscope, which consists of an objective lens and an eyepiece lens. (b) A compound microscope.

to the focal point of the objective: $p_1 \approx f_o$. Therefore, the lateral magnification by the objective is

$$M_o \approx -\frac{L}{f_o}$$

The angular magnification by the eyepiece for an object (corresponding to the image at I_1) placed at the focal point of the eyepiece is, from Equation 36.25,

$$m_e = \frac{25 \text{ cm}}{f_e}$$

The overall magnification of the image formed by a compound microscope is defined as the product of the lateral and angular magnifications:

$$M = M_o m_e = -\frac{L}{f_o}\left(\frac{25 \text{ cm}}{f_e}\right) \qquad \text{(36.26)}$$

The negative sign indicates that the image is inverted.

The microscope has extended human vision to the point where we can view previously unknown details of incredibly small objects. The capabilities of this instrument have steadily increased with improved techniques for precision grinding of lenses. A question often asked about microscopes is, "If one were extremely patient and careful, would it be possible to construct a microscope that would enable the human eye to see an atom?" The answer is no, as long as light is used to illuminate the object. For an object under an optical microscope (one that uses visible light) to be seen, the object must be at least as large as a wavelength of light. Because the diameter of any atom is many times smaller than the wavelengths of visible light, the mysteries of the atom must be probed using other types of "microscopes."

36.10 The Telescope

Two fundamentally different types of **telescopes** exist; both are designed to aid in viewing distant objects such as the planets in our solar system. The first type, the **refracting telescope,** uses a combination of lenses to form an image.

Like the compound microscope, the refracting telescope shown in Figure 36.42a has an objective and an eyepiece. The two lenses are arranged so that the objective

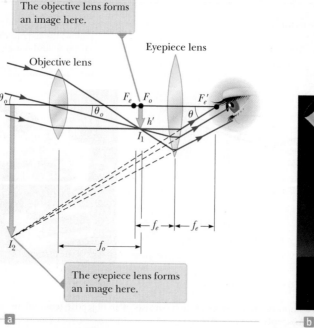

The objective lens forms an image here.

Eyepiece lens

Objective lens

The eyepiece lens forms an image here.

Figure 36.42 (a) Lens arrangement in a refracting telescope, with the object at infinity. (b) A refracting telescope.

©Tony Freeman/Photo Edit

forms a real, inverted image of a distant object very near the focal point of the eyepiece. Because the object is essentially at infinity, this point at which I_1 forms is the focal point of the objective. The eyepiece then forms, at I_2, an enlarged, inverted image of the image at I_1. To provide the largest possible magnification, the image distance for the eyepiece is infinite. Therefore, the image due to the objective lens, which acts as the object for the eyepiece lens, must be located at the focal point of the eyepiece. Hence, the two lenses are separated by a distance $f_o + f_e$, which corresponds to the length of the telescope tube.

The angular magnification of the telescope is given by θ/θ_o, where θ_o is the angle subtended by the object at the objective and θ is the angle subtended by the final image at the viewer's eye. Consider Figure 36.42a, in which the object is a very great distance to the left of the figure. The angle θ_o (to the *left* of the objective) subtended by the object at the objective is the same as the angle (to the *right* of the objective) subtended by the first image at the objective. Therefore,

$$\tan \theta_o \approx \theta_o \approx -\frac{h'}{f_o}$$

where the negative sign indicates that the image is inverted.

The angle θ subtended by the final image at the eye is the same as the angle that a ray coming from the tip of I_1 and traveling parallel to the principal axis makes with the principal axis after it passes through the lens. Therefore,

$$\tan \theta \approx \theta \approx \frac{h'}{f_e}$$

We have not used a negative sign in this equation because the final image is not inverted; the object creating this final image I_2 is I_1, and both it and I_2 point in the same direction. Therefore, the angular magnification of the telescope can be expressed as

$$m = \frac{\theta}{\theta_o} = \frac{h'/f_e}{-h'/f_o} = -\frac{f_o}{f_e} \tag{36.27}$$

This result shows that the angular magnification of a telescope equals the ratio of the objective focal length to the eyepiece focal length. The negative sign indicates that the image is inverted.

When you look through a telescope at such relatively nearby objects as the Moon and the planets, magnification is important. Individual stars in our galaxy, however, are so far away that they always appear as small points of light no matter how great the magnification. To gather as much light as possible, large research telescopes used to study very distant objects must have a large diameter. It is difficult and expensive to manufacture large lenses for refracting telescopes. Another difficulty with large lenses is that their weight leads to sagging, which is an additional source of aberration.

These problems associated with large lenses can be partially overcome by replacing the objective with a concave mirror, which results in the second type of telescope, the **reflecting telescope**. Because light is reflected from the mirror and does not pass through a lens, the mirror can have rigid supports on the back side. Such supports eliminate the problem of sagging.

Figure 36.43a shows the design for a typical reflecting telescope. The incoming light rays are reflected by a parabolic mirror at the base. These reflected rays converge toward point A in the figure, where an image would be formed. Before this image is formed, however, a small, flat mirror M reflects the light toward an opening in the tube's side and it passes into an eyepiece. This particular design is said to have a Newtonian focus because Newton developed it. Figure 36.43b shows such a telescope. Notice that the light never passes through glass (except through the small eyepiece) in the reflecting telescope. As a result, problems associated with chromatic aberration are virtually eliminated. The reflecting telescope can be made even shorter by orienting the flat mirror so that it reflects the light back

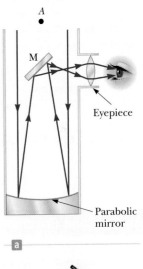

a

Orion® Sky View Pro

b

Figure 36.43 (a) A Newtonian-focus reflecting telescope. (b) A reflecting telescope. This type of telescope is shorter than that in Figure 36.42b.

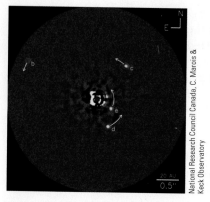

National Research Council Canada, C. Marois & Keck Observatory

Figure 36.44 A direct optical image of a solar system around the star HR8799, developed at the Keck Observatory in Hawaii.

toward the objective mirror and the light enters an eyepiece in a hole in the middle of the mirror.

The largest reflecting telescopes in the world are at the Keck Observatory on Mauna Kea, Hawaii. The site includes two telescopes with diameters of 10 m, each containing 36 hexagonally shaped, computer-controlled mirrors that work together to form a large reflecting surface. In addition, the two telescopes can work together to provide a telescope with an effective diameter of 85 m. In contrast, the largest refracting telescope in the world, at the Yerkes Observatory in Williams Bay, Wisconsin, has a diameter of only 1 m.

Figure 36.44 shows a remarkable optical image from the Keck Observatory of a solar system around the star HR8799, located 129 light-years from the Earth. The planets labeled b, c, and d were seen in 2008 and the innermost planet, labeled e, was observed in December 2010. This photograph represents the first direct image of another solar system and was made possible by the adaptive optics technology used in the Keck Observatory.

Summary

Definitions

The **lateral magnification** M of the image due to a mirror or lens is defined as the ratio of the image height h' to the object height h. It is equal to the negative of the ratio of the image distance q to the object distance p:

$$M \equiv \frac{\text{image height}}{\text{object height}} = \frac{h'}{h} = -\frac{q}{p} \quad \text{(36.1, 36.2, 36.17)}$$

The **angular magnification** m is the ratio of the angle subtended by an object with a lens in use (angle θ in Fig. 36.40b) to the angle subtended by the object placed at the near point with no lens in use (angle θ_0 in Fig. 36.40a):

$$m \equiv \frac{\theta}{\theta_0} \quad \text{(36.22)}$$

The ratio of the focal length of a camera lens to the diameter of the lens is called the *f*-**number** of the lens:

$$f\text{-number} \equiv \frac{f}{D} \quad \text{(36.20)}$$

Concepts and Principles

In the paraxial ray approximation, the object distance p and image distance q for a spherical mirror of radius R are related by the **mirror equation:**

$$\frac{1}{p} + \frac{1}{q} = \frac{2}{R} = \frac{1}{f} \quad \text{(36.4, 36.6)}$$

where $f = R/2$ is the **focal length** of the mirror.

An image can be formed by refraction from a spherical surface of radius R. The object and image distances for refraction from such a surface are related by

$$\frac{n_1}{p} + \frac{n_2}{q} = \frac{n_2 - n_1}{R} \quad \text{(36.8)}$$

where the light is incident in the medium for which the index of refraction is n_1 and is refracted in the medium for which the index of refraction is n_2.

The inverse of the **focal length** f of a thin lens surrounded by air is given by the **lens-makers' equation:**

$$\frac{1}{f} = (n - 1)\left(\frac{1}{R_1} - \frac{1}{R_2}\right) \quad \text{(36.15)}$$

Converging lenses have positive focal lengths, and **diverging lenses** have negative focal lengths.

For a thin lens, and in the paraxial ray approximation, the object and image distances are related by the **thin lens equation:**

$$\frac{1}{p} + \frac{1}{q} = \frac{1}{f} \quad \text{(36.16)}$$

The maximum magnification of a single lens of focal length f used as a simple magnifier is

$$m_{max} = 1 + \frac{25 \text{ cm}}{f} \qquad \textbf{(36.24)}$$

The overall magnification of the image formed by a compound microscope is

$$M = -\frac{L}{f_o}\left(\frac{25 \text{ cm}}{f_e}\right) \qquad \textbf{(36.26)}$$

where f_o and f_e are the focal lengths of the objective and eyepiece lenses, respectively, and L is the distance between the lenses.

The angular magnification of a refracting telescope can be expressed as

$$m = -\frac{f_o}{f_e} \qquad \textbf{(36.27)}$$

where f_o and f_e are the focal lengths of the objective and eyepiece lenses, respectively. The angular magnification of a reflecting telescope is given by the same expression where f_o is the focal length of the objective mirror.

Objective Questions   **1.** denotes answer available in *Student Solutions Manual/Study Guide*

1. The faceplate of a diving mask can be ground into a corrective lens for a diver who does not have perfect vision. The proper design allows the person to see clearly both under water and in the air. Normal eyeglasses have lenses with both the front and back surfaces curved. Should the lenses of a diving mask be curved (a) on the outer surface only, (b) on the inner surface only, or (c) on both surfaces?

2. Lulu looks at her image in a makeup mirror. It is enlarged when she is close to the mirror. As she backs away, the image becomes larger, then impossible to identify when she is 30.0 cm from the mirror, then upside down when she is beyond 30.0 cm, and finally small, clear, and upside down when she is much farther from the mirror. (i) Is the mirror (a) convex, (b) plane, or (c) concave? (ii) Is the magnitude of its focal length (a) 0, (b) 15.0 cm, (c) 30.0 cm, (d) 60.0 cm, or (e) ∞?

3. An object is located 50.0 cm from a converging lens having a focal length of 15.0 cm. Which of the following statements is true regarding the image formed by the lens? (a) It is virtual, upright, and larger than the object. (b) It is real, inverted, and smaller than the object. (c) It is virtual, inverted, and smaller than the object. (d) It is real, inverted, and larger than the object. (e) It is real, upright, and larger than the object.

4. **(i)** When an image of an object is formed by a converging lens, which of the following statements is *always* true? More than one statement may be correct. (a) The image is virtual. (b) The image is real. (c) The image is upright. (d) The image is inverted. (e) None of those statements is always true. **(ii)** When the image of an object is formed by a diverging lens, which of the statements is *always* true?

5. A converging lens in a vertical plane receives light from an object and forms an inverted image on a screen. An opaque card is then placed next to the lens, covering only the upper half of the lens. What happens to the image on the screen? (a) The upper half of the image disappears. (b) The lower half of the image disappears. (c) The entire image disappears. (d) The entire image is still visible, but is dimmer. (e) No change in the image occurs.

6. If Josh's face is 30.0 cm in front of a concave shaving mirror creating an upright image 1.50 times as large as the object, what is the mirror's focal length? (a) 12.0 cm (b) 20.0 cm (c) 70.0 cm (d) 90.0 cm (e) none of those answers

7. Two thin lenses of focal lengths $f_1 = 15.0$ and $f_2 = 10.0$ cm, respectively, are separated by 35.0 cm along a common axis. The f_1 lens is located to the left of the f_2 lens. An object is now placed 50.0 cm to the left of the f_1 lens, and a final image due to light passing though both lenses forms. By what factor is the final image different in size from the object? (a) 0.600 (b) 1.20 (c) 2.40 (d) 3.60 (e) none of those answers

8. If you increase the aperture diameter of a camera by a factor of 3, how is the intensity of the light striking the film affected? (a) It increases by factor of 3. (b) It decreases by a factor of 3. (c) It increases by a factor of 9. (d) It decreases by a factor of 9. (e) Increasing the aperture size doesn't affect the intensity.

9. A person spearfishing from a boat sees a stationary fish a few meters away in a direction about 30° below the horizontal. To spear the fish, and assuming the spear does not change direction when it enters the water, should the person (a) aim above where he sees the fish, (b) aim below the fish, or (c) aim precisely at the fish?

10. Model each of the following devices in use as consisting of a single converging lens. Rank the cases according to the ratio of the distance from the object to the lens to the focal length of the lens, from the largest ratio to the smallest. (a) a film-based movie projector showing a movie (b) a magnifying glass being used to examine a postage stamp (c) an astronomical refracting telescope being used to make a sharp image of stars on an electronic detector (d) a searchlight being used to produce a beam of parallel rays from a point source (e) a camera lens being used to photograph a soccer game

11. A converging lens made of crown glass has a focal length of 15.0 cm when used in air. If the lens is immersed in water, what is its focal length? (a) negative

(b) less than 15.0 cm (c) equal to 15.0 cm (d) greater than 15.0 cm (e) none of those answers

12. A converging lens of focal length 8 cm forms a sharp image of an object on a screen. What is the smallest possible distance between the object and the screen? (a) 0 (b) 4 cm (c) 8 cm (d) 16 cm (e) 32 cm

13. (i) When an image of an object is formed by a plane mirror, which of the following statements is *always* true? More than one statement may be correct. (a) The image is virtual. (b) The image is real. (c) The image is upright. (d) The image is inverted. (e) None of those statements is always true. **(ii)** When the image of an object is formed by a concave mirror, which of the preceding statements are *always* true? **(iii)** When the image of an object is formed by a convex mirror, which of the preceding statements are *always* true?

14. An object, represented by a gray arrow, is placed in front of a plane mirror. Which of the diagrams in Figure OQ36.14 correctly describes the image, represented by the pink arrow?

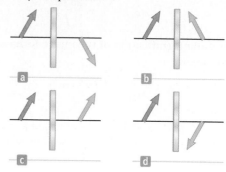

Figure OQ36.14

Conceptual Questions

1. denotes answer available in *Student Solutions Manual/Study Guide*

1. A converging lens of short focal length can take light diverging from a small source and refract it into a beam of parallel rays. A Fresnel lens as shown in Figure 36.27 is used in a lighthouse for this purpose. A concave mirror can take light diverging from a small source and reflect it into a beam of parallel rays. (a) Is it possible to make a Fresnel mirror? (b) Is this idea original, or has it already been done?

2. Explain this statement: "The focal point of a lens is the location of the image of a point object at infinity." (a) Discuss the notion of infinity in real terms as it applies to object distances. (b) Based on this statement, can you think of a simple method for determining the focal length of a converging lens?

3. Why do some emergency vehicles have the symbol ƎƆИA⅃UᗺMA written on the front?

4. Explain why a mirror cannot give rise to chromatic aberration.

5. (a) Can a converging lens be made to diverge light if it is placed into a liquid? (b) **What If?** What about a converging mirror?

6. Explain why a fish in a spherical goldfish bowl appears larger than it really is.

7. In Figure 36.26a, assume the gray object arrow is replaced by one that is much taller than the lens. (a) How many rays from the top of the object will strike the lens? (b) How many principal rays can be drawn in a ray diagram?

8. Lenses used in eyeglasses, whether converging or diverging, are always designed so that the middle of the lens curves away from the eye like the center lenses of Figures 36.25a and 36.25b. Why?

9. Suppose you want to use a converging lens to project the image of two trees onto a screen. As shown in Figure CQ36.9, one tree is a distance *x* from the lens and the other is at 2*x*. You adjust the screen so that the near tree is in focus. If you now want the far tree to be in focus, do you move the screen toward or away from the lens?

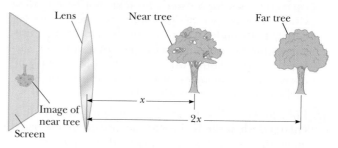

Figure CQ36.9

10. Consider a spherical concave mirror with the object located to the left of the mirror beyond the focal point. Using ray diagrams, show that the image moves to the left as the object approaches the focal point.

11. In Figures CQ36.11a and CQ36.11b, which glasses correct nearsightedness and which correct farsightedness?

Figure CQ36.11 Conceptual Questions 11 and 12.

12. Bethany tries on either her hyperopic grandfather's or her myopic brother's glasses and complains, "Everything looks blurry." Why do the eyes of a person wearing glasses not look blurry? (See Fig. CQ36.11.)

13. In a Jules Verne novel, a piece of ice is shaped to form a magnifying lens, focusing sunlight to start a fire. Is that possible?

14. A solar furnace can be constructed by using a concave mirror to reflect and focus sunlight into a furnace enclosure. What factors in the design of the reflecting mirror would guarantee very high temperatures?

15. Figure CQ36.15 shows a lithograph by M. C. Escher titled *Hand with Reflection Sphere (Self-Portrait in Spherical Mirror)*. Escher said about the work: "The picture shows a spherical mirror, resting on a left hand. But as a print is the reverse of the original drawing on stone, it was my right hand that you see depicted. (Being left-handed, I needed my left hand to make the drawing.) Such a globe reflection collects almost one's whole surroundings in one disk-shaped image. The whole room, four walls,

Figure CQ36.15

the floor, and the ceiling, everything, albeit distorted, is compressed into that one small circle. Your own head, or more exactly the point between your eyes, is the absolute center. No matter how you turn or twist yourself, you can't get out of that central point. You are immovably the focus, the unshakable core, of your world." Comment on the accuracy of Escher's description.

16. If a cylinder of solid glass or clear plastic is placed above the words LEAD OXIDE and viewed from the side as shown in Figure CQ36.16, the word LEAD appears inverted, but the word OXIDE does not. Explain.

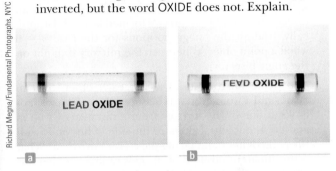

Figure CQ36.16

17. Do the equations $1/p + 1/q = 1/f$ and $M = -q/p$ apply to the image formed by a flat mirror? Explain your answer.

Problems

The problems found in this chapter may be assigned online in Enhanced WebAssign

1. straightforward; **2.** intermediate;
3. challenging

|1.| full solution available in the *Student Solutions Manual/Study Guide*

AMT Analysis Model tutorial available in Enhanced WebAssign

GP Guided Problem

M Master It tutorial available in Enhanced WebAssign

W Watch It video solution available in Enhanced WebAssign

Section 36.1 Images Formed by Flat Mirrors

1. Determine the minimum height of a vertical flat mirror in which a person 178 cm tall can see his or her full image. *Suggestion:* Drawing a ray diagram would be helpful.

2. In a choir practice room, two parallel walls are 5.30 m apart. The singers stand against the north wall. The organist faces the south wall, sitting 0.800 m away from it. To enable her to see the choir, a flat mirror 0.600 m wide is mounted on the south wall, straight in front of her. What width of the north wall can the organist see? *Suggestion:* Draw a top-view diagram to justify your answer.

3. (a) Does your bathroom mirror show you older or younger than you actually are? (b) Compute an order-of-magnitude estimate for the age difference based on data you specify.

4. A person walks into a room that has two flat mirrors on opposite walls. The mirrors produce multiple images of the person. Consider *only* the images formed in the

mirror on the left. When the person is 2.00 m from the mirror on the left wall and 4.00 m from the mirror on the right wall, find the distance from the person to the first three images seen in the mirror on the left wall.

5. A periscope (Fig. P36.5) is useful for viewing objects that cannot be seen directly. It can be used in submarines and when watching golf matches or parades from behind a crowd of people. Suppose the object is a distance p_1 from the upper mirror and the centers

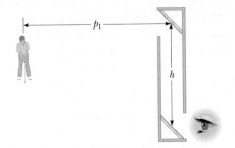

Figure P36.5

of the two flat mirrors are separated by a distance h. (a) What is the distance of the final image from the lower mirror? (b) Is the final image real or virtual? (c) Is it upright or inverted? (d) What is its magnification? (e) Does it appear to be left–right reversed?

6. Two flat mirrors have their reflecting surfaces facing each other, with the edge of one mirror in contact with an edge of the other, so that the angle between the mirrors is α. When an object is placed between the mirrors, a number of images are formed. In general, if the angle α is such that $n\alpha = 360°$, where n is an integer, the number of images formed is $n - 1$. Graphically, find all the image positions for the case $n = 6$ when a point object is between the mirrors (but not on the angle bisector).

7. Two plane mirrors stand facing each other, 3.00 m apart, and a woman stands between them. The woman looks at one of the mirrors from a distance of 1.00 m and holds her left arm out to the side of her body with the palm of her hand facing the closer mirror. (a) What is the apparent position of the closest image of her left hand, measured perpendicularly from the surface of the mirror in front of her? (b) Does it show the palm of her hand or the back of her hand? (c) What is the position of the next closest image? (d) Does it show the palm of her hand or the back of her hand? (e) What is the position of the third closest image? (f) Does it show the palm of her hand or the back of her hand? (g) Which of the images are real and which are virtual?

Section 36.2 Images Formed by Spherical Mirrors

8. An object is placed 50.0 cm from a concave spherical mirror with focal length of magnitude 20.0 cm. (a) Find the location of the image. (b) What is the magnification of the image? (c) Is the image real or virtual? (d) Is the image upright or inverted?

9. A concave spherical mirror has a radius of curvature of magnitude 20.0 cm. (a) Find the location of the image for object distances of (i) 40.0 cm, (ii) 20.0 cm, and (iii) 10.0 cm. For each case, state whether the image is (b) real or virtual and (c) upright or inverted. (d) Find the magnification in each case.

10. An object is placed 20.0 cm from a concave spherical mirror having a focal length of magnitude 40.0 cm. (a) Use graph paper to construct an accurate ray diagram for this situation. (b) From your ray diagram, determine the location of the image. (c) What is the magnification of the image? (d) Check your answers to parts (b) and (c) using the mirror equation.

11. A convex spherical mirror has a radius of curvature of magnitude 40.0 cm. Determine the position of the virtual image and the magnification for object distances of (a) 30.0 cm and (b) 60.0 cm. (c) Are the images in parts (a) and (b) upright or inverted?

12. At an intersection of hospital hallways, a convex spherical mirror is mounted high on a wall to help people avoid collisions. The magnitude of the mirror's radius of curvature is 0.550 m. (a) Locate the image of a patient 10.0 m from the mirror. (b) Indicate whether the image is upright or inverted. (c) Determine the magnification of the image.

13. An object of height 2.00 cm is placed 30.0 cm from a convex spherical mirror of focal length of magnitude 10.0 cm. (a) Find the location of the image. (b) Indicate whether the image is upright or inverted. (c) Determine the height of the image.

14. A dentist uses a spherical mirror to examine a tooth. The tooth is 1.00 cm in front of the mirror, and the image is formed 10.0 cm behind the mirror. Determine (a) the mirror's radius of curvature and (b) the magnification of the image.

15. A large hall in a museum has a niche in one wall. On the floor plan, the niche appears as a semicircular indentation of radius 2.50 m. A tourist stands on the centerline of the niche, 2.00 m out from its deepest point, and whispers "Hello." Where is the sound concentrated after reflection from the niche?

16. *Why is the following situation impossible?* At a blind corner in an outdoor shopping mall, a convex mirror is mounted so pedestrians can see around the corner before arriving there and bumping into someone traveling in the perpendicular direction. The installers of the mirror failed to take into account the position of the Sun, and the mirror focuses the Sun's rays on a nearby bush and sets it on fire.

17. To fit a contact lens to a patient's eye, a *keratometer* can be used to measure the curvature of the eye's front surface, the cornea. This instrument places an illuminated object of known size at a known distance p from the cornea. The cornea reflects some light from the object, forming an image of the object. The magnification M of the image is measured by using a small viewing telescope that allows comparison of the image formed by the cornea with a second calibrated image projected into the field of view by a prism arrangement. Determine the radius of curvature of the cornea for the case $p = 30.0$ cm and $M = 0.013\,0$.

18. A certain Christmas tree ornament is a silver sphere having a diameter of 8.50 cm. (a) If the size of an image created by reflection in the ornament is three-fourths the reflected object's actual size, determine the object's location. (b) Use a principal-ray diagram to determine whether the image is upright or inverted.

19. (a) A concave spherical mirror forms an inverted image 4.00 times larger than the object. Assuming the distance between object and image is 0.600 m, find the focal length of the mirror. (b) **What If?** Suppose the mirror is convex. The distance between the image and the object is the same as in part (a), but the image is 0.500 the size of the object. Determine the focal length of the mirror.

20. (a) A concave spherical mirror forms an inverted image different in size from the object by a factor $a > 1$. The distance between object and image is d. Find the focal length of the mirror. (b) **What If?** Suppose the mirror is convex, an upright image is formed, and $a < 1$. Determine the focal length of the mirror.

21. An object 10.0 cm tall is placed at the zero mark of a **W** meterstick. A spherical mirror located at some point on the meterstick creates an image of the object that is upright, 4.00 cm tall, and located at the 42.0-cm mark of the meterstick. (a) Is the mirror convex or concave? (b) Where is the mirror? (c) What is the mirror's focal length?

22. A concave spherical mirror has a radius of curvature of magnitude 24.0 cm. (a) Determine the object position for which the resulting image is upright and larger than the object by a factor of 3.00. (b) Draw a ray diagram to determine the position of the image. (c) Is the image real or virtual?

23. A dedicated sports car enthusiast polishes the inside **W** and outside surfaces of a hubcap that is a thin section of a sphere. When she looks into one side of the hubcap, she sees an image of her face 30.0 cm in back of the hubcap. She then flips the hubcap over and sees another image of her face 10.0 cm in back of the hubcap. (a) How far is her face from the hubcap? (b) What is the radius of curvature of the hubcap?

24. A convex spherical mirror has a focal length of magnitude 8.00 cm. (a) What is the location of an object for which the magnitude of the image distance is one-third the magnitude of the object distance? (b) Find the magnification of the image and (c) state whether it is upright or inverted.

25. A spherical mirror is to be used to form an image 5.00 **M** times the size of an object on a screen located 5.00 m from the object. (a) Is the mirror required concave or convex? (b) What is the required radius of curvature of the mirror? (c) Where should the mirror be positioned relative to the object?

26. **Review.** A ball is dropped at $t = 0$ from rest 3.00 m **AMT** directly above the vertex of a concave spherical mirror that has a radius of curvature of magnitude 1.00 m and lies in a horizontal plane. (a) Describe the motion of the ball's image in the mirror. (b) At what instant or instants do the ball and its image coincide?

27. You unconsciously estimate the distance to an object from the angle it subtends in your field of view. This angle θ in radians is related to the linear height of the object h and to the distance d by $\theta = h/d$. Assume you are driving a car and another car, 1.50 m high, is 24.0 m behind you. (a) Suppose your car has a flat passenger-side rearview mirror, 1.55 m from your eyes. How far from your eyes is the image of the car following you? (b) What angle does the image subtend in your field of view? (c) **What If?** Now suppose your car has a convex rearview mirror with a radius of curvature of magnitude 2.00 m (as suggested in Fig. 36.15). How far from your eyes is the image of the car behind you? (d) What angle does the image subtend at your eyes? (e) Based on its angular size, how far away does the following car appear to be?

28. A man standing 1.52 m in front of a shaving mirror pro-**M** duces an inverted image 18.0 cm in front of it. How close to the mirror should he stand if he wants to form an upright image of his chin that is twice the chin's actual size?

Section 36.3 Images Formed by Refraction

29. One end of a long glass rod ($n = 1.50$) is formed into a convex surface with a radius of curvature of magnitude 6.00 cm. An object is located in air along the axis of the rod. Find the image positions corresponding to object distances of (a) 20.0 cm, (b) 10.0 cm, and (c) 3.00 cm from the convex end of the rod.

30. A cubical block of ice 50.0 cm on a side is placed over a speck of dust on a level floor. Find the location of the image of the speck as viewed from above. The index of refraction of ice is 1.309.

31. The top of a swimming pool is at ground level. If the pool is 2.00 m deep, how far below ground level does the bottom of the pool appear to be located when (a) the pool is completely filled with water? (b) When it is filled halfway with water?

32. The magnification of the image formed by a refracting surface is given by

$$M = -\frac{n_1 q}{n_2 p}$$

where n_1, n_2, p, and q are defined as they are for Figure 36.17 and Equation 36.8. A paperweight is made of a solid glass hemisphere with index of refraction 1.50. The radius of the circular cross section is 4.00 cm. The hemisphere is placed on its flat surface, with the center directly over a 2.50-mm-long line drawn on a sheet of paper. What is the length of this line as seen by someone looking vertically down on the hemisphere?

33. A flint glass plate rests on the bottom of an aquarium tank. The plate is 8.00 cm thick (vertical dimension) and is covered with a layer of water 12.0 cm deep. Calculate the apparent thickness of the plate as viewed from straight above the water.

34. Figure P36.34 shows a curved surface separating a material with index of refraction n_1 from a material with index n_2. The surface forms an image I of object O. The ray shown in red passes through the surface along a radial line. Its angles of incidence and refraction are both zero, so its direction does not change at the surface. For the ray shown in blue, the direction changes according to Snell's law, $n_1 \sin \theta_1 = n_2 \sin \theta_2$. For paraxial rays, we assume θ_1 and θ_2 are small, so we may write $n_1 \tan \theta_1 = n_2 \tan \theta_2$. The magnification is defined as $M = h'/h$. Prove that the magnification is given by $M = -n_1 q/n_2 p$.

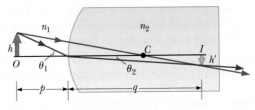

Figure P36.34

35. A glass sphere ($n = 1.50$) with a radius of 15.0 cm has a **M** tiny air bubble 5.00 cm above its center. The sphere is viewed looking down along the extended radius containing the bubble. What is the apparent depth of the bubble below the surface of the sphere?

36. As shown in Figure P36.36, Ben and Jacob check out an aquarium that has a curved front made of plastic with uniform thickness and a radius of curvature of magnitude $R = 2.25$ m. (a) Locate the images of fish that are located (i) 5.00 cm and (ii) 25.0 cm from the front wall of the aquarium. (b) Find the magnification of images (i) and (ii) from the previous part. (See Problem 32 to find an expression for the magnification of an image formed by a refracting surface.) (c) Explain why you don't need to know the refractive index of the plastic to solve this problem. (d) If this aquarium were very long from front to back, could the image of a fish ever be farther from the front surface than the fish itself is? (e) If not, explain why not. If so, give an example and find the magnification.

Chris Candela

Figure P36.36

37. A goldfish is swimming at 2.00 cm/s toward the front wall of a rectangular aquarium. What is the apparent speed of the fish measured by an observer looking in from outside the front wall of the tank?

Section 36.4 Images Formed by Thin Lenses

38. A thin lens has a focal length of 25.0 cm. Locate and describe the image when the object is placed (a) 26.0 cm and (b) 24.0 cm in front of the lens.

39. An object located 32.0 cm in front of a lens forms an image on a screen 8.00 cm behind the lens. (a) Find the focal length of the lens. (b) Determine the magnification. (c) Is the lens converging or diverging?

40. An object is located 20.0 cm to the left of a diverging lens having a focal length $f = -32.0$ cm. Determine (a) the location and (b) the magnification of the image. (c) Construct a ray diagram for this arrangement.

41. The projection lens in a certain slide projector is a single thin lens. A slide 24.0 mm high is to be projected so that its image fills a screen 1.80 m high. The slide-to-screen distance is 3.00 m. (a) Determine the focal length of the projection lens. (b) How far from the slide should the lens of the projector be placed so as to form the image on the screen?

42. An object's distance from a converging lens is 5.00 times the focal length. (a) Determine the location of the image. Express the answer as a fraction of the focal length. (b) Find the magnification of the image and indicate whether it is (c) upright or inverted and (d) real or virtual.

43. A contact lens is made of plastic with an index of refraction of 1.50. The lens has an outer radius of curvature of +2.00 cm and an inner radius of curvature of +2.50 cm. What is the focal length of the lens?

44. A converging lens has a focal length of 10.0 cm. Construct accurate ray diagrams for object distances of **(i)** 20.0 cm and **(ii)** 5.00 cm. (a) From your ray diagrams, determine the location of each image. (b) Is the image real or virtual? (c) Is the image upright or inverted? (d) What is the magnification of the image? (e) Compare your results with the values found algebraically. (f) Comment on difficulties in constructing the graph that could lead to differences between the graphical and algebraic answers.

45. A converging lens has a focal length of 10.0 cm. Locate the object if a real image is located at a distance from the lens of (a) 20.0 cm and (b) 50.0 cm. **What If?** Redo the calculations if the images are virtual and located at a distance from the lens of (c) 20.0 cm and (d) 50.0 cm.

46. A diverging lens has a focal length of magnitude 20.0 cm. (a) Locate the image for object distances of **(i)** 40.0 cm, **(ii)** 20.0 cm, and **(iii)** 10.0 cm. For each case, state whether the image is (b) real or virtual and (c) upright or inverted.(d) For each case, find the magnification.

47. The nickel's image in Figure P36.47 has twice the diameter of the nickel and is 2.84 cm from the lens. Determine the focal length of the lens.

Figure P36.47

48. Suppose an object has thickness dp so that it extends from object distance p to $p + dp$. (a) Prove that the thickness dq of its image is given by $(-q^2/p^2)dp$. (b) The longitudinal magnification of the object is $M_{long} = dq/dp$. How is the longitudinal magnification related to the lateral magnification M?

49. The left face of a biconvex lens has a radius of curvature of magnitude 12.0 cm, and the right face has a radius of curvature of magnitude 18.0 cm. The index of refraction of the glass is 1.44. (a) Calculate the focal length of the lens for light incident from the left. (b) **What If?** After the lens is turned around to interchange the radii of curvature of the two faces, calculate the focal length of the lens for light incident from the left.

50. In Figure P36.50, a thin converging lens of focal length 14.0 cm forms an image of the square *abcd*, which is $h_c = h_b = 10.0$ cm high and lies between distances of $p_d = 20.0$ cm and $p_a = 30.0$ cm from the lens. Let a', b', c', and d' represent the respective corners of the image. Let q_a represent the image distance for points a' and b', q_d represent the image distance for points c' and d',

h'_b represent the distance from point b' to the axis, and h'_c represent the height of c'. (a) Find q_a, q_d, h'_b, and h'_c. (b) Make a sketch of the image. (c) The area of the object is 100 cm². By carrying out the following steps, you will evaluate

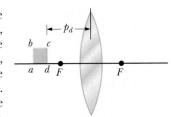

Figure P36.50

the area of the image. Let q represent the image distance of any point between a' and d', for which the object distance is p. Let h' represent the distance from the axis to the point at the edge of the image between b' and c' at image distance q. Demonstrate that

$$|h'| = 10.0q\left(\frac{1}{14.0} - \frac{1}{q}\right)$$

where h' and q are in centimeters. (d) Explain why the geometric area of the image is given by

$$\int_{q_a}^{q_d} |h'| \, dq$$

(e) Carry out the integration to find the area of the image.

51. An antelope is at a distance of 20.0 m from a converging lens of focal length 30.0 cm. The lens forms an image of the animal. (a) If the antelope runs away from the lens at a speed of 5.00 m/s, how fast does the image move? (b) Does the image move toward or away from the lens?

52. *Why is the following situation impossible?* An illuminated object is placed a distance $d = 2.00$ m from a screen. By placing a converging lens of focal length $f = 60.0$ cm at two locations between the object and the screen, a sharp, real image of the object can be formed on the screen. In one location of the lens, the image is larger than the object, and in the other, the image is smaller.

53. A 1.00-cm-high object is placed 4.00 cm to the left of a converging lens of focal length 8.00 cm. A diverging lens of focal length −16.00 cm is 6.00 cm to the right of the converging lens. Find the position and height of the final image. Is the image inverted or upright? Real or virtual?

Section 36.5 **Lens Aberrations**

54. The magnitudes of the radii of curvature are 32.5 cm and 42.5 cm for the two faces of a biconcave lens. The glass has index of refraction 1.53 for violet light and 1.51 for red light. For a very distant object, locate (a) the image formed by violet light and (b) the image formed by red light.

55. Two rays traveling parallel to the principal axis strike a large plano-convex lens having a refractive index of 1.60 (Fig. P36.55). If the convex face is spherical, a ray near the edge does not pass through the focal point (spherical aberration occurs). Assume this face has a radius of curvature of $R = 20.0$ cm and the two rays are at distances $h_1 = 0.500$ cm and $h_2 = 12.0$ cm from the

principal axis. Find the difference Δx in the positions where each crosses the principal axis.

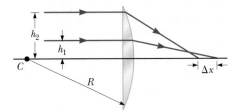

Figure P36.55

Section 36.6 **The Camera**

56. A camera is being used with a correct exposure at $f/4$ and a shutter speed of $\frac{1}{15}$ s. In addition to the f-numbers listed in Section 36.6, this camera has f-numbers $f/1$, $f/1.4$, and $f/2$. To photograph a rapidly moving subject, the shutter speed is changed to $\frac{1}{125}$ s. Find the new f-number setting needed on this camera to maintain satisfactory exposure.

57. Figure 36.33 diagrams a cross section of a camera. It has a single lens of focal length 65.0 mm, which is to form an image on the CCD at the back of the camera. Suppose the position of the lens has been adjusted to focus the image of a distant object. How far and in what direction must the lens be moved to form a sharp image of an object that is 2.00 m away?

Section 36.7 **The Eye**

58. A nearsighted person cannot see objects clearly beyond 25.0 cm (her far point). If she has no astigmatism and contact lenses are prescribed for her, what (a) power and (b) type of lens are required to correct her vision?

59. The near point of a person's eye is 60.0 cm. To see objects clearly at a distance of 25.0 cm, what should be the (a) focal length and (b) power of the appropriate corrective lens? (Neglect the distance from the lens to the eye.)

60. A person sees clearly wearing eyeglasses that have a power of −4.00 diopters when the lenses are 2.00 cm in front of the eyes. (a) What is the focal length of the lens? (b) Is the person nearsighted or farsighted? (c) If the person wants to switch to contact lenses placed directly on the eyes, what lens power should be prescribed?

61. The accommodation limits for a nearsighted person's eyes are 18.0 cm and 80.0 cm. When he wears his glasses, he can see faraway objects clearly. At what minimum distance is he able to see objects clearly?

62. A certain child's near point is 10.0 cm; her far point (with eyes relaxed) is 125 cm. Each eye lens is 2.00 cm from the retina. (a) Between what limits, measured in diopters, does the power of this lens–cornea combination vary? (b) Calculate the power of the eyeglass lens the child should use for relaxed distance vision. Is the lens converging or diverging?

63. A person is to be fitted with bifocals. She can see clearly when the object is between 30 cm and 1.5 m

from the eye. (a) The upper portions of the bifocals (Fig. P36.63) should be designed to enable her to see distant objects clearly. What power should they have? (b) The lower portions of the bifocals should enable her to see objects located 25 cm in front of the eye. What power should they have?

Far vision

Near vision

Figure P36.63

64. A simple model of the human eye ignores its lens entirely. Most of what the eye does to light happens at the outer surface of the transparent cornea. Assume that this surface has a radius of curvature of 6.00 mm and that the eyeball contains just one fluid with a refractive index of 1.40. Prove that a very distant object will be imaged on the retina, 21.0 mm behind the cornea. Describe the image.

65. A patient has a near point of 45.0 cm and far point of 85.0 cm. (a) Can a single pair of glasses correct the patient's vision? Explain the patient's options. (b) Calculate the power lens needed to correct the near point so that the patient can see objects 25.0 cm away. Neglect the eye–lens distance. (c) Calculate the power lens needed to correct the patient's far point, again neglecting the eye–lens distance.

Section 36.8 The Simple Magnifier

66. A lens that has a focal length of 5.00 cm is used as a magnifying glass. (a) To obtain maximum magnification and an image that can be seen clearly by a normal eye, where should the object be placed? (b) What is the magnification?

Section 36.9 The Compound Microscope

67. The distance between the eyepiece and the objective lens in a certain compound microscope is 23.0 cm. The focal length of the eyepiece is 2.50 cm and that of the objective is 0.400 cm. What is the overall magnification of the microscope?

Section 36.10 The Telescope

68. The refracting telescope at the Yerkes Observatory has a 1.00-m diameter objective lens of focal length 20.0 m. Assume it is used with an eyepiece of focal length 2.50 cm. (a) Determine the magnification of Mars as seen through this telescope. (b) Are the Martian polar caps right side up or upside down?

69. A certain telescope has an objective mirror with an aperture diameter of 200 mm and a focal length of 2 000 mm. It captures the image of a nebula on photographic film at its prime focus with an exposure time of 1.50 min. To produce the same light energy per unit area on the film, what is the required exposure time to photograph the same nebula with a smaller telescope that has an objective with a 60.0-mm diameter and a 900-mm focal length?

70. Astronomers often take photographs with the objective lens or mirror of a telescope alone, without an eyepiece. (a) Show that the image size h' for such a telescope is given by $h' = fh/(f - p)$, where f is the objective focal length, h is the object size, and p is the object distance. (b) **What If?** Simplify the expression in part (a) for the case in which the object distance is much greater than objective focal length. (c) The "wingspan" of the International Space Station is 108.6 m, the overall width of its solar panel configuration. When the station is orbiting at an altitude of 407 km, find the width of the image formed by a telescope objective of focal length 4.00 m.

Additional Problems

71. The lens-makers' equation applies to a lens immersed in a liquid if n in the equation is replaced by n_2/n_1. Here n_2 refers to the index of refraction of the lens material and n_1 is that of the medium surrounding the lens. (a) A certain lens has focal length 79.0 cm in air and index of refraction 1.55. Find its focal length in water. (b) A certain mirror has focal length 79.0 cm in air. Find its focal length in water.

72. A real object is located at the zero end of a meterstick. A large concave spherical mirror at the 100-cm end of the meterstick forms an image of the object at the 70.0-cm position. A small convex spherical mirror placed at the 20.0-cm position forms a final image at the 10.0-cm point. What is the radius of curvature of the convex mirror?

73. The distance between an object and its upright image is 20.0 cm. If the magnification is 0.500, what is the focal length of the lens being used to form the image?

74. The distance between an object and its upright image is d. If the magnification is M, what is the focal length of the lens being used to form the image?

75. A person decides to use an old pair of eyeglasses to make some optical instruments. He knows that the near point in his left eye is 50.0 cm and the near point in his right eye is 100 cm. (a) What is the maximum angular magnification he can produce in a telescope? (b) If he places the lenses 10.0 cm apart, what is the maximum overall magnification he can produce in a microscope? *Hint:* Go back to basics and use the thin lens equation to solve part (b).

76. You are designing an endoscope for use inside an air-filled body cavity. A lens at the end of the endoscope will form an image covering the end of a bundle of optical fibers. This image will then be carried by the optical fibers to an eyepiece lens at the outside end of the fiberscope. The radius of the bundle is 1.00 mm. The scene within the body that is to appear within the image fills a circle of radius 6.00 cm. The lens will be located 5.00 cm from the tissues you wish to observe. (a) How far should the lens be located from the end of an optical fiber bundle? (b) What is the focal length of the lens required?

77. The lens and mirror in Figure P36.77 are separated by $d = 1.00$ m and have focal lengths of +80.0 cm and

−50.0 cm, respectively. An object is placed $p = 1.00$ m to the left of the lens as shown. (a) Locate the final image, formed by light that has gone through the lens twice. (b) Determine the overall magnification of the image and (c) state whether the image is upright or inverted.

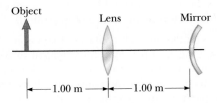

Figure P36.77

78. Two converging lenses having focal lengths of $f_1 = 10.0$ cm and $f_2 = 20.0$ cm are placed a distance $d = 50.0$ cm apart as shown in Figure P36.78. The image due to light passing through both lenses is to be located between the lenses at the position $x = 31.0$ cm indicated. (a) At what value of p should the object be positioned to the left of the first lens? (b) What is the magnification of the final image? (c) Is the final image upright or inverted? (d) Is the final image real or virtual?

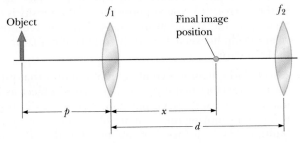

Figure P36.78

79. Figure P36.79 shows a piece of glass with index of refraction $n = 1.50$ surrounded by air. The ends are hemispheres with radii $R_1 = 2.00$ cm and $R_2 = 4.00$ cm, and the centers of the hemispherical ends are separated by a distance of $d = 8.00$ cm. A point object is in air, a distance $p = 1.00$ cm from the left end of the glass. (a) Locate the image of the object due to refraction at the two spherical surfaces. (b) Is the final image real or virtual?

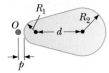

Figure P36.79

80. An object is originally at the $x_i = 0$ cm position of a meterstick located on the x axis. A converging lens of focal length 26.0 cm is fixed at the position 32.0 cm. Then we gradually slide the object to the position $x_f = 12.0$ cm. (a) Find the location x' of the object's image as a function of the object position x. (b) Describe the pattern of the image's motion with reference to a graph or a table of values. (c) As the object moves 12.0 cm to the right, how far does the image move? (d) In what direction or directions?

81. The object in Figure P36.81 is midway between the lens and the mirror, which are separated by a distance

$d = 25.0$ cm. The magnitude of the mirror's radius of curvature is 20.0 cm, and the lens has a focal length of −16.7 cm. (a) Considering only the light that leaves the object and travels first toward the mirror, locate the final image formed by this system. (b) Is this image real or virtual? (c) Is it upright or inverted? (d) What is the overall magnification?

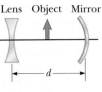

Figure P36.81

82. In many applications, it is necessary to expand or decrease the diameter of a beam of parallel rays of light, which can be accomplished by using a converging lens and a diverging lens in combination. Suppose you have a converging lens of focal length 21.0 cm and a diverging lens of focal length −12.0 cm. (a) How can you arrange these lenses to increase the diameter of a beam of parallel rays? (b) By what factor will the diameter increase?

83. **Review.** A spherical lightbulb of diameter 3.20 cm radiates light equally in all directions, with power 4.50 W. (a) Find the light intensity at the surface of the lightbulb. (b) Find the light intensity 7.20 m away from the center of the lightbulb. (c) At this 7.20-m distance, a lens is set up with its axis pointing toward the lightbulb. The lens has a circular face with a diameter of 15.0 cm and has a focal length of 35.0 cm. Find the diameter of the lightbulb's image. (d) Find the light intensity at the image.

84. A parallel beam of light enters a glass hemisphere perpendicular to the flat face as shown in Figure P36.84. The magnitude of the radius of the hemisphere is $R = 6.00$ cm, and its index of refraction is $n = 1.560$. Assuming paraxial rays, determine the point at which the beam is focused.

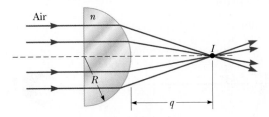

Figure P36.84

85. Two lenses made of kinds of glass having different indices of refraction n_1 and n_2 are cemented together to form an *optical doublet*. Optical doublets are often used to correct chromatic aberrations in optical devices. The first lens of a certain doublet has index of refraction n_1, one flat side, and one concave side with a radius of curvature of magnitude R. The second lens has index of refraction n_2 and two convex sides with radii of curvature also of magnitude R. Show that the doublet can be modeled as a single thin lens with a focal length described by

$$\frac{1}{f} = \frac{2n_2 - n_1 - 1}{R}$$

86. *Why is the following situation impossible?* Consider the lens–mirror combination shown in Figure P36.86 on page 1132. The lens has a focal length of $f_L = 0.200$ m,

and the mirror has a focal length of $f_M = 0.500$ m. The lens and mirror are placed a distance $d = 1.30$ m apart, and an object is placed at $p = 0.300$ m from the lens. By moving a screen to various positions to the left of the lens, a student finds two different positions of the screen that produce a sharp image of the object. One of these positions corresponds to light leaving the object and traveling to the left through the lens. The other position corresponds to light traveling to the right from the object, reflecting from the mirror and then passing through the lens.

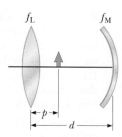

Figure P36.86
Problems 86 and 97.

87. An object is placed 12.0 cm to the left of a diverging **M** lens of focal length −6.00 cm. A converging lens of focal length 12.0 cm is placed a distance d to the right of the diverging lens. Find the distance d so that the final image is infinitely far away to the right.

88. An object is placed a distance p to the left of a diverging lens of focal length f_1. A converging lens of focal length f_2 is placed a distance d to the right of the diverging lens. Find the distance d so that the final image is infinitely far away to the right.

89. An observer to the right of the mirror–lens combination shown in Figure P36.89 (not to scale) sees two real images that are the same size and in the same location. One image is upright, and the other is inverted. Both images are 1.50 times larger than the object. The lens has a focal length of 10.0 cm. The lens and mirror are separated by 40.0 cm. Determine the focal length of the mirror.

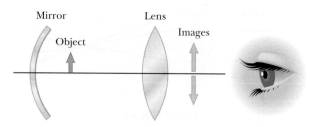

Figure P36.89

90. In a darkened room, a burning candle is placed 1.50 m **GP** from a white wall. A lens is placed between the candle **M** and the wall at a location that causes a larger, inverted image to form on the wall. When the lens is in this position, the object distance is p_1. When the lens is moved 90.0 cm toward the wall, another image of the candle is formed on the wall. From this information, we wish to find p_1 and the focal length of the lens. (a) From the lens equation for the first position of the lens, write an equation relating the focal length f of the lens to the object distance p_1, with no other variables in the equation. (b) From the lens equation for the second position of the lens, write another equation relat-

ing the focal length f of the lens to the object distance p_1. (c) Solve the equations in parts (a) and (b) simultaneously to find p_1. (d) Use the value in part (c) to find the focal length f of the lens.

91. The disk of the Sun subtends an angle of 0.533° at the Earth. What are (a) the position and (b) the diameter of the solar image formed by a concave spherical mirror with a radius of curvature of magnitude 3.00 m?

92. An object 2.00 cm high is placed 40.0 cm to the left of a converging lens having a focal length of 30.0 cm. A diverging lens with a focal length of −20.0 cm is placed 110 cm to the right of the converging lens. Determine (a) the position and (b) the magnification of the final image. (c) Is the image upright or inverted? (d) **What If?** Repeat parts (a) through (c) for the case in which the second lens is a converging lens having a focal length of 20.0 cm.

Challenge Problems

93. Assume the intensity of sunlight is 1.00 kW/m² at a particular location. A highly reflecting concave mirror is to be pointed toward the Sun to produce a power of at least 350 W at the image point. (a) Assuming the disk of the Sun subtends an angle of 0.533° at the Earth, find the required radius R_a of the circular face area of the mirror. (b) Now suppose the light intensity is to be at least 120 kW/m² at the image. Find the required relationship between R_a and the radius of curvature R of the mirror.

94. A *zoom lens* system is a combination of lenses that produces a variable magnification of a fixed object as it maintains a fixed image position. The magnification is varied by moving one or more lenses along the axis. Multiple lenses are used in practice, but the effect of zooming in on an object can be demonstrated with a simple two-lens system. An object, two converging lenses, and a screen are mounted on an optical bench. Lens 1, which is to the right of the object, has a focal length of $f_1 = 5.00$ cm, and lens 2, which is to the right of the first lens, has a focal length of $f_2 = 10.0$ cm. The screen is to the right of lens 2. Initially, an object is situated at a distance of 7.50 cm to the left of lens 1, and the image formed on the screen has a magnification of +1.00. (a) Find the distance between the object and the screen. (b) Both lenses are now moved along their common axis while the object and the screen maintain fixed positions until the image formed on the screen has a magnification of +3.00. Find the displacement of each lens from its initial position in part (a). (c) Can the lenses be displaced in more than one way?

95. Figure P36.95 shows a thin converging lens for which the radii of curvature of its surfaces have magnitudes of 9.00 cm and 11.0 cm. The lens is in front of a concave spherical mirror with the radius of curvature $R = 8.00$ cm. Assume the focal points F_1 and F_2 of the lens are 5.00 cm from the center of the lens. (a) Determine the index of refraction of the lens material. The lens and mirror are 20.0 cm apart, and an object is placed

8.00 cm to the left of the lens. Determine (b) the position of the final image and (c) its magnification as seen by the eye in the figure. (d) Is the final image inverted or upright? Explain.

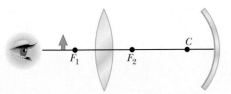

Figure P36.95

96. A floating strawberry illusion is achieved with two parabolic mirrors, each having a focal length 7.50 cm, facing each other as shown in Figure P36.96. If a strawberry is placed on the lower mirror, an image of the strawberry is formed at the small opening at the center of the top mirror, 7.50 cm above the lowest point of the bottom mirror. The position of the eye in Figure P36.96a corresponds to the view of the apparatus in Figure P36.96b. Consider the light path marked *A*. Notice that this light path is blocked by the upper mirror so that the strawberry itself is not directly observable. The light path marked *B* corresponds to the eye viewing the image of the strawberry that is formed at the opening at the top of the apparatus. (a) Show that the final image is formed at that location and describe its characteristics. (b) A very startling effect is to shine a flashlight beam on this image. Even at a glancing angle, the incoming light beam is seemingly reflected from the image! Explain.

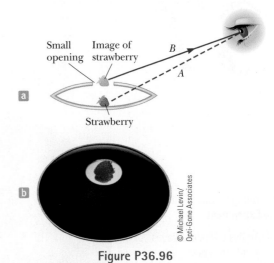

Figure P36.96

97. Consider the lens–mirror arrangement shown in Figure P36.86. There are two final image positions to the left of the lens of focal length f_L. One image position is due to light traveling from the object to the left and passing through the lens. The other image position is due to light traveling to the right from the object, reflecting from the mirror of focal length f_M and then passing through the lens. For a given object position p between the lens and the mirror and measured with respect to the lens, there are two separation distances d between the lens and mirror that will cause the two images described above to be at the same location. Find both positions.

Wave Optics

The colors in many of a hummingbird's feathers are not due to pigment. The *iridescence* that makes the brilliant colors that often appear on the bird's throat and belly is due to an interference effect caused by structures in the feathers. The colors will vary with the viewing angle. *(Dec Hogan/ Shutterstock.com)*

In Chapter 36, we studied light rays passing through a lens or reflecting from a mirror to describe the formation of images. This discussion completed our study of *ray optics.* In this chapter and in Chapter 38, we are concerned with *wave optics*, sometimes called *physical optics*, the study of interference, diffraction, and polarization of light. These phenomena cannot be adequately explained with the ray optics used in Chapters 35 and 36. We now learn how treating light as waves rather than as rays leads to a satisfying description of such phenomena.

37.1 Young's Double-Slit Experiment

In Chapter 18, we studied the waves in interference model and found that the superposition of two mechanical waves can be constructive or destructive. In constructive interference, the amplitude of the resultant wave is greater than that of either individual wave, whereas in destructive interference, the resultant amplitude is less than that of the larger wave. Light waves also interfere with one another. Fundamentally, all interference associated with light waves arises when the electromagnetic fields that constitute the individual waves combine.

Interference in light waves from two sources was first demonstrated by Thomas Young in 1801. A schematic diagram of the apparatus Young used is shown in Figure 37.1a. Plane light waves arrive at a barrier that contains two slits S_1 and S_2. The light from S_1 and S_2 produces on a viewing screen a visible pattern of bright and dark parallel bands called **fringes** (Fig. 37.1b). When the light from S_1 and that from S_2 both arrive at a point on the screen such that constructive interference occurs at

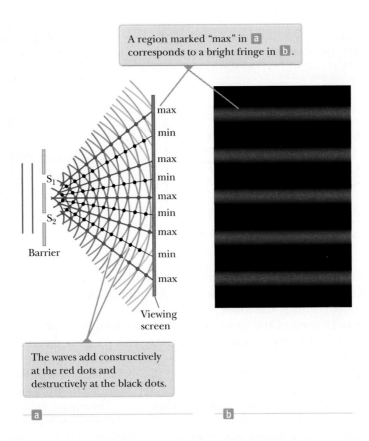

A region marked "max" in **a** corresponds to a bright fringe in **b**.

S$_1$

S$_2$

Barrier

max
min
max
min
max
min
max
min
max

Viewing screen

The waves add constructively at the red dots and destructively at the black dots.

a

b

Figure 37.1 (a) Schematic diagram of Young's double-slit experiment. Slits S$_1$ and S$_2$ behave as coherent sources of light waves that produce an interference pattern on the viewing screen (drawing not to scale). (b) A simulation of an enlargement of the center of a fringe pattern formed on the viewing screen.

that location, a bright fringe appears. When the light from the two slits combines destructively at any location on the screen, a dark fringe results.

Figure 37.2 is a photograph looking down on an interference pattern produced on the surface of a water tank by two vibrating sources. The linear regions of constructive interference, such as at *A*, and destructive interference, such as at *B*, radiating from the area between the sources are analogous to the red and black lines in Figure 37.1a.

Figure 37.3 on page 1136 shows some of the ways in which two waves can combine at the screen. In Figure 37.3a, the two waves, which leave the two slits in phase, strike the screen at the central point *O*. Because both waves travel the same distance, they arrive at *O* in phase. As a result, constructive interference occurs at this location and a bright fringe is observed. In Figure 37.3b, the two waves also start in phase, but here the lower wave has to travel one wavelength farther than the upper wave to reach point *P*. Because the lower wave falls behind

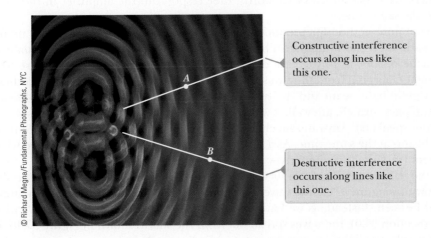

A

B

Constructive interference occurs along lines like this one.

Destructive interference occurs along lines like this one.

Figure 37.2 An interference pattern involving water waves is produced by two vibrating sources at the water's surface.

Figure 37.3 Waves leave the slits and combine at various points on the viewing screen. (All figures not to scale.)

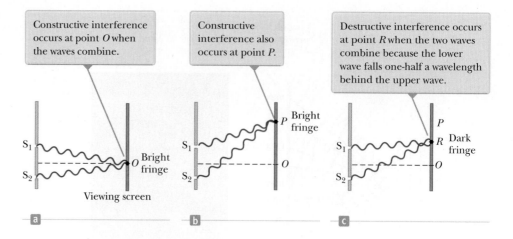

Constructive interference occurs at point O when the waves combine.

Constructive interference also occurs at point P.

Destructive interference occurs at point R when the two waves combine because the lower wave falls one-half a wavelength behind the upper wave.

the upper one by exactly one wavelength, they still arrive in phase at P and a second bright fringe appears at this location. At point R in Figure 37.3c, however, between points O and P, the lower wave has fallen half a wavelength behind the upper wave and a crest of the upper wave overlaps a trough of the lower wave, giving rise to destructive interference at point R. A dark fringe is therefore observed at this location.

If two lightbulbs are placed side by side so that light from both bulbs combines, no interference effects are observed because the light waves from one bulb are emitted independently of those from the other bulb. The emissions from the two lightbulbs do not maintain a constant phase relationship with each other over time. Light waves from an ordinary source such as a lightbulb undergo random phase changes in time intervals of less than a nanosecond. Therefore, the conditions for constructive interference, destructive interference, or some intermediate state are maintained only for such short time intervals. Because the eye cannot follow such rapid changes, no interference effects are observed. Such light sources are said to be **incoherent.**

To observe interference of waves from two sources, the following conditions must be met:

Conditions for interference ▶

- The sources must be **coherent;** that is, they must maintain a constant phase with respect to each other.
- The sources should be **monochromatic;** that is, they should be of a single wavelength.

As an example, single-frequency sound waves emitted by two side-by-side loudspeakers driven by a single amplifier can interfere with each other because the two speakers are coherent. In other words, they respond to the amplifier in the same way at the same time.

A common method for producing two coherent light sources is to use a monochromatic source to illuminate a barrier containing two small openings, usually in the shape of slits, as in the case of Young's experiment illustrated in Figure 37.1. The light emerging from the two slits is coherent because a single source produces the original light beam and the two slits serve only to separate the original beam into two parts (which, after all, is what is done to the sound signal from two side-by-side loudspeakers). Any random change in the light emitted by the source occurs in both beams at the same time. As a result, interference effects can be observed when the light from the two slits arrives at a viewing screen.

If the light traveled only in its original direction after passing through the slits as shown in Figure 37.4a, the waves would not overlap and no interference pattern would be seen. Instead, as we have discussed in our treatment of Huygens's principle (Section 35.6), the waves spread out from the slits as shown in Figure 37.4b. In other words, the light deviates from a straight-line path and enters the region that

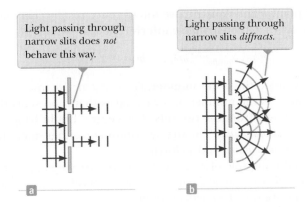

Figure 37.4 (a) If light waves did not spread out after passing through the slits, no interference would occur. (b) The light waves from the two slits overlap as they spread out, filling what we expect to be shadowed regions with light and producing interference fringes on a screen placed to the right of the slits.

would otherwise be shadowed. As noted in Section 35.3, this divergence of light from its initial line of travel is called **diffraction.**

37.2 Analysis Model: Waves in Interference

We discussed the superposition principle for waves on strings in Section 18.1, leading to a one-dimensional version of the waves in interference analysis model. In Example 18.1 on page 537, we briefly discussed a two-dimensional interference phenomenon for sound from two loudspeakers. In walking from point O to point P in Figure 18.5, the listener experienced a maximum in sound intensity at O and a minimum at P. This experience is exactly analogous to an observer looking at point O in Figure 37.3 and seeing a bright fringe and then sweeping his eyes upward to point R, where there is a minimum in light intensity.

Let's look in more detail at the two-dimensional nature of Young's experiment with the help of Figure 37.5. The viewing screen is located a perpendicular distance L from the barrier containing two slits, S_1 and S_2 (Fig. 37.5a). These slits are separated by a distance d, and the source is monochromatic. To reach any arbitrary point P in the upper half of the screen, a wave from the lower slit must travel farther than a wave from the upper slit. The extra distance traveled from the lower slit is the **path difference** δ (Greek letter delta). If we assume the rays labeled r_1 and r_2 are parallel (Fig. 37.5b), which is approximately true if L is much greater than d, then δ is given by

$$\delta = r_2 - r_1 = d \sin \theta \tag{37.1}$$

The value of δ determines whether the two waves are in phase when they arrive at point P. If δ is either zero or some integer multiple of the wavelength, the two waves

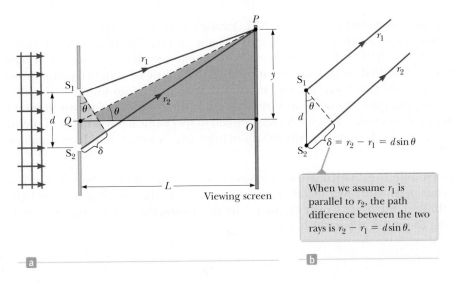

When we assume r_1 is parallel to r_2, the path difference between the two rays is $r_2 - r_1 = d \sin \theta$.

Figure 37.5 (a) Geometric construction for describing Young's double-slit experiment (not to scale). (b) The slits are represented as sources, and the outgoing light rays are assumed to be parallel as they travel to P. To achieve that in practice, it is essential that $L \gg d$.

are in phase at point P and constructive interference results. Therefore, the condition for bright fringes, or **constructive interference,** at point P is

Condition for constructive ▶
interference

$$d \sin \theta_{\text{bright}} = m\lambda \qquad m = 0, \pm 1, \pm 2, \ldots \tag{37.2}$$

The number m is called the **order number.** For constructive interference, the order number is the same as the number of wavelengths that represents the path difference between the waves from the two slits. The central bright fringe at $\theta_{\text{bright}} = 0$ is called the *zeroth-order maximum.* The first maximum on either side, where $m = \pm 1$, is called the *first-order maximum,* and so forth.

When δ is an odd multiple of $\lambda/2$, the two waves arriving at point P are 180° out of phase and give rise to destructive interference. Therefore, the condition for dark fringes, or **destructive interference,** at point P is

Condition for destructive ▶
interference

$$d \sin \theta_{\text{dark}} = (m + \tfrac{1}{2})\lambda \qquad m = 0, \pm 1, \pm 2, \ldots \tag{37.3}$$

These equations provide the *angular* positions of the fringes. It is also useful to obtain expressions for the *linear* positions measured along the screen from O to P. From the triangle OPQ in Figure 37.5a, we see that

$$\tan \theta = \frac{y}{L} \tag{37.4}$$

Using this result, the linear positions of bright and dark fringes are given by

$$y_{\text{bright}} = L \tan \theta_{\text{bright}} \tag{37.5}$$

$$y_{\text{dark}} = L \tan \theta_{\text{dark}} \tag{37.6}$$

where θ_{bright} and θ_{dark} are given by Equations 37.2 and 37.3.

When the angles to the fringes are small, the positions of the fringes are linear near the center of the pattern. That can be verified by noting that for small angles, $\tan \theta \approx \sin \theta$, so Equation 37.5 gives the positions of the bright fringes as $y_{\text{bright}} = L \sin \theta_{\text{bright}}$. Incorporating Equation 37.2 gives

$$y_{\text{bright}} = L \frac{m\lambda}{d} \quad \text{(small angles)} \tag{37.7}$$

This result shows that y_{bright} is linear in the order number m, so the fringes are equally spaced for small angles. Similarly, for dark fringes,

$$y_{\text{dark}} = L \frac{(m + \tfrac{1}{2})\lambda}{d} \quad \text{(small angles)} \tag{37.8}$$

As demonstrated in Example 37.1, Young's double-slit experiment provides a method for measuring the wavelength of light. In fact, Young used this technique to do precisely that. In addition, his experiment gave the wave model of light a great deal of credibility. It was inconceivable that particles of light coming through the slits could cancel one another in a way that would explain the dark fringes.

The principles discussed in this section are the basis of the **waves in interference** analysis model. This model was applied to mechanical waves in one dimension in Chapter 18. Here we see the details of applying this model in three dimensions to light.

Quick Quiz 37.1 Which of the following causes the fringes in a two-slit interference pattern to move farther apart? **(a)** decreasing the wavelength of the light **(b)** decreasing the screen distance L **(c)** decreasing the slit spacing d **(d)** immersing the entire apparatus in water

| Analysis Model | Waves in Interference |

Imagine a broad beam of light that illuminates a double slit in an otherwise opaque material. An interference pattern of bright and dark fringes is created on a distant screen. The condition for bright fringes (**constructive interference**) is

$$d \sin \theta_{\text{bright}} = m\lambda \quad m = 0, \pm 1, \pm 2, \ldots \quad \textbf{(37.2)}$$

The condition for dark fringes (**destructive interference**) is

$$d \sin \theta_{\text{dark}} = (m + \tfrac{1}{2})\lambda \quad m = 0, \pm 1, \pm 2, \ldots \quad \textbf{(37.3)}$$

The number m is called the **order number** of the fringe.

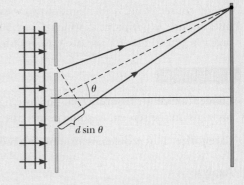

Examples:

- a thin film of oil on top of water shows swirls of color (Section 37.5)
- x-rays passing through a crystalline solid combine to form a Laue pattern (Chapter 38)
- a Michelson interferometer (Section 37.6) is used to search for the ether representing the medium through which light travels (Chapter 39)
- electrons exhibit interference just like light waves when they pass through a double slit (Chapter 40)

| Example 37.1 | **Measuring the Wavelength of a Light Source** | AM |

A viewing screen is separated from a double slit by 4.80 m. The distance between the two slits is 0.030 0 mm. Monochromatic light is directed toward the double slit and forms an interference pattern on the screen. The first dark fringe is 4.50 cm from the center line on the screen.

(A) Determine the wavelength of the light.

SOLUTION

Conceptualize Study Figure 37.5 to be sure you understand the phenomenon of interference of light waves. The distance of 4.50 cm is y in Figure 37.5. Because $L \gg y$, the angles for the fringes are small.

Categorize This problem is a simple application of the *waves in interference* model.

··

Analyze

Solve Equation 37.8 for the wavelength and substitute numerical values, taking $m = 0$ for the first dark fringe:

$$\lambda = \frac{y_{\text{dark}} d}{(m + \tfrac{1}{2})L} = \frac{(4.50 \times 10^{-2}\ \text{m})(3.00 \times 10^{-5}\ \text{m})}{(0 + \tfrac{1}{2})(4.80\ \text{m})}$$

$$= 5.62 \times 10^{-7}\ \text{m} = \boxed{562\ \text{nm}}$$

(B) Calculate the distance between adjacent bright fringes.

SOLUTION

Find the distance between adjacent bright fringes from Equation 37.7 and the results of part (A):

$$y_{m+1} - y_m = L\frac{(m+1)\lambda}{d} - L\frac{m\lambda}{d}$$

$$= L\frac{\lambda}{d} = 4.80\ \text{m}\left(\frac{5.62 \times 10^{-7}\ \text{m}}{3.00 \times 10^{-5}\ \text{m}}\right)$$

$$= 9.00 \times 10^{-2}\ \text{m} = \boxed{9.00\ \text{cm}}$$

Finalize For practice, find the wavelength of the sound in Example 18.1 using the procedure in part (A) of this example.

| **Example 37.2** | **Separating Double-Slit Fringes of Two Wavelengths**   **AM** |

A light source emits visible light of two wavelengths: $\lambda = 430$ nm and $\lambda' = 510$ nm. The source is used in a double-slit interference experiment in which $L = 1.50$ m and $d = 0.025\ 0$ mm. Find the separation distance between the third-order bright fringes for the two wavelengths.

SOLUTION

Conceptualize In Figure 37.5a, imagine light of two wavelengths incident on the slits and forming two interference patterns on the screen. At some points, the fringes of the two colors might overlap, but at most points, they will not.

Categorize This problem is an application of the mathematical representation of the *waves in interference* analysis model.

Analyze

Use Equation 37.7 to find the fringe positions corresponding to these two wavelengths and subtract them:

$$\Delta y = y'_{bright} - y_{bright} = L\frac{m\lambda'}{d} - L\frac{m\lambda}{d} = \frac{Lm}{d}(\lambda' - \lambda)$$

Substitute numerical values:

$$\Delta y = \frac{(1.50\ \text{m})(3)}{0.025\ 0 \times 10^{-3}\ \text{m}}(510 \times 10^{-9}\ \text{m} - 430 \times 10^{-9}\ \text{m})$$

$$= 0.014\ 4\ \text{m} = \boxed{1.44\ \text{cm}}$$

Finalize Let's explore further details of the interference pattern in the following **What If?**

WHAT IF? What if we examine the entire interference pattern due to the two wavelengths and look for overlapping fringes? Are there any locations on the screen where the bright fringes from the two wavelengths overlap exactly?

Answer Find such a location by setting the location of any bright fringe due to λ equal to one due to λ', using Equation 37.7:

$$L\frac{m\lambda}{d} = L\frac{m'\lambda'}{d} \quad \rightarrow \quad \frac{m'}{m} = \frac{\lambda}{\lambda'}$$

Substitute the wavelengths:

$$\frac{m'}{m} = \frac{430\ \text{nm}}{510\ \text{nm}} = \frac{43}{51}$$

Therefore, the 51st fringe of the 430-nm light overlaps with the 43rd fringe of the 510-nm light.

Use Equation 37.7 to find the value of y for these fringes:

$$y = (1.50\ \text{m})\left[\frac{51(430 \times 10^{-9}\ \text{m})}{0.025\ 0 \times 10^{-3}\ \text{m}}\right] = 1.32\ \text{m}$$

This value of y is comparable to L, so the small-angle approximation used for Equation 37.7 is *not* valid. This conclusion suggests we should not expect Equation 37.7 to give us the correct result. If you use Equation 37.5, you can show that the bright fringes do indeed overlap when the same condition, $m'/m = \lambda/\lambda'$, is met (see Problem 48). Therefore, the 51st fringe of the 430-nm light does overlap with the 43rd fringe of the 510-nm light, but not at the location of 1.32 m. You are asked to find the correct location as part of Problem 48.

37.3 Intensity Distribution of the Double-Slit Interference Pattern

Notice that the edges of the bright fringes in Figure 37.1b are not sharp; rather, there is a gradual change from bright to dark. So far, we have discussed the locations of only the centers of the bright and dark fringes on a distant screen. Let's now direct our attention to the intensity of the light at other points between the positions of maximum constructive and destructive interference. In other words, we now calculate the distribution of light intensity associated with the double-slit interference pattern.

Again, suppose the two slits represent coherent sources of sinusoidal waves such that the two waves from the slits have the same angular frequency ω and are in phase. The total magnitude of the electric field at point P on the screen in Figure 37.5 is the superposition of the two waves. Assuming the two waves have the same amplitude E_0, we can write the magnitude of the electric field at point P due to each wave separately as

$$E_1 = E_0 \sin \omega t \quad \text{and} \quad E_2 = E_0 \sin (\omega t + \phi) \tag{37.9}$$

Although the waves are in phase at the slits, their phase difference ϕ at P depends on the path difference $\delta = r_2 - r_1 = d \sin \theta$. A path difference of λ (for constructive interference) corresponds to a phase difference of 2π rad. A path difference of δ is the same fraction of λ as the phase difference ϕ is of 2π. We can describe this fraction mathematically with the ratio

$$\frac{\delta}{\lambda} = \frac{\phi}{2\pi}$$

which gives

$$\phi = \frac{2\pi}{\lambda} \delta = \frac{2\pi}{\lambda} d \sin \theta \tag{37.10}$$ ◀ **Phase difference**

This equation shows how the phase difference ϕ depends on the angle θ in Figure 37.5.

Using the superposition principle and Equation 37.9, we obtain the following expression for the magnitude of the resultant electric field at point P:

$$E_P = E_1 + E_2 = E_0[\sin \omega t + \sin (\omega t + \phi)] \tag{37.11}$$

We can simplify this expression by using the trigonometric identity

$$\sin A + \sin B = 2 \sin \left(\frac{A + B}{2}\right) \cos \left(\frac{A - B}{2}\right)$$

Taking $A = \omega t + \phi$ and $B = \omega t$, Equation 37.11 becomes

$$E_P = 2E_0 \cos \left(\frac{\phi}{2}\right) \sin \left(\omega t + \frac{\phi}{2}\right) \tag{37.12}$$

This result indicates that the electric field at point P has the same frequency ω as the light at the slits but that the amplitude of the field is multiplied by the factor $2 \cos (\phi/2)$. To check the consistency of this result, note that if $\phi = 0, 2\pi, 4\pi, \ldots$, the magnitude of the electric field at point P is $2E_0$, corresponding to the condition for maximum constructive interference. These values of ϕ are consistent with Equation 37.2 for constructive interference. Likewise, if $\phi = \pi, 3\pi, 5\pi, \ldots$, the magnitude of the electric field at point P is zero, which is consistent with Equation 37.3 for total destructive interference.

Finally, to obtain an expression for the light intensity at point P, recall from Section 34.4 that the intensity of a wave is proportional to the square of the resultant electric field magnitude at that point (Eq. 34.24). Using Equation 37.12, we can therefore express the light intensity at point P as

$$I \propto E_P^2 = 4E_0^2 \cos^2 \left(\frac{\phi}{2}\right) \sin^2 \left(\omega t + \frac{\phi}{2}\right)$$

Most light-detecting instruments measure time-averaged light intensity, and the time-averaged value of $\sin^2 (\omega t + \phi/2)$ over one cycle is $\frac{1}{2}$. (See Fig. 33.5.) Therefore, we can write the average light intensity at point P as

$$I = I_{max} \cos^2 \left(\frac{\phi}{2}\right) \tag{37.13}$$

where I_{max} is the maximum intensity on the screen and the expression represents the time average. Substituting the value for ϕ given by Equation 37.10 into this expression gives

$$I = I_{max} \cos^2\left(\frac{\pi d \sin \theta}{\lambda}\right) \qquad (37.14)$$

Alternatively, because $\sin \theta \approx y/L$ for small values of θ in Figure 37.5, we can write Equation 37.14 in the form

$$I = I_{max} \cos^2\left(\frac{\pi d}{\lambda L} y\right) \quad \text{(small angles)} \qquad (37.15)$$

Constructive interference, which produces light intensity maxima, occurs when the quantity $\pi dy/\lambda L$ is an integral multiple of π, corresponding to $y = (\lambda L/d)m$, where m is the order number. This result is consistent with Equation 37.7.

A plot of light intensity versus $d \sin \theta$ is given in Figure 37.6. The interference pattern consists of equally spaced fringes of equal intensity.

Figure 37.7 shows similar plots of light intensity versus $d \sin \theta$ for light passing through multiple slits. For more than two slits, we would add together more electric field magnitudes than the two in Equation 37.9. In this case, the pattern contains primary and secondary maxima. For three slits, notice that the primary maxima are nine times more intense than the secondary maxima as measured by the height of the curve because the intensity varies as E^2. For N slits, the intensity of the primary maxima is N^2 times greater than that for the secondary maxima. As the number of slits increases, the primary maxima increase in intensity and become narrower, while the secondary maxima decrease in intensity relative to the primary maxima. Figure 37.7 also shows that as the number of slits increases, the number of secondary maxima also increases. In fact, the number of secondary maxima is always $N - 2$, where N is the number of slits. In Section 38.4, we shall investigate the pattern for a very large number of slits in a device called a *diffraction grating*.

Quick **Quiz** 37.2 Using Figure 37.7 as a model, sketch the interference pattern from six slits.

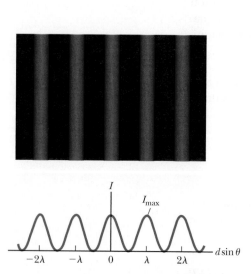

Figure 37.6 Light intensity versus $d \sin \theta$ for a double-slit interference pattern when the screen is far from the two slits ($L \gg d$).

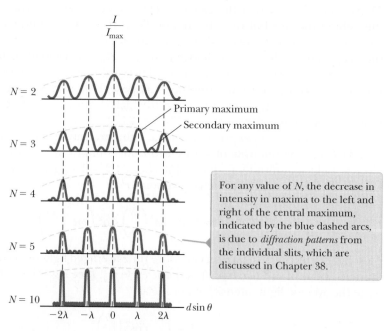

For any value of N, the decrease in intensity in maxima to the left and right of the central maximum, indicated by the blue dashed arcs, is due to *diffraction patterns* from the individual slits, which are discussed in Chapter 38.

Figure 37.7 Multiple-slit interference patterns. As N, the number of slits, is increased, the primary maxima (the tallest peaks in each graph) become narrower but remain fixed in position and the number of secondary maxima increases.

37.4 Change of Phase Due to Reflection

Young's method for producing two coherent light sources involves illuminating a pair of slits with a single source. Another simple, yet ingenious, arrangement for producing an interference pattern with a single light source is known as *Lloyd's mirror*[1] (Fig. 37.8). A point light source S is placed close to a mirror, and a viewing screen is positioned some distance away and perpendicular to the mirror. Light waves can reach point P on the screen either directly from S to P or by the path involving reflection from the mirror. The reflected ray can be treated as a ray originating from a virtual source S'. As a result, we can think of this arrangement as a double-slit source where the distance d between sources S and S' in Figure 37.8 is analogous to length d in Figure 37.5. Hence, at observation points far from the source ($L \gg d$), we expect waves from S and S' to form an interference pattern exactly like the one formed by two real coherent sources. An interference pattern is indeed observed. The positions of the dark and bright fringes, however, are reversed relative to the pattern created by two real coherent sources (Young's experiment). Such a reversal can only occur if the coherent sources S and S' differ in phase by 180°.

To illustrate further, consider point P', the point where the mirror intersects the screen. This point is equidistant from sources S and S'. If path difference alone were responsible for the phase difference, we would see a bright fringe at P' (because the path difference is zero for this point), corresponding to the central bright fringe of the two-slit interference pattern. Instead, a dark fringe is observed at P'. We therefore conclude that a 180° phase change must be produced by reflection from the mirror. In general, an electromagnetic wave undergoes a phase change of 180° upon reflection from a medium that has a higher index of refraction than the one in which the wave is traveling.

It is useful to draw an analogy between reflected light waves and the reflections of a transverse pulse on a stretched string (Section 16.4). The reflected pulse on a string undergoes a phase change of 180° when reflected from the boundary of a denser string or a rigid support, but no phase change occurs when the pulse is reflected from the boundary of a less dense string or a freely-supported end. Similarly, an electromagnetic wave undergoes a 180° phase change when reflected from a boundary leading to an optically denser medium (defined as a medium with a higher index of refraction), but no phase change occurs when the wave is reflected from a boundary leading to a less dense medium. These rules, summarized in Figure 37.9, can be deduced from Maxwell's equations, but the treatment is beyond the scope of this text.

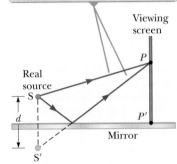

An interference pattern is produced on the screen as a result of the combination of the direct ray (red) and the reflected ray (blue).

Figure 37.8 Lloyd's mirror. The reflected ray undergoes a phase change of 180°.

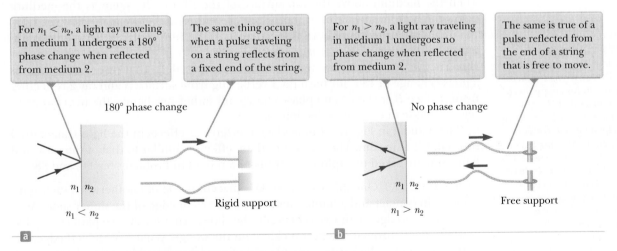

For $n_1 < n_2$, a light ray traveling in medium 1 undergoes a 180° phase change when reflected from medium 2.

The same thing occurs when a pulse traveling on a string reflects from a fixed end of the string.

For $n_1 > n_2$, a light ray traveling in medium 1 undergoes no phase change when reflected from medium 2.

The same is true of a pulse reflected from the end of a string that is free to move.

180° phase change

No phase change

n_1 n_2

$n_1 < n_2$

Rigid support

n_1 n_2

$n_1 > n_2$

Free support

a

b

Figure 37.9 Comparisons of reflections of light waves and waves on strings.

[1]Developed in 1834 by Humphrey Lloyd (1800–1881), Professor of Natural and Experimental Philosophy, Trinity College, Dublin.

37.5 Interference in Thin Films

Interference effects are commonly observed in thin films, such as thin layers of oil on water or the thin surface of a soap bubble. The varied colors observed when white light is incident on such films result from the interference of waves reflected from the two surfaces of the film.

Consider a film of uniform thickness t and index of refraction n. The wavelength of light λ_n in the film (see Section 35.5) is

$$\lambda_n = \frac{\lambda}{n}$$

where λ is the wavelength of the light in free space and n is the index of refraction of the film material. Let's assume light rays traveling in air are nearly normal to the two surfaces of the film as shown in Figure 37.10.

Reflected ray 1, which is reflected from the upper surface (A) in Figure 37.10, undergoes a phase change of 180° with respect to the incident wave. Reflected ray 2, which is reflected from the lower film surface (B), undergoes no phase change because it is reflected from a medium (air) that has a lower index of refraction. Therefore, ray 1 is 180° out of phase with ray 2, which is equivalent to a path difference of $\lambda_n/2$. We must also consider, however, that ray 2 travels an extra distance $2t$ before the waves recombine in the air above surface A. (Remember that we are considering light rays that are close to normal to the surface. If the rays are not close to normal, the path difference is larger than $2t$.) If $2t = \lambda_n/2$, rays 1 and 2 recombine in phase and the result is constructive interference. In general, the condition for *constructive* interference in thin films is[2]

$$2t = \left(m + \tfrac{1}{2}\right)\lambda_n \quad m = 0, 1, 2, \ldots \tag{37.16}$$

This condition takes into account two factors: (1) the difference in path length for the two rays (the term $m\lambda_n$) and (2) the 180° phase change upon reflection (the term $\tfrac{1}{2}\lambda_n$). Because $\lambda_n = \lambda/n$, we can write Equation 37.16 as

$$2nt = \left(m + \tfrac{1}{2}\right)\lambda \quad m = 0, 1, 2, \ldots \tag{37.17}$$

If the extra distance $2t$ traveled by ray 2 corresponds to a multiple of λ_n, the two waves combine out of phase and the result is destructive interference. The general equation for *destructive* interference in thin films is

$$2nt = m\lambda \quad m = 0, 1, 2, \ldots \tag{37.18}$$

The foregoing conditions for constructive and destructive interference are valid when the medium above the top surface of the film is the same as the medium below the bottom surface or, if there are different media above and below the film, the index of refraction of both is less than n. If the film is placed between two different media, one with $n < n_{\text{film}}$ and the other with $n > n_{\text{film}}$, the conditions for constructive and destructive interference are reversed. In that case, either there is a phase change of 180° for both ray 1 reflecting from surface A and ray 2 reflecting from surface B or there is no phase change for either ray; hence, the net change in relative phase due to the reflections is zero.

Rays 3 and 4 in Figure 37.10 lead to interference effects in the light transmitted through the thin film. The analysis of these effects is similar to that of the reflected light. You are asked to explore the transmitted light in Problems 35, 36, and 38.

Quick Quiz 37.3 One microscope slide is placed on top of another with their left edges in contact and a human hair under the right edge of the upper slide. As a result, a wedge of air exists between the slides. An interference pattern results when monochromatic light is incident on the wedge. What is at the left edges of the slides? **(a)** a dark fringe **(b)** a bright fringe **(c)** impossible to determine

---

Interference in light reflected from a thin film is due to a combination of rays 1 and 2 reflected from the upper and lower surfaces of the film.

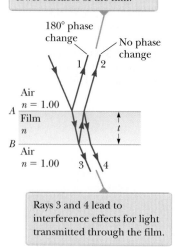

Rays 3 and 4 lead to interference effects for light transmitted through the film.

Figure 37.10 Light paths through a thin film.

Pitfall Prevention 37.1

Be Careful with Thin Films Be sure to include *both* effects—path length and phase change—when analyzing an interference pattern resulting from a thin film. The possible phase change is a new feature we did not need to consider for double-slit interference. Also think carefully about the material on either side of the film. If there are different materials on either side of the film, you may have a situation in which there is a 180° phase change at *both* surfaces or at *neither* surface.

---

[2]The full interference effect in a thin film requires an analysis of an infinite number of reflections back and forth between the top and bottom surfaces of the film. We focus here only on a single reflection from the bottom of the film, which provides the largest contribution to the interference effect.

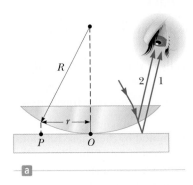

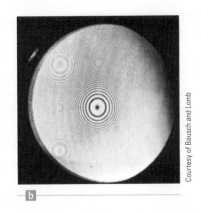

Figure 37.11 (a) The combination of rays reflected from the flat plate and the curved lens surface gives rise to an interference pattern known as Newton's rings. (b) Photograph of Newton's rings.

Newton's Rings

Another method for observing interference in light waves is to place a plano-convex lens on top of a flat glass surface as shown in Figure 37.11a. With this arrangement, the air film between the glass surfaces varies in thickness from zero at the point of contact to some nonzero value at point P. If the radius of curvature R of the lens is much greater than the distance r and the system is viewed from above, a pattern of light and dark rings is observed as shown in Figure 37.11b. These circular fringes, discovered by Newton, are called **Newton's rings.**

The interference effect is due to the combination of ray 1, reflected from the flat plate, with ray 2, reflected from the curved surface of the lens. Ray 1 undergoes a phase change of 180° upon reflection (because it is reflected from a medium of higher index of refraction), whereas ray 2 undergoes no phase change (because it is reflected from a medium of lower index of refraction). Hence, the conditions for constructive and destructive interference are given by Equations 37.17 and 37.18, respectively, with $n = 1$ because the film is air. Because there is no path difference and the total phase change is due only to the 180° phase change upon reflection, the contact point at O is dark as seen in Figure 37.11b.

Using the geometry shown in Figure 37.11a, we can obtain expressions for the radii of the bright and dark bands in terms of the radius of curvature R and wavelength λ. For example, the dark rings have radii given by the expression $r \approx \sqrt{m\lambda R/n}$. The details are left as a problem (see Problem 66). We can obtain the wavelength of the light causing the interference pattern by measuring the radii of the rings, provided R is known. Conversely, we can use a known wavelength to obtain R.

One important use of Newton's rings is in the testing of optical lenses. A circular pattern like that pictured in Figure 37.11b is obtained only when the lens is ground to a perfectly symmetric curvature. Variations from such symmetry produce a pattern with fringes that vary from a smooth, circular shape. These variations indicate how the lens must be reground and repolished to remove imperfections.

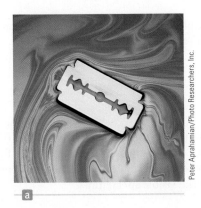

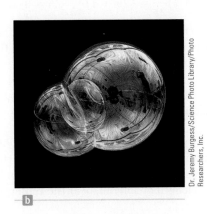

(a) A thin film of oil floating on water displays interference, shown by the pattern of colors when white light is incident on the film. Variations in film thickness produce the interesting color pattern. The razor blade gives you an idea of the size of the colored bands. (b) Interference in soap bubbles. The colors are due to interference between light rays reflected from the inner and outer surfaces of the thin film of soap making up the bubble. The color depends on the thickness of the film, ranging from black, where the film is thinnest, to magenta, where it is thickest.

Problem-Solving Strategy Thin-Film Interference

The following features should be kept in mind when working thin-film interference problems.

1. Conceptualize. Think about what is going on physically in the problem. Identify the light source and the location of the observer.

2. Categorize. Confirm that you should use the techniques for thin-film interference by identifying the thin film causing the interference.

3. Analyze. The type of interference that occurs is determined by the phase relationship between the portion of the wave reflected at the upper surface of the film and the portion reflected at the lower surface. Phase differences between the two portions of the wave have two causes: differences in the distances traveled by the two portions and phase changes occurring on reflection. *Both* causes must be considered when determining which type of interference occurs. If the media above and below the film both have index of refraction larger than that of the film or if both indices are smaller, use Equation 37.17 for constructive interference and Equation 37.18 for destructive interference. If the film is located between two different media, one with $n < n_{film}$ and the other with $n > n_{film}$, reverse these two equations for constructive and destructive interference.

4. Finalize. Inspect your final results to see if they make sense physically and are of an appropriate size.

Example 37.3 Interference in a Soap Film

Calculate the minimum thickness of a soap-bubble film that results in constructive interference in the reflected light if the film is illuminated with light whose wavelength in free space is $\lambda = 600$ nm. The index of refraction of the soap film is 1.33.

SOLUTION

Conceptualize Imagine that the film in Figure 37.10 is soap, with air on both sides.

Categorize We determine the result using an equation from this section, so we categorize this example as a substitution problem.

The minimum film thickness for constructive interference in the reflected light corresponds to $m = 0$ in Equation 37.17. Solve this equation for t and substitute numerical values:

$$t = \frac{(0 + \frac{1}{2})\lambda}{2n} = \frac{\lambda}{4n} = \frac{(600 \text{ nm})}{4(1.33)} = \boxed{113 \text{ nm}}$$

WHAT IF? What if the film is twice as thick? Does this situation produce constructive interference?

Answer Using Equation 37.17, we can solve for the thicknesses at which constructive interference occurs:

$$t = \left(m + \frac{1}{2}\right)\frac{\lambda}{2n} = (2m + 1)\frac{\lambda}{4n} \quad m = 0, 1, 2, \dots$$

The allowed values of m show that constructive interference occurs for *odd* multiples of the thickness corresponding to $m = 0$, $t = 113$ nm. Therefore, constructive interference does *not* occur for a film that is twice as thick.

Example 37.4 Nonreflective Coatings for Solar Cells

Solar cells—devices that generate electricity when exposed to sunlight—are often coated with a transparent, thin film of silicon monoxide (SiO, $n = 1.45$) to minimize reflective losses from the surface. Suppose a silicon solar cell ($n = 3.5$) is coated with a thin film of silicon monoxide for this purpose (Fig. 37.12a). Determine the minimum film thickness that produces the least reflection at a wavelength of 550 nm, near the center of the visible spectrum.

▶ **37.4** continued

SOLUTION

Conceptualize Figure 37.12a helps us visualize the path of the rays in the SiO film that result in interference in the reflected light.

Categorize Based on the geometry of the SiO layer, we categorize this example as a thin-film interference problem.

Analyze The reflected light is a minimum when rays 1 and 2 in Figure 37.12a meet the condition of destructive interference. In this situation, *both* rays undergo a 180° phase change upon reflection: ray 1 from the upper SiO surface and ray 2 from the lower SiO surface. The net change in phase due to reflection is therefore zero, and the condition for a reflection minimum requires a path difference of $\lambda_n/2$, where λ_n is the wavelength of the light in SiO. Hence, $2nt = \lambda/2$, where λ is the wavelength in air and n is the index of refraction of SiO.

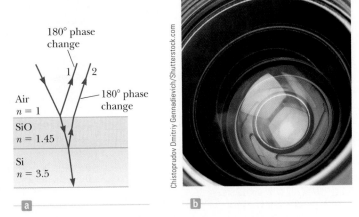

Figure 37.12 (Example 37.4) (a) Reflective losses from a silicon solar cell are minimized by coating the surface of the cell with a thin film of silicon monoxide. (b) The reflected light from a coated camera lens often has a reddish-violet appearance.

Solve the equation $2nt = \lambda/2$ for t and substitute numerical values:

$$t = \frac{\lambda}{4n} = \frac{550 \text{ nm}}{4(1.45)} = \boxed{94.8 \text{ nm}}$$

Finalize A typical uncoated solar cell has reflective losses as high as 30%, but a coating of SiO can reduce this value to about 10%. This significant decrease in reflective losses increases the cell's efficiency because less reflection means that more sunlight enters the silicon to create charge carriers in the cell. No coating can ever be made perfectly nonreflecting because the required thickness is wavelength-dependent and the incident light covers a wide range of wavelengths.

Glass lenses used in cameras and other optical instruments are usually coated with a transparent thin film to reduce or eliminate unwanted reflection and to enhance the transmission of light through the lenses. The camera lens in Figure 37.12b has several coatings (of different thicknesses) to minimize reflection of light waves having wavelengths near the center of the visible spectrum. As a result, the small amount of light that is reflected by the lens has a greater proportion of the far ends of the spectrum and often appears reddish violet.

37.6 The Michelson Interferometer

The **interferometer,** invented by American physicist A. A. Michelson (1852–1931), splits a light beam into two parts and then recombines the parts to form an interference pattern. The device can be used to measure wavelengths or other lengths with great precision because a large and precisely measurable displacement of one of the mirrors is related to an exactly countable number of wavelengths of light.

A schematic diagram of the interferometer is shown in Figure 37.13 (page 1148). A ray of light from a monochromatic source is split into two rays by mirror M_0, which is inclined at 45° to the incident light beam. Mirror M_0, called a *beam splitter,* transmits half the light incident on it and reflects the rest. One ray is reflected from M_0 to the right toward mirror M_1, and the second ray is transmitted vertically through M_0 toward mirror M_2. Hence, the two rays travel separate paths L_1 and L_2. After reflecting from M_1 and M_2, the two rays eventually recombine at M_0 to produce an interference pattern, which can be viewed through a telescope.

The interference condition for the two rays is determined by the difference in their path length. When the two mirrors are exactly perpendicular to each other, the interference pattern is a target pattern of bright and dark circular fringes. As M_1 is moved, the fringe pattern collapses or expands, depending on the direction in which M_1 is moved. For example, if a dark circle appears at the center of the

Figure 37.13 Diagram of the Michelson interferometer.

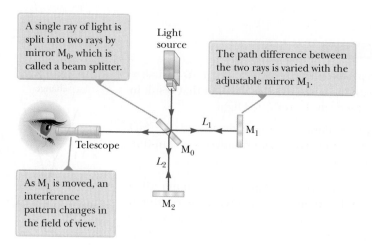

A single ray of light is split into two rays by mirror M_0, which is called a beam splitter.

Light source

The path difference between the two rays is varied with the adjustable mirror M_1.

L_1

M_1

Telescope

M_0

L_2

As M_1 is moved, an interference pattern changes in the field of view.

M_2

target pattern (corresponding to destructive interference) and M_1 is then moved a distance $\lambda/4$ toward M_0, the path difference changes by $\lambda/2$. What was a dark circle at the center now becomes a bright circle. As M_1 is moved an additional distance $\lambda/4$ toward M_0, the bright circle becomes a dark circle again. Therefore, the fringe pattern shifts by one-half fringe each time M_1 is moved a distance $\lambda/4$. The wavelength of light is then measured by counting the number of fringe shifts for a given displacement of M_1. If the wavelength is accurately known, mirror displacements can be measured to within a fraction of the wavelength.

We will see an important historical use of the Michelson interferometer in our discussion of relativity in Chapter 39. Modern uses include the following two applications, Fourier transform infrared spectroscopy and the laser interferometer gravitational-wave observatory.

Fourier Transform Infrared Spectroscopy

Spectroscopy is the study of the wavelength distribution of radiation from a sample that can be used to identify the characteristics of atoms or molecules in the sample. Infrared spectroscopy is particularly important to organic chemists when analyzing organic molecules. Traditional spectroscopy involves the use of an optical element, such as a prism (Section 35.5) or a diffraction grating (Section 38.4), which spreads out various wavelengths in a complex optical signal from the sample into different angles. In this way, the various wavelengths of radiation and their intensities in the signal can be determined. These types of devices are limited in their resolution and effectiveness because they must be scanned through the various angular deviations of the radiation.

The technique of *Fourier transform infrared* (FTIR) *spectroscopy* is used to create a higher-resolution spectrum in a time interval of 1 second that may have required 30 minutes with a standard spectrometer. In this technique, the radiation from a sample enters a Michelson interferometer. The movable mirror is swept through the zero-path-difference condition, and the intensity of radiation at the viewing position is recorded. The result is a complex set of data relating light intensity as a function of mirror position, called an *interferogram*. Because there is a relationship between mirror position and light intensity for a given wavelength, the interferogram contains information about all wavelengths in the signal.

In Section 18.8, we discussed Fourier analysis of a waveform. The waveform is a function that contains information about all the individual frequency components that make up the waveform.[3] Equation 18.13 shows how the waveform is generated from the individual frequency components. Similarly, the interferogram can be

---

[3]In acoustics, it is common to talk about the components of a complex signal in terms of frequency. In optics, it is more common to identify the components by wavelength.

Figure 37.14 The Laser Interferometer Gravitational-Wave Observatory (LIGO) near Richland, Washington. Notice the two perpendicular arms of the Michelson interferometer.

analyzed by computer, in a process called a *Fourier transform,* to provide all the wavelength components. This information is the same as that generated by traditional spectroscopy, but the resolution of FTIR spectroscopy is much higher.

Laser Interferometer Gravitational-Wave Observatory

Einstein's general theory of relativity (Section 39.9) predicts the existence of *gravitational waves.* These waves propagate from the site of any gravitational disturbance, which could be periodic and predictable, such as the rotation of a double star around a center of mass, or unpredictable, such as the supernova explosion of a massive star.

In Einstein's theory, gravitation is equivalent to a distortion of space. Therefore, a gravitational disturbance causes an additional distortion that propagates through space in a manner similar to mechanical or electromagnetic waves. When gravitational waves from a disturbance pass by the Earth, they create a distortion of the local space. The laser interferometer gravitational-wave observatory (LIGO) apparatus is designed to detect this distortion. The apparatus employs a Michelson interferometer that uses laser beams with an effective path length of several kilometers. At the end of an arm of the interferometer, a mirror is mounted on a massive pendulum. When a gravitational wave passes by, the pendulum and the attached mirror move and the interference pattern due to the laser beams from the two arms changes.

Two sites for interferometers have been developed in the United States—in Richland, Washington, and in Livingston, Louisiana—to allow coincidence studies of gravitational waves. Figure 37.14 shows the Washington site. The two arms of the Michelson interferometer are evident in the photograph. Six data runs have been performed as of 2010. These runs have been coordinated with other gravitational wave detectors, such as GEO in Hannover, Germany, TAMA in Mitaka, Japan, and VIRGO in Cascina, Italy. So far, gravitational waves have not yet been detected, but the data runs have provided critical information for modifications and design features for the next generation of detectors. The original detectors are currently being dismantled, in preparation for the installation of Advanced LIGO, an upgrade that should increase the sensitivity of the observatory by a factor of 10. The target date for the beginning of scientific operation of Advanced LIGO is 2014.

Summary

Concepts and Principles

▌ **Interference** in light waves occurs whenever two or more waves overlap at a given point. An interference pattern is observed if (1) the sources are coherent and (2) the sources have identical wavelengths.

▌ The **intensity** at a point in a double-slit interference pattern is

$$I = I_{max} \cos^2\left(\frac{\pi d \sin \theta}{\lambda}\right) \qquad \textbf{(37.14)}$$

where I_{max} is the maximum intensity on the screen and the expression represents the time average.

continued

A wave traveling from a medium of index of refraction n_1 toward a medium of index of refraction n_2 undergoes a 180° phase change upon reflection when $n_2 > n_1$ and undergoes no phase change when $n_2 < n_1$.

The condition for constructive interference in a film of thickness t and index of refraction n surrounded by air is

$$2nt = \left(m + \tfrac{1}{2}\right)\lambda \quad m = 0, 1, 2, \ldots \quad \textbf{(37.17)}$$

where λ is the wavelength of the light in free space.

Similarly, the condition for destructive interference in a thin film surrounded by air is

$$2nt = m\lambda \quad m = 0, 1, 2, \ldots \quad \textbf{(37.18)}$$

Analysis Models for Problem Solving

Waves in Interference. Young's double-slit experiment serves as a prototype for interference phenomena involving electromagnetic radiation. In this experiment, two slits separated by a distance d are illuminated by a single-wavelength light source. The condition for bright fringes (**constructive interference**) is

$$d \sin \theta_{\text{bright}} = m\lambda \quad m = 0, \pm 1, \pm 2, \ldots \quad \textbf{(37.2)}$$

The condition for dark fringes (**destructive interference**) is

$$d \sin \theta_{\text{dark}} = \left(m + \tfrac{1}{2}\right)\lambda \quad m = 0, \pm 1, \pm 2, \ldots \quad \textbf{(37.3)}$$

The number m is called the **order number** of the fringe.

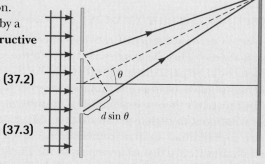

Objective Questions

1. denotes answer available in *Student Solutions Manual/Study Guide*

1. While using a Michelson interferometer (shown in Fig. 37.13), you see a dark circle at the center of the interference pattern. **(i)** As you gradually move the light source toward the central mirror M_0, through a distance $\lambda/2$, what do you see? (a) There is no change in the pattern. (b) The dark circle changes into a bright circle. (c) The dark circle changes into a bright circle and then back into a dark circle. (d) The dark circle changes into a bright circle, then into a dark circle, and then into a bright circle. **(ii)** As you gradually move the moving mirror toward the central mirror M_0, through a distance $\lambda/2$, what do you see? Choose from the same possibilities.

2. Four trials of Young's double-slit experiment are conducted. (a) In the first trial, blue light passes through two fine slits 400 μm apart and forms an interference pattern on a screen 4 m away. (b) In a second trial, red light passes through the same slits and falls on the same screen. (c) A third trial is performed with red light and the same screen, but with slits 800 μm apart. (d) A final trial is performed with red light, slits 800 μm apart, and a screen 8 m away. **(i)** Rank the trials (a) through (d) from the largest to the smallest value of the angle between the central maximum and the first-order side maximum. In your ranking, note any cases of equality. **(ii)** Rank the same trials according to the distance between the central maximum and the first-order side maximum on the screen.

3. Suppose Young's double-slit experiment is performed in air using red light and then the apparatus is immersed in water. What happens to the interference pattern on the screen? (a) It disappears. (b) The bright and dark fringes stay in the same locations, but the contrast is reduced. (c) The bright fringes are closer together. (d) The bright fringes are farther apart. (e) No change happens in the interference pattern.

4. Green light has a wavelength of 500 nm in air. **(i)** Assume green light is reflected from a mirror with angle of incidence 0°. The incident and reflected waves together constitute a standing wave with what distance from one node to the next node? (a) 1 000 nm (b) 500 nm (c) 250 nm (d) 125 nm (e) 62.5 nm **(ii).** The green light is sent into a Michelson interferometer that is adjusted to produce a central bright circle. How far must the interferometer's moving mirror be shifted to change the center of the pattern into a dark circle? Choose from the same possibilities as in part (i). **(iii).** The green light is reflected perpendicularly from a thin film of a plastic with an index of refraction 2.00. The film appears bright in the reflected light. How much additional thickness would make the film appear dark?

5. A thin layer of oil ($n = 1.25$) is floating on water ($n = 1.33$). What is the minimum nonzero thickness of the oil in the region that strongly reflects green light ($\lambda = 530$ nm)? (a) 500 nm (b) 313 nm (c) 404 nm (d) 212 nm (e) 285 nm

6. A monochromatic beam of light of wavelength 500 nm illuminates a double slit having a slit separation of 2.00×10^{-5} m. What is the angle of the second-order bright fringe? (a) 0.050 0 rad (b) 0.025 0 rad (c) 0.100 rad (d) 0.250 rad (e) 0.010 0 rad

7. According to Table 35.1, the index of refraction of flint glass is 1.66 and the index of refraction of crown glass is 1.52. (i) A film formed by one drop of sassafras oil, on a horizontal surface of a flint glass block, is viewed by reflected light. The film appears brightest at its outer margin, where it is thinnest. A film of the same oil on crown glass appears dark at its outer margin. What can you say about the index of refraction of the oil? (a) It must be less than 1.52. (b) It must be between 1.52 and 1.66. (c) It must be greater than 1.66. (d) None of those statements is necessarily true. (ii) Could a very thin film of some other liquid appear bright by reflected light on both of the glass blocks? (iii) Could it appear dark on both? (iv) Could it appear dark on crown glass and bright on flint glass? Experiments described by Thomas Young suggested this question.

8. Suppose you perform Young's double-slit experiment with the slit separation slightly smaller than the wavelength of the light. As a screen, you use a large half-cylinder with its axis along the midline between the slits. What interference pattern will you see on the interior surface of the cylinder? (a) bright and dark fringes so closely spaced as to be indistinguishable (b) one central bright fringe and two dark fringes only (c) a completely bright screen with no dark fringes (d) one central dark fringe and two bright fringes only (e) a completely dark screen with no bright fringes

9. A plane monochromatic light wave is incident on a double slit as illustrated in Figure 37.1. (i) As the viewing screen is moved away from the double slit, what happens to the separation between the interference fringes on the screen? (a) It increases. (b) It decreases. (c) It remains the same. (d) It may increase or decrease, depending on the wavelength of the light. (e) More information is required. (ii) As the slit separation increases, what happens to the separation between the interference fringes on the screen? Select from the same choices.

10. A film of oil on a puddle in a parking lot shows a variety of bright colors in swirled patches. What can you say about the thickness of the oil film? (a) It is much less than the wavelength of visible light. (b) It is on the same order of magnitude as the wavelength of visible light. (c) It is much greater than the wavelength of visible light. (d) It might have any relationship to the wavelength of visible light.

Conceptual Questions   1. denotes answer available in *Student Solutions Manual/Study Guide*

1. Why is the lens on a good-quality camera coated with a thin film?

2. A soap film is held vertically in air and is viewed in reflected light as in Figure CQ37.2. Explain why the film appears to be dark at the top.

3. Explain why two flashlights held close together do not produce an interference pattern on a distant screen.

4. A lens with outer radius of curvature R and index of refraction n rests on a flat glass plate. The combination is illuminated with white light from above and observed from above. (a) Is there a dark spot or a light spot at the center of the lens? (b) What does it mean if the observed rings are noncircular?

5. Consider a dark fringe in a double-slit interference pattern at which almost no light energy is arriving. Light from both slits is arriving at the location of the dark fringe, but the waves cancel. Where does the energy at the positions of dark fringes go?

6. (a) In Young's double-slit experiment, why do we use monochromatic light? (b) If white light is used, how would the pattern change?

7. What is the necessary condition on the path length difference between two waves that interfere (a) constructively and (b) destructively?

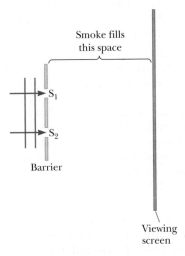

Figure CQ37.2
Conceptual Question 2 and Problem 70.

© Richard Megna/Fundamental Photographs, NYC

8. In a laboratory accident, you spill two liquids onto different parts of a water surface. Neither of the liquids mixes with the water. Both liquids form thin films on the water surface. As the films spread and become very thin, you notice that one film becomes brighter and the other darker in reflected light. Why?

9. A theatrical smoke machine fills the space between the barrier and the viewing screen in the Young's double-slit experiment shown in Figure CQ37.9. Would the smoke show evidence of interference within this space? Explain your answer.

Smoke fills this space

S_1

S_2

Barrier

Viewing screen

Figure CQ37.9

Problems

Section 37.1 Young's Double-Slit Experiment
Section 37.2 Analysis Model: Waves in Interference

Problems 3, 5, 8, 10, and 13 in Chapter 18 can be assigned with this section.

1. Two slits are separated by 0.320 mm. A beam of 500-nm light strikes the slits, producing an interference pattern. Determine the number of maxima observed in the angular range $-30.0° < \theta < 30.0°$.

2. Light of wavelength 530 nm illuminates a pair of slits separated by 0.300 mm. If a screen is placed 2.00 m from the slits, determine the distance between the first and second dark fringes.

3. A laser beam is incident on two slits with a separation of 0.200 mm, and a screen is placed 5.00 m from the slits. An interference pattern appears on the screen. If the angle from the center fringe to the first bright fringe to the side is 0.181°, what is the wavelength of the laser light?

4. A Young's interference experiment is performed with blue-green argon laser light. The separation between the slits is 0.500 mm, and the screen is located 3.30 m from the slits. The first bright fringe is located 3.40 mm from the center of the interference pattern. What is the wavelength of the argon laser light?

5. Young's double-slit experiment is performed with 589-nm light and a distance of 2.00 m between the slits and the screen. The tenth interference minimum is observed 7.26 mm from the central maximum. Determine the spacing of the slits.

6. *Why is the following situation impossible?* Two narrow slits are separated by 8.00 mm in a piece of metal. A beam of microwaves strikes the metal perpendicularly, passes through the two slits, and then proceeds toward a wall some distance away. You know that the wavelength of the radiation is 1.00 cm ±5%, but you wish to measure it more precisely. Moving a microwave detector along the wall to study the interference pattern, you measure the position of the $m = 1$ bright fringe, which leads to a successful measurement of the wavelength of the radiation.

7. Light of wavelength 620 nm falls on a double slit, and the first bright fringe of the interference pattern is seen at an angle of 15.0° with the horizontal. Find the separation between the slits.

8. In a Young's double-slit experiment, two parallel slits with a slit separation of 0.100 mm are illuminated by light of wavelength 589 nm, and the interference pattern is observed on a screen located 4.00 m from the slits. (a) What is the difference in path lengths from each of the slits to the location of the center of a third-order bright fringe on the screen? (b) What is the difference in path lengths from the two slits to the location of the center of the third dark fringe away from the center of the pattern?

9. A pair of narrow, parallel slits separated by 0.250 mm is illuminated by green light ($\lambda = 546.1$ nm). The interference pattern is observed on a screen 1.20 m away from the plane of the parallel slits. Calculate the distance (a) from the central maximum to the first bright region on either side of the central maximum and (b) between the first and second dark bands in the interference pattern.

10. Light with wavelength 442 nm passes through a double-slit system that has a slit separation $d = 0.400$ mm. Determine how far away a screen must be placed so that dark fringes appear directly opposite both slits, with only one bright fringe between them.

11. The two speakers of a boom box are 35.0 cm apart. A single oscillator makes the speakers vibrate in phase at a frequency of 2.00 kHz. At what angles, measured from the perpendicular bisector of the line joining the speakers, would a distant observer hear maximum sound intensity? Minimum sound intensity? (Take the speed of sound as 340 m/s.)

12. In a location where the speed of sound is 343 m/s, a 2 000-Hz sound wave impinges on two slits 30.0 cm apart. (a) At what angle is the first maximum of sound intensity located? (b) **What If?** If the sound wave is replaced by 3.00-cm microwaves, what slit separation gives the same angle for the first maximum of microwave intensity? (c) **What If?** If the slit separation is 1.00 μm, what frequency of light gives the same angle to the first maximum of light intensity?

13. Two radio antennas separated by $d = 300$ m as shown in Figure P37.13 simultaneously broadcast identical signals at the same wavelength. A car travels due north along a straight line at position $x = 1\ 000$ m from the center point between the antennas, and its radio receives the signals. (a) If the car is at the position of the second maximum after that at point O when it has

traveled a distance $y = 400$ m northward, what is the wavelength of the signals? (b) How much farther must the car travel from this position to encounter the next minimum in reception? *Note:* Do not use the small-angle approximation in this problem.

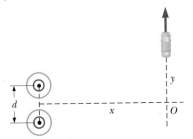

Figure P37.13

14. A riverside warehouse has several small doors facing
[W] the river. Two of these doors are open as shown in Figure P37.14. The walls of the warehouse are lined with sound-absorbing material. Two people stand at a distance $L = 150$ m from the wall with the open doors. Person A stands along a line passing through the midpoint between the open doors, and person B stands a distance $y = 20$ m to his side. A boat on the river sounds its horn. To person A, the sound is loud and clear. To person B, the sound is barely audible. The principal wavelength of the sound waves is 3.00 m. Assuming person B is at the position of the first minimum, determine the distance d between the doors, center to center.

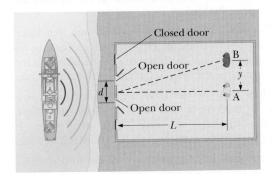

Figure P37.14

15. A student holds a laser that emits light of wavelength 632.8 nm. The laser beam passes though a pair of slits separated by 0.300 mm, in a glass plate attached to the front of the laser. The beam then falls perpendicularly on a screen, creating an interference pattern on it. The student begins to walk directly toward the screen at 3.00 m/s. The central maximum on the screen is stationary. Find the speed of the 50th-order maxima on the screen.

16. A student holds a laser that emits light of wavelength λ. The laser beam passes though a pair of slits separated by a distance d, in a glass plate attached to the front of the laser. The beam then falls perpendicularly on a screen, creating an interference pattern on it. The student begins to walk directly toward the screen at speed v. The central maximum on the screen is sta-

tionary. Find the speed of the mth-order maxima on the screen, where m can be very large.

17. Radio waves of wavelength 125 m from a galaxy reach a radio telescope by two separate paths as shown in Figure P37.17. One is a direct path to the receiver, which is situated on the edge of a tall cliff by the ocean, and the second is by reflection off the water. As the galaxy rises in the east over the water, the first minimum of destructive interference occurs when the galaxy is $\theta = 25.0°$ above the horizon. Find the height of the radio telescope dish above the water.

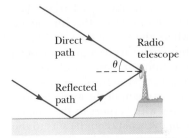

Figure P37.17 Problems 17 and 69.

18. In Figure P37.18 (not to scale), let $L = 1.20$ m and $d =$
[M] 0.120 mm and assume the slit system is illuminated with monochromatic 500-nm light. Calculate the phase difference between the two wave fronts arriving at P when (a) $\theta = 0.500°$ and (b) $y = 5.00$ mm. (c) What is the value of θ for which the phase difference is 0.333 rad? (d) What is the value of θ for which the path difference is $\lambda/4$?

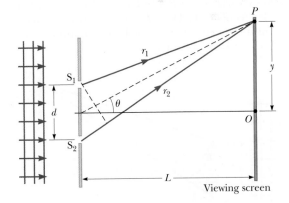

Figure P37.18 Problems 18 and 25.

19. Coherent light rays of wavelength λ strike a pair of slits separated by distance d at an angle θ_1 with respect to the normal to the plane containing the slits as shown in Figure P37.19. The rays leaving the slits make an

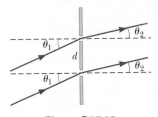

Figure P37.19

angle θ_2 with respect to the normal, and an interference maximum is formed by those rays on a screen that is a great distance from the slits. Show that the angle θ_2 is given by

$$\theta_2 = \sin^{-1}\left(\sin\theta_1 - \frac{m\lambda}{d}\right)$$

where m is an integer.

20. Monochromatic light of wavelength λ is incident on a **GP** pair of slits separated by 2.40×10^{-4} m and forms an interference pattern on a screen placed 1.80 m from the slits. The first-order bright fringe is at a position $y_{bright} = 4.52$ mm measured from the center of the central maximum. From this information, we wish to predict where the fringe for $n = 50$ would be located. (a) Assuming the fringes are laid out linearly along the screen, find the position of the $n = 50$ fringe by multiplying the position of the $n = 1$ fringe by 50.0. (b) Find the tangent of the angle the first-order bright fringe makes with respect to the line extending from the point midway between the slits to the center of the central maximum. (c) Using the result of part (b) and Equation 37.2, calculate the wavelength of the light. (d) Compute the angle for the 50th-order bright fringe from Equation 37.2. (e) Find the position of the 50th-order bright fringe on the screen from Equation 37.5. (f) Comment on the agreement between the answers to parts (a) and (e).

21. In the double-slit arrangement of Figure P37.21, $d =$ **W** 0.150 mm, $L = 140$ cm, $\lambda = 643$ nm, and $y = 1.80$ cm. (a) What is the path difference δ for the rays from the two slits arriving at P? (b) Express this path difference in terms of λ. (c) Does P correspond to a maximum, a minimum, or an intermediate condition? Give evidence for your answer.

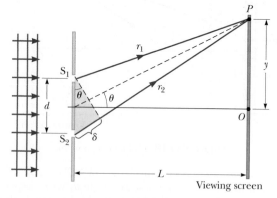

Figure P37.21

22. Young's double-slit experiment underlies the *instrument landing system* used to guide aircraft to safe landings at some airports when the visibility is poor. Although real systems are more complicated than the example described here, they operate on the same principles. A pilot is trying to align her plane with a runway as suggested in Figure P37.22. Two radio antennas (the black dots in the figure) are positioned adjacent to the runway, separated by $d = 40.0$ m. The antennas broadcast unmodulated coherent radio waves at 30.0 MHz.

The red lines in Figure P37.22 represent paths along which maxima in the interference pattern of the radio waves exist. (a) Find the wavelength of the waves. The pilot "locks onto" the strong signal radiated along an interference maximum and steers the plane to keep the received signal strong. If she has found the central maximum, the plane will have precisely the correct heading to land when it reaches the runway as exhibited by plane A. (b) **What If?** Suppose the plane is flying along the first side maximum instead as is the case for plane B. How far to the side of the runway centerline will the plane be when it is 2.00 km from the antennas, measured along its direction of travel? (c) It is possible to tell the pilot that she is on the wrong maximum by sending out two signals from each antenna and equipping the aircraft with a two-channel receiver. The ratio of the two frequencies must not be the ratio of small integers (such as $\frac{3}{4}$). Explain how this two-frequency system would work and why it would not necessarily work if the frequencies were related by an integer ratio.

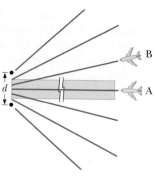

Figure P37.22

Section 37.3 Intensity Distribution of the Double-Slit Interference Pattern

23. Two slits are separated by 0.180 mm. An interference pattern is formed on a screen 80.0 cm away by 656.3-nm light. Calculate the fraction of the maximum intensity a distance $y = 0.600$ cm away from the central maximum.

24. Show that the two waves with wave functions given by $E_1 = 6.00 \sin(100\pi t)$ and $E_2 = 8.00 \sin(100\pi t + \pi/2)$ add to give a wave with the wave function $E_R \sin(100\pi t + \phi)$. Find the required values for E_R and ϕ.

25. In Figure 37.18, let $L = 120$ cm and $d = 0.250$ cm. **M** The slits are illuminated with coherent 600-nm light. Calculate the distance y from the central maximum for which the average intensity on the screen is 75.0% of the maximum.

26. Monochromatic coherent light of amplitude E_0 and angular frequency ω passes through three parallel slits, each separated by a distance d from its neighbor. (a) Show that the time-averaged intensity as a function of the angle θ is

$$I(\theta) = I_{max}\left[1 + 2\cos\left(\frac{2\pi d\sin\theta}{\lambda}\right)\right]^2$$

(b) Explain how this expression describes both the primary and the secondary maxima. (c) Determine the ratio of the intensities of the primary and secondary maxima.

27. The intensity on the screen at a certain point in a double-slit interference pattern is 64.0% of the maximum value. (a) What minimum phase difference (in radians) between sources produces this result? (b) Express

this phase difference as a path difference for 486.1-nm light.

28. Green light (λ = 546 nm) illuminates a pair of narrow, parallel slits separated by 0.250 mm. Make a graph of I/I_{max} as a function of θ for the interference pattern observed on a screen 1.20 m away from the plane of the parallel slits. Let θ range over the interval from $-0.3°$ to $+0.3°$.

29. Two narrow, parallel slits separated by 0.850 mm are **W** illuminated by 600-nm light, and the viewing screen is 2.80 m away from the slits. (a) What is the phase difference between the two interfering waves on a screen at a point 2.50 mm from the central bright fringe? (b) What is the ratio of the intensity at this point to the intensity at the center of a bright fringe?

Section 37.4 Change of Phase Due to Reflection

Section 37.5 Interference in Thin Films

30. A soap bubble (n = 1.33) floating in air has the shape of a spherical shell with a wall thickness of 120 nm. (a) What is the wavelength of the visible light that is most strongly reflected? (b) Explain how a bubble of different thickness could also strongly reflect light of this same wavelength. (c) Find the two smallest film thicknesses larger than 120 nm that can produce strongly reflected light of the same wavelength.

31. A thin film of oil (n = 1.25) is located on smooth, **W** wet pavement. When viewed perpendicular to the pavement, the film reflects most strongly red light at 640 nm and reflects no green light at 512 nm. How thick is the oil film?

32. A material having an index of refraction of 1.30 is used as an antireflective coating on a piece of glass (n = 1.50). What should the minimum thickness of this film be to minimize reflection of 500-nm light?

33. A possible means for making an airplane invisible to **M** radar is to coat the plane with an antireflective polymer. If radar waves have a wavelength of 3.00 cm and the index of refraction of the polymer is n = 1.50, how thick would you make the coating?

34. A film of MgF_2 (n = 1.38) having thickness 1.00×10^{-5} cm is used to coat a camera lens. (a) What are the three longest wavelengths that are intensified in the reflected light? (b) Are any of these wavelengths in the visible spectrum?

35. A beam of 580-nm light passes through two closely **W** spaced glass plates at close to normal incidence as shown in Figure P37.35. For what minimum nonzero value of the plate separation d is the transmitted light bright?

36. An oil film (n = 1.45) floating on water is illumi-**M** nated by white light at normal incidence. The film is 280 nm thick. Find (a) the wavelength and color of the light in the visible spectrum most strongly reflected and (b) the wavelength and color of the light in the spectrum most strongly transmitted. Explain your reasoning.

37. An air wedge is formed between two glass plates sepa-**M** rated at one edge by a very fine wire of circular cross section as shown in Figure P37.37. When the wedge is illuminated from above by 600-nm light and viewed from above, 30 dark fringes are observed. Calculate the diameter d of the wire.

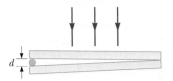

Figure P37.37 Problems 37, 41, 49, and 59.

38. Astronomers observe the chromosphere of the Sun with a filter that passes the red hydrogen spectral line of wavelength 656.3 nm, called the H_α line. The filter consists of a transparent dielectric of thickness d held between two partially aluminized glass plates. The filter is held at a constant temperature. (a) Find the minimum value of d that produces maximum transmission of perpendicular H_α light if the dielectric has an index of refraction of 1.378. (b) **What If?** If the temperature of the filter increases above the normal value, increasing its thickness, what happens to the transmitted wavelength? (c) The dielectric will also pass what near-visible wavelength? One of the glass plates is colored red to absorb this light.

39. When a liquid is introduced into the air space between **W** the lens and the plate in a Newton's-rings apparatus, the diameter of the tenth ring changes from 1.50 to 1.31 cm. Find the index of refraction of the liquid.

40. A lens made of glass (n_g = 1.52) is coated with a thin film of MgF_2 (n_s = 1.38) of thickness t. Visible light is incident normally on the coated lens as in Figure P37.40. (a) For what minimum value of t will the

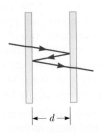

Figure P37.35

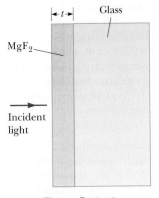

Figure P37.40

reflected light of wavelength 540 nm (in air) be missing? (b) Are there other values of t that will minimize the reflected light at this wavelength? Explain.

41. Two glass plates 10.0 cm long are in contact at one end and separated at the other end by a thread with a diameter $d = 0.050\,0$ mm (Fig. P37.37). Light containing the two wavelengths 400 nm and 600 nm is incident perpendicularly and viewed by reflection. At what distance from the contact point is the next dark fringe?

Section 37.6 The Michelson Interferometer

42. Mirror M_1 in Figure 37.13 is moved through a displacement ΔL. During this displacement, 250 fringe reversals (formation of successive dark or bright bands) are counted. The light being used has a wavelength of 632.8 nm. Calculate the displacement ΔL.

43. The Michelson interferometer can be used to measure the index of refraction of a gas by placing an evacuated transparent tube in the light path along one arm of the device. Fringe shifts occur as the gas is slowly added to the tube. Assume 600-nm light is used, the tube is 5.00 cm long, and 160 bright fringes pass on the screen as the pressure of the gas in the tube increases to atmospheric pressure. What is the index of refraction of the gas? *Hint:* The fringe shifts occur because the wavelength of the light changes inside the gas-filled tube.

44. One leg of a Michelson interferometer contains an evacuated cylinder of length L, having glass plates on each end. A gas is slowly leaked into the cylinder until a pressure of 1 atm is reached. If N bright fringes pass on the screen during this process when light of wavelength λ is used, what is the index of refraction of the gas? *Hint:* The fringe shifts occur because the wavelength of the light changes inside the gas-filled tube.

Additional Problems

45. Radio transmitter A operating at 60.0 MHz is 10.0 m from another similar transmitter B that is 180° out of phase with A. How far must an observer move from A toward B along the line connecting the two transmitters to reach the nearest point where the two beams are in phase?

46. A room is 6.0 m long and 3.0 m wide. At the front of the room, along one of the 3.0-m-wide walls, two loudspeakers are set 1.0 m apart, with the center point between them coinciding with the center point of the wall. The speakers emit a sound wave of a single frequency and a maximum in sound intensity is heard at the center of the back wall, 6.0 m from the speakers. What is the highest possible frequency of the sound from the speakers if no other maxima are heard anywhere along the back wall?

47. In an experiment similar to that of Example 37.1, green light with wavelength 560 nm, sent through a pair of slits 30.0 μm apart, produces bright fringes 2.24 cm apart on a screen 1.20 m away. If the apparatus is now submerged in a tank containing a sugar solution

with index of refraction 1.38, calculate the fringe separation for this same arrangement.

48. In the What If? section of Example 37.2, it was claimed that overlapping fringes in a two-slit interference pattern for two different wavelengths obey the following relationship even for large values of the angle θ:

$$\frac{m'}{m} = \frac{\lambda}{\lambda'}$$

(a) Prove this assertion. (b) Using the data in Example 37.2, find the nonzero value of y on the screen at which the fringes from the two wavelengths first coincide.

49. An investigator finds a fiber at a crime scene that he wishes to use as evidence against a suspect. He gives the fiber to a technician to test the properties of the fiber. To measure the diameter d of the fiber, the technician places it between two flat glass plates at their ends as in Figure P37.37. When the plates, of length 14.0 cm, are illuminated from above with light of wavelength 650 nm, she observes interference bands separated by 0.580 mm. What is the diameter of the fiber?

50. Raise your hand and hold it flat. Think of the space between your index finger and your middle finger as one slit and think of the space between middle finger and ring finger as a second slit. (a) Consider the interference resulting from sending coherent visible light perpendicularly through this pair of openings. Compute an order-of-magnitude estimate for the angle between adjacent zones of constructive interference. (b) To make the angles in the interference pattern easy to measure with a plastic protractor, you should use an electromagnetic wave with frequency of what order of magnitude? (c) How is this wave classified on the electromagnetic spectrum?

51. Two coherent waves, coming from sources at different locations, move along the x axis. Their wave functions are

$$E_1 = 860 \sin\left[\frac{2\pi x_1}{650} - 924\pi t + \frac{\pi}{6}\right]$$

and

$$E_2 = 860 \sin\left[\frac{2\pi x_2}{650} - 924\pi t + \frac{\pi}{8}\right]$$

where E_1 and E_2 are in volts per meter, x_1 and x_2 are in nanometers, and t is in picoseconds. When the two waves are superposed, determine the relationship between x_1 and x_2 that produces constructive interference.

52. In a Young's interference experiment, the two slits are separated by 0.150 mm and the incident light includes two wavelengths: $\lambda_1 = 540$ nm (green) and $\lambda_2 = 450$ nm (blue). The overlapping interference patterns are observed on a screen 1.40 m from the slits. Calculate the minimum distance from the center of the screen to a point where a bright fringe of the green light coincides with a bright fringe of the blue light.

53. In a Young's double-slit experiment using light of wavelength λ, a thin piece of Plexiglas having index of refraction n covers one of the slits. If the center point

on the screen is a dark spot instead of a bright spot, what is the minimum thickness of the Plexiglas?

54. **Review.** A flat piece of glass is held stationary and
AMT horizontal above the highly polished, flat top end of a 10.0-cm-long vertical metal rod that has its lower end rigidly fixed. The thin film of air between the rod and glass is observed to be bright by reflected light when it is illuminated by light of wavelength 500 nm. As the temperature is slowly increased by 25.0°C, the film changes from bright to dark and back to bright 200 times. What is the coefficient of linear expansion of the metal?

55. A certain grade of crude oil has an index of refraction of 1.25. A ship accidentally spills $1.00 \ m^3$ of this oil into the ocean, and the oil spreads into a thin, uniform slick. If the film produces a first-order maximum of light of wavelength 500 nm normally incident on it, how much surface area of the ocean does the oil slick cover? Assume the index of refraction of the ocean water is 1.34.

56. The waves from a radio station can reach a home receiver by two paths. One is a straight-line path from transmitter to home, a distance of 30.0 km. The second is by reflection from the ionosphere (a layer of ionized air molecules high in the atmosphere). Assume this reflection takes place at a point midway between receiver and transmitter, the wavelength broadcast by the radio station is 350 m, and no phase change occurs on reflection. Find the minimum height of the ionospheric layer that could produce destructive interference between the direct and reflected beams.

57. Interference effects are produced at point P on a screen as a result of direct rays from a 500-nm source and reflected rays from the mirror as shown in Figure P37.57. Assume the source is 100 m to the left of the screen and 1.00 cm above the mirror. Find the distance y to the first dark band above the mirror.

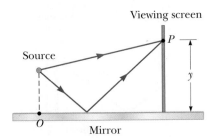

Figure P37.57

58. Measurements are made of the intensity distribution within the central bright fringe in a Young's interference pattern (see Fig. 37.6). At a particular value of y, it is found that $I/I_{max} = 0.810$ when 600-nm light is used. What wavelength of light should be used to reduce the relative intensity at the same location to 64.0% of the maximum intensity?

59. Many cells are transparent and colorless. Structures of great interest in biology and medicine can be practically invisible to ordinary microscopy. To indicate the size and shape of cell structures, an *interference micro-*

scope reveals a difference in index of refraction as a shift in interference fringes. The idea is exemplified in the following problem. An air wedge is formed between two glass plates in contact along one edge and slightly separated at the opposite edge as in Figure P37.37. When the plates are illuminated with monochromatic light from above, the reflected light has 85 dark fringes. Calculate the number of dark fringes that appear if water ($n = 1.33$) replaces the air between the plates.

60. Consider the double-slit arrangement shown in Figure
W P37.60, where the slit separation is d and the distance from the slit to the screen is L. A sheet of transparent plastic having an index of refraction n and thickness t is placed over the upper slit. As a result, the central maximum of the interference pattern moves upward a distance y'. Find y'.

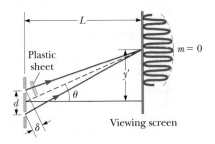

Figure P37.60

61. Figure P37.61 shows a radio-wave transmitter and a receiver separated by a distance $d = 50.0$ m and both a distance $h = 35.0$ m above the ground. The receiver can receive signals both directly from the transmitter and indirectly from signals that reflect from the ground. Assume the ground is level between the transmitter and receiver and a 180° phase shift occurs upon reflection. Determine the longest wavelengths that interfere (a) constructively and (b) destructively.

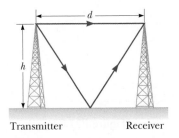

Figure P37.61 Problems 61 and 62.

62. Figure P37.61 shows a radio-wave transmitter and a receiver separated by a distance d and both a distance h above the ground. The receiver can receive signals both directly from the transmitter and indirectly from signals that reflect from the ground. Assume the ground is level between the transmitter and receiver and a 180° phase shift occurs upon reflection. Determine the longest wavelengths that interfere (a) constructively and (b) destructively.

63. In a Newton's-rings experiment, a plano-convex glass ($n = 1.52$) lens having radius $r = 5.00$ cm is placed on a flat plate as shown in Figure P37.63 (page 1158). When

light of wavelength $\lambda = 650$ nm is incident normally, 55 bright rings are observed, with the last one precisely on the edge of the lens. (a) What is the radius R of curvature of the convex surface of the lens? (b) What is the focal length of the lens?

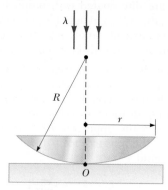

Figure P37.63

64. *Why is the following situation impossible?* A piece of transparent material having an index of refraction $n = 1.50$ is cut into the shape of a wedge as shown in Figure P37.64. Both the top and bottom surfaces of the wedge are in contact with air. Monochromatic light of wavelength $\lambda = 632.8$ nm is normally incident from above, and the wedge is viewed from above. Let $h = 1.00$ mm represent the height of the wedge and $\ell = 0.500$ m its length. A thin-film interference pattern appears in the wedge due to reflection from the top and bottom surfaces. You have been given the task of counting the number of bright fringes that appear in the entire length ℓ of the wedge. You find this task tedious, and your concentration is broken by a noisy distraction after accurately counting 5 000 bright fringes.

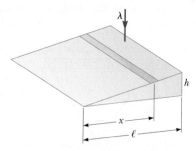

Figure P37.64

65. A plano-concave lens having index of refraction 1.50 is placed on a flat glass plate as shown in Figure P37.65. Its curved surface, with radius of curvature 8.00 m, is on the bottom. The lens is illuminated from above with yellow sodium light of wavelength 589 nm, and a series of concentric bright and dark rings is observed by reflection. The interference pattern has a dark spot

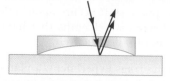

Figure P37.65

at the center that is surrounded by 50 dark rings, the largest of which is at the outer edge of the lens. (a) What is the thickness of the air layer at the center of the interference pattern? (b) Calculate the radius of the outermost dark ring. (c) Find the focal length of the lens.

66. A plano-convex lens has index of refraction n. The curved side of the lens has radius of curvature R and rests on a flat glass surface of the same index of refraction, with a film of index n_{film} between them, as shown in Figure 37.66. The lens is illuminated from above by light of wavelength λ. Show that the dark Newton's rings have radii given approximately by

$$r \approx \sqrt{\frac{m\lambda R}{n_{film}}}$$

where $r << R$ and m is an integer.

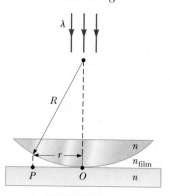

Figure P37.66

67. Interference fringes are produced using Lloyd's mirror and a source S of wavelength $\lambda = 606$ nm as shown in Figure P37.67. Fringes separated by $\Delta y = 1.20$ mm are formed on a screen a distance $L = 2.00$ m from the source. Find the vertical distance h of the source above the reflecting surface.

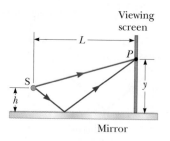

Figure P37.67

68. The quantity nt in Equations 37.17 and 37.18 is called the *optical path length* corresponding to the geometrical distance t and is analogous to the quantity δ in Equation 37.1, the path difference. The optical path length is proportional to n because a larger index of refraction shortens the wavelength, so more cycles of a wave fit into a particular geometrical distance. (a) Assume a mixture of corn syrup and water is prepared in a tank, with its index of refraction n increasing uniformly from 1.33 at $y = 20.0$ cm at the top to 1.90 at $y = 0$. Write the index of refraction $n(y)$ as a function of y.

(b) Compute the optical path length corresponding to the 20.0-cm height of the tank by calculating

$$\int_0^{20\,\text{cm}} n(y)\,dy$$

(c) Suppose a narrow beam of light is directed into the mixture at a nonzero angle with respect to the normal to the surface of the mixture. Qualitatively describe its path.

69. Astronomers observe a 60.0-MHz radio source both **M** directly and by reflection from the sea as shown in Figure P37.17. If the receiving dish is 20.0 m above sea level, what is the angle of the radio source above the horizon at first maximum?

70. Figure CQ37.2 shows an unbroken soap film in a circular frame. The film thickness increases from top to bottom, slowly at first and then rapidly. As a simpler model, consider a soap film ($n = 1.33$) contained within a rectangular wire frame. The frame is held vertically so that the film drains downward and forms a wedge with flat faces. The thickness of the film at the top is essentially zero. The film is viewed in reflected white light with near-normal incidence, and the first violet ($\lambda = 420$ nm) interference band is observed 3.00 cm from the top edge of the film. (a) Locate the first red ($\lambda = 680$ nm) interference band. (b) Determine the film thickness at the positions of the violet and red bands. (c) What is the wedge angle of the film?

Challenge Problems

71. Our discussion of the techniques for determining constructive and destructive interference by reflection from a thin film in air has been confined to rays striking the film at nearly normal incidence. **What If?** Assume a ray is incident at an angle of 30.0° (relative to the normal) on a film with index of refraction 1.38 surrounded by vacuum. Calculate the minimum thickness for constructive interference of sodium light with a wavelength of 590 nm.

72. The condition for constructive interference by reflection from a thin film in air as developed in Section 37.5 assumes nearly normal incidence. **What If?** Suppose the light is incident on the film at a nonzero angle θ_1 (relative to the normal). The index of refraction of the film is n, and the film is surrounded by vacuum. Find the condition for constructive interference that relates the thickness t of the film, the index of refraction n of the film, the wavelength λ of the light, and the angle of incidence θ_1.

73. Both sides of a uniform film that has index of refraction n and thickness d are in contact with air. For normal incidence of light, an intensity minimum is observed in the reflected light at λ_2 and an intensity maximum is observed at λ_1, where $\lambda_1 > \lambda_2$. (a) Assuming no intensity minima are observed between λ_1 and λ_2, find an expression for the integer m in Equations 37.17 and 37.18 in terms of the wavelengths λ_1 and λ_2. (b) Assuming $n = 1.40$, $\lambda_1 = 500$ nm, and $\lambda_2 = 370$ nm, determine the best estimate for the thickness of the film.

74. Slit 1 of a double slit is wider than slit 2 so that the light from slit 1 has an amplitude 3.00 times that of the light from slit 2. Show that Equation 37.13 is replaced by the equation $I = I_{\text{max}}(1 + 3\cos^2 \phi/2)$ for this situation.

75. Monochromatic light of wavelength 620 nm passes through a very narrow slit S and then strikes a screen in which are two parallel slits, S_1 and S_2, as shown in Figure P37.75. Slit S_1 is directly in line with S and at a distance of $L = 1.20$ m away from S, whereas S_2 is displaced a distance d to one side. The light is detected at point P on a second screen, equidistant from S_1 and S_2. When either slit S_1 or S_2 is open, equal light intensities are measured at point P. When both slits are open, the intensity is three times larger. Find the minimum possible value for the slit separation d.

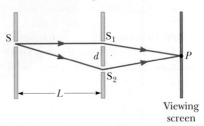

Figure P37.75

76. A plano-convex lens having a radius of curvature of $r = 4.00$ m is placed on a concave glass surface whose radius of curvature is $R = 12.0$ m as shown in Figure P37.76. Assuming 500-nm light is incident normal to the flat surface of the lens, determine the radius of the 100th bright ring.

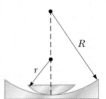

Figure P37.76

Diffraction Patterns and Polarization

The Hubble Space Telescope does its viewing above the atmosphere and does not suffer from the atmospheric blurring, caused by air turbulence, that plagues ground-based telescopes. Despite this advantage, it does have limitations due to diffraction effects. In this chapter, we show how the wave nature of light limits the ability of any optical system to distinguish between closely spaced objects. *(NASA Hubble Space Telescope Collection)*

When plane light waves pass through a small aperture in an opaque barrier, the aperture acts as if it were a point source of light, with waves entering the shadow region behind the barrier. This phenomenon, known as diffraction, was first mentioned in Section 35.3, and can be described only with a wave model for light. In this chapter, we investigate the features of the *diffraction pattern* that occurs when the light from the aperture is allowed to fall upon a screen.

In Chapter 34, we learned that electromagnetic waves are transverse. That is, the electric and magnetic field vectors associated with electromagnetic waves are perpendicular to the direction of wave propagation. In this chapter, we show that under certain conditions these transverse waves with electric field vectors in all possible transverse directions can be *polarized* in various ways. In other words, only certain directions of the electric field vectors are present in the polarized wave.

38.1 Introduction to Diffraction Patterns

In Sections 35.3 and 37.1, we discussed that light of wavelength comparable to or larger than the width of a slit spreads out in all forward directions upon passing through the slit. This phenomenon is called *diffraction*. When light passes through a narrow slit, it spreads beyond the narrow path defined by the slit into regions that would be in shadow if light traveled in straight lines. Other waves, such as sound waves and water waves, also have this property of spreading when passing through apertures or by sharp edges.

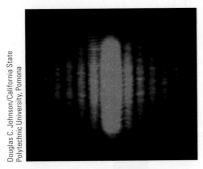

Figure 38.1 The diffraction pattern that appears on a screen when light passes through a narrow vertical slit. The pattern consists of a broad central fringe and a series of less intense and narrower side fringes.

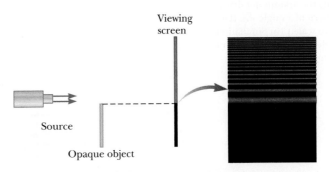

Figure 38.2 Light from a small source passes by the edge of an opaque object and continues on to a screen. A diffraction pattern consisting of bright and dark fringes appears on the screen in the region above the edge of the object.

You might expect that the light passing through a small opening would simply result in a broad region of light on a screen due to the spreading of the light as it passes through the opening. We find something more interesting, however. A **diffraction pattern** consisting of light and dark areas is observed, somewhat similar to the interference patterns discussed earlier. For example, when a narrow slit is placed between a distant light source (or a laser beam) and a screen, the light produces a diffraction pattern like that shown in Figure 38.1. The pattern consists of a broad, intense central band (called the **central maximum**) flanked by a series of narrower, less intense additional bands (called **side maxima** or **secondary maxima**) and a series of intervening dark bands (or **minima**). Figure 38.2 shows a diffraction pattern associated with light passing by the edge of an object. Again we see bright and dark fringes, which is reminiscent of an interference pattern.

Figure 38.3 shows a diffraction pattern associated with the shadow of a penny. A bright spot occurs at the center, and circular fringes extend outward from the shadow's edge. We can explain the central bright spot by using the wave theory of light, which predicts constructive interference at this point. From the viewpoint of ray optics (in which light is viewed as rays traveling in straight lines), we expect the center of the shadow to be dark because that part of the viewing screen is completely shielded by the penny.

Shortly before the central bright spot was first observed, one of the supporters of ray optics, Simeon Poisson, argued that if Augustin Fresnel's wave theory of light were valid, a central bright spot should be observed in the shadow of a circular object illuminated by a point source of light. To Poisson's astonishment, the spot was observed by Dominique Arago shortly thereafter. Therefore, Poisson's prediction reinforced the wave theory rather than disproving it.

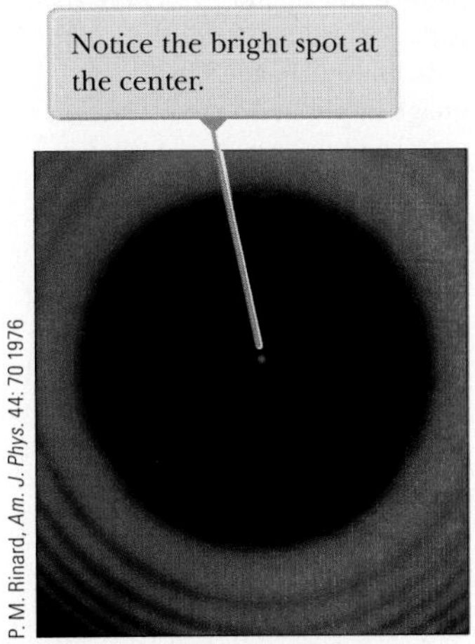

Notice the bright spot at the center.

Figure 38.3 Diffraction pattern created by the illumination of a penny, with the penny positioned midway between the screen and light source.

38.2 Diffraction Patterns from Narrow Slits

Let's consider a common situation, that of light passing through a narrow opening modeled as a slit and projected onto a screen. To simplify our analysis, we assume the observing screen is far from the slit and the rays reaching the screen are approximately parallel. (This situation can also be achieved experimentally by using a converging lens to focus the parallel rays on a nearby screen.) In this model, the pattern on the screen is called a **Fraunhofer diffraction pattern**.[1]

Figure 38.4a (page 1162) shows light entering a single slit from the left and diffracting as it propagates toward a screen. Figure 38.4b shows the fringe structure of

---

[1]If the screen is brought close to the slit (and no lens is used), the pattern is a *Fresnel* diffraction pattern. The Fresnel pattern is more difficult to analyze, so we shall restrict our discussion to Fraunhofer diffraction.

Figure 38.4 (a) Geometry for analyzing the Fraunhofer diffraction pattern of a single slit. (Drawing not to scale.) (b) Simulation of a single-slit Fraunhofer diffraction pattern.

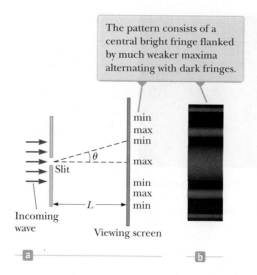

The pattern consists of a central bright fringe flanked by much weaker maxima alternating with dark fringes.

Pitfall Prevention 38.1

Diffraction Versus Diffraction Pattern *Diffraction* refers to the general behavior of waves spreading out as they pass through a slit. We used diffraction in explaining the existence of an interference pattern in Chapter 37. A *diffraction pattern* is actually a misnomer, but is deeply entrenched in the language of physics. The diffraction pattern seen on a screen when a single slit is illuminated is actually another interference pattern. The interference is between parts of the incident light illuminating different regions of the slit.

a Fraunhofer diffraction pattern. A bright fringe is observed along the axis at $\theta = 0$, with alternating dark and bright fringes on each side of the central bright fringe.

Until now, we have assumed slits are point sources of light. In this section, we abandon that assumption and see how the finite width of slits is the basis for understanding Fraunhofer diffraction. We can explain some important features of this phenomenon by examining waves coming from various portions of the slit as shown in Figure 38.5. According to Huygens's principle, each portion of the slit acts as a source of light waves. Hence, light from one portion of the slit can interfere with light from another portion, and the resultant light intensity on a viewing screen depends on the direction θ. Based on this analysis, we recognize that a diffraction pattern is actually an interference pattern in which the different sources of light are different portions of the single slit! Therefore, the diffraction patterns we discuss in this chapter are applications of the waves in interference analysis model.

To analyze the diffraction pattern, let's divide the slit into two halves as shown in Figure 38.5. Keeping in mind that all the waves are in phase as they leave the slit, consider rays 1 and 3. As these two rays travel toward a viewing screen far to the right of the figure, ray 1 travels farther than ray 3 by an amount equal to the path difference $(a/2) \sin \theta$, where a is the width of the slit. Similarly, the path difference between rays 2 and 4 is also $(a/2) \sin \theta$, as is that between rays 3 and 5. If this path difference is exactly half a wavelength (corresponding to a phase difference of 180°), the pairs of waves cancel each other and destructive interference results. This cancellation occurs for any two rays that originate at points separated by half the slit width because the phase difference between two such points is 180°. Therefore, waves from the upper half of the slit interfere destructively with waves from the lower half when

$$\frac{a}{2} \sin \theta = \frac{\lambda}{2}$$

or, if we consider waves at angle θ both above the dashed line in Figure 38.5 and below,

$$\sin \theta = \pm \frac{\lambda}{a}$$

Dividing the slit into four equal parts and using similar reasoning, we find that the viewing screen is also dark when

$$\sin \theta = \pm 2 \frac{\lambda}{a}$$

Likewise, dividing the slit into six equal parts shows that darkness occurs on the screen when

$$\sin \theta = \pm 3 \frac{\lambda}{a}$$

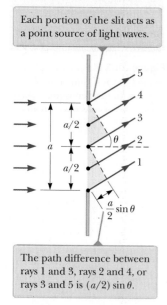

Each portion of the slit acts as a point source of light waves.

The path difference between rays 1 and 3, rays 2 and 4, or rays 3 and 5 is $(a/2) \sin \theta$.

Figure 38.5 Paths of light rays that encounter a narrow slit of width a and diffract toward a screen in the direction described by angle θ (not to scale).

Therefore, the general condition for destructive interference is

$$\sin \theta_{\text{dark}} = m\frac{\lambda}{a} \qquad m = \pm 1, \pm 2, \pm 3, \ldots \qquad (38.1)$$

◀ Condition for destructive interference for a single slit

This equation gives the values of θ_{dark} for which the diffraction pattern has zero light intensity, that is, when a dark fringe is formed. It tells us nothing, however, about the variation in light intensity along the screen. The general features of the intensity distribution are shown in Figure 38.4. A broad, central bright fringe is observed; this fringe is flanked by much weaker bright fringes alternating with dark fringes. The various dark fringes occur at the values of θ_{dark} that satisfy Equation 38.1. Each bright-fringe peak lies approximately halfway between its bordering dark-fringe minima. Notice that the central bright maximum is twice as wide as the secondary maxima. There is no central dark fringe, represented by the absence of $m = 0$ in Equation 38.1.

> **Pitfall Prevention 38.2**
> **Similar Equation Warning!** Equation 38.1 has exactly the same form as Equation 37.2, with d, the slit separation, used in Equation 37.2 and a, the slit width, used in Equation 38.1. Equation 37.2, however, describes the *bright* regions in a two-slit interference pattern, whereas Equation 38.1 describes the *dark* regions in a single-slit diffraction pattern.

Quick Quiz 38.1 Suppose the slit width in Figure 38.4 is made half as wide. Does the central bright fringe (a) become wider, (b) remain the same, or (c) become narrower?

Example 38.1   Where Are the Dark Fringes?   AM

Light of wavelength 580 nm is incident on a slit having a width of 0.300 mm. The viewing screen is 2.00 m from the slit. Find the width of the central bright fringe.

SOLUTION

Conceptualize Based on the problem statement, we imagine a single-slit diffraction pattern similar to that in Figure 38.4.

Categorize We categorize this example as a straightforward application of our discussion of single-slit diffraction patterns, which comes from the *waves in interference* analysis model.

Analyze Evaluate Equation 38.1 for the two dark fringes that flank the central bright fringe, which correspond to $m = \pm 1$:

$$\sin \theta_{\text{dark}} = \pm\frac{\lambda}{a}$$

Let y represent the vertical position along the viewing screen in Figure 38.4a, measured from the point on the screen directly behind the slit. Then, $\tan \theta_{\text{dark}} = y_1/L$, where the subscript 1 refers to the first dark fringe. Because θ_{dark} is very small, we can use the approximation $\sin \theta_{\text{dark}} \approx \tan \theta_{\text{dark}}$; therefore, $y_1 = L \sin \theta_{\text{dark}}$.

The width of the central bright fringe is twice the absolute value of y_1:

$$2|y_1| = 2|L \sin \theta_{\text{dark}}| = 2\left|\pm L\frac{\lambda}{a}\right| = 2L\frac{\lambda}{a} = 2(2.00 \text{ m})\frac{580 \times 10^{-9} \text{ m}}{0.300 \times 10^{-3} \text{ m}}$$

$$= 7.73 \times 10^{-3} \text{ m} = \boxed{7.73 \text{ mm}}$$

Finalize Notice that this value is much greater than the width of the slit. Let's explore below what happens if we change the slit width.

WHAT IF? What if the slit width is increased by an order of magnitude to 3.00 mm? What happens to the diffraction pattern?

Answer Based on Equation 38.1, we expect that the angles at which the dark bands appear will decrease as a increases. Therefore, the diffraction pattern narrows.

Repeat the calculation with the larger slit width:

$$2|y_1| = 2L\frac{\lambda}{a} = 2(2.00 \text{ m})\frac{580 \times 10^{-9} \text{ m}}{3.00 \times 10^{-3} \text{ m}} = 7.73 \times 10^{-4} \text{ m} = \boxed{0.773 \text{ mm}}$$

Notice that this result is *smaller* than the width of the slit. In general, for large values of a, the various maxima and minima are so closely spaced that only a large, central bright area resembling the geometric image of the slit is observed. This concept is very important in the performance of optical instruments such as telescopes.

Intensity of Single-Slit Diffraction Patterns

Analysis of the intensity variation in a diffraction pattern from a single slit of width a shows that the intensity is given by

◄ **Intensity of a single-slit Fraunhofer diffraction pattern**

$$I = I_{max} \left[\frac{\sin(\pi a \sin \theta / \lambda)}{\pi a \sin \theta / \lambda} \right]^2 \qquad \text{(38.2)}$$

where I_{max} is the intensity at $\theta = 0$ (the central maximum) and λ is the wavelength of light used to illuminate the slit. This expression shows that *minima* occur when

$$\frac{\pi a \sin \theta_{dark}}{\lambda} = m\pi$$

or

◄ **Condition for intensity minima for a single slit**

$$\sin \theta_{dark} = m \frac{\lambda}{a} \quad m = \pm 1, \pm 2, \pm 3, \ldots$$

in agreement with Equation 38.1.

Figure 38.6a represents a plot of the intensity in the single-slit pattern as given by Equation 38.2, and Figure 38.6b is a simulation of a single-slit Fraunhofer diffraction pattern. Notice that most of the light intensity is concentrated in the central bright fringe.

Intensity of Two-Slit Diffraction Patterns

When more than one slit is present, we must consider not only diffraction patterns due to the individual slits but also the interference patterns due to the waves coming from different slits. Notice the curved dashed lines in Figure 37.7 in Chapter 37, which indicate a decrease in intensity of the interference maxima as θ increases. This decrease is due to a diffraction pattern. The interference patterns in that figure are located entirely within the central bright fringe of the diffraction pattern, so the only hint of the diffraction pattern we see is the falloff in intensity toward the outside of the pattern. To determine the effects of both two-slit interference and a single-slit diffraction pattern from each slit from a wider viewpoint than that in Figure 37.7, we combine Equations 37.14 and 38.2:

$$I = I_{max} \cos^2 \left(\frac{\pi d \sin \theta}{\lambda} \right) \left[\frac{\sin(\pi a \sin \theta / \lambda)}{\pi a \sin \theta / \lambda} \right]^2 \qquad \text{(38.3)}$$

Although this expression looks complicated, it merely represents the single-slit diffraction pattern (the factor in square brackets) acting as an "envelope" for a two-slit interference pattern (the cosine-squared factor) as shown in Figure 38.7. The broken

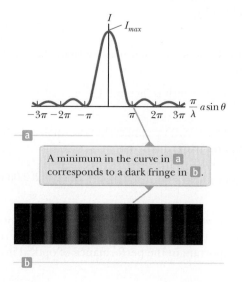

A minimum in the curve in **a** corresponds to a dark fringe in **b**.

Figure 38.6 (a) A plot of light intensity I versus $(\pi/\lambda)a \sin \theta$ for the single-slit Fraunhofer diffraction pattern. (b) Simulation of a single-slit Fraunhofer diffraction pattern.

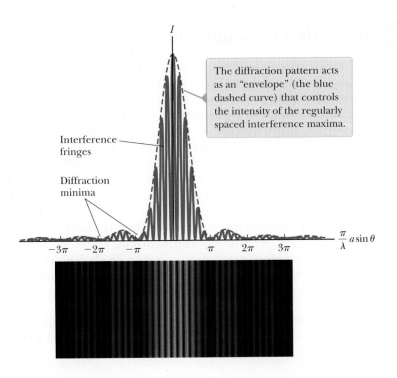

Figure 38.7 The combined effects of two-slit and single-slit interference. This pattern is produced when 650-nm light waves pass through two 3.0-μm slits that are 18 μm apart.

blue curve in Figure 38.7 represents the factor in square brackets in Equation 38.3. The cosine-squared factor by itself would give a series of peaks all with the same height as the highest peak of the red-brown curve in Figure 38.7. Because of the effect of the square-bracket factor, however, these peaks vary in height as shown.

Equation 37.2 indicates the conditions for interference maxima as $d \sin \theta = m\lambda$, where d is the distance between the two slits. Equation 38.1 specifies that the first diffraction minimum occurs when $a \sin \theta = \lambda$, where a is the slit width. Dividing Equation 37.2 by Equation 38.1 (with $m = 1$) allows us to determine which interference maximum coincides with the first diffraction minimum:

$$\frac{d \sin \theta}{a \sin \theta} = \frac{m\lambda}{\lambda}$$

$$\frac{d}{a} = m \tag{38.4}$$

In Figure 38.7, $d/a = 18 \ \mu\text{m}/3.0 \ \mu\text{m} = 6$. Therefore, the sixth interference maximum (if we count the central maximum as $m = 0$) is aligned with the first diffraction minimum and is dark.

Quick Quiz 38.2 Consider the central peak in the diffraction envelope in Figure 38.7 and look closely at the horizontal scale. Suppose the wavelength of the light is changed to 450 nm. What happens to this central peak? **(a)** The width of the peak decreases, and the number of interference fringes it encloses decreases. **(b)** The width of the peak decreases, and the number of interference fringes it encloses increases. **(c)** The width of the peak decreases, and the number of interference fringes it encloses remains the same. **(d)** The width of the peak increases, and the number of interference fringes it encloses decreases. **(e)** The width of the peak increases, and the number of interference fringes it encloses increases. **(f)** The width of the peak increases, and the number of interference fringes it encloses remains the same. **(g)** The width of the peak remains the same, and the number of interference fringes it encloses decreases. **(h)** The width of the peak remains the same, and the number of interference fringes it encloses increases. **(i)** The width of the peak remains the same, and the number of interference fringes it encloses remains the same.

38.3 Resolution of Single-Slit and Circular Apertures

The ability of optical systems to distinguish between closely spaced objects is limited because of the wave nature of light. To understand this limitation, consider Figure 38.8, which shows two light sources far from a narrow slit of width a. The sources can be two noncoherent point sources S_1 and S_2; for example, they could be two distant stars. If no interference occurred between light passing through different parts of the slit, two distinct bright spots (or images) would be observed on the viewing screen. Because of such interference, however, each source is imaged as a bright central region flanked by weaker bright and dark fringes, a diffraction pattern. What is observed on the screen is the sum of two diffraction patterns: one from S_1 and the other from S_2.

If the two sources are far enough apart to keep their central maxima from overlapping as in Figure 38.8a, their images can be distinguished and are said to be *resolved*. If the sources are close together as in Figure 38.8b, however, the two central maxima overlap and the images are not resolved. To determine whether two images are resolved, the following condition is often used:

> When the central maximum of one image falls on the first minimum of another image, the images are said to be just resolved. This limiting condition of resolution is known as **Rayleigh's criterion.**

From Rayleigh's criterion, we can determine the minimum angular separation θ_{min} subtended by the sources at the slit in Figure 38.8 for which the images are just resolved. Equation 38.1 indicates that the first minimum in a single-slit diffraction pattern occurs at the angle for which

$$\sin \theta = \frac{\lambda}{a}$$

where a is the width of the slit. According to Rayleigh's criterion, this expression gives the smallest angular separation for which the two images are resolved. Because $\lambda << a$ in most situations, $\sin \theta$ is small and we can use the approximation $\sin \theta \approx \theta$. Therefore, the limiting angle of resolution for a slit of width a is

$$\theta_{min} = \frac{\lambda}{a} \tag{38.5}$$

where θ_{min} is expressed in radians. Hence, the angle subtended by the two sources at the slit must be greater than λ/a if the images are to be resolved.

Many optical systems use circular apertures rather than slits. The diffraction pattern of a circular aperture as shown in the photographs of Figure 38.9 consists of

Figure 38.8 Two point sources far from a narrow slit each produce a diffraction pattern. (a) The sources are separated by a large angle. (b) The sources are separated by a small angle. (Notice that the angles are greatly exaggerated. The drawing is not to scale.)

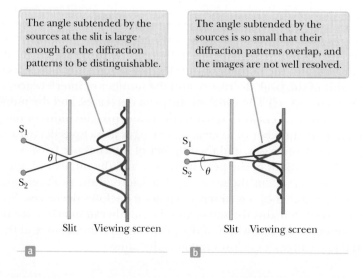

The angle subtended by the sources at the slit is large enough for the diffraction patterns to be distinguishable.

The angle subtended by the sources is so small that their diffraction patterns overlap, and the images are not well resolved.

S_1

S_2

θ

Slit   Viewing screen

a

S_1

S_2

θ

Slit   Viewing screen

b

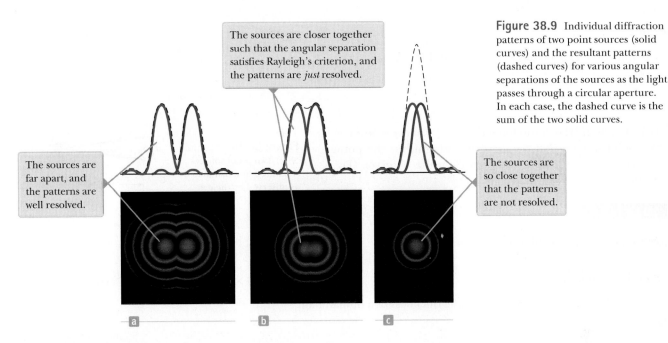

The sources are closer together such that the angular separation satisfies Rayleigh's criterion, and the patterns are *just* resolved.

The sources are far apart, and the patterns are well resolved.

The sources are so close together that the patterns are not resolved.

Figure 38.9 Individual diffraction patterns of two point sources (solid curves) and the resultant patterns (dashed curves) for various angular separations of the sources as the light passes through a circular aperture. In each case, the dashed curve is the sum of the two solid curves.

a central circular bright disk surrounded by progressively fainter bright and dark rings. Figure 38.9 shows diffraction patterns for three situations in which light from two point sources passes through a circular aperture. When the sources are far apart, their images are well resolved (Fig. 38.9a). When the angular separation of the sources satisfies Rayleigh's criterion, the images are just resolved (Fig. 38.9b). Finally, when the sources are close together, the images are said to be unresolved (Fig. 38.9c) and the pattern looks like that of a single source.

Analysis shows that the limiting angle of resolution of the circular aperture is

$$\theta_{min} = 1.22 \frac{\lambda}{D} \qquad (38.6)$$

◀ **Limiting angle of resolution for a circular aperture**

where D is the diameter of the aperture. This expression is similar to Equation 38.5 except for the factor 1.22, which arises from a mathematical analysis of diffraction from the circular aperture.

Quick Quiz 38.3 Cat's eyes have pupils that can be modeled as vertical slits. At night, would cats be more successful in resolving **(a)** headlights on a distant car or **(b)** vertically separated lights on the mast of a distant boat?

Quick Quiz 38.4 Suppose you are observing a binary star with a telescope and are having difficulty resolving the two stars. You decide to use a colored filter to maximize the resolution. (A filter of a given color transmits only that color of light.) What color filter should you choose? **(a)** blue **(b)** green **(c)** yellow **(d)** red

---

Example 38.2   **Resolution of the Eye**

Light of wavelength 500 nm, near the center of the visible spectrum, enters a human eye. Although pupil diameter varies from person to person, let's estimate a daytime diameter of 2 mm.

(A) Estimate the limiting angle of resolution for this eye, assuming its resolution is limited only by diffraction.

SOLUTION

Conceptualize Identify the pupil of the eye as the aperture through which the light travels. Light passing through this small aperture causes diffraction patterns to occur on the retina.

Categorize We determine the result using equations developed in this section, so we categorize this example as a substitution problem.

continued

▶ **38.2** continued

Use Equation 38.6, taking $\lambda = 500$ nm and $D = 2$ mm:

$$\theta_{min} = 1.22\frac{\lambda}{D} = 1.22\left(\frac{5.00 \times 10^{-7}\text{ m}}{2 \times 10^{-3}\text{ m}}\right)$$

$$= 3 \times 10^{-4}\text{ rad} \approx 1\text{ min of arc}$$

(B) Determine the minimum separation distance d between two point sources that the eye can distinguish if the point sources are a distance $L = 25$ cm from the observer (Fig. 38.10).

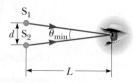

Figure 38.10 (Example 38.2) Two point sources separated by a distance d as observed by the eye.

SOLUTION

Noting that θ_{min} is small, find d:

$$\sin\theta_{min} \approx \theta_{min} \approx \frac{d}{L} \rightarrow d = L\theta_{min}$$

Substitute numerical values:

$$d = (25\text{ cm})(3 \times 10^{-4}\text{ rad}) = 8 \times 10^{-3}\text{ cm}$$

This result is approximately equal to the thickness of a human hair.

Example 38.3 **Resolution of a Telescope**

Each of the two telescopes at the Keck Observatory on the dormant Mauna Kea volcano in Hawaii has an effective diameter of 10 m. What is its limiting angle of resolution for 600-nm light?

SOLUTION

Conceptualize Identify the aperture through which the light travels as the opening of the telescope. Light passing through this aperture causes diffraction patterns to occur in the final image.

Categorize We determine the result using equations developed in this section, so we categorize this example as a substitution problem.

Use Equation 38.6, taking $\lambda = 6.00 \times 10^{-7}$ m and $D = 10$ m:

$$\theta_{min} = 1.22\frac{\lambda}{D} = 1.22\left(\frac{6.00 \times 10^{-7}\text{ m}}{10\text{ m}}\right)$$

$$= 7.3 \times 10^{-8}\text{ rad} \approx 0.015\text{ s of arc}$$

Any two stars that subtend an angle greater than or equal to this value are resolved (if atmospheric conditions are ideal).

WHAT IF? What if we consider radio telescopes? They are much larger in diameter than optical telescopes, but do they have better angular resolutions than optical telescopes? For example, the radio telescope at Arecibo, Puerto Rico, has a diameter of 305 m and is designed to detect radio waves of 0.75-m wavelength. How does its resolution compare with that of one of the Keck telescopes?

Answer The increase in diameter might suggest that radio telescopes would have better resolution than a Keck telescope, but Equation 38.6 shows that θ_{min} depends on *both* diameter and wavelength. Calculating the minimum angle of resolution for the radio telescope, we find

$$\theta_{min} = 1.22\frac{\lambda}{D} = 1.22\left(\frac{0.75\text{ m}}{305\text{ m}}\right)$$

$$= 3.0 \times 10^{-3}\text{ rad} \approx 10\text{ min of arc}$$

This limiting angle of resolution is measured in *minutes* of arc rather than the *seconds* of arc for the optical telescope. Therefore, the change in wavelength more than compensates for the increase in diameter. The limiting angle of resolution for the Arecibo radio telescope is more than 40 000 times larger (that is, *worse*) than the Keck minimum.

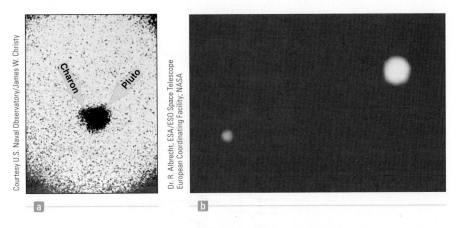

Figure 38.11 (a) The photograph on which Charon, the moon of Pluto, was discovered in 1978. From an Earth-based telescope, atmospheric blurring results in Charon appearing only as a subtle bump on the edge of Pluto. (b) A Hubble Space Telescope photo of Pluto and Charon, clearly resolving the two objects.

A telescope such as the one discussed in Example 38.3 can never reach its diffraction limit because the limiting angle of resolution is always set by atmospheric blurring at optical wavelengths. This seeing limit is usually about 1 s of arc and is never smaller than about 0.1 s of arc. The atmospheric blurring is caused by variations in index of refraction with temperature variations in the air. This blurring is one reason for the superiority of photographs from orbiting telescopes, which view celestial objects from a position above the atmosphere.

As an example of the effects of atmospheric blurring, consider telescopic images of Pluto and its moon, Charon. Figure 38.11a, an image taken in 1978, represents the discovery of Charon. In this photograph, taken from an Earth-based telescope, atmospheric turbulence causes the image of Charon to appear only as a bump on the edge of Pluto. In comparison, Figure 38.11b shows a photograph taken from the Hubble Space Telescope. Without the problems of atmospheric turbulence, Pluto and its moon are clearly resolved.

38.4 The Diffraction Grating

The **diffraction grating,** a useful device for analyzing light sources, consists of a large number of equally spaced parallel slits. A *transmission grating* can be made by cutting parallel grooves on a glass plate with a precision ruling machine. The spaces between the grooves are transparent to the light and hence act as separate slits. A *reflection grating* can be made by cutting parallel grooves on the surface of a reflective material. The reflection of light from the spaces between the grooves is specular, and the reflection from the grooves cut into the material is diffuse. Therefore, the spaces between the grooves act as parallel sources of reflected light like the slits in a transmission grating. Current technology can produce gratings that have very small slit spacings. For example, a typical grating ruled with 5 000 grooves/cm has a slit spacing $d = (1/5\,000)$ cm $= 2.00 \times 10^{-4}$ cm.

A section of a diffraction grating is illustrated in Figure 38.12 (page 1170). A plane wave is incident from the left, normal to the plane of the grating. The pattern observed on the screen far to the right of the grating is the result of the combined effects of interference and diffraction. Each slit produces diffraction, and the diffracted beams interfere with one another to produce the final pattern.

The waves from all slits are in phase as they leave the slits. For an arbitrary direction θ measured from the horizontal, however, the waves must travel different path lengths before reaching the screen. Notice in Figure 38.12 that the path difference δ between rays from any two adjacent slits is equal to $d \sin \theta$. If this path difference equals one wavelength or some integral multiple of a wavelength, waves from all slits are in phase at the screen and a bright fringe is observed. Therefore, the condition for *maxima* in the interference pattern at the angle θ_{bright} is

$$d \sin \theta_{bright} = m\lambda \quad m = 0, \pm 1, \pm 2, \pm 3, \ldots \quad (38.7)$$

◀ Condition for interference maxima for a grating

Pitfall Prevention 38.3
A Diffraction Grating Is an Interference Grating As with *diffraction pattern, diffraction grating* is a misnomer, but is deeply entrenched in the language of physics. The diffraction grating depends on diffraction in the same way as the double slit, spreading the light so that light from different slits can interfere. It would be more correct to call it an *interference grating,* but *diffraction grating* is the name in use.

Figure 38.12 Side view of a diffraction grating. The slit separation is d, and the path difference between adjacent slits is $d \sin \theta$.

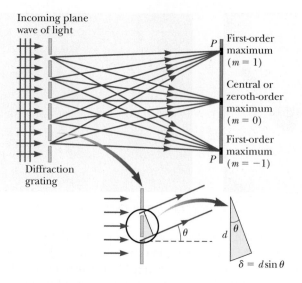

Incoming plane wave of light

Diffraction grating

P First-order maximum ($m = 1$)

Central or zeroth-order maximum ($m = 0$)

P First-order maximum ($m = -1$)

$\delta = d \sin \theta$

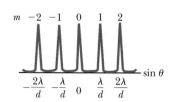

Figure 38.13 Intensity versus $\sin \theta$ for a diffraction grating. The zeroth-, first-, and second-order maxima are shown.

We can use this expression to calculate the wavelength if we know the grating spacing d and the angle θ_{bright}. If the incident radiation contains several wavelengths, the mth-order maximum for each wavelength occurs at a specific angle. All wavelengths are seen at $\theta = 0$, corresponding to $m = 0$, the zeroth-order maximum. The first-order maximum ($m = 1$) is observed at an angle that satisfies the relationship $\sin \theta_{\text{bright}} = \lambda/d$, the second-order maximum ($m = 2$) is observed at a larger angle θ_{bright}, and so on. For the small values of d typical in a diffraction grating, the angles θ_{bright} are large, as we see in Example 38.5.

The intensity distribution for a diffraction grating obtained with the use of a monochromatic source is shown in Figure 38.13. Notice the sharpness of the principal maxima and the broadness of the dark areas compared with the broad bright fringes characteristic of the two-slit interference pattern (see Fig. 37.6). You should also review Figure 37.7, which shows that the width of the intensity maxima decreases as the number of slits increases. Because the principal maxima are so sharp, they are much brighter than two-slit interference maxima.

Quick **Quiz** 38.5 Ultraviolet light of wavelength 350 nm is incident on a diffraction grating with slit spacing d and forms an interference pattern on a screen a distance L away. The angular positions θ_{bright} of the interference maxima are large. The locations of the bright fringes are marked on the screen. Now red light of wavelength 700 nm is used with a diffraction grating to form another diffraction pattern on the screen. Will the bright fringes of this pattern be located at the marks on the screen if **(a)** the screen is moved to a distance $2L$ from the grating, **(b)** the screen is moved to a distance $L/2$ from the grating, **(c)** the grating is replaced with one of slit spacing $2d$, **(d)** the grating is replaced with one of slit spacing $d/2$, or **(e)** nothing is changed?

Conceptual Example 38.4 **A Compact Disc Is a Diffraction Grating**

Light reflected from the surface of a compact disc is multicolored as shown in Figure 38.14. The colors and their intensities depend on the orientation of the CD relative to the eye and relative to the light source. Explain how that works.

Figure 38.14 (Conceptual Example 38.4) A compact disc observed under white light. The colors observed in the reflected light and their intensities depend on the orientation of the CD relative to the eye and relative to the light source.

SOLUTION

The surface of a CD has a spiral grooved track (with adjacent grooves having a separation on

▶ **38.4** continued

the order of 1 μm). Therefore, the surface acts as a reflection grating. The light reflecting from the regions between these closely spaced grooves interferes constructively only in certain directions that depend on the wavelength and the direction of the incident light. Any section of the CD serves as a diffraction grating for white light, sending different colors in different directions. The different colors you see upon viewing one section change when the light source, the CD, or you change position. This change in position causes the angle of incidence or the angle of the diffracted light to be altered.

Example 38.5   The Orders of a Diffraction Grating

Monochromatic light from a helium–neon laser ($\lambda = 632.8$ nm) is incident normally on a diffraction grating containing 6 000 grooves per centimeter. Find the angles at which the first- and second-order maxima are observed.

SOLUTION

Conceptualize Study Figure 38.12 and imagine that the light coming from the left originates from the helium–neon laser. Let's evaluate the possible values of the angle θ for constructive interference.

Categorize We determine results using equations developed in this section, so we categorize this example as a substitution problem.

Calculate the slit separation as the inverse of the number of grooves per centimeter:

$$d = \frac{1}{6\,000}\ \text{cm} = 1.667 \times 10^{-4}\ \text{cm} = 1\,667\ \text{nm}$$

Solve Equation 38.7 for $\sin\theta$ and substitute numerical values for the first-order maximum ($m = 1$) to find θ_1:

$$\sin\theta_1 = \frac{(1)\lambda}{d} = \frac{632.8\ \text{nm}}{1\,667\ \text{nm}} = 0.379\,7$$

$$\theta_1 = \boxed{22.31°}$$

Repeat for the second-order maximum ($m = 2$):

$$\sin\theta_2 = \frac{(2)\lambda}{d} = \frac{2(632.8\ \text{nm})}{1\,667\ \text{nm}} = 0.759\,4$$

$$\theta_2 = \boxed{49.41°}$$

WHAT IF? What if you looked for the third-order maximum? Would you find it?

Answer For $m = 3$, we find $\sin\theta_3 = 1.139$. Because $\sin\theta$ cannot exceed unity, this result does not represent a realistic solution. Hence, only zeroth-, first-, and second-order maxima can be observed for this situation.

Applications of Diffraction Gratings

A schematic drawing of a simple apparatus used to measure angles in a diffraction pattern is shown in Figure 38.15 (page 1172). This apparatus is a *diffraction grating spectrometer*. The light to be analyzed passes through a slit, and a collimated beam of light is incident on the grating. The diffracted light leaves the grating at angles that satisfy Equation 38.7, and a telescope is used to view the image of the slit. The wavelength can be determined by measuring the precise angles at which the images of the slit appear for the various orders.

   The spectrometer is a useful tool in *atomic spectroscopy,* in which the light from an atom is analyzed to find the wavelength components. These wavelength components can be used to identify the atom. We shall investigate atomic spectra in Chapter 42 of the extended version of this text.

   Another application of diffraction gratings is the *grating light valve* (GLV), which competes in some video display applications with the digital micromirror devices (DMDs) discussed in Section 35.4. A GLV is a silicon microchip fitted with an array

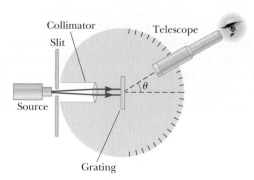

Figure 38.15 Diagram of a diffraction grating spectrometer. The collimated beam incident on the grating is spread into its various wavelength components with constructive interference for a particular wavelength occurring at the angles θ_{bright} that satisfy the equation $d \sin \theta_{bright} = m\lambda$, where $m = 0$, ± 1, ± 2, ...

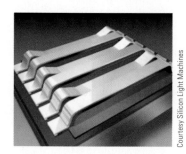

Figure 38.16 A small portion of a grating light valve. The alternating reflective ribbons at different levels act as a diffraction grating, offering very high-speed control of the direction of light toward a digital display device.

of parallel silicon nitride ribbons coated with a thin layer of aluminum (Fig. 38.16). Each ribbon is approximately 20 μm long and 5 μm wide and is separated from the silicon substrate by an air gap on the order of 100 nm. With no voltage applied, all ribbons are at the same level. In this situation, the array of ribbons acts as a flat surface, specularly reflecting incident light.

When a voltage is applied between a ribbon and the electrode on the silicon substrate, an electric force pulls the ribbon downward, closer to the substrate. Alternate ribbons can be pulled down, while those in between remain in an elevated configuration. As a result, the array of ribbons acts as a diffraction grating such that the constructive interference for a particular wavelength of light can be directed toward a screen or other optical display system. If one uses three such devices—one each for red, blue, and green light—full-color display is possible.

In addition to its use in video display, the GLV has found applications in laser optical navigation sensor technology, computer-to-plate commercial printing, and other types of imaging.

Another interesting application of diffraction gratings is **holography,** the production of three-dimensional images of objects. The physics of holography was developed by Dennis Gabor (1900–1979) in 1948 and resulted in the Nobel Prize in Physics for Gabor in 1971. The requirement of coherent light for holography delayed the realization of holographic images from Gabor's work until the development of lasers in the 1960s. Figure 38.17 shows a single hologram viewed from two different positions and the three-dimensional character of its image. Notice in particular the difference in the view through the magnifying glass in Figures 38.17a and 38.17b.

Figure 38.18 shows how a hologram is made. Light from the laser is split into two parts by a half-silvered mirror at B. One part of the beam reflects off the object to be photographed and strikes an ordinary photographic film. The other half of the beam is diverged by lens L_2, reflects from mirrors M_1 and M_2, and finally

Figure 38.17 In this hologram, a circuit board is shown from two different views. Notice the difference in the appearance of the measuring tape and the view through the magnifying lens in (a) and (b).

a

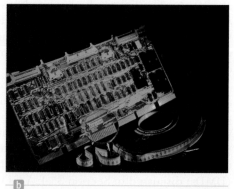

b

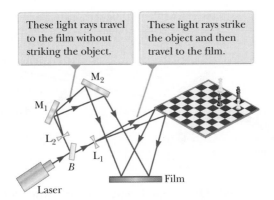

These light rays travel to the film without striking the object.

These light rays strike the object and then travel to the film.

M_2

M_1

L_2

L_1

B

Laser

Film

Figure 38.18 Experimental arrangement for producing a hologram.

strikes the film. The two beams overlap to form an extremely complicated interference pattern on the film. Such an interference pattern can be produced only if the phase relationship of the two waves is constant throughout the exposure of the film. This condition is met by illuminating the scene with light coming through a pinhole or with coherent laser radiation. The hologram records not only the intensity of the light scattered from the object (as in a conventional photograph), but also the phase difference between the reference beam and the beam scattered from the object. Because of this phase difference, an interference pattern is formed that produces an image in which all three-dimensional information available from the perspective of any point on the hologram is preserved.

In a normal photographic image, a lens is used to focus the image so that each point on the object corresponds to a single point on the photograph. Notice that there is no lens used in Figure 38.18 to focus the light onto the film. Therefore, light from each point on the object reaches *all* points on the film. As a result, each region of the photographic film on which the hologram is recorded contains information about all illuminated points on the object, which leads to a remarkable result: if a small section of the hologram is cut from the film, the complete image can be formed from the small piece! (The quality of the image is reduced, but the entire image is present.)

A hologram is best viewed by allowing coherent light to pass through the developed film as one looks back along the direction from which the beam comes. The interference pattern on the film acts as a diffraction grating. Figure 38.19 shows two rays of light striking and passing through the film. For each ray, the $m = 0$ and $m = \pm 1$ rays in the diffraction pattern are shown emerging from the right side of the film. The $m = +1$ rays converge to form a real image of the scene, which is not the image that is normally viewed. By extending the light rays corresponding to $m = -1$ behind the film, we see that there is a virtual image located there, with light coming from it in exactly the same way that light came from the actual object

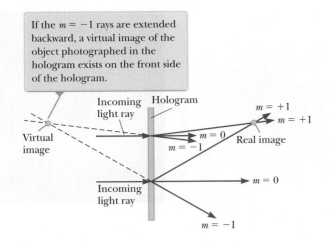

If the $m = -1$ rays are extended backward, a virtual image of the object photographed in the hologram exists on the front side of the hologram.

Virtual image

Incoming light ray

Hologram

$m = +1$

$m = +1$

$m = 0$

$m = -1$

Real image

$m = 0$

Incoming light ray

$m = -1$

Figure 38.19 Two light rays strike a hologram at normal incidence. For each ray, outgoing rays corresponding to $m = 0$ and $m = \pm 1$ are shown.

Figure 38.20 Schematic diagram of the technique used to observe the diffraction of x-rays by a crystal. The array of spots formed on the film is called a Laue pattern.

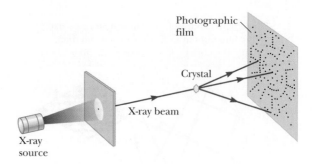

when the film was exposed. This image is what one sees when looking through the holographic film.

Holograms are finding a number of applications. You may have a hologram on your credit card. This special type of hologram is called a *rainbow hologram* and is designed to be viewed in reflected white light.

38.5 Diffraction of X-Rays by Crystals

In principle, the wavelength of any electromagnetic wave can be determined if a grating of the proper spacing (on the order of λ) is available. X-rays, discovered by Wilhelm Roentgen (1845–1923) in 1895, are electromagnetic waves of very short wavelength (on the order of 0.1 nm). It would be impossible to construct a grating having such a small spacing by the cutting process described at the beginning of Section 38.4. The atomic spacing in a solid is known to be about 0.1 nm, however. In 1913, Max von Laue (1879–1960) suggested that the regular array of atoms in a crystal could act as a three-dimensional diffraction grating for x-rays. Subsequent experiments confirmed this prediction. The diffraction patterns from crystals are complex because of the three-dimensional nature of the crystal structure. Nevertheless, x-ray diffraction has proved to be an invaluable technique for elucidating these structures and for understanding the structure of matter.

Figure 38.20 shows one experimental arrangement for observing x-ray diffraction from a crystal. A collimated beam of monochromatic x-rays is incident on a crystal. The diffracted beams are very intense in certain directions, corresponding to constructive interference from waves reflected from layers of atoms in the crystal. The diffracted beams, which can be detected by a photographic film, form an array of spots known as a *Laue pattern* as in Figure 38.21a. One can deduce the crystalline structure by analyzing the positions and intensities of the various spots in the pattern. Figure 38.21b shows a Laue pattern from a crystalline enzyme, using a wide range of wavelengths so that a swirling pattern results.

Figure 38.21 (a) A Laue pattern of a single crystal of the mineral beryl (beryllium aluminum silicate). Each dot represents a point of constructive interference. (b) A Laue pattern of the enzyme Rubisco, produced with a wide-band x-ray spectrum. This enzyme is present in plants and takes part in the process of photosynthesis. The Laue pattern is used to determine the crystal structure of Rubisco.

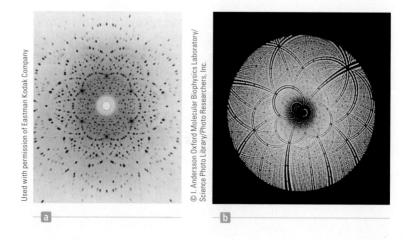

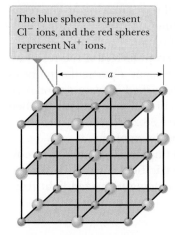

The blue spheres represent Cl⁻ ions, and the red spheres represent Na⁺ ions.

Figure 38.22 Crystalline structure of sodium chloride (NaCl). The length of the cube edge is $a = 0.562\,737$ nm.

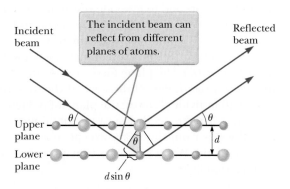

The incident beam can reflect from different planes of atoms.

Figure 38.23 A two-dimensional description of the reflection of an x-ray beam from two parallel crystalline planes separated by a distance d. The beam reflected from the lower plane travels farther than the beam reflected from the upper plane by a distance $2d \sin \theta$.

The arrangement of atoms in a crystal of sodium chloride (NaCl) is shown in Figure 38.22. Each unit cell (the geometric solid that repeats throughout the crystal) is a cube having an edge length a. A careful examination of the NaCl structure shows that the ions lie in discrete planes (the shaded areas in Fig. 38.22). Now suppose an incident x-ray beam makes an angle θ with one of the planes as in Figure 38.23. The beam can be reflected from both the upper plane and the lower one, but the beam reflected from the lower plane travels farther than the beam reflected from the upper plane. The effective path difference is $2d \sin \theta$. The two beams reinforce each other (constructive interference) when this path difference equals some integer multiple of λ. The same is true for reflection from the entire family of parallel planes. Hence, the condition for *constructive* interference (maxima in the reflected beam) is

$$2d \sin \theta = m\lambda \quad m = 1, 2, 3, \ldots \tag{38.8}$$

◄ Bragg's law

This condition is known as **Bragg's law,** after W. L. Bragg (1890–1971), who first derived the relationship. If the wavelength and diffraction angle are measured, Equation 38.8 can be used to calculate the spacing between atomic planes.

Pitfall Prevention 38.4
Different Angles Notice in Figure 38.23 that the angle θ is measured from the reflecting surface rather than from the normal as in the case of the law of reflection in Chapter 35. With slits and diffraction gratings, we also measured the angle θ from the normal to the array of slits. Because of historical tradition, the angle is measured differently in Bragg diffraction, so interpret Equation 38.8 with care.

38.6 Polarization of Light Waves

In Chapter 34, we described the transverse nature of light and all other electromagnetic waves. Polarization, discussed in this section, is firm evidence of this transverse nature.

An ordinary beam of light consists of a large number of waves emitted by the atoms of the light source. Each atom produces a wave having some particular orientation of the electric field vector $\vec{E}$, corresponding to the direction of atomic vibration. The *direction of polarization* of each individual wave is defined to be the direction in which the electric field is vibrating. In Figure 38.24, this direction happens to lie along the y axis. All individual electromagnetic waves traveling in the x direction have an $\vec{E}$ vector parallel to the yz plane, but this vector could be at any possible angle with respect to the y axis. Because all directions of vibration from a wave source are possible, the resultant electromagnetic wave is a superposition of waves vibrating in many different directions. The result is an **unpolarized** light beam, represented in Figure 38.25a (page 1176). The direction of wave propagation in this figure is perpendicular to the page. The arrows show a few possible

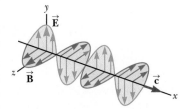

Figure 38.24 Schematic diagram of an electromagnetic wave propagating at velocity $\vec{c}$ in the x direction. The electric field vibrates in the xy plane, and the magnetic field vibrates in the xz plane.

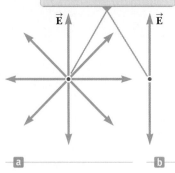

The red dot signifies the velocity vector for the wave coming out of the page.

$\vec{\mathbf{E}}$ $\vec{\mathbf{E}}$

ⓐ   ⓑ

Figure 38.25 (a) A representation of an unpolarized light beam viewed along the direction of propagation. The transverse electric field can vibrate in any direction in the plane of the page with equal probability. (b) A linearly polarized light beam with the electric field vibrating in the vertical direction.

directions of the electric field vectors for the individual waves making up the resultant beam. At any given point and at some instant of time, all these individual electric field vectors add to give one resultant electric field vector.

As noted in Section 34.3, a wave is said to be **linearly polarized** if the resultant electric field $\vec{\mathbf{E}}$ vibrates in the same direction *at all times* at a particular point as shown in Figure 38.25b. (Sometimes, such a wave is described as *plane-polarized*, or simply *polarized*.) The plane formed by $\vec{\mathbf{E}}$ and the direction of propagation is called the *plane of polarization* of the wave. If the wave in Figure 38.24 represents the resultant of all individual waves, the plane of polarization is the *xy* plane.

A linearly polarized beam can be obtained from an unpolarized beam by removing all waves from the beam except those whose electric field vectors oscillate in a single plane. We now discuss four processes for producing polarized light from unpolarized light.

Polarization by Selective Absorption

The most common technique for producing polarized light is to use a material that transmits waves whose electric fields vibrate in a plane parallel to a certain direction and that absorbs waves whose electric fields vibrate in all other directions.

In 1938, E. H. Land (1909–1991) discovered a material, which he called *Polaroid*, that polarizes light through selective absorption. This material is fabricated in thin sheets of long-chain hydrocarbons. The sheets are stretched during manufacture so that the long-chain molecules align. After a sheet is dipped into a solution containing iodine, the molecules become good electrical conductors. Conduction takes place primarily along the hydrocarbon chains because electrons can move easily only along the chains. If light whose electric field vector is parallel to the chains is incident on the material, the electric field accelerates electrons along the chains and energy is absorbed from the radiation. Therefore, the light does not pass through the material. Light whose electric field vector is perpendicular to the chains passes through the material because electrons cannot move from one molecule to the next. As a result, when unpolarized light is incident on the material, the exiting light is polarized perpendicular to the molecular chains.

It is common to refer to the direction perpendicular to the molecular chains as the *transmission axis*. In an ideal polarizer, all light with $\vec{\mathbf{E}}$ parallel to the transmission axis is transmitted and all light with $\vec{\mathbf{E}}$ perpendicular to the transmission axis is absorbed.

Figure 38.26 represents an unpolarized light beam incident on a first polarizing sheet, called the *polarizer*. Because the transmission axis is oriented vertically in the figure, the light transmitted through this sheet is polarized vertically. A second polarizing sheet, called the *analyzer*, intercepts the beam. In Figure 38.26, the analyzer transmission axis is set at an angle θ to the polarizer axis. We call the electric field vector of the first transmitted beam $\vec{\mathbf{E}}_0$. The component of $\vec{\mathbf{E}}_0$ perpendicular to the analyzer axis is completely absorbed. The component of $\vec{\mathbf{E}}_0$ parallel to the

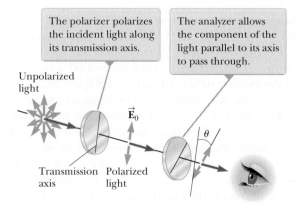

The polarizer polarizes the incident light along its transmission axis.

The analyzer allows the component of the light parallel to its axis to pass through.

Unpolarized light

$\vec{\mathbf{E}}_0$

θ

Transmission axis   Polarized light

Figure 38.26 Two polarizing sheets whose transmission axes make an angle θ with each other. Only a fraction of the polarized light incident on the analyzer is transmitted through it.

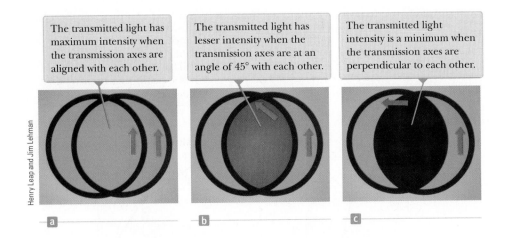

The transmitted light has maximum intensity when the transmission axes are aligned with each other.

The transmitted light has lesser intensity when the transmission axes are at an angle of 45° with each other.

The transmitted light intensity is a minimum when the transmission axes are perpendicular to each other.

Figure 38.27 The intensity of light transmitted through two polarizers depends on the relative orientation of their transmission axes. The red arrows indicate the transmission axes of the polarizers.

analyzer axis, which is transmitted through the analyzer, is $E_0 \cos \theta$. Because the intensity of the transmitted beam varies as the square of its magnitude, we conclude that the intensity I of the (polarized) beam transmitted through the analyzer varies as

$$I = I_{\max} \cos^2 \theta \qquad (38.9)$$

◀ Malus's law

where $I_{\max}$ is the intensity of the polarized beam incident on the analyzer. This expression, known as **Malus's law,**[2] applies to any two polarizing materials whose transmission axes are at an angle θ to each other. This expression shows that the intensity of the transmitted beam is maximum when the transmission axes are parallel ($\theta = 0$ or $180°$) and is zero (complete absorption by the analyzer) when the transmission axes are perpendicular to each other. This variation in transmitted intensity through a pair of polarizing sheets is illustrated in Figure 38.27. Because the average value of $\cos^2 \theta$ is $\frac{1}{2}$, the intensity of initially unpolarized light is reduced by a factor of one-half as the light passes through a single ideal polarizer.

Polarization by Reflection

When an unpolarized light beam is reflected from a surface, the polarization of the reflected light depends on the angle of incidence. If the angle of incidence is $0°$, the reflected beam is unpolarized. For other angles of incidence, the reflected light is polarized to some extent, and for one particular angle of incidence, the reflected light is completely polarized. Let's now investigate reflection at that special angle.

Suppose an unpolarized light beam is incident on a surface as in Figure 38.28a (page 1178). Each individual electric field vector can be resolved into two components: one parallel to the surface (and perpendicular to the page in Fig. 38.28, represented by the dots) and the other (represented by the orange arrows) perpendicular both to the first component and to the direction of propagation. Therefore, the polarization of the entire beam can be described by two electric field components in these directions. It is found that the parallel component represented by the dots reflects more strongly than the other component represented by the arrows, resulting in a partially polarized reflected beam. Furthermore, the refracted beam is also partially polarized.

Now suppose the angle of incidence θ_1 is varied until the angle between the reflected and refracted beams is $90°$ as in Figure 38.28b. At this particular angle of incidence, the reflected beam is completely polarized (with its electric field vector parallel to the surface) and the refracted beam is still only partially polarized. The angle of incidence at which this polarization occurs is called the **polarizing angle** θ_p.

---

[2]Named after its discoverer, E. L. Malus (1775–1812). Malus discovered that reflected light was polarized by viewing it through a calcite ($CaCO_3$) crystal.

Figure 38.28 (a) When unpolarized light is incident on a reflecting surface, the reflected and refracted beams are partially polarized. (b) The reflected beam is completely polarized when the angle of incidence equals the polarizing angle θ_p, which satisfies the equation $n_2/n_1 = \tan \theta_p$. At this incident angle, the reflected and refracted rays are perpendicular to each other.

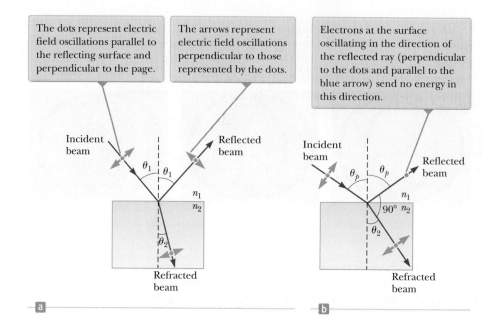

We can obtain an expression relating the polarizing angle to the index of refraction of the reflecting substance by using Figure 38.28b. From this figure, we see that $\theta_p + 90° + \theta_2 = 180°$; therefore, $\theta_2 = 90° - \theta_p$. Using Snell's law of refraction (Eq. 35.8) gives

$$\frac{n_2}{n_1} = \frac{\sin \theta_1}{\sin \theta_2} = \frac{\sin \theta_p}{\sin \theta_2}$$

Because $\sin \theta_2 = \sin (90° - \theta_p) = \cos \theta_p$, we can write this expression as $n_2/n_1 = \sin \theta_p/\cos \theta_p$, which means that

Brewster's law ▶

$$\tan \theta_p = \frac{n_2}{n_1} \tag{38.10}$$

This expression is called **Brewster's law,** and the polarizing angle θ_p is sometimes called **Brewster's angle,** after its discoverer, David Brewster (1781–1868). Because n varies with wavelength for a given substance, Brewster's angle is also a function of wavelength.

We can understand polarization by reflection by imagining that the electric field in the incident light sets electrons at the surface of the material in Figure 38.28b into oscillation. The component directions of oscillation are (1) parallel to the arrows shown on the refracted beam of light and therefore parallel to the reflected beam and (2) perpendicular to the page. The oscillating electrons act as dipole antennas radiating light with a polarization parallel to the direction of oscillation. Consult Figure 34.12, which shows the pattern of radiation from a dipole antenna. Notice that there is no radiation at an angle of $\theta = 0$, that is, along the oscillation direction of the antenna. Therefore, for the oscillations in direction 1, there is no radiation in the direction along the reflected ray. For oscillations in direction 2, the electrons radiate light with a polarization perpendicular to the page. Therefore, the light reflected from the surface at this angle is completely polarized parallel to the surface.

Polarization by reflection is a common phenomenon. Sunlight reflected from water, glass, and snow is partially polarized. If the surface is horizontal, the electric field vector of the reflected light has a strong horizontal component. Sunglasses made of polarizing material reduce the glare of reflected light. The transmission axes of such lenses are oriented vertically so that they absorb the strong horizontal component of the reflected light. If you rotate sunglasses through 90°, they are not as effective at blocking the glare from shiny horizontal surfaces.

Polarization by Double Refraction

Solids can be classified on the basis of internal structure. Those in which the atoms are arranged in a specific order are called *crystalline;* the NaCl structure of Figure 38.22 is one example of a crystalline solid. Those solids in which the atoms are distributed randomly are called *amorphous.* When light travels through an amorphous material such as glass, it travels with a speed that is the same in all directions. That is, glass has a single index of refraction. In certain crystalline materials such as calcite and quartz, however, the speed of light is not the same in all directions. In these materials, the speed of light depends on the direction of propagation *and* on the plane of polarization of the light. Such materials are characterized by two indices of refraction. Hence, they are often referred to as **double-refracting** or **birefringent** materials.

When unpolarized light enters a birefringent material, it may split into an **ordinary (O) ray** and an **extraordinary (E) ray.** These two rays have mutually perpendicular polarizations and travel at different speeds through the material. The two speeds correspond to two indices of refraction, n_O for the ordinary ray and n_E for the extraordinary ray.

There is one direction, called the **optic axis,** along which the ordinary and extraordinary rays have the same speed. If light enters a birefringent material at an angle to the optic axis, however, the different indices of refraction will cause the two polarized rays to split and travel in different directions as shown in Figure 38.29.

The index of refraction n_O for the ordinary ray is the same in all directions. If one could place a point source of light inside the crystal as in Figure 38.30, the ordinary waves would spread out from the source as spheres. The index of refraction n_E varies with the direction of propagation. A point source sends out an extraordinary wave having wave fronts that are elliptical in cross section. The difference in speed for the two rays is a maximum in the direction perpendicular to the optic axis. For example, in calcite, $n_O = 1.658$ at a wavelength of 589.3 nm and n_E varies from 1.658 along the optic axis to 1.486 perpendicular to the optic axis. Values for n_O and the extreme value of n_E for various double-refracting crystals are given in Table 38.1.

If you place a calcite crystal on a sheet of paper and then look through the crystal at any writing on the paper, you would see two images as shown in Figure 38.31. As can be seen from Figure 38.29, these two images correspond to one formed by the ordinary ray and one formed by the extraordinary ray. If the two images are viewed through a sheet of rotating polarizing glass, they alternately appear and disappear because the ordinary and extraordinary rays are plane-polarized along mutually perpendicular directions.

Some materials such as glass and plastic become birefringent when stressed. Suppose an unstressed piece of plastic is placed between a polarizer and an analyzer so that light passes from polarizer to plastic to analyzer. When the plastic is unstressed and the analyzer axis is perpendicular to the polarizer axis, none of the polarized light passes through the analyzer. In other words, the unstressed plastic has no effect on the light passing through it. If the plastic is stressed, however, regions of greatest stress become birefringent and the polarization of the light passing through the plastic changes. Hence, a series of bright and dark bands is observed in the transmitted light, with the bright bands corresponding to regions of greatest stress.

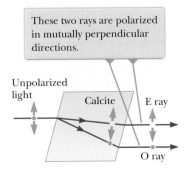

Figure 38.29 Unpolarized light incident at an angle to the optic axis in a calcite crystal splits into an ordinary (O) ray and an extraordinary (E) ray (not to scale).

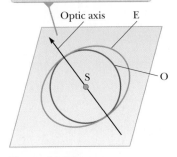

Figure 38.30 A point source S inside a double-refracting crystal produces a spherical wave front corresponding to the ordinary (O) ray and an elliptical wave front corresponding to the extraordinary (E) ray.

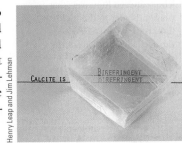

Figure 38.31 A calcite crystal produces a double image because it is a birefringent (double-refracting) material.

Table 38.1	Indices of Refraction for Some Double-Refracting Crystals at a Wavelength of 589.3 nm		
Crystal	n_O	n_E	n_O/n_E
Calcite (CaCO$_3$)	1.658	1.486	1.116
Quartz (SiO$_2$)	1.544	1.553	0.994
Sodium nitrate (NaNO$_3$)	1.587	1.336	1.188
Sodium sulfite (NaSO$_3$)	1.565	1.515	1.033
Zinc chloride (ZnCl$_2$)	1.687	1.713	0.985
Zinc sulfide (ZnS)	2.356	2.378	0.991

Figure 38.32 A plastic model of an arch structure under load conditions. The pattern is produced when the plastic model is viewed between a polarizer and analyzer oriented perpendicular to each other. Such patterns are useful in the optimal design of architectural components.

Engineers often use this technique, called *optical stress analysis*, in designing structures ranging from bridges to small tools. They build a plastic model and analyze it under different load conditions to determine regions of potential weakness and failure under stress. An example of a plastic model under stress is shown in Figure 38.32.

Polarization by Scattering

When light is incident on any material, the electrons in the material can absorb and reradiate part of the light. Such absorption and reradiation of light by electrons in the gas molecules that make up air is what causes sunlight reaching an observer on the Earth to be partially polarized. You can observe this effect—called **scattering**—by looking directly up at the sky through a pair of sunglasses whose lenses are made of polarizing material. Less light passes through at certain orientations of the lenses than at others.

Figure 38.33 illustrates how sunlight becomes polarized when it is scattered. The phenomenon is similar to that creating completely polarized light upon reflection from a surface at Brewster's angle. An unpolarized beam of sunlight traveling in the horizontal direction (parallel to the ground) strikes a molecule of one of the gases that make up air, setting the electrons of the molecule into vibration. These vibrating charges act like the vibrating charges in an antenna. The horizontal component of the electric field vector in the incident wave results in a horizontal component of the vibration of the charges, and the vertical component of the vector results in a vertical component of vibration. If the observer in Figure 38.33 is looking straight up (perpendicular to the original direction of propagation of the light), the vertical oscillations of the charges send no radiation toward the observer. Therefore, the observer sees light that is completely polarized in the horizontal direction as indicated by the orange arrows. If the observer looks in other directions, the light is partially polarized in the horizontal direction.

Variations in the color of scattered light in the atmosphere can be understood as follows. When light of various wavelengths λ is incident on gas molecules of diameter d, where $d \ll \lambda$, the relative intensity of the scattered light varies as $1/\lambda^4$. The condition $d \ll \lambda$ is satisfied for scattering from oxygen (O_2) and nitrogen (N_2) molecules in the atmosphere, whose diameters are about 0.2 nm. Hence, short wavelengths (violet light) are scattered more efficiently than long wavelengths (red light). Therefore, when sunlight is scattered by gas molecules in the air, the short-wavelength radiation (violet) is scattered more intensely than the long-wavelength radiation (red).

When you look up into the sky in a direction that is not toward the Sun, you see the scattered light, which is predominantly violet. Your eyes, however, are not very sensitive to violet light. Light of the next color in the spectrum, blue, is scattered with less intensity than violet, but your eyes are far more sensitive to blue light than to violet light. Hence, you see a blue sky. If you look toward the west at sunset (or toward the east at sunrise), you are looking in a direction toward the Sun and are seeing light that has passed through a large distance of air. Most of the blue light has been scattered by the air between you and the Sun. The light that survives this

The scattered light traveling perpendicular to the incident light is plane-polarized because the vertical vibrations of the charges in the air molecule send no light in this direction.

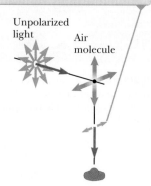

Unpolarized light

Air molecule

Figure 38.33 The scattering of unpolarized sunlight by air molecules.

trip through the air to you has had much of its blue component scattered and is therefore heavily weighted toward the red end of the spectrum; as a result, you see the red and orange colors of sunset (or sunrise).

Optical Activity

Many important applications of polarized light involve materials that display **optical activity.** A material is said to be optically active if it rotates the plane of polarization of any light transmitted through the material. The angle through which the light is rotated by a specific material depends on the length of the path through the material and on concentration if the material is in solution. One optically active material is a solution of the common sugar dextrose. A standard method for determining the concentration of sugar solutions is to measure the rotation produced by a fixed length of the solution.

Molecular asymmetry determines whether a material is optically active. For example, some proteins are optically active because of their spiral shape.

The liquid crystal displays found in most calculators have their optical activity changed by the application of electric potential across different parts of the display. Try using a pair of polarizing sunglasses to investigate the polarization used in the display of your calculator.

Quick Quiz 38.6 A polarizer for microwaves can be made as a grid of parallel metal wires approximately 1 cm apart. Is the electric field vector for microwaves transmitted through this polarizer **(a)** parallel or **(b)** perpendicular to the metal wires?

Quick Quiz 38.7 You are walking down a long hallway that has many light fixtures in the ceiling and a very shiny, newly waxed floor. When looking at the floor, you see reflections of every light fixture. Now you put on sunglasses that are polarized. Some of the reflections of the light fixtures can no longer be seen. (Try it!) Are the reflections that disappear those **(a)** nearest to you, **(b)** farthest from you, or **(c)** at an intermediate distance from you?

Summary

Concepts and Principles

Diffraction is the deviation of light from a straight-line path when the light passes through an aperture or around an obstacle. Diffraction is due to the wave nature of light.

The **Fraunhofer diffraction pattern** produced by a single slit of width a on a distant screen consists of a central bright fringe and alternating bright and dark fringes of much lower intensities. The angles θ_{dark} at which the diffraction pattern has zero intensity, corresponding to destructive interference, are given by

$$\sin \theta_{dark} = m \frac{\lambda}{a} \quad m = \pm 1, \pm 2, \pm 3, \ldots \quad (38.1)$$

Rayleigh's criterion, which is a limiting condition of resolution, states that two images formed by an aperture are just distinguishable if the central maximum of the diffraction pattern for one image falls on the first minimum of the diffraction pattern for the other image. The limiting angle of resolution for a slit of width a is $\theta_{min} = \lambda/a$, and the limiting angle of resolution for a circular aperture of diameter D is given by $\theta_{min} = 1.22\lambda/D$.

A **diffraction grating** consists of a large number of equally spaced, identical slits. The condition for intensity maxima in the interference pattern of a diffraction grating for normal incidence is

$$d \sin \theta_{bright} = m\lambda \quad m = 0, \pm 1, \pm 2, \pm 3, \ldots \quad (38.7)$$

where d is the spacing between adjacent slits and m is the order number of the intensity maximum.

continued

When polarized light of intensity I_{max} is emitted by a polarizer and then is incident on an analyzer, the light transmitted through the analyzer has an intensity equal to $I_{max} \cos^2 \theta$, where θ is the angle between the polarizer and analyzer transmission axes.

In general, reflected light is partially polarized. Reflected light, however, is completely polarized when the angle of incidence is such that the angle between the reflected and refracted beams is 90°. This angle of incidence, called the **polarizing angle** θ_p, satisfies **Brewster's law:**

$$\tan \theta_p = \frac{n_2}{n_1} \qquad \text{(38.10)}$$

where n_1 is the index of refraction of the medium in which the light initially travels and n_2 is the index of refraction of the reflecting medium.

Objective Questions   1. denotes answer available in *Student Solutions Manual/Study Guide*

1. Certain sunglasses use a polarizing material to reduce the intensity of light reflected as glare from water or automobile windshields. What orientation should the polarizing filters have to be most effective? (a) The polarizers should absorb light with its electric field horizontal. (b) The polarizers should absorb light with its electric field vertical. (c) The polarizers should absorb both horizontal and vertical electric fields. (d) The polarizers should not absorb either horizontal or vertical electric fields.

2. What is most likely to happen to a beam of light when it reflects from a shiny metallic surface at an arbitrary angle? Choose the best answer. (a) It is totally absorbed by the surface. (b) It is totally polarized. (c) It is unpolarized. (d) It is partially polarized. (e) More information is required.

3. In Figure 38.4, assume the slit is in a barrier that is opaque to x-rays as well as to visible light. The photograph in Figure 38.4b shows the diffraction pattern produced with visible light. What will happen if the experiment is repeated with x-rays as the incoming wave and with no other changes? (a) The diffraction pattern is similar. (b) There is no noticeable diffraction pattern but rather a projected shadow of high intensity on the screen, having the same width as the slit. (c) The central maximum is much wider, and the minima occur at larger angles than with visible light. (d) No x-rays reach the screen.

4. A Fraunhofer diffraction pattern is produced on a screen located 1.00 m from a single slit. If a light source of wavelength 5.00×10^{-7} m is used and the distance from the center of the central bright fringe to the first dark fringe is 5.00×10^{-3} m, what is the slit width? (a) 0.010 0 mm (b) 0.100 mm (c) 0.200 mm (d) 1.00 mm (e) 0.005 00 mm

5. Consider a wave passing through a single slit. What happens to the width of the central maximum of its diffraction pattern as the slit is made half as wide? (a) It becomes one-fourth as wide. (b) It becomes one-half as wide. (c) Its width does not change. (d) It becomes twice as wide. (e) It becomes four times as wide.

6. Assume Figure 38.1 was photographed with red light of a single wavelength λ_0. The light passed through a single slit of width a and traveled distance L to the screen where the photograph was made. Consider the width of the central bright fringe, measured between the centers of the dark fringes on both sides of it. Rank from largest to smallest the widths of the central fringe in the following situations and note any cases of equality. (a) The experiment is performed as photographed. (b) The experiment is performed with light whose frequency is increased by 50%. (c) The experiment is performed with light whose wavelength is increased by 50%. (d) The experiment is performed with the original light and with a slit of width $2a$. (e) The experiment is performed with the original light and slit and with distance $2L$ to the screen.

7. If plane polarized light is sent through two polarizers, the first at 45° to the original plane of polarization and the second at 90° to the original plane of polarization, what fraction of the original polarized intensity passes through the last polarizer? (a) 0 (b) $\frac{1}{4}$ (c) $\frac{1}{2}$ (d) $\frac{1}{8}$ (e) $\frac{1}{10}$

8. Why is it advantageous to use a large-diameter objective lens in a telescope? (a) It diffracts the light more effectively than smaller-diameter objective lenses. (b) It increases its magnification. (c) It enables you to see more objects in the field of view. (d) It reflects unwanted wavelengths. (e) It increases its resolution.

9. What combination of optical phenomena causes the bright colored patterns sometimes seen on wet streets covered with a layer of oil? Choose the best answer. (a) diffraction and polarization (b) interference and diffraction (c) polarization and reflection (d) refraction and diffraction (e) reflection and interference

10. When you receive a chest x-ray at a hospital, the x-rays pass through a set of parallel ribs in your chest. Do your ribs act as a diffraction grating for x-rays? (a) Yes. They produce diffracted beams that can be observed separately. (b) Not to a measurable extent. The ribs are too far apart. (c) Essentially not. The ribs are too close together. (d) Essentially not. The ribs are too few in number. (e) Absolutely not. X-rays cannot diffract.

11. When unpolarized light passes through a diffraction grating, does it become polarized? (a) No, it does not. (b) Yes, it does, with the transmission axis parallel to the slits or grooves in the grating. (c) Yes, it does, with the transmission axis perpendicular to the slits or grooves in the grating. (d) It possibly does because an electric field above some threshold is blocked out by the grating if the field is perpendicular to the slits.

12. Off in the distance, you see the headlights of a car, but they are indistinguishable from the single headlight of a motorcycle. Assume the car's headlights are now switched from low beam to high beam so that the light intensity you receive becomes three times greater. What then happens to your ability to resolve the two light sources? (a) It increases by a factor of 9. (b) It increases by a factor of 3. (c) It remains the same. (d) It becomes one-third as good. (e) It becomes one-ninth as good.

Conceptual Questions

1. denotes answer available in *Student Solutions Manual/Study Guide*

1. The atoms in a crystal lie in planes separated by a few tenths of a nanometer. Can they produce a diffraction pattern for visible light as they do for x-rays? Explain your answer with reference to Bragg's law.

2. Holding your hand at arm's length, you can readily block sunlight from reaching your eyes. Why can you not block sound from reaching your ears this way?

3. How could the index of refraction of a flat piece of opaque obsidian glass be determined?

4. (a) Is light from the sky polarized? (b) Why is it that clouds seen through Polaroid glasses stand out in bold contrast to the sky?

5. A laser beam is incident at a shallow angle on a horizontal machinist's ruler that has a finely calibrated scale. The engraved rulings on the scale give rise to a diffraction pattern on a vertical screen. Discuss how you can use this technique to obtain a measure of the wavelength of the laser light.

6. If a coin is glued to a glass sheet and this arrangement is held in front of a laser beam, the projected shadow has diffraction rings around its edge and a bright spot in the center. How are these effects possible?

7. Fingerprints left on a piece of glass such as a windowpane often show colored spectra like that from a diffraction grating. Why?

8. A laser produces a beam a few millimeters wide, with uniform intensity across its width. A hair is stretched vertically across the front of the laser to cross the beam. (a) How is the diffraction pattern it produces on a distant screen related to that of a vertical slit equal in width to the hair? (b) How could you determine the width of the hair from measurements of its diffraction pattern?

9. A radio station serves listeners in a city to the northeast of its broadcast site. It broadcasts from three adjacent towers on a mountain ridge, along a line running east to west, in what's called a *phased array*. Show that by introducing time delays among the signals the individual towers radiate, the station can maximize net intensity in the direction toward the city (and in the opposite direction) and minimize the signal transmitted in other directions.

10. John William Strutt, Lord Rayleigh (1842–1919), invented an improved foghorn. To warn ships of a coastline, a foghorn should radiate sound in a wide horizontal sheet over the ocean's surface. It should not waste energy by broadcasting sound upward or downward. Rayleigh's foghorn trumpet is shown in two possible configurations, horizontal and vertical, in Figure CQ38.10. Which is the correct orientation? Decide whether the long dimension of the rectangular opening should be horizontal or vertical and argue for your decision.

Figure CQ38.10

11. Why can you hear around corners, but not see around corners?

12. Figure CQ38.12 shows a megaphone in use. Construct a theoretical description of how a megaphone works. You may assume the sound of your voice radiates just through the opening of your mouth. Most of the information in speech is carried not in a signal at the fundamental frequency, but in noises and in harmonics, with frequencies of a few thousand hertz. Does your theory allow any prediction that is simple to test?

Doug Pensinger/Getty Images

Figure CQ38.12

Problems

WebAssign The problems found in this chapter may be assigned online in Enhanced WebAssign

1. straightforward; **2.** intermediate; **3.** challenging

1. full solution available in the *Student Solutions Manual/Study Guide*

AMT Analysis Model tutorial available in Enhanced WebAssign

GP Guided Problem

M Master It tutorial available in Enhanced WebAssign

W Watch It video solution available in Enhanced WebAssign

Section 38.2 Diffraction Patterns from Narrow Slits

1. Light of wavelength 587.5 nm illuminates a slit of width 0.75 mm. (a) At what distance from the slit should a screen be placed if the first minimum in the diffraction pattern is to be 0.85 mm from the central maximum? (b) Calculate the width of the central maximum.

2. Helium–neon laser light (λ = 632.8 nm) is sent through a 0.300-mm-wide single slit. What is the width of the central maximum on a screen 1.00 m from the slit?

3. Sound with a frequency 650 Hz from a distant source passes through a doorway 1.10 m wide in a sound-absorbing wall. Find (a) the number and (b) the angular directions of the diffraction minima at listening positions along a line parallel to the wall.

4. A horizontal laser beam of wavelength 632.8 nm has a circular cross section 2.00 mm in diameter. A rectangular aperture is to be placed in the center of the beam so that when the light falls perpendicularly on a wall 4.50 m away, the central maximum fills a rectangle 110 mm wide and 6.00 mm high. The dimensions are measured between the minima bracketing the central maximum. Find the required (a) width and (b) height of the aperture. (c) Is the longer dimension of the central bright patch in the diffraction pattern horizontal or vertical? (d) Is the longer dimension of the aperture horizontal or vertical? (e) Explain the relationship between these two rectangles, using a diagram.

5. Coherent microwaves of wavelength 5.00 cm enter a tall, narrow window in a building otherwise essentially opaque to the microwaves. If the window is 36.0 cm wide, what is the distance from the central maximum to the first-order minimum along a wall 6.50 m from the window?

6. Light of wavelength 540 nm passes through a slit of width 0.200 mm. (a) The width of the central maximum on a screen is 8.10 mm. How far is the screen from the slit? (b) Determine the width of the first bright fringe to the side of the central maximum.

7. A screen is placed 50.0 cm from a single slit, which is illuminated with light of wavelength 690 nm. If the distance between the first and third minima in the diffraction pattern is 3.00 mm, what is the width of the slit?

8. A screen is placed a distance L from a single slit of width a, which is illuminated with light of wavelength λ. Assume $L \gg a$. If the distance between the minima for $m = m_1$ and $m = m_2$ in the diffraction pattern is Δy, what is the width of the slit?

9. Assume light of wavelength 650 nm passes through two slits 3.00 μm wide, with their centers 9.00 μm apart. Make a sketch of the combined diffraction and interference pattern in the form of a graph of intensity versus $\phi = (\pi a \sin \theta)/\lambda$. You may use Figure 38.7 as a starting point.

10. What If? Suppose light strikes a single slit of width a at an angle β from the perpendicular direction as shown in Figure P38.10. Show that Equation 38.1, the condition for destructive interference, must be modified to read

$$\sin \theta_{\text{dark}} = m\frac{\lambda}{a} - \sin \beta$$
$$m = \pm 1, \pm 2, \pm 3, \ldots$$

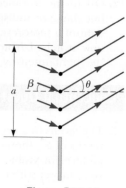

Figure P38.10

11. A diffraction pattern is formed on a screen 120 cm away from a 0.400-mm-wide slit. Monochromatic 546.1-nm light is used. Calculate the fractional intensity $I/I_{\max}$ at a point on the screen 4.10 mm from the center of the principal maximum.

12. Coherent light of wavelength 501.5 nm is sent through two parallel slits in an opaque material. Each slit is 0.700 μm wide. Their centers are 2.80 μm apart. The light then falls on a semicylindrical screen, with its axis at the midline between the slits. We would like to describe the appearance of the pattern of light visible on the screen. (a) Find the direction for each two-slit interference maximum on the screen as an angle away from the bisector of the line joining the slits. (b) How many angles are there that represent two-slit interference maxima? (c) Find the direction for each single-slit interference minimum on the screen as an angle away from the bisector of the line joining the slits. (d) How many angles are there that represent single-slit interference minima? (e) How many of the angles in part (d) are identical to those in part (a)? (f) How many bright fringes are visible on the screen? (g) If the intensity of the central fringe is $I_{\max}$, what is the intensity of the last fringe visible on the screen?

13. A beam of monochromatic light is incident on a single slit of width 0.600 mm. A diffraction pattern forms on a wall 1.30 m beyond the slit. The distance between the positions of zero intensity on both sides of the central maximum is 2.00 mm. Calculate the wavelength of the light.

Section 38.3 Resolution of Single–Slit and Circular Apertures

Note: In Problems 14, 19, 22, 23, and 67, you may use the Rayleigh criterion for the limiting angle of resolution of an eye. The standard may be overly optimistic for human vision.

14. The pupil of a cat's eye narrows to a vertical slit of width 0.500 mm in daylight. Assume the average wavelength of the light is 500 nm. What is the angular resolution for horizontally separated mice?

15. The angular resolution of a radio telescope is to be 0.100° when the incident waves have a wavelength of 3.00 mm. What minimum diameter is required for the telescope's receiving dish?

16. A *pinhole camera* has a small circular aperture of diameter D. Light from distant objects passes through the aperture into an otherwise dark box, falling on a screen at the other end of the box. The aperture in a pinhole camera has diameter $D = 0.600$ mm. Two point sources of light of wavelength 550 nm are at a distance L from the hole. The separation between the sources is 2.80 cm, and they are just resolved by the camera. What is L?

17. The objective lens of a certain refracting telescope has a diameter of 58.0 cm. The telescope is mounted in a satellite that orbits the Earth at an altitude of 270 km to view objects on the Earth's surface. Assuming an average wavelength of 500 nm, find the minimum distance between two objects on the ground if their images are to be resolved by this lens.

18. Yellow light of wavelength 589 nm is used to view an object under a microscope. The objective lens diameter is 9.00 mm. (a) What is the limiting angle of resolution? (b) Suppose it is possible to use visible light of any wavelength. What color should you choose to give the smallest possible angle of resolution, and what is this angle? (c) Suppose water fills the space between the object and the objective. What effect does this change have on the resolving power when 589-nm light is used?

19. What is the approximate size of the smallest object on the Earth that astronauts can resolve by eye when they are orbiting 250 km above the Earth? Assume $\lambda = 500$ nm and a pupil diameter of 5.00 mm.

20. A helium–neon laser emits light that has a wavelength of 632.8 nm. The circular aperture through which the beam emerges has a diameter of 0.500 cm. Estimate the diameter of the beam 10.0 km from the laser.

21. To increase the resolving power of a microscope, the object and the objective are immersed in oil ($n = 1.5$). If the limiting angle of resolution without the oil is 0.60 μrad, what is the limiting angle of resolution with the oil? *Hint:* The oil changes the wavelength of the light.

22. Narrow, parallel, glowing gas-filled tubes in a variety of colors form block letters to spell out the name of a nightclub. Adjacent tubes are all 2.80 cm apart. The tubes forming one letter are filled with neon and radiate predominantly red light with a wavelength of 640 nm. For another letter, the tubes emit predominantly blue light at 440 nm. The pupil of a dark-adapted viewer's eye is 5.20 mm in diameter. (a) Which color is easier to resolve? State how you decide. (b) If she is in a certain range of distances away, the viewer can resolve the separate tubes of one color but not the other. The viewer's distance must be in what range for her to resolve the tubes of only one of these two colors?

23. Impressionist painter Georges Seurat created paintings with an enormous number of dots of pure pigment, each of which was approximately 2.00 mm in diameter. The idea was to have colors such as red and green next to each other to form a scintillating canvas, such as in his masterpiece, *A Sunday Afternoon on the Island of La Grande Jatte* (Fig. P38.23). Assume $\lambda = 500$ nm and a pupil diameter of 5.00 mm. Beyond what distance would a viewer be unable to discern individual dots on the canvas?

© SuperStock/SuperStock

Figure P38.23

24. A circular radar antenna on a Coast Guard ship has a diameter of 2.10 m and radiates at a frequency of 15.0 GHz. Two small boats are located 9.00 km away from the ship. How close together could the boats be and still be detected as two objects?

Section 38.4 The Diffraction Grating

Note: In the following problems, assume the light is incident normally on the gratings.

25. A helium–neon laser ($\lambda = 632.8$ nm) is used to calibrate a diffraction grating. If the first-order maximum occurs at 20.5°, what is the spacing between adjacent grooves in the grating?

26. White light is spread out into its spectral components by a diffraction grating. If the grating has 2 000 grooves per centimeter, at what angle does red light of wavelength 640 nm appear in first order?

27. Consider an array of parallel wires with uniform spacing of 1.30 cm between centers. In air at 20.0°C, ultrasound with a frequency of 37.2 kHz from a distant source is incident perpendicular to the array. (a) Find the number of directions on the other side of the array in which there is a maximum of intensity. (b) Find the angle for each of these directions relative to the direction of the incident beam.

28. Three discrete spectral lines occur at angles of 10.1°, 13.7°, and 14.8° in the first-order spectrum of a grating spectrometer. (a) If the grating has 3 660 slits/cm, what are the wavelengths of the light? (b) At what angles are these lines found in the second-order spectrum?

29. The laser in a compact disc player must precisely follow the spiral track on the CD, along which the distance between one loop of the spiral and the next is only about 1.25 μm. Figure P38.29 (page 1186) shows how a diffraction grating is used to provide information to keep the beam on track. The laser light passes through a diffraction grating before it reaches the CD. The strong central maximum of the diffraction pattern is used to read the information in the track of pits. The two first-order side maxima are designed to fall on the flat surfaces on both sides of the information track and are used for steering. As long as both beams are reflecting from smooth,

nonpitted surfaces, they are detected with constant high intensity. If the main beam wanders off the track, however, one of the side beams begins to strike pits on the information track and the reflected light diminishes. This change is used with an electronic circuit to guide the beam back to the desired location. Assume the laser light has a wavelength of 780 nm and the diffraction grating is positioned 6.90 μm from the disk. Assume the first-order beams are to fall on the CD 0.400 μm on either side of the information track. What should be the number of grooves per millimeter in the grating?

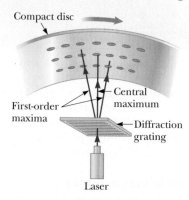

Figure P38.29

30. A grating with 250 grooves/mm is used with an incandescent light source. Assume the visible spectrum to range in wavelength from 400 nm to 700 nm. In how many orders can one see (a) the entire visible spectrum and (b) the short-wavelength region of the visible spectrum?

31. A diffraction grating has 4 200 rulings/cm. On a screen 2.00 m from the grating, it is found that for a particular order m, the maxima corresponding to two closely spaced wavelengths of sodium (589.0 nm and 589.6 nm) are separated by 1.54 mm. Determine the value of m.

32. The hydrogen spectrum includes a red line at 656 nm and a blue-violet line at 434 nm. What are the angular separations between these two spectral lines for all visible orders obtained with a diffraction grating that has 4 500 grooves/cm?

33. Light from an argon laser strikes a diffraction grating that has 5 310 grooves per centimeter. The central and first-order principal maxima are separated by 0.488 m on a wall 1.72 m from the grating. Determine the wavelength of the laser light.

34. Show that whenever white light is passed through a diffraction grating of any spacing size, the violet end of the spectrum in the third order on a screen always overlaps the red end of the spectrum in the second order.

35. Light of wavelength 500 nm is incident normally on a diffraction grating. If the third-order maximum of the diffraction pattern is observed at 32.0°, (a) what is the number of rulings per centimeter for the grating? (b) Determine the total number of primary maxima that can be observed in this situation.

36. A wide beam of laser light with a wavelength of 632.8 nm is directed through several narrow parallel slits, separated by 1.20 mm, and falls on a sheet of photographic film 1.40 m away. The exposure time is chosen so that the film stays unexposed everywhere except at the central region of each bright fringe. (a) Find the distance between these interference maxima. The film is printed as a transparency; it is opaque everywhere except at the exposed lines. Next, the same beam of laser light is directed through the transparency and allowed to fall on a screen 1.40 m beyond. (b) Argue that several narrow, parallel, bright regions, separated by 1.20 mm, appear on the screen as real images of the original slits. (A similar train of thought, at a soccer game, led Dennis Gabor to invent holography.)

37. A beam of bright red light of wavelength 654 nm passes through a diffraction grating. Enclosing the space beyond the grating is a large semicylindrical screen centered on the grating, with its axis parallel to the slits in the grating. Fifteen bright spots appear on the screen. Find (a) the maximum and (b) the minimum possible values for the slit separation in the diffraction grating.

Section 38.5 Diffraction of X-Rays by Crystals

38. If the spacing between planes of atoms in a NaCl crystal is 0.281 nm, what is the predicted angle at which 0.140-nm x-rays are diffracted in a first-order maximum?

39. Potassium iodide (KI) has the same crystalline structure as NaCl, with atomic planes separated by 0.353 nm. A monochromatic x-ray beam shows a first-order diffraction maximum when the grazing angle is 7.60°. Calculate the x-ray wavelength.

40. Monochromatic x-rays ($\lambda = 0.166$ nm) from a nickel target are incident on a potassium chloride (KCl) crystal surface. The spacing between planes of atoms in KCl is 0.314 nm. At what angle (relative to the surface) should the beam be directed for a second-order maximum to be observed?

41. The first-order diffraction maximum is observed at 12.6° for a crystal having a spacing between planes of atoms of 0.250 nm. (a) What wavelength x-ray is used to observe this first-order pattern? (b) How many orders can be observed for this crystal at this wavelength?

Section 38.6 Polarization of Light Waves

Problem 62 in Chapter 34 can be assigned with this section.

42. *Why is the following situation impossible?* A technician is measuring the index of refraction of a solid material by observing the polarization of light reflected from its surface. She notices that when a light beam is projected from air onto the material surface, the reflected light is totally polarized parallel to the surface when the incident angle is 41.0°.

43. Plane-polarized light is incident on a single polarizing disk with the direction of $\vec{E}_0$ parallel to the direction of the transmission axis. Through what angle should the disk be rotated so that the intensity in the transmitted beam is reduced by a factor of (a) 3.00, (b) 5.00, and (c) 10.0?

44. The angle of incidence of a light beam onto a reflect-
AMT ing surface is continuously variable. The reflected ray
in air is completely polarized when the angle of incidence is 48.0°. What is the index of refraction of the
reflecting material?

45. Unpolarized light passes through two ideal Polaroid
sheets. The axis of the first is vertical, and the axis of
the second is at 30.0° to the vertical. What fraction of
the incident light is transmitted?

46. Two handheld radio transceivers with dipole antennas
are separated by a large fixed distance. If the transmitting antenna is vertical, what fraction of the maximum
received power will appear in the receiving antenna
when it is inclined from the vertical by (a) 15.0°,
(b) 45.0°, and (c) 90.0°?

47. You use a sequence of ideal polarizing filters, each
with its axis making the same angle with the axis of
the previous filter, to rotate the plane of polarization
of a polarized light beam by a total of 45.0°. You wish
to have an intensity reduction no larger than 10.0%.
(a) How many polarizers do you need to achieve
your goal? (b) What is the angle between adjacent
polarizers?

48. An unpolarized beam of light is incident on a stack of
ideal polarizing filters. The axis of the first filter is perpendicular to the axis of the last filter in the stack. Find
the fraction by which the transmitted beam's intensity
is reduced in the three following cases. (a) Three filters
are in the stack, each with its transmission axis at 45.0°
relative to the preceding filter. (b) Four filters are in
the stack, each with its transmission axis at 30.0° relative to the preceding filter. (c) Seven filters are in the
stack, each with its transmission axis at 15.0° relative to
the preceding filter. (d) Comment on comparing the
answers to parts (a), (b), and (c).

49. The critical angle for total internal reflection for sapphire surrounded by air is 34.4°. Calculate the polarizing angle for sapphire.

50. For a particular transparent medium surrounded by
air, find the polarizing angle θ_p in terms of the critical
angle for total internal reflection θ_c.

51. Three polarizing plates whose planes are parallel are
M centered on a common axis. The directions of the
transmission axes relative to the common vertical
direction are shown in Figure P38.51. A linearly polarized beam of light with plane of polarization parallel
to the vertical reference direction is incident from
the left onto the first disk with intensity $I_i = 10.0$ units

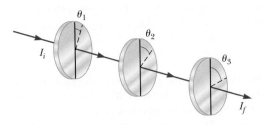

Figure P38.51

(arbitrary). Calculate the transmitted intensity I_f when
$\theta_1 = 20.0°$, $\theta_2 = 40.0°$, and $\theta_3 = 60.0°$. *Hint:* Make
repeated use of Malus's law.

52. Two polarizing sheets are placed together with their
transmission axes crossed so that no light is transmitted. A third sheet is inserted between them with
its transmission axis at an angle of 45.0° with respect
to each of the other axes. Find the fraction of incident unpolarized light intensity transmitted by the
three-sheet combination. (Assume each polarizing
sheet is ideal.)

Additional Problems

53. In a single-slit diffraction pattern, assuming each side
maximum is halfway between the adjacent minima,
find the ratio of the intensity of (a) the first-order side
maximum and (b) the second-order side maximum to
the intensity of the central maximum.

54. Laser light with a wavelength of 632.8 nm is directed
through one slit or two slits and allowed to fall on a
screen 2.60 m beyond. Figure P38.54 shows the pattern on the screen, with a centimeter ruler below it.
(a) Did the light pass through one slit or two slits?
Explain how you can determine the answer. (b) If one
slit, find its width. If two slits, find the distance between
their centers.

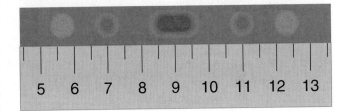

Figure P38.54

55. In water of uniform depth, a wide pier is supported on
pilings in several parallel rows 2.80 m apart. Ocean
waves of uniform wavelength roll in, moving in a direction that makes an angle of 80.0° with the rows of pilings. Find the three longest wavelengths of waves that
are strongly reflected by the pilings.

56. The second-order dark fringe in a single-slit diffraction pattern is 1.40 mm from the center of the central
maximum. Assuming the screen is 85.0 cm from a slit
of width 0.800 mm and assuming monochromatic
incident light, calculate the wavelength of the incident
light.

57. Light from a helium–neon laser ($\lambda = 632.8$ nm) is incident on a single slit. What is the maximum width of the
slit for which no diffraction minima are observed?

58. Two motorcycles separated laterally by 2.30 m are
W approaching an observer wearing night-vision gog-
gles sensitive to infrared light of wavelength 885 nm.
(a) Assume the light propagates through perfectly
steady and uniform air. What aperture diameter
is required if the motorcycles' headlights are to be
resolved at a distance of 12.0 km? (b) Comment on
how realistic the assumption in part (a) is.

59. The *Very Large Array* (VLA) is a set of 27 radio telescope dishes in Catron and Socorro counties, New Mexico (Fig. P38.59). The antennas can be moved apart on railroad tracks, and their combined signals give the resolving power of a synthetic aperture 36.0 km in diameter. (a) If the detectors are tuned to a frequency of 1.40 GHz, what is the angular resolution of the VLA? (b) Clouds of interstellar hydrogen radiate at the frequency used in part (a). What must be the separation distance of two clouds at the center of the galaxy, 26 000 light-years away, if they are to be resolved? (c) **What If?** As the telescope looks up, a circling hawk looks down. Assume the hawk is most sensitive to green light having a wavelength of 500 nm and has a pupil of diameter 12.0 mm. Find the angular resolution of the hawk's eye. (d) A mouse is on the ground 30.0 m below. By what distance must the mouse's whiskers be separated if the hawk can resolve them?

Figure P38.59

60. Two wavelengths λ and $\lambda + \Delta\lambda$ (with $\Delta\lambda << \lambda$) are incident on a diffraction grating. Show that the angular separation between the spectral lines in the mth-order spectrum is

$$\Delta\theta = \frac{\Delta\lambda}{\sqrt{(d/m)^2 - \lambda^2}}$$

where d is the slit spacing and m is the order number.

61. Review. A beam of 541-nm light is incident on a diffraction grating that has 400 grooves/mm. (a) Determine the angle of the second-order ray. (b) **What If?** If the entire apparatus is immersed in water, what is the new second-order angle of diffraction? (c) Show that the two diffracted rays of parts (a) and (b) are related through the law of refraction.

62. *Why is the following situation impossible?* A technician is sending laser light of wavelength 632.8 nm through a pair of slits separated by 30.0 μm. Each slit is of width 2.00 μm. The screen on which he projects the pattern is not wide enough, so light from the $m = 15$ interference maximum misses the edge of the screen and passes into the next lab station, startling a coworker.

63. A 750-nm light beam in air hits the flat surface of a certain liquid, and the beam is split into a reflected ray and a refracted ray. If the reflected ray is completely polarized when it is at 36.0° with respect to the surface, what is the wavelength of the refracted ray?

64. Iridescent peacock feathers are shown in Figure P38.64a. The surface of one microscopic barbule is composed of transparent keratin that supports rods of dark brown melanin in a regular lattice, represented in Figure P38.64b. (Your fingernails are made of keratin, and melanin is the dark pigment giving color to human skin.) In a portion of the feather that can appear turquoise (blue-green), assume the melanin rods are uniformly separated by 0.25 μm, with air between them. (a) Explain how this structure can appear turquoise when it contains no blue or green pigment. (b) Explain how it can also appear violet if light falls on it in a different direction. (c) Explain how it can present different colors to your two eyes simultaneously, which is a characteristic of iridescence. (d) A compact disc can appear to be any color of the rainbow. Explain why the portion of the feather in Figure P38.64b cannot appear yellow or red. (e) What could be different about the array of melanin rods in a portion of the feather that does appear to be red?

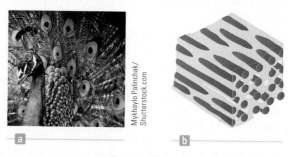

Mykhaylo Palinchak/ Shutterstock.com

a   b

Figure P38.64

65. Light in air strikes a water surface at the polarizing
_{AMT} angle. The part of the beam refracted into the water strikes a submerged slab of material with refractive index $n = 1.62$ as shown in Figure P38.65. The light reflected from the upper surface of the slab is completely polarized. Find the angle θ between the water surface and the surface of the slab.

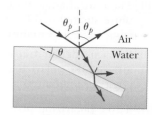

Figure P38.65
Problems 65 and 66.

66. Light in air (assume $n = 1$) strikes the surface of a liquid of index of refraction n_ℓ at the polarizing angle. The part of the beam refracted into the liquid strikes a submerged slab of material with refractive index n as shown in Figure P38.65. The light reflected from the upper surface of the slab is completely polarized. Find the angle θ between the water surface and the surface of the slab as a function of n and n_ℓ.

67. An American standard analog television picture (non-HDTV), also known as NTSC, is composed of approximately 485 visible horizontal lines of varying light intensity. Assume your ability to resolve the lines is limited only by the Rayleigh criterion, the pupils of your eyes are 5.00 mm in diameter, and the average wavelength of the light coming from the screen is 550 nm. Calculate the ratio of the minimum viewing distance to the vertical dimension of the picture such that you will not be able to resolve the lines.

68. A pinhole camera has a small circular aperture of diameter D. Light from distant objects passes through the aperture into an otherwise dark box, falling on a screen located a distance L away. If D is too large, the display on the screen will be fuzzy because a bright point in the field of view will send light onto a circle of diameter slightly larger than D. On the other hand, if D is too small, diffraction will blur the display on the screen. The screen shows a reasonably sharp image if the diameter of the central disk of the diffraction pattern, specified by Equation 38.6, is equal to D at the screen. (a) Show that for monochromatic light with plane wave fronts and $L \gg D$, the condition for a sharp view is fulfilled if $D^2 = 2.44\lambda L$. (b) Find the optimum pinhole diameter for 500-nm light projected onto a screen 15.0 cm away.

69. The *scale* of a map is a number of kilometers per centimeter specifying the distance on the ground that any distance on the map represents. The scale of a spectrum is its *dispersion*, a number of nanometers per centimeter, specifying the change in wavelength that a distance across the spectrum represents. You must know the dispersion if you want to compare one spectrum with another or make a measurement of, for example, a Doppler shift. Let y represent the position relative to the center of a diffraction pattern projected onto a flat screen at distance L by a diffraction grating with slit spacing d. The dispersion is $d\lambda/dy$. (a) Prove that the dispersion is given by

$$\frac{d\lambda}{dy} = \frac{L^2 d}{m(L^2 + y^2)^{3/2}}$$

(b) A light with a mean wavelength of 550 nm is analyzed with a grating having 8 000 rulings/cm and projected onto a screen 2.40 m away. Calculate the dispersion in first order.

70. (a) Light traveling in a medium of index of refraction n_1 is incident at an angle θ on the surface of a medium of index n_2. The angle between reflected and refracted rays is β. Show that

$$\tan\theta = \frac{n_2 \sin\beta}{n_1 - n_2 \cos\beta}$$

(b) **What If?** Show that this expression for $\tan\theta$ reduces to Brewster's law when $\beta = 90°$.

71. The intensity of light in a diffraction pattern of a single slit is described by the equation

$$I = I_{max}\frac{\sin^2\phi}{\phi^2}$$

where $\phi = (\pi a \sin\theta)/\lambda$. The central maximum is at $\phi = 0$, and the side maxima are *approximately* at $\phi = (m + \frac{1}{2})\pi$ for $m = 1, 2, 3, \ldots$. Determine more precisely (a) the location of the first side maximum, where $m = 1$, and (b) the location of the second side maximum. *Suggestion:* Observe in Figure 38.6a that the graph of intensity versus ϕ has a horizontal tangent at maxima and also at minima.

72. How much diffraction spreading does a light beam undergo? One quantitative answer is the *full width at half maximum* of the central maximum of the single-slit Fraunhofer diffraction pattern. You can evaluate this angle of spreading in this problem. (a) In Equation 38.2, define $\phi = \pi a \sin\theta/\lambda$ and show that at the point where $I = 0.5I_{max}$ we must have $\phi = \sqrt{2}\sin\phi$. (b) Let $y_1 = \sin\phi$ and $y_2 = \phi/\sqrt{2}$. Plot y_1 and y_2 on the same set of axes over a range from $\phi = 1$ rad to $\phi = \pi/2$ rad. Determine ϕ from the point of intersection of the two curves. (c) Then show that if the fraction λ/a is not large, the angular full width at half maximum of the central diffraction maximum is $\theta = 0.885\lambda/a$. (d) **What If?** Another method to solve the transcendental equation $\phi = \sqrt{2}\sin\phi$ in part (a) is to guess a first value of ϕ, use a computer or calculator to see how nearly it fits, and continue to update your estimate until the equation balances. How many steps (iterations) does this process take?

73. Two closely spaced wavelengths of light are incident on a diffraction grating. (a) Starting with Equation 38.7, show that the angular dispersion of the grating is given by

$$\frac{d\theta}{d\lambda} = \frac{m}{d\cos\theta}$$

(b) A square grating 2.00 cm on each side containing 8 000 equally spaced slits is used to analyze the spectrum of mercury. Two closely spaced lines emitted by this element have wavelengths of 579.065 nm and 576.959 nm. What is the angular separation of these two wavelengths in the second-order spectrum?

74. Light of wavelength 632.8 nm illuminates a single slit, and a diffraction pattern is formed on a screen 1.00 m from the slit. (a) Using the data in the following table, plot relative intensity versus position. Choose an appropriate value for the slit width a and, on the same graph used for the experimental data, plot the theoretical expression for the relative intensity

$$\frac{I}{I_{max}} = \frac{\sin^2\phi}{\phi^2}$$

where $\phi = (\pi a \sin\theta)/\lambda$. (b) What value of a gives the best fit of theory and experiment?

Position Relative to Central Maximum (mm)	Relative Intensity
0	1.00
0.8	0.95
1.6	0.80
3.2	0.39
4.8	0.079
6.5	0.003
8.1	0.036
9.7	0.043
11.3	0.013
12.9	0.000 3
14.5	0.012
16.1	0.015
17.7	0.004 4
19.3	0.000 3

Challenge Problems

75. Figure P38.75a is a three-dimensional sketch of a bire-fringent crystal. The dotted lines illustrate how a thin, parallel-faced slab of material could be cut from the larger specimen with the crystal's optic axis parallel to the faces of the plate. A section cut from the crystal in this manner is known as a *retardation plate*. When a beam of light is incident on the plate perpendicular to the direction of the optic axis as shown in Figure P38.75b, the O ray and the E ray travel along a single straight line, but with different speeds. The figure shows the wave fronts for the two rays. (a) Let the thickness of the plate be d. Show that the phase difference between the O ray and the E ray after traveling the thickness of the plate is

$$\theta = \frac{2\pi d}{\lambda} |n_O - n_E|$$

where λ is the wavelength in air. (b) In a particular case, the incident light has a wavelength of 550 nm. Find the minimum value of d for a quartz plate for which $\theta = \pi/2$. Such a plate is called a *quarter-wave plate*. Use values of n_O and n_E from Table 38.1.

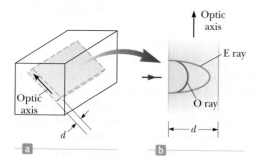

Figure P38.75

76. A spy satellite can consist of a large-diameter concave mirror forming an image on a digital-camera detector and sending the picture to a ground receiver by radio waves. In effect, it is an astronomical telescope in orbit, looking down instead of up. (a) Can a spy satellite read a license plate? (b) Can it read the date on a dime? Argue for your answers by making an order-of-magnitude calculation, specifying the data you estimate.

77. Suppose the single slit in Figure 38.4 is 6.00 cm wide and in front of a microwave source operating at 7.50 GHz. (a) Calculate the angle for the first minimum in the diffraction pattern. (b) What is the relative intensity I/I_{max} at $\theta = 15.0°$? (c) Assume two such sources, separated laterally by 20.0 cm, are behind the slit. What must be the maximum distance between the plane of the sources and the slit if the diffraction patterns are to be resolved? In this case, the approximation $\sin \theta \approx \tan \theta$ is not valid because of the relatively small value of a/λ.

78. In Figure P38.78, suppose the transmission axes of the left and right polarizing disks are perpendicular to each other. Also, let the center disk be rotated on the common axis with an angular speed ω. Show that if unpolarized light is incident on the left disk with an intensity I_{max}, the intensity of the beam emerging from the right disk is

$$I = \tfrac{1}{16} I_{max} (1 - \cos 4\omega t)$$

This result means that the intensity of the emerging beam is modulated at a rate four times the rate of rotation of the center disk. *Suggestion:* Use the trigonometric identities $\cos^2 \theta = \tfrac{1}{2}(1 + \cos 2\theta)$ and $\sin^2 \theta = \tfrac{1}{2}(1 - \cos 2\theta)$.

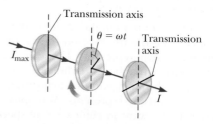

Figure P38.78

79. Consider a light wave passing through a slit and propagating toward a distant screen. Figure P38.79 shows the intensity variation for the pattern on the screen. Give a mathematical argument that more than 90% of the transmitted energy is in the central maximum of the diffraction pattern. *Suggestion:* You are not expected to calculate the precise percentage, but explain the steps of your reasoning. You may use the identification

$$\frac{1}{1^2} + \frac{1}{3^2} + \frac{1}{5^2} + \cdots = \frac{\pi^2}{8}$$

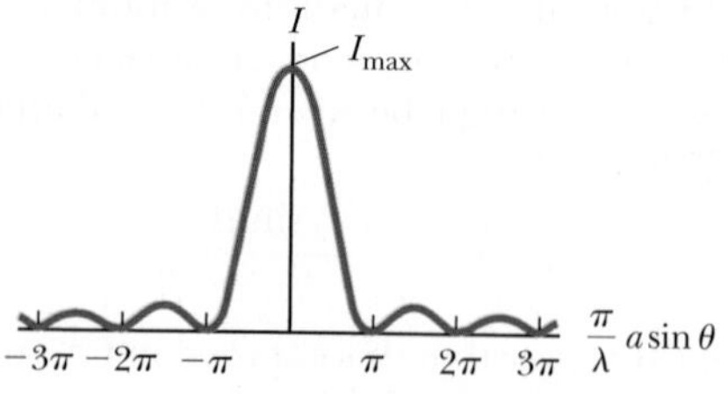

Figure P38.79

Modern Physics

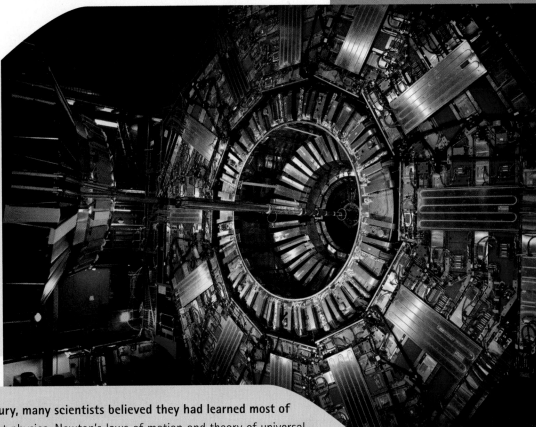

The Compact Muon Solenoid (CMS) Detector is part of the Large Hadron Collider at the European Laboratory for Particle Physics operated by CERN. It is one of several detectors that search for elementary particles. For a sense of scale, the green structure to the left of the detector and extending to the top is five stories high. *(CERN)*

At the end of the 19th century, many scientists believed they had learned most of what there was to know about physics. Newton's laws of motion and theory of universal gravitation, Maxwell's theoretical work in unifying electricity and magnetism, the laws of thermodynamics and kinetic theory, and the principles of optics were highly successful in explaining a variety of phenomena.

At the turn of the 20th century, however, a major revolution shook the world of physics. In 1900, Max Planck provided the basic ideas that led to the formulation of the quantum theory, and in 1905, Albert Einstein formulated his special theory of relativity. The excitement of the times is captured in Einstein's own words: "It was a marvelous time to be alive." Both theories were to have a profound effect on our understanding of nature. Within a few decades, they inspired new developments in the fields of atomic physics, nuclear physics, and condensed-matter physics.

In Chapter 39, we shall introduce the special theory of relativity. The theory provides us with a new and deeper view of physical laws. Although the predictions of this theory often violate our common sense, the theory correctly describes the results of experiments involving speeds near the speed of light. The extended version of this textbook, *Physics for Scientists and Engineers with Modern Physics*, covers the basic concepts of quantum mechanics and their application to atomic and molecular physics. In addition, we introduce condensed matter physics, nuclear physics, particle physics, and cosmology in the extended version.

Even though the physics that was developed during the 20th century has led to a multitude of important technological achievements, the story is still incomplete. Discoveries will continue to evolve during our lifetimes, and many of these discoveries will deepen or refine our understanding of nature and the Universe around us. It is still a "marvelous time to be alive." ∎

Relativity

Standing on the shoulders of a giant. David Serway, son of one of the authors, watches over two of his children, Nathan and Kaitlyn, as they frolic in the arms of Albert Einstein's statue at the Einstein memorial in Washington, D.C. It is well known that Einstein, the principal architect of relativity, was very fond of children. *(Emily Serway)*

Our everyday experiences and observations involve objects that move at speeds much less than the speed of light. Newtonian mechanics was formulated by observing and describing the motion of such objects, and this formalism is very successful in describing a wide range of phenomena that occur at low speeds. Nonetheless, it fails to describe properly the motion of objects whose speeds approach that of light.

Experimentally, the predictions of Newtonian theory can be tested at high speeds by accelerating electrons or other charged particles through a large electric potential difference. For example, it is possible to accelerate an electron to a speed of $0.99c$ (where c is the speed of light) by using a potential difference of several million volts. According to Newtonian mechanics, if the potential difference is increased by a factor of 4, the electron's kinetic energy is four times greater and its speed should double to $1.98c$. Experiments show, however, that the speed of the electron—as well as the speed of any other object in the Universe—always remains less than the speed of light, regardless of the size of the accelerating voltage. Because it places no upper limit on speed, Newtonian mechanics is contrary to modern experimental results and is clearly a limited theory.

In 1905, at the age of only 26, Einstein published his special theory of relativity. Regarding the theory, Einstein wrote:

The relativity theory arose from necessity, from serious and deep contradictions in the old theory from which there seemed no escape. The strength of the new theory lies in the consistency and simplicity with which it solves all these difficulties.[1]

Although Einstein made many other important contributions to science, the special theory of relativity alone represents one of the greatest intellectual achievements of all time. With this theory, experimental observations can be correctly predicted over the range of speeds from $v = 0$ to speeds approaching the speed of light. At low speeds, Einstein's theory reduces to Newtonian mechanics as a limiting situation. It is important to recognize that Einstein was working on electromagnetism when he developed the special theory of relativity. He was convinced that Maxwell's equations were correct, and to reconcile them with one of his postulates, he was forced into the revolutionary notion of assuming that space and time are not absolute.

This chapter gives an introduction to the special theory of relativity, with emphasis on some of its predictions. In addition to its well-known and essential role in theoretical physics, the special theory of relativity has practical applications, including the design of nuclear power plants and modern global positioning system (GPS) units. These devices depend on relativistic principles for proper design and operation.

39.1 The Principle of Galilean Relativity

To describe a physical event, we must establish a frame of reference. You should recall from Chapter 5 that an inertial frame of reference is one in which an object is observed to have no acceleration when no forces act on it. Furthermore, any frame moving with constant velocity with respect to an inertial frame must also be an inertial frame.

There is no absolute inertial reference frame. Therefore, the results of an experiment performed in a vehicle moving with uniform velocity must be identical to the results of the same experiment performed in a stationary vehicle. The formal statement of this result is called the **principle of Galilean relativity:**

> The laws of mechanics must be the same in all inertial frames of reference.

◀ **Principle of Galilean relativity**

Let's consider an observation that illustrates the equivalence of the laws of mechanics in different inertial frames. The pickup truck in Figure 39.1a moves with a

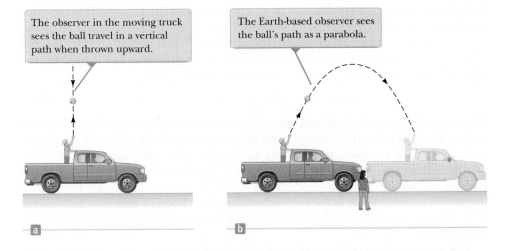

The observer in the moving truck sees the ball travel in a vertical path when thrown upward.

The Earth-based observer sees the ball's path as a parabola.

a

b

Figure 39.1 Two observers watch the path of a thrown ball and obtain different results.

---

[1]A. Einstein and L. Infield, *The Evolution of Physics* (New York: Simon and Schuster, 1961).

constant velocity with respect to the ground. If a passenger in the truck throws a ball straight up and if air resistance is neglected, the passenger observes that the ball moves in a vertical path. The motion of the ball appears to be precisely the same as if the ball were thrown by a person at rest on the Earth. The law of universal gravitation and the equations of motion under constant acceleration are obeyed whether the truck is at rest or in uniform motion.

Consider also an observer on the ground as in Figure 39.1b. Both observers agree on the laws of physics: the observer in the truck throws a ball straight up, and it rises and falls back into his hand according to the particle under constant acceleration model. Do the observers agree on the path of the ball thrown by the observer in the truck? The observer on the ground sees the path of the ball as a parabola as illustrated in Figure 39.1b, whereas, as mentioned earlier, the observer in the truck sees the ball move in a vertical path. Furthermore, according to the observer on the ground, the ball has a horizontal component of velocity equal to the velocity of the truck, and the horizontal motion of the ball is described by the particle under constant velocity model. Although the two observers disagree on certain aspects of the situation, they agree on the validity of Newton's laws and on the results of applying appropriate analysis models that we have learned. This agreement implies that no mechanical experiment can detect any difference between the two inertial frames. The only thing that can be detected is the relative motion of one frame with respect to the other.

Quick Quiz 39.1 Which observer in Figure 39.1 sees the ball's *correct* path? **(a)** the observer in the truck **(b)** the observer on the ground **(c)** both observers

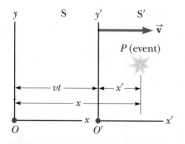

Figure 39.2 An event occurs at a point *P*. The event is seen by two observers in inertial frames S and S', where S' moves with a velocity $\vec{v}$ relative to S.

Suppose some physical phenomenon, which we call an *event*, occurs and is observed by an observer at rest in an inertial reference frame. The wording "in a frame" means that the observer is at rest with respect to the origin of that frame. The event's location and time of occurrence can be specified by the four coordinates (x, y, z, t). We would like to be able to transform these coordinates from those of an observer in one inertial frame to those of another observer in a frame moving with uniform relative velocity compared with the first frame.

Consider two inertial frames S and S' (Fig. 39.2). The S' frame moves with a constant velocity $\vec{v}$ along the common x and x' axes, where $\vec{v}$ is measured relative to S. We assume the origins of S and S' coincide at $t = 0$ and an event occurs at point P in space at some instant of time. For simplicity, we show the observer O in the S frame and the observer O' in the S' frame as blue dots at the origins of their coordinate frames in Figure 39.2, but that is not necessary: either observer could be at any fixed location in his or her frame. Observer O describes the event with space–time coordinates (x, y, z, t), whereas observer O' in S' uses the coordinates (x', y', z', t') to describe the same event. Model the origin of S' as a particle under constant velocity relative to the origin of S. As we see from the geometry in Figure 39.2, the relationships among these various coordinates can be written

Galilean transformation ▶
equations

$$x' = x - vt \qquad y' = y \qquad z' = z \qquad t' = t \qquad \text{(39.1)}$$

These equations are the **Galilean space–time transformation equations.** Note that time is assumed to be the same in both inertial frames. That is, within the framework of classical mechanics, all clocks run at the same rate, regardless of their velocity, so the time at which an event occurs for an observer in S is the same as the time for the same event in S'. Consequently, the time interval between two successive events should be the same for both observers. Although this assumption may seem obvious, it turns out to be incorrect in situations where v is comparable to the speed of light.

Now suppose a particle moves through a displacement of magnitude dx along the x axis in a time interval dt as measured by an observer in S. It follows from Equations 39.1 that the corresponding displacement dx' measured by an observer in S' is

$dx' = dx - v\,dt$, where frame S′ is moving with speed v in the x direction relative to frame S. Because $dt = dt'$, we find that

$$\frac{dx'}{dt'} = \frac{dx}{dt} - v$$

or

$$u'_x = u_x - v \qquad \qquad \textbf{(39.2)}$$

where u_x and u'_x are the x components of the velocity of the particle measured by observers in S and S′, respectively. (We use the symbol $\vec{\mathbf{u}}$ rather than $\vec{\mathbf{v}}$ for particle velocity because $\vec{\mathbf{v}}$ is already used for the relative velocity of two reference frames.) Equation 39.2 is the **Galilean velocity transformation equation.** It is consistent with our intuitive notion of time and space as well as with our discussions in Section 4.6. As we shall soon see, however, it leads to serious contradictions when applied to electromagnetic waves.

Quick Quiz 39.2 A baseball pitcher with a 90-mi/h fastball throws a ball while standing on a railroad flatcar moving at 110 mi/h. The ball is thrown in the same direction as that of the velocity of the train. If you apply the Galilean velocity transformation equation to this situation, is the speed of the ball relative to the Earth **(a)** 90 mi/h, **(b)** 110 mi/h, **(c)** 20 mi/h, **(d)** 200 mi/h, or **(e)** impossible to determine?

The Speed of Light

It is quite natural to ask whether the principle of Galilean relativity also applies to electricity, magnetism, and optics. Experiments indicate that the answer is no. Recall from Chapter 34 that Maxwell showed that the speed of light in free space is $c = 3.00 \times 10^8$ m/s. Physicists of the late 1800s thought light waves move through a medium called the *ether* and the speed of light is c only in a special, absolute frame at rest with respect to the ether. The Galilean velocity transformation equation was expected to hold for observations of light made by an observer in any frame moving at speed v relative to the absolute ether frame. That is, if light travels along the x axis and an observer moves with velocity $\vec{\mathbf{v}}$ along the x axis, the observer measures the light to have speed $c \pm v$, depending on the directions of travel of the observer and the light.

Because the existence of a preferred, absolute ether frame would show that light is similar to other classical waves and that Newtonian ideas of an absolute frame are true, considerable importance was attached to establishing the existence of the ether frame. Prior to the late 1800s, experiments involving light traveling in media moving at the highest laboratory speeds attainable at that time were not capable of detecting differences as small as that between c and $c \pm v$. Starting in about 1880, scientists decided to use the Earth as the moving frame in an attempt to improve their chances of detecting these small changes in the speed of light.

Observers fixed on the Earth can take the view that they are stationary and that the absolute ether frame containing the medium for light propagation moves past them with speed v. Determining the speed of light under these circumstances is similar to determining the speed of an aircraft traveling in a moving air current, or wind; consequently, we speak of an "ether wind" blowing through our apparatus fixed to the Earth.

A direct method for detecting an ether wind would use an apparatus fixed to the Earth to measure the ether wind's influence on the speed of light. If v is the speed of the ether relative to the Earth, light should have its maximum speed $c + v$ when propagating downwind as in Figure 39.3a. Likewise, the speed of light should have its minimum value $c - v$ when the light is propagating upwind as in Figure 39.3b and an intermediate value $(c^2 - v^2)^{1/2}$ when the light is directed such that it travels perpendicular to the ether wind as in Figure 39.3c. In this latter case, the vector $\vec{\mathbf{c}}$

> **Pitfall Prevention 39.1**
> **The Relationship Between the S and S′ Frames** Many of the mathematical representations in this chapter are true *only* for the specified relationship between the S and S′ frames. The x and x' axes coincide, except their origins are different. The y and y' axes (and the z and z' axes) are parallel, but they only coincide at one instant due to the time-varying position of the origin of S′ with respect to that of S. We choose the time $t = 0$ to be the instant at which the origins of the two coordinate systems coincide. If the S′ frame is moving in the positive x direction relative to S, then v is positive; otherwise, it is negative.

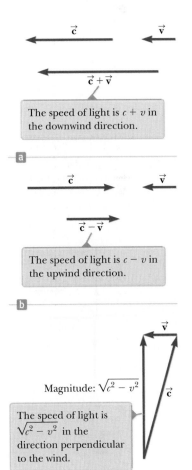

The speed of light is $c + v$ in the downwind direction.

a

The speed of light is $c - v$ in the upwind direction.

b

Magnitude: $\sqrt{c^2 - v^2}$

The speed of light is $\sqrt{c^2 - v^2}$ in the direction perpendicular to the wind.

c

Figure 39.3 If the velocity of the ether wind relative to the Earth is $\vec{\mathbf{v}}$ and the velocity of light relative to the ether is $\vec{\mathbf{c}}$, the speed of light relative to the Earth depends on the direction of the Earth's velocity.

must be aimed upstream so that the resultant velocity is perpendicular to the wind, like the boat in Figure 4.21b. If the Sun is assumed to be at rest in the ether, the velocity of the ether wind would be equal to the orbital velocity of the Earth around the Sun, which has a magnitude of approximately 30 km/s or 3×10^4 m/s. Because $c = 3 \times 10^8$ m/s, it is necessary to detect a change in speed of approximately 1 part in 10^4 for measurements in the upwind or downwind directions. Although such a change is experimentally measurable, all attempts to detect such changes and establish the existence of the ether wind (and hence the absolute frame) proved futile! We shall discuss the classic experimental search for the ether in Section 39.2.

The principle of Galilean relativity refers only to the laws of mechanics. If it is assumed the laws of electricity and magnetism are the same in all inertial frames, a paradox concerning the speed of light immediately arises. That can be understood by recognizing that Maxwell's equations imply that the speed of light always has the fixed value 3.00×10^8 m/s in all inertial frames, a result in direct contradiction to what is expected based on the Galilean velocity transformation equation. According to Galilean relativity, the speed of light should *not* be the same in all inertial frames.

To resolve this contradiction in theories, we must conclude that either (1) the laws of electricity and magnetism are not the same in all inertial frames or (2) the Galilean velocity transformation equation is incorrect. If we assume the first alternative, a preferred reference frame in which the speed of light has the value c must exist and the measured speed must be greater or less than this value in any other reference frame, in accordance with the Galilean velocity transformation equation. If we assume the second alternative, we must abandon the notions of absolute time and absolute length that form the basis of the Galilean space–time transformation equations.

39.2 The Michelson–Morley Experiment

The most famous experiment designed to detect small changes in the speed of light was first performed in 1881 by A. A. Michelson (see Section 37.6) and later repeated under various conditions by Michelson and Edward W. Morley (1838–1923). As we shall see, the outcome of the experiment contradicted the ether hypothesis.

The experiment was designed to determine the velocity of the Earth relative to that of the hypothetical ether. The experimental tool used was the Michelson interferometer, which was discussed in Section 37.6 and is shown again in Figure 39.4. Arm 2 is aligned along the direction of the Earth's motion through space. The Earth moving through the ether at speed v is equivalent to the ether flowing past the Earth in the opposite direction with speed v. This ether wind blowing in the direction opposite the direction of the Earth's motion should cause the speed of light measured in the Earth frame to be $c - v$ as the light approaches mirror M_2 and $c + v$ after reflection, where c is the speed of light in the ether frame.

The two light beams reflect from M_1 and M_2 and recombine, and an interference pattern is formed as discussed in Section 37.6. The interference pattern is then observed while the interferometer is rotated through an angle of 90°. This rotation interchanges the speed of the ether wind between the arms of the interferometer. The rotation should cause the fringe pattern to shift slightly but measurably. Measurements failed, however, to show any change in the interference pattern! The Michelson–Morley experiment was repeated at different times of the year when the ether wind was expected to change direction and magnitude, but the results were always the same: no fringe shift of the magnitude required was *ever* observed.[2]

The negative results of the Michelson–Morley experiment not only contradicted the ether hypothesis, but also showed that it is impossible to measure the absolute

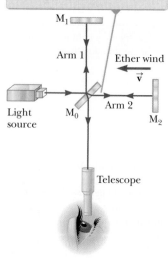

According to the ether wind theory, the speed of light should be $c - v$ as the beam approaches mirror M_2 and $c + v$ after reflection.

Figure 39.4 A Michelson interferometer is used in an attempt to detect the ether wind.

---

[2]From an Earth-based observer's point of view, changes in the Earth's speed and direction of motion in the course of a year are viewed as ether wind shifts. Even if the speed of the Earth with respect to the ether were zero at some time, six months later the speed of the Earth would be 60 km/s with respect to the ether and as a result a fringe shift should be noticed. No shift has ever been observed, however.

velocity of the Earth with respect to the ether frame. Einstein, however, offered a postulate for his special theory of relativity that places quite a different interpretation on these null results. In later years, when more was known about the nature of light, the idea of an ether that permeates all of space was abandoned. Light is now understood to be an electromagnetic wave, which requires no medium for its propagation. As a result, the idea of an ether in which these waves travel became unnecessary.

Details of the Michelson–Morley Experiment

To understand the outcome of the Michelson–Morley experiment, let's assume the two arms of the interferometer in Figure 39.4 are of equal length L. We shall analyze the situation as if there were an ether wind because that is what Michelson and Morley expected to find. As noted above, the speed of the light beam along arm 2 should be $c - v$ as the beam approaches M_2 and $c + v$ after the beam is reflected. We model a pulse of light as a particle under constant speed. Therefore, the time interval for travel to the right for the pulse is $\Delta t = L/(c - v)$ and the time interval for travel to the left is $\Delta t = L/(c + v)$. The total time interval for the round trip along arm 2 is

$$\Delta t_{arm\,2} = \frac{L}{c + v} + \frac{L}{c - v} = \frac{2Lc}{c^2 - v^2} = \frac{2L}{c}\left(1 - \frac{v^2}{c^2}\right)^{-1}$$

Now consider the light beam traveling along arm 1, perpendicular to the ether wind. Because the speed of the beam relative to the Earth is $(c^2 - v^2)^{1/2}$ in this case (see Fig. 39.3c), the time interval for travel for each half of the trip is $\Delta t = L/(c^2 - v^2)^{1/2}$ and the total time interval for the round trip is

$$\Delta t_{arm\,1} = \frac{2L}{(c^2 - v^2)^{1/2}} = \frac{2L}{c}\left(1 - \frac{v^2}{c^2}\right)^{-1/2}$$

The time difference Δt between the horizontal round trip (arm 2) and the vertical round trip (arm 1) is

$$\Delta t = \Delta t_{arm\,2} - \Delta t_{arm\,1} = \frac{2L}{c}\left[\left(1 - \frac{v^2}{c^2}\right)^{-1} - \left(1 - \frac{v^2}{c^2}\right)^{-1/2}\right]$$

Because $v^2/c^2 \ll 1$, we can simplify this expression by using the following binomial expansion after dropping all terms higher than second order:

$$(1 - x)^n \approx 1 - nx \quad (\text{for } x \ll 1)$$

In our case, $x = v^2/c^2$, and we find that

$$\Delta t = \Delta t_{arm\,2} - \Delta t_{arm\,1} \approx \frac{Lv^2}{c^3} \qquad \textbf{(39.3)}$$

This time difference between the two instants at which the reflected beams arrive at the viewing telescope gives rise to a phase difference between the beams, producing an interference pattern when they combine at the position of the telescope. A shift in the interference pattern should be detected when the interferometer is rotated through 90° in a horizontal plane so that the two beams exchange roles. This rotation results in a time difference twice that given by Equation 39.3. Therefore, the path difference that corresponds to this time difference is

$$\Delta d = c(2\,\Delta t) = \frac{2Lv^2}{c^2}$$

Because a change in path length of one wavelength corresponds to a shift of one fringe, the corresponding fringe shift is equal to this path difference divided by the wavelength of the light:

$$\text{Shift} = \frac{2Lv^2}{\lambda c^2} \qquad \textbf{(39.4)}$$

In the experiments by Michelson and Morley, each light beam was reflected by mirrors many times to give an effective path length L of approximately 11 m. Using this value, taking v to be equal to 3.0×10^4 m/s (the speed of the Earth around the Sun), and using 500 nm for the wavelength of the light, we expect a fringe shift of

$$\text{Shift} = \frac{2(11 \text{ m})(3.0 \times 10^4 \text{ m/s})^2}{(5.0 \times 10^{-7} \text{ m})(3.0 \times 10^8 \text{ m/s})^2} = 0.44$$

The instrument used by Michelson and Morley could detect shifts as small as 0.01 fringe, but it detected no shift whatsoever in the fringe pattern! The experiment has been repeated many times since by different scientists under a wide variety of conditions, and no fringe shift has ever been detected. Therefore, it was concluded that the motion of the Earth with respect to the postulated ether cannot be detected.

Many efforts were made to explain the null results of the Michelson–Morley experiment and to save the ether frame concept and the Galilean velocity transformation equation for light. All proposals resulting from these efforts have been shown to be wrong. No experiment in the history of physics received such valiant efforts to explain the absence of an expected result as did the Michelson–Morley experiment. The stage was set for Einstein, who solved the problem in 1905 with his special theory of relativity.

39.3 Einstein's Principle of Relativity

In the previous section, we noted the impossibility of measuring the speed of the ether with respect to the Earth and the failure of the Galilean velocity transformation equation in the case of light. Einstein proposed a theory that boldly removed these difficulties and at the same time completely altered our notion of space and time.[3] He based his special theory of relativity on two postulates:

1. **The principle of relativity:** The laws of physics must be the same in all inertial reference frames.
2. **The constancy of the speed of light:** The speed of light in vacuum has the same value, $c = 3.00 \times 10^8$ m/s, in all inertial frames, regardless of the velocity of the observer or the velocity of the source emitting the light.

The first postulate asserts that *all* the laws of physics—those dealing with mechanics, electricity and magnetism, optics, thermodynamics, and so on—are the same in all reference frames moving with constant velocity relative to one another. This postulate is a generalization of the principle of Galilean relativity, which refers only to the laws of mechanics. From an experimental point of view, Einstein's principle of relativity means that any kind of experiment (measuring the speed of light, for example) performed in a laboratory at rest must give the same result when performed in a laboratory moving at a constant velocity with respect to the first one. Hence, no preferred inertial reference frame exists, and it is impossible to detect absolute motion.

Note that postulate 2 is required by postulate 1: if the speed of light were not the same in all inertial frames, measurements of different speeds would make it possible to distinguish between inertial frames. As a result, a preferred, absolute frame could be identified, in contradiction to postulate 1.

Although the Michelson–Morley experiment was performed before Einstein published his work on relativity, it is not clear whether or not Einstein was aware of the details of the experiment. Nonetheless, the null result of the experiment can be readily understood within the framework of Einstein's theory. According to

Albert Einstein
German-American Physicist
(1879–1955)
Einstein, one of the greatest physicists of all time, was born in Ulm, Germany. In 1905, at age 26, he published four scientific papers that revolutionized physics. Two of these papers were concerned with what is now considered his most important contribution: the special theory of relativity.

In 1916, Einstein published his work on the general theory of relativity. The most dramatic prediction of this theory is the degree to which light is deflected by a gravitational field. Measurements made by astronomers on bright stars in the vicinity of the eclipsed Sun in 1919 confirmed Einstein's prediction, and Einstein became a world celebrity as a result. Einstein was deeply disturbed by the development of quantum mechanics in the 1920s despite his own role as a scientific revolutionary. In particular, he could never accept the probabilistic view of events in nature that is a central feature of quantum theory. The last few decades of his life were devoted to an unsuccessful search for a unified theory that would combine gravitation and electromagnetism.

---

[3]A. Einstein, "On the Electrodynamics of Moving Bodies," *Ann. Physik* **17**:891, 1905. For an English translation of this article and other publications by Einstein, see the book by H. Lorentz, A. Einstein, H. Minkowski, and H. Weyl, *The Principle of Relativity* (New York: Dover, 1958).

his principle of relativity, the premises of the Michelson–Morley experiment were incorrect. In the process of trying to explain the expected results, we stated that when light traveled against the ether wind, its speed was $c - v$, in accordance with the Galilean velocity transformation equation. If the state of motion of the observer or of the source has no influence on the value found for the speed of light, however, one always measures the value to be c. Likewise, the light makes the return trip after reflection from the mirror at speed c, not at speed $c + v$. Therefore, the motion of the Earth does not influence the interference pattern observed in the Michelson–Morley experiment, and a null result should be expected.

If we accept Einstein's theory of relativity, we must conclude that relative motion is unimportant when measuring the speed of light. At the same time, we must alter our commonsense notion of space and time and be prepared for some surprising consequences. As you read the pages ahead, keep in mind that our commonsense ideas are based on a lifetime of everyday experiences and not on observations of objects moving at hundreds of thousands of kilometers per second. Therefore, these results may seem strange, but that is only because we have no experience with them.

39.4 Consequences of the Special Theory of Relativity

As we examine some of the consequences of relativity in this section, we restrict our discussion to the concepts of simultaneity, time intervals, and lengths, all three of which are quite different in relativistic mechanics from what they are in Newtonian mechanics. In relativistic mechanics, for example, the distance between two points and the time interval between two events depend on the frame of reference in which they are measured.

Simultaneity and the Relativity of Time

A basic premise of Newtonian mechanics is that a universal time scale exists that is the same for all observers. Newton and his followers took simultaneity for granted. In his special theory of relativity, Einstein abandoned this assumption.

Einstein devised the following thought experiment to illustrate this point. A boxcar moves with uniform velocity, and two bolts of lightning strike its ends as illustrated in Figure 39.5a, leaving marks on the boxcar and on the ground. The marks on the boxcar are labeled A' and B', and those on the ground are labeled A and B. An observer O' moving with the boxcar is midway between A' and B', and a ground observer O is midway between A and B. The events recorded by the observers are the striking of the boxcar by the two lightning bolts.

The light signals emitted from A and B at the instant at which the two bolts strike later reach observer O at the same time as indicated in Figure 39.5b. This observer

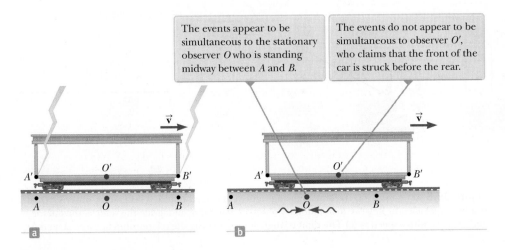

The events appear to be simultaneous to the stationary observer O who is standing midway between A and B.

The events do not appear to be simultaneous to observer O', who claims that the front of the car is struck before the rear.

Figure 39.5 (a) Two lightning bolts strike the ends of a moving boxcar. (b) The leftward-traveling light signal has already passed O', but the rightward-traveling signal has not yet reached O'.

realizes that the signals traveled at the same speed over equal distances and so concludes that the events at *A* and *B* occurred simultaneously. Now consider the same events as viewed by observer *O'*. By the time the signals have reached observer *O*, observer *O'* has moved as indicated in Figure 39.5b. Therefore, the signal from *B'* has already swept past *O'*, but the signal from *A'* has not yet reached *O'*. In other words, *O'* sees the signal from *B'* before seeing the signal from *A'*. According to Einstein, *the two observers must find that light travels at the same speed.* Therefore, observer *O'* concludes that one lightning bolt strikes the front of the boxcar *before* the other one strikes the back.

This thought experiment clearly demonstrates that the two events that appear to be simultaneous to observer *O* do *not* appear to be simultaneous to observer *O'*. Simultaneity is not an absolute concept but rather one that depends on the state of motion of the observer. Einstein's thought experiment demonstrates that two observers can disagree on the simultaneity of two events. This disagreement, however, depends on the transit time of light to the observers and therefore does *not* demonstrate the deeper meaning of relativity. In relativistic analyses of high-speed situations, simultaneity is relative even when the transit time is subtracted out. In fact, in all the relativistic effects we discuss, we ignore differences caused by the transit time of light to the observers.

Time Dilation

To illustrate that observers in different inertial frames can measure different time intervals between a pair of events, consider a vehicle moving to the right with a speed *v* such as the boxcar shown in Figure 39.6a. A mirror is fixed to the ceiling of the vehicle, and observer *O'* at rest in the frame attached to the vehicle holds a flashlight a distance *d* below the mirror. At some instant, the flashlight emits a pulse of light directed toward the mirror (event 1), and at some later time after reflecting from the mirror, the pulse arrives back at the flashlight (event 2). Observer *O'* carries a clock and uses it to measure the time interval Δt_p between these two events. (The subscript *p* stands for *proper*, as we shall see in a moment.) We model the pulse of light as a particle under constant speed. Because the light pulse has a speed *c*, the time interval required for the pulse to travel from *O'* to the mirror and back is

$$\Delta t_p = \frac{\text{distance traveled}}{\text{speed}} = \frac{2d}{c} \tag{39.5}$$

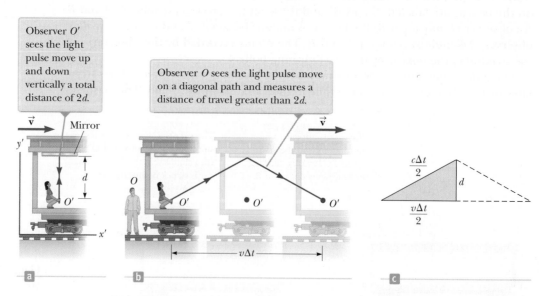

Figure 39.6 (a) A mirror is fixed to a moving vehicle, and a light pulse is sent out by observer *O'* at rest in the vehicle. (b) Relative to a stationary observer *O* standing alongside the vehicle, the mirror and *O'* move with a speed *v*. (c) The right triangle for calculating the relationship between Δt and Δt_p.

Now consider the same pair of events as viewed by observer O in a second frame at rest with respect to the ground as shown in Figure 39.6b. According to this observer, the mirror and the flashlight are moving to the right with a speed v, and as a result, the sequence of events appears entirely different. By the time the light from the flashlight reaches the mirror, the mirror has moved to the right a distance $v\,\Delta t/2$, where Δt is the time interval required for the light to travel from O' to the mirror and back to O' as measured by O. Observer O concludes that because of the motion of the vehicle, if the light is to hit the mirror, it must leave the flashlight at an angle with respect to the vertical direction. Comparing Figure 39.6a with Figure 39.6b, we see that the light must travel farther in part (b) than in part (a). (Notice that neither observer "knows" that he or she is moving. Each is at rest in his or her own inertial frame.)

According to the second postulate of the special theory of relativity, both observers must measure c for the speed of light. Because the light travels farther according to O, the time interval Δt measured by O is longer than the time interval Δt_p measured by O'. To obtain a relationship between these two time intervals, let's use the right triangle shown in Figure 39.6c. The Pythagorean theorem gives

$$\left(\frac{c\,\Delta t}{2}\right)^2 = \left(\frac{v\,\Delta t}{2}\right)^2 + d^2$$

Solving for Δt gives

$$\Delta t = \frac{2d}{\sqrt{c^2 - v^2}} = \frac{2d}{c\sqrt{1 - \dfrac{v^2}{c^2}}} \qquad \textbf{(39.6)}$$

Because $\Delta t_p = 2d/c$, we can express this result as

$$\Delta t = \frac{\Delta t_p}{\sqrt{1 - \dfrac{v^2}{c^2}}} = \gamma\,\Delta t_p \qquad \textbf{(39.7)} \qquad \blacktriangleleft \text{ Time dilation}$$

where

$$\gamma = \frac{1}{\sqrt{1 - \dfrac{v^2}{c^2}}} \qquad \textbf{(39.8)}$$

Because γ is always greater than unity, Equation 39.7 shows that the time interval Δt measured by an observer moving with respect to a clock is longer than the time interval Δt_p measured by an observer at rest with respect to the clock. This effect is known as **time dilation.**

Time dilation is not observed in our everyday lives, which can be understood by considering the factor γ. This factor deviates significantly from a value of 1 only for very high speeds as shown in Figure 39.7 and Table 39.1. For example, for a speed of $0.1c$, the value of γ is 1.005. Therefore, there is a time dilation of only 0.5% at

Table 39.1	Approximate Values for γ at Various Speeds
v/c	γ
0	1
0.001 0	1.000 000 5
0.010	1.000 05
0.10	1.005
0.20	1.021
0.30	1.048
0.40	1.091
0.50	1.155
0.60	1.250
0.70	1.400
0.80	1.667
0.90	2.294
0.92	2.552
0.94	2.931
0.96	3.571
0.98	5.025
0.99	7.089
0.995	10.01
0.999	22.37

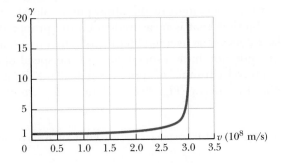

Figure 39.7 Graph of γ versus v. As the speed approaches that of light, γ increases rapidly.

one-tenth the speed of light. Speeds encountered on an everyday basis are far slower than $0.1c$, so we do not experience time dilation in normal situations.

The time interval Δt_p in Equations 39.5 and 39.7 is called the **proper time interval.** (Einstein used the German term *Eigenzeit,* which means "own-time.") In general, the proper time interval is the time interval between two events measured by an observer *who sees the events occur at the same point in space.*

If a clock is moving with respect to you, the time interval between ticks of the moving clock is observed to be longer than the time interval between ticks of an identical clock in your reference frame. Therefore, it is often said that a moving clock is measured to run more slowly than a clock in your reference frame by a factor γ. We can generalize this result by stating that all physical processes, including mechanical, chemical, and biological ones, are measured to slow down when those processes occur in a frame moving with respect to the observer. For example, the heartbeat of an astronaut moving through space keeps time with a clock inside the spacecraft. Both the astronaut's clock and heartbeat are measured to slow down relative to a clock back on the Earth (although the astronaut would have no sensation of life slowing down in the spacecraft).

Quick Quiz 39.3 Suppose the observer O' on the train in Figure 39.6 aims her flashlight at the far wall of the boxcar and turns it on and off, sending a pulse of light toward the far wall. Both O' and O measure the time interval between when the pulse leaves the flashlight and when it hits the far wall. Which observer measures the proper time interval between these two events? **(a)** O' **(b)** O **(c)** both observers **(d)** neither observer

Quick Quiz 39.4 A crew on a spacecraft watches a movie that is two hours long. The spacecraft is moving at high speed through space. Does an Earth-based observer watching the movie screen on the spacecraft through a powerful telescope measure the duration of the movie to be **(a)** longer than, **(b)** shorter than, or **(c)** equal to two hours?

Pitfall Prevention 39.3

The Proper Time Interval It is *very* important in relativistic calculations to correctly identify the observer who measures the proper time interval. The proper time interval between two events is always the time interval measured by an observer for whom the two events take place at the same position.

Time dilation is a very real phenomenon that has been verified by various experiments involving natural clocks. One experiment reported by J. C. Hafele and R. E. Keating provided direct evidence of time dilation.[4] Time intervals measured with four cesium atomic clocks in jet flight were compared with time intervals measured by Earth-based reference atomic clocks. To compare these results with theory, many factors had to be considered, including periods of speeding up and slowing down relative to the Earth, variations in direction of travel, and the weaker gravitational field experienced by the flying clocks than that experienced by the Earth-based clock. The results were in good agreement with the predictions of the special theory of relativity and were explained in terms of the relative motion between the Earth and the jet aircraft. In their paper, Hafele and Keating stated that "relative to the atomic time scale of the U.S. Naval Observatory, the flying clocks lost 59 ± 10 ns during the eastward trip and gained 273 ± 7 ns during the westward trip."

Another interesting example of time dilation involves the observation of *muons,* unstable elementary particles that have a charge equal to that of the electron and a mass 207 times that of the electron. Muons can be produced by the collision of cosmic radiation with atoms high in the atmosphere. Slow-moving muons in the laboratory have a lifetime that is measured to be the proper time interval $\Delta t_p = 2.2 \ \mu s$. If we take 2.2 μs as the average lifetime of a muon and assume that muons created by cosmic radiation have a speed close to the speed of light, we find that these particles can travel a distance of approximately $(3.0 \times 10^8 \ \text{m/s})(2.2 \times 10^{-6} \ \text{s}) \approx 6.6 \times 10^2$ m before they decay (Fig. 39.8a). Hence, they are unlikely to reach the

---

[4]J. C. Hafele and R. E. Keating, "Around the World Atomic Clocks: Relativistic Time Gains Observed," *Science* **177**:168, 1972.

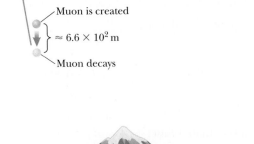

Without relativistic considerations, according to an observer on the Earth, muons created in the atmosphere and traveling downward with a speed close to c travel only about 6.6×10^2 m before decaying with an average lifetime of 2.2 μs. Therefore, very few muons would reach the surface of the Earth.

Muon is created

$\approx 6.6 \times 10^2$ m

Muon decays

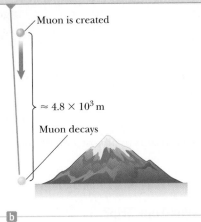

With relativistic considerations, the muon's lifetime is dilated according to an observer on the Earth. Hence, according to this observer, the muon can travel about 4.8×10^3 m before decaying. The result is many of them arriving at the surface.

Muon is created

$\approx 4.8 \times 10^3$ m

Muon decays

Figure 39.8 Travel of muons according to an Earth-based observer.

surface of the Earth from high in the atmosphere where they are produced. Experiments show, however, that a large number of muons *do* reach the surface. The phenomenon of time dilation explains this effect. As measured by an observer on the Earth, the muons have a dilated lifetime equal to $\gamma \, \Delta t_p$. For example, for $v = 0.99c$, $\gamma \approx 7.1$, and $\gamma \, \Delta t_p \approx 16$ μs. Hence, the average distance traveled by the muons in this time interval as measured by an observer on the Earth is approximately $(0.99)(3.0 \times 10^8 \text{ m/s})(16 \times 10^{-6} \text{ s}) \approx 4.8 \times 10^3$ m as indicated in Figure 39.8b.

In 1976, at the laboratory of the European Council for Nuclear Research (CERN) in Geneva, muons injected into a large storage ring reached speeds of approximately $0.999\,4c$. Electrons produced by the decaying muons were detected by counters around the ring, enabling scientists to measure the decay rate and hence the muon lifetime. The lifetime of the moving muons was measured to be approximately 30 times as long as that of the stationary muon, in agreement with the prediction of relativity to within two parts in a thousand.

Example 39.1 **What Is the Period of the Pendulum?**

The period of a pendulum is measured to be 3.00 s in the reference frame of the pendulum. What is the period when measured by an observer moving at a speed of $0.960c$ relative to the pendulum?

SOLUTION

Conceptualize Let's change frames of reference. Instead of the observer moving at $0.960c$, we can take the equivalent point of view that the observer is at rest and the pendulum is moving at $0.960c$ past the stationary observer. Hence, the pendulum is an example of a clock moving at high speed with respect to an observer.

Categorize Based on the Conceptualize step, we can categorize this example as a substitution problem involving relativistic time dilation.

The proper time interval, measured in the rest frame of the pendulum, is $\Delta t_p = 3.00$ s.

Use Equation 39.7 to find the dilated time interval:

$$\Delta t = \gamma \, \Delta t_p = \frac{1}{\sqrt{1 - \dfrac{(0.960c)^2}{c^2}}} \, \Delta t_p = \frac{1}{\sqrt{1 - 0.921\,6}} \, \Delta t_p$$

$$= 3.57(3.00 \text{ s}) = \boxed{10.7 \text{ s}}$$

continued

▶ **39.1** c o n t i n u e d

This result shows that a moving pendulum is indeed measured to take longer to complete a period than a pendulum at rest does. The period increases by a factor of $\gamma = 3.57$.

WHAT IF?   What if the speed of the observer increases by 4.00%? Does the dilated time interval increase by 4.00%?

Answer Based on the highly nonlinear behavior of γ as a function of v in Figure 39.7, we would guess that the increase in Δt would be different from 4.00%.

Find the new speed if it increases by 4.00%:

$$v_{new} = (1.040\,0)(0.960c) = 0.998\,4c$$

Perform the time dilation calculation again:

$$\Delta t = \gamma\,\Delta t_p = \frac{1}{\sqrt{1 - \dfrac{(0.998\,4c)^2}{c^2}}}\,\Delta t_p = \frac{1}{\sqrt{1 - 0.996\,8}}\,\Delta t_p$$

$$= 17.68(3.00\text{ s}) = 53.1\text{ s}$$

Therefore, the 4.00% increase in speed results in almost a 400% increase in the dilated time!

Example 39.2   **How Long Was Your Trip?**

Suppose you are driving your car on a business trip and are traveling at 30 m/s. Your boss, who is waiting at your destination, expects the trip to take 5.0 h. When you arrive late, your excuse is that the clock in your car registered the passage of 5.0 h but that you were driving fast and so your clock ran more slowly than the clock in your boss's office. If your car clock actually did indicate a 5.0-h trip, how much time passed on your boss's clock, which was at rest on the Earth?

S O L U T I O N

Conceptualize The observer is your boss standing stationary on the Earth. The clock is in your car, moving at 30 m/s with respect to your boss.

Categorize The low speed of 30 m/s suggests we might categorize this problem as one in which we use classical concepts and equations. Based on the problem statement that the moving clock runs more slowly than a stationary clock, however, we categorize this problem as one involving time dilation.

· ·

Analyze The proper time interval, measured in the rest frame of the car, is $\Delta t_p = 5.0$ h.

Use Equation 39.8 to evaluate γ:

$$\gamma = \frac{1}{\sqrt{1 - \dfrac{v^2}{c^2}}} = \frac{1}{\sqrt{1 - \dfrac{(3.0 \times 10^1\text{ m/s})^2}{(3.0 \times 10^8\text{ m/s})^2}}} = \frac{1}{\sqrt{1 - 10^{-14}}}$$

If you try to determine this value on your calculator, you will probably obtain $\gamma = 1$. Instead, perform a binomial expansion:

$$\gamma = (1 - 10^{-14})^{-1/2} \approx 1 + \tfrac{1}{2}(10^{-14}) = 1 + 5.0 \times 10^{-15}$$

Use Equation 39.7 to find the dilated time interval measured by your boss:

$$\Delta t = \gamma\,\Delta t_p = (1 + 5.0 \times 10^{-15})(5.0\text{ h})$$

$$= 5.0\text{ h} + 2.5 \times 10^{-14}\text{ h} = \boxed{5.0\text{ h} + 0.090\text{ ns}}$$

· ·

Finalize Your boss's clock would be only 0.090 ns ahead of your car clock. You might want to think of another excuse!

The Twin Paradox

An intriguing consequence of time dilation is the *twin paradox* (Fig. 39.9). Consider an experiment involving a set of twins named Speedo and Goslo. When they are 20 years old, Speedo, the more adventuresome of the two, sets out on an epic journey from the Earth to Planet X, located 20 light-years away. One light-year (ly) is the distance light travels through free space in 1 year. Furthermore, Speedo's

As Speedo (on the left) leaves his brother on Earth, both twins are the same age.

When Speedo returns from his journey, Goslo (on the right) is much older than Speedo.

a

b

Figure 39.9 The twin paradox. Speedo takes a journey to a star 20 light-years away and returns to the Earth.

spacecraft is capable of reaching a speed of $0.95c$ relative to the inertial frame of his twin brother back home on the Earth. After reaching Planet X, Speedo becomes homesick and immediately returns to the Earth at the same speed $0.95c$. Upon his return, Speedo is shocked to discover that Goslo has aged 42 years and is now 62 years old. Speedo, on the other hand, has aged only 13 years.

The paradox is *not* that the twins have aged at different rates. Here is the apparent paradox. From Goslo's frame of reference, he was at rest while his brother traveled at a high speed away from him and then came back. According to Speedo, however, he himself remained stationary while Goslo and the Earth raced away from him and then headed back. Therefore, we might expect Speedo to claim that Goslo ages more slowly than himself. The situation appears to be symmetrical from either twin's point of view. Which twin *actually* ages more slowly?

The situation is actually not symmetrical. Consider a third observer moving at a constant speed relative to Goslo. According to the third observer, Goslo never changes inertial frames. Goslo's speed relative to the third observer is always the same. The third observer notes, however, that Speedo accelerates during his journey when he slows down and starts moving back toward the Earth, *changing reference frames in the process*. From the third observer's perspective, there is something very different about the motion of Goslo when compared to Speedo. Therefore, there is no paradox: only Goslo, who is always in a single inertial frame, can make correct predictions based on special relativity. Goslo finds that instead of aging 42 years, Speedo ages only $(1 - v^2/c^2)^{1/2}(42 \text{ years}) = 13$ years. Of these 13 years, Speedo spends 6.5 years traveling to Planet X and 6.5 years returning.

Quick Quiz 39.5 Suppose astronauts are paid according to the amount of time they spend traveling in space. After a long voyage traveling at a speed approaching c, would a crew rather be paid according to **(a)** an Earth-based clock, **(b)** their spacecraft's clock, or **(c)** either clock?

Length Contraction

The measured distance between two points in space also depends on the frame of reference of the observer. The **proper length** L_p of an object is the length measured by an observer *at rest relative to the object*. The length of an object measured by someone in a reference frame that is moving with respect to the object is always less than the proper length. This effect is known as **length contraction.**

To understand length contraction, consider a spacecraft traveling with a speed v from one star to another. There are two observers: one on the Earth and the other in the spacecraft. The observer at rest on the Earth (and also assumed to be at rest with respect to the two stars) measures the distance between the stars to be the proper length L_p. According to this observer, the time interval required for the spacecraft to complete the voyage is given by the particle under constant velocity model as $\Delta t = L_p/v$. The passages of the two stars by the spacecraft occur at the same position for the space traveler. Therefore, the space traveler measures the proper time interval Δt_p. Because of time dilation, the proper time interval is related to the Earth-measured time interval by $\Delta t_p = \Delta t/\gamma$. Because the space traveler reaches the second star in the time Δt_p, he or she concludes that the distance L between the stars is

$$L = v\,\Delta t_p = v\,\frac{\Delta t}{\gamma}$$

Because the proper length is $L_p = v\,\Delta t$, we see that

Length contraction ▶

$$L = \frac{L_p}{\gamma} = L_p\sqrt{1 - \frac{v^2}{c^2}} \tag{39.9}$$

where $\sqrt{1 - v^2/c^2}$ is a factor less than unity. If an object has a proper length L_p when it is measured by an observer at rest with respect to the object, its length L when it moves with speed v in a direction parallel to its length is measured to be shorter according to Equation 39.9.

For example, suppose a meterstick moves past a stationary Earth-based observer with speed v as in Figure 39.10. The length of the meterstick as measured by an observer in a frame attached to the stick is the proper length L_p shown in Figure 39.10a. The length of the stick L measured by the Earth observer is shorter than L_p by the factor $(1 - v^2/c^2)^{1/2}$ as suggested in Figure 39.10b. Notice that length contraction takes place only along the direction of motion.

The proper length and the proper time interval are defined differently. The proper length is measured by an observer for whom the endpoints of the length remain fixed in space. The proper time interval is measured by someone for whom the two events take place at the same position in space. As an example of this point, let's return to the decaying muons moving at speeds close to the speed of light. An observer in the muon's reference frame measures the proper lifetime, whereas an Earth-based observer measures the proper length (the distance between the creation point and the decay point in Fig. 39.8b). In the muon's reference frame, there is no time dilation, but the distance of travel to the surface is shorter when measured in this frame. Likewise, in the Earth observer's reference frame, there is time dilation, but the distance of travel is measured to be the proper length. Therefore, when calculations on the muon are performed in both frames, the outcome of the experiment in one frame is the same as the outcome in the other frame: more muons reach the surface than would be predicted without relativistic effects.

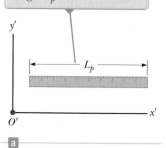

A meterstick measured by an observer in a frame attached to the stick has its proper length L_p.

L_p

y′

O′

x′

a

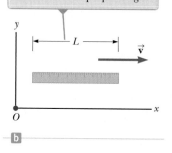

A meterstick measured by an observer in a frame in which the stick has a velocity relative to the frame is measured to be shorter than its proper length.

L

$\vec{v}$

y

O

x

b

Figure 39.10 The length of a meterstick is measured by two observers.

Quick **Quiz 39.6** You are packing for a trip to another star. During the journey, you will be traveling at $0.99c$. You are trying to decide whether you should buy smaller sizes of your clothing because you will be thinner on your trip due to length contraction. You also plan to save money by reserving a smaller cabin to sleep in because you will be shorter when you lie down. Should you **(a)** buy smaller sizes of clothing, **(b)** reserve a smaller cabin, **(c)** do neither of these things, or **(d)** do both of these things?

Quick **Quiz 39.7** You are observing a spacecraft moving away from you. You measure it to be shorter than when it was at rest on the ground next to you. You also see a clock through the spacecraft window, and you observe that the passage of time on the clock is measured to be slower than that of the watch on your wrist. Compared with when the spacecraft was on the ground, what do you measure if the spacecraft turns around and comes *toward* you at the same speed? **(a)** The spacecraft is measured to be longer, and the clock runs faster. **(b)** The spacecraft is measured to be longer, and the clock runs slower. **(c)** The spacecraft is measured to be shorter, and the clock runs faster. **(d)** The spacecraft is measured to be shorter, and the clock runs slower.

Space–Time Graphs

It is sometimes helpful to represent a physical situation with a **space–time graph,** in which ct is the ordinate and position x is the abscissa. The twin paradox is displayed in such a graph in Figure 39.11 from Goslo's point of view. A path through space–time is called a **world-line.** At the origin, the world-lines of Speedo (blue) and Goslo (green) coincide because the twins are in the same location at the same time. After Speedo leaves on his trip, his world-line diverges from that of his brother. Goslo's world-line is vertical because he remains fixed in location. At Goslo and Speedo's reunion, the two world-lines again come together. It would be impossible for Speedo to have a world-line that crossed the path of a light beam that left the Earth when he did. To do so would require him to have a speed greater than c (which, as shown in Sections 39.6 and 39.7, is not possible).

World-lines for light beams are diagonal lines on space–time graphs, typically drawn at 45° to the right or left of vertical (assuming the x and ct axes have the same scales), depending on whether the light beam is traveling in the direction of increasing or decreasing x. All possible future events for Goslo and Speedo lie above the x axis and between the red-brown lines in Figure 39.11 because neither twin can travel faster than light. The only past events that Goslo and Speedo could have experienced occur between two similar 45° world-lines that approach the origin from below the x axis.

If Figure 39.11 is rotated about the ct axis, the red-brown lines sweep out a cone, called the *light cone,* which generalizes Figure 39.11 to two space dimensions. The y axis can be imagined coming out of the page. All future events for an observer at the origin must lie within the light cone. We can imagine another rotation that would generalize the light cone to three space dimensions to include z, but because of the requirement for four dimensions (three space dimensions and time), we cannot represent this situation in a two-dimensional drawing on paper.

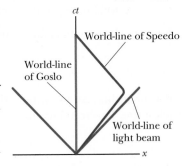

Figure 39.11 The twin paradox on a space–time graph. The twin who stays on the Earth has a world-line along the ct axis (green). The path of the traveling twin through space–time is represented by a world-line that changes direction (blue). The red-brown lines are world-lines for light beams traveling in the positive x direction (on the right) or the negative x direction (on the left).

Example 39.3 **A Voyage to Sirius**

An astronaut takes a trip to Sirius, which is located a distance of 8 light-years from the Earth. The astronaut measures the time of the one-way journey to be 6 years. If the spaceship moves at a constant speed of 0.8c, how can the 8-ly distance be reconciled with the 6-year trip time measured by the astronaut?

SOLUTION

Conceptualize An observer on the Earth measures light to require 8 years to travel between Sirius and the Earth. The astronaut measures a time interval for his travel of only 6 years. Is the astronaut traveling faster than light?

Categorize Because the astronaut is measuring a length of space between the Earth and Sirius that is in motion with respect to her, we categorize this example as a length contraction problem. We also model the astronaut as a *particle under constant velocity.*

Analyze The distance of 8 ly represents the proper length from the Earth to Sirius measured by an observer on the Earth seeing both objects nearly at rest.

Calculate the contracted length measured by the astronaut using Equation 39.9:

$$L = \frac{8 \text{ ly}}{\gamma} = (8 \text{ ly})\sqrt{1 - \frac{v^2}{c^2}} = (8 \text{ ly})\sqrt{1 - \frac{(0.8c)^2}{c^2}} = 5 \text{ ly}$$

Use the particle under constant velocity model to find the travel time measured on the astronaut's clock:

$$\Delta t = \frac{L}{v} = \frac{5 \text{ ly}}{0.8c} = \frac{5 \text{ ly}}{0.8(1 \text{ ly/yr})} = 6 \text{ yr}$$

continued

▶ **39.3** continued

Finalize Notice that we have used the value for the speed of light as $c = 1$ ly/yr. The trip takes a time interval shorter than 8 years for the astronaut because, to her, the distance between the Earth and Sirius is measured to be shorter.

WHAT IF? What if this trip is observed with a very powerful telescope by a technician in Mission Control on the Earth? At what time will this technician *see* that the astronaut has arrived at Sirius?

Answer The time interval the technician measures for the astronaut to arrive is

$$\Delta t = \frac{L_p}{v} = \frac{8 \text{ ly}}{0.8c} = 10 \text{ yr}$$

For the technician to *see* the arrival, the light from the scene of the arrival must travel back to the Earth and enter the telescope. This travel requires a time interval of

$$\Delta t = \frac{L_p}{v} = \frac{8 \text{ ly}}{c} = 8 \text{ yr}$$

Therefore, the technician sees the arrival after 10 yr + 8 yr = 18 yr. If the astronaut immediately turns around and comes back home, she arrives, according to the technician, 20 years after leaving, only 2 years *after the technician saw her arrive!* In addition, the astronaut would have aged by only 12 years.

Example 39.4   **The Pole-in-the-Barn Paradox**   **AM**

The twin paradox, discussed earlier, is a classic "paradox" in relativity. Another classic "paradox" is as follows. Suppose a runner moving at 0.75c carries a horizontal pole 15 m long toward a barn that is 10 m long. The barn has front and rear doors that are initially open. An observer on the ground can instantly and simultaneously close and open the two doors by remote control. When the runner and the pole are inside the barn, the ground observer closes and then opens both doors so that the runner and pole are momentarily captured inside the barn and then proceed to exit the barn from the back doorway. Do both the runner and the ground observer agree that the runner makes it safely through the barn?

SOLUTION

Conceptualize From your everyday experience, you would be surprised to see a 15-m pole fit inside a 10-m barn, but we are becoming used to surprising results in relativistic situations.

Categorize The pole is in motion with respect to the ground observer so that the observer measures its length to be contracted, whereas the stationary barn has a proper length of 10 m. We categorize this example as a length contraction problem. The runner carrying the pole is modeled as a *particle under constant velocity*.

Analyze Use Equation 39.9 to find the contracted length of the pole according to the ground observer:

$$L_{\text{pole}} = L_p \sqrt{1 - \frac{v^2}{c^2}} = (15 \text{ m}) \sqrt{1 - (0.75)^2} = 9.9 \text{ m}$$

Therefore, the ground observer measures the pole to be slightly shorter than the barn and there is no problem with momentarily capturing the pole inside it. The "paradox" arises when we consider the runner's point of view.

Use Equation 39.9 to find the contracted length of the barn according to the running observer:

$$L_{\text{barn}} = L_p \sqrt{1 - \frac{v^2}{c^2}} = (10 \text{ m}) \sqrt{1 - (0.75)^2} = 6.6 \text{ m}$$

Because the pole is in the rest frame of the runner, the runner measures it to have its proper length of 15 m. Now the situation looks even worse: How can a 15-m pole fit inside a *6.6-m* barn? Although this question is the classic one that is often asked, it is not the question we have asked because it is not the important one. We asked, "*Does the runner make it safely through the barn?*"

The resolution of the "paradox" lies in the relativity of simultaneity. The closing of the two doors is measured to be simultaneous by the ground observer. Because the doors are at different positions, however, they do not close simulta-

▶ **39.4** continued

neously as measured by the runner. The rear door closes and then opens first, allowing the leading end of the pole to exit. The front door of the barn does not close until the trailing end of the pole passes by.

We can analyze this "paradox" using a space–time graph. Figure 39.12a is a space–time graph from the ground observer's point of view. We choose $x = 0$ as the position of the front doorway of the barn and $t = 0$ as the instant at which the leading end of the pole is located at the front doorway of the barn. The world-lines for the two doorways of the barn are separated by 10 m and are vertical because the barn is not moving relative to this observer. For the pole, we follow two tilted world-lines, one for each end of the moving pole. These world-lines are 9.9 m apart horizontally, which is the contracted length seen by the ground observer. As seen in Figure 39.12a, the pole is entirely within the barn at some time.

Figure 39.12b shows the space–time graph according to the runner. Here, the world-lines for the pole are separated by 15 m and are vertical because the pole is at rest in the runner's frame of reference. The barn is hurtling *toward* the runner, so the world-lines for the front and rear doorways of the barn are tilted to the left. The world-lines for the barn are separated by 6.6 m, the contracted length as seen by the runner. The leading end of the pole leaves the rear doorway of the barn long before the trailing end of the pole enters the barn. Therefore, the opening of the rear door occurs before the closing of the front door.

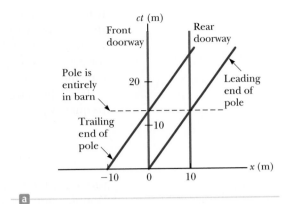

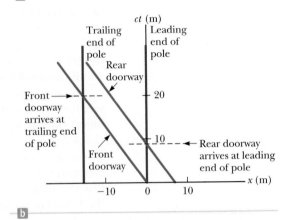

Figure 39.12 (Example 39.4) Space–time graphs for the pole-in-the-barn paradox (a) from the ground observer's point of view and (b) from the runner's point of view.

From the ground observer's point of view, use the particle under constant velocity model to find the time after $t = 0$ at which the trailing end of the pole enters the barn:

$$(1) \quad t = \frac{\Delta x}{v} = \frac{9.9 \text{ m}}{0.75c} = \frac{13.2 \text{ m}}{c}$$

From the runner's point of view, use the particle under constant velocity model to find the time at which the leading end of the pole leaves the barn:

$$(2) \quad t = \frac{\Delta x}{v} = \frac{6.6 \text{ m}}{0.75c} = \frac{8.8 \text{ m}}{c}$$

Find the time at which the trailing end of the pole enters the front door of the barn:

$$(3) \quad t = \frac{\Delta x}{v} = \frac{15 \text{ m}}{0.75c} = \frac{20 \text{ m}}{c}$$

Finalize From Equation (1), the pole should be completely inside the barn at a time corresponding to $ct = 13.2$ m. This situation is consistent with the point on the ct axis in Figure 39.12a where the pole is inside the barn. From Equation (2), the leading end of the pole leaves the barn at $ct = 8.8$ m. This situation is consistent with the point on the ct axis in Figure 39.12b where the rear doorway of the barn arrives at the leading end of the pole. Equation (3) gives $ct = 20$ m, which agrees with the instant shown in Figure 39.12b at which the front doorway of the barn arrives at the trailing end of the pole.

The Relativistic Doppler Effect

Another important consequence of time dilation is the shift in frequency observed for light emitted by atoms in motion as opposed to light emitted by atoms at rest. This phenomenon, known as the Doppler effect, was introduced in Chapter 17 as it pertains to sound waves. In the case of sound, the velocity v_S of the source with respect to the medium of propagation can be distinguished from the velocity v_O of the observer with respect to the medium (the air). Light waves must be analyzed differently, however, because *they require no medium of propagation,* and no method exists for distinguishing the velocity of a light source from the velocity of the observer. The only measurable velocity is the *relative velocity* v between the source and the observer.

If a light source and an observer approach each other with a relative speed v, the frequency f' measured by the observer is

$$f' = \frac{\sqrt{1 + v/c}}{\sqrt{1 - v/c}} f \tag{39.10}$$

where f is the frequency of the source measured in its rest frame. This relativistic Doppler shift equation, unlike the Doppler shift equation for sound, depends only on the relative speed v of the source and observer and holds for relative speeds as great as c. As you might expect, the equation predicts that $f' > f$ when the source and observer approach each other. We obtain the expression for the case in which the source and observer recede from each other by substituting negative values for v in Equation 39.10.

The most spectacular and dramatic use of the relativistic Doppler effect is the measurement of shifts in the frequency of light emitted by a moving astronomical object such as a galaxy. Light emitted by atoms and normally found in the extreme violet region of the spectrum is shifted toward the red end of the spectrum for atoms in other galaxies, indicating that these galaxies are *receding* from us. American astronomer Edwin Hubble (1889–1953) performed extensive measurements of this *red shift* to confirm that most galaxies are moving away from us, indicating that the Universe is expanding.

39.5 The Lorentz Transformation Equations

Suppose two events occur at points P and Q and are reported by two observers, one at rest in a frame S and another in a frame S′ that is moving to the right with speed v as in Figure 39.13. The observer in S reports the events with space–time coordinates (x, y, z, t), and the observer in S′ reports the same events using the coordinates (x', y', z', t'). Equation 39.1 predicts that the distance between the two points in space at which the events occur does not depend on motion of the observer: $\Delta x = \Delta x'$. Because this prediction is contradictory to the notion of length contraction, the Galilean transformation is not valid when v approaches the speed of light. In this section, we present the correct transformation equations that apply for all speeds in the range $0 < v < c$.

The equations that are valid for all speeds and that enable us to transform coordinates from S to S′ are the **Lorentz transformation equations:**

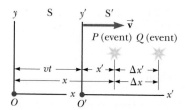

Figure 39.13 Events occur at points P and Q and are observed by an observer at rest in the S frame and another in the S′ frame, which is moving to the right with a speed v.

Lorentz transformation for S → S′ ▶

$$x' = \gamma(x - vt) \qquad y' = y \qquad z' = z \qquad t' = \gamma\left(t - \frac{v}{c^2}x\right) \tag{39.11}$$

These transformation equations were developed by Hendrik A. Lorentz (1853–1928) in 1890 in connection with electromagnetism. It was Einstein, however, who recognized their physical significance and took the bold step of interpreting them within the framework of the special theory of relativity.

Notice the difference between the Galilean and Lorentz time equations. In the Galilean case, $t = t'$. In the Lorentz case, however, the value for t' assigned to an event by an observer O' in the S′ frame in Figure 39.13 depends both on the time t and on the coordinate x as measured by an observer O in the S frame, which is consistent with the notion that an event is characterized by four space–time coordinates (x, y, z, t). In other words, in relativity, space and time are *not* separate concepts but rather are closely interwoven with each other.

If you wish to transform coordinates in the S′ frame to coordinates in the S frame, simply replace v by $-v$ and interchange the primed and unprimed coordinates in Equations 39.11:

$$x = \gamma(x' + vt') \qquad y = y' \qquad z = z' \qquad t = \gamma\left(t' + \frac{v}{c^2}x'\right) \qquad \textbf{(39.12)}$$

◀ **Inverse Lorentz transformation for S′ → S**

When $v \ll c$, the Lorentz transformation equations should reduce to the Galilean equations. As v approaches zero, $v/c \ll 1$; therefore, $\gamma \to 1$ and Equations 39.11 indeed reduce to the Galilean space–time transformation equations in Equation 39.1.

In many situations, we would like to know the difference in coordinates between two events or the time interval between two events as seen by observers O and O'. From Equations 39.11 and 39.12, we can express the differences between the four variables x, x', t, and t' in the form

$$\left.\begin{array}{l} \Delta x' = \gamma(\Delta x - v\,\Delta t) \\[2mm] \Delta t' = \gamma\left(\Delta t - \dfrac{v}{c^2}\Delta x\right) \end{array}\right\}\text{S} \rightarrow \text{S}' \qquad \textbf{(39.13)}$$

$$\left.\begin{array}{l} \Delta x = \gamma(\Delta x' + v\,\Delta t') \\[2mm] \Delta t = \gamma\left(\Delta t' + \dfrac{v}{c^2}\Delta x'\right) \end{array}\right\}\text{S}' \rightarrow \text{S} \qquad \textbf{(39.14)}$$

where $\Delta x' = x_2' - x_1'$ and $\Delta t' = t_2' - t_1'$ are the differences measured by observer O' and $\Delta x = x_2 - x_1$ and $\Delta t = t_2 - t_1$ are the differences measured by observer O. (We have not included the expressions for relating the y and z coordinates because they are unaffected by motion along the x direction.[5])

| Example 39.5 | **Simultaneity and Time Dilation Revisited** |

(A) Use the Lorentz transformation equations in difference form to show that simultaneity is not an absolute concept.

SOLUTION

Conceptualize Imagine two events that are simultaneous and separated in space as measured in the S′ frame such that $\Delta t' = 0$ and $\Delta x' \neq 0$. These measurements are made by an observer O' who is moving with speed v relative to O.

Categorize The statement of the problem tells us to categorize this example as one involving the use of the Lorentz transformation.

. . . .

Analyze From the expression for Δt given in Equation 39.14, find the time interval Δt measured by observer O:

$$\Delta t = \gamma\left(\Delta t' + \frac{v}{c^2}\Delta x'\right) = \gamma\left(0 + \frac{v}{c^2}\Delta x'\right) = \gamma\frac{v}{c^2}\Delta x'$$

. . . .

Finalize The time interval for the same two events as measured by O is nonzero, so the events do not appear to be simultaneous to O.

(B) Use the Lorentz transformation equations in difference form to show that a moving clock is measured to run more slowly than a clock that is at rest with respect to an observer.

SOLUTION

Conceptualize Imagine that observer O' carries a clock that he uses to measure a time interval $\Delta t'$. He finds that two events occur at the same place in his reference frame ($\Delta x' = 0$) but at different times ($\Delta t' \neq 0$). Observer O' is moving with speed v relative to O.

continued

---

[5]Although relative motion of the two frames along the x axis does not change the y and z coordinates of an object, it does change the y and z velocity components of an object moving in either frame as noted in Section 39.6.

◀ **39.5** continued

Categorize The statement of the problem tells us to categorize this example as one involving the use of the Lorentz transformation.

Analyze From the expression for Δt given in Equation 39.14, find the time interval Δt measured by observer O:

$$\Delta t = \gamma\left(\Delta t' + \frac{v}{c^2}\Delta x'\right) = \gamma\left[\Delta t' + \frac{v}{c^2}(0)\right] = \gamma\,\Delta t'$$

Finalize This result is the equation for time dilation found earlier (Eq. 39.7), where $\Delta t' = \Delta t_p$ is the proper time interval measured by the clock carried by observer O'. Therefore, O measures the moving clock to run slow.

39.6 The Lorentz Velocity Transformation Equations

Suppose two observers in relative motion with respect to each other are both observing an object's motion. Previously, we defined an event as occurring at an instant of time. Now let's interpret the "event" as the object's motion. We know that the Galilean velocity transformation (Eq. 39.2) is valid for low speeds. How do the observers' measurements of the velocity of the object relate to each other if the speed of the object or the relative speed of the observers is close to that of light? Once again, S′ is our frame moving at a speed v relative to S. Suppose an object has a velocity component u'_x measured in the S′ frame, where

$$u'_x = \frac{dx'}{dt'} \tag{39.15}$$

Using Equation 39.11, we have

$$dx' = \gamma(dx - v\,dt)$$

$$dt' = \gamma\left(dt - \frac{v}{c^2}\,dx\right)$$

Substituting these values into Equation 39.15 gives

$$u'_x = \frac{dx - v\,dt}{dt - \frac{v}{c^2}\,dx} = \frac{\dfrac{dx}{dt} - v}{1 - \dfrac{v}{c^2}\dfrac{dx}{dt}}$$

The term dx/dt, however, is simply the velocity component u_x of the object measured by an observer in S, so this expression becomes

Lorentz velocity trans- ▶
formation for S → S′

$$u'_x = \frac{u_x - v}{1 - \dfrac{u_x v}{c^2}} \tag{39.16}$$

If the object has velocity components along the y and z axes, the components as measured by an observer in S′ are

$$u'_y = \frac{u_y}{\gamma\left(1 - \dfrac{u_x v}{c^2}\right)} \quad \text{and} \quad u'_z = \frac{u_z}{\gamma\left(1 - \dfrac{u_x v}{c^2}\right)} \tag{39.17}$$

Notice that u'_y and u'_z do not contain the parameter v in the numerator because the relative velocity is along the x axis.

When v is much smaller than c (the nonrelativistic case), the denominator of Equation 39.16 approaches unity and so $u'_x \approx u_x - v$, which is the Galilean velocity transformation equation. In another extreme, when $u_x = c$, Equation 39.16 becomes

$$u'_x = \frac{c - v}{1 - \dfrac{cv}{c^2}} = \frac{c\left(1 - \dfrac{v}{c}\right)}{1 - \dfrac{v}{c}} = c$$

This result shows that a speed measured as c by an observer in S is also measured as c by an observer in S′, independent of the relative motion of S and S′. This conclusion is consistent with Einstein's second postulate: the speed of light must be c relative to all inertial reference frames. Furthermore, we find that the speed of an object can never be measured as larger than c. That is, the speed of light is the ultimate speed. We shall return to this point later.

To obtain u_x in terms of u'_x, we replace v by $-v$ in Equation 39.16 and interchange the roles of u_x and u'_x:

$$u_x = \frac{u'_x + v}{1 + \dfrac{u'_x v}{c^2}} \qquad \textbf{(39.18)}$$

Quick Quiz 39.8 You are driving on a freeway at a relativistic speed. **(i)** Straight ahead of you, a technician standing on the ground turns on a searchlight and a beam of light moves exactly vertically upward as seen by the technician. As you observe the beam of light, do you measure the magnitude of the vertical component of its velocity as (a) equal to c, (b) greater than c, or (c) less than c? **(ii)** If the technician aims the searchlight directly at you instead of upward, do you measure the magnitude of the horizontal component of its velocity as (a) equal to c, (b) greater than c, or (c) less than c?

> **Pitfall Prevention 39.5**
> **What Can the Observers Agree On?** We have seen several measurements that the two observers O and O' do *not* agree on: (1) the time interval between events that take place in the same position in one of their frames, (2) the distance between two points that remain fixed in one of their frames, (3) the velocity components of a moving particle, and (4) whether two events occurring at different locations in both frames are simultaneous or not. The two observers *can* agree on (1) their relative speed of motion v with respect to each other, (2) the speed c of any ray of light, and (3) the simultaneity of two events that take place at the same position *and* time in some frame.

Example 39.6   Relative Velocity of Two Spacecraft

Two spacecraft A and B are moving in opposite directions as shown in Figure 39.14. An observer on the Earth measures the speed of spacecraft A to be $0.750c$ and the speed of spacecraft B to be $0.850c$. Find the velocity of spacecraft B as observed by the crew on spacecraft A.

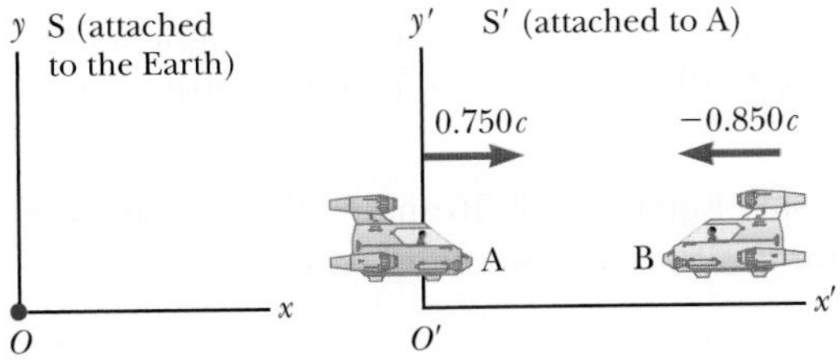

Figure 39.14 (Example 39.6) Two spacecraft A and B move in opposite directions. The speed of spacecraft B relative to spacecraft A is *less* than c and is obtained from the relativistic velocity transformation equation.

SOLUTION

Conceptualize There are two observers, one (O) on the Earth and one (O') on spacecraft A. The event is the motion of spacecraft B.

Categorize Because the problem asks to find an observed velocity, we categorize this example as one requiring the Lorentz velocity transformation.

Analyze The Earth-based observer at rest in the S frame makes two measurements, one of each spacecraft. We want to find the velocity of spacecraft B as measured by the crew on spacecraft A. Therefore, $u_x = -0.850c$. The velocity of spacecraft A is also the velocity of the observer at rest in spacecraft A (the S′ frame) relative to the observer at rest on the Earth. Therefore, $v = 0.750c$.

Obtain the velocity u'_x of spacecraft B relative to spacecraft A using Equation 39.16:

$$u'_x = \frac{u_x - v}{1 - \dfrac{u_x v}{c^2}} = \frac{-0.850c - 0.750c}{1 - \dfrac{(-0.850c)(0.750c)}{c^2}} = \boxed{-0.977c}$$

Finalize The negative sign indicates that spacecraft B is moving in the negative x direction as observed by the crew on spacecraft A. Is that consistent with your expectation from Figure 39.14? Notice that the speed is less than c. That is, an

continued

▶ **39.6** c o n t i n u e d

object whose speed is less than c in one frame of reference must have a speed less than c in any other frame. (Had you used the Galilean velocity transformation equation in this example, you would have found that $u'_x = u_x - v = -0.850c - 0.750c = -1.60c$, which is impossible. The Galilean transformation equation does not work in relativistic situations.)

WHAT IF?   What if the two spacecraft pass each other? What is their relative speed now?

Answer   The calculation using Equation 39.16 involves only the velocities of the two spacecraft and does not depend on their locations. After they pass each other, they have the same velocities, so the velocity of spacecraft B as observed by the crew on spacecraft A is the same, $-0.977c$. The only difference after they pass is that spacecraft B is receding from spacecraft A, whereas it was approaching spacecraft A before it passed.

Example 39.7   **Relativistic Leaders of the Pack**

Two motorcycle pack leaders named David and Emily are racing at relativistic speeds along perpendicular paths as shown in Figure 39.15. How fast does Emily recede as seen by David over his right shoulder?

SOLUTION

Conceptualize   The two observers are David and the police officer in Figure 39.15. The event is the motion of Emily. Figure 39.15 represents the situation as seen by the police officer at rest in frame S. Frame S′ moves along with David.

Categorize   Because the problem asks to find an observed velocity, we categorize this problem as one requiring the Lorentz velocity transformation. The motion takes place in two dimensions.

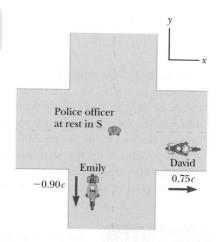

Figure 39.15 (Example 39.7) David moves east with a speed $0.75c$ relative to the police officer, and Emily travels south at a speed $0.90c$ relative to the officer.

Analyze   Identify the velocity components for David and Emily according to the police officer:

David: $v_x = v = 0.75c$ $v_y = 0$

Emily: $u_x = 0$ $u_y = -0.90c$

Using Equations 39.16 and 39.17, calculate u'_x and u'_y for Emily as measured by David:

$$u'_x = \frac{u_x - v}{1 - \dfrac{u_x v}{c^2}} = \frac{0 - 0.75c}{1 - \dfrac{(0)(0.75c)}{c^2}} = -0.75c$$

$$u'_y = \frac{u_y}{\gamma\left(1 - \dfrac{u_x v}{c^2}\right)} = \frac{\sqrt{1 - \dfrac{(0.75c)^2}{c^2}}\,(-0.90c)}{1 - \dfrac{(0)(0.75c)}{c^2}} = -0.60c$$

Using the Pythagorean theorem, find the speed of Emily as measured by David:

$$u' = \sqrt{(u'_x)^2 + (u'_y)^2} = \sqrt{(-0.75c)^2 + (-0.60c)^2} = \boxed{0.96c}$$

Finalize   This speed is less than c, as required by the special theory of relativity.

39.7 Relativistic Linear Momentum

To describe the motion of particles within the framework of the special theory of relativity properly, you must replace the Galilean transformation equations by the Lorentz transformation equations. Because the laws of physics must remain unchanged under the Lorentz transformation, we must generalize Newton's laws and the definitions of linear momentum and energy to conform to the Lorentz transformation equations and the principle of relativity. These generalized definitions should reduce to the classical (nonrelativistic) definitions for $v \ll c$.

First, recall from the isolated system model that when two particles (or objects that can be modeled as particles) collide, the total momentum of the isolated system of the two particles remains constant. Suppose we observe this collision in a reference frame S and confirm that the momentum of the system is conserved. Now imagine that the momenta of the particles are measured by an observer in a second reference frame S′ moving with velocity $\vec{v}$ relative to the first frame. Using the Lorentz velocity transformation equation and the classical definition of linear momentum, $\vec{p} = m\vec{u}$ (where $\vec{u}$ is the velocity of a particle), we find that linear momentum of the system is *not* measured to be conserved by the observer in S′. Because the laws of physics are the same in all inertial frames, however, linear momentum of the system must be conserved in all frames. We have a contradiction. In view of this contradiction and assuming the Lorentz velocity transformation equation is correct, we must modify the definition of linear momentum so that the momentum of an isolated system is conserved for all observers. For any particle, the correct relativistic equation for linear momentum that satisfies this condition is

$$\vec{p} \equiv \frac{m\vec{u}}{\sqrt{1 - \dfrac{u^2}{c^2}}} = \gamma m\vec{u} \qquad (39.19)$$

◀ **Definition of relativistic linear momentum**

where m is the mass of the particle and $\vec{u}$ is the velocity of the particle. When u is much less than c, $\gamma = (1 - u^2/c^2)^{-1/2}$ approaches unity and $\vec{p}$ approaches $m\vec{u}$. Therefore, the relativistic equation for $\vec{p}$ reduces to the classical expression when u is much smaller than c, as it should.

The relativistic force $\vec{F}$ acting on a particle whose linear momentum is $\vec{p}$ is defined as

$$\vec{F} \equiv \frac{d\vec{p}}{dt} \qquad (39.20)$$

where $\vec{p}$ is given by Equation 39.19. This expression, which is the relativistic form of Newton's second law, is reasonable because it preserves classical mechanics in the limit of low velocities and is consistent with conservation of linear momentum for an isolated system ($\vec{F}_{ext} = 0$) both relativistically and classically.

It is left as an end-of-chapter problem (Problem 88) to show that under relativistic conditions, the acceleration $\vec{a}$ of a particle decreases under the action of a constant force, in which case $a \propto (1 - u^2/c^2)^{3/2}$. This proportionality shows that as the particle's speed approaches c, the acceleration caused by any finite force approaches zero. Hence, it is impossible to accelerate a particle from rest to a speed $u \geq c$. This argument reinforces that the speed of light is the ultimate speed, the speed limit of the Universe. It is the maximum possible speed for energy transfer and for information transfer. Any object with mass must move at a lower speed.

Pitfall Prevention 39.6
Watch Out for "Relativistic Mass" Some older treatments of relativity maintained the conservation of momentum principle at high speeds by using a model in which a particle's mass increases with speed. You might still encounter this notion of "relativistic mass" in your outside reading, especially in older books. Be aware that this notion is no longer widely accepted; today, mass is considered as *invariant*, independent of speed. The mass of an object in all frames is considered to be the mass as measured by an observer at rest with respect to the object.

Example 39.8 Linear Momentum of an Electron

An electron, which has a mass of 9.11×10^{-31} kg, moves with a speed of $0.750c$. Find the magnitude of its relativistic momentum and compare this value with the momentum calculated from the classical expression.

SOLUTION

Conceptualize Imagine an electron moving with high speed. The electron carries momentum, but the magnitude of its momentum is not given by $p = mu$ because the speed is relativistic.

Categorize We categorize this example as a substitution problem involving a relativistic equation.

continued

▶ **39.8** continued

Use Equation 39.19 with $u = 0.750c$ to find the magnitude of the momentum:

$$p = \frac{m_e u}{\sqrt{1 - \dfrac{u^2}{c^2}}}$$

$$p = \frac{(9.11 \times 10^{-31} \text{ kg})(0.750)(3.00 \times 10^8 \text{ m/s})}{\sqrt{1 - \dfrac{(0.750c)^2}{c^2}}}$$

$$= 3.10 \times 10^{-22} \text{ kg} \cdot \text{m/s}$$

The classical expression (used incorrectly here) gives $p_{\text{classical}} = m_e u = 2.05 \times 10^{-22}$ kg · m/s. Hence, the correct relativistic result is 50% greater than the classical result!

39.8 Relativistic Energy

We have seen that the definition of linear momentum requires generalization to make it compatible with Einstein's postulates. This conclusion implies that the definition of kinetic energy must most likely be modified also.

To derive the relativistic form of the work–kinetic energy theorem, imagine a particle moving in one dimension along the x axis. A force in the x direction causes the momentum of the particle to change according to Equation 39.20. In what follows, we assume the particle is accelerated from rest to some final speed u. The work done by the force F on the particle is

$$W = \int_{x_1}^{x_2} F \, dx = \int_{x_1}^{x_2} \frac{dp}{dt} \, dx \tag{39.21}$$

To perform this integration and find the work done on the particle and the relativistic kinetic energy as a function of u, we first evaluate dp/dt:

$$\frac{dp}{dt} = \frac{d}{dt} \frac{mu}{\sqrt{1 - \dfrac{u^2}{c^2}}} = \frac{m}{\left(1 - \dfrac{u^2}{c^2}\right)^{3/2}} \frac{du}{dt}$$

Substituting this expression for dp/dt and $dx = u \, dt$ into Equation 39.21 gives

$$W = \int_0^t \frac{m}{\left(1 - \dfrac{u^2}{c^2}\right)^{3/2}} \frac{du}{dt}(u \, dt) = m \int_0^u \frac{u}{\left(1 - \dfrac{u^2}{c^2}\right)^{3/2}} \, du$$

where we use the limits 0 and u in the integral because the integration variable has been changed from t to u. Evaluating the integral gives

$$W = \frac{mc^2}{\sqrt{1 - \dfrac{u^2}{c^2}}} - mc^2 \tag{39.22}$$

Recall from Chapter 7 that the work done by a force acting on a system consisting of a single particle equals the change in kinetic energy of the particle: $W = \Delta K$. Because we assumed the initial speed of the particle is zero, its initial kinetic energy is zero, so $W = K - K_i = K - 0 = K$. Therefore, the work W in Equation 39.22 is equivalent to the relativistic kinetic energy K:

Relativistic kinetic energy ▶

$$K = \frac{mc^2}{\sqrt{1 - \dfrac{u^2}{c^2}}} - mc^2 = \gamma mc^2 - mc^2 = (\gamma - 1)mc^2 \tag{39.23}$$

This equation is routinely confirmed by experiments using high-energy particle accelerators.

At low speeds, where $u/c \ll 1$, Equation 39.23 should reduce to the classical expression $K = \frac{1}{2}mu^2$. We can check that by using the binomial expansion $(1 - \beta^2)^{-1/2} \approx 1 + \frac{1}{2}\beta^2 + \cdots$ for $\beta \ll 1$, where the higher-order powers of β are neglected in the expansion. (In treatments of relativity, β is a common symbol used to represent u/c or v/c.) In our case, $\beta = u/c$, so

$$\gamma = \frac{1}{\sqrt{1 - \dfrac{u^2}{c^2}}} = \left(1 - \frac{u^2}{c^2}\right)^{-1/2} \approx 1 + \frac{1}{2}\frac{u^2}{c^2}$$

Substituting this result into Equation 39.23 gives

$$K \approx \left[\left(1 + \frac{1}{2}\frac{u^2}{c^2}\right) - 1\right]mc^2 = \frac{1}{2}mu^2 \quad \text{(for } u/c \ll 1\text{)}$$

which is the classical expression for kinetic energy. A graph comparing the relativistic and nonrelativistic expressions is given in Figure 39.16. In the relativistic case, the particle speed never exceeds c, regardless of the kinetic energy. The two curves are in good agreement when $u \ll c$.

The constant term mc^2 in Equation 39.23, which is independent of the speed of the particle, is called the **rest energy** E_R of the particle:

$$E_R = mc^2 \tag{39.24}$$

Equation 39.24 shows that **mass is a form of energy**, where c^2 is simply a constant conversion factor. This expression also shows that a small mass corresponds to an enormous amount of energy, a concept fundamental to nuclear and elementary-particle physics.

The term γmc^2 in Equation 39.23, which depends on the particle speed, is the sum of the kinetic and rest energies. It is called the **total energy** E:

$$\text{Total energy} = \text{kinetic energy} + \text{rest energy}$$

$$E = K + mc^2 \tag{39.25}$$

or

$$E = \frac{mc^2}{\sqrt{1 - \dfrac{u^2}{c^2}}} = \gamma mc^2 \tag{39.26}$$

◀ **Total energy of a relativistic particle**

In many situations, the linear momentum or energy of a particle rather than its speed is measured. It is therefore useful to have an expression relating the total energy E to the relativistic linear momentum p, which is accomplished by using the expressions $E = \gamma mc^2$ and $p = \gamma mu$. By squaring these equations and subtracting, we can eliminate u (Problem 58). The result, after some algebra, is[6]

$$E^2 = p^2c^2 + (mc^2)^2 \tag{39.27}$$

◀ **Energy–momentum relationship for a relativistic particle**

When the particle is at rest, $p = 0$, so $E = E_R = mc^2$.

In Section 35.1, we introduced the concept of a particle of light, called a **photon.** For particles that have zero mass, such as photons, we set $m = 0$ in Equation 39.27 and find that

$$E = pc \tag{39.28}$$

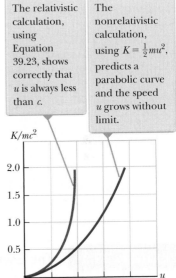

The relativistic calculation, using Equation 39.23, shows correctly that u is always less than c.

The nonrelativistic calculation, using $K = \frac{1}{2}mu^2$, predicts a parabolic curve and the speed u grows without limit.

Figure 39.16 A graph comparing relativistic and nonrelativistic kinetic energy of a moving particle. The energies are plotted as a function of particle speed u.

---

[6]One way to remember this relationship is to draw a right triangle having a hypotenuse of length E and legs of lengths pc and mc^2.

This equation is an exact expression relating total energy and linear momentum for photons, which always travel at the speed of light (in vacuum).

Finally, because the mass m of a particle is independent of its motion, m must have the same value in all reference frames. For this reason, m is often called the **invariant mass.** On the other hand, because the total energy and linear momentum of a particle both depend on velocity, these quantities depend on the reference frame in which they are measured.

When dealing with subatomic particles, it is convenient to express their energy in electron volts (Section 25.1) because the particles are usually given this energy by acceleration through a potential difference. The conversion factor, as you recall from Equation 25.5, is

$$1 \text{ eV} = 1.602 \times 10^{-19} \text{ J}$$

For example, the mass of an electron is 9.109×10^{-31} kg. Hence, the rest energy of the electron is

$$m_e c^2 = (9.109 \times 10^{-31} \text{ kg})(2.998 \times 10^8 \text{ m/s})^2 = 8.187 \times 10^{-14} \text{ J}$$

$$= (8.187 \times 10^{-14} \text{ J})(1 \text{ eV}/1.602 \times 10^{-19} \text{ J}) = 0.511 \text{ MeV}$$

Quick Quiz 39.9 The following *pairs* of energies—particle 1: E, $2E$; particle 2: E, $3E$; particle 3: $2E$, $4E$—represent the rest energy and total energy of three different particles. Rank the particles from greatest to least according to their **(a)** mass, **(b)** kinetic energy, and **(c)** speed.

Example 39.9 **The Energy of a Speedy Proton**

(A) Find the rest energy of a proton in units of electron volts.

SOLUTION

Conceptualize Even if the proton is not moving, it has energy associated with its mass. If it moves, the proton possesses more energy, with the total energy being the sum of its rest energy and its kinetic energy.

Categorize The phrase "rest energy" suggests we must take a relativistic rather than a classical approach to this problem.

Analyze Use Equation 39.24 to find the rest energy:

$$E_R = m_p c^2 = (1.672\,6 \times 10^{-27} \text{ kg})(2.998 \times 10^8 \text{ m/s})^2$$

$$= (1.504 \times 10^{-10} \text{ J})\left(\frac{1.00 \text{ eV}}{1.602 \times 10^{-19} \text{ J}}\right) = \boxed{938 \text{ MeV}}$$

(B) If the total energy of a proton is three times its rest energy, what is the speed of the proton?

SOLUTION

Use Equation 39.26 to relate the total energy of the proton to the rest energy:

$$E = 3m_p c^2 = \frac{m_p c^2}{\sqrt{1 - \dfrac{u^2}{c^2}}} \;\rightarrow\; 3 = \frac{1}{\sqrt{1 - \dfrac{u^2}{c^2}}}$$

Solve for u:

$$1 - \frac{u^2}{c^2} = \tfrac{1}{9} \;\rightarrow\; \frac{u^2}{c^2} = \tfrac{8}{9}$$

$$u = \frac{\sqrt{8}}{3} c = 0.943c = \boxed{2.83 \times 10^8 \text{ m/s}}$$

(C) Determine the kinetic energy of the proton in units of electron volts.

▶ **39.9** continued

SOLUTION

Use Equation 39.25 to find the kinetic energy of the proton:

$$K = E - m_p c^2 = 3m_p c^2 - m_p c^2 = 2m_p c^2$$
$$= 2(938 \text{ MeV}) = \boxed{1.88 \times 10^3 \text{ MeV}}$$

(D) What is the proton's momentum?

SOLUTION

Use Equation 39.27 to calculate the momentum:

$$E^2 = p^2 c^2 + (m_p c^2)^2 = (3m_p c^2)^2$$
$$p^2 c^2 = 9(m_p c^2)^2 - (m_p c^2)^2 = 8(m_p c^2)^2$$
$$p = \sqrt{8} \, \frac{m_p c^2}{c} = \sqrt{8} \, \frac{938 \text{ MeV}}{c} = \boxed{2.65 \times 10^3 \text{ MeV}/c}$$

Finalize The unit of momentum in part (D) is written MeV/c, which is a common unit in particle physics. For comparison, you might want to solve this example using classical equations.

WHAT IF?   In classical physics, if the momentum of a particle doubles, the kinetic energy increases by a factor of 4. What happens to the kinetic energy of the proton in this example if its momentum doubles?

Answer Based on what we have seen so far in relativity, it is likely you would predict that its kinetic energy does not increase by a factor of 4.

Find the new doubled momentum:

$$p_{\text{new}} = 2\left(\sqrt{8} \, \frac{m_p c^2}{c}\right) = 4\sqrt{2} \, \frac{m_p c^2}{c}$$

Use this result in Equation 39.27 to find the new total energy:

$$E^2_{\text{new}} = p^2_{\text{new}} c^2 + (m_p c^2)^2$$
$$E^2_{\text{new}} = \left(4\sqrt{2} \, \frac{m_p c^2}{c}\right)^2 c^2 + (m_p c^2)^2 = 33(m_p c^2)^2$$
$$E_{\text{new}} = \sqrt{33} \, m_p c^2 = 5.7 m_p c^2$$

Use Equation 39.25 to find the new kinetic energy:

$$K_{\text{new}} = E_{\text{new}} - m_p c^2 = 5.7 m_p c^2 - m_p c^2 = 4.7 m_p c^2$$

This value is a little more than twice the kinetic energy found in part (C), not four times. In general, the factor by which the kinetic energy increases if the momentum doubles depends on the initial momentum, but it approaches 4 as the momentum approaches zero. In this latter situation, classical physics correctly describes the situation.

Equation 39.26, $E = \gamma mc^2$, represents the total energy of a particle. This important equation suggests that even when a particle is at rest ($\gamma = 1$), it still possesses enormous energy through its mass. The clearest experimental proof of the equivalence of mass and energy occurs in nuclear and elementary-particle interactions in which the conversion of mass into kinetic energy takes place. Consequently, we cannot use the principle of conservation of energy in relativistic situations as it was outlined in Chapter 8. We must modify the principle by including rest energy as another form of energy storage.

This concept is important in atomic and nuclear processes, in which the change in mass is a relatively large fraction of the initial mass. In a conventional nuclear reactor, for example, the uranium nucleus undergoes *fission*, a reaction that results in several lighter fragments having considerable kinetic energy. In the case of ^{235}U, which is used as fuel in nuclear power plants, the fragments are two lighter nuclei and a few neutrons. The total mass of the fragments is less than that of the ^{235}U by an amount Δm. The corresponding energy Δmc^2 associated with this mass difference is exactly equal to the sum of the kinetic energies of the fragments. The kinetic energy is absorbed as the fragments move through water, raising the internal energy of the water. This internal energy is used to produce steam for the generation of electricity.

Next, consider a basic *fusion* reaction in which two deuterium atoms combine to form one helium atom. The decrease in mass that results from the creation of one helium atom from two deuterium atoms is $\Delta m = 4.25 \times 10^{-29}$ kg. Hence, the corresponding energy that results from one fusion reaction is $\Delta mc^2 = 3.83 \times 10^{-12}$ J = 23.9 MeV. To appreciate the magnitude of this result, consider that if only 1 g of deuterium were converted to helium, the energy released would be on the order of 10^{12} J! In 2013's cost of electrical energy, this energy would be worth approximately $35 000. We shall present more details of these nuclear processes in Chapter 45 of the extended version of this textbook.

Example 39.10   Mass Change in a Radioactive Decay

The ^{216}Po nucleus is unstable and exhibits radioactivity (Chapter 44). It decays to ^{212}Pb by emitting an alpha particle, which is a helium nucleus, ^{4}He. The relevant masses, in atomic mass units (see Table A.1 in Appendix A), are $m_i = m(^{216}\text{Po}) = 216.001\ 915$ u and $m_f = m(^{212}\text{Pb}) + m(^4\text{He}) = 211.991\ 898$ u + 4.002 603 u.

(A) Find the mass change of the system in this decay.

SOLUTION

Conceptualize The initial system is the ^{216}Po nucleus. Imagine the mass of the system decreasing during the decay and transforming to kinetic energy of the alpha particle and the ^{212}Pb nucleus after the decay.

Categorize We use concepts discussed in this section, so we categorize this example as a substitution problem.

Calculate the change in mass using the mass values given in the problem statement.

$$\Delta m = 216.001\ 915 \text{ u} - (211.991\ 898 \text{ u} + 4.002\ 603 \text{ u})$$

$$= 0.007\ 414 \text{ u} = \boxed{1.23 \times 10^{-29} \text{ kg}}$$

(B) Find the energy this mass change represents.

SOLUTION

Use Equation 39.24 to find the energy associated with this mass change:

$$E = \Delta mc^2 = (1.23 \times 10^{-29} \text{ kg})(3.00 \times 10^8 \text{ m/s})^2$$

$$= 1.11 \times 10^{-12} \text{ J} = \boxed{6.92 \text{ MeV}}$$

39.9 The General Theory of Relativity

Up to this point, we have sidestepped a curious puzzle. Mass has two seemingly different properties: a *gravitational attraction* for other masses and an *inertial* property that represents a resistance to acceleration. We first discussed these two attributes for mass in Section 5.5. To designate these two attributes, we use the subscripts g and i and write

Gravitational property: $F_g = m_g g$

Inertial property: $\sum F = m_i a$

The value for the gravitational constant G was chosen to make the magnitudes of m_g and m_i numerically equal. Regardless of how G is chosen, however, the strict proportionality of m_g and m_i has been established experimentally to an extremely high degree: a few parts in 10^{12}. Therefore, it appears that gravitational mass and inertial mass may indeed be exactly proportional.

Why, though? They seem to involve two entirely different concepts: a force of mutual gravitational attraction between two masses and the resistance of a single mass to being accelerated. This question, which puzzled Newton and many other physicists over the years, was answered by Einstein in 1916 when he published his theory of gravitation, known as the *general theory of relativity*. Because it is a mathematically complex theory, we offer merely a hint of its elegance and insight.

In Einstein's view, the dual behavior of mass was evidence for a very intimate and basic connection between the two behaviors. He pointed out that no mechanical experiment (such as dropping an object) could distinguish between the two situations illustrated in Figures 39.17a and 39.17b. In Figure 39.17a, a person standing in an elevator on the surface of a planet feels pressed into the floor due to the gravitational force. If he releases his briefcase, he observes it moving toward the floor with acceleration $\vec{\mathbf{g}} = -g\hat{\mathbf{j}}$. In Figure 39.17b, the person is in an elevator in empty space accelerating upward with $\vec{\mathbf{a}}_{el} = +g\hat{\mathbf{j}}$. The person feels pressed into the floor with the same force as in Figure 39.17a. If he releases his briefcase, he observes it moving toward the floor with acceleration g, exactly as in the previous situation. In each situation, an object released by the observer undergoes a downward acceleration of magnitude g relative to the floor. In Figure 39.17a, the person is at rest in an inertial frame in a gravitational field due to the planet. In Figure 39.17b, the person is in a noninertial frame accelerating in gravity-free space. Einstein's claim is that these two situations are completely equivalent.

Einstein carried this idea further and proposed that *no* experiment, mechanical or otherwise, could distinguish between the two situations. This extension to include all phenomena (not just mechanical ones) has interesting consequences. For example, suppose a light pulse is sent horizontally across the elevator as in Figure 39.17c, in which the elevator is accelerating upward in empty space. From the point of view of an observer in an inertial frame outside the elevator, the light travels in a straight line while the floor of the elevator accelerates upward. According to the observer on the elevator, however, the trajectory of the light pulse bends downward as the floor of the elevator (and the observer) accelerates upward. Therefore, based on the equality of parts (a) and (b) of the figure, Einstein proposed that a

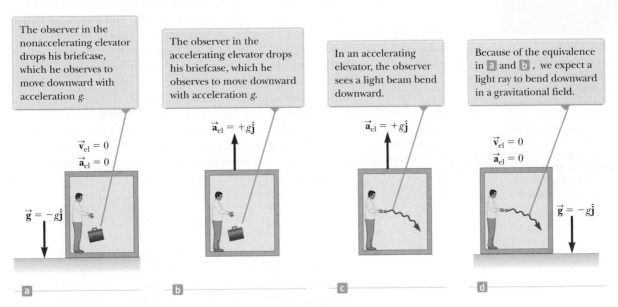

The observer in the nonaccelerating elevator drops his briefcase, which he observes to move downward with acceleration g.

The observer in the accelerating elevator drops his briefcase, which he observes to move downward with acceleration g.

In an accelerating elevator, the observer sees a light beam bend downward.

Because of the equivalence in **a** and **b**, we expect a light ray to bend downward in a gravitational field.

$\vec{\mathbf{a}}_{el} = +g\hat{\mathbf{j}}$

$\vec{\mathbf{a}}_{el} = +g\hat{\mathbf{j}}$

$\vec{\mathbf{v}}_{el} = 0$
$\vec{\mathbf{a}}_{el} = 0$

$\vec{\mathbf{v}}_{el} = 0$
$\vec{\mathbf{a}}_{el} = 0$

$\vec{\mathbf{g}} = -g\hat{\mathbf{j}}$

$\vec{\mathbf{g}} = -g\hat{\mathbf{j}}$

a   **b**   **c**   **d**

Figure 39.17 (a) The observer is at rest in an elevator in a uniform gravitational field $\vec{\mathbf{g}} = -g\hat{\mathbf{j}}$, directed downward. (b) The observer is in a region where gravity is negligible, but the elevator moves upward with an acceleration $\vec{\mathbf{a}}_{el} = +g\hat{\mathbf{j}}$. According to Einstein, the frames of reference in (a) and (b) are equivalent in every way. No local experiment can distinguish any difference between the two frames. (c) An observer watches a beam of light in an accelerating elevator. (d) Einstein's prediction of the behavior of a beam of light in a gravitational field.

beam of light should also be bent downward by a gravitational field as in Figure 39.17d. Experiments have verified the effect, although the bending is small. A laser aimed at the horizon falls less than 1 cm after traveling 6 000 km. (No such bending is predicted in Newton's theory of gravitation.)

Einstein's **general theory of relativity** has two postulates:

- All the laws of nature have the same form for observers in any frame of reference, whether accelerated or not.
- In the vicinity of any point, a gravitational field is equivalent to an accelerated frame of reference in gravity-free space (the **principle of equivalence**).

One interesting effect predicted by the general theory is that time is altered by gravity. A clock in the presence of gravity runs slower than one located where gravity is negligible. Consequently, the frequencies of radiation emitted by atoms in the presence of a strong gravitational field are *redshifted* to lower frequencies when compared with the same emissions in the presence of a weak field. This gravitational redshift has been detected in spectral lines emitted by atoms in massive stars. It has also been verified on the Earth by comparing the frequencies of gamma rays emitted from nuclei separated vertically by about 20 m.

The second postulate suggests a gravitational field may be "transformed away" at any point if we choose an appropriate accelerated frame of reference, a freely falling one. Einstein developed an ingenious method of describing the acceleration necessary to make the gravitational field "disappear." He specified a concept, the *curvature of space–time,* that describes the gravitational effect at every point. In fact, the curvature of space–time completely replaces Newton's gravitational theory. According to Einstein, there is no such thing as a gravitational force. Rather, the presence of a mass causes a curvature of space–time in the vicinity of the mass, and this curvature dictates the space–time path that all freely moving objects must follow.

As an example of the effects of curved space–time, imagine two travelers moving on parallel paths a few meters apart on the surface of the Earth and maintaining an exact northward heading along two longitude lines. As they observe each other near the equator, they will claim that their paths are exactly parallel. As they approach the North Pole, however, they notice that they are moving closer together and will meet at the North Pole. Therefore, they claim that they moved along parallel paths, but moved toward each other, *as if there were an attractive force between them.* The travelers make this conclusion based on their everyday experience of moving on flat surfaces. From our mental representation, however, we realize they are walking on a curved surface, and it is the geometry of the curved surface, rather than an attractive force, that causes them to converge. In a similar way, general relativity replaces the notion of forces with the movement of objects through curved space–time.

One prediction of the general theory of relativity is that a light ray passing near the Sun should be deflected in the curved space–time created by the Sun's mass. This prediction was confirmed when astronomers detected the bending of starlight near the Sun during a total solar eclipse that occurred shortly after World War I (Fig. 39.18). When this discovery was announced, Einstein became an international celebrity.

Einstein's cross. The four outer bright spots are images of the same galaxy that have been bent around a massive object located between the galaxy and the Earth. The massive object acts like a lens, causing the rays of light that were diverging from the distant galaxy to converge on the Earth. (If the intervening massive object had a uniform mass distribution, we would see a bright ring instead of four spots.)

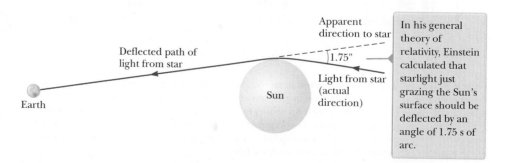

Figure 39.18 Deflection of starlight passing near the Sun. Because of this effect, the Sun or some other remote object can act as a *gravitational lens.*

Apparent direction to star

Deflected path of light from star

1.75"

Light from star (actual direction)

Earth

Sun

In his general theory of relativity, Einstein calculated that starlight just grazing the Sun's surface should be deflected by an angle of 1.75 s of arc.

If the concentration of mass becomes very great as is believed to occur when a large star exhausts its nuclear fuel and collapses to a very small volume, a **black hole** may form as discussed in Chapter 13. Here, the curvature of space–time is so extreme that within a certain distance from the center of the black hole all matter and light become trapped as discussed in Section 13.6.

Summary

Definitions

The relativistic expression for the **linear momentum** of a particle moving with a velocity $\vec{\mathbf{u}}$ is

$$\vec{\mathbf{p}} \equiv \frac{m\vec{\mathbf{u}}}{\sqrt{1 - \dfrac{u^2}{c^2}}} = \gamma m \vec{\mathbf{u}} \qquad \text{(39.19)}$$

The relativistic force $\vec{\mathbf{F}}$ acting on a particle whose linear momentum is $\vec{\mathbf{p}}$ is defined as

$$\vec{\mathbf{F}} \equiv \frac{d\vec{\mathbf{p}}}{dt} \qquad \text{(39.20)}$$

Concepts and Principles

The two basic postulates of the special theory of relativity are as follows:

- The laws of physics must be the same in all inertial reference frames.
- The speed of light in vacuum has the same value, $c = 3.00 \times 10^8$ m/s, in all inertial frames, regardless of the velocity of the observer or the velocity of the source emitting the light.

Three consequences of the special theory of relativity are as follows:

- Events that are measured to be simultaneous for one observer are not necessarily measured to be simultaneous for another observer who is in motion relative to the first.
- Clocks in motion relative to an observer are measured to run slower by a factor $\gamma = (1 - v^2/c^2)^{-1/2}$. This phenomenon is known as **time dilation.**
- The lengths of objects in motion are measured to be shorter in the direction of motion by a factor $1/\gamma = (1 - v^2/c^2)^{1/2}$. This phenomenon is known as **length contraction.**

To satisfy the postulates of special relativity, the Galilean transformation equations must be replaced by the **Lorentz transformation equations:**

$$x' = \gamma(x - vt) \quad y' = y \quad z' = z \quad t' = \gamma\left(t - \frac{v}{c^2}x\right) \qquad \text{(39.11)}$$

where $\gamma = (1 - v^2/c^2)^{-1/2}$ and the S′ frame moves in the x direction at speed v relative to the S frame.

The relativistic form of the **Lorentz velocity transformation equation** is

$$u'_x = \frac{u_x - v}{1 - \dfrac{u_x v}{c^2}} \qquad \text{(39.16)}$$

where u'_x is the x component of the velocity of an object as measured in the S′ frame and u_x is its component as measured in the S frame.

The relativistic expression for the **kinetic energy** of a particle is

$$K = \frac{mc^2}{\sqrt{1 - \dfrac{u^2}{c^2}}} - mc^2 = (\gamma - 1)mc^2 \qquad \text{(39.23)}$$

The constant term mc^2 in Equation 39.23 is called the **rest energy** E_R of the particle:

$$E_R = mc^2 \qquad \text{(39.24)}$$

The **total energy** E of a particle is given by

$$E = \frac{mc^2}{\sqrt{1 - \dfrac{u^2}{c^2}}} = \gamma mc^2 \qquad \text{(39.26)}$$

The relativistic linear momentum of a particle is related to its total energy through the equation

$$E^2 = p^2c^2 + (mc^2)^2 \qquad \text{(39.27)}$$

Objective Questions

1. denotes answer available in *Student Solutions Manual/Study Guide*

1. **(i)** Does the speed of an electron have an upper limit? (a) yes, the speed of light c (b) yes, with another value (c) no **(ii)** Does the magnitude of an electron's momentum have an upper limit? (a) yes, $m_e c$ (b) yes, with another value (c) no **(iii)** Does the electron's kinetic energy have an upper limit? (a) yes, $m_e c^2$ (b) yes, $\frac{1}{2} m_e c^2$ (c) yes, with another value (d) no

2. A spacecraft zooms past the Earth with a constant velocity. An observer on the Earth measures that an undamaged clock on the spacecraft is ticking at one-third the rate of an identical clock on the Earth. What does an observer on the spacecraft measure about the Earth-based clock's ticking rate? (a) It runs more than three times faster than his own clock. (b) It runs three times faster than his own. (c) It runs at the same rate as his own. (d) It runs at one-third the rate of his own. (e) It runs at less than one-third the rate of his own.

3. As a car heads down a highway traveling at a speed v away from a ground observer, which of the following statements are true about the measured speed of the light beam from the car's headlights? More than one statement may be correct. (a) The ground observer measures the light speed to be $c + v$. (b) The driver measures the light speed to be c. (c) The ground observer measures the light speed to be c. (d) The driver measures the light speed to be $c - v$. (e) The ground observer measures the light speed to be $c - v$.

4. A spacecraft built in the shape of a sphere moves past an observer on the Earth with a speed of $0.500c$. What shape does the observer measure for the spacecraft as it goes by? (a) a sphere (b) a cigar shape, elongated along the direction of motion (c) a round pillow shape, flattened along the direction of motion (d) a conical shape, pointing in the direction of motion

5. An astronaut is traveling in a spacecraft in outer space in a straight line at a constant speed of $0.500c$. Which of the following effects would she experience? (a) She would feel heavier. (b) She would find it harder to breathe. (c) Her heart rate would change. (d) Some

of the dimensions of her spacecraft would be shorter. (e) None of those answers is correct.

6. You measure the volume of a cube at rest to be V_0. You then measure the volume of the same cube as it passes you in a direction parallel to one side of the cube. The speed of the cube is $0.980c$, so $\gamma \approx 5$. Is the volume you measure close to (a) $V_0/25$, (b) $V_0/5$, (c) V_0, (d) $5V_0$, or (e) $25V_0$?

7. Two identical clocks are set side by side and synchronized. One remains on the Earth. The other is put into orbit around the Earth moving rapidly toward the east. **(i)** As measured by an observer on the Earth, does the orbiting clock (a) run faster than the Earth-based clock, (b) run at the same rate, or (c) run slower? **(ii)** The orbiting clock is returned to its original location and brought to rest relative to the Earth-based clock. Thereafter, what happens? (a) Its reading lags farther and farther behind the Earth-based clock. (b) It lags behind the Earth-based clock by a constant amount. (c) It is synchronous with the Earth-based clock. (d) It is ahead of the Earth-based clock by a constant amount. (e) It gets farther and farther ahead of the Earth-based clock.

8. The following three particles all have the same total energy E: (a) a photon, (b) a proton, and (c) an electron. Rank the magnitudes of the particles' momenta from greatest to smallest.

9. Which of the following statements are fundamental postulates of the special theory of relativity? More than one statement may be correct. (a) Light moves through a substance called the ether. (b) The speed of light depends on the inertial reference frame in which it is measured. (c) The laws of physics depend on the inertial reference frame in which they are used. (d) The laws of physics are the same in all inertial reference frames. (e) The speed of light is independent of the inertial reference frame in which it is measured.

10. A distant astronomical object (a quasar) is moving away from us at half the speed of light. What is the speed of the light we receive from this quasar? (a) greater than c (b) c (c) between $c/2$ and c (d) $c/2$ (e) between 0 and $c/2$

Conceptual Questions

1. denotes answer available in *Student Solutions Manual/Study Guide*

1. In several cases, a nearby star has been found to have a large planet orbiting about it, although light from the planet could not be seen separately from the starlight. Using the ideas of a system rotating about its center of mass and of the Doppler shift for light, explain how an astronomer could determine the presence of the invisible planet.

2. Explain why, when defining the length of a rod, it is necessary to specify that the positions of the ends of the rod are to be measured simultaneously.

3. A train is approaching you at very high speed as you stand next to the tracks. Just as an observer on the train passes you, you both begin to play the same

recorded version of a Beethoven symphony on identical iPods. (a) According to you, whose iPod finishes the symphony first? (b) **What If?** According to the observer on the train, whose iPod finishes the symphony first? (c) Whose iPod actually finishes the symphony first?

4. List three ways our day-to-day lives would change if the speed of light were only 50 m/s.

5. How is acceleration indicated on a space–time graph?

6. (a) "Newtonian mechanics correctly describes objects moving at ordinary speeds, and relativistic mechanics correctly describes objects moving very fast." (b) "Relativistic mechanics must make a smooth transition as

it reduces to Newtonian mechanics in a case in which the speed of an object becomes small compared with the speed of light." Argue for or against statements (a) and (b).

7. The speed of light in water is 230 Mm/s. Suppose an electron is moving through water at 250 Mm/s. Does that violate the principle of relativity? Explain.

8. A particle is moving at a speed less than $c/2$. If the speed of the particle is doubled, what happens to its momentum?

9. Give a physical argument that shows it is impossible to accelerate an object of mass m to the speed of light, even with a continuous force acting on it.

10. Explain how the Doppler effect with microwaves is used to determine the speed of an automobile.

11. It is said that Einstein, in his teenage years, asked the question, "What would I see in a mirror if I carried it in my hands and ran at a speed near that of light?" How would you answer this question?

12. (i) An object is placed at a position $p > f$ from a concave mirror as shown in Figure CQ39.12a, where f is the focal length of the mirror. In a finite time interval, the object is moved to the right to a position at the focal point F of the mirror. Show that the image of the object moves at a speed greater than the speed of

light. (ii) A laser pointer is suspended in a horizontal plane and set into rapid rotation as shown in Figure CQ39.12b. Show that the spot of light it produces on a distant screen can move across the screen at a speed greater than the speed of light. (If you carry out this experiment, make sure the direct laser light cannot enter a person's eyes.) (iii) Argue that the experiments in parts (i) and (ii) do not invalidate the principle that no material, no energy, and no information can move faster than light moves in a vacuum.

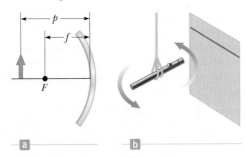

a   b

Figure CQ39.12

13. With regard to reference frames, how does general relativity differ from special relativity?

14. Two identical clocks are in the same house, one upstairs in a bedroom and the other downstairs in the kitchen. Which clock runs slower? Explain.

Problems

WebAssign The problems found in this chapter may be assigned online in Enhanced WebAssign

1. straightforward; 2. intermediate; 3. challenging

1. full solution available in the *Student Solutions Manual/Study Guide*

AMT Analysis Model tutorial available in Enhanced WebAssign

GP Guided Problem

M Master It tutorial available in Enhanced WebAssign

W Watch It video solution available in Enhanced WebAssign

Section 39.1 The Principle of Galilean Relativity

Problems 46–48, 50, 51, 53–54, and 79 in Chapter 4 can be assigned with this section.

1. The truck in Figure P39.1 is moving at a speed of 10.0 m/s relative to the ground. The person on the truck throws a baseball in the backward direction at a speed of 20.0 m/s relative to the truck. What is the velocity of the baseball as measured by the observer on the ground?

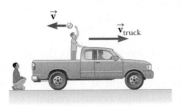

Figure P39.1

2. In a laboratory frame of reference, an observer notes that Newton's second law is valid. Assume forces and masses are measured to be the same in any reference frame for speeds small compared with the speed of light. (a) Show that Newton's second law is also valid for an observer moving at a constant speed, small compared with the speed of light, relative to the laboratory frame. (b) Show that Newton's second law is *not* valid in a reference frame moving past the laboratory frame with a constant acceleration.

3. The speed of the Earth in its orbit is 29.8 km/s. If that is the magnitude of the velocity $\vec{v}$ of the ether wind in Figure P39.3, find the angle ϕ between the velocity of light $\vec{c}$ in vacuum and the resultant velocity of light if there were an ether.

Magnitude: $\sqrt{c^2 - v^2}$

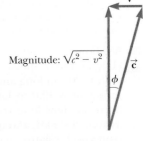

Figure P39.3

4. A car of mass 2 000 kg moving with a speed of 20.0 m/s
AMT collides and locks together with a 1 500-kg car at rest at a stop sign. Show that momentum is conserved in a reference frame moving at 10.0 m/s in the direction of the moving car.

Section 39.2 The Michelson–Morley Experiment
Section 39.3 Einstein's Principle of Relativity
Section 39.4 Consequences of the Special Theory of Relativity

Problem 82 in Chapter 4 can be assigned with this section.

5. A star is 5.00 ly from the Earth. At what speed must a spacecraft travel on its journey to the star such that the Earth–star distance measured in the frame of the spacecraft is 2.00 ly?

6. A meterstick moving at $0.900c$ relative to the Earth's surface approaches an observer at rest with respect to the Earth's surface. (a) What is the meterstick's length as measured by the observer? (b) Qualitatively, how would the answer to part (a) change if the observer started running toward the meterstick?

7. At what speed does a clock move if it is measured to
W run at a rate one-half the rate of a clock at rest with respect to an observer?

8. A muon formed high in the Earth's atmosphere is measured by an observer on the Earth's surface to travel at speed $v = 0.990c$ for a distance of 4.60 km before it decays into an electron, a neutrino, and an antineutrino ($\mu^- \rightarrow e^- + \nu + \bar{\nu}$). (a) For what time interval does the muon live as measured in its reference frame? (b) How far does the Earth travel as measured in the frame of the muon?

9. How fast must a meterstick be moving if its length is
W measured to shrink to 0.500 m?

10. An astronaut is traveling in a space vehicle moving at $0.500c$ relative to the Earth. The astronaut measures her pulse rate at 75.0 beats per minute. Signals generated by the astronaut's pulse are radioed to the Earth when the vehicle is moving in a direction perpendicular to the line that connects the vehicle with an observer on the Earth. (a) What pulse rate does the Earth-based observer measure? (b) **What If?** What would be the pulse rate if the speed of the space vehicle were increased to $0.990c$?

11. A physicist drives through a stop light. When he is pulled over, he tells the police officer that the Doppler shift made the red light of wavelength 650 nm appear green to him, with a wavelength of 520 nm. The police officer writes out a traffic citation for speeding. How fast was the physicist traveling, according to his own testimony?

12. A fellow astronaut passes by you in a spacecraft trav-
W eling at a high speed. The astronaut tells you that his craft is 20.0 m long and that the identical craft you are sitting in is 19.0 m long. According to your observations, (a) how long is your craft, (b) how long is the astronaut's craft, and (c) what is the speed of the astronaut's craft relative to your craft?

13. A deep-space vehicle moves away from the Earth with a speed of $0.800c$. An astronaut on the vehicle measures a time interval of 3.00 s to rotate her body through 1.00 rev as she floats in the vehicle. What time interval is required for this rotation according to an observer on the Earth?

14. For what value of v does $\gamma = 1.010\ 0$? Observe that for speeds lower than this value, time dilation and length contraction are effects amounting to less than 1%.

15. A supertrain with a proper length of 100 m travels at a speed of $0.950c$ as it passes through a tunnel having a proper length of 50.0 m. As seen by a trackside observer, is the train ever completely within the tunnel? If so, by how much do the train's ends clear the ends of the tunnel?

16. The average lifetime of a pi meson in its own frame of
M reference (i.e., the proper lifetime) is 2.6×10^{-8} s. If the meson moves with a speed of $0.98c$, what is (a) its mean lifetime as measured by an observer on Earth, and (b) the average distance it travels before decaying, as measured by an observer on Earth? (c) What distance would it travel if time dilation did not occur?

17. An astronomer on the Earth observes a meteoroid in the southern sky approaching the Earth at a speed of $0.800c$. At the time of its discovery the meteoroid is 20.0 ly from the Earth. Calculate (a) the time interval required for the meteoroid to reach the Earth as measured by the Earthbound astronomer, (b) this time interval as measured by a tourist on the meteoroid, and (c) the distance to the Earth as measured by the tourist.

18. A cube of steel has a volume of 1.00 cm³ and a mass of 8.00 g when at rest on the Earth. If this cube is now given a speed $u = 0.900c$, what is its density as measured by a stationary observer? Note that relativistic density is defined as E_R/c^2V.

19. A spacecraft with a proper length of 300 m passes by
AMT an observer on the Earth. According to this observer, it
M takes 0.750 µs for the spacecraft to pass a fixed point. Determine the speed of the spacecraft as measured by the Earth-based observer.

20. A spacecraft with a proper length of L_p passes by an observer on the Earth. According to this observer, it takes a time interval Δt for the spacecraft to pass a fixed point. Determine the speed of the object as measured by the Earth-based observer.

21. A light source recedes from an observer with a speed v_S that is small compared with c. (a) Show that the fractional shift in the measured wavelength is given by the approximate expression

$$\frac{\Delta \lambda}{\lambda} \approx \frac{v_S}{c}$$

This phenomenon is known as the *redshift* because the visible light is shifted toward the red. (b) Spectroscopic measurements of light at $\lambda = 397$ nm coming from a galaxy in Ursa Major reveal a redshift of 20.0 nm. What is the recessional speed of the galaxy?

22. **Review.** In 1963, astronaut Gordon Cooper orbited the Earth 22 times. The press stated that for each orbit, he aged two-millionths of a second less than he would have had he remained on the Earth. (a) Assuming Cooper was 160 km above the Earth in a circular orbit, determine the difference in elapsed time between someone on the Earth and the orbiting astronaut for the 22 orbits. You may use the approximation

$$\frac{1}{\sqrt{1-x}} \approx 1 + \frac{x}{2}$$

for small x. (b) Did the press report accurate information? Explain.

23. Police radar detects the speed of a car (Fig. P39.23) as follows. Microwaves of a precisely known frequency are broadcast toward the car. The moving car reflects the microwaves with a Doppler shift. The reflected waves are received and combined with an attenuated version of the transmitted wave. Beats occur between the two microwave signals. The beat frequency is measured. (a) For an electromagnetic wave reflected back to its source from a mirror approaching at speed v, show that the reflected wave has frequency

$$f' = \frac{c+v}{c-v}f$$

where f is the source frequency. (b) Noting that v is much less than c, show that the beat frequency can be written as $f_{beat} = 2v/\lambda$. (c) What beat frequency is measured for a car speed of 30.0 m/s if the microwaves have frequency 10.0 GHz? (d) If the beat frequency measurement in part (c) is accurate to ±5.0 Hz, how accurate is the speed measurement?

Figure P39.23

24. The identical twins Speedo and Goslo join a migration from the Earth to Planet X, 20.0 ly away in a reference frame in which both planets are at rest. The twins, of the same age, depart at the same moment on different spacecraft. Speedo's spacecraft travels steadily at $0.950c$ and Goslo's at $0.750c$. (a) Calculate the age difference between the twins after Goslo's spacecraft lands on Planet X. (b) Which twin is older?

25. An atomic clock moves at 1 000 km/h for 1.00 h as measured by an identical clock on the Earth. At the end of the 1.00-h interval, how many nanoseconds slow will the moving clock be compared with the Earth-based clock?

26. **Review.** An alien civilization occupies a planet circling a brown dwarf, several light-years away. The plane of the planet's orbit is perpendicular to a line from the brown dwarf to the Sun, so the planet is at nearly a fixed position relative to the Sun. The extraterrestrials have come to love broadcasts of *MacGyver*, on television channel 2, at carrier frequency 57.0 MHz. Their line of sight to us is in the plane of the Earth's orbit. Find the difference between the highest and lowest frequencies they receive due to the Earth's orbital motion around the Sun.

Section 39.5 The Lorentz Transformation Equations

27. A red light flashes at position $x_R = 3.00$ m and time $t_R = 1.00 \times 10^{-9}$ s, and a blue light flashes at $x_B = 5.00$ m and $t_B = 9.00 \times 10^{-9}$ s, all measured in the S reference frame. Reference frame S′ moves uniformly to the right and has its origin at the same point as S at $t = t' = 0$. Both flashes are observed to occur at the same place in S′. (a) Find the relative speed between S and S′. (b) Find the location of the two flashes in frame S′. (c) At what time does the red flash occur in the S′ frame?

28. Shannon observes two light pulses to be emitted from the same location, but separated in time by 3.00 μs. Kimmie observes the emission of the same two pulses to be separated in time by 9.00 μs. (a) How fast is Kimmie moving relative to Shannon? (b) According to Kimmie, what is the separation in space of the two pulses?

29. A moving rod is observed to have a length of $\ell = 2.00$ m and to be oriented at an angle of $\theta = 30.0°$ with respect to the direction of motion as shown in Figure P39.29. The rod has a speed of $0.995c$. (a) What is the proper length of the rod? (b) What is the orientation angle in the proper frame?

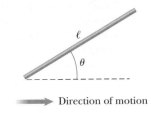

Direction of motion

Figure P39.29

30. A rod moving with a speed v along the horizontal direction is observed to have length ℓ and to make an angle θ with respect to the direction of motion as shown in Figure P39.29. (a) Show that the length of the rod as measured by an observer at rest with respect to the rod is $\ell_p = \ell[1 - (v^2/c^2) \cos^2 \theta]^{1/2}$. (b) Show that the angle θ_p that the rod makes with the x axis according to an observer at rest with respect to the rod can be found from $\tan \theta_p = \gamma \tan \theta$. These results show that the rod is observed to be both contracted and rotated. (Take the lower end of the rod to be at the origin of the coordinate system in which the rod is at rest.)

31. Keilah, in reference frame S, measures two events to be simultaneous. Event A occurs at the point (50.0 m, 0, 0) at the instant 9:00:00 Universal time on January 15,

2013. Event B occurs at the point (150 m, 0, 0) at the same moment. Torrey, moving past with a velocity of $0.800c\hat{\mathbf{i}}$, also observes the two events. In her reference frame S′, which event occurred first and what time interval elapsed between the events?

Section 39.6 The Lorentz Velocity Transformation Equations

32. Figure P39.32 shows a jet of material (at the upper right) being ejected by galaxy M87 (at the lower left). Such jets are believed to be evidence of supermassive black holes at the center of a galaxy. Suppose two jets of material from the center of a galaxy are ejected in opposite directions. Both jets move at $0.750c$ relative to the galaxy center. Determine the speed of one jet relative to the other.

NASA/STScI

Figure P39.32

33. An enemy spacecraft moves away from the Earth at a speed of $v = 0.800c$ (Fig. P39.33). A galactic patrol spacecraft pursues at a speed of $u = 0.900c$ relative to the Earth. Observers on the Earth measure the patrol craft to be overtaking the enemy craft at a relative speed of $0.100c$. With what speed is the patrol craft overtaking the enemy craft as measured by the patrol craft's crew?

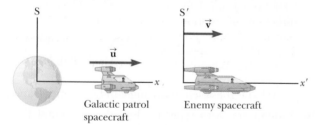

Galactic patrol spacecraft   Enemy spacecraft

Figure P39.33

34. A spacecraft is launched from the surface of the Earth with a velocity of $0.600c$ at an angle of $50.0°$ above the horizontal positive x axis. Another spacecraft is moving past with a velocity of $0.700c$ in the negative x direction. Determine the magnitude and direction of the velocity of the first spacecraft as measured by the pilot of the second spacecraft.

35. A rocket moves with a velocity of $0.92c$ to the right with respect to a stationary observer A. An observer B moving relative to observer A finds that the rocket is moving with a velocity of $0.95c$ to the left. What is the velocity of observer B relative to observer A? (*Hint:*

Consider observer B's velocity in the frame of reference of the rocket.)

Section 39.7 Relativistic Linear Momentum

36. Calculate the momentum of an electron moving with a speed of (a) $0.010\ 0c$, (b) $0.500c$, and (c) $0.900c$.

37. An electron has a momentum that is three times larger than its classical momentum. (a) Find the speed of the electron. (b) **What If?** How would your result change if the particle were a proton?

38. Show that the speed of an object having momentum of magnitude p and mass m is

$$u = \frac{c}{\sqrt{1 + (mc/p)^2}}$$

39. (a) Calculate the classical momentum of a proton traveling at $0.990c$, neglecting relativistic effects. (b) Repeat the calculation while including relativistic effects. (c) Does it make sense to neglect relativity at such speeds?

40. The speed limit on a certain roadway is 90.0 km/h. Suppose speeding fines are made proportional to the amount by which a vehicle's momentum exceeds the momentum it would have when traveling at the speed limit. The fine for driving at 190 km/h (that is, 100 km/h over the speed limit) is $80.0. What, then, is the fine for traveling (a) at 1 090 km/h? (b) At 1 000 000 090 km/h?

41. A golf ball travels with a speed of 90.0 m/s. By what fraction does its relativistic momentum magnitude p differ from its classical value mu? That is, find the ratio $(p - mu)/mu$.

42. The nonrelativistic expression for the momentum of a particle, $p = mu$, agrees with experiment if $u \ll c$. For what speed does the use of this equation give an error in the measured momentum of (a) 1.00% and (b) 10.0%?

43. An unstable particle at rest spontaneously breaks into two fragments of unequal mass. The mass of the first fragment is 2.50×10^{-28} kg, and that of the other is 1.67×10^{-27} kg. If the lighter fragment has a speed of $0.893c$ after the breakup, what is the speed of the heavier fragment?

Section 39.8 Relativistic Energy

44. Determine the energy required to accelerate an electron from (a) $0.500c$ to $0.900c$ and (b) $0.900c$ to $0.990c$.

45. An electron has a kinetic energy five times greater than its rest energy. Find (a) its total energy and (b) its speed.

46. Protons in an accelerator at the Fermi National Laboratory near Chicago are accelerated to a total energy that is 400 times their rest energy. (a) What is the speed of these protons in terms of c? (b) What is their kinetic energy in MeV?

47. A proton moves at $0.950c$. Calculate its (a) rest energy, (b) total energy, and (c) kinetic energy.

48. (a) Find the kinetic energy of a 78.0-kg spacecraft launched out of the solar system with speed 106 km/s by using the classical equation $K = \frac{1}{2}mu^2$. (b) **What If?** Calculate its kinetic energy using the relativistic equation. (c) Explain the result of comparing the answers of parts (a) and (b).

49. A proton in a high-energy accelerator moves with a speed of $c/2$. Use the work–kinetic energy theorem to find the work required to increase its speed to (a) $0.750c$ and (b) $0.995c$.

50. Show that for any object moving at less than one-tenth the speed of light, the relativistic kinetic energy agrees with the result of the classical equation $K = \frac{1}{2}mu^2$ to within less than 1%. Therefore, for most purposes, the classical equation is sufficient to describe these objects.

51. The total energy of a proton is twice its rest energy. Find the momentum of the proton in MeV/c units.

52. Consider electrons accelerated to a total energy of 20.0 GeV in the 3.00-km-long Stanford Linear Accelerator. (a) What is the factor γ for the electrons? (b) What is the electrons' speed at the given energy? (c) What is the length of the accelerator in the electrons' frame of reference when they are moving at their highest speed?

53. When 1.00 g of hydrogen combines with 8.00 g of oxygen, 9.00 g of water is formed. During this chemical reaction, 2.86×10^5 J of energy is released. (a) Is the mass of the water larger or smaller than the mass of the reactants? (b) What is the difference in mass? (c) Explain whether the change in mass is likely to be detectable.

54. In a nuclear power plant, the fuel rods last 3 yr before they are replaced. The plant can transform energy at a maximum possible rate of 1.00 GW. Supposing it operates at 80.0% capacity for 3.00 yr, what is the loss of mass of the fuel?

55. The power output of the Sun is 3.85×10^{26} W. By how much does the mass of the Sun decrease each second?

56. A gamma ray (a high-energy photon) can produce an electron (e^-) and a positron (e^+) of equal mass when it enters the electric field of a heavy nucleus: $\gamma \rightarrow e^+ + e^-$. What minimum gamma-ray energy is required to accomplish this task?

57. A spaceship of mass 2.40×10^6 kg is to be accelerated to a speed of $0.700c$. (a) What minimum amount of energy does this acceleration require from the spaceship's fuel, assuming perfect efficiency? (b) How much fuel would it take to provide this much energy if all the rest energy of the fuel could be transformed to kinetic energy of the spaceship?

58. Show that the energy–momentum relationship in Equation 39.27, $E^2 = p^2c^2 + (mc^2)^2$, follows from the expressions $E = \gamma mc^2$ and $p = \gamma mu$.

59. The rest energy of an electron is 0.511 MeV. The rest energy of a proton is 938 MeV. Assume both particles have kinetic energies of 2.00 MeV. Find the speed of (a) the electron and (b) the proton. (c) By what factor does the speed of the electron exceed that of the proton? (d) Repeat the calculations in parts (a) through

(c) assuming both particles have kinetic energies of 2 000 MeV.

60. Consider a car moving at highway speed u. Is its actual kinetic energy larger or smaller than $\frac{1}{2}mu^2$? Make an order-of-magnitude estimate of the amount by which its actual kinetic energy differs from $\frac{1}{2}mu^2$. In your solution, state the quantities you take as data and the values you measure or estimate for them. You may find Appendix B.5 useful.

61. A pion at rest ($m_\pi = 273m_e$) decays to a muon ($m_\mu = 207m_e$) and an antineutrino ($m_{\bar\nu} \approx 0$). The reaction is written $\pi^- \rightarrow \mu^- + \bar\nu$. Find (a) the kinetic energy of the muon and (b) the energy of the antineutrino in electron volts.

62. An unstable particle with mass $m = 3.34 \times 10^{-27}$ kg is initially at rest. The particle decays into two fragments that fly off along the x axis with velocity components $u_1 = 0.987c$ and $u_2 = -0.868c$. From this information, we wish to determine the masses of fragments 1 and 2. (a) Is the initial system of the unstable particle, which becomes the system of the two fragments, isolated or nonisolated? (b) Based on your answer to part (a), what two analysis models are appropriate for this situation? (c) Find the values of γ for the two fragments after the decay. (d) Using one of the analysis models in part (b), find a relationship between the masses m_1 and m_2 of the fragments. (e) Using the second analysis model in part (b), find a second relationship between the masses m_1 and m_2. (f) Solve the relationships in parts (d) and (e) simultaneously for the masses m_1 and m_2.

63. Massive stars ending their lives in supernova explosions produce the nuclei of all the atoms in the bottom half of the periodic table by fusion of smaller nuclei. This problem roughly models that process. A particle of mass $m = 1.99 \times 10^{-26}$ kg moving with a velocity $\vec{u} = 0.500c\hat{i}$ collides head-on and sticks to a particle of mass $m' = m/3$ moving with the velocity $\vec{u} = -0.500c\hat{i}$. What is the mass of the resulting particle?

64. Massive stars ending their lives in supernova explosions produce the nuclei of all the atoms in the bottom half of the periodic table by fusion of smaller nuclei. This problem roughly models that process. A particle of mass m moving along the x axis with a velocity component $+u$ collides head-on and sticks to a particle of mass $m/3$ moving along the x axis with the velocity component $-u$. (a) What is the mass M of the resulting particle? (b) Evaluate the expression from part (a) in the limit $u \rightarrow 0$. (c) Explain whether the result agrees with what you should expect from nonrelativistic physics.

Section 39.9 The General Theory of Relativity

65. **Review.** A global positioning system (GPS) satellite moves in a circular orbit with period 11 h 58 min. (a) Determine the radius of its orbit. (b) Determine its speed. (c) The nonmilitary GPS signal is broadcast at a frequency of 1 575.42 MHz in the reference frame of the satellite. When it is received on the Earth's surface by a GPS receiver (Fig. P39.65 on page 1230), what is the fractional change in this frequency due to time dilation as described by special relativity? (d) The gravitational "blueshift" of the frequency according to general relativity is a separate effect. It is called a blue-shift to indicate a change to a higher frequency. The magnitude of that fractional change is given by

$$\frac{\Delta f}{f} = \frac{\Delta U_g}{mc^2}$$

where U_g is the change in gravitational potential energy of an object–Earth system when the object of mass m is moved between the two points where the signal is observed. Calculate this fractional change in frequency due to the change in position of the satellite from the Earth's surface to its orbital position. (e) What is the overall fractional change in frequency due to both time dilation and gravitational blueshift?

Figure P39.65

Additional Problems

66. An electron has a speed of 0.750*c*. (a) Find the speed of a proton that has the same kinetic energy as the electron. (b) **What If?** Find the speed of a proton that has the same momentum as the electron.

67. The net nuclear fusion reaction inside the Sun can **M** be written as $4^1\text{H} \rightarrow {}^4\text{He} + E$. The rest energy of each hydrogen atom is 938.78 MeV, and the rest energy of the helium-4 atom is 3 728.4 MeV. Calculate the percentage of the starting mass that is transformed to other forms of energy.

68. *Why is the following situation impossible?* On their 40th birthday, twins Speedo and Goslo say good-bye as Speedo takes off for a planet that is 50 ly away. He travels at a constant speed of 0.85*c* and immediately turns around and comes back to the Earth after arriving at the planet. Upon arriving back at the Earth, Speedo has a joyous reunion with Goslo.

69. A Doppler weather radar station broadcasts a pulse of radio waves at frequency 2.85 GHz. From a relatively small batch of raindrops at bearing 38.6° east of north, the station receives a reflected pulse after 180 μs with a frequency shifted upward by 254 Hz. From a similar batch of raindrops at bearing 39.6° east of north, the station receives a reflected pulse after the same time

delay, with a frequency shifted downward by 254 Hz. These pulses have the highest and lowest frequencies the station receives. (a) Calculate the radial velocity components of both batches of raindrops. (b) Assume that these raindrops are swirling in a uniformly rotating vortex. Find the angular speed of their rotation.

70. An object having mass 900 kg and traveling at speed 0.850*c* collides with a stationary object having mass 1 400 kg. The two objects stick together. Find (a) the speed and (b) the mass of the composite object.

71. An astronaut wishes to visit the Andromeda galaxy, **M** making a one-way trip that will take 30.0 years in the spaceship's frame of reference. Assume the galaxy is 2.00 million light-years away and his speed is constant. (a) How fast must he travel relative to Earth? (b) What will be the kinetic energy of his spacecraft, which has mass of 1.00×10^6 kg? (c) What is the cost of this energy if it is purchased at a typical consumer price for electric energy, 13.0¢ per kWh? The following approximation will prove useful:

$$\frac{1}{\sqrt{1 + x}} \approx 1 - \frac{x}{2} \text{ for } x \ll 1$$

72. A physics professor on the Earth gives an exam to her students, who are in a spacecraft traveling at speed *v* relative to the Earth. The moment the craft passes the professor, she signals the start of the exam. She wishes her students to have a time interval T_0 (spacecraft time) to complete the exam. Show that she should wait a time interval (Earth time) of

$$T = T_0 \sqrt{\frac{1 - v/c}{1 + v/c}}$$

before sending a light signal telling them to stop. (*Suggestion:* Remember that it takes some time for the second light signal to travel from the professor to the students.)

73. An interstellar space probe is launched from Earth. **M** After a brief period of acceleration, it moves with a constant velocity, 70.0% of the speed of light. Its nuclear-powered batteries supply the energy to keep its data transmitter active continuously. The batteries have a lifetime of 15.0 years as measured in a rest frame. (a) How long do the batteries on the space probe last as measured by mission control on Earth? (b) How far is the probe from Earth when its batteries fail as measured by mission control? (c) How far is the probe from Earth as measured by its built-in trip odometer when its batteries fail? (d) For what total time after launch are data received from the probe by mission control? Note that radio waves travel at the speed of light and fill the space between the probe and Earth at the time the battery fails.

74. The equation

$$K = \left(\frac{1}{\sqrt{1 - u^2/c^2}} - 1 \right) mc^2$$

gives the kinetic energy of a particle moving at speed *u*. (a) Solve the equation for *u*. (b) From the equation for *u*, identify the minimum possible value of speed and the corresponding kinetic energy. (c) Identify the maximum possible speed and the corresponding kinetic energy. (d) Differentiate the equation for u with respect to time to obtain an equation describing the acceleration of a particle as a function of its kinetic energy and the power input to the particle. (e) Observe that for a nonrelativistic particle we have $u = (2K/m)^{1/2}$ and that differentiating this equation with respect to time gives $a = P/(2mK)^{1/2}$. State the limiting form of the expression in part (d) at low energy. State how it compares with the nonrelativistic expression. (f) State the limiting form of the expression in part (d) at high energy. (g) Consider a particle with constant input power. Explain how the answer to part (f) helps account for the answer to part (c).

75. Consider the astronaut planning the trip to Andromeda in Problem 71. (a) To three significant figures, what is the value for γ for the speed found in part (a) of Problem 71? (b) Just as the astronaut leaves on his constant-speed trip, a light beam is also sent in the direction of Andromeda. According to the Earth observer, how much later does the astronaut arrive at Andromeda after the arrival of the light beam?

76. An object disintegrates into two fragments. One fragment has mass 1.00 MeV/c^2 and momentum 1.75 MeV/c in the positive x direction, and the other has mass 1.50 MeV/c^2 and momentum 2.00 MeV/c in the positive y direction. Find (a) the mass and (b) the speed of the original object.

77. The cosmic rays of highest energy are protons that [M] have kinetic energy on the order of 10^{13} MeV. (a) As measured in the proton's frame, what time interval would a proton of this energy require to travel across the Milky Way galaxy, which has a proper diameter $\sim 10^5$ ly? (b) From the point of view of the proton, how many kilometers across is the galaxy?

78. Spacecraft I, containing students taking a physics [M] exam, approaches the Earth with a speed of $0.600c$ (relative to the Earth), while spacecraft II, containing professors proctoring the exam, moves at $0.280c$ (relative to the Earth) directly toward the students. If the professors stop the exam after 50.0 min have passed on their clock, for what time interval does the exam last as measured by (a) the students and (b) an observer on the Earth?

79. **Review.** Around the core of a nuclear reactor shielded by a large pool of water, Cerenkov radiation appears as a blue glow. (See Fig. P17.38 on page 528.) Cerenkov radiation occurs when a particle travels faster through a medium than the speed of light in that medium. It is the electromagnetic equivalent of a bow wave or a sonic boom. An electron is traveling through water at a speed 10.0% faster than the speed of light in water. Determine the electron's (a) total energy, (b) kinetic energy, and (c) momentum. (d) Find the angle between the shock wave and the electron's direction of motion.

80. The motion of a transparent medium influences the speed of light. This effect was first observed by Fizeau in 1851. Consider a light beam in water. The water moves with speed v in a horizontal pipe. Assume the light travels in the same direction as the water moves. The speed of light with respect to the water is c/n, where $n = 1.33$ is the index of refraction of water. (a) Use the velocity transformation equation to show that the speed of the light measured in the laboratory frame is

$$u = \frac{c}{n}\left(\frac{1 + nv/c}{1 + v/nc}\right)$$

(b) Show that for $v \ll c$, the expression from part (a) becomes, to a good approximation,

$$u \approx \frac{c}{n} + v - \frac{v}{n^2}$$

(c) Argue for or against the view that we should expect the result to be $u = (c/n) + v$ according to the Galilean transformation and that the presence of the term $-v/n^2$ represents a relativistic effect appearing even at "nonrelativistic" speeds. (d) Evaluate u in the limit as the speed of the water approaches c.

81. Imagine that the entire Sun, of mass M_S, collapses to a sphere of radius R_g such that the work required to remove a small mass m from the surface would be equal to its rest energy mc^2. This radius is called the *gravitational radius* for the Sun. (a) Use this approach to show that $R_g = GM_S/c^2$. (b) Find a numerical value for R_g.

82. *Why is the following situation impossible?* An experimenter is accelerating electrons for use in probing a material. She finds that when she accelerates them through a potential difference of 84.0 kV, the electrons have half the speed she wishes. She quadruples the potential difference to 336 kV, and the electrons accelerated through this potential difference have her desired speed.

83. An alien spaceship traveling at $0.600c$ toward the Earth launches a landing craft. The landing craft travels in the same direction with a speed of $0.800c$ relative to the mother ship. As measured on the Earth, the spaceship is 0.200 ly from the Earth when the landing craft is launched. (a) What speed do the Earth-based observers measure for the approaching landing craft? (b) What is the distance to the Earth at the moment of the landing craft's launch as measured by the aliens? (c) What travel time is required for the landing craft to reach the Earth as measured by the aliens on the mother ship? (d) If the landing craft has a mass of 4.00×10^5 kg, what is its kinetic energy as measured in the Earth reference frame?

84. (a) Prepare a graph of the relativistic kinetic energy and the classical kinetic energy, both as a function of speed, for an object with a mass of your choice. (b) At what speed does the classical kinetic energy underestimate the experimental value by 1%? (c) By 5%? (d) By 50%?

85. An observer in a coasting spacecraft moves toward a [AMT] mirror at speed $v = 0.650c$ relative to the reference frame labeled S in Figure P39.85 (page 1232). The mirror is stationary with respect to S. A light pulse emitted by the spacecraft travels toward the mirror and is

reflected back to the spacecraft. The spacecraft is a distance $d = 5.66 \times 10^{10}$ m from the mirror (as measured by observers in S) at the moment the light pulse leaves the spacecraft. What is the total travel time of the pulse as measured by observers in (a) the S frame and (b) the spacecraft?

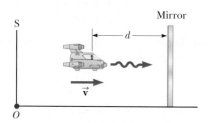

Figure P39.85 Problems 85 and 86.

86. An observer in a coasting spacecraft moves toward a mirror at speed v relative to the reference frame labeled S in Figure P39.85. The mirror is stationary with respect to S. A light pulse emitted by the spacecraft travels toward the mirror and is reflected back to the spacecraft. The spacecraft is a distance d from the mirror (as measured by observers in S) at the moment the light pulse leaves the spacecraft. What is the total travel time of the pulse as measured by observers in (a) the S frame and (b) the spacecraft?

87. A ^{57}Fe nucleus at rest emits a 14.0-keV photon. Use conservation of energy and momentum to find the kinetic energy of the recoiling nucleus in electron volts. Use $Mc^2 = 8.60 \times 10^{-9}$ J for the final state of the ^{57}Fe nucleus.

Challenge Problems

88. A particle with electric charge q moves along a straight line in a uniform electric field $\vec{E}$ with speed u. The electric force exerted on the charge is $q\vec{E}$. The velocity of the particle and the electric field are both in the x direction. (a) Show that the acceleration of the particle in the x direction is given by

$$a = \frac{du}{dt} = \frac{qE}{m}\left(1 - \frac{u^2}{c^2}\right)^{3/2}$$

(b) Discuss the significance of the dependence of the acceleration on the speed. (c) **What If?** If the particle starts from rest at $x = 0$ at $t = 0$, how would you proceed to find the speed of the particle and its position at time t?

89. The creation and study of new and very massive elementary particles is an important part of contemporary physics. To create a particle of mass M requires an energy Mc^2. With enough energy, an exotic particle can be created by allowing a fast-moving proton to collide with a similar target particle. Consider a perfectly inelastic collision between two protons: an incident proton with mass m_p, kinetic energy K, and momentum magnitude p joins with an originally stationary target proton to form a single product particle of mass M. Not all the kinetic energy of the incoming proton is available to create the product particle because conservation of

momentum requires that the system as a whole still must have some kinetic energy after the collision. Therefore, only a fraction of the energy of the incident particle is available to create a new particle. (a) Show that the energy available to create a product particle is given by

$$Mc^2 = 2m_pc^2\sqrt{1 + \frac{K}{2m_pc^2}}$$

This result shows that when the kinetic energy K of the incident proton is large compared with its rest energy m_pc^2, then M approaches $(2m_pK)^{1/2}/c$. Therefore, if the energy of the incoming proton is increased by a factor of 9, the mass you can create increases only by a factor of 3, not by a factor of 9 as would be expected. (b) This problem can be alleviated by using *colliding beams* as is the case in most modern accelerators. Here the total momentum of a pair of interacting particles can be zero. The center of mass can be at rest after the collision, so, in principle, all the initial kinetic energy can be used for particle creation. Show that

$$Mc^2 = 2mc^2\left(1 + \frac{K}{mc^2}\right)$$

where K is the kinetic energy of each of the two identical colliding particles. Here, if $K \gg mc^2$, we have M directly proportional to K as we would desire.

90. Suppose our Sun is about to explode. In an effort to escape, we depart in a spacecraft at $v = 0.800c$ and head toward the star Tau Ceti, 12.0 ly away. When we reach the midpoint of our journey from the Earth, we see our Sun explode, and, unfortunately, at the same instant, we see Tau Ceti explode as well. (a) In the spacecraft's frame of reference, should we conclude that the two explosions occurred simultaneously? If not, which occurred first? (b) **What If?** In a frame of reference in which the Sun and Tau Ceti are at rest, did they explode simultaneously? If not, which exploded first?

91. Owen and Dina are at rest in frame S′, which is moving at $0.600c$ with respect to frame S. They play a game of catch while Ed, at rest in frame S, watches the action (Fig. P39.91). Owen throws the ball to Dina at $0.800c$ (according to Owen), and their separation (measured in S′) is equal to 1.80×10^{12} m. (a) According to Dina, how fast is the ball moving? (b) According to Dina, what time interval is required for the ball to reach her? According to Ed, (c) how far apart are Owen and Dina, (d) how fast is the ball moving, and (e) what time interval is required for the ball to reach Dina?

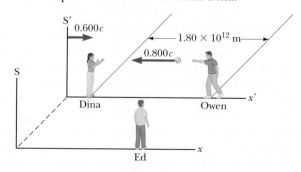

Figure P39.91

Introduction to Quantum Physics

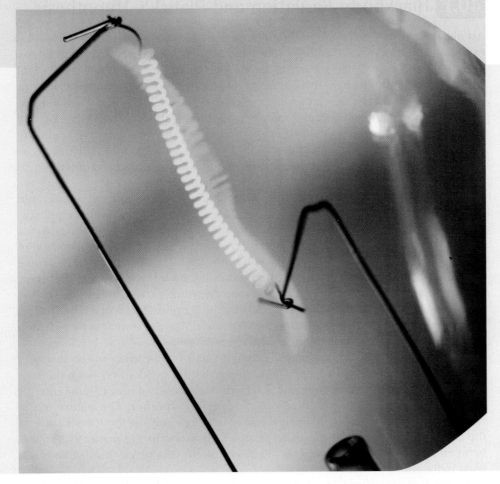

In Chapter 39, we discussed that Newtonian mechanics must be replaced by Einstein's special theory of relativity when dealing with particle speeds comparable to the speed of light. As the 20th century progressed, many experimental and theoretical problems were resolved by the special theory of relativity. For many other problems, however, neither relativity nor classical physics could provide a theoretical answer. Attempts to apply the laws of classical physics to explain the behavior of matter on the atomic scale were consistently unsuccessful. For example, the emission of discrete wavelengths of light from atoms in a high-temperature gas could not be explained within the framework of classical physics.

As physicists sought new ways to solve these puzzles, another revolution took place in physics between 1900 and 1930. A new theory called *quantum mechanics* was highly successful in explaining the behavior of particles of microscopic size. Like the special theory of relativity, the quantum theory requires a modification of our ideas concerning the physical world.

The first explanation of a phenomenon using quantum theory was introduced by Max Planck. Many subsequent mathematical developments and interpretations were made by a number of distinguished physicists, including Einstein, Bohr, de Broglie, Schrödinger, and

This lightbulb filament glows with an orange color. Why? Classical physics is unable to explain the experimentally observed wavelength distribution of electromagnetic radiation from a hot object. A theory proposed in 1900 and describing the radiation from such objects represents the dawn of quantum physics. *(Steve Cole/Getty Images)*

Heisenberg. Despite the great success of the quantum theory, Einstein frequently played the role of its critic, especially with regard to the manner in which the theory was interpreted.

Because an extensive study of quantum theory is beyond the scope of this book, this chapter is simply an introduction to its underlying principles.

40.1 Blackbody Radiation and Planck's Hypothesis

An object at any temperature emits electromagnetic waves in the form of **thermal radiation** from its surface as discussed in Section 20.7. The characteristics of this radiation depend on the temperature and properties of the object's surface. Careful study shows that the radiation consists of a continuous distribution of wavelengths from all portions of the electromagnetic spectrum. If the object is at room temperature, the wavelengths of thermal radiation are mainly in the infrared region and hence the radiation is not detected by the human eye. As the surface temperature of the object increases, the object eventually begins to glow visibly red, like the coils of a toaster. At sufficiently high temperatures, the glowing object appears white, as in the hot tungsten filament of an incandescent lightbulb.

From a classical viewpoint, thermal radiation originates from accelerated charged particles in the atoms near the surface of the object; those charged particles emit radiation much as small antennas do. The thermally agitated particles can have a distribution of energies, which accounts for the continuous spectrum of radiation emitted by the object. By the end of the 19th century, however, it became apparent that the classical theory of thermal radiation was inadequate. The basic problem was in understanding the observed distribution of wavelengths in the radiation emitted by a black body. As defined in Section 20.7, a **black body** is an ideal system that absorbs all radiation incident on it. The electromagnetic radiation emitted by the black body is called **blackbody radiation.**

A good approximation of a black body is a small hole leading to the inside of a hollow object as shown in Figure 40.1. Any radiation incident on the hole from outside the cavity enters the hole and is reflected a number of times on the interior walls of the cavity; hence, the hole acts as a perfect absorber. The nature of the radiation leaving the cavity through the hole depends only on the temperature of the cavity walls and not on the material of which the walls are made. The spaces between lumps of hot charcoal (Fig. 40.2) emit light that is very much like blackbody radiation.

The radiation emitted by oscillators in the cavity walls in Figure 40.1 experiences boundary conditions and can be analyzed using the waves under boundary conditions analysis model. As the radiation reflects from the cavity's walls, standing electromagnetic waves are established within the three-dimensional interior of the cavity. Many standing-wave modes are possible, and the distribution of the energy in the cavity among these modes determines the wavelength distribution of the radiation leaving the cavity through the hole.

The wavelength distribution of radiation from cavities was studied experimentally in the late 19th century. Figure 40.3 shows how the intensity of blackbody radiation varies with temperature and wavelength. The following two consistent experimental findings were seen as especially significant:

1. **The total power of the emitted radiation increases with temperature.**
   We discussed this behavior briefly in Chapter 20, where we introduced Stefan's law:

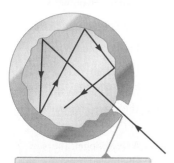

The opening to a cavity inside a hollow object is a good approximation of a black body: the hole acts as a perfect absorber.

Figure 40.1 A physical model of a black body.

SOMMAI/Shutterstock.com

Figure 40.2 The glow emanating from the spaces between these hot charcoal briquettes is, to a close approximation, blackbody radiation. The color of the light depends only on the temperature of the briquettes.

Stefan's law ▶

$$P = \sigma A e T^4 \tag{40.1}$$

where P is the power in watts radiated at all wavelengths from the surface of an object, $\sigma = 5.670 \times 10^{-8}$ W/m$^2 \cdot$ K^4 is the Stefan–Boltzmann constant, A is the surface area of the object in square meters, e is the emissivity of the surface, and T is the surface temperature in kelvins. For a black body, the emissivity is $e = 1$ exactly.

2. **The peak of the wavelength distribution shifts to shorter wavelengths as the temperature increases.** This behavior is described by the following relationship, called **Wien's displacement law:**

$$\lambda_{\max} T = 2.898 \times 10^{-3} \text{ m} \cdot \text{K} \qquad (40.2)$$

◀ Wien's displacement law

where $\lambda_{\max}$ is the wavelength at which the curve peaks and T is the absolute temperature of the surface of the object emitting the radiation. The wavelength at the curve's peak is inversely proportional to the absolute temperature; that is, as the temperature increases, the peak is "displaced" to shorter wavelengths (Fig. 40.3).

Wien's displacement law is consistent with the behavior of the object mentioned at the beginning of this section. At room temperature, the object does not appear to glow because the peak is in the infrared region of the electromagnetic spectrum. At higher temperatures, it glows red because the peak is in the near infrared with some radiation at the red end of the visible spectrum, and at still higher temperatures, it glows white because the peak is in the visible so that all colors are emitted.

Quick Quiz 40.1 Figure 40.4 shows two stars in the constellation Orion. Betelgeuse appears to glow red, whereas Rigel looks blue in color. Which star has a higher surface temperature? **(a)** Betelgeuse **(b)** Rigel **(c)** both the same **(d)** impossible to determine

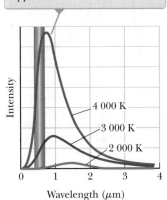

The 4 000-K curve has a peak near the visible range. This curve represents an object that would glow with a yellowish-white appearance.

Figure 40.3 Intensity of blackbody radiation versus wavelength at three temperatures. The visible range of wavelengths is between 0.4 μm and 0.7 μm. At approximately 6 000 K, the peak is in the center of the visible wavelengths and the object appears white.

John Chumack/Photo Researchers, Inc.

Figure 40.4 (Quick Quiz 40.1) Which star is hotter, Betelgeuse or Rigel?

A successful theory for blackbody radiation must predict the shape of the curves in Figure 40.3, the temperature dependence expressed in Stefan's law, and the shift of the peak with temperature described by Wien's displacement law. Early attempts to use classical ideas to explain the shapes of the curves in Figure 40.3 failed.

Let's consider one of these early attempts. To describe the distribution of energy from a black body, we define $I(\lambda,T)\, d\lambda$ to be the intensity, or power per unit area, emitted in the wavelength interval $d\lambda$. The result of a calculation based on a classical theory of blackbody radiation known as the **Rayleigh–Jeans law** is

$$I(\lambda,T) = \frac{2\pi c k_{\text{B}} T}{\lambda^4} \qquad (40.3)$$

◀ Rayleigh–Jeans law

where k_{B} is Boltzmann's constant. The black body is modeled as the hole leading into a cavity (Fig. 40.1), resulting in many modes of oscillation of the electromagnetic field caused by accelerated charges in the cavity walls and the emission of electromagnetic waves at all wavelengths. In the classical theory used to derive

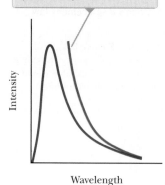

The classical theory (red-brown curve) shows intensity growing without bound for short wavelengths, unlike the experimental data (blue curve).

Intensity

Wavelength

Figure 40.5 Comparison of experimental results and the curve predicted by the Rayleigh–Jeans law for the distribution of blackbody radiation.

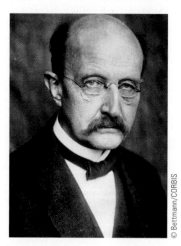

© Bettmann/CORBIS

Max Planck
German Physicist (1858–1947)
Planck introduced the concept of "quantum of action" (Planck's constant, *h*) in an attempt to explain the spectral distribution of blackbody radiation, which laid the foundations for quantum theory. In 1918, he was awarded the Nobel Prize in Physics for this discovery of the quantized nature of energy.

Equation 40.3, the average energy for each wavelength of the standing-wave modes is assumed to be proportional to $k_B T$, based on the theorem of equipartition of energy discussed in Section 21.1.

An experimental plot of the blackbody radiation spectrum, together with the theoretical prediction of the Rayleigh–Jeans law, is shown in Figure 40.5. At long wavelengths, the Rayleigh–Jeans law is in reasonable agreement with experimental data, but at short wavelengths, major disagreement is apparent.

As λ approaches zero, the function $I(\lambda,T)$ given by Equation 40.3 approaches infinity. Hence, according to classical theory, not only should short wavelengths predominate in a blackbody spectrum, but also the energy emitted by any black body should become infinite in the limit of zero wavelength. In contrast to this prediction, the experimental data plotted in Figure 40.5 show that as λ approaches zero, $I(\lambda,T)$ also approaches zero. This mismatch of theory and experiment was so disconcerting that scientists called it the *ultraviolet catastrophe*. (This "catastrophe"—infinite energy—occurs as the wavelength approaches zero; the word *ultraviolet* was applied because ultraviolet wavelengths are short.)

In 1900, Max Planck developed a theory of blackbody radiation that leads to an equation for $I(\lambda,T)$ that is in complete agreement with experimental results at all wavelengths. In discussing this theory, we use the outline of properties of structural models introduced in Chapter 21:

1. *Physical components:*
   Planck assumed the cavity radiation came from atomic oscillators in the cavity walls in Figure 40.1.
2. *Behavior of the components:*
   (a) The energy of an oscillator can have only certain *discrete* values E_n:

$$E_n = nhf \tag{40.4}$$

where *n* is a positive integer called a **quantum number,**[1] *f* is the oscillator's frequency, and *h* is a parameter Planck introduced that is now called **Planck's constant.** Because the energy of each oscillator can have only discrete values given by Equation 40.4, we say the energy is **quantized.** Each discrete energy value corresponds to a different **quantum state,** represented by the quantum number *n*. When the oscillator is in the $n = 1$ quantum state, its energy is hf; when it is in the $n = 2$ quantum state, its energy is $2hf$; and so on.

   (b) The oscillators emit or absorb energy when making a transition from one quantum state to another. The entire energy difference between the initial and final states in the transition is emitted or absorbed as a single quantum of radiation. If the transition is from one state to a lower adjacent state—say, from the $n = 3$ state to the $n = 2$ state—Equation 40.4 shows that the amount of energy emitted by the oscillator and carried by the quantum of radiation is

$$E = hf \tag{40.5}$$

According to property 2(b), an oscillator emits or absorbs energy only when it changes quantum states. If it remains in one quantum state, no energy is absorbed or emitted. Figure 40.6 is an **energy-level diagram** showing the quantized energy levels and allowed transitions proposed by Planck. This important semigraphical representation is used often in quantum physics.[2] The vertical axis is linear in energy, and the allowed energy levels are represented as horizontal lines. The quantized system can have only the energies represented by the horizontal lines.

---

[1] A quantum number is generally an integer (although half-integer quantum numbers can occur) that describes an allowed state of a system, such as the values of *n* describing the normal modes of oscillation of a string fixed at both ends, as discussed in Section 18.3.

[2] We first saw an energy-level diagram in Section 21.3.

The key point in Planck's theory is the radical assumption of quantized energy states. This development—a clear deviation from classical physics—marked the birth of the quantum theory.

In the Rayleigh–Jeans model, the average energy associated with a particular wavelength of standing waves in the cavity is the same for all wavelengths and is equal to $k_B T$. Planck used the same classical ideas as in the Rayleigh–Jeans model to arrive at the energy density as a product of constants and the average energy for a given wavelength, but the average energy is not given by the equipartition theorem. A wave's average energy is the average energy difference between levels of the oscillator, *weighted according to the probability of the wave being emitted*. This weighting is based on the occupation of higher-energy states as described by the Boltzmann distribution law, which was discussed in Section 21.5. According to this law, the probability of a state being occupied is proportional to the factor $e^{-E/k_B T}$, where E is the energy of the state.

At low frequencies (long wavelengths), according to property 2(a), the energy levels are close together as on the right in Figure 40.7, and many of the energy states are excited because the Boltzmann factor $e^{-E/k_B T}$ is relatively large for these states. Therefore, there are many contributions to the outgoing radiation, although each contribution has very low energy. Now, consider high-frequency radiation, that is, radiation with short wavelength. To obtain this radiation, the allowed energies are very far apart as on the left in Figure 40.7. The probability of thermal agitation exciting these high energy levels is small because of the small value of the Boltzmann factor for large values of E. At high frequencies, the low probability of excitation results in very little contribution to the total energy, even though each quantum is of large energy. This low probability "turns the curve over" and brings it down to zero again at short wavelengths.

Using this approach, Planck generated a theoretical expression for the wavelength distribution that agreed remarkably well with the experimental curves in Figure 40.3:

$$I(\lambda, T) = \frac{2\pi hc^2}{\lambda^5 \left(e^{hc/\lambda k_B T} - 1 \right)} \tag{40.6}$$

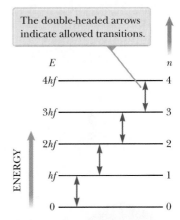

The double-headed arrows indicate allowed transitions.

Figure 40.6 Allowed energy levels for an oscillator with frequency f.

◀ **Planck's wavelength distribution function**

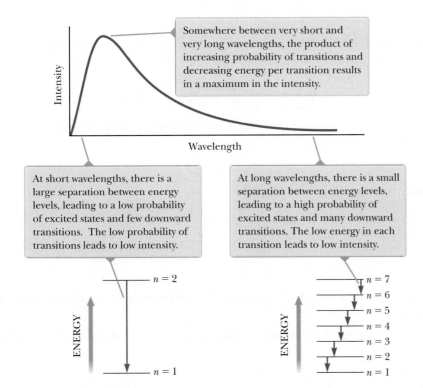

Somewhere between very short and very long wavelengths, the product of increasing probability of transitions and decreasing energy per transition results in a maximum in the intensity.

At short wavelengths, there is a large separation between energy levels, leading to a low probability of excited states and few downward transitions. The low probability of transitions leads to low intensity.

At long wavelengths, there is a small separation between energy levels, leading to a high probability of excited states and many downward transitions. The low energy in each transition leads to low intensity.

Figure 40.7 In Planck's model, the average energy associated with a given wavelength is the product of the energy of a transition and a factor related to the probability of the transition occurring.

This function includes the parameter h, which Planck adjusted so that his curve matched the experimental data at all wavelengths. The value of this parameter is found to be independent of the material of which the black body is made and independent of the temperature; it is a fundamental constant of nature. The value of h, Planck's constant, which was first introduced in Chapter 35, is

◀ Planck's constant

$$h = 6.626 \times 10^{-34}\,\text{J} \cdot \text{s} \qquad \text{(40.7)}$$

At long wavelengths, Equation 40.6 reduces to the Rayleigh–Jeans expression, Equation 40.3 (see Problem 14), and at short wavelengths, it predicts an exponential decrease in $I(\lambda,T)$ with decreasing wavelength, in agreement with experimental results.

When Planck presented his theory, most scientists (including Planck!) did not consider the quantum concept to be realistic. They believed it was a mathematical trick that happened to predict the correct results. Hence, Planck and others continued to search for a more "rational" explanation of blackbody radiation. Subsequent developments, however, showed that a theory based on the quantum concept (rather than on classical concepts) had to be used to explain not only blackbody radiation but also a number of other phenomena at the atomic level.

In 1905, Einstein rederived Planck's results by assuming the oscillations of the electromagnetic field were themselves quantized. In other words, he proposed that quantization is a fundamental property of light and other electromagnetic radiation, which led to the concept of photons as shall be discussed in Section 40.2. Critical to the success of the quantum or photon theory was the relation between energy and frequency, which classical theory completely failed to predict.

You may have had your body temperature measured at the doctor's office by an *ear thermometer*, which can read your temperature very quickly (Fig. 40.8). In a fraction of a second, this type of thermometer measures the amount of infrared radiation emitted by the eardrum. It then converts the amount of radiation into a temperature reading. This thermometer is very sensitive because temperature is raised to the fourth power in Stefan's law. Suppose you have a fever 1°C above normal. Because absolute temperatures are found by adding 273 to Celsius temperatures, the ratio of your fever temperature to normal body temperature of 37°C is

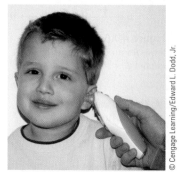

$$\frac{T_{\text{fever}}}{T_{\text{normal}}} = \frac{38°\text{C} + 273°\text{C}}{37°\text{C} + 273°\text{C}} = 1.003\,2$$

Figure 40.8 An ear thermometer measures a patient's temperature by detecting the intensity of infrared radiation leaving the eardrum.

which is only a 0.32% increase in temperature. The increase in radiated power, however, is proportional to the fourth power of temperature, so

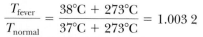

$$\frac{P_{\text{fever}}}{P_{\text{normal}}} = \left(\frac{38°\text{C} + 273°\text{C}}{37°\text{C} + 273°\text{C}}\right)^4 = 1.013$$

The result is a 1.3% increase in radiated power, which is easily measured by modern infrared radiation sensors.

Example 40.1   Thermal Radiation from Different Objects

(A) Find the peak wavelength of the blackbody radiation emitted by the human body when the skin temperature is 35°C.

SOLUTION

Conceptualize Thermal radiation is emitted from the surface of any object. The peak wavelength is related to the surface temperature through Wien's displacement law (Eq. 40.2).

Categorize We evaluate results using an equation developed in this section, so we categorize this example as a substitution problem.

© Cengage Learning/Edward L. Dodd, Jr.

▶ **40.1** continued

Solve Equation 40.2 for λ_{max}:

$$(1) \quad \lambda_{max} = \frac{2.898 \times 10^{-3}\ \text{m} \cdot \text{K}}{T}$$

Substitute the surface temperature:

$$\lambda_{max} = \frac{2.898 \times 10^{-3}\ \text{m} \cdot \text{K}}{308\ \text{K}} = \boxed{9.41\ \mu\text{m}}$$

This radiation is in the infrared region of the spectrum and is invisible to the human eye. Some animals (pit vipers, for instance) are able to detect radiation of this wavelength and therefore can locate warm-blooded prey even in the dark.

(B) Find the peak wavelength of the blackbody radiation emitted by the tungsten filament of a lightbulb, which operates at 2 000 K.

SOLUTION

Substitute the filament temperature into Equation (1):

$$\lambda_{max} = \frac{2.898 \times 10^{-3}\ \text{m} \cdot \text{K}}{2\ 000\ \text{K}} = \boxed{1.45\ \mu\text{m}}$$

This radiation is also in the infrared, meaning that most of the energy emitted by a lightbulb is not visible to us.

(C) Find the peak wavelength of the blackbody radiation emitted by the Sun, which has a surface temperature of approximately 5 800 K.

SOLUTION

Substitute the surface temperature into Equation (1):

$$\lambda_{max} = \frac{2.898 \times 10^{-3}\ \text{m} \cdot \text{K}}{5\ 800\ \text{K}} = \boxed{0.500\ \mu\text{m}}$$

This radiation is near the center of the visible spectrum, near the color of a yellow-green tennis ball. Because it is the most prevalent color in sunlight, our eyes have evolved to be most sensitive to light of approximately this wavelength.

Example 40.2   **The Quantized Oscillator**

A 2.00-kg block is attached to a massless spring that has a force constant of $k = 25.0$ N/m. The spring is stretched 0.400 m from its equilibrium position and released from rest.

(A) Find the total energy of the system and the frequency of oscillation according to classical calculations.

SOLUTION

Conceptualize We understand the details of the block's motion from our study of simple harmonic motion in Chapter 15. Review that material if you need to.

Categorize The phrase "according to classical calculations" tells us to categorize this part of the problem as a classical analysis of the oscillator. We model the block as a *particle in simple harmonic motion*.

....

Analyze Based on the way the block is set into motion, its amplitude is 0.400 m.

Evaluate the total energy of the block–spring system using Equation 15.21:

$$E = \tfrac{1}{2}kA^2 = \tfrac{1}{2}(25.0\ \text{N/m})(0.400\ \text{m})^2 = \boxed{2.00\ \text{J}}$$

Evaluate the frequency of oscillation from Equation 15.14:

$$f = \frac{1}{2\pi}\sqrt{\frac{k}{m}} = \frac{1}{2\pi}\sqrt{\frac{25.0\ \text{N/m}}{2.00\ \text{kg}}} = \boxed{0.563\ \text{Hz}}$$

(B) Assuming the energy of the oscillator is quantized, find the quantum number n for the system oscillating with this amplitude.

continued

▶ **40.2** c o n t i n u e d

SOLUTION

Categorize This part of the problem is categorized as a quantum analysis of the oscillator. We model the block–spring system as a Planck oscillator.

⋯⋯⋯

Analyze Solve Equation 40.4 for the quantum number n:
$$n = \frac{E_n}{hf}$$

Substitute numerical values:
$$n = \frac{2.00 \text{ J}}{(6.626 \times 10^{-34} \text{ J} \cdot \text{s})(0.563 \text{ Hz})} = \boxed{5.36 \times 10^{33}}$$

⋯⋯⋯

Finalize Notice that 5.36×10^{33} is a very large quantum number, which is typical for macroscopic systems. Changes between quantum states for the oscillator are explored next.

WHAT IF? Suppose the oscillator makes a transition from the $n = 5.36 \times 10^{33}$ state to the state corresponding to $n = 5.36 \times 10^{33} - 1$. By how much does the energy of the oscillator change in this one-quantum change?

Answer From Equation 40.5 and the result to part (A), the energy carried away due to the transition between states differing in n by 1 is

$$E = hf = (6.626 \times 10^{-34} \text{ J} \cdot \text{s})(0.563 \text{ Hz}) = 3.73 \times 10^{-34} \text{ J}$$

This energy change due to a one-quantum change is fractionally equal to 3.73×10^{-34} J/2.00 J, or on the order of one part in 10^{34}! It is such a small fraction of the total energy of the oscillator that it cannot be detected. Therefore, even though the energy of a macroscopic block–spring system is quantized and does indeed decrease by small quantum jumps, our senses perceive the decrease as continuous. Quantum effects become important and detectable only on the submicroscopic level of atoms and molecules.

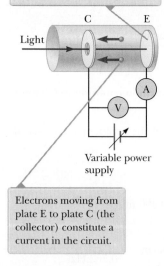

When light strikes plate E (the emitter), photoelectrons are ejected from the plate.

Electrons moving from plate E to plate C (the collector) constitute a current in the circuit.

Figure 40.9 A circuit diagram for studying the photoelectric effect.

40.2 The Photoelectric Effect

Blackbody radiation was the first phenomenon to be explained with a quantum model. In the latter part of the 19th century, at the same time that data were taken on thermal radiation, experiments showed that light incident on certain metallic surfaces causes electrons to be emitted from those surfaces. This phenomenon, which was first discussed in Section 35.1, is known as the **photoelectric effect,** and the emitted electrons are called **photoelectrons.**[3]

Figure 40.9 is a diagram of an apparatus for studying the photoelectric effect. An evacuated glass or quartz tube contains a metallic plate E (the emitter) connected to the negative terminal of a battery and another metallic plate C (the collector) that is connected to the positive terminal of the battery. When the tube is kept in the dark, the ammeter reads zero, indicating no current in the circuit. However, when plate E is illuminated by light having an appropriate wavelength, a current is detected by the ammeter, indicating a flow of charges across the gap between plates E and C. This current arises from photoelectrons emitted from plate E and collected at plate C.

Figure 40.10 is a plot of photoelectric current versus potential difference ΔV applied between plates E and C for two light intensities. At large values of ΔV, the current reaches a maximum value; all the electrons emitted from E are collected at C, and the current cannot increase further. In addition, the maximum current increases as the intensity of the incident light increases, as you might expect,

---

[3]Photoelectrons are not different from other electrons. They are given this name solely because of their ejection from a metal by light in the photoelectric effect.

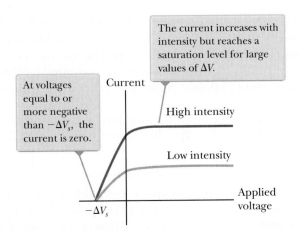

Figure 40.10 Photoelectric current versus applied potential difference for two light intensities.

The current increases with intensity but reaches a saturation level for large values of ΔV.

At voltages equal to or more negative than $-\Delta V_s$, the current is zero.

Current

High intensity

Low intensity

Applied voltage

$-\Delta V_s$

because more electrons are ejected by the higher-intensity light. Finally, when ΔV is negative—that is, when the battery in the circuit is reversed to make plate E positive and plate C negative—the current drops because many of the photoelectrons emitted from E are repelled by the now negative plate C. In this situation, only those photoelectrons having a kinetic energy greater than $e|\Delta V|$ reach plate C, where e is the magnitude of the charge on the electron. When ΔV is equal to or more negative than $-\Delta V_s$, where ΔV_s is the **stopping potential,** no photoelectrons reach C and the current is zero.

Let's model the combination of the electric field between the plates and an electron ejected from plate E as an isolated system. Suppose this electron stops just as it reaches plate C. Because the system is isolated, the appropriate reduction of Equation 8.2 is

$$\Delta K + \Delta U = 0$$

where the initial configuration is at the instant the electron leaves the metal with kinetic energy K_i and the final configuration is when the electron stops just before touching plate C. If we define the electric potential energy of the system in the initial configuration to be zero, we have

$$(0 - K_i) + [(q)(\Delta V) - 0] = 0 \quad \rightarrow \quad K_i = q\Delta V = -e\Delta V$$

Now suppose the potential difference ΔV is increased in the negative direction just until the current is zero at $\Delta V = -\Delta V_s$. In this case, the electron that stops immediately before reaching plate C has the maximum possible kinetic energy upon leaving the metal surface. The previous equation can then be written as

$$K_{\max} = e\,\Delta V_s \qquad\qquad \textbf{(40.8)}$$

This equation allows us to measure $K_{\max}$ experimentally by determining the magnitude of the voltage ΔV_s at which the current drops to zero.

Several features of the photoelectric effect are listed below. For each feature, we compare the predictions made by a classical approach, using the wave model for light, with the experimental results.

1. Dependence of photoelectron kinetic energy on light intensity

   Classical prediction: Electrons should absorb energy continuously from the electromagnetic waves. As the light intensity incident on a metal is increased, energy should be transferred into the metal at a higher rate and the electrons should be ejected with more kinetic energy.

   Experimental result: The maximum kinetic energy of photoelectrons is *independent* of light intensity as shown in Figure 40.10 with both curves falling to zero at the *same* negative voltage. (According to Equation 40.8, the maximum kinetic energy is proportional to the stopping potential.)

2. Time interval between incidence of light and ejection of photoelectrons
   Classical prediction: At low light intensities, a measurable time interval should pass between the instant the light is turned on and the time an electron is ejected from the metal. This time interval is required for the electron to absorb the incident radiation before it acquires enough energy to escape from the metal.
   Experimental result: Electrons are emitted from the surface of the metal almost *instantaneously* (less than 10^{-9} s after the surface is illuminated), even at very low light intensities.

3. Dependence of ejection of electrons on light frequency
   Classical prediction: Electrons should be ejected from the metal at any incident light frequency, as long as the light intensity is high enough, because energy is transferred to the metal regardless of the incident light frequency.
   Experimental result: No electrons are emitted if the incident light frequency falls below some **cutoff frequency** f_c, whose value is characteristic of the material being illuminated. No electrons are ejected below this cutoff frequency *regardless* of the light intensity.

4. Dependence of photoelectron kinetic energy on light frequency
   Classical prediction: There should be *no* relationship between the frequency of the light and the electron kinetic energy. The kinetic energy should be related to the intensity of the light.
   Experimental result: The maximum kinetic energy of the photoelectrons increases with increasing light frequency.

For these features, experimental results contradict *all four* classical predictions. A successful explanation of the photoelectric effect was given by Einstein in 1905, the same year he published his special theory of relativity. As part of a general paper on electromagnetic radiation, for which he received a Nobel Prize in Physics in 1921, Einstein extended Planck's concept of quantization to electromagnetic waves as mentioned in Section 40.1. Einstein assumed light (or any other electromagnetic wave) of frequency f from *any* source can be considered a stream of quanta. Today we call these quanta **photons.** Each photon has an energy E given by Equation 40.5, $E = hf$, and each moves in a vacuum at the speed of light c, where $c = 3.00 \times 10^8$ m/s.

Quick Quiz 40.2 While standing outdoors one evening, you are exposed to the following four types of electromagnetic radiation: yellow light from a sodium street lamp, radio waves from an AM radio station, radio waves from an FM radio station, and microwaves from an antenna of a communications system. Rank these types of waves in terms of photon energy from highest to lowest.

Let us organize Einstein's model for the photoelectric effect using the properties of structural models:

1. *Physical components:*
   We imagine the system to consist of two physical components: (1) an electron that is to be ejected by an incoming photon and (2) the remainder of the metal.

2. *Behavior of the components:*
   (a) In Einstein's model, a photon of the incident light gives *all* its energy hf to a *single* electron in the metal. Therefore, the absorption of energy by the electrons is not a continuous process as envisioned in the wave model, but rather a discontinuous process in which energy is delivered to the electrons in bundles. The energy transfer is accomplished via a one-photon/one-electron event.[4]

---

[4]In principle, two photons could combine to provide an electron with their combined energy. That is highly improbable, however, without the high intensity of radiation available from very strong lasers.

(b) We can describe the time evolution of the system by applying the non-isolated system model for energy over a time interval that includes the absorption of one photon and the ejection of the corresponding electron. Energy is transferred into the system by electromagnetic radiation, the photon. The system has two types of energy: the potential energy of the metal–electron system and the kinetic energy of the ejected electron. Therefore, we can write the conservation of energy equation (Eq. 8.2) as

$$\Delta K + \Delta U = T_{ER} \tag{40.9}$$

The energy transfer into the system is that of the photon, $T_{ER} = hf$. During the process, the kinetic energy of the electron increases from zero to its final value, which we assume to be the maximum possible value K_{max}. The potential energy of the system increases because the electron is pulled away from the metal to which it is attracted. We define the potential energy of the system when the electron is outside the metal as zero. The potential energy of the system when the electron is in the metal is $U = -\phi$, where ϕ is called the **work function** of the metal. The work function represents the minimum energy with which an electron is bound in the metal and is on the order of a few electron volts. Table 40.1 lists selected values. The increase in potential energy of the system when the electron is removed from the metal is the work function ϕ. Substituting these energies into Equation 40.9, we have

$$(K_{max} - 0) + [0 - (-\phi)] = hf$$
$$K_{max} + \phi = hf \tag{40.10}$$

If the electron makes collisions with other electrons or metal ions as it is being ejected, some of the incoming energy is transferred to the metal and the electron is ejected with less kinetic energy than K_{max}.

The prediction made by Einstein is an equation for the maximum kinetic energy of an ejected electron as a function of frequency of the illuminating radiation. This equation can be found by rearranging Equation 40.10:

$$K_{max} = hf - \phi \tag{40.11}$$

◀ Photoelectric effect equation

With Einstein's structural model, one can explain the observed features of the photoelectric effect that cannot be understood using classical concepts:

1. **Dependence of photoelectron kinetic energy on light intensity**

   Equation 40.11 shows that K_{max} is independent of the light intensity. The maximum kinetic energy of any one electron, which equals $hf - \phi$, depends only on the light frequency and the work function. If the light intensity is doubled, the number of photons arriving per unit time is doubled, which doubles the rate at which photoelectrons are emitted. The maximum kinetic energy of any one photoelectron, however, is unchanged.

2. **Time interval between incidence of light and ejection of photoelectrons**

   Near-instantaneous emission of electrons is consistent with the photon model of light. The incident energy appears in small packets, and there is a one-to-one interaction between photons and electrons. If the incident light has very low intensity, there are very few photons arriving per unit time interval; each photon, however, can have sufficient energy to eject an electron immediately.

3. **Dependence of ejection of electrons on light frequency**

   Because the photon must have energy greater than the work function ϕ to eject an electron, the photoelectric effect cannot be observed below a

Table 40.1 Work Functions of Selected Metals

Metal	ϕ (eV)
Na	2.46
Al	4.08
Fe	4.50
Cu	4.70
Zn	4.31
Ag	4.73
Pt	6.35
Pb	4.14

Note: Values are typical for metals listed. Actual values may vary depending on whether the metal is a single crystal or polycrystalline. Values may also depend on the face from which electrons are ejected from crystalline metals. Furthermore, different experimental procedures may produce differing values.

Figure 40.11 A plot of K_{max} for photoelectrons versus frequency of incident light in a typical photoelectric effect experiment.

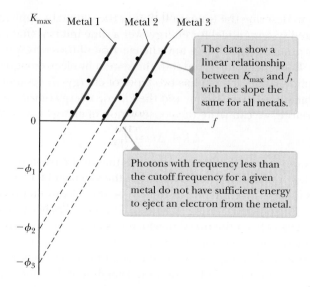

certain cutoff frequency. If the energy of an incoming photon does not satisfy this requirement, an electron cannot be ejected from the surface, even though many photons per unit time are incident on the metal in a very intense light beam.

4. Dependence of photoelectron kinetic energy on light frequency

A photon of higher frequency carries more energy and therefore ejects a photoelectron with more kinetic energy than does a photon of lower frequency.

Einstein's model predicts a linear relationship (Eq. 40.11) between the maximum electron kinetic energy K_{max} and the light frequency f. Experimental observation of a linear relationship between K_{max} and f would be a final confirmation of Einstein's theory. Indeed, such a linear relationship was observed experimentally within a few years of Einstein's theory and is sketched in Figure 40.11. The slope of the lines in such a plot is Planck's constant h. The intercept on the horizontal axis gives the cutoff frequency below which no photoelectrons are emitted. The cutoff frequency is related to the work function through the relationship $f_c = \phi/h$. The cutoff frequency corresponds to a **cutoff wavelength** λ_c, where

Cutoff wavelength ▶

$$\lambda_c = \frac{c}{f_c} = \frac{c}{\phi/h} = \frac{hc}{\phi} \tag{40.12}$$

and c is the speed of light. Wavelengths greater than λ_c incident on a material having a work function ϕ do not result in the emission of photoelectrons.

The combination hc in Equation 40.12 often occurs when relating a photon's energy to its wavelength. A common shortcut when solving problems is to express this combination in useful units according to the following approximation:

$$hc = 1\,240 \text{ eV} \cdot \text{nm}$$

One of the first practical uses of the photoelectric effect was as the detector in a camera's light meter. Light reflected from the object to be photographed strikes a photoelectric surface in the meter, causing it to emit photoelectrons that then pass through a sensitive ammeter. The magnitude of the current in the ammeter depends on the light intensity.

The phototube, another early application of the photoelectric effect, acts much like a switch in an electric circuit. It produces a current in the circuit when light of sufficiently high frequency falls on a metal plate in the phototube, but produces no current in the dark. Phototubes were used in burglar alarms and in the detection of the soundtrack on motion picture film. Modern semiconductor devices have now replaced older devices based on the photoelectric effect.

Today, the photoelectric effect is used in the operation of photomultiplier tubes. Figure 40.12 shows the structure of such a device. A photon striking the photocathode ejects an electron by means of the photoelectric effect. This electron accelerates across the potential difference between the photocathode and the first *dynode*, shown as being at +200 V relative to the photocathode in Figure 40.12. This high-energy electron strikes the dynode and ejects several more electrons. The same process is repeated through a series of dynodes at ever higher potentials until an electrical pulse is produced as millions of electrons strike the last dynode. The tube is therefore called a *multiplier*: one photon at the input has resulted in millions of electrons at the output.

The photomultiplier tube is used in nuclear detectors to detect photons produced by the interaction of energetic charged particles or gamma rays with certain materials. It is also used in astronomy in a technique called *photoelectric photometry*. In that technique, the light collected by a telescope from a single star is allowed to fall on a photomultiplier tube for a time interval. The tube measures the total energy transferred by light during the time interval, which can then be converted to a luminosity of the star.

The photomultiplier tube is being replaced in many astronomical observations with a *charge-coupled device* (CCD), which is the same device used in a digital camera (Section 36.6). Half of the 2009 Nobel Prize in Physics was awarded to Willard S. Boyle (b. 1924) and George E. Smith (b. 1930) for their 1969 invention of the charge-coupled device. In a CCD, an array of pixels is formed on the silicon surface of an integrated circuit (Section 43.7). When the surface is exposed to light from an astronomical scene through a telescope or a terrestrial scene through a digital camera, electrons generated by the photoelectric effect are caught in "traps" beneath the surface. The number of electrons is related to the intensity of the light striking the surface. A signal processor measures the number of electrons associated with each pixel and converts this information into a digital code that a computer can use to reconstruct and display the scene.

The *electron bombardment CCD camera* allows higher sensitivity than a conventional CCD. In this device, electrons ejected from a photocathode by the photoelectric effect are accelerated through a high voltage before striking a CCD array. The higher energy of the electrons results in a very sensitive detector of low-intensity radiation.

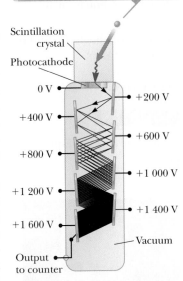

An incoming particle enters the scintillation crystal, where a collision results in a photon. The photon strikes the photocathode, which emits an electron by the photoelectric effect.

Figure 40.12 The multiplication of electrons in a photomultiplier tube.

Quick Quiz 40.3 Consider one of the curves in Figure 40.10. Suppose the intensity of the incident light is held fixed but its frequency is increased. Does the stopping potential in Figure 40.10 **(a)** remain fixed, **(b)** move to the right, or **(c)** move to the left?

Quick Quiz 40.4 Suppose classical physicists had the idea of plotting K_{max} versus f as in Figure 40.11. Draw a graph of what the expected plot would look like, based on the wave model for light.

Example 40.3	**The Photoelectric Effect for Sodium**

A sodium surface is illuminated with light having a wavelength of 300 nm. As indicated in Table 40.1, the work function for sodium metal is 2.46 eV.

(A) Find the maximum kinetic energy of the ejected photoelectrons.

SOLUTION

Conceptualize Imagine a photon striking the metal surface and ejecting an electron. The electron with the maximum energy is one near the surface that experiences no interactions with other particles in the metal that would reduce its energy on its way out of the metal.

continued

▶ **40.3** c o n t i n u e d

Categorize We evaluate the results using equations developed in this section, so we categorize this example as a substitution problem.

Find the energy of each photon in the illuminating light beam from Equation 40.5:

$$E = hf = \frac{hc}{\lambda}$$

From Equation 40.11, find the maximum kinetic energy of an electron:

$$K_{max} = \frac{hc}{\lambda} - \phi = \frac{1\ 240\ \text{eV} \cdot \text{nm}}{300\ \text{nm}} - 2.46\ \text{eV} = \boxed{1.67\ \text{eV}}$$

(B) Find the cutoff wavelength λ_c for sodium.

SOLUTION

Calculate λ_c using Equation 40.12:

$$\lambda_c = \frac{hc}{\phi} = \frac{1\ 240\ \text{eV} \cdot \text{nm}}{2.46\ \text{eV}} = \boxed{504\ \text{nm}}$$

Arthur Holly Compton
American Physicist (1892–1962)
Compton was born in Wooster, Ohio, and attended Wooster College and Princeton University. He became the director of the laboratory at the University of Chicago, where experimental work concerned with sustained nuclear chain reactions was conducted. This work was of central importance to the construction of the first nuclear weapon. His discovery of the Compton effect led to his sharing of the 1927 Nobel Prize in Physics with Charles Wilson.

© Mary Evans Picture Library/Alamy

40.3 The Compton Effect

In 1919, Einstein concluded that a photon of energy E travels in a single direction and carries a momentum equal to $E/c = hf/c$. In 1923, Arthur Holly Compton (1892–1962) and Peter Debye (1884–1966) independently carried Einstein's idea of photon momentum further.

Prior to 1922, Compton and his coworkers had accumulated evidence showing that the classical wave theory of light failed to explain the scattering of x-rays from electrons. According to classical theory, electromagnetic waves of frequency f incident on electrons should have two effects: (1) radiation pressure (see Section 34.5) should cause the electrons to accelerate in the direction of propagation of the waves, and (2) the oscillating electric field of the incident radiation should set the electrons into oscillation at the apparent frequency f', where f' is the frequency in the frame of the moving electrons. This apparent frequency is different from the frequency f of the incident radiation because of the Doppler effect (see Section 17.4). Each electron first absorbs radiation as a moving particle and then reradiates as a moving particle, thereby exhibiting two Doppler shifts in the frequency of radiation.

Because different electrons move at different speeds after the interaction, depending on the amount of energy absorbed from the electromagnetic waves, the scattered wave frequency at a given angle to the incoming radiation should show a distribution of Doppler-shifted values. Contrary to this prediction, Compton's experiments showed that at a given angle only *one* frequency of radiation is observed. Compton and his coworkers explained these experiments by treating photons not as waves but rather as point-like particles having energy hf and momentum hf/c and by assuming the energy and momentum of the isolated system of the colliding photon–electron pair are conserved. Compton adopted a particle model for something that was well known as a wave, and today this scattering phenomenon is known as the **Compton effect.** Figure 40.13 shows the quantum picture of the collision between an individual x-ray photon of frequency f_0 and an electron. In the quantum model, the electron is scattered through an angle ϕ with respect to this direction as in a billiard-ball type of collision. (The symbol ϕ used here is an angle and is not to be confused with the work function, which was discussed in the preceding section.) Compare Figure 40.13 with the two-dimensional collision shown in Figure 9.11.

Figure 40.14 is a schematic diagram of the apparatus used by Compton. The x-rays, scattered from a carbon target, were diffracted by a rotating crystal spectrometer, and the intensity was measured with an ionization chamber that generated a current proportional to the intensity. The incident beam consisted of monochromatic x-rays of wavelength $\lambda_0 = 0.071$ nm. The experimental intensity-

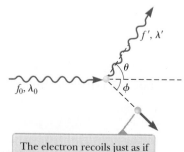

The electron recoils just as if struck by a classical particle, revealing the particle-like nature of the photon.

Figure 40.13 The quantum model for x-ray scattering from an electron.

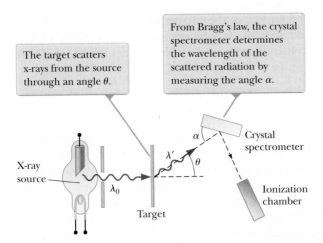

Figure 40.14 Schematic diagram of Compton's apparatus.

versus-wavelength plots observed by Compton for four scattering angles (corresponding to θ in Fig. 40.13) are shown in Figure 40.15. The graphs for the three nonzero angles show two peaks, one at λ_0 and one at $\lambda' > \lambda_0$. The shifted peak at λ' is caused by the scattering of x-rays from free electrons, which was predicted by Compton to depend on scattering angle as

$$\lambda' - \lambda_0 = \frac{h}{m_e c}(1 - \cos\theta) \qquad (40.13)$$

◀ **Compton shift equation**

where m_e is the mass of the electron. This expression is known as the **Compton shift equation** and correctly describes the positions of the peaks in Figure 40.15. The factor $h/m_e c$, called the **Compton wavelength** of the electron, has a currently accepted value of

$$\lambda_C = \frac{h}{m_e c} = 0.002\ 43\ \text{nm}$$

◀ **Compton wavelength**

The unshifted peak at λ_0 in Figure 40.15 is caused by x-rays scattered from electrons tightly bound to the target atoms. This unshifted peak also is predicted by Equation 40.13 if the electron mass is replaced with the mass of a carbon atom, which is approximately 23 000 times the mass of the electron. Therefore, there is a wavelength shift for scattering from an electron bound to an atom, but it is so small that it was undetectable in Compton's experiment.

Compton's measurements were in excellent agreement with the predictions of Equation 40.13. These results were the first to convince many physicists of the fundamental validity of quantum theory.

Quick Quiz 40.5 For any given scattering angle θ, Equation 40.13 gives the same value for the Compton shift for any wavelength. Keeping that in mind, for which of the following types of radiation is the fractional shift in wavelength at a given scattering angle the largest? **(a)** radio waves **(b)** microwaves **(c)** visible light **(d)** x-rays

Derivation of the Compton Shift Equation

We can derive the Compton shift equation by assuming the photon behaves like a particle and collides elastically with a free electron initially at rest as shown in Figure 40.13. The photon is treated as a particle having energy $E = hf = hc/\lambda$ and zero rest energy. We apply the isolated system analysis models for energy and momentum to the photon and the electron. In the scattering process, the total energy and total linear momentum of the system are conserved. Applying the isolated system model for energy to this process gives

$$\Delta K_{\text{photon}} + \Delta K_e = 0 \quad \rightarrow \quad \frac{hc}{\lambda_0} = \frac{hc}{\lambda'} + K_e$$

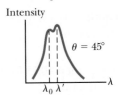
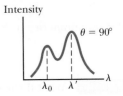
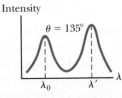

Figure 40.15 Scattered x-ray intensity versus wavelength for Compton scattering at $\theta = 0°$, 45°, 90°, and 135°.

where hc/λ_0 is the energy of the incident photon, hc/λ' is the energy of the scattered photon, and K_e is the kinetic energy of the recoiling electron. Because the electron may recoil at a speed comparable to that of light, we must use the relativistic expression $K_e = (\gamma - 1)m_e c^2$ (Eq. 39.23). Therefore,

$$\frac{hc}{\lambda_0} = \frac{hc}{\lambda'} + (\gamma - 1)m_e c^2 \tag{40.14}$$

where $\gamma = 1/\sqrt{1 - (u^2/c^2)}$ and u is the speed of the electron.

Next, let's apply the isolated system model for momentum to this collision, noting that the x and y components of momentum are each conserved independently. Equation 39.28 shows that the momentum of a photon has a magnitude $p = E/c$, and we know from Equation 40.5 that $E = hf$. Therefore, $p = hf/c$. Substituting λf for c (Eq. 34.20) in this expression gives $p = h/\lambda$. Because the relativistic expression for the momentum of the recoiling electron is $p_e = \gamma m_e u$ (Eq. 39.19), we obtain the following expressions for the x and y components of linear momentum, where the angles are as described in Figure 40.13:

$$x \text{ component: } \quad \frac{h}{\lambda_0} = \frac{h}{\lambda'} \cos\theta + \gamma m_e u \cos\phi \tag{40.15}$$

$$y \text{ component: } \quad 0 = \frac{h}{\lambda'} \sin\theta - \gamma m_e u \sin\phi \tag{40.16}$$

Eliminating u and ϕ from Equations 40.14 through 40.16 gives a single expression that relates the remaining three variables (λ', λ_0, and θ). After some algebra (see Problem 64), we obtain Equation 40.13.

Example 40.4   Compton Scattering at 45°

X-rays of wavelength $\lambda_0 = 0.200\,000$ nm are scattered from a block of material. The scattered x-rays are observed at an angle of 45.0° to the incident beam. Calculate their wavelength.

SOLUTION

Conceptualize Imagine the process in Figure 40.13, with the photon scattered at 45° to its original direction.

Categorize We evaluate the result using an equation developed in this section, so we categorize this example as a substitution problem.

Solve Equation 40.13 for the wavelength of the scattered x-ray:

$$(1) \quad \lambda' = \lambda_0 + \frac{h(1 - \cos\theta)}{m_e c}$$

Substitute numerical values:

$$\lambda' = 0.200\,000 \times 10^{-9}\,\text{m} + \frac{(6.626 \times 10^{-34}\,\text{J} \cdot \text{s})(1 - \cos 45.0°)}{(9.11 \times 10^{-31}\,\text{kg})(3.00 \times 10^{8}\,\text{m/s})}$$

$$= 0.200\,000 \times 10^{-9}\,\text{m} + 7.10 \times 10^{-13}\,\text{m} = \boxed{0.200\,710\,\text{nm}}$$

WHAT IF?   What if the detector is moved so that scattered x-rays are detected at an angle larger than 45°? Does the wavelength of the scattered x-rays increase or decrease as the angle θ increases?

Answer In Equation (1), if the angle θ increases, $\cos\theta$ decreases. Consequently, the factor $(1 - \cos\theta)$ increases. Therefore, the scattered wavelength increases.

We could also apply an energy argument to achieve this same result. As the scattering angle increases, more energy is transferred from the incident photon to the electron. As a result, the energy of the scattered photon decreases with increasing scattering angle. Because $E = hf$, the frequency of the scattered photon decreases, and because $\lambda = c/f$, the wavelength increases.

40.4 The Nature of Electromagnetic Waves

In Section 35.1, we introduced the notion of competing models of light: particles and waves. Let's expand on that earlier discussion. Phenomena such as the photoelectric effect and the Compton effect offer ironclad evidence that when light (or other forms of electromagnetic radiation) and matter interact, the light behaves as if it were composed of particles having energy hf and momentum h/λ. How can light be considered a photon (in other words, a particle) when we know it is a wave? On the one hand, we describe light in terms of photons having energy and momentum. On the other hand, light and other electromagnetic waves exhibit interference and diffraction effects, which are consistent only with a wave interpretation.

Which model is correct? Is light a wave or a particle? The answer depends on the phenomenon being observed. Some experiments can be explained either better or solely with the photon model, whereas others are explained either better or solely with the wave model. We must accept both models and admit that the true nature of light is not describable in terms of any single classical picture. The same light beam that can eject photoelectrons from a metal (meaning that the beam consists of photons) can also be diffracted by a grating (meaning that the beam is a wave). In other words, the particle model and the wave model of light complement each other.

The success of the particle model of light in explaining the photoelectric effect and the Compton effect raises many other questions. If light is a particle, what is the meaning of the "frequency" and "wavelength" of the particle, and which of these two properties determines its energy and momentum? Is light *simultaneously* a wave and a particle? Although photons have no rest energy (a nonobservable quantity because a photon cannot be at rest), is there a simple expression for the *effective mass* of a moving photon? If photons have effective mass, do they experience gravitational attraction? What is the spatial extent of a photon, and how does an electron absorb or scatter one photon? Some of these questions can be answered, but others demand a view of atomic processes that is too pictorial and literal. Many of them stem from classical analogies such as colliding billiard balls and ocean waves breaking on a seashore. Quantum mechanics gives light a more flexible nature by treating the particle model and the wave model of light as both necessary and complementary. Neither model can be used exclusively to describe all properties of light. A complete understanding of the observed behavior of light can be attained only if the two models are combined in a complementary manner.

40.5 The Wave Properties of Particles

Students introduced to the dual nature of light often find the concept difficult to accept. In the world around us, we are accustomed to regarding such things as baseballs solely as particles and other things such as sound waves solely as forms of wave motion. Every large-scale observation can be interpreted by considering either a wave explanation or a particle explanation, but in the world of photons and electrons, such distinctions are not as sharply drawn.

Even more disconcerting is that, under certain conditions, the things we unambiguously call "particles" exhibit wave characteristics. In his 1923 doctoral dissertation, Louis de Broglie postulated that because photons have both wave and particle characteristics, perhaps all forms of matter have both properties. This highly revolutionary idea had no experimental confirmation at the time. According to de Broglie, electrons, just like light, have a dual particle–wave nature.

In Section 40.3, we found that the momentum of a photon can be expressed as

$$p = \frac{h}{\lambda}$$

SPL/Getty Images

Louis de Broglie
French Physicist (1892–1987)
De Broglie was born in Dieppe, France. At the Sorbonne in Paris, he studied history in preparation for what he hoped would be a career in the diplomatic service. The world of science is lucky he changed his career path to become a theoretical physicist. De Broglie was awarded the Nobel Prize in Physics in 1929 for his prediction of the wave nature of electrons.

This equation shows that the photon wavelength can be specified by its momentum: $\lambda = h/p$. De Broglie suggested that material particles of momentum p have a characteristic wavelength that is given by the *same expression*. Because the magnitude of the momentum of a particle of mass m and speed u is $p = mu$, the **de Broglie wavelength** of that particle is[5]

$$\lambda = \frac{h}{p} = \frac{h}{mu} \tag{40.17}$$

Furthermore, in analogy with photons, de Broglie postulated that particles obey the Einstein relation $E = hf$, where E is the total energy of the particle. The frequency of a particle is then

$$f = \frac{E}{h} \tag{40.18}$$

The dual nature of matter is apparent in Equations 40.17 and 40.18 because each contains both particle quantities (p and E) and wave quantities (λ and f).

The problem of understanding the dual nature of matter and radiation is conceptually difficult because the two models seem to contradict each other. This problem as it applies to light was discussed earlier. The **principle of complementarity** states that

> the wave and particle models of either matter or radiation complement each other.

Neither model can be used exclusively to describe matter or radiation adequately. Because humans tend to generate mental images based on their experiences from the everyday world, we use both descriptions in a complementary manner to explain any given set of data from the quantum world.

The Davisson–Germer Experiment

De Broglie's 1923 proposal that matter exhibits both wave and particle properties was regarded as pure speculation. If particles such as electrons had wave properties, under the correct conditions they should exhibit diffraction effects. Only three years later, C. J. Davisson (1881–1958) and L. H. Germer (1896–1971) succeeded in observing electron diffraction and measuring the wavelength of electrons. Their important discovery provided the first experimental confirmation of the waves proposed by de Broglie.

Interestingly, the intent of the initial Davisson–Germer experiment was not to confirm the de Broglie hypothesis. In fact, their discovery was made by accident (as is often the case). The experiment involved the scattering of low-energy electrons (approximately 54 eV) from a nickel target in a vacuum. During one experiment, the nickel surface was badly oxidized because of an accidental break in the vacuum system. After the target was heated in a flowing stream of hydrogen to remove the oxide coating, electrons scattered by it exhibited intensity maxima and minima at specific angles. The experimenters finally realized that the nickel had formed large crystalline regions upon heating and that the regularly spaced planes of atoms in these regions served as a diffraction grating for electrons. (See the discussion of diffraction of x-rays by crystals in Section 38.5.)

Shortly thereafter, Davisson and Germer performed more extensive diffraction measurements on electrons scattered from single-crystal targets. Their results showed conclusively the wave nature of electrons and confirmed the de Broglie relationship $p = h/\lambda$. In the same year, G. P. Thomson (1892–1975) of Scotland also observed electron diffraction patterns by passing electrons through very thin gold

Pitfall Prevention 40.3

What's Waving? If particles have wave properties, what's waving? You are familiar with waves on strings, which are very concrete. Sound waves are more abstract, but you are likely comfortable with them. Electromagnetic waves are even more abstract, but at least they can be described in terms of physical variables and electric and magnetic fields. In contrast, waves associated with particles are completely abstract and cannot be associated with a physical variable. In Chapter 41, we describe the wave associated with a particle in terms of probability.

---

[5]The de Broglie wavelength for a particle moving at *any* speed u is $\lambda = h/\gamma mu$, where $\gamma = [1 - (u^2/c^2)]^{-1/2}$.

foils. Diffraction patterns were subsequently observed in the scattering of helium atoms, hydrogen atoms, and neutrons. Hence, the wave nature of particles has been established in various ways.

Quick Quiz 40.6 An electron and a proton both moving at nonrelativistic speeds have the same de Broglie wavelength. Which of the following quantities are also the same for the two particles? **(a)** speed **(b)** kinetic energy **(c)** momentum **(d)** frequency

---

Example 40.5   Wavelengths for Microscopic and Macroscopic Objects

(A) Calculate the de Broglie wavelength for an electron ($m_e = 9.11 \times 10^{-31}$ kg) moving at 1.00×10^7 m/s.

SOLUTION

Conceptualize Imagine the electron moving through space. From a classical viewpoint, it is a particle under constant velocity. From the quantum viewpoint, the electron has a wavelength associated with it.

Categorize We evaluate the result using an equation developed in this section, so we categorize this example as a substitution problem.

Evaluate the de Broglie wavelength using Equation 40.17:

$$\lambda = \frac{h}{m_e u} = \frac{6.626 \times 10^{-34}\,\text{J}\cdot\text{s}}{(9.11 \times 10^{-31}\,\text{kg})(1.00 \times 10^7\,\text{m/s})} = \boxed{7.27 \times 10^{-11}\,\text{m}}$$

The wave nature of this electron could be detected by diffraction techniques such as those in the Davisson–Germer experiment.

(B) A rock of mass 50 g is thrown with a speed of 40 m/s. What is its de Broglie wavelength?

SOLUTION

Evaluate the de Broglie wavelength using Equation 40.17:

$$\lambda = \frac{h}{mu} = \frac{6.626 \times 10^{-34}\,\text{J}\cdot\text{s}}{(50 \times 10^{-3}\,\text{kg})(40\,\text{m/s})} = \boxed{3.3 \times 10^{-34}\,\text{m}}$$

This wavelength is much smaller than any aperture through which the rock could possibly pass. Hence, we could not observe diffraction effects, and as a result, the wave properties of large-scale objects cannot be observed.

The Electron Microscope

A practical device that relies on the wave characteristics of electrons is the **electron microscope.** A *transmission* electron microscope, used for viewing flat, thin samples, is shown in Figure 40.16 on page 1252. In many respects, it is similar to an optical microscope; the electron microscope, however, has a much greater resolving power because it can accelerate electrons to very high kinetic energies, giving them very short wavelengths. No microscope can resolve details that are significantly smaller than the wavelength of the waves used to illuminate the object. The shorter wavelengths of electrons gives an electron microscope a resolution that can be 1 000 times better than that from the visible light used in optical microscopes. As a result, an electron microscope with ideal lenses would be able to distinguish details approximately 1 000 times smaller than those distinguished by an optical microscope. (Electromagnetic radiation of the same wavelength as the electrons in an electron microscope is in the x-ray region of the spectrum.)

The electron beam in an electron microscope is controlled by electrostatic or magnetic deflection, which acts on the electrons to focus the beam and form an image. Rather than examining the image through an eyepiece as in an optical microscope, the viewer looks at an image formed on a monitor or other type of

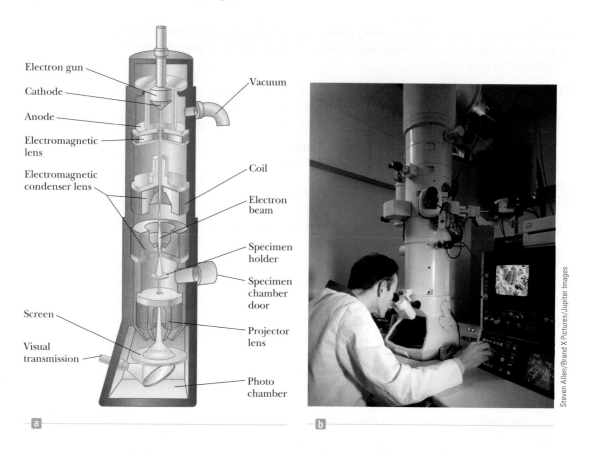

Figure 40.16 (a) Diagram of a transmission electron microscope for viewing a thinly sectioned sample. The "lenses" that control the electron beam are magnetic deflection coils. (b) An electron microscope in use.

Figure 40.17 A scanning electron microscope photograph shows significant detail of a cheese mite, *Tyrolichus casei*. The mite is so small, with a maximum length of 0.70 mm, that ordinary microscopes do not reveal minute anatomical details.

display screen. Figure 40.17 shows the amazing detail available with an electron microscope.

40.6 A New Model: The Quantum Particle

Because in the past we considered the particle and wave models to be distinct, the discussions presented in previous sections may be quite disturbing. The notion that both light and material particles have both particle and wave properties does not fit with this distinction. Experimental evidence shows, however, that this conclusion is exactly what we must accept. The recognition of this dual nature leads to a new model, the **quantum particle,** which is a combination of the particle model introduced in Chapter 2 and the wave model discussed in Chapter 16. In this new model, entities have both particle and wave characteristics, and we must choose one appropriate behavior—particle or wave—to understand a particular phenomenon.

In this section, we shall explore this model in a way that might make you more comfortable with this idea. We shall do so by demonstrating that an entity that exhibits properties of a particle can be constructed from waves.

Let's first recall some characteristics of ideal particles and ideal waves. An ideal particle has zero size. Therefore, an essential feature of a particle is that it is *localized* in space. An ideal wave has a single frequency and is infinitely long as suggested by Figure 40.18a. Therefore, an ideal wave is *unlocalized* in space. A localized entity can be built from infinitely long waves as follows. Imagine drawing one wave along the *x* axis, with one of its crests located at $x = 0$, as at the top of Figure 40.18b. Now draw a second wave, of the same amplitude but a different frequency, with one of its crests also at $x = 0$. As a result of the superposition of these two waves, *beats* exist as the waves are alternately in phase and out of phase. (Beats were discussed in Section 18.7.) The bottom curve in Figure 40.18b shows the results of superposing these two waves.

Notice that we have already introduced some localization by superposing the two waves. A single wave has the same amplitude everywhere in space; no point in space is any different from any other point. By adding a second wave, however, there is something different about the in-phase points compared with the out-of-phase points.

Now imagine that more and more waves are added to our original two, each new wave having a new frequency. Each new wave is added so that one of its crests is at $x = 0$ with the result that all the waves add constructively at $x = 0$. When we add a large number of waves, the probability of a positive value of a wave function at any point $x \neq 0$ is equal to the probability of a negative value, and there is destructive interference *everywhere* except near $x = 0$, where all the crests are superposed. The result is shown in Figure 40.19. The small region of constructive interference is called a **wave packet.** This localized region of space is different from all other regions. We can identify the wave packet as a particle because it has the localized nature of a particle! The location of the wave packet corresponds to the particle's position.

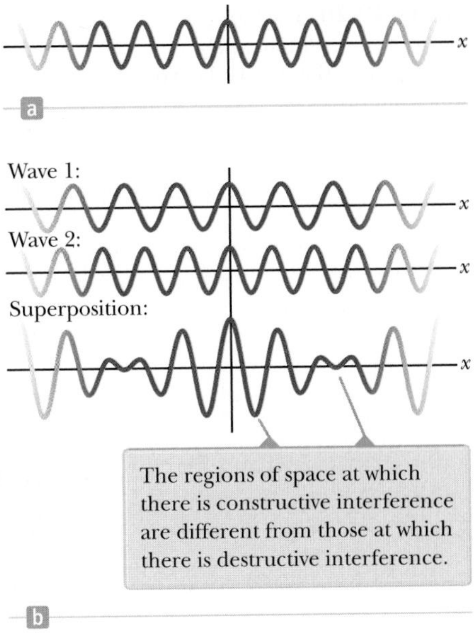

The localized nature of this entity is the *only* characteristic of a particle that was generated with this process. We have not addressed how the wave packet might achieve such particle characteristics as mass, electric charge, and spin. Therefore, you may not be completely convinced that we have built a particle. As further evidence that the wave packet can represent the particle, let's show that the wave packet has another characteristic of a particle.

To simplify the mathematical representation, we return to our combination of two waves. Consider two waves with equal amplitudes but different angular frequencies ω_1 and ω_2. We can represent the waves mathematically as

$$y_1 = A \cos (k_1 x - \omega_1 t) \quad \text{and} \quad y_2 = A \cos (k_2 x - \omega_2 t)$$

where, as in Chapter 16, $k = 2\pi/\lambda$ and $\omega = 2\pi f$. Using the superposition principle, let's add the waves:

$$y = y_1 + y_2 = A \cos (k_1 x - \omega_1 t) + A \cos (k_2 x - \omega_2 t)$$

It is convenient to write this expression in a form that uses the trigonometric identity

$$\cos a + \cos b = 2 \cos \left(\frac{a - b}{2} \right) \cos \left(\frac{a + b}{2} \right)$$

Letting $a = k_1 x - \omega_1 t$ and $b = k_2 x - \omega_2 t$ gives

$$y = 2A \cos \left[\frac{(k_1 x - \omega_1 t) - (k_2 x - \omega_2 t)}{2} \right] \cos \left[\frac{(k_1 x - \omega_1 t) + (k_2 x - \omega_2 t)}{2} \right]$$

$$y = \left[2A \cos \left(\frac{\Delta k}{2} x - \frac{\Delta \omega}{2} t \right) \right] \cos \left(\frac{k_1 + k_2}{2} x - \frac{\omega_1 + \omega_2}{2} t \right) \quad \textbf{(40.19)}$$

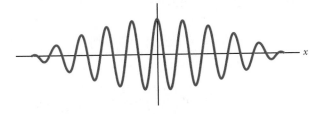

Figure 40.18 (a) An idealized wave of an exact single frequency is the same throughout space and time. (b) If two ideal waves with slightly different frequencies are combined, beats result (Section 18.7).

a

Wave 1:

Wave 2:

Superposition:

The regions of space at which there is constructive interference are different from those at which there is destructive interference.

b

Figure 40.19 If a large number of waves are combined, the result is a wave packet, which represents a particle.

Figure 40.20 The beat pattern of Figure 40.18b, with an envelope function (dashed curve) superimposed.

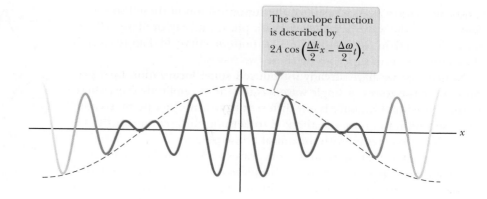

The envelope function is described by $2A \cos\left(\frac{\Delta k}{2}x - \frac{\Delta\omega}{2}t\right)$.

where $\Delta k = k_1 - k_2$ and $\Delta\omega = \omega_1 - \omega_2$. The second cosine factor represents a wave with a wave number and frequency that are equal to the averages of the values for the individual waves.

In Equation 40.19, the factor in square brackets represents the envelope of the wave as shown by the dashed curve in Figure 40.20. This factor also has the mathematical form of a wave. This envelope of the combination can travel through space with a different speed than the individual waves. As an extreme example of this possibility, imagine combining two identical waves moving in opposite directions. The two waves move with the same speed, but the envelope has a speed of *zero* because we have built a standing wave, which we studied in Section 18.2.

For an individual wave, the speed is given by Equation 16.11,

Phase speed of a wave ▶ in a wave packet

$$v_{\text{phase}} = \frac{\omega}{k} \tag{40.20}$$

This speed is called the **phase speed** because it is the rate of advance of a crest on a single wave, which is a point of fixed phase. Equation 40.20 can be interpreted as follows: the phase speed of a wave is the ratio of the coefficient of the time variable t to the coefficient of the space variable x in the equation representing the wave, $y = A \cos(kx - \omega t)$.

The factor in brackets in Equation 40.19 is of the form of a wave, so it moves with a speed given by this same ratio:

$$v_g = \frac{\text{coefficient of time variable } t}{\text{coefficient of space variable } x} = \frac{(\Delta\omega/2)}{(\Delta k/2)} = \frac{\Delta\omega}{\Delta k}$$

The subscript g on the speed indicates that it is commonly called the **group speed**, or the speed of the wave packet (the *group* of waves) we have built. We have generated this expression for a simple addition of two waves. When a large number of waves are superposed to form a wave packet, this ratio becomes a derivative:

Group speed of a wave packet ▶

$$v_g = \frac{d\omega}{dk} \tag{40.21}$$

Multiplying the numerator and the denominator by $\hbar$, where $\hbar = h/2\pi$, gives

$$v_g = \frac{\hbar\, d\omega}{\hbar\, dk} = \frac{d(\hbar\omega)}{d(\hbar k)} \tag{40.22}$$

Let's look at the terms in the parentheses of Equation 40.22 separately. For the numerator,

$$\hbar\omega = \frac{h}{2\pi}(2\pi f) = hf = E$$

For the denominator,

$$\hbar k = \frac{h}{2\pi}\left(\frac{2\pi}{\lambda}\right) = \frac{h}{\lambda} = p$$

Therefore, Equation 40.22 can be written as

$$v_g = \frac{d(\hbar\omega)}{d(\hbar k)} = \frac{dE}{dp} \qquad \textbf{(40.23)}$$

Because we are exploring the possibility that the envelope of the combined waves represents the particle, consider a free particle moving with a speed u that is small compared with the speed of light. The energy of the particle is its kinetic energy:

$$E = \tfrac{1}{2}mu^2 = \frac{p^2}{2m}$$

Differentiating this equation with respect to p gives

$$v_g = \frac{dE}{dp} = \frac{d}{dp}\left(\frac{p^2}{2m}\right) = \frac{1}{2m}(2p) = u \qquad \textbf{(40.24)}$$

Therefore, the group speed of the wave packet is identical to the speed of the particle that it is modeled to represent, giving us further confidence that the wave packet is a reasonable way to build a particle.

Quick Quiz 40.7 As an analogy to wave packets, consider an "automobile packet" that occurs near the scene of an accident on a freeway. The phase speed is analogous to the speed of individual automobiles as they move through the backup caused by the accident. The group speed can be identified as the speed of the leading edge of the packet of cars. For the automobile packet, is the group speed **(a)** the same as the phase speed, **(b)** less than the phase speed, or **(c)** greater than the phase speed?

40.7 The Double-Slit Experiment Revisited

Wave–particle duality is now a firmly accepted concept reinforced by experimental results, including those of the Davisson–Germer experiment. As with the postulates of special relativity, however, this concept often leads to clashes with familiar thought patterns we hold from everyday experience.

One way to crystallize our ideas about the electron's wave–particle duality is through an experiment in which electrons are fired at a double slit. Consider a parallel beam of mono-energetic electrons incident on a double slit as in Figure 40.21. Let's assume the slit widths are small compared with the electron wavelength so that we need not worry about diffraction maxima and minima as discussed for light in Section 38.2. An electron detector screen is positioned far from the slits at a distance much greater than d, the separation distance of the slits. If the detector screen collects electrons for a long enough time, we find a typical wave interference pattern for the counts per minute, or probability of arrival of electrons. Such an interference

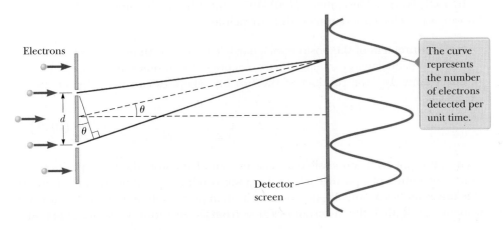

Figure 40.21 Electron interference. The slit separation d is much greater than the individual slit widths and much less than the distance between the slit and the detector screen.

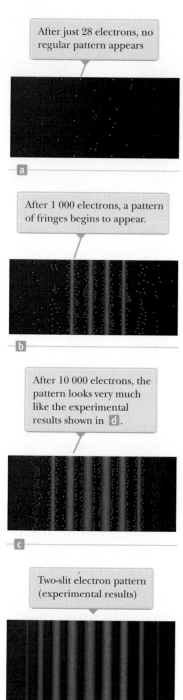

After just 28 electrons, no regular pattern appears

a

After 1 000 electrons, a pattern of fringes begins to appear.

b

After 10 000 electrons, the pattern looks very much like the experimental results shown in **d**.

c

Two-slit electron pattern (experimental results)

d

Figure 40.22 (a)–(c) Computer-simulated interference patterns for a small number of electrons incident on a double slit. (d) Computer simulation of a double-slit interference pattern produced by many electrons.

pattern would not be expected if the electrons behaved as classical particles, giving clear evidence that electrons are interfering, a distinct wave-like behavior.

If we measure the angles θ at which the maximum intensity of electrons arrives at the detector screen in Figure 40.21, we find they are described by exactly the same equation as that for light, $d \sin \theta = m\lambda$ (Eq. 37.2), where m is the order number and λ is the electron wavelength. Therefore, the dual nature of the electron is clearly shown in this experiment: the electrons are detected as particles at a localized spot on the detector screen at some instant of time, but the probability of arrival at that spot is determined by finding the intensity of two interfering waves.

Now imagine that we lower the beam intensity so that one electron at a time arrives at the double slit. It is tempting to assume the electron goes through either slit 1 or slit 2. You might argue that there are no interference effects because there is not a second electron going through the other slit to interfere with the first. This assumption places too much emphasis on the particle model of the electron, however. The interference pattern is still observed if the time interval for the measurement is sufficiently long for many electrons to pass one at a time through the slits and arrive at the detector screen! This situation is illustrated by the computer-simulated patterns in Figure 40.22 where the interference pattern becomes clearer as the number of electrons reaching the detector screen increases. Hence, our assumption that the electron is localized and goes through only one slit when both slits are open must be wrong (a painful conclusion!).

To interpret these results, we are forced to conclude that an electron interacts with both slits *simultaneously*. If you try to determine experimentally which slit the electron goes through, the act of measuring destroys the interference pattern. It is impossible to determine which slit the electron goes through. In effect, we can say only that the electron passes through *both* slits! The same arguments apply to photons.

If we restrict ourselves to a pure particle model, it is an uncomfortable notion that the electron can be present at both slits at once. From the quantum particle model, however, the particle can be considered to be built from waves that exist throughout space. Therefore, the wave components of the electron are present at both slits at the same time, and this model leads to a more comfortable interpretation of this experiment.

40.8 The Uncertainty Principle

Whenever one measures the position or velocity of a particle at any instant, experimental uncertainties are built into the measurements. According to classical mechanics, there is no fundamental barrier to an ultimate refinement of the apparatus or experimental procedures. In other words, it is possible, in principle, to make such measurements with arbitrarily small uncertainty. Quantum theory predicts, however, that it is fundamentally impossible to make simultaneous measurements of a particle's position and momentum with infinite accuracy.

In 1927, Werner Heisenberg (1901–1976) introduced this notion, which is now known as the **Heisenberg uncertainty principle:**

> If a measurement of the position of a particle is made with uncertainty Δx and a simultaneous measurement of its x component of momentum is made with uncertainty Δp_x, the product of the two uncertainties can never be smaller than $\hbar/2$:
>
> $$\Delta x \, \Delta p_x \geq \frac{\hbar}{2} \qquad \text{(40.25)}$$

That is, it is physically impossible to measure simultaneously the exact position and exact momentum of a particle. Heisenberg was careful to point out that the inescapable uncertainties Δx and Δp_x do not arise from imperfections in practical measuring instruments. Rather, the uncertainties arise from the quantum structure of matter.

To understand the uncertainty principle, imagine that a particle has a single wavelength that is known *exactly*. According to the de Broglie relation, $\lambda = h/p$, we would therefore know the momentum to be precisely $p = h/\lambda$. In reality, a single-wavelength wave would exist throughout space. Any region along this wave is the same as any other region (Fig. 40.18a). Suppose we ask, Where is the particle this wave represents? No special location in space along the wave could be identified with the particle; all points along the wave are the same. Therefore, we have *infinite* uncertainty in the position of the particle, and we know nothing about its location. Perfect knowledge of the particle's momentum has cost us all information about its location.

In comparison, now consider a particle whose momentum is uncertain so that it has a range of possible values of momentum. According to the de Broglie relation, the result is a range of wavelengths. Therefore, the particle is not represented by a single wavelength, but rather by a combination of wavelengths within this range. This combination forms a wave packet as we discussed in Section 40.6 and illustrated in Figure 40.19. If you were asked to determine the location of the particle, you could only say that it is somewhere in the region defined by the wave packet because there is a distinct difference between this region and the rest of space. Therefore, by losing some information about the momentum of the particle, we have gained information about its position.

If you were to lose *all* information about the momentum, you would be adding together waves of all possible wavelengths, resulting in a wave packet of zero length. Therefore, if you know nothing about the momentum, you know exactly where the particle is.

The mathematical form of the uncertainty principle states that the product of the uncertainties in position and momentum is always larger than some minimum value. This value can be calculated from the types of arguments discussed above, and the result is the value of $\hbar/2$ in Equation 40.25.

Another form of the uncertainty principle can be generated by reconsidering Figure 40.19. Imagine that the horizontal axis is time rather than spatial position x. We can then make the same arguments that were made about knowledge of wavelength and position in the time domain. The corresponding variables would be frequency and time. Because frequency is related to the energy of the particle by $E = hf$, the uncertainty principle in this form is

$$\Delta E\,\Delta t \geq \frac{\hbar}{2} \tag{40.26}$$

The form of the uncertainty principle given in Equation 40.26 suggests that energy conservation can appear to be violated by an amount ΔE as long as it is only for a short time interval Δt consistent with that equation. We shall use this notion to estimate the rest energies of particles in Chapter 46.

Quick Quiz 40.8 A particle's location is measured and specified as being exactly at $x = 0$, with *zero* uncertainty in the x direction. How does that location affect the uncertainty of its velocity component in the y direction? **(a)** It does not affect it. **(b)** It makes it infinite. **(c)** It makes it zero.

Werner Heisenberg
German Theoretical Physicist
(1901–1976)
Heisenberg obtained his Ph.D. in 1923 at the University of Munich. While other physicists tried to develop physical models of quantum phenomena, Heisenberg developed an abstract mathematical model called *matrix mechanics*. The more widely accepted physical models were shown to be equivalent to matrix mechanics. Heisenberg made many other significant contributions to physics, including his famous uncertainty principle for which he received a Nobel Prize in Physics in 1932, the prediction of two forms of molecular hydrogen, and theoretical models of the nucleus.

Pitfall Prevention 40.4
The Uncertainty Principle Some students incorrectly interpret the uncertainty principle as meaning that a measurement interferes with the system. For example, if an electron is observed in a hypothetical experiment using an optical microscope, the photon used to see the electron collides with it and makes it move, giving it an uncertainty in momentum. This scenario does *not* represent the basis of the uncertainty principle. The uncertainty principle is independent of the measurement process and is based on the wave nature of matter.

Example 40.6   **Locating an Electron**

The speed of an electron is measured to be 5.00×10^3 m/s to an accuracy of 0.003 00%. Find the minimum uncertainty in determining the position of this electron.

SOLUTION

Conceptualize The fractional value given for the accuracy of the electron's speed can be interpreted as the fractional uncertainty in its momentum. This uncertainty corresponds to a minimum uncertainty in the electron's position through the uncertainty principle.

continued

▶ **40.6** continued

Categorize We evaluate the result using concepts developed in this section, so we categorize this example as a substitution problem.

Assume the electron is moving along the x axis and find the uncertainty in p_x, letting f represent the accuracy of the measurement of its speed:

$$\Delta p_x = m\,\Delta v_x = mf v_x$$

Solve Equation 40.25 for the uncertainty in the electron's position and substitute numerical values:

$$\Delta x \geq \frac{\hbar}{2\,\Delta p_x} = \frac{\hbar}{2mf v_x} = \frac{1.055 \times 10^{-34}\,\text{J} \cdot \text{s}}{2(9.11 \times 10^{-31}\,\text{kg})(0.000\,030\,0)(5.00 \times 10^3\,\text{m/s})}$$

$$= 3.86 \times 10^{-4}\,\text{m} = \boxed{0.386\,\text{mm}}$$

Example 40.7   The Line Width of Atomic Emissions

Atoms have quantized energy levels similar to those of Planck's oscillators, although the energy levels of an atom are usually not evenly spaced. When an atom makes a transition between states separated in energy by ΔE, energy is emitted in the form of a photon of frequency $f = \Delta E/h$. Although an excited atom can radiate at any time from $t = 0$ to $t = \infty$, the average time interval after excitation during which an atom radiates is called the **lifetime** τ. If $\tau = 1.0 \times 10^{-8}$ s, use the uncertainty principle to compute the line width Δf produced by this finite lifetime.

SOLUTION

Conceptualize The lifetime τ given for the excited state can be interpreted as the uncertainty Δt in the time at which the transition occurs. This uncertainty corresponds to a minimum uncertainty in the frequency of the radiated photon through the uncertainty principle.

Categorize We evaluate the result using concepts developed in this section, so we categorize this example as a substitution problem.

Use Equation 40.5 to relate the uncertainty in the photon's frequency to the uncertainty in its energy:

$$E = hf \;\rightarrow\; \Delta E = h\,\Delta f \;\rightarrow\; \Delta f = \frac{\Delta E}{h}$$

Use Equation 40.26 to substitute for the uncertainty in the photon's energy, giving the minimum value of Δf:

$$\Delta f \geq \frac{1}{h}\frac{\hbar}{2\,\Delta t} = \frac{1}{h}\frac{h/2\pi}{2\,\Delta t} = \frac{1}{4\pi\,\Delta t} = \frac{1}{4\pi\tau}$$

Substitute for the lifetime of the excited state:

$$\Delta f \geq \frac{1}{4\pi(1.0 \times 10^{-8}\,\text{s})} = \boxed{8.0 \times 10^6\,\text{Hz}}$$

WHAT IF? What if this same lifetime were associated with a transition that emits a radio wave rather than a visible light wave from an atom? Is the fractional line width $\Delta f/f$ larger or smaller than for the visible light?

Answer Because we are assuming the same lifetime for both transitions, Δf is independent of the frequency of radiation. Radio waves have lower frequencies than light waves, so the ratio $\Delta f/f$ will be larger for the radio waves. Assuming a light-wave frequency f of 6.00×10^{14} Hz, the fractional line width is

$$\frac{\Delta f}{f} = \frac{8.0 \times 10^6\,\text{Hz}}{6.00 \times 10^{14}\,\text{Hz}} = 1.3 \times 10^{-8}$$

This narrow fractional line width can be measured with a sensitive interferometer. Usually, however, temperature and pressure effects overshadow the natural line width and broaden the line through mechanisms associated with the Doppler effect and collisions.

Assuming a radio-wave frequency f of 94.7×10^6 Hz, the fractional line width is

$$\frac{\Delta f}{f} = \frac{8.0 \times 10^6\,\text{Hz}}{94.7 \times 10^6\,\text{Hz}} = 8.4 \times 10^{-2}$$

Therefore, for the radio wave, this same absolute line width corresponds to a fractional line width of more than 8%.

Summary

Concepts and Principles

The characteristics of **blackbody radiation** cannot be explained using classical concepts. Planck introduced the quantum concept and Planck's constant h when he assumed atomic oscillators existing only in discrete energy states were responsible for this radiation. In Planck's model, radiation is emitted in single quantized packets whenever an oscillator makes a transition between discrete energy states. The energy of a packet is

$$E = hf \qquad \textbf{(40.5)}$$

where f is the frequency of the oscillator. Einstein successfully extended Planck's quantum hypothesis to the standing waves of electromagnetic radiation in a cavity used in the blackbody radiation model.

The **photoelectric effect** is a process whereby electrons are ejected from a metal surface when light is incident on that surface. In Einstein's model, light is viewed as a stream of particles, or **photons,** each having energy $E = hf$, where h is Planck's constant and f is the frequency. The maximum kinetic energy of the ejected photoelectron is

$$K_{max} = hf - \phi \qquad \textbf{(40.11)}$$

where ϕ is the **work function** of the metal.

X-rays are scattered at various angles by electrons in a target. In such a scattering event, a shift in wavelength is observed for the scattered x-rays, a phenomenon known as the **Compton effect.** Classical physics does not predict the correct behavior in this effect. If the x-ray is treated as a photon, conservation of energy and linear momentum applied to the photon–electron collisions yields, for the Compton shift,

$$\lambda' - \lambda_0 = \frac{h}{m_e c}(1 - \cos\theta) \qquad \textbf{(40.13)}$$

where m_e is the mass of the electron, c is the speed of light, and θ is the scattering angle.

Light has a dual nature in that it has both wave and particle characteristics. Some experiments can be explained either better or solely by the particle model, whereas others can be explained either better or solely by the wave model.

Every object of mass m and momentum $p = mu$ has wave properties, with a **de Broglie wavelength** given by

$$\lambda = \frac{h}{p} = \frac{h}{mu} \qquad \textbf{(40.17)}$$

By combining a large number of waves, a single region of constructive interference, called a **wave packet,** can be created. The wave packet carries the characteristic of localization like a particle does, but it has wave properties because it is built from waves. For an individual wave in the wave packet, the **phase speed** is

$$v_{phase} = \frac{\omega}{k} \qquad \textbf{(40.20)}$$

For the wave packet as a whole, the **group speed** is

$$v_g = \frac{d\omega}{dk} \qquad \textbf{(40.21)}$$

For a wave packet representing a particle, the group speed can be shown to be the same as the speed of the particle.

The **Heisenberg uncertainty principle** states that if a measurement of the position of a particle is made with uncertainty Δx and a simultaneous measurement of its linear momentum is made with uncertainty Δp_x, the product of the two uncertainties is restricted to

$$\Delta x \, \Delta p_x \geq \frac{\hbar}{2} \qquad \textbf{(40.25)}$$

Another form of the uncertainty principle relates measurements of energy and time:

$$\Delta E \, \Delta t \geq \frac{\hbar}{2} \qquad \textbf{(40.26)}$$

Objective Questions   1. denotes answer available in *Student Solutions Manual/Study Guide*

1. Rank the wavelengths of the following quantum particles from the largest to the smallest. If any have equal wavelengths, display the equality in your ranking. (a) a photon with energy 3 eV (b) an electron with

kinetic energy 3 eV (c) a proton with kinetic energy 3 eV (d) a photon with energy 0.3 eV (e) an electron with momentum 3 eV/c

2. An x-ray photon is scattered by an originally stationary electron. Relative to the frequency of the incident photon, is the frequency of the scattered photon (a) lower, (b) higher, or (c) unchanged?

3. In a Compton scattering experiment, a photon of energy E is scattered from an electron at rest. After the scattering event occurs, which of the following statements is true? (a) The frequency of the photon is greater than E/h. (b) The energy of the photon is less than E. (c) The wavelength of the photon is less than hc/E. (d) The momentum of the photon increases. (e) None of those statements is true.

4. In a certain experiment, a filament in an evacuated lightbulb carries a current I_1 and you measure the spectrum of light emitted by the filament, which behaves as a black body at temperature T_1. The wavelength emitted with highest intensity (symbolized by λ_{max}) has the value λ_1. You then increase the potential difference across the filament by a factor of 8, and the current increases by a factor of 2. (i) After this change, what is the new value of the temperature of the filament? (a) $16T_1$ (b) $8T_1$ (c) $4T_1$ (d) $2T_1$ (e) still T_1 (ii) What is the new value of the wavelength emitted with highest intensity? (a) $4\lambda_1$ (b) $2\lambda_1$ (c) λ_1 (d) $\frac{1}{2}\lambda_1$ (e) $\frac{1}{4}\lambda_1$

5. Which of the following statements are true according to the uncertainty principle? More than one statement may be correct. (a) It is impossible to simultaneously determine both the position and the momentum of a particle along the same axis with arbitrary accuracy. (b) It is impossible to simultaneously determine both the energy and momentum of a particle with arbitrary accuracy. (c) It is impossible to determine a particle's energy with arbitrary accuracy in a finite amount of time. (d) It is impossible to measure the position of a particle with arbitrary accuracy in a finite amount of time. (e) It is impossible to simultaneously measure both the energy and position of a particle with arbitrary accuracy.

6. A monochromatic light beam is incident on a barium target that has a work function of 2.50 eV. If a potential difference of 1.00 V is required to turn back all the ejected electrons, what is the wavelength of the light

beam? (a) 355 nm (b) 497 nm (c) 744 nm (d) 1.42 pm (e) none of those answers

7. Which of the following is most likely to cause sunburn by delivering more energy to individual molecules in skin cells? (a) infrared light (b) visible light (c) ultraviolet light (d) microwaves (e) Choices (a) through (d) are equally likely.

8. Which of the following phenomena most clearly demonstrates the wave nature of electrons? (a) the photoelectric effect (b) blackbody radiation (c) the Compton effect (d) diffraction of electrons by crystals (e) none of those answers

9. What is the de Broglie wavelength of an electron accelerated from rest through a potential difference of 50.0 V? (a) 0.100 nm (b) 0.139 nm (c) 0.174 nm (d) 0.834 nm (e) none of those answers

10. A proton, an electron, and a helium nucleus all move at speed v. Rank their de Broglie wavelengths from largest to smallest.

11. Consider (a) an electron, (b) a photon, and (c) a proton, all moving in vacuum. Choose all correct answers for each question. (i) Which of the three possess rest energy? (ii) Which have charge? (iii) Which carry energy? (iv) Which carry momentum? (v) Which move at the speed of light? (vi) Which have a wavelength characterizing their motion?

12. An electron and a proton, moving in opposite directions, are accelerated from rest through the same potential difference. Which particle has the longer wavelength? (a) The electron does. (b) The proton does. (c) Both are the same. (d) Neither has a wavelength.

13. Which of the following phenomena most clearly demonstrates the particle nature of light? (a) diffraction (b) the photoelectric effect (c) polarization (d) interference (e) refraction

14. Both an electron and a proton are accelerated to the same speed, and the experimental uncertainty in the speed is the same for the two particles. The positions of the two particles are also measured. Is the minimum possible uncertainty in the electron's position (a) less than the minimum possible uncertainty in the proton's position, (b) the same as that for the proton, (c) more than that for the proton, or (d) impossible to tell from the given information?

Conceptual Questions

1. denotes answer available in *Student Solutions Manual/Study Guide*

1. The opening photograph for this chapter shows a filament of a lightbulb in operation. Look carefully at the last turns of wire at the upper and lower ends of the filament. Why are these turns dimmer than the others?

2. How does the Compton effect differ from the photoelectric effect?

3. If matter has a wave nature, why is this wave-like characteristic not observable in our daily experiences?

4. If the photoelectric effect is observed for one metal, can you conclude that the effect will also be observed

for another metal under the same conditions? Explain.

5. In the photoelectric effect, explain why the stopping potential depends on the frequency of light but not on the intensity.

6. Why does the existence of a cutoff frequency in the photoelectric effect favor a particle theory for light over a wave theory?

7. Which has more energy, a photon of ultraviolet radiation or a photon of yellow light? Explain.

8. All objects radiate energy. Why, then, are we not able to see all objects in a dark room?

9. Is an electron a wave or a particle? Support your answer by citing some experimental results.

10. Suppose a photograph were made of a person's face using only a few photons. Would the result be simply a very faint image of the face? Explain your answer.

11. Why is an electron microscope more suitable than an optical microscope for "seeing" objects less than 1 μm in size?

12. Is light a wave or a particle? Support your answer by citing specific experimental evidence.

13. (a) What does the slope of the lines in Figure 40.11 represent? (b) What does the y intercept represent? (c) How would such graphs for different metals compare with one another?

14. Why was the demonstration of electron diffraction by Davisson and Germer an important experiment?

15. *Iridescence* is the phenomenon that gives shining colors to the feathers of peacocks, hummingbirds (see page 1134), resplendent quetzals, and even ducks and grackles. Without pigments, it colors Morpho butterflies (Fig. CQ40.15), Urania moths, some beetles and flies, rainbow trout, and mother-of-pearl in abalone shells. Iridescent colors change as you turn an object. They are produced by a wide variety of intricate structures in different species. Problem 64 in Chapter 38 describes the structures that produce iridescence in a peacock feather. These structures were all unknown until the invention of the electron microscope. Explain why light microscopes cannot reveal them.

Figure CQ40.15

16. In describing the passage of electrons through a slit and arriving at a screen, physicist Richard Feynman said that "electrons arrive in lumps, like particles, but the probability of arrival of these lumps is determined as the intensity of the waves would be. It is in this sense that the electron behaves sometimes like a particle and sometimes like a wave." Elaborate on this point in your own words. For further discussion, see R. Feynman, *The Character of Physical Law* (Cambridge, MA: MIT Press, 1980), chap. 6.

17. The classical model of blackbody radiation given by the Rayleigh–Jeans law has two major flaws. (a) Identify the flaws and (b) explain how Planck's law deals with them.

Problems

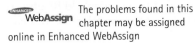 The problems found in this chapter may be assigned online in Enhanced WebAssign

1. straightforward; **2.** intermediate; **3.** challenging

1. full solution available in the *Student Solutions Manual/Study Guide*

AMT Analysis Model tutorial available in Enhanced WebAssign

GP Guided Problem

M Master It tutorial available in Enhanced WebAssign

W Watch It video solution available in Enhanced WebAssign

Section 40.1 Blackbody Radiation and Planck's Hypothesis

1. The temperature of an electric heating element is 150°C. At what wavelength does the radiation emitted from the heating element reach its peak?

2. Model the tungsten filament of a lightbulb as a black body at temperature 2 900 K. (a) Determine the wavelength of light it emits most strongly. (b) Explain why the answer to part (a) suggests that more energy from the lightbulb goes into infrared radiation than into visible light.

3. Lightning produces a maximum air temperature on the order of 10^4 K, whereas a nuclear explosion produces a temperature on the order of 10^7 K. (a) Use Wien's displacement law to find the order of magnitude of the wavelength of the thermally produced photons radiated with greatest intensity by each of these sources. (b) Name the part of the electromagnetic spectrum where you would expect each to radiate most strongly.

4. Figure P40.4 shows the spectrum of light emitted by a firefly. (a) Determine the temperature of a black body that would emit radiation peaked at the same wavelength. (b) Based on your result, explain whether firefly radiation is blackbody radiation.

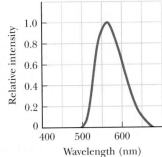

Figure P40.4

5. The average threshold of dark-adapted (scotopic) [W] vision is 4.00×10^{-11} W/m² at a central wavelength of 500 nm. If light with this intensity and wavelength enters the eye and the pupil is open to its maximum diameter of 8.50 mm, how many photons per second enter the eye?

6. (i) Calculate the energy, in electron volts, of a pho- [M] ton whose frequency is (a) 620 THz, (b) 3.10 GHz, and (c) 46.0 MHz. (ii) Determine the corresponding wavelengths for the photons listed in part (i) and (iii) state the classification of each on the electromagnetic spectrum.

7. (a) What is the surface temperature of Betelgeuse, a red giant star in the constellation Orion (Fig. 40.4), which radiates with a peak wavelength of about 970 nm? (b) Rigel, a bluish-white star in Orion, radiates with a peak wavelength of 145 nm. Find the temperature of Rigel's surface.

8. An FM radio transmitter has a power output of 150 kW [M] and operates at a frequency of 99.7 MHz. How many photons per second does the transmitter emit?

9. The human eye is most sensitive to 560-nm (green) [W] light. What is the temperature of a black body that would radiate most intensely at this wavelength?

10. The radius of our Sun is 6.96×10^{8} m, and its total [W] power output is 3.85×10^{26} W. (a) Assuming the Sun's surface emits as a black body, calculate its surface temperature. (b) Using the result of part (a), find λ_{max} for the Sun.

11. A black body at 7 500 K consists of an opening of diam- [W] eter 0.050 0 mm, looking into an oven. Find the number of photons per second escaping the opening and having wavelengths between 500 nm and 501 nm.

12. Consider a black body of surface area 20.0 cm² and temperature 5 000 K. (a) How much power does it radiate? (b) At what wavelength does it radiate most intensely? Find the spectral power per wavelength interval at (c) this wavelength and at wavelengths of (d) 1.00 nm (an x- or gamma ray), (e) 5.00 nm (ultraviolet light or an x-ray), (f) 400 nm (at the boundary between UV and visible light), (g) 700 nm (at the boundary between visible and infrared light), (h) 1.00 mm (infrared light or a microwave), and (i) 10.0 cm (a microwave or radio wave). (j) Approximately how much power does the object radiate as visible light?

13. **Review.** This problem is about how strongly matter is coupled to radiation, the subject with which quantum mechanics began. For a simple model, consider a solid iron sphere 2.00 cm in radius. Assume its temperature is always uniform throughout its volume. (a) Find the mass of the sphere. (b) Assume the sphere is at 20.0°C and has emissivity 0.860. Find the power with which it radiates electromagnetic waves. (c) If it were alone in the Universe, at what rate would the sphere's temperature be changing? (d) Assume Wien's law describes the sphere. Find the wavelength λ_{max} of electromagnetic radiation it emits most strongly. Although it emits a spectrum of waves having all different wavelengths,

assume its power output is carried by photons of wavelength λ_{max}. Find (e) the energy of one photon and (f) the number of photons it emits each second.

14. Show that at long wavelengths, Planck's radiation law (Eq. 40.6) reduces to the Rayleigh–Jeans law (Eq. 40.3).

15. A simple pendulum has a length of 1.00 m and a mass of 1.00 kg. The maximum horizontal displacement of the pendulum bob from equilibrium is 3.00 cm. Calculate the quantum number n for the pendulum.

16. A pulsed ruby laser emits light at 694.3 nm. For a 14.0-ps pulse containing 3.00 J of energy, find (a) the physical length of the pulse as it travels through space and (b) the number of photons in it. (c) Assuming that the beam has a circular cross-section of 0.600 cm diameter, find the number of photons per cubic millimeter.

Section 40.2 The Photoelectric Effect

17. Molybdenum has a work function of 4.20 eV. (a) Find the cutoff wavelength and cutoff frequency for the photoelectric effect. (b) What is the stopping potential if the incident light has a wavelength of 180 nm?

18. The work function for zinc is 4.31 eV. (a) Find the cutoff wavelength for zinc. (b) What is the lowest frequency of light incident on zinc that releases photoelectrons from its surface? (c) If photons of energy 5.50 eV are incident on zinc, what is the maximum kinetic energy of the ejected photoelectrons?

19. Two light sources are used in a photoelectric experiment to determine the work function for a particular metal surface. When green light from a mercury lamp ($\lambda = 546.1$ nm) is used, a stopping potential of 0.376 V reduces the photocurrent to zero. (a) Based on this measurement, what is the work function for this metal? (b) What stopping potential would be observed when using the yellow light from a helium discharge tube ($\lambda = 587.5$ nm)?

20. Lithium, beryllium, and mercury have work functions [W] of 2.30 eV, 3.90 eV, and 4.50 eV, respectively. Light with a wavelength of 400 nm is incident on each of these metals. (a) Determine which of these metals exhibit the photoelectric effect for this incident light. Explain your reasoning. (b) Find the maximum kinetic energy for the photoelectrons in each case.

21. Electrons are ejected from a metallic surface with speeds of up to 4.60×10^{5} m/s when light with a wavelength of 625 nm is used. (a) What is the work function of the surface? (b) What is the cutoff frequency for this surface?

22. From the scattering of sunlight, J. J. Thomson calculated the classical radius of the electron as having the value 2.82×10^{-15} m. Sunlight with an intensity of 500 W/m² falls on a disk with this radius. Assume light is a classical wave and the light striking the disk is completely absorbed. (a) Calculate the time interval required to accumulate 1.00 eV of energy. (b) Explain how your result for part (a) compares with the observation that photoelectrons are emitted promptly (within 10^{-9} s).

23. Review. An isolated copper sphere of radius 5.00 cm, **AMT** initially uncharged, is illuminated by ultraviolet light of wavelength 200 nm. The work function for copper is 4.70 eV. What charge does the photoelectric effect induce on the sphere?

24. The work function for platinum is 6.35 eV. Ultraviolet **GP** light of wavelength 150 nm is incident on the clean surface of a platinum sample. We wish to predict the stopping voltage we will need for electrons ejected from the surface. (a) What is the photon energy of the ultraviolet light? (b) How do you know that these photons will eject electrons from platinum? (c) What is the maximum kinetic energy of the ejected photoelectrons? (d) What stopping voltage would be required to arrest the current of photoelectrons?

Section 40.3 The Compton Effect

25. X-rays are scattered from a target at an angle of 55.0° with the direction of the incident beam. Find the wavelength shift of the scattered x-rays.

26. A photon having wavelength λ scatters off a free electron at A (Fig. P40.26), producing a second photon having wavelength λ'. This photon then scatters off another free electron at B, producing a third photon having wavelength λ'' and moving in a direction directly opposite the original photon as shown in the figure. Determine the value of $\Delta\lambda = \lambda'' - \lambda$.

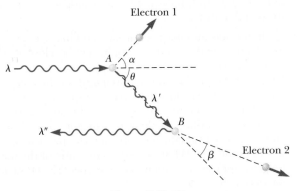

Electron 1

Electron 2

Figure P40.26

27. A 0.110-nm photon collides with a stationary electron. After the collision, the electron moves forward and the photon recoils backward. Find the momentum and the kinetic energy of the electron.

28. X-rays with a wavelength of 120.0 pm undergo Compton scattering. (a) Find the wavelengths of the photons scattered at angles of 30.0°, 60.0°, 90.0°, 120°, 150°, and 180°. (b) Find the energy of the scattered electron in each case. (c) Which of the scattering angles provides the electron with the greatest energy? Explain whether you could answer this question without doing any calculations.

29. A 0.001 60-nm photon scatters from a free electron. **M** For what (photon) scattering angle does the recoiling electron have kinetic energy equal to the energy of the scattered photon?

30. After a 0.800-nm x-ray photon scatters from a free electron, the electron recoils at 1.40×10^6 m/s. (a) What

is the Compton shift in the photon's wavelength? (b) Through what angle is the photon scattered?

31. A photon having energy $E_0 = 0.880$ MeV is scattered **W** by a free electron initially at rest such that the scattering angle of the scattered electron is equal to that of the scattered photon as shown in Figure P40.31. (a) Determine the scattering angle of the photon and the electron. (b) Determine the energy and momentum of the scattered photon. (c) Determine the kinetic energy and momentum of the scattered electron.

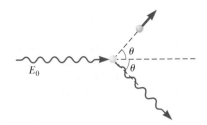

Figure P40.31 Problems 31 and 32.

32. A photon having energy E_0 is scattered by a free electron initially at rest such that the scattering angle of the scattered electron is equal to that of the scattered photon as shown in Figure P40.31. (a) Determine the angle θ. (b) Determine the energy and momentum of the scattered photon. (c) Determine the kinetic energy and momentum of the scattered electron.

33. X-rays having an energy of 300 keV undergo Compton scattering from a target. The scattered rays are detected at 37.0° relative to the incident rays. Find (a) the Compton shift at this angle, (b) the energy of the scattered x-ray, and (c) the energy of the recoiling electron.

34. In a Compton scattering experiment, a photon is scattered through an angle of 90.0° and the electron is set into motion in a direction at an angle of 20.0° to the original direction of the photon. (a) Explain how this information is sufficient to determine uniquely the wavelength of the scattered photon and (b) find this wavelength.

35. In a Compton scattering experiment, an x-ray photon scatters through an angle of 17.4° from a free electron that is initially at rest. The electron recoils with a speed of 2 180 km/s. Calculate (a) the wavelength of the incident photon and (b) the angle through which the electron scatters.

36. Find the maximum fractional energy loss for a 0.511-MeV gamma ray that is Compton scattered from (a) a free electron and (b) a free proton.

Section 40.4 The Nature of Electromagnetic Waves

37. An electromagnetic wave is called *ionizing radiation* if its photon energy is larger than, say, 10.0 eV so that a single photon has enough energy to break apart an atom. With reference to Figure P40.37 (page 1264), explain what region or regions of the electromagnetic spectrum fit this definition of ionizing radiation and what do not. (If you wish to consult a larger version of Fig. P40.37, see Fig. 34.13.)

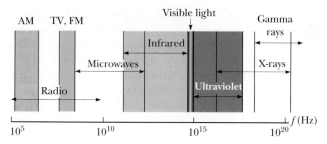

Figure P40.37

38. **Review.** A helium–neon laser produces a beam of diameter 1.75 mm, delivering 2.00×10^{18} photons/s. Each photon has a wavelength of 633 nm. Calculate the amplitudes of (a) the electric fields and (b) the magnetic fields inside the beam. (c) If the beam shines perpendicularly onto a perfectly reflecting surface, what force does it exert on the surface? (d) If the beam is absorbed by a block of ice at 0°C for 1.50 h, what mass of ice is melted?

Section 40.5 The Wave Properties of Particles

39. (a) Calculate the momentum of a photon whose wavelength is 4.00×10^{-7} m. (b) Find the speed of an electron having the same momentum as the photon in part (a).

40. (a) An electron has a kinetic energy of 3.00 eV. Find its wavelength. (b) **What If?** A photon has energy 3.00 eV. Find its wavelength.

41. The resolving power of a microscope depends on the wavelength used. If you wanted to "see" an atom, a wavelength of approximately 1.00×10^{-11} m would be required. (a) If electrons are used (in an electron microscope), what minimum kinetic energy is required for the electrons? (b) **What If?** If photons are used, what minimum photon energy is needed to obtain the required resolution?

42. Calculate the de Broglie wavelength for a proton moving with a speed of 1.00×10^6 m/s.

43. In the Davisson–Germer experiment, 54.0-eV electrons were diffracted from a nickel lattice. If the first maximum in the diffraction pattern was observed at $\phi = 50.0°$ (Fig. P40.43), what was the lattice spacing a between the vertical columns of atoms in the figure?

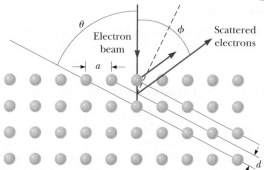

Figure P40.43

44. The nucleus of an atom is on the order of 10^{-14} m in diameter. For an electron to be confined to a nucleus, its de Broglie wavelength would have to be on this order of magnitude or smaller. (a) What would be the kinetic energy of an electron confined to this region? (b) Make an order-of-magnitude estimate of the electric potential energy of a system of an electron inside an atomic nucleus. (c) Would you expect to find an electron in a nucleus? Explain.

45. Robert Hofstadter won the 1961 Nobel Prize in Physics for his pioneering work in studying the scattering of 20-GeV electrons from nuclei. (a) What is the γ factor for an electron with total energy 20.0 GeV, defined by $\gamma = 1/\sqrt{1 - u^2/c^2}$? (b) Find the momentum of the electron. (c) Find the wavelength of the electron. (d) State how the wavelength compares with the diameter of an atomic nucleus, typically on the order of 10^{-14} m.

46. *Why is the following situation impossible?* After learning about de Broglie's hypothesis that material particles of momentum p move as waves with wavelength $\lambda = h/p$, an 80-kg student has grown concerned about being diffracted when passing through a doorway of width $w = 75$ cm. Assume significant diffraction occurs when the width of the diffraction aperture is less than ten times the wavelength of the wave being diffracted. Together with his classmates, the student performs precision experiments and finds that he does indeed experience measurable diffraction.

47. A photon has an energy equal to the kinetic energy of an electron with speed u, which may be close to the speed of light c. (a) Calculate the ratio of the wavelength of the photon to the wavelength of the electron. (b) Evaluate the ratio for the particle speed $u = 0.900c$. (c) **What If?** What would happen to the answer to part (b) if the material particle were a proton instead of an electron? (d) Evaluate the ratio for the particle speed $u = 0.001\,00c$. (e) What value does the ratio of the wavelengths approach at high particle speeds? (f) At low particle speeds?

48. (a) Show that the frequency f and wavelength λ of a freely moving quantum particle with mass are related by the expression

$$\left(\frac{f}{c}\right)^2 = \frac{1}{\lambda^2} + \frac{1}{\lambda_C^2}$$

where $\lambda_C = h/mc$ is the Compton wavelength of the particle. (b) Is it ever possible for a particle having nonzero mass to have the same wavelength *and* frequency as a photon? Explain.

Section 40.6 A New Model: The Quantum Particle

49. Consider a freely moving quantum particle with mass m and speed u. Its energy is $E = K = \frac{1}{2}mu^2$. (a) Determine the phase speed of the quantum wave representing the particle and (b) show that it is different from the speed at which the particle transports mass and energy.

50. For a free relativistic quantum particle moving with speed u, the total energy of the particle is $E = hf = \hbar\omega = \sqrt{p^2c^2 + m^2c^4}$ and the momentum is $p = h/\lambda = \hbar k =$

γmu. For the quantum wave representing the particle, the group speed is $v_g = d\omega/dk$. Prove that the group speed of the wave is the same as the speed of the particle.

Section 40.7 The Double-Slit Experiment Revisited

51. Neutrons traveling at 0.400 m/s are directed through
M a pair of slits separated by 1.00 mm. An array of detectors is placed 10.0 m from the slits. (a) What is the de Broglie wavelength of the neutrons? (b) How far off axis is the first zero-intensity point on the detector array? (c) When a neutron reaches a detector, can we say which slit the neutron passed through? Explain.

52. In a certain vacuum tube, electrons evaporate from a hot cathode at a slow, steady rate and accelerate from rest through a potential difference of 45.0 V. Then they travel 28.0 cm as they pass through an array of slits and fall on a screen to produce an interference pattern. If the beam current is below a certain value, only one electron at a time will be in flight in the tube. In this situation, the interference pattern still appears, showing that each individual electron can interfere with itself. What is the maximum value for the beam current that will result in only one electron at a time in flight in the tube?

53. A modified oscilloscope is used to perform an elec-
W tron interference experiment. Electrons are incident on a pair of narrow slits 0.060 0 μm apart. The bright bands in the interference pattern are separated by 0.400 mm on a screen 20.0 cm from the slits. Determine the potential difference through which the electrons were accelerated to give this pattern.

Section 40.8 The Uncertainty Principle

54. Suppose a duck lives in a universe in which $h = 2\pi$ J · s. The duck has a mass of 2.00 kg and is initially known to be within a pond 1.00 m wide. (a) What is the minimum uncertainty in the component of the duck's velocity parallel to the pond's width? (b) Assuming this uncertainty in speed prevails for 5.00 s, determine the uncertainty in the duck's position after this time interval.

55. An electron and a 0.020 0-kg bullet each have a
M velocity of magnitude 500 m/s, accurate to within 0.010 0%. Within what lower limit could we determine the position of each object along the direction of the velocity?

56. A 0.500-kg block rests on the frictionless, icy surface of a frozen pond. If the location of the block is measured to a precision of 0.150 cm and its mass is known exactly, what is the minimum uncertainty in the block's speed?

57. The average lifetime of a muon is about 2 μs. Estimate the minimum uncertainty in the rest energy of a muon.

58. *Why is the following situation impossible?* An air rifle is used to shoot 1.00-g particles at a speed of $v_x = 100$ m/s. The rifle's barrel has a diameter of 2.00 mm. The rifle is mounted on a perfectly rigid support so that it is fired in exactly the same way each time. Because of the uncertainty principle, however, after many firings, the diameter of the spray of pellets on a paper target is 1.00 cm.

59. Use the uncertainty principle to show that if an electron were confined inside an atomic nucleus of diameter on the order of 10^{-14} m, it would have to be moving relativistically, whereas a proton confined to the same nucleus can be moving nonrelativistically.

Additional Problems

60. The accompanying table shows data obtained in a photoelectric experiment. (a) Using these data, make a graph similar to Figure 40.11 that plots as a straight line. From the graph, determine (b) an experimental value for Planck's constant (in joule-seconds) and (c) the work function (in electron volts) for the surface. (Two significant figures for each answer are sufficient.)

Wavelength (nm)	Maximum Kinetic Energy of Photoelectrons (eV)
588	0.67
505	0.98
445	1.35
399	1.63

61. Photons of wavelength 450 nm are incident on a metal.
AMT The most energetic electrons ejected from the metal
M are bent into a circular arc of radius 20.0 cm by a magnetic field with a magnitude of 2.00×10^{-5} T. What is the work function of the metal?

62. Review. Photons of wavelength λ are incident on a metal. The most energetic electrons ejected from the metal are bent into a circular arc of radius R by a magnetic field having a magnitude B. What is the work function of the metal?

63. Review. Design an incandescent lamp filament. A tungsten wire radiates electromagnetic waves with power 75.0 W when its ends are connected across a 120-V power supply. Assume its constant operating temperature is 2 900 K and its emissivity is 0.450. Also assume it takes in energy only by electric transmission and emits energy only by electromagnetic radiation. You may take the resistivity of tungsten at 2 900 K as 7.13×10^{-7} Ω · m. Specify (a) the radius and (b) the length of the filament.

64. Derive the equation for the Compton shift (Eq. 40.13) from Equations 40.14 through 40.16.

65. Figure P40.65 shows the stopping potential versus the incident photon frequency for the photoelectric effect

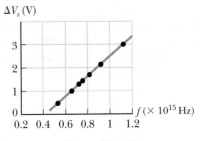

Figure P40.65

for sodium. Use the graph to find (a) the work function of sodium, (b) the ratio h/e, and (c) the cutoff wavelength. The data are taken from R. A. Millikan, *Physical Review* **7**:362 (1916).

66. A photon of initial energy E_0 undergoes Compton scattering at an angle θ from a free electron (mass m_e) initially at rest. Derive the following relationship for the final energy E' of the scattered photon:

$$E' = \frac{E_0}{1 + \left(\dfrac{E_0}{m_e c^2}\right)(1 - \cos\theta)}$$

67. A daredevil's favorite trick is to step out of a 16th-story window and fall 50.0 m into a pool. A news reporter takes a picture of the 75.0-kg daredevil just before he makes a splash, using an exposure time of 5.00 ms. Find (a) the daredevil's de Broglie wavelength at this moment, (b) the uncertainty of his kinetic energy measurement during the 5.00-ms time interval, and (c) the percent error caused by such an uncertainty.

68. Show that the ratio of the Compton wavelength λ_C to the de Broglie wavelength $\lambda = h/p$ for a relativistic electron is

$$\frac{\lambda_C}{\lambda} = \left[\left(\frac{E}{m_e c^2}\right)^2 - 1\right]^{1/2}$$

where E is the total energy of the electron and m_e is its mass.

69. Monochromatic ultraviolet light with intensity 550 W/m^2 is incident normally on the surface of a metal that has a work function of 3.44 eV. Photoelectrons are emitted with a maximum speed of 420 km/s. (a) Find the maximum possible rate of photoelectron emission from 1.00 cm^2 of the surface by imagining that every photon produces one photoelectron. (b) Find the electric current these electrons constitute. (c) How do you suppose the actual current compares with this maximum possible current?

70. A π^0 meson is an unstable particle produced in high-energy particle collisions. Its rest energy is approximately 135 MeV, and it exists for a lifetime of only 8.70×10^{-17} s before decaying into two gamma rays. Using the uncertainty principle, estimate the fractional uncertainty $\Delta m/m$ in its mass determination.

71. The neutron has a mass of 1.67×10^{-27} kg. Neutrons emitted in nuclear reactions can be slowed down by collisions with matter. They are referred to as thermal neutrons after they come into thermal equilibrium with the environment. The average kinetic energy $\left(\frac{3}{2}k_B T\right)$ of a thermal neutron is approximately 0.04 eV. (a) Calculate the de Broglie wavelength of a neutron with a kinetic energy of 0.040 0 eV. (b) How does your answer compare with the characteristic atomic spacing in a crystal? (c) Explain whether you expect thermal neutrons to exhibit diffraction effects when scattered by a crystal.

Challenge Problems

72. A woman on a ladder drops small pellets toward a point target on the floor. (a) Show that, according to the uncertainty principle, the average miss distance must be at least

$$\Delta x_f = \left(\frac{2\hbar}{m}\right)^{1/2}\left(\frac{2H}{g}\right)^{1/4}$$

where H is the initial height of each pellet above the floor and m is the mass of each pellet. Assume that the spread in impact points is given by $\Delta x_f = \Delta x_i + (\Delta v_x)t$. (b) If $H = 2.00$ m and $m = 0.500$ g, what is Δx_f?

73. **Review.** A light source emitting radiation at frequency 7.00×10^{14} Hz is incapable of ejecting photoelectrons from a certain metal. In an attempt to use this source to eject photoelectrons from the metal, the source is given a velocity toward the metal. (a) Explain how this procedure can produce photoelectrons. (b) When the speed of the light source is equal to $0.280c$, photoelectrons just begin to be ejected from the metal. What is the work function of the metal? (c) When the speed of the light source is increased to $0.900c$, determine the maximum kinetic energy of the photoelectrons.

74. Using conservation principles, prove that a photon cannot transfer all its energy to a free electron.

75. The total power per unit area radiated by a black body at a temperature T is the area under the $I(\lambda, T)$-versus-λ curve as shown in Figure 40.3. (a) Show that this power per unit area is

$$\int_0^\infty I(\lambda, T)\, d\lambda = \sigma T^4$$

where $I(\lambda, T)$ is given by Planck's radiation law and σ is a constant independent of T. This result is known as Stefan's law. (See Section 20.7.) To carry out the integration, you should make the change of variable $x = hc/\lambda k_B T$ and use

$$\int_0^\infty \frac{x^3\, dx}{e^x - 1} = \frac{\pi^4}{15}$$

(b) Show that the Stefan–Boltzmann constant σ has the value

$$\sigma = \frac{2\pi^5 k_B{}^4}{15 c^2 h^3} = 5.67 \times 10^{-8}\ \text{W/m}^2 \cdot \text{K}^4$$

76. (a) Derive Wien's displacement law from Planck's law. Proceed as follows. In Figure 40.3, notice that the wavelength at which a black body radiates with greatest intensity is the wavelength for which the graph of $I(\lambda, T)$ versus λ has a horizontal tangent. From Equation 40.6, evaluate the derivative $dI/d\lambda$. Set it equal to zero. Solve the resulting transcendental equation numerically to prove that $hc/\lambda_{max} k_B T = 4.965 \ldots$ or $\lambda_{max} T = hc/4.965 k_B$. (b) Evaluate the constant as precisely as possible and compare it with Wien's experimental value.

Quantum Mechanics

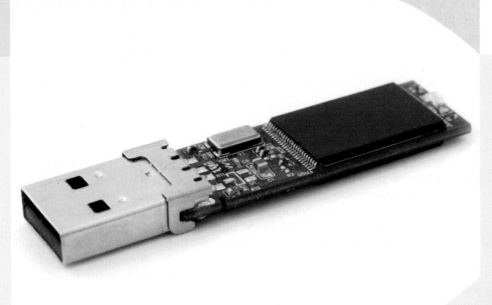

In this chapter, we introduce quantum mechanics, an extremely successful theory for explaining the behavior of microscopic particles. This theory, developed in the 1920s by Erwin Schrödinger, Werner Heisenberg, and others, enables us to understand a host of phenomena involving atoms, molecules, nuclei, and solids. The discussion in this chapter follows from the quantum particle model that was developed in Chapter 40 and incorporates some of the features of the waves under boundary conditions model that was explored in Chapter 18. We also discuss practical applications of quantum mechanics, including the scanning tunneling microscope and nanoscale devices that may be used in future quantum computers. Finally, we shall return to the simple harmonic oscillator that was introduced in Chapter 15 and examine it from a quantum mechanical point of view.

An opened flash drive of the type used as an external data storage device for a computer. Flash drives are employed extensively in computers, digital cameras, cell phones, and other devices. Writing data to and erasing data from flash drives incorporate the phenomenon of quantum tunneling, which we explore in this chapter. (Image copyright Vasilius, 2009. Used under license from Shutterstock.com)

41.1 The Wave Function

In Chapter 40, we introduced some new and strange ideas. In particular, we concluded on the basis of experimental evidence that both matter and electromagnetic radiation are sometimes best modeled as particles and sometimes as waves, depending on the phenomenon being observed. We can improve our understanding of quantum physics by making another connection between particles and waves using the notion of probability, a concept that was introduced in Chapter 40.

We begin by discussing electromagnetic radiation using the particle model. The probability per unit volume of finding a photon in a given region of space at an instant of time is proportional to the number of photons per unit volume at that time:

$$\frac{\text{Probability}}{V} \propto \frac{N}{V}$$

The number of photons per unit volume is proportional to the intensity of the radiation:

$$\frac{N}{V} \propto I$$

Now, let's form a connection between the particle model and the wave model by recalling that the intensity of electromagnetic radiation is proportional to the square of the electric field amplitude E for the electromagnetic wave (Eq. 34.24):

$$I \propto E^2$$

Equating the beginning and the end of this series of proportionalities gives

$$\frac{\text{Probability}}{V} \propto E^2 \qquad (41.1)$$

Therefore, for electromagnetic radiation, the probability per unit volume of finding a particle associated with this radiation (the photon) is proportional to the square of the amplitude of the associated electromagnetic wave.

Recognizing the wave–particle duality of both electromagnetic radiation and matter, we should suspect a parallel proportionality for a material particle: the probability per unit volume of finding the particle is proportional to the square of the amplitude of a wave representing the particle. In Chapter 40, we learned that there is a de Broglie wave associated with every particle. The amplitude of the de Broglie wave associated with a particle is not a measurable quantity because the wave function representing a particle is generally a complex function as we discuss below. In contrast, the electric field for an electromagnetic wave is a real function. The matter analog to Equation 41.1 relates the square of the amplitude of the wave to the probability per unit volume of finding the particle. Hence, the amplitude of the wave associated with the particle is called the **probability amplitude,** or the **wave function,** and it has the symbol Ψ.

In general, the complete wave function Ψ for a system depends on the positions of all the particles in the system and on time; therefore, it can be written $\Psi(\vec{\mathbf{r}}_1, \vec{\mathbf{r}}_2, \vec{\mathbf{r}}_3, \ldots, \vec{\mathbf{r}}_j, \ldots, t)$, where $\vec{\mathbf{r}}_j$ is the position vector of the jth particle in the system. For many systems of interest, including all those we study in this text, the wave function Ψ is mathematically separable in space and time and can be written as a product of a space function ψ for one particle of the system and a complex time function:[1]

$$\Psi(\vec{\mathbf{r}}_1, \vec{\mathbf{r}}_2, \vec{\mathbf{r}}_3, \ldots, \vec{\mathbf{r}}_j, \ldots, t) = \psi(\vec{\mathbf{r}}_j) e^{-i\omega t} \qquad (41.2)$$

◀ Space- and time-dependent wave function Ψ

where ω ($= 2\pi f$) is the angular frequency of the wave function and $i = \sqrt{-1}$.

For any system in which the potential energy is time-independent and depends only on the positions of particles within the system, the important information about the system is contained within the space part of the wave function. The time part is simply the factor $e^{-i\omega t}$. Therefore, an understanding of ψ is the critical aspect of a given problem.

The wave function ψ is often complex-valued. The absolute square $|\psi|^2 = \psi^*\psi$, where ψ^* is the complex conjugate[2] of ψ, is always real and positive and is proportional to the probability per unit volume of finding a particle at a given point at some instant. The wave function contains within it all the information that can be known about the particle.

---

[1] The standard form of a complex number is $a + ib$. The notation $e^{i\theta}$ is equivalent to the standard form as follows:

$$e^{i\theta} = \cos\theta + i\sin\theta$$

Therefore, the notation $e^{-i\omega t}$ in Equation 41.2 is equivalent to $\cos(-\omega t) + i\sin(-\omega t) = \cos\omega t - i\sin\omega t$.

[2] For a complex number $z = a + ib$, the complex conjugate is found by changing i to $-i$: $z^* = a - ib$. The product of a complex number and its complex conjugate is always real and positive. That is, $z^*z = (a - ib)(a + ib) = a^2 - (ib)^2 = a^2 - (i)^2 b^2 = a^2 + b^2$.

Although ψ cannot be measured, we can measure the real quantity $|\psi|^2$, which can be interpreted as follows. If ψ represents a single particle, then $|\psi|^2$—called the **probability density**—is the relative probability per unit volume that the particle will be found at any given point in the volume. This interpretation can also be stated in the following manner. If dV is a small volume element surrounding some point, the probability of finding the particle in that volume element is

◀ Probability density $|\psi|^2$

$$P(x, y, z)\ dV = |\psi|^2\ dV \tag{41.3}$$

This probabilistic interpretation of the wave function was first suggested by Max Born (1882–1970) in 1928. In 1926, Erwin Schrödinger proposed a wave equation that describes the manner in which the wave function changes in space and time. The *Schrödinger wave equation*, which we shall examine in Section 41.3, represents a key element in the theory of quantum mechanics.

The concepts of quantum mechanics, strange as they sometimes may seem, developed from classical ideas. In fact, when the techniques of quantum mechanics are applied to macroscopic systems, the results are essentially identical to those of classical physics. This blending of the two approaches occurs when the de Broglie wavelength is small compared with the dimensions of the system. The situation is similar to the agreement between relativistic mechanics and classical mechanics when $v \ll c$.

In Section 40.5, we found that the de Broglie equation relates the momentum of a particle to its wavelength through the relation $p = h/\lambda$. If an ideal free particle has a precisely known momentum p_x, its wave function is an infinitely long sinusoidal wave of wavelength $\lambda = h/p_x$ and the particle has equal probability of being at any point along the x axis (Fig. 40.18a). The wave function ψ for such a free particle moving along the x axis can be written as

$$\psi(x) = Ae^{ikx} \tag{41.4}$$

◀ Wave function for a free particle

where A is a constant amplitude and $k = 2\pi/\lambda$ is the angular wave number (Eq. 16.8) of the wave representing the particle.[3]

One-Dimensional Wave Functions and Expectation Values

This section discusses only one-dimensional systems, where the particle must be located along the x axis, so the probability $|\psi|^2\ dV$ in Equation 41.3 is modified to become $|\psi|^2\ dx$. The probability that the particle will be found in the infinitesimal interval dx around the point x is

$$P(x)\ dx = |\psi|^2\ dx \tag{41.5}$$

Although it is not possible to specify the position of a particle with complete certainty, it is possible through $|\psi|^2$ to specify the probability of observing it in a region surrounding a given point x. The probability of finding the particle in the arbitrary interval $a \le x \le b$ is

$$P_{ab} = \int_a^b |\psi|^2\ dx \tag{41.6}$$

The probability P_{ab} is the area under the curve of $|\psi|^2$ versus x between the points $x = a$ and $x = b$ as in Figure 41.1.

Experimentally, there is a finite probability of finding a particle in an interval near some point at some instant. The value of that probability must lie between the

Pitfall Prevention 41.1
The Wave Function Belongs to a System The common language in quantum mechanics is to associate a wave function with a particle. The wave function, however, is determined by the particle *and* its interaction with its environment, so it more rightfully belongs to a system. In many cases, the particle is the only part of the system that experiences a change, which is why the common language has developed. You will see examples in the future in which it is more proper to think of the system wave function rather than the particle wave function.

The probability of a particle being in the interval $a \le x \le b$ is the area under the probability density curve from a to b.

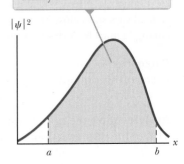

Figure 41.1 An arbitrary probability density curve for a particle.

---

[3]For the free particle, the full wave function, based on Equation 41.2, is

$$\Psi(x, t) = Ae^{ikx}e^{-i\omega t} = Ae^{i(kx - \omega t)} = A[\cos(kx - \omega t) + i\sin(kx - \omega t)]$$

The real part of this wave function has the same form as the waves we added together to form wave packets in Section 40.6.

limits 0 and 1. For example, if the probability is 0.30, there is a 30% chance of finding the particle in the interval.

Because the particle must be somewhere along the x axis, the sum of the probabilities over all values of x must be 1:

Normalization condition on ψ ▶

$$\int_{-\infty}^{\infty} |\psi|^2 \, dx = 1 \tag{41.7}$$

Any wave function satisfying Equation 41.7 is said to be **normalized.** Normalization is simply a statement that the particle exists at some point in space.

Once the wave function for a particle is known, it is possible to calculate the average position at which you would expect to find the particle after many measurements. This average position is called the **expectation value** of x and is defined by the equation

Expectation value ▶
for position x

$$\langle x \rangle \equiv \int_{-\infty}^{\infty} \psi^* x \psi \, dx \tag{41.8}$$

(Brackets, $\langle \ldots \rangle$, are used to denote expectation values.) Furthermore, one can find the expectation value of any function $f(x)$ associated with the particle by using the following equation:[4]

Expectation value for ▶
a function $f(x)$

$$\langle f(x) \rangle \equiv \int_{-\infty}^{\infty} \psi^* f(x) \psi \, dx \tag{41.9}$$

Quick Quiz 41.1 Consider the wave function for the free particle, Equation 41.4. At what value of x is the particle most likely to be found at a given time? **(a)** at $x = 0$ **(b)** at small nonzero values of x **(c)** at large values of x **(d)** anywhere along the x axis

---

Example 41.1   **A Wave Function for a Particle**

Consider a particle whose wave function is graphed in Figure 41.2 and is given by

$$\psi(x) = Ae^{-ax^2}$$

(A) What is the value of A if this wave function is normalized?

SOLUTION

Conceptualize The particle is not a free particle because the wave function is not a sinusoidal function. Figure 41.2 indicates that the particle is constrained to remain close to $x = 0$ at all times. Think of a physical system in which the particle always stays close to a given point. Examples of such systems are a block on a spring, a marble at the bottom of a bowl, and the bob of a simple pendulum.

Categorize Because the statement of the problem describes the wave nature of a particle, this example requires a quantum approach rather than a classical approach.

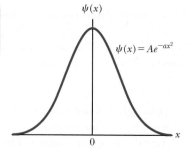

Figure 41.2 (Example 41.1) A symmetric wave function for a particle, given by $\psi(x) = Ae^{-ax^2}$.

....

Analyze Apply the normalization condition, Equation 41.7, to the wave function:

$$\int_{-\infty}^{\infty} |\psi|^2 \, dx = \int_{-\infty}^{\infty} (Ae^{-ax^2})^2 \, dx = A^2 \int_{-\infty}^{\infty} e^{-2ax^2} \, dx = 1$$

---

[4]Expectation values are analogous to "weighted averages," in which each possible value of a function is multiplied by the probability of the occurrence of that value before summing over all possible values. We write the expectation value as $\int_{-\infty}^{\infty} \psi^* f(x) \psi \, dx$ rather than $\int_{-\infty}^{\infty} f(x) \psi^2 \, dx$ because $f(x)$ may be represented by an operator (such as a derivative) rather than a simple multiplicative function in more advanced treatments of quantum mechanics. In these situations, the operator is applied only to ψ and not to ψ^*.

▶ **41.1** c o n t i n u e d

Express the integral as the sum of two integrals:

$$(1) \quad A^2 \int_{-\infty}^{\infty} e^{-2ax^2}\, dx = A^2 \left(\int_{0}^{\infty} e^{-2ax^2}\, dx + \int_{-\infty}^{0} e^{-2ax^2}\, dx \right) = 1$$

Change the integration variable from x to $-x$ in the second integral:

$$\int_{-\infty}^{0} e^{-2ax^2}\, dx = \int_{\infty}^{0} e^{-2a(-x)^2}\, (-dx) = -\int_{\infty}^{0} e^{-2ax^2}\, dx$$

Reverse the order of the limits, which introduces a negative sign:

$$-\int_{\infty}^{0} e^{-2ax^2}\, dx = \int_{0}^{\infty} e^{-2ax^2}\, dx$$

Substitute this expression for the second integral in Equation (1):

$$A^2 \left(\int_{0}^{\infty} e^{-2ax^2}\, dx + \int_{0}^{\infty} e^{-2ax^2}\, dx \right) = 1$$

$$(2) \quad 2A^2 \int_{0}^{\infty} e^{-2ax^2}\, dx = 1$$

Evaluate the integral with the help of Table B.6 in Appendix B:

$$\int_{0}^{\infty} e^{-2ax^2}\, dx = \frac{1}{2}\sqrt{\frac{\pi}{2a}}$$

Substitute this result into Equation (2) and solve for A:

$$2A^2 \left(\frac{1}{2}\sqrt{\frac{\pi}{2a}} \right) = 1 \quad \rightarrow \quad A = \boxed{\left(\frac{2a}{\pi}\right)^{1/4}}$$

(B) What is the expectation value of x for this particle?

S O L U T I O N

Evaluate the expectation value using Equation 41.8:

$$\langle x \rangle \equiv \int_{-\infty}^{\infty} \psi^* x \psi \, dx = \int_{-\infty}^{\infty} \left(Ae^{-ax^2}\right) x \left(Ae^{-ax^2}\right) dx$$

$$= A^2 \int_{-\infty}^{\infty} x e^{-2ax^2}\, dx$$

As in part (A), express the integral as a sum of two integrals:

$$(3) \quad \langle x \rangle = A^2 \left(\int_{0}^{\infty} x e^{-2ax^2}\, dx + \int_{-\infty}^{0} x e^{-2ax^2}\, dx \right)$$

Change the integration variable from x to $-x$ in the second integral:

$$\int_{-\infty}^{0} x e^{-2ax^2}\, dx = \int_{\infty}^{0} -x e^{-2a(-x)^2}\, (-dx) = \int_{\infty}^{0} x e^{-2ax^2}\, dx$$

Reverse the order of the limits, which introduces a negative sign:

$$\int_{\infty}^{0} x e^{-2ax^2}\, dx = -\int_{0}^{\infty} x e^{-2ax^2}\, dx$$

Substitute this expression for the second integral in Equation (3):

$$\langle x \rangle = A^2 \left(\int_{0}^{\infty} x e^{-2ax^2}\, dx - \int_{0}^{\infty} x e^{-2ax^2}\, dx \right) = \boxed{0}$$

Finalize Given the symmetry of the wave function around $x = 0$ in Figure 41.2, it is not surprising that the average position of the particle is at $x = 0$. In Section 41.7, we show that the wave function studied in this example represents the lowest-energy state of the quantum harmonic oscillator.

41.2 Analysis Model: Quantum Particle Under Boundary Conditions

The free particle discussed in Section 41.1 has no boundary conditions; it can be anywhere in space. The particle in Example 41.1 is not a free particle. Figure 41.2 shows that the particle is always restricted to positions near $x = 0$. In this section, we shall investigate the effects of restrictions on the motion of a quantum particle.

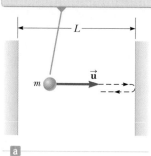

a

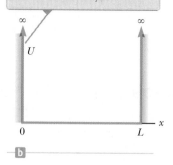

b

Figure 41.3 (a) The particle in a box. (b) The potential energy function for the system.

A Particle in a Box

We begin by applying some of the ideas we have developed to a simple physical problem, a particle confined to a one-dimensional region of space, called the *particle-in-a-box* problem (even though the "box" is one-dimensional!). From a classical viewpoint, if a particle is bouncing elastically back and forth along the x axis between two impenetrable walls separated by a distance L as in Figure 41.3a, it can be modeled as a particle under constant speed. If the speed of the particle is u, the magnitude of its momentum mu remains constant as does its kinetic energy. (Recall that in Chapter 39 we used u for particle speed to distinguish it from v, the speed of a reference frame.) Classical physics places no restrictions on the values of a particle's momentum and energy. The quantum-mechanical approach to this problem is quite different and requires that we find the appropriate wave function consistent with the conditions of the situation.

Because the walls are impenetrable, there is zero probability of finding the particle outside the box, so the wave function $\psi(x)$ must be zero for $x < 0$ and $x > L$. To be a mathematically well-behaved function, $\psi(x)$ must be continuous in space. There must be no discontinuous jumps in the value of the wave function at any point.[5] Therefore, if ψ is zero outside the walls, it must also be zero *at* the walls; that is, $\psi(0) = 0$ and $\psi(L) = 0$. Only those wave functions that satisfy these boundary conditions are allowed.

Figure 41.3b, a graphical representation of the particle-in-a-box problem, shows the potential energy of the particle–environment system as a function of the position of the particle. As long as the particle is inside the box, the potential energy of the system does not depend on the location of the particle and we can choose its constant value to be zero. Outside the box, we must ensure that the wave function is zero. We can do so by defining the system's potential energy as infinitely large if the particle were outside the box. Therefore, the only way a particle could be outside the box is if the system has an infinite amount of energy, which is impossible.

The wave function for a particle in the box can be expressed as a real sinusoidal function:[6]

$$\psi(x) = A \sin\left(\frac{2\pi x}{\lambda}\right) \tag{41.10}$$

where λ is the de Broglie wavelength associated with the particle. This wave function must satisfy the boundary conditions at the walls. The boundary condition $\psi(0) = 0$ is satisfied already because the sine function is zero when $x = 0$. The boundary condition $\psi(L) = 0$ gives

$$\psi(L) = 0 = A \sin\left(\frac{2\pi L}{\lambda}\right)$$

which can only be true if

$$\frac{2\pi L}{\lambda} = n\pi \quad \rightarrow \quad \lambda = \frac{2L}{n} \tag{41.11}$$

where $n = 1, 2, 3, \ldots$. Therefore, only certain wavelengths for the particle are allowed! Each of the allowed wavelengths corresponds to a quantum state for the system, and n is the quantum number. Incorporating Equation 41.11 in Equation 41.10 gives

Wave functions for ▶
a particle in a box

$$\psi_n(x) = A \sin\left(\frac{2\pi x}{2L/n}\right) = A \sin\left(\frac{n\pi x}{L}\right) \tag{41.12}$$

---

[5]If the wave function were not continuous at a point, the derivative of the wave function at that point would be infinite. This result leads to difficulties in the Schrödinger equation, for which the wave function is a solution as discussed in Section 41.3.

[6]We shall show this result explicitly in Section 41.3.

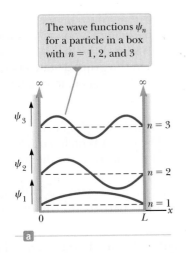

The wave functions ψ_n for a particle in a box with $n = 1, 2$, and 3

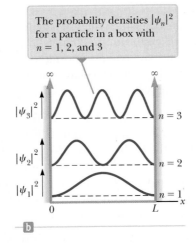

The probability densities $|\psi_n|^2$ for a particle in a box with $n = 1, 2$, and 3

Figure 41.4 The first three allowed states for a particle confined to a one-dimensional box. The states are shown superimposed on the potential energy function of Figure 41.3b. The wave functions and probability densities are plotted vertically from separate axes that are offset vertically for clarity. The positions of these axes on the potential energy function suggest the relative energies of the states.

Normalizing this wave function shows that $A = \sqrt{2/L}$. (See Problem 18.) Therefore, the normalized wave function for the particle in a box is

$$\psi_n(x) = \sqrt{\frac{2}{L}} \sin\left(\frac{n\pi x}{L}\right) \tag{41.13}$$

◀ **Normalized wave function for a particle in a box**

Figures 41.4a and b are graphical representations of ψ_n versus x and $|\psi_n|^2$ versus x for $n = 1, 2$, and 3 for the particle in a box.[7] Although a general wave function ψ can have positive and negative values, $|\psi|^2$ is always positive. Because $|\psi|^2$ represents a probability density, a negative value for $|\psi|^2$ would be meaningless.

Further inspection of Figure 41.4b shows that $|\psi|^2$ is zero at the boundaries, satisfying our boundary conditions. In addition, $|\psi|^2$ is zero at other points, depending on the value of n. For $n = 2$, $|\psi_2|^2 = 0$ at $x = L/2$; for $n = 3$, $|\psi_3|^2 = 0$ at $x = L/3$ and at $x = 2L/3$. The number of zero points increases by one each time the quantum number increases by one.

Because the wavelengths of the particle are restricted by the condition $\lambda = 2L/n$, the magnitude of the momentum of the particle is also restricted to specific values, which can be found from the expression for the de Broglie wavelength, Equation 40.17:

$$p = \frac{h}{\lambda} = \frac{h}{2L/n} = \frac{nh}{2L}$$

We have chosen the potential energy of the system to be zero when the particle is inside the box. Therefore, the energy of the system is simply the kinetic energy of the particle and the allowed values are given by

$$E_n = \tfrac{1}{2}mu^2 = \frac{p^2}{2m} = \frac{(nh/2L)^2}{2m}$$

$$E_n = \left(\frac{h^2}{8mL^2}\right)n^2 \quad n = 1, 2, 3, \ldots \tag{41.14}$$

◀ **Quantized energies for a particle in a box**

This expression shows that the energy of the particle is quantized. The lowest allowed energy corresponds to the **ground state,** which is the lowest energy state for any system. For the particle in a box, the ground state corresponds to $n = 1$, for which $E_1 = h^2/8mL^2$. Because $E_n = n^2E_1$, the **excited states** corresponding to $n = 2, 3, 4, \ldots$ have energies given by $4E_1, 9E_1, 16E_1, \ldots$.

Pitfall Prevention 41.2
Reminder: Energy Belongs to a System We often refer to the energy of a particle in commonly used language. As in Pitfall Prevention 41.1, we are actually describing the energy of the *system* of the particle and whatever environment is establishing the impenetrable walls. For the particle in a box, the only type of energy is kinetic energy belonging to the particle, which is the origin of the common description.

---

[7]Note that $n = 0$ is not allowed because, according to Equation 41.12, the wave function would be $\psi = 0$, which is not a physically reasonable wave function. For example, it cannot be normalized because $\int_{-\infty}^{\infty} |\psi|^2\, dx = \int_{-\infty}^{\infty} (0)\, dx = 0$, but Equation 41.7 tells us that this integral must equal 1.

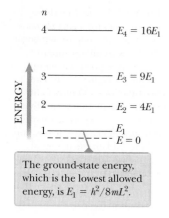

The ground-state energy, which is the lowest allowed energy, is $E_1 = h^2/8mL^2$.

Figure 41.5 Energy-level diagram for a particle confined to a one-dimensional box of length L.

Figure 41.5 is an energy-level diagram describing the energy values of the allowed states. Because the lowest energy of the particle in a box is not zero, then, according to quantum mechanics, the particle can never be at rest! The smallest energy it can have, corresponding to $n = 1$, is called the **ground-state energy.** This result contradicts the classical viewpoint, in which $E = 0$ is an acceptable state, as are *all* positive values of E.

Quick Quiz 41.2 Consider an electron, a proton, and an alpha particle (a helium nucleus), each trapped separately in identical boxes. **(i)** Which particle corresponds to the highest ground-state energy? (a) the electron (b) the proton (c) the alpha particle (d) The ground-state energy is the same in all three cases. **(ii)** Which particle has the longest wavelength when the system is in the ground state? (a) the electron (b) the proton (c) the alpha particle (d) All three particles have the same wavelength.

Quick Quiz 41.3 A particle is in a box of length L. Suddenly, the length of the box is increased to $2L$. What happens to the energy levels shown in Figure 41.5? **(a)** nothing; they are unaffected. **(b)** They move farther apart. **(c)** They move closer together.

Example 41.2 Microscopic and Macroscopic Particles in Boxes

(A) An electron is confined between two impenetrable walls 0.200 nm apart. Determine the energy levels for the states $n = 1, 2,$ and 3.

SOLUTION

Conceptualize In Figure 41.3a, imagine that the particle is an electron and the walls are very close together.

Categorize We evaluate the energy levels using an equation developed in this section, so we categorize this example as a substitution problem.

Use Equation 41.14 for the $n = 1$ state:

$$E_1 = \frac{h^2}{8m_e L^2}(1)^2 = \frac{(6.63 \times 10^{-34}\,\text{J} \cdot \text{s})^2}{8(9.11 \times 10^{-31}\,\text{kg})(2.00 \times 10^{-10}\,\text{m})^2}$$

$$= 1.51 \times 10^{-18}\,\text{J} = \boxed{9.42\,\text{eV}}$$

Using $E_n = n^2 E_1$, find the energies of the $n = 2$ and $n = 3$ states:

$$E_2 = (2)^2 E_1 = 4(9.42\,\text{eV}) = \boxed{37.7\,\text{eV}}$$

$$E_3 = (3)^2 E_1 = 9(9.42\,\text{eV}) = \boxed{84.8\,\text{eV}}$$

(B) Find the speed of the electron in the $n = 1$ state.

SOLUTION

Solve the classical expression for kinetic energy for the particle speed:

$$K = \tfrac{1}{2}m_e u^2 \quad \rightarrow \quad u = \sqrt{\frac{2K}{m_e}}$$

Recognize that the kinetic energy of the particle is equal to the system energy and substitute E_n for K:

$$(1) \quad u = \sqrt{\frac{2E_n}{m_e}}$$

Substitute numerical values from part (A):

$$u = \sqrt{\frac{2(1.51 \times 10^{-18}\,\text{J})}{9.11 \times 10^{-31}\,\text{kg}}} = \boxed{1.82 \times 10^6\,\text{m/s}}$$

Simply placing the electron in the box results in a *minimum* speed of the electron equal to 0.6% of the speed of light!

(C) A 0.500-kg baseball is confined between two rigid walls of a stadium that can be modeled as a box of length 100 m. Calculate the minimum speed of the baseball.

▶ **41.2** c o n t i n u e d

SOLUTION

Conceptualize In Figure 41.3a, imagine that the particle is a baseball and the walls are those of the stadium.

Categorize This part of the example is a substitution problem in which we apply a quantum approach to a macroscopic object.

Use Equation 41.14 for the $n = 1$ state:

$$E_1 = \frac{h^2}{8mL^2}(1)^2 = \frac{(6.63 \times 10^{-34}\,\text{J}\cdot\text{s})^2}{8(0.500\,\text{kg})(100\,\text{m})^2} = 1.10 \times 10^{-71}\,\text{J}$$

Use Equation (1) to find the speed:

$$u = \sqrt{\frac{2(1.10 \times 10^{-71}\,\text{J})}{0.500\,\text{kg}}} = \boxed{6.63 \times 10^{-36}\,\text{m/s}}$$

This speed is so small that the object can be considered to be at rest, which is what one would expect for the minimum speed of a macroscopic object.

WHAT IF? What if a sharp line drive is hit so that the baseball is moving with a speed of 150 m/s? What is the quantum number of the state in which the baseball now resides?

Answer We expect the quantum number to be very large because the baseball is a macroscopic object.

Evaluate the kinetic energy of the baseball: $\qquad \frac{1}{2}mu^2 = \frac{1}{2}(0.500\,\text{kg})(150\,\text{m/s})^2 = 5.62 \times 10^3\,\text{J}$

From Equation 41.14, calculate the quantum number n:

$$n = \sqrt{\frac{8mL^2E_n}{h^2}} = \sqrt{\frac{8(0.500\,\text{kg})(100\,\text{m})^2(5.62 \times 10^3\,\text{J})}{(6.63 \times 10^{-34}\,\text{J}\cdot\text{s})^2}} = 2.26 \times 10^{37}$$

This result is a tremendously large quantum number. As the baseball pushes air out of the way, hits the ground, and rolls to a stop, it moves through more than 10^{37} quantum states. These states are so close together in energy that we cannot observe the transitions from one state to the next. Rather, we see what appears to be a smooth variation in the speed of the ball. The quantum nature of the universe is simply not evident in the motion of macroscopic objects.

Example 41.3 The Expectation Values for the Particle in a Box

A particle of mass m is confined to a one-dimensional box between $x = 0$ and $x = L$. Find the expectation value of the position x of the particle in the state characterized by quantum number n.

SOLUTION

Conceptualize Figure 41.4b shows that the probability for the particle to be at a given location varies with position within the box. Can you predict what the expectation value of x will be from the symmetry of the wave functions?

Categorize The statement of the example categorizes the problem for us: we focus on a quantum particle in a box and on the calculation of its expectation value of x.

Analyze In Equation 41.8, the integration from $-\infty$ to ∞ reduces to the limits 0 to L because $\psi = 0$ everywhere except in the box.

Substitute Equation 41.13 into Equation 41.8 to find the expectation value for x:

$$\langle x \rangle = \int_{-\infty}^{\infty} \psi_n{}^* x \psi_n \, dx = \int_0^L x \left[\sqrt{\frac{2}{L}} \sin\left(\frac{n\pi x}{L}\right) \right]^2 dx$$

$$= \frac{2}{L} \int_0^L x \sin^2\left(\frac{n\pi x}{L}\right) dx$$

continued

▶ **41.3** continued

Evaluate the integral by consulting an integral table or by mathematical integration:[8]

$$\langle x \rangle = \frac{2}{L}\left[\frac{x^2}{4} - \frac{x\sin\left(2\frac{n\pi x}{L}\right)}{4\frac{n\pi}{L}} - \frac{\cos\left(2\frac{n\pi x}{L}\right)}{8\left(\frac{n\pi}{L}\right)^2} \right]_0^L$$

$$= \frac{2}{L}\left[\frac{L^2}{4}\right] = \boxed{\frac{L}{2}}$$

Finalize This result shows that the expectation value of x is at the center of the box for all values of n, which you would expect from the symmetry of the square of the wave functions (the probability density) about the center (Fig. 41.4b).

The $n = 2$ wave function in Figure 41.4b has a value of zero at the midpoint of the box. Can the expectation value of the particle be at a position at which the particle has zero probability of existing? Remember that the expectation value is the *average* position. Therefore, the particle is as likely to be found to the right of the midpoint as to the left, so its average position is at the midpoint even though its probability of being there is zero. As an analogy, consider a group of students for whom the average final examination score is 50%. There is no requirement that some student achieve a score of exactly 50% for the average of all students to be 50%.

Boundary Conditions on Particles in General

The discussion of the particle in a box is very similar to the discussion in Chapter 18 of standing waves on strings:

- Because the ends of the string must be nodes, the wave functions for allowed waves must be zero at the boundaries of the string. Because the particle in a box cannot exist outside the box, the allowed wave functions for the particle must be zero at the boundaries.
- The boundary conditions on the string waves lead to quantized wavelengths and frequencies of the waves. The boundary conditions on the wave function for the particle in a box lead to quantized wavelengths and frequencies of the particle.

In quantum mechanics, it is very common for particles to be subject to boundary conditions. We therefore introduce a new analysis model, the **quantum particle under boundary conditions.** In many ways, this model is similar to the waves under boundary conditions model studied in Section 18.3. In fact, the allowed wavelengths for the wave function of a particle in a box (Eq. 41.11) are identical in form to the allowed wavelengths for mechanical waves on a string fixed at both ends (Eq. 18.4).

The quantum particle under boundary conditions model *differs* in some ways from the waves under boundary conditions model:

- In most cases of quantum particles, the wave function is *not* a simple sinusoidal function like the wave function for waves on strings. Furthermore, the wave function for a quantum particle may be a complex function.
- For a quantum particle, frequency is related to energy through $E = hf$, so the quantized frequencies lead to quantized energies.
- There may be no stationary "nodes" associated with the wave function of a quantum particle under boundary conditions. Systems more complicated than the particle in a box have more complicated wave functions, and some boundary conditions may not lead to zeroes of the wave function at fixed points.

---

[8]To integrate this function, first replace $\sin^2(n\pi x/L)$ with $\frac{1}{2}(1 - \cos 2n\pi x/L)$ (refer to Table B.3 in Appendix B), which allows $\langle x \rangle$ to be expressed as two integrals. The second integral can then be evaluated by partial integration (Section B.7 in Appendix B).

In general,

an interaction of a quantum particle with its environment represents one or more boundary conditions, and, if the interaction restricts the particle to a finite region of space, results in quantization of the energy of the system.

Boundary conditions on quantum wave functions are related to the coordinates describing the problem. For the particle in a box, the wave function must be zero at two values of x. In the case of a three-dimensional system such as the hydrogen atom we shall discuss in Chapter 42, the problem is best presented in *spherical coordinates*. These coordinates, an extension of the plane polar coordinates introduced in Section 3.1, consist of a radial coordinate r and two angular coordinates. The generation of the wave function and application of the boundary conditions for the hydrogen atom are beyond the scope of this book. We shall, however, examine the behavior of some of the hydrogen-atom wave functions in Chapter 42.

Boundary conditions on wave functions that exist for all values of x require that the wave function approach zero as $x \to \infty$ (so that the wave function can be normalized) and remain finite as $x \to 0$. One boundary condition on any angular parts of wave functions is that adding 2π radians to the angle must return the wave function to the same value because an addition of 2π results in the same angular position.

Analysis Model | **Quantum Particle Under Boundary Conditions**

Imagine a particle described by quantum physics that is subject to one or more boundary conditions. If the particle is restricted to a finite region of space by the boundary conditions, the energy of the system is quantized. Associated with each quantized energy is a quantum state characterized by a wave function and a quantum number.

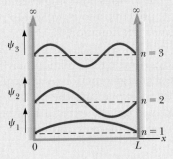

Examples:

- an electron in a quantum dot cannot escape, quantizing the energies of the electron (Section 41.4)
- an electron in a hydrogen atom is restricted to stay near the nucleus of the atom, quantizing the energies of the atom (Chapter 42)
- two atoms are bound to form a diatomic molecule, quantizing the energies of vibration and rotation of the molecule (Chapter 43)
- a proton is trapped in a nucleus, quantizing its energy levels (Chapter 44)

41.3 The Schrödinger Equation

In Section 34.3, we discussed a linear wave equation for electromagnetic radiation that follows from Maxwell's equations. The waves associated with particles also satisfy a wave equation. The wave equation for material particles is different from that associated with photons because material particles have a nonzero rest energy. The appropriate wave equation was developed by Schrödinger in 1926. In analyzing the behavior of a quantum system, the approach is to determine a solution to this equation and then apply the appropriate boundary conditions to the solution. This process yields the allowed wave functions and energy levels of the system under consideration. Proper manipulation of the wave function then enables one to calculate all measurable features of the system.

The Schrödinger equation as it applies to a particle of mass m confined to moving along the x axis and interacting with its environment through a potential energy function $U(x)$ is

$$-\frac{\hbar^2}{2m}\frac{d^2\psi}{dx^2} + U\psi = E\psi \qquad (41.15)$$

◀ Time-independent Schrödinger equation

Erwin Schrödinger
Austrian Theoretical Physicist
(1887–1961)
Schrödinger is best known as one of the creators of quantum mechanics. His approach to quantum mechanics was demonstrated to be mathematically equivalent to the more abstract matrix mechanics developed by Heisenberg. Schrödinger also produced important papers in the fields of statistical mechanics, color vision, and general relativity.

© INTERFOTO/Alamy

where E is a constant equal to the total energy of the system (the particle and its environment). Because this equation is independent of time, it is commonly referred to as the **time-independent Schrödinger equation.** (We shall not discuss the time-dependent Schrödinger equation in this book.)

The Schrödinger equation is consistent with the principle of conservation of mechanical energy for an isolated system with no nonconservative forces acting. Problem 44 shows, both for a free particle and a particle in a box, that the first term in the Schrödinger equation reduces to the kinetic energy of the particle multiplied by the wave function. Therefore, Equation 41.15 indicates that the total energy of the system is the sum of the kinetic energy and the potential energy and that the total energy is a constant: $K + U = E =$ constant.

In principle, if the potential energy function U for a system is known, one can solve Equation 41.15 and obtain the wave functions and energies for the allowed states of the system. In addition, in many cases, the wave function ψ must satisfy boundary conditions. Therefore, once we have a preliminary solution to the Schrödinger equation, we impose the following conditions to find the exact solution and the allowed energies:

- ψ must be normalizable. That is, Equation 41.7 must be satisfied.
- ψ must go to 0 as $x \rightarrow \pm\infty$ and remain finite as $x \rightarrow 0$.
- ψ must be continuous in x and be single-valued everywhere; solutions to Equation 41.15 in different regions must join smoothly at the boundaries between the regions.
- $d\psi/dx$ must be finite, continuous, and single-valued everywhere for finite values of U. If $d\psi/dx$ were not continuous, we would not be able to evaluate the second derivative $d^2\psi/dx^2$ in Equation 41.15 at the point of discontinuity.

The task of solving the Schrödinger equation may be very difficult, depending on the form of the potential energy function. As it turns out, the Schrödinger equation is extremely successful in explaining the behavior of atomic and nuclear systems, whereas classical physics fails to explain this behavior. Furthermore, when quantum mechanics is applied to macroscopic objects, the results agree with classical physics.

The Particle in a Box Revisited

To see how the quantum particle under boundary conditions model is applied to a problem, let's return to our particle in a one-dimensional box of length L (see Fig. 41.3) and analyze it with the Schrödinger equation. Figure 41.3b is the potential-energy diagram that describes this problem. Potential-energy diagrams are a useful representation for understanding and solving problems with the Schrödinger equation.

Because of the shape of the curve in Figure 41.3b, the particle in a box is sometimes said to be in a **square well,**[9] where a **well** is an upward-facing region of the curve in a potential-energy diagram. (A downward-facing region is called a *barrier,* which we investigate in Section 41.5.) Figure 41.3b shows an infinite square well.

In the region $0 < x < L$, where $U = 0$, we can express the Schrödinger equation in the form

$$\frac{d^2\psi}{dx^2} = -\frac{2mE}{\hbar^2}\psi = -k^2\psi \tag{41.16}$$

where

$$k = \frac{\sqrt{2mE}}{\hbar}$$

---

[9]It is called a square well even if it has a rectangular shape in a potential-energy diagram.

The solution to Equation 41.16 is a function ψ whose second derivative is the negative of the same function multiplied by a constant k^2. Both the sine and cosine functions satisfy this requirement. Therefore, the most general solution to the equation is a linear combination of both solutions:

$$\psi(x) = A \sin kx + B \cos kx$$

where A and B are constants that are determined by the boundary and normalization conditions.

The first boundary condition on the wave function is that $\psi(0) = 0$:

$$\psi(0) = A \sin 0 + B \cos 0 = 0 + B = 0$$

which means that $B = 0$. Therefore, our solution reduces to

$$\psi(x) = A \sin kx$$

The second boundary condition, $\psi(L) = 0$, when applied to the reduced solution gives

$$\psi(L) = A \sin kL = 0$$

This equation could be satisfied by setting $A = 0$, but that would mean that $\psi = 0$ everywhere, which is not a valid wave function. The boundary condition is also satisfied if kL is an integral multiple of π, that is, if $kL = n\pi$, where n is an integer. Substituting $k = \sqrt{2mE}/\hbar$ into this expression gives

$$kL = \frac{\sqrt{2mE}}{\hbar} L = n\pi$$

Each value of the integer n corresponds to a quantized energy that we call E_n. Solving for the allowed energies E_n gives

$$E_n = \left(\frac{h^2}{8mL^2}\right) n^2 \tag{41.17}$$

which are identical to the allowed energies in Equation 41.14.

Substituting the values of k in the wave function, the allowed wave functions $\psi_n(x)$ are given by

$$\psi_n(x) = A \sin\left(\frac{n\pi x}{L}\right) \tag{41.18}$$

which is the wave function (Eq. 41.12) used in our initial discussion of the particle in a box.

41.4 A Particle in a Well of Finite Height

Now consider a particle in a *finite* potential well, that is, a system having a potential energy that is zero when the particle is in the region $0 < x < L$ and a finite value U when the particle is outside this region as in Figure 41.6. Classically, if the total energy E of the system is less than U, the particle is permanently bound in the potential well. If the particle were outside the well, its kinetic energy would have to be negative, which is an impossibility. According to quantum mechanics, however, a finite probability exists that the particle can be found outside the well even if $E < U$. That is, the wave function ψ is generally nonzero outside the well—regions I and III in Figure 41.6—so the probability density $|\psi|^2$ is also nonzero in these regions. Although this notion may be uncomfortable to accept, the uncertainty principle indicates that the energy of the system is uncertain. This uncertainty allows the particle to be outside the well as long as the apparent violation of conservation of energy does not exist in any measurable way.

In region II, where $U = 0$, the allowed wave functions are again sinusoidal because they represent solutions of Equation 41.16. The boundary conditions, however,

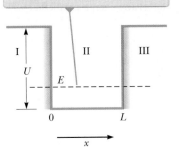

If the total energy E of the particle–well system is less than U, the particle is trapped in the well.

Figure 41.6 Potential-energy diagram of a well of finite height U and length L.

The wave functions ψ_n for a particle in a potential well of finite height with $n = 1, 2,$ and 3

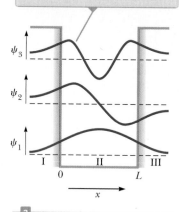

a

The probability densities $|\psi_n|^2$ for a particle in a potential well of finite height with $n = 1, 2,$ and 3

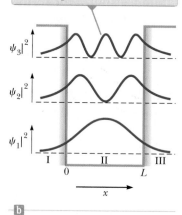

b

Figure 41.7 The first three allowed states for a particle in a potential well of finite height. The states are shown superimposed on the potential energy function of Figure 41.6. The wave functions and probability densities are plotted vertically from separate axes that are offset vertically for clarity. The positions of these axes on the potential energy function suggest the relative energies of the states.

no longer require that ψ be zero at the ends of the well, as was the case with the infinite square well.

The Schrödinger equation for regions I and III may be written

$$\frac{d^2\psi}{dx^2} = \frac{2m(U - E)}{\hbar^2}\psi \qquad \textbf{(41.19)}$$

Because $U > E$, the coefficient of ψ on the right-hand side is necessarily positive. Therefore, we can express Equation 41.19 as

$$\frac{d^2\psi}{dx^2} = C^2\psi \qquad \textbf{(41.20)}$$

where $C^2 = 2m(U - E)/\hbar^2$ is a positive constant in regions I and III. As you can verify by substitution, the general solution of Equation 41.20 is

$$\psi = Ae^{Cx} + Be^{-Cx} \qquad \textbf{(41.21)}$$

where A and B are constants.

We can use this general solution as a starting point for determining the appropriate solution for regions I and III. The solution must remain finite as $x \to \pm\infty$. Therefore, in region I, where $x < 0$, the function ψ cannot contain the term Be^{-Cx}. This requirement is handled by taking $B = 0$ in this region to avoid an infinite value for ψ for large negative values of x. Likewise, in region III, where $x > L$, the function ψ cannot contain the term Ae^{Cx}. This requirement is handled by taking $A = 0$ in this region to avoid an infinite value for ψ for large positive x values. Hence, the solutions in regions I and III are

$$\psi_I = Ae^{Cx} \qquad \text{for } x < 0$$

$$\psi_{III} = Be^{-Cx} \qquad \text{for } x > L$$

In region II, the wave function is sinusoidal and has the general form

$$\psi_{II}(x) = F\sin kx + G\cos kx$$

where F and G are constants.

These results show that the wave functions outside the potential well (where classical physics forbids the presence of the particle) decay exponentially with distance. At large negative x values, ψ_I approaches zero; at large positive x values, ψ_{III} approaches zero. These functions, together with the sinusoidal solution in region II, are shown in Figure 41.7a for the first three energy states. In evaluating the complete wave function, we impose the following boundary conditions:

$$\psi_I = \psi_{II} \quad \text{and} \quad \frac{d\psi_I}{dx} = \frac{d\psi_{II}}{dx} \qquad \text{at } x = 0$$

$$\psi_{II} = \psi_{III} \quad \text{and} \quad \frac{d\psi_{II}}{dx} = \frac{d\psi_{III}}{dx} \qquad \text{at } x = L$$

These four boundary conditions and the normalization condition (Eq. 41.7) are sufficient to determine the four constants A, B, F, and G and the allowed values of the energy E. Figure 41.7b plots the probability densities for these states. In each case, the wave functions inside and outside the potential well join smoothly at the boundaries.

The notion of trapping particles in potential wells is used in the burgeoning field of **nanotechnology,** which refers to the design and application of devices having dimensions ranging from 1 to 100 nm. The fabrication of these devices often involves manipulating single atoms or small groups of atoms to form very tiny structures or mechanisms.

One area of nanotechnology of interest to researchers is the **quantum dot,** a small region that is grown in a silicon crystal and acts as a potential well. This region can trap electrons into states with quantized energies. The wave functions for a particle in a quantum dot look similar to those in Figure 41.7a if L is on the order of nanometers. The storage of binary information using quantum dots is an active field of research. A simple binary scheme would involve associating a one with a quantum dot containing an electron and a zero with an empty dot. Other schemes involve cells of multiple dots such that arrangements of electrons among the dots correspond to ones and zeroes. Several research laboratories are studying the properties and potential applications of quantum dots. Information should be forthcoming from these laboratories at a steady rate in the next few years.

41.5 Tunneling Through a Potential Energy Barrier

Consider the potential energy function shown in Figure 41.8. In this situation, the potential energy has a constant value of U in the region of width L and is zero in all other regions.[10] A potential energy function of this shape is called a **square barrier,** and U is called the **barrier height.** A very interesting and peculiar phenomenon occurs when a moving particle encounters such a barrier of finite height and width. Suppose a particle of energy $E < U$ is incident on the barrier from the left (Fig. 41.8). Classically, the particle is reflected by the barrier. If the particle were located in region II, its kinetic energy would be negative, which is not classically allowed. Consequently, region II and therefore region III are both classically *forbidden* to the particle incident from the left. According to quantum mechanics, however, all regions are accessible to the particle, regardless of its energy. (Although all regions are accessible, the probability of the particle being in a classically forbidden region is very low.) According to the uncertainty principle, the particle could be within the barrier as long as the time interval during which it is in the barrier is short and consistent with Equation 40.26. If the barrier is relatively narrow, this short time interval can allow the particle to pass through the barrier.

Let's approach this situation using a mathematical representation. The Schrödinger equation has valid solutions in all three regions. The solutions in regions I and III are sinusoidal like Equation 41.18. In region II, the solution is exponential like Equation 41.21. Applying the boundary conditions that the wave functions in the three regions and their derivatives must join smoothly at the boundaries, a full solution, such as the one represented by the curve in Figure 41.8, can be found. Because the probability of locating the particle is proportional to $|\psi|^2$, the probability of finding the particle beyond the barrier in region III is nonzero. This result is in complete disagreement with classical physics. The movement of the particle to the far side of the barrier is called **tunneling** or **barrier penetration.**

The probability of tunneling can be described with a **transmission coefficient** T and a **reflection coefficient** R. The transmission coefficient represents the probability that the particle penetrates to the other side of the barrier, and the reflection coefficient is the probability that the particle is reflected by the barrier. Because the incident particle is either reflected or transmitted, we require that $T + R = 1$. An approximate expression for the transmission coefficient that is obtained in the case of $T \ll 1$ (a very wide barrier or a very high barrier, that is, $U \gg E$) is

$$T \approx e^{-2CL} \tag{41.22}$$

where

$$C = \frac{\sqrt{2m(U - E)}}{\hbar} \tag{41.23}$$

This quantum model of barrier penetration and specifically Equation 41.22 show that T can be nonzero. That the phenomenon of tunneling is observed experimentally provides further confidence in the principles of quantum physics.

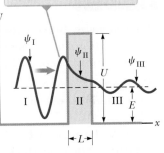

The wave function is sinusoidal in regions I and III, but is exponentially decaying in region II.

Figure 41.8 Wave function ψ for a particle incident from the left on a barrier of height U and width L. The wave function is plotted vertically from an axis positioned at the energy of the particle.

Pitfall Prevention 41.4
"Height" on an Energy Diagram The word *height* (as in *barrier height*) refers to an energy in discussions of barriers in potential-energy diagrams. For example, we might say the height of the barrier is 10 eV. On the other hand, the barrier *width* refers to the traditional usage of such a word and is an actual physical length measurement between the locations of the two vertical sides of the barrier.

---

[10]It is common in physics to refer to L as the *length* of a well but the *width* of a barrier.

Quick Quiz 41.4 Which of the following changes would increase the probability of transmission of a particle through a potential barrier? (You may choose more than one answer.) **(a)** decreasing the width of the barrier **(b)** increasing the width of the barrier **(c)** decreasing the height of the barrier **(d)** increasing the height of the barrier **(e)** decreasing the kinetic energy of the incident particle **(f)** increasing the kinetic energy of the incident particle

---

Example 41.4 **Transmission Coefficient for an Electron**

A 30-eV electron is incident on a square barrier of height 40 eV.

(A) What is the probability that the electron tunnels through the barrier if its width is 1.0 nm?

SOLUTION

Conceptualize Because the particle energy is smaller than the height of the potential barrier, we expect the electron to reflect from the barrier with a probability of 100% according to classical physics. Because of the tunneling phenomenon, however, there is a finite probability that the particle can appear on the other side of the barrier.

Categorize We evaluate the probability using an equation developed in this section, so we categorize this example as a substitution problem.

Evaluate the quantity $U - E$ that appears in Equation 41.23:

$$U - E = 40 \text{ eV} - 30 \text{ eV} = 10 \text{ eV} \left(\frac{1.6 \times 10^{-19} \text{J}}{1 \text{ eV}} \right) = 1.6 \times 10^{-18} \text{J}$$

Evaluate the quantity $2CL$ using Equation 41.23:

$$(1) \quad 2CL = 2 \frac{\sqrt{2(9.11 \times 10^{-31} \text{ kg})(1.6 \times 10^{-18} \text{J})}}{1.055 \times 10^{-34} \text{J} \cdot \text{s}} (1.0 \times 10^{-9} \text{ m}) = 32.4$$

From Equation 41.22, find the probability of tunneling through the barrier:

$$T \approx e^{-2CL} = e^{-32.4} = \boxed{8.5 \times 10^{-15}}$$

(B) What is the probability that the electron tunnels through the barrier if its width is 0.10 nm?

SOLUTION

In this case, the width L in Equation (1) is one-tenth as large, so evaluate the new value of $2CL$:

$$2CL = (0.1)(32.4) = 3.24$$

From Equation 41.22, find the new probability of tunneling through the barrier:

$$T \approx e^{-2CL} = e^{-3.24} = \boxed{0.039}$$

In part (A), the electron has approximately 1 chance in 10^{14} of tunneling through the barrier. In part (B), however, the electron has a much higher probability (3.9%) of penetrating the barrier. Therefore, reducing the width of the barrier by only one order of magnitude increases the probability of tunneling by about 12 orders of magnitude!

41.6 Applications of Tunneling

As we have seen, tunneling is a quantum phenomenon, a manifestation of the wave nature of matter. Many examples exist (on the atomic and nuclear scales) for which tunneling is very important.

Alpha Decay

One form of radioactive decay is the emission of alpha particles (the nuclei of helium atoms) by unstable, heavy nuclei (Chapter 44). To escape from the nucleus, an alpha particle must penetrate a barrier whose height is several times larger than the energy of the nucleus–alpha particle system as shown in Figure 41.9. The barrier results from a combination of the attractive nuclear force (discussed in Chapter 44) and the Coulomb repulsion (discussed in Chapter 23) between the alpha particle and the rest of the nucleus. Occasionally, an alpha particle tunnels through the barrier, which explains the basic mechanism for this type of decay and the large variations in the mean lifetimes of various radioactive nuclei.

Figure 41.8 shows the wave function of a particle tunneling through a barrier in one dimension. A similar wave function having spherical symmetry describes the barrier penetration of an alpha particle leaving a radioactive nucleus. The wave function exists both inside and outside the nucleus, and its amplitude is constant in time. In this way, the wave function correctly describes the small but constant probability that the nucleus will decay. The moment of decay cannot be predicted. In general, quantum mechanics implies that the future is indeterminate. This feature is in contrast to classical mechanics, from which the trajectory of an object can be calculated to arbitrarily high precision from precise knowledge of its initial position and velocity and of the forces exerted on it. Do not think that the future is undetermined simply because we have incomplete information about the present. The wave function contains all the information about the state of a system. Sometimes precise predictions can be made, such as the energy of a bound system, but sometimes only probabilities can be calculated about the future. The fundamental laws of nature are probabilistic. Therefore, it appears that Einstein's famous statement about quantum mechanics, "God does not roll dice," was wrong.

A radiation detector can be used to show that a nucleus decays by emitting a particle at a particular moment and in a particular direction. To point out the contrast between this experimental result and the wave function describing it, Schrödinger imagined a box containing a cat, a radioactive sample, a radiation counter, and a vial of poison. When a nucleus in the sample decays, the counter triggers the administration of lethal poison to the cat. Quantum mechanics correctly predicts the probability of finding the cat dead when the box is opened. Before the box is opened, does the cat have a wave function describing it as fractionally dead, with some chance of being alive?

This question is under continuing investigation, never with actual cats but sometimes with interference experiments building upon the experiment described in Section 40.7. Does the act of measurement change the system from a probabilistic to a definite state? When a particle emitted by a radioactive nucleus is detected at one particular location, does the wave function describing the particle drop instantaneously to zero everywhere else in the Universe? (Einstein called such a state change a "spooky action at a distance.") Is there a fundamental difference between a quantum system and a macroscopic system? The answers to these questions are unknown.

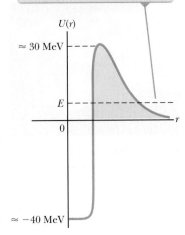

The alpha particle can tunnel through the barrier and escape from the nucleus even though its energy is lower than the height of the well.

$U(r)$

≈ 30 MeV

E

0

r

≈ -40 MeV

Figure 41.9 The potential well for an alpha particle in a nucleus.

Nuclear Fusion

The basic reaction that powers the Sun and, indirectly, almost everything else in the solar system is fusion, which we shall study in Chapter 45. In one step of the process that occurs at the core of the Sun, protons must approach one another to within such a small distance that they fuse and form a deuterium nucleus. (See Section 45.4.) According to classical physics, these protons cannot overcome and penetrate the barrier caused by their mutual electrical repulsion. Quantum mechanically, however, the protons are able to tunnel through the barrier and fuse together.

Scanning Tunneling Microscopes

The scanning tunneling microscope (STM) enables scientists to obtain highly detailed images of surfaces at resolutions comparable to the size of a *single atom*. Figure 41.10 (page 1284), showing the surface of a piece of graphite, demonstrates what STMs can do. What makes this image so remarkable is that its resolution is

The contours seen here represent the ring-like arrangement of individual carbon atoms on the crystal surface.

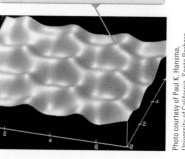

Photo courtesy of Paul K. Hansma, University of California, Santa Barbara

Figure 41.10 The surface of graphite as "viewed" with a scanning tunneling microscope. This type of microscope enables scientists to see details with a lateral resolution of about 0.2 nm and a vertical resolution of 0.001 nm.

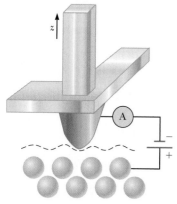

Figure 41.11 Schematic view of a scanning tunneling microscope. A scan of the tip over the sample can reveal surface contours down to the atomic level. An STM image is composed of a series of scans displaced laterally from one another. (Based on a drawing from P. K. Hansma, V. B. Elings, O. Marti, and C. Bracker, *Science* **242**:209, 1988. © 1988 by the AAAS.)

approximately 0.2 nm. For an optical microscope, the resolution is limited by the wavelength of the light used to make the image. Therefore, an optical microscope has a resolution no better than 200 nm, about half the wavelength of visible light, and so could never show the detail displayed in Figure 41.10.

Scanning tunneling microscopes achieve such high resolution by using the basic idea shown in Figure 41.11. An electrically conducting probe with a very sharp tip is brought near the surface to be studied. The empty space between tip and surface represents the "barrier" we have been discussing, and the tip and surface are the two walls of the "potential well." Because electrons obey quantum rules rather than Newtonian rules, they can "tunnel" across the barrier of empty space. If a voltage is applied between surface and tip, electrons in the atoms of the surface material can tunnel preferentially from surface to tip to produce a tunneling current. In this way, the tip samples the distribution of electrons immediately above the surface.

In the empty space between tip and surface, the electron wave function falls off exponentially (see region II in Fig. 41.8 and Example 41.4). For tip-to-surface distances $z > 1$ nm (that is, beyond a few atomic diameters), essentially no tunneling takes place. This exponential behavior causes the current of electrons tunneling from surface to tip to depend very strongly on z. By monitoring the tunneling current as the tip is scanned over the surface, scientists obtain a sensitive measure of the topography of the electron distribution on the surface. The result of this scan is used to make images like that in Figure 41.10. In this way, the STM can measure the height of surface features to within 0.001 nm, approximately 1/100 of an atomic diameter!

You can appreciate the sensitivity of STMs by examining Figure 41.10. Of the six carbon atoms in each ring, three appear lower than the other three. In fact, all six atoms are at the same height, but all have slightly different electron distributions. The three atoms that appear lower are bonded to other carbon atoms directly beneath them in the underlying atomic layer; as a result, their electron distributions, which are responsible for the bonding, extend downward beneath the surface. The atoms in the surface layer that appear higher do not lie directly over subsurface atoms and hence are not bonded to any underlying atoms. For these higher-appearing atoms, the electron distribution extends upward into the space above the surface. Because STMs map the topography of the electron distribution, this extra electron density makes these atoms appear higher in Figure 41.10.

The STM has one serious limitation: Its operation depends on the electrical conductivity of the sample and the tip. Unfortunately, most materials are not electrically conductive at their surfaces. Even metals, which are usually excellent electrical conductors, are covered with nonconductive oxides. A newer microscope, the atomic force microscope, or AFM, overcomes this limitation.

Resonant Tunneling Devices

Let's expand on the quantum-dot discussion in Section 41.4 by exploring the **resonant tunneling device.** Figure 41.12a shows the physical construction of such a device. The island of gallium arsenide in the center is a quantum dot located between two barriers formed from the thin extensions of aluminum arsenide. Figure 41.12b shows both the potential barriers encountered by electrons incident from the left and the quantized energy levels in the quantum dot. This situation differs from the one shown in Figure 41.8 in that there are quantized energy levels on the right of the first barrier. In Figure 41.8, an electron that tunnels through the barrier is considered a free particle and can have any energy. In contrast, the second barrier in Figure 41.12b imposes boundary conditions on the particle and quantizes its energy in the quantum dot. In Figure 41.12b, as the electron with the energy shown encounters the first barrier, it has no matching energy levels available on the right side of the barrier, which greatly reduces the probability of tunneling.

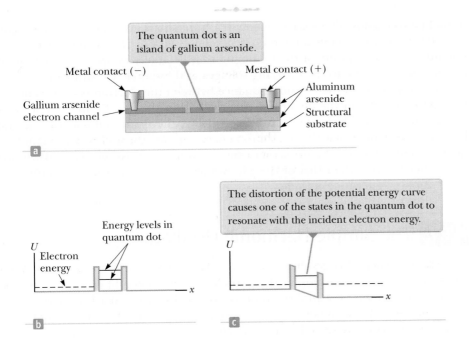

Figure 41.12 (a) The physical structure of a resonant tunneling device. (b) A potential-energy diagram showing the double barrier representing the walls of the quantum dot. (c) A voltage is applied across the device.

Figure 41.12c shows the effect of applying a voltage: the potential decreases with position as we move to the right across the device. The deformation of the potential barrier results in an energy level in the quantum dot coinciding with the energy of the incident electrons. This "resonance" of energies gives the device its name. When the voltage is applied, the probability of tunneling increases tremendously and the device carries current. In this manner, the device can be used as a very fast switch on a nanotechnological scale.

Resonant Tunneling Transistors

Figure 41.13a shows the addition of a gate electrode at the top of the resonant tunneling device over the quantum dot. This electrode turns the device into a **resonant**

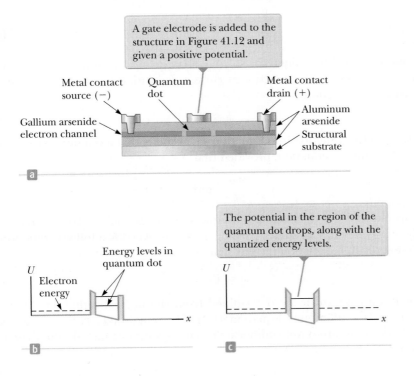

Figure 41.13 (a) A resonant tunneling transistor. (b) A potential-energy diagram showing the double barrier representing the walls of the quantum dot. (c) A voltage is applied to the gate electrode.

tunneling transistor. The basic function of a transistor is amplification, converting a small varying voltage into a large varying voltage. Figure 41.13b, representing the potential-energy diagram for the tunneling transistor, has a slope at the bottom of the quantum dot due to the differing voltages at the source and drain electrodes. In this configuration, there is no resonance between the electron energies outside the quantum dot and the quantized energies within the dot. By applying a small voltage to the gate electrode as in Figure 41.13c, the quantized energies can be brought into resonance with the electron energy outside the well and resonant tunneling occurs. The resulting current causes a voltage across an external resistor that is much larger than that of the gate voltage; hence, the device amplifies the input signal to the gate electrode.

41.7 The Simple Harmonic Oscillator

Consider a particle that is subject to a linear restoring force $F = -kx$, where k is a constant and x is the position of the particle relative to equilibrium ($x = 0$). The classical description of such a situation is provided by the particle in simple harmonic motion analysis model, which was discussed in Chapter 15. The potential energy of the system is, from Equation 15.20,

$$U = \tfrac{1}{2}kx^2 = \tfrac{1}{2}m\omega^2 x^2$$

where the angular frequency of vibration is $\omega = \sqrt{k/m}$. Classically, if the particle is displaced from its equilibrium position and released, it oscillates between the points $x = -A$ and $x = A$, where A is the amplitude of the motion. Furthermore, its total energy E is, from Equation 15.21,

$$E = K + U = \tfrac{1}{2}kA^2 = \tfrac{1}{2}m\omega^2 A^2$$

In the classical model, any value of E is allowed, including $E = 0$, which is the total energy when the particle is at rest at $x = 0$.

Let's investigate how the simple harmonic oscillator is treated from a quantum point of view. The Schrödinger equation for this problem is obtained by substituting $U = \tfrac{1}{2}m\omega^2 x^2$ into Equation 41.15:

$$-\frac{\hbar^2}{2m}\frac{d^2\psi}{dx^2} + \tfrac{1}{2}m\omega^2 x^2\psi = E\psi \tag{41.24}$$

The mathematical technique for solving this equation is beyond the level of this book; nonetheless, it is instructive to guess at a solution. We take as our guess the following wave function:

$$\psi = Be^{-Cx^2} \tag{41.25}$$

Substituting this function into Equation 41.24 shows that it is a satisfactory solution to the Schrödinger equation, provided that

$$C = \frac{m\omega}{2\hbar} \quad \text{and} \quad E = \tfrac{1}{2}\hbar\omega$$

It turns out that the solution we have guessed corresponds to the ground state of the system, which has an energy $\tfrac{1}{2}\hbar\omega$. Because $C = m\omega/2\hbar$, it follows from Equation 41.25 that the wave function for this state is

◄ Wave function for the ground state of a simple harmonic oscillator

$$\psi = Be^{-(m\omega/2\hbar)x^2} \tag{41.26}$$

where B is a constant to be determined from the normalization condition. This result is but one solution to Equation 41.24. The remaining solutions that describe the excited states are more complicated, but all solutions include the exponential factor e^{-Cx^2}.

The energy levels of a harmonic oscillator are quantized as we would expect because the oscillating particle is bound to stay near $x = 0$. The energy of a state having an arbitrary quantum number n is

$$E_n = \left(n + \tfrac{1}{2}\right)\hbar\omega \quad n = 0, 1, 2, \ldots \qquad \text{(41.27)}$$

The state $n = 0$ corresponds to the ground state, whose energy is $E_0 = \tfrac{1}{2}\hbar\omega$; the state $n = 1$ corresponds to the first excited state, whose energy is $E_1 = \tfrac{3}{2}\hbar\omega$; and so on. The energy-level diagram for this system is shown in Figure 41.14. The separations between adjacent levels are equal and given by

$$\Delta E = \hbar\omega \qquad \text{(41.28)}$$

Notice that the energy levels for the harmonic oscillator in Figure 41.14 are equally spaced, just as Planck proposed for the oscillators in the walls of the cavity that was used in the model for blackbody radiation in Section 40.1. Planck's Equation 40.4 for the energy levels of the oscillators differs from Equation 41.27 only in the term $\tfrac{1}{2}$ added to n. This additional term does not affect the energy emitted in a transition, given by Equation 40.5, which is equivalent to Equation 41.28. That Planck generated these concepts without the benefit of the Schrödinger equation is testimony to his genius.

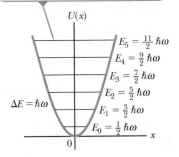

The levels are equally spaced, with separation $\hbar\omega$. The ground-state energy is $E_0 = \tfrac{1}{2}\hbar\omega$.

Figure 41.14 Energy-level diagram for a simple harmonic oscillator, superimposed on the potential energy function.

Example 41.5 | Molar Specific Heat of Hydrogen Gas

In Figure 21.6 (Section 21.3), which shows the molar specific heat of hydrogen as a function of temperature, vibration does not contribute to the molar specific heat at room temperature. Explain why, modeling the hydrogen molecule as a simple harmonic oscillator. The effective spring constant for the bond in the hydrogen molecule is 573 N/m.

SOLUTION

Conceptualize Imagine the only mode of vibration available to a diatomic molecule. This mode (shown in Fig. 21.5c) consists of the two atoms always moving in opposite directions with equal speeds.

Categorize We categorize this example as a quantum harmonic oscillator problem, with the molecule modeled as a two-particle system.

Analyze The motion of the particles relative to the center of mass can be analyzed by considering the oscillation of a single particle with reduced mass μ. (See Problem 40.)

Use the result of Problem 40 to evaluate the reduced mass of the hydrogen molecule, in which the masses of the two particles are the same:

$$\mu = \frac{m_1 m_2}{m_1 + m_2} = \frac{m^2}{2m} = \tfrac{1}{2}m$$

Using Equation 41.28, calculate the energy necessary to excite the molecule from its ground vibrational state to its first excited vibrational state:

$$\Delta E = \hbar\omega = \hbar\sqrt{\frac{k}{\mu}} = \hbar\sqrt{\frac{k}{\tfrac{1}{2}m}} = \hbar\sqrt{\frac{2k}{m}}$$

Substitute numerical values, noting that m is the mass of a hydrogen atom:

$$\Delta E = (1.055 \times 10^{-34}\,\text{J} \cdot \text{s})\sqrt{\frac{2(573\,\text{N/m})}{1.67 \times 10^{-27}\,\text{kg}}} = 8.74 \times 10^{-20}\,\text{J}$$

Set this energy equal to $\tfrac{3}{2}k_B T$ from Equation 21.19 and find the temperature at which the average molecular translational kinetic energy is equal to that required to excite the first vibrational state of the molecule:

$$\tfrac{3}{2}k_B T = \Delta E$$

$$T = \tfrac{2}{3}\left(\frac{\Delta E}{k_B}\right) = \tfrac{2}{3}\left(\frac{8.74 \times 10^{-20}\,\text{J}}{1.38 \times 10^{-23}\,\text{J/K}}\right) = 4.22 \times 10^3\,\text{K}$$

Finalize The temperature of the gas must be more than 4 000 K for the translational kinetic energy to be comparable to the energy required to excite the first vibrational state. This excitation energy must come from collisions between

continued

▶ **41.5** continued

molecules, so if the molecules do not have sufficient translational kinetic energy, they cannot be excited to the first vibrational state and vibration does not contribute to the molar specific heat. Hence, the curve in Figure 21.6 does not rise to a value corresponding to the contribution of vibration until the hydrogen gas has been raised to thousands of kelvins.

Figure 21.6 shows that rotational energy levels must be more closely spaced in energy than vibrational levels because they are excited at a lower temperature than the vibrational levels. The translational energy levels are those of a particle in a three-dimensional box, where the box is the container holding the gas. These levels are given by an expression similar to Equation 41.14. Because the box is macroscopic in size, L is very large and the energy levels are very close together. In fact, they are so close together that translational energy levels are excited at the temperature at which liquid hydrogen becomes a gas shown in Figure 21.6.

Summary

Definitions

The **wave function** Ψ for a system is a mathematical function that can be written as a product of a space function ψ for one particle of the system and a complex time function:

$$\Psi(\vec{\mathbf{r}}_1, \vec{\mathbf{r}}_2, \vec{\mathbf{r}}_3, \ldots, \vec{\mathbf{r}}_j, \ldots, t) = \psi(\vec{\mathbf{r}}_j)e^{-i\omega t} \tag{41.2}$$

where ω ($= 2\pi f$) is the angular frequency of the wave function and $i = \sqrt{-1}$. The wave function contains within it all the information that can be known about the particle.

The measured position x of a particle, averaged over many trials, is called the **expectation value** of x and is defined by

$$\langle x \rangle \equiv \int_{-\infty}^{\infty} \psi^* x \psi \, dx \tag{41.8}$$

Concepts and Principles

In quantum mechanics, a particle in a system can be represented by a wave function $\psi(x, y, z)$. The probability per unit volume (or probability density) that a particle will be found at a point is $|\psi|^2 = \psi^*\psi$, where ψ^* is the complex conjugate of ψ. If the particle is confined to moving along the x axis, the probability that it is located in an interval dx is $|\psi|^2 \, dx$. Furthermore, the sum of all these probabilities over all values of x must be 1:

$$\int_{-\infty}^{\infty} |\psi|^2 \, dx = 1 \tag{41.7}$$

This expression is called the **normalization condition**.

If a particle of mass m is confined to moving in a one-dimensional box of length L whose walls are impenetrable, then ψ must be zero at the walls and outside the box. The wave functions for this system are given by

$$\psi(x) = A \sin\left(\frac{n\pi x}{L}\right) \quad n = 1, 2, 3, \ldots \tag{41.12}$$

where A is the maximum value of ψ. The allowed states of a particle in a box have quantized energies given by

$$E_n = \left(\frac{h^2}{8mL^2}\right)n^2 \quad n = 1, 2, 3, \ldots \tag{41.14}$$

The wave function for a system must satisfy the **Schrödinger equation**. The time-independent Schrödinger equation for a particle confined to moving along the x axis is

$$-\frac{\hbar^2}{2m}\frac{d^2\psi}{dx^2} + U\psi = E\psi \tag{41.15}$$

where U is the potential energy of the system and E is the total energy.

Quantum Particle Under Boundary Conditions. An interaction of a quantum particle with its environment represents one or more boundary conditions. If the interaction restricts the particle to a finite region of space, the energy of the system is quantized. All wave functions must satisfy the following four boundary conditions: (1) $\psi(x)$ must remain finite as x approaches 0, (2) $\psi(x)$ must approach zero as x approaches $\pm\infty$, (3) $\psi(x)$ must be continuous for all values of x, and (4) $d\psi/dx$ must be continuous for all finite values of $U(x)$. If the solution to Equation 41.15 is piecewise, conditions (3) and (4) must be applied at the boundaries between regions of x in which Equation 41.15 has been solved.

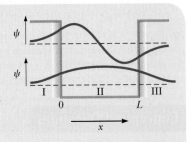

1. A beam of quantum particles with kinetic energy 2.00 eV is reflected from a potential barrier of small width and original height 3.00 eV. How does the fraction of the particles that are reflected change as the barrier height is reduced to 2.01 eV? (a) It increases. (b) It decreases. (c) It stays constant at zero. (d) It stays constant at 1. (e) It stays constant with some other value.

2. A quantum particle of mass m_1 is in a square well with infinitely high walls and length 3 nm. Rank the situations (a) through (e) according to the particle's energy from highest to lowest, noting any cases of equality. (a) The particle of mass m_1 is in the ground state of the well. (b) The same particle is in the $n = 2$ excited state of the same well. (c) A particle with mass $2m_1$ is in the ground state of the same well. (d) A particle of mass m_1 in the ground state of the same well, and the uncertainty principle has become inoperative; that is, Planck's constant has been reduced to zero. (e) A particle of mass m_1 is in the ground state of a well of length 6 nm.

3. Is each one of the following statements (a) through (e) true or false for an electron? (a) It is a quantum particle, behaving in some experiments like a classical particle and in some experiments like a classical wave. (b) Its rest energy is zero. (c) It carries energy in its motion. (d) It carries momentum in its motion. (e) Its motion is described by a wave function that has a wavelength and satisfies a wave equation.

4. Is each one of the following statements (a) through (e) true or false for a photon? (a) It is a quantum particle, behaving in some experiments like a classical particle and in some experiments like a classical wave. (b) Its rest energy is zero. (c) It carries energy in its motion. (d) It carries momentum in its motion. (e) Its motion is described by a wave function that has a wavelength and satisfies a wave equation.

5. A particle in a rigid box of length L is in the first excited state for which $n = 2$ (Fig. OQ41.5). Where is the particle most likely to be found? (a) At the center of the box. (b) At either end of the box. (c) All points in the box are equally likely. (d) One-fourth of the way

from either end of the box. (e) None of those answers is correct.

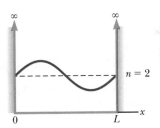

Figure OQ41.5

6. Two square wells have the same length. Well 1 has walls of finite height, and well 2 has walls of infinite height. Both wells contain identical quantum particles, one in each well. (i) Is the wavelength of the ground-state wave function (a) greater for well 1, (b) greater for well 2, or (c) equal for both wells? (ii) Is the magnitude of the ground-state momentum (a) greater for well 1, (b) greater for well 2, or (c) equal for both wells? (iii) Is the ground-state energy of the particle (a) greater for well 1, (b) greater for well 2, or (c) equal for both wells?

7. The probability of finding a certain quantum particle in the section of the x axis between $x = 4$ nm and $x = 7$ nm is 48%. The particle's wave function $\psi(x)$ is constant over this range. What numerical value can be attributed to $\psi(x)$, in units of $\text{nm}^{-1/2}$? (a) 0.48 (b) 0.16 (c) 0.12 (d) 0.69 (e) 0.40

8. Suppose a tunneling current in an electronic device goes through a potential-energy barrier. The tunneling current is small because the width of the barrier is large and the barrier is high. To increase the current most effectively, what should you do? (a) Reduce the width of the barrier. (b) Reduce the height of the barrier. (c) Either choice (a) or choice (b) is equally effective. (d) Neither choice (a) nor choice (b) increases the current.

9. Unlike the idealized diagram of Figure 41.11, a typical tip used for a scanning tunneling microscope is rather jagged on the atomic scale, with several irregularly spaced points. For such a tip, does most of the

tunneling current occur between the sample and (a) all the points of the tip equally, (b) the most centrally located point, (c) the point closest to the sample, or (d) the point farthest from the sample?

10. Figure OQ41.10 represents the wave function for a hypothetical quantum particle in a given region. From the choices *a* through *e*, at what value of *x* is the particle most likely to be found?

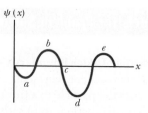

Figure OQ41.10

Conceptual Questions

1. denotes answer available in *Student Solutions Manual/Study Guide*

1. Richard Feynman said, "A philosopher once said that 'it is necessary for the very existence of science that the same conditions always produce the same results.' Well, they don't!" In view of what has been discussed in this chapter, present an argument showing that the philosopher's statement is false. How might the statement be reworded to make it true?

2. Discuss the relationship between ground-state energy and the uncertainty principle.

3. For a quantum particle in a box, the probability density at certain points is zero as seen in Figure CQ41.3. Does this value imply that the particle cannot move across these points? Explain.

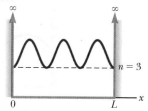

Figure CQ41.3

4. Why are the following wave functions not physically possible for all values of *x*? (a) $\psi(x) = Ae^x$ (b) $\psi(x) = A \tan x$

5. What is the significance of the wave function ψ?

6. In quantum mechanics, it is possible for the energy *E* of a particle to be less than the potential energy, but classically this condition is not possible. Explain.

7. Consider the wave functions in Figure CQ41.7. Which of them are not physically significant in the interval shown? For those that are not, state why they fail to qualify.

8. How is the Schrödinger equation useful in describing quantum phenomena?

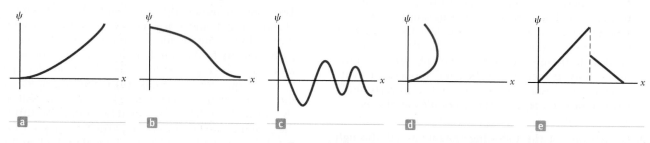

Figure CQ41.7

Problems

WebAssign The problems found in this chapter may be assigned online in Enhanced WebAssign

1. straightforward; 2. intermediate; 3. challenging

1. full solution available in the *Student Solutions Manual/Study Guide*

AMT Analysis Model tutorial available in Enhanced WebAssign

GP Guided Problem

M Master It tutorial available in Enhanced WebAssign

W Watch It video solution available in Enhanced WebAssign

Section 41.1 The Wave Function

1. A free electron has a wave function

$$\psi(x) = Ae^{i(5.00 \times 10^{10} x)}$$

where *x* is in meters. Find its (a) de Broglie wavelength, (b) momentum, and (c) kinetic energy in electron volts.

2. The wave function for a particle is given by $\psi(x) = Ae^{-|x|/a}$, where *A* and *a* are constants. (a) Sketch this function for values of *x* in the interval $-3a < x < 3a$. (b) Determine the value of *A*. (c) Find the probability that the particle will be found in the interval $-a < x < a$.

3. The wave function for a quantum particle is given by $\psi(x) = Ax$ between $x = 0$ and $x = 1.00$, and $\psi(x) = 0$ elsewhere. Find (a) the value of the normalization constant A, (b) the probability that the particle will be found between $x = 0.300$ and $x = 0.400$, and (c) the expectation value of the particle's position.

4. The wave function for a quantum particle is

$$\psi(x) = \sqrt{\frac{a}{\pi(x^2 + a^2)}}$$

for $a > 0$ and $-\infty < x < +\infty$. Determine the probability that the particle is located somewhere between $x = -a$ and $x = +a$.

Section 41.2 Analysis Model: Quantum Particle Under Boundary Conditions

5. (a) Use the quantum-particle-in-a-box model to calculate the first three energy levels of a neutron trapped in an atomic nucleus of diameter 20.0 fm. (b) Explain whether the energy-level differences have a realistic order of magnitude.

6. An electron that has an energy of approximately 6 eV moves between infinitely high walls 1.00 nm apart. Find (a) the quantum number n for the energy state the electron occupies and (b) the precise energy of the electron.

7. An electron is contained in a one-dimensional box of length 0.100 nm. (a) Draw an energy-level diagram for the electron for levels up to $n = 4$. (b) Photons are emitted by the electron making downward transitions that could eventually carry it from the $n = 4$ state to the $n = 1$ state. Find the wavelengths of all such photons.

8. *Why is the following situation impossible?* A proton is in an infinitely deep potential well of length 1.00 nm. It absorbs a microwave photon of wavelength 6.06 mm and is excited into the next available quantum state.

9. A ruby laser emits 694.3-nm light. Assume light of this wavelength is due to a transition of an electron in a box from its $n = 2$ state to its $n = 1$ state. Find the length of the box.

10. A laser emits light of wavelength λ. Assume this light is due to a transition of an electron in a box from its $n = 2$ state to its $n = 1$ state. Find the length of the box.

11. The nuclear potential energy that binds protons and neutrons in a nucleus is often approximated by a square well. Imagine a proton confined in an infinitely high square well of length 10.0 fm, a typical nuclear diameter. Assuming the proton makes a transition from the $n = 2$ state to the ground state, calculate (a) the energy and (b) the wavelength of the emitted photon. (c) Identify the region of the electromagnetic spectrum to which this wavelength belongs.

12. A proton is confined to move in a one-dimensional box of length 0.200 nm. (a) Find the lowest possible energy of the proton. (b) **What If?** What is the lowest possible energy of an electron confined to the same box?

(c) How do you account for the great difference in your results for parts (a) and (b)?

13. An electron is confined to a one-dimensional region in which its ground-state ($n = 1$) energy is 2.00 eV. (a) What is the length L of the region? (b) What energy input is required to promote the electron to its first excited state?

14. A 4.00-g particle confined to a box of length L has a speed of 1.00 mm/s. (a) What is the classical kinetic energy of the particle? (b) If the energy of the first excited state ($n = 2$) is equal to the kinetic energy found in part (a), what is the value of L? (c) Is the result found in part (b) realistic? Explain.

15. A photon with wavelength λ is absorbed by an electron confined to a box. As a result, the electron moves from state $n = 1$ to $n = 4$. (a) Find the length of the box. (b) What is the wavelength λ' of the photon emitted in the transition of that electron from the state $n = 4$ to the state $n = 2$?

16. For a quantum particle of mass m in the ground state of a square well with length L and infinitely high walls, the uncertainty in position is $\Delta x \approx L$. (a) Use the uncertainty principle to estimate the uncertainty in its momentum. (b) Because the particle stays inside the box, its average momentum must be zero. Its average squared momentum is then $\langle p^2 \rangle \approx (\Delta p)^2$. Estimate the energy of the particle. (c) State how the result of part (b) compares with the actual ground-state energy.

17. A quantum particle is described by the wave function

$$\psi(x) = \begin{cases} A \cos\left(\dfrac{2\pi x}{L}\right) & \text{for } -\dfrac{L}{4} \le x \le \dfrac{L}{4} \\ 0 & \text{elsewhere} \end{cases}$$

(a) Determine the normalization constant A. (b) What is the probability that the particle will be found between $x = 0$ and $x = L/8$ if its position is measured?

18. The wave function for a quantum particle confined to moving in a one-dimensional box located between $x = 0$ and $x = L$ is

$$\psi(x) = A \sin\left(\frac{n\pi x}{L}\right)$$

Use the normalization condition on ψ to show that

$$A = \sqrt{\frac{2}{L}}$$

19. A quantum particle in an infinitely deep square well has a wave function given by

$$\psi_2(x) = \sqrt{\frac{2}{L}} \sin\left(\frac{2\pi x}{L}\right)$$

for $0 \le x \le L$ and zero otherwise. (a) Determine the expectation value of x. (b) Determine the probability of finding the particle near $\frac{1}{2}L$ by calculating the probability that the particle lies in the range $0.490L \le x \le 0.510L$. (c) **What If?** Determine the probability of finding the particle near $\frac{1}{4}L$ by calculating the probability that the particle lies in the range $0.240L \le x \le 0.260L$.

(d) Argue that the result of part (a) does not contradict the results of parts (b) and (c).

20. An electron in an infinitely deep square well has a wave function that is given by

$$\psi_3(x) = \sqrt{\frac{2}{L}} \sin\left(\frac{3\pi x}{L}\right)$$

for $0 \le x \le L$ and is zero otherwise. (a) What are the most probable positions of the electron? (b) Explain how you identify them.

21. An electron is trapped in an infinitely deep potential well 0.300 nm in length. (a) If the electron is in its ground state, what is the probability of finding it within 0.100 nm of the left-hand wall? (b) Identify the classical probability of finding the electron in this interval and state how it compares with the answer to part (a). (c) Repeat parts (a) and (b) assuming the particle is in the 99th energy state.

22. A quantum particle is in the $n = 1$ state of an infinitely deep square well with walls at $x = 0$ and $x = L$. Let ℓ be an arbitrary value of x between $x = 0$ and $x = L$. (a) Find an expression for the probability, as a function of ℓ, that the particle will be found between $x = 0$ and $x = \ell$. (b) Sketch the probability as a function of the variable ℓ/L. Choose values of ℓ/L ranging from 0 to 1.00 in steps of 0.100. (c) Explain why the probability function must have particular values at $\ell/L = 0$ and at $\ell/L = 1$. (d) Find the value of ℓ for which the probability of finding the particle between $x = 0$ and $x = \ell$ is twice the probability of finding the particle between $x = \ell$ and $x = L$. *Suggestion:* Solve the transcendental equation for ℓ/L numerically.

23. A quantum particle in an infinitely deep square well has a wave function that is given by

$$\psi_1(x) = \sqrt{\frac{2}{L}} \sin\left(\frac{\pi x}{L}\right)$$

for $0 \le x \le L$ and is zero otherwise. (a) Determine the probability of finding the particle between $x = 0$ and $x = \frac{1}{3}L$. (b) Use the result of this calculation and a symmetry argument to find the probability of finding the particle between $x = \frac{1}{3}L$ and $x = \frac{2}{3}L$. Do not re-evaluate the integral.

Section 41.3 The Schrödinger Equation

24. Show that the wave function $\psi = Ae^{i(kx-\omega t)}$ is a solution to the Schrödinger equation (Eq. 41.15), where $k = 2\pi/\lambda$ and $U = 0$.

25. The wave function of a quantum particle of mass m is

$$\psi(x) = A \cos(kx) + B \sin(kx)$$

where A, B, and k are constants. (a) Assuming the particle is free ($U = 0$), show that $\psi(x)$ is a solution of the Schrödinger equation (Eq. 41.15). (b) Find the corresponding energy E of the particle.

26. Consider a quantum particle moving in a one-dimensional box for which the walls are at $x = -L/2$ and $x = L/2$. (a) Write the wave functions and probability

densities for $n = 1$, $n = 2$, and $n = 3$. (b) Sketch the wave functions and probability densities.

27. In a region of space, a quantum particle with zero total energy has a wave function

$$\psi(x) = Axe^{-x^2/L^2}$$

(a) Find the potential energy U as a function of x. (b) Make a sketch of $U(x)$ versus x.

28. A quantum particle of mass m moves in a potential well of length $2L$. Its potential energy is infinite for $x < -L$ and for $x > +L$. In the region $-L < x < L$, its potential energy is given by

$$U(x) = \frac{-\hbar^2 x^2}{mL^2(L^2 - x^2)}$$

In addition, the particle is in a stationary state that is described by the wave function $\psi(x) = A(1 - x^2/L^2)$ for $-L < x < +L$ and by $\psi(x) = 0$ elsewhere. (a) Determine the energy of the particle in terms of $\hbar$, m, and L. (b) Determine the normalization constant A. (c) Determine the probability that the particle is located between $x = -L/3$ and $x = +L/3$.

Section 41.4 A Particle in a Well of Finite Height

29. Sketch (a) the wave function $\psi(x)$ and (b) the probability density $|\psi(x)|^2$ for the $n = 4$ state of a quantum particle in a finite potential well. (See Fig. 41.7.)

30. Suppose a quantum particle is in its ground state in a box that has infinitely high walls (see Fig. 41.4a). Now suppose the left-hand wall is suddenly lowered to a finite height and width. (a) Qualitatively sketch the wave function for the particle a short time later. (b) If the box has a length L, what is the wavelength of the wave that penetrates the left-hand wall?

Section 41.5 Tunneling Through a Potential Energy Barrier

31. An electron with kinetic energy $E = 5.00$ eV is incident on a barrier of width $L = 0.200$ nm and height $U = 10.0$ eV (Fig. P41.31). What is the probability that the electron (a) tunnels through the barrier? (b) Is reflected?

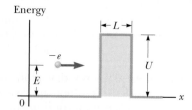

Figure P41.31 Problems 31 and 32.

32. An electron having total energy $E = 4.50$ eV approaches a rectangular energy barrier with $U = 5.00$ eV and $L = 950$ pm as shown in Figure P41.31. Classically, the electron cannot pass through the barrier because $E < U$. Quantum-mechanically, however, the probability of tunneling is not zero. (a) Calculate this probability, which is the transmission coefficient. (b) To what value would the width L of the potential barrier have to be increased for the chance of an inci-

dent 4.50-eV electron tunneling through the barrier to be one in one million?

33. An electron has a kinetic energy of 12.0 eV. The elec-
[W] tron is incident upon a rectangular barrier of height 20.0 eV and width 1.00 nm. If the electron absorbed all the energy of a photon of green light (with wavelength 546 nm) at the instant it reached the barrier, by what factor would the electron's probability of tunneling through the barrier increase?

Section 41.6 Applications of Tunneling

34. A scanning tunneling microscope (STM) can precisely
[W] determine the depths of surface features because the current through its tip is very sensitive to differences in the width of the gap between the tip and the sample surface. Assume the electron wave function falls off exponentially in this direction with a decay length of 0.100 nm, that is, with $C = 10.0$ nm^{-1}. Determine the ratio of the current when the STM tip is 0.500 nm above a surface feature to the current when the tip is 0.515 nm above the surface.

35. The design criterion for a typical scanning tunnel-
[M] ing microscope (STM) specifies that it must be able to detect, on the sample below its tip, surface features that differ in height by only 0.002 00 nm. Assuming the electron transmission coefficient is e^{-2CL} with $C = 10.0$ nm^{-1}, what percentage change in electron transmission must the electronics of the STM be able to detect to achieve this resolution?

Section 41.7 The Simple Harmonic Oscillator

36. A one-dimensional harmonic oscillator wave function is

$$\psi = Axe^{-bx^2}$$

(a) Show that ψ satisfies Equation 41.24. (b) Find b and the total energy E. (c) Is this wave function for the ground state or for the first excited state?

37. A quantum simple harmonic oscillator consists of an
[W] electron bound by a restoring force proportional to its position relative to a certain equilibrium point. The proportionality constant is 8.99 N/m. What is the longest wavelength of light that can excite the oscillator?

38. A quantum simple harmonic oscillator consists of a particle of mass m bound by a restoring force proportional to its position relative to a certain equilibrium point. The proportionality constant is k. What is the longest wavelength of light that can excite the oscillator?

39. (a) Normalize the wave function for the ground state of a simple harmonic oscillator. That is, apply Equation 41.7 to Equation 41.26 and find the required value for the constant B in terms of m, ω, and fundamental constants. (b) Determine the probability of finding the oscillator in a narrow interval $-\delta/2 < x < \delta/2$ around its equilibrium position.

40. Two particles with masses m_1 and m_2 are joined by a light spring of force constant k. They vibrate along a straight line with their center of mass fixed. (a) Show that the total energy

$$\tfrac{1}{2}m_1u_1^2 + \tfrac{1}{2}m_2u_2^2 + \tfrac{1}{2}kx^2$$

can be written as $\tfrac{1}{2}\mu u^2 + \tfrac{1}{2}kx^2$, where $u = |u_1| + |u_2|$ is the *relative* speed of the particles and $\mu = m_1m_2/(m_1 + m_2)$ is the reduced mass of the system. This result demonstrates that the pair of freely vibrating particles can be precisely modeled as a single particle vibrating on one end of a spring that has its other end fixed. (b) Differentiate the equation

$$\tfrac{1}{2}\mu u^2 + \tfrac{1}{2}kx^2 = \text{constant}$$

with respect to x. Proceed to show that the system executes simple harmonic motion. (c) Find its frequency.

41. The total energy of a particle–spring system in which the particle moves with simple harmonic motion along the x axis is

$$E = \frac{p_x^2}{2m} + \frac{kx^2}{2}$$

where p_x is the momentum of the quantum particle and k is the spring constant. (a) Using the uncertainty principle, show that this expression can also be written as

$$E \geq \frac{p_x^2}{2m} + \frac{k\hbar^2}{8p_x^2}$$

(b) Show that the minimum energy of the harmonic oscillator is

$$E_{\min} = K + U = \tfrac{1}{4}\hbar\sqrt{\frac{k}{m}} + \frac{\hbar\omega}{4} = \frac{\hbar\omega}{2}$$

42. Show that Equation 41.26 is a solution of Equation 41.24 with energy $E = \tfrac{1}{2}\hbar\omega$.

Additional Problems

43. A particle of mass 2.00×10^{-28} kg is confined to a one-
[W] dimensional box of length 1.00×10^{-10} m. For $n = 1$, what are (a) the particle's wavelength, (b) its momentum, and (c) its ground-state energy?

44. Prove that the first term in the Schrödinger equation, $-(\hbar^2/2m)(d^2\psi/dx^2)$, reduces to the kinetic energy of the quantum particle multiplied by the wave function (a) for a freely moving particle, with the wave function given by Equation 41.4, and (b) for a particle in a box, with the wave function given by Equation 41.13.

45. A particle in a one-dimensional box of length L is in its first excited state, corresponding to $n = 2$. Determine the probability of finding the particle between $x = 0$ and $x = L/4$.

46. Prove that assuming $n = 0$ for a quantum particle in an infinitely deep potential well leads to a violation of the uncertainty principle $\Delta p_x \Delta x \geq \hbar/2$.

47. Calculate the transmission probability for quantum-mechanical tunneling in each of the following cases. (a) An electron with an energy deficit of $U - E = 0.010\ 0$ eV is incident on a square barrier of width $L = 0.100$ nm. (b) An electron with an energy deficit of 1.00 eV is incident on the same barrier. (c) An alpha particle (mass 6.64×10^{-27} kg) with an energy deficit

of 1.00 MeV is incident on a square barrier of width 1.00 fm. (d) An 8.00-kg bowling ball with an energy deficit of 1.00 J is incident on a square barrier of width 2.00 cm.

48. An electron in an infinitely deep potential well has a ground-state energy of 0.300 eV. (a) Show that the photon emitted in a transition from the $n = 3$ state to the $n = 1$ state has a wavelength of 517 nm, which makes it green visible light. (b) Find the wavelength and the spectral region for each of the other five transitions that take place among the four lowest energy levels.

49. An atom in an excited state 1.80 eV above the ground **W** state remains in that excited state 2.00 μs before moving to the ground state. Find (a) the frequency and (b) the wavelength of the emitted photon. (c) Find the approximate uncertainty in energy of the photon.

50. A marble rolls back and forth across a shoebox at a constant speed of 0.8 m/s. Make an order-of-magnitude estimate of the probability of it escaping through the wall of the box by quantum tunneling. State the quantities you take as data and the values you measure or estimate for them.

51. An electron confined to a box absorbs a photon with wavelength λ. As a result, the electron makes a transition from the $n = 1$ state to the $n = 3$ state. (a) Find the length of the box. (b) What is the wavelength λ' of the photon emitted when the electron makes a transition from the $n = 3$ state to the $n = 2$ state?

52. For a quantum particle described by a wave function $\psi(x)$, the expectation value of a physical quantity $f(x)$ associated with the particle is defined by

$$\langle f(x) \rangle \equiv \int_{-\infty}^{\infty} \psi^* f(x) \psi \, dx$$

For a particle in an infinitely deep one-dimensional box extending from $x = 0$ to $x = L$, show that

$$\langle x^2 \rangle = \frac{L^2}{3} - \frac{L^2}{2n^2\pi^2}$$

53. A quantum particle of mass m is placed in a one-dimensional box of length L. Assume the box is so small that the particle's motion is relativistic and $K = p^2/2m$ is not valid. (a) Derive an expression for the kinetic energy levels of the particle. (b) Assume the particle is an electron in a box of length $L = 1.00 \times 10^{-12}$ m. Find its lowest possible kinetic energy. (c) By what percent is the nonrelativistic equation in error? *Suggestion:* See Equation 39.23.

54. *Why is the following situation impossible?* A particle is in the ground state of an infinite square well of length L. A light source is adjusted so that the photons of wavelength λ are absorbed by the particle as it makes a transition to the first excited state. An identical particle is in the ground state of a finite square well of length L. The light source sends photons of the same wavelength λ toward this particle. The photons are not absorbed because the allowed energies of the finite square well are different from those of the infinite square well. To cause the photons to be absorbed, you move the light source at a high speed toward the particle in the finite square well. You are able to find a speed at which the Doppler-shifted photons are absorbed as the particle makes a transition to the first excited state.

55. A quantum particle has a wave function

$$\psi(x) = \begin{cases} \sqrt{\dfrac{2}{a}}\, e^{-x/a} & \text{for } x > 0 \\ 0 & \text{for } x < 0 \end{cases}$$

(a) Find and sketch the probability density. (b) Find the probability that the particle will be at any point where $x < 0$. (c) Show that ψ is normalized and then (d) find the probability of finding the particle between $x = 0$ and $x = a$.

56. An electron is confined to move in the xy plane in a **GP** rectangle whose dimensions are L_x and L_y. That is, the electron is trapped in a two-dimensional potential well having lengths of L_x and L_y. In this situation, the allowed energies of the electron depend on two quantum numbers n_x and n_y and are given by

$$E = \frac{h^2}{8m_e}\left(\frac{n_x^2}{L_x^2} + \frac{n_y^2}{L_y^2}\right)$$

Using this information, we wish to find the wavelength of a photon needed to excite the electron from the ground state to the second excited state, assuming $L_x = L_y = L$. (a) Using the assumption on the lengths, write an expression for the allowed energies of the electron in terms of the quantum numbers n_x and n_y. (b) What values of n_x and n_y correspond to the ground state? (c) Find the energy of the ground state. (d) What are the possible values of n_x and n_y for the first excited state, that is, the next-highest state in terms of energy? (e) What are the possible values of n_x and n_y for the second excited state? (f) Using the values in part (e), what is the energy of the second excited state? (g) What is the energy difference between the ground state and the second excited state? (h) What is the wavelength of a photon that will cause the transition between the ground state and the second excited state?

57. The normalized wave functions for the ground state, $\psi_0(x)$, and the first excited state, $\psi_1(x)$, of a quantum harmonic oscillator are

$$\psi_0(x) = \left(\frac{a}{\pi}\right)^{1/4} e^{-ax^2/2} \qquad \psi_1(x) = \left(\frac{4a^3}{\pi}\right)^{1/4} x e^{-ax^2/2}$$

where $a = m\omega/\hbar$. A mixed state, $\psi_{01}(x)$, is constructed from these states:

$$\psi_{01}(x) = \frac{1}{\sqrt{2}}\left[\psi_0(x) + \psi_1(x)\right]$$

The symbol $\langle q \rangle_s$ denotes the expectation value of the quantity q for the state $\psi_s(x)$. Calculate the expectation values (a) $\langle x \rangle_0$, (b) $\langle x \rangle_1$, and (c) $\langle x \rangle_{01}$.

58. A two-slit electron diffraction experiment is done with slits of *unequal* widths. When only slit 1 is open, the

number of electrons reaching the screen per second is 25.0 times the number of electrons reaching the screen per second when only slit 2 is open. When both slits are open, an interference pattern results in which the destructive interference is not complete. Find the ratio of the probability of an electron arriving at an interference maximum to the probability of an electron arriving at an adjacent interference minimum. *Suggestion:* Use the superposition principle.

Challenge Problems

59. Particles incident from the left in Figure P41.59 are confronted with a step in potential energy. The step has a height U at $x = 0$. The particles have energy $E > U$. Classically, all the particles would continue moving forward with reduced speed. According to quantum mechanics, however, a fraction of the particles are reflected at the step. (a) Prove that the reflection coefficient R for this case is

$$R = \frac{(k_1 - k_2)^2}{(k_1 + k_2)^2}$$

where $k_1 = 2\pi/\lambda_1$ and $k_2 = 2\pi/\lambda_2$ are the wave numbers for the incident and transmitted particles, respectively. Proceed as follows. Show that the wave function $\psi_1 = Ae^{ik_1x} + Be^{-ik_1x}$ satisfies the Schrödinger equation in region 1, for $x < 0$. Here Ae^{ik_1x} represents the incident beam and Be^{-ik_1x} represents the reflected particles. Show that $\psi_2 = Ce^{ik_2x}$ satisfies the Schrödinger equation in region 2, for $x > 0$. Impose the boundary conditions $\psi_1 = \psi_2$ and $d\psi_1/dx = d\psi_2/dx$, at $x = 0$, to find the relationship between B and A. Then evaluate $R = B^2/A^2$. A particle that has kinetic energy $E = 7.00$ eV is incident from a region where the potential energy is zero onto one where $U = 5.00$ eV. Find (b) its probability of being reflected and (c) its probability of being transmitted.

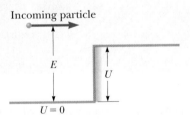

Incoming particle

E U

$U = 0$

Figure P41.59

60. Consider a "crystal" consisting of two fixed ions of charge $+e$ and two electrons as shown in Figure P41.60. (a) Taking into account all the pairs of interactions, find the potential energy of the system as a function of d. (b) Assuming the electrons to be restricted to a one-

$\leftarrow d \rightarrow \leftarrow d \rightarrow \leftarrow d \rightarrow$

Figure P41.60

dimensional box of length $3d$, find the minimum kinetic energy of the two electrons. (c) Find the value of d for which the total energy is a minimum. (d) State how this value of d compares with the spacing of atoms in lithium, which has a density of 0.530 g/cm³ and a molar mass of 6.94 g/mol.

61. An electron is trapped in a quantum dot. The quantum dot may be modeled as a one-dimensional, rigid-walled box of length 1.00 nm. (a) Taking $x = 0$ as the left side of the box, calculate the probability of finding the electron between $x_1 = 0.150$ nm and $x_2 = 0.350$ nm for the $n = 1$ state. (b) Repeat part (a) for the $n = 2$ state. Calculate the energies in electron volts of (c) the $n = 1$ state and (d) the $n = 2$ state.

62. An electron is represented by the time-independent wave function

$$\psi(x) = \begin{cases} Ae^{-\alpha x} & \text{for } x > 0 \\ Ae^{+\alpha x} & \text{for } x < 0 \end{cases}$$

(a) Sketch the wave function as a function of x. (b) Sketch the probability density representing the likelihood that the electron is found between x and $x + dx$. (c) Only an infinite value of potential energy could produce the discontinuity in the derivative of the wave function at $x = 0$. Aside from this feature, argue that $\psi(x)$ can be a physically reasonable wave function. (d) Normalize the wave function. (e) Determine the probability of finding the electron somewhere in the range

$$-\frac{1}{2\alpha} \le x \le \frac{1}{2\alpha}$$

63. The wave function

$$\psi(x) = Bxe^{-(m\omega/2\hbar)x^2}$$

is a solution to the simple harmonic oscillator problem. (a) Find the energy of this state. (b) At what position are you least likely to find the particle? (c) At what positions are you most likely to find the particle? (d) Determine the value of B required to normalize the wave function. (e) **What If?** Determine the classical probability of finding the particle in an interval of small length δ centered at the position $x = 2(\hbar/m\omega)^{1/2}$. (f) What is the actual probability of finding the particle in this interval?

64. (a) Find the normalization constant A for a wave function made up of the two lowest states of a quantum particle in a box extending from $x = 0$ to $x = L$:

$$\psi(x) = A\left[\sin\left(\frac{\pi x}{L}\right) + 4\sin\left(\frac{2\pi x}{L}\right)\right]$$

(b) A particle is described in the space $-a \le x \le a$ by the wave function

$$\psi(x) = A\cos\left(\frac{\pi x}{2a}\right) + B\sin\left(\frac{\pi x}{a}\right)$$

Determine the relationship between the values of A and B required for normalization.

Atomic Physics

This street in the Ginza district in Tokyo displays many signs formed from neon lamps of varying bright colors. The light from these lamps has its origin in transitions between quantized energy states in the atoms contained in the lamps. In this chapter, we investigate those transitions. (© Ken Straiton/Corbis)

In Chapter 41, we introduced some basic concepts and techniques used in quantum mechanics along with their applications to various one-dimensional systems. In this chapter, we apply quantum mechanics to atomic systems. A large portion of the chapter is focused on the application of quantum mechanics to the study of the hydrogen atom. Understanding the hydrogen atom, the simplest atomic system, is important for several reasons:

- The hydrogen atom is the only atomic system that can be solved exactly.
- Much of what was learned in the 20th century about the hydrogen atom, with its single electron, can be extended to such single-electron ions as He^+ and Li^{2+}.
- The hydrogen atom is an ideal system for performing precise tests of theory against experiment and for improving our overall understanding of atomic structure.

- The quantum numbers that are used to characterize the allowed states of hydrogen can also be used to investigate more complex atoms, and such a description enables us to understand the periodic table of the elements. This understanding is one of the greatest triumphs of quantum mechanics.
- The basic ideas about atomic structure must be well understood before we attempt to deal with the complexities of molecular structures and the electronic structure of solids.

The full mathematical solution of the Schrödinger equation applied to the hydrogen atom gives a complete and beautiful description of the atom's properties. Because the mathematical procedures involved are beyond the scope of this text, however, many details are omitted. The solutions for some states of hydrogen are discussed, together with the quantum numbers used to characterize various allowed states. We also discuss the physical significance of the quantum numbers and the effect of a magnetic field on certain quantum states.

A new physical idea, the *exclusion principle,* is presented in this chapter. This principle is extremely important for understanding the properties of multielectron atoms and the arrangement of elements in the periodic table.

Finally, we apply our knowledge of atomic structure to describe the mechanisms involved in the production of x-rays and in the operation of a laser.

42.1 Atomic Spectra of Gases

As pointed out in Section 40.1, all objects emit thermal radiation characterized by a *continuous* distribution of wavelengths. In sharp contrast to this continuous-distribution spectrum is the *discrete* **line spectrum** observed when a low-pressure gas undergoes an electric discharge. (Electric discharge occurs when the gas is subject to a potential difference that creates an electric field greater than the dielectric strength of the gas.) Observation and analysis of these spectral lines is called **emission spectroscopy.**

When the light from a gas discharge is examined using a spectrometer (see Fig. 38.15), it is found to consist of a few bright lines of color on a generally dark background. This discrete line spectrum contrasts sharply with the continuous rainbow of colors seen when a glowing solid is viewed through the same instrument. Figure 42.1a (page 1298) shows that the wavelengths contained in a given line spectrum are characteristic of the element emitting the light. The simplest line spectrum is that for atomic hydrogen, and we describe this spectrum in detail. Because no two elements have the same line spectrum, this phenomenon represents a practical and sensitive technique for identifying the elements present in unknown samples.

Another form of spectroscopy very useful in analyzing substances is **absorption spectroscopy.** An absorption spectrum is obtained by passing white light from a continuous source through a gas or a dilute solution of the element being analyzed. The absorption spectrum consists of a series of dark lines superimposed on the continuous spectrum of the light source as shown in Figure 42.1b for atomic hydrogen.

The absorption spectrum of an element has many practical applications. For example, the continuous spectrum of radiation emitted by the Sun must pass through the cooler gases of the solar atmosphere. The various absorption lines observed in the solar spectrum have been used to identify elements in the solar atmosphere. In early studies of the solar spectrum, experimenters found some lines that did not correspond to any known element. A new element had been discovered!

Pitfall Prevention 42.1
Why Lines? The phrase "spectral lines" is often used when discussing the radiation from atoms. Lines are seen because the light passes through a long and very narrow slit before being separated by wavelength. You will see many references to these "lines" in both physics and chemistry.

Figure 42.1 (a) Emission line spectra for hydrogen, mercury, and neon. (b) The absorption spectrum for hydrogen. Notice that the dark absorption lines occur at the same wavelengths as the hydrogen emission lines in (a). (K. W. Whitten, R. E. Davis, M. L. Peck, and G. G. Stanley, *General Chemistry*, 7th ed., Belmont, CA, Brooks/Cole, 2004.)

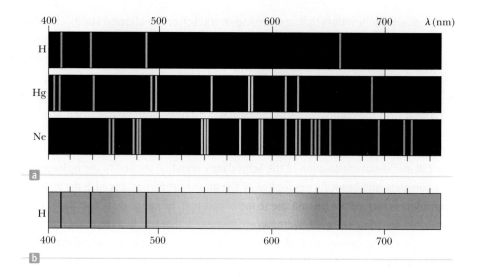

The new element was named helium, after the Greek word for Sun, *helios*. Helium was subsequently isolated from subterranean gas on the Earth.

Using this technique, scientists have examined the light from stars other than our Sun and have never detected elements other than those present on the Earth. Absorption spectroscopy has also been useful in analyzing heavy-metal contamination of the food chain. For example, the first determination of high levels of mercury in tuna was made with the use of atomic absorption spectroscopy.

The discrete emissions of light from gas discharges are used in "neon" signs such as those in the opening photograph of this chapter. Neon, the first gas used in these types of signs and the gas after which these signs are named, emits strongly in the red region. As a result, a glass tube filled with neon gas emits bright red light when an applied voltage causes a continuous discharge. Early signs used different gases to provide different colors, although the brightness of these signs was generally very low. Many present-day "neon" signs contain mercury vapor, which emits strongly in the ultraviolet range of the electromagnetic spectrum. The inside of a present-day sign's glass tube is coated with a material that emits a particular color when it absorbs ultraviolet radiation from the mercury. The color of the light from the tube results from the particular material chosen. A household fluorescent light operates in the same manner, with a white-emitting material coating the inside of the glass tube.

The lines shown in color are in the visible range of wavelengths.

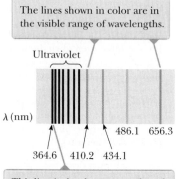

Figure 42.2 The Balmer series of spectral lines for atomic hydrogen, with several lines marked with the wavelength in nanometers. (The horizontal wavelength axis is not to scale.)

This line is the shortest wavelength line and is in the ultraviolet region of the electromagnetic spectrum.

From 1860 to 1885, scientists accumulated a great deal of data on atomic emissions using spectroscopic measurements. In 1885, a Swiss schoolteacher, Johann Jacob Balmer (1825–1898), found an empirical equation that correctly predicted the wavelengths of four visible emission lines of hydrogen: H_α (red), H_β (blue-green), H_γ (blue-violet), and H_δ (violet). Figure 42.2 shows these and other lines (in the ultraviolet) in the emission spectrum of hydrogen. The four visible lines occur at the wavelengths 656.3 nm, 486.1 nm, 434.1 nm, and 410.2 nm. The complete set of lines is called the **Balmer series.** The wavelengths of these lines can be described by the following equation, which is a modification made by Johannes Rydberg (1854–1919) of Balmer's original equation:

Balmer series ▶

$$\frac{1}{\lambda} = R_H \left(\frac{1}{2^2} - \frac{1}{n^2} \right) \qquad n = 3, 4, 5, \ldots \qquad \textbf{(42.1)}$$

where R_H is a constant now called the **Rydberg constant** with a value of $1.097\,373\,2 \times 10^7$ m^{-1}. The integer values of n from 3 to 6 give the four visible lines from 656.3 nm (red) down to 410.2 nm (violet). Equation 42.1 also describes the ultraviolet spectral lines in the Balmer series if n is carried out beyond $n = 6$. The **series limit** is the shortest wavelength in the series and corresponds to $n \to \infty$, with a wavelength of 364.6 nm as in Figure 42.2. The measured spectral lines agree with the empirical equation, Equation 42.1, to within 0.1%.

Other lines in the spectrum of hydrogen were found following Balmer's discovery. These spectra are called the Lyman, Paschen, and Brackett series after their discoverers. The wavelengths of the lines in these series can be calculated through the use of the following empirical equations:

$$\frac{1}{\lambda} = R_{\mathrm{H}}\left(1 - \frac{1}{n^2}\right) \quad n = 2, 3, 4, \ldots \qquad (42.2)$$ ◀ Lyman series

$$\frac{1}{\lambda} = R_{\mathrm{H}}\left(\frac{1}{3^2} - \frac{1}{n^2}\right) \quad n = 4, 5, 6, \ldots \qquad (42.3)$$ ◀ Paschen series

$$\frac{1}{\lambda} = R_{\mathrm{H}}\left(\frac{1}{4^2} - \frac{1}{n^2}\right) \quad n = 5, 6, 7, \ldots \qquad (42.4)$$ ◀ Brackett series

No theoretical basis existed for these equations; they simply worked. The same constant R_{H} appears in each equation, and all equations involve small integers. In Section 42.3, we shall discuss the remarkable achievement of a theory for the hydrogen atom that provided an explanation for these equations.

42.2 Early Models of the Atom

The model of the atom in the days of Newton was a tiny, hard, indestructible sphere. Although this model provided a good basis for the kinetic theory of gases (Chapter 21), new models had to be devised when experiments revealed the electrical nature of atoms. In 1897, J. J. Thomson established the charge-to-mass ratio for electrons. (See Fig. 29.15 in Section 29.3.) The following year, he suggested a model that describes the atom as a region in which positive charge is spread out in space with electrons embedded throughout the region, much like the seeds in a watermelon or raisins in thick pudding (Fig. 42.3). The atom as a whole would then be electrically neutral.

In 1911, Ernest Rutherford (1871–1937) and his students Hans Geiger and Ernest Marsden performed a critical experiment that showed that Thomson's model could not be correct. In this experiment, a beam of positively charged alpha particles (helium nuclei) was projected into a thin metallic foil such as the target in Figure 42.4a (page 1300). Most of the particles passed through the foil as if it were empty space, but some of the results of the experiment were astounding. Many of the particles deflected from their original direction of travel were scattered through *large* angles. Some particles were even deflected backward, completely reversing their direction of travel! When Geiger informed Rutherford that some alpha particles were scattered backward, Rutherford wrote, "It was quite the most incredible event that has ever happened to me in my life. It was almost as incredible as if you fired a 15-inch [artillery] shell at a piece of tissue paper and it came back and hit you."

Such large deflections were not expected on the basis of Thomson's model. According to that model, the positive charge of an atom in the foil is spread out over such a great volume (the entire atom) that there is no concentration of positive charge strong enough to cause any large-angle deflections of the positively charged alpha particles. Furthermore, the electrons are so much less massive than the alpha particles that they would not cause large-angle scattering either. Rutherford explained his astonishing results by developing a new atomic model, one that assumed the positive charge in the atom was concentrated in a region that was small relative to the size of the atom. He called this concentration of positive charge the **nucleus** of the atom. Any electrons belonging to the atom were assumed to be in the relatively large volume outside the nucleus. To explain why these electrons were not pulled into the nucleus by the attractive electric force, Rutherford modeled them as moving in orbits around the nucleus in the same manner as the planets orbit the Sun (Fig. 42.4b). For this reason, this model is often referred to as the planetary model of the atom.

Two basic difficulties exist with Rutherford's planetary model. As we saw in Section 42.1, an atom emits (and absorbs) certain characteristic frequencies of

Joseph John Thomson
English physicist (1856–1940)
The recipient of a Nobel Prize in Physics in 1906, Thomson is usually considered the discoverer of the electron. He opened up the field of subatomic particle physics with his extensive work on the deflection of cathode rays (electrons) in an electric field.

The electrons are small negative charges at various locations within the atom.

The positive charge of the atom is distributed continuously in a spherical volume.

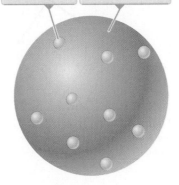

Figure 42.3 Thomson's model of the atom.

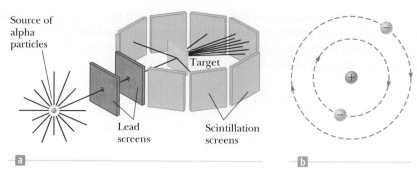

Figure 42.4 (a) Rutherford's technique for observing the scattering of alpha particles from a thin foil target. The source is a naturally occurring radioactive substance, such as radium. (b) Rutherford's planetary model of the atom.

Source of alpha particles

Target

Lead screens

Scintillation screens

a

b

Because the accelerating electron radiates energy, the size of the orbit decreases until the electron falls into the nucleus.

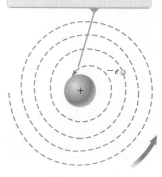

Figure 42.5 The classical model of the nuclear atom predicts that the atom decays.

electromagnetic radiation and no others, but the Rutherford model cannot explain this phenomenon. A second difficulty is that Rutherford's electrons are described by the particle in uniform circular motion model; they have a centripetal acceleration. According to Maxwell's theory of electromagnetism, centripetally accelerated charges revolving with frequency f should radiate electromagnetic waves of frequency f. Unfortunately, this classical model leads to a prediction of self-destruction when applied to the atom. Identifying the electron and the proton as a nonisolated system for energy, Equation 8.2 becomes $\Delta K + \Delta U = T_{ER}$, where K is the kinetic energy of the electron, U is the electric potential energy of the electron–nucleus system, and T_{ER} represents the outgoing electromagnetic radiation. As energy leaves the system, the radius of the electron's orbit steadily decreases (Fig. 42.5). The system is an isolated system for angular momentum because there is no torque on the system. Therefore, as the electron moves closer to the nucleus, the angular speed of the electron will increase, just like the spinning skater in Figure 11.10 in Section 11.4. This process leads to an ever-increasing frequency of emitted radiation and an ultimate collapse of the atom as the electron plunges into the nucleus.

42.3 Bohr's Model of the Hydrogen Atom

Given the situation described at the end of Section 42.2, the stage was set for Niels Bohr in 1913 when he presented a new model of the hydrogen atom that circumvented the difficulties of Rutherford's planetary model. Bohr applied Planck's ideas of quantized energy levels (Section 40.1) to Rutherford's orbiting atomic electrons. Bohr's theory was historically important to the development of quantum physics, and it appeared to explain the spectral line series described by Equations 42.1 through 42.4. Although Bohr's model is now considered obsolete and has been completely replaced by a probabilistic quantum-mechanical theory, we can use the Bohr model to develop the notions of energy quantization and angular momentum quantization as applied to atomic-sized systems.

Bohr combined ideas from Planck's original quantum theory, Einstein's concept of the photon, Rutherford's planetary model of the atom, and Newtonian mechanics to arrive at a semiclassical structural model based on some revolutionary ideas. The structural model of the Bohr theory as it applies to the hydrogen atom has the following properties:

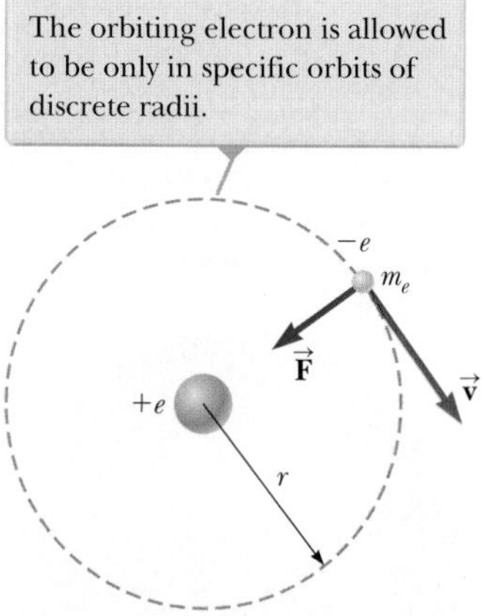

The orbiting electron is allowed to be only in specific orbits of discrete radii.

Figure 42.6 Diagram representing Bohr's model of the hydrogen atom.

1. *Physical components:*
   The electron moves in circular orbits around the proton under the influence of the electric force of attraction as shown in Figure 42.6.

2. *Behavior of the components:*

(a) Only certain electron orbits are stable. When in one of these **stationary states,** as Bohr called them, the electron does not emit energy in the form of radiation, even though it is accelerating. Hence, the total energy of the atom remains constant and classical mechanics can be used to describe the electron's motion. Bohr's model claims that the centripetally accelerated electron does not continuously emit radiation, losing energy and eventually spiraling into the nucleus, as predicted by classical physics in the form of Rutherford's planetary model.

(b) The atom emits radiation when the electron makes a transition from a more energetic initial stationary state to a lower-energy stationary state. This transition cannot be visualized or treated classically. In particular, the frequency f of the photon emitted in the transition is related to the change in the atom's energy and is not equal to the frequency of the electron's orbital motion. The frequency of the emitted radiation is found from the energy-conservation expression

$$E_i - E_f = hf \qquad \text{(42.5)}$$

where E_i is the energy of the initial state, E_f is the energy of the final state, and $E_i > E_f$. In addition, energy of an incident photon can be absorbed by the atom, but only if the photon has an energy that exactly matches the difference in energy between an allowed state of the atom and a higher-energy state. Upon absorption, the photon disappears and the atom makes a transition to the higher-energy state.

(c) The size of an allowed electron orbit is determined by a condition imposed on the electron's orbital angular momentum: the allowed orbits are those for which the electron's orbital angular momentum about the nucleus is quantized and equal to an integral multiple of $\hbar = h/2\pi$,

$$m_e v r = n\hbar \qquad n = 1, 2, 3, \ldots \qquad \text{(42.6)}$$

where m_e is the electron mass, v is the electron's speed in its orbit, and r is the orbital radius.

Niels Bohr
Danish Physicist (1885–1962)
Bohr was an active participant in the early development of quantum mechanics and provided much of its philosophical framework. During the 1920s and 1930s, he headed the Institute for Advanced Studies in Copenhagen. The institute was a magnet for many of the world's best physicists and provided a forum for the exchange of ideas. Bohr was awarded the 1922 Nobel Prize in Physics for his investigation of the structure of atoms and the radiation emanating from them. When Bohr visited the United States in 1939 to attend a scientific conference, he brought news that the fission of uranium had been observed by Hahn and Strassman in Berlin. The results were the foundations of the nuclear weapon developed in the United States during World War II.

These postulates are a mixture of established principles and completely new and untested ideas at the time. Property 1, from classical mechanics, treats the electron in orbit around the nucleus in the same way we treat a planet in a circular orbit around a star, using the particle in uniform circular motion analysis model. Property 2(a) was a radical new idea in 1913 that was completely at odds with the understanding of electromagnetism at the time. Property 2(b) represents the principle of conservation of energy as described by the nonisolated system model for energy. Property 2(c) is another new idea that had no basis in classical physics.

Property 2(b) implies qualitatively the existence of a characteristic discrete emission line spectrum *and also* a corresponding absorption line spectrum of the kind shown in Figure 42.1 for hydrogen. Using these postulates, let's calculate the allowed energy levels and find quantitative values of the emission wavelengths of the hydrogen atom.

The electric potential energy of the system shown in Figure 42.6 is given by Equation 25.13, $U = k_e q_1 q_2 / r = -k_e e^2 / r$, where k_e is the Coulomb constant and the negative sign arises from the charge $-e$ on the electron. Therefore, the *total* energy of the atom, which consists of the electron's kinetic energy and the system's potential energy, is

$$E = K + U = \tfrac{1}{2} m_e v^2 - k_e \frac{e^2}{r} \qquad \text{(42.7)}$$

The electron is modeled as a particle in uniform circular motion, so the electric force $k_e e^2/r^2$ exerted on the electron must equal the product of its mass and its centripetal acceleration ($a_c = v^2/r$):

$$\frac{k_e e^2}{r^2} = \frac{m_e v^2}{r}$$

$$v^2 = \frac{k_e e^2}{m_e r} \tag{42.8}$$

From Equation 42.8, we find that the kinetic energy of the electron is

$$K = \tfrac{1}{2} m_e v^2 = \frac{k_e e^2}{2r}$$

Substituting this value of K into Equation 42.7 gives the following expression for the total energy of the atom:[1]

$$E = -\frac{k_e e^2}{2r} \tag{42.9}$$

Because the total energy is *negative*, which indicates a bound electron–proton system, energy in the amount of $k_e e^2/2r$ must be added to the atom to remove the electron and make the total energy of the system zero.

We can obtain an expression for r, the radius of the allowed orbits, by solving Equation 42.6 for v^2 and equating it to Equation 42.8:

$$v^2 = \frac{n^2 \hbar^2}{m_e^2 r^2} = \frac{k_e e^2}{m_e r}$$

$$r_n = \frac{n^2 \hbar^2}{m_e k_e e^2} \quad n = 1, 2, 3, \ldots \tag{42.10}$$

Equation 42.10 shows that the radii of the allowed orbits have discrete values: they are quantized. The result is based on the *assumption* that the electron can exist only in certain allowed orbits determined by the integer n (Bohr's Property 2(c)).

The orbit with the smallest radius, called the **Bohr radius** a_0, corresponds to $n = 1$ and has the value

$$a_0 = \frac{\hbar^2}{m_e k_e e^2} = 0.052\ 9 \text{ nm} \tag{42.11}$$

Substituting Equation 42.11 into Equation 42.10 gives a general expression for the radius of any orbit in the hydrogen atom:

$$r_n = n^2 a_0 = n^2(0.052\ 9 \text{ nm}) \quad n = 1, 2, 3, \ldots \tag{42.12}$$

Bohr's theory predicts a value for the radius of a hydrogen atom on the right order of magnitude, based on experimental measurements. This result was a striking triumph for Bohr's theory. The first three Bohr orbits are shown to scale in Figure 42.7.

The quantization of orbit radii leads to energy quantization. Substituting $r_n = n^2 a_0$ into Equation 42.9 gives

$$E_n = -\frac{k_e e^2}{2a_0}\left(\frac{1}{n^2}\right) \quad n = 1, 2, 3, \ldots \tag{42.13}$$

Inserting numerical values into this expression, we find that

$$E_n = -\frac{13.606 \text{ eV}}{n^2} \quad n = 1, 2, 3, \ldots \tag{42.14}$$

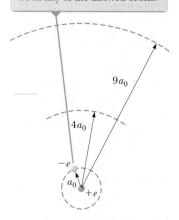

The electron is shown in the lowest-energy orbit, but it could be in any of the allowed orbits.

$9a_0$

$4a_0$

$-e$

a_0

$+e$

Figure 42.7 The first three circular orbits predicted by the Bohr model of the hydrogen atom.

▶ **Bohr radius**

▶ **Radii of Bohr orbits in hydrogen**

---

[1]Compare Equation 42.9 with its gravitational counterpart, Equation 13.19.

Only energies satisfying this equation are permitted. The lowest allowed energy level, the ground state, has $n = 1$ and energy $E_1 = -13.606$ eV. The next energy level, the first excited state, has $n = 2$ and energy $E_2 = E_1/2^2 = -3.401$ eV. Figure 42.8 is an energy-level diagram showing the energies of these discrete energy states and the corresponding quantum numbers n. The uppermost level corresponds to $n = \infty$ (or $r = \infty$) and $E = 0$.

Notice how the allowed energies of the hydrogen atom differ from those of the particle in a box. The particle-in-a-box energies (Eq. 41.14) increase as n^2, so they become farther apart in energy as n increases. On the other hand, the energies of the hydrogen atom (Eq. 42.14) are inversely proportional to n^2, so their separation in energy becomes smaller as n increases. The separation between energy levels approaches zero as n approaches infinity and the energy approaches zero.

Zero energy represents the boundary between a bound system of an electron and a proton and an unbound system. If the energy of the atom is raised from that of the ground state to any energy larger than zero, the atom is **ionized.** The minimum energy required to ionize the atom in its ground state is called the **ionization energy.** As can be seen from Figure 42.8, the ionization energy for hydrogen in the ground state, based on Bohr's calculation, is 13.6 eV. This finding constituted another major achievement for the Bohr theory because the ionization energy for hydrogen had already been measured to be 13.6 eV.

Equations 42.5 and 42.13 can be used to calculate the frequency of the photon emitted when the electron makes a transition from an outer orbit to an inner orbit:

$$f = \frac{E_i - E_f}{h} = \frac{k_e e^2}{2a_0 h}\left(\frac{1}{n_f^2} - \frac{1}{n_i^2}\right) \tag{42.15}$$

Because the quantity measured experimentally is wavelength, it is convenient to use $c = f\lambda$ to express Equation 42.15 in terms of wavelength:

$$\frac{1}{\lambda} = \frac{f}{c} = \frac{k_e e^2}{2a_0 hc}\left(\frac{1}{n_f^2} - \frac{1}{n_i^2}\right) \tag{42.16}$$

Remarkably, this expression, which is purely *theoretical*, is *identical* to the general form of the empirical relationships discovered by Balmer and Rydberg and given by Equations 42.1 to 42.4:

$$\frac{1}{\lambda} = R_H\left(\frac{1}{n_f^2} - \frac{1}{n_i^2}\right) \tag{42.17}$$

provided the constant $k_e e^2/2a_0 hc$ is equal to the experimentally determined Rydberg constant. Soon after Bohr demonstrated that these two quantities agree to within approximately 1%, this work was recognized as the crowning achievement of his new quantum theory of the hydrogen atom. Furthermore, Bohr showed that all the spectral series for hydrogen have a natural interpretation in his theory. The different series correspond to transitions to different final states characterized by the quantum number n_f. Figure 42.8 shows the origin of these spectral series as transitions between energy levels.

Bohr extended his model for hydrogen to other elements in which all but one electron had been removed. These systems have the same structure as the hydrogen atom except that the nuclear charge is larger. Ionized elements such as He^+, Li^{2+}, and Be^{3+} were suspected to exist in hot stellar atmospheres, where atomic collisions frequently have enough energy to completely remove one or more atomic electrons. Bohr showed that many mysterious lines observed in the spectra of the Sun and several other stars could not be due to hydrogen but were correctly predicted by his theory if attributed to singly ionized helium. In general, the number of protons in the nucleus of an atom is called the **atomic number** of the element

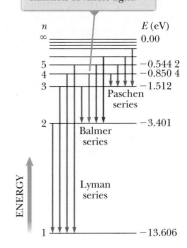

The colored arrows for the Balmer series indicate that this series results in the emission of visible light.

Figure 42.8 An energy-level diagram for the hydrogen atom. Quantum numbers are given on the left, and energies (in electron volts) are given on the right. Vertical arrows represent the four lowest-energy transitions for each of the spectral series shown.

and is given the symbol Z. To describe a single electron orbiting a fixed nucleus of charge $+Ze$, Bohr's theory gives

$$r_n = (n^2)\frac{a_0}{Z} \tag{42.18}$$

$$E_n = -\frac{k_e e^2}{2a_0}\left(\frac{Z^2}{n^2}\right) \quad n = 1, 2, 3, \ldots \tag{42.19}$$

Although the Bohr theory was triumphant in its agreement with some experimental results on the hydrogen atom, it suffered from some difficulties. One of the first indications that the Bohr theory needed to be modified arose when improved spectroscopic techniques were used to examine the spectral lines of hydrogen. It was found that many of the lines in the Balmer and other series were not single lines at all. Instead, each was a group of lines spaced very close together. An additional difficulty arose when it was observed that in some situations certain single spectral lines were split into three closely spaced lines when the atoms were placed in a strong magnetic field. Efforts to explain these and other deviations from the Bohr model led to modifications in the theory and ultimately to a replacement theory that will be discussed in Section 42.4.

Bohr's Correspondence Principle

In our study of relativity, we found that Newtonian mechanics is a special case of relativistic mechanics and is usable only for speeds much less than c. Similarly,

> quantum physics agrees with classical physics when the difference between quantized levels becomes vanishingly small.

This principle, first set forth by Bohr, is called the **correspondence principle.**[2]

For example, consider an electron orbiting the hydrogen atom with $n > 10\ 000$. For such large values of n, the energy differences between adjacent levels approach zero; therefore, the levels are nearly continuous. Consequently, the classical model is reasonably accurate in describing the system for large values of n. According to the classical picture, the frequency of the light emitted by the atom is equal to the frequency of revolution of the electron in its orbit about the nucleus. Calculations show that for $n > 10\ 000$, this frequency is different from that predicted by quantum mechanics by less than 0.015%.

Quick Quiz 42.1 A hydrogen atom is in its ground state. Incident on the atom is a photon having an energy of 10.5 eV. What is the result? **(a)** The atom is excited to a higher allowed state. **(b)** The atom is ionized. **(c)** The photon passes by the atom without interaction.

Quick Quiz 42.2 A hydrogen atom makes a transition from the $n = 3$ level to the $n = 2$ level. It then makes a transition from the $n = 2$ level to the $n = 1$ level. Which transition results in emission of the longer-wavelength photon? **(a)** the first transition **(b)** the second transition **(c)** neither transition because the wavelengths are the same for both

---

Example 42.1   **Electronic Transitions in Hydrogen**

(A) The electron in a hydrogen atom makes a transition from the $n = 2$ energy level to the ground level $(n = 1)$. Find the wavelength and frequency of the emitted photon.

---

[2]In reality, the correspondence principle is the starting point for Bohr's property 2(c) on angular momentum quantization. To see how property 2(c) arises from the correspondence principle, see J. W. Jewett Jr., *Physics Begins with Another M . . . Mysteries, Magic, Myth, and Modern Physics* (Boston: Allyn & Bacon, 1996), pp. 353–356.

▶ **42.1** c o n t i n u e d

SOLUTION

Conceptualize Imagine the electron in a circular orbit about the nucleus as in the Bohr model in Figure 42.6. When the electron makes a transition to a lower stationary state, it emits a photon with a given frequency and drops to a circular orbit of smaller radius.

Categorize We evaluate the results using equations developed in this section, so we categorize this example as a substitution problem.

Use Equation 42.17 to obtain λ, with $n_i = 2$ and $n_f = 1$:

$$\frac{1}{\lambda} = R_H\left(\frac{1}{1^2} - \frac{1}{2^2}\right) = \frac{3R_H}{4}$$

$$\lambda = \frac{4}{3R_H} = \frac{4}{3(1.097 \times 10^7\text{ m}^{-1})} = 1.22 \times 10^{-7}\text{ m} = \boxed{122\text{ nm}}$$

Use Equation 34.20 to find the frequency of the photon:

$$f = \frac{c}{\lambda} = \frac{3.00 \times 10^8\text{ m/s}}{1.22 \times 10^{-7}\text{ m}} = \boxed{2.47 \times 10^{15}\text{ Hz}}$$

(B) In interstellar space, highly excited hydrogen atoms called Rydberg atoms have been observed. Find the wavelength to which radio astronomers must tune to detect signals from electrons dropping from the $n = 273$ level to the $n = 272$ level.

SOLUTION

Use Equation 42.17, this time with $n_i = 273$ and $n_f = 272$:

$$\frac{1}{\lambda} = R_H\left(\frac{1}{n_f^2} - \frac{1}{n_i^2}\right) = R_H\left(\frac{1}{(272)^2} - \frac{1}{(273)^2}\right) = 9.88 \times 10^{-8}\, R_H$$

Solve for λ:

$$\lambda = \frac{1}{9.88 \times 10^{-8}R_H} = \frac{1}{(9.88 \times 10^{-8})(1.097 \times 10^7\text{ m}^{-1})} = \boxed{0.922\text{ m}}$$

(C) What is the radius of the electron orbit for a Rydberg atom for which $n = 273$?

SOLUTION

Use Equation 42.12 to find the radius of the orbit:

$$r_{273} = (273)^2\,(0.052\,9\text{ nm}) = \boxed{3.94\ \mu\text{m}}$$

This radius is large enough that the atom is on the verge of becoming macroscopic!

(D) How fast is the electron moving in a Rydberg atom for which $n = 273$?

SOLUTION

Solve Equation 42.8 for the electron's speed:

$$v = \sqrt{\frac{k_e e^2}{m_e r}} = \sqrt{\frac{(8.99 \times 10^9\text{ N} \cdot \text{m}^2/\text{C}^2)(1.60 \times 10^{-19}\text{ C})^2}{(9.11 \times 10^{-31}\text{ kg})(3.94 \times 10^{-6}\text{ m})}}$$

$$= \boxed{8.01 \times 10^3\text{ m/s}}$$

WHAT IF? What if radiation from the Rydberg atom in part (B) is treated classically? What is the wavelength of radiation emitted by the atom in the $n = 273$ level?

Answer Classically, the frequency of the emitted radiation is that of the rotation of the electron around the nucleus.

Calculate this frequency using the period defined in Equation 4.15:

$$f = \frac{1}{T} = \frac{v}{2\pi r}$$

Substitute the radius and speed from parts (C) and (D):

$$f = \frac{v}{2\pi r} = \frac{8.02 \times 10^3\text{ m/s}}{2\pi(3.94 \times 10^{-6}\text{ m})} = 3.24 \times 10^8\text{ Hz}$$

Find the wavelength of the radiation from Equation 34.20:

$$\lambda = \frac{c}{f} = \frac{3.00 \times 10^8\text{ m/s}}{3.24 \times 10^8\text{ Hz}} = 0.927\text{ m}$$

This value is about 0.5% different from the wavelength calculated in part (B). As indicated in the discussion of Bohr's correspondence principle, this difference becomes even smaller for higher values of n.

42.4 The Quantum Model of the Hydrogen Atom

In the preceding section, we described how the Bohr model views the electron as a particle orbiting the nucleus in nonradiating, quantized energy levels. This model combines both classical and quantum concepts. Although the model demonstrates excellent agreement with some experimental results, it cannot explain others. These difficulties are removed when a full quantum model involving the Schrödinger equation is used to describe the hydrogen atom.

The formal procedure for solving the problem of the hydrogen atom is to substitute the appropriate potential energy function into the Schrödinger equation, find solutions to the equation, and apply boundary conditions as we did for the particle in a box in Chapter 41. The potential energy function for the hydrogen atom is that due to the electrical interaction between the electron and the proton (see Section 25.3):

$$U(r) = -k_e \frac{e^2}{r} \tag{42.20}$$

where k_e is the Coulomb constant and r is the radial distance from the proton (situated at $r = 0$) to the electron.

The mathematics for the hydrogen atom is more complicated than that for the particle in a box for two primary reasons: (1) the atom is three-dimensional, and (2) U is not constant, but rather depends on the radial coordinate r. If the time-independent Schrödinger equation (Eq. 41.15) is extended to three-dimensional rectangular coordinates, the result is

$$-\frac{\hbar^2}{2m}\left(\frac{\partial^2 \psi}{\partial x^2} + \frac{\partial^2 \psi}{\partial y^2} + \frac{\partial^2 \psi}{\partial z^2}\right) + U\psi = E\psi$$

It is easier to solve this equation for the hydrogen atom if rectangular coordinates are converted to spherical polar coordinates, an extension of the plane polar coordinates introduced in Section 3.1. In spherical polar coordinates, a point in space is represented by the three variables r, θ, and ϕ, where r is the radial distance from the origin, $r = \sqrt{x^2 + y^2 + z^2}$. With the point represented at the end of a position vector $\vec{r}$ as shown in Figure 42.9, the angular coordinate θ specifies its angular position relative to the z axis. Once that position vector is projected onto the xy plane, the angular coordinate ϕ specifies the projection's (and therefore the point's) angular position relative to the x axis.

The conversion of the three-dimensional time-independent Schrödinger equation for $\psi(x, y, z)$ to the equivalent form for $\psi(r, \theta, \phi)$ is straightforward but very tedious, so we omit the details.[3] In Chapter 41, we separated the time dependence from the space dependence in the general wave function Ψ. In this case of the hydrogen atom, the three space variables in $\psi(r, \theta, \phi)$ can be similarly separated by writing the wave function as a product of functions of each single variable:

$$\psi(r, \theta, \phi) = R(r)f(\theta)g(\phi)$$

In this way, Schrödinger's equation, which is a three-dimensional partial differential equation, can be transformed into three separate ordinary differential equations: one for $R(r)$, one for $f(\theta)$, and one for $g(\phi)$. Each of these functions is subject to boundary conditions. For example, $R(r)$ must remain finite as $r \to 0$ and $r \to \infty$; furthermore, $g(\phi)$ must have the same value as $g(\phi + 2\pi)$.

The potential energy function given in Equation 42.20 depends *only* on the radial coordinate r and not on either of the angular coordinates; therefore, it appears only in the equation for $R(r)$. As a result, the equations for θ and ϕ are independent of the particular system and their solutions are valid for *any* system exhibiting rotation.

When the full set of boundary conditions is applied to all three functions, three different quantum numbers are found for each allowed state of the hydrogen atom,

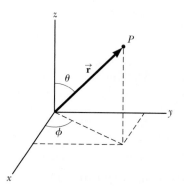

Figure 42.9 A point P in space is located by means of a position vector $\vec{r}$. In Cartesian coordinates, the components of this vector are x, y, and z. In spherical polar coordinates, the point is described by r, the distance from the origin; θ, the angle between $\vec{r}$ and the z axis; and ϕ, the angle between the x axis and a projection of $\vec{r}$ onto the xy plane.

---

[3]Descriptions of the solutions to the Schrödinger equation for the hydrogen atom are available in modern physics textbooks such as R. A. Serway, C. Moses, and C. A. Moyer, *Modern Physics*, 3rd ed. (Belmont, CA: Brooks/Cole, 2005).

one for each of the separate differential equations. These quantum numbers are restricted to integer values and correspond to the three independent degrees of freedom (three space dimensions).

The first quantum number, associated with the radial function $R(r)$ of the full wave function, is called the **principal quantum number** and is assigned the symbol n. The differential equation for $R(r)$ leads to functions giving the probability of finding the electron at a certain radial distance from the nucleus. In Section 42.5, we will describe two of these radial wave functions. From the boundary conditions, the energies of the allowed states for the hydrogen atom are found to be related to n as follows:

$$E_n = -\left(\frac{k_e e^2}{2a_0}\right)\frac{1}{n^2} = -\frac{13.606 \text{ eV}}{n^2} \quad n = 1, 2, 3, \ldots \tag{42.21}$$

◀ Allowed energies of the quantum hydrogen atom

This result is in exact agreement with that obtained in the Bohr theory (Eqs. 42.13 and 42.14)! This agreement is *remarkable* because the Bohr theory and the full quantum theory arrive at the result from completely different starting points.

The **orbital quantum number,** symbolized ℓ, comes from the differential equation for $f(\theta)$ and is associated with the orbital angular momentum of the electron. The **orbital magnetic quantum number** m_ℓ arises from the differential equation for $g(\phi)$. Both ℓ and m_ℓ are integers. We will expand our discussion of these two quantum numbers in Section 42.6, where we also introduce a fourth (nonintegral) quantum number, resulting from a relativistic treatment of the hydrogen atom.

The application of boundary conditions on the three parts of the full wave function leads to important relationships among the three quantum numbers as well as certain restrictions on their values:

The values of n are integers that can range from 1 to ∞.

The values of ℓ are integers that can range from 0 to $n - 1$.

The values of m_ℓ are integers that can range from $-\ell$ to ℓ.

◀ Restrictions on the values of hydrogen-atom quantum numbers

For example, if $n = 1$, only $\ell = 0$ and $m_\ell = 0$ are permitted. If $n = 2$, then ℓ may be 0 or 1; if $\ell = 0$, then $m_\ell = 0$; but if $\ell = 1$, then m_ℓ may be 1, 0, or -1. Table 42.1 summarizes the rules for determining the allowed values of ℓ and m_ℓ for a given n.

For historical reasons, all states having the same principal quantum number are said to form a **shell.** Shells are identified by the letters K, L, M, . . . , which designate the states for which $n = 1, 2, 3, \ldots$. Likewise, all states having the same values of n and ℓ are said to form a **subshell.** The letters[4] s, p, d, f, g, h, . . . are used to designate the subshells for which $\ell = 0, 1, 2, 3, \ldots$. The state designated by $3p$, for example, has the quantum numbers $n = 3$ and $\ell = 1$; the $2s$ state has the quantum numbers $n = 2$ and $\ell = 0$. These notations are summarized in Tables 42.2 and 42.3 (page 1308).

States that violate the rules given in Table 42.1 do not exist. (They do not satisfy the boundary conditions on the wave function.) For instance, the $2d$ state, which

Pitfall Prevention 42.3

Energy Depends on n Only for Hydrogen The implication in Equation 42.21 that the energy depends only on the quantum number n is true only for the hydrogen atom. For more complicated atoms, we will use the same quantum numbers developed here for hydrogen. The energy levels for these atoms depend primarily on n, but they also depend to a lesser degree on other quantum numbers.

Pitfall Prevention 42.4

Quantum Numbers Describe a System It is common to assign the quantum numbers to an electron. Remember, however, that these quantum numbers arise from the Schrödinger equation, which involves a potential energy function for the *system* of the electron and the nucleus. Therefore, it is more *proper* to assign the quantum numbers to the atom, but it is more *popular* to assign them to an electron. We follow this latter usage because it is so common.

Table 42.1	**Three Quantum Numbers for the Hydrogen Atom**		
Quantum Number	Name	Allowed Values	Number of Allowed States
n	Principal quantum number	$1, 2, 3, \ldots$	Any number
ℓ	Orbital quantum number	$0, 1, 2, \ldots, n - 1$	n
m_ℓ	Orbital magnetic quantum number	$-\ell, -\ell + 1, \ldots, 0, \ldots, \ell - 1, \ell$	$2\ell + 1$

Table 42.2	**Atomic Shell Notations**
n	Shell Symbol
1	K
2	L
3	M
4	N
5	O
6	P

[4]The first four of these letters come from early classifications of spectral lines: sharp, principal, diffuse, and fundamental. The remaining letters are in alphabetical order.

Table 42.3   **Atomic Subshell Notations**

ℓ	Subshell Symbol
0	s
1	p
2	d
3	f
4	g
5	h

would have $n = 2$ and $\ell = 2$, cannot exist because the highest allowed value of ℓ is $n - 1$, which in this case is 1. Therefore, for $n = 2$, the $2s$ and $2p$ states are allowed but $2d$, $2f$, ... are not. For $n = 3$, the allowed subshells are $3s$, $3p$, and $3d$.

Quick Quiz 42.3 How many possible subshells are there for the $n = 4$ level of hydrogen? **(a)** 5 **(b)** 4 **(c)** 3 **(d)** 2 **(e)** 1

Quick Quiz 42.4 When the principal quantum number is $n = 5$, how many different values of **(a)** ℓ and **(b)** m_ℓ are possible?

Example 42.2   **The $n = 2$ Level of Hydrogen**

For a hydrogen atom, determine the allowed states corresponding to the principal quantum number $n = 2$ and calculate the energies of these states.

SOLUTION

Conceptualize Think about the atom in the $n = 2$ quantum state. There is only one such state in the Bohr theory, but our discussion of the quantum theory allows for more states because of the possible values of ℓ and m_ℓ.

Categorize We evaluate the results using rules discussed in this section, so we categorize this example as a substitution problem.

From Table 42.1, we find that when $n = 2$, ℓ can be 0 or 1. Find the possible values of m_ℓ from Table 42.1:

$$\ell = 0 \quad \rightarrow \quad m_\ell = 0$$
$$\ell = 1 \quad \rightarrow \quad m_\ell = -1, 0, \text{ or } 1$$

Hence, we have one state, designated as the $2s$ state, that is associated with the quantum numbers $n = 2$, $\ell = 0$, and $m_\ell = 0$, and we have three states, designated as $2p$ states, for which the quantum numbers are $n = 2$, $\ell = 1$, and $m_\ell = -1$; $n = 2$, $\ell = 1$, and $m_\ell = 0$; and $n = 2$, $\ell = 1$, and $m_\ell = 1$.

Find the energy for all four of these states with $n = 2$ from Equation 42.21:

$$E_2 = -\frac{13.606 \text{ eV}}{2^2} = \boxed{-3.401 \text{ eV}}$$

42.5 The Wave Functions for Hydrogen

Because the potential energy of the hydrogen atom depends only on the radial distance r between nucleus and electron, some of the allowed states for this atom can be represented by wave functions that depend only on r. For these states, $f(\theta)$ and $g(\phi)$ are constants. The simplest wave function for hydrogen is the one that describes the $1s$ state and is designated $\psi_{1s}(r)$:

Wave function for hydrogen in its ground state ▶

$$\psi_{1s}(r) = \frac{1}{\sqrt{\pi a_0^3}} e^{-r/a_0} \tag{42.22}$$

where a_0 is the Bohr radius. (In Problem 26, you can verify that this function satisfies the Schrödinger equation.) Note that ψ_{1s} approaches zero as r approaches ∞ and is normalized as presented (see Eq. 41.7). Furthermore, because ψ_{1s} depends only on r, it is *spherically symmetric*. This symmetry exists for all s states.

Recall that the probability of finding a particle in any region is equal to an integral of the probability density $|\psi|^2$ for the particle over the region. The probability density for the $1s$ state is

$$|\psi_{1s}|^2 = \left(\frac{1}{\pi a_0^3}\right) e^{-2r/a_0} \tag{42.23}$$

Because we imagine the nucleus to be fixed in space at $r = 0$, we can assign this probability density to the question of locating the electron. According to Equation 41.3, the probability of finding the electron in a volume element dV is $|\psi|^2 \, dV$. It is convenient to define the *radial probability density function* $P(r)$ as the probability per unit radial length of finding the electron in a spherical shell of radius r and thickness dr. Therefore, $P(r) \, dr$ is the probability of finding the electron in this shell. The volume dV of such an infinitesimally thin shell equals its surface area $4\pi r^2$ multiplied by the shell thickness dr (Fig. 42.10), so we can write this probability as

$$P(r) \, dr = |\psi|^2 \, dV = |\psi|^2 \, 4\pi r^2 \, dr$$

Therefore, the radial probability density function for an s state is

$$P(r) = 4\pi r^2 |\psi|^2 \qquad \textbf{(42.24)}$$

Substituting Equation 42.23 into Equation 42.24 gives the radial probability density function for the hydrogen atom in its ground state:

$$P_{1s}(r) = \left(\frac{4r^2}{a_0{}^3}\right) e^{-2r/a_0} \qquad \textbf{(42.25)}$$

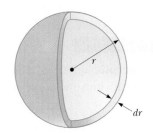

Figure 42.10 A spherical shell of radius r and infinitesimal thickness dr has a volume equal to $4\pi r^2 \, dr$.

◀ **Radial probability density for the 1s state of hydrogen**

A plot of the function $P_{1s}(r)$ versus r is presented in Figure 42.11a. The peak of the curve corresponds to the most probable value of r for this particular state. We show in Example 42.3 that this peak occurs at the Bohr radius, the radial position of the electron when the hydrogen atom is in its ground state in the Bohr theory, another remarkable agreement between the Bohr theory and the quantum theory.

According to quantum mechanics, the atom has no sharply defined boundary as suggested by the Bohr theory. The probability distribution in Figure 42.11a suggests that the charge of the electron can be modeled as being extended throughout a region of space, commonly referred to as an *electron cloud*. Figure 42.11b shows the probability density of the electron in a hydrogen atom in the $1s$ state as a function of position in the xy plane. The darkness of the blue color corresponds to the value of the probability density. The darkest portion of the distribution appears at $r = a_0$, corresponding to the most probable value of r for the electron.

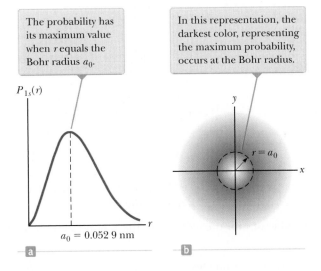

The probability has its maximum value when r equals the Bohr radius a_0.

In this representation, the darkest color, representing the maximum probability, occurs at the Bohr radius.

$P_{1s}(r)$

$a_0 = 0.052\,9$ nm

$r = a_0$

ⓐ ⓑ

Figure 42.11 (a) The probability of finding the electron as a function of distance from the nucleus for the hydrogen atom in the $1s$ (ground) state. (b) The cross section in the xy plane of the spherical electronic charge distribution for the hydrogen atom in its $1s$ state.

| Example 42.3 | **The Ground State of Hydrogen** |

(A) Calculate the most probable value of r for an electron in the ground state of the hydrogen atom.

continued

▶ **42.3** continued

SOLUTION

Conceptualize Do not imagine the electron in orbit around the proton as in the Bohr theory of the hydrogen atom. Instead, imagine the charge of the electron spread out in space around the proton in an electron cloud with spherical symmetry.

Categorize Because the statement of the problem asks for the "most probable value of r," we categorize this example as a problem in which the quantum approach is used. (In the Bohr atom, the electron moves in an orbit with an *exact* value of r.)

Analyze The most probable value of r corresponds to the maximum in the plot of $P_{1s}(r)$ versus r. We can evaluate the most probable value of r by setting $dP_{1s}/dr = 0$ and solving for r.

Differentiate Equation 42.25 and set the result equal to zero:

$$\frac{dP_{1s}}{dr} = \frac{d}{dr}\left[\left(\frac{4r^2}{a_0{}^3}\right)e^{-2r/a_0}\right] = 0$$

$$e^{-2r/a_0}\frac{d}{dr}(r^2) + r^2\frac{d}{dr}(e^{-2r/a_0}) = 0$$

$$2re^{-2r/a_0} + r^2(-2/a_0)e^{-2r/a_0} = 0$$

$$(1)\quad 2r[1 - (r/a_0)]e^{-2r/a_0} = 0$$

Set the bracketed expression equal to zero and solve for r:

$$1 - \frac{r}{a_0} = 0 \quad \rightarrow \quad r = \boxed{a_0}$$

Finalize The most probable value of r is the Bohr radius! Equation (1) is also satisfied at $r = 0$ and as $r \rightarrow \infty$. These points are locations of the *minimum* probability, which is equal to zero as seen in Figure 42.11a.

(B) Calculate the probability that the electron in the ground state of hydrogen will be found outside the Bohr radius.

SOLUTION

Analyze The probability is found by integrating the radial probability density function $P_{1s}(r)$ for this state from the Bohr radius a_0 to ∞.

Set up this integral using Equation 42.25:

$$P = \int_{a_0}^{\infty} P_{1s}(r)\, dr = \frac{4}{a_0{}^3}\int_{a_0}^{\infty} r^2 e^{-2r/a_0}\, dr$$

Put the integral in dimensionless form by changing variables from r to $z = 2r/a_0$, noting that $z = 2$ when $r = a_0$ and that $dr = (a_0/2)\, dz$:

$$P = \frac{4}{a_0{}^3}\int_2^{\infty}\left(\frac{za_0}{2}\right)^2 e^{-z}\left(\frac{a_0}{2}\right)dz = \tfrac{1}{2}\int_2^{\infty} z^2 e^{-z}\, dz$$

Evaluate the integral using partial integration (see Appendix B.7):

$$P = -\tfrac{1}{2}(z^2 + 2z + 2)e^{-z}\Big|_2^{\infty}$$

Evaluate between the limits:

$$P = 0 - [-\tfrac{1}{2}(4 + 4 + 2)e^{-2}] = 5e^{-2} = \boxed{0.677 \text{ or } 67.7\%}$$

Finalize This probability is larger than 50%. The reason for this value is the asymmetry in the radial probability density function (Fig. 42.11a), which has more area to the right of the peak than to the left.

WHAT IF?   What if you were asked for the *average* value of r for the electron in the ground state rather than the most probable value?

Answer The average value of r is the same as the expectation value for r.

Use Equation 42.25 to evaluate the average value of r:

$$r_{avg} = \langle r \rangle = \int_0^{\infty} rP(r)\, dr = \int_0^{\infty} r\left(\frac{4r^2}{a_0{}^3}\right)e^{-2r/a_0}\, dr$$

$$= \left(\frac{4}{a_0{}^3}\right)\int_0^{\infty} r^3 e^{-2r/a_0}\, dr$$

▶ **42.3** continued

Evaluate the integral with the help of the first integral listed in Table B.6 in Appendix B:

$$r_{avg} = \left(\frac{4}{a_0^{\,3}}\right)\left(\frac{3!}{(2/a_0)^4}\right) = \tfrac{3}{2}a_0$$

Again, the average value is larger than the most probable value because of the asymmetry in the wave function as seen in Figure 42.11a.

The next-simplest wave function for the hydrogen atom is the one corresponding to the 2s state ($n = 2$, $\ell = 0$). The normalized wave function for this state is

$$\psi_{2s}(r) = \frac{1}{4\sqrt{2\pi}}\left(\frac{1}{a_0}\right)^{3/2}\left(2 - \frac{r}{a_0}\right)e^{-r/2a_0} \qquad \text{(42.26)}$$

◀ **Wave function for hydrogen in the 2s state**

Again notice that ψ_{2s} depends only on r and is spherically symmetric. The energy corresponding to this state is $E_2 = -(13.606/4)$ eV $= -3.401$ eV. This energy level represents the first excited state of hydrogen. A plot of the radial probability density function for this state in comparison to the 1s state is shown in Figure 42.12. The plot for the 2s state has two peaks. In this case, the most probable value corresponds to that value of r that has the highest value of P ($\approx 5a_0$). An electron in the 2s state would be much farther from the nucleus (on the average) than an electron in the 1s state.

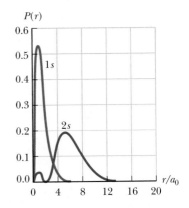

Figure 42.12 The radial probability density function versus r/a_0 for the 1s and 2s states of the hydrogen atom.

42.6 Physical Interpretation of the Quantum Numbers

The principal quantum number n of a particular state in the hydrogen atom determines the energy of the atom according to Equation 42.21. Now let's see what the other quantum numbers in our atomic model correspond to physically.

The Orbital Quantum Number ℓ

We begin this discussion by returning briefly to the Bohr model of the atom. If the electron moves in a circle of radius r, the magnitude of its angular momentum relative to the center of the circle is $L = m_e vr$. The direction of $\vec{L}$ is perpendicular to the plane of the circle and is given by a right-hand rule. According to classical physics, the magnitude L of the orbital angular momentum can have any value. The Bohr model of hydrogen, however, postulates that the magnitude of the angular momentum of the electron is restricted to multiples of $\hbar$; that is, $L = n\hbar$. This model must be modified because it predicts (incorrectly) that the ground state of hydrogen has one unit of angular momentum. Furthermore, if L is taken to be zero in the Bohr model, the electron must be pictured as a particle oscillating along a straight line through the nucleus, which is a physically unacceptable situation.

These difficulties are resolved with the quantum-mechanical model of the atom, although we must give up the convenient mental representation of an electron orbiting in a well-defined circular path. Despite the absence of this representation, the atom does indeed possess an angular momentum and it is still called orbital angular momentum. According to quantum mechanics, an atom in a state whose principal quantum number is n can take on the following *discrete* values of the magnitude of the orbital angular momentum:[5]

$$L = \sqrt{\ell(\ell + 1)}\,\hbar \qquad \ell = 0, 1, 2, \ldots, n - 1 \qquad \text{(42.27)}$$

◀ **Allowed values of L**

[5]Equation 42.27 is a direct result of the mathematical solution of the Schrödinger equation and the application of angular boundary conditions. This development, however, is beyond the scope of this book.

Given these allowed values of ℓ, we see that $L = 0$ (corresponding to $\ell = 0$) is an acceptable value of the magnitude of the angular momentum. That L can be zero in this model serves to point out the inherent difficulties in any attempt to describe results based on quantum mechanics in terms of a purely particle-like (classical) model. In the quantum-mechanical interpretation, the electron cloud for the $L = 0$ state is spherically symmetric and has no fundamental rotation axis.

The Orbital Magnetic Quantum Number m_ℓ

Because angular momentum is a vector, its direction must be specified. Recall from Chapter 29 that a current loop has a corresponding magnetic moment $\vec{\mu} = I\vec{A}$ (Eq. 29.15), where I is the current in the loop and $\vec{A}$ is a vector perpendicular to the loop whose magnitude is the area of the loop. Such a moment placed in a magnetic field $\vec{B}$ interacts with the field. Suppose a weak magnetic field applied along the z axis defines a direction in space. According to classical physics, the energy of the loop–field system depends on the direction of the magnetic moment of the loop with respect to the magnetic field as described by Equation 29.18, $U_B = -\vec{\mu} \cdot \vec{B}$. Any energy between $-\mu B$ and $+\mu B$ is allowed by classical physics.

In the Bohr theory, the circulating electron represents a current loop. In the quantum-mechanical approach to the hydrogen atom, we abandon the circular orbit viewpoint of the Bohr theory, but the atom still possesses an orbital angular momentum. Therefore, there is some sense of rotation of the electron around the nucleus and a magnetic moment is present due to this angular momentum.

As mentioned in Section 42.3, spectral lines from some atoms are observed to split into groups of three closely spaced lines when the atoms are placed in a magnetic field. Suppose the hydrogen atom is located in a magnetic field. According to quantum mechanics, there are *discrete* directions allowed for the magnetic moment vector $\vec{\mu}$ with respect to the magnetic field vector $\vec{B}$. This situation is very different from that in classical physics, in which all directions are allowed.

Because the magnetic moment $\vec{\mu}$ of the atom can be related[6] to the angular momentum vector $\vec{L}$, the discrete directions of $\vec{\mu}$ translate to the direction of $\vec{L}$ being quantized. This quantization means that L_z (the projection of $\vec{L}$ along the z axis) can have only discrete values. The orbital magnetic quantum number m_ℓ specifies the allowed values of the z component of the orbital angular momentum according to the expression[7]

Allowed values of L_z ▶

$$L_z = m_\ell \hbar \qquad (42.28)$$

The quantization of the possible orientations of $\vec{L}$ with respect to an external magnetic field is often referred to as **space quantization.**

Let's look at the possible magnitudes and orientations of $\vec{L}$ for a given value of ℓ. Recall that m_ℓ can have values ranging from $-\ell$ to ℓ. If $\ell = 0$, then $L = 0$; the only allowed value of m_ℓ is $m_\ell = 0$ and $L_z = 0$. If $\ell = 1$, then $L = \sqrt{2}\hbar$ from Equation 42.27. The possible values of m_ℓ are -1, 0, and 1, so Equation 42.28 tells us that L_z may be $-\hbar$, 0, or $\hbar$. If $\ell = 2$, the magnitude of the orbital angular momentum is $\sqrt{6}\hbar$. The value of m_ℓ can be -2, -1, 0, 1, or 2, corresponding to L_z values of $-2\hbar$, $-\hbar$, 0, $\hbar$, or $2\hbar$, and so on.

Figure 42.13a shows a **vector model** that describes space quantization for the case $\ell = 2$. Notice that $\vec{L}$ can never be aligned parallel or antiparallel to $\vec{B}$ because the maximum value of L_z is $\ell\hbar$, which is less than the magnitude of the angular momentum $L = \sqrt{\ell(\ell + 1)}\hbar$. The angular momentum vector $\vec{L}$ is allowed to be perpendicular to $\vec{B}$, which corresponds to the case of $L_z = 0$ and $\ell = 0$.

---

[6]See Equation 30.22 for this relationship as derived from a classical viewpoint. Quantum mechanics arrives at the same result.

[7]As with Equation 42.27, the relationship expressed in Equation 42.28 arises from the solution to the Schrödinger equation and application of boundary conditions.

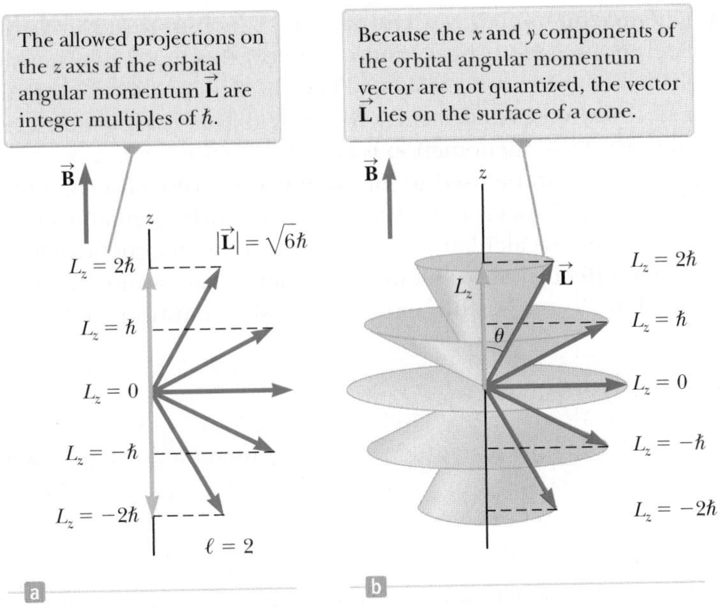

Figure 42.13 A vector model for $\ell = 2$.

The allowed projections on the z axis af the orbital angular momentum $\vec{\mathbf{L}}$ are integer multiples of $\hbar$.

Because the x and y components of the orbital angular momentum vector are not quantized, the vector $\vec{\mathbf{L}}$ lies on the surface of a cone.

$\vec{\mathbf{B}}$

$L_z = 2\hbar$ $|\vec{\mathbf{L}}| = \sqrt{6}\hbar$

$L_z = \hbar$

$L_z = 0$

$L_z = -\hbar$

$L_z = -2\hbar$

$\ell = 2$

$\vec{\mathbf{B}}$

$L_z = 2\hbar$

$L_z = \hbar$

$L_z = 0$

$L_z = -\hbar$

$L_z = -2\hbar$

a

b

The vector $\vec{\mathbf{L}}$ does not point in one specific direction. If $\vec{\mathbf{L}}$ were known exactly, all three components L_x, L_y, and L_z would be specified, which is inconsistent with an angular momentum version of the uncertainty principle. How can the magnitude and z component of a vector be specified, but the vector not be completely specified? To answer, imagine that L_x and L_y are completely unspecified so that $\vec{\mathbf{L}}$ lies anywhere on the surface of a cone that makes an angle θ with the z axis as shown in Figure 42.13b. From the figure, we see that θ is also quantized and that its values are specified through the relationship

$$\cos \theta = \frac{L_z}{L} = \frac{m_\ell}{\sqrt{\ell(\ell + 1)}} \qquad (42.29)$$

◀ **Allowed directions of the orbital angular momentum vector**

If the atom is placed in a magnetic field, the energy $U_B = -\vec{\boldsymbol{\mu}} \cdot \vec{\mathbf{B}}$ is additional energy for the atom–field system beyond that described in Equation 42.21. Because the directions of $\vec{\boldsymbol{\mu}}$ are quantized, there are discrete total energies for the system corresponding to different values of m_ℓ. Figure 42.14a shows a transition between two atomic levels in the absence of a magnetic field. In Figure 42.14b, a magnetic

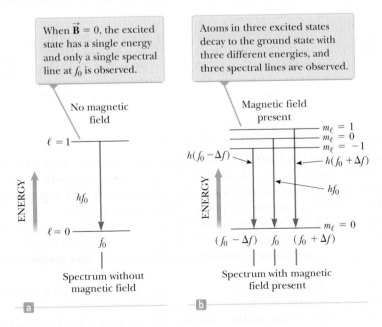

When $\vec{\mathbf{B}} = 0$, the excited state has a single energy and only a single spectral line at f_0 is observed.

Atoms in three excited states decay to the ground state with three different energies, and three spectral lines are observed.

No magnetic field

$\ell = 1$

ENERGY

hf_0

$\ell = 0$

f_0

Spectrum without magnetic field

Magnetic field present

$m_\ell = 1$
$m_\ell = 0$
$m_\ell = -1$

$h(f_0 - \Delta f)$ $h(f_0 + \Delta f)$

hf_0

ENERGY

$m_\ell = 0$

$(f_0 - \Delta f)$ f_0 $(f_0 + \Delta f)$

Spectrum with magnetic field present

a

b

Figure 42.14 The Zeeman effect. (a) Energy levels for the ground and first excited states of a hydrogen atom. (b) When the atom is immersed in a magnetic field $\vec{\mathbf{B}}$, the state with $\ell = 1$ splits into three states, giving rise to emission lines at f_0, $f_0 + \Delta f$, and $f_0 - \Delta f$, where Δf is the frequency shift of the emission caused by the magnetic field.

field is applied and the upper level, with $\ell = 1$, splits into three levels corresponding to the different directions of $\vec{\boldsymbol{\mu}}$. There are now three possible transitions from the $\ell = 1$ subshell to the $\ell = 0$ subshell. Therefore, in a collection of atoms, there are atoms in all three states and the single spectral line in Figure 42.14a splits into three spectral lines. This phenomenon is called the *Zeeman effect*.

The Zeeman effect can be used to measure extraterrestrial magnetic fields. For example, the splitting of spectral lines in light from hydrogen atoms in the surface of the Sun can be used to calculate the magnitude of the magnetic field at that location. The Zeeman effect is one of many phenomena that cannot be explained with the Bohr model but are successfully explained by the quantum model of the atom.

Example 42.4   Space Quantization for Hydrogen

Consider the hydrogen atom in the $\ell = 3$ state. Calculate the magnitude of $\vec{\mathbf{L}}$, the allowed values of L_z, and the corresponding angles θ that $\vec{\mathbf{L}}$ makes with the z axis.

SOLUTION

Conceptualize Consider Figure 42.13a, which is a vector model for $\ell = 2$. Draw such a vector model for $\ell = 3$ to help with this problem.

Categorize We evaluate results using equations developed in this section, so we categorize this example as a substitution problem.

Calculate the magnitude of the orbital angular momentum using Equation 42.27:

$$L = \sqrt{\ell(\ell + 1)}\,\hbar = \sqrt{3(3 + 1)}\,\hbar = \boxed{2\sqrt{3}\,\hbar}$$

Calculate the allowed values of L_z using Equation 42.28 with $m_\ell = -3, -2, -1, 0, 1, 2,$ and 3:

$$L_z = \boxed{-3\hbar, -2\hbar, -\hbar, 0, \hbar, 2\hbar, 3\hbar}$$

Calculate the allowed values of $\cos \theta$ using Equation 42.29:

$$\cos \theta = \frac{\pm 3}{2\sqrt{3}} = \pm 0.866 \qquad \cos \theta = \frac{\pm 2}{2\sqrt{3}} = \pm 0.577$$

$$\cos \theta = \frac{\pm 1}{2\sqrt{3}} = \pm 0.289 \qquad \cos \theta = \frac{0}{2\sqrt{3}} = 0$$

Find the angles corresponding to these values of $\cos \theta$:

$$\theta = \boxed{30.0°, 54.7°, 73.2°, 90.0°, 107°, 125°, 150°}$$

WHAT IF?   What if the value of ℓ is an arbitrary integer? For an arbitrary value of ℓ, how many values of m_ℓ are allowed?

Answer For a given value of ℓ, the values of m_ℓ range from $-\ell$ to $+\ell$ in steps of 1. Therefore, there are 2ℓ nonzero values of m_ℓ (specifically, $\pm 1, \pm 2, \ldots, \pm \ell$). In addition, one more value of $m_\ell = 0$ is possible, for a total of $2\ell + 1$ values of m_ℓ. This result is critical in understanding the results of the Stern–Gerlach experiment described below with regard to spin.

Wolfgang Pauli and Niels Bohr watch a spinning top. The spin of the electron is analogous to the spin of the top but is different in many ways.

The Spin Magnetic Quantum Number m_s

The three quantum numbers n, ℓ, and m_ℓ discussed so far are generated by applying boundary conditions to solutions of the Schrödinger equation, and we can assign a physical interpretation to each quantum number. Let's now consider **electron spin**, which does *not* come from the Schrödinger equation.

In Example 42.2, we found four quantum states corresponding to $n = 2$. In reality, however, eight such states occur. The additional four states can be explained by requiring a fourth quantum number for each state, the **spin magnetic quantum number m_s**.

The need for this new quantum number arises because of an unusual feature observed in the spectra of certain gases, such as sodium vapor. Close examination of one prominent line in the emission spectrum of sodium reveals that the line is, in fact, two closely spaced lines called a *doublet*.[8] The wavelengths of these lines occur in the yellow region of the electromagnetic spectrum at 589.0 nm and 589.6 nm. In 1925, when this doublet was first observed, it could not be explained with the existing atomic theory. To resolve this dilemma, Samuel Goudsmit (1902–1978) and George Uhlenbeck (1900–1988), following a suggestion made by Austrian physicist Wolfgang Pauli, proposed the spin quantum number.

To describe this new quantum number, it is convenient (but technically incorrect) to imagine the electron spinning about its axis as it orbits the nucleus as described in Section 30.6. As illustrated in Figure 42.15, only two directions exist for the electron spin. If the direction of spin is as shown in Figure 42.15a, the electron is said to have *spin up*. If the direction of spin is as shown in Figure 42.15b, the electron is said to have *spin down*. In the presence of a magnetic field, the energy associated with the electron is slightly different for the two spin directions. This energy difference accounts for the sodium doublet.

The classical description of electron spin—as resulting from a spinning electron—is incorrect. More recent theory indicates that the electron is a point particle, without spatial extent. Therefore, the electron is not modeled as a rigid object and cannot be considered to be spinning. Despite this conceptual difficulty, all experimental evidence supports the idea that an electron does have some intrinsic angular momentum that can be described by the spin magnetic quantum number. Paul Dirac (1902–1984) showed that this fourth quantum number originates in the relativistic properties of the electron.

In 1921, Otto Stern (1888–1969) and Walter Gerlach (1889–1979) performed an experiment that demonstrated space quantization. Their results, however, were not in quantitative agreement with the atomic theory that existed at that time. In their experiment, a beam of silver atoms sent through a nonuniform magnetic field was split into two discrete components (Fig. 42.16). Stern and Gerlach repeated the experiment using other atoms, and in each case the beam split into two or more components. The classical argument is as follows. If the z direction is chosen to be the direction of the maximum nonuniformity of $\vec{\mathbf{B}}$, the net magnetic force on the atoms is along the z axis and is proportional to the component of the magnetic moment $\vec{\boldsymbol{\mu}}$ of the atom in the z direction. Classically, $\vec{\boldsymbol{\mu}}$ can have any orientation, so the deflected beam should be spread out continuously. According to quantum mechanics, however, the deflected beam has an integral number of discrete components and the number of components determines the number of possible values of μ_z. Therefore, because the Stern–Gerlach experiment showed split beams, space quantization was at least qualitatively verified.

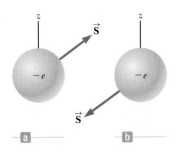

Figure 42.15 The spin of an electron can be either (a) up or (b) down relative to a specified z axis. As in the case of orbital angular momentum, the x and y components of the spin angular momentum vector are not quantized.

Pitfall Prevention 42.5
The Electron Is Not Spinning Although the concept of a spinning electron is conceptually useful, it should not be taken literally. The spin of the Earth is a mechanical rotation. On the other hand, electron spin is a purely quantum effect that gives the electron an angular momentum as if it were physically spinning.

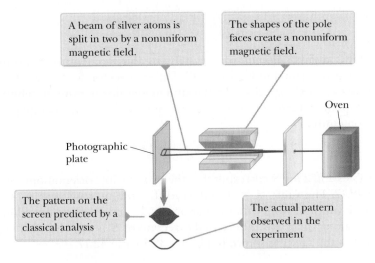

A beam of silver atoms is split in two by a nonuniform magnetic field.

The shapes of the pole faces create a nonuniform magnetic field.

Oven

Photographic plate

The pattern on the screen predicted by a classical analysis

The actual pattern observed in the experiment

Figure 42.16 The technique used by Stern and Gerlach to verify space quantization.

---

[8]This phenomenon is a Zeeman effect for spin and is identical in nature to the Zeeman effect for orbital angular momentum discussed before Example 42.4 except that no external magnetic field is required. The magnetic field for this Zeeman effect is internal to the atom and arises from the relative motion of the electron and the nucleus.

For the moment, let's assume the magnetic moment of the atom is due to the orbital angular momentum. Because μ_z is proportional to m_ℓ, the number of possible values of μ_z is $2\ell + 1$ as found in the What If? section of Example 42.4. Furthermore, because ℓ is an integer, the number of values of μ_z is always odd. This prediction is not consistent with Stern and Gerlach's observation of two components (an *even* number) in the deflected beam of silver atoms. Hence, either quantum mechanics is incorrect or the model is in need of refinement.

In 1927, T. E. Phipps and J. B. Taylor repeated the Stern–Gerlach experiment using a beam of hydrogen atoms. Their experiment was important because it involved an atom containing a single electron in its ground state, for which the quantum theory makes reliable predictions. Recall that $\ell = 0$ for hydrogen in its ground state, so $m_\ell = 0$. Therefore, we would not expect the beam to be deflected by the magnetic field at all because the magnetic moment $\vec{\mu}$ of the atom is zero. The beam in the Phipps–Taylor experiment, however, was again split into two components! On the basis of that result, we must conclude that something other than the electron's orbital motion is contributing to the atomic magnetic moment.

As we learned earlier, Goudsmit and Uhlenbeck had proposed that the electron has an intrinsic angular momentum, spin, apart from its orbital angular momentum. In other words, the total angular momentum of the electron in a particular electronic state contains both an orbital contribution $\vec{L}$ and a spin contribution $\vec{S}$. The Phipps–Taylor result confirmed the hypothesis of Goudsmit and Uhlenbeck.

In 1929, Dirac used the relativistic form of the total energy of a system to solve the relativistic wave equation for the electron in a potential well. His analysis confirmed the fundamental nature of electron spin. (Spin, like mass and charge, is an *intrinsic* property of a particle, independent of its surroundings.) Furthermore, the analysis showed that electron spin[9] can be described by a single quantum number s, whose value can be only $s = \frac{1}{2}$. The spin angular momentum of the electron *never changes*. This notion contradicts classical laws, which dictate that a rotating charge slows down in the presence of an applied magnetic field because of the Faraday emf that accompanies the changing field (Chapter 31). Furthermore, if the electron is viewed as a spinning ball of charge subject to classical laws, parts of the electron near its surface would be rotating with speeds exceeding the speed of light. Therefore, the classical picture must not be pressed too far; ultimately, spin of an electron is a quantum entity defying any simple classical description.

Because spin is a form of angular momentum, it must follow the same quantum rules as orbital angular momentum. In accordance with Equation 42.27, the magnitude of the **spin angular momentum $\vec{S}$** for the electron is

▶ Magnitude of the spin angular momentum of an electron

$$S = \sqrt{s(s + 1)}\,\hbar = \frac{\sqrt{3}}{2}\hbar \qquad (42.30)$$

Like orbital angular momentum $\vec{L}$, spin angular momentum $\vec{S}$ exhibits space quantization as described in Figure 42.17. The spin vector $\vec{S}$ can have two orientations relative to a z axis, specified by the **spin magnetic quantum number** $m_s = \pm\frac{1}{2}$. Similar to Equation 42.28 for orbital angular momentum, the z component of spin angular momentum is

▶ Allowed values of S_z

$$S_z = m_s \hbar = \pm\tfrac{1}{2}\hbar \qquad (42.31)$$

The two values $\pm\hbar/2$ for S_z correspond to the two possible orientations for $\vec{S}$ shown in Figure 42.17. The value $m_s = +\frac{1}{2}$ refers to the spin-up case, and $m_s = -\frac{1}{2}$ refers to the spin-down case. Notice that Equations 42.30 and 42.31 do not allow the spin vector to lie along the z axis. The actual direction of $\vec{S}$ is at a relatively large angle with respect to the z axis as shown in Figures 42.15 and 42.17.

---

[9]Scientists often use the word *spin* when referring to the spin angular momentum quantum number. For example, it is common to say, "The electron has a spin of one half."

As discussed in the What If? feature of Example 42.4, there are $2\ell + 1$ possible values of m_ℓ for orbital angular momentum. Similarly, for spin angular momentum, there are $2s + 1$ values of m_s. For a spin of $s = \frac{1}{2}$, the number of values of m_s is $2s + 1 = 2$. These two possibilities for m_s lead to the splitting of the beams into two components in the Stern–Gerlach and Phipps–Taylor experiments.

The spin magnetic moment $\vec{\boldsymbol{\mu}}_{\text{spin}}$ of the electron is related to its spin angular momentum $\vec{\mathbf{S}}$ by the expression

$$\vec{\boldsymbol{\mu}}_{\text{spin}} = -\frac{e}{m_e}\,\vec{\mathbf{S}} \tag{42.32}$$

where e is the electronic charge and m_e is the mass of the electron. Because $S_z = \pm\frac{1}{2}\hbar$, the z component of the spin magnetic moment can have the values

$$\vec{\boldsymbol{\mu}}_{\text{spin},z} = \pm\frac{e\hbar}{2m_e} \tag{42.33}$$

As we learned in Section 30.6, the quantity $e\hbar/2m_e$ is the Bohr magneton $\mu_B = 9.27 \times 10^{-24}$ J/T. The ratio of magnetic moment to angular momentum is twice as great for spin angular momentum (Eq. 42.32) as it is for orbital angular momentum (Eq. 30.22). The factor of 2 is explained in a relativistic treatment first carried out by Dirac.

Today, physicists explain the Stern–Gerlach and Phipps–Taylor experiments as follows. The observed magnetic moments for both silver and hydrogen are due to spin angular momentum only, with no contribution from orbital angular momentum. In the Phipps–Taylor experiment, the single electron in the hydrogen atom has its electron spin quantized in the magnetic field in such a way that the z component of spin angular momentum is either $\frac{1}{2}\hbar$ or $-\frac{1}{2}\hbar$, corresponding to $m_s = \pm\frac{1}{2}$. Electrons with spin $+\frac{1}{2}$ are deflected downward, and those with spin $-\frac{1}{2}$ are deflected upward. In the Stern–Gerlach experiment, 46 of a silver atom's 47 electrons are in filled subshells with paired spins. Therefore, these 46 electrons have a net zero contribution to both orbital and spin angular momentum for the atom. The angular momentum of the atom is due to only the 47th electron. This electron lies in the $5s$ subshell, so there is no contribution from orbital angular momentum. As a result, the silver atoms have angular momentum due to just the spin of one electron and behave in the same way in a nonuniform magnetic field as the hydrogen atoms in the Phipps–Taylor experiment.

The Stern–Gerlach experiment provided two important results. First, it verified the concept of space quantization. Second, it showed that spin angular momentum exists, even though this property was not recognized until four years after the experiments were performed.

As mentioned earlier, there are eight quantum states corresponding to $n = 2$ in the hydrogen atom, not four as found in Example 42.2. Each of the four states in Example 42.2 is actually two states because of the two possible values of m_s. Table 42.4 shows the quantum numbers corresponding to these eight states.

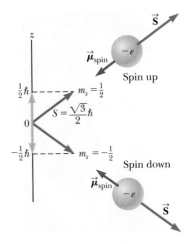

Figure 42.17 Spin angular momentum $\vec{\mathbf{S}}$ exhibits space quantization. This figure shows the two allowed orientations of the spin angular momentum vector $\vec{\mathbf{S}}$ and the spin magnetic moment $\vec{\boldsymbol{\mu}}_{\text{spin}}$ for a spin-$\frac{1}{2}$ particle, such as the electron.

Table 42.4   Quantum Numbers for the $n = 2$ State of Hydrogen

n	ℓ	m_ℓ	m_s	Subshell	Shell	Number of States in Subshell
2	0	0	$\frac{1}{2}$	2s	L	2
2	0	0	$-\frac{1}{2}$			
2	1	1	$\frac{1}{2}$			
2	1	1	$-\frac{1}{2}$			
2	1	0	$\frac{1}{2}$	2p	L	6
2	1	0	$-\frac{1}{2}$			
2	1	-1	$\frac{1}{2}$			
2	1	-1	$-\frac{1}{2}$			

Wolfgang Pauli
Austrian Theoretical Physicist
(1900–1958)
An extremely talented theoretician who made important contributions in many areas of modern physics, Pauli gained public recognition at the age of 21 with a masterful review article on relativity that is still considered one of the finest and most comprehensive introductions to the subject. His other major contributions were the discovery of the exclusion principle, the explanation of the connection between particle spin and statistics, theories of relativistic quantum electrodynamics, the neutrino hypothesis, and the hypothesis of nuclear spin.

Pitfall Prevention 42.6

The Exclusion Principle Is More General A more general form of the exclusion principle, discussed in Chapter 46, states that no two *fermions* can be in the same quantum state. Fermions are particles with half-integral spin ($\frac{1}{2}, \frac{3}{2}, \frac{5}{2}$, and so on).

42.7 The Exclusion Principle and the Periodic Table

We have found that the state of a hydrogen atom is specified by four quantum numbers: n, ℓ, m_ℓ, and m_s. As it turns out, the number of states available to other atoms may also be predicted by this same set of quantum numbers. In fact, these four quantum numbers can be used to describe all the electronic states of an atom, regardless of the number of electrons in its structure.

For our discussion of atoms with many electrons, it is often easiest to assign the quantum numbers to the electrons in the atom as opposed to the entire atom. An obvious question that arises here is, "How many electrons can be in a particular quantum state?" Pauli answered this important question in 1925, in a statement known as the **exclusion principle**:

> No two electrons can ever be in the same quantum state; therefore, no two electrons in the same atom can have the same set of quantum numbers.

If this principle were not valid, an atom could radiate energy until every electron in the atom is in the lowest possible energy state and therefore the chemical behavior of the elements would be grossly modified. Nature as we know it would not exist.

In reality, we can view the electronic structure of complex atoms as a succession of filled levels increasing in energy. As a general rule, the order of filling of an atom's subshells is as follows. Once a subshell is filled, the next electron goes into the lowest-energy vacant subshell. We can understand this behavior by recognizing that if the atom were not in the lowest energy state available to it, it would radiate energy until it reached this state. This tendency of a quantum system to achieve the lowest energy state is consistent with the second law of thermodynamics discussed in Chapter 22. The entropy of the Universe is increased by the system emitting photons, so that energy is spread out over a larger volume of space.

Before we discuss the electronic configuration of various elements, it is convenient to define an *orbital* as the atomic state characterized by the quantum numbers n, ℓ, and m_ℓ. The exclusion principle tells us that only two electrons can be present in any orbital. One of these electrons has a spin magnetic quantum number $m_s = +\frac{1}{2}$, and the other has $m_s = -\frac{1}{2}$. Because each orbital is limited to two electrons, the number of electrons that can occupy the various shells is also limited.

Table 42.5 shows the allowed quantum states for an atom up to $n = 3$. The arrows pointing upward indicate an electron described by $m_s = +\frac{1}{2}$, and those pointing downward indicate that $m_s = -\frac{1}{2}$. The $n = 1$ shell can accommodate only two electrons because $m_\ell = 0$ means that only one orbital is allowed. (The three quantum numbers describing this orbital are $n = 1$, $\ell = 0$, and $m_\ell = 0$.) The $n = 2$ shell has two subshells, one for $\ell = 0$ and one for $\ell = 1$. The $\ell = 0$ subshell is limited to two electrons because $m_\ell = 0$. The $\ell = 1$ subshell has three allowed orbitals, corresponding to $m_\ell = 1$, 0, and -1. Because each orbital can accommodate two electrons, the $\ell = 1$ subshell can hold six electrons. Therefore, the $n = 2$ shell can contain eight electrons as shown in Table 42.4. The $n = 3$ shell has three subshells ($\ell = 0, 1, 2$) and nine orbitals, accommodating up to 18 electrons. In general, each shell can accommodate up to $2n^2$ electrons.

Table 42.5   Allowed Quantum States for an Atom Up to $n = 3$

Shell	n	1		2				3								
Subshell	ℓ	0	0		1		0		1			2				
Orbital	m_ℓ	0	0	1	0	−1	0	1	0	−1	2	1	0	−1	−2		
	m_s	↑↓	↑↓	↑↓	↑↓	↑↓	↑↓	↑↓	↑↓	↑↓	↑↓	↑↓	↑↓	↑↓	↑↓		

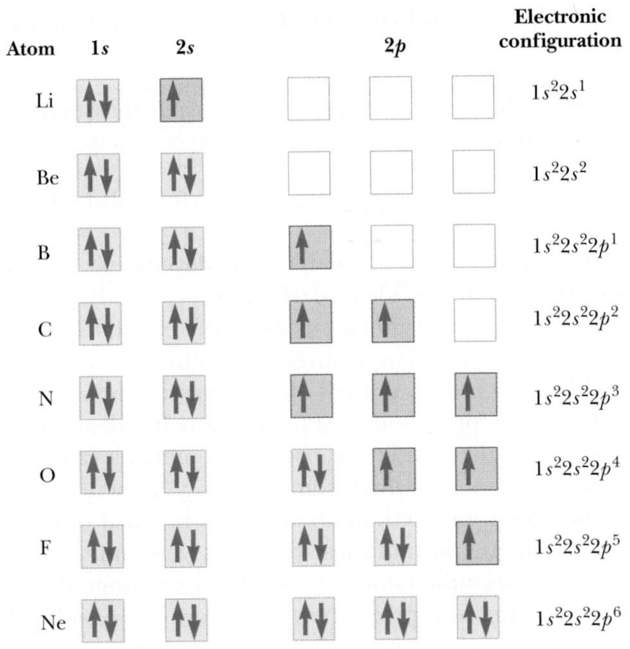

Atom	1s	2s		2p		Electronic configuration
Li	↑↓	↑				$1s^2 2s^1$
Be	↑↓	↑↓				$1s^2 2s^2$
B	↑↓	↑↓	↑			$1s^2 2s^2 2p^1$
C	↑↓	↑↓	↑	↑		$1s^2 2s^2 2p^2$
N	↑↓	↑↓	↑	↑	↑	$1s^2 2s^2 2p^3$
O	↑↓	↑↓	↑↓	↑	↑	$1s^2 2s^2 2p^4$
F	↑↓	↑↓	↑↓	↑↓	↑	$1s^2 2s^2 2p^5$
Ne	↑↓	↑↓	↑↓	↑↓	↑↓	$1s^2 2s^2 2p^6$

Figure 42.18 The filling of electronic states must obey both the exclusion principle and Hund's rule (page 1320).

The exclusion principle can be illustrated by examining the electronic arrangement in a few of the lighter atoms. The atomic number Z of any element is the number of protons in the nucleus of an atom of that element. A neutral atom of that element has Z electrons. Hydrogen ($Z = 1$) has only one electron, which, in the ground state of the atom, can be described by either of two sets of quantum numbers n, ℓ, m_ℓ, m_s: $1, 0, 0, \frac{1}{2}$ or $1, 0, 0, -\frac{1}{2}$. This electronic configuration is often written $1s^1$. The notation $1s$ refers to a state for which $n = 1$ and $\ell = 0$, and the superscript indicates that one electron is present in the s subshell.

Helium ($Z = 2$) has two electrons. In the ground state, their quantum numbers are $1, 0, 0, \frac{1}{2}$ and $1, 0, 0, -\frac{1}{2}$. No other possible combinations of quantum numbers exist for this level, and we say that the K shell is filled. This electronic configuration is written $1s^2$.

Lithium ($Z = 3$) has three electrons. In the ground state, two of them are in the $1s$ subshell. The third is in the $2s$ subshell because this subshell is slightly lower in energy than the $2p$ subshell.[10] Hence, the electronic configuration for lithium is $1s^2 2s^1$.

The electronic configurations of lithium and the next several elements are provided in Figure 42.18. The electronic configuration of beryllium ($Z = 4$), with its four electrons, is $1s^2 2s^2$, and boron ($Z = 5$) has a configuration of $1s^2 2s^2 2p^1$. The $2p$ electron in boron may be described by any of the six equally probable sets of quantum numbers listed in Table 42.4. In Figure 42.18, we show this electron in the leftmost $2p$ box with spin up, but it is equally likely to be in any $2p$ box with spin either up or down.

Carbon ($Z = 6$) has six electrons, giving rise to a question concerning how to assign the two $2p$ electrons. Do they go into the same orbital with paired spins (↑ ↓), or do they occupy different orbitals with unpaired spins (↑ ↑)? Experimental data show that the most stable configuration (that is, the one with the lowest energy) is the latter, in which the spins are unpaired. Hence, the two $2p$ electrons in carbon and the three $2p$ electrons in nitrogen ($Z = 7$) have unpaired spins as

---

[10]To a first approximation, energy depends only on the quantum number n, as we have discussed. Because of the effect of the electronic charge shielding the nuclear charge, however, energy depends on ℓ also in multielectron atoms. We shall discuss these shielding effects in Section 42.8.

Figure 42.18 shows. The general rule that governs such situations, called **Hund's rule,** states that

> when an atom has orbitals of equal energy, the order in which they are filled by electrons is such that a maximum number of electrons have unpaired spins.

Some exceptions to this rule occur in elements having subshells that are close to being filled or half-filled.

In 1871, long before quantum mechanics was developed, the Russian chemist Dmitri Mendeleev (1834–1907) made an early attempt at finding some order among the chemical elements. He was trying to organize the elements for the table of contents of a book he was writing. He arranged the atoms in a table similar to that shown in Figure 42.19, according to their atomic masses and chemical similarities. The first table Mendeleev proposed contained many blank spaces, and he boldly stated that the gaps were there only because the elements had not yet been discovered. By noting the columns in which some missing elements should be located, he was able to make rough predictions about their chemical properties. Within 20 years of this announcement, most of these elements were indeed discovered.

The elements in the **periodic table** (Fig. 42.19) are arranged so that all those in a column have similar chemical properties. For example, consider the elements in the last column, which are all gases at room temperature: He (helium), Ne (neon), Ar (argon), Kr (krypton), Xe (xenon), and Rn (radon). The outstanding characteristic of all these elements is that they do not normally take part in chemical reactions; that is, they do not readily join with other atoms to form molecules. They are therefore called *inert gases* or *noble gases*.

Group I	Group II	Transition elements										Group III	Group IV	Group V	Group VI	Group VII	Group 0
H 1 $1s^1$																H 1 $1s^1$	He 2 $1s^2$
Li 3 $2s^1$	Be 4 $2s^2$											B 5 $2p^1$	C 6 $2p^2$	N 7 $2p^3$	O 8 $2p^4$	F 9 $2p^5$	Ne 10 $2p^6$
Na 11 $3s^1$	Mg 12 $3s^2$											Al 13 $3p^1$	Si 14 $3p^2$	P 15 $3p^3$	S 16 $3p^4$	Cl 17 $3p^5$	Ar 18 $3p^6$
K 19 $4s^1$	Ca 20 $4s^2$	Sc 21 $3d^14s^2$	Ti 22 $3d^24s^2$	V 23 $3d^34s^2$	Cr 24 $3d^54s^1$	Mn 25 $3d^54s^2$	Fe 26 $3d^64s^2$	Co 27 $3d^74s^2$	Ni 28 $3d^84s^2$	Cu 29 $3d^{10}4s^1$	Zn 30 $3d^{10}4s^2$	Ga 31 $4p^1$	Ge 32 $4p^2$	As 33 $4p^3$	Se 34 $4p^4$	Br 35 $4p^5$	Kr 36 $4p^6$
Rb 37 $5s^1$	Sr 38 $5s^2$	Y 39 $4d^15s^2$	Zr 40 $4d^25s^2$	Nb 41 $4d^45s^1$	Mo 42 $4d^55s^1$	Tc 43 $4d^55s^2$	Ru 44 $4d^75s^1$	Rh 45 $4d^85s^1$	Pd 46 $4d^{10}$	Ag 47 $4d^{10}5s^1$	Cd 48 $4d^{10}5s^2$	In 49 $5p^1$	Sn 50 $5p^2$	Sb 51 $5p^3$	Te 52 $5p^4$	I 53 $5p^5$	Xe 54 $5p^6$
Cs 55 $6s^1$	Ba 56 $6s^2$	57–71*	Hf 72 $5d^26s^2$	Ta 73 $5d^36s^2$	W 74 $5d^46s^2$	Re 75 $5d^56s^2$	Os 76 $5d^66s^2$	Ir 77 $5d^76s^2$	Pt 78 $5d^96s^1$	Au 79 $5d^{10}6s^1$	Hg 80 $5d^{10}6s^2$	Tl 81 $6p^1$	Pb 82 $6p^2$	Bi 83 $6p^3$	Po 84 $6p^4$	At 85 $6p^5$	Rn 86 $6p^6$
Fr 87 $7s^1$	Ra 88 $7s^2$	89–103**	Rf 104 $6d^27s^2$	Db 105 $6d^37s^2$	Sg 106 $6d^47s^2$	Bh 107 $6d^57s^2$	Hs 108 $6d^67s^2$	Mt 109 $6d^77s^2$	Ds 110 $6d^97s^1$	Rg 111	Cn 112	113	Fl 114	115	Lv 116	117	118

*Lanthanide series	La 57 $5d^16s^2$	Ce 58 $5d^14f^16s^2$	Pr 59 $4f^36s^2$	Nd 60 $4f^46s^2$	Pm 61 $4f^56s^2$	Sm 62 $4f^66s^2$	Eu 63 $4f^76s^2$	Gd 64 $5d^14f^76s^2$	Tb 65 $5d^14f^86s^2$	Dy 66 $4f^{10}6s^2$	Ho 67 $4f^{11}6s^2$	Er 68 $4f^{12}6s^2$	Tm 69 $4f^{13}6s^2$	Yb 70 $4f^{14}6s^2$	Lu 71 $5d^14f^{14}6s^2$
**Actinide series	Ac 89 $6d^17s^2$	Th 90 $6d^27s^2$	Pa 91 $5f^26d^17s^2$	U 92 $5f^36d^17s^2$	Np 93 $5f^46d^17s^2$	Pu 94 $5f^67s^2$	Am 95 $5f^77s^2$	Cm 96 $5f^76d^17s^2$	Bk 97 $5f^86d^17s^2$	Cf 98 $5f^{10}7s^2$	Es 99 $5f^{11}7s^2$	Fm 100 $5f^{12}7s^2$	Md 101 $5f^{13}7s^2$	No 102 $5f^{14}7s^2$	Lr 103 $5f^{14}6d^17s^2$

Figure 42.19 The periodic table of the elements is an organized tabular representation of the elements that shows their periodic chemical behavior. Elements in a given column have similar chemical behavior. This table shows the chemical symbol for the element, the atomic number, and the electron configuration. A more complete periodic table is available in Appendix C.

We can partially understand this behavior by looking at the electronic configurations in Figure 42.19. The chemical behavior of an element depends on the outermost shell that contains electrons. The electronic configuration for helium is $1s^2$, and the $n = 1$ shell (which is the outermost shell because it is the only shell) is filled. Also, the energy of the atom in this configuration is considerably lower than the energy for the configuration in which an electron is in the next available level, the $2s$ subshell. Next, look at the electronic configuration for neon, $1s^2 2s^2 2p^6$. Again, the outermost shell ($n = 2$ in this case) is filled and a wide gap in energy occurs between the filled $2p$ subshell and the next available one, the $3s$ subshell. Argon has the configuration $1s^2 2s^2 2p^6 3s^2 3p^6$. Here, it is only the $3p$ subshell that is filled, but again a wide gap in energy occurs between the filled $3p$ subshell and the next available one, the $3d$ subshell. This pattern continues through all the noble gases. Krypton has a filled $4p$ subshell, xenon a filled $5p$ subshell, and radon a filled $6p$ subshell.

The column to the left of the noble gases in the periodic table consists of a group of elements called the *halogens:* fluorine, chlorine, bromine, iodine, and astatine. At room temperature, fluorine and chlorine are gases, bromine is a liquid, and iodine and astatine are solids. In each of these atoms, the outer subshell is one electron short of being filled. As a result, the halogens are chemically very active, readily accepting an electron from another atom to form a closed shell. The halogens tend to form strong ionic bonds with atoms at the other side of the periodic table. (We shall discuss ionic bonds in Chapter 43.) In a halogen lightbulb, bromine or iodine atoms combine with tungsten atoms evaporated from the filament and return them to the filament, resulting in a longer-lasting lightbulb. In addition, the filament can be operated at a higher temperature than in ordinary lightbulbs, giving a brighter and whiter light.

At the left side of the periodic table, the Group I elements consist of hydrogen and the *alkali metals:* lithium, sodium, potassium, rubidium, cesium, and francium. Each of these atoms contains one electron in a subshell outside of a closed subshell. Therefore, these elements easily form positive ions because the lone electron is bound with a relatively low energy and is easily removed. Therefore, the alkali metal atoms are chemically active and form very strong bonds with halogen atoms. For example, table salt, NaCl, is a combination of an alkali metal and a halogen. Because the outer electron is weakly bound, pure alkali metals tend to be good electrical conductors. Because of their high chemical activity, however, they are not generally found in nature in pure form.

It is interesting to plot ionization energy versus atomic number Z as in Figure 42.20. Notice the pattern of $\Delta Z = 2, 8, 8, 18, 18, 32$ for the various peaks. This pattern follows from the exclusion principle and helps explain why the elements repeat their chemical properties in groups. For example, the peaks at $Z = 2, 10, 18,$

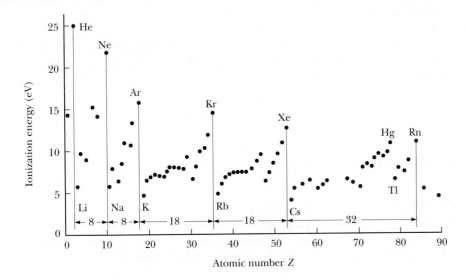

Figure 42.20 Ionization energy of the elements versus atomic number.

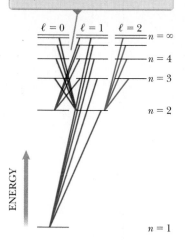

Figure 42.21 Some allowed electronic transitions for hydrogen, represented by the colored lines.

Selection rules for allowed ▶
atomic transitions

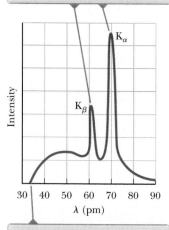

Figure 42.22 The x-ray spectrum of a metal target. The data shown were obtained when 37-keV electrons bombarded a molybdenum target.

and 36 correspond to the noble gases helium, neon, argon, and krypton, respectively, which, as we have mentioned, all have filled outermost shells. These elements have relatively high ionization energies and similar chemical behavior.

42.8 More on Atomic Spectra: Visible and X-Ray

In Section 42.1, we discussed the observation and early interpretation of visible spectral lines from gases. These spectral lines have their origin in transitions between quantized atomic states. We shall investigate these transitions more deeply in these final three sections of this chapter.

A modified energy-level diagram for hydrogen is shown in Figure 42.21. In this diagram, the allowed values of ℓ for each shell are separated horizontally. Figure 42.21 shows only those states up to $\ell = 2$; the shells from $n = 4$ upward would have more sets of states to the right, which are not shown. Transitions for which ℓ does not change are very unlikely to occur and are called *forbidden transitions*. (Such transitions actually can occur, but their probability is very low relative to the probability of "allowed" transitions.) The various diagonal lines represent allowed transitions between stationary states. Whenever an atom makes a transition from a higher energy state to a lower one, a photon of light is emitted. The frequency of this photon is $f = \Delta E/h$, where ΔE is the energy difference between the two states and h is Planck's constant. The **selection rules** for the *allowed transitions* are

$$\Delta \ell = \pm 1 \quad \text{and} \quad \Delta m_\ell = 0, \pm 1 \tag{42.34}$$

Figure 42.21 shows that the orbital angular momentum of an atom *changes* when it makes a transition to a lower energy state. Therefore, the atom alone is a *non-isolated* system for angular momentum. If we consider the atom–photon system, however, it must be an *isolated* system for angular momentum because nothing else is interacting with this system. The photon involved in the process must carry angular momentum away from the atom when the transition occurs. In fact, the photon has an angular momentum equivalent to that of a particle having a spin of 1. We have now determined over several chapters that a photon has energy, linear momentum, and angular momentum, and each of these is conserved in atomic processes.

Recall from Equation 42.19 that the allowed energies for one-electron atoms and ions, such as hydrogen and He^+, are

$$E_n = -\frac{k_e e^2}{2a_0}\left(\frac{Z^2}{n^2}\right) = -\frac{(13.6 \text{ eV})Z^2}{n^2} \tag{42.35}$$

This equation was developed from the Bohr theory, but it serves as a good first approximation in quantum theory as well. For multielectron atoms, the positive nuclear charge Ze is largely shielded by the negative charge of the inner-shell electrons. Therefore, the outer electrons interact with a net charge that is smaller than the nuclear charge. The expression for the allowed energies for multielectron atoms has the same form as Equation 42.35 with Z replaced by an effective atomic number Z_{eff}:

$$E_n = -\frac{(13.6 \text{ eV})Z_{\text{eff}}^2}{n^2} \tag{42.36}$$

where Z_{eff} depends on n and ℓ.

X-Ray Spectra

X-rays are emitted when high-energy electrons or any other charged particles bombard a metal target. The x-ray spectrum typically consists of a broad continuous band containing a series of sharp lines as shown in Figure 42.22. In Section 34.6, we mentioned that an accelerated electric charge emits electromagnetic radiation. The x-rays in Figure 42.22 are the result of the slowing down of high-energy electrons as they strike the target. It may take several interactions with the atoms of the target before the electron gives up all its kinetic energy. The amount of kinetic energy given up in any interaction can vary from zero up to the entire kinetic energy of the electron. Therefore, the wavelength of radiation from these interactions lies in a continuous range from some minimum value up to infinity. It is this general slowing down of the electrons that provides the continuous curve in Figure 42.22, which shows the cutoff of x-rays below a minimum wavelength value that depends on the kinetic energy of the incoming electrons. X-ray radiation with its origin in the slowing down of electrons is called **bremsstrahlung,** the German word for "braking radiation."

Extremely high-energy bremsstrahlung can be used for the treatment of cancerous tissues. Figure 42.23 shows a machine that uses a linear accelerator to accelerate electrons up to 18 MeV and smash them into a tungsten target. The result is a beam of photons, up to a maximum energy of 18 MeV, which is actually in the gamma-ray range in Figure 34.13. This radiation is directed at the tumor in the patient.

The discrete lines in Figure 42.22, called **characteristic x-rays** and discovered in 1908, have a different origin. Their origin remained unexplained until the details of atomic structure were understood. The first step in the production of characteristic x-rays occurs when a bombarding electron collides with a target atom. The electron must have sufficient energy to remove an inner-shell electron from the atom. The vacancy created in the shell is filled when an electron in a higher level drops down into the level containing the vacancy. The existence of characteristic lines in an x-ray spectrum is further direct evidence of the quantization of energy in atomic systems.

The time interval for atomic transitions to happen is very short, less than 10^{-9} s. This transition is accompanied by the emission of a photon whose energy equals the difference in energy between the two levels. Typically, the energy of such transitions is greater than 1 000 eV and the emitted x-ray photons have wavelengths in the range of 0.01 nm to 1 nm.

Let's assume the incoming electron has dislodged an atomic electron from the innermost shell, the K shell. If the vacancy is filled by an electron dropping from the next higher shell—the L shell—the photon emitted has an energy corresponding to the K_α characteristic x-ray line on the curve of Figure 42.22. In this notation, K refers to the final level of the electron and the subscript α, as the *first* letter of the Greek alphabet, refers to the initial level as the *first* one above the final level. Figure 42.24 shows this transition as well as others discussed below. If the vacancy in the K shell is filled by an electron dropping from the M shell, the K_β line in Figure 42.22 is produced.

Other characteristic x-ray lines are formed when electrons drop from upper levels to vacancies other than those in the K shell. For example, L lines are produced when vacancies in the L shell are filled by electrons dropping from higher shells. An L_α line is produced as an electron drops from the M shell to the L shell, and an L_β line is produced by a transition from the N shell to the L shell.

Although multielectron atoms cannot be analyzed exactly with either the Bohr model or the Schrödinger equation, we can apply Gauss's law from Chapter 24 to make some surprisingly accurate estimates of expected x-ray energies and wavelengths. Consider an atom of atomic number Z in which one of the two electrons in the K shell has been ejected. Imagine drawing a gaussian sphere immediately inside the most probable radius of the L electrons. The electric field at the position of the L electrons is a combination of the fields created by the nucleus, the single K electron, the other L electrons, and the outer electrons. The wave functions of the outer electrons are such that the electrons have a very high probability of being farther from the nucleus than the L electrons are. Therefore, the outer

Figure 42.23 Bremsstrahlung is created by this machine and used to treat cancer in a patient.

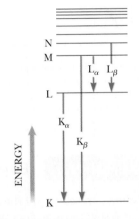

Figure 42.24 Transitions between higher and lower atomic energy levels that give rise to x-ray photons from heavy atoms when they are bombarded with high-energy electrons.

electrons are much more likely to be outside the gaussian surface than inside and, on average, do not contribute significantly to the electric field at the position of the L electrons. The effective charge inside the gaussian surface is the positive nuclear charge and one negative charge due to the single K electron. Ignoring the interactions between L electrons, a single L electron behaves as if it experiences an electric field due to a charge $(Z - 1)e$ enclosed by the gaussian surface. The nuclear charge is shielded by the electron in the K shell such that Z_{eff} in Equation 42.36 is $Z - 1$. For higher-level shells, the nuclear charge is shielded by electrons in all the inner shells.

We can now use Equation 42.36 to estimate the energy associated with an electron in the L shell:

$$E_{\text{L}} = -(Z - 1)^2 \frac{13.6 \text{ eV}}{2^2}$$

After the atom makes the transition, there are two electrons in the K shell. We can approximate the energy associated with one of these electrons as that of a one-electron atom. (In reality, the nuclear charge is reduced somewhat by the negative charge of the other electron, but let's ignore this effect.) Therefore,

$$E_{\text{K}} \approx -Z^2(13.6 \text{ eV}) \tag{42.37}$$

As Example 42.5 shows, the energy of the atom with an electron in an M shell can be estimated in a similar fashion. Taking the energy difference between the initial and final levels, we can then calculate the energy and wavelength of the emitted photon.

In 1914, Henry G. J. Moseley (1887–1915) plotted $\sqrt{1/\lambda}$ versus the Z values for a number of elements where λ is the wavelength of the K_α line of each element. He found that the plot is a straight line as in Figure 42.25, which is consistent with rough calculations of the energy levels given by Equation 42.37. From this plot, Moseley determined the Z values of elements that had not yet been discovered and produced a periodic table in excellent agreement with the known chemical properties of the elements. Until that experiment, atomic numbers had been merely placeholders for the elements that appeared in the periodic table, the elements being ordered according to mass.

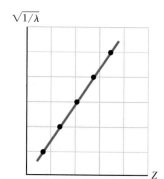

Figure 42.25 A Moseley plot of $\sqrt{1/\lambda}$ versus Z, where λ is the wavelength of the K_α x-ray line of the element of atomic number Z.

Quick Quiz 42.5 In an x-ray tube, as you increase the energy of the electrons striking the metal target, do the wavelengths of the characteristic x-rays **(a)** increase, **(b)** decrease, or **(c)** remain constant?

Quick Quiz 42.6 True or False: It is possible for an x-ray spectrum to show the continuous spectrum of x-rays without the presence of the characteristic x-rays.

Example 42.5   **Estimating the Energy of an X-Ray**

Estimate the energy of the characteristic x-ray emitted from a tungsten target when an electron drops from an M shell ($n = 3$ state) to a vacancy in the K shell ($n = 1$ state). The atomic number for tungsten is $Z = 74$.

SOLUTION

Conceptualize Imagine an accelerated electron striking a tungsten atom and ejecting an electron from the K shell ($n = 1$). Subsequently, an electron in the M shell ($n = 3$) drops down to fill the vacancy and the energy difference between the states is emitted as an x-ray photon.

Categorize We estimate the results using equations developed in this section, so we categorize this example as a substitution problem.

Use Equation 42.37 and $Z = 74$ for tungsten to estimate the energy associated with the electron in the K shell:

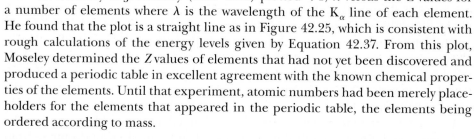

$E_{\text{K}} \approx -(74)^2(13.6 \text{ eV}) = -7.4 \times 10^4 \text{ eV}$

▶ **42.5** continued

Use Equation 42.36 and that nine electrons shield the nuclear charge (eight electrons in the $n = 2$ state and one electron in the $n = 1$ state) to estimate the energy of the M shell:

$$E_M \approx -\frac{(13.6 \text{ eV})(74 - 9)^2}{(3)^2} \approx -6.4 \times 10^3 \text{ eV}$$

Find the energy of the emitted x-ray photon:

$$hf = E_M - E_K \approx -6.4 \times 10^3 \text{ eV} - (-7.4 \times 10^4 \text{ eV})$$
$$\approx 6.8 \times 10^4 \text{ eV} = \boxed{68 \text{ keV}}$$

Consultation of x-ray tables shows that the M–K transition energies in tungsten vary from 66.9 keV to 67.7 keV, where the range of energies is due to slightly different energy values for states of different ℓ. Therefore, our estimate differs from the midpoint of this experimentally measured range by approximately 1%.

42.9 Spontaneous and Stimulated Transitions

We have seen that an atom absorbs and emits electromagnetic radiation only at frequencies that correspond to the energy differences between allowed states. Let's now examine more details of these processes. Consider an atom having the allowed energy levels labeled E_1, E_2, E_3, ... When radiation is incident on the atom, only those photons whose energy hf matches the energy separation ΔE between two energy levels can be absorbed by the atom as represented in Figure 42.26. This process is called **stimulated absorption** because the photon stimulates the atom to make the upward transition. At ordinary temperatures, most of the atoms in a sample are in the ground state. If a vessel containing many atoms of a gaseous element is illuminated with radiation of all possible photon frequencies (that is, a continuous spectrum), only those photons having energy $E_2 - E_1$, $E_3 - E_1$, $E_4 - E_1$, and so on are absorbed by the atoms. As a result of this absorption, some of the atoms are raised to excited states.

Once an atom is in an excited state, the excited atom can make a transition back to a lower energy level, emitting a photon in the process as in Figure 42.27. This process is known as **spontaneous emission** because it happens naturally, without requiring an event to trigger the transition. Typically, an atom remains in an excited state for only about 10^{-8} s.

In addition to spontaneous emission, **stimulated emission** occurs. Suppose an atom is in an excited state E_2 as in Figure 42.28 (page 1326). If the excited state is a *metastable state*—that is, if its lifetime is much longer than the typical 10^{-8} s lifetime of

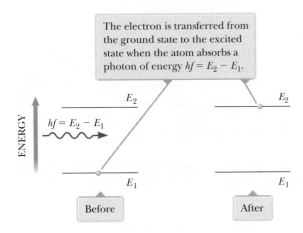

Figure 42.26 Stimulated absorption of a photon.

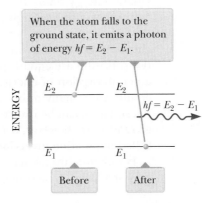

Figure 42.27 Spontaneous emission of a photon by an atom that is initially in the excited state E_2.

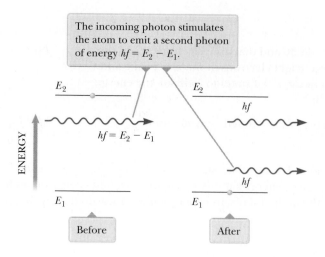

The incoming photon stimulates the atom to emit a second photon of energy $hf = E_2 - E_1$.

E_2

$hf = E_2 - E_1$

E_1

Before

E_2

hf

hf

E_1

After

excited states—the time interval until spontaneous emission occurs is relatively long. Let's imagine that during that interval a photon of energy $hf = E_2 - E_1$ is incident on the atom. One possibility is that the photon energy is sufficient for the photon to ionize the atom. Another possibility is that the interaction between the incoming photon and the atom causes the atom to return to the ground state[11] and thereby emit a second photon with energy $hf = E_2 - E_1$. In this process, the incident photon is not absorbed; therefore, after the stimulated emission, two photons with identical energy exist: the incident photon and the emitted photon. The two are in phase and travel in the same direction, which is an important consideration in lasers, discussed next.

42.10 Lasers

In this section, we explore the nature of laser light and a variety of applications of lasers in our technological society. The primary properties of laser light that make it useful in these technological applications are the following:

- Laser light is coherent. The individual rays of light in a laser beam maintain a fixed phase relationship with one another.
- Laser light is monochromatic. Light in a laser beam has a very narrow range of wavelengths.
- Laser light has a small angle of divergence. The beam spreads out very little, even over large distances.

To understand the origin of these properties, let's combine our knowledge of atomic energy levels from this chapter with some special requirements for the atoms that emit laser light.

We have described how an incident photon can cause atomic energy transitions either upward (stimulated absorption) or downward (stimulated emission). The two processes are equally probable. When light is incident on a collection of atoms, a net absorption of energy usually occurs because when the system is in thermal equilibrium, many more atoms are in the ground state than in excited states. If the situation can be inverted so that more atoms are in an excited state than in the ground state, however, a net emission of photons can result. Such a condition is called **population inversion.**

Population inversion is, in fact, the fundamental principle involved in the operation of a **laser** (an acronym for *light amplification by stimulated emission of radia-*

---

[11]This phenomenon is fundamentally due to *resonance*. The incoming photon has a frequency and drives the system of the atom at that frequency. Because the driving frequency matches that associated with a transition between states—one of the natural frequencies of the atom—there is a large response: the atom makes the transition.

tion). The full name indicates one of the requirements for laser light: to achieve laser action, the process of stimulated emission must occur.

Suppose an atom is in the excited state E_2 as in Figure 42.28 and a photon with energy $hf = E_2 - E_1$ is incident on it. As described in Section 42.9, the incoming photon can stimulate the excited atom to return to the ground state and thereby emit a second photon having the same energy hf and traveling in the same direction. The incident photon is not absorbed, so after the stimulated emission, there are two identical photons: the incident photon and the emitted photon. The emitted photon is in phase with the incident photon. These photons can stimulate other atoms to emit photons in a chain of similar processes. The many photons produced in this fashion are the source of the intense, coherent light in a laser.

For the stimulated emission to result in laser light, there must be a buildup of photons in the system. The following three conditions must be satisfied to achieve this buildup:

- The system must be in a state of population inversion: there must be more atoms in an excited state than in the ground state. That must be true because the number of photons emitted must be greater than the number absorbed.
- The excited state of the system must be a *metastable state*, meaning that its lifetime must be long compared with the usually short lifetimes of excited states, which are typically 10^{-8} s. In this case, the population inversion can be established and stimulated emission is likely to occur before spontaneous emission.
- The emitted photons must be confined in the system long enough to enable them to stimulate further emission from other excited atoms. That is achieved by using reflecting mirrors at the ends of the system. One end is made totally reflecting, and the other is partially reflecting. A fraction of the light intensity passes through the partially reflecting end, forming the beam of laser light (Fig. 42.29).

One device that exhibits stimulated emission of radiation is the helium–neon gas laser. Figure 42.30 is an energy-level diagram for the neon atom in this system. The mixture of helium and neon is confined to a glass tube that is sealed at the ends by mirrors. A voltage applied across the tube causes electrons to sweep through the tube, colliding with the atoms of the gases and raising them into excited states. Neon atoms are excited to state E_3^* through this process (the asterisk indicates a metastable state) and also as a result of collisions with excited helium atoms. Stimulated emission occurs, causing neon atoms to make transitions to state E_2. Neighboring excited atoms are also stimulated. The result is the production of coherent light at a wavelength of 632.8 nm.

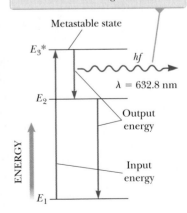

The atom emits 632.8-nm photons through stimulated emission in the transition $E_3^* - E_2$. That is the source of coherent light in the laser.

Metastable state

E_3^*

hf

$\lambda = 632.8$ nm

E_2

Output energy

Input energy

E_1

ENERGY

Figure 42.30 Energy-level diagram for a neon atom in a helium–neon laser.

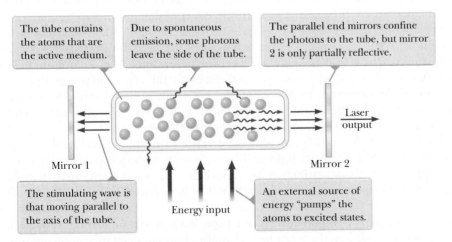

The tube contains the atoms that are the active medium.

Due to spontaneous emission, some photons leave the side of the tube.

The parallel end mirrors confine the photons to the tube, but mirror 2 is only partially reflective.

Laser output

Mirror 1

Mirror 2

The stimulating wave is that moving parallel to the axis of the tube.

Energy input

An external source of energy "pumps" the atoms to excited states.

Figure 42.29 Schematic diagram of a laser design.

Figure 42.31 This robot carrying laser scissors, which can cut up to 50 layers of fabric at a time, is one of the many applications of laser technology.

Applications

Since the development of the first laser in 1960, tremendous growth has occurred in laser technology. Lasers that cover wavelengths in the infrared, visible, and ultraviolet regions are now available. *Laser diodes* are used as laser pointers, and in surveying and construction rangefinders, fiber optic communication, DVD and Blu-ray players, and bar code readers. *Carbon dioxide lasers* are used in industry for welding and cutting, such as the process shown to cut fabric in Figure 42.31. *Excimer lasers* are used in Lasik eye surgery. A variety of other types of lasers exist and are used in various applications. These applications are possible because of the unique characteristics of laser light. In addition to being highly monochromatic, laser light is also highly directional and can be sharply focused to produce regions of extremely intense light energy (with energy densities 10^{12} times the density in the flame of a typical cutting torch).

Lasers are used in precision long-range distance measurement (range finding). In recent years, it has become important in astronomy and geophysics to measure as precisely as possible the distances from various points on the surface of the Earth to a point on the Moon's surface. To facilitate these measurements, the *Apollo* astronauts set up a 0.5-m square of reflector prisms on the Moon, which enables laser pulses directed from an Earth-based station to be retroreflected to the same station (see Fig. 35.8a). Using the known speed of light and the measured round-trip travel time of a laser pulse, the Earth–Moon distance can be determined to a precision of better than 10 cm.

Because various laser wavelengths can be absorbed in specific biological tissues, lasers have a number of medical applications. For example, certain laser procedures have greatly reduced blindness in patients with glaucoma and diabetes. Glaucoma is a widespread eye condition characterized by a high fluid pressure in the eye, a condition that can lead to destruction of the optic nerve. A simple laser operation (iridectomy) can "burn" open a tiny hole in a clogged membrane, relieving the destructive pressure. A serious side effect of diabetes is neovascularization, the proliferation of weak blood vessels, which often leak blood. When neovascularization occurs in the retina, vision deteriorates (diabetic retinopathy) and finally is destroyed. Today, it is possible to direct the green light from an argon ion laser through the clear eye lens and eye fluid, focus on the retina edges, and photocoagulate the leaky vessels. Even people who have only minor vision defects such as nearsightedness are benefiting from the use of lasers to reshape the cornea, changing its focal length and reducing the need for eyeglasses.

Laser surgery is now an everyday occurrence at hospitals and medical clinics around the world. Infrared light at 10 μm from a carbon dioxide laser can cut through muscle tissue, primarily by vaporizing the water contained in cellular material. Laser power of approximately 100 W is required in this technique. The advantage of the "laser knife" over conventional methods is that laser radiation cuts tissue and coagulates blood at the same time, leading to a substantial reduction in blood loss. In addition, the technique virtually eliminates cell migration, an important consideration when tumors are being removed.

A laser beam can be trapped in fine optical fiber light guides (endoscopes) by means of total internal reflection. An endoscope can be introduced through natural orifices, conducted around internal organs, and directed to specific interior body locations, eliminating the need for invasive surgery. For example, bleeding in the gastrointestinal tract can be optically cauterized by endoscopes inserted through the patient's mouth.

In biological and medical research, it is often important to isolate and collect unusual cells for study and growth. A laser cell separator exploits the tagging of specific cells with fluorescent dyes. All cells are then dropped from a tiny charged nozzle and laser-scanned for the dye tag. If triggered by the correct light-emitting tag, a small voltage applied to parallel plates deflects the falling electrically charged cell into a collection beaker.

An exciting area of research and technological applications arose in the 1990s with the development of *laser trapping* of atoms. One scheme, called *optical molasses* and developed by Steven Chu of Stanford University and his colleagues, involves focusing six laser beams onto a small region in which atoms are to be trapped. Each pair of lasers is along one of the x, y, and z axes and emits light in opposite directions (Fig. 42.32). The frequency of the laser light is tuned to be slightly below the absorption frequency of the subject atom. Imagine that an atom has been placed into the trap region and moves along the positive x axis toward the laser that is emitting light toward it (the rightmost laser on the x axis in Fig. 42.32). Because the atom is moving, the light from the laser appears Doppler-shifted upward in frequency in the reference frame of the atom. Therefore, a match between the Doppler-shifted laser frequency and the absorption frequency of the atom exists and the atom absorbs photons.[12] The momentum carried by these photons results in the atom being pushed back to the center of the trap. By incorporating six lasers, the atoms are pushed back into the trap regardless of which way they move along any axis.

In 1986, Chu developed *optical tweezers*, a device that uses a single tightly focused laser beam to trap and manipulate small particles. In combination with microscopes, optical tweezers have opened up many new possibilities for biologists. Optical tweezers have been used to manipulate live bacteria without damage, move chromosomes within a cell nucleus, and measure the elastic properties of a single DNA molecule. Chu shared the 1997 Nobel Prize in Physics with two of his colleagues for the development of the techniques of optical trapping.

An extension of laser trapping, *laser cooling*, is possible because the normal high speeds of the atoms are reduced when they are restricted to the region of the trap. As a result, the temperature of the collection of atoms can be reduced to a few microkelvins. The technique of laser cooling allows scientists to study the behavior of atoms at extremely low temperatures (Fig. 42.33).

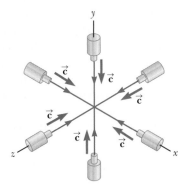

Figure 42.32 An optical trap for atoms is formed at the intersection point of six counterpropagating laser beams along mutually perpendicular axes.

The orange dot is the sample of trapped sodium atoms.

Courtesy of National Institute of Standards and Technology, U.S. Dept. of Commerce

Figure 42.33 A staff member of the National Institute of Standards and Technology views a sample of trapped sodium atoms cooled to a temperature of less than 1 mK.

Summary

Concepts and Principles

The wavelengths of spectral lines from hydrogen, called the **Balmer series**, can be described by the equation

$$\frac{1}{\lambda} = R_{\text{H}}\left(\frac{1}{2^2} - \frac{1}{n^2}\right) \quad n = 3, 4, 5, \ldots \tag{42.1}$$

where R_{H} is the **Rydberg constant.** The spectral lines corresponding to values of n from 3 to 6 are in the visible range of the electromagnetic spectrum. Values of n higher than 6 correspond to spectral lines in the ultraviolet region of the spectrum.

continued

---

[12]The laser light traveling in the same direction as the atom is Doppler-shifted further downward in frequency, so there is no absorption. Therefore, the atom is not pushed out of the trap by the diametrically opposed laser.

The Bohr model of the atom is successful in describing some details of the spectra of atomic hydrogen and hydrogen-like ions. One basic assumption of the model is that the electron can exist only in discrete orbits such that the angular momentum of the electron is an integral multiple of $h/2\pi = \hbar$. When we assume circular orbits and a simple Coulomb attraction between electron and proton, the energies of the quantum states for hydrogen are calculated to be

$$E_n = -\frac{k_e e^2}{2a_0}\left(\frac{1}{n^2}\right) \quad n = 1, 2, 3, \ldots \quad \textbf{(42.13)}$$

where n is an integer called the **quantum number,** k_e is the Coulomb constant, e is the electronic charge, and $a_0 = 0.0529$ nm is the **Bohr radius.**

If the electron in a hydrogen atom makes a transition from an orbit whose quantum number is n_i to one whose quantum number is n_f, where $n_f < n_i$, a photon is emitted by the atom. The frequency of this photon is

$$f = \frac{k_e e^2}{2a_0 h}\left(\frac{1}{n_f^2} - \frac{1}{n_i^2}\right) \quad \textbf{(42.15)}$$

Quantum mechanics can be applied to the hydrogen atom by the use of the potential energy function $U(r) = -k_e e^2/r$ in the Schrödinger equation. The solution to this equation yields wave functions for allowed states and allowed energies:

$$E_n = -\left(\frac{k_e e^2}{2a_0}\right)\frac{1}{n^2} = -\frac{13.606 \text{ eV}}{n^2} \quad n = 1, 2, 3, \ldots \quad \textbf{(42.21)}$$

where n is the **principal quantum number.** The allowed wave functions depend on three quantum numbers: n, ℓ, and m_ℓ, where ℓ is the **orbital quantum number** and m_ℓ is the **orbital magnetic quantum number.** The restrictions on the quantum numbers are

$$n = 1, 2, 3, \ldots$$
$$\ell = 0, 1, 2, \ldots, n - 1$$
$$m_\ell = -\ell, -\ell + 1, \ldots \ell - 1, \ell$$

All states having the same principal quantum number n form a **shell,** identified by the letters K, L, M, . . . (corresponding to $n = 1, 2, 3, \ldots$). All states having the same values of n and ℓ form a **subshell,** designated by the letters $s, p, d, f, \ldots$ (corresponding to $\ell = 0, 1, 2, 3, \ldots$).

An atom in a state characterized by a specific value of n can have the following values of L, the magnitude of the atom's orbital angular momentum $\vec{\mathbf{L}}$:

$$L = \sqrt{\ell(\ell + 1)}\,\hbar$$
$$\ell = 0, 1, 2, \ldots, n - 1 \quad \textbf{(42.27)}$$

The allowed values of the projection of $\vec{\mathbf{L}}$ along the z axis are

$$L_z = m_\ell \hbar \quad \textbf{(42.28)}$$

Only discrete values of L_z are allowed as determined by the restrictions on m_ℓ. This quantization of L_z is referred to as **space quantization.**

The electron has an intrinsic angular momentum called the **spin angular momentum.** Electron spin can be described by a single quantum number $s = \frac{1}{2}$. To describe a quantum state completely, it is necessary to include a fourth quantum number m_s, called the **spin magnetic quantum number.** This quantum number can have only two values, $\pm\frac{1}{2}$. The magnitude of the spin angular momentum is

$$S = \frac{\sqrt{3}}{2}\hbar \quad \textbf{(42.30)}$$

and the z component of $\vec{\mathbf{S}}$ is

$$S_z = m_s \hbar = \pm\frac{1}{2}\hbar \quad \textbf{(42.31)}$$

That is, the spin angular momentum is also quantized in space, as specified by the spin magnetic quantum number $m_s = \pm\frac{1}{2}$.

The **exclusion principle** states that **no two electrons in an atom can be in the same quantum state.** In other words, no two electrons can have the same set of quantum numbers n, ℓ, m_ℓ, and m_s. Using this principle, the electronic configurations of the elements can be determined. This principle serves as a basis for understanding atomic structure and the chemical properties of the elements.

The magnetic moment $\vec{\boldsymbol{\mu}}_{\text{spin}}$ associated with the spin angular momentum of an electron is

$$\vec{\boldsymbol{\mu}}_{\text{spin}} = -\frac{e}{m_e}\vec{\mathbf{S}} \quad \textbf{(42.32)}$$

The z component of $\vec{\boldsymbol{\mu}}_{\text{spin}}$ can have the values

$$\mu_{\text{spin},z} = \pm\frac{e\hbar}{2m_e} \quad \textbf{(42.33)}$$

The x-ray spectrum of a metal target consists of a set of sharp characteristic lines superimposed on a broad continuous spectrum. **Bremsstrahlung** is x-radiation with its origin in the slowing down of high-energy electrons as they encounter the target. **Characteristic x-rays** are emitted by atoms when an electron undergoes a transition from an outer shell to a vacancy in an inner shell.

Atomic transitions can be described with three processes: **stimulated absorption,** in which an incoming photon raises the atom to a higher energy state; **spontaneous emission,** in which the atom makes a transition to a lower energy state, emitting a photon; and **stimulated emission,** in which an incident photon causes an excited atom to make a downward transition, emitting a photon identical to the incident one.

Objective Questions   **1.** denotes answer available in *Student Solutions Manual/Study Guide*

1. **(i)** What is the principal quantum number of the initial state of an atom as it emits an M_β line in an x-ray spectrum? (a) 1 (b) 2 (c) 3 (d) 4 (e) 5 **(ii)** What is the principal quantum number of the final state for this transition? Choose from the same possibilities as in part (i).

2. If an electron in an atom has the quantum numbers $n = 3$, $\ell = 2$, $m_\ell = 1$, and $m_s = \frac{1}{2}$, what state is it in? (a) $3s$ (b) $3p$ (c) $3d$ (d) $4d$ (e) $3f$

3. An electron in the $n = 5$ energy level of hydrogen undergoes a transition to the $n = 3$ energy level. What is the wavelength of the photon the atom emits in this process? (a) 2.28×10^{-6} m (b) 8.20×10^{-7} m (c) 3.64×10^{-7} m (d) 1.28×10^{-6} m (e) 5.92×10^{-5} m

4. Consider the $n = 3$ energy level in a hydrogen atom. How many electrons can be placed in this level? (a) 1 (b) 2 (c) 8 (d) 9 (e) 18

5. Which of the following is *not* one of the basic assumptions of the Bohr model of hydrogen? (a) Only certain electron orbits are stable and allowed. (b) The electron moves in circular orbits about the proton under the influence of the Coulomb force. (c) The charge on the electron is quantized. (d) Radiation is emitted by the atom when the electron moves from a higher energy state to a lower energy state. (e) The angular momentum associated with the electron's orbital motion is quantized.

6. Let $-E$ represent the energy of a hydrogen atom. **(i)** What is the kinetic energy of the electron? (a) $2E$ (b) E (c) 0 (d) $-E$ (e) $-2E$ **(ii)** What is the potential energy of the atom? Choose from the same possibilities (a) through (e).

7. The periodic table is based on which of the following principles? (a) The uncertainty principle. (b) All electrons in an atom must have the same set of quantum numbers. (c) Energy is conserved in all interactions. (d) All electrons in an atom are in orbitals having the same energy. (e) No two electrons in an atom can have the same set of quantum numbers.

8. (a) Can a hydrogen atom in the ground state absorb a photon of energy less than 13.6 eV? (b) Can this atom absorb a photon of energy greater than 13.6 eV?

9. Which of the following electronic configurations are *not* allowed for an atom? Choose all correct answers. (a) $2s^2 2p^6$ (b) $3s^2 3p^7$ (c) $3d^7 4s^2$ (d) $3d^{10} 4s^2 4p^6$ (e) $1s^2 2s^2 2d^1$

10. What can be concluded about a hydrogen atom with its electron in the d state? (a) The atom is ionized. (b) The orbital quantum number is $\ell = 1$. (c) The principal quantum number is $n = 2$. (d) The atom is in its ground state. (e) The orbital angular momentum of the atom is not zero.

11. **(i)** Rank the following transitions for a hydrogen atom from the transition with the greatest gain in energy to that with the greatest loss, showing any cases of equality. (a) $n_i = 2$; $n_f = 5$ (b) $n_i = 5$; $n_f = 3$ (c) $n_i = 7$; $n_f = 4$ (d) $n_i = 4$; $n_f = 7$ **(ii)** Rank the same transitions as in part (i) according to the wavelength of the photon absorbed or emitted by an otherwise isolated atom from greatest wavelength to smallest.

12. When an atom emits a photon, what happens? (a) One of its electrons leaves the atom. (b) The atom moves to a state of higher energy. (c) The atom moves to a state of lower energy. (d) One of its electrons collides with another particle. (e) None of those events occur.

13. (a) In the hydrogen atom, can the quantum number n increase without limit? (b) Can the frequency of possible discrete lines in the spectrum of hydrogen increase without limit? (c) Can the wavelength of possible discrete lines in the spectrum of hydrogen increase without limit?

14. Consider the quantum numbers (a) n, (b) ℓ, (c) m_ℓ, and (d) m_s. **(i)** Which of these quantum numbers are fractional as opposed to being integers? **(ii)** Which can sometimes attain negative values? **(iii)** Which can be zero?

15. When an electron collides with an atom, it can transfer all or some of its energy to the atom. A hydrogen atom is in its ground state. Incident on the atom are several electrons, each having a kinetic energy of 10.5 eV. What is the result? (a) The atom can be excited to a higher allowed state. (b) The atom is ionized. (c) The electrons pass by the atom without interaction.

1. Why is stimulated emission so important in the operation of a laser?

2. An energy of about 21 eV is required to excite an electron in a helium atom from the 1s state to the 2s state. The same transition for the He$^+$ ion requires approximately twice as much energy. Explain.

3. Why are three quantum numbers needed to describe the state of a one-electron atom (ignoring spin)?

4. Compare the Bohr theory and the Schrödinger treatment of the hydrogen atom, specifically commenting on their treatment of total energy and orbital angular momentum of the atom.

5. Could the Stern–Gerlach experiment be performed with ions rather than neutral atoms? Explain.

6. Why is a *nonuniform* magnetic field used in the Stern–Gerlach experiment?

7. Discuss some consequences of the exclusion principle.

8. (a) According to Bohr's model of the hydrogen atom, what is the uncertainty in the radial coordinate of the electron? (b) What is the uncertainty in the radial component of the velocity of the electron? (c) In what way does the model violate the uncertainty principle?

9. Why do lithium, potassium, and sodium exhibit similar chemical properties?

10. It is easy to understand how two electrons (one spin up, one spin down) fill the $n = 1$ or K shell for a helium atom. How is it possible that eight more electrons are allowed in the $n = 2$ shell, filling the K and L shells for a neon atom?

11. Suppose the electron in the hydrogen atom obeyed classical mechanics rather than quantum mechanics. Why should a gas of such hypothetical atoms emit a continuous spectrum rather than the observed line spectrum?

12. Does the intensity of light from a laser fall off as $1/r^2$? Explain.

Problems

Section 42.1 Atomic Spectra of Gases

1. The wavelengths of the Lyman series for hydrogen are given by

$$\frac{1}{\lambda} = R_H\left(1 - \frac{1}{n^2}\right) \quad n = 2, 3, 4, \ldots$$

(a) Calculate the wavelengths of the first three lines in this series. (b) Identify the region of the electromagnetic spectrum in which these lines appear.

2. The wavelengths of the Paschen series for hydrogen are given by

$$\frac{1}{\lambda} = R_H\left(\frac{1}{3^2} - \frac{1}{n^2}\right) \quad n = 4, 5, 6, \ldots$$

(a) Calculate the wavelengths of the first three lines in this series. (b) Identify the region of the electromagnetic spectrum in which these lines appear.

3. An isolated atom of a certain element emits light of wavelength 520 nm when the atom falls from its fifth excited state into its second excited state. The atom emits a photon of wavelength 410 nm when it drops from its sixth excited state into its second excited state. Find the wavelength of the light radiated when the atom makes a transition from its sixth to its fifth excited state.

4. An isolated atom of a certain element emits light of wavelength λ_{m1} when the atom falls from its state with quantum number m into its ground state of quantum number 1. The atom emits a photon of wavelength λ_{n1} when the atom falls from its state with quantum number n into its ground state. (a) Find the wavelength of the light radiated when the atom makes a transition from the m state to the n state. (b) Show that $k_{mn} = |k_{m1} - k_{n1}|$, where $k_{ij} = 2\pi/\lambda_{ij}$ is the wave number of the photon. This problem exemplifies the *Ritz combination principle*, an empirical rule formulated in 1908.

5. (a) What value of n_i is associated with the 94.96-nm spectral line in the Lyman series of hydrogen? (b) **What If?** Could this wavelength be associated with the Paschen series? (c) Could this wavelength be associated with the Balmer series?

Section 42.2 Early Models of the Atom

6. According to classical physics, a charge e moving with an acceleration a radiates energy at a rate

$$\frac{dE}{dt} = -\frac{1}{6\pi\epsilon_0}\frac{e^2 a^2}{c^3}$$

(a) Show that an electron in a classical hydrogen atom (see Fig. 42.5) spirals into the nucleus at a rate

$$\frac{dr}{dt} = -\frac{e^4}{12\pi^2\epsilon_0^2 m_e^2 c^3}\left(\frac{1}{r^2}\right)$$

(b) Find the time interval over which the electron reaches $r = 0$, starting from $r_0 = 2.00 \times 10^{-10}$ m.

7. **Review.** In the Rutherford scattering experiment, 4.00-MeV alpha particles scatter off gold nuclei (containing 79 protons and 118 neutrons). Assume a particular alpha particle moves directly toward the gold nucleus and scatters backward at 180°, and that the gold nucleus remains fixed throughout the entire process. Determine (a) the distance of closest approach of the alpha particle to the gold nucleus and (b) the maximum force exerted on the alpha particle.

Section 42.3 Bohr's Model of the Hydrogen Atom

Note: In this section, unless otherwise indicated, assume the hydrogen atom is treated with the Bohr model.

8. Show that the speed of the electron in the nth Bohr orbit in hydrogen is given by
$$v_n = \frac{k_e e^2}{n\hbar}$$

9. How much energy is required to ionize hydrogen (a) when it is in the ground state and (b) when it is in the state for which $n = 3$?

10. What is the energy of a photon that, when absorbed by **M** a hydrogen atom, could cause an electronic transition from (a) the $n = 2$ state to the $n = 5$ state and (b) the $n = 4$ state to the $n = 6$ state?

11. A photon is emitted when a hydrogen atom undergoes a transition from the $n = 5$ state to the $n = 3$ state. Calculate (a) the energy (in electron volts), (b) the wavelength, and (c) the frequency of the emitted photon.

12. The Balmer series for the hydrogen atom corresponds to electronic transitions that terminate in the state with quantum number $n = 2$ as shown in Figure P42.12. Consider the photon of longest wavelength corresponding to a transition shown in the figure. Determine (a) its energy and (b) its wavelength. Consider the spectral line of shortest wavelength corresponding to a transition shown in the figure. Find (c) its photon energy and (d) its wavelength. (e) What is the shortest possible wavelength in the Balmer series?

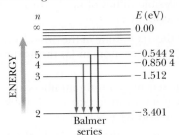

Figure P42.12

13. For a hydrogen atom in its ground state, compute (a) the orbital speed of the electron, (b) the kinetic energy of the electron, and (c) the electric potential energy of the atom.

14. Two hydrogen atoms collide head-on and end up with **AMT** zero kinetic energy. Each atom then emits light with a wavelength of 121.6 nm ($n = 2$ to $n = 1$ transition). At what speed were the atoms moving before the collision?

15. (a) Calculate the angular momentum of the Moon due to its orbital motion about the Earth. In your calculation, use 3.84×10^8 m as the average Earth–Moon distance and 2.36×10^6 s as the period of the Moon in its orbit. (b) Assume that the Moon's angular momentum is described by Bohr's assumption $mvr = n\hbar$. Determine the corresponding quantum number. (c) By what fraction would the Earth–Moon distance have to be increased to raise the quantum number by 1?

16. A monochromatic beam of light is absorbed by a col- **W** lection of ground-state hydrogen atoms in such a way that six different wavelengths are observed when the hydrogen relaxes back to the ground state. (a) What is the wavelength of the incident beam? Explain the steps in your solution. (b) What is the longest wavelength in the emission spectrum of these atoms? (c) To what portion of the electromagnetic spectrum and (d) to what series does it belong? (e) What is the shortest wavelength? (f) To what portion of the electromagnetic spectrum and (g) to what series does it belong?

17. A hydrogen atom is in its second excited state, corresponding to $n = 3$. Find (a) the radius of the electron's Bohr orbit and (b) the de Broglie wavelength of the electron in this orbit.

18. A hydrogen atom is in its first excited state ($n = 2$). Cal- **M** culate (a) the radius of the orbit, (b) the linear momen- **W** tum of the electron, (c) the angular momentum of the electron, (d) the kinetic energy of the electron, (e) the potential energy of the system, and (f) the total energy of the system.

19. A photon with energy 2.28 eV is absorbed by a hydrogen atom. Find (a) the minimum n for a hydrogen atom that can be ionized by such a photon and (b) the speed of the electron released from the state in part (a) when it is far from the nucleus.

20. An electron is in the nth Bohr orbit of the hydrogen atom. (a) Show that the period of the electron is $T = n^3 t_0$ and determine the numerical value of t_0. (b) On average, an electron remains in the $n = 2$ orbit for approximately 10 μs before it jumps down to the $n = 1$ (ground-state) orbit. How many revolutions does the electron make in the excited state? (c) Define the period of one revolution as an electron year, analogous to an Earth year being the period of the Earth's motion around the Sun. Explain whether we should think of the electron in the $n = 2$ orbit as "living for a long time."

21. (a) Construct an energy-level diagram for the He^+ ion, for which $Z = 2$, using the Bohr model. (b) What is the ionization energy for He^+?

Section 42.4 The Quantum Model of the Hydrogen Atom

22. A general expression for the energy levels of one-electron atoms and ions is
$$E_n = -\frac{\mu k_e^2 q_1^2 q_2^2}{2\hbar^2 n^2}$$

Here μ is the reduced mass of the atom, given by $\mu = m_1m_2/(m_1 + m_2)$, where m_1 is the mass of the electron and m_2 is the mass of the nucleus; k_e is the Coulomb constant; and q_1 and q_2 are the charges of the electron and the nucleus, respectively. The wavelength for the $n = 3$ to $n = 2$ transition of the hydrogen atom is 656.3 nm (visible red light). What are the wavelengths for this same transition in (a) positronium, which consists of an electron and a positron, and (b) singly ionized helium? *Note:* A positron is a positively charged electron.

23. Atoms of the same element but with different numbers of neutrons in the nucleus are called *isotopes*. Ordinary hydrogen gas is a mixture of two isotopes containing either one- or two-particle nuclei. These isotopes are hydrogen-1, with a proton nucleus, and hydrogen-2, called deuterium, with a deuteron nucleus. A deuteron is one proton and one neutron bound together. Hydrogen-1 and deuterium have identical chemical properties, but they can be separated via an ultracentrifuge or by other methods. Their emission spectra show lines of the same colors at very slightly different wavelengths. (a) Use the equation given in Problem 22 to show that the difference in wavelength between the hydrogen-1 and deuterium spectral lines associated with a particular electron transition is given by

$$\lambda_H - \lambda_D = \left(1 - \frac{\mu_H}{\mu_D}\right)\lambda_H$$

(b) Find the wavelength difference for the Balmer alpha line of hydrogen, with wavelength 656.3 nm, emitted by an atom making a transition from an $n = 3$ state to an $n = 2$ state. Harold Urey observed this wavelength difference in 1931 and so confirmed his discovery of deuterium.

24. An electron of momentum p is at a distance r from a stationary proton. The electron has kinetic energy $K = p^2/2m_e$. The atom has potential energy $U = -k_e e^2/r$ and total energy $E = K + U$. If the electron is bound to the proton to form a hydrogen atom, its average position is at the proton but the uncertainty in its position is approximately equal to the radius r of its orbit. The electron's average vector momentum is zero, but its average squared momentum is approximately equal to the squared uncertainty in its momentum as given by the uncertainty principle. Treating the atom as a one-dimensional system, (a) estimate the uncertainty in the electron's momentum in terms of r. Estimate the electron's (b) kinetic energy and (c) total energy in terms of r. The actual value of r is the one that *minimizes the total energy*, resulting in a stable atom. Find (d) that value of r and (e) the resulting total energy. (f) State how your answers compare with the predictions of the Bohr theory.

Section 42.5 The Wave Functions for Hydrogen

25. Plot the wave function $\psi_{1s}(r)$ versus r (see Eq. 42.22) and the radial probability density function $P_{1s}(r)$ versus r (see Eq. 42.25) for hydrogen. Let r range from 0 to $1.5a_0$, where a_0 is the Bohr radius.

26. For a spherically symmetric state of a hydrogen atom, the Schrödinger equation in spherical coordinates is

$$-\frac{\hbar^2}{2m_e}\left(\frac{d^2\psi}{dr^2} + \frac{2}{r}\frac{d\psi}{dr}\right) - \frac{k_e e^2}{r}\psi = E\psi$$

(a) Show that the 1s wave function for an electron in hydrogen,

$$\psi_{1s}(r) = \frac{1}{\sqrt{\pi a_0^3}}e^{-r/a_0}$$

satisfies the Schrödinger equation. (b) What is the energy of the atom for this state?

27. The radial function $R(r)$ of the wave function for a hydrogen atom in the 2p state is

$$\psi_{2p} = \frac{1}{\sqrt{3}(2a_0)^{3/2}}\frac{r}{a_0}e^{-r/2a_0}$$

What is the most likely distance from the nucleus to find an electron in the 2p state?

28. The ground-state wave function for the electron in a hydrogen atom is

$$\psi_{1s}(r) = \frac{1}{\sqrt{\pi a_0^3}}e^{-r/a_0}$$

where r is the radial coordinate of the electron and a_0 is the Bohr radius. (a) Show that the wave function as given is normalized. (b) Find the probability of locating the electron between $r_1 = a_0/2$ and $r_2 = 3a_0/2$.

29. In an experiment, a large number of electrons are fired at a sample of neutral hydrogen atoms and observations are made of how the incident particles scatter. The electron in the ground state of a hydrogen atom is found to be momentarily at a distance $a_0/2$ from the nucleus in 1 000 of the observations. In this set of trials, how many times is the atomic electron observed at a distance $2a_0$ from the nucleus?

Section 42.6 Physical Interpretation of the Quantum Numbers

30. List the possible sets of quantum numbers for the hydrogen atom associated with (a) the 3d subshell and (b) the 3p subshell.

31. If a hydrogen atom has orbital angular momentum 4.714×10^{-34} J · s, what is the orbital quantum number for the state of the atom?

32. Find all possible values of (a) L, (b) L_z, and (c) θ for a hydrogen atom in a 3d state.

33. Calculate the magnitude of the orbital angular momentum for a hydrogen atom in (a) the 4d state and (b) the 6f state.

34. How many sets of quantum numbers are possible for a hydrogen atom for which (a) $n = 1$, (b) $n = 2$, (c) $n = 3$, (d) $n = 4$, and (e) $n = 5$?

35. An electron in a sodium atom is in the N shell. Determine the maximum value the z component of its angular momentum could have.

36. (a) Find the mass density of a proton, modeling it as a solid sphere of radius 1.00×10^{-15} m. (b) **What If?** Consider a classical model of an electron as a uniform

solid sphere with the same density as the proton. Find its radius. (c) Imagine that this electron possesses spin angular momentum $I\omega = \hbar/2$ because of classical rotation about the z axis. Determine the speed of a point on the equator of the electron. (d) State how this speed compares with the speed of light.

37. A hydrogen atom is in its fifth excited state, with principal quantum number 6. The atom emits a photon with a wavelength of 1 090 nm. Determine the maximum possible magnitude of the orbital angular momentum of the atom after emission.

38. *Why is the following situation impossible?* A photon of wavelength 88.0 nm strikes a clean aluminum surface, ejecting a photoelectron. The photoelectron then strikes a hydrogen atom in its ground state, transferring energy to it and exciting the atom to a higher quantum state.

39. The ρ^- meson has a charge of $-e$, a spin quantum number of 1, and a mass 1 507 times that of the electron. The possible values for its spin magnetic quantum number are -1, 0, and 1. **What If?** Imagine that the electrons in atoms are replaced by ρ^- mesons. List the possible sets of quantum numbers for ρ^- mesons in the $3d$ subshell.

Section 42.7 The Exclusion Principle and the Periodic Table

40. (a) As we go down the periodic table, which subshell is filled first, the $3d$ or the $4s$ subshell? (b) Which electronic configuration has a lower energy, $[Ar]3d^44s^2$ or $[Ar]3d^54s^1$? *Note:* The notation $[Ar]$ represents the filled configuration for argon. *Suggestion:* Which has the greater number of unpaired spins? (c) Identify the element with the electronic configuration in part (b).

41. (a) Write out the electronic configuration of the ground state for nitrogen ($Z = 7$). (b) Write out the values for the possible set of quantum numbers n, ℓ, m_ℓ, and m_s for the electrons in nitrogen.

42. Devise a table similar to that shown in Figure 42.18 for atoms containing 11 through 19 electrons. Use Hund's rule and educated guesswork.

43. A certain element has its outermost electron in a $3p$ subshell. It has valence $+3$ because it has three more electrons than a certain noble gas. What element is it?

44. Scanning through Figure 42.19 in order of increasing atomic number, notice that the electrons usually fill the subshells in such a way that those subshells with the lowest values of $n + \ell$ are filled first. If two subshells have the same value of $n + \ell$, the one with the lower value of n is generally filled first. Using these two rules, write the order in which the subshells are filled through $n + \ell = 7$.

45. Two electrons in the same atom both have $n = 3$ and $\ell = 1$. Assume the electrons are distinguishable, so that interchanging them defines a new state. (a) How many states of the atom are possible considering the quantum numbers these two electrons can have? (b) **What If?** How many states would be possible if the exclusion principle were inoperative?

46. For a neutral atom of element 110, what would be the probable ground-state electronic configuration?

47. **Review.** For an electron with magnetic moment $\vec{\boldsymbol{\mu}}_s$ in a magnetic field $\vec{\mathbf{B}}$, Section 29.5 showed the following. The electron–field system can be in a higher energy state with the z component of the electron's magnetic moment opposite the field or a lower energy state with the z component of the magnetic moment in the direction of the field. The difference in energy between the two states is $2\mu_B B$.

Under high resolution, many spectral lines are observed to be doublets. The most famous doublet is the pair of two yellow lines in the spectrum of sodium (the D lines), with wavelengths of 588.995 nm and 589.592 nm. Their existence was explained in 1925 by Goudsmit and Uhlenbeck, who postulated that an electron has intrinsic spin angular momentum. When the sodium atom is excited with its outermost electron in a $3p$ state, the orbital motion of the outermost electron creates a magnetic field. The atom's energy is somewhat different depending on whether the electron is itself spin-up or spin-down in this field. Then the photon energy the atom radiates as it falls back into its ground state depends on the energy of the excited state. Calculate the magnitude of the internal magnetic field, mediating this so-called spin-orbit coupling.

Section 42.8 More on Atomic Spectra: Visible and X–Ray

48. In x-ray production, electrons are accelerated through a high voltage ΔV and then decelerated by striking a target. Show that the shortest wavelength of an x-ray that can be produced is

$$\lambda_{min} = \frac{1\,240\,\text{nm} \cdot \text{V}}{\Delta V}$$

49. What minimum accelerating voltage would be required to produce an x-ray with a wavelength of 70.0 pm?

50. A tungsten target is struck by electrons that have been accelerated from rest through a 40.0-keV potential difference. Find the shortest wavelength of the radiation emitted.

51. A bismuth target is struck by electrons, and x-rays are emitted. Estimate (a) the M- to L-shell transitional energy for bismuth and (b) the wavelength of the x-ray emitted when an electron falls from the M shell to the L shell.

52. The $3p$ level of sodium has an energy of -3.0 eV, and the $3d$ level has an energy of -1.5 eV. (a) Determine Z_{eff} for each of these states. (b) Explain the difference.

53. (a) Determine the possible values of the quantum numbers ℓ and m_ℓ for the He^+ ion in the state corresponding to $n = 3$. (b) What is the energy of this state?

54. The K series of the discrete spectrum of tungsten contains wavelengths of 0.018 5 nm, 0.020 9 nm, and 0.021 5 nm. The K-shell ionization energy is 69.5 keV. Determine the ionization energies of the L, M, and N shells.

55. Use the method illustrated in Example 42.5 to calculate the wavelength of the x-ray emitted from a

molybdenum target ($Z = 42$) when an electron moves from the L shell ($n = 2$) to the K shell ($n = 1$).

56. In x-ray production, electrons are accelerated through a high voltage and then decelerated by striking a target. (a) To make possible the production of x-rays of wavelength λ, what is the minimum potential difference ΔV through which the electrons must be accelerated? (b) State in words how the required potential difference depends on the wavelength. (c) Explain whether your result predicts the correct minimum wavelength in Figure 42.22. (d) Does the relationship from part (a) apply to other kinds of electromagnetic radiation besides x-rays? (e) What does the potential difference approach as λ goes to zero? (f) What does the potential difference approach as λ increases without limit?

57. When an electron drops from the M shell ($n = 3$) to a vacancy in the K shell ($n = 1$), the measured wavelength of the emitted x-ray is found to be 0.101 nm. Identify the element.

Section 42.9 Spontaneous and Stimulated Transitions

Section 42.10 Lasers

58. Figure P42.58 shows portions of the energy-level diagrams of the helium and neon atoms. An electrical discharge excites the He atom from its ground state (arbitrarily assigned the energy $E_1 = 0$) to its excited state of 20.61 eV. The excited He atom collides with a Ne atom in its ground state and excites this atom to the state at 20.66 eV. Lasing action takes place for electron transitions from E_3^* to E_2 in the Ne atoms. From the data in the figure, show that the wavelength of the red He–Ne laser light is approximately 633 nm.

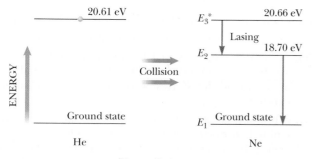

Figure P42.58

59. The carbon dioxide laser is one of the most powerful developed. The energy difference between the two laser levels is 0.117 eV. Determine (a) the frequency and (b) the wavelength of the radiation emitted by this laser. (c) In what portion of the electromagnetic spectrum is this radiation?

60. **Review.** A helium–neon laser can produce a green laser beam instead of a red one. Figure P42.60 shows the transitions involved to form the red beam and the green beam. After a population inversion is established, neon atoms make a variety of downward transitions in falling from the state labeled E_4^* down eventually to level E_1 (arbitrarily assigned the energy $E_1 = 0$). The atoms emit both red light with a wavelength

of 632.8 nm in a transition $E_4^* - E_3$ and green light with a wavelength of 543 nm in a competing transition $E_4^* - E_2$. (a) What is the energy E_2? Assume the atoms are in a cavity between mirrors designed to reflect the green light with high efficiency but to allow the red light to leave the cavity immediately. Then stimulated emission can lead to the buildup of a collimated beam of green light between the mirrors having a greater intensity than that of the red light. To constitute the radiated laser beam, a small fraction of the green light is permitted to escape by transmission through one mirror. The mirrors forming the resonant cavity can be made of layers of silicon dioxide (index of refraction $n = 1.458$) and titanium dioxide (index of refraction varies between 1.9 and 2.6). (b) How thick a layer of silicon dioxide, between layers of titanium dioxide, would minimize reflection of the red light? (c) What should be the thickness of a similar but separate layer of silicon dioxide to maximize reflection of the green light?

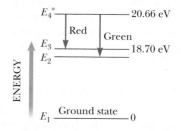

Figure P42.60 Problems 60 and 62.

61. A ruby laser delivers a 10.0-ns pulse of 1.00-MW average power. If the photons have a wavelength of 694.3 nm, how many are contained in the pulse?

62. The number N of atoms in a particular state is called the population of that state. This number depends on the energy of that state and the temperature. In thermal equilibrium, the population of atoms in a state of energy E_n is given by a Boltzmann distribution expression

$$N = N_g e^{-(E_n - E_g)/k_B T}$$

where N_g is the population of the ground state of energy E_g, k_B is Boltzmann's constant, and T is the absolute temperature. For simplicity, assume each energy level has only one quantum state associated with it. (a) Before the power is switched on, the neon atoms in a laser are in thermal equilibrium at 27.0°C. Find the equilibrium ratio of the populations of the states E_4^* and E_3 shown for the red transition in Figure P42.60. Lasers operate by a clever artificial production of a "population inversion" between the upper and lower atomic energy states involved in the lasing transition. This term means that more atoms are in the upper excited state than in the lower one. Consider the $E_4^* - E_3$ transition in Figure P42.60. Assume 2% more atoms occur in the upper state than in the lower. (b) To demonstrate how unnatural such a situation is, find the temperature for which the Boltzmann distribution describes a 2.00% population inversion. (c) Why does such a situation not occur naturally?

63. A neodymium–yttrium–aluminum garnet laser used in **M** eye surgery emits a 3.00-mJ pulse in 1.00 ns, focused to **W** a spot 30.0 μm in diameter on the retina. (a) Find (in SI units) the power per unit area at the retina. (In the optics industry, this quantity is called the *irradiance*.) (b) What energy is delivered by the pulse to an area of molecular size, taken as a circular area 0.600 nm in diameter?

64. **Review.** Figure 42.29 represents the light bouncing **AMT** between two mirrors in a laser cavity as two traveling waves. These traveling waves moving in opposite directions constitute a standing wave. If the reflecting surfaces are metallic films, the electric field has nodes at both ends. The electromagnetic standing wave is analogous to the standing string wave represented in Figure 18.10. (a) Assume that a helium–neon laser has precisely flat and parallel mirrors 35.124 103 cm apart. Assume that the active medium can efficiently amplify only light with wavelengths between 632.808 40 nm and 632.809 80 nm. Find the number of components that constitute the laser light, and the wavelength of each component, precise to eight digits. (b) Find the root-mean-square speed for a neon atom at 120°C. (c) Show that at this temperature the Doppler effect for light emission by moving neon atoms should realistically make the bandwidth of the light amplifier larger than the 0.001 40 nm assumed in part (a).

Additional Problems

65. How much energy is required to ionize a hydrogen atom when it is in (a) the $n = 2$ state and (b) the $n = 10$ state?

66. The force on a magnetic moment μ_z in a nonuniform magnetic field B_z is given by $F_z = \mu_z(dB_z/dz)$. If a beam of silver atoms travels a horizontal distance of 1.00 m through such a field and each atom has a speed of 100 m/s, how strong must be the field gradient dB_z/dz to deflect the beam 1.00 mm?

67. Suppose a hydrogen atom is in the $2s$ state, with its wave function given by Equation 42.26. Taking $r = a_0$, calculate values for (a) $\psi_{2s}(a_0)$, (b) $|\psi_{2s}(a_0)|^2$, and (c) $P_{2s}(a_0)$.

68. **Review.** (a) How much energy is required to cause an **W** electron in hydrogen to move from the $n = 1$ state to the $n = 2$ state? (b) Suppose the atom gains this energy through collisions among hydrogen atoms at a high temperature. At what temperature would the average atomic kinetic energy $\frac{3}{2}k_BT$ be great enough to excite the electron? Here k_B is Boltzmann's constant.

69. In the technique known as electron spin resonance **M** (ESR), a sample containing unpaired electrons is placed in a magnetic field. Consider a situation in which a single electron (*not* contained in an atom) is immersed in a magnetic field. In this simple situation, only two energy states are possible, corresponding to $m_s = \pm\frac{1}{2}$. In ESR, the absorption of a photon causes the electron's spin magnetic moment to flip from the lower energy state to the higher energy state. According to Section 29.5, the change in energy is $2\mu_BB$. (The

lower energy state corresponds to the case in which the z component of the magnetic moment $\vec{\mu}_{spin}$ is aligned with the magnetic field, and the higher energy state corresponds to the case in which the z component of $\vec{\mu}_{spin}$ is aligned opposite to the field.) What is the photon frequency required to excite an ESR transition in a 0.350-T magnetic field?

70. An electron in chromium moves from the $n = 2$ state to the $n = 1$ state without emitting a photon. Instead, the excess energy is transferred to an outer electron (one in the $n = 4$ state), which is then ejected by the atom. In this Auger (pronounced "ohjay") process, the ejected electron is referred to as an Auger electron. Use the Bohr theory to find the kinetic energy of the Auger electron.

71. The states of matter are solid, liquid, gas, and plasma. Plasma can be described as a gas of charged particles or a gas of ionized atoms. Most of the matter in the Solar System is plasma (throughout the interior of the Sun). In fact, most of the matter in the Universe is plasma; so is a candle flame. Use the information in Figure 42.20 to make an order-of-magnitude estimate for the temperature to which a typical chemical element must be raised to turn into plasma by ionizing most of the atoms in a sample. Explain your reasoning.

72. Show that the wave function for a hydrogen atom in the $2s$ state

$$\psi_{2s}(r) = \frac{1}{4\sqrt{2\pi}}\left(\frac{1}{a_0}\right)^{3/2}\left(2 - \frac{r}{a_0}\right)e^{-r/2a_0}$$

satisfies the spherically symmetric Schrödinger equation given in Problem 26.

73. Find the average (expectation) value of $1/r$ in the $1s$ state of hydrogen. Note that the general expression is given by

$$\langle 1/r \rangle = \int_{\text{all space}} |\psi|^2(1/r)\,dV = \int_0^\infty P(r)(1/r)\,dr$$

Is the result equal to the inverse of the average value of r?

74. *Why is the following situation impossible?* An experiment is performed on an atom. Measurements of the atom when it is in a particular excited state show five possible values of the z component of orbital angular momentum, ranging between 3.16×10^{-34} kg · m²/s and -3.16×10^{-34} kg · m²/s.

75. In the Bohr model of the hydrogen atom, an electron travels in a circular path. Consider another case in which an electron travels in a circular path: a single electron moving perpendicular to a magnetic field $\vec{B}$. Lev Davidovich Landau (1908–1968) solved the Schrödinger equation for such an electron. The electron can be considered as a model atom without a nucleus or as the irreducible quantum limit of the cyclotron. Landau proved its energy is quantized in uniform steps of $e\hbar B/m_e$. In 1999, a single electron was trapped by a Harvard University research team in an evacuated centimeter-size metal can cooled to a temperature of 80 mK. In a magnetic field of magnitude

5.26 T, the electron circulated for hours in its lowest energy level. (a) Evaluate the size of a quantum jump in the electron's energy. (b) For comparison, evaluate $k_B T$ as a measure of the energy available to the electron in blackbody radiation from the walls of its container. Microwave radiation was introduced to excite the electron. Calculate (c) the frequency and (d) the wavelength of the photon the electron absorbed as it jumped to its second energy level. Measurement of the resonant absorption frequency verified the theory and permitted precise determination of properties of the electron.

76. As the Earth moves around the Sun, its orbits are quantized. (a) Follow the steps of Bohr's analysis of the hydrogen atom to show that the allowed radii of the Earth's orbit are given by

$$r = \frac{n^2 \hbar^2}{GM_S M_E^2}$$

where n is an integer quantum number, M_S is the mass of the Sun, and M_E is the mass of the Earth. (b) Calculate the numerical value of n for the Sun–Earth system. (c) Find the distance between the orbit for quantum number n and the next orbit out from the Sun corresponding to the quantum number $n + 1$. (d) Discuss the significance of your results from parts (b) and (c).

77. An elementary theorem in statistics states that the root-mean-square uncertainty in a quantity r is given by $\Delta r = \sqrt{\langle r^2 \rangle - \langle r \rangle^2}$. Determine the uncertainty in the radial position of the electron in the ground state of the hydrogen atom. Use the average value of r found in Example 42.3: $\langle r \rangle = 3a_0/2$. The average value of the squared distance between the electron and the proton is given by

$$\langle r^2 \rangle = \int\limits_{\text{all space}} |\psi|^2 r^2 \, dV = \int_0^\infty P(r) r^2 \, dr$$

78. Example 42.3 calculates the most probable value and the average value for the radial coordinate r of the electron in the ground state of a hydrogen atom. For comparison with these modal and mean values, find the median value of r. Proceed as follows. (a) Derive an expression for the probability, as a function of r, that the electron in the ground state of hydrogen will be found outside a sphere of radius r centered on the nucleus. (b) Make a graph of the probability as a function of r/a_0. Choose values of r/a_0 ranging from 0 to 4.00 in steps of 0.250. (c) Find the value of r for which the probability of finding the electron outside a sphere of radius r is equal to the probability of finding the electron inside this sphere. You must solve a transcendental equation numerically, and your graph is a good starting point.

79. (a) For a hydrogen atom making a transition from the $n = 4$ state to the $n = 2$ state, determine the wavelength of the photon created in the process. (b) Assuming the atom was initially at rest, determine the recoil speed of the hydrogen atom when it emits this photon.

80. Astronomers observe a series of spectral lines in the light from a distant galaxy. On the hypothesis that the lines form the Lyman series for a (new?) one-electron atom, they start to construct the energy-level diagram shown in Figure P42.80, which gives the wavelengths of the first four lines and the short-wavelength limit of this series. Based on this information, calculate (a) the energies of the ground state and first four excited states for this one-electron atom and (b) the wavelengths of the first three lines and the short-wavelength limit in the Balmer series for this atom. (c) Show that the wavelengths of the first four lines and the short-wavelength limit of the Lyman series for the hydrogen atom are all 60.0% of the wavelengths for the Lyman series in the one-electron atom in the distant galaxy. (d) Based on this observation, explain why this atom could be hydrogen.

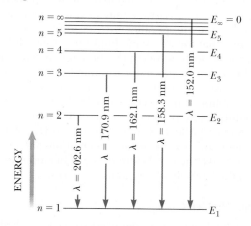

Figure P42.80

81. We wish to show that the most probable radial position for an electron in the 2s state of hydrogen is $r = 5.236a_0$. (a) Use Equations 42.24 and 42.26 to find the radial probability density for the 2s state of hydrogen. (b) Calculate the derivative of the radial probability density with respect to r. (c) Set the derivative in part (b) equal to zero and identify three values of r that represent minima in the function. (d) Find two values of r that represent maxima in the function. (e) Identify which of the values in part (c) represents the highest probability.

82. All atoms have the same size, to an order of magnitude. (a) To demonstrate this fact, estimate the atomic diameters for aluminum (with molar mass 27.0 g/mol and density 2.70 g/cm³) and uranium (molar mass 238 g/mol and density 18.9 g/cm³). (b) What do the results of part (a) imply about the wave functions for inner-shell electrons as we progress to higher and higher atomic mass atoms?

83. A pulsed ruby laser emits light at 694.3 nm. For a 14.0-ps pulse containing 3.00 J of energy, find (a) the physical length of the pulse as it travels through space and (b) the number of photons in it. (c) The beam has a circular cross section of diameter 0.600 cm. Find the number of photons per cubic millimeter.

84. A pulsed laser emits light of wavelength λ. For a pulse of duration Δt having energy T_{ER}, find (a) the physical length of the pulse as it travels through space and (b) the number of photons in it. (c) The beam has a

circular cross section having diameter d. Find the number of photons per unit volume.

85. Assume three identical uncharged particles of mass m and spin $\frac{1}{2}$ are contained in a one-dimensional box of length L. What is the ground-state energy of this system?

86. Suppose the ionization energy of an atom is 4.10 eV. In the spectrum of this same atom, we observe emission lines with wavelengths 310 nm, 400 nm, and 1 377.8 nm. Use this information to construct the energy-level diagram with the fewest levels. Assume the higher levels are closer together.

87. For hydrogen in the 1s state, what is the probability of finding the electron farther than $2.50a_0$ from the nucleus?

88. For hydrogen in the 1s state, what is the probability of finding the electron farther than βa_0 from the nucleus, where β is an arbitrary number?

Challenge Problems

89. The positron is the antiparticle to the electron. It has the same mass and a positive electric charge of the same magnitude as that of the electron. Positronium is a hydrogen-like atom consisting of a positron and an electron revolving around each other. Using the Bohr model, find (a) the allowed distances between the two particles and (b) the allowed energies of the system.

90. **Review.** Steven Chu, Claude Cohen-Tannoudji, and William Phillips received the 1997 Nobel Prize in Physics for "the development of methods to cool and trap atoms with laser light." One part of their work was with a beam of atoms (mass $\sim 10^{-25}$ kg) that move at a speed on the order of 1 km/s, similar to the speed of molecules in air at room temperature. An intense laser light beam tuned to a visible atomic transition (assume 500 nm) is directed straight into the atomic beam; that is, the atomic beam and the light beam are traveling in opposite directions. An atom in the ground state immediately absorbs a photon. Total system momentum is conserved in the absorption process. After a lifetime on the order of 10^{-8} s, the excited atom radiates by spontaneous emission. It has an equal probability of emitting a photon in any direction. Therefore, the average "recoil" of the atom is zero over many absorption and emission cycles. (a) Estimate the average deceleration of the atomic beam. (b) What is the order of magnitude of the distance over which the atoms in the beam are brought to a halt?

91. (a) Use Bohr's model of the hydrogen atom to show that when the electron moves from the n state to the $n-1$ state, the frequency of the emitted light is

$$f = \left(\frac{2\pi^2 m_e k_e^2 e^4}{h^3} \right) \frac{2n-1}{n^2(n-1)^2}$$

(b) Bohr's correspondence principle claims that quantum results should reduce to classical results in the limit of large quantum numbers. Show that as $n \to \infty$, this expression varies as $1/n^3$ and reduces to the classical frequency one expects the atom to emit. *Suggestion:* To calculate the classical frequency, note that the frequency of revolution is $v/2\pi r$, where v is the speed of the electron and r is given by Equation 42.10.

CHAPTER

43

Molecules and Solids

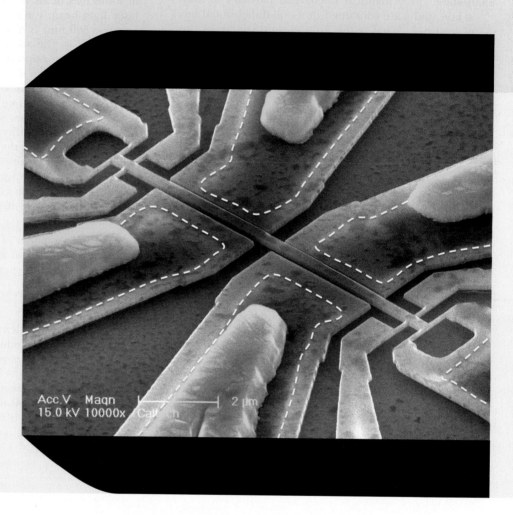

The photograph shows a *NEMS resonator*, where NEMS is an acronym for *nanoelectromechanical system*. The device employs a semiconductor bridge vibrating in a standing wave like the strings in Chapter 18. When a single molecule or other particle adheres to the bridge, the resonance frequencies of the normal modes shift in a measurable way. Scientists can determine the mass of the particle from the shifts in the frequencies. The new device shows promise in allowing the masses of molecules and many biological particles to be measured with great accuracy. *(Caltech/Scott Kelberg and Michael Roukes)*

The most random atomic arrangement, that of a gas, was well understood in the 1800s as discussed in our study of kinetic theory in Chapter 21. In a crystalline solid, the atoms are not randomly arranged; rather, they form a regular array. The symmetry of the arrangement of atoms both stimulated and allowed rapid progress in the field of solid-state physics in the 20th century. Recently, our understanding of liquids and amorphous solids has advanced. (In an amorphous solid such as glass or paraffin, the atoms do not form a regular array.) The recent interest in the physics of low-cost amorphous materials has been driven by their use in such devices as solar cells, memory elements, and fiber-optic waveguides. With the addition of liquids, amorphous solids, and some more exotic forms of matter, such as Bose–Einstein condensates, solid-state physics expanded in the middle of the 20th century to become known as *condensed matter physics.*

We begin this chapter by studying the aggregates of atoms known as molecules. We describe the bonding mechanisms in molecules, the various modes of molecular excitation, and the radiation emitted or absorbed by molecules. Next, we show how molecules combine to form solids. Then, by examining their energy-level structure, we explain the differences between insulating, conducting, semiconducting, and superconducting materials. The chapter also includes discussions of semiconducting junctions and several semiconductor devices.

43.1 Molecular Bonds

The bonding mechanisms in a molecule are fundamentally due to electric forces between atoms (or ions). Because the electric force is conservative, the forces between atoms in the system of a molecule are related to a potential energy function. A stable molecule is expected at a configuration for which the potential energy function for the molecule has its minimum value. (See Section 7.9.)

A potential energy function that can be used to model a molecule should account for two known features of molecular bonding:

1. The force between atoms is repulsive at very small separation distances. When two atoms are brought close to each other, some of their electron shells overlap, resulting in repulsion between the shells. This repulsion is partly electrostatic in origin and partly the result of the exclusion principle. Because all electrons must obey the exclusion principle, some electrons in the overlapping shells are forced into higher energy states and the system energy increases as if a repulsive force existed between the atoms.

2. At somewhat larger separations, the force between atoms is attractive. If that were not true, the atoms in a molecule would not be bound together.

Taking into account these two features, the potential energy for a system of two atoms can be represented by an expression of the form

$$U(r) = -\frac{A}{r^n} + \frac{B}{r^m} \qquad (43.1)$$

where r is the internuclear separation distance between the two atoms and n and m are small integers. The parameter A is associated with the attractive force and B with the repulsive force. Example 7.9 gives one common model for such a potential energy function, the Lennard–Jones potential.

Potential energy versus internuclear separation distance for a two-atom system is graphed in Figure 43.1. At large separation distances between the two atoms, the slope of the curve is positive, corresponding to a net attractive force. At the equilibrium separation distance, the attractive and repulsive forces just balance. At this point, the potential energy has its minimum value and the slope of the curve is zero.

A complete description of the bonding mechanisms in molecules is highly complex because bonding involves the mutual interactions of many particles. In this section, we discuss only some simplified models.

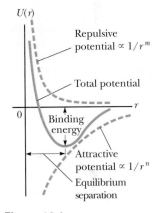

Figure 43.1 Total potential energy as a function of internuclear separation distance for a system of two atoms.

Ionic Bonding

When two atoms combine in such a way that one or more outer electrons are transferred from one atom to the other, the bond formed is called an **ionic bond.** Ionic bonds are fundamentally caused by the Coulomb attraction between oppositely charged ions.

A familiar example of an ionically bonded solid is sodium chloride, NaCl, which is common table salt. Sodium, which has the electronic configuration $1s^2 2s^2 2p^6 3s^1$, is ionized relatively easily, giving up its $3s$ electron to form a Na^+ ion. The energy required to ionize the atom to form Na^+ is 5.1 eV. Chlorine, which has the electronic configuration $1s^2 2s^2 2p^5$, is one electron short of the filled-shell structure of argon. If we compare the energy of a system of a free electron and a Cl atom with one in which the electron joins the atom to make the Cl^- ion, we find that the energy of the ion is lower. When the electron makes a transition from the $E = 0$ state to the negative energy state associated with the available shell in the atom, energy is released. This amount of energy is called the **electron affinity** of the atom. For chlorine, the electron affinity is 3.6 eV. Therefore, the energy required to form Na^+ and Cl^- from isolated atoms is $5.1 - 3.6 = 1.5$ eV. It costs 5.1 eV to remove

Figure 43.2 Total energy versus internuclear separation distance for Na^+ and Cl^- ions.

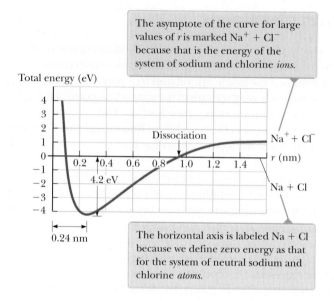

The asymptote of the curve for large values of r is marked $Na^+ + Cl^-$ because that is the energy of the system of sodium and chlorine *ions*.

The horizontal axis is labeled Na + Cl because we define zero energy as that for the system of neutral sodium and chlorine *atoms*.

Pitfall Prevention 43.1

Ionic and Covalent Bonds In practice, these descriptions of ionic and covalent bonds represent extreme ends of a spectrum of bonds involving electron transfer. In a real bond, the electron may not be *completely* transferred as in an ionic bond or *equally* shared as in a covalent bond. Therefore, real bonds lie somewhere between these extremes.

the electron from the Na atom, but 3.6 eV of it is gained back when that electron is allowed to join with the Cl atom.

Now imagine that these two charged ions interact with one another to form a NaCl "molecule."[1] The total energy of the NaCl molecule versus internuclear separation distance is graphed in Figure 43.2. At very large separation distances, the energy of the system of ions is 1.5 eV as calculated above. The total energy has a minimum value of -4.2 eV at the equilibrium separation distance, which is approximately 0.24 nm. Hence, the energy required to break the Na^+-Cl^- bond and form neutral sodium and chlorine atoms, called the **dissociation energy,** is 4.2 eV. The energy of the molecule is lower than that of the system of two neutral atoms. Consequently, it is **energetically favorable** for the molecule to form: if a lower energy state of a system exists, the system tends to emit energy to achieve this lower energy state. The system of neutral sodium and chlorine atoms can reduce its total energy by transferring energy out of the system (by electromagnetic radiation, for example) and forming the NaCl molecule.

Covalent Bonding

A **covalent bond** between two atoms is one in which electrons supplied by either one or both atoms are shared by the two atoms. Many diatomic molecules—such as H_2, F_2, and CO—owe their stability to covalent bonds. The bond between two hydrogen atoms can be described by using atomic wave functions. The ground-state wave function for a hydrogen atom (Chapter 42) is

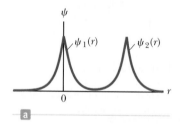

$$\psi_{1s}(r) = \frac{1}{\sqrt{\pi a_0^3}} e^{-r/a_0}$$

This wave function is graphed in Figure 43.3a for two hydrogen atoms that are far apart. There is very little overlap of the wave functions $\psi_1(r)$ for atom 1, located at $r = 0$, and $\psi_2(r)$ for atom 2, located some distance away. Suppose now the two atoms are brought close together. As that happens, their wave functions overlap and form the compound wave function $\psi_1(r) + \psi_2(r)$ shown in Figure 43.3b. Notice that the probability amplitude is larger between the atoms than it is on either side of the combination of atoms. As a result, the probability is higher that the electrons associated with the atoms will be located between the atoms than on the outer regions

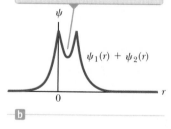

The probability amplitude for an electron to be between the atoms is high.

Figure 43.3 Ground-state wave functions $\psi_1(r)$ and $\psi_2(r)$ for two atoms making a covalent bond. (a) The atoms are far apart, and their wave functions overlap minimally. (b) The atoms are close together, forming a composite wave function $\psi_1(r) + \psi_2(r)$ for the system.

[1]NaCl does not tend to form as an isolated molecule at room temperature. In the solid state, NaCl forms a crystalline array of ions as described in Section 43.3. In the liquid state or in solution with water, the Na^+ and Cl^- ions dissociate and are free to move relative to each other.

of the system. Consequently, the average position of negative charge in the system is halfway between the atoms. This scenario can be modeled as if there were a fixed negative charge between the atoms, exerting attractive Coulomb forces on both nuclei. Therefore, there is an overall attractive force between the atoms, resulting in a covalent bond.

Because of the exclusion principle, the two electrons in the ground state of H_2 must have antiparallel spins. Also because of the exclusion principle, if a third H atom is brought near the H_2 molecule, the third electron would have to occupy a higher energy level, which is not an energetically favorable situation. For this reason, the H_3 molecule is not stable and does not form.

Van der Waals Bonding

Ionic and covalent bonds occur between atoms to form molecules or ionic solids, so they can be described as bonds *within* molecules. Two additional types of bonds, van der Waals bonds and hydrogen bonds, can occur *between* molecules.

You might think that two neutral molecules would not interact by means of the electric force because they each have zero net charge. They are attracted to each other, however, by weak electrostatic forces called **van der Waals forces.** Likewise, atoms that do not form ionic or covalent bonds are attracted to each other by van der Waals forces. Noble gas atoms, for example, because of their filled shell structure, do not generally form molecules or bond to each other to form a liquid. Because of van der Waals forces, however, at sufficiently low temperatures at which thermal excitations are negligible, noble gases first condense to liquids and then solidify. (The exception is helium, which does not solidify at atmospheric pressure.)

The van der Waals force results from the following situation. While being electrically neutral, a molecule has a charge distribution with positive and negative centers at different positions in the molecule. As a result, the molecule may act as an electric dipole. (See Section 23.4.) Because of the dipole electric fields, two molecules can interact such that there is an attractive force between them.

There are three types of van der Waals forces. The first type, called the *dipole–dipole force*, is an interaction between two molecules each having a permanent electric dipole moment. For example, polar molecules such as HCl have permanent electric dipole moments and attract other polar molecules.

The second type, the *dipole–induced dipole force*, results when a polar molecule having a permanent electric dipole moment induces a dipole moment in a nonpolar molecule. In this case, the electric field of the polar molecule creates the dipole moment in the nonpolar molecule, which then results in an attractive force between the molecules.

The third type is called the *dispersion force*, an attractive force that occurs between two nonpolar molecules. In this case, although the average dipole moment of a nonpolar molecule is zero, the average of the square of the dipole moment is nonzero because of charge fluctuations. Two nonpolar molecules near each other tend to have dipole moments that are correlated in time so as to produce an attractive van der Waals force.

Hydrogen Bonding

Because hydrogen has only one electron, it is expected to form a covalent bond with only one other atom within a molecule. A hydrogen atom in a given molecule can also form a second type of bond between molecules called a **hydrogen bond.** Let's use the water molecule H_2O as an example. In the two covalent bonds in this molecule, the electrons from the hydrogen atoms are more likely to be found near the oxygen atom than near the hydrogen atoms, leaving essentially bare protons at the positions of the hydrogen atoms. This unshielded positive charge can be attracted to the negative end of another polar molecule. Because the proton is unshielded by electrons, the negative end of the other molecule can come very close to the proton to form a bond strong enough to form a solid crystalline structure, such as

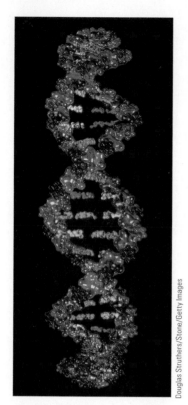

that of ordinary ice. The bonds within a water molecule are covalent, but the bonds between water molecules in ice are hydrogen bonds.

The hydrogen bond is relatively weak compared with other chemical bonds and can be broken with an input energy of approximately 0.1 eV. Because of this weakness, ice melts at the low temperature of 0°C. Even though this bond is weak, however, hydrogen bonding is a critical mechanism responsible for the linking of biological molecules and polymers. For example, in the case of the DNA (deoxyribonucleic acid) molecule, which has a double-helix structure (Fig. 43.4), hydrogen bonds form by the sharing of a proton between two atoms and create linkages between the turns of the helix.

Quick Quiz 43.1 For each of the following atoms or molecules, identify the most likely type of bonding that occurs between the atoms or between the molecules. Choose from the following list: ionic, covalent, van der Waals, hydrogen. **(a)** atoms of krypton **(b)** potassium and chlorine atoms **(c)** hydrogen fluoride (HF) molecules **(d)** chlorine and oxygen atoms in a hypochlorite ion (ClO⁻)

Figure 43.4 DNA molecules are held together by hydrogen bonds.

Douglas Struthers/Stone/Getty Images

▶ Total energy of a molecule

43.2 Energy States and Spectra of Molecules

Consider an individual molecule in the gaseous phase of a substance. The energy E of the molecule can be divided into four categories: (1) electronic energy, due to the interactions between the molecule's electrons and nuclei; (2) translational energy, due to the motion of the molecule's center of mass through space; (3) rotational energy, due to the rotation of the molecule about its center of mass; and (4) vibrational energy, due to the vibration of the molecule's constituent atoms:

$$E = E_{el} + E_{trans} + E_{rot} + E_{vib}$$

We explored the roles of translational, rotational, and vibrational energy of molecules in determining the molar specific heats of gases in Sections 21.2 and 21.3. The translational energy is important in kinetic theory, but it is unrelated to internal structure of the molecule, so this molecular energy is unimportant in interpreting molecular spectra. The electronic energy of a molecule is very complex because it involves the interaction of many charged particles, but various techniques have been developed to approximate its values. Although the electronic energies can be studied, significant information about a molecule can be determined by analyzing its quantized rotational and vibrational energy states. Transitions between these states give spectral lines in the microwave and infrared regions of the electromagnetic spectrum, respectively.

Rotational Motion of Molecules

Let's consider the rotation of a molecule around its center of mass, confining our discussion to the diatomic molecule (Fig. 43.5a) but noting that the same ideas can be extended to polyatomic molecules. A diatomic molecule aligned along a y axis has only two rotational degrees of freedom, corresponding to rotations about the x and z axes passing through the molecule's center of mass. We discussed the rotation of such a molecule and its contribution to the specific heat of a gas in Section 21.3. If ω is the angular frequency of rotation about one of these axes, the rotational kinetic energy of the molecule about that axis can be expressed with Equation 10.24:

$$E_{rot} = \tfrac{1}{2}I\omega^2 \qquad \textbf{(43.2)}$$

In this equation, I is the moment of inertia of the molecule about its center of mass, given by

◀ Moment of inertia for a diatomic molecule

$$I = \left(\frac{m_1 m_2}{m_1 + m_2}\right)r^2 = \mu r^2 \qquad \textbf{(43.3)}$$

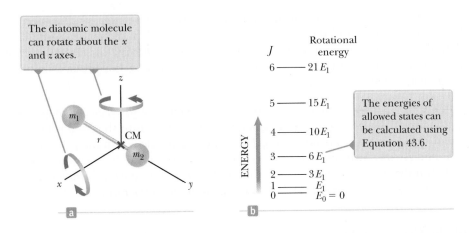

The diatomic molecule can rotate about the x and z axes.

The energies of allowed states can be calculated using Equation 43.6.

Figure 43.5 Rotation of a diatomic molecule around its center of mass. (a) A diatomic molecule oriented along the y axis. (b) Allowed rotational energies of a diatomic molecule expressed as multiples of $E_1 = \hbar^2/I$.

where m_1 and m_2 are the masses of the atoms that form the molecule, r is the atomic separation, and μ is the **reduced mass** of the molecule (see Example 41.5 and Problem 40 in Chapter 41):

$$\mu = \frac{m_1\, m_2}{m_1 + m_2} \qquad \text{(43.4)}$$

◀ **Reduced mass of a diatomic molecule**

The magnitude of the molecule's angular momentum about its center of mass is given by Equation 11.14, $L = I\omega$, which classically can have any value. Quantum mechanics, however, restricts the molecule to certain quantized rotational frequencies such that the angular momentum of the molecule has the values[2]

$$L = \sqrt{J(J + 1)}\,\hbar \qquad J = 0, 1, 2, \ldots \qquad \text{(43.5)}$$

◀ **Allowed values of rotational angular momentum**

where J is an integer called the **rotational quantum number.** Combining Equations 43.5 and 43.2, we obtain an expression for the allowed values of the rotational kinetic energy of the molecule:

$$E_{\text{rot}} = \tfrac{1}{2}I\omega^2 = \frac{1}{2I}(I\omega)^2 = \frac{L^2}{2I} = \frac{(\sqrt{J(J+1)}\,\hbar)^2}{2I}$$

$$E_{\text{rot}} = E_J = \frac{\hbar^2}{2I}J(J + 1) \qquad J = 0, 1, 2, \ldots \qquad \text{(43.6)}$$

◀ **Allowed values of rotational energy**

The allowed rotational energies of a diatomic molecule are plotted in Figure 43.5b. As the quantum number J goes up, the states become farther apart as displayed earlier for rotational energy levels in Figure 21.7.

For most molecules, transitions between adjacent rotational energy levels result in radiation that lies in the microwave range of frequencies ($f \sim 10^{11}$ Hz). When a molecule absorbs a microwave photon, the molecule jumps from a lower rotational energy level to a higher one. The allowed rotational transitions of linear molecules are regulated by the selection rule $\Delta J = \pm 1$. Given this selection rule, all absorption lines in the spectrum of a linear molecule correspond to energy separations equal to $E_J - E_{J-1}$, where $J = 1, 2, 3, \ldots$. From Equation 43.6, we see that the energies of the absorbed photons are given by

$$E_{\text{photon}} = \Delta E_{\text{rot}} = E_J - E_{J-1} = \frac{\hbar^2}{2I}[J(J + 1) - (J - 1)J]$$

$$E_{\text{photon}} = \frac{\hbar^2}{I}J = \frac{h^2}{4\pi^2 I}J \qquad J = 1, 2, 3, \ldots \qquad \text{(43.7)}$$

◀ **Energy of a photon absorbed in a transition between adjacent rotational levels**

[2]Equation 43.5 is similar to Equation 42.27 for orbital angular momentum in an atom. The relationship between the magnitude of the angular momentum of a system and the associated quantum number is the same as it is in these equations for *any* system that exhibits rotation as long as the potential energy function for the system is spherically symmetric.

where J is the rotational quantum number of the higher energy state. Because $E_{photon} = hf$, where f is the frequency of the absorbed photon, we see that the allowed frequency for the transition $J = 0$ to $J = 1$ is $f_1 = h/4\pi^2 I$. The frequency corresponding to the $J = 1$ to $J = 2$ transition is $2f_1$, and so on. These predictions are in excellent agreement with the observed frequencies.

Quick Quiz 43.2 A gas of identical diatomic molecules absorbs electromagnetic radiation over a wide range of frequencies. Molecule 1 is in the $J = 0$ rotation state and makes a transition to the $J = 1$ state. Molecule 2 is in the $J = 2$ state and makes a transition to the $J = 3$ state. Is the ratio of the frequency of the photon that excited molecule 2 to that of the photon that excited molecule 1 equal to (a) 1, (b) 2, (c) 3, (d) 4, or (e) impossible to determine?

Example 43.1   Rotation of the CO Molecule

The $J = 0$ to $J = 1$ rotational transition of the CO molecule occurs at a frequency of 1.15×10^{11} Hz.

(A) Use this information to calculate the moment of inertia of the molecule.

SOLUTION

Conceptualize Imagine that the two atoms in Figure 43.5a are carbon and oxygen. The center of mass of the molecule is not midway between the atoms because of the difference in masses of the C and O atoms.

Categorize The statement of the problem tells us to categorize this example as one involving a quantum-mechanical treatment and to restrict our investigation to the rotational motion of a diatomic molecule.

Analyze Use Equation 43.7 to find the energy of a photon that excites the molecule from the $J = 0$ to the $J = 1$ rotational level:

$$E_{photon} = \frac{h^2}{4\pi^2 I}(1) = \frac{h^2}{4\pi^2 I}$$

Equate this energy to $E = hf$ for the absorbed photon and solve for I:

$$\frac{h^2}{4\pi^2 I} = hf \rightarrow I = \frac{h}{4\pi^2 f}$$

Substitute the frequency given in the problem statement:

$$I = \frac{6.626 \times 10^{-34} \, J \cdot s}{4\pi^2 (1.15 \times 10^{11} \, s^{-1})} = \boxed{1.46 \times 10^{-46} \, kg \cdot m^2}$$

(B) Calculate the bond length of the molecule.

SOLUTION

Find the reduced mass μ of the CO molecule:

$$\mu = \frac{m_1 m_2}{m_1 + m_2} = \frac{(12 \, u)(16 \, u)}{12 \, u + 16 \, u} = 6.86 \, u$$

$$= (6.86 \, u)\left(\frac{1.66 \times 10^{-27} \, kg}{1 \, u}\right) = 1.14 \times 10^{-26} \, kg$$

Solve Equation 43.3 for r and substitute for the reduced mass and the moment of inertia from part (A):

$$r = \sqrt{\frac{I}{\mu}} = \sqrt{\frac{1.46 \times 10^{-46} \, kg \cdot m^2}{1.14 \times 10^{-26} \, kg}}$$

$$= 1.13 \times 10^{-10} \, m = \boxed{0.113 \, nm}$$

Finalize The moment of inertia of the molecule and the separation distance between the atoms are both very small, as expected for a microscopic system.

WHAT IF? What if another photon of frequency 1.15×10^{11} Hz is incident on the CO molecule while that molecule is in the $J = 1$ state? What happens?

Answer Because the rotational quantum states are not equally spaced in energy, the $J = 1$ to $J = 2$ transition does not have the same energy as the $J = 0$ to $J = 1$ transition. Therefore, the molecule will *not* be excited to the $J = 2$ state. Two

▶ **43.1** continued

possibilities exist. The photon could pass by the molecule with no interaction, or the photon could induce a stimulated emission, similar to that for atoms and discussed in Section 42.9. In this case, the molecule makes a transition back to the $J = 0$ state and the original photon and a second identical photon leave the scene of the interaction.

Vibrational Motion of Molecules

If we consider a molecule to be a flexible structure in which the atoms are bonded together by "effective springs" as shown in Figure 43.6a, we can apply the particle in simple harmonic motion analysis model to the molecule as long as the atoms in the molecule are not too far from their equilibrium positions. Recall from Section 15.3 that the potential energy function for a simple harmonic oscillator is parabolic, varying as the square of the position of the particle relative to the equilibrium position. (See Eq. 15.20 and Fig. 15.9b.) Figure 43.6b shows a plot of potential energy versus atomic separation for a diatomic molecule, where r_0 is the equilibrium atomic separation. For separations close to r_0, the shape of the potential energy curve closely resembles the parabolic shape of the potential energy function in the particle in simple harmonic motion model.

According to classical mechanics, the frequency of vibration for the system shown in Figure 43.6a is given by Equation 15.14:

$$f = \frac{1}{2\pi}\sqrt{\frac{k}{\mu}} \tag{43.8}$$

where k is the effective spring constant and μ is the reduced mass given by Equation 43.4. In Section 21.3, we studied the contribution of a molecule's vibration to the specific heats of gases.

Quantum mechanics predicts that a molecule vibrates in quantized states as described in Section 41.7. The vibrational motion and quantized vibrational energy can be altered if the molecule acquires energy of the proper value to cause a transition between quantized vibrational states. As discussed in Section 41.7, the allowed vibrational energies are

$$E_{\text{vib}} = (v + \tfrac{1}{2})hf \qquad v = 0, 1, 2, \ldots \tag{43.9}$$

where v is an integer called the **vibrational quantum number.** (We used n in Section 41.7 for a general harmonic oscillator, but v is often used for the quantum number when discussing molecular vibrations.) If the system is in the lowest vibrational state, for which $v = 0$, its ground-state energy is $\tfrac{1}{2}hf$. In the first excited vibrational state, $v = 1$ and the energy is $\tfrac{3}{2}hf$, and so on.

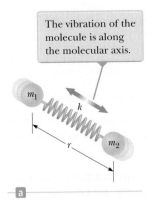

The vibration of the molecule is along the molecular axis.

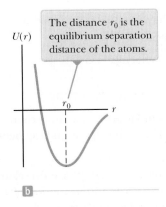

The distance r_0 is the equilibrium separation distance of the atoms.

Figure 43.6 (a) Effective-spring model of a diatomic molecule. (b) Plot of the potential energy of a diatomic molecule versus atomic separation distance. Compare with Figure 15.11a.

Figure 43.7 Allowed vibrational energies of a diatomic molecule, where f is the frequency of vibration of the molecule, given by Equation 43.8.

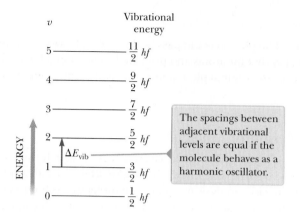

The spacings between adjacent vibrational levels are equal if the molecule behaves as a harmonic oscillator.

Substituting Equation 43.8 into Equation 43.9 gives the following expression for the allowed vibrational energies:

Allowed values of ▶
vibrational energy

$$E_{vib} = \left(v + \tfrac{1}{2}\right)\frac{h}{2\pi}\sqrt{\frac{k}{\mu}} \qquad v = 0, 1, 2, \ldots \tag{43.10}$$

The selection rule for the allowed vibrational transitions is $\Delta v = \pm 1$. Transitions between vibrational levels are caused by absorption of photons in the infrared region of the spectrum. The energy of an absorbed photon is equal to the energy difference between any two successive vibrational levels. Therefore, the photon energy is given by

$$E_{photon} = \Delta E_{vib} = \frac{h}{2\pi}\sqrt{\frac{k}{\mu}} \tag{43.11}$$

The vibrational energies of a diatomic molecule are plotted in Figure 43.7. At ordinary temperatures, most molecules have vibrational energies corresponding to the $v = 0$ state because the spacing between vibrational states is much greater than $k_B T$, where k_B is Boltzmann's constant and T is the temperature.

Quick Quiz 43.3 A gas of identical diatomic molecules absorbs electromagnetic radiation over a wide range of frequencies. Molecule 1, initially in the $v = 0$ vibrational state, makes a transition to the $v = 1$ state. Molecule 2, initially in the $v = 2$ state, makes a transition to the $v = 3$ state. What is the ratio of the frequency of the photon that excited molecule 2 to that of the photon that excited molecule 1? **(a)** 1 **(b)** 2 **(c)** 3 **(d)** 4 **(e)** impossible to determine

Example 43.2 Vibration of the CO Molecule AM

The frequency of the photon that causes the $v = 0$ to $v = 1$ transition in the CO molecule is 6.42×10^{13} Hz. We ignore any changes in the rotational energy for this example.

(A) Calculate the force constant k for this molecule.

SOLUTION

Conceptualize Imagine that the two atoms in Figure 43.6a are carbon and oxygen. As the molecule vibrates, a given point on the imaginary spring is at rest. This point is not midway between the atoms because of the difference in masses of the C and O atoms.

Categorize The statement of the problem tells us to categorize this example as one involving a quantum-mechanical treatment and to restrict our investigation to the vibrational motion of a diatomic molecule. The molecule is analyzed with portions of the *particle in simple harmonic motion* analysis model.

▶ **43.2** continued

Analyze Set Equation 43.11 equal to the photon energy hf and solve for the force constant:

$$\frac{h}{2\pi}\sqrt{\frac{k}{\mu}} = hf \quad \rightarrow \quad k = 4\pi^2 \mu f^2$$

Substitute the frequency given in the problem statement and the reduced mass from Example 43.1:

$$k = 4\pi^2(1.14 \times 10^{-26}\ \text{kg})(6.42 \times 10^{13}\ \text{s}^{-1})^2 = \boxed{1.85 \times 10^3\ \text{N/m}}$$

(B) What is the classical amplitude A of vibration for this molecule in the $v = 0$ vibrational state?

SOLUTION

Equate the maximum elastic potential energy $\frac{1}{2}kA^2$ in the molecule (Eq. 15.21) to the vibrational energy given by Equation 43.10 with $v = 0$ and solve for A:

$$\frac{1}{2}kA^2 = \frac{h}{4\pi}\sqrt{\frac{k}{\mu}} \quad \rightarrow \quad A = \sqrt{\frac{h}{2\pi}}\left(\frac{1}{\mu k}\right)^{1/4}$$

Substitute the value for k from part (A) and the value for μ:

$$A = \sqrt{\frac{6.626 \times 10^{-34}\ \text{J} \cdot \text{s}}{2\pi}}\left[\frac{1}{(1.14 \times 10^{-26}\ \text{kg})(1.85 \times 10^3\ \text{N/m})}\right]^{1/4}$$

$$= 4.79 \times 10^{-12}\ \text{m} = \boxed{0.004\ 79\ \text{nm}}$$

Finalize Comparing this result with the bond length of 0.113 nm we calculated in Example 43.1 shows that the classical amplitude of vibration is approximately 4% of the bond length.

Molecular Spectra

In general, a molecule vibrates and rotates simultaneously. To a first approximation, these motions are independent of each other, so the total energy of the molecule for these motions is the sum of Equations 43.6 and 43.9:

$$E = (v + \tfrac{1}{2})hf + \frac{\hbar^2}{2I}J(J + 1) \tag{43.12}$$

The energy levels of any molecule can be calculated from this expression, and each level is indexed by the two quantum numbers v and J. From these calculations, an energy-level diagram like the one shown in Figure 43.8a (page 1350) can be constructed. For each allowed value of the vibrational quantum number v, there is a complete set of rotational levels corresponding to $J = 0, 1, 2, \ldots$. The energy separation between successive rotational levels is much smaller than the separation between successive vibrational levels. As noted earlier, most molecules at ordinary temperatures are in the $v = 0$ vibrational state; these molecules can be in various rotational states as Figure 43.8a shows.

When a molecule absorbs a photon with the appropriate energy, the vibrational quantum number v increases by one unit while the rotational quantum number J either increases or decreases by one unit as can be seen in Figure 43.8. Therefore, the molecular absorption spectrum in Figure 43.8b consists of two groups of lines: one group to the right of center and satisfying the selection rules $\Delta J = +1$ and $\Delta v = +1$, and the other group to the left of center and satisfying the selection rules $\Delta J = -1$ and $\Delta v = +1$.

The energies of the absorbed photons can be calculated from Equation 43.12:

$$E_{\text{photon}} = \Delta E = hf + \frac{\hbar^2}{I}(J + 1) \qquad J = 0, 1, 2, \ldots \quad (\Delta J = +1) \tag{43.13}$$

$$E_{\text{photon}} = \Delta E = hf - \frac{\hbar^2}{I}J \qquad J = 1, 2, 3, \ldots \quad (\Delta J = -1) \tag{43.14}$$

Figure 43.8 (a) Absorptive transitions between the $v = 0$ and $v = 1$ vibrational states of a diatomic molecule. Compare the energy levels in this figure with those in Figure 21.7. (b) Expected lines in the absorption spectrum of a molecule. These same lines appear in the emission spectrum.

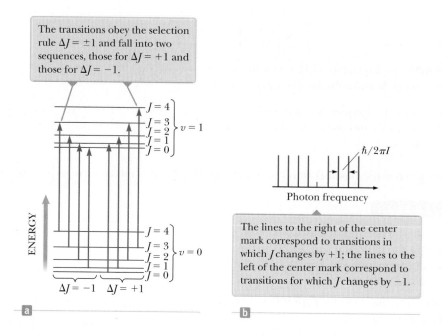

The transitions obey the selection rule $\Delta J = \pm 1$ and fall into two sequences, those for $\Delta J = +1$ and those for $\Delta J = -1$.

The lines to the right of the center mark correspond to transitions in which J changes by $+1$; the lines to the left of the center mark correspond to transitions for which J changes by -1.

where J is the rotational quantum number of the *initial* state. Equation 43.13 generates the series of equally spaced lines *higher* than the frequency f, whereas Equation 43.14 generates the series *lower* than this frequency. Adjacent lines are separated in frequency by the fundamental unit $\hbar/2\pi I$. Figure 43.8b shows the expected frequencies in the absorption spectrum of the molecule; these same frequencies appear in the emission spectrum.

The experimental absorption spectrum of the HCl molecule shown in Figure 43.9 follows this pattern very well and reinforces our model. One peculiarity is apparent, however: each line is split into a doublet. This doubling occurs because two chlorine isotopes (Cl-35 and Cl-37; see Section 44.1) were present in the sample used to obtain this spectrum. Because the isotopes have different masses, the two HCl molecules have different values of I.

The intensity of the spectral lines in Figure 43.9 follows an interesting pattern, rising first as one moves away from the central gap (located at about 8.65×10^{13} Hz, corresponding to the forbidden $J = 0$ to $J = 0$ transition) and then falling. This intensity is determined by a product of two functions of J. The first function corresponds to the number of available states for a given value of J. This function is $2J + 1$, corresponding to the number of values of m_J, the molecular rotation analog to m_ℓ for atomic states. For example, the $J = 2$ state has five substates with five values of m_J ($m_J = -2, -1, 0, 1, 2$), whereas the $J = 1$ state has only three substates ($m_J = -1, 0, 1$). Therefore, on average and without regard for the second function described below, five-thirds as many molecules make the transition from the $J = 2$ state as from the $J = 1$ state.

The second function determining the envelope of the intensity of the spectral lines is the Boltzmann factor, introduced in Section 21.5. The number of molecules in an excited rotational state is given by

$$n = n_0 e^{-\hbar^2 J(J+1)/(2Ik_B T)}$$

where n_0 is the number of molecules in the $J = 0$ state.

Multiplying these factors together indicates that the intensity of spectral lines should be described by a function of J as follows:

Intensity variation in the vibration–rotation spectrum of a molecule ▶

$$I \propto (2J + 1)e^{-\hbar^2 J(J+1)/(2Ik_B T)} \tag{43.15}$$

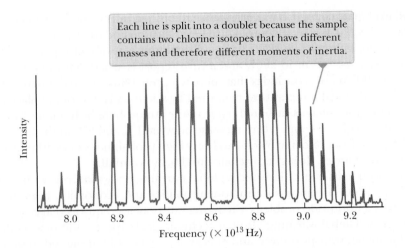

Each line is split into a doublet because the sample contains two chlorine isotopes that have different masses and therefore different moments of inertia.

Figure 43.9 Experimental absorption spectrum of the HCl molecule.

The factor $(2J + 1)$ increases with J while the exponential second factor decreases. The product of the two factors gives a behavior that closely describes the envelope of the spectral lines in Figure 43.9.

The excitation of rotational and vibrational energy levels is an important consideration in current models of global warming. Most of the absorption lines for CO_2 are in the infrared portion of the spectrum. Therefore, visible light from the Sun is not absorbed by atmospheric CO_2 but instead strikes the Earth's surface, warming it. In turn, the surface of the Earth, being at a much lower temperature than the Sun, emits thermal radiation that peaks in the infrared portion of the electromagnetic spectrum (Section 40.1). This infrared radiation is absorbed by the CO_2 molecules in the air instead of radiating out into space. Atmospheric CO_2 acts like a one-way valve for energy from the Sun and is responsible, along with some other atmospheric molecules, for raising the temperature of the Earth's surface above its value in the absence of an atmosphere. This phenomenon is commonly called the "greenhouse effect." The burning of fossil fuels in today's industrialized society adds more CO_2 to the atmosphere. This addition of CO_2 increases the absorption of infrared radiation, raising the Earth's temperature further. In turn, this increase in temperature causes substantial climatic changes.

As seen in Figure 43.10, the amount of carbon dioxide in the atmosphere has been steadily increasing since the middle of the 20th century. This graph shows hard data that indicate that the atmosphere is undergoing a distinct change, although not all scientists agree on the interpretation of what that change means in terms of global temperatures.

The Intergovernmental Panel on Climate Change (IPCC) is a scientific body that assesses the available information related to global warming and associated effects

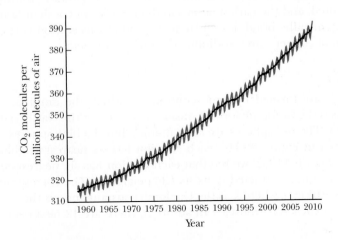

Figure 43.10 The concentration of atmospheric carbon dioxide in parts per million (ppm) of dry air as a function of time. These data were recorded at the Mauna Loa Observatory in Hawaii. The yearly variations (red-brown curve) coincide with growing seasons because vegetation absorbs carbon dioxide from the air. The steady increase in the average concentration (black curve) is of concern to scientists.

related to climate change. It was originally established in 1988 by two United Nations organizations, the World Meteorological Organization and the United Nations Environment Programme. The IPCC has published four assessment reports on climate change, the most recent in 2007, and a fifth report is scheduled to be released in 2014. The 2007 report concludes that there is a probability of greater than 90% that the increased global temperature measured by scientists is due to the placement of greenhouse gases such as carbon dioxide in the atmosphere by humans. The report also predicts a global temperature increase between 1°C and 6°C in the 21st century, a sea level rise from 18 cm to 59 cm, and very high probabilities of weather extremes, including heat waves, droughts, cyclones, and heavy rainfall.

In addition to its scientific aspects, global warming is a social issue with many facets. These facets encompass international politics and economics, because global warming is a worldwide problem. Changing our policies requires real costs to solve the problem. Global warming also has technological aspects, and new methods of manufacturing, transportation, and energy supply must be designed to slow down or reverse the increase in temperature.

Conceptual Example 43.3   Comparing Figures 43.8 and 43.9

In Figure 43.8a, the transitions indicated correspond to spectral lines that are equally spaced as shown in Figure 43.8b. The actual spectrum in Figure 43.9, however, shows lines that move closer together as the frequency increases. Why does the spacing of the actual spectral lines differ from the diagram in Figure 43.8?

SOLUTION

In Figure 43.8, we modeled the rotating diatomic molecule as a rigid object (Chapter 10). In reality, however, as the molecule rotates faster and faster, the effective spring in Figure 43.6a stretches and provides the increased force associated with the larger centripetal acceleration of each atom. As the molecule stretches along its length, its moment of inertia I increases. Therefore, the rotational part of the energy expression in Equation 43.12 has an extra dependence on J in the moment of inertia I. Because the increasing moment of inertia is in the denominator, as J increases, the energies do not increase as rapidly with J as indicated in Equation 43.12. With each higher energy level being lower than indicated by Equation 43.12, the energy associated with a transition to that level is smaller, as is the frequency of the absorbed photon, destroying the even spacing of the spectral lines and giving the spacing that decreases with increasing frequency seen in Figure 43.9.

43.3 Bonding in Solids

A crystalline solid consists of a large number of atoms arranged in a regular array, forming a periodic structure. The ions in the NaCl crystal are ionically bonded, as already noted, and the carbon atoms in diamond form covalent bonds with one another. The metallic bond described at the end of this section is responsible for the cohesion of copper, silver, sodium, and other solid metals.

Ionic Solids

Many crystals are formed by ionic bonding, in which the dominant interaction between ions is the Coulomb force. Consider a portion of the NaCl crystal shown in Figure 43.11a. The red spheres are sodium ions, and the blue spheres are chlorine ions. As shown in Figure 43.11b, each Na^+ ion has six nearest-neighbor Cl^- ions. Similarly, in Figure 43.11c, we see that each Cl^- ion has six nearest-neighbor Na^+ ions. Each Na^+ ion is attracted to its six Cl^- neighbors. The corresponding potential energy is $-6k_e e^2/r$, where k_e is the Coulomb constant and r is the separation distance between each Na^+ and Cl^-. In addition, there are 12 next-nearest-neighbor

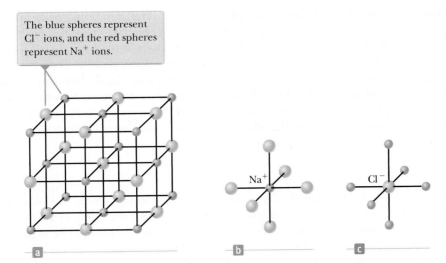

The blue spheres represent Cl^- ions, and the red spheres represent Na^+ ions.

Na^+

Cl^-

a   b   c

Figure 43.11 (a) Crystalline structure of NaCl. (b) Each positive sodium ion is surrounded by six negative chlorine ions. (c) Each chlorine ion is surrounded by six sodium ions.

Na^+ ions at a distance of $\sqrt{2}r$ from the Na^+ ion, and these 12 positive ions exert weaker repulsive forces on the central Na^+. Furthermore, beyond these 12 Na^+ ions are more Cl^- ions that exert an attractive force, and so on. The net effect of all these interactions is a resultant negative electric potential energy

$$U_{\text{attractive}} = -\alpha k_e \frac{e^2}{r} \qquad (43.16)$$

where α is a dimensionless number known as the **Madelung constant.** The value of α depends only on the particular crystalline structure of the solid. For example, $\alpha = 1.747\,6$ for the NaCl structure. When the constituent ions of a crystal are brought close together, a repulsive force exists because of electrostatic forces and the exclusion principle as discussed in Section 43.1. The potential energy term B/r^m in Equation 43.1 accounts for this repulsive force. We do not include neighbors other than nearest neighbors here because the repulsive forces occur only for ions that are very close together. (Electron shells must overlap for exclusion-principle effects to become important.) Therefore, we can express the total potential energy of the crystal as

$$U_{\text{total}} = -\alpha k_e \frac{e^2}{r} + \frac{B}{r^m} \qquad (43.17)$$

where m in this expression is some small integer.

A plot of total potential energy versus ion separation distance is shown in Figure 43.12. The potential energy has its minimum value U_0 at the equilibrium separation, when $r = r_0$. It is left as a problem (Problem 59) to show that

$$U_0 = -\alpha k_e \frac{e^2}{r_0}\left(1 - \frac{1}{m}\right) \qquad (43.18)$$

This minimum energy U_0 is called the **ionic cohesive energy** of the solid, and its absolute value represents the energy required to separate the solid into a collection of isolated positive and negative ions. Its value for NaCl is -7.84 eV per ion pair.

To calculate the **atomic cohesive energy,** which is the binding energy relative to the energy of the neutral atoms, 5.14 eV must be added to the ionic cohesive energy value to account for the transition from Na^+ to Na and 3.62 eV must be subtracted to account for the conversion of Cl^- to Cl. Therefore, the atomic cohesive energy of NaCl is

$$-7.84 \text{ eV} + 5.14 \text{ eV} - 3.62 \text{ eV} = -6.32 \text{ eV}$$

In other words, 6.32 eV of energy per ion pair is needed to separate the solid into isolated neutral atoms of Na and Cl.

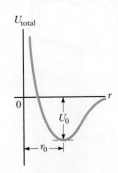

U_{total}

U_0

r_0

Figure 43.12 Total potential energy versus ion separation distance for an ionic solid, where U_0 is the ionic cohesive energy and r_0 is the equilibrium separation distance between ions.

Ionic crystals form relatively stable, hard crystals. They are poor electrical conductors because they contain no free electrons; each electron in the solid is bound tightly to one of the ions, so it is not sufficiently mobile to carry current. Ionic crystals have high melting points; for example, the melting point of NaCl is 801°C. Ionic crystals are transparent to visible radiation because the shells formed by the electrons in ionic solids are so tightly bound that visible radiation does not possess sufficient energy to promote electrons to the next allowed shell. Infrared radiation is absorbed strongly because the vibrations of the ions have natural resonant frequencies in the low-energy infrared region.

Covalent Solids

Solid carbon, in the form of diamond, is a crystal whose atoms are covalently bonded. Because atomic carbon has the electronic configuration $1s^22s^22p^2$, it is four electrons short of filling its $n = 2$ shell, which can accommodate eight electrons. Because of this electron structure, two carbon atoms have a strong attraction for each other, with a cohesive energy of 7.37 eV. In the diamond structure, each carbon atom is covalently bonded to four other carbon atoms located at four corners of a cube as shown in Figure 43.13a.

The crystalline structure of diamond is shown in Figure 43.13b. Notice that each carbon atom forms covalent bonds with four nearest-neighbor atoms. The basic structure of diamond is called tetrahedral (each carbon atom is at the center of a regular tetrahedron), and the angle between the bonds is 109.5°. Other crystals such as silicon and germanium have the same structure.

Carbon is interesting in that it can form several different types of structures. In addition to the diamond structure, it forms graphite, with completely different properties. In this form, the carbon atoms form flat layers with hexagonal arrays of atoms. A very weak interaction between the layers allows the layers to be removed easily under friction, as occurs in the graphite used in pencil lead.

Carbon atoms can also form a large hollow structure; in this case, the compound is called **buckminsterfullerene** after the famous architect R. Buckminster Fuller, who invented the geodesic dome. The unique shape of this molecule (Fig. 43.14) provides a "cage" to hold other atoms or molecules. Related structures, called "buckytubes" because of their long, narrow cylindrical arrangements of carbon atoms, may provide the basis for extremely strong, yet lightweight, materials.

A current area of active research is in the properties and applications of **graphene.** Graphene consists of a monolayer of carbon atoms, with the atoms arranged in hexagons so that the monolayer looks like chicken wire. Graphite flakes that are shed from a pencil while writing contain small fragments of graphene. Pioneers in graphene research include Andre Geim (b. 1958) and Konstantin Novoselov (b. 1974) of the University of Manchester, who received the Nobel Prize in Physics in 2010 for their experiments. Graphene has interesting electronic, thermal, and optical properties that are currently under investigation. Its mechanical properties include a breaking strength 200 times that of steel. Potential applications under

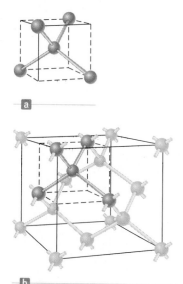

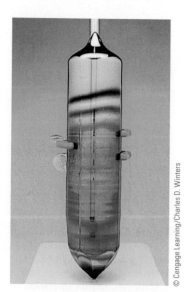

Figure 43.13 (a) Each carbon atom in a diamond crystal is covalently bonded to four other carbon atoms so that a tetrahedral structure is formed. (b) The crystal structure of diamond, showing the tetrahedral bond arrangement.

A cylinder of nearly pure crystalline silicon (Si), approximately 25 cm long. Such crystals are cut into wafers and processed to make various semiconductor devices.

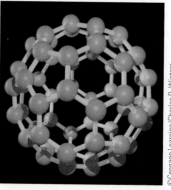

Figure 43.14 Computer rendering of a "buckyball," short for the molecule buckminsterfullerene. These nearly spherical molecular structures that look like soccer balls were named for the inventor of the geodesic dome. This form of carbon, C_{60}, was discovered by astrophysicists investigating the carbon gas that exists between stars. Scientists are actively studying the properties and potential uses of buckminsterfullerene and related molecules.

Table 43.1	Atomic Cohesive Energies of Some Covalent Solids
Solid	Cohesive Energy (eV per ion pair)
C (diamond)	7.37
Si	4.63
Ge	3.85
InAs	5.70
SiC	6.15
ZnS	6.32
CuCl	9.24

study include graphene nanoribbons, quantum dots, transistors, optical modulators, and integrated circuits.

The atomic cohesive energies of some covalent solids are given in Table 43.1. The large energies account for the hardness of covalent solids. Diamond is particularly hard and has an extremely high melting point (about 4 000 K). Covalently bonded solids usually have high bond energies and high melting points, and are good electrical insulators.

Metallic Solids

Metallic bonds are generally weaker than ionic or covalent bonds. The outer electrons in the atoms of a metal are relatively free to move throughout the material, and the number of such mobile electrons in a metal is large. The metallic structure can be viewed as a "sea" or a "gas" of nearly free electrons surrounding a lattice of positive ions (Fig. 43.15). The bonding mechanism in a metal is the attractive force between the entire collection of positive ions and the electron gas. Metals have a cohesive energy in the range of 1 to 3 eV per atom, which is less than the cohesive energies of ionic or covalent solids.

Light interacts strongly with the free electrons in metals. Hence, visible light is absorbed and re-emitted quite close to the surface of a metal, which accounts for the shiny nature of metal surfaces. In addition to the high electrical conductivity of metals produced by the free electrons, the nondirectional nature of the metallic bond allows many different types of metal atoms to be dissolved in a host metal in varying amounts. The resulting *solid solutions*, or *alloys* (steel, bronze, brass, etc.), may be designed to have particular properties, such as tensile strength, ductility, electrical and thermal conductivity, and resistance to corrosion.

Because the bonding in metals is between all the electrons and all the positive ions, metals tend to bend when stressed. This bending is in contrast to nonmetallic solids, which tend to fracture when stressed. Fracturing results because bonding in nonmetallic solids is primarily with nearest-neighbor ions or atoms. When the distortion causes sufficient stress between some set of nearest neighbors, fracture occurs.

> The blue area represents the electron gas, and the red spheres represent the positive metal ions.

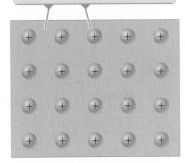

Figure 43.15 Highly schematic diagram of a metal.

43.4 Free-Electron Theory of Metals

In Section 27.3, we described a classical free-electron theory of electrical conduction in metals, a structural model that led to Ohm's law. According to this theory, a metal is modeled as a classical gas of conduction electrons moving through a fixed lattice of ions. Although this theory predicts the correct functional form of Ohm's law, it does not predict the correct values of electrical and thermal conductivities.

A quantum-based free-electron theory of metals remedies the shortcomings of the classical model by taking into account the wave nature of the electrons. In this model, based on the quantum particle under boundary conditions analysis model, the outer-shell electrons are free to move through the metal but are trapped within a three-dimensional box formed by the metal surfaces. Therefore, each electron is represented as a particle in a box. As discussed in Section 41.2, particles in a box are restricted to quantized energy levels.

Statistical physics can be applied to a collection of particles in an effort to relate microscopic properties to macroscopic properties as we saw with kinetic theory of gases in Chapter 21. In the case of electrons, it is necessary to use *quantum statistics*, with the requirement that each state of the system can be occupied by only two electrons (one with spin up and the other with spin down) as a consequence of the exclusion principle. The probability that a particular state having energy E is occupied by one of the electrons in a solid is

Fermi–Dirac distribution function ▶

$$f(E) = \frac{1}{e^{(E - E_F)/k_B T} + 1} \tag{43.19}$$

where $f(E)$ is called the **Fermi–Dirac distribution function** and E_F is called the **Fermi energy**. A plot of $f(E)$ versus E at $T = 0$ K is shown in Figure 43.16a. Notice that $f(E) = 1$ for $E < E_F$ and $f(E) = 0$ for $E > E_F$. That is, at 0 K, all states having energies less than the Fermi energy are occupied and all states having energies greater than the Fermi energy are vacant. A plot of $f(E)$ versus E at some temperature $T > 0$ K is shown in Figure 43.16b. This curve shows that as T increases, the distribution rounds off slightly. Because of thermal excitation, states near and below E_F lose population and states near and above E_F gain population. The Fermi energy E_F also depends on temperature, but the dependence is weak in metals.

Let's now follow up on our discussion of the particle in a box in Chapter 41 to generalize the results to a three-dimensional box. Recall that if a particle of mass m is confined to move in a one-dimensional box of length L, the allowed states have quantized energy levels given by Equation 41.14:

$$E_n = \left(\frac{h^2}{8mL^2}\right)n^2 = \left(\frac{\hbar^2 \pi^2}{2mL^2}\right)n^2 \quad n = 1, 2, 3, \ldots$$

Now imagine a piece of metal in the shape of a solid cube of sides L and volume L^3 and focus on one electron that is free to move anywhere in this volume. Therefore, the electron is modeled as a particle in a three-dimensional box. In this model, we require that $\psi(x, y, z) = 0$ at the boundaries of the metal. It can be shown (see Problem 37) that the energy for such an electron is

$$E = \frac{\hbar^2 \pi^2}{2m_e L^2}\left(n_x^2 + n_y^2 + n_z^2\right) \tag{43.20}$$

where m_e is the mass of the electron and n_x, n_y, and n_z are quantum numbers. As we expect, the energies are quantized, and each allowed value of the energy is characterized by this set of three quantum numbers (one for each degree of freedom) and the spin quantum number m_s. For example, the ground state, corresponding to

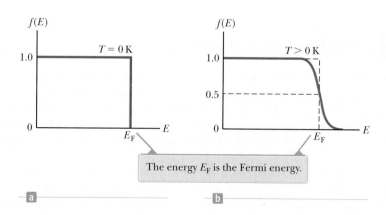

Figure 43.16 Plot of the Fermi–Dirac distribution function $f(E)$ versus energy at (a) $T = 0$ K and (b) $T > 0$ K.

The energy E_F is the Fermi energy.

$n_x = n_y = n_z = 1$, has an energy equal to $3\hbar^2\pi^2/2m_eL^2$ and can be occupied by two electrons, corresponding to spin up and spin down.

Because of the macroscopic size L of the box, the energy levels for the electrons are very close together. As a result, we can treat the quantum numbers as continuous variables. Under this assumption, the number of allowed states per unit volume that have energies between E and $E + dE$ is

$$g(E)\, dE = \frac{8\sqrt{2}\,\pi m_e^{3/2}}{h^3} E^{1/2}\, dE \qquad \textbf{(43.21)}$$

(See Example 43.5.) The function $g(E)$ is called the **density-of-states function.**

If a metal is in thermal equilibrium, the number of electrons per unit volume $N(E)\, dE$ that have energy between E and $E + dE$ is equal to the product of the number of allowed states per unit volume and the probability that a state is occupied; that is, $N(E)\, dE = g(E)f(E)\, dE$:

$$N(E)\, dE = \left(\frac{8\sqrt{2}\,\pi m_e^{3/2}}{h^3} E^{1/2}\right)\left(\frac{1}{e^{(E-E_F)/k_BT} + 1}\right) dE \qquad \textbf{(43.22)}$$

Plots of $N(E)$ versus E for two temperatures are given in Figure 43.17.

If n_e is the total number of electrons per unit volume, we require that

$$n_e = \int_0^\infty N(E)\, dE = \frac{8\sqrt{2}\,\pi m_e^{3/2}}{h^3}\int_0^\infty \frac{E^{1/2}\, dE}{e^{(E-E_F)/k_BT} + 1} \qquad \textbf{(43.23)}$$

We can use this condition to calculate the Fermi energy. At $T = 0$ K, the Fermi–Dirac distribution function $f(E) = 1$ for $E < E_F$ and $f(E) = 0$ for $E > E_F$. Therefore, at $T = 0$ K, Equation 43.23 becomes

$$n_e = \frac{8\sqrt{2}\,\pi m_e^{3/2}}{h^3}\int_0^{E_F} E^{1/2}\, dE = \tfrac{2}{3}\frac{8\sqrt{2}\,\pi m_e^{3/2}}{h^3} E_F^{3/2} \qquad \textbf{(43.24)}$$

Solving for the Fermi energy at 0 K gives

$$E_F(0) = \frac{h^2}{2m_e}\left(\frac{3n_e}{8\pi}\right)^{2/3} \qquad \textbf{(43.25)}$$

The Fermi energies for metals are in the range of a few electron volts. Representative values for various metals are given in Table 43.2. It is left as a problem (Problem 39) to show that the average energy of a free electron in a metal at 0 K is

$$E_{avg} = \tfrac{3}{5}E_F \qquad \textbf{(43.26)}$$

In summary, we can consider a metal to be a system comprising a very large number of energy levels available to the free electrons. These electrons fill the levels in accordance with the Pauli exclusion principle, beginning with $E = 0$ and ending with E_F. At $T = 0$ K, all levels below the Fermi energy are filled and all levels above the Fermi energy are empty. At 300 K, a small fraction of the free electrons are excited above the Fermi energy.

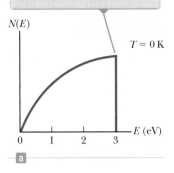

To provide a sense of scale, imagine that the Fermi energy E_F of the metal is 3 eV.

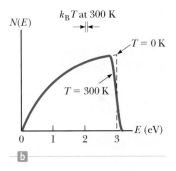

Figure 43.17 Plot of the electron distribution function versus energy in a metal at (a) $T = 0$ K and (b) $T = 300$ K.

◀ Fermi energy at $T = 0$ K

Table 43.2	Calculated Values of the Fermi Energy for Metals at 300 K Based on the Free-Electron Theory	
Metal	**Electron Concentration (m^{-3})**	**Fermi Energy (eV)**
Li	4.70×10^{28}	4.72
Na	2.65×10^{28}	3.23
K	1.40×10^{28}	2.12
Cu	8.46×10^{28}	7.05
Ag	5.85×10^{28}	5.48
Au	5.90×10^{28}	5.53

Example 43.4	The Fermi Energy of Gold

Each atom of gold (Au) contributes one free electron to the metal. Compute the Fermi energy for gold.

SOLUTION

Conceptualize Imagine electrons filling available levels at $T = 0$ K in gold until the solid is neutral. The highest energy filled is the Fermi energy.

Categorize We evaluate the result using a result from this section, so we categorize this example as a substitution problem.

Substitute the concentration of free electrons in gold from Table 43.2 into Equation 43.25 to calculate the Fermi energy at 0 K:

$$E_F(0) = \frac{(6.626 \times 10^{-34} \, \text{J} \cdot \text{s})^2}{2(9.11 \times 10^{-31} \, \text{kg})} \left[\frac{3(5.90 \times 10^{28} \, \text{m}^{-3})}{8\pi} \right]^{2/3}$$

$$= 8.85 \times 10^{-19} \, \text{J} = \boxed{5.53 \, \text{eV}}$$

Example 43.5	Deriving Equation 43.21

Based on the allowed states of a particle in a three-dimensional box, derive Equation 43.21.

SOLUTION

Conceptualize Imagine a particle confined to a three-dimensional box, subject to boundary conditions in three dimensions. Imagine also a three-dimensional *quantum number space* whose axes represent n_x, n_y, and n_z. The allowed states in this space can be repre- sented as dots located at integral values of the three quantum numbers as in Figure 43.18. This space is not traditional space in which a location is specified by coordinates x, y, and z; rather, it is a space in which allowed states can be specified by coordinates representing the quantum numbers. The number of allowed states having energies between E and $E + dE$ corresponds to the number of dots in the spherical shell of radius n and thickness dn.

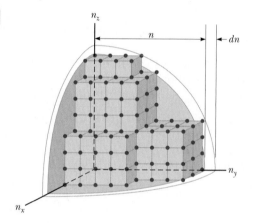

Figure 43.18 The dots rep- resenting the allowed states are located at integer values of n_x, n_y, and n_z and are therefore at the corners of cubes with sides of "length" 1.

Categorize We categorize this problem as that of a quantum system in which the energies of the particle are quantized. Furthermore, we can base the solution to the problem on our understanding of the particle in a one-dimensional box.

Analyze As noted previously, the allowed states of the particle in a three-dimensional box are described by three quan- tum numbers n_x, n_y, and n_z. For a macroscopic sample of metal, the number of allowed values of these quantum num- bers is tremendous, so on a macroscopic scale, the allowed states in the number space can be modeled as continuous.

Defining $E_0 = \hbar^2 \pi^2 / 2 m_e L^2$ and $n = (E/E_0)^{1/2}$, rewrite Equation 43.20:

$$(1) \quad n_x^2 + n_y^2 + n_z^2 = \frac{2 m_e L^2}{\hbar^2 \pi^2} E = \frac{E}{E_0} = n^2$$

In the quantum number space, Equation (1) is the equation of a sphere of radius n. Therefore, the number of allowed states having energies between E and $E + dE$ is equal to the number of points in a spherical shell of radius n and thick- ness dn.

Find the "volume" of this shell, which represents the total number of states $G(E) \, dE$:

$$(2) \quad G(E) \, dE = \tfrac{1}{8}(4\pi n^2 \, dn) = \tfrac{1}{2}\pi n^2 \, dn$$

We have taken one-eighth of the total volume because we are restricted to the octant of a three-dimensional space in which all three quantum numbers are positive.

Replace n in Equation (2) with its equivalent in terms of E using the relation $n^2 = E/E_0$ from Equation (1):

$$G(E) \, dE = \tfrac{1}{2}\pi \left(\frac{E}{E_0} \right) d\left[\left(\frac{E}{E_0} \right)^{1/2} \right] = \tfrac{1}{2}\pi \frac{E}{(E_0)^{3/2}} d[(E)^{1/2}]$$

▶ **43.5** c o n t i n u e d

Evaluate the differential:

$$G(E)\, dE = \tfrac{1}{2}\pi\left[\frac{E}{(E_0)^{3/2}}\right]\!\left(\tfrac{1}{2}E^{-1/2}\, dE\right) = \tfrac{1}{4}\pi E_0^{-3/2}E^{1/2}\, dE$$

Substitute for E_0 from its definition above:

$$G(E)\, dE = \tfrac{1}{4}\pi\left(\frac{\hbar^2\pi^2}{2m_eL^2}\right)^{-3/2}E^{1/2}\, dE$$

$$= \frac{\sqrt{2}}{2}\,\frac{m_e^{3/2}L^3}{\hbar^3\pi^2}\,E^{1/2}\, dE$$

Letting $g(E)$ represent the number of states per unit volume, where L^3 is the volume V of the cubical box in normal space, find $g(E) = G(E)/V$:

$$g(E)\, dE = \frac{G(E)}{V}\, dE = \frac{\sqrt{2}}{2}\,\frac{m_e^{3/2}}{\hbar^3\pi^2}\,E^{1/2}\, dE$$

Substitute $\hbar = h/2\pi$:

$$g(E)\, dE = \frac{4\sqrt{2}\,\pi m_e^{3/2}}{h^3}\,E^{1/2}\, dE$$

Multiply by 2 for the two possible spin states in each particle-in-a-box state:

$$g(E)\, dE = \frac{8\sqrt{2}\,\pi m_e^{3/2}}{h^3}\,E^{1/2}\, dE$$

Finalize This result is Equation 43.21, which is what we set out to derive.

43.5 Band Theory of Solids

In Section 43.4, the electrons in a metal were modeled as particles free to move around inside a three-dimensional box and we ignored the influence of the parent atoms. In this section, we make the model more sophisticated by incorporating the contribution of the parent atoms that form the crystal.

Recall from Section 41.1 that the probability density $|\psi|^2$ for a system is physically significant, but the probability amplitude ψ is not. Let's consider as an example an atom that has a single s electron outside of a closed shell. Both of the following wave functions are valid for such an atom with atomic number Z:

$$\psi_s^{+}(r) = +Af(r)e^{-Zr/na_0} \qquad \psi_s^{-}(r) = -Af(r)e^{-Zr/na_0}$$

where A is the normalization constant and $f(r)$ is a function[3] of r that varies with the value of n. Choosing either of these wave functions leads to the same value of $|\psi|^2$, so both choices are equivalent. A difference arises, however, when two atoms are combined.

If two identical atoms are very far apart, they do not interact and their electronic energy levels can be considered to be those of isolated atoms. Suppose the two atoms are sodium, each having a lone $3s$ electron that is in a well-defined quantum state. As the two sodium atoms are brought closer together, their wave functions begin to overlap as we discussed for covalent bonding in Section 43.1. The properties of the combined system differ depending on whether the two atoms are combined with wave functions $\psi_s^{+}(r)$ as in Figure 43.19a or whether they are combined with one having wave function $\psi_s^{+}(r)$ and the other $\psi_s^{-}(r)$ as in Figure 43.19b. The choice of two atoms with wave function $\psi_s^{-}(r)$ is physically equivalent to that with two positive wave functions, so we do not consider it separately. When two wave functions $\psi_s^{+}(r)$ are combined, the result is a composite wave function in which the probability amplitudes add between the atoms. If $\psi_s^{+}(r)$ combines with $\psi_s^{-}(r)$,

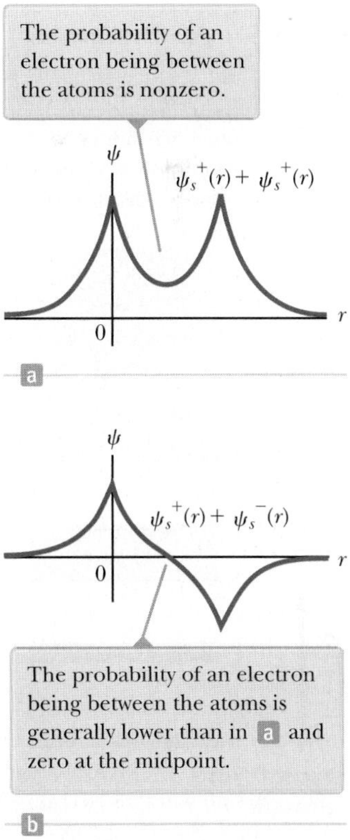

The probability of an electron being between the atoms is nonzero.

The probability of an electron being between the atoms is generally lower than in **a** and zero at the midpoint.

Figure 43.19 The wave functions of two atoms combine to form a composite wave function for the two-atom system when the atoms are close together. (a) Two atoms with wave functions $\psi_s^{+}(r)$ combine. (b) Two atoms with wave functions $\psi_s^{+}(r)$ and $\psi_s^{-}(r)$ combine.

[3]The functions $f(r)$ are called *Laguerre polynomials*. They can be found in the quantum treatment of the hydrogen atom in modern physics textbooks.

Figure 43.20 Energies of the 1s and 2s levels in sodium as a function of the separation distance *r* between atoms.

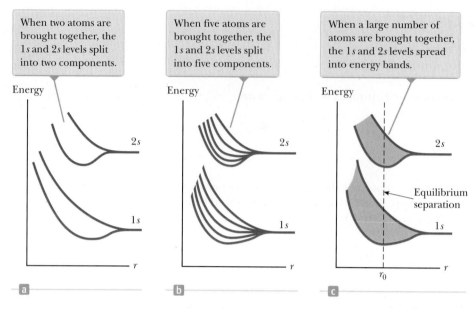

When two atoms are brought together, the 1s and 2s levels split into two components.

When five atoms are brought together, the 1s and 2s levels split into five components.

When a large number of atoms are brought together, the 1s and 2s levels spread into energy bands.

however, the wave functions between the nuclei subtract. Therefore, the composite probability amplitudes for the two possibilities are different. These two possible combinations of wave functions represent two possible states of the two-atom system. We interpret these curves as representing the probability amplitude of finding an electron. The positive–positive curve shows some probability of finding the electron at the midpoint between the atoms. The positive–negative function shows no such probability. A state with a high probability of an electron *between* two positive nuclei must have a different energy than a state with a high probability of the electron being elsewhere! Therefore, the states are *split* into two energy levels due to the two ways of combining the wave functions. The energy difference is relatively small, so the two states are close together on an energy scale.

Figure 43.20a shows this splitting effect as a function of separation distance. For large separations *r*, the electron clouds do not overlap and there is no splitting. As the atoms are brought closer so that *r* decreases, the electron clouds overlap and we need to consider the system of two atoms.

When a large number of atoms are brought together to form a solid, a similar phenomenon occurs. The individual wave functions can be brought together in various combinations of $\psi_s^{+}(r)$ and $\psi_s^{-}(r)$, each possible combination corresponding to a different energy. As the atoms are brought close together, the various isolated-atom energy levels split into multiple energy levels for the composite system. This splitting in levels for five atoms in close proximity is shown in Figure 43.20b. In this case, there are five energy levels corresponding to five different combinations of isolated-atom wave functions.

As the number of atoms grows, the number of combinations of wave functions grows, as does the number of possible energies. If we extend this argument to the large number of atoms found in solids (on the order of 10^{23} atoms per cubic centimeter), we obtain a huge number of levels of varying energy so closely spaced that they may be regarded as a continuous **band** of energy levels as shown in Figure 43.20c. In the case of sodium, it is customary to refer to the continuous distributions of allowed energy levels as *s* bands because the bands originate from the *s* levels of the individual sodium atoms.

Each energy level in the atom can spread into a band when the atoms are combined into a solid. Figure 43.21 shows the allowed energy bands of sodium at a fixed separation distance between the atoms. Notice that energy gaps, corresponding to *forbidden energies,* occur between the allowed bands. In addition, some bands exhibit sufficient spreading in energy that there is an overlap between bands arising from different quantum states (3s and 3p).

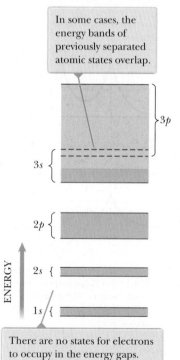

In some cases, the energy bands of previously separated atomic states overlap.

There are no states for electrons to occupy in the energy gaps.

Figure 43.21 Energy bands of a sodium crystal. Blue represents energy bands occupied by the sodium electrons when the atom is in its ground state. Gold represents energy bands that are empty.

As indicated by the blue-shaded areas in Figure 43.21, the 1s, 2s, and 2p bands of sodium are each full of electrons because the 1s, 2s, and 2p states of each atom are full. An energy level in which the orbital angular momentum is ℓ can hold $2(2\ell + 1)$ electrons. The factor 2 arises from the two possible electron spin orientations, and the factor $2\ell + 1$ corresponds to the number of possible orientations of the orbital angular momentum. The capacity of each band for a system of N atoms is $2(2\ell + 1)N$ electrons. Therefore, the 1s and 2s bands each contain 2N electrons ($\ell = 0$), and the 2p band contains 6N electrons ($\ell = 1$). Because sodium has only one 3s electron and there are a total of N atoms in the solid, the 3s band contains only N electrons and is partially full as indicated by the blue coloring in Figure 43.21. The 3p band, which is the higher region of the overlapping bands, is completely empty (all gold in the figure).

Band theory allows us to build simple models to understand the behavior of conductors, insulators, and semiconductors as well as that of semiconductor devices, as we shall discuss in the following sections.

43.6 Electrical Conduction in Metals, Insulators, and Semiconductors

Good electrical conductors contain a high density of free charge carriers, and the density of free charge carriers in insulators is nearly zero. Semiconductors, first introduced in Section 23.2, are a class of technologically important materials in which charge-carrier densities are intermediate between those of insulators and those of conductors. In this section, we discuss the mechanisms of conduction in these three classes of materials in terms of a model based on energy bands.

Metals

If a material is to be a good electrical conductor, the charge carriers in the material must be free to move in response to an applied electric field. Let's consider the electrons in a metal as the charge carriers. The motion of the electrons in response to an electric field represents an increase in energy of the system (the metal lattice and the free electrons) corresponding to the additional kinetic energy of the moving electrons. The system is described by the nonisolated system model for energy. Equation 8.2 becomes $W = \Delta K$, where the work is done on the electrons by the electric field. Therefore, when an electric field is applied to a conductor, electrons must move upward to an available higher energy state on an energy-level diagram to represent the additional kinetic energy.

Figure 43.22 shows a half-filled band in a metal at $T = 0$ K, where the blue region represents levels filled with electrons. Because electrons obey Fermi–Dirac statistics, all levels below the Fermi energy are filled with electrons and all levels above the Fermi energy are empty. The Fermi energy lies in the band at the highest filled state. At temperatures slightly greater than 0 K, some electrons are thermally excited to levels above E_F, but overall there is little change from the 0 K case. If a potential difference is applied to the metal, however, electrons having energies near the Fermi energy require only a small amount of additional energy from the applied electric field to reach nearby empty energy states above the Fermi energy. Therefore, electrons in a metal experiencing only a weak applied electric field are free to move because many empty levels are available close to the occupied energy levels. The model of metals based on band theory demonstrates that metals are excellent electrical conductors.

Insulators

Now consider the two outermost energy bands of a material in which the lower band is filled with electrons and the higher band is empty at 0 K (Fig. 43.23). The

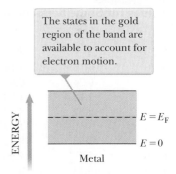

The states in the gold region of the band are available to account for electron motion.

$E = E_F$

$E = 0$

Metal

Figure 43.22 Half-filled band of a metal, an electrical conductor. At $T = 0$ K, the Fermi energy lies in the middle of the band.

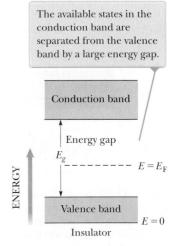

The available states in the conduction band are separated from the valence band by a large energy gap.

Conduction band

Energy gap

E_g

$E = E_F$

Valence band

$E = 0$

Insulator

Figure 43.23 An electrical insulator at $T = 0$ K has a filled valence band and an empty conduction band. The Fermi level lies somewhere between these bands in the region known as the energy gap.

lower, filled band is called the **valence band,** and the upper, empty band is the **conduction band.** (The conduction band is the one that is partially filled in a metal.) It is common to refer to the energy separation between the valence and conduction bands as the **energy gap** E_g of the material. The Fermi energy lies somewhere in the energy gap[4] as shown in Figure 43.23.

Suppose a material has a relatively large energy gap of, for example, approximately 5 eV. At 300 K (room temperature), $k_BT = 0.025$ eV, which is much smaller than the energy gap. At such temperatures, the Fermi–Dirac distribution predicts that very few electrons are thermally excited into the conduction band. There are no available states that lie close in energy above the valence band and into which electrons can move upward to account for the extra kinetic energy associated with motion through the material in response to an electric field. Consequently, the electrons do not move; the material is an insulator. Although an insulator has many vacant states in its conduction band that can accept electrons, these states are separated from the filled states by a large energy gap. Only a few electrons occupy these states, so the overall electrical conductivity of insulators is very small.

Semiconductors

Semiconductors have the same type of band structure as an insulator, but the energy gap is much smaller, on the order of 1 eV. Table 43.3 shows the energy gaps for some representative materials. The band structure of a semiconductor is shown in Figure 43.24. Because the Fermi level is located near the middle of the gap for a semiconductor and E_g is small, appreciable numbers of electrons are thermally excited from the valence band to the conduction band. Because of the many empty levels above the thermally filled levels in the conduction band, a small applied potential difference can easily raise the electrons in the conduction band into available energy states, resulting in a moderate current.

At $T = 0$ K, all electrons in these materials are in the valence band and no energy is available to excite them across the energy gap. Therefore, semiconductors are poor conductors at very low temperatures. Because the thermal excitation of electrons across the narrow gap is more probable at higher temperatures, the conductivity of semiconductors increases rapidly with temperature, contrasting sharply with the conductivity of metals, which decreases slowly with increasing temperature.

Charge carriers in a semiconductor can be negative, positive, or both. When an electron moves from the valence band into the conduction band, it leaves behind a vacant site, called a **hole,** in the otherwise filled valence band. This hole (electron-deficient site) acts as a charge carrier in the sense that a free electron from a nearby site can transfer into the hole. Whenever an electron does so, it creates a new hole at the site it abandoned. Therefore, the net effect can be viewed as the hole migrating through the material in the direction opposite the direction of electron movement. The hole behaves as if it were a particle with a positive charge $+e$.

A pure semiconductor crystal containing only one element or one compound is called an **intrinsic semiconductor.** In these semiconductors, there are equal numbers of conduction electrons and holes. Such combinations of charges are called **electron–hole pairs.** In the presence of an external electric field, the holes move in the direction of the field and the conduction electrons move in the direction opposite the field (Fig. 43.25). Because the electrons and holes have opposite signs, both motions correspond to a current in the same direction.

Table 43.3 **Energy-Gap Values for Some Semiconductors**

Crystal	E_g (eV) 0 K	E_g (eV) 300 K
Si	1.17	1.14
Ge	0.74	0.67
InP	1.42	1.34
GaP	2.32	2.26
GaAs	1.52	1.42
CdS	2.58	2.42
CdTe	1.61	1.56
ZnO	3.44	3.2
ZnS	3.91	3.6

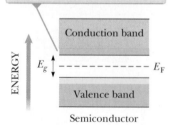

The small energy gap allows electrons to be thermally excited into the conduction band.

Figure 43.24 Band structure of a semiconductor at ordinary temperatures ($T \approx 300$ K). The energy gap is much smaller than in an insulator.

---

[4]We defined the Fermi energy as the energy of the highest filled state at $T = 0$, which might suggest that the Fermi energy should be at the top of the valence band in Figure 43.23. A more sophisticated general treatment of the Fermi energy, however, shows that it is located at that energy at which the probability of occupation is one-half (see Fig. 43.16b). According to this definition, the Fermi energy lies in the energy gap between the bands.

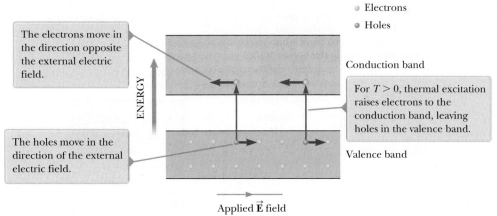

● Electrons
● Holes

The electrons move in the direction opposite the external electric field.

For $T > 0$, thermal excitation raises electrons to the conduction band, leaving holes in the valence band.

The holes move in the direction of the external electric field.

Conduction band

Valence band

Applied $\vec{E}$ field

Figure 43.25 Movement of charges (holes and electrons) in an intrinsic semiconductor.

Quick Quiz 43.4 Consider the data on three materials given in the table.

Material	Conduction Band	E_g
A	Empty	1.2 eV
B	Half full	1.2 eV
C	Empty	8.0 eV

Identify each material as a conductor, an insulator, or a semiconductor.

Doped Semiconductors

When impurities are added to a semiconductor, both the band structure of the semiconductor and its resistivity are modified. The process of adding impurities, called **doping,** is important in controlling the conductivity of semiconductors. For example, when an atom containing five outer-shell electrons, such as arsenic, is added to a Group IV semiconductor, four of the electrons form covalent bonds with atoms of the semiconductor and one is left over (Fig. 43.26a). This extra electron is nearly free of its parent atom and can be modeled as having an energy level that lies in the energy gap, immediately below the conduction band (Fig. 43.26b). Such a pentavalent atom in effect donates an electron to the structure and hence is referred to as a **donor atom.** Because the spacing between the energy level of the electron of the donor atom and the bottom of the conduction band is very

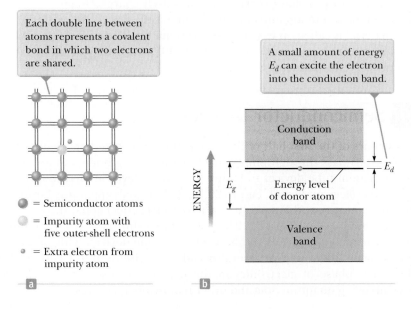

Each double line between atoms represents a covalent bond in which two electrons are shared.

A small amount of energy E_d can excite the electron into the conduction band.

Conduction band

E_g

E_d

Energy level of donor atom

Valence band

ENERGY

● = Semiconductor atoms

● = Impurity atom with five outer-shell electrons

● = Extra electron from impurity atom

a

b

Figure 43.26 (a) Two-dimensional representation of a semiconductor consisting of Group IV atoms (gray) and an impurity atom (yellow) that has five outer-shell electrons. (b) Energy-band diagram for a semiconductor in which the nearly free electron of the impurity atom lies in the energy gap, immediately below the bottom of the conduction band.

Figure 43.27 (a) Two-dimensional representation of a semiconductor consisting of Group IV atoms (gray) and an impurity atom (yellow) having three outer-shell electrons. (b) Energy-band diagram for a semiconductor in which the energy level associated with the trivalent impurity atom lies in the energy gap, immediately above the top of the valence band.

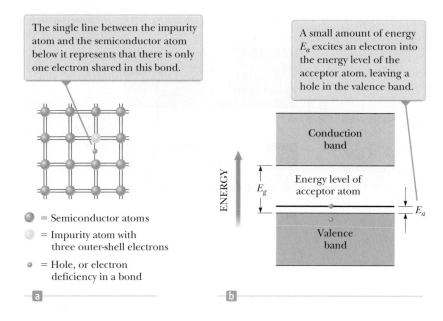

The single line between the impurity atom and the semiconductor atom below it represents that there is only one electron shared in this bond.

A small amount of energy E_a excites an electron into the energy level of the acceptor atom, leaving a hole in the valence band.

= Semiconductor atoms

= Impurity atom with three outer-shell electrons

= Hole, or electron deficiency in a bond

small (typically, approximately 0.05 eV), only a small amount of thermal excitation is needed to cause this electron to move into the conduction band. (Recall that the average energy of an electron at room temperature is approximately $k_B T \approx$ 0.025 eV.) Semiconductors doped with donor atoms are called **n-type semiconductors** because the majority of charge carriers are electrons, which are *n*egatively charged.

If a Group IV semiconductor is doped with atoms containing three outer-shell electrons, such as indium and aluminum, the three electrons form covalent bonds with neighboring semiconductor atoms, leaving an electron deficiency—a hole—where the fourth bond would be if an impurity-atom electron were available to form it (Fig. 43.27a). This situation can be modeled by placing an energy level in the energy gap, immediately above the valence band, as in Figure 43.27b. An electron from the valence band has enough energy at room temperature to fill this impurity level, leaving behind a hole in the valence band. This hole can carry current in the presence of an electric field. Because a trivalent atom accepts an electron from the valence band, such impurities are referred to as **acceptor atoms.** A semiconductor doped with trivalent (acceptor) impurities is known as a **p-type semiconductor** because the majority of charge carriers are *p*ositively charged holes.

When conduction in a semiconductor is the result of acceptor or donor impurities, the material is called an **extrinsic semiconductor.** The typical range of doping densities for extrinsic semiconductors is 10^{13} to 10^{19} cm^{-3}, whereas the electron density in a typical semiconductor is roughly 10^{21} cm^{-3}.

43.7 Semiconductor Devices

The electronics of the first half of the 20th century was based on vacuum tubes, in which electrons pass through empty space between a cathode and an anode. We have seen vacuum tube devices in Figure 29.6 (the television picture tube), Figure 29.10 (circular electron beam), Figure 29.15a (Thomson's apparatus for measuring e/m_e for the electron), and Figure 40.9 (photoelectric effect apparatus).

The transistor was invented in 1948, leading to a shift away from vacuum tubes and toward semiconductors as the basis of electronic devices. This phase of electronics has been under way for several decades. As discussed in Chapter 41, there may be a new phase of electronics in the near future using nanotechnological devices employing quantum dots and other nanoscale structures.

In this section, we discuss electronic devices based on semiconductors, which are still in wide use and will be for many years to come.

The Junction Diode

A fundamental unit of a semiconductor device is formed when a *p*-type semiconductor is joined to an *n*-type semiconductor to form a **p–n junction.** A **junction diode** is a device that is based on a single *p–n* junction. The role of a diode of any type is to pass current in one direction but not the other. Therefore, it acts as a one-way valve for current.

The *p–n* junction shown in Figure 43.28a consists of three distinct regions: a *p* region, an *n* region, and a small area that extends several micrometers to either side of the interface, called a *depletion region.*

The depletion region may be visualized as arising when the two halves of the junction are brought together. The mobile *n*-side donor electrons nearest the junction (deep-blue area in Fig. 43.28a) diffuse to the *p* side and fill holes located there, leaving behind immobile positive ions. While this process occurs, we can model the holes that are being filled as diffusing to the *n* side, leaving behind a region (brown area in Fig. 43.28a) of fixed negative ions.

Because the two sides of the depletion region each carry a net charge, an internal electric field on the order of 10^4 to 10^6 V/cm exists in the depletion region (see Fig. 43.28b). This field produces an electric force on any remaining mobile charge carriers that sweeps them out of the depletion region, so named because it is a region depleted of mobile charge carriers. This internal electric field creates an internal potential difference ΔV_0 that prevents further diffusion of holes and electrons across the junction and thereby ensures zero current in the junction when no potential difference is applied.

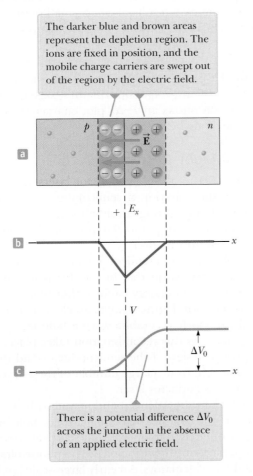

The darker blue and brown areas represent the depletion region. The ions are fixed in position, and the mobile charge carriers are swept out of the region by the electric field.

There is a potential difference ΔV_0 across the junction in the absence of an applied electric field.

Figure 43.28 (a) Physical arrangement of a *p–n* junction. (b) Component E_x of the internal electric field versus *x* for the *p–n* junction. (c) Internal electric potential difference ΔV versus *x* for the *p–n* junction.

Figure 43.29 (a) A p–n junction under forward bias. The middle diagram shows the potentials applied at the ends of the junction. Below that is a circuit diagram showing a battery with an adjustable voltage. The upper diagram shows how the potential varies across the junction. The dashed line shows the potential difference across the unbiased junction. (b) When the battery is reversed and the p–n junction is under reverse bias, the current is very small. (c) The characteristic curve for a real p–n junction.

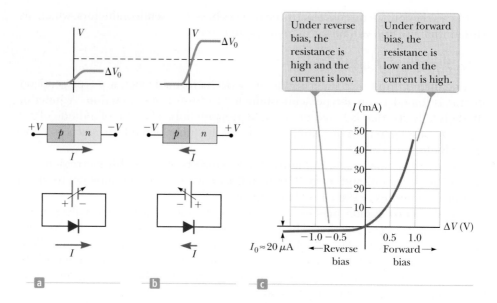

Under reverse bias, the resistance is high and the current is low.

Under forward bias, the resistance is low and the current is high.

$I_0 \approx 20\ \mu A$

The operation of the junction as a diode is easiest to understand in terms of the potential difference graph shown in Figure 43.28c. If a voltage ΔV is applied to the junction such that the p side is connected to the positive terminal of a voltage source as shown in Figure 43.29a, the internal potential difference ΔV_0 across the junction decreases as shown at the top of the figure; the decrease results in a current that increases exponentially with increasing forward voltage, or *forward bias*. For *reverse bias* (where the n side of the junction is connected to the positive terminal of a voltage source), the internal potential difference ΔV_0 increases with increasing reverse bias as in Figure 43.29b; the increase results in a very small reverse current that quickly reaches a saturation value I_0. The current–voltage relationship for an ideal diode is

$$I = I_0 \left(e^{e\,\Delta V / k_B T} - 1 \right) \qquad (43.27)$$

where the first e is the base of the natural logarithm, the second e represents the magnitude of the electron charge, k_B is Boltzmann's constant, and T is the absolute temperature. Figure 43.29c shows an I–ΔV plot characteristic of a real p–n junction, demonstrating the diode behavior.

Light-Emitting and Light-Absorbing Diodes

Light-emitting diodes (LEDs) and semiconductor lasers are common examples of devices that depend on the behavior of semiconductors. LEDs are used in LCD television displays, household lighting, flashlights, and camera flash units. Semiconductor lasers are often used for pointers in presentations and in playback equipment for digitally recorded information.

Light emission and absorption in semiconductors is similar to light emission and absorption by gaseous atoms except that in the discussion of semiconductors we must incorporate the concept of energy bands rather than the discrete energy levels in single atoms. As shown in Figure 43.30a, an electron excited electrically into the conduction band can easily recombine with a hole (especially if the electron is injected into a p region). As this recombination takes place, a photon of energy E_g is emitted. With proper design of the semiconductor and the associated plastic envelope or mirrors, the light from a large number of these transitions serves as the source of an LED or a semiconductor laser.

Conversely, an electron in the valence band may absorb an incoming photon of light and be promoted to the conduction band, leaving a hole behind (Fig. 43.30b). This absorbed energy can be used to operate an electrical circuit.

One device that operates on this principle is the **photovoltaic solar cell,** which appears in many handheld calculators. An early large-scale application of arrays of

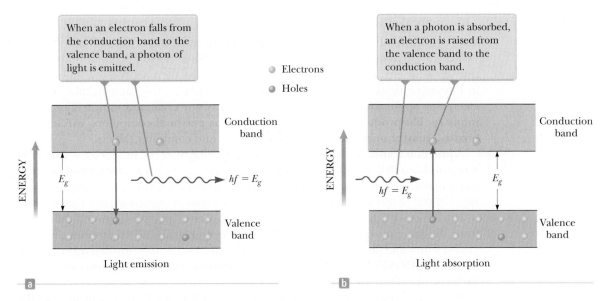

Figure 43.30 (a) Light emission from a semiconductor. (b) Light absorption by a semiconductor.

photovoltaic cells is the energy supply for orbiting spacecraft. The solar panels of the Hubble Space Telescope can be seen in the chapter-opening photograph for Chapter 38 on page 1160.

During the early years of the current century, application of photovoltaics for ground-based generation of electricity has been one of the world's fastest-growing energy technologies. At the time of this printing, the global generation of energy by means of photovoltaics is over 70 GW. A homeowner can install arrays of photovoltaic panels on the roof of his or her house and generate enough energy to operate the home as well as feed excess energy back into the electrical grid. Several photovoltaic power plants have recently been completed, including the Agua Caliente Solar Project in Arizona (200 MW completed in 2012, and 397 MW projected at completion), the Golmud Solar Park in China (200 MW), and the Charanka Solar Park in India (214 MW completed in 2012, and 500 MW projected at completion), the latter of which will be one location in the Gujarat Solar Park, a collection of several sites that is hoped to eventually supply close to 1 GW of power.

Example 43.6 | **Where's the Remote?**

Estimate the band gap of the semiconductor in the infrared LED of a typical television remote control.

SOLUTION

Conceptualize Imagine electrons in Figure 43.30a falling from the conduction band to the valence band, emitting infrared photons in the process.

Categorize We use concepts discussed in this section, so we categorize this example as a substitution problem.

In Chapter 34, we learned that the wavelength of infrared light ranges from 700 nm to 1 mm. Let's pick a number that is easy to work with, such as 1 000 nm (which is not a bad estimate because remote controls typically operate in the range of 880 to 950 nm.)

Estimate the energy hf of the photons from the remote control:

$$E = hf = \frac{hc}{\lambda} = \frac{1\,240 \text{ eV} \cdot \text{nm}}{1\,000 \text{ nm}} = \boxed{1.2 \text{ eV}}$$

This value corresponds to an energy gap E_g of approximately 1.2 eV in the LED's semiconductor.

The Transistor

The invention of the transistor by John Bardeen (1908–1991), Walter Brattain (1902–1987), and William Shockley (1910–1989) in 1948 totally revolutionized the world of electronics. For this work, these three men shared the Nobel Prize in Physics in 1956. By 1960, the transistor had replaced the vacuum tube in many electronic applications. The advent of the transistor created a multitrillion-dollar industry that produces such popular devices as personal computers, wireless keyboards, smartphones, electronic book readers, and computer tablets.

A **junction transistor** consists of a semiconducting material in which a very narrow *n* region is sandwiched between two *p* regions or a *p* region is sandwiched between two *n* regions. In either case, the transistor is formed from two *p–n* junctions. These types of transistors were used widely in the early days of semiconductor electronics.

During the 1960s, the electronics industry converted many electronic applications from the junction transistor to the **field-effect transistor,** which is much easier to manufacture and just as effective. Figure 43.31a shows the structure of a very common device, the **MOSFET,** or **metal-oxide-semiconductor field-effect transistor.** You are likely using millions of MOSFET devices when you are working on your computer.

There are three metal connections (the M in MOSFET) to the transistor: the *source, drain,* and *gate.* The source and drain are connected to *n*-type semiconductor regions (the S in MOSFET) at either end of the structure. These regions are connected by a narrow channel of additional *n*-type material, the *n* channel. The source and drain regions and the *n* channel are embedded in a *p*-type substrate material, which forms a depletion region, as in the junction diode, along the bottom of the *n* channel. (Depletion regions also exist at the junctions underneath the source and drain regions, but we will ignore them because the operation of the device depends primarily on the behavior in the channel.)

The gate is separated from the *n* channel by a layer of insulating silicon dioxide (the O in MOSFET, for oxide). Therefore, it does not make electrical contact with the rest of the semiconducting material.

Imagine that a voltage source ΔV_{SD} is applied across the source and drain as shown in Figure 43.31b. In this situation, electrons flow through the upper region of the *n* channel. Electrons cannot flow through the depletion region in the lower part of the *n* channel because this region is depleted of charge carriers. Now a second voltage ΔV_{SG} is applied across the source and gate as in Figure 43.31c. The positive potential on the gate electrode results in an electric field below the gate that is directed downward in the *n* channel (the field in "field-effect"). This electric field exerts upward forces on electrons in the region below the gate, causing them to move into the *n* channel. Consequently, the depletion region becomes smaller,

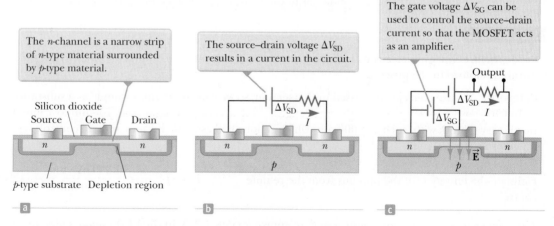

Figure 43.31   (a) The structure of a metal-oxide-semiconductor field-effect transistor (MOSFET). (b) A source–drain voltage is applied. (c) A gate voltage is applied.

widening the area through which there is current between the top of the n channel and the depletion region. As the area becomes wider, the current increases.

If a varying voltage, such as that generated from music stored in the memory of a smartphone, is applied to the gate, the area through which the source–drain current exists varies in size according to the varying gate voltage. A small variation in gate voltage results in a large variation in current and a correspondingly large voltage across the resistor in Figure 43.31c. Therefore, the MOSFET acts as a voltage amplifier. A circuit consisting of a chain of such transistors can result in a very small initial signal from a microphone being amplified enough to drive powerful speakers at an outdoor concert.

The Integrated Circuit

Invented independently by Jack Kilby (1923–2005, Nobel Prize in Physics, 2000) at Texas Instruments in late 1958 and by Robert Noyce (1927–1990) at Fairchild Camera and Instrument in early 1959, the integrated circuit has been justly called "the most remarkable technology ever to hit mankind." Kilby's first device is shown in Figure 43.32. Integrated circuits have indeed started a "second industrial revolution" and are found at the heart of computers, watches, cameras, automobiles, aircraft, robots, space vehicles, and all sorts of communication and switching networks.

In simplest terms, an **integrated circuit** is a collection of interconnected transistors, diodes, resistors, and capacitors fabricated on a single piece of silicon known as a *chip*. Contemporary electronic devices often contain many integrated circuits (Fig. 43.33). State-of-the-art chips easily contain several million components within a 1-cm² area, and the number of components per square inch has increased steadily since the integrated circuit was invented. The dramatic advances in chip technology can be seen by looking at microchips manufactured by Intel. The 4004 chip, introduced in 1971, contained 2 300 transistors. This number increased to 3.2 million 24 years later in 1995 with the Pentium processor. Sixteen years later, the Core i7 Sandy Bridge processor introduced in November 2011 contained 2 270 million transistors.

Integrated circuits were invented partly to solve the interconnection problem spawned by the transistor. In the era of vacuum tubes, power and size considerations of individual components set modest limits on the number of components that could be interconnected in a given circuit. With the advent of the tiny, low-power, highly reliable transistor, design limits on the number of components disappeared and were replaced by the problem of wiring together hundreds of thousands of components. The magnitude of this problem can be appreciated when we consider that second-generation computers (consisting of discrete transistors rather than integrated circuits) contained several hundred thousand components requiring more than a million joints that had to be hand-soldered and tested.

In addition to solving the interconnection problem, integrated circuits possess the advantages of miniaturization and fast response, two attributes critical for high-speed computers. Because the response time of a circuit depends on the time

Figure 43.32 Jack Kilby's first integrated circuit, tested on September 12, 1958.

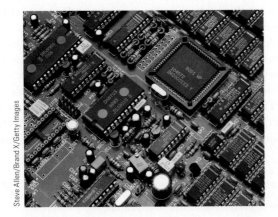

Figure 43.33 Integrated circuits are prevalent in many electronic devices. All the flat circuit elements with black-topped surfaces in this photograph are integrated circuits.

interval required for electrical signals traveling at the speed of light to pass from one component to another, miniaturization and close packing of components result in fast response times.

43.8 Superconductivity

We learned in Section 27.5 that there is a class of metals and compounds known as **superconductors** whose electrical resistance decreases to virtually zero below a certain temperature T_c called the *critical temperature* (Table 27.3). Let's now look at these amazing materials in greater detail, using what we know about the properties of solids to help us understand the behavior of superconductors.

Let's start by examining the Meissner effect, introduced in Section 30.6 as the exclusion of magnetic flux from the interior of superconductors. The Meissner effect is illustrated in Figure 43.34 for a superconducting material in the shape of a long cylinder. Notice that the magnetic field penetrates the cylinder when its temperature is greater than T_c (Fig. 43.34a). As the temperature is lowered to below T_c, however, the field lines are spontaneously expelled from the interior of the superconductor (Fig. 43.34b). Therefore, a superconductor is more than a perfect conductor (resistivity $\rho = 0$); it is also a perfect diamagnet ($\vec{B} = 0$). The property that $\vec{B} = 0$ in the interior of a superconductor is as fundamental as the property of zero resistance. If the magnitude of the applied magnetic field exceeds a critical value B_c, defined as the value of B that destroys a material's superconducting properties, the field again penetrates the sample.

Because a superconductor is a perfect diamagnet, it repels a permanent magnet. In fact, one can perform a demonstration of the Meissner effect by floating a small permanent magnet above a superconductor and achieving magnetic levitation as seen in Figure 30.27 in Section 30.6.

Recall from our study of electricity that a good conductor expels static electric fields by moving charges to its surface. In effect, the surface charges produce an electric field that exactly cancels the externally applied field inside the conductor. In a similar manner, a superconductor expels magnetic fields by forming surface currents. To see why that happens, consider again the superconductor shown in Figure 43.34. Let's assume the sample is initially at a temperature $T > T_c$ as illustrated in Figure 43.34a so that the magnetic field penetrates the cylinder. As the cylinder is cooled to a temperature $T < T_c$, the field is expelled as shown in Figure 43.34b. Surface currents induced on the superconductor's surface produce a magnetic field that exactly cancels the externally applied field inside the superconductor. As you would expect, the surface currents disappear when the external magnetic field is removed.

A successful theory for superconductivity in metals was published in 1957 by John Bardeen, L. N. Cooper (b. 1930), and J. R. Schrieffer (b. 1931); it is generally called BCS theory, based on the first letters of their last names. This theory led to a Nobel Prize in Physics for the three scientists in 1972. In this theory, two electrons can interact via distortions in the array of lattice ions so that there is a net attractive force between the electrons.[5] As a result, the two electrons are bound into an entity called a *Cooper pair*, which behaves like a particle with integral spin. Particles with integral spin are called *bosons*. (As noted in Pitfall Prevention 42.6, *fermions* make up another class of particles, those with half-integral spin.) An important feature of bosons is that they do not obey the Pauli exclusion principle. Consequently, at very low temperatures, it is possible for all bosons in a collection of such particles to be in the lowest quantum state. The entire collection of Cooper pairs in the metal is described by a single wave function. Above the energy level associated with this wave function is an energy gap equal to the binding energy of a Cooper pair. Under

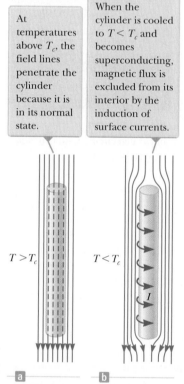

At temperatures above T_c, the field lines penetrate the cylinder because it is in its normal state.

When the cylinder is cooled to $T < T_c$ and becomes superconducting, magnetic flux is excluded from its interior by the induction of surface currents.

$T > T_c$

$T < T_c$

I

a   **b**

Figure 43.34 A superconductor in the form of a long cylinder in the presence of an external magnetic field.

---

[5]A highly simplified explanation of this attraction between electrons is as follows. The attractive Coulomb force between one electron and the surrounding positively charged lattice ions causes the ions to move inward slightly toward the electron. As a result, there is a higher concentration of positive charge in this region than elsewhere in the lattice. A second electron is attracted to the higher concentration of positive charge.

the action of an applied electric field, the Cooper pairs experience an electric force and move through the metal. A random scattering event of a Cooper pair from a lattice ion would represent resistance to the electric current. Such a collision would change the energy of the Cooper pair because some energy would be transferred to the lattice ion. There are no available energy levels below that of the Cooper pair (it is already in the lowest state), however, and none available above because of the energy gap. As a result, collisions do not occur and there is no resistance to the movement of Cooper pairs.

An important development in physics that elicited much excitement in the scientific community was the discovery of high-temperature copper oxide-based superconductors. The excitement began with a 1986 publication by J. Georg Bednorz (b. 1950) and K. Alex Müller (b. 1927), scientists at the IBM Zurich Research Laboratory in Switzerland. In their seminal paper,[6] Bednorz and Müller reported strong evidence for superconductivity at 30 K in an oxide of barium, lanthanum, and copper. They were awarded the Nobel Prize in Physics in 1987 for their remarkable discovery. Shortly thereafter, a new family of compounds was open for investigation and research activity in the field of superconductivity proceeded vigorously. In early 1987, groups at the University of Alabama at Huntsville and the University of Houston announced superconductivity at approximately 92 K in an oxide of yttrium, barium, and copper ($YBa_2Cu_3O_7$). Later that year, teams of scientists from Japan and the United States reported superconductivity at 105 K in an oxide of bismuth, strontium, calcium, and copper. Superconductivity at temperatures as high as 150 K have been reported in an oxide containing mercury. In 2006, Japanese scientists discovered superconductivity for the first time in iron-based materials, beginning with LaFePO, with a critical temperature of 4 K. The highest critical temperature that has been reported so far in the iron-based materials is 55 K, a milestone held by fluorine-doped SmFeAsO. These newly discovered materials have rejuvenated the field of high-T_c superconductivity. Today, one cannot rule out the possibility of room-temperature superconductivity, and the mechanisms responsible for the behavior of high-temperature superconductors are still under investigation. The search for novel superconducting materials continues both for scientific reasons and because practical applications become more probable and widespread as the critical temperature is raised.

Although BCS theory was very successful in explaining superconductivity in metals, there is currently no widely accepted theory for high-temperature superconductivity. It remains an area of active research.

Summary

Concepts and Principles

Two or more atoms combine to form molecules because of a net attractive force between the atoms. The mechanisms responsible for molecular bonding can be classified as follows:

- **Ionic bonds** form primarily because of the Coulomb attraction between oppositely charged ions. Sodium chloride (NaCl) is one example.
- **Covalent bonds** form when the constituent atoms of a molecule share electrons. For example, the two electrons of the H_2 molecule are equally shared between the two nuclei.
- **Van der Waals bonds** are weak electrostatic bonds between molecules or between atoms that do not form ionic or covalent bonds. These bonds are responsible for the condensation of noble gas atoms and nonpolar molecules into the liquid phase.
- **Hydrogen bonds** form between the center of positive charge in a polar molecule that includes one or more hydrogen atoms and the center of negative charge in another polar molecule.

continued

[6]J. G. Bednorz and K. A. Müller, *Z. Phys. B* **64:**189, 1986.

The allowed values of the rotational energy of a diatomic molecule are

$$E_{rot} = E_J = \frac{\hbar^2}{2I} J(J+1) \quad J = 0, 1, 2, \ldots \quad \textbf{(43.6)}$$

where I is the moment of inertia of the molecule and J is an integer called the **rotational quantum number.** The selection rule for transitions between rotational states is $\Delta J = \pm 1$.

The allowed values of the vibrational energy of a diatomic molecule are

$$E_{vib} = \left(v + \tfrac{1}{2}\right) \frac{h}{2\pi} \sqrt{\frac{k}{\mu}} \quad v = 0, 1, 2, \ldots \quad \textbf{(43.10)}$$

where v is the **vibrational quantum number,** k is the force constant of the "effective spring" bonding the molecule, and μ is the **reduced mass** of the molecule. The selection rule for allowed vibrational transitions is $\Delta v = \pm 1$, and the energy difference between any two adjacent levels is the same, regardless of which two levels are involved.

Bonding mechanisms in solids can be classified in a manner similar to the schemes for molecules. For example, the Na$^+$ and Cl$^-$ ions in NaCl form **ionic bonds,** whereas the carbon atoms in diamond form **covalent bonds.** The **metallic bond** is characterized by a net attractive force between positive ion cores and the mobile free electrons of a metal.

In the **free-electron theory of metals,** the free electrons fill the quantized levels in accordance with the Pauli exclusion principle. The number of states per unit volume available to the conduction electrons having energies between E and $E + dE$ is

$$N(E)\, dE = \left(\frac{8\sqrt{2}\,\pi m_e^{3/2}}{h^3} E^{1/2} \right) \left(\frac{1}{e^{(E - E_F)/k_B T} + 1} \right) dE \quad \textbf{(43.22)}$$

where E_F is the **Fermi energy.** At $T = 0$ K, all levels below E_F are filled, all levels above E_F are empty, and

$$E_F(0) = \frac{h^2}{2m_e} \left(\frac{3n_e}{8\pi} \right)^{2/3} \quad \textbf{(43.25)}$$

where n_e is the total number of conduction electrons per unit volume. Only those electrons having energies near E_F can contribute to the electrical conductivity of the metal.

In a crystalline solid, the energy levels of the system form a set of **bands.** Electrons occupy the lowest energy states, with no more than one electron per state. Energy gaps are present between the bands of allowed states.

A **semiconductor** is a material having an energy gap of approximately 1 eV and a valence band that is filled at $T = 0$ K. Because of the small energy gap, a significant number of electrons can be thermally excited from the valence band into the conduction band. The band structures and electrical properties of a Group IV semiconductor can be modified by the addition of either donor atoms containing five outer-shell electrons or acceptor atoms containing three outer-shell electrons. A semiconductor **doped** with donor impurity atoms is called an **n-type semiconductor,** and one doped with acceptor impurity atoms is called a **p-type semiconductor.**

Objective Questions   **1.** denotes answer available in *Student Solutions Manual/Study Guide*

1. Is each one of the following statements true or false for a superconductor below its critical temperature? (a) It can carry infinite current. (b) It must carry some nonzero current. (c) Its interior electric field must be zero. (d) Its internal magnetic field must be zero. (e) No internal energy appears when it carries electric current.

2. An infrared absorption spectrum of a molecule is shown in Figure OQ43.2. Notice that the highest peak on either side of the gap is the third peak from the gap. After this spectrum is taken, the temperature of the sample of molecules is raised to a much higher value. Compared with Figure OQ43.2, in this new spectrum is the highest absorption peak (a) at the same frequency, (b) farther from the gap, or (c) closer to the gap?

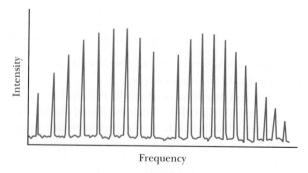

Figure OQ43.2

3. What kind of bonding likely holds the atoms together in the following solids (i), (ii), and (iii)? Choose your

answers from these possibilities: (a) ionic bonding, (b) covalent bonding, and (c) metallic bonding. **(i)** The solid is opaque, shiny, flexible, and a good electric conductor. **(ii)** The crystal is transparent, brittle, and soluble in water. It is a poor conductor of electricity. **(iii)** The crystal is opaque, brittle, very hard, and a good electric insulator.

4. The Fermi energy for silver is 5.48 eV. In a piece of solid silver, free-electron energy levels are measured near 2 eV and near 6 eV. **(i)** Near which of these energies are the energy levels closer together? (a) 2 eV (b) 6 eV (c) The spacing is the same. **(ii)** Near which of these energies are more electrons occupying energy levels? (a) 2 eV (b) 6 eV (c) The number of electrons is the same.

5. As discussed in Chapter 27, the conductivity of metals decreases with increasing temperature due to electron collisions with vibrating atoms. In contrast, the conductivity of semiconductors increases with increasing temperature. What property of a semicon-ductor is responsible for this behavior? (a) Atomic vibrations decrease as temperature increases. (b) The number of conduction electrons and the number of holes increase steeply with increasing temperature. (c) The energy gap decreases with increasing temperature. (d) Electrons do not collide with atoms in a semiconductor.

6. **(i)** Should you expect an *n*-type doped semiconductor to have (a) higher, (b) lower, or (c) the same conductivity as an intrinsic (pure) semiconductor? **(ii)** Should you expect a *p*-type doped semiconductor to have (a) higher, (b) lower, or (c) the same conductivity as an intrinsic (pure) semiconductor?

7. Consider a typical material composed of covalently bonded diatomic molecules. Rank the following energies from the largest in magnitude to the smallest in magnitude. (a) the latent heat of fusion per molecule (b) the molecular binding energy (c) the energy of the first excited state of molecular rotation (d) the energy of the first excited state of molecular vibration

Conceptual Questions

1. denotes answer available in *Student Solutions Manual/Study Guide*

Conceptual Questions 5 and 6 in Chapter 27 can be assigned with this chapter.

1. The energies of photons of visible light range between the approximate values 1.8 eV and 3.1 eV. Explain why silicon, with an energy gap of 1.14 eV at room temperature (see Table 43.3), appears opaque, whereas diamond, with an energy gap of 5.47 eV, appears transparent.

2. Discuss the three major forms of excitation of a molecule (other than translational motion) and the relative energies associated with these three forms.

3. How can the analysis of the rotational spectrum of a molecule lead to an estimate of the size of that molecule?

4. Pentavalent atoms such as arsenic are donor atoms in a semiconductor such as silicon, whereas trivalent atoms such as indium are acceptors. Inspect the periodic table in Appendix C and determine what other elements might make good donors or acceptors.

5. When a photon is absorbed by a semiconductor, an electron–hole pair is created. Give a physical explanation of this statement using the energy-band model as the basis for your description.

6. (a) Discuss the differences in the band structures of metals, insulators, and semiconductors. (b) How does the band-structure model enable you to understand the electrical properties of these materials better?

7. (a) What essential assumptions are made in the free-electron theory of metals? (b) How does the energy-band model differ from the free-electron theory in describing the properties of metals?

8. How do the vibrational and rotational levels of heavy hydrogen (D₂) molecules compare with those of H₂ molecules?

9. Discuss models for the different types of bonds that form stable molecules.

10. Discuss the differences between crystalline solids, amorphous solids, and gases.

Problems

WebAssign The problems found in this chapter may be assigned online in Enhanced WebAssign

1. straightforward; **2.** intermediate; **3.** challenging

1. full solution available in the *Student Solutions Manual/Study Guide*

AMT Analysis Model tutorial available in Enhanced WebAssign

GP Guided Problem

M Master It tutorial available in Enhanced WebAssign

W Watch It video solution available in Enhanced WebAssign

Section 43.1 **Molecular Bonds**

1. A van der Waals dispersion force between helium atoms produces a very shallow potential well, with a depth on the order of 1 meV. At approximately what temperature would you expect helium to condense?

2. **Review.** A K^+ ion and a Cl^- ion are separated by a distance of 5.00×10^{-10} m. Assuming the two ions act like charged particles, determine (a) the force each ion exerts on the other and (b) the potential energy of the two-ion system in electron volts.

3. Potassium chloride is an ionically bonded molecule that is sold as a salt substitute for use in a low-sodium diet. The electron affinity of chlorine is 3.6 eV. An energy input of 0.70 eV is required to form separate K^+ and Cl^- ions from separate K and Cl atoms. What is the ionization energy of K?

4. In the potassium iodide (KI) molecule, assume the K and I atoms bond ionically by the transfer of one electron from K to I. (a) The ionization energy of K is 4.34 eV, and the electron affinity of I is 3.06 eV. What energy is needed to transfer an electron from K to I, to form K^+ and I^- ions from neutral atoms? This quantity is sometimes called the activation energy E_a. (b) A model potential energy function for the KI molecule is the Lennard–Jones potential:

$$U(r) = 4\epsilon \left[\left(\frac{\sigma}{r} \right)^{12} - \left(\frac{\sigma}{r} \right)^{6} \right] + E_a$$

where r is the internuclear separation distance and ϵ and σ are adjustable parameters. The E_a term is added to ensure the correct asymptotic behavior at large r. At the equilibrium separation distance, $r = r_0 = 0.305$ nm, $U(r)$ is a minimum, and $dU/dr = 0$. In addition, $U(r_0)$ is the negative of the dissociation energy: $U(r_0) = -3.37$ eV. Find σ and ϵ. (c) Calculate the force needed to break up a KI molecule. (d) Calculate the force constant for small oscillations about $r = r_0$. *Suggestion:* Set $r = r_0 + s$, where $s/r_0 \ll 1$, and expand $U(r)$ in powers of s/r_0 up to second-order terms.

5. One description of the potential energy of a diatomic molecule is given by the Lennard–Jones potential,

$$U = \frac{A}{r^{12}} - \frac{B}{r^{6}}$$

where A and B are constants and r is the separation distance between the atoms. For the H_2 molecule, take $A = 0.124 \times 10^{-120}$ eV · m^{12} and $B = 1.488 \times 10^{-60}$ eV · m^6. Find (a) the separation distance r_0 at which the energy of the molecule is a minimum and (b) the energy E required to break up the H_2 molecule.

6. One description of the potential energy of a diatomic molecule is given by the Lennard–Jones potential,

$$U = \frac{A}{r^{12}} - \frac{B}{r^{6}}$$

where A and B are constants and r is the separation distance between the atoms. Find, in terms of A and B, (a) the value r_0 at which the energy is a minimum and (b) the energy E required to break up a diatomic molecule.

Section 43.2 **Energy States and Spectra of Molecules**

7. The CO molecule makes a transition from the $J = 1$ to the $J = 2$ rotational state when it absorbs a photon of frequency 2.30×10^{11} Hz. (a) Find the moment of inertia of this molecule from these data. (b) Compare your answer with that obtained in Example 43.1 and comment on the significance of the two results.

8. The cesium iodide (CsI) molecule has an atomic separation of 0.127 nm. (a) Determine the energy of the second excited rotational state, with $J = 2$. (b) Find the frequency of the photon absorbed in the $J = 1$ to $J = 2$ transition.

9. An HCl molecule is excited to its second rotational energy level, corresponding to $J = 2$. If the distance between its nuclei is 0.127 5 nm, what is the angular speed of the molecule about its center of mass?

10. The photon frequency that would be absorbed by the NO molecule in a transition from vibration state $v = 0$ to $v = 1$, with no change in rotation state, is 56.3 THz. The bond between the atoms has an effective spring constant of 1 530 N/m. (a) Use this information to calculate the reduced mass of the NO molecule. (b) Compute a value for μ using Equation 43.4. (c) Compare your results to parts (a) and (b) and explain their difference, if any.

11. Assume the distance between the protons in the H_2 molecule is 0.750×10^{-10} m. (a) Find the energy of the first excited rotational state, with $J = 1$. (b) Find the wavelength of radiation emitted in the transition from $J = 1$ to $J = 0$.

12. *Why is the following situation impossible?* The effective force constant of a vibrating HCl molecule is $k = 480$ N/m. A beam of infrared radiation of wavelength 6.20×10^3 nm is directed through a gas of HCl molecules. As a result, the molecules are excited from the ground vibrational state to the first excited vibrational state.

13. The effective spring constant describing the potential energy of the HI molecule is 320 N/m and that for the HF molecule is 970 N/m. Calculate the minimum amplitude of vibration for (a) the HI molecule and (b) the HF molecule.

14. The rotational spectrum of the HCl molecule contains lines with wavelengths of 0.060 4, 0.069 0, 0.080 4, 0.096 4, and 0.120 4 mm. What is the moment of inertia of the molecule?

15. The atoms of an NaCl molecule are separated by a distance $r = 0.280$ nm. Calculate (a) the reduced mass

of an NaCl molecule, (b) the moment of inertia of an NaCl molecule, and (c) the wavelength of radiation emitted when an NaCl molecule undergoes a transition from the $J = 2$ state to the $J = 1$ state.

16. A diatomic molecule consists of two atoms having masses m_1 and m_2 separated by a distance r. Show that the moment of inertia about an axis through the center of mass of the molecule is given by Equation 43.3, $I = \mu r^2$.

17. The nuclei of the O_2 molecule are separated by a distance 1.20×10^{-10} m. The mass of each oxygen atom in the molecule is 2.66×10^{-26} kg. (a) Determine the rotational energies of an oxygen molecule in electron volts for the levels corresponding to $J = 0$, 1, and 2. (b) The effective force constant k between the atoms in the oxygen molecule is 1 177 N/m. Determine the vibrational energies (in electron volts) corresponding to $v = 0$, 1, and 2.

18. Figure P43.18 is a model of a benzene molecule. All atoms lie in a plane, and the carbon atoms ($m_C = 1.99 \times 10^{-26}$ kg) form a regular hexagon, as do the hydrogen atoms ($m_H = 1.67 \times 10^{-27}$ kg). The carbon atoms are 0.110 nm apart center to center, and the adjacent carbon and hydrogen atoms are 0.100 nm apart center to center. (a) Calculate the moment of inertia of the molecule about an axis perpendicular to the plane of the paper through the center point O. (b) Determine the allowed rotational energies about this axis.

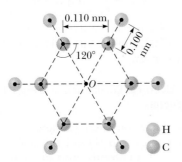

Figure P43.18

19. (a) In an HCl molecule, take the Cl atom to be the isotope ^{35}Cl. The equilibrium separation of the H and Cl atoms is 0.127 46 nm. The atomic mass of the H atom is 1.007 825 u and that of the ^{35}Cl atom is 34.968 853 u. Calculate the longest wavelength in the rotational spectrum of this molecule. (b) **What If?** Repeat the calculation in part (a), but take the Cl atom to be the isotope ^{37}Cl, which has atomic mass 36.965 903 u. The equilibrium separation distance is the same as in part (a). (c) Naturally occurring chlorine contains approximately three parts of ^{35}Cl to one part of ^{37}Cl. Because of the two different Cl masses, each line in the microwave rotational spectrum of HCl is split into a doublet as shown in Figure P43.19. Calculate the separation in wavelength between the doublet lines for the longest wavelength.

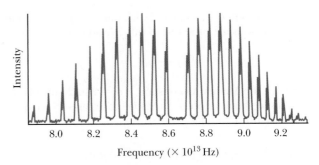

Figure P43.19 Problems 19 and 20.

20. Estimate the moment of inertia of an HCl molecule from its infrared absorption spectrum shown in Figure P43.19.

21. An H_2 molecule is in its vibrational and rotational ground states. It absorbs a photon of wavelength 2.211 2 μm and makes a transition to the $v = 1$, $J = 1$ energy level. It then drops to the $v = 0$, $J = 2$ energy level while emitting a photon of wavelength 2.405 4 μm. Calculate (a) the moment of inertia of the H_2 molecule about an axis through its center of mass and perpendicular to the H–H bond, (b) the vibrational frequency of the H_2 molecule, and (c) the equilibrium separation distance for this molecule.

22. Photons of what frequencies can be spontaneously emitted by CO molecules in the state with $v = 1$ and $J = 0$?

23. Most of the mass of an atom is in its nucleus. Model the mass distribution in a diatomic molecule as two spheres of uniform density, each of radius 2.00×10^{-15} m and mass 1.00×10^{-26} kg, located at points along the y axis as in Figure 43.5a, and separated by 2.00×10^{-10} m. Rotation about the axis joining the nuclei in the diatomic molecule is ordinarily ignored because the first excited state would have an energy that is too high to access. To see why, calculate the ratio of the energy of the first excited state for rotation about the y axis to the energy of the first excited state for rotation about the x axis.

Section 43.3 Bonding in Solids

24. Use a magnifying glass to look at the grains of table salt that come out of a salt shaker. Compare what you see with Figure 43.11a. The distance between a sodium ion and a nearest-neighbor chlorine ion is 0.261 nm. (a) Make an order-of-magnitude estimate of the number N of atoms in a typical grain of salt. (b) **What If?** Suppose you had a number of grains of salt equal to this number N. What would be the volume of this quantity of salt?

25. Use Equation 43.18 to calculate the ionic cohesive energy for NaCl. Take $\alpha = 1.747$ 6, $r_0 = 0.281$ nm, and $m = 8$.

26. Consider a one-dimensional chain of alternating singly-ionized positive and negative ions. Show that the

potential energy associated with one of the ions and its interactions with the rest of this hypothetical crystal is

$$U(r) = -\alpha k_e \frac{e^2}{r}$$

where the Madelung constant is $\alpha = 2 \ln 2$ and r is the distance between ions. *Suggestion:* Use the series expansion for $\ln (1 + x)$.

Section 43.4 Free-Electron Theory of Metals

Section 43.5 Band Theory of Solids

27. Calculate the energy of a conduction electron in silver at 800 K, assuming the probability of finding an electron in that state is 0.950. The Fermi energy of silver is 5.48 eV at this temperature.

28. (a) State what the Fermi energy depends on according to the free-electron theory of metals and how the Fermi energy depends on that quantity. (b) Show that Equation 43.25 can be expressed as $E_F = (3.65 \times 10^{-19}) n_e^{2/3}$, where E_F is in electron volts when n_e is in electrons per cubic meter. (c) According to Table 43.2, by what factor does the free-electron concentration in copper exceed that in potassium? (d) Which of these metals has the larger Fermi energy? (e) By what factor is the Fermi energy larger? (f) Explain whether this behavior is predicted by Equation 43.25.

29. When solid silver starts to melt, what is the approximate fraction of the conduction electrons that are thermally excited above the Fermi level?

30. (a) Find the typical speed of a conduction electron in copper, taking its kinetic energy as equal to the Fermi energy, 7.05 eV. (b) Suppose the copper is a current-carrying wire. How does the speed found in part (a) compare with a typical drift speed (see Section 27.1) of electrons in the wire of 0.1 mm/s?

31. The Fermi energy of copper at 300 K is 7.05 eV. (a) What is the average energy of a conduction electron in copper at 300 K? (b) At what temperature would the average translational energy of a molecule in an ideal gas be equal to the energy calculated in part (a)?

32. Consider a cube of gold 1.00 mm on an edge. Calculate the approximate number of conduction electrons in this cube whose energies lie in the range 4.000 to 4.025 eV.

33. Sodium is a monovalent metal having a density of 0.971 g/cm³ and a molar mass of 23.0 g/mol. Use this information to calculate (a) the density of charge carriers and (b) the Fermi energy of sodium.

34. *Why is the following situation impossible?* A hypothetical metal has the following properties: its Fermi energy is 5.48 eV, its density is 4.90×10^3 kg/m³, its molar mass is 100 g/mol, and it has one free electron per atom.

35. For copper at 300 K, calculate the probability that a state with an energy equal to 99.0% of the Fermi energy is occupied.

36. For a metal at temperature T, calculate the probability that a state with an energy equal to βE_F is occupied where β is a fraction between 0 and 1.

37. **Review.** An electron moves in a three-dimensional box of edge length L and volume L^3. The wave function of the particle is $\psi = A \sin (k_x x) \sin (k_y y) \sin (k_z z)$. Show that its energy is given by Equation 43.20,

$$E = \frac{\hbar^2 \pi^2}{2 m_e L^2} (n_x^2 + n_y^2 + n_z^2)$$

where the quantum numbers (n_x, n_y, n_z) are integers ≥ 1. *Suggestion:* The Schrödinger equation in three dimensions may be written

$$\frac{\hbar^2}{2m} \left(\frac{\partial^2 \psi}{\partial x^2} + \frac{\partial^2 \psi}{\partial y^2} + \frac{\partial^2 \psi}{\partial z^2} \right) = (U - E)\psi$$

38. (a) Consider a system of electrons confined to a three-dimensional box. Calculate the ratio of the number of allowed energy levels at 8.50 eV to the number at 7.05 eV. (b) **What If?** Copper has a Fermi energy of 7.05 eV at 300 K. Calculate the ratio of the number of occupied levels in copper at an energy of 8.50 eV to the number at the Fermi energy. (c) How does your answer to part (b) compare with that obtained in part (a)?

39. Show that the average kinetic energy of a conduction electron in a metal at 0 K is $E_{avg} = \frac{3}{5} E_F$. *Suggestion:* In general, the average kinetic energy is

$$E_{avg} = \frac{1}{n_e} \int_0^\infty E N(E) \, dE$$

where n_e is the density of particles, $N(E) \, dE$ is given by Equation 43.22, and the integral is over all possible values of the energy.

Section 43.6 Electrical Conduction in Metals, Insulators, and Semiconductors

40. Most solar radiation has a wavelength of 1 μm or less. (a) What energy gap should the material in a solar cell have if it is to absorb this radiation? (b) Is silicon an appropriate solar cell material (see Table 43.3)? Explain your answer.

41. The energy gap for silicon at 300 K is 1.14 eV. (a) Find the lowest-frequency photon that can promote an electron from the valence band to the conduction band. (b) What is the wavelength of this photon?

42. Light from a hydrogen discharge tube is incident on a CdS crystal. (a) Which spectral lines from the Balmer series are absorbed and (b) which are transmitted?

43. A light-emitting diode (LED) made of the semiconductor GaAsP emits red light ($\lambda = 650$ nm). Determine the energy-band gap E_g for this semiconductor.

44. The longest wavelength of radiation absorbed by a certain semiconductor is 0.512 μm. Calculate the energy gap for this semiconductor.

45. You are asked to build a scientific instrument that is thermally isolated from its surroundings. The isolation container may be a calorimeter, but these design criteria could apply to other containers as well. You wish to use a laser external to the container to raise the temperature of a target inside the instrument. You decide to use a diamond window in the container. Diamond has an energy gap of 5.47 eV. What is the shortest laser wavelength you can use to warm the sample inside the instrument?

46. **Review.** When a phosphorus atom is substituted for a silicon atom in a crystal, four of the phosphorus valence electrons form bonds with neighboring atoms and the remaining electron is much more loosely bound. You can model the electron as free to move through the crystal lattice. The phosphorus nucleus has one more positive charge than does the silicon nucleus, however, so the extra electron provided by the phosphorus atom is attracted to this single nuclear charge $+e$. The energy levels of the extra electron are similar to those of of the electron in the Bohr hydrogen atom with two important exceptions. First, the Coulomb attraction between the electron and the positive charge on the phosphorus nucleus is reduced by a factor of $1/\kappa$ from what it would be in free space (see Eq. 26.21), where κ is the dielectric constant of the crystal. As a result, the orbit radii are greatly increased over those of the hydrogen atom. Second, the influence of the periodic electric potential of the lattice causes the electron to move as if it had an effective mass m^*, which is quite different from the mass m_e of a free electron. You can use the Bohr model of hydrogen to obtain relatively accurate values for the allowed energy levels of the extra electron. We wish to find the typical energy of these donor states, which play an important role in semiconductor devices. Assume $\kappa = 11.7$ for silicon and $m^* = 0.220 m_e$. (a) Find a symbolic expression for the smallest radius of the electron orbit in terms of a_0, the Bohr radius. (b) Substitute numerical values to find the numerical value of the smallest radius. (c) Find a symbolic expression for the energy levels E_n' of the electron in the Bohr orbits around the donor atom in terms of m_e, m^*, κ, and E_n, the energy of the hydrogen atom in the Bohr model. (d) Find the numerical value of the energy for the ground state of the electron.

Section 43.7 Semiconductor Devices

47. Assuming $T = 300$ K, (a) for what value of the bias voltage ΔV in Equation 43.27 does $I = 9.00 I_0$? (b) **What If?** What if $I = -0.900 I_0$?

48. A diode, a resistor, and a battery are connected in a series circuit. The diode is at a temperature for which $k_B T = 25.0$ meV, and the saturation value of the current is $I_0 = 1.00 \ \mu$A. The resistance of the resistor is $R = 745 \ \Omega$, and the battery maintains a constant potential difference of $\varepsilon = 2.42$ V between its terminals. (a) Use Kirchhoff's loop rule to show that

$$\varepsilon - \Delta V = I_0 R(e^{e \Delta V / k_B T} - 1)$$

where ΔV is the voltage across the diode. (b) To solve this transcendental equation for the voltage ΔV, graph the left-hand side of the above equation and the right-hand side as functions of ΔV and find the value of ΔV at which the curves cross. (c) Find the current I in the circuit. (d) Find the ohmic resistance of the diode, defined as the ratio $\Delta V/I$, at the voltage in part (b). (e) Find the dynamic resistance of the diode, which is defined as the derivative $d(\Delta V)/dI$, at the voltage in part (b).

49. You put a diode in a microelectronic circuit to protect the system in case an untrained person installs the battery backward. In the correct forward-bias situation, the current is 200 mA with a potential difference of 100 mV across the diode at room temperature (300 K). If the battery were reversed, so that the potential difference across the diode is still 100 mV but with the opposite sign, what would be the magnitude of the current in the diode?

50. A diode is at room temperature so that $k_B T = 0.025\ 0$ eV. Taking the applied voltages across the diode to be $+ 1.00$ V (under forward bias) and -1.00 V (under reverse bias), calculate the ratio of the forward current to the reverse current if the diode is described by Equation 43.27.

Section 43.8 Superconductivity

Problem 30 in Chapter 30 and Problems 73 through 76 in Chapter 32 can also be assigned with this section.

51. A thin rod of superconducting material 2.50 cm long is placed into a 0.540-T magnetic field with its cylindrical axis along the magnetic field lines. (a) Sketch the directions of the applied field and the induced surface current. (b) Find the magnitude of the surface current on the curved surface of the rod.

52. A direct and relatively simple demonstration of zero DC resistance can be carried out using the four-point probe method. The probe shown in Figure P43.52 consists of a disk of $YBa_2Cu_3O_7$ (a high-T_c superconductor) to which four wires are attached. Current is maintained through the sample by applying a DC voltage between points a and b, and it is measured with a DC ammeter. The current can be varied with the variable resistance R. The potential difference ΔV_{cd} between c and d is measured with a digital voltmeter. When the probe is immersed in liquid nitrogen, the sample quickly cools

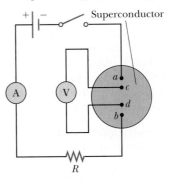

Figure P43.52

to 77 K, below the critical temperature of the material, 92 K. The current remains approximately constant, but ΔV_{cd} *drops abruptly to zero.* (a) Explain this observation on the basis of what you know about superconductors. (b) The data in the accompanying table represent actual values of ΔV_{cd} for different values of I taken on the sample at room temperature in the senior author's laboratory. A 6-V battery in series with a variable resistor R supplied the current. The values of R ranged from 10 Ω to 100 Ω. Make an I–ΔV plot of the data and determine whether the sample behaves in a linear manner. (c) From the data, obtain a value for the DC resistance of the sample at room temperature. (d) At room temperature, it was found that $\Delta V_{cd} = 2.234$ mV for $I = 100.3$ mA, but after the sample was cooled to 77 K, $\Delta V_{cd} = 0$ and $I = 98.1$ mA. What do you think might have caused the slight decrease in current?

Current Versus Potential Difference ΔV_{cd} Measured in a Bulk Ceramic Sample of $YBa_2Cu_3O_{7-\delta}$ at Room Temperature

I (mA)	ΔV_{cd} (mV)
57.8	1.356
61.5	1.441
68.3	1.602
76.8	1.802
87.5	2.053
102.2	2.398
123.7	2.904
155	3.61

53. A superconducting ring of niobium metal 2.00 cm in diameter is immersed in a uniform 0.020 0-T magnetic field directed perpendicular to the ring and carries no current. Determine the current generated in the ring when the magnetic field is suddenly decreased to zero. The inductance of the ring is 3.10×10^{-8} H.

Additional Problems

54. The effective spring constant associated with bonding in the N_2 molecule is 2 297 N/m. The nitrogen atoms each have a mass of 2.32×10^{-26} kg, and their nuclei are 0.120 nm apart. Assume the molecule is rigid. The first excited vibrational state of the molecule is above the vibrational ground state by an energy difference ΔE. Calculate the J value of the rotational state that is above the rotational ground state by the same energy difference ΔE.

55. The hydrogen molecule comes apart (dissociates) when it is excited internally by 4.48 eV. Assuming this molecule behaves like a harmonic oscillator having classical angular frequency $\omega = 8.28 \times 10^{14}$ rad/s, find the highest vibrational quantum number for a state below the 4.48-eV dissociation energy.

56. (a) Starting with Equation 43.17, show that the force exerted on an ion in an ionic solid can be written as

$$F = -\alpha k_e \frac{e^2}{r^2} \left[1 - \left(\frac{r_0}{r} \right)^{m-1} \right]$$

where α is the Madelung constant and r_0 is the equilibrium separation. (b) Imagine that an ion in the solid is displaced a small distance s from r_0. Show that the ion experiences a restoring force $F = -Ks$, where

$$K = \frac{\alpha k_e e^2}{r_0^3} (m - 1)$$

(c) Use the result of part (b) to find the frequency of vibration of a Na^+ ion in NaCl. Take $m = 8$ and use the value $\alpha = 1.747$ 6.

57. Under pressure, liquid helium can solidify as each atom bonds with four others, and each bond has an average energy of 1.74×10^{-23} J. Find the latent heat of fusion for helium in joules per gram. (The molar mass of He is 4.00 g/mol.)

58. The dissociation energy of ground-state molecular hydrogen is 4.48 eV, but it only takes 3.96 eV to dissociate it when it starts in the first excited vibrational state with $J = 0$. Using this information, determine the depth of the H_2 molecular potential-energy function.

59. Starting with Equation 43.17, show that the ionic cohesive energy of an ionically bonded solid is given by Equation 43.18.

60. The Fermi–Dirac distribution function can be written as

$$f(E) = \frac{1}{e^{(E-E_F)/k_B T} + 1} = \frac{1}{e^{(E/E_F - 1)T_F/T} + 1}$$

where T_F is the *Fermi temperature*, defined according to

$$k_B T_F \equiv E_F$$

(a) Write a spreadsheet to calculate and plot $f(E)$ versus E/E_F at a fixed temperature T. (b) Describe the curves obtained for $T = 0.1T_F$, $0.2T_F$, and $0.5T_F$.

61. A particle moves in one-dimensional motion through a field for which the potential energy of the particle–field system is

$$U(x) = \frac{A}{x^3} - \frac{B}{x}$$

where $A = 0.150$ eV $\cdot$ nm^3 and $B = 3.68$ eV $\cdot$ nm. The shape of this function is shown in Figure P43.61. (a) Find the equilibrium position x_0 of the particle. (b) Determine the depth U_0 of this potential well.

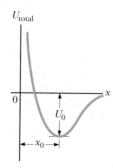

Figure P43.61 Problems 61 and 62.

(c) In moving along the x axis, what maximum force toward the negative x direction does the particle experience?

62. A particle of mass m moves in one-dimensional motion through a field for which the potential energy of the particle–field system is

$$U(x) = \frac{A}{x^3} - \frac{B}{x}$$

where A and B are constants. The general shape of this function is shown in Figure P43.61. (a) Find the equilibrium position x_0 of the particle in terms of m, A, and B. (b) Determine the depth U_0 of this potential well. (c) In moving along the x axis, what maximum force toward the negative x direction does the particle experience?

Challenge Problems

63. As you will learn in Chapter 44, carbon-14 (^{14}C) is an
W   unstable isotope of carbon. It has the same chemical properties and electronic structure as the much more abundant isotope carbon-12 (^{12}C), but it has different nuclear properties. Its mass is 14 u, greater than that of carbon-12 because of the two extra neutrons in the carbon-14 nucleus. Assume the CO molecular potential energy is the same for both isotopes of carbon and the examples in Section 43.2 contain accurate data and results for carbon monoxide with carbon-12 atoms. (a) What is the vibrational frequency of ^{14}CO? (b) What is the moment of inertia of ^{14}CO? (c) What wavelengths of light can be absorbed by ^{14}CO in the ($v = 0, J = 10$) state that cause it to end up in the $v = 1$ state?

64. As an alternative to Equation 43.1, another useful model for the potential energy of a diatomic molecule is the Morse potential

$$U(r) = B\left[e^{-a(r - r_0)} - 1\right]^2$$

where B, a, and r_0 are parameters used to adjust the shape of the potential and its depth. (a) What is the equilibrium separation of the nuclei? (b) What is the depth of the potential well, defined as the difference in energy between the potential's minimum value and its asymptote as r approaches infinity? (c) If μ is the reduced mass of the system of two nuclei and assuming the potential is nearly parabolic about the well minimum, what is the vibrational frequency of the diatomic molecule in its ground state? (d) What amount of energy needs to be supplied to the ground-state molecule to separate the two nuclei to infinity?

Nuclear Structure

Ötzi the Iceman, a Copper Age man, was discovered by German tourists in the Italian Alps in 1991 when a glacier melted enough to expose his remains. Analysis of his corpse has exposed his last meal, illnesses he suffered, and places he lived. Radioactivity was used to determine that he lived in about 3300 BC. Ötzi can be seen today in the Südtiroler Archäologiemuseum (South Tyrol Museum of Archaeology) in Bolzano, Italy. (©Vienna Report Agency/Sygma/Corbis)

The year 1896 marks the birth of nuclear physics when French physicist Antoine-Henri Becquerel (1852–1908) discovered radioactivity in uranium compounds. This discovery prompted scientists to investigate the details of radioactivity and, ultimately, the structure of the nucleus. Pioneering work by Ernest Rutherford showed that the radiation emitted from radioactive substances is of three types—alpha, beta, and gamma rays—classified according to the nature of their electric charge and their ability to penetrate matter and ionize air. Later experiments showed that alpha rays are helium nuclei, beta rays are electrons, and gamma rays are high-energy photons.

In 1911, Rutherford, Hans Geiger, and Ernest Marsden performed the alpha-particle scattering experiments described in Section 42.2. These experiments established that the nucleus of an atom can be modeled as a point mass and point charge and that most of the atomic mass is contained in the nucleus. Subsequent studies revealed the presence of a new type of force, the short-range nuclear force, which is predominant at particle separation distances less than approximately 10^{-14} m and is zero for large distances.

In this chapter, we discuss the properties and structure of the atomic nucleus. We start by describing the basic properties of nuclei, followed by a discussion of nuclear forces and binding energy, nuclear models, and the phenomenon of radioactivity. Finally, we explore the various processes by which nuclei decay and the ways that nuclei can react with each other.

44.1 Some Properties of Nuclei

All nuclei are composed of two types of particles: protons and neutrons. The only exception is the ordinary hydrogen nucleus, which is a single proton. We describe the atomic nucleus by the number of protons and neutrons it contains, using the following quantities:

- the **atomic number** Z, which equals the number of protons in the nucleus (sometimes called the *charge number*)
- the **neutron number** N, which equals the number of neutrons in the nucleus
- the **mass number** $A = Z + N$, which equals the number of **nucleons** (neutrons plus protons) in the nucleus

A **nuclide** is a specific combination of atomic number and mass number that represents a nucleus. In representing nuclides, it is convenient to use the symbol $^A_Z X$ to convey the numbers of protons and neutrons, where X represents the chemical symbol of the element. For example, $^{56}_{26}Fe$ (iron) has mass number 56 and atomic number 26; therefore, it contains 26 protons and 30 neutrons. When no confusion is likely to arise, we omit the subscript Z because the chemical symbol can always be used to determine Z. Therefore, $^{56}_{26}Fe$ is the same as ^{56}Fe and can also be expressed as "iron-56" or "Fe-56."

The nuclei of all atoms of a particular element contain the same number of protons but often contain different numbers of neutrons. Nuclei related in this way are called **isotopes.** The isotopes of an element have the same Z value but different N and A values.

The natural abundance of isotopes can differ substantially. For example $^{11}_6C$, $^{12}_6C$, $^{13}_6C$, and $^{14}_6C$ are four isotopes of carbon. The natural abundance of the $^{12}_6C$ isotope is approximately 98.9%, whereas that of the $^{13}_6C$ isotope is only about 1.1%. Some isotopes, such as $^{11}_6C$ and $^{14}_6C$, do not occur naturally but can be produced by nuclear reactions in the laboratory or by cosmic rays.

Even the simplest element, hydrogen, has isotopes: ^1_1H, the ordinary hydrogen nucleus; ^2_1H, deuterium; and ^3_1H, tritium.

Quick Quiz 44.1 For each part of this Quick Quiz, choose from the following answers: (a) protons (b) neutrons (c) nucleons. **(i)** The three nuclei ^{12}C, ^{13}N, and ^{14}O have the same number of what type of particle? **(ii)** The three nuclei ^{12}N, ^{13}N, and ^{14}N have the same number of what type of particle? **(iii)** The three nuclei ^{14}C, ^{14}N, and ^{14}O have the same number of what type of particle?

Charge and Mass

The proton carries a single positive charge e, equal in magnitude to the charge $-e$ on the electron ($e = 1.6 \times 10^{-19}$ C). The neutron is electrically neutral as its name implies. Because the neutron has no charge, it was difficult to detect with early experimental apparatus and techniques. Today, neutrons are easily detected with devices such as plastic scintillators.

Nuclear masses can be measured with great precision using a mass spectrometer (see Section 29.3) and by the analysis of nuclear reactions. The proton is approximately 1 836 times as massive as the electron, and the masses of the proton and the neutron are almost equal. The **atomic mass unit** u is defined in such a way that the mass of one atom of the isotope ^{12}C is exactly 12 u, where 1 u is equal to $1.660\,539 \times 10^{-27}$ kg. According to this definition, the proton and neutron each have a mass of approximately 1 u and the electron has a mass that is only a small fraction of this value. The masses of these particles and others important to the phenomena discussed in this chapter are given in Table 44.1 (page 1382).

Pitfall Prevention 44.1
Mass Number Is Not Atomic Mass
The mass number A should not be confused with the atomic mass. Mass number is an integer specific to an isotope and has no units; it is simply a count of the number of nucleons. Atomic mass has units and is generally not an integer because it is an average of the masses of a given element's naturally occurring isotopes.

Table 44.1	**Masses of Selected Particles in Various Units**		
		Mass	
Particle	**kg**	**u**	**MeV/c²**
Proton	$1.672\,62 \times 10^{-27}$	$1.007\,276$	938.27
Neutron	$1.674\,93 \times 10^{-27}$	$1.008\,665$	939.57
Electron (β particle)	$9.109\,38 \times 10^{-31}$	$5.485\,79 \times 10^{-4}$	$0.510\,999$
$^{1}_{1}$H atom	$1.673\,53 \times 10^{-27}$	$1.007\,825$	938.783
$^{4}_{2}$He nucleus (α particle)	$6.644\,66 \times 10^{-27}$	$4.001\,506$	$3\,727.38$
$^{4}_{2}$He atom	$6.646\,48 \times 10^{-27}$	$4.002\,603$	$3\,728.40$
$^{12}_{6}$C atom	$1.992\,65 \times 10^{-27}$	$12.000\,000$	$11\,177.9$

You might wonder how six protons and six neutrons, each having a mass larger than 1 u, can be combined with six electrons to form a carbon-12 atom having a mass of exactly 12 u. The bound system of ^{12}C has a lower rest energy (Section 39.8) than that of six separate protons and six separate neutrons. According to Equation 39.24, $E_R = mc^2$, this lower rest energy corresponds to a smaller mass for the bound system. The difference in mass accounts for the binding energy when the particles are combined to form the nucleus. We shall discuss this point in more detail in Section 44.2.

It is often convenient to express the atomic mass unit in terms of its *rest-energy equivalent*. For one atomic mass unit,

$$E_R = mc^2 = (1.660\,539 \times 10^{-27}\text{ kg})(2.997\,92 \times 10^8\text{ m/s})^2 = 931.494\text{ MeV}$$

where we have used the conversion 1 eV = $1.602\,176 \times 10^{-19}$ J.

Based on the rest-energy expression in Equation 39.24, nuclear physicists often express mass in terms of the unit MeV/c^2.

Example 44.1 The Atomic Mass Unit

Use Avogadro's number to show that 1 u = 1.66×10^{-27} kg.

SOLUTION

Conceptualize From the definition of the mole given in Section 19.5, we know that exactly 12 g (= 1 mol) of ^{12}C contains Avogadro's number of atoms.

Categorize We evaluate the atomic mass unit that was introduced in this section, so we categorize this example as a substitution problem.

Find the mass m of one ^{12}C atom:

$$m = \frac{0.012\text{ kg}}{6.02 \times 10^{23}\text{ atoms}} = 1.99 \times 10^{-26}\text{ kg}$$

Because one atom of ^{12}C is defined to have a mass of 12.0 u, divide by 12.0 to find the mass equivalent to 1 u:

$$1\text{ u} = \frac{1.99 \times 10^{-26}\text{ kg}}{12.0} = 1.66 \times 10^{-27}\text{ kg}$$

The Size and Structure of Nuclei

In Rutherford's scattering experiments, positively charged nuclei of helium atoms (alpha particles) were directed at a thin piece of metallic foil. As the alpha particles moved through the foil, they often passed near a metal nucleus. Because of the positive charge on both the incident particles and the nuclei, the particles were deflected from their straight-line paths by the Coulomb repulsive force.

Rutherford used the isolated system (energy) analysis model to find an expression for the separation distance d at which an alpha particle approaching a nucleus head-on is turned around by Coulomb repulsion. In such a head-on collision, the mechanical energy of the nucleus–alpha particle system is conserved. The initial kinetic energy of the incoming particle is transformed completely to electric potential energy of the system when the alpha particle stops momentarily at the point of closest approach (the final configuration of the system) before moving back along the same path (Fig. 44.1). Applying Equation 8.2, the conservation of energy principle, to the system gives

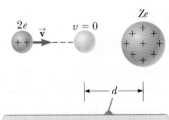

$$\Delta K + \Delta U = 0$$

$$\left(0 - \tfrac{1}{2}mv^2\right) + \left(k_e \frac{q_1 q_2}{d} - 0\right) = 0$$

Because of the Coulomb repulsion between the charges of the same sign, the alpha particle approaches to a distance d from the nucleus, called the distance of closest approach.

where m is the mass of the alpha particle and v is its initial speed. Solving for d gives

$$d = 2k_e \frac{q_1 q_2}{mv^2} = 2k_e \frac{(2e)(Ze)}{mv^2} = 4k_e \frac{Ze^2}{mv^2}$$

Figure 44.1 An alpha particle on a head-on collision course with a nucleus of charge Ze.

where Z is the atomic number of the target nucleus. From this expression, Rutherford found that the alpha particles approached nuclei to within 3.2×10^{-14} m when the foil was made of gold. Therefore, the radius of the gold nucleus must be less than this value. From the results of his scattering experiments, Rutherford concluded that the positive charge in an atom is concentrated in a small sphere, which he called the nucleus, whose radius is no greater than approximately 10^{-14} m.

Because such small lengths are common in nuclear physics, an often-used convenient length unit is the femtometer (fm), which is sometimes called the **fermi** and is defined as

$$1 \text{ fm} \equiv 10^{-15} \text{ m}$$

In the early 1920s, it was known that the nucleus of an atom contains Z protons and has a mass nearly equivalent to that of A protons, where on average $A \approx 2Z$ for lighter nuclei ($Z \le 20$) and $A > 2Z$ for heavier nuclei. To account for the nuclear mass, Rutherford proposed that each nucleus must also contain $A - Z$ neutral particles that he called neutrons. In 1932, British physicist James Chadwick (1891–1974) discovered the neutron, and he was awarded the Nobel Prize in Physics in 1935 for this important work.

Since the time of Rutherford's scattering experiments, a multitude of other experiments have shown that most nuclei are approximately spherical and have an average radius given by

$$r = aA^{1/3} \tag{44.1}$$

◀ **Nuclear radius**

where a is a constant equal to 1.2×10^{-15} m and A is the mass number. Because the volume of a sphere is proportional to the cube of its radius, it follows from Equation 44.1 that the volume of a nucleus (assumed to be spherical) is directly proportional to A, the total number of nucleons. This proportionality suggests that *all nuclei have nearly the same density.* When nucleons combine to form a nucleus, they combine as though they were tightly packed spheres (Fig. 44.2). This fact has led to an analogy between the nucleus and a drop of liquid, in which the density of the drop is independent of its size. We shall discuss the liquid-drop model of the nucleus in Section 44.3.

Figure 44.2 A nucleus can be modeled as a cluster of tightly packed spheres, where each sphere is a nucleon.

Example 44.2 **The Volume and Density of a Nucleus**

Consider a nucleus of mass number A.

(A) Find an approximate expression for the mass of the nucleus.

continued

▶ **44.2** continued

SOLUTION

Conceptualize Imagine the nucleus to be a collection of protons and neutrons as shown in Figure 44.2. The mass number A counts *both* protons and neutrons.

Categorize Let's assume A is large enough that we can imagine the nucleus to be spherical.

....

Analyze The mass of the proton is approximately equal to that of the neutron. Therefore, if the mass of one of these particles is m, the mass of the nucleus is approximately Am.

(B) Find an expression for the volume of this nucleus in terms of A.

SOLUTION

Assume the nucleus is spherical and use Equation 44.1:

$$(1) \quad V_{\text{nucleus}} = \tfrac{4}{3}\pi r^3 = \tfrac{4}{3}\pi a^3 A$$

(C) Find a numerical value for the density of this nucleus.

SOLUTION

Use Equation 1.1 and substitute Equation (1):

$$\rho = \frac{m_{\text{nucleus}}}{V_{\text{nucleus}}} = \frac{Am}{\tfrac{4}{3}\pi a^3 A} = \frac{3m}{4\pi a^3}$$

Substitute numerical values:

$$\rho = \frac{3(1.67 \times 10^{-27}\ \text{kg})}{4\pi(1.2 \times 10^{-15}\ \text{m})^3} = 2.3 \times 10^{17}\ \text{kg/m}^3$$

Finalize The nuclear density is approximately 2.3×10^{14} times the density of water ($\rho_{\text{water}} = 1.0 \times 10^3\ \text{kg/m}^3$).

WHAT IF? What if the Earth could be compressed until it had this density? How large would it be?

Answer Because this density is so large, we predict that an Earth of this density would be very small.

Use Equation 1.1 and the mass of the Earth to find the volume of the compressed Earth:

$$V = \frac{M_E}{\rho} = \frac{5.97 \times 10^{24}\ \text{kg}}{2.3 \times 10^{17}\ \text{kg/m}^3} = 2.6 \times 10^7\ \text{m}^3$$

From this volume, find the radius:

$$V = \tfrac{4}{3}\pi r^3 \quad \rightarrow \quad r = \left(\frac{3V}{4\pi}\right)^{1/3} = \left[\frac{3(2.6 \times 10^7\ \text{m}^3)}{4\pi}\right]^{1/3}$$

$$r = 1.8 \times 10^2\ \text{m}$$

An Earth of this radius is indeed a small Earth!

Nuclear Stability

You might expect that the very large repulsive Coulomb forces between the close-packed protons in a nucleus should cause the nucleus to fly apart. Because that does not happen, there must be a counteracting attractive force. The **nuclear force** is a very short range (about 2 fm) attractive force that acts between all nuclear particles. The protons attract each other by means of the nuclear force, and, at the same time, they repel each other through the Coulomb force. The nuclear force also acts between pairs of neutrons and between neutrons and protons. The nuclear force dominates the Coulomb repulsive force within the nucleus (at short ranges), so stable nuclei can exist.

The nuclear force is independent of charge. In other words, the forces associated with the proton–proton, proton–neutron, and neutron–neutron interactions are the same, apart from the additional repulsive Coulomb force for the proton–proton interaction.

Evidence for the limited range of nuclear forces comes from scattering experiments and from studies of nuclear binding energies. The short range of the nuclear force is shown in the neutron–proton (n–p) potential energy plot of Figure 44.3a obtained by scattering neutrons from a target containing hydrogen. The depth of the n–p potential energy well is 40 to 50 MeV, and there is a strong repulsive component that prevents the nucleons from approaching much closer than 0.4 fm.

The nuclear force does not affect electrons, enabling energetic electrons to serve as point-like probes of nuclei. The charge independence of the nuclear force also means that the main difference between the n–p and p–p interactions is that the p–p potential energy consists of a *superposition* of nuclear and Coulomb interactions as shown in Figure 44.3b. At distances less than 2 fm, both p–p and n–p potential energies are nearly identical, but for distances of 2 fm or greater, the p–p potential has a positive energy barrier with a maximum at 4 fm.

The existence of the nuclear force results in approximately 270 stable nuclei; hundreds of other nuclei have been observed, but they are unstable. A plot of neutron number N versus atomic number Z for a number of stable nuclei is given in Figure 44.4. The stable nuclei are represented by the black dots, which lie in a narrow range called the *line of stability*. Notice that the light stable nuclei contain an equal number of protons and neutrons; that is, $N = Z$. Also notice that in heavy stable nuclei, the number of neutrons exceeds the number of protons: above $Z = 20$, the line of stability deviates upward from the line representing $N = Z$. This deviation can be understood by recognizing that as the number of protons increases, the strength of the Coulomb force increases, which tends to break the nucleus apart. As a result, more neutrons are needed to keep the nucleus stable because neutrons experience only the attractive nuclear force. Eventually, the repulsive Coulomb forces between protons cannot be compensated by the addition of more neutrons. This point occurs at $Z = 83$, meaning that elements that contain more than 83 protons do not have stable nuclei.

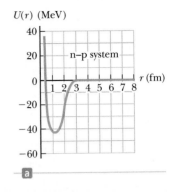

a

The difference in the two curves is due to the large Coulomb repulsion in the case of the proton–proton interaction.

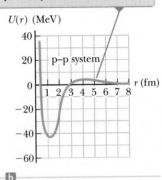

b

Figure 44.3 (a) Potential energy versus separation distance for a neutron–proton system. (b) Potential energy versus separation distance for a proton–proton system. To display the difference in the curves on this scale, the height of the peak for the proton–proton curve has been exaggerated by a factor of 10.

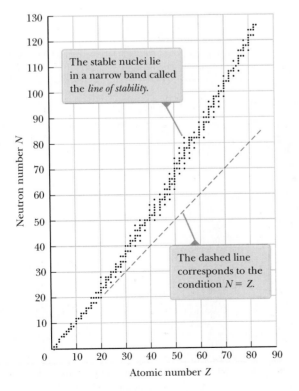

Figure 44.4 Neutron number N versus atomic number Z for stable nuclei (black dots).

44.2 Nuclear Binding Energy

As mentioned in the discussion of ^{12}C in Section 44.1, the total mass of a nucleus is less than the sum of the masses of its individual nucleons. Therefore, the rest energy of the bound system (the nucleus) is less than the combined rest energy of the separated nucleons. This difference in energy is called the **binding energy** of the nucleus and can be interpreted as the energy that must be added to a nucleus to break it apart into its components. Therefore, to separate a nucleus into protons and neutrons, energy must be delivered to the system.

Conservation of energy and the Einstein mass–energy equivalence relationship show that the binding energy E_b in MeV of any nucleus is

◀ Binding energy of a nucleus

$$E_b = [ZM(H) + Nm_n - M(^A_Z X)] \times 931.494 \text{ MeV/u} \qquad (44.2)$$

where $M(H)$ is the atomic mass of the neutral hydrogen atom, m_n is the mass of the neutron, $M(^A_Z X)$ represents the atomic mass of an atom of the isotope $^A_Z X$, and the masses are all in atomic mass units. The mass of the Z electrons included in $M(H)$ cancels with the mass of the Z electrons included in the term $M(^A_Z X)$ within a small difference associated with the atomic binding energy of the electrons. Because atomic binding energies are typically several electron volts and nuclear binding energies are several million electron volts, this difference is negligible.

A plot of binding energy per nucleon E_b/A as a function of mass number A for various stable nuclei is shown in Figure 44.5. Notice that the binding energy in Figure 44.5 peaks in the vicinity of $A = 60$. That is, nuclei having mass numbers either greater or less than 60 are not as strongly bound as those near the middle of the periodic table. The decrease in binding energy per nucleon for $A > 60$ implies that energy is released when a heavy nucleus splits, or *fissions*, into two lighter nuclei. Energy is released in fission because the nucleons in each product nucleus are more tightly bound to one another than are the nucleons in the original nucleus. The important process of fission and a second important process of *fusion*, in which energy is released as light nuclei combine, shall be considered in detail in Chapter 45.

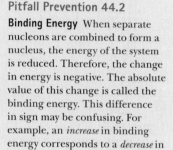

Pitfall Prevention 44.2

Binding Energy When separate nucleons are combined to form a nucleus, the energy of the system is reduced. Therefore, the change in energy is negative. The absolute value of this change is called the binding energy. This difference in sign may be confusing. For example, an *increase* in binding energy corresponds to a *decrease* in the energy of the system.

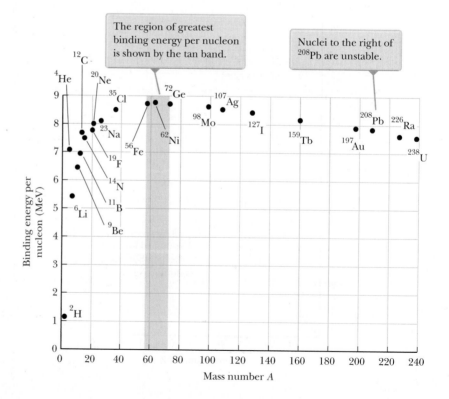

Figure 44.5 Binding energy per nucleon versus mass number for nuclides that lie along the line of stability in Figure 44.4. Some representative nuclides appear as black dots with labels.

Another important feature of Figure 44.5 is that the binding energy per nucleon is approximately constant at around 8 MeV per nucleon for all nuclei with $A > 50$. For these nuclei, the nuclear forces are said to be *saturated*, meaning that in the closely packed structure shown in Figure 44.2, a particular nucleon can form attractive bonds with only a limited number of other nucleons.

Figure 44.5 provides insight into fundamental questions about the origin of the chemical elements. In the early life of the Universe, the only elements that existed were hydrogen and helium. Clouds of cosmic gas coalesced under gravitational forces to form stars. As a star ages, it produces heavier elements from the lighter elements contained within it, beginning by fusing hydrogen atoms to form helium. This process continues as the star becomes older, generating atoms having larger and larger atomic numbers, up to the tan band shown in Figure 44.5.

The nucleus $^{63}_{28}\text{Ni}$ has the largest binding energy per nucleon of 8.794 5 MeV. It takes additional energy to create elements with mass numbers larger than 63 because of their lower binding energies per nucleon. This energy comes from the supernova explosion that occurs at the end of some large stars' lives. Therefore, all the heavy atoms in your body were produced from the explosions of ancient stars. You are literally made of stardust!

44.3 Nuclear Models

The details of the nuclear force are still an area of active research. Several nuclear models have been proposed that are useful in understanding general features of nuclear experimental data and the mechanisms responsible for binding energy. Two such models, the liquid-drop model and the shell model, are discussed below.

The Liquid–Drop Model

In 1936, Bohr proposed treating nucleons like molecules in a drop of liquid. In this **liquid-drop model,** the nucleons interact strongly with one another and undergo frequent collisions as they jiggle around within the nucleus. This jiggling motion is analogous to the thermally agitated motion of molecules in a drop of liquid.

Four major effects influence the binding energy of the nucleus in the liquid-drop model:

- **The volume effect.** Figure 44.5 shows that for $A > 50$, the binding energy per nucleon is approximately constant, which indicates that the nuclear force on a given nucleon is due only to a few nearest neighbors and not to all the other nucleons in the nucleus. On average, then, the binding energy associated with the nuclear force for each nucleon is the same in all nuclei: that associated with an interaction with a few neighbors. This property indicates that the total binding energy of the nucleus is proportional to A and therefore proportional to the nuclear volume. The contribution to the binding energy of the entire nucleus is C_1A, where C_1 is an adjustable constant that can be determined by fitting the prediction of the model to experimental results.
- **The surface effect.** Because nucleons on the surface of the drop have fewer neighbors than those in the interior, surface nucleons reduce the binding energy by an amount proportional to their number. Because the number of surface nucleons is proportional to the surface area $4\pi r^2$ of the nucleus (modeled as a sphere) and because $r^2 \propto A^{2/3}$ (Eq. 44.1), the surface term can be expressed as $-C_2A^{2/3}$, where C_2 is a second adjustable constant.
- **The Coulomb repulsion effect.** Each proton repels every other proton in the nucleus. The corresponding potential energy per pair of interacting protons is $k_e e^2/r$, where k_e is the Coulomb constant. The total electric potential energy is equivalent to the work required to assemble Z protons, initially infinitely far apart, into a sphere of volume V. This energy is proportional to the number

of proton pairs $Z(Z-1)/2$ and inversely proportional to the nuclear radius. Consequently, the reduction in binding energy that results from the Coulomb effect is $-C_3 Z(Z-1)/A^{1/3}$, where C_3 is yet another adjustable constant.

- **The symmetry effect.** Another effect that lowers the binding energy is related to the symmetry of the nucleus in terms of values of N and Z. For small values of A, stable nuclei tend to have $N \approx Z$. Any large asymmetry between N and Z for light nuclei reduces the binding energy and makes the nucleus less stable. For larger A, the value of N for stable nuclei is naturally larger than Z. This effect can be described by a binding-energy term of the form $-C_4(N-Z)^2/A$, where C_4 is another adjustable constant.[1] For small A, any large asymmetry between values of N and Z makes this term relatively large and reduces the binding energy. For large A, this term is small and has little effect on the overall binding energy.

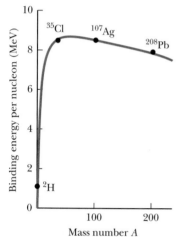

Figure 44.6 The binding-energy curve plotted by using the semiempirical binding-energy formula (red-brown). For comparison to the theoretical curve, experimental values for four sample nuclei are shown.

Adding these contributions gives the following expression for the total binding energy:

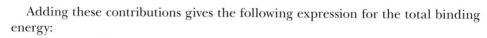

$$E_b = C_1 A - C_2 A^{2/3} - C_3 \frac{Z(Z-1)}{A^{1/3}} - C_4 \frac{(N-Z)^2}{A} \qquad \textbf{(44.3)}$$

This equation, often referred to as the **semiempirical binding-energy formula,** contains four constants that are adjusted to fit the theoretical expression to experimental data. For nuclei having $A \geq 15$, the constants have the values

$$C_1 = 15.7 \text{ MeV} \qquad C_2 = 17.8 \text{ MeV}$$

$$C_3 = 0.71 \text{ MeV} \qquad C_4 = 23.6 \text{ MeV}$$

Equation 44.3, together with these constants, fits the known nuclear mass values very well as shown by the theoretical curve and sample experimental values in Figure 44.6. The liquid-drop model does not, however, account for some finer details of nuclear structure, such as stability rules and angular momentum. Equation 44.3 is a *theoretical* equation for the binding energy, based on the liquid-drop model, whereas binding energies calculated from Equation 44.2 are *experimental* values based on mass measurements.

Example 44.3	Applying the Semiempirical Binding–Energy Formula

The nucleus ^{64}Zn has a tabulated binding energy of 559.09 MeV. Use the semiempirical binding-energy formula to generate a theoretical estimate of the binding energy for this nucleus.

SOLUTION

Conceptualize Imagine bringing the separate protons and neutrons together to form a ^{64}Zn nucleus. The rest energy of the nucleus is smaller than the rest energy of the individual particles. The difference in rest energy is the binding energy.

Categorize From the text of the problem, we know to apply the liquid-drop model. This example is a substitution problem.

For the ^{64}Zn nucleus, $Z = 30$, $N = 34$, and $A = 64$. Evaluate the four terms of the semiempirical binding-energy formula:

$$C_1 A = (15.7 \text{ MeV})(64) = 1\,005 \text{ MeV}$$

$$C_2 A^{2/3} = (17.8 \text{ MeV})(64)^{2/3} = 285 \text{ MeV}$$

$$C_3 \frac{Z(Z-1)}{A^{1/3}} = (0.71 \text{ MeV})\frac{(30)(29)}{(64)^{1/3}} = 154 \text{ MeV}$$

$$C_4 \frac{(N-Z)^2}{A} = (23.6 \text{ MeV})\frac{(34-30)^2}{64} = 5.90 \text{ MeV}$$

---

[1]The liquid-drop model *describes* that heavy nuclei have $N > Z$. The shell model, as we shall see shortly, *explains* why that is true with a physical argument.

▶ **44.3** continued

Substitute these values into Equation 44.3: $E_b = 1\,005\text{ MeV} - 285\text{ MeV} - 154\text{ MeV} - 5.90\text{ MeV} = \boxed{560\text{ MeV}}$

This value differs from the tabulated value by less than 0.2%. Notice how the sizes of the terms decrease from the first to the fourth term. The fourth term is particularly small for this nucleus, which does not have an excessive number of neutrons.

The Shell Model

The liquid-drop model describes the general behavior of nuclear binding energies relatively well. When the binding energies are studied more closely, however, we find the following features:

- Most stable nuclei have an even value of A. Furthermore, only eight stable nuclei have odd values for both Z and N.
- Figure 44.7 shows a graph of the difference between the binding energy per nucleon calculated by Equation 44.3 and the measured binding energy. There is evidence for regularly spaced peaks in the data that are not described by the semiempirical binding-energy formula. The peaks occur at values of N or Z that have become known as **magic numbers:**

$$Z \text{ or } N = 2, 8, 20, 28, 50, 82 \qquad\qquad (44.4)$$ ◀ **Magic numbers**

- High-precision studies of nuclear radii show deviations from the simple expression in Equation 44.1. Graphs of experimental data show peaks in the curve of radius versus N at values of N equal to the magic numbers.
- A group of *isotones* is a collection of nuclei having the same value of N and varying values of Z. When the number of stable isotones is graphed as function of N, there are peaks in the graph, again at the magic numbers in Equation 44.4.
- Several other nuclear measurements show anomalous behavior at the magic numbers.[2]

These peaks in graphs of experimental data are reminiscent of the peaks in Figure 42.20 for the ionization energy of atoms, which arose because of the shell structure of the atom. The **shell model** of the nucleus, also called the **independent-particle model,** was developed independently by two German scientists: Maria Goeppert-Mayer in 1949 and Hans Jensen (1907–1973) in 1950. Goeppert-Mayer and Jensen

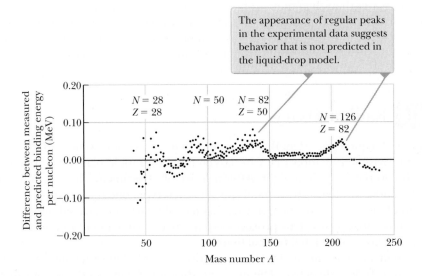

Figure 44.7 The difference between measured binding energies and those calculated from the liquid-drop model as a function of A. (Adapted from R. A. Dunlap, *The Physics of Nuclei and Particles,* Brooks/Cole, Belmont, CA, 2004.)

---

[2]For further details, see chapter 5 of R. A. Dunlap, *The Physics of Nuclei and Particles,* Brooks/Cole, Belmont, CA, 2004.

Maria Goeppert-Mayer
German Scientist (1906–1972)
Goeppert-Mayer was born and educated in Germany. She is best known for her development of the shell model (independent-particle model) of the nucleus, published in 1950. A similar model was simultaneously developed by Hans Jensen, another German scientist. Goeppert-Mayer and Jensen were awarded the Nobel Prize in Physics in 1963 for their extraordinary work in understanding the structure of the nucleus.

The energy levels for the protons are slightly higher than those for the neutrons because of the electric potential energy associated with the system of protons.

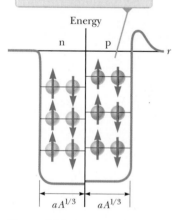

Figure 44.8 A square potential well containing 12 nucleons. The red spheres represent protons, and the gray spheres represent neutrons.

shared the 1963 Nobel Prize in Physics for their work. In this model, each nucleon is assumed to exist in a shell, similar to an atomic shell for an electron. The nucleons exist in quantized energy states, and there are few collisions between nucleons. Obviously, the assumptions of this model differ greatly from those made in the liquid-drop model.

The quantized states occupied by the nucleons can be described by a set of quantum numbers. Because both the proton and the neutron have spin $\frac{1}{2}$, the exclusion principle can be applied to describe the allowed states (as it was for electrons in Chapter 42). That is, each state can contain only two protons (or two neutrons) having *opposite* spins (Fig. 44.8). The proton states differ from those of the neutrons because the two species move in different potential wells. The proton energy levels are farther apart than the neutron levels because the protons experience a superposition of the Coulomb force and the nuclear force, whereas the neutrons experience only the nuclear force.

One factor influencing the observed characteristics of nuclear ground states is *nuclear spin–orbit* effects. The atomic spin–orbit interaction between the spin of an electron and its orbital motion in an atom gives rise to the sodium doublet discussed in Section 42.6 and is magnetic in origin. In contrast, the nuclear spin–orbit effect for nucleons is due to the nuclear force. It is much stronger than in the atomic case, and it has opposite sign. When these effects are taken into account, the shell model is able to account for the observed magic numbers.

The shell model helps us understand why nuclei containing an even number of protons and neutrons are more stable than other nuclei. (There are 160 stable even–even isotopes.) Any particular state is filled when it contains two protons (or two neutrons) having opposite spins. An extra proton or neutron can be added to the nucleus only at the expense of increasing the energy of the nucleus. This increase in energy leads to a nucleus that is less stable than the original nucleus. A careful inspection of the stable nuclei shows that the majority have a special stability when their nucleons combine in pairs, which results in a total angular momentum of zero.

The shell model also helps us understand why nuclei tend to have more neutrons than protons. As in Figure 44.8, the proton energy levels are higher than those for neutrons due to the extra energy associated with Coulomb repulsion. This effect becomes more pronounced as Z increases. Consequently, as Z increases and higher states are filled, a proton level for a given quantum number will be much higher in energy than the neutron level for the same quantum number. In fact, it will be even higher in energy than neutron levels for higher quantum numbers. Hence, it is more energetically favorable for the nucleus to form with neutrons in the lower energy levels rather than protons in the higher energy levels, so the number of neutrons is greater than the number of protons.

More sophisticated models of the nucleus have been and continue to be developed. For example, the *collective model* combines features of the liquid-drop and shell models. The development of theoretical models of the nucleus continues to be an active area of research.

44.4 Radioactivity

In 1896, Becquerel accidentally discovered that uranyl potassium sulfate crystals emit an invisible radiation that can darken a photographic plate even though the plate is covered to exclude light. After a series of experiments, he concluded that the radiation emitted by the crystals was of a new type, one that requires no external stimulation and was so penetrating that it could darken protected photographic plates and ionize gases. This process of spontaneous emission of radiation by uranium was soon to be called **radioactivity.**

Subsequent experiments by other scientists showed that other substances were more powerfully radioactive. The most significant early investigations of this type were conducted by Marie and Pierre Curie (1859–1906). After several years of careful and laborious chemical separation processes on tons of pitchblende, a radioactive ore, the Curies reported the discovery of two previously unknown elements, both radioactive, named polonium and radium. Additional experiments, including Rutherford's famous work on alpha-particle scattering, suggested that radioactivity is the result of the *decay*, or disintegration, of unstable nuclei.

Three types of radioactive decay occur in radioactive substances: alpha (α) decay, in which the emitted particles are ^{4}He nuclei; beta (β) decay, in which the emitted particles are either electrons or positrons; and gamma (γ) decay, in which the emitted particles are high-energy photons. A **positron** is a particle like the electron in all respects except that the positron has a charge of $+e$. (The positron is the *antiparticle* of the electron; see Section 46.2.) The symbol e^- is used to designate an electron, and e^+ designates a positron.

We can distinguish among these three forms of radiation by using the scheme described in Figure 44.9. The radiation from radioactive samples that emit all three types of particles is directed into a region in which there is a magnetic field. Following the particle in a field (magnetic) analysis model, the radiation beam splits into three components, two bending in opposite directions and the third experiencing no change in direction. This simple observation shows that the radiation of the undeflected beam carries no charge (the gamma ray), the component deflected upward corresponds to positively charged particles (alpha particles), and the component deflected downward corresponds to negatively charged particles (e^-). If the beam includes a positron (e^+), it is deflected upward like the alpha particle, but it follows a different trajectory due to its smaller mass.

The three types of radiation have quite different penetrating powers. Alpha particles barely penetrate a sheet of paper, beta particles can penetrate a few millimeters of aluminum, and gamma rays can penetrate several centimeters of lead.

The decay process is probabilistic in nature and can be described with statistical calculations for a radioactive substance of macroscopic size containing a large number of radioactive nuclei. For such large numbers, the rate at which a particular decay process occurs in a sample is proportional to the number of radioactive nuclei present (that is, the number of nuclei that have not yet decayed). If N is the number of undecayed radioactive nuclei present at some instant, the rate of change of N with time is

$$\frac{dN}{dt} = -\lambda N \tag{44.5}$$

where λ, called the **decay constant,** is the probability of decay per nucleus per second. The negative sign indicates that dN/dt is negative; that is, N decreases in time.

Equation 44.5 can be written in the form

$$\frac{dN}{N} = -\lambda \, dt$$

Marie Curie
Polish Scientist (1867–1934)
In 1903, Marie Curie shared the Nobel Prize in Physics with her husband, Pierre, and with Becquerel for their studies of radioactive substances. In 1911, she was awarded a Nobel Prize in Chemistry for the discovery of radium and polonium.

Pitfall Prevention 44.3
Rays or Particles? Early in the history of nuclear physics, the term *radiation* was used to describe the emanations from radioactive nuclei. We now know that alpha radiation and beta radiation involve the emission of particles with nonzero rest energy. Even though they are not examples of electromagnetic radiation, the use of the term *radiation* for all three types of emission is deeply entrenched in our language and in the physics community.

Pitfall Prevention 44.4
Notation Warning In Section 44.1, we introduced the symbol N as an integer representing the number of neutrons in a nucleus. In this discussion, the symbol N represents the number of undecayed nuclei in a radioactive sample remaining after some time interval. As you read further, be sure to consider the context to determine the appropriate meaning for the symbol N.

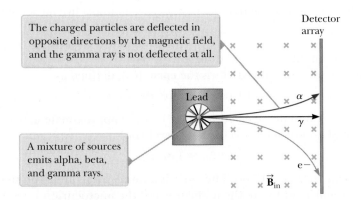

The charged particles are deflected in opposite directions by the magnetic field, and the gamma ray is not deflected at all.

A mixture of sources emits alpha, beta, and gamma rays.

Detector array

Lead

α

γ

e^-

$\vec{B}_{in}$

Figure 44.9 The radiation from radioactive sources can be separated into three components by using a magnetic field to deflect the charged particles. The detector array at the right records the events.

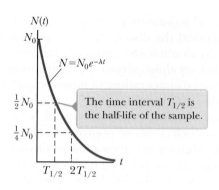

which, upon integration, gives

◀ **Exponential behavior of the number of undecayed nuclei**

$$N = N_0 e^{-\lambda t} \tag{44.6}$$

where the constant N_0 represents the number of undecayed radioactive nuclei at $t = 0$. Equation 44.6 shows that the number of undecayed radioactive nuclei in a sample decreases exponentially with time. The plot of N versus t shown in Figure 44.10 illustrates the exponential nature of the decay. The curve is similar to that for the time variation of electric charge on a discharging capacitor in an RC circuit, as studied in Section 28.4.

The **decay rate** R, which is the number of decays per second, can be obtained by combining Equations 44.5 and 44.6:

◀ **Exponential behavior of the decay rate**

$$R = \left| \frac{dN}{dt} \right| = \lambda N = \lambda N_0 e^{-\lambda t} = R_0 e^{-\lambda t} \tag{44.7}$$

where $R_0 = \lambda N_0$ is the decay rate at $t = 0$. The decay rate R of a sample is often referred to as its **activity.** Note that both N and R decrease exponentially with time.

Another parameter useful in characterizing nuclear decay is the **half-life** $T_{1/2}$:

> The **half-life** of a radioactive substance is the time interval during which half of a given number of radioactive nuclei decay.

Pitfall Prevention 44.5

Half-life It is *not* true that all the original nuclei have decayed after two half-lives! In one half-life, half of the original nuclei will decay. In the second half-life, half of those remaining will decay, leaving $\frac{1}{4}$ of the original number.

To find an expression for the half-life, we first set $N = N_0/2$ and $t = T_{1/2}$ in Equation 44.6 to give

$$\frac{N_0}{2} = N_0 e^{-\lambda T_{1/2}}$$

Canceling the N_0 factors and then taking the reciprocal of both sides, we obtain $e^{\lambda T_{1/2}} = 2$. Taking the natural logarithm of both sides gives

Half-life ▶

$$T_{1/2} = \frac{\ln 2}{\lambda} = \frac{0.693}{\lambda} \tag{44.8}$$

After a time interval equal to one half-life, there are $N_0/2$ radioactive nuclei remaining (by definition); after two half-lives, half of these remaining nuclei have decayed and $N_0/4$ radioactive nuclei are left; after three half-lives, $N_0/8$ are left; and so on. In general, after n half-lives, the number of undecayed radioactive nuclei remaining is

$$N = N_0 \left(\tfrac{1}{2} \right)^n \tag{44.9}$$

where n can be an integer or a noninteger.

A frequently used unit of activity is the **curie** (Ci), defined as

The curie ▶

$$1 \text{ Ci} \equiv 3.7 \times 10^{10} \text{ decays/s}$$

This value was originally selected because it is the approximate activity of 1 g of radium. The SI unit of activity is the **becquerel** (Bq):

The becquerel ▶

$$1 \text{ Bq} \equiv 1 \text{ decay/s}$$

Therefore, 1 Ci = 3.7×10^{10} Bq. The curie is a rather large unit, and the more frequently used activity units are the millicurie and the microcurie.

Quick Quiz 44.2 On your birthday, you measure the activity of a sample of ^{210}Bi, which has a half-life of 5.01 days. The activity you measure is 1.000 μCi. What is the activity of this sample on your next birthday? **(a)** 1.000 μCi **(b)** 0 **(c)** ~ 0.2 μCi **(d)** ~ 0.01 μCi **(e)** $\sim 10^{-22}$ μCi

---

Example 44.4   **How Many Nuclei Are Left?**

The isotope carbon-14, $^{14}_{6}\text{C}$, is radioactive and has a half-life of 5 730 years. If you start with a sample of 1 000 carbon-14 nuclei, how many nuclei will still be undecayed in 25 000 years?

SOLUTION

Conceptualize The time interval of 25 000 years is much longer than the half-life, so only a small fraction of the originally undecayed nuclei will remain.

Categorize The text of the problem allows us to categorize this example as a substitution problem involving radioactive decay.

Analyze Divide the time interval by the half-life to determine the number of half-lives:

$$n = \frac{25\ 000\ \text{yr}}{5\ 730\ \text{yr}} = 4.363$$

Determine how many undecayed nuclei are left after this many half-lives using Equation 44.9:

$$N = N_0(\tfrac{1}{2})^n = 1\ 000(\tfrac{1}{2})^{4.363} = \boxed{49}$$

Finalize As we have mentioned, radioactive decay is a probabilistic process and accurate statistical predictions are possible only with a very large number of atoms. The original sample in this example contains only 1 000 nuclei, which is certainly not a very large number. Therefore, if you counted the number of undecayed nuclei remaining after 25 000 years, it might not be exactly 49.

Example 44.5   **The Activity of Carbon**

At time $t = 0$, a radioactive sample contains 3.50 μg of pure $^{11}_{6}\text{C}$, which has a half-life of 20.4 min.

(A) Determine the number N_0 of nuclei in the sample at $t = 0$.

SOLUTION

Conceptualize The half-life is relatively short, so the number of undecayed nuclei drops rapidly. The molar mass of $^{11}_{6}\text{C}$ is approximately 11.0 g/mol.

Categorize We evaluate results using equations developed in this section, so we categorize this example as a substitution problem.

Find the number of moles in 3.50 μg of pure $^{11}_{6}\text{C}$:

$$n = \frac{3.50 \times 10^{-6}\ \text{g}}{11.0\ \text{g/mol}} = 3.18 \times 10^{-7}\ \text{mol}$$

Find the number of undecayed nuclei in this amount of pure $^{11}_{6}\text{C}$:

$$N_0 = (3.18 \times 10^{-7}\ \text{mol})(6.02 \times 10^{23}\ \text{nuclei/mol}) = \boxed{1.92 \times 10^{17}\ \text{nuclei}}$$

(B) What is the activity of the sample initially and after 8.00 h?

SOLUTION

Find the initial activity of the sample using Equations 44.7 and 44.8:

$$R_0 = \lambda N_0 = \frac{0.693}{T_{1/2}} N_0 = \frac{0.693}{20.4\ \text{min}} \left(\frac{1\ \text{min}}{60\ \text{s}}\right)(1.92 \times 10^{17})$$

$$= (5.66 \times 10^{-4}\ \text{s}^{-1})(1.92 \times 10^{17}) = \boxed{1.09 \times 10^{14}\ \text{Bq}}$$

continued

▶ **44.5** continued

Use Equation 44.7 to find the activity at
$t = 8.00 \text{ h} = 2.88 \times 10^4 \text{ s}$:

$$R = R_0 e^{-\lambda t} = (1.09 \times 10^{14} \text{ Bq})e^{-(5.66 \times 10^{-4} \text{ s}^{-1})(2.88 \times 10^4 \text{ s})} = \boxed{8.96 \times 10^6 \text{ Bq}}$$

Example 44.6   A Radioactive Isotope of Iodine

A sample of the isotope ^{131}I, which has a half-life of 8.04 days, has an activity of 5.0 mCi at the time of shipment. Upon receipt of the sample at a medical laboratory, the activity is 2.1 mCi. How much time has elapsed between the two measurements?

SOLUTION

Conceptualize The sample is continuously decaying as it is in transit. The decrease in the activity is 58% during the time interval between shipment and receipt, so we expect the elapsed time to be greater than the half-life of 8.04 d.

Categorize The stated activity corresponds to many decays per second, so N is large and we can categorize this problem as one in which we can use our statistical analysis of radioactivity.

Analyze Solve Equation 44.7 for the ratio of the final activity to the initial activity:

$$\frac{R}{R_0} = e^{-\lambda t}$$

Take the natural logarithm of both sides:

$$\ln\left(\frac{R}{R_0}\right) = -\lambda t$$

Solve for the time t:

$$(1) \quad t = -\frac{1}{\lambda}\ln\left(\frac{R}{R_0}\right)$$

Use Equation 44.8 to substitute for λ:

$$t = -\frac{T_{1/2}}{\ln 2}\ln\left(\frac{R}{R_0}\right)$$

Substitute numerical values:

$$t = -\frac{8.04 \text{ d}}{0.693}\ln\left(\frac{2.1 \text{ mCi}}{5.0 \text{ mCi}}\right) = \boxed{10 \text{ d}}$$

Finalize This result is indeed greater than the half-life, as expected. This example demonstrates the difficulty in shipping radioactive samples with short half-lives. If the shipment is delayed by several days, only a small fraction of the sample might remain upon receipt. This difficulty can be addressed by shipping a combination of isotopes in which the desired isotope is the product of a decay occurring within the sample. It is possible for the desired isotope to be in *equilibrium*, in which case it is created at the same rate as it decays. Therefore, the amount of the desired isotope remains constant during the shipping process and subsequent storage. When needed, the desired isotope can be separated from the rest of the sample; its decay from the initial activity begins at this point rather than upon shipment.

44.5 The Decay Processes

As we stated in Section 44.4, a radioactive nucleus spontaneously decays by one of three processes: alpha decay, beta decay, or gamma decay. Figure 44.11 shows a close-up view of a portion of Figure 44.4 from $Z = 65$ to $Z = 80$. The black circles are the stable nuclei seen in Figure 44.4. In addition, unstable nuclei above and below the line of stability for each value of Z are shown. Above the line of stability, the blue circles show unstable nuclei that are neutron-rich and undergo a beta decay process in which an electron is emitted. Below the black circles are red circles corresponding to proton-rich unstable nuclei that primarily undergo a beta-decay process in which a positron is emitted or a competing process called electron capture. Beta decay and electron capture are described in more detail below. Further below the line of stability (with a few exceptions) are tan circles that represent very proton-rich nuclei for which the primary decay mechanism is alpha decay, which we discuss first.

Alpha Decay

A nucleus emitting an alpha particle ($^4_2$He) loses two protons and two neutrons. Therefore, the atomic number Z decreases by 2, the mass number A decreases by 4, and the neutron number decreases by 2. The decay can be written

$$^A_Z X \rightarrow {}^{A-4}_{Z-2} Y + {}^4_2 He \tag{44.10}$$

where X is called the **parent nucleus** and Y the **daughter nucleus.** As a general rule in any decay expression such as this one, (1) the sum of the mass numbers A must be the same on both sides of the decay and (2) the sum of the atomic numbers Z must be the same on both sides of the decay. As examples, ^{238}U and ^{226}Ra are both alpha emitters and decay according to the schemes

$$^{238}_{92} U \rightarrow {}^{234}_{90} Th + {}^4_2 He \tag{44.11}$$

$$^{226}_{88} Ra \rightarrow {}^{222}_{86} Rn + {}^4_2 He \tag{44.12}$$

The decay of ^{226}Ra is shown in Figure 44.12.

When the nucleus of one element changes into the nucleus of another as happens in alpha decay, the process is called **spontaneous decay.** In any spontaneous decay, relativistic energy and momentum of the parent nucleus as an isolated system must be conserved. The final components of the system are the daughter nucleus and the alpha particle. If we call M_X the mass of the parent nucleus, M_Y the mass of the daughter nucleus, and M_α the mass of the alpha particle, we can define the **disintegration energy** Q of the system as

$$Q = (M_X - M_Y - M_\alpha)c^2 \tag{44.13}$$

The energy Q is in joules when the masses are in kilograms and c is the speed of light, 3.00×10^8 m/s. When the masses are expressed in atomic mass units u, however, Q can be calculated in MeV using the expression

$$Q = (M_X - M_Y - M_\alpha) \times 931.494 \text{ MeV/u} \tag{44.14}$$

Table 44.2 (page 1396) contains information on selected isotopes, including masses of neutral atoms that can be used in Equation 44.14 and similar equations.

The disintegration energy Q is the amount of rest energy transformed and appears in the form of kinetic energy in the daughter nucleus and the alpha particle and is sometimes referred to as the Q value of the nuclear decay. Consider the case of the ^{226}Ra decay described in Figure 44.12. If the parent nucleus is at rest before the decay, the total kinetic energy of the products is 4.87 MeV. (See Example 44.7.) Most of this kinetic energy is associated with the alpha particle because this particle is much less massive than the daughter nucleus ^{222}Rn. That is, because the system is also isolated in terms of momentum, the lighter alpha particle recoils with a much higher speed than does the daughter nucleus. Generally, less massive particles carry off most of the energy in nuclear decays.

Experimental observations of alpha-particle energies show a number of discrete energies rather than a single energy because the daughter nucleus may be left in an

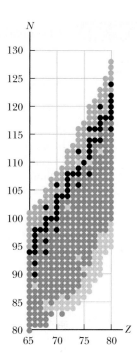

- ● Beta (electron)
- ● Stable
- ● Beta (positron) or electron capture
- ● Alpha

Figure 44.11 A close-up view of the line of stability in Figure 44.4 from $Z = 65$ to $Z = 80$. The black dots represent stable nuclei as in Figure 44.4. The other colored dots represent unstable isotopes above and below the line of stability, with the color of the dot indicating the primary means of decay.

Pitfall Prevention 44.6

Another Q We have seen the symbol Q before, but this use is a brand-new meaning for this symbol: the disintegration energy. In this context, it is not heat, charge, or quality factor for a resonance, for which we have used Q before.

Figure 44.12 The alpha decay of radium-226. The radium nucleus is initially at rest. After the decay, the radon nucleus has kinetic energy K_{Rn} and momentum $\vec{p}_{Rn}$ and the alpha particle has kinetic energy K_α and momentum $\vec{p}_\alpha$.

Table 44.2   Chemical and Nuclear Information for Selected Isotopes

Atomic Number Z	Element	Chemical Symbol	Mass Number A (* means radioactive)	Mass of Neutral Atom (u)	Percent Abundance	Half-life, if Radioactive $T_{1/2}$
−1	electron	e^-	0	0.000 549		
0	neutron	n	1*	1.008 665		614 s
1	hydrogen	$^1H = p$	1	1.007 825	99.988 5	
	[deuterium	$^2H = D$]	2	2.014 102	0.011 5	
	[tritium	$^3H = T$]	3*	3.016 049		12.33 yr
2	helium	He	3	3.016 029	0.000 137	
	[alpha particle	$\alpha = {}^4He$]	4	4.002 603	99.999 863	
			6*	6.018 889		0.81 s
3	lithium	Li	6	6.015 123	7.5	
			7	7.016 005	92.5	
4	beryllium	Be	7*	7.016 930		53.3 d
			8*	8.005 305		10^{-17} s
			9	9.012 182	100	
5	boron	B	10	10.012 937	19.9	
			11	11.009 305	80.1	
6	carbon	C	11*	11.011 434		20.4 min
			12	12.000 000	98.93	
			13	13.003 355	1.07	
			14*	14.003 242		5 730 yr
7	nitrogen	N	13*	13.005 739		9.96 min
			14	14.003 074	99.632	
			15	15.000 109	0.368	
8	oxygen	O	14*	14.008 596		70.6 s
			15*	15.003 066		122 s
			16	15.994 915	99.757	
			17	16.999 132	0.038	
			18	17.999 161	0.205	
9	fluorine	F	18*	18.000 938		109.8 min
			19	18.998 403	100	
10	neon	Ne	20	19.992 440	90.48	
11	sodium	Na	23	22.989 769	100	
12	magnesium	Mg	23*	22.994 124		11.3 s
			24	23.985 042	78.99	
13	aluminum	Al	27	26.981 539	100	
14	silicon	Si	27*	26.986 705		4.2 s
15	phosphorus	P	30*	29.978 314		2.50 min
			31	30.973 762	100	
			32*	31.973 907		14.26 d
16	sulfur	S	32	31.972 071	94.93	
19	potassium	K	39	38.963 707	93.258 1	
			40*	39.963 998	0.011 7	1.28×10^9 yr
20	calcium	Ca	40	39.962 591	96.941	
			42	41.958 618	0.647	
			43	42.958 767	0.135	
25	manganese	Mn	55	54.938 045	100	
26	iron	Fe	56	55.934 938	91.754	
			57	56.935 394	2.119	

continued

Table 44.2   **Chemical and Nuclear Information for Selected Isotopes** *(continued)*

Atomic Number Z	Element	Chemical Symbol	Mass Number A (* means radioactive)	Mass of Neutral Atom (u)	Percent Abundance	Half-life, if Radioactive $T_{1/2}$
27	cobalt	Co	57*	56.936 291		272 d
			59	58.933 195	100	
			60*	59.933 817		5.27 yr
28	nickel	Ni	58	57.935 343	68.076 9	
			60	59.930 786	26.223 1	
29	copper	Cu	63	62.929 598	69.17	
			64*	63.929 764		12.7 h
			65	64.927 789	30.83	
30	zinc	Zn	64	63.929 142	48.63	
37	rubidium	Rb	87*	86.909 181	27.83	
38	strontium	Sr	87	86.908 877	7.00	
			88	87.905 612	82.58	
			90*	89.907 738		29.1 yr
41	niobium	Nb	93	92.906 378	100	
42	molybdenum	Mo	94	93.905 088	9.25	
44	ruthenium	Ru	98	97.905 287	1.87	
54	xenon	Xe	136*	135.907 219		2.4×10^{21} yr
55	cesium	Cs	137*	136.907 090		30 yr
56	barium	Ba	137	136.905 827	11.232	
58	cerium	Ce	140	139.905 439	88.450	
59	praseodymium	Pr	141	140.907 653	100	
60	neodymium	Nd	144*	143.910 087	23.8	2.3×10^{15} yr
61	promethium	Pm	145*	144.912 749		17.7 yr
79	gold	Au	197	196.966 569	100	
80	mercury	Hg	198	197.966 769	9.97	
			202	201.970 643	29.86	
82	lead	Pb	206	205.974 465	24.1	
			207	206.975 897	22.1	
			208	207.976 652	52.4	
			214*	213.999 805		26.8 min
83	bismuth	Bi	209	208.980 399	100	
84	polonium	Po	210*	209.982 874		138.38 d
			216*	216.001 915		0.145 s
			218*	218.008 973		3.10 min
86	radon	Rn	220*	220.011 394		55.6 s
			222*	222.017 578		3.823 d
88	radium	Ra	226*	226.025 410		1 600 yr
90	thorium	Th	232*	232.038 055	100	1.40×10^{10} yr
			234*	234.043 601		24.1 d
92	uranium	U	234*	234.040 952		2.45×10^{5} yr
			235*	235.043 930	0.720 0	7.04×10^{8} yr
			236*	236.045 568		2.34×10^{7} yr
			238*	238.050 788	99.274 5	4.47×10^{9} yr
93	neptunium	Np	236*	236.046 570		1.15×10^{5} yr
			237*	237.048 173		2.14×10^{6} yr
94	plutonium	Pu	239*	239.052 163		24 120 yr

Source: G. Audi, A. H. Wapstra, and C. Thibault, "The AME2003 Atomic Mass Evaluation," *Nuclear Physics A* **729**:337–676, 2003.

excited quantum state after the decay. As a result, not all the disintegration energy is available as kinetic energy of the alpha particle and daughter nucleus. The emission of an alpha particle is followed by one or more gamma-ray photons (discussed shortly) as the excited nucleus decays to the ground state. The observed discrete alpha-particle energies represent evidence of the quantized nature of the nucleus and allow a determination of the energies of the quantum states.

If one assumes ^{238}U (or any other alpha emitter) decays by emitting either a proton or a neutron, the mass of the decay products would exceed that of the parent nucleus, corresponding to a negative Q value. A negative Q value indicates that such a proposed decay does not occur spontaneously.

> Quick Quiz 44.3 Which of the following is the correct daughter nucleus associated with the alpha decay of $^{157}_{72}$Hf?   (a) $^{153}_{72}$Hf   (b) $^{153}_{70}$Yb   (c) $^{157}_{70}$Yb

---

Example 44.7   **The Energy Liberated When Radium Decays**   **AM**

The ^{226}Ra nucleus undergoes alpha decay according to Equation 44.12.

(A) Calculate the Q value for this process. From Table 44.2, the masses are 226.025 410 u for ^{226}Ra, 222.017 578 u for ^{222}Rn, and 4.002 603 u for $^4_2$He.

SOLUTION

Conceptualize Study Figure 44.12 to understand the process of alpha decay in this nucleus.

Categorize The parent nucleus is an *isolated system* that decays into an alpha particle and a daughter nucleus. The system is isolated in terms of both *energy* and *momentum*.

Analyze Evaluate Q using Equation 44.14:

$$Q = (M_X - M_Y - M_\alpha) \times 931.494 \text{ MeV/u}$$

$$= (226.025\ 410 \text{ u} - 222.017\ 578 \text{ u} - 4.002\ 603 \text{ u}) \times 931.494 \text{ MeV/u}$$

$$= (0.005\ 229 \text{ u}) \times 931.494 \text{ MeV/u} = \boxed{4.87 \text{ MeV}}$$

(B) What is the kinetic energy of the alpha particle after the decay?

Analyze The value of 4.87 MeV is the disintegration energy for the decay. It includes the kinetic energy of both the alpha particle and the daughter nucleus after the decay. Therefore, the kinetic energy of the alpha particle would be *less* than 4.87 MeV.

Set up a conservation of momentum equation, noting that the initial momentum of the system is zero:

$$(1) \quad 0 = M_Y v_Y - M_\alpha v_\alpha$$

Set the disintegration energy equal to the sum of the kinetic energies of the alpha particle and the daughter nucleus (assuming the daughter nucleus is left in the ground state):

$$(2) \quad Q = \tfrac{1}{2}M_\alpha v_\alpha^2 + \tfrac{1}{2}M_Y v_Y^2$$

Solve Equation (1) for v_Y and substitute into Equation (2):

$$Q = \tfrac{1}{2}M_\alpha v_\alpha^2 + \tfrac{1}{2}M_Y\left(\frac{M_\alpha v_\alpha}{M_Y}\right)^2 = \tfrac{1}{2}M_\alpha v_\alpha^2\left(1 + \frac{M_\alpha}{M_Y}\right)$$

$$Q = K_\alpha\left(\frac{M_Y + M_\alpha}{M_Y}\right)$$

Solve for the kinetic energy of the alpha particle:

$$K_\alpha = Q\left(\frac{M_Y}{M_Y + M_\alpha}\right)$$

Evaluate this kinetic energy for the specific decay of ^{226}Ra that we are exploring in this example:

$$K_\alpha = (4.87 \text{ MeV})\left(\frac{222}{222 + 4}\right) = \boxed{4.78 \text{ MeV}}$$

Finalize The kinetic energy of the alpha particle is indeed less than the disintegration energy, but notice that the alpha particle carries away *most* of the energy available in the decay.

To understand the mechanism of alpha decay, let's model the parent nucleus as a system consisting of (1) the alpha particle, already formed as an entity within the nucleus, and (2) the daughter nucleus that will result when the alpha particle is emitted. Figure 44.13 shows a plot of potential energy versus separation distance r between the alpha particle and the daughter nucleus, where the distance marked R is the range of the nuclear force. The curve represents the combined effects of (1) the repulsive Coulomb force, which gives the positive part of the curve for $r > R$, and (2) the attractive nuclear force, which causes the curve to be negative for $r < R$. As shown in Example 44.7, a typical disintegration energy Q is approximately 5 MeV, which is the approximate kinetic energy of the alpha particle, represented by the lower dashed line in Figure 44.13.

According to classical physics, the alpha particle is trapped in a potential well. How, then, does it ever escape from the nucleus? The answer to this question was first provided by George Gamow (1904–1968) in 1928 and independently by R. W. Gurney (1898–1953) and E. U. Condon (1902–1974) in 1929, using quantum mechanics. In the view of quantum mechanics, there is always some probability that a particle can tunnel through a barrier (Section 41.5). That is exactly how we can describe alpha decay: the alpha particle tunnels through the barrier in Figure 44.13, escaping the nucleus. Furthermore, this model agrees with the observation that higher-energy alpha particles come from nuclei with shorter half-lives. For higher-energy alpha particles in Figure 44.13, the barrier is narrower and the probability is higher that tunneling occurs. The higher probability translates to a shorter half-life.

As an example, consider the decays of ^{238}U and ^{226}Ra in Equations 44.11 and 44.12, along with the corresponding half-lives and alpha-particle energies:

$$^{238}\text{U}: \quad T_{1/2} = 4.47 \times 10^9 \text{ yr} \quad K_\alpha = 4.20 \text{ MeV}$$

$$^{226}\text{Ra}: \quad T_{1/2} = 1.60 \times 10^3 \text{ yr} \quad K_\alpha = 4.78 \text{ MeV}$$

Notice that a relatively small difference in alpha-particle energy is associated with a tremendous difference of six orders of magnitude in the half-life. The origin of this effect can be understood as follows. Figure 44.13 shows that the curve below an alpha-particle energy of 5 MeV has a slope with a relatively small magnitude. Therefore, a small difference in energy on the vertical axis has a relatively large effect on the width of the potential barrier. Second, recall Equation 41.22, which describes the exponential dependence of the probability of transmission on the barrier width. These two factors combine to give the very sensitive relationship between half-life and alpha-particle energy that the data above suggest.

A life-saving application of alpha decay is the household smoke detector, shown in Figure 44.14. The detector consists of an ionization chamber, a sensitive current detector, and an alarm. A weak radioactive source (usually $^{241}_{95}$Am) ionizes the air in the chamber of the detector, creating charged particles. A voltage is maintained between the plates inside the chamber, setting up a small but detectable current in the external circuit due to the ions acting as charge carriers between the plates. As long as the current is maintained, the alarm is deactivated. If smoke drifts into the chamber, however, the ions become attached to the smoke particles. These heavier particles do not drift as readily as do the lighter ions, which causes a decrease in the detector current. The external circuit senses this decrease in current and sets off the alarm.

Beta Decay

When a radioactive nucleus undergoes beta decay, the daughter nucleus contains the same number of nucleons as the parent nucleus but the atomic number is changed by 1, which means that the number of protons changes:

$$^A_Z\text{X} \rightarrow ^{A}_{Z+1}\text{Y} + e^- \quad \text{(incomplete expression)} \tag{44.15}$$

$$^A_Z\text{X} \rightarrow ^{A}_{Z-1}\text{Y} + e^+ \quad \text{(incomplete expression)} \tag{44.16}$$

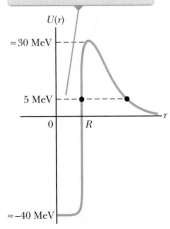

Classically, the 5-MeV energy of the alpha particle is not sufficiently large to overcome the energy barrier, so the particle should not be able to escape from the nucleus.

$U(r)$

≈ 30 MeV

5 MeV

0 R r

≈ -40 MeV

Figure 44.13 Potential energy versus separation distance for a system consisting of an alpha particle and a daughter nucleus. The alpha particle escapes by tunneling through the barrier.

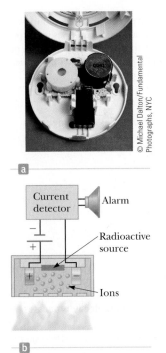

a

Current detector — Alarm

Radioactive source

Ions

b

Figure 44.14 (a) A smoke detector uses alpha decay to determine whether smoke is in the air. The alpha source is in the black cylinder at the right. (b) Smoke entering the chamber reduces the detected current, causing the alarm to sound.

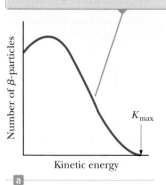

The observed energies of beta particles are continuous, having all values up to a maximum value.

Number of β-particles

$K_{\max}$

Kinetic energy

a

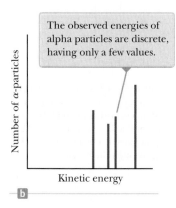

The observed energies of alpha particles are discrete, having only a few values.

Number of α-particles

Kinetic energy

b

Figure 44.15 (a) Distribution of beta-particle energies in a typical beta decay. (b) Distribution of alpha-particle energies in a typical alpha decay.

where, as mentioned in Section 44.4, e⁻ designates an electron and e⁺ designates a positron, with *beta particle* being the general term referring to either. *Beta decay is not described completely by these expressions.* We shall give reasons for this statement shortly.

As with alpha decay, the nucleon number and total charge are both conserved in beta decays. Because *A* does not change but *Z* does, we conclude that in beta decay, either a neutron changes to a proton (Eq. 44.15) or a proton changes to a neutron (Eq. 44.16). Note that the electron or positron emitted in these decays is not present beforehand in the nucleus; it is created in the process of the decay from the rest energy of the decaying nucleus. Two typical beta-decay processes are

$$^{14}_{6}\text{C} \;\rightarrow\; ^{14}_{7}\text{N} + \text{e}^- \qquad \text{(incomplete expression)} \qquad \textbf{(44.17)}$$

$$^{12}_{7}\text{N} \;\rightarrow\; ^{12}_{6}\text{C} + \text{e}^+ \qquad \text{(incomplete expression)} \qquad \textbf{(44.18)}$$

Let's consider the energy of the system undergoing beta decay before and after the decay. As with alpha decay, energy of the isolated system must be conserved. Experimentally, it is found that beta particles from a single type of nucleus are emitted over a continuous range of energies (Fig. 44.15a), as opposed to alpha decay, in which the alpha particles are emitted with discrete energies (Fig. 44.15b). The kinetic energy of the system after the decay is equal to the decrease in rest energy of the system, that is, the *Q* value. Because all decaying nuclei in the sample have the same initial mass, however, *the Q value must be the same for each decay.* So, why do the emitted particles have the range of kinetic energies shown in Figure 44.15a? The isolated system model and the law of conservation of energy seem to be violated! It becomes worse: further analysis of the decay processes described by Equations 44.15 and 44.16 shows that the laws of conservation of angular momentum (spin) and linear momentum are also violated!

After a great deal of experimental and theoretical study, Pauli in 1930 proposed that a third particle must be present in the decay products to carry away the "missing" energy and momentum. Fermi later named this particle the **neutrino** (little neutral one) because it had to be electrically neutral and have little or no mass. Although it eluded detection for many years, the neutrino (symbol ν, Greek nu) was finally detected experimentally in 1956 by Frederick Reines (1918–1998), who received the Nobel Prize in Physics for this work in 1995. The neutrino has the following properties:

Properties of the neutrino ▶

- It has zero electric charge.
- Its mass is either zero (in which case it travels at the speed of light) or very small; much recent persuasive experimental evidence suggests that the neutrino mass is not zero. Current experiments place the upper bound of the mass of the neutrino at approximately 7 eV/c^2.
- It has a spin of $\frac{1}{2}$, which allows the law of conservation of angular momentum to be satisfied in beta decay.
- It interacts very weakly with matter and is therefore very difficult to detect.

We can now write the beta-decay processes (Eqs. 44.15 and 44.16) in their correct and complete form:

Beta decay processes ▶

$$^{A}_{Z}\text{X} \;\rightarrow\; ^{A}_{Z+1}\text{Y} + \text{e}^- + \bar{\nu} \qquad \text{(complete expression)} \qquad \textbf{(44.19)}$$

$$^{A}_{Z}\text{X} \;\rightarrow\; ^{A}_{Z-1}\text{Y} + \text{e}^+ + \nu \qquad \text{(complete expression)} \qquad \textbf{(44.20)}$$

as well as those for carbon-14 and nitrogen-12 (Eqs. 44.17 and 44.18):

$$^{14}_{6}\text{C} \;\rightarrow\; ^{14}_{7}\text{N} + \text{e}^- + \bar{\nu} \qquad \text{(complete expression)} \qquad \textbf{(44.21)}$$

$$^{12}_{7}\text{N} \;\rightarrow\; ^{12}_{6}\text{C} + \text{e}^+ + \nu \qquad \text{(complete expression)} \qquad \textbf{(44.22)}$$

where the symbol $\bar{\nu}$ represents the **antineutrino,** the antiparticle to the neutrino. We shall discuss antiparticles further in Chapter 46. For now, it suffices to say that a neutrino is emitted in positron decay and an antineutrino is emitted in electron

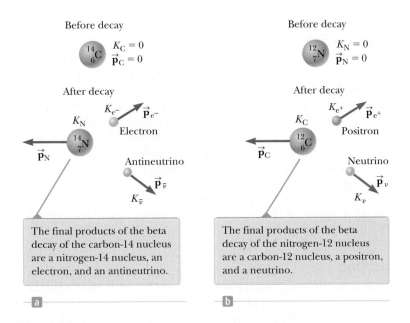

Figure 44.16 (a) The beta decay of carbon-14. (b) The beta decay of nitrogen-12.

decay. As with alpha decay, the decays listed above are analyzed by applying conservation laws, but relativistic expressions must be used for beta particles because their kinetic energy is large (typically 1 MeV) compared with their rest energy of 0.511 MeV. Figure 44.16 shows a pictorial representation of the decays described by Equations 44.21 and 44.22.

In Equation 44.19, the number of protons has increased by one and the number of neutrons has decreased by one. We can write the fundamental process of e^- decay in terms of a neutron changing into a proton as follows:

$$\text{n} \rightarrow p + e^- + \bar{\nu} \tag{44.23}$$

The electron and the antineutrino are ejected from the nucleus, with the net result that there is one more proton and one fewer neutron, consistent with the changes in Z and $A - Z$. A similar process occurs in e^+ decay, with a proton changing into a neutron, a positron, and a neutrino. This latter process can only occur within the nucleus, with the result that the nuclear mass decreases. It cannot occur for an isolated proton because its mass is less than that of the neutron.

A process that competes with e^+ decay is **electron capture,** which occurs when a parent nucleus captures one of its own orbital electrons and emits a neutrino. The final product after decay is a nucleus whose charge is $Z - 1$:

$$^A_Z\text{X} + ^{\ 0}_{-1}\text{e} \rightarrow ^{\ A}_{Z-1}\text{Y} + \nu \tag{44.24}$$

◀ **Electron capture**

In most cases, it is a K-shell electron that is captured and the process is therefore referred to as **K capture.** One example is the capture of an electron by ^7_4Be:

$$^7_4\text{Be} + ^{\ 0}_{-1}\text{e} \rightarrow ^7_3\text{Li} + \nu$$

Because the neutrino is very difficult to detect, electron capture is usually observed by the x-rays given off as higher-shell electrons cascade downward to fill the vacancy created in the K shell.

Finally, we specify Q values for the beta-decay processes. The Q values for e^- decay and electron capture are given by $Q = (M_X - M_Y)c^2$, where M_X and M_Y are the masses of neutral atoms. In e^- decay, the parent nucleus experiences an increase in atomic number and, for the atom to become neutral, an electron must be absorbed by the atom. If the neutral parent atom and an electron (which will eventually combine with the daughter to form a neutral atom) is the initial system and the final system is the neutral daughter atom and the beta-ejected electron, the system contains a free electron both before and after the decay. Therefore, in subtracting the initial and final masses of the system, this electron mass cancels.

Pitfall Prevention 44.7

Mass Number of the Electron An alternative notation for an electron, as we see in Equation 44.24, is the symbol $^{\ 0}_{-1}\text{e}$, which does not imply that the electron has zero rest energy. The mass of the electron is so much smaller than that of the lightest nucleon, however, that we approximate it as zero in the context of nuclear decays and reactions.

The Q values for e^+ decay are given by $Q = (M_X - M_Y - 2m_e)c^2$. The extra term $-2m_ec^2$ in this expression is necessary because the atomic number of the parent decreases by one when the daughter is formed. After it is formed by the decay, the daughter atom sheds one electron to form a neutral atom. Therefore, the final products are the daughter atom, the shed electron, and the ejected positron.

These relationships are useful in determining whether or not a process is energetically possible. For example, the Q value for proposed e^+ decay for a particular parent nucleus may turn out to be negative. In that case, this decay does not occur. The Q value for electron capture for this parent nucleus, however, may be a positive number, so electron capture can occur even though e^+ decay is not possible. Such is the case for the decay of ^7_4Be shown above.

Quick **Quiz** 44.4 Which of the following is the correct daughter nucleus associated with the beta decay of $^{184}_{72}\text{Hf}$? **(a)** $^{183}_{72}\text{Hf}$ **(b)** $^{183}_{73}\text{Ta}$ **(c)** $^{184}_{73}\text{Ta}$

Carbon Dating

The beta decay of ^{14}C (Eq. 44.21) is commonly used to date organic samples. Cosmic rays in the upper atmosphere cause nuclear reactions (Section 44.7) that create ^{14}C. The ratio of ^{14}C to ^{12}C in the carbon dioxide molecules of our atmosphere has a constant value of approximately $r_0 = 1.3 \times 10^{-12}$. The carbon atoms in all living organisms have this same $^{14}\text{C}/^{12}\text{C}$ ratio r_0 because the organisms continuously exchange carbon dioxide with their surroundings. When an organism dies, however, it no longer absorbs ^{14}C from the atmosphere, and so the $^{14}\text{C}/^{12}\text{C}$ ratio decreases as the ^{14}C decays with a half-life of 5 730 yr. It is therefore possible to measure the age of a material by measuring its ^{14}C activity. Using this technique, scientists have been able to identify samples of wood, charcoal, bone, and shell as having lived from 1 000 to 25 000 years ago. This knowledge has helped us reconstruct the history of living organisms—including humans—during this time span.

A particularly interesting example is the dating of the Dead Sea Scrolls. This group of manuscripts was discovered by a shepherd in 1947. Translation showed them to be religious documents, including most of the books of the Old Testament. Because of their historical and religious significance, scholars wanted to know their age. Carbon dating applied to the material in which they were wrapped established their age at approximately 1 950 yr.

Conceptual Example 44.8 **The Age of Iceman**

In 1991, German tourists discovered the well-preserved remains of a man, now called "Ötzi the Iceman," trapped in a glacier in the Italian Alps. (See the photograph at the opening of this chapter.) Radioactive dating with ^{14}C revealed that this person was alive approximately 5 300 years ago. Why did scientists date a sample of Ötzi using ^{14}C rather than ^{11}C, which is a beta emitter having a half-life of 20.4 min?

SOLUTION

Because ^{14}C has a half-life of 5 730 yr, the fraction of ^{14}C nuclei remaining after thousands of years is high enough to allow accurate measurements of changes in the sample's activity. Because ^{11}C has a very short half-life, it is not useful; its activity decreases to a vanishingly small value over the age of the sample, making it impossible to detect.

An isotope used to date a sample must be present in a known amount in the sample when it is formed. As a general rule, the isotope chosen to date a sample should also have a half-life that is on the same order of magnitude as the age of the sample. If the half-life is much less than the age of the sample, there won't be enough activity left to measure because almost all the original radioactive nuclei will have decayed. If the half-life is much greater than the age of the sample, the amount of decay that has taken place since the sample died will be too small to measure. For example, if you have a specimen estimated

▶ **44.8** continued

to have died 50 years ago, neither ^{14}C (5 730 yr) nor ^{11}C (20 min) is suitable. If you know your sample contains

hydrogen, however, you can measure the activity of ^3H (tritium), a beta emitter that has a half-life of 12.3 yr.

Example 44.9 Radioactive Dating

A piece of charcoal containing 25.0 g of carbon is found in some ruins of an ancient city. The sample shows a ^{14}C activity R of 250 decays/min. How long has the tree from which this charcoal came been dead?

SOLUTION

Conceptualize Because the charcoal was found in ancient ruins, we expect the current activity to be smaller than the initial activity. If we can determine the initial activity, we can find out how long the wood has been dead.

Categorize The text of the question helps us categorize this example as a carbon dating problem.

Analyze Solve Equation 44.7 for t:

$$(1) \quad t = -\frac{1}{\lambda} \ln\left(\frac{R}{R_0}\right)$$

Evaluate the ratio R/R_0 using Equation 44.7, the initial value of the $^{14}C/^{12}C$ ratio r_0, the number of moles n of carbon, and Avogadro's number N_A:

$$\frac{R}{R_0} = \frac{R}{\lambda N_0(^{14}C)} = \frac{R}{\lambda r_0 N_0(^{12}C)} = \frac{R}{\lambda r_0 n N_A}$$

Replace the number of moles in terms of the molar mass M of carbon and the mass m of the sample and substitute for the decay constant λ:

$$\frac{R}{R_0} = \frac{R}{(\ln 2/T_{1/2}) r_0 (m/M) N_A} = \frac{R M T_{1/2}}{r_0 m N_A \ln 2}$$

Substitute numerical values:

$$\frac{R}{R_0} = \frac{(250 \text{ min}^{-1})(12.0 \text{ g/mol})(5\,730 \text{ yr})}{(1.3 \times 10^{-12})(25.0 \text{ g})(6.022 \times 10^{23} \text{ mol}^{-1}) \ln 2}\left(\frac{3.156 \times 10^7 \text{ s}}{1 \text{ yr}}\right)\left(\frac{1 \text{ min}}{60 \text{ s}}\right)$$

$$= 0.667$$

Substitute this ratio into Equation (1) and substitute for the decay constant λ:

$$t = -\frac{1}{\lambda} \ln\left(\frac{R}{R_0}\right) = -\frac{T_{1/2}}{\ln 2} \ln\left(\frac{R}{R_0}\right)$$

$$= -\frac{5\,730 \text{ yr}}{\ln 2} \ln (0.667) = \boxed{3.4 \times 10^3 \text{ yr}}$$

Finalize Note that the time interval found here is on the same order of magnitude as the half-life, so ^{14}C is a valid isotope to use for this sample, as discussed in Conceptual Example 44.8.

Gamma Decay

Very often, a nucleus that undergoes radioactive decay is left in an excited energy state. The nucleus can then undergo a second decay to a lower-energy state, perhaps to the ground state, by emitting a high-energy photon:

$$_Z^A X^* \rightarrow {}_Z^A X + \gamma \qquad\qquad (44.25) \qquad ◀ \text{ Gamma decay}$$

where X* indicates a nucleus in an excited state. The typical half-life of an excited nuclear state is 10^{-10} s. Photons emitted in such a de-excitation process are called gamma rays. Such photons have very high energy (1 MeV to 1 GeV) relative to the energy of visible light (approximately 1 eV). Recall from Section 42.3 that the energy of a photon emitted or absorbed by an atom equals the difference in energy

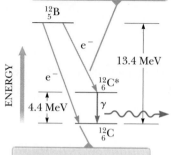

In this decay process, the daughter nucleus is in an excited state, denoted by $^{12}_{6}C^*$, and the beta decay is followed by a gamma decay.

In this decay process, the daughter nucleus $^{12}_{6}C$ is left in the ground state.

Figure 44.17 An energy-level diagram showing the initial nuclear state of a ^{12}B nucleus and two possible lower-energy states of the ^{12}C nucleus.

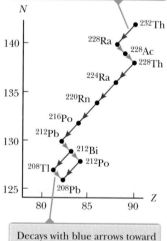

Decays with violet arrows toward the lower left are alpha decays, in which A changes by 4.

Decays with blue arrows toward the lower right are beta decays, in which A does not change.

Figure 44.18 Successive decays for the ^{232}Th series.

between the two electronic states involved in the transition. Similarly, a gamma-ray photon has an energy hf that equals the energy difference ΔE between two nuclear energy levels. When a nucleus decays by emitting a gamma ray, the only change in the nucleus is that it ends up in a lower-energy state. There are no changes in Z, N, or A.

A nucleus may reach an excited state as the result of a violent collision with another particle. More common, however, is for a nucleus to be in an excited state after it has undergone alpha or beta decay. The following sequence of events represents a typical situation in which gamma decay occurs:

$$^{12}_{5}B \;\rightarrow\; ^{12}_{6}C^* + e^- + \overline{\nu} \tag{44.26}$$

$$^{12}_{6}C^* \;\rightarrow\; ^{12}_{6}C + \gamma \tag{44.27}$$

Figure 44.17 shows the decay scheme for ^{12}B, which undergoes beta decay to either of two levels of ^{12}C. It can either (1) decay directly to the ground state of ^{12}C by emitting a 13.4-MeV electron or (2) undergo beta decay to an excited state of $^{12}C^*$ followed by gamma decay to the ground state. The latter process results in the emission of a 9.0-MeV electron and a 4.4-MeV photon.

The various pathways by which a radioactive nucleus can undergo decay are summarized in Table 44.3.

44.6 Natural Radioactivity

Radioactive nuclei are generally classified into two groups: (1) unstable nuclei found in nature, which give rise to **natural radioactivity,** and (2) unstable nuclei produced in the laboratory through nuclear reactions, which exhibit **artificial radioactivity.**

As Table 44.4 shows, there are three series of naturally occurring radioactive nuclei. Each series starts with a specific long-lived radioactive isotope whose half-life exceeds that of any of its unstable descendants. The three natural series begin with the isotopes ^{238}U, ^{235}U, and ^{232}Th, and the corresponding stable end products are three isotopes of lead: ^{206}Pb, ^{207}Pb, and ^{208}Pb. The fourth series in Table 44.4 begins with ^{237}Np and has as its stable end product ^{209}Bi. The element ^{237}Np is a *transuranic* element (one having an atomic number greater than that of uranium) not found in nature. This element has a half-life of "only" 2.14×10^6 years.

Figure 44.18 shows the successive decays for the ^{232}Th series. First, ^{232}Th undergoes alpha decay to ^{228}Ra. Next, ^{228}Ra undergoes two successive beta decays to ^{228}Th. The series continues and finally branches when it reaches ^{212}Bi. At this point, there are two decay possibilities. The sequence shown in Figure 44.18 is characterized by a mass-number decrease of either 4 (for alpha decays) or 0 (for beta or gamma decays). The two uranium series are more complex than the ^{232}Th series. In addition, several naturally occurring radioactive isotopes, such as ^{14}C and ^{40}K, are not part of any decay series.

Because of these radioactive series, our environment is constantly replenished with radioactive elements that would otherwise have disappeared long ago. For example, because our solar system is approximately 5×10^9 years old, the supply of

Table 44.3	Various Decay Pathways
Alpha decay	$^{A}_{Z}X \rightarrow \;^{A-4}_{Z-2}Y + \;^{4}_{2}He$
Beta decay (e^-)	$^{A}_{Z}X \rightarrow \;^{A}_{Z+1}Y + e^- + \overline{\nu}$
Beta decay (e^+)	$^{A}_{Z}X \rightarrow \;^{A}_{Z-1}Y + e^+ + \nu$
Electron capture	$^{A}_{Z}X + e^- \rightarrow \;^{A}_{Z-1}Y + \nu$
Gamma decay	$^{A}_{Z}X^* \rightarrow \;^{A}_{Z}X + \gamma$

Table 44.4   The Four Radioactive Series

Series		Starting Isotope	Half-life (years)	Stable End Product
Uranium	⎫	$^{238}_{92}U$	4.47×10^9	$^{206}_{82}Pb$
Actinium	⎬ Natural	$^{235}_{92}U$	7.04×10^8	$^{207}_{82}Pb$
Thorium	⎭	$^{232}_{90}Th$	1.41×10^{10}	$^{208}_{82}Pb$
Neptunium		$^{237}_{93}Np$	2.14×10^6	$^{209}_{83}Bi$

^{226}Ra (whose half-life is only 1 600 years) would have been depleted by radioactive decay long ago if it were not for the radioactive series starting with ^{238}U.

44.7 Nuclear Reactions

We have studied radioactivity, which is a spontaneous process in which the structure of a nucleus changes. It is also possible to stimulate changes in the structure of nuclei by bombarding them with energetic particles. Such collisions, which change the identity of the target nuclei, are called **nuclear reactions.** Rutherford was the first to observe them, in 1919, using naturally occurring radioactive sources for the bombarding particles. Since then, a wide variety of nuclear reactions has been observed following the development of charged-particle accelerators in the 1930s. With today's advanced technology in particle accelerators and particle detectors, the Large Hadron Collider (see Section 46.10) in Europe can achieve particle energies of 14 000 GeV = 14 TeV. These high-energy particles are used to create new particles whose properties are helping to solve the mysteries of the nucleus.

Consider a reaction in which a target nucleus X is bombarded by a particle a, resulting in a daughter nucleus Y and an outgoing particle b:

$$a + X \rightarrow Y + b \qquad \text{(44.28)}$$

◀ Nuclear reaction

Sometimes this reaction is written in the more compact form

$$X(a, b)Y$$

In Section 44.5, the Q value, or disintegration energy, of a radioactive decay was defined as the rest energy transformed to kinetic energy as a result of the decay process. Likewise, we define the **reaction energy** Q associated with a nuclear reaction as *the difference between the initial and final rest energies resulting from the reaction:*

$$Q = (M_a + M_X - M_Y - M_b)c^2 \qquad \text{(44.29)}$$

◀ Reaction energy Q

As an example, consider the reaction ^{7}Li(p, α)^{4}He. The notation p indicates a proton, which is a hydrogen nucleus. Therefore, we can write this reaction in the expanded form

$$^1_1\text{H} + {}^7_3\text{Li} \rightarrow {}^4_2\text{He} + {}^4_2\text{He}$$

The Q value for this reaction is 17.3 MeV. A reaction such as this one, for which Q is positive, is called **exothermic.** A reaction for which Q is negative is called **endothermic.** To satisfy conservation of momentum for the isolated system, an endothermic reaction does not occur unless the bombarding particle has a kinetic energy greater than Q. (See Problem 74.) The minimum energy necessary for such a reaction to occur is called the **threshold energy.**

If particles a and b in a nuclear reaction are identical so that X and Y are also necessarily identical, the reaction is called a **scattering event.** If the kinetic energy of the system (a and X) before the event is the same as that of the system (b and Y) after the event, it is classified as *elastic scattering.* If the kinetic energy of the system after the event is less than that before the event, the reaction is described as *inelastic scattering.* In this case, the target nucleus has been raised to an excited state by the event, which accounts for the difference in energy. The final system now consists of b and an excited nucleus Y*, and eventually it will become b, Y, and γ, where γ is the gamma-ray photon that is emitted when the system returns to the ground state. This elastic and inelastic terminology is identical to that used in describing collisions between macroscopic objects as discussed in Section 9.4.

In addition to energy and momentum, the total charge and total number of nucleons must be conserved in any nuclear reaction. For example, consider the reaction $^{19}\text{F}(\text{p}, \alpha)^{16}\text{O}$, which has a Q value of 8.11 MeV. We can show this reaction more completely as

$$^{1}_{1}\text{H} + ^{19}_{9}\text{F} \rightarrow ^{16}_{8}\text{O} + ^{4}_{2}\text{He} \tag{44.30}$$

The total number of nucleons before the reaction ($1 + 19 = 20$) is equal to the total number after the reaction ($16 + 4 = 20$). Furthermore, the total charge is the same before ($1 + 9$) and after ($8 + 2$) the reaction.

44.8 Nuclear Magnetic Resonance and Magnetic Resonance Imaging

In this section, we describe an important application of nuclear physics in medicine called *magnetic resonance imaging*. To understand this application, we first discuss the spin angular momentum of the nucleus. This discussion has parallels with the discussion of spin for atomic electrons.

In Chapter 42, we discussed that the electron has an intrinsic angular momentum, called spin. Nuclei also have spin because their component particles—neutrons and protons—each have spin $\frac{1}{2}$ as well as orbital angular momentum within the nucleus. All types of angular momentum obey the quantum rules that were outlined for orbital and spin angular momentum in Chapter 42. In particular, two quantum numbers associated with the angular momentum determine the allowed values of the magnitude of the angular momentum vector and its direction in space. The magnitude of the nuclear angular momentum is $\sqrt{I(I + 1)}\,\hbar$, where I is called the **nuclear spin quantum number** and may be an integer or a half-integer, depending on how the individual proton and neutron spins combine. The quantum number I is the analog to ℓ for the electron in an atom as discussed in Section 42.6. Furthermore, there is a quantum number m_I that is the analog to m_ℓ, in that the allowed projections of the nuclear spin angular momentum vector on the z axis are $m_I \hbar$. The values of m_I range from $-I$ to $+I$ in steps of 1. (In fact, for *any* type of spin with a quantum number S, there is a quantum number m_S that ranges in value from $-S$ to $+S$ in steps of 1.) Therefore, the maximum value of the z component of the spin angular momentum vector is $I\hbar$. Figure 44.19 is a vector model (see Section 42.6) illustrating the possible orientations of the nuclear spin vector and its projections along the z axis for the case in which $I = \frac{3}{2}$.

Nuclear spin has an associated nuclear magnetic moment, similar to that of the electron. The spin magnetic moment of a nucleus is measured in terms of the **nuclear magneton** μ_n, a unit of moment defined as

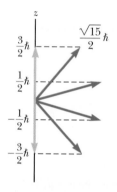

Figure 44.19 A vector model showing possible orientations of the nuclear spin angular momentum vector and its projections along the z axis for the case $I = \frac{3}{2}$.

◀ Nuclear magneton

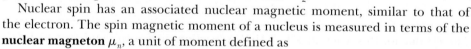

$$\mu_n \equiv \frac{e\hbar}{2m_p} = 5.05 \times 10^{-27}\,\text{J/T} \tag{44.31}$$

where m_p is the mass of the proton. This definition is analogous to that of the Bohr magneton μ_B, which corresponds to the spin magnetic moment of a free electron (see Section 42.6). Note that μ_n is smaller than μ_B ($= 9.274 \times 10^{-24}\,\text{J/T}$) by a factor of 1 836 because of the large difference between the proton mass and the electron mass.

The magnetic moment of a free proton is $2.792\,8\mu_n$. Unfortunately, there is no general theory of nuclear magnetism that explains this value. The neutron also has a magnetic moment, which has a value of $-1.913\,5\mu_n$. The negative sign indicates that this moment is opposite the spin angular momentum of the neutron. The existence of a magnetic moment for the neutron is surprising in view of the neutron being uncharged. That suggests that the neutron is not a fundamental particle but rather has an underlying structure consisting of charged constituents. We shall explore this structure in Chapter 46.

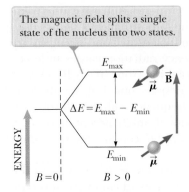

The magnetic field splits a single state of the nucleus into two states.

Figure 44.20 A nucleus with spin $\frac{1}{2}$ is placed in a magnetic field.

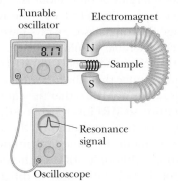

Tunable oscillator   Electromagnet

Sample

Resonance signal

Oscilloscope

Figure 44.21 Experimental arrangement for nuclear magnetic resonance. The radio-frequency magnetic field created by the coil surrounding the sample and provided by the variable-frequency oscillator is perpendicular to the constant magnetic field created by the electromagnet. When the nuclei in the sample meet the resonance condition, the nuclei absorb energy from the radio-frequency field of the coil; this absorption changes the characteristics of the circuit in which the coil is included. Most modern NMR spectrometers use superconducting magnets at fixed field strengths and operate at frequencies of approximately 200 MHz.

The potential energy associated with a magnetic dipole moment $\vec{\mu}$ in an external magnetic field $\vec{B}$ is given by $-\vec{\mu} \cdot \vec{B}$ (Eq. 29.18). When the magnetic moment $\vec{\mu}$ is lined up with the field as closely as quantum physics allows, the potential energy of the dipole–field system has its minimum value E_{min}. When $\vec{\mu}$ is as antiparallel to the field as possible, the potential energy has its maximum value E_{max}. In general, there are other energy states between these values corresponding to the quantized directions of the magnetic moment with respect to the field. For a nucleus with spin $\frac{1}{2}$, there are only two allowed states, with energies E_{min} and E_{max}. These two energy states are shown in Figure 44.20.

It is possible to observe transitions between these two spin states using a technique called **NMR**, for **nuclear magnetic resonance.** A constant magnetic field ($\vec{B}$ in Fig. 44.20) is introduced to define a z axis and split the energies of the spin states. A second, weaker, oscillating magnetic field is then applied perpendicular to $\vec{B}$, creating a cloud of radio-frequency photons around the sample. When the frequency of the oscillating field is adjusted so that the photon energy matches the energy difference between the spin states, there is a net absorption of photons by the nuclei that can be detected electronically.

Figure 44.21 is a simplified diagram of the apparatus used in nuclear magnetic resonance. The energy absorbed by the nuclei is supplied by the tunable oscillator producing the oscillating magnetic field. Nuclear magnetic resonance and a related technique called *electron spin resonance* are extremely important methods for studying nuclear and atomic systems and the ways in which these systems interact with their surroundings.

A widely used medical diagnostic technique called **MRI**, for **magnetic resonance imaging,** is based on nuclear magnetic resonance. Because nearly two-thirds of the atoms in the human body are hydrogen (which gives a strong NMR signal), MRI works exceptionally well for viewing internal tissues. The patient is placed inside a large solenoid that supplies a magnetic field that is constant in time but whose magnitude varies spatially across the body. Because of the variation in the field, hydrogen atoms in different parts of the body have different energy splittings between spin states, so the resonance signal can be used to provide information about the positions of the protons. A computer is used to analyze the position information to provide data for constructing a final image. Contrast in the final image among different types of tissues is created by computer analysis of the time intervals for the nuclei to return to the lower-energy spin state between pulses of radio-frequency photons. Contrast can be enhanced with the use of contrast agents such as gadolinium compounds or iron oxide nanoparticles taken orally or injected intravenously. An MRI scan showing incredible detail in internal body structure is shown in Figure 44.22.

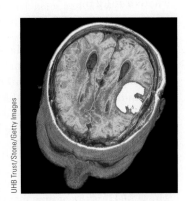

UHB Trust/Stone/Getty Images

Figure 44.22 A color-enhanced MRI scan of a human brain, showing a tumor in white.

The main advantage of MRI over other imaging techniques is that it causes minimal cellular damage. The photons associated with the radio-frequency signals used in MRI have energies of only about 10^{-7} eV. Because molecular bond strengths are much larger (approximately 1 eV), the radio-frequency radiation causes little cellular damage. In comparison, x-rays have energies ranging from 10^4 to 10^6 eV and can cause considerable cellular damage. Therefore, despite some individuals' fears of the word *nuclear* associated with MRI, the radio-frequency radiation involved is overwhelmingly safer than the x-rays that these individuals might accept more readily. A disadvantage of MRI is that the equipment required to conduct the procedure is very expensive, so MRI images are costly.

The magnetic field produced by the solenoid is sufficient to lift a car, and the radio signal is about the same magnitude as that from a small commercial broadcasting station. Although MRI is inherently safe in normal use, the strong magnetic field of the solenoid requires diligent care to ensure that no ferromagnetic materials are located in the room near the MRI apparatus. Several accidents have occurred, such as a 2000 incident in which a gun pulled from a police officer's hand discharged upon striking the machine.

Summary

Definitions

A nucleus is represented by the symbol ^A_ZX, where A is the **mass number** (the total number of nucleons) and Z is the **atomic number** (the total number of protons). The total number of neutrons in a nucleus is the **neutron number** N, where $A = N + Z$. Nuclei having the same Z value but different A and N values are **isotopes** of each other.

The magnetic moment of a nucleus is measured in terms of the **nuclear magneton** μ_n, where

$$\mu_n \equiv \frac{e\hbar}{2m_p} = 5.05 \times 10^{-27}\,\text{J/T} \quad \textbf{(44.31)}$$

Concepts and Principles

Assuming nuclei are spherical, their radius is given by

$$r = aA^{1/3} \quad \textbf{(44.1)}$$

where $a = 1.2$ fm.

Nuclei are stable because of the **nuclear force** between nucleons. This short-range force dominates the Coulomb repulsive force at distances of less than about 2 fm and is independent of charge. Light stable nuclei have equal numbers of protons and neutrons. Heavy stable nuclei have more neutrons than protons. The most stable nuclei have Z and N values that are both even.

The difference between the sum of the masses of a group of separate nucleons and the mass of the compound nucleus containing these nucleons, when multiplied by c^2, gives the **binding energy** E_b of the nucleus. The binding energy of a nucleus can be calculated in MeV using the expression

$$E_b = [ZM(\text{H}) + Nm_n - M(^A_Z\text{X})] \times 931.494 \text{ MeV/u}$$
$$\textbf{(44.2)}$$

where $M(\text{H})$ is the atomic mass of the neutral hydrogen atom, $M(^A_Z\text{X})$ represents the atomic mass of an atom of the isotope ^A_ZX, and m_n is the mass of the neutron.

The **liquid-drop model** of nuclear structure treats the nucleons as molecules in a drop of liquid. The four main contributions influencing binding energy are the volume effect, the surface effect, the Coulomb repulsion effect, and the symmetry effect. Summing such contributions results in the **semiempirical binding-energy formula:**

$$E_b = C_1 A - C_2 A^{2/3} - C_3 \frac{Z(Z-1)}{A^{1/3}} - C_4 \frac{(N-Z)^2}{A} \quad \textbf{(44.3)}$$

The **shell model,** or **independent-particle model,** assumes each nucleon exists in a shell and can only have discrete energy values. The stability of certain nuclei can be explained with this model.

A radioactive substance decays by **alpha decay, beta decay,** or **gamma decay.** An alpha particle is the ^{4}He nucleus, a beta particle is either an electron (e^-) or a positron (e^+), and a gamma particle is a high-energy photon.

If a radioactive material contains N_0 radioactive nuclei at $t = 0$, the number N of nuclei remaining after a time t has elapsed is

$$N = N_0 e^{-\lambda t} \tag{44.6}$$

where λ is the **decay constant,** a number equal to the probability per second that a nucleus will decay. The **decay rate,** or **activity,** of a radioactive substance is

$$R = \left| \frac{dN}{dt} \right| = R_0 e^{-\lambda t} \tag{44.7}$$

where $R_0 = \lambda N_0$ is the activity at $t = 0$. The **half-life** $T_{1/2}$ is the time interval required for half of a given number of radioactive nuclei to decay, where

$$T_{1/2} = \frac{0.693}{\lambda} \tag{44.8}$$

In alpha decay, a helium nucleus is ejected from the parent nucleus with a discrete set of kinetic energies. A nucleus undergoing beta decay emits either an electron (e^-) and an antineutrino ($\bar{\nu}$) or a positron (e^+) and a neutrino (ν). The electron or positron is ejected with a continuous range of energies. In **electron capture,** the nucleus of an atom absorbs one of its own electrons and emits a neutrino. In gamma decay, a nucleus in an excited state decays to its ground state and emits a gamma ray.

Nuclear reactions can occur when a target nucleus X is bombarded by a particle a, resulting in a daughter nucleus Y and an outgoing particle b:

$$a + X \rightarrow Y + b \tag{44.28}$$

The mass–energy conversion in such a reaction, called the **reaction energy** Q, is

$$Q = (M_a + M_X - M_Y - M_b)c^2 \tag{44.29}$$

Objective Questions

1. denotes answer available in *Student Solutions Manual/Study Guide*

1. In nuclear magnetic resonance, suppose we increase the value of the constant magnetic field. As a result, the frequency of the photons that are absorbed in a particular transition changes. How is the frequency of the photons absorbed related to the magnetic field? (a) The frequency is proportional to the square of the magnetic field. (b) The frequency is directly proportional to the magnetic field. (c) The frequency is independent of the magnetic field. (d) The frequency is inversely proportional to the magnetic field. (e) The frequency is proportional to the reciprocal of the square of the magnetic field.

2. When the $^{95}_{36}$Kr nucleus undergoes beta decay by emitting an electron and an antineutrino, does the daughter nucleus (Rb) contain (a) 58 neutrons and 37 protons, (b) 58 protons and 37 neutrons, (c) 54 neutrons and 41 protons, or (d) 55 neutrons and 40 protons?

3. When $^{32}_{15}$P decays to $^{32}_{16}$S, which of the following particles is emitted? (a) a proton (b) an alpha particle (c) an electron (d) a gamma ray (e) an antineutrino

4. The half-life of radium-224 is about 3.6 days. What approximate fraction of a sample remains undecayed after two weeks? (a) $\frac{1}{2}$ (b) $\frac{1}{4}$ (c) $\frac{1}{8}$ (d) $\frac{1}{16}$ (e) $\frac{1}{32}$

5. Two samples of the same radioactive nuclide are prepared. Sample G has twice the initial activity of

sample H. **(i)** How does the half-life of G compare with the half-life of H? (a) It is two times larger. (b) It is the same. (c) It is half as large. **(ii)** After each has passed through five half-lives, how do their activities compare? (a) G has more than twice the activity of H. (b) G has twice the activity of H. (c) G and H have the same activity. (d) G has lower activity than H.

6. If a radioactive nuclide ^A_ZX decays by emitting a gamma ray, what happens? (a) The resulting nuclide has a different Z value. (b) The resulting nuclide has the same A and Z values. (c) The resulting nuclide has a different A value. (d) Both A and Z decrease by one. (e) None of those statements is correct.

7. Does a nucleus designated as $^{40}_{18}$X contain (a) 20 neutrons and 20 protons, (b) 22 protons and 18 neutrons, (c) 18 protons and 22 neutrons, (d) 18 protons and 40 neutrons, or (e) 40 protons and 18 neutrons?

8. When $^{144}_{60}$Nd decays to $^{140}_{58}$Ce, identify the particle that is released. (a) a proton (b) an alpha particle (c) an electron (d) a neutron (e) a neutrino

9. What is the Q value for the reaction ^{9}Be $+ \alpha \rightarrow {}^{12}$C $+$ n? (a) 8.4 MeV (b) 7.3 MeV (c) 6.2 MeV (d) 5.7 MeV (e) 4.2 MeV

10. (i) To predict the behavior of a nucleus in a fission reaction, which model would be more appropriate,

(a) the liquid-drop model or (b) the shell model? **(ii)** Which model would be more successful in predicting the magnetic moment of a given nucleus? Choose from the same answers as in part (i). **(iii)** Which could better explain the gamma-ray spectrum of an excited nucleus? Choose from the same answers as in part (i).

11. A free neutron has a half-life of 614 s. It undergoes beta decay by emitting an electron. Can a free proton undergo a similar decay? (a) yes, the same decay (b) yes, but by emitting a positron (c) yes, but with a very different half-life (d) no

12. Which of the following quantities represents the reaction energy of a nuclear reaction? (a) (final mass − initial mass)$/c^2$ (b) (initial mass − final mass)$/c^2$ (c) (final mass − initial mass)c^2 (d) (initial mass − final mass)c^2 (e) none of those quantities

13. In the decay $^{234}_{90}\text{Th} \rightarrow {}^{A}_{Z}\text{Ra} + {}^{4}_{2}\text{He}$, identify the mass number and the atomic number of the Ra nucleus: (a) $A = 230$, $Z = 92$ (b) $A = 238$, $Z = 88$ (c) $A = 230$, $Z = 88$ (d) $A = 234$, $Z = 88$ (e) $A = 238$, $Z = 86$

Conceptual Questions

1. denotes answer available in *Student Solutions Manual/Study Guide*

1. If a nucleus such as ^{226}Ra initially at rest undergoes alpha decay, which has more kinetic energy after the decay, the alpha particle or the daughter nucleus? Explain your answer.

2. "If no more people were to be born, the law of population growth would strongly resemble the radioactive decay law." Discuss this statement.

3. A student claims that a heavy form of hydrogen decays by alpha emission. How do you respond?

4. In beta decay, the energy of the electron or positron emitted from the nucleus lies somewhere in a relatively large range of possibilities. In alpha decay, however, the alpha-particle energy can only have discrete values. Explain this difference.

5. Can carbon-14 dating be used to measure the age of a rock? Explain.

6. In positron decay, a proton in the nucleus becomes a neutron and its positive charge is carried away by the positron. A neutron, though, has a larger rest energy than a proton. How is that possible?

7. (a) How many values of I_z are possible for $I = \frac{5}{2}$? (b) For $I = 3$?

8. Why do nearly all the naturally occurring isotopes lie above the $N = Z$ line in Figure 44.4?

9. Why are very heavy nuclei unstable?

10. Explain why nuclei that are well off the line of stability in Figure 44.4 tend to be unstable.

11. Consider two heavy nuclei X and Y having similar mass numbers. If X has the higher binding energy, which nucleus tends to be more unstable? Explain your answer.

12. What fraction of a radioactive sample has decayed after two half-lives have elapsed?

13. Figure CQ44.13 shows a watch from the early 20th century. The numbers and the hands of the watch are painted with a paint that contains a small amount of natural radium $^{226}_{88}\text{Ra}$ mixed with a phosphorescent material. The decay of the radium causes the phosphorescent material to glow continuously. The radioactive nuclide $^{226}_{88}\text{Ra}$ has a half-life of approximately 1.60×10^3 years. Being that the solar system is approximately 5 billion years old, why was this isotope still available in the 20th century for use on this watch?

© Richard Megna/Fundamental Photographs

Figure CQ44.13

14. Can a nucleus emit alpha particles that have different energies? Explain.

15. In Rutherford's experiment, assume an alpha particle is headed directly toward the nucleus of an atom. Why doesn't the alpha particle make physical contact with the nucleus?

16. Suppose it could be shown that the cosmic-ray intensity at the Earth's surface was much greater 10 000 years ago. How would this difference affect what we accept as valid carbon-dated values of the age of ancient samples of once-living matter? Explain your answer.

17. Compare and contrast the properties of a photon and a neutrino.

Problems

Section 44.1 Some Properties of Nuclei

1. Find the nuclear radii of (a) $_1^2$H, (b) $_{27}^{60}$Co, (c) $_{79}^{197}$Au, and (d) $_{94}^{239}$Pu.

2. (a) Determine the mass number of a nucleus whose radius is approximately equal to two-thirds the radius of $_{88}^{230}$Ra. (b) Identify the element. (c) Are any other answers possible? Explain.

3. (a) Use energy methods to calculate the distance of **M** closest approach for a head-on collision between an alpha particle having an initial energy of 0.500 MeV and a gold nucleus (^{197}Au) at rest. Assume the gold nucleus remains at rest during the collision. (b) What minimum initial speed must the alpha particle have to approach as close as 300 fm to the gold nucleus?

4. (a) What is the order of magnitude of the number of protons in your body? (b) Of the number of neutrons? (c) Of the number of electrons?

5. Consider the $_{29}^{65}$Cu nucleus. Find approximate values for its (a) radius, (b) volume, and (c) density.

6. Using 2.30×10^{17} kg/m^3 as the density of nuclear matter, find the radius of a sphere of such matter that would have a mass equal to that of a baseball, 0.145 kg.

7. A star ending its life with a mass of four to eight times the Sun's mass is expected to collapse and then undergo a supernova event. In the remnant that is not carried away by the supernova explosion, protons and electrons combine to form a neutron star with approximately twice the mass of the Sun. Such a star can be thought of as a gigantic atomic nucleus. Assume $r = aA^{1/3}$ (Eq. 44.1). If a star of mass 3.98×10^{30} kg is composed entirely of neutrons ($m_n = 1.67 \times 10^{-27}$ kg), what would its radius be?

8. Figure P44.8 shows the potential energy for two protons as a function of separation distance. In the text, it was claimed that, to be visible on such a graph, the peak in the curve is exaggerated by a factor of ten. (a) Find the electric potential energy of a pair of protons separated by 4.00 fm. (b) Verify that the peak in Figure P44.8 is exaggerated by a factor of ten.

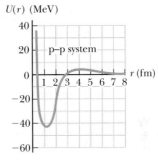

Figure P44.8

9. **Review.** Singly ionized carbon is accelerated through **AMT** 1 000 V and passed into a mass spectrometer to determine the isotopes present (see Chapter 29). The magnitude of the magnetic field in the spectrometer is 0.200 T. The orbit radius for a ^{12}C isotope as it passes through the field is $r = 7.89$ cm. Find the radius of the orbit of a ^{13}C isotope.

10. **Review.** Singly ionized carbon is accelerated through a potential difference ΔV and passed into a mass spectrometer to determine the isotopes present (see Chapter 29). The magnitude of the magnetic field in the spectrometer is B. The orbit radius for an isotope of mass m_1 as it passes through the field is r_1. Find the radius of the orbit of an isotope of mass m_2.

11. An alpha particle ($Z = 2$, mass $= 6.64 \times 10^{-27}$ kg) **M** approaches to within 1.00×10^{-14} m of a carbon nucleus ($Z = 6$). What are (a) the magnitude of the maximum Coulomb force on the alpha particle, (b) the magnitude of the acceleration of the alpha particle at the time of the maximum force, and (c) the potential energy of the system of the alpha particle and the carbon nucleus at this time?

12. In a Rutherford scattering experiment, alpha particles having kinetic energy of 7.70 MeV are fired toward a gold nucleus that remains at rest during the collision. The alpha particles come as close as 29.5 fm to the gold nucleus before turning around. (a) Calculate the de Broglie wavelength for the 7.70-MeV alpha particle and compare it with the distance of closest approach,

29.5 fm. (b) Based on this comparison, why is it proper to treat the alpha particle as a particle and not as a wave in the Rutherford scattering experiment?

13. **Review.** Two golf balls each have a 4.30-cm diameter and are 1.00 m apart. What would be the gravitational force exerted by each ball on the other if the balls were made of nuclear matter?

14. Assume a hydrogen atom is a sphere with diameter 0.100 nm and a hydrogen molecule consists of two such spheres in contact. (a) What fraction of the space in a tank of hydrogen gas at 0°C and 1.00 atm is occupied by the hydrogen molecules themselves? (b) What fraction of the space within one hydrogen atom is occupied by its nucleus, of radius 1.20 fm?

Section 44.2 Nuclear Binding Energy

15. Calculate the binding energy per nucleon for (a) ^{2}H, (b) ^{4}He, (c) ^{56}Fe, and (d) ^{238}U.

16. (a) Calculate the difference in binding energy per nucleon for the nuclei $^{23}_{11}$Na and $^{23}_{12}$Mg. (b) How do you account for the difference?

17. A pair of nuclei for which $Z_1 = N_2$ and $Z_2 = N_1$ are called *mirror isobars* (the atomic and neutron numbers are interchanged). Binding-energy measurements on these nuclei can be used to obtain evidence of the charge independence of nuclear forces (that is, proton–proton, proton–neutron, and neutron–neutron nuclear forces are equal). Calculate the difference in binding energy for the two mirror isobars $^{15}_8$O and $^{15}_7$N. The electric repulsion among eight protons rather than seven accounts for the difference.

18. The peak of the graph of nuclear binding energy per nucleon occurs near ^{56}Fe, which is why iron is prominent in the spectrum of the Sun and stars. Show that ^{56}Fe has a higher binding energy per nucleon than its neighbors ^{55}Mn and ^{59}Co.

19. Nuclei having the same mass numbers are called *isobars*. The isotope $^{139}_{57}$La is stable. A radioactive isobar, $^{139}_{59}$Pr, is located below the line of stable nuclei as shown in Figure P44.19 and decays by e^+ emission. Another

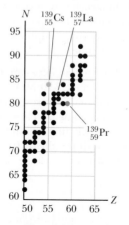

Figure P44.19

radioactive isobar of $^{139}_{57}$La, $^{139}_{55}$Cs, decays by e^- emission and is located above the line of stable nuclei in Figure P44.19. (a) Which of these three isobars has the highest neutron-to-proton ratio? (b) Which has the greatest binding energy per nucleon? (c) Which do you expect to be heavier, $^{139}_{59}$Pr or $^{139}_{55}$Cs?

20. The energy required to construct a uniformly charged sphere of total charge Q and radius R is $U = 3k_eQ^2/5R$, where k_e is the Coulomb constant (see Problem 77). Assume a ^{40}Ca nucleus contains 20 protons uniformly distributed in a spherical volume. (a) How much energy is required to counter their electrical repulsion according to the above equation? (b) Calculate the binding energy of ^{40}Ca. (c) Explain what you can conclude from comparing the result of part (b) with that of part (a).

21. Calculate the minimum energy required to remove a neutron from the $^{43}_{20}$Ca nucleus.

Section 44.3 Nuclear Models

22. Using the graph in Figure 44.5, estimate how much energy is released when a nucleus of mass number 200 fissions into two nuclei each of mass number 100.

23. (a) Use the semiempirical binding-energy formula (Eq. 44.3) to compute the binding energy for $^{56}_{26}$Fe. (b) What percentage is contributed to the binding energy by each of the four terms?

24. (a) In the liquid-drop model of nuclear structure, why does the surface-effect term $-C_2A^{2/3}$ have a negative sign? (b) **What If?** The binding energy of the nucleus increases as the volume-to-surface area ratio increases. Calculate this ratio for both spherical and cubical shapes and explain which is more plausible for nuclei.

Section 44.4 Radioactivity

25. What time interval is required for the activity of a sample of the radioactive isotope $^{72}_{33}$As to decrease by 90.0% from its original value? The half-life of $^{72}_{33}$As is 26 h.

26. A freshly prepared sample of a certain radioactive isotope has an activity of 10.0 mCi. After 4.00 h, its activity is 8.00 mCi. Find (a) the decay constant and (b) the half-life. (c) How many atoms of the isotope were contained in the freshly prepared sample? (d) What is the sample's activity 30.0 h after it is prepared?

27. A sample of radioactive material contains 1.00×10^{15} atoms and has an activity of 6.00×10^{11} Bq. What is its half-life?

28. From the equation expressing the law of radioactive decay, derive the following useful expressions for the decay constant and the half-life, in terms of the time interval Δt during which the decay rate decreases from R_0 to R:

$$\lambda = \frac{1}{\Delta t}\ln\left(\frac{R_0}{R}\right) \qquad T_{1/2} = \frac{(\ln 2)\,\Delta t}{\ln(R_0/R)}$$

29. The radioactive isotope ^{198}Au has a half-life of 64.8 h. **M** A sample containing this isotope has an initial activity ($t = 0$) of 40.0 μCi. Calculate the number of nuclei that decay in the time interval between $t_1 = 10.0$ h and $t_2 = 12.0$ h.

30. A radioactive nucleus has half-life $T_{1/2}$. A sample containing these nuclei has initial activity R_0 at $t = 0$. Calculate the number of nuclei that decay during the interval between the later times t_1 and t_2.

31. The half-life of ^{131}I is 8.04 days. (a) Calculate the decay constant for this nuclide. (b) Find the number of ^{131}I nuclei necessary to produce a sample with an activity of 6.40 mCi. (c) A sample of ^{131}I with this initial activity decays for 40.2 d. What is the activity at the end of that period?

32. Tritium has a half-life of 12.33 years. What fraction of the nuclei in a tritium sample will remain (a) after 5.00 yr? (b) After 10.0 yr? (c) After 123.3 yr? (d) According to Equation 44.6, an infinite amount of time is required for the entire sample to decay. Discuss whether that is realistic.

33. Consider a radioactive sample. Determine the ratio of the number of nuclei decaying during the first half of its half-life to the number of nuclei decaying during the second half of its half-life.

34. (a) The daughter nucleus formed in radioactive decay is often radioactive. Let N_{10} represent the number of parent nuclei at time $t = 0$, $N_1(t)$ the number of parent nuclei at time t, and λ_1 the decay constant of the parent. Suppose the number of daughter nuclei at time $t = 0$ is zero. Let $N_2(t)$ be the number of daughter nuclei at time t and let λ_2 be the decay constant of the daughter. Show that $N_2(t)$ satisfies the differential equation

$$\frac{dN_2}{dt} = \lambda_1 N_1 - \lambda_2 N_2$$

(b) Verify by substitution that this differential equation has the solution

$$N_2(t) = \frac{N_{10}\lambda_1}{\lambda_1 - \lambda_2}\left(e^{-\lambda_2 t} - e^{-\lambda_1 t}\right)$$

This equation is the law of successive radioactive decays. (c) ^{218}Po decays into ^{214}Pb with a half-life of 3.10 min, and ^{214}Pb decays into ^{214}Bi with a half-life of 26.8 min. On the same axes, plot graphs of $N_1(t)$ for ^{218}Po and $N_2(t)$ for ^{214}Pb. Let $N_{10} = 1\,000$ nuclei and choose values of t from 0 to 36 min in 2-min intervals. (d) The curve for ^{214}Pb obtained in part (c) at first rises to a maximum and then starts to decay. At what instant t_m is the number of ^{214}Pb nuclei a maximum? (e) By applying the condition for a maximum $dN_2/dt = 0$, derive a symbolic equation for t_m in terms of λ_1 and λ_2. (f) Explain whether the value obtained in part (c) agrees with this equation.

Section 44.5 The Decay Processes

35. Determine which decays can occur spontaneously.
(a) $^{40}_{20}$Ca $\rightarrow$ e$^+$ + $^{40}_{19}$K (b) $^{98}_{44}$Ru $\rightarrow$ $^{4}_{2}$He + $^{94}_{42}$Mo
(c) $^{144}_{60}$Nd $\rightarrow$ $^{4}_{2}$He + $^{140}_{58}$Ce

36. A ^{3}H nucleus beta decays into ^{3}He by creating an electron and an antineutrino according to the reaction

$$^{3}_{1}\text{H} \rightarrow ^{3}_{2}\text{He} + \text{e}^- + \bar{\nu}$$

Determine the total energy released in this decay.

37. The ^{14}C isotope undergoes beta decay according to the process given by Equation 44.21. Find the Q value for this process.

38. Identify the unknown nuclide or particle (X).
(a) X $\rightarrow$ $^{65}_{28}$Ni + γ (b) $^{215}_{84}$Po $\rightarrow$ X + α
(c) X $\rightarrow$ $^{55}_{26}$Fe + e$^+$ + ν

39. Find the energy released in the alpha decay

$$^{238}_{92}\text{U} \rightarrow ^{234}_{90}\text{Th} + ^{4}_{2}\text{He}$$

40. A sample consists of 1.00×10^6 radioactive nuclei with a half-life of 10.0 h. No other nuclei are present at time $t = 0$. The stable daughter nuclei accumulate in the sample as time goes on. (a) Derive an equation giving the number of daughter nuclei N_d as a function of time. (b) Sketch or describe a graph of the number of daughter nuclei as a function of time. (c) What are the maximum and minimum numbers of daughter nuclei, and when do they occur? (d) What are the maximum and minimum rates of change in the number of daughter nuclei, and when do they occur?

41. The nucleus $^{15}_{8}$O decays by electron capture. The nuclear reaction is written

$$^{15}_{8}\text{O} + \text{e}^- \rightarrow ^{15}_{7}\text{N} + \nu$$

(a) Write the process going on for a single particle within the nucleus. (b) Disregarding the daughter's recoil, determine the energy of the neutrino.

42. A living specimen in equilibrium with the atmosphere **GP** contains one atom of ^{14}C (half-life = 5 730 yr) for every **W** 7.70×10^{11} stable carbon atoms. An archeological sample of wood (cellulose, $C_{12}H_{22}O_{11}$) contains 21.0 mg of carbon. When the sample is placed inside a shielded beta counter with 88.0% counting efficiency, 837 counts are accumulated in one week. We wish to find the age of the sample. (a) Find the number of carbon atoms in the sample. (b) Find the number of carbon-14 atoms in the sample. (c) Find the decay constant for carbon-14 in inverse seconds. (d) Find the initial number of decays per week just after the specimen died. (e) Find the corrected number of decays per week from the current sample. (f) From the answers to parts (d) and (e), find the time interval in years since the specimen died.

Section 44.6 Natural Radioactivity

43. Uranium is naturally present in rock and soil. At one step in its series of radioactive decays, ^{238}U produces the chemically inert gas radon-222, with a half-life of 3.82 days. The radon seeps out of the ground to mix into the atmosphere, typically making open air radioactive with activity 0.3 pCi/L. In homes, ^{222}Rn can be a serious pollutant, accumulating to reach much higher activities in enclosed spaces, sometimes reaching 4.00 pCi/L. If the radon radioactivity exceeds 4.00 pCi/L, the U.S. Environmental Protection Agency suggests taking action to reduce it such as by reducing infiltration of air from the ground. (a) Convert the activity 4.00 pCi/L to units of becquerels per cubic meter. (b) How many ^{222}Rn atoms are in 1 m^3 of air displaying this activity? (c) What fraction of the mass of the air does the radon constitute?

44. The most common isotope of radon is ^{222}Rn, which has half-life 3.82 days. (a) What fraction of the nuclei that were on the Earth one week ago are now undecayed? (b) Of those that existed one year ago? (c) In view of these results, explain why radon remains a problem, contributing significantly to our background radiation exposure.

45. Enter the correct nuclide symbol in each open tan rectangle in Figure P44.45, which shows the sequences of decays in the natural radioactive series starting with the long-lived isotope uranium-235 and ending with the stable nucleus lead-207.

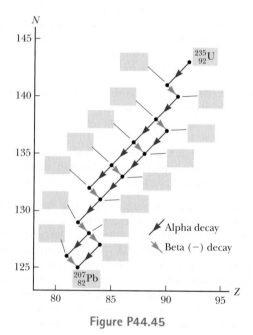

Figure P44.45

46. A rock sample contains traces of ^{238}U, ^{235}U, ^{232}Th, ^{208}Pb, ^{207}Pb, and ^{206}Pb. Analysis shows that the ratio of the amount of ^{238}U to ^{206}Pb is 1.164. (a) Assuming the rock originally contained no lead, determine the age of the rock. (b) What should be the ratios of ^{235}U to ^{207}Pb and of ^{232}Th to ^{208}Pb so that they would yield the same age for the rock? Ignore the minute amounts of the intermediate decay products in the decay chains. *Note:* This form of multiple dating gives reliable geological dates.

Section 44.7 Nuclear Reactions

47. A beam of 6.61-MeV protons is incident on a target of $^{27}_{13}$Al. Those that collide produce the reaction

$$p + {}^{27}_{13}\text{Al} \rightarrow {}^{27}_{14}\text{Si} + n$$

Ignoring any recoil of the product nucleus, determine the kinetic energy of the emerging neutrons.

48. (a) One method of producing neutrons for experimental use is bombardment of light nuclei with alpha particles. In the method used by James Chadwick in 1932, alpha particles emitted by polonium are incident on beryllium nuclei:

$$^{4}_{2}\text{He} + {}^{9}_{4}\text{Be} \rightarrow {}^{12}_{6}\text{C} + {}^{1}_{0}\text{n}$$

What is the Q value of this reaction? (b) Neutrons are also often produced by small-particle accelerators. In one design, deuterons accelerated in a Van de Graaff generator bombard other deuterium nuclei and cause the reaction

$$^{2}_{1}\text{H} + {}^{2}_{1}\text{H} \rightarrow {}^{3}_{2}\text{He} + {}^{1}_{0}\text{n}$$

Calculate the Q value of the reaction. (c) Is the reaction in part (b) exothermic or endothermic?

49. Identify the unknown nuclides and particles X and X′ in the nuclear reactions (a) X + $^{4}_{2}$He → $^{24}_{12}$Mg + $^{1}_{0}$n, (b) $^{235}_{92}$U + $^{1}_{0}$n → $^{90}_{38}$Sr + X + 2($^{1}_{0}$n), and (c) 2($^{1}_{1}$H) → $^{2}_{1}$H + X + X′.

50. Natural gold has only one isotope, $^{197}_{79}$Au. If natural gold is irradiated by a flux of slow neutrons, electrons are emitted. (a) Write the reaction equation. (b) Calculate the maximum energy of the emitted electrons.

51. The following reactions are observed:

$$^{9}_{4}\text{Be} + n \rightarrow {}^{10}_{4}\text{Be} + \gamma \quad Q = 6.812 \text{ MeV}$$

$$^{9}_{4}\text{Be} + \gamma \rightarrow {}^{8}_{4}\text{Be} + n \quad Q = -1.665 \text{ MeV}$$

Calculate the masses of ^{8}Be and ^{10}Be in unified mass units to four decimal places from these data.

Section 44.8 Nuclear Magnetic Resonance and Magnetic Resonance Imaging

52. Construct a diagram like that of Figure 44.19 for the cases when *I* equals (a) $\frac{5}{2}$ and (b) 4.

53. The radio frequency at which a nucleus having a magnetic moment of magnitude μ displays resonance absorption between spin states is called the Larmor frequency and is given by

$$f = \frac{\Delta E}{h} = \frac{2\mu B}{h}$$

Calculate the Larmor frequency for (a) free neutrons in a magnetic field of 1.00 T, (b) free protons in a magnetic field of 1.00 T, and (c) free protons in the Earth's magnetic field at a location where the magnitude of the field is 50.0 μT.

Additional Problems

54. A wooden artifact is found in an ancient tomb. Its carbon-14 ($^{14}_{6}$C) activity is measured to be 60.0% of that in a fresh sample of wood from the same region.

Assuming the same amount of ^{14}C was initially present in the artifact as is now contained in the fresh sample, determine the age of the artifact.

55. A 200.0-mCi sample of a radioactive isotope is purchased by a medical supply house. If the sample has a half-life of 14.0 days, how long will it be before its activity is reduced to 20.0 mCi?

56. *Why is the following situation impossible?* A ^{10}B nucleus is struck by an incoming alpha particle. As a result, a proton and a ^{12}C nucleus leave the site after the reaction.

57. (a) Find the radius of the $^{12}_{6}$C nucleus. (b) Find the force of repulsion between a proton at the surface of a $^{12}_{6}$C nucleus and the remaining five protons. (c) How much work (in MeV) has to be done to overcome this electric repulsion in transporting the last proton from a large distance up to the surface of the nucleus? (d) Repeat parts (a), (b), and (c) for $^{238}_{92}$U.

58. (a) Why is the beta decay $p \rightarrow n + e^+ + \nu$ forbidden for a free proton? (b) **What If?** Why is the same reaction possible if the proton is bound in a nucleus? For example, the following reaction occurs:

$$^{13}_{7}\text{N} \rightarrow {}^{13}_{6}\text{C} + e^+ + \nu$$

(c) How much energy is released in the reaction given in part (b)?

59. **Review.** Consider the Bohr model of the hydrogen atom, with the electron in the ground state. The magnetic field at the nucleus produced by the orbiting electron has a value of 12.5 T. (See Problem 6 in Chapter 30.) The proton can have its magnetic moment aligned in either of two directions perpendicular to the plane of the electron's orbit. The interaction of the proton's magnetic moment with the electron's magnetic field causes a difference in energy between the states with the two different orientations of the proton's magnetic moment. Find that energy difference in electron volts.

60. Show that the ^{238}U isotope cannot spontaneously emit a proton by analyzing the hypothetical process

$$^{238}_{92}\text{U} \rightarrow {}^{237}_{91}\text{Pa} + {}^{1}_{1}\text{H}$$

Note: The ^{237}Pa isotope has a mass of 237.051 144 u.

61. **Review.** (a) Is the mass of a hydrogen atom in its ground state larger or smaller than the sum of the masses of a proton and an electron? (b) What is the mass difference? (c) How large is the difference as a percentage of the total mass? (d) Is it large enough to affect the value of the atomic mass listed to six decimal places in Table 44.2?

62. *Why is the following situation impossible?* In an effort to study positronium, a scientist places ^{57}Co and ^{14}C in proximity. The ^{57}Co nuclei decay by e^+ emission, and the ^{14}C nuclei decay by e^- emission. Some of the positrons and electrons from these decays combine to form sufficient amounts of positronium for the scientist to gather data.

63. A by-product of some fission reactors is the isotope $^{239}_{94}$Pu, an alpha emitter having a half-life of 24 120 yr:

$$^{239}_{94}\text{Pu} \rightarrow {}^{235}_{92}\text{U} + \alpha$$

Consider a sample of 1.00 kg of pure $^{239}_{94}$Pu at $t = 0$. Calculate (a) the number of $^{239}_{94}$Pu nuclei present at $t = 0$ and (b) the initial activity in the sample. (c) **What If?** For what time interval does the sample have to be stored if a "safe" activity level is 0.100 Bq?

64. After the sudden release of radioactivity from the Chernobyl nuclear reactor accident in 1986, the radioactivity of milk in Poland rose to 2 000 Bq/L due to iodine-131 present in the grass eaten by dairy cattle. Radioactive iodine, with half-life 8.04 days, is particularly hazardous because the thyroid gland concentrates iodine. The Chernobyl accident caused a measurable increase in thyroid cancers among children in Poland and many other Eastern European countries. (a) For comparison, find the activity of milk due to potassium. Assume 1.00 liter of milk contains 2.00 g of potassium, of which 0.011 7% is the isotope ^{40}K with half-life 1.28×10^9 yr. (b) After what elapsed time would the activity due to iodine fall below that due to potassium?

65. A theory of nuclear astrophysics proposes that all the elements heavier than iron are formed in supernova explosions ending the lives of massive stars. Assume equal amounts of ^{235}U and ^{238}U were created at the time of the explosion and the present ^{235}U/^{238}U ratio on the Earth is 0.007 25. The half-lives of ^{235}U and ^{238}U are 0.704×10^9 yr and 4.47×10^9 yr, respectively. How long ago did the star(s) explode that released the elements that formed the Earth?

66. The activity of a radioactive sample was measured over 12 h, with the net count rates shown in the accompanying table. (a) Plot the logarithm of the counting rate as a function of time. (b) Determine the decay constant and half-life of the radioactive nuclei in the sample. (c) What counting rate would you expect for the sample at $t = 0$? (d) Assuming the efficiency of the counting instrument is 10.0%, calculate the number of radioactive atoms in the sample at $t = 0$.

Time (h)	Counting Rate (counts/min)
1.00	3 100
2.00	2 450
4.00	1 480
6.00	910
8.00	545
10.0	330
12.0	200

67. When, after a reaction or disturbance of any kind, a nucleus is left in an excited state, it can return to its normal (ground) state by emission of a gamma-ray photon (or several photons). This process is illustrated by Equation 44.25. The emitting nucleus must recoil to

conserve both energy and momentum. (a) Show that the recoil energy of the nucleus is

$$E_r = \frac{(\Delta E)^2}{2Mc^2}$$

where ΔE is the difference in energy between the excited and ground states of a nucleus of mass M. (b) Calculate the recoil energy of the ^{57}Fe nucleus when it decays by gamma emission from the 14.4-keV excited state. For this calculation, take the mass to be 57 u. *Suggestion:* Assume $hf \ll Mc^2$.

68. In a piece of rock from the Moon, the ^{87}Rb content is assayed to be 1.82×10^{10} atoms per gram of material and the ^{87}Sr content is found to be 1.07×10^9 atoms per gram. The relevant decay relating these nuclides is $^{87}\text{Rb} \rightarrow {}^{87}\text{Sr} + e^- + \bar{\nu}$. The half-life of the decay is 4.75×10^{10} yr. (a) Calculate the age of the rock. (b) **What If?** Could the material in the rock actually be much older? What assumption is implicit in using the radioactive dating method?

69. Free neutrons have a characteristic half-life of 10.4 min. What fraction of a group of free neutrons with kinetic energy 0.040 0 eV decays before traveling a distance of 10.0 km?

70. On July 4, 1054, a brilliant light appeared in the constellation Taurus the Bull. The supernova, which could be seen in daylight for some days, was recorded by Arab and Chinese astronomers. As it faded, it remained visible for years, dimming for a time with the 77.1-day half-life of the radioactive cobalt-56 that had been created in the explosion. (a) The remains of the star now form the Crab nebula (see the photograph opening Chapter 34). In it, the cobalt-56 has now decreased to what fraction of its original activity? (b) Suppose that an American, of the people called the Anasazi, made a charcoal drawing of the supernova. The carbon-14 in the charcoal has now decayed to what fraction of its original activity?

71. When a nucleus decays, it can leave the daughter nucleus in an excited state. The $^{93}_{43}$Tc nucleus (molar mass 92.910 2 g/mol) in the ground state decays by electron capture and e^+ emission to energy levels of the daughter (molar mass 92.906 8 g/mol in the ground state) at 2.44 MeV, 2.03 MeV, 1.48 MeV, and 1.35 MeV. (a) Identify the daughter nuclide. (b) To which of the listed levels of the daughter are electron capture and e^+ decay of $^{93}_{43}$Tc allowed?

72. The radioactive isotope ^{137}Ba has a relatively short half-life and can be easily extracted from a solution containing its parent ^{137}Cs. This barium isotope is commonly used in an undergraduate laboratory exercise for demonstrating the radioactive decay law. Undergraduate students using modest experimental equipment took the data presented in Figure P44.72. Determine the half-life for the decay of ^{137}Ba using their data.

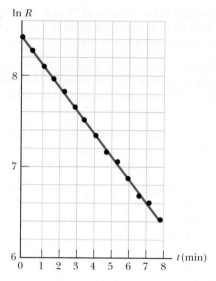

Figure P44.72

73. As part of his discovery of the neutron in 1932, James Chadwick determined the mass of the newly identified particle by firing a beam of fast neutrons, all having the same speed, at two different targets and measuring the maximum recoil speeds of the target nuclei. The maximum speeds arise when an elastic head-on collision occurs between a neutron and a stationary target nucleus. (a) Represent the masses and final speeds of the two target nuclei as m_1, v_1, m_2, and v_2 and assume Newtonian mechanics applies. Show that the neutron mass can be calculated from the equation

$$m_n = \frac{m_1 v_1 - m_2 v_2}{v_2 - v_1}$$

(b) Chadwick directed a beam of neutrons (produced from a nuclear reaction) on paraffin, which contains hydrogen. The maximum speed of the protons ejected was found to be 3.30×10^7 m/s. Because the velocity of the neutrons could not be determined directly, a second experiment was performed using neutrons from the same source and nitrogen nuclei as the target. The maximum recoil speed of the nitrogen nuclei was found to be 4.70×10^6 m/s. The masses of a proton and a nitrogen nucleus were taken as 1.00 u and 14.0 u, respectively. What was Chadwick's value for the neutron mass?

74. When the nuclear reaction represented by Equation 44.28 is endothermic, the reaction energy Q is negative. For the reaction to proceed, the incoming particle must have a minimum energy called the threshold energy, E_{th}. Some fraction of the energy of the incident particle is transferred to the compound nucleus to conserve momentum. Therefore, E_{th} must be greater than Q. (a) Show that

$$E_{th} = -Q\left(1 + \frac{M_a}{M_X}\right)$$

(b) Calculate the threshold energy of the incident alpha particle in the reaction

$$^{4}_{2}\text{He} + ^{14}_{7}\text{N} \rightarrow ^{17}_{8}\text{O} + ^{1}_{1}\text{H}$$

75. In an experiment on the transport of nutrients in a plant's root structure, two radioactive nuclides X and Y are used. Initially, 2.50 times more nuclei of type X are present than of type Y. At a time 3.00 d later, there are 4.20 times more nuclei of type X than of type Y. Isotope Y has a half-life of 1.60 d. What is the half-life of isotope X?

76. In an experiment on the transport of nutrients in a plant's root structure, two radioactive nuclides X and Y are used. Initially, the ratio of the number of nuclei of type X present to that of type Y is r_1. After a time interval Δt, the ratio of the number of nuclei of type X present to that of type Y is r_2. Isotope Y has a half-life of T_Y. What is the half-life of isotope X?

Challenge Problems

77. **Review.** Consider a model of the nucleus in which the positive charge (Ze) is uniformly distributed throughout a sphere of radius R. By integrating the energy density $\frac{1}{2}\epsilon_0 E^2$ over all space, show that the electric potential energy may be written

$$U = \frac{3Z^2 e^2}{20\pi\epsilon_0 R} = \frac{3k_e Z^2 e^2}{5R}$$

Problem 72 in Chapter 25 derived the same result by a different method.

78. After determining that the Sun has existed for hundreds of millions of years, but before the discovery of nuclear physics, scientists could not explain why the Sun has continued to burn for such a long time interval. For example, if it were a coal fire, it would have burned up in about 3 000 yr. Assume the Sun, whose mass is equal to 1.99×10^{30} kg, originally consisted entirely of hydrogen and its total power output is 3.85×10^{26} W. (a) Assuming the energy-generating mechanism of the Sun is the fusion of hydrogen into helium via the net reaction

$$4(^{1}_{1}\text{H}) + 2(e^{-}) \rightarrow ^{4}_{2}\text{He} + 2\nu + \gamma$$

calculate the energy (in joules) given off by this reaction. (b) Take the mass of one hydrogen atom to be equal to 1.67×10^{-27} kg. Determine how many hydrogen atoms constitute the Sun. (c) If the total power output remains constant, after what time interval will all the hydrogen be converted into helium, making the Sun die? (d) How does your answer to part (c) compare with current estimates of the expected life of the Sun, which are 4 billion to 7 billion years?

Applications of Nuclear Physics

In this chapter, we study both nuclear fission and nuclear fusion. The structure above is the target assembly for the inertial confinement procedure for initiating fusion by laser at the National Ignition Facility in Livermore, California. The triangle-shaped shrouds protect the fuel pellets and then open a few seconds before very powerful lasers bombard the target. *(Courtesy of Lawrence Livermore National Library)*

In this chapter, we study two means for deriving energy from nuclear reactions: fission, in which a large nucleus splits into two smaller nuclei, and fusion, in which two small nuclei fuse to form a larger one. In both cases, the released energy can be used either constructively (as in electric power plants) or destructively (as in nuclear weapons). We also examine the ways in which radiation interacts with matter and discuss the structure of fission and fusion reactors. The chapter concludes with a discussion of some industrial and biological applications of radiation.

45.1 Interactions Involving Neutrons

Nuclear fission is the process that occurs in present-day nuclear reactors and ultimately results in energy supplied to a community by electrical transmission. Nuclear fusion is an area of active research, but it has not yet been commercially developed for the supply of energy. We will discuss fission first and then explore fusion in Section 45.4.

To understand nuclear fission and the physics of nuclear reactors, we must first understand how neutrons interact with nuclei. Because of their charge neutrality, neutrons are not subject to Coulomb forces and as a result do not interact electrically with electrons or the nucleus. Therefore, neutrons can easily penetrate deep into an atom and collide with the nucleus.

A fast neutron (energy greater than approximately 1 MeV) traveling through matter undergoes many collisions with nuclei, giving up some of its kinetic energy in each collision. For fast neutrons in some materials, elastic collisions dominate. Materials for which that occurs are called **moderators** because they slow down (or moderate) the originally energetic neutrons very effectively. Moderator nuclei should be of low mass so that a large amount of kinetic energy is transferred to them when struck by neutrons. For this reason, materials that are abundant in hydrogen, such as paraffin and water, are good moderators for neutrons.

Eventually, most neutrons bombarding a moderator become **thermal neutrons,** which means they have given up so much of their energy that they are in thermal equilibrium with the moderator material. Their average kinetic energy at room temperature is, from Equation 21.19,

$$K_{\text{avg}} = \tfrac{3}{2}k_{\text{B}}T \approx \tfrac{3}{2}(1.38 \times 10^{-23}\,\text{J/K})(300\,\text{K}) = 6.21 \times 10^{-21}\,\text{J} \approx 0.04\,\text{eV}$$

which corresponds to a neutron root-mean-square speed of approximately 2 800 m/s. Thermal neutrons have a distribution of speeds, just as the molecules in a container of gas do (see Chapter 21). High-energy neutrons, those with energy of several MeV, *thermalize* (that is, their average energy reaches K_{avg}) in less than 1 ms when they are incident on a moderator.

Once the neutrons have thermalized and the energy of a particular neutron is sufficiently low, there is a high probability the neutron will be captured by a nucleus, an event that is accompanied by the emission of a gamma ray. This **neutron capture** reaction can be written

$$\tfrac{1}{0}\text{n} + {}^{A}_{Z}\text{X} \;\rightarrow\; {}^{A+1}_{Z}\text{X}^{*} \;\rightarrow\; {}^{A+1}_{Z}\text{X} + \gamma \qquad (45.1)$$

◄ Neutron capture reaction

Once the neutron is captured, the nucleus ${}^{A+1}_{Z}\text{X}^{*}$ is in an excited state for a very short time before it undergoes gamma decay. The product nucleus ${}^{A+1}_{Z}\text{X}$ is usually radioactive and decays by beta emission.

The neutron-capture rate for neutrons passing through any sample depends on the type of atoms in the sample and on the energy of the incident neutrons. The interaction of neutrons with matter increases with decreasing neutron energy because a slow neutron spends a larger time interval in the vicinity of target nuclei.

45.2 Nuclear Fission

As mentioned in Section 44.2, nuclear **fission** occurs when a heavy nucleus, such as ^{235}U, splits into two smaller nuclei. Fission is initiated when a heavy nucleus captures a thermal neutron as described by the first step in Equation 45.1. The absorption of the neutron creates a nucleus that is unstable and can change to a lower-energy configuration by splitting into two smaller nuclei. In such a reaction, the combined mass of the daughter nuclei is less than the mass of the parent nucleus, and the difference in mass is called the **mass defect.** Multiplying the mass defect by c^2 gives the numerical value of the released energy. This energy is in the form of kinetic energy associated with the motion of the neutrons and the daughter nuclei after the fission event. Energy is released because the binding energy per nucleon of the daughter nuclei is approximately 1 MeV greater than that of the parent nucleus (see Fig. 44.5).

Nuclear fission was first observed in 1938 by Otto Hahn (1879–1968) and Fritz Strassmann (1902–1980) following some basic studies by Fermi. After bombarding uranium with neutrons, Hahn and Strassmann discovered among the reaction products two medium-mass elements, barium and lanthanum. Shortly thereafter, Lise Meitner (1878–1968) and her nephew Otto Frisch (1904–1979) explained what had happened. After absorbing a neutron, the uranium nucleus had split into two

Pitfall Prevention 45.1
Binding Energy Reminder Remember from Chapter 44 that binding energy is the absolute value of the system energy and is related to the system mass. Therefore, when considering Figure 44.5, imagine flipping it upside down for a graph representing system mass. In a fission reaction, the system mass decreases. This decrease in mass appears in the system as kinetic energy of the fission products.

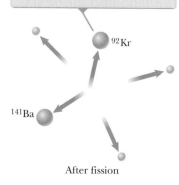

Before the event, a slow neutron approaches a ^{235}U nucleus.

^{235}U

Before fission

After the event, there are two lighter nuclei and three neutrons.

^{92}Kr

^{141}Ba

After fission

Figure 45.1 A nuclear fission event.

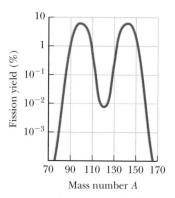

Figure 45.2 Distribution of fission products versus mass number for the fission of ^{235}U bombarded with thermal neutrons. Notice that the vertical axis is logarithmic.

nearly equal fragments plus several neutrons. Such an occurrence was of considerable interest to physicists attempting to understand the nucleus, but it was to have even more far-reaching consequences. Measurements showed that approximately 200 MeV of energy was released in each fission event, and this fact was to affect the course of history in World War II.

The fission of ^{235}U by thermal neutrons can be represented by the reaction

$$^{1}_{0}n + {}^{235}_{92}U \rightarrow {}^{236}_{92}U^* \rightarrow X + Y + \text{neutrons} \qquad (45.2)$$

where ^{236}U* is an intermediate excited state that lasts for approximately 10^{-12} s before splitting into medium-mass nuclei X and Y, which are called **fission fragments.** In any fission reaction, there are many combinations of X and Y that satisfy the requirements of conservation of energy and charge. In the case of uranium, for example, approximately 90 daughter nuclei can be formed.

Fission also results in the production of several neutrons, typically two or three. On average, approximately 2.5 neutrons are released per event. A typical fission reaction for uranium is

$$^{1}_{0}n + {}^{235}_{92}U \rightarrow {}^{141}_{56}Ba + {}^{92}_{36}Kr + 3({}^{1}_{0}n) \qquad (45.3)$$

Figure 45.1 shows a pictorial representation of the fission event in Equation 45.3.

Figure 45.2 is a graph of the distribution of fission products versus mass number A. The most probable products have mass numbers $A \approx 95$ and $A \approx 140$. Suppose these products are $^{95}_{39}Y$ (with 56 neutrons) and $^{140}_{53}I$ (with 87 neutrons). If these nuclei are located on the graph of Figure 44.4, it is seen that both are well above the line of stability. Because these fragments are very unstable owing to their unusually high number of neutrons, they almost instantaneously release two or three neutrons.

Let's estimate the disintegration energy Q released in a typical fission process. From Figure 44.5, we see that the binding energy per nucleon is approximately 7.2 MeV for heavy nuclei ($A \approx 240$) and approximately 8.2 MeV for nuclei of intermediate mass. The amount of energy released is 8.2 MeV − 7.2 MeV = 1 MeV per nucleon. Because there are a total of 235 nucleons in $^{235}_{92}U$, the energy released per fission event is approximately 235 MeV, a large amount of energy relative to the amount released in chemical processes. For example, the energy released in the combustion of one molecule of octane used in gasoline engines is about one-millionth of the energy released in a single fission event!

Quick Quiz 45.1 When a nucleus undergoes fission, the two daughter nuclei are generally radioactive. By which process are they most likely to decay? **(a)** alpha decay **(b)** beta decay (e^-) **(c)** beta decay (e^+)

Quick Quiz 45.2 Which of the following are possible fission reactions?

(a) $\quad ^{1}_{0}n + {}^{235}_{92}U \rightarrow {}^{140}_{54}Xe + {}^{94}_{38}Sr + 2({}^{1}_{0}n)$

(b) $\quad ^{1}_{0}n + {}^{235}_{92}U \rightarrow {}^{132}_{50}Sn + {}^{101}_{42}Mo + 3({}^{1}_{0}n)$

(c) $\quad ^{1}_{0}n + {}^{239}_{94}Pu \rightarrow {}^{137}_{53}I + {}^{97}_{41}Nb + 3({}^{1}_{0}n)$

| Example 45.1 | The Energy Released in the Fission of ^{235}U |

Calculate the energy released when 1.00 kg of ^{235}U fissions, taking the disintegration energy per event to be $Q = 208$ MeV.

SOLUTION

Conceptualize Imagine a nucleus of ^{235}U absorbing a neutron and then splitting into two smaller nuclei and several neutrons as in Figure 45.1.

▶ **45.1** c o n t i n u e d

Categorize The problem statement tells us to categorize this example as one involving an energy analysis of nuclear fission.

Analyze Because $A = 235$ for uranium, one mole of this isotope has a mass of $M = 235$ g.

Find the number of nuclei in our sample in terms of the number of moles n and Avogadro's number, and then in terms of the sample mass m and the molar mass M of ^{235}U:

$$N = nN_A = \frac{m}{M} N_A$$

Find the total energy released when all nuclei undergo fission:

$$E = NQ = \frac{m}{M} N_A Q = \frac{1.00 \times 10^3 \text{ g}}{235 \text{ g/mol}} (6.02 \times 10^{23} \text{ mol}^{-1})(208 \text{ MeV})$$

$$= 5.33 \times 10^{26} \text{ MeV}$$

Finalize Convert this energy to kWh:

$$E = (5.33 \times 10^{26} \text{ MeV})\left(\frac{1.60 \times 10^{-13} \text{ J}}{1 \text{ MeV}}\right)\left(\frac{1 \text{ kWh}}{3.60 \times 10^6 \text{ J}}\right) = 2.37 \times 10^7 \text{ kWh}$$

which, if released slowly, is enough energy to keep a 100-W lightbulb operating for 30 000 years! If the available fission energy in 1 kg of ^{235}U were suddenly released, it would be equivalent to detonating about 20 000 tons of TNT.

45.3 Nuclear Reactors

In Section 45.2, we learned that when ^{235}U fissions, one incoming neutron results in an average of 2.5 neutrons emitted per event. These neutrons can trigger other nuclei to fission. Because more neutrons are produced by the event than are absorbed, there is the possibility of an ever-building chain reaction (Fig. 45.3). Experience shows that if the chain reaction is not controlled (that is, if it does not

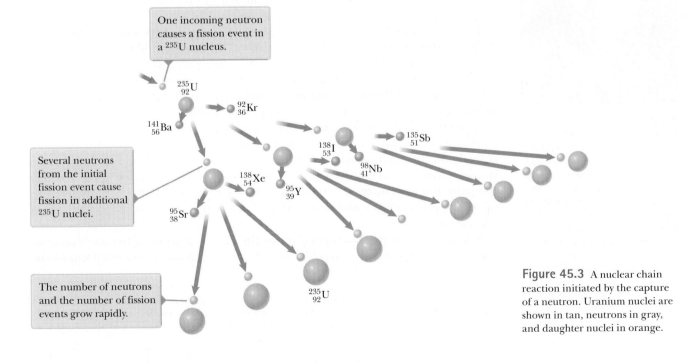

One incoming neutron causes a fission event in a $^{235}_{92}$U nucleus.

Several neutrons from the initial fission event cause fission in additional ^{235}U nuclei.

The number of neutrons and the number of fission events grow rapidly.

$^{235}_{92}$U $^{92}_{36}$Kr $^{141}_{56}$Ba $^{138}_{53}$I $^{135}_{51}$Sb $^{98}_{41}$Nb $^{138}_{54}$Xe $^{95}_{39}$Y $^{95}_{38}$Sr $^{235}_{92}$U

Figure 45.3 A nuclear chain reaction initiated by the capture of a neutron. Uranium nuclei are shown in tan, neutrons in gray, and daughter nuclei in orange.

Figure 45.4 Artist's rendition of the world's first nuclear reactor. Because of wartime secrecy, there are few photographs of the completed reactor, which was composed of layers of moderating graphite interspersed with uranium. A self-sustained chain reaction was first achieved on December 2, 1942. Word of the success was telephoned immediately to Washington, D.C., with this message: "The Italian navigator has landed in the New World and found the natives very friendly." The historic event took place in an improvised laboratory in the racquet court under the stands of the University of Chicago's Stagg Field, and the Italian navigator was Enrico Fermi.

Courtesy of Chicago Historical Society

© World History Archive/Alamy

Enrico Fermi
Italian Physicist (1901–1954)
Fermi was awarded the Nobel Prize in Physics in 1938 for producing transuranic elements by neutron irradiation and for his discovery of nuclear reactions brought about by thermal neutrons. He made many other outstanding contributions to physics, including his theory of beta decay, the free-electron theory of metals, and the development of the world's first fission reactor in 1942. Fermi was truly a gifted theoretical and experimental physicist. He was also well known for his ability to present physics in a clear and exciting manner.

proceed slowly), it can result in a violent explosion, with the sudden release of an enormous amount of energy. When the reaction is controlled, however, the energy released can be put to constructive use. In the United States, for example, nearly 20% of the electricity generated each year comes from nuclear power plants, and nuclear power is used extensively in many other countries, including France, Russia, and India.

A nuclear reactor is a system designed to maintain what is called a **self-sustained chain reaction.** This important process was first achieved in 1942 by Enrico Fermi and his team at the University of Chicago, using naturally occurring uranium as the fuel.[1] In the first nuclear reactor (Fig. 45.4), Fermi placed bricks of graphite (carbon) between the fuel elements. Carbon nuclei are about 12 times more massive than neutrons, but after several collisions with carbon nuclei, a neutron is slowed sufficiently to increase its likelihood of fission with ^{235}U. In this design, carbon is the moderator; most modern reactors use water as the moderator.

Most reactors in operation today also use uranium as fuel. Naturally occurring uranium contains only 0.7% of the ^{235}U isotope, however, with the remaining 99.3% being ^{238}U. This fact is important to the operation of a reactor because ^{238}U almost never fissions. Instead, it tends to absorb neutrons without a subsequent fission event, producing neptunium and plutonium. For this reason, reactor fuels must be artificially *enriched* to contain at least a few percent ^{235}U.

To achieve a self-sustained chain reaction, an average of one neutron emitted in each ^{235}U fission must be captured by another ^{235}U nucleus and cause that nucleus to undergo fission. A useful parameter for describing the level of reactor operation is the **reproduction constant K,** defined as **the average number of neutrons from each fission event that cause another fission event.** As we have seen, K has an average value of 2.5 in the uncontrolled fission of uranium.

A self-sustained and controlled chain reaction is achieved when $K = 1$. When in this condition, the reactor is said to be **critical.** When $K < 1$, the reactor is subcritical and the reaction dies out. When $K > 1$, the reactor is supercritical and a runaway reaction occurs. In a nuclear reactor used to furnish power to a utility company, it is necessary to maintain a value of K close to 1. If K rises above this value, the rest energy transformed to internal energy in the reaction could melt the reactor.

Several types of reactor systems allow the kinetic energy of fission fragments to be transformed to other types of energy and eventually transferred out of the

---

[1]Although Fermi's reactor was the first manufactured nuclear reactor, there is evidence that a natural fission reaction may have sustained itself for perhaps hundreds of thousands of years in a deposit of uranium in Gabon, West Africa. See G. Cowan, "A Natural Fission Reactor," *Scientific American* **235**(5): 36, 1976.

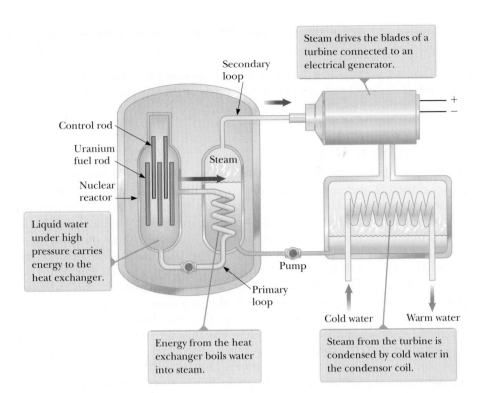

 labels:
Secondary loop

Steam drives the blades of a turbine connected to an electrical generator.

Control rod

Uranium fuel rod

Steam

Nuclear reactor

Liquid water under high pressure carries energy to the heat exchanger.

Pump

Primary loop

Cold water   Warm water

Energy from the heat exchanger boils water into steam.

Steam from the turbine is condensed by cold water in the condensor coil.

Figure 45.5 Main components of a pressurized-water nuclear reactor.

reactor plant by electrical transmission. The most common reactor in use in the United States is the pressurized-water reactor (Fig. 45.5). We shall examine this type because its main parts are common to all reactor designs. Fission events in the uranium **fuel elements** in the reactor core raise the temperature of the water contained in the primary loop, which is maintained at high pressure to keep the water from boiling. (This water also serves as the moderator to slow down the neutrons released in the fission events with energy of approximately 2 MeV.) The hot water is pumped through a heat exchanger, where the internal energy of the water is transferred by conduction to the water contained in the secondary loop. The hot water in the secondary loop is converted to steam, which does work to drive a turbine–generator system to create electric power. The water in the secondary loop is isolated from the water in the primary loop to avoid contamination of the secondary water and the steam by radioactive nuclei from the reactor core.

In any reactor, a fraction of the neutrons produced in fission leak out of the uranium fuel elements before inducing other fission events. If the fraction leaking out is too large, the reactor will not operate. The percentage lost is large if the fuel elements are very small because leakage is a function of the ratio of surface area to volume. Therefore, a critical feature of the reactor design is an optimal surface area–to–volume ratio of the fuel elements.

Control of Power Level

Safety is of critical importance in the operation of a nuclear reactor. The reproduction constant K must not be allowed to rise above 1, lest a runaway reaction occur. Consequently, reactor design must include a means of controlling the value of K.

The basic design of a nuclear reactor core is shown in Figure 45.6. The fuel elements consist of uranium that has been enriched in the ^{235}U isotope. To control the power level, **control rods** are inserted into the reactor core. These rods are made of materials such as cadmium that are very efficient in absorbing neutrons. By adjusting the number and position of the control rods in the reactor core, the K value

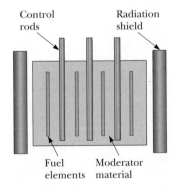

Labels: Control rods   Radiation shield   Fuel elements   Moderator material

Figure 45.6 Cross section of a reactor core showing the control rods, fuel elements containing enriched fuel, and moderating material, all surrounded by a radiation shield.

can be varied and any power level within the design range of the reactor can be achieved.

Quick Quiz 45.3 To reduce the value of the reproduction constant K, do you **(a)** push the control rods deeper into the core or **(b)** pull the control rods farther out of the core?

Safety and Waste Disposal

The 1986 accident at the Chernobyl reactor in Ukraine and the 2011 nuclear disaster caused by the earthquake and tsunami in Japan rightfully focused attention on reactor safety. Unfortunately, at Chernobyl the activity of the materials released immediately after the accident totaled approximately 1.2×10^{19} Bq and resulted in the evacuation of 135 000 people. Thirty individuals died during the accident or shortly thereafter, and data from the Ukraine Radiological Institute suggest that more than 2 500 deaths could be attributed to the Chernobyl accident. In the period 1986–1997, there was a tenfold increase in the number of children contracting thyroid cancer from the ingestion of radioactive iodine in milk from cows that ate contaminated grass. One conclusion of an international conference studying the Ukraine accident was that the main causes of the Chernobyl accident were the coincidence of severe deficiencies in the reactor physical design and a violation of safety procedures. Most of these deficiencies have since been addressed at plants of similar design in Russia and neighboring countries of the former Soviet Union.

The March 2011 accident in Japan was caused by an unfortunate combination of a massive earthquake and subsequent tsunami. The most hard-hit power plant, Fukushima I, shut down automatically after the earthquake. Shutting down a nuclear power plant, however, is not an instantaneous process. Cooling water must continue to be circulated to carry the energy generated by beta decay of the fission by-products out of the reactor core. Unfortunately, the water from the tsunami broke the connection to the power grid, leaving the plant without outside electrical support for circulating the water. While the plant had emergency generators to take over in such a situation, the tsunami inundated the generator rooms, making the generators inoperable. Three of the six reactors at Fukushima experienced meltdown, and there were several explosions. Significant radiation was released into the environment. At the time of this printing, all 54 of Japan's nuclear power plants have been taken offline, and the Japanese public has expressed strong reluctance to continue with nuclear power.

Commercial reactors achieve safety through careful design and rigid operating protocol, and only when these variables are compromised do reactors pose a danger. Radiation exposure and the potential health risks associated with such exposure are controlled by three layers of containment. The fuel and radioactive fission products are contained inside the reactor vessel. Should this vessel rupture, the reactor building acts as a second containment structure to prevent radioactive material from contaminating the environment. Finally, the reactor facilities must be in a remote location to protect the general public from exposure should radiation escape the reactor building.

A continuing concern about nuclear fission reactors is the safe disposal of radioactive material when the reactor core is replaced. This waste material contains long-lived, highly radioactive isotopes and must be stored over long time intervals in such a way that there is no chance of environmental contamination. At present, sealing radioactive wastes in waterproof containers and burying them in deep geologic repositories seems to be the most promising solution.

Transport of reactor fuel and reactor wastes poses additional safety risks. Accidents during transport of nuclear fuel could expose the public to harmful levels of radiation. The U.S. Department of Energy requires stringent crash tests of all containers used to transport nuclear materials. Container manufacturers must demonstrate that their containers will not rupture even in high-speed collisions.

Despite these risks, there are advantages to the use of nuclear power to be weighed against the risks. For example, nuclear power plants do not produce air pollution and greenhouse gases as do fossil fuel plants, and the supply of uranium on the Earth is predicted to last longer than the supply of fossil fuels. For each source of energy—whether nuclear, hydroelectric, fossil fuel, wind, solar, or other—the risks must be weighed against the benefits and the availability of the energy source.

45.4 Nuclear Fusion

In Chapter 44, we found that the binding energy for light nuclei ($A < 20$) is much smaller than the binding energy for heavier nuclei, which suggests a process that is the reverse of fission. As mentioned in Section 39.8, when two light nuclei combine to form a heavier nucleus, the process is called nuclear **fusion.** Because the mass of the final nucleus is less than the combined masses of the original nuclei, there is a loss of mass accompanied by a release of energy.

Two examples of such energy-liberating fusion reactions are as follows:

$$^1_1\text{H} + ^1_1\text{H} \rightarrow ^2_1\text{H} + e^+ + \nu$$

$$^1_1\text{H} + ^2_1\text{H} \rightarrow ^3_2\text{He} + \gamma$$

These reactions occur in the core of a star and are responsible for the outpouring of energy from the star. The second reaction is followed by either hydrogen–helium fusion or helium–helium fusion:

$$^1_1\text{H} + ^3_2\text{He} \rightarrow ^4_2\text{He} + e^+ + \nu$$

$$^3_2\text{He} + ^3_2\text{He} \rightarrow ^4_2\text{He} + ^1_1\text{H} + ^1_1\text{H}$$

These fusion reactions are the basic reactions in the **proton–proton cycle,** believed to be one of the basic cycles by which energy is generated in the Sun and other stars that contain an abundance of hydrogen. Most of the energy production takes place in the Sun's interior, where the temperature is approximately 1.5×10^7 K. Because such high temperatures are required to drive these reactions, they are called **thermonuclear fusion reactions.** All the reactions in the proton–proton cycle are exothermic. An overview of the cycle is that four protons combine to generate an alpha particle, positrons, gamma rays, and neutrinos.

Quick Quiz 45.4 In the core of a star, hydrogen nuclei combine in fusion reactions. Once the hydrogen has been exhausted, fusion of helium nuclei can occur. If the star is sufficiently massive, fusion of heavier and heavier nuclei can occur once the helium is used up. Consider a fusion reaction involving two nuclei with the same value of A. For this reaction to be exothermic, which of the following values of A are impossible? **(a)** 12 **(b)** 20 **(c)** 28 **(d)** 64

> **Pitfall Prevention 45.2**
> **Fission and Fusion** The words *fission* and *fusion* sound similar, but they correspond to different processes. Consider the binding-energy graph in Figure 44.5. There are two directions from which you can approach the peak of the graph so that energy is released: combining two light nuclei, or fusion, and separating a heavy nucleus into two lighter nuclei, or fission.

Example 45.2   Energy Released in Fusion

Find the total energy released in the fusion reactions in the proton–proton cycle.

SOLUTION

Conceptualize The net nuclear result of the proton–proton cycle is to fuse four protons to form an alpha particle. Study the reactions above for the proton–proton cycle to be sure you understand how four protons become an alpha particle.

Categorize We use concepts discussed in this section, so we categorize this example as a substitution problem.

continued

▶ **45.2** c o n t i n u e d

Find the initial mass of the system using the atomic mass of hydrogen from Table 44.2:

$4(1.007\ 825\ \text{u}) = 4.031\ 300\ \text{u}$

Find the change in mass of the system as this value minus the mass of a ^{4}He atom:

$4.031\ 300\ \text{u} - 4.002\ 603\ \text{u} = 0.028\ 697\ \text{u}$

Convert this mass change into energy units:

$E = 0.028\ 697\ \text{u} \times 931.494\ \text{MeV/u} = \boxed{26.7\ \text{MeV}}$

This energy is shared among the alpha particle and other particles such as positrons, gamma rays, and neutrinos.

Terrestrial Fusion Reactions

The enormous amount of energy released in fusion reactions suggests the possibility of harnessing this energy for useful purposes. A great deal of effort is currently under way to develop a sustained and controllable thermonuclear reactor, a fusion power reactor. Controlled fusion is often called the ultimate energy source because of the availability of its fuel source: water. For example, if deuterium were used as the fuel, 0.12 g of it could be extracted from 1 gal of water at a cost of about four cents. This amount of deuterium would release approximately 10^{10} J if all nuclei underwent fusion. By comparison, 1 gal of gasoline releases approximately 10^8 J upon burning and costs far more than four cents.

An additional advantage of fusion reactors is that comparatively few radioactive by-products are formed. For the proton–proton cycle, for instance, the end product is safe, nonradioactive helium. Unfortunately, a thermonuclear reactor that can deliver a net power output spread over a reasonable time interval is not yet a reality, and many difficulties must be resolved before a successful device is constructed.

The Sun's energy is based in part on a set of reactions in which hydrogen is converted to helium. The proton–proton interaction is not suitable for use in a fusion reactor, however, because the event requires very high temperatures and densities. The process works in the Sun only because of the extremely high density of protons in the Sun's interior.

The reactions that appear most promising for a fusion power reactor involve deuterium ($^2_1$H) and tritium ($^3_1$H):

$$^2_1\text{H} + {}^2_1\text{H} \rightarrow {}^3_2\text{He} + {}^1_0\text{n} \quad Q = 3.27\ \text{MeV}$$
$$^2_1\text{H} + {}^2_1\text{H} \rightarrow {}^3_1\text{H} + {}^1_1\text{H} \quad Q = 4.03\ \text{MeV} \tag{45.4}$$
$$^2_1\text{H} + {}^3_1\text{H} \rightarrow {}^4_2\text{He} + {}^1_0\text{n} \quad Q = 17.59\ \text{MeV}$$

As noted earlier, deuterium is available in almost unlimited quantities from our lakes and oceans and is very inexpensive to extract. Tritium, however, is radioactive ($T_{1/2} = 12.3$ yr) and undergoes beta decay to ^{3}He. For this reason, tritium does not occur naturally to any great extent and must be artificially produced.

One major problem in obtaining energy from nuclear fusion is that the Coulomb repulsive force between two nuclei, which carry positive charges, must be overcome before they can fuse. Figure 45.7 is a graph of potential energy as a function of the separation distance between two deuterons (deuterium nuclei, each having charge $+e$). The potential energy is positive in the region $r > R$, where the Coulomb repulsive force dominates ($R \approx 1$ fm), and negative in the region $r < R$, where the nuclear force dominates. The fundamental problem then is to give the two nuclei enough kinetic energy to overcome this repulsive force. This requirement can be accomplished by raising the fuel to extremely high temperatures (to approximately 10^8 K). At these high temperatures, the atoms are ionized and the system consists of a collection of electrons and nuclei, commonly referred to as a *plasma*.

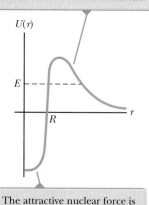

The Coulomb repulsive force is dominant for large separation distances between the deuterons.

The attractive nuclear force is dominant when the deuterons are close together.

Figure 45.7 Potential energy as a function of separation distance between two deuterons. R is on the order of 1 fm. If we neglect tunneling, the two deuterons require an energy E greater than the height of the barrier to undergo fusion.

Example 45.3 **The Fusion of Two Deuterons**

For the nuclear force to overcome the repulsive Coulomb force, the separation distance between two deuterons must be approximately 1.0×10^{-14} m.

(A) Calculate the height of the potential barrier due to the repulsive force.

SOLUTION

Conceptualize Imagine moving two deuterons toward each other. As they move closer together, the Coulomb repulsion force becomes stronger. Work must be done on the system to push against this force, and this work appears in the system of two deuterons as electric potential energy.

Categorize We categorize this problem as one involving the electric potential energy of a system of two charged particles.

Analyze Evaluate the potential energy associated with two charges separated by a distance r (Eq. 25.13) for two deuterons:

$$U = k_e \frac{q_1 q_2}{r} = k_e \frac{(+e)^2}{r} = (8.99 \times 10^9 \text{ N} \cdot \text{m}^2/\text{C}^2) \frac{(1.60 \times 10^{-19} \text{ C})^2}{1.0 \times 10^{-14} \text{ m}}$$

$$= 2.3 \times 10^{-14} \text{ J} = \boxed{0.14 \text{ MeV}}$$

(B) Estimate the temperature required for a deuteron to overcome the potential barrier, assuming an energy of $\frac{3}{2} k_B T$ per deuteron (where k_B is Boltzmann's constant).

SOLUTION

Because the total Coulomb energy of the pair is 0.14 MeV, the Coulomb energy per deuteron is equal to 0.07 MeV = 1.1×10^{-14} J.

Set this energy equal to the average energy per deuteron:

$$\tfrac{3}{2} k_B T = 1.1 \times 10^{-14} \text{ J}$$

Solve for T:

$$T = \frac{2(1.1 \times 10^{-14} \text{ J})}{3(1.38 \times 10^{-23} \text{ J/K})} = \boxed{5.6 \times 10^8 \text{ K}}$$

(C) Find the energy released in the deuterium–deuterium reaction

$$^2_1\text{H} + \,^2_1\text{H} \;\rightarrow\; ^3_1\text{H} + \,^1_1\text{H}$$

SOLUTION

The mass of a single deuterium atom is equal to 2.014 102 u. Therefore, the total mass of the system before the reaction is 4.028 204 u.

Find the sum of the masses after the reaction: 3.016 049 u + 1.007 825 u = 4.023 874 u

Find the change in mass and convert to energy units: 4.028 204 u − 4.023 874 u = 0.004 33 u

$$= 0.004\ 33 \text{ u} \times 931.494 \text{ MeV/u} = \boxed{4.03 \text{ MeV}}$$

Finalize The calculated temperature in part (B) is too high because the particles in the plasma have a Maxwellian speed distribution (Section 21.5) and therefore some fusion reactions are caused by particles in the high-energy tail of this distribution. Furthermore, even those particles that do not have enough energy to overcome the barrier have some probability of tunneling through (Section 41.5). When these effects are taken into account, a temperature of "only" 4×10^8 K appears adequate to fuse two deuterons in a plasma. In part (C), notice that the energy value is consistent with that already given in Equation 45.4.

WHAT IF? Suppose the tritium resulting from the reaction in part (C) reacts with another deuterium in the reaction

$$^2_1\text{H} + \,^3_1\text{H} \;\rightarrow\; ^4_2\text{He} + \,^1_0\text{n}$$

How much energy is released in the sequence of two reactions?

continued

▶ **45.3** continued

Answer The overall effect of the sequence of two reactions is that three deuterium nuclei have combined to form a helium nucleus, a hydrogen nucleus, and a neutron. The initial mass is 3(2.014 102 u) = 6.042 306 u. After the reaction, the sum of the masses is 4.002 603 u + 1.007 825 u + 1.008 665 = 6.019 093 u. The excess mass is equal to 0.023 213 u, equivalent to an energy of 21.6 MeV. Notice that this value is the sum of the Q values for the second and third reactions in Equation 45.4.

The temperature at which the power generation rate in any fusion reaction exceeds the loss rate is called the **critical ignition temperature** T_{ignit}. This temperature for the deuterium–deuterium (D–D) reaction is 4×10^8 K. From the relationship $E \approx \frac{3}{2} k_B T$, the ignition temperature is equivalent to approximately 52 keV. The critical ignition temperature for the deuterium–tritium (D–T) reaction is approximately 4.5×10^7 K, or only 6 keV. A plot of the power P_{gen} generated by fusion versus temperature for the two reactions is shown in Figure 45.8. The straight green line represents the power P_{lost} lost via the radiation mechanism known as bremsstrahlung (Section 42.8). In this principal mechanism of energy loss, radiation (primarily x-rays) is emitted as the result of electron–ion collisions within the plasma. The intersections of the P_{lost} line with the P_{gen} curves give the critical ignition temperatures.

In addition to the high-temperature requirements, two other critical parameters determine whether or not a thermonuclear reactor is successful: the **ion density** n and **confinement time** τ, which is the time interval during which energy injected into the plasma remains within the plasma. British physicist J. D. Lawson (1923–2008) showed that both the ion density and confinement time must be large enough to ensure that more fusion energy is released than the amount required to raise the temperature of the plasma. For a given value of n, the probability of fusion between two particles increases as τ increases. For a given value of τ, the collision rate between nuclei increases as n increases. The product $n\tau$ is referred to as the **Lawson number** of a reaction. A graph of the value of $n\tau$ necessary to achieve a net energy output for the D–T and D–D reactions at different temperatures is shown in Figure 45.9. In particular, **Lawson's criterion** states that a net energy output is possible for values of $n\tau$ that meet the following conditions:

$$n\tau \geq 10^{14} \text{ s/cm}^3 \quad \text{(D–T)} \tag{45.5}$$
$$n\tau \geq 10^{16} \text{ s/cm}^3 \quad \text{(D–D)}$$

These values represent the minima of the curves in Figure 45.9.

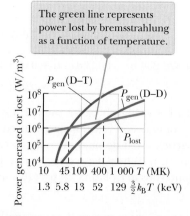

Figure 45.8 Power generated versus temperature for deuterium–deuterium (D–D) and deuterium–tritium (D–T) fusion. When the generation rate exceeds the loss rate, ignition takes place.

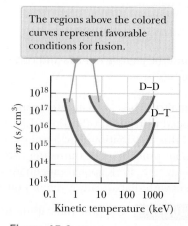

Figure 45.9 The Lawson number $n\tau$ at which net energy output is possible versus temperature for the D–T and D–D fusion reactions.

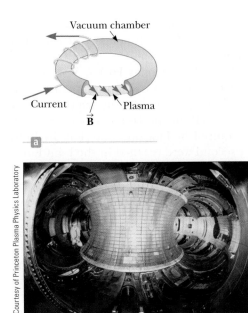

Vacuum chamber

Current   Plasma

$\vec{\mathbf{B}}$

Courtesy of Princeton Plasma Physics Laboratory

a

b

Courtesy of Princeton University

c

Figure 45.10 (a) Diagram of a tokamak used in the magnetic confinement scheme. (b) Interior view of the closed Tokamak Fusion Test Reactor (TFTR) vacuum vessel at the Princeton Plasma Physics Laboratory. (c) The National Spherical Torus Experiment (NSTX) that began operation in March 1999.

Lawson's criterion was arrived at by comparing the energy required to raise the temperature of a given plasma with the energy generated by the fusion process.[2] The energy E_{in} required to raise the temperature of the plasma is proportional to the ion density n, which we can express as $E_{in} = C_1 n$, where C_1 is some constant. The energy generated by the fusion process is proportional to $n^2 \tau$, or $E_{gen} = C_2 n^2 \tau$. This dependence may be understood by realizing that the fusion energy released is proportional to both the rate at which interacting ions collide ($\propto n^2$) and the confinement time τ. Net energy is produced when $E_{gen} > E_{in}$. When the constants C_1 and C_2 are calculated for different reactions, the condition that $E_{gen} \geq E_{in}$ leads to Lawson's criterion.

Current efforts are aimed at meeting Lawson's criterion at temperatures exceeding T_{ignit}. Although the minimum required plasma densities have been achieved, the problem of confinement time is more difficult. The two basic techniques under investigation for solving this problem are *magnetic confinement* and *inertial confinement*.

Magnetic Confinement

Many fusion-related plasma experiments use **magnetic confinement** to contain the plasma. A toroidal device called a **tokamak,** first developed in Russia, is shown in Figure 45.10a. A combination of two magnetic fields is used to confine and stabilize the plasma: (1) a strong toroidal field produced by the current in the toroidal windings surrounding a doughnut-shaped vacuum chamber and (2) a weaker "poloidal" field produced by the toroidal current. In addition to confining the plasma, the toroidal current is used to raise its temperature. The resultant helical magnetic field lines spiral around the plasma and keep it from touching the walls of the vacuum chamber. (If the plasma touches the walls, its temperature is reduced and heavy impurities sputtered from the walls "poison" it, leading to large power losses.)

One major breakthrough in magnetic confinement in the 1980s was in the area of auxiliary energy input to reach ignition temperatures. Experiments have shown

---

[2]Lawson's criterion neglects the energy needed to set up the strong magnetic field used to confine the hot plasma in a magnetic confinement approach. This energy is expected to be about 20 times greater than the energy required to raise the temperature of the plasma. It is therefore necessary either to have a magnetic energy recovery system or to use superconducting magnets.

that injecting a beam of energetic neutral particles into the plasma is a very efficient method of raising it to ignition temperatures. Radio-frequency energy input will probably be needed for reactor-size plasmas.

When it was in operation from 1982 to 1997, the Tokamak Fusion Test Reactor (TFTR, Fig. 45.10b) at Princeton University reported central ion temperatures of 510 million degrees Celsius, more than 30 times greater than the temperature at the center of the Sun. The $n\tau$ values in the TFTR for the D–T reaction were well above 10^{13} s/cm^3 and close to the value required by Lawson's criterion. In 1991, reaction rates of 6×10^{17} D–T fusions per second were reached in the Joint European Torus (JET) tokamak at Abington, England.

One of the new generation of fusion experiments is the National Spherical Torus Experiment (NSTX) at the Princeton Plasma Physics Laboratory and shown in Figure 45.10c. This reactor was brought on line in February 1999 and has been running fusion experiments since then. Rather than the doughnut-shaped plasma of a tokamak, the NSTX produces a spherical plasma that has a hole through its center. The major advantage of the spherical configuration is its ability to confine the plasma at a higher pressure in a given magnetic field. This approach could lead to development of smaller, more economical fusion reactors.

An international collaborative effort involving the United States, the European Union, Japan, China, South Korea, India, and Russia is currently under way to build a fusion reactor called ITER. This acronym stands for International Thermonuclear Experimental Reactor, although recently the emphasis has shifted to interpreting "iter" in terms of its Latin meaning, "the way." One reason proposed for this change is to avoid public misunderstanding and negative connotations toward the word *thermonuclear*. This facility will address the remaining technological and scientific issues concerning the feasibility of fusion power. The design is completed, and Cadarache, France, was chosen in June 2005 as the reactor site. Construction began in 2007 and will require about 10 years, with fusion operation projected to begin in 2019. If the planned device works as expected, the Lawson number for ITER will be about six times greater than the current record holder, the JT-60U tokamak in Japan. ITER is expected to produce ten times as much output power as input power, and the energy content of the alpha particles inside the reactor will be so intense that they will sustain the fusion reaction, allowing the auxiliary energy sources to be turned off once the reaction is initiated.

Example 45.4 Inside a Fusion Reactor

In 1998, the JT-60U tokamak in Japan operated with a D–T plasma density of 4.8×10^{13} cm^{-3} at a temperature (in energy units) of 24.1 keV. It confined this plasma inside a magnetic field for 1.1 s.

(A) Do these data meet Lawson's criterion?

SOLUTION

Conceptualize With the help of the third of Equations 45.4, imagine many such reactions occurring in a plasma of high temperature and high density.

Categorize We use the concept of the Lawson number discussed in this section, so we categorize this example as a substitution problem.

Evaluate the Lawson number for the JT-60U:

$$n\tau = (4.8 \times 10^{13}\ \text{cm}^{-3})(1.1\ \text{s}) = 5.3 \times 10^{13}\ \text{s/cm}^3$$

This value is close to meeting Lawson's criterion of 10^{14} s/cm^3 for a D–T plasma given in Equation 45.5. In fact, scientists recorded a power gain of 1.25, indicating that the reactor operated slightly past the break-even point and produced more energy than it required to maintain the plasma.

(B) How does the plasma density compare with the density of atoms in an ideal gas when the gas is under standard conditions ($T = 0°C$ and $P = 1$ atm)?

▶ **45.4** continued

Find the density of atoms in a sample of ideal gas by evaluating N_A/V_{mol}, where N_A is Avogadro's number and V_{mol} is the molar volume of an ideal gas under standard conditions, 2.24×10^{-2} m³/mol:

$$\frac{N_A}{V_{mol}} = \frac{6.02 \times 10^{23}\ \text{atoms/mol}}{2.24 \times 10^{-2}\ \text{m}^3/\text{mol}} = 2.7 \times 10^{25}\ \text{atoms/m}^3$$
$$= 2.7 \times 10^{19}\ \text{atoms/cm}^3$$

This value is more than 500 000 times greater than the plasma density in the reactor.

Inertial Confinement

The second technique for confining a plasma, called **inertial confinement,** makes use of a D–T target that has a very high particle density. In this scheme, the confinement time is very short (typically 10^{-11} to 10^{-9} s), and, because of their own inertia, the particles do not have a chance to move appreciably from their initial positions. Therefore, Lawson's criterion can be satisfied by combining a high particle density with a short confinement time.

Laser fusion is the most common form of inertial confinement. A small D–T pellet, approximately 1 mm in diameter, is struck simultaneously by several focused, high-intensity laser beams, resulting in a large pulse of input energy that causes the surface of the fuel pellet to evaporate (Fig. 45.11). The escaping particles exert a third-law reaction force on the core of the pellet, resulting in a strong, inwardly moving compressive shock wave. This shock wave increases the pressure and density of the core and produces a corresponding increase in temperature. When the temperature of the core reaches ignition temperature, fusion reactions occur.

One of the leading laser fusion laboratories in the United States is the Omega facility at the University of Rochester in New York. This facility focuses 24 laser beams on the target. Currently under operation at the Lawrence Livermore National Laboratory in Livermore, California, is the National Ignition Facility. The research apparatus there includes 192 laser beams that can be focused on a deuterium–tritium pellet. Construction was completed in early 2009, and a test firing of the lasers in March 2012 broke the record for lasers, delivering 1.87 MJ to a target. This energy is delivered in such a short time interval that the power is immense: 500 trillion watts, more than 1 000 times the power used in the United States at any moment.

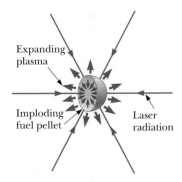

Figure 45.11 In inertial confinement, a D–T fuel pellet fuses when struck by several high-intensity laser beams simultaneously.

Fusion Reactor Design

In the D–T fusion reaction

$$^2_1\text{H} + {}^3_1\text{H} \rightarrow {}^4_2\text{He} + {}^1_0\text{n} \qquad Q = 17.59\ \text{MeV}$$

the alpha particle carries 20% of the energy and the neutron carries 80%, or approximately 14 MeV. A diagram of the deuterium–tritium fusion reaction is shown in Figure 45.12. Because the alpha particles are charged, they are primarily absorbed by the plasma, causing the plasma's temperature to increase. In contrast, the 14-MeV neutrons, being electrically neutral, pass through the plasma and are absorbed by a surrounding blanket material, where their large kinetic energy is extracted and used to generate electric power.

One scheme is to use molten lithium metal as the neutron-absorbing material and to circulate the lithium in a closed heat-exchange loop, thereby producing steam and driving turbines as in a conventional power plant. Figure 45.13 (page 1432) shows a diagram of such a reactor. It is estimated that a blanket of lithium approximately 1 m thick will capture nearly 100% of the neutrons from the fusion of a small D–T pellet.

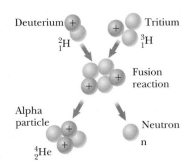

Figure 45.12 Deuterium–tritium fusion. Eighty percent of the energy released is in the 14-MeV neutron.

Figure 45.13 Diagram of a fusion reactor.

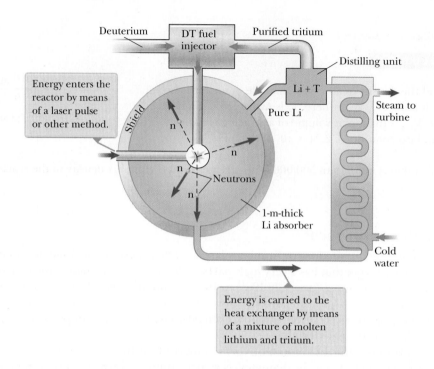

The capture of neutrons by lithium is described by the reaction

$$\,_0^1n + \,_3^6Li \;\rightarrow\; \,_1^3H + \,_2^4He$$

where the kinetic energies of the charged tritium $\,_1^3H$ and alpha particle are transformed to internal energy in the molten lithium. An extra advantage of using lithium as the energy-transfer medium is that the tritium produced can be separated from the lithium and returned as fuel to the reactor.

Advantages and Problems of Fusion

If fusion power can ever be harnessed, it will offer several advantages over fission-generated power: (1) low cost and abundance of fuel (deuterium), (2) impossibility of runaway accidents, and (3) decreased radiation hazard. Some of the anticipated problems and disadvantages include (1) scarcity of lithium, (2) limited supply of helium, which is needed for cooling the superconducting magnets used to produce strong confining fields, and (3) structural damage and induced radioactivity caused by neutron bombardment. If such problems and the engineering design factors can be resolved, nuclear fusion may become a feasible source of energy in the twenty-first century.

45.5 Radiation Damage

In Chapter 34, we learned that electromagnetic radiation is all around us in the form of radio waves, microwaves, light waves, and so on. In this section, we describe forms of radiation that can cause severe damage as they pass through matter, such as radiation resulting from radioactive processes and radiation in the form of energetic particles such as neutrons and protons.

The degree and type of damage depend on several factors, including the type and energy of the radiation and the properties of the matter. The metals used in nuclear reactor structures can be severely weakened by high fluxes of energetic neutrons because these high fluxes often lead to metal fatigue. The damage in such situations is in the form of atomic displacements, often resulting in major alterations in the properties of the material.

Radiation damage in biological organisms is primarily due to ionization effects in cells. A cell's normal operation may be disrupted when highly reactive ions are formed as the result of ionizing radiation. For example, hydrogen and the hydroxyl radical OH^- produced from water molecules can induce chemical reactions that may break bonds in proteins and other vital molecules. Furthermore, the ionizing radiation may affect vital molecules directly by removing electrons from their structure. Large doses of radiation are especially dangerous because damage to a great number of molecules in a cell may cause the cell to die. Although the death of a single cell is usually not a problem, the death of many cells may result in irreversible damage to the organism. Cells that divide rapidly, such as those of the digestive tract, reproductive organs, and hair follicles, are especially susceptible. In addition, cells that survive the radiation may become defective. These defective cells can produce more defective cells and can lead to cancer.

In biological systems, it is common to separate radiation damage into two categories: somatic damage and genetic damage. *Somatic damage* is that associated with any body cell except the reproductive cells. Somatic damage can lead to cancer or can seriously alter the characteristics of specific organisms. *Genetic damage* affects only reproductive cells. Damage to the genes in reproductive cells can lead to defective offspring. It is important to be aware of the effect of diagnostic treatments, such as x-rays and other forms of radiation exposure, and to balance the significant benefits of treatment with the damaging effects.

Damage caused by radiation also depends on the radiation's penetrating power. Alpha particles cause extensive damage, but penetrate only to a shallow depth in a material due to the strong interaction with other charged particles. Neutrons do not interact via the electric force and hence penetrate deeper, causing significant damage. Gamma rays are high-energy photons that can cause severe damage, but often pass through matter without interaction.

Several units have been used historically to quantify the amount, or dose, of any radiation that interacts with a substance.

> The **roentgen** (R) is that amount of ionizing radiation that produces an electric charge of 3.33×10^{-10} C in 1 cm^3 of air under standard conditions.

Equivalently, the roentgen is that amount of radiation that increases the energy of 1 kg of air by 8.76×10^{-3} J.

For most applications, the roentgen has been replaced by the rad (an acronym for *radiation absorbed dose*):

> One **rad** is that amount of radiation that increases the energy of 1 kg of absorbing material by 1×10^{-2} J.

Although the rad is a perfectly good physical unit, it is not the best unit for measuring the degree of biological damage produced by radiation because damage depends not only on the dose but also on the type of the radiation. For example, a given dose of alpha particles causes about ten times more biological damage than an equal dose of x-rays. The **RBE** (relative biological effectiveness) factor for a given type of radiation is **the number of rads of x-radiation or gamma radiation that produces the same biological damage as 1 rad of the radiation being used.** The RBE factors for different types of radiation are given in Table 45.1 (page 1434). The values are only approximate because they vary with particle energy and with the form of the damage. The RBE factor should be considered only a first-approximation guide to the actual effects of radiation.

Finally, the **rem** (radiation equivalent in man) is the product of the dose in rad and the RBE factor:

$$\text{Dose in rem} \equiv \text{dose in rad} \times \text{RBE} \qquad \textbf{(45.6)} \qquad \blacktriangleleft \textbf{ Radiation dose in rem}$$

Table 45.1 **RBE Factors for Several Types of Radiation**

Radiation	RBE Factor
X-rays and gamma rays	1.0
Beta particles	1.0–1.7
Alpha particles	10–20
Thermal neutrons	4–5
Fast neutrons and protons	10
Heavy ions	20

Note: RBE = relative biological effectiveness.

Table 45.2 **Units for Radiation Dosage**

Quantity	SI Unit	Symbol	Relations to Other SI Units	Older Unit	Conversion
Absorbed dose	gray	Gy	= 1 J/kg	rad	1 Gy = 100 rad
Dose equivalent	sievert	Sv	= 1 J/kg	rem	1 Sv = 100 rem

According to this definition, 1 rem of any two types of radiation produces the same amount of biological damage. Table 45.1 shows that a dose of 1 rad of fast neutrons represents an effective dose of 10 rem, but 1 rad of gamma radiation is equivalent to a dose of only 1 rem.

This discussion has focused on measurements of radiation dosage in units such as rads and rems because these units are still widely used. They have, however, been formally replaced with new SI units. The rad has been replaced with the *gray* (Gy), equal to 100 rad, and the rem has been replaced with the *sievert* (Sv), equal to 100 rem. Table 45.2 summarizes the older and the current SI units of radiation dosage.

Low-level radiation from natural sources such as cosmic rays and radioactive rocks and soil delivers to each of us a dose of approximately 2.4 mSv/yr. This radiation, called *background radiation,* varies with geography, with the main factors being altitude (exposure to cosmic rays) and geology (radon gas released by some rock formations, deposits of naturally radioactive minerals).

The upper limit of radiation dose rate recommended by the U.S. government (apart from background radiation) is approximately 5 mSv/yr. Many occupations involve much higher radiation exposures, so an upper limit of 50 mSv/yr has been set for combined whole-body exposure. Higher upper limits are permissible for certain parts of the body, such as the hands and the forearms. A dose of 4 to 5 Sv results in a mortality rate of approximately 50% (which means that half the people exposed to this radiation level die). The most dangerous form of exposure for most people is either ingestion or inhalation of radioactive isotopes, especially isotopes of those elements the body retains and concentrates, such as ^{90}Sr.

45.6 Uses of Radiation

Nuclear physics applications are extremely widespread in manufacturing, medicine, and biology. In this section, we present a few of these applications and the underlying theories supporting them.

Tracing

Radioactive tracers are used to track chemicals participating in various reactions. One of the most valuable uses of radioactive tracers is in medicine. For example, iodine, a nutrient needed by the human body, is obtained largely through the

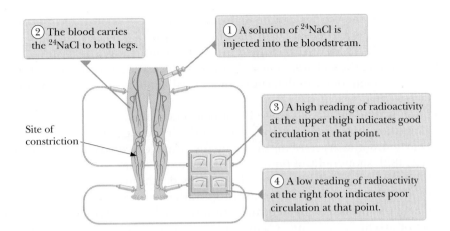

② The blood carries the ^{24}NaCl to both legs.

① A solution of ^{24}NaCl is injected into the bloodstream.

Site of constriction

③ A high reading of radioactivity at the upper thigh indicates good circulation at that point.

④ A low reading of radioactivity at the right foot indicates poor circulation at that point.

Figure 45.14 A tracer technique for determining the condition of the human circulatory system.

intake of iodized salt and seafood. To evaluate the performance of the thyroid, the patient drinks a very small amount of radioactive sodium iodide containing ^{131}I, an artificially produced isotope of iodine (the natural, nonradioactive isotope is ^{127}I). The amount of iodine in the thyroid gland is determined as a function of time by measuring the radiation intensity at the neck area. How much of the isotope ^{131}I remains in the thyroid is a measure of how well that gland is functioning.

A second medical application is indicated in Figure 45.14. A solution containing radioactive sodium is injected into a vein in the leg, and the time at which the radioisotope arrives at another part of the body is detected with a radiation counter. The elapsed time is a good indication of the presence or absence of constrictions in the circulatory system.

Tracers are also useful in agricultural research. Suppose the best method of fertilizing a plant is to be determined. A certain element in a fertilizer, such as nitrogen, can be *tagged* (identified) with one of its radioactive isotopes. The fertilizer is then sprayed on one group of plants, sprinkled on the ground for a second group, and raked into the soil for a third. A Geiger counter is then used to track the nitrogen through each of the three groups.

Tracing techniques are as wide ranging as human ingenuity can devise. Today, applications range from checking how teeth absorb fluoride to monitoring how cleansers contaminate food-processing equipment to studying deterioration inside an automobile engine. In this last case, a radioactive material is used in the manufacture of the car's piston rings and the oil is checked for radioactivity to determine the amount of wear on the rings.

Materials Analysis

For centuries, a standard method of identifying the elements in a sample of material has been chemical analysis, which involves determining how the material reacts with various chemicals. A second method is spectral analysis, which works because each element, when excited, emits its own characteristic set of electromagnetic wavelengths. These methods are now supplemented by a third technique, **neutron activation analysis.** A disadvantage of both chemical and spectral methods is that a fairly large sample of the material must be destroyed for the analysis. In addition, extremely small quantities of an element may go undetected by either method. Neutron activation analysis has an advantage over chemical analysis and spectral analysis in both respects.

When a material is irradiated with neutrons, nuclei in the material absorb the neutrons and are changed to different isotopes, most of which are radioactive. For example, ^{65}Cu absorbs a neutron to become ^{66}Cu, which undergoes beta decay:

$$\,^{1}_{0}n + \,^{65}_{29}Cu \;\rightarrow\; \,^{66}_{29}Cu \;\rightarrow\; \,^{66}_{30}Zn + e^- + \bar{\nu}$$

The presence of the copper can be deduced because it is known that ^{66}Cu has a half-life of 5.1 min and decays with the emission of beta particles having a maximum energy of 2.63 MeV. Also emitted in the decay of ^{66}Cu is a 1.04-MeV gamma ray. By examining the radiation emitted by a substance after it has been exposed to neutron irradiation, one can detect extremely small amounts of an element in that substance.

Neutron activation analysis is used routinely in a number of industries. In commercial aviation, for example, it is used to check airline luggage for hidden explosives. One nonroutine use is of historical interest. Napoleon died on the island of St. Helena in 1821, supposedly of natural causes. Over the years, suspicion has existed that his death was not all that natural. After his death, his head was shaved and locks of his hair were sold as souvenirs. In 1961, the amount of arsenic in a sample of this hair was measured by neutron activation analysis, and an unusually large quantity of arsenic was found. (Activation analysis is so sensitive that very small pieces of a single hair could be analyzed.) Results showed that the arsenic was fed to him irregularly. In fact, the arsenic concentration pattern corresponded to the fluctuations in the severity of Napoleon's illness as determined from historical records.

Art historians use neutron activation analysis to detect forgeries. The pigments used in paints have changed throughout history, and old and new pigments react differently to neutron activation. The method can even reveal hidden works of art behind existing paintings because an older, hidden layer of paint reacts differently than the surface layer to neutron activation.

Radiation Therapy

Radiation causes much damage to rapidly dividing cells. Therefore, it is useful in cancer treatment because tumor cells divide extremely rapidly. Several mechanisms can be used to deliver radiation to a tumor. In Section 42.8, we discussed the use of high-energy x-rays in the treatment of cancerous tissue. Other treatment protocols include the use of narrow beams of radiation from a radioactive source. As an example, Figure 45.15 shows a machine that uses ^{60}Co as a source. The ^{60}Co isotope emits gamma rays with photon energies higher than 1 MeV.

In other situations, a technique called *brachytherapy* is used. In this treatment plan, thin radioactive needles called *seeds* are implanted in the cancerous tissue. The energy emitted from the seeds is delivered directly to the tumor, reducing the exposure of surrounding tissue to radiation damage. In the case of prostate cancer, the active isotopes used in brachytherapy include ^{125}I and ^{103}Pd.

Food Preservation

Radiation is finding increasing use as a means of preserving food because exposure to high levels of radiation can destroy or incapacitate bacteria and mold spores (Fig. 45.16). Techniques include exposing foods to gamma rays, high-energy electron beams, and x-rays. Food preserved by such exposure can be placed in a sealed container (to keep out new spoiling agents) and stored for long periods of time. There is little or no evidence of adverse effect on the taste or nutritional value of food

Figure 45.15 This large machine is being set to deliver a dose of radiation from ^{60}Co in an effort to destroy a cancerous tumor. Cancer cells are especially susceptible to this type of therapy because they tend to divide more often than cells of healthy tissue nearby.

Figure 45.16 The strawberries on the left are untreated and have become moldy. The unspoiled strawberries on the right have been irradiated. The radiation has killed or incapacitated the mold spores that have spoiled the strawberries on the left.

NON - IRRADIATED -

IRRADIATED - (0.2 M RAD)

Council for Agricultural Science & Technology

from irradiation. The safety of irradiated foods has been endorsed by the World Health Organization, the Centers for Disease Control and Prevention, the U.S. Department of Agriculture, and the Food and Drug Administration. Irradiation of food is presently permitted in more than 50 countries. Some estimates place the amount of irradiated food in the world as high as 500 000 metric tons each year.

Summary

Concepts and Principles

The probability that neutrons are captured as they move through matter generally increases with decreasing neutron energy. A **thermal neutron** is a slow-moving neutron that has a high probability of being captured by a nucleus in a **neutron capture event:**

$$\ce{^{1}_{0}n} + \ce{^{A}_{Z}X} \rightarrow \ce{^{A+1}_{Z}X^{*}} \rightarrow \ce{^{A+1}_{Z}X} + \gamma \tag{45.1}$$

where $^{A+1}_{Z}X^{*}$ is an excited intermediate nucleus that rapidly emits a photon.

Nuclear fission occurs when a very heavy nucleus, such as ^{235}U, splits into two smaller **fission fragments.** Thermal neutrons can create fission in ^{235}U:

$$\ce{^{1}_{0}n} + \ce{^{235}_{92}U} \rightarrow \ce{^{236}_{92}U^{*}} \rightarrow X + Y + \text{neutrons} \tag{45.2}$$

where $^{236}U^{*}$ is an intermediate excited state and X and Y are the fission fragments. On average, 2.5 neutrons are released per fission event. The fragments then undergo a series of beta and gamma decays to various stable isotopes. The energy released per fission event is approximately 200 MeV.

The **reproduction constant** K is the average number of neutrons released from each fission event that cause another event. In a fission reactor, it is necessary to maintain $K \approx 1$. The value of K is affected by such factors as reactor geometry, mean neutron energy, and probability of neutron capture.

In **nuclear fusion,** two light nuclei fuse to form a heavier nucleus and release energy. The major obstacle in obtaining useful energy from fusion is the large Coulomb repulsive force between the charged nuclei at small separation distances. The temperature required to produce fusion is on the order of 10^8 K, and at this temperature, all matter occurs as a plasma.

In a fusion reactor, the plasma temperature must reach the **critical ignition temperature,** the temperature at which the power generated by the fusion reactions exceeds the power lost in the system. The most promising fusion reaction is the D–T reaction, which has a critical ignition temperature of approximately 4.5×10^7 K. Two critical parameters in fusion reactor design are **ion density** n and **confinement time** τ, the time interval during which the interacting particles must be maintained at $T > T_{ignit}$. **Lawson's criterion** states that for the D–T reaction, $n\tau \geq 10^{14}$ s/cm^3.

1. In a certain fission reaction, a ^{235}U nucleus captures a neutron. This process results in the creation of the products ^{137}I and ^{96}Y along with how many neutrons? (a) 1 (b) 2 (c) 3 (d) 4 (e) 5

2. Which particle is most likely to be captured by a ^{235}U nucleus and cause it to undergo fission? (a) an energetic proton (b) an energetic neutron (c) a slow-moving alpha particle (d) a slow-moving neutron (e) a fast-moving electron

3. In the first nuclear weapon test carried out in New Mexico, the energy released was equivalent to approximately 17 kilotons of TNT. Estimate the mass decrease in the nuclear fuel representing the energy converted from rest energy into other forms in this event. *Note:* One ton of TNT has the energy equivalent of 4.2×10^9 J. (a) 1 μg (b) 1 mg (c) 1 g (d) 1 kg (e) 20 kg

4. Working with radioactive materials at a laboratory over one year, (a) Tom received 1 rem of alpha radiation, (b) Karen received 1 rad of fast neutrons, (c) Paul received 1 rad of thermal neutrons as a whole-body dose, and (d) Ingrid received 1 rad of thermal neutrons to her hands only. Rank these four doses according to the likely amount of biological damage from the greatest to the least, noting any cases of equality.

5. If the moderator were suddenly removed from a nuclear reactor in an electric generating station, what is the most likely consequence? (a) The reactor would go supercritical, and a runaway reaction would occur. (b) The nuclear reaction would proceed in the same way, but the reactor would overheat. (c) The reactor would become subcritical, and the reaction would die out. (d) No change would occur in the reactor's operation.

6. You may use Figure 44.5 to answer this question. Three nuclear reactions take place, each involving 108 nucleons: (1) eighteen ^{6}Li nuclei fuse in pairs to form nine ^{12}C nuclei, (2) four nuclei each with 27 nucleons fuse in pairs to form two nuclei with 54 nucleons, and (3) one nucleus with 108 nucleons fissions to form two nuclei with 54 nucleons. Rank these three reactions from the largest positive Q value (representing energy output) to the largest negative

value (representing energy input). Also include $Q = 0$ in your ranking to make clear which of the reactions put out energy and which absorb energy. Note any cases of equality in your ranking.

7. A device called a *bubble chamber* uses a liquid (usually liquid hydrogen) maintained near its boiling point. Ions produced by incoming charged particles from nuclear decays leave bubble tracks, which can be photographed. Figure OQ45.7 shows particle tracks in a bubble chamber immersed in a magnetic field. The tracks are generally spirals rather than sections of circles. What is the primary reason for this shape? (a) The magnetic field is not perpendicular to the velocity of the particles. (b) The magnetic field is not uniform in space. (c) The forces on the particles increase with time. (d) The speeds of the particles decrease with time.

Figure OQ45.7

8. If an alpha particle and an electron have the same kinetic energy, which undergoes the greater deflection when passed through a magnetic field? (a) The alpha particle does. (b) The electron does. (c) They undergo the same deflection. (d) Neither is deflected.

9. Which of the following fuel conditions is *not* necessary to operate a self-sustained controlled fusion reactor? (a) The fuel must be at a sufficiently high temperature. (b) The fuel must be radioactive. (c) The fuel must be at a sufficiently high density. (d) The fuel must be confined for a sufficiently long period of time. (e) Conditions (a) through (d) are all necessary.

1. What factors make a terrestrial fusion reaction difficult to achieve?

2. Lawson's criterion states that the product of ion density and confinement time must exceed a certain number before a break-even fusion reaction can occur. Why should these two parameters determine the outcome?

3. Why would a fusion reactor produce less radioactive waste than a fission reactor?

4. Discuss the advantages and disadvantages of fission reactors from the point of view of safety, pollution, and resources. Make a comparison with power generated from the burning of fossil fuels.

5. Discuss the similarities and differences between fusion and fission.

6. If a nucleus captures a slow-moving neutron, the product is left in a highly excited state, with an energy approximately 8 MeV above the ground state. Explain the source of the excitation energy.

7. Discuss the advantages and disadvantages of fusion power from the viewpoint of safety, pollution, and resources.

8. A scintillation crystal can be a detector of radiation when combined with a photomultiplier tube (Section 40.2). The scintillator is usually a solid or liquid material whose atoms are easily excited by radiation. The excited atoms then emit photons when they return to their ground state. The design of the radiation detector in Figure CQ45.8 might suggest that any number of dynodes may be used to amplify a weak signal. What factors do you suppose would limit the amplification in this device?

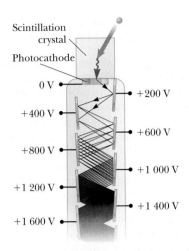

Figure CQ45.8

9. Why is water a better shield against neutrons than lead or steel?

Problems

Section 45.1 Interactions Involving Neutrons

Section 45.2 Nuclear Fission

Problem 57 in Chapter 25 and Problems 22 and 78 in Chapter 44 can be assigned with this chapter.

1. If the average energy released in a fission event is **M** 208 MeV, find the total number of fission events required to operate a 100-W lightbulb for 1.0 h.

2. Burning one metric ton (1 000 kg) of coal can yield an energy of 3.30×10^{10} J. Fission of one nucleus of uranium-235 yields an average of approximately 200 MeV. What mass of uranium produces the same energy in fission as burning one metric ton of coal?

3. Strontium-90 is a particularly dangerous fission product of ^{235}U because it is radioactive and it substitutes for calcium in bones. What other direct fission products would accompany it in the neutron-induced fission of ^{235}U? *Note:* This reaction may release two, three, or four free neutrons.

4. A typical nuclear fission power plant produces approximately 1.00 GW of electrical power. Assume the plant has an overall efficiency of 40.0% and each fission reaction produces 200 MeV of energy. Calculate the mass of ^{235}U consumed each day.

5. List the nuclear reactions required to produce ^{233}U from ^{232}Th under fast neutron bombardment.

6. The following fission reaction is typical of those occurring in a nuclear electric generating station:

$$^{1}_{0}n + {}^{235}_{92}U \rightarrow {}^{141}_{56}Ba + {}^{92}_{36}Kr + 3({}^{1}_{0}n)$$

(a) Find the energy released in the reaction. The masses of the products are 140.914 411 u for $^{141}_{56}$Ba and 91.926 156 u for $^{92}_{36}$Kr. (b) What fraction of the initial rest energy of the system is transformed to other forms?

7. Find the energy released in the fission reaction

$$^{1}_{0}n + {}^{235}_{92}U \rightarrow {}^{88}_{38}Sr + {}^{136}_{54}Xe + 12({}^{1}_{0}n)$$

8. A 2.00-MeV neutron is emitted in a fission reactor. If it loses half its kinetic energy in each collision with a moderator atom, how many collisions does it undergo as it becomes a thermal neutron, with energy 0.039 eV?

9. Find the energy released in the fission reaction

$$^{1}_{0}n + {}^{235}_{92}U \rightarrow {}^{98}_{40}Zr + {}^{135}_{52}Te + 3({}^{1}_{0}n)$$

The atomic masses of the fission products are 97.912 735 u for $^{98}_{40}$Zr and 134.916 450 u for $^{135}_{52}$Te.

10. Seawater contains 3.00 mg of uranium per cubic meter. (a) Given that the average ocean depth is about 4.00 km and water covers two-thirds of the Earth's surface, estimate the amount of uranium dissolved in the ocean. (b) About 0.700% of naturally occurring uranium is the fissionable isotope ^{235}U. Estimate how long the uranium in the oceans could supply the world's energy needs at the current usage of 1.50×10^{13} J/s. (c) Where does the dissolved uranium come from? (d) Is it a renewable energy source?.

11. **Review.** Suppose seawater exerts an average frictional drag force of 1.00×10^5 N on a nuclear-powered ship. The fuel consists of enriched uranium containing 3.40% of the fissionable isotope $^{235}_{92}$U, and the ship's reactor has an efficiency of 20.0%. Assuming 200 MeV is released per fission event, how far can the ship travel per kilogram of fuel?

Section 45.3 Nuclear Reactors

12. Assume ordinary soil contains natural uranium in an amount of 1 part per million by mass. (a) How much uranium is in the top 1.00 m of soil on a 1-acre (43 560-ft^2) plot of ground, assuming the specific gravity of soil is 4.00? (b) How much of the isotope ^{235}U, appropriate for nuclear reactor fuel, is in this soil? *Hint:* See Table 44.2 for the percent abundance of $^{235}_{92}$U.

13. If the reproduction constant is 1.000 25 for a chain reaction in a fission reactor and the average time interval between successive fissions is 1.20 ms, by what factor does the reaction rate increase in one minute?

14. To minimize neutron leakage from a reactor, the ratio of the surface area to the volume should be a minimum. For a given volume V, calculate this ratio for (a) a sphere, (b) a cube, and (c) a parallelepiped of dimensions $a \times a \times 2a$. (d) Which of these shapes would have minimum leakage? Which would have maximum leakage? Explain your answers.

15. The probability of a nuclear reaction increases dramatically when the incident particle is given energy above the "Coulomb barrier," which is the electric potential energy of the two nuclei when their surfaces barely touch. Compute the Coulomb barrier for the absorption of an alpha particle by a gold nucleus.

16. A large nuclear power reactor produces approximately 3 000 MW of power in its core. Three months after a reactor is shut down, the core power from radioactive by-products is 10.0 MW. Assuming each emission delivers 1.00 MeV of energy to the power, find the activity in becquerels three months after the reactor is shut down.

17. According to one estimate, there are 4.40×10^6 metric tons of world uranium reserves extractable at \$130/kg or less. We wish to determine if these reserves

are sufficient to supply all the world's energy needs. About 0.700% of naturally occurring uranium is the fissionable isotope ^{235}U. (a) Calculate the mass of ^{235}U in the reserve in grams. (b) Find the number of moles of ^{235}U in the reserve. (c) Find the number of ^{235}U nuclei in the reserve. (d) Assuming 200 MeV is obtained from each fission reaction and all this energy is captured, calculate the total energy in joules that can be extracted from the reserve. (e) Assuming the rate of world power consumption remains constant at 1.5×10^{13} J/s, how many years could the uranium reserve provide for all the world's energy needs? (f) What conclusion can be drawn?

18. *Why is the following situation impossible?* An engineer working on nuclear power makes a breakthrough so that he is able to control what daughter nuclei are created in a fission reaction. By carefully controlling the process, he is able to restrict the fission reactions to just this single possibility: the uranium-235 nucleus absorbs a slow neutron and splits into lanthanum-141 and bromine-94. Using this breakthrough, he is able to design and build a successful nuclear reactor in which only this single process occurs.

19. An all-electric home uses approximately 2 000 kWh of electric energy per month. How much uranium-235 would be required to provide this house with its energy needs for one year? Assume 100% conversion efficiency and 208 MeV released per fission.

20. A particle cannot generally be localized to distances much smaller than its de Broglie wavelength. This fact can be taken to mean that a slow neutron appears to be larger to a target particle than does a fast neutron in the sense that the slow neutron has probabilities of being found over a larger volume of space. For a thermal neutron at room temperature of 300 K, find (a) the linear momentum and (b) the de Broglie wavelength. (c) State how this effective size compares with both nuclear and atomic dimensions.

Section 45.4 Nuclear Fusion

21. When a star has exhausted its hydrogen fuel, it may fuse other nuclear fuels. At temperatures above 1.00×10^8 K, helium fusion can occur. Consider the following processes. (a) Two alpha particles fuse to produce a nucleus A and a gamma ray. What is nucleus A? (b) Nucleus A from part (a) absorbs an alpha particle to produce nucleus B and a gamma ray. What is nucleus B? (c) Find the total energy released in the sequence of reactions given in parts (a) and (b).

22. An all-electric home uses 2 000 kWh of electric energy per month. Assuming all energy released from fusion could be captured, how many fusion events described by the reaction $^2_1\text{H} + ^3_1\text{H} \rightarrow ^4_2\text{He} + ^1_0\text{n}$ would be required to keep this home running for one year?

23. Find the energy released in the fusion reaction

$$^1_1\text{H} + ^2_1\text{H} \rightarrow ^3_2\text{He} + \gamma$$

24. Two nuclei having atomic numbers Z_1 and Z_2 approach each other with a total energy E. (a) When they are far apart, they interact only by electric repulsion. If they approach to a distance of 1.00×10^{-14} m, the nuclear force suddenly takes over to make them fuse. Find the minimum value of E, in terms of Z_1 and Z_2, required to produce fusion. (b) State how E depends on the atomic numbers. (c) If $Z_1 + Z_2$ is to have a certain target value such as 60, would it be energetically favorable to take $Z_1 = 1$ and $Z_2 = 59$, or $Z_1 = Z_2 = 30$, or some other choice? Explain your answer. (d) Evaluate from your expression the minimum energy for fusion for the D–D and D–T reactions (the first and third reactions in Eq. 45.4).

25. (a) Consider a fusion generator built to create 3.00 GW of power. Determine the rate of fuel burning in grams per hour if the D–T reaction is used. (b) Do the same for the D–D reaction, assuming the reaction products are split evenly between (n, ^{3}He) and (p, ^{3}H).

26. **Review.** Consider the deuterium–tritium fusion reaction with the tritium nucleus at rest:

$$^2_1\text{H} + ^3_1\text{H} \rightarrow ^4_2\text{He} + ^1_0\text{n}$$

(a) Suppose the reactant nuclei will spontaneously fuse if their surfaces touch. From Equation 44.1, determine the required distance of closest approach between their centers. (b) What is the electric potential energy (in electron volts) at this distance? (c) Suppose the deuteron is fired straight at an originally stationary tritium nucleus with just enough energy to reach the required distance of closest approach. What is the common speed of the deuterium and tritium nuclei, in terms of the initial deuteron speed v_i, as they touch? (d) Use energy methods to find the minimum initial deuteron energy required to achieve fusion. (e) Why does the fusion reaction actually occur at much lower deuteron energies than the energy calculated in part (d)?

27. Of all the hydrogen in the oceans, 0.030 0% of the mass is deuterium. The oceans have a volume of 317 million mi^3. (a) If nuclear fusion were controlled and all the deuterium in the oceans were fused to $^4_2$He, how many joules of energy would be released? (b) **What If?** World power consumption is approximately 1.50×10^{13} W. If consumption were 100 times greater, how many years would the energy calculated in part (a) last?

28. It has been suggested that fusion reactors are safe from explosion because the plasma never contains enough energy to do much damage. (a) In 1992, the TFTR reactor, with a plasma volume of approximately 50.0 m^3, achieved an ion temperature of 4.00×10^8 K, an ion density of 2.00×10^{13} cm^{-3}, and a confinement time of 1.40 s. Calculate the amount of energy stored in the plasma of the TFTR reactor. (b) How many kilograms of water at 27.0°C could be boiled away by this much energy?

29. To understand why plasma containment is necessary, consider the rate at which an unconfined plasma would be lost. (a) Estimate the rms speed of deuterons in a plasma at a temperature of 4.00×10^8 K. (b) **What If?** Estimate the order of magnitude of the time interval during which such a plasma would remain in a 10.0-cm cube if no steps were taken to contain it.

30. Another series of nuclear reactions that can produce energy in the interior of stars is the carbon cycle first proposed by Hans Bethe in 1939, leading to his Nobel Prize in Physics in 1967. This cycle is most efficient when the central temperature in a star is above 1.6×10^7 K. Because the temperature at the center of the Sun is only 1.5×10^7 K, the following cycle produces less than 10% of the Sun's energy. (a) A high-energy proton is absorbed by ^{12}C. Another nucleus, A, is produced in the reaction, along with a gamma ray. Identify nucleus A. (b) Nucleus A decays through positron emission to form nucleus B. Identify nucleus B. (c) Nucleus B absorbs a proton to produce nucleus C and a gamma ray. Identify nucleus C. (d) Nucleus C absorbs a proton to produce nucleus D and a gamma ray. Identify nucleus D. (e) Nucleus D decays through positron emission to produce nucleus E. Identify nucleus E. (f) Nucleus E absorbs a proton to produce nucleus F plus an alpha particle. Identify nucleus F. (g) What is the significance of the final nucleus in the last step of the cycle outlined in part (f)?

31. **Review.** To confine a stable plasma, the magnetic energy density in the magnetic field (Eq. 32.14) must exceed the pressure $2nk_\text{B}T$ of the plasma by a factor of at least 10. In this problem, assume a confinement time $\tau = 1.00$ s. (a) Using Lawson's criterion, determine the ion density required for the D–T reaction. (b) From the ignition-temperature criterion, determine the required plasma pressure. (c) Determine the magnitude of the magnetic field required to contain the plasma.

Section 45.5 Radiation Damage

32. Assume an x-ray technician takes an average of eight x-rays per workday and receives a dose of 5.0 rem/yr as a result. (a) Estimate the dose in rem per x-ray taken. (b) Explain how the technician's exposure compares with low-level background radiation.

33. When gamma rays are incident on matter, the intensity of the gamma rays passing through the material varies with depth x as $I(x) = I_0 e^{-\mu x}$, where I_0 is the intensity of the radiation at the surface of the material (at $x = 0$) and μ is the linear absorption coefficient. For 0.400-MeV gamma rays in lead, the linear absorption coefficient is 1.59 cm^{-1}. (a) Determine the "half-thickness" for lead, that is, the thickness of lead that would absorb half the incident gamma rays. (b) What thickness reduces the radiation by a factor of 10^4?

34. When gamma rays are incident on matter, the intensity of the gamma rays passing through the material varies with depth x as $I(x) = I_0 e^{-\mu x}$, where I_0 is the intensity of the radiation at the surface of the material (at $x = 0$) and μ is the linear absorption coefficient. (a) Determine

the "half-thickness" for a material with linear absorption coefficient μ, that is, the thickness of the material that would absorb half the incident gamma rays. (b) What thickness changes the radiation by a factor of f?

35. Review. A particular radioactive source produces 100 mrad of 2.00-MeV gamma rays per hour at a distance of 1.00 m from the source. (a) How long could a person stand at this distance before accumulating an intolerable dose of 1.00 rem? (b) **What If?** Assuming the radioactive source is a point source, at what distance would a person receive a dose of 10.0 mrad/h?

36. A person whose mass is 75.0 kg is exposed to a whole-
M body dose of 0.250 Gy. How many joules of energy are deposited in the person's body?

37. Review. The danger to the body from a high dose of gamma rays is not due to the amount of energy absorbed; rather, it is due to the ionizing nature of the radiation. As an illustration, calculate the rise in body temperature that results if a "lethal" dose of 1 000 rad is absorbed strictly as internal energy. Take the specific heat of living tissue as 4 186 J/kg · °C.

38. Review. *Why is the following situation impossible?* A "clever" technician takes his 20-min coffee break and boils some water for his coffee with an x-ray machine. The machine produces 10.0 rad/s, and the temperature of the water in an insulated cup is initially 50.0°C.

39. A small building has become accidentally contami-
W nated with radioactivity. The longest-lived material in the building is strontium-90. ($^{90}_{38}$Sr has an atomic mass 89.907 7 u, and its half-life is 29.1 yr. It is particularly dangerous because it substitutes for calcium in bones.) Assume the building initially contained 5.00 kg of this substance uniformly distributed throughout the building and the safe level is defined as less than 10.0 decays/min (which is small compared with background radiation). How long will the building be unsafe?

40. Technetium-99 is used in certain medical diagnostic procedures. Assume 1.00×10^{-8} g of ^{99}Tc is injected into a 60.0-kg patient and half of the 0.140-MeV gamma rays are absorbed in the body. Determine the total radiation dose received by the patient.

41. To destroy a cancerous tumor, a dose of gamma radia-
W tion with a total energy of 2.12 J is to be delivered in 30.0 days from implanted sealed capsules containing palladium-103. Assume this isotope has a half-life of 17.0 d and emits gamma rays of energy 21.0 keV, which are entirely absorbed within the tumor. (a) Find the initial activity of the set of capsules. (b) Find the total mass of radioactive palladium these "seeds" should contain.

42. Strontium-90 from the testing of nuclear bombs can still be found in the atmosphere. Each decay of ^{90}Sr releases 1.10 MeV of energy into the bones of a person who has had strontium replace his or her body's calcium. Assume a 70.0-kg person receives 1.00 ng of ^{90}Sr from contaminated milk. Take the half-life of ^{90}Sr to

be 29.1 yr. Calculate the absorbed dose rate (in joules per kilogram) in one year.

Section 45.6 Uses of Radiation

43. When gamma rays are incident on matter, the inten-
M sity of the gamma rays passing through the material varies with depth x as $I(x) = I_0 e^{-\mu x}$, where I_0 is the intensity of the radiation at the surface of the material (at $x = 0$) and μ is the linear absorption coefficient. For low-energy gamma rays in steel, take the absorption coefficient to be 0.720 mm^{-1}. (a) Determine the "half-thickness" for steel, that is, the thickness of steel that would absorb half the incident gamma rays. (b) In a steel mill, the thickness of sheet steel passing into a roller is measured by monitoring the intensity of gamma radiation reaching a detector below the rapidly moving metal from a small source immediately above the metal. If the thickness of the sheet changes from 0.800 mm to 0.700 mm, by what percentage does the gamma-ray intensity change?

44. A method called *neutron activation analysis* can be used for chemical analysis at the level of isotopes. When a sample is irradiated by neutrons, radioactive atoms are produced continuously and then decay according to their characteristic half-lives. (a) Assume one species of radioactive nuclei is produced at a constant rate R and its decay is described by the conventional radioactive decay law. Assuming irradiation begins at time $t = 0$, show that the number of radioactive atoms accumulated at time t is

$$N = \frac{R}{\lambda}(1 - e^{-\lambda t})$$

(b) What is the maximum number of radioactive atoms that can be produced?

45. You want to find out how many atoms of the isotope ^{65}Cu are in a small sample of material. You bombard the sample with neutrons to ensure that on the order of 1% of these copper nuclei absorb a neutron. After activation, you turn off the neutron flux and then use a highly efficient detector to monitor the gamma radiation that comes out of the sample. Assume half of the ^{66}Cu nuclei emit a 1.04-MeV gamma ray in their decay. (The other half of the activated nuclei decay directly to the ground state of ^{66}Ni.) If after 10 min (two half-lives) you have detected 1.00×10^4 MeV of photon energy at 1.04 MeV, (a) approximately how many ^{65}Cu atoms are in the sample? (b) Assume the sample contains natural copper. Refer to the isotopic abundances listed in Table 44.2 and estimate the total mass of copper in the sample.

Additional Problems

46. A fusion reaction that has been considered as a source of energy is the absorption of a proton by a boron-11 nucleus to produce three alpha particles:

$$^1_1\text{H} + ^{11}_5\text{B} \rightarrow 3(^4_2\text{He})$$

This reaction is an attractive possibility because boron is easily obtained from the Earth's crust. A disadvantage is that the protons and boron nuclei must have large kinetic energies for the reaction to take place. This requirement contrasts with the initiation of uranium fission by slow neutrons. (a) How much energy is released in each reaction? (b) Why must the reactant particles have high kinetic energies?

47. **Review.** A very slow neutron (with speed approximately equal to zero) can initiate the reaction

$$\,^1_0n + \,^{10}_5B \;\rightarrow\; \,^7_3Li + \,^4_2He$$

The alpha particle moves away with speed 9.25×10^6 m/s. Calculate the kinetic energy of the lithium nucleus. Use nonrelativistic equations.

48. **Review.** The first nuclear bomb was a fissioning mass of plutonium-239 that exploded in the Trinity test before dawn on July 16, 1945, at Alamogordo, New Mexico. Enrico Fermi was 14 km away, lying on the ground facing away from the bomb. After the whole sky had flashed with unbelievable brightness, Fermi stood up and began dropping bits of paper to the ground. They first fell at his feet in the calm and silent air. As the shock wave passed, about 40 s after the explosion, the paper then in flight jumped approximately 2.5 m away from ground zero. (a) Equation 17.10 describes the relationship between the pressure amplitude ΔP_{max} of a sinusoidal air compression wave and its displacement amplitude s_{max}. The compression pulse produced by the bomb explosion was not a sinusoidal wave, but let's use the same equation to compute an estimate for the pressure amplitude, taking $\omega \sim 1 \ s^{-1}$ as an estimate for the angular frequency at which the pulse ramps up and down. (b) Find the change in volume ΔV of a sphere of radius 14 km when its radius increases by 2.5 m. (c) The energy carried by the blast wave is the work done by one layer of air on the next as the wave crest passes. An extension of the logic used to derive Equation 20.8 shows that this work is given by $(\Delta P_{max})(\Delta V)$. Compute an estimate for this energy. (d) Assume the blast wave carried on the order of one-tenth of the explosion's energy. Make an order-of-magnitude estimate of the bomb yield. (e) One ton of exploding TNT releases 4.2 GJ of energy. What was the order of magnitude of the energy of the Trinity test in equivalent tons of TNT? Fermi's immediate knowledge of the bomb yield agreed with that determined days later by analysis of elaborate measurements.

49. On August 6, 1945, the United States dropped on Hiroshima a nuclear bomb that released 5×10^{13} J of energy, equivalent to that from 12 000 tons of TNT. The fission of one $^{235}_{92}U$ nucleus releases an average of 208 MeV. Estimate (a) the number of nuclei fissioned and (b) the mass of this $^{235}_{92}U$.

50. (a) A student wishes to measure the half-life of a radioactive substance using a small sample. Consecutive clicks of her radiation counter are randomly spaced in time. The counter registers 372 counts during one 5.00-min

interval and 337 counts during the next 5.00 min. The average background rate is 15 counts per minute. Find the most probable value for the half-life. (b) Estimate the uncertainty in the half-life determination in part (a). Explain your reasoning.

51. In a Geiger–Mueller tube for detecting radiation (see Problem 68 in Chapter 25), the voltage between the electrodes is typically 1.00 kV and the current pulse discharges a 5.00-pF capacitor. (a) What is the energy amplification of this device for a 0.500-MeV electron? (b) How many electrons participate in the avalanche caused by the single initial electron?

52. **Review.** Consider a nucleus at rest, which then spontaneously splits into two fragments of masses m_1 and m_2. (a) Show that the fraction of the total kinetic energy carried by fragment m_1 is

$$\frac{K_1}{K_{\text{tot}}} = \frac{m_2}{m_1 + m_2}$$

and the fraction carried by m_2 is

$$\frac{K_2}{K_{\text{tot}}} = \frac{m_1}{m_1 + m_2}$$

assuming relativistic corrections can be ignored. A stationary $^{236}_{92}U$ nucleus fissions spontaneously into two primary fragments, $^{87}_{35}Br$ and $^{149}_{57}La$. (b) Calculate the disintegration energy. The required atomic masses are 86.920 711 u for $^{87}_{35}Br$, 148.934 370 u for $^{149}_{57}La$, and 236.045 562 u for $^{236}_{92}U$. (c) How is the disintegration energy split between the two primary fragments? (d) Calculate the speed of each fragment immediately after the fission.

53. Consider the carbon cycle in Problem 30. (a) Calculate the Q value for each of the six steps in the carbon cycle listed in Problem 30. (b) In the second and fifth steps of the cycle, the positron that is ejected combines with an electron to form two photons. The energies of these photons must be included in the energy released in the cycle. How much energy is released by these annihilations in each of the two steps? (c) What is the overall energy released in the carbon cycle? (d) Do you think that the energy carried off by the neutrinos is deposited in the star? Explain.

54. A fission reactor is hit by a missile, and 5.00×10^6 Ci of ^{90}Sr, with half-life 29.1 yr, evaporates into the air. The strontium falls out over an area of 10^4 km^2. After what time interval will the activity of the ^{90}Sr reach the agriculturally "safe" level of 2.00 μCi/m^2?

55. The alpha-emitter plutonium-238 ($^{238}_{94}Pu$, atomic mass 238.049 560 u, half-life 87.7 yr) was used in a nuclear energy source on the Apollo Lunar Surface Experiments Package (Fig. P45.55, page 1444). The energy source, called the Radioisotope Thermoelectric Generator, is the small gray object to the left of the gold-shrouded Central Station in the photograph. Assume the source contains 3.80 kg of ^{238}Pu and the efficiency

for conversion of radioactive decay energy to energy transferred by electrical transmission is 3.20%. Determine the initial power output of the source.

NASA Johnson Space Center Collection

Figure P45.55

56. The half-life of tritium is 12.3 yr. (a) If the TFTR fusion reactor contained 50.0 m³ of tritium at a density equal to 2.00×10^{14} ions/cm³, how many curies of tritium were in the plasma? (b) State how this value compares with a fission inventory (the estimated supply of fissionable material) of 4.00×10^{10} Ci.

57. **Review.** A nuclear power plant operates by using the energy released in nuclear fission to convert 20°C water into 400°C steam. How much water could theoretically be converted to steam by the complete fissioning of 1.00 g of ^{235}U at 200 MeV/fission?

58. **Review.** A nuclear power plant operates by using the energy released in nuclear fission to convert liquid water at T_c into steam at T_h. How much water could theoretically be converted to steam by the complete fissioning of a mass m of ^{235}U if the energy released per fission event is E?

59. Consider the two nuclear reactions

$$A + B \rightarrow C + E$$
$$C + D \rightarrow F + G$$

(a) Show that the net disintegration energy for these two reactions ($Q_{net} = Q_I + Q_{II}$) is identical to the disintegration energy for the net reaction

$$A + B + D \rightarrow E + F + G$$

(b) One chain of reactions in the proton–proton cycle in the Sun's core is

$$\begin{aligned}
{}^1_1\text{H} + {}^1_1\text{H} &\rightarrow {}^2_1\text{H} + {}^0_1\text{e} + \nu \\
{}^0_1\text{e} + {}^0_{-1}\text{e} &\rightarrow 2\gamma \\
{}^1_1\text{H} + {}^2_1\text{H} &\rightarrow {}^3_2\text{He} + \gamma \\
{}^1_1\text{H} + {}^3_2\text{He} &\rightarrow {}^4_2\text{He} + {}^0_1\text{e} + \nu \\
{}^0_1\text{e} + {}^0_{-1}\text{e} &\rightarrow 2\gamma
\end{aligned}$$

Based on part (a), what is Q_{net} for this sequence?

60. Natural uranium must be processed to produce uranium enriched in ^{235}U for weapons and power plants. The processing yields a large quantity of nearly pure ^{238}U as a by-product, called "depleted uranium." Because of its high mass density, ^{238}U is used in armor-piercing artillery shells. (a) Find the edge dimension of a 70.0-kg cube of ^{238}U ($\rho = 19.1 \times 10^3$ kg/m³). (b) The isotope ^{238}U has a long half-life of 4.47×10^9 yr. As soon as one nucleus decays, a relatively rapid series of 14 steps begins that together constitute the net reaction

$$^{238}_{92}\text{U} \rightarrow 8({}^4_2\text{He}) + 6({}^0_{-1}\text{e}) + {}^{206}_{82}\text{Pb} + 6\bar{\nu} + Q_{net}$$

Find the net decay energy. (Refer to Table 44.2.) (c) Argue that a radioactive sample with decay rate R and decay energy Q has power output $P = QR$. (d) Consider an artillery shell with a jacket of 70.0 kg of ^{238}U. Find its power output due to the radioactivity of the uranium and its daughters. Assume the shell is old enough that the daughters have reached steady-state amounts. Express the power in joules per year. (e) **What If?** A 17-year-old soldier of mass 70.0 kg works in an arsenal where many such artillery shells are stored. Assume his radiation exposure is limited to 5.00 rem per year. Find the rate in joules per year at which he can absorb energy of radiation. Assume an average RBE factor of 1.10.

61. Suppose the target in a laser fusion reactor is a sphere of solid hydrogen that has a diameter of 1.50×10^{-4} m and a density of 0.200 g/cm³. Assume half of the nuclei are ^{2}H and half are ^{3}H. (a) If 1.00% of a 200-kJ laser pulse is delivered to this sphere, what temperature does the sphere reach? (b) If all the hydrogen fuses according to the D–T reaction, how many joules of energy are released?

62. When photons pass through matter, the intensity I of the beam (measured in watts per square meter) decreases exponentially according to

$$I = I_0 e^{-\mu x}$$

where I is the intensity of the beam that just passed through a thickness x of material and I_0 is the intensity of the incident beam. The constant μ is known as the linear absorption coefficient, and its value depends on the absorbing material and the wavelength of the photon beam. This wavelength (or energy) dependence allows us to filter out unwanted wavelengths from a broad-spectrum x-ray beam. (a) Two x-ray beams of wavelengths λ_1 and λ_2 and equal incident intensities pass through the same metal plate. Show that the ratio of the emergent beam intensities is

$$\frac{I_2}{I_1} = e^{-(\mu_2 - \mu_1)x}$$

(b) Compute the ratio of intensities emerging from an aluminum plate 1.00 mm thick if the incident beam contains equal intensities of 50 pm and 100 pm x-rays. The values of μ for aluminum at these two wavelengths are $\mu_1 = 5.40$ cm^{-1} at 50 pm and $\mu_2 = 41.0$ cm^{-1} at 100 pm. (c) Repeat part (b) for an aluminum plate 10.0 mm thick.

63. Assume a deuteron and a triton are at rest when they fuse according to the reaction

$$^2_1\text{H} + ^3_1\text{H} \rightarrow ^4_2\text{He} + ^1_0\text{n}$$

Determine the kinetic energy acquired by the neutron.

64. (a) Calculate the energy (in kilowatt-hours) released if 1.00 kg of ^{239}Pu undergoes complete fission and the energy released per fission event is 200 MeV. (b) Calculate the energy (in electron volts) released in the deuterium–tritium fusion reaction

$$^2_1\text{H} + ^3_1\text{H} \rightarrow ^4_2\text{He} + ^1_0\text{n}$$

(c) Calculate the energy (in kilowatt-hours) released if 1.00 kg of deuterium undergoes fusion according to this reaction. (d) **What If?** Calculate the energy (in kilowatt-hours) released by the combustion of 1.00 kg of carbon in coal if each C + $O_2 \rightarrow CO_2$ reaction yields 4.20 eV. (e) List advantages and disadvantages of each of these methods of energy generation.

65. Consider a 1.00-kg sample of natural uranium composed primarily of ^{238}U, a smaller amount (0.720% by mass) of ^{235}U, and a trace (0.005 00%) of ^{234}U, which has a half-life of 2.44×10^5 yr. (a) Find the activity in curies due to each of the isotopes. (b) What fraction of the total activity is due to each isotope? (c) Explain whether the activity of this sample is dangerous.

66. Approximately 1 of every 3 300 water molecules contains one deuterium atom. (a) If all the deuterium nuclei in 1 L of water are fused in pairs according to the D–D fusion reaction $^2\text{H} + ^2\text{H} \rightarrow ^3\text{He} + \text{n} + 3.27$ MeV, how much energy in joules is liberated? (b) **What If?** Burning gasoline produces approximately 3.40×10^7 J/L. State how the energy obtainable from the fusion of the deuterium in 1 L of water compares with the energy liberated from the burning of 1 L of gasoline.

67. Carbon detonations are powerful nuclear reactions that temporarily tear apart the cores inside massive stars late in their lives. These blasts are produced by carbon fusion, which requires a temperature of approximately 6×10^8 K to overcome the strong Coulomb repulsion between carbon nuclei. (a) Estimate the repulsive energy barrier to fusion, using the temperature required for carbon fusion. (In other words, what is the average kinetic energy of a carbon nucleus at 6×10^8 K?) (b) Calculate the energy (in MeV) released in each of these "carbon-burning" reactions:

$$^{12}\text{C} + ^{12}\text{C} \rightarrow ^{20}\text{Ne} + ^4\text{He}$$
$$^{12}\text{C} + ^{12}\text{C} \rightarrow ^{24}\text{Mg} + \gamma$$

(c) Calculate the energy in kilowatt-hours given off when 2.00 kg of carbon completely fuse according to the first reaction.

68. A sealed capsule containing the radiopharmaceutical phosphorus-32, an e^- emitter, is implanted into a patient's tumor. The average kinetic energy of the beta particles is 700 keV. The initial activity is 5.22 MBq. Assume the beta particles are completely absorbed in 100 g of tissue. Determine the absorbed dose during a 10.0-day period.

69. A certain nuclear plant generates internal energy at a rate of 3.065 GW and transfers energy out of the plant by electrical transmission at a rate of 1.000 GW. Of the waste energy, 3.0% is ejected to the atmosphere and the remainder is passed into a river. A state law requires that the river water be warmed by no more than 3.50°C when it is returned to the river. (a) Determine the amount of cooling water necessary (in kilograms per hour and cubic meters per hour) to cool the plant. (b) Assume fission generates 7.80×10^{10} J/g of ^{235}U. Determine the rate of fuel burning (in kilograms per hour) of ^{235}U.

70. The Sun radiates energy at the rate of 3.85×10^{26} W. Suppose the net reaction $4(^1_1\text{H}) + 2(^0_{-1}\text{e}) \rightarrow ^4_2\text{He} + 2\nu + \gamma$ accounts for all the energy released. Calculate the number of protons fused per second.

Challenge Problems

71. During the manufacture of a steel engine component, radioactive iron (^{59}Fe) with a half-life of 45.1 d is included in the total mass of 0.200 kg. The component is placed in a test engine when the activity due to this isotope is 20.0 μCi. After a 1 000-h test period, some of the lubricating oil is removed from the engine and found to contain enough ^{59}Fe to produce 800 disintegrations/min/L of oil. The total volume of oil in the engine is 6.50 L. Calculate the total mass worn from the engine component per hour of operation.

72. (a) At time $t = 0$, a sample of uranium is exposed to a neutron source that causes N_0 nuclei to undergo fission. The sample is in a supercritical state, with a reproduction constant $K > 1$. A chain reaction occurs that

proliferates fission throughout the mass of uranium. The chain reaction can be thought of as a succession of *generations*. The N_0 fissions produced initially are the zeroth generation of fissions. From this generation, N_0K neutrons go off to produce fission of new uranium nuclei. The N_0K fissions that occur subsequently are the first generation of fissions, and from this generation N_0K^2 neutrons go in search of uranium nuclei in which to cause fission. The subsequent N_0K^2 fissions are the second generation of fissions. This process can continue until all the uranium nuclei have fissioned. Show that the cumulative total of fissions N that have occurred up to and including the nth generation after the zeroth generation is given by

$$N = N_0\left(\frac{K^{n+1} - 1}{K - 1}\right)$$

(b) Consider a hypothetical uranium weapon made from 5.50 kg of isotopically pure ^{235}U. The chain reaction has a reproduction constant of 1.10 and starts with a zeroth generation of 1.00×10^{20} fissions. The average time interval between one fission generation and the next is 10.0 ns. How long after the zeroth generation does it take the uranium in this weapon to fission completely? (c) Assume the bulk modulus of uranium is 150 GPa. Find the speed of sound in uranium. You may ignore the density difference between ^{235}U and natural uranium. (d) Find the time interval required for a compressional wave to cross the radius of a 5.50-kg sphere of uranium. This time interval indicates how quickly the motion of explosion begins. (e) Fission must occur in a time interval that is short compared with that in part (d); otherwise, most of the uranium will disperse in small chunks without

having fissioned. Can the weapon considered in part (b) release the explosive energy of all its uranium? If so, how much energy does it release in equivalent tons of TNT? Assume one ton of TNT releases 4.20 GJ and each uranium fission releases 200 MeV of energy.

73. Assume a photomultiplier tube for detecting radiation has seven dynodes with potentials of 100, 200, 300, ..., 700 V as shown in Figure P45.73. The average energy required to free an electron from the dynode surface is 10.0 eV. Assume only one electron is incident and the tube functions with 100% efficiency. (a) How many electrons are freed at the first dynode at 100 V? (b) How many electrons are collected at the last dynode? (c) What is the energy available to the counter for all the electrons arriving at the last dynode?

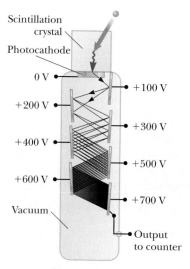

Figure P45.73

Particle Physics and Cosmology

The word *atom* comes from the Greek *atomos,* which means "indivisible." The early Greeks believed that atoms were the indivisible constituents of matter; that is, they regarded them as elementary particles. After 1932, physicists viewed all matter as consisting of three constituent particles: electrons, protons, and neutrons. Beginning in the 1940s, many "new" particles were discovered in experiments involving high-energy collisions between known particles. The new particles are characteristically very unstable and have very short half-lives, ranging between 10^{-6} s and 10^{-23} s. So far, more than 300 of these particles have been catalogued.

Until the 1960s, physicists were bewildered by the great number and variety of subatomic particles that were being discovered. They wondered whether the particles had no systematic relationship connecting them or whether a pattern was emerging that would provide a better understanding of the elaborate structure in the subatomic world. For example, that the neutron has a magnetic moment despite having zero electric charge (Section 44.8) suggests an underlying structure to the neutron. The periodic table explains how more than 100 elements can be formed from three types of particles (electrons, protons, and

One of the most intense areas of current research is the hunt for the Higgs boson, discussed in Section 46.10. The photo shows an event recorded at the Large Hadron Collider in July 2012 that shows particles consistent with the creation of a Higgs boson. The data is not entirely conclusive, however, and the hunt continues. *(CERN)*

neutrons), which suggests there is, perhaps, a means of forming more than 300 subatomic particles from a small number of basic building blocks.

Recall Figure 1.2, which illustrated the various levels of structure in matter. We studied the atomic structure of matter in Chapter 42. In Chapter 44, we investigated the substructure of the atom by describing the structure of the nucleus. As mentioned in Section 1.2, the protons and neutrons in the nucleus, and a host of other exotic particles, are now known to be composed of six different varieties of particles called *quarks*. In this concluding chapter, we examine the current theory of elementary particles, in which all matter is constructed from only two families of particles, quarks and leptons. We also discuss how clarifications of such models might help scientists understand the birth and evolution of the Universe.

46.1 The Fundamental Forces in Nature

As noted in Section 5.1, all natural phenomena can be described by four fundamental forces acting between particles. In order of decreasing strength, they are the nuclear force, the electromagnetic force, the weak force, and the gravitational force.

The nuclear force discussed in Chapter 44 is an attractive force between nucleons. It has a very short range and is negligible for separation distances between nucleons greater than approximately 10^{-15} m (about the size of the nucleus). The electromagnetic force, which binds atoms and molecules together to form ordinary matter, has a strength of approximately 10^{-2} times that of the nuclear force. This long-range force decreases in magnitude as the inverse square of the separation between interacting particles. The weak force is a short-range force that tends to produce instability in certain nuclei. It is responsible for decay processes, and its strength is only about 10^{-5} times that of the nuclear force. Finally, the gravitational force is a long-range force that has a strength of only about 10^{-39} times that of the nuclear force. Although this familiar interaction is the force that holds the planets, stars, and galaxies together, its effect on elementary particles is negligible.

In Section 13.3, we discussed the difficulty early scientists had with the notion of the gravitational force acting at a distance, with no physical contact between the interacting objects. To resolve this difficulty, the concept of the gravitational field was introduced. Similarly, in Chapter 23, we introduced the electric field to describe the electric force acting between charged objects, and we followed that with a discussion of the magnetic field in Chapter 29. For each of these types of fields, we developed a particle in a field analysis model. In modern physics, the nature of the interaction between particles is carried a step further. These interactions are described in terms of the exchange of entities called **field particles** or **exchange particles.** Field particles are also called **gauge bosons.**[1] The interacting particles continuously emit and absorb field particles. The emission of a field particle by one particle and its absorption by another manifests as a force between the two interacting particles. In the case of the electromagnetic interaction, for instance, the field particles are photons. In the language of modern physics, the electromagnetic force is said to be *mediated* by photons, and photons are the field particles of the electromagnetic field. Likewise, the nuclear force is mediated by field particles called *gluons*. The weak force is mediated by field particles called *W* and *Z bosons*, and the gravitational force is proposed to be mediated by field particles called *gravitons*. These interactions, their ranges, and their relative strengths are summarized in Table 46.1.

---

[1]The word *bosons* suggests that the field particles have integral spin as discussed in Section 43.8. The word *gauge* comes from *gauge theory*, which is a sophisticated mathematical analysis that is beyond the scope of this book.

Table 46.1	Particle Interactions			
Interactions	Relative Strength	Range of Force	Mediating Field Particle	Mass of Field Particle (GeV/c^2)
Nuclear	1	Short (≈ 1 fm)	Gluon	0
Electromagnetic	10^{-2}	∞	Photon	0
Weak	10^{-5}	Short ($\approx 10^{-3}$ fm)	$W^{\pm}$, Z^0 bosons	80.4, 80.4, 91.2
Gravitational	10^{-39}	∞	Graviton	0

© INTERFOTO/Alamy

Paul Adrien Maurice Dirac
British Physicist (1902–1984)
Dirac was instrumental in the understanding of antimatter and the unification of quantum mechanics and relativity. He made many contributions to the development of quantum physics and cosmology. In 1933, Dirac won a Nobel Prize in Physics.

46.2 Positrons and Other Antiparticles

In the 1920s, Paul Dirac developed a relativistic quantum-mechanical description of the electron that successfully explained the origin of the electron's spin and its magnetic moment. His theory had one major problem, however: its relativistic wave equation required solutions corresponding to negative energy states, and if negative energy states existed, an electron in a state of positive energy would be expected to make a rapid transition to one of these states, emitting a photon in the process.

Dirac circumvented this difficulty by postulating that all negative energy states are filled. The electrons occupying these negative energy states are collectively called the *Dirac sea*. Electrons in the Dirac sea (the blue area in Fig. 46.1) are not directly observable because the Pauli exclusion principle does not allow them to react to external forces; there are no available states to which an electron can make a transition in response to an external force. Therefore, an electron in such a state acts as an isolated system unless an interaction with the environment is strong enough to excite the electron to a positive energy state. Such an excitation causes one of the negative energy states to be vacant as in Figure 46.1, leaving a hole in the sea of filled states. This process is described by the nonisolated system model: as energy enters the system by some transfer mechanism, the system energy increases and the electron is excited to a higher energy level. *The hole can react to external forces and is observable.* The hole reacts in a way similar to that of the electron except that it has a positive charge: it is the *antiparticle* to the electron.

This theory strongly suggested that *an antiparticle exists for every particle*, not only for fermions such as electrons but also for bosons. It has subsequently been verified that practically every known elementary particle has a distinct antiparticle. Among the exceptions are the photon and the neutral pion (π^0; see Section 46.3). Following the construction of high-energy accelerators in the 1950s, many other antiparticles were revealed. They included the antiproton, discovered by Emilio Segré (1905–1989) and Owen Chamberlain (1920–2006) in 1955, and the antineutron, discovered shortly thereafter. The antiparticle for a charged particle has the same mass as the particle but opposite charge.[2] For example, the electron's antiparticle (the *positron* mentioned in Section 44.4) has a rest energy of 0.511 MeV and a positive charge of $+1.60 \times 10^{-19}$ C.

Carl Anderson (1905–1991) observed the positron experimentally in 1932 and was awarded a Nobel Prize in Physics in 1936 for this achievement. Anderson discovered the positron while examining tracks created in a cloud chamber by electron-like particles of positive charge. (These early experiments used cosmic rays—mostly energetic protons passing through interstellar space—to initiate high-energy reactions on the order of several GeV.) To discriminate between positive and negative charges, Anderson placed the cloud chamber in a magnetic field,

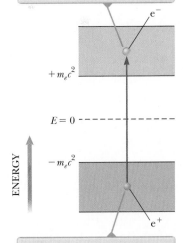

An electron can make a transition out of the Dirac sea only if it is provided with energy equal to or larger than $2m_ec^2$.

$+m_ec^2$

e^-

$E = 0$

$-m_ec^2$

e^+

An upward transition of an electron leaves a vacancy in the Dirac sea, which can behave as a particle identical to the electron except for its positive charge.

Figure 46.1 Dirac's model for the existence of antielectrons (positrons). The minimum energy for an electron to exist in the gold band is its rest energy m_ec^2. The blue band of negative energies is filled with electrons.

---

[2]Antiparticles for uncharged particles, such as the neutron, are a little more difficult to describe. One basic process that can detect the existence of an antiparticle is pair annihilation. For example, a neutron and an antineutron can annihilate to form two gamma rays. Because the photon and the neutral pion do not have distinct antiparticles, pair annihilation is not observed with either of these particles.

Figure 46.2 (a) Bubble-chamber tracks of electron–positron pairs produced by 300-MeV gamma rays striking a lead sheet from the left. (b) The pertinent pair-production events. The positrons deflect upward and the electrons downward in an applied magnetic field.

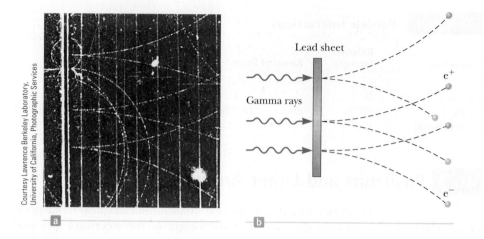

Courtesy Lawrence Berkeley Laboratory, University of California, Photographic Services

Lead sheet

Gamma rays

e⁺

e⁻

a

b

causing moving charges to follow curved paths. He noted that some of the electron-like tracks deflected in a direction corresponding to a positively charged particle.

Since Anderson's discovery, positrons have been observed in a number of experiments. A common source of positrons is **pair production.** In this process, a gamma-ray photon with sufficiently high energy interacts with a nucleus and an electron–positron pair is created from the photon. (The presence of the nucleus allows the principle of conservation of momentum to be satisfied.) Because the total rest energy of the electron–positron pair is $2m_e c^2 = 1.02$ MeV (where m_e is the mass of the electron), the photon must have at least this much energy to create an electron–positron pair. The energy of a photon is converted to rest energy of the electron and positron in accordance with Einstein's relationship $E_R = mc^2$. If the gamma-ray photon has energy in excess of the rest energy of the electron–positron pair, the excess appears as kinetic energy of the two particles. Figure 46.2 shows early observations of tracks of electron–positron pairs in a bubble chamber created by 300-MeV gamma rays striking a lead sheet.

Quick Quiz 46.1 Given the identification of the particles in Figure 46.2b, is the direction of the external magnetic field in Figure 46.2a **(a)** into the page, **(b)** out of the page, or **(c)** impossible to determine?

The reverse process can also occur. Under the proper conditions, an electron and a positron can annihilate each other to produce two gamma-ray photons that have a combined energy of at least 1.02 MeV:

$$e^- + e^+ \rightarrow 2\gamma$$

Because the initial momentum of the electron–positron system is approximately zero, the two gamma rays travel in opposite directions after the annihilation, satisfying the principle of conservation of momentum for the isolated system.

Electron–positron annihilation is used in the medical diagnostic technique called *positron-emission tomography* (PET). The patient is injected with a glucose solution containing a radioactive substance that decays by positron emission, and the material is carried throughout the body by the blood. A positron emitted during a decay event in one of the radioactive nuclei in the glucose solution annihilates with an electron in the surrounding tissue, resulting in two gamma-ray photons emitted in opposite directions. A gamma detector surrounding the patient pinpoints the source of the photons and, with the assistance of a computer, displays an image of the sites at which the glucose accumulates. (Glucose metabolizes rapidly in cancerous tumors and accumulates at those sites, providing a strong signal for a PET detector system.) The images from a PET scan can indicate a wide variety of disorders in the brain, including Alzheimer's disease (Fig. 46.3). In addition, because glucose metabolizes more rapidly in active areas

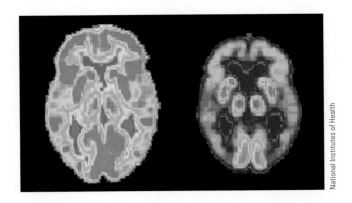

Figure 46.3 PET scans of the brain of a healthy older person *(left)* and that of a patient suffering from Alzheimer's disease *(right)*. Lighter regions contain higher concentrations of radioactive glucose, indicating higher metabolism rates and therefore increased brain activity.

of the brain, a PET scan can indicate areas of the brain involved in the activities in which the patient is engaging at the time of the scan, such as language use, music, and vision.

46.3 Mesons and the Beginning of Particle Physics

Physicists in the mid-1930s had a fairly simple view of the structure of matter. The building blocks were the proton, the electron, and the neutron. Three other particles were either known or postulated at the time: the photon, the neutrino, and the positron. Together these six particles were considered the fundamental constituents of matter. With this simple picture, however, no one was able to answer the following important question: the protons in any nucleus should strongly repel one another due to their charges of the same sign, so what is the nature of the force that holds the nucleus together? Scientists recognized that this mysterious force must be much stronger than anything encountered in nature up to that time. This force is the nuclear force discussed in Section 44.1 and examined in historical perspective in the following paragraphs.

The first theory to explain the nature of the nuclear force was proposed in 1935 by Japanese physicist Hideki Yukawa, an effort that earned him a Nobel Prize in Physics in 1949. To understand Yukawa's theory, recall the introduction of field particles in Section 46.1, which stated that each fundamental force is mediated by a field particle exchanged between the interacting particles. Yukawa used this idea to explain the nuclear force, proposing the existence of a new particle whose exchange between nucleons in the nucleus causes the nuclear force. He established that the range of the force is inversely proportional to the mass of this particle and predicted the mass to be approximately 200 times the mass of the electron. (Yukawa's predicted particle is *not* the gluon mentioned in Section 46.1, which is massless and is today considered to be the field particle for the nuclear force.) Because the new particle would have a mass between that of the electron and that of the proton, it was called a **meson** (from the Greek *meso*, "middle").

In efforts to substantiate Yukawa's predictions, physicists began experimental searches for the meson by studying cosmic rays entering the Earth's atmosphere. In 1937, Carl Anderson and his collaborators discovered a particle of mass 106 MeV/c^2, approximately 207 times the mass of the electron. This particle was thought to be Yukawa's meson. Subsequent experiments, however, showed that the particle interacted very weakly with matter and hence could not be the field particle for the nuclear force. That puzzling situation inspired several theoreticians to propose two mesons having slightly different masses equal to approximately 200 times that of the electron, one having been discovered by Anderson and the other, still undiscovered, predicted by Yukawa. This idea was confirmed in 1947 with the discovery of the **pi meson** (π), or simply **pion**. The particle discovered by Anderson in 1937, the one initially thought to be Yukawa's meson, is not really a

Hideki Yukawa
Japanese Physicist (1907–1981)
Yukawa was awarded the Nobel Prize in Physics in 1949 for predicting the existence of mesons. This photograph of him at work was taken in 1950 in his office at Columbia University. Yukawa came to Columbia in 1949 after spending the early part of his career in Japan.

Figure 46.4 Feynman diagram representing a photon mediating the electromagnetic force between two electrons.

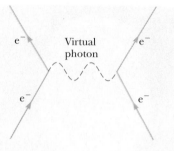

meson. (We shall discuss the characteristics of mesons in Section 46.4.) Instead, it takes part in the weak and electromagnetic interactions only and is now called the **muon** (μ).

The pion comes in three varieties, corresponding to three charge states: π^+, π^-, and π^0. The π^+ and π^- particles (π^- is the antiparticle of π^+) each have a mass of 139.6 MeV/c^2, and the π^0 mass is 135.0 MeV/c^2. Two muons exist: μ^- and its antiparticle μ^+.

Pions and muons are very unstable particles. For example, the π^-, which has a mean lifetime of 2.6×10^{-8} s, decays to a muon and an antineutrino.[3] The muon, which has a mean lifetime of 2.2 μs, then decays to an electron, a neutrino, and an antineutrino:

$$\pi^- \rightarrow \mu^- + \bar{\nu}$$
$$\mu^- \rightarrow e^- + \nu + \bar{\nu} \tag{46.1}$$

For chargeless particles (as well as some charged particles, such as the proton), a bar over the symbol indicates an antiparticle, as for the neutrino in beta decay (see Section 44.5). Other antiparticles, such as e^+ and μ^+, use a different notation.

The interaction between two particles can be represented in a simple diagram called a **Feynman diagram,** developed by American physicist Richard P. Feynman. Figure 46.4 is such a diagram for the electromagnetic interaction between two electrons. A Feynman diagram is a qualitative graph of time on the vertical axis versus space on the horizontal axis. It is qualitative in the sense that the actual values of time and space are not important, but the overall appearance of the graph provides a pictorial representation of the process.

In the simple case of the electron–electron interaction in Figure 46.4, a photon (the field particle) mediates the electromagnetic force between the electrons. Notice that the entire interaction is represented in the diagram as occurring at a single point in time. Therefore, the paths of the electrons appear to undergo a discontinuous change in direction at the moment of interaction. The electron paths shown in Figure 46.4 are different from the *actual* paths, which would be curved due to the continuous exchange of large numbers of field particles.

In the electron–electron interaction, the photon, which transfers energy and momentum from one electron to the other, is called a *virtual photon* because it vanishes during the interaction without having been detected. In Chapter 40, we discussed that a photon has energy $E = hf$, where f is its frequency. Consequently, for a system of two electrons initially at rest, the system has energy $2m_ec^2$ before a virtual photon is released and energy $2m_ec^2 + hf$ after the virtual photon is released (plus any kinetic energy of the electron resulting from the emission of the photon). Is that a violation of the law of conservation of energy for an isolated system? No; this process does *not* violate the law of conservation of energy because the virtual

Richard Feynman
American Physicist (1918–1988)
Inspired by Dirac, Feynman developed quantum electrodynamics, the theory of the interaction of light and matter on a relativistic and quantum basis. In 1965, Feynman won the Nobel Prize in Physics. The prize was shared by Feynman, Julian Schwinger, and Sin Itiro Tomonaga. Early in Feynman's career, he was a leading member of the team developing the first nuclear weapon in the Manhattan Project. Toward the end of his career, he worked on the commission investigating the 1986 *Challenger* tragedy and demonstrated the effects of cold temperatures on the rubber O-rings used in the space shuttle.

© Shelly Gazin/CORBIS

---

[3]The antineutrino is another zero-charge particle for which the identification of the antiparticle is more difficult than that for a charged particle. Although the details are beyond the scope of this book, the neutrino and antineutrino can be differentiated by means of the relationship between the linear momentum and the spin angular momentum of the particles.

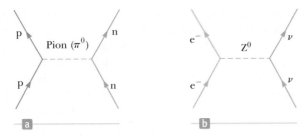

Figure 46.5 (a) Feynman diagram representing a proton and a neutron interacting via the nuclear force with a neutral pion mediating the force. (This model is *not* the current model for nucleon interaction.) (b) Feynman diagram for an electron and a neutrino interacting via the weak force, with a Z^0 boson mediating the force.

photon has a very short lifetime Δt that makes the uncertainty in the energy $\Delta E \approx \hbar/2 \, \Delta t$ of the system greater than the photon energy. Therefore, within the constraints of the uncertainty principle, the energy of the system is conserved.

Now consider a pion exchange between a proton and a neutron according to Yukawa's model (Fig. 46.5a). The energy ΔE_R needed to create a pion of mass m_π is given by Einstein's equation $\Delta E_R = m_\pi c^2$. As with the photon in Figure 46.4, the very existence of the pion would appear to violate the law of conservation of energy if the particle existed for a time interval greater than $\Delta t \approx \hbar/2 \, \Delta E_R$ (from the uncertainty principle), where Δt is the time interval required for the pion to transfer from one nucleon to the other. Therefore,

$$\Delta t \approx \frac{\hbar}{2 \, \Delta E_R} = \frac{\hbar}{2 m_\pi c^2}$$

and the rest energy of the pion is

$$m_\pi c^2 = \frac{\hbar}{2 \, \Delta t} \tag{46.2}$$

Because the pion cannot travel faster than the speed of light, the maximum distance d it can travel in a time interval Δt is $c \, \Delta t$. Therefore, using Equation 46.2 and $d = c \, \Delta t$, we find

$$m_\pi c^2 = \frac{\hbar c}{2d} \tag{46.3}$$

From Table 46.1, we know that the range of the nuclear force is on the order of 10^{-15} fm. Using this value for d in Equation 46.3, we estimate the rest energy of the pion to be

$$m_p c^2 \approx \frac{(1.055 \times 10^{-34} \, \text{J} \cdot \text{s})(3.00 \times 10^8 \, \text{m/s})}{2(1 \times 10^{-15} \, \text{m})}$$

$$= 1.6 \times 10^{-11} \, \text{J} \approx 100 \, \text{MeV}$$

which corresponds to a mass of 100 MeV/c^2 (approximately 200 times the mass of the electron). This value is in reasonable agreement with the observed pion mass.

The concept just described is quite revolutionary. In effect, it says that a system of two nucleons can change into two nucleons plus a pion as long as it returns to its original state in a very short time interval. (Remember that this description is the older historical model, which assumes the pion is the field particle for the nuclear force; the gluon is the actual field particle in current models.) Physicists often say that a nucleon undergoes *fluctuations* as it emits and absorbs field particles. These fluctuations are a consequence of a combination of quantum mechanics (through the uncertainty principle) and special relativity (through Einstein's energy–mass relationship $E_R = mc^2$).

In this section, we discussed the field particles that were originally proposed to mediate the nuclear force (pions) and those that mediate the electromagnetic force (photons). The graviton, the field particle for the gravitational force, has yet to be observed. In 1983, $W^{\pm}$ and Z^0 particles, which mediate the weak force, were discovered by Italian physicist Carlo Rubbia (b. 1934) and his associates, using a proton–antiproton collider. Rubbia and Simon van der Meer (1925–2011), both at CERN,[4] shared the 1984 Nobel Prize in Physics for the discovery of the $W^{\pm}$ and Z^0 particles and the development of the proton–antiproton collider. Figure 46.5b shows a Feynman diagram for a weak interaction mediated by a Z^0 boson.

46.4 Classification of Particles

All particles other than field particles can be classified into two broad categories, *hadrons* and *leptons*. The criterion for separating these particles into categories is whether or not they interact via the strong force. The nuclear force between nucleons in a nucleus is a particular manifestation of the strong force, but we will use the term *strong force* to refer to any interaction between particles made up of quarks. (For more detail on quarks and the strong force, see Section 46.8.) Table 46.2 provides a summary of the properties of hadrons and leptons.

Table 46.2   Some Particles and Their Properties

Category	Particle Name	Symbol	Anti-particle	Mass (MeV/c^2)	B	L_e	L_μ	L_τ	S	Lifetime(s)	Spin
Leptons	Electron	e^-	e^+	0.511	0	+1	0	0	0	Stable	$\frac{1}{2}$
	Electron–neutrino	ν_e	$\bar{\nu}_e$	< 2 eV/c^2	0	+1	0	0	0	Stable	$\frac{1}{2}$
	Muon	μ^-	μ^+	105.7	0	0	+1	0	0	2.20×10^{-6}	$\frac{1}{2}$
	Muon–neutrino	ν_μ	$\bar{\nu}_\mu$	< 0.17	0	0	+1	0	0	Stable	$\frac{1}{2}$
	Tau	τ^-	τ^+	1 784	0	0	0	+1	0	$< 4 \times 10^{-13}$	$\frac{1}{2}$
	Tau–neutrino	ν_τ	$\bar{\nu}_\tau$	< 18	0	0	0	+1	0	Stable	$\frac{1}{2}$
Hadrons											
Mesons	Pion	π^+	π^-	139.6	0	0	0	0	0	2.60×10^{-8}	0
		π^0	Self	135.0	0	0	0	0	0	0.83×10^{-16}	0
	Kaon	K^+	K^-	493.7	0	0	0	0	+1	1.24×10^{-8}	0
		K_S^0	$\bar{K}_S^0$	497.7	0	0	0	0	+1	0.89×10^{-10}	0
		K_L^0	$\bar{K}_L^0$	497.7	0	0	0	0	+1	5.2×10^{-8}	0
	Eta	η	Self	548.8	0	0	0	0	0	$< 10^{-18}$	0
		η'	Self	958	0	0	0	0	0	2.2×10^{-21}	0
Baryons	Proton	p	$\bar{p}$	938.3	+1	0	0	0	0	Stable	$\frac{1}{2}$
	Neutron	n	$\bar{n}$	939.6	+1	0	0	0	0	614	$\frac{1}{2}$
	Lambda	Λ^0	$\bar{\Lambda}^0$	1 115.6	+1	0	0	0	−1	2.6×10^{-10}	$\frac{1}{2}$
	Sigma	Σ^+	$\bar{\Sigma}^-$	1 189.4	+1	0	0	0	−1	0.80×10^{-10}	$\frac{1}{2}$
		Σ^0	$\bar{\Sigma}^0$	1 192.5	+1	0	0	0	−1	6×10^{-20}	$\frac{1}{2}$
		Σ^-	$\bar{\Sigma}^+$	1 197.3	+1	0	0	0	−1	1.5×10^{-10}	$\frac{1}{2}$
	Delta	Δ^{++}	$\bar{\Delta}^{--}$	1 230	+1	0	0	0	0	6×10^{-24}	$\frac{3}{2}$
		Δ^+	$\bar{\Delta}^-$	1 231	+1	0	0	0	0	6×10^{-24}	$\frac{3}{2}$
		Δ^0	$\bar{\Delta}^0$	1 232	+1	0	0	0	0	6×10^{-24}	$\frac{3}{2}$
		Δ^-	$\bar{\Delta}^+$	1 234	+1	0	0	0	0	6×10^{-24}	$\frac{3}{2}$
	Xi	Ξ^0	$\bar{\Xi}^0$	1 315	+1	0	0	0	−2	2.9×10^{-10}	$\frac{1}{2}$
		Ξ^-	$\bar{\Xi}^+$	1 321	+1	0	0	0	−2	1.64×10^{-10}	$\frac{1}{2}$
	Omega	Ω^-	Ω^+	1 672	+1	0	0	0	−3	0.82×10^{-10}	$\frac{3}{2}$

[4]CERN was originally the Conseil Européen pour la Recherche Nucléaire; the name has been altered to the European Organization for Nuclear Research, and the laboratory operated by CERN is called the European Laboratory for Particle Physics. The CERN acronym has been retained and is commonly used to refer to both the organization and the laboratory.

Hadrons

Particles that interact through the strong force (as well as through the other fundamental forces) are called **hadrons.** The two classes of hadrons, *mesons* and *baryons,* are distinguished by their masses and spins.

Mesons all have zero or integer spin (0 or 1). As indicated in Section 46.3, the name comes from the expectation that Yukawa's proposed meson mass would lie between the masses of the electron and the proton. Several meson masses do lie in this range, although mesons having masses greater than that of the proton have been found to exist.

All mesons decay finally into electrons, positrons, neutrinos, and photons. The pions are the lightest known mesons and have masses of approximately 1.4×10^2 MeV/c^2, and all three pions—π^+, π^-, and π^0—have a spin of 0. (This spin-0 characteristic indicates that the particle discovered by Anderson in 1937, the muon, is not a meson. The muon has spin $\frac{1}{2}$ and belongs in the *lepton* classification, described below.)

Baryons, the second class of hadrons, have masses equal to or greater than the proton mass (the name *baryon* means "heavy" in Greek), and their spin is always a half-integer value ($\frac{1}{2}$, $\frac{3}{2}$, . . .). Protons and neutrons are baryons, as are many other particles. With the exception of the proton, all baryons decay in such a way that the end products include a proton. For example, the baryon called the Ξ^0 hyperon (Greek letter xi) decays to the Λ^0 baryon (Greek letter lambda) in approximately 10^{-10} s. The Λ^0 then decays to a proton and a π^- in approximately 3×10^{-10} s.

Today it is believed that hadrons are not elementary particles but instead are composed of more elementary units called quarks, per Section 46.8.

Leptons

Leptons (from the Greek *leptos,* meaning "small" or "light") are particles that do not interact by means of the strong force. All leptons have spin $\frac{1}{2}$. Unlike hadrons, which have size and structure, leptons appear to be truly elementary, meaning that they have no structure and are point-like.

Quite unlike the case with hadrons, the number of known leptons is small. Currently, scientists believe that only six leptons exist: the electron, the muon, the tau, and a neutrino associated with each: e^-, μ^-, τ^-, ν_e, ν_μ, and ν_τ. The tau lepton, discovered in 1975, has a mass about twice that of the proton. Direct experimental evidence for the neutrino associated with the tau was announced by the Fermi National Accelerator Laboratory (Fermilab) in July 2000. Each of the six leptons has an antiparticle.

Current studies indicate that neutrinos have a small but nonzero mass. If they do have mass, they cannot travel at the speed of light. In addition, because so many neutrinos exist, their combined mass may be sufficient to cause all the matter in the Universe to eventually collapse into a single point, which might then explode and create a completely new Universe! We shall discuss this possibility in more detail in Section 46.11.

46.5 Conservation Laws

The laws of conservation of energy, linear momentum, angular momentum, and electric charge for an isolated system provide us with a set of rules that all processes must follow. In Chapter 44, we learned that conservation laws are important for understanding why certain radioactive decays and nuclear reactions occur and others do not. In the study of elementary particles, a number of additional conservation laws are important. Although the two described here have no theoretical foundation, they are supported by abundant empirical evidence.

Baryon Number

Experimental results show that whenever a baryon is created in a decay or nuclear reaction, an antibaryon is also created. This scheme can be quantified by assigning every particle a quantum number, the **baryon number,** as follows: $B = +1$ for all baryons, $B = -1$ for all antibaryons, and $B = 0$ for all other particles. (See Table 46.2.) The **law of conservation of baryon number** states that

◄ Conservation of baryon number

> whenever a nuclear reaction or decay occurs, the sum of the baryon numbers before the process must equal the sum of the baryon numbers after the process.

If baryon number is conserved, the proton must be absolutely stable. For example, a decay of the proton to a positron and a neutral pion would satisfy conservation of energy, momentum, and electric charge. Such a decay has never been observed, however. The law of conservation of baryon number would be consistent with the absence of this decay because the proposed decay would involve the loss of a baryon. Based on experimental observations as pointed out in Example 46.2, all we can say at present is that protons have a half-life of at least 10^{33} years (the estimated age of the Universe is only 10^{10} years). Some recent theories, however, predict that the proton is unstable. According to this theory, baryon number is not absolutely conserved.

Quick Quiz 46.2 Consider the decays (i) $n \rightarrow \pi^+ + \pi^- + \mu^+ + \mu^-$ and (ii) $n \rightarrow p + \pi^-$. From the following choices, which conservation laws are violated by each decay? (a) energy (b) electric charge (c) baryon number (d) angular momentum (e) no conservation laws

Example 46.1   Checking Baryon Numbers

Use the law of conservation of baryon number to determine whether each of the following reactions can occur:

(A) $p + n \rightarrow p + p + n + \bar{p}$

SOLUTION

Conceptualize The mass on the right is larger than the mass on the left. Therefore, one might be tempted to claim that the reaction violates energy conservation. The reaction can indeed occur, however, if the initial particles have sufficient kinetic energy to allow for the increase in rest energy of the system.

Categorize We use a conservation law developed in this section, so we categorize this example as a substitution problem.

Evaluate the total baryon number for the left side of the reaction: $1 + 1 = 2$

Evaluate the total baryon number for the right side of the reaction: $1 + 1 + 1 + (-1) = 2$

Therefore, baryon number is conserved and the reaction can occur.

(B) $p + n \rightarrow p + p + \bar{p}$

SOLUTION

Evaluate the total baryon number for the left side of the reaction: $1 + 1 = 2$

Evaluate the total baryon number for the right side of the reaction: $1 + 1 + (-1) = 1$

Because baryon number is not conserved, the reaction cannot occur.

Example 46.2 **Detecting Proton Decay**

Measurements taken at two neutrino detection facilities, the Irvine–Michigan–Brookhaven detector (Fig. 46.6) and the Super Kamiokande in Japan, indicate that the half-life of protons is at least 10^{33} yr.

(A) Estimate how long we would have to watch, on average, to see a proton in a glass of water decay.

SOLUTION

Conceptualize Imagine the number of protons in a glass of water. Although this number is huge, the probability of a single proton undergoing decay is small, so we would expect to wait for a long time interval before observing a decay.

Figure 46.6 (Example 46.2) A diver swims through ultrapure water in the Irvine–Michigan–Brookhaven neutrino detector. This detector holds almost 7 000 metric tons of water and is lined with over 2 000 photomultiplier tubes, many of which are visible in the photograph.

Categorize Because a half-life is provided in the problem, we categorize this problem as one in which we can apply our statistical analysis techniques from Section 44.4.

Analyze Let's estimate that a drinking glass contains a number of moles n of water, with a mass of $m = 250$ g and a molar mass $M = 18$ g/mol.

Find the number of molecules of water in the glass:

$$N_{\text{molecules}} = nN_{\text{A}} = \frac{m}{M} N_{\text{A}}$$

Each water molecule contains one proton in each of its two hydrogen atoms plus eight protons in its oxygen atom, for a total of ten protons. Therefore, there are $N = 10 N_{\text{molecules}}$ protons in the glass of water.

Find the activity of the protons from Equation 44.7:

$$(1) \quad R = \lambda N = \frac{\ln 2}{T_{1/2}} \left(10 \frac{m}{M} N_{\text{A}} \right) = \frac{\ln 2}{10^{33} \text{ yr}} (10) \left(\frac{250 \text{ g}}{18 \text{ g/mol}} \right) (6.02 \times 10^{23} \text{ mol}^{-1})$$

$$= 5.8 \times 10^{-8} \text{ yr}^{-1}$$

Finalize The decay constant represents the probability that *one* proton decays in one year. The probability that *any* proton in our glass of water decays in the one-year interval is given by Equation (1). Therefore, we must watch our glass of water for $1/R \approx$ 17 million years! That indeed is a long time interval, as expected.

(B) The Super Kamiokande neutrino facility contains 50 000 metric tons of water. Estimate the average time interval between detected proton decays in this much water if the half-life of a proton is 10^{33} yr.

SOLUTION

Analyze The proton decay rate R in a sample of water is proportional to the number N of protons. Set up a ratio of the decay rate in the Super Kamiokande facility to that in a glass of water:

$$\frac{R_{\text{Kamiokande}}}{R_{\text{glass}}} = \frac{N_{\text{Kamiokande}}}{N_{\text{glass}}} \rightarrow R_{\text{Kamiokande}} = \frac{N_{\text{Kamiokande}}}{N_{\text{glass}}} R_{\text{glass}}$$

The number of protons is proportional to the mass of the sample, so express the decay rate in terms of mass:

$$R_{\text{Kamiokande}} = \frac{m_{\text{Kamiokande}}}{m_{\text{glass}}} R_{\text{glass}}$$

Substitute numerical values:

$$R_{\text{Kamiokande}} = \left(\frac{50\ 000 \text{ metric tons}}{0.250 \text{ kg}} \right) \left(\frac{1\ 000 \text{ kg}}{1 \text{ metric ton}} \right) (5.8 \times 10^{-8} \text{ yr}^{-1}) \approx 12 \text{ yr}^{-1}$$

Finalize The average time interval between decays is about one-twelfth of a year, or approximately one month. That is much shorter than the time interval in part (A) due to the tremendous amount of water in the detector facility. Despite this rosy prediction of one proton decay per month, a proton decay has never been observed. This suggests that the half-life of the proton may be larger than 10^{33} years or that proton decay simply does not occur.

Lepton Number

There are three conservation laws involving lepton numbers, one for each variety of lepton. The **law of conservation of electron lepton number** states that

Conservation of electron ▶
lepton number

whenever a nuclear reaction or decay occurs, the sum of the electron lepton numbers before the process must equal the sum of the electron lepton numbers after the process.

The electron and the electron neutrino are assigned an electron lepton number $L_e = +1$, and the antileptons e^+ and $\bar{\nu}_e$ are assigned an electron lepton number $L_e = -1$. All other particles have $L_e = 0$. For example, consider the decay of the neutron:

$$n \rightarrow p + e^- + \bar{\nu}_e$$

Before the decay, the electron lepton number is $L_e = 0$; after the decay, it is $0 + 1 + (-1) = 0$. Therefore, electron lepton number is conserved. (Baryon number must also be conserved, of course, and it is: before the decay, $B = +1$, and after the decay, $B = +1 + 0 + 0 = +1$.)

Similarly, when a decay involves muons, the muon lepton number L_μ is conserved. The μ^- and the ν_μ are assigned a muon lepton number $L_\mu = +1$, and the antimuons μ^+ and $\bar{\nu}_\mu$ are assigned a muon lepton number $L_\mu = -1$. All other particles have $L_\mu = 0$.

Finally, tau lepton number L_τ is conserved with similar assignments made for the tau lepton, its neutrino, and their two antiparticles.

Quick Quiz 46.3 Consider the following decay: $\pi^0 \rightarrow \mu^- + e^+ + \nu_\mu$. What conservation laws are violated by this decay? **(a)** energy **(b)** angular momentum **(c)** electric charge **(d)** baryon number **(e)** electron lepton number **(f)** muon lepton number **(g)** tau lepton number **(h)** no conservation laws

Quick Quiz 46.4 Suppose a claim is made that the decay of the neutron is given by $n \rightarrow p + e^-$. What conservation laws are violated by this decay? **(a)** energy **(b)** angular momentum **(c)** electric charge **(d)** baryon number **(e)** electron lepton number **(f)** muon lepton number **(g)** tau lepton number **(h)** no conservation laws

---

Example 46.3   **Checking Lepton Numbers**

Use the law of conservation of lepton numbers to determine whether each of the following decay schemes (A) and (B) can occur:

(A) $\mu^- \rightarrow e^- + \bar{\nu}_e + \nu_\mu$

SOLUTION

Conceptualize Because this decay involves a muon and an electron, L_μ and L_e must each be conserved separately if the decay is to occur.

Categorize We use a conservation law developed in this section, so we categorize this example as a substitution problem.

Evaluate the lepton numbers before the decay: $L_\mu = +1$ $L_e = 0$

Evaluate the total lepton numbers after the decay: $L_\mu = 0 + 0 + 1 = +1$ $L_e = +1 + (-1) + 0 = 0$

Therefore, both numbers are conserved and on this basis the decay is possible.

(B) $\pi^+ \rightarrow \mu^+ + \nu_\mu + \nu_e$

▶ **46.3** c o n t i n u e d

Evaluate the lepton numbers before the decay: $\qquad L_\mu = 0 \qquad L_e = 0$

Evaluate the total lepton numbers after the decay: $\qquad L_\mu = -1 + 1 + 0 = 0 \qquad L_e = 0 + 0 + 1 = 1$

Therefore, the decay is not possible because electron lepton number is not conserved.

46.6 Strange Particles and Strangeness

Many particles discovered in the 1950s were produced by the interaction of pions with protons and neutrons in the atmosphere. A group of these—the kaon (K), lambda (Λ), and sigma (Σ) particles—exhibited unusual properties both as they were created and as they decayed; hence, they were called *strange particles.*

One unusual property of strange particles is that they are always produced in pairs. For example, when a pion collides with a proton, a highly probable result is the production of two neutral strange particles (Fig. 46.7):

$$\pi^- + p \ \rightarrow \ K^0 + \Lambda^0$$

The reaction $\pi^- + p \rightarrow K^0 + n$, where only one final particle is strange, never occurs, however, even though no previously known conservation laws would be violated and even though the energy of the pion is sufficient to initiate the reaction.

The second peculiar feature of strange particles is that although they are produced in reactions involving the strong interaction at a high rate, they do not decay into particles that interact via the strong force at a high rate. Instead, they decay very slowly, which is characteristic of the weak interaction. Their half-lives are in

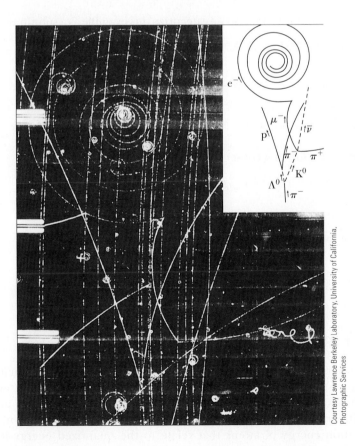

Courtesy Lawrence Berkeley Laboratory, University of California, Photographic Services

Figure 46.7 This bubble-chamber photograph shows many events, and the inset is a drawing of identified tracks. The strange particles Λ^0 and K^0 are formed at the bottom as a π^- particle interacts with a proton in the reaction $\pi^- + p \rightarrow K^0 + \Lambda^0$. (Notice that the neutral particles leave no tracks, as indicated by the dashed lines in the inset.) The Λ^0 then decays in the reaction $\Lambda^0 \rightarrow \pi^- + p$ and the K^0 in the reaction $K^0 \rightarrow \pi^+ + \mu^- + \bar{\nu}_\mu$.

the range 10^{-10} s to 10^{-8} s, whereas most other particles that interact via the strong force have much shorter lifetimes on the order of 10^{-23} s.

To explain these unusual properties of strange particles, a new quantum number S, called **strangeness,** was introduced, together with a conservation law. The strangeness numbers for some particles are given in Table 46.2. The production of strange particles in pairs is handled mathematically by assigning $S = +1$ to one of the particles, $S = -1$ to the other, and $S = 0$ to all nonstrange particles. The **law of conservation of strangeness** states that

Conservation of strangeness ▶

> in a nuclear reaction or decay that occurs via the strong force, strangeness is conserved; that is, the sum of the strangeness numbers before the process must equal the sum of the strangeness numbers after the process. In processes that occur via the weak interaction, strangeness may not be conserved.

The low decay rate of strange particles can be explained by assuming the strong and electromagnetic interactions obey the law of conservation of strangeness but the weak interaction does not. Because the decay of a strange particle involves the loss of one strange particle, it violates strangeness conservation and hence proceeds slowly via the weak interaction.

Example 46.4   **Is Strangeness Conserved?**

(A) Use the law of strangeness conservation to determine whether the reaction $\pi^0 + n \rightarrow K^+ + \Sigma^-$ occurs.

SOLUTION

Conceptualize We recognize that there are strange particles appearing in this reaction, so we see that we will need to investigate conservation of strangeness.

Categorize We use a conservation law developed in this section, so we categorize this example as a substitution problem.

Evaluate the strangeness for the left side of the reaction using Table 46.2:

$$S = 0 + 0 = 0$$

Evaluate the strangeness for the right side of the reaction:

$$S = +1 - 1 = 0$$

Therefore, strangeness is conserved and the reaction is allowed.

(B) Show that the reaction $\pi^- + p \rightarrow \pi^- + \Sigma^+$ does not conserve strangeness.

SOLUTION

Evaluate the strangeness for the left side of the reaction: $S = 0 + 0 = 0$

Evaluate the strangeness for the right side of the reaction: $S = 0 + (-1) = -1$

Therefore, strangeness is not conserved.

46.7 Finding Patterns in the Particles

One tool scientists use is the detection of patterns in data, patterns that contribute to our understanding of nature. For example, Table 21.2 shows a pattern of molar specific heats of gases that allows us to understand the differences among monatomic, diatomic, and polyatomic gases. Figure 42.20 shows a pattern of peaks in the ionization energy of atoms that relate to the quantized energy levels in the

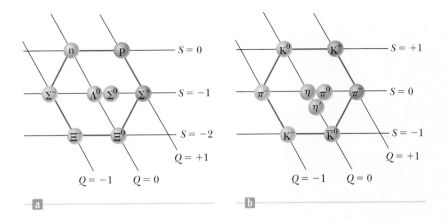

© Linh Hassel/Age Fotostock

Figure 46.8 (a) The hexagonal eightfold-way pattern for the eight spin-$\frac{1}{2}$ baryons. This strangeness-versus-charge plot uses a sloping axis for charge number Q and a horizontal axis for strangeness S. (b) The eightfold-way pattern for the nine spin-zero mesons.

atoms. Figure 44.7 shows a pattern of peaks in the binding energy that suggest a shell structure within the nucleus. One of the best examples of this tool's use is the development of the periodic table, which provides a fundamental understanding of the chemical behavior of the elements. As mentioned in the introduction, the periodic table explains how more than 100 elements can be formed from three particles, the electron, the proton, and the neutron. The table of nuclides, part of which is shown in Table 44.2, contains hundreds of nuclides, but all can be built from protons and neutrons.

The number of particles observed by particle physicists is in the hundreds. Is it possible that a small number of entities exist from which all these particles can be built? Taking a hint from the success of the periodic table and the table of nuclides, let explore the historical search for patterns among the particles.

Many classification schemes have been proposed for grouping particles into families. Consider, for instance, the baryons listed in Table 46.2 that have spins of $\frac{1}{2}$: p, n, Λ^0, Σ^+, Σ^0, Σ^-, Ξ^0, and Ξ^-. If we plot strangeness versus charge for these baryons using a sloping coordinate system as in Figure 46.8a, a fascinating pattern is observed: six of the baryons form a hexagon, and the remaining two are at the hexagon's center.

As a second example, consider the following nine spin-zero mesons listed in Table 46.2: π^+, π^0, π^-, K^+, K^0, K^-, η, η', and the antiparticle $\overline{K}^0$. Figure 46.8b is a plot of strangeness versus charge for this family. Again, a hexagonal pattern emerges. In this case, each particle on the perimeter of the hexagon lies opposite its antiparticle and the remaining three (which form their own antiparticles) are at the center of the hexagon. These and related symmetric patterns were developed independently in 1961 by Murray Gell-Mann and Yuval Ne'eman (1925–2006). Gell-Mann called the patterns the **eightfold way,** after the eightfold path to nirvana in Buddhism.

Groups of baryons and mesons can be displayed in many other symmetric patterns within the framework of the eightfold way. For example, the family of spin-$\frac{3}{2}$ baryons known in 1961 contains nine particles arranged in a pattern like that of the pins in a bowling alley as in Figure 46.9. (The particles Σ^{*+}, Σ^{*0}, Σ^{*-}, Ξ^{*0},

Murray Gell-Mann
American Physicist (b. 1929)
In 1969, Murray Gell-Mann was awarded the Nobel Prize in Physics for his theoretical studies dealing with subatomic particles.

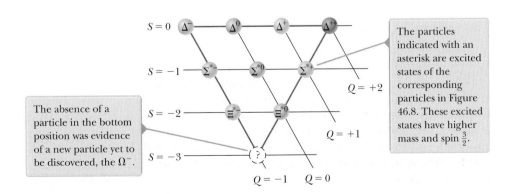

The particles indicated with an asterisk are excited states of the corresponding particles in Figure 46.8. These excited states have higher mass and spin $\frac{3}{2}$.

The absence of a particle in the bottom position was evidence of a new particle yet to be discovered, the Ω^-.

Figure 46.9 The pattern for the higher-mass, spin-$\frac{3}{2}$ baryons known at the time the pattern was proposed.

Figure 46.10 Discovery of the Ω⁻ particle. The photograph on the left shows the original bubble-chamber tracks. The drawing on the right isolates the tracks of the important events.

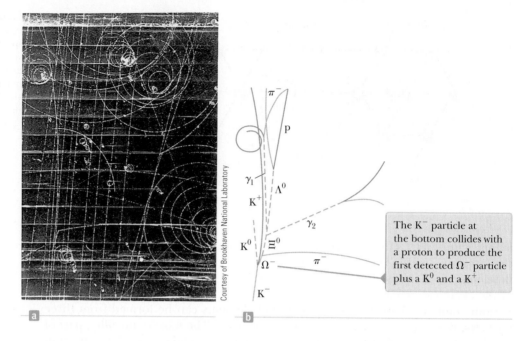

Courtesy of Brookhaven National Laboratory

The K⁻ particle at the bottom collides with a proton to produce the first detected Ω⁻ particle plus a K⁰ and a K⁺.

and Ξ^{*-} are excited states of the particles Σ^{+}, Σ^{0}, Σ^{-}, Ξ^{0}, and Ξ^{-}. In these higher-energy states, the spins of the three quarks—see Section 46.8—making up the particle are aligned so that the total spin of the particle is $\frac{3}{2}$.) When this pattern was proposed, an empty spot occurred in it (at the bottom position), corresponding to a particle that had never been observed. Gell-Mann predicted that the missing particle, which he called the omega minus (Ω^{-}), should have spin $\frac{3}{2}$, charge -1, strangeness -3, and rest energy of approximately 1 680 MeV. Shortly thereafter, in 1964, scientists at the Brookhaven National Laboratory found the missing particle through careful analyses of bubble-chamber photographs (Fig. 46.10) and confirmed all its predicted properties.

The prediction of the missing particle in the eightfold way has much in common with the prediction of missing elements in the periodic table. Whenever a vacancy occurs in an organized pattern of information, experimentalists have a guide for their investigations.

46.8 Quarks

As mentioned earlier, leptons appear to be truly elementary particles because there are only a few types of them, and experiments indicate that they have no measurable size or internal structure. Hadrons, on the other hand, are complex particles having size and structure. The existence of the strangeness–charge patterns of the eightfold way suggests that hadrons have substructure. Furthermore, hundreds of types of hadrons exist and many decay into other hadrons.

The Original Quark Model

In 1963, Gell-Mann and George Zweig (b. 1937) independently proposed a model for the substructure of hadrons. According to their model, all hadrons are composed of two or three elementary constituents called **quarks.** (Gell-Mann borrowed the word *quark* from the passage "Three quarks for Muster Mark" in James Joyce's *Finnegans Wake.* In Zweig's model, he called the constituents "aces.") The model has three types of quarks, designated by the symbols u, d, and s, that are given the arbitrary names **up, down,** and **strange.** The various types of quarks are called **flavors.** Figure 46.11 is a pictorial representation of the quark compositions of several hadrons.

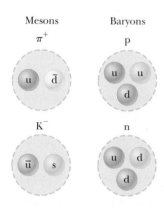

Figure 46.11 Quark composition of two mesons and two baryons.

Table 46.3 Properties of Quarks and Antiquarks

Quarks

Name	Symbol	Spin	Charge	Baryon Number	Strangeness	Charm	Bottomness	Topness
Up	u	$\frac{1}{2}$	$+\frac{2}{3}e$	$\frac{1}{3}$	0	0	0	0
Down	d	$\frac{1}{2}$	$-\frac{1}{3}e$	$\frac{1}{3}$	0	0	0	0
Strange	s	$\frac{1}{2}$	$-\frac{1}{3}e$	$\frac{1}{3}$	−1	0	0	0
Charmed	c	$\frac{1}{2}$	$+\frac{2}{3}e$	$\frac{1}{3}$	0	+1	0	0
Bottom	b	$\frac{1}{2}$	$-\frac{1}{3}e$	$\frac{1}{3}$	0	0	+1	0
Top	t	$\frac{1}{2}$	$+\frac{2}{3}e$	$\frac{1}{3}$	0	0	0	+1

Antiquarks

Name	Symbol	Spin	Charge	Baryon Number	Strangeness	Charm	Bottomness	Topness
Anti-up	$\bar{u}$	$\frac{1}{2}$	$-\frac{2}{3}e$	$-\frac{1}{3}$	0	0	0	0
Anti-down	$\bar{d}$	$\frac{1}{2}$	$+\frac{1}{3}e$	$-\frac{1}{3}$	0	0	0	0
Anti-strange	$\bar{s}$	$\frac{1}{2}$	$+\frac{1}{3}e$	$-\frac{1}{3}$	+1	0	0	0
Anti-charmed	$\bar{c}$	$\frac{1}{2}$	$-\frac{2}{3}e$	$-\frac{1}{3}$	0	−1	0	0
Anti-bottom	$\bar{b}$	$\frac{1}{2}$	$+\frac{1}{3}e$	$-\frac{1}{3}$	0	0	−1	0
Anti-top	$\bar{t}$	$\frac{1}{2}$	$-\frac{2}{3}e$	$-\frac{1}{3}$	0	0	0	−1

An unusual property of quarks is that they carry a fractional electric charge. The u, d, and s quarks have charges of $+2e/3$, $-e/3$, and $-e/3$, respectively, where e is the elementary charge 1.60×10^{-19} C. These and other properties of quarks and antiquarks are given in Table 46.3. Quarks have spin $\frac{1}{2}$, which means that all quarks are fermions, defined as any particle having half-integral spin, as pointed out in Section 43.8. As Table 46.3 shows, associated with each quark is an antiquark of opposite charge, baryon number, and strangeness.

The compositions of all hadrons known when Gell-Mann and Zweig presented their model can be completely specified by three simple rules:

- A meson consists of one quark and one antiquark, giving it a baryon number of 0, as required.
- A baryon consists of three quarks.
- An antibaryon consists of three antiquarks.

The theory put forth by Gell-Mann and Zweig is referred to as the *original quark model*.

Quick Quiz 46.5 Using a coordinate system like that in Figure 46.8, draw an eightfold-way diagram for the three quarks in the original quark model.

Charm and Other Developments

Although the original quark model was highly successful in classifying particles into families, some discrepancies occurred between its predictions and certain experimental decay rates. Consequently, several physicists proposed a fourth quark flavor in 1967. They argued that if four types of leptons exist (as was thought at the time), there should also be four flavors of quarks because of an underlying symmetry in nature. The fourth quark, designated c, was assigned a property called **charm.** A *charmed* quark has charge $+2e/3$, just as the up quark does, but its charm distinguishes it from the other three quarks. This introduces a new quantum number C, representing charm. The new quark has charm $C = +1$, its antiquark has charm of $C = -1$, and all other quarks have $C = 0$. Charm, like strangeness, is conserved in strong and electromagnetic interactions but not in weak interactions.

Table 46.4 **Quark Composition of Mesons**

		Antiquarks									
		$\overline{b}$		$\overline{c}$		$\overline{s}$		$\overline{d}$		$\overline{u}$	
Quarks	**b**	Y	$(b\overline{b})$	B_c^-	$(\overline{c}b)$	$\overline{B}_s^0$	$(\overline{s}b)$	$\overline{B}_d^0$	$(\overline{d}b)$	B^-	$(\overline{u}b)$
	c	B_c^+	$(\overline{b}c)$	J/Ψ	$(c\overline{c})$	D_s^+	$(\overline{s}c)$	D^+	$(\overline{d}c)$	D^0	$(\overline{u}c)$
	s	B_s^0	$(\overline{b}s)$	D_s^-	$(\overline{c}s)$	η, η'	$(\overline{s}s)$	$\overline{K}^0$	$(\overline{d}s)$	K^-	$(\overline{u}s)$
	d	B_d^0	$(\overline{b}d)$	D^-	$(\overline{c}d)$	K^0	$(\overline{s}d)$	π^0, η, η'	$(\overline{d}d)$	π^-	$(\overline{u}d)$
	u	B^+	$(\overline{b}u)$	$\overline{D}^0$	$(\overline{c}u)$	K^+	$(\overline{s}u)$	π^+	$(\overline{d}u)$	π^0, η, η'	$(\overline{u}u)$

Note: The top quark does not form mesons because it decays too quickly.

Table 46.5 **Quark Composition of Several Baryons**

Particle	Quark Composition
p	uud
n	udd
Λ^0	uds
Σ^+	uus
Σ^0	uds
Σ^-	dds
Δ^{++}	uuu
Δ^+	uud
Δ^0	udd
Δ^-	ddd
Ξ^0	uss
Ξ^-	dss
Ω^-	sss

Note: Some baryons have the same quark composition, such as the p and the Δ^+ and the n and the Δ^0. In these cases, the Δ particles are considered to be excited states of the proton and neutron.

Evidence that the charmed quark exists began to accumulate in 1974, when a heavy meson called the J/Ψ particle (or simply Ψ, Greek letter psi) was discovered independently by two groups, one led by Burton Richter (b. 1931) at the Stanford Linear Accelerator (SLAC), and the other led by Samuel Ting (b. 1936) at the Brookhaven National Laboratory. In 1976, Richter and Ting were awarded the Nobel Prize in Physics for this work. The J/Ψ particle does not fit into the three-quark model; instead, it has properties of a combination of the proposed charmed quark and its antiquark ($c\overline{c}$). It is much more massive than the other known mesons (~3 100 MeV/c^2), and its lifetime is much longer than the lifetimes of particles that interact via the strong force. Soon, related mesons were discovered, corresponding to such quark combinations as $\overline{c}d$ and $c\overline{d}$, all of which have great masses and long lifetimes. The existence of these new mesons provided firm evidence for the fourth quark flavor.

In 1975, researchers at Stanford University reported strong evidence for the tau (τ) lepton, mass 1 784 MeV/c^2. This fifth type of lepton led physicists to propose that more flavors of quarks might exist, on the basis of symmetry arguments similar to those leading to the proposal of the charmed quark. These proposals led to more elaborate quark models and the prediction of two new quarks, **top** (t) and **bottom** (b). (Some physicists prefer *truth* and *beauty*.) To distinguish these quarks from the others, quantum numbers called *topness* and *bottomness* (with allowed values +1, 0, −1) were assigned to all quarks and antiquarks (see Table 46.3). In 1977, researchers at the Fermi National Laboratory, under the direction of Leon Lederman (b. 1922), reported the discovery of a very massive new meson Y (Greek letter upsilon), whose composition is considered to be bb, providing evidence for the bottom quark. In March 1995, researchers at Fermilab announced the discovery of the top quark (supposedly the last of the quarks to be found), which has a mass of 173 GeV/c^2.

Table 46.4 lists the quark compositions of mesons formed from the up, down, strange, charmed, and bottom quarks. Table 46.5 shows the quark combinations for the baryons listed in Table 46.2. Notice that only two flavors of quarks, u and d, are contained in all hadrons encountered in ordinary matter (protons and neutrons).

Will the discoveries of elementary particles ever end? How many "building blocks" of matter actually exist? At present, physicists believe that the elementary particles in nature are six quarks and six leptons, together with their antiparticles, and the four field particles listed in Table 46.1. Table 46.6 lists the rest energies and charges of the quarks and leptons.

Despite extensive experimental effort, no isolated quark has ever been observed. Physicists now believe that at ordinary temperatures, quarks are permanently confined inside ordinary particles because of an exceptionally strong force that prevents them from escaping, called (appropriately) the **strong force**[5] (which we

---

[5]As a reminder, the original meaning of the term *strong force* was the short-range attractive force between nucleons, which we have called the *nuclear force*. The nuclear force between nucleons is a secondary effect of the strong force between quarks.

Table 46.6	The Elementary Particles and Their Rest Energies and Charges	
Particle	Approximate Rest Energy	Charge
Quarks		
u	2.4 MeV	$+\frac{2}{3}e$
d	4.8 MeV	$-\frac{1}{3}e$
s	104 MeV	$-\frac{1}{3}e$
c	1.27 GeV	$+\frac{2}{3}e$
b	4.2 GeV	$-\frac{1}{3}e$
t	173 GeV	$+\frac{2}{3}e$
Leptons		
e^-	511 keV	$-e$
μ^-	105.7 MeV	$-e$
τ^-	1.78 GeV	$-e$
ν_e	< 2 eV	0
ν_μ	< 0.17 MeV	0
ν_τ	< 18 MeV	0

introduced at the beginning of Section 46.4 and will discuss further in Section 46.10). This force increases with separation distance, similar to the force exerted by a stretched spring. Current efforts are under way to form a **quark–gluon plasma,** a state of matter in which the quarks are freed from neutrons and protons. In 2000, scientists at CERN announced evidence for a quark–gluon plasma formed by colliding lead nuclei. In 2005, experiments at the Relativistic Heavy Ion Collider (RHIC) at Brookhaven suggested the creation of a quark–gluon plasma. Neither laboratory has provided definitive data to verify the existence of a quark–gluon plasma. Experiments continue, and the ALICE project (A Large Ion Collider Experiment) at the Large Hadron Collider at CERN has joined the search.

Quick Quiz 46.6 Doubly charged baryons, such as the Δ^{++}, are known to exist. True or False: Doubly charged mesons also exist.

46.9 Multicolored Quarks

Shortly after the concept of quarks was proposed, scientists recognized that certain particles had quark compositions that violated the exclusion principle. In Section 42.7, we applied the exclusion principle to electrons in atoms. The principle is more general, however, and applies to all particles with half-integral spin ($\frac{1}{2}$, $\frac{3}{2}$, etc.), which are collectively called fermions. Because all quarks are fermions having spin $\frac{1}{2}$, they are expected to follow the exclusion principle. One example of a particle that appears to violate the exclusion principle is the Ω^- (sss) baryon, which contains three strange quarks having parallel spins, giving it a total spin of $\frac{3}{2}$. All three quarks have the same spin quantum number, in violation of the exclusion principle. Other examples of baryons made up of identical quarks having parallel spins are the Δ^{++} (uuu) and the Δ^- (ddd).

To resolve this problem, it was suggested that quarks possess an additional property called **color charge.** This property is similar in many respects to electric charge except that it occurs in six varieties rather than two. The colors assigned to quarks are red, green, and blue, and antiquarks have the colors antired, antigreen, and antiblue. Therefore, the colors red, green, and blue serve as the "quantum numbers" for the color of the quark. To satisfy the exclusion principle, the three quarks in any baryon must all have different colors. Look again at the quarks in the baryons in Figure 46.11 and notice the colors. The three colors "neutralize" to white.

Pitfall Prevention 46.3
Color Charge Is Not Really Color
The description of color for a quark has nothing to do with visual sensation from light. It is simply a convenient name for a property that is analogous to electric charge.

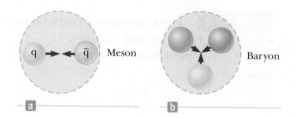

A quark and an antiquark in a meson must be of a color and the corresponding anticolor and will consequently neutralize to white, similar to the way electric charges $+$ and $-$ neutralize to zero net charge. (See the mesons in Fig. 46.11.) The apparent violation of the exclusion principle in the Ω^- baryon is removed because the three quarks in the particle have different colors.

The new property of color increases the number of quarks by a factor of 3 because each of the six quarks comes in three colors. Although the concept of color in the quark model was originally conceived to satisfy the exclusion principle, it also provided a better theory for explaining certain experimental results. For example, the modified theory correctly predicts the lifetime of the π^0 meson.

The theory of how quarks interact with each other is called **quantum chromodynamics,** or QCD, to parallel the name *quantum electrodynamics* (the theory of the electrical interaction between light and matter). In QCD, each quark is said to carry a color charge, in analogy to electric charge. The strong force between quarks is often called the **color force.** Therefore, the terms *strong force* and *color force* are used interchangeably.

In Section 46.1, we stated that the nuclear interaction between hadrons is mediated by massless field particles called **gluons.** As mentioned earlier, the nuclear force is actually a secondary effect of the strong force between quarks. The gluons are the mediators of the strong force. When a quark emits or absorbs a gluon, the quark's color may change. For example, a blue quark that emits a gluon may become a red quark and a red quark that absorbs this gluon becomes a blue quark.

The color force between quarks is analogous to the electric force between charges: particles with the same color repel, and those with opposite colors attract. Therefore, two green quarks repel each other, but a green quark is attracted to an antigreen quark. The attraction between quarks of opposite color to form a meson (q$\bar{q}$) is indicated in Figure 46.12a. Differently colored quarks also attract one another, although with less intensity than the oppositely colored quark and antiquark. For example, a cluster of red, blue, and green quarks all attract one another to form a baryon as in Figure 46.12b. Therefore, every baryon contains three quarks of three different colors.

Although the nuclear force between two colorless hadrons is negligible at large separations, the net strong force between their constituent quarks is not exactly zero at small separations. This residual strong force is the nuclear force that binds protons and neutrons to form nuclei. It is similar to the force between two electric dipoles. Each dipole is electrically neutral. An electric field surrounds the dipoles, however, because of the separation of the positive and negative charges (see Section 23.6). As a result, an electric interaction occurs between the dipoles that is weaker than the force between single charges. In Section 43.1, we explored how this interaction results in the Van der Waals force between neutral molecules.

According to QCD, a more basic explanation of the nuclear force can be given in terms of quarks and gluons. Figure 46.13a shows the nuclear interaction between a neutron and a proton by means of Yukawa's pion, in this case a π^-. This drawing differs from Figure 46.5a, in which the field particle is a π^0; there is no transfer of charge from one nucleon to the other in Figure 46.5a. In Figure 46.13a, the charged pion carries charge from one nucleon to the other, so the nucleons change identities, with the proton becoming a neutron and the neutron becoming a proton.

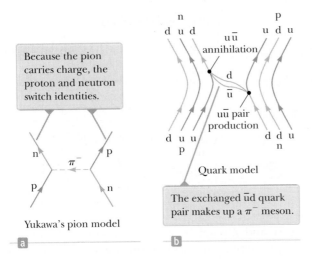

Because the pion carries charge, the proton and neutron switch identities.

Yukawa's pion model

uū annihilation

uū pair production

The exchanged ūd quark pair makes up a π^- meson.

Quark model

Figure 46.13 (a) A nuclear interaction between a proton and a neutron explained in terms of Yukawa's pion-exchange model. (b) The same interaction, explained in terms of quarks and gluons.

Let's look at the same interaction from the viewpoint of the quark model, shown in Figure 46.13b. In this Feynman diagram, the proton and neutron are represented by their quark constituents. Each quark in the neutron and proton is continuously emitting and absorbing gluons. The energy of a gluon can result in the creation of quark–antiquark pairs. This process is similar to the creation of electron–positron pairs in pair production, which we investigated in Section 46.2. When the neutron and proton approach to within 1 fm of each other, these gluons and quarks can be exchanged between the two nucleons, and such exchanges produce the nuclear force. Figure 46.13b depicts one possibility for the process shown in Figure 46.13a. A down quark in the neutron on the right emits a gluon. The energy of the gluon is then transformed to create a uū pair. The u quark stays within the nucleon (which has now changed to a proton), and the recoiling d quark and the ū antiquark are transmitted to the proton on the left side of the diagram. Here the ū annihilates a u quark within the proton and the d is captured. The net effect is to change a u quark to a d quark, and the proton on the left has changed to a neutron.

As the d quark and ū antiquark in Figure 46.13b transfer between the nucleons, the d and ū exchange gluons with each other and can be considered to be bound to each other by means of the strong force. Looking back at Table 46.4, we see that this combination is a π^-, or Yukawa's field particle! Therefore, the quark model of interactions between nucleons is consistent with the pion-exchange model.

46.10 The Standard Model

Scientists now believe there are three classifications of truly elementary particles: leptons, quarks, and field particles. These three types of particles are further classified as either fermions or bosons. Quarks and leptons have spin $\frac{1}{2}$ and hence are fermions, whereas the field particles have integral spin of 1 or higher and are bosons.

Recall from Section 46.1 that the weak force is believed to be mediated by the W^+, W^-, and Z^0 bosons. These particles are said to have *weak charge,* just as quarks have color charge. Therefore, each elementary particle can have mass, electric charge, color charge, and weak charge. Of course, one or more of these could be zero.

In 1979, Sheldon Glashow (b. 1932), Abdus Salam (1926–1996), and Steven Weinberg (b. 1933) won the Nobel Prize in Physics for developing a theory that unifies the electromagnetic and weak interactions. This **electroweak theory** postulates that the weak and electromagnetic interactions have the same strength when the particles involved have very high energies. The two interactions are viewed as different manifestations of a single unifying electroweak interaction. The theory makes many concrete predictions, but perhaps the most spectacular is the prediction of

Figure 46.14 The Standard Model of particle physics.

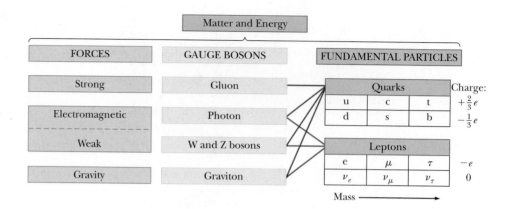

the masses of the W and Z particles at approximately 82 GeV/c^2 and 93 GeV/c^2, respectively. These predictions are close to the masses in Table 46.1 determined by experiment.

The combination of the electroweak theory and QCD for the strong interaction is referred to in high-energy physics as the **Standard Model.** Although the details of the Standard Model are complex, its essential ingredients can be summarized with the help of Fig. 46.14. (Although the Standard Model does not include the gravitational force at present, we include gravity in Fig. 46.14 because physicists hope to eventually incorporate this force into a unified theory.) This diagram shows that quarks participate in all the fundamental forces and that leptons participate in all except the strong force.

The Standard Model does not answer all questions. A major question still unanswered is why, of the two mediators of the electroweak interaction, the photon has no mass but the W and Z bosons do. Because of this mass difference, the electromagnetic and weak forces are quite distinct at low energies but become similar at very high energies, when the rest energy is negligible relative to the total energy. The behavior as one goes from high to low energies is called *symmetry breaking* because the forces are similar, or symmetric, at high energies but are very different at low energies. The nonzero rest energies of the W and Z bosons raise the question of the origin of particle masses. To resolve this problem, a hypothetical particle called the **Higgs boson,** which provides a mechanism for breaking the electroweak symmetry, has been proposed. The Standard Model modified to include the Higgs boson provides a logically consistent explanation of the massive nature of the W and Z bosons. In July 2012, announcements from the ATLAS (A Toroidal LHC Apparatus) and CMS (Compact Muon Solenoid) experiments at the Large Hadron Collider (LHC) at CERN claimed the discovery of a new particle having properties consistent with that of a Higgs boson. The mass of the particle is 125–127 GeV, within the range of predictions made from theoretical considerations using the Standard Model.

Because of the limited energy available in conventional accelerators using fixed targets, it is necessary to employ colliding-beam accelerators called **colliders.** The concept of colliders is straightforward. Particles that have equal masses and equal kinetic energies, traveling in opposite directions in an accelerator ring, collide head-on to produce the required reaction and form new particles. Because the total momentum of the interacting particles is zero, all their kinetic energy is available for the reaction.

Several colliders provided important data for understanding the Standard Model in the latter part of the 20th century and the first decade of the 21st century: the Large Electron–Positron (LEP) Collider and the Super Proton Synchrotron at CERN, the Stanford Linear Collider, and the Tevatron at the Fermi National Laboratory in Illinois. The Relativistic Heavy Ion Collider at Brookhaven National Laboratory is the sole remaining collider in operation in the United States. The Large Hadron Collider at CERN, which began collision operations in March 2010, has

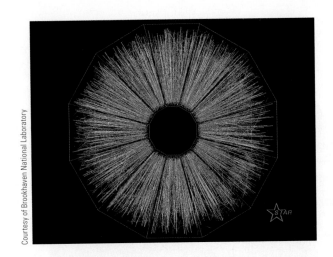

Figure 46.15 A shower of particle tracks from a head-on collision of gold nuclei, each moving with energy 100 GeV. This collision occurred at the Relativistic Heavy Ion Collider (RHIC) at Brookhaven National Laboratory and was recorded with the STAR (Solenoidal Tracker at RHIC) detector. The tracks represent many fundamental particles arising from the energy of the collision.

Courtesy of Brookhaven National Laboratory

taken the lead in particle studies due to its extremely high energy capabilities. The expected upper limit for the LHC is a center-of-mass energy of 14 TeV. (See page 868 for a photo of a magnet used by the LHC.)

In addition to increasing energies in modern accelerators, detection techniques have become increasingly sophisticated. We saw simple bubble-chamber photographs earlier in this chapter that required hours of analysis by hand. Figure 46.15 shows a complex set of tracks from a collision of gold nuclei.

46.11 The Cosmic Connection

In this section, we describe one of the most fascinating theories in all science—the Big Bang theory of the creation of the Universe—and the experimental evidence that supports it. This theory of cosmology states that the Universe had a beginning and furthermore that the beginning was so cataclysmic that it is impossible to look back beyond it. According to this theory, the Universe erupted from an infinitely dense singularity about 14 billion years ago. The first few moments after the Big Bang saw such extremely high energy that it is believed that all four interactions of physics were unified and all matter was contained in a quark–gluon plasma.

The evolution of the four fundamental forces from the Big Bang to the present is shown in Figure 46.16 (page 1470). During the first 10^{-43} s (the ultrahot epoch, $T \sim 10^{32}$ K), it is presumed the strong, electroweak, and gravitational forces were joined to form a completely unified force. In the first 10^{-35} s following the Big Bang (the hot epoch, $T \sim 10^{29}$ K), symmetry breaking occurred for gravity while the strong and electroweak forces remained unified. It was a period when particle energies were so great ($> 10^{16}$ GeV) that very massive particles as well as quarks, leptons, and their antiparticles existed. Then, after 10^{-35} s, the Universe rapidly expanded and cooled (the warm epoch, $T \sim 10^{29}$ to 10^{15} K) and the strong and electroweak forces parted company. As the Universe continued to cool, the electroweak force split into the weak force and the electromagnetic force approximately 10^{-10} s after the Big Bang.

After a few minutes, protons and neutrons condensed out of the plasma. For half an hour, the Universe underwent thermonuclear fusion, exploding as a hydrogen bomb and producing most of the helium nuclei that now exist. The Universe continued to expand, and its temperature dropped. Until about 700 000 years after the Big Bang, the Universe was dominated by radiation. Energetic radiation prevented matter from forming single hydrogen atoms because photons would instantly ionize any atoms that happened to form. Photons experienced continuous Compton scattering from the vast numbers of free electrons, resulting in a Universe that was opaque to radiation. By the time the Universe was about 700 000 years old, it had

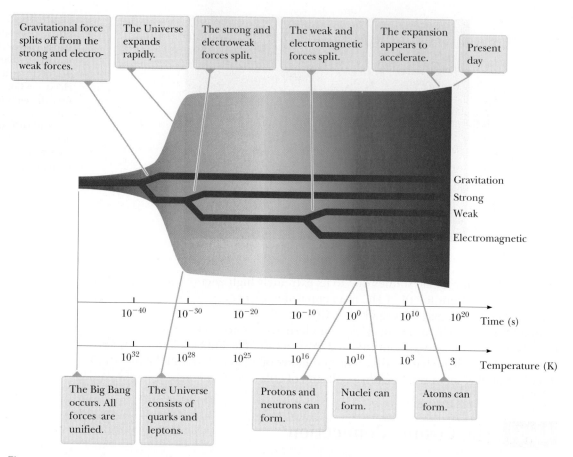

Gravitational force splits off from the strong and electro-weak forces.

The Universe expands rapidly.

The strong and electroweak forces split.

The weak and electromagnetic forces split.

The expansion appears to accelerate.

Present day

Gravitation

Strong

Weak

Electromagnetic

10^{-40} 10^{-30} 10^{-20} 10^{-10} 10^{0} 10^{10} 10^{20} Time (s)

10^{32} 10^{28} 10^{25} 10^{16} 10^{10} 10^{3} 3 Temperature (K)

The Big Bang occurs. All forces are unified.

The Universe consists of quarks and leptons.

Protons and neutrons can form.

Nuclei can form.

Atoms can form.

Figure 46.16 A brief history of the Universe from the Big Bang to the present. The four forces became distinguishable during the first nanosecond. Following that, all the quarks combined to form particles that interact via the nuclear force. The leptons, however, remained separate and to this day exist as individual, observable particles.

expanded and cooled to approximately 3 000 K and protons could bind to electrons to form neutral hydrogen atoms. Because of the quantized energies of the atoms, far more wavelengths of radiation were not absorbed by atoms than were absorbed, and the Universe suddenly became transparent to photons. Radiation no longer dominated the Universe, and clumps of neutral matter steadily grew: first atoms, then molecules, gas clouds, stars, and finally galaxies.

Observation of Radiation from the Primordial Fireball

In 1965, Arno A. Penzias (b. 1933) and Robert W. Wilson (b. 1936) of Bell Laboratories were testing a sensitive microwave receiver and made an amazing discovery. A pesky signal producing a faint background hiss was interfering with their satellite communications experiments. The microwave horn that served as their receiving antenna is shown in Figure 46.17. Evicting a flock of pigeons from the 20-ft horn and cooling the microwave detector both failed to remove the signal.

The intensity of the detected signal remained unchanged as the antenna was pointed in different directions. That the radiation had equal strengths in all directions suggested that the entire Universe was the source of this radiation. Ultimately, it became clear that they were detecting microwave background radiation (at a wavelength of 7.35 cm), which represented the leftover "glow" from the Big Bang. Through a casual conversation, Penzias and Wilson discovered that a group at Princeton University had predicted the residual radiation from the Big Bang and were planning an experiment to attempt to confirm the theory. The excitement in the scientific community was high when Penzias and Wilson announced that they had already observed an excess microwave background compatible with a 3-K

Figure 46.17 Robert W. Wilson (*left*) and Arno A. Penzias with the Bell Telephone Laboratories horn-reflector antenna.

AT&T Bell Laboratories

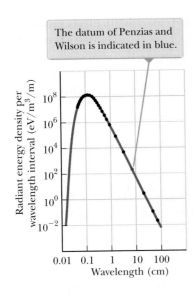

The datum of Penzias and Wilson is indicated in blue.

Figure 46.18 Theoretical black-body (brown curve) and measured radiation spectra (black points) of the Big Bang. Most of the data were collected from the COsmic Background Explorer, or COBE, satellite.

blackbody source, which was consistent with the predicted temperature of the Universe at this time after the Big Bang.

Because Penzias and Wilson made their measurements at a single wavelength, they did not completely confirm the radiation as 3-K blackbody radiation. Subsequent experiments by other groups added intensity data at different wavelengths as shown in Figure 46.18. The results confirm that the radiation is that of a black body at 2.7 K. This figure is perhaps the most clear-cut evidence for the Big Bang theory. The 1978 Nobel Prize in Physics was awarded to Penzias and Wilson for this most important discovery.

In the years following Penzias and Wilson's discovery, other researchers made measurements at different wavelengths. In 1989, the COBE (COsmic Background Explorer) satellite was launched by NASA and added critical measurements at wavelengths below 0.1 cm. The results of these measurements led to a Nobel Prize in Physics for the principal investigators in 2006. Several data points from COBE are shown in Figure 46.18. The Wilkinson Microwave Anisotropy Probe, launched in June 2001, exhibits data that allow observation of temperature differences in the cosmos in the microkelvin range. Ongoing observations are also being made from Earth-based facilities, associated with projects such as QUaD, Qubic, and the South Pole Telescope. In addition, the Planck satellite was launched in May 2009 by the European Space Agency. This space-based observatory has been measuring the cosmic background radiation with higher sensitivity than the Wilkinson probe. The series of measurements taken since 1965 are consistent with thermal radiation associated with a temperature of 2.7 K. The whole story of the cosmic temperature is a remarkable example of science at work: building a model, making a prediction, taking measurements, and testing the measurements against the predictions.

Other Evidence for an Expanding Universe

The Big Bang theory of cosmology predicts that the Universe is expanding. Most of the key discoveries supporting the theory of an expanding Universe were made in the 20th century. Vesto Melvin Slipher (1875–1969), an American astronomer, reported in 1912 that most galaxies are receding from the Earth at speeds up to several million miles per hour. Slipher was one of the first scientists to use Doppler shifts (see Section 17.4) in spectral lines to measure galaxy velocities.

In the late 1920s, Edwin P. Hubble (1889–1953) made the bold assertion that the whole Universe is expanding. From 1928 to 1936, until they reached the limits of the 100-inch telescope, Hubble and Milton Humason (1891–1972) worked at Mount Wilson in California to prove this assertion. The results of that work and of its continuation with the use of a 200-inch telescope in the 1940s showed that the speeds at which galaxies are receding from the Earth increase in direct proportion to their distance R from us. This linear relationship, known as **Hubble's law,** may be written

◀ Hubble's law

$$v = HR \qquad (46.4)$$

where H, called the **Hubble constant,** has the approximate value

$$H \approx 22 \times 10^{-3} \text{ m/(s} \cdot \text{ly)}$$

Example 46.5   Recession of a Quasar

A quasar is an object that appears similar to a star and is very distant from the Earth. Its speed can be determined from Doppler-shift measurements in the light it emits. A certain quasar recedes from the Earth at a speed of $0.55c$. How far away is it?

SOLUTION

Conceptualize A common mental representation for the Hubble law is that of raisin bread cooking in an oven. Imagine yourself at the center of the loaf of bread. As the entire loaf of bread expands upon heating, raisins near you move slowly with respect to you. Raisins far away from you on the edge of the loaf move at a higher speed.

Categorize We use a concept developed in this section, so we categorize this example as a substitution problem.

Find the distance through Hubble's law:

$$R = \frac{v}{H} = \frac{(0.55)(3.00 \times 10^{8} \text{ m/s})}{22 \times 10^{-3} \text{ m/(s} \cdot \text{ly)}} = \boxed{7.5 \times 10^{9} \text{ ly}}$$

WHAT IF? Suppose the quasar has moved at this speed ever since the Big Bang. With this assumption, estimate the age of the Universe.

Answer Let's approximate the distance from the Earth to the quasar as the distance the quasar has moved from the singularity since the Big Bang. We can then find the time interval from the *particle under constant speed* model: $\Delta t = d/v = R/v = 1/H \approx 14$ billion years, which is in approximate agreement with other calculations.

Will the Universe Expand Forever?

In the 1950s and 1960s, Allan R. Sandage (1926–2010) used the 200-inch telescope at Mount Palomar to measure the speeds of galaxies at distances of up to 6 billion light-years away from the Earth. These measurements showed that these very distant galaxies were moving approximately 10 000 km/s faster than Hubble's law predicted. According to this result, the Universe must have been expanding more rapidly 1 billion years ago, and consequently we conclude from these data that the expansion rate is slowing.[6] Today, astronomers and physicists are trying to determine the rate of expansion. If the average mass density of the Universe is less than some critical value ρ_c, the galaxies will slow in their outward rush but still escape to infinity. If the average density exceeds the critical value, the expansion will eventually stop and contraction will begin, possibly leading to a superdense state followed by another expansion. In this scenario, we have an oscillating Universe.

Example 46.6   The Critical Density of the Universe

(A) Starting from energy conservation, derive an expression for the critical mass density of the Universe ρ_c in terms of the Hubble constant H and the universal gravitational constant G.

---

[6]The data at large distances have large observational uncertainties and may be systematically in error from effects such as abnormal brightness in the most distant visible clusters.

▶ **46.6** continued

SOLUTION

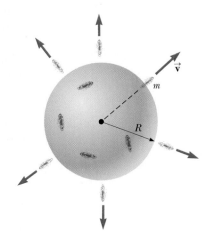

Figure 46.19 (Example 46.6) The galaxy marked with mass m is escaping from a large cluster of galaxies contained within a spherical volume of radius R. Only the mass within R slows the escaping galaxy.

Conceptualize Figure 46.19 shows a large section of the Universe, contained within a sphere of radius R. The total mass in this volume is M. A galaxy of mass $m \ll M$ that has a speed v at a distance R from the center of the sphere escapes to infinity (at which its speed approaches zero) if the sum of its kinetic energy and the gravitational potential energy of the system is zero.

Categorize The Universe may be infinite in spatial extent, but Gauss's law for gravitation (an analog to Gauss's law for electric fields in Chapter 24) implies that only the mass M inside the sphere contributes to the gravitational potential energy of the galaxy–sphere system. Therefore, we categorize this problem as one in which we apply Gauss's law for gravitation. We model the sphere in Figure 46.19 and the escaping galaxy as an *isolated system* for *energy*.

Analyze Write the appropriate reduction of Equation 8.2, assuming that the galaxy leaves the spherical volume while moving at the escape speed:

$$\Delta K + \Delta U = 0$$

$$(0 - \tfrac{1}{2}mv^2) + \left[0 - \left(-\frac{GmM}{R} \right) \right] = 0$$

Substitute for the mass M contained within the sphere the product of the critical density and the volume of the sphere:

$$\tfrac{1}{2}mv^2 = \frac{Gm(\tfrac{4}{3}\pi R^3 \rho_c)}{R}$$

Solve for the critical density:

$$\rho_c = \frac{3v^2}{8\pi GR^2}$$

From Hubble's law, substitute for the ratio $v/R = H$:

$$(1) \quad \rho_c = \frac{3}{8\pi G} \left(\frac{v}{R} \right)^2 = \frac{3H^2}{8\pi G}$$

(B) Estimate a numerical value for the critical density in grams per cubic centimeter.

SOLUTION

In Equation (1), substitute numerical values for H and G:

$$\rho_c = \frac{3H^2}{8\pi G} = \frac{3[22 \times 10^{-3} \text{ m}/(\text{s} \cdot \text{ly})]^2}{8\pi (6.67 \times 10^{-11} \text{ N} \cdot \text{m}^2/\text{kg}^2)} = 8.7 \times 10^5 \text{ kg/m} \cdot (\text{ly})^2$$

Reconcile the units by converting light-years to meters:

$$\rho_c = 8.7 \times 10^5 \text{ kg/m} \cdot (\text{ly})^2 \left(\frac{1 \text{ ly}}{9.46 \times 10^{15} \text{ m}} \right)^2$$

$$= 9.7 \times 10^{-27} \text{ kg/m}^3 = 9.7 \times 10^{-30} \text{ g/cm}^3$$

Finalize Because the mass of a hydrogen atom is 1.67×10^{-24} g, this value of ρ_c corresponds to 6×10^{-6} hydrogen atoms per cubic centimeter or 6 atoms per cubic meter.

Missing Mass in the Universe?

The luminous matter in galaxies averages out to a Universe density of about 5×10^{-33} g/cm^3. The radiation in the Universe has a mass equivalent of approximately 2% of the luminous matter. The total mass of all nonluminous matter (such as interstellar gas and black holes) may be estimated from the speeds of galaxies orbiting each other in a cluster. The higher the galaxy speeds, the more mass in the cluster. Measurements on the Coma cluster of galaxies indicate, surprisingly, that the amount of nonluminous matter is 20 to 30 times the amount of luminous matter present in stars and luminous gas clouds. Yet even this large, invisible component of *dark matter* (see Section 13.6), if extrapolated to the Universe as a whole, leaves the observed mass density a factor of 10 less than ρ_c calculated in Example 46.6. The deficit, called *missing mass*, has been the subject of intense theoretical and experimental work, with exotic particles such as axions, photinos, and superstring particles suggested as candidates for the missing mass. Some researchers have made the more mundane proposal that the missing mass is present in neutrinos. In fact, neutrinos are so abundant that a tiny neutrino rest energy on the order of only 20 eV would furnish the missing mass and "close" the Universe. Current experiments designed to measure the rest energy of the neutrino will have an effect on predictions for the future of the Universe.

Mysterious Energy in the Universe?

A surprising twist in the story of the Universe arose in 1998 with the observation of a class of supernovae that have a fixed absolute brightness. By combining the apparent brightness and the redshift of light from these explosions, their distance and speed of recession from the Earth can be determined. These observations led to the conclusion that the expansion of the Universe is not slowing down, but is accelerating! Observations by other groups also led to the same interpretation.

To explain this acceleration, physicists have proposed *dark energy*, which is energy possessed by the vacuum of space. In the early life of the Universe, gravity dominated over the dark energy. As the Universe expanded and the gravitational force between galaxies became smaller because of the great distances between them, the dark energy became more important. The dark energy results in an effective repulsive force that causes the expansion rate to increase.[7]

Although there is some degree of certainty about the beginning of the Universe, we are uncertain about how the story will end. Will the Universe keep on expanding forever, or will it someday collapse and then expand again, perhaps in an endless series of oscillations? Results and answers to these questions remain inconclusive, and the exciting controversy continues.

46.12 Problems and Perspectives

While particle physicists have been exploring the realm of the very small, cosmologists have been exploring cosmic history back to the first microsecond of the Big Bang. Observation of the events that occur when two particles collide in an accelerator is essential for reconstructing the early moments in cosmic history. For this reason, perhaps the key to understanding the early Universe is to first understand the world of elementary particles. Cosmologists and physicists now find that they have many common goals and are joining hands in an attempt to understand the physical world at its most fundamental level.

Our understanding of physics at short distances is far from complete. Particle physics is faced with many questions. Why does so little antimatter exist in the Universe? Is it possible to unify the strong and electroweak theories in a logical and consistent manner? Why do quarks and leptons form three similar but distinct families? Are muons the same as electrons apart from their difference in mass, or do they have other subtle differences that have not been detected? Why are some particles charged and others neutral? Why do quarks carry a fractional charge? What determines the masses of the elementary constituents of matter? Can isolated quarks exist? Why do electrons and protons have *exactly* the same magnitude of

---

[7]For an overview of dark energy, see S. Perlmutter, "Supernovae, Dark Energy, and the Accelerating Universe," *Physics Today* **56**(4): 53–60, April 2003.

charge when one is a truly fundamental particle and the other is built from smaller particles?

An important and obvious question that remains is whether leptons and quarks have an underlying structure. If they do, we can envision an infinite number of deeper structure levels. If leptons and quarks are indeed the ultimate constituents of matter, however, scientists hope to construct a final theory of the structure of matter, just as Einstein dreamed of doing. This theory, whimsically called the Theory of Everything, is a combination of the Standard Model and a quantum theory of gravity.

String Theory: A New Perspective

Let's briefly discuss one current effort at answering some of these questions by proposing a new perspective on particles. While reading this book, you may recall starting off with the *particle* model in Chapter 2 and doing quite a bit of physics with it. In Chapter 16, we introduced the *wave* model, and there was more physics to be investigated via the properties of waves. We used a *wave* model for light in Chapter 35; in Chapter 40, however, we saw the need to return to the *particle* model for light. Furthermore, we found that material particles had wave-like characteristics. The quantum particle model discussed in Chapter 40 allowed us to build particles out of waves, suggesting that a *wave* is the fundamental entity. In the current Chapter 46, however, we introduced elementary *particles* as the fundamental entities. It seems as if we cannot make up our mind! In this final section, we discuss a current research effort to build particles out of waves and vibrations on strings!

String theory is an effort to unify the four fundamental forces by modeling all particles as various quantized vibrational modes of a single entity, an incredibly small string. The typical length of such a string is on the order of 10^{-35} m, called the **Planck length.** We have seen quantized modes before in the frequencies of vibrating guitar strings in Chapter 18 and the quantized energy levels of atoms in Chapter 42. In string theory, each quantized mode of vibration of the string corresponds to a different elementary particle in the Standard Model.

One complicating factor in string theory is that it requires space–time to have ten dimensions. Despite the theoretical and conceptual difficulties in dealing with ten dimensions, string theory holds promise in incorporating gravity with the other forces. Four of the ten dimensions—three space dimensions and one time dimension—are visible to us. The other six are said to be *compactified;* that is, the six dimensions are curled up so tightly that they are not visible in the macroscopic world.

As an analogy, consider a soda straw. You can build a soda straw by cutting a rectangular piece of paper (Fig. 46.20a), which clearly has two dimensions, and rolling it into a small tube (Fig. 46.20b). From far away, the soda straw looks like a one-dimensional straight line. The second dimension has been curled up and is not visible. String theory claims that six space–time dimensions are curled up in an analogous way, with the curling being on the size of the Planck length and impossible to see from our viewpoint.

Another complicating factor with string theory is that it is difficult for string theorists to guide experimentalists as to what to look for in an experiment. The

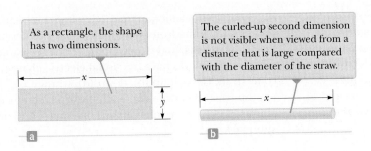

As a rectangle, the shape has two dimensions.

The curled-up second dimension is not visible when viewed from a distance that is large compared with the diameter of the straw.

Figure 46.20 (a) A piece of paper is cut into a rectangular shape. (b) The paper is rolled up into a soda straw.

Planck length is so small that direct experimentation on strings is impossible. Until the theory has been further developed, string theorists are restricted to applying the theory to known results and testing for consistency.

One of the predictions of string theory, called **supersymmetry,** or SUSY, suggests that every elementary particle has a superpartner that has not yet been observed. It is believed that supersymmetry is a broken symmetry (like the broken electroweak symmetry at low energies) and the masses of the superpartners are above our current capabilities of detection by accelerators. Some theorists claim that the mass of superpartners is the missing mass discussed in Section 46.11. Keeping with the whimsical trend in naming particles and their properties, superpartners are given names such as the *squark* (the superpartner to a quark), the *selectron* (electron), and the *gluino* (gluon).

Other theorists are working on **M-theory,** which is an eleven-dimensional theory based on membranes rather than strings. In a way reminiscent of the correspondence principle, M-theory is claimed to reduce to string theory if one compactifies from eleven dimensions to ten dimensions.

The questions listed at the beginning of this section go on and on. Because of the rapid advances and new discoveries in the field of particle physics, many of these questions may be resolved in the next decade and other new questions may emerge.

Summary

Concepts and Principles

Before quark theory was developed, the four fundamental forces in nature were identified as nuclear, electromagnetic, weak, and gravitational. All the interactions in which these forces take part are mediated by **field particles.** The electromagnetic interaction is mediated by photons; the weak interaction is mediated by the $W^\pm$ and Z^0 bosons; the gravitational interaction is mediated by gravitons; and the nuclear interaction is mediated by gluons.

A charged particle and its **antiparticle** have the same mass but opposite charge, and other properties will have opposite values, such as lepton number and baryon number. It is possible to produce particle–antiparticle pairs in nuclear reactions if the available energy is greater than $2mc^2$, where m is the mass of the particle (or antiparticle).

Particles other than field particles are classified as hadrons or leptons. **Hadrons** interact via all four fundamental forces. They have size and structure and are not elementary particles. There are two types, **baryons** and **mesons.** Baryons, which generally are the most massive particles, have nonzero **baryon number** and a spin of $\frac{1}{2}$ or $\frac{3}{2}$. Mesons have baryon number zero and either zero or integral spin.

Leptons have no structure or size and are considered truly elementary. They interact only via the weak, gravitational, and electromagnetic forces. Six types of leptons exist: the electron e^-, the muon μ^-, and the tau τ^-, and their neutrinos ν_e, ν_μ, and ν_τ.

In all reactions and decays, quantities such as energy, linear momentum, angular momentum, electric charge, baryon number, and lepton number are strictly conserved. Certain particles have properties called **strangeness** and **charm.** These unusual properties are conserved in all decays and nuclear reactions except those that occur via the weak force.

Theorists in elementary particle physics have postulated that all hadrons are composed of smaller units known as **quarks,** and experimental evidence agrees with this model. Quarks have fractional electric charge and come in six **flavors:** up (u), down (d), strange (s), charmed (c), top (t), and bottom (b). Each baryon contains three quarks, and each meson contains one quark and one antiquark.

According to the theory of **quantum chromodynamics,** quarks have a property called **color;** the force between quarks is referred to as the **strong force** or the **color force.** The strong force is now considered to be a fundamental force. The nuclear force, which was originally considered to be fundamental, is now understood to be a secondary effect of the strong force due to gluon exchanges between hadrons.

The electromagnetic and weak forces are now considered to be manifestations of a single force called the **electroweak force.** The combination of quantum chromodynamics and the electroweak theory is called the **Standard Model.**

The background microwave radiation discovered by Penzias and Wilson strongly suggests that the Universe started with a Big Bang about 14 billion years ago. The background radiation is equivalent to that of a black body at 3 K. Various astronomical measurements strongly suggest that the Universe is expanding. According to **Hubble's law,** distant galaxies are receding from the Earth at a speed $v = HR$, where H is the **Hubble constant,** $H \approx 22 \times 10^{-3}$ m/(s · ly), and R is the distance from the Earth to the galaxy.

Objective Questions   1. denotes answer available in *Student Solutions Manual/Study Guide*

1. What interactions affect protons in an atomic nucleus? More than one answer may be correct. (a) the nuclear interaction (b) the weak interaction (c) the electromagnetic interaction (d) the gravitational interaction

2. In one experiment, two balls of clay of the same mass travel with the same speed v toward each other. They collide head-on and come to rest. In a second experiment, two clay balls of the same mass are again used. One ball hangs at rest, suspended from the ceiling by a thread. The second ball is fired toward the first at speed v, to collide, stick to the first ball, and continue to move forward. Is the kinetic energy that is transformed into internal energy in the first experiment (a) one-fourth as much as in the second experiment, (b) one-half as much as in the second experiment, (c) the same as in the second experiment, (d) twice as much as in the second experiment, or (e) four times as much as in the second experiment?

3. The Ω^- particle is a baryon with spin $\frac{3}{2}$. Does the Ω^- particle have (a) three possible spin states in a magnetic field, (b) four possible spin states, (c) three times the charge of a spin $-\frac{1}{2}$ particle, or (d) three times the mass of a spin $-\frac{1}{2}$ particle, or (e) are none of those choices correct?

4. Which of the following field particles mediates the strong force? (a) photon (b) gluon (c) graviton (d) $W^\pm$ and Z bosons (e) none of those field particles

5. An isolated stationary muon decays into an electron, an electron antineutrino, and a muon neutrino. Is the total kinetic energy of these three particles (a) zero, (b) small, or (c) large compared to their rest energies, or (d) none of those choices are possible?

6. Define the average density of the solar system ρ_{SS} as the total mass of the Sun, planets, satellites, rings, asteroids, icy outliers, and comets, divided by the volume of a sphere around the Sun large enough to contain all these objects. The sphere extends about halfway to the nearest star, with a radius of approximately 2×10^{16} m, about two light-years. How does this average density of the solar system compare with the critical density ρ_c required for the Universe to stop its Hubble's-law expansion? (a) ρ_{SS} is much greater than ρ_c. (b) ρ_{SS} is approximately or precisely equal to ρ_c. (c) ρ_{SS} is much less than ρ_c. (d) It is impossible to determine.

7. When an electron and a positron meet at low speed in empty space, they annihilate each other to produce two 0.511-MeV gamma rays. What law would be violated if they produced one gamma ray with an energy of 1.02 MeV? (a) conservation of energy (b) conservation of momentum (c) conservation of charge (d) conservation of baryon number (e) conservation of electron lepton number

8. Place the following events into the correct sequence from the earliest in the history of the Universe to the latest. (a) Neutral atoms form. (b) Protons and neutrons are no longer annihilated as fast as they form. (c) The Universe is a quark–gluon soup. (d) The Universe is like the core of a normal star today, forming helium by nuclear fusion. (e) The Universe is like the surface of a hot star today, consisting of a plasma of ionized atoms. (f) Polyatomic molecules form. (g) Solid materials form.

Conceptual Questions   1. denotes answer available in *Student Solutions Manual/Study Guide*

1. The W and Z bosons were first produced at CERN in 1983 by causing a beam of protons and a beam of antiprotons to meet at high energy. Why was this discovery important?

2. What are the differences between hadrons and leptons?

3. Neutral atoms did not exist until hundreds of thousands of years after the Big Bang. Why?

4. Describe the properties of baryons and mesons and the important differences between them.

5. The Ξ^0 particle decays by the weak interaction according to the decay mode $\Xi^0 \rightarrow \Lambda^0 + \pi^0$. Would you expect this decay to be fast or slow? Explain.

6. In the theory of quantum chromodynamics, quarks come in three colors. How would you justify the statement that "all baryons and mesons are colorless"?

7. An antibaryon interacts with a meson. Can a baryon be produced in such an interaction? Explain.

8. Describe the essential features of the Standard Model of particle physics.

9. How many quarks are in each of the following: (a) a baryon, (b) an antibaryon, (c) a meson, (d) an antimeson? (e) How do you explain that baryons have half-integral spins, whereas mesons have spins of 0 or 1?

10. Are the laws of conservation of baryon number, lepton number, and strangeness based on fundamental properties of nature (as are the laws of conservation of momentum and energy, for example)? Explain.

11. Name the four fundamental interactions and the field particle that mediates each.

12. How did Edwin Hubble determine in 1928 that the Universe is expanding?

13. Kaons all decay into final states that contain no protons or neutrons. What is the baryon number for kaons?

Problems

WebAssign The problems found in this chapter may be assigned online in Enhanced WebAssign

1. straightforward; 2. intermediate; 3. challenging

1. full solution available in the *Student Solutions Manual/Study Guide*

AMT Analysis Model tutorial available in Enhanced WebAssign

GP Guided Problem

M Master It tutorial available in Enhanced WebAssign

W Watch It video solution available in Enhanced WebAssign

Section 46.1 The Fundamental Forces in Nature

Section 46.2 Positrons and Other Antiparticles

1. Model a penny as 3.10 g of pure copper. Consider an anti-penny minted from 3.10 g of copper anti-atoms, each with 29 positrons in orbit around a nucleus comprising 29 antiprotons and 34 or 36 antineutrons. (a) Find the energy released if the two coins collide. (b) Find the value of this energy at the unit price of $0.11/kWh, a representative retail rate for energy from the electric company.

2. Two photons are produced when a proton and an antiproton annihilate each other. In the reference frame in which the center of mass of the proton–antiproton system is stationary, what are (a) the minimum frequency and (b) the corresponding wavelength of each photon?

3. A photon produces a proton–antiproton pair according to the reaction $\gamma \rightarrow p + \bar{p}$. (a) What is the minimum possible frequency of the photon? (b) What is its wavelength?

4. At some time in your life, you may find yourself in a hospital to have a PET, or positron-emission tomography, scan. In the procedure, a radioactive element that undergoes e^+ decay is introduced into your body. The equipment detects the gamma rays that result from pair annihilation when the emitted positron encounters an electron in your body's tissue. During such a scan, suppose you receive an injection of glucose containing on the order of 10^{10} atoms of ^{14}O, with half-life 70.6 s. Assume the oxygen remaining after 5 min is uniformly distributed through 2 L of blood. What is then the order of magnitude of the oxygen atoms' activity in 1 cm^3 of the blood?

5. A photon with an energy $E_\gamma = 2.09$ GeV creates a proton–antiproton pair in which the proton has a kinetic energy of 95.0 MeV. What is the kinetic energy of the antiproton? *Note:* $m_p c^2 = 938.3$ MeV.

Section 46.3 Mesons and the Beginning of Particle Physics

6. One mediator of the weak interaction is the Z^0 boson, with mass 91 GeV/c^2. Use this information to find the order of magnitude of the range of the weak interaction.

7. (a) Prove that the exchange of a virtual particle of mass m can be associated with a force with a range given by

$$d \approx \frac{1\,240}{4\pi mc^2} = \frac{98.7}{mc^2}$$

where d is in nanometers and mc^2 is in electron volts. (b) State the pattern of dependence of the range on the mass. (c) What is the range of the force that might be produced by the virtual exchange of a proton?

Section 46.4 **Classification of Particles**

Section 46.5 **Conservation Laws**

8. The first of the following two reactions can occur, but the second cannot. Explain.

$$K_S^0 \rightarrow \pi^+ + \pi^- \quad \text{(can occur)}$$

$$\Lambda^0 \rightarrow \pi^+ + \pi^- \quad \text{(cannot occur)}$$

9. A neutral pion at rest decays into two photons according to $\pi^0 \rightarrow \gamma + \gamma$. Find the (a) energy, (b) momentum, and (c) frequency of each photon.

10. When a high-energy proton or pion traveling near the speed of light collides with a nucleus, it travels an average distance of 3×10^{-15} m before interacting. From this information, find the order of magnitude of the time interval required for the strong interaction to occur.

11. Each of the following reactions is forbidden. Determine what conservation laws are violated for each reaction.

(a) $p + \bar{p} \rightarrow \mu^+ + e^-$

(b) $\pi^- + p \rightarrow p + \pi^+$

(c) $p + p \rightarrow p + p + n$

(d) $\gamma + p \rightarrow n + \pi^0$

(e) $\nu_e + p \rightarrow n + e^+$

12. (a) Show that baryon number and charge are conserved in the following reactions of a pion with a proton:

$$(1) \quad \pi^+ + p \rightarrow K^+ + \Sigma^+$$

$$(2) \quad \pi^+ + p \rightarrow \pi^+ + \Sigma^+$$

(b) The first reaction is observed, but the second never occurs. Explain.

13. The following reactions or decays involve one or more neutrinos. In each case, supply the missing neutrino (ν_e, ν_μ, or ν_τ) or antineutrino.

(a) $\pi^- \rightarrow \mu^- + ?$ (b) $K^+ \rightarrow \mu^+ + ?$

(c) $? + p \rightarrow n + e^+$ (d) $? + n \rightarrow p + e^-$

(e) $? + n \rightarrow p + \mu^-$ (f) $\mu^- \rightarrow e^- + ? + ?$

14. Determine the type of neutrino or antineutrino involved in each of the following processes.

(a) $\pi^+ \rightarrow \pi^0 + e^+ + ?$ (b) $? + p \rightarrow \mu^- + p + \pi^+$

(c) $\Lambda^0 \rightarrow p + \mu^- + ?$ (d) $\tau^+ \rightarrow \mu^+ + ? + ?$

15. Determine which of the following reactions can occur. For those that cannot occur, determine the conservation law (or laws) violated.

(a) $p \rightarrow \pi^+ + \pi^0$ (b) $p + p \rightarrow p + p + \pi^0$

(c) $p + p \rightarrow p + \pi^+$ (d) $\pi^+ \rightarrow \mu^+ + \nu_\mu$

(e) $n \rightarrow p + e^- + \bar{\nu}_e$ (f) $\pi^+ \rightarrow \mu^+ + n$

16. Occasionally, high-energy muons collide with electrons and produce two neutrinos according to the reaction $\mu^+ + e^- \rightarrow 2\nu$. What kind of neutrinos are they?

17. A K_S^0 particle at rest decays into a π^+ and a π^-. The mass of the K_S^0 is 497.7 MeV/c^2, and the mass of each π meson is 139.6 MeV/c^2. What is the speed of each pion?

18. (a) Show that the proton-decay $p \rightarrow e^+ + \gamma$ cannot occur because it violates the conservation of baryon number. (b) **What If?** Imagine that this reaction does occur and the proton is initially at rest. Determine the energies and magnitudes of the momentum of the positron and photon after the reaction. (c) Determine the speed of the positron after the reaction.

19. A Λ^0 particle at rest decays into a proton and a π^- meson. (a) Use the data in Table 46.2 to find the Q value for this decay in MeV. (b) What is the total kinetic energy shared by the proton and the π^- meson after the decay? (c) What is the total momentum shared by the proton and the π^- meson? (d) The proton and the π^- meson have momenta with the same magnitude after the decay. Do they have equal kinetic energies? Explain.

Section 46.6 **Strange Particles and Strangeness**

20. The neutral meson ρ^0 decays by the strong interaction into two pions:

$$\rho^0 \rightarrow \pi^+ + \pi^- \quad (T_{1/2} \sim 10^{-23} \text{ s})$$

The neutral kaon also decays into two pions:

$$K_S^0 \rightarrow \pi^+ + \pi^- \quad (T_{1/2} \sim 10^{-10} \text{ s})$$

How do you explain the difference in half-lives?

21. Which of the following processes are allowed by the strong interaction, the electromagnetic interaction, the weak interaction, or no interaction at all?

(a) $\pi^- + p \rightarrow 2\eta$ (b) $K^- + n \rightarrow \Lambda^0 + \pi^-$

(c) $K^- \rightarrow \pi^- + \pi^0$ (d) $\Omega^- \rightarrow \Xi^- + \pi^0$

(e) $\eta \rightarrow 2\gamma$

22. For each of the following forbidden decays, determine what conservation laws are violated.

(a) $\mu^- \rightarrow e^- + \gamma$ (b) $n \rightarrow p + e^- + \nu_e$

(c) $\Lambda^0 \rightarrow p + \pi^0$ (d) $p \rightarrow e^+ + \pi^0$

(e) $\Xi^0 \rightarrow n + \pi^0$

23. Fill in the missing particle. Assume reaction (a) occurs via the strong interaction and reactions (b) and (c) involve the weak interaction. Assume also the total strangeness changes by one unit if strangeness is not conserved.

(a) $K^+ + p \rightarrow ? + p$ (c) $K^+ \rightarrow ? + \mu^+ + \nu_\mu$

(b) $\Omega^- \rightarrow ? + \pi^-$

24. Identify the conserved quantities in the following processes.

(a) $\Xi^- \rightarrow \Lambda^0 + \mu^- + \nu_\mu$ (b) $K_S^0 \rightarrow 2\pi^0$

(c) $K^- + p \rightarrow \Sigma^0 + n$ (d) $\Sigma^0 \rightarrow \Lambda^0 + \gamma$

(e) $e^+ + e^- \rightarrow \mu^+ + \mu^-$ (f) $\bar{p} + n \rightarrow \overline{\Lambda^0} + \Sigma^-$

(g) Which reactions cannot occur? Why not?

25. Determine whether or not strangeness is conserved in the following decays and reactions.

(a) $\Lambda^0 \rightarrow p + \pi^-$ (b) $\pi^- + p \rightarrow \Lambda^0 + K^0$

(c) $\bar{p} + p \rightarrow \overline{\Lambda^0} + \Lambda^0$ (d) $\pi^- + p \rightarrow \pi^- + \Sigma^+$

(e) $\Xi^- \rightarrow \Lambda^0 + \pi^-$ (f) $\Xi^0 \rightarrow p + \pi^-$

26. The particle decay $\Sigma^+ \rightarrow \pi^+ + n$ is observed in a
GP bubble chamber. Figure P46.26 represents the curved tracks of the particles Σ^+ and π^+ and the invisible track of the neutron in the presence of a uniform magnetic field of 1.15 T directed out of the page. The measured radii of curvature are 1.99 m for the Σ^+ particle and 0.580 m for the π^+ particle. From this information, we wish to determine the mass of the Σ^+ particle. (a) Find the magnitudes of the momenta of the Σ^+ and the π^+ particles in units of MeV/c. (b) The angle between the momenta of the Σ^+ and the π^+ particles at the moment of decay is $\theta = 64.5°$. Find the magnitude of the momentum of the neutron. (c) Calculate the total energy of the π^+ particle and of the neutron from their known masses ($m_\pi = 139.6$ MeV/c^2, $m_n = 939.6$ MeV/c^2) and the relativistic energy–momentum relation. (d) What is the total energy of the Σ^+ particle? (e) Calculate the mass of the Σ^+ particle. (f) Compare the mass with the value in Table 46.2.

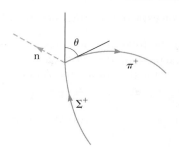

Figure P46.26

27. If a K_S^0 meson at rest decays in 0.900×10^{-10} s, how far
M does a K_S^0 meson travel if it is moving at $0.960c$?

Section 46.7 Finding Patterns in the Particles

Section 46.8 Quarks

Section 46.9 Multicolored Quarks

Section 46.10 The Standard Model

Problem 89 in Chapter 39 can be assigned with Section 46.10.

28. The quark compositions of the K^0 and Λ^0 particles are $d\bar{s}$ and uds, respectively. Show that the charge, baryon number, and strangeness of these particles equal the sums of these numbers for the quark constituents.

29. The reaction $\pi^- + p \rightarrow K^0 + \Lambda^0$ occurs with high probability, whereas the reaction $\pi^- + p \rightarrow K^0 + n$ never occurs. Analyze these reactions at the quark level. Show that the first reaction conserves the total number of each type of quark and the second reaction does not.

30. Identify the particles corresponding to the quark states (a) suu, (b) $\bar{u}d$, (c) $\bar{s}d$, and (d) ssd.

31. The quark composition of the proton is uud, whereas that of the neutron is udd. Show that the charge, baryon number, and strangeness of these particles equal the sums of these numbers for their quark constituents.

32. Analyze each of the following reactions in terms of constituent quarks and show that each type of quark is conserved. (a) $\pi^+ + p \rightarrow K^+ + \Sigma^+$ (b) $K^- + p \rightarrow K^+ + K^0 + \Omega^-$ (c) Determine the quarks in the final particle for this reaction: $p + p \rightarrow K^0 + p + \pi^+ + ?$ (d) In the reaction in part (c), identify the mystery particle.

33. What is the electrical charge of the baryons with the quark compositions (a) $\bar{u}\bar{u}\bar{d}$ and (b) $\bar{u}d\bar{d}$ (c) What are these baryons called?

34. Find the number of electrons, and of each species of
M quark, in 1 L of water.

35. A Σ^0 particle traveling through matter strikes a proton; then a Σ^+ and a gamma ray as well as a third particle emerge. Use the quark model of each to determine the identity of the third particle.

36. **What If?** Imagine that binding energies could be ignored. Find the masses of the u and d quarks from the masses of the proton and neutron.

Section 46.11 The Cosmic Connection

Problem 21 in Chapter 39 can be assigned with this section.

37. **Review.** Refer to Section 39.4. Prove that the Doppler shift in wavelength of electromagnetic waves is described by

$$\lambda' = \lambda \sqrt{\frac{1 + v/c}{1 - v/c}}$$

where λ' is the wavelength measured by an observer moving at speed v away from a source radiating waves of wavelength λ.

38. Gravitation and other forces prevent Hubble's-law expansion from taking place except in systems larger than clusters of galaxies. **What If?** Imagine that these forces could be ignored and all distances expanded at a rate described by the Hubble constant of 22×10^{-3} m/(s · ly). (a) At what rate would the 1.85-m height of a basketball player be increasing? (b) At what rate would the distance between the Earth and the Moon be increasing?

39. **Review.** The cosmic background radiation is blackbody radiation from a source at a temperature of 2.73 K.

(a) Use Wien's law to determine the wavelength at which this radiation has its maximum intensity. (b) In what part of the electromagnetic spectrum is the peak of the distribution?

40. Assume dark matter exists throughout space with a uniform density of 6.00×10^{-28} kg/m^3. (a) Find the amount of such dark matter inside a sphere centered on the Sun, having the Earth's orbit as its equator. (b) Explain whether the gravitational field of this dark matter would have a measurable effect on the Earth's revolution.

41. The early Universe was dense with gamma-ray photons of energy $\sim k_B T$ and at such a high temperature that protons and antiprotons were created by the process $\gamma \rightarrow p + \bar{p}$ as rapidly as they annihilated each other. As the Universe cooled in adiabatic expansion, its temperature fell below a certain value and proton pair production became rare. At that time, slightly more protons than antiprotons existed, and essentially all the protons in the Universe today date from that time. (a) Estimate the order of magnitude of the temperature of the Universe when protons condensed out. (b) Estimate the order of magnitude of the temperature of the Universe when electrons condensed out.

42. If the average density of the Universe is small compared with the critical density, the expansion of the Universe described by Hubble's law proceeds with speeds that are nearly constant over time. (a) Prove that in this case the age of the Universe is given by the inverse of the Hubble constant. (b) Calculate $1/H$ and express it in years.

43. **Review.** A star moving away from the Earth at $0.280c$ emits radiation that we measure to be most intense at the wavelength 500 nm. Determine the surface temperature of this star.

44. **Review.** Use Stefan's law to find the intensity of the cosmic background radiation emitted by the fireball of the Big Bang at a temperature of 2.73 K.

45. The first quasar to be identified and the brightest found to date, 3C 273 in the constellation Virgo, was observed to be moving away from the Earth at such high speed that the observed blue 434-nm H_γ line of hydrogen is Doppler-shifted to 510 nm, in the green portion of the spectrum. (a) How fast is the quasar receding? (b) Edwin Hubble discovered that all objects outside the local group of galaxies are moving away from us, with speeds v proportional to their distances R. Hubble's law is expressed as $v = HR$, where the Hubble constant has the approximate value $H \approx 22 \times 10^{-3}$ m/(s $\cdot$ ly). Determine the distance from the Earth to this quasar.

46. The various spectral lines observed in the light from a distant quasar have longer wavelengths λ'_n than the wavelengths λ_n measured in light from a stationary source. Here n is an index taking different values for different spectral lines. The fractional change in wavelength toward the red is the same for all spectral lines. That is, the Doppler redshift parameter Z defined by

$$Z = \frac{\lambda'_n - \lambda_n}{\lambda_n}$$

is common to all spectral lines for one object. In terms of Z, use Hubble's law to determine (a) the speed of recession of the quasar and (b) the distance from the Earth to this quasar.

47. Using Hubble's law, find the wavelength of the 590-nm sodium line emitted from galaxies (a) 2.00×10^6 ly, (b) 2.00×10^8 ly, and (c) 2.00×10^9 ly away from the Earth.

48. The visible section of the Universe is a sphere centered on the bridge of your nose, with radius 13.7 billion light-years. (a) Explain why the visible Universe is getting larger, with its radius increasing by one light-year in every year. (b) Find the rate at which the volume of the visible section of the Universe is increasing.

49. In Section 13.6, we discussed dark matter along with one proposal for the origin of dark matter: WIMPs, or *weakly interacting massive particles*. Another proposal is that dark matter consists of large planet-sized objects, called MACHOs, or *massive astrophysical compact halo objects,* that drift through interstellar space and are not bound to a solar system. Whether WIMPs or MACHOs, suppose astronomers perform theoretical calculations and determine the average density of the observable Universe to be $1.20\rho_c$. If this value were correct, how many times larger will the Universe become before it begins to collapse? That is, by what factor will the distance between remote galaxies increase in the future?

Section 46.12 Problems and Perspectives

50. Classical general relativity views the structure of space–time as deterministic and well defined down to arbitrarily small distances. On the other hand, quantum general relativity forbids distances smaller than the Planck length given by $L = (\hbar G/c^3)^{1/2}$. (a) Calculate the value of the Planck length. The quantum limitation suggests that after the Big Bang, when all the presently observable section of the Universe was contained within a point-like singularity, nothing could be observed until that singularity grew larger than the Planck length. Because the size of the singularity grew at the speed of light, we can infer that no observations were possible during the time interval required for light to travel the Planck length. (b) Calculate this time interval, known as the Planck time T, and state how it compares with the ultrahot epoch mentioned in the text.

Additional Problems

51. For each of the following decays or reactions, name at least one conservation law that prevents it from occurring.

(a) $\pi^- + p \rightarrow \Sigma^+ + \pi^0$

(b) $\mu^- \rightarrow \pi^- + \nu_e$

(c) $p \rightarrow \pi^+ + \pi^+ + \pi^-$

52. Identify the unknown particle on the left side of the following reaction:

$$? + p \rightarrow n + \mu^+$$

53. Assume that the half-life of free neutrons is 614 s. What fraction of a group of free thermal neutrons with kinetic energy 0.040 0 eV will decay before traveling a distance of 10.0 km?

54. *Why is the following situation impossible?* A gamma-ray photon with energy 1.05 MeV strikes a stationary electron, causing the following reaction to occur:

$$\gamma^- + e^- \rightarrow e^- + e^- + e^+$$

Assume all three final particles move with the same speed in the same direction after the reaction.

55. **Review.** Supernova Shelton 1987A, located approximately 170 000 ly from the Earth, is estimated to have emitted a burst of neutrinos carrying energy $\sim 10^{46}$ J (Fig. P46.55). Suppose the average neutrino energy was 6 MeV and your mother's body presented cross-sectional area 5 000 cm^2. To an order of magnitude, how many of these neutrinos passed through her?

Australian Astronomical Observatory, photography by David Malin from AAT Plates

Figure P46.55 Problems 55 and 72.

56. The energy flux carried by neutrinos from the Sun is estimated to be on the order of 0.400 W/m^2 at the Earth's surface. Estimate the fractional mass loss of the Sun over 10^9 yr due to the emission of neutrinos. The mass of the Sun is 1.989×10^{30} kg. The Earth–Sun distance is equal to 1.496×10^{11} m.

57. Hubble's law can be stated in vector form as $\vec{v} = H\vec{R}$. Outside the local group of galaxies, all objects are moving away from us with velocities proportional to their positions relative to us. In this form, it sounds as if our location in the Universe is specially privileged. Prove that Hubble's law is equally true for an observer elsewhere in the Universe. Proceed as follows. Assume we are at the origin of coordinates, one galaxy cluster is at location $\vec{R}_1$ and has velocity $\vec{v}_1 = H\vec{R}_1$ relative to us, and another galaxy cluster has position vector $\vec{R}_2$ and velocity $\vec{v}_2 = H\vec{R}_2$. Suppose the speeds are nonrelativistic. Consider the frame of reference of an observer in the first of these galaxy clusters. (a) Show that our velocity relative to her, together with the position vector of our galaxy cluster from hers, satisfies Hubble's law. (b) Show that the position and velocity of cluster 2 relative to cluster 1 satisfy Hubble's law.

58. A π^- meson at rest decays according to $\pi^- \rightarrow \mu^- + \bar{\nu}_\mu$. Assume the antineutrino has no mass and moves off with the speed of light. Take $m_\pi c^2 = 139.6$ MeV and $m_\mu c^2 = 105.7$ MeV. What is the energy carried off by the neutrino?

59. **AMT** An unstable particle, initially at rest, decays into a proton (rest energy 938.3 MeV) and a negative pion (rest energy 139.6 MeV). A uniform magnetic field of 0.250 T exists perpendicular to the velocities of the created particles. The radius of curvature of each track is found to be 1.33 m. What is the mass of the original unstable particle?

60. An unstable particle, initially at rest, decays into a positively charged particle of charge $+e$ and rest energy E_+ and a negatively charged particle of charge $-e$ and rest energy E_-. A uniform magnetic field of magnitude B exists perpendicular to the velocities of the created particles. The radius of curvature of each track is r. What is the mass of the original unstable particle?

61. (a) What processes are described by the Feynman diagrams in Figure P46.61? (b) What is the exchanged particle in each process?

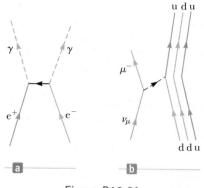

Figure P46.61

62. Identify the mediators for the two interactions described in the Feynman diagrams shown in Figure P46.62.

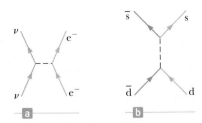

Figure P46.62

63. **Review.** The energy required to excite an atom is on the order of 1 eV. As the temperature of the Universe dropped below a threshold, neutral atoms could form from plasma and the Universe became transparent. Use the Boltzmann distribution function e^{-E/k_BT} to find the order of magnitude of the threshold temperature at which 1.00% of a population of photons has energy greater than 1.00 eV.

64. A Σ^0 particle at rest decays according to $\Sigma^0 \rightarrow \Lambda^0 + \gamma$. Find the gamma-ray energy.

65. Two protons approach each other head-on, each with 70.4 MeV of kinetic energy, and engage in a reaction in which a proton and positive pion emerge at rest. What third particle, obviously uncharged and therefore difficult to detect, must have been created?

66. Two protons approach each other with velocities of equal magnitude in opposite directions. What is the minimum kinetic energy of each proton if the two are to produce a π^+ meson at rest in the reaction $p + p \rightarrow p + n + \pi^+$?

Challenge Problems

67. Determine the kinetic energies of the proton and pion resulting from the decay of a Λ^0 at rest:

$$\Lambda^0 \rightarrow p + \pi^-$$

68. A particle of mass m_1 is fired at a stationary particle of mass m_2, and a reaction takes place in which new particles are created out of the incident kinetic energy. Taken together, the product particles have total mass m_3. The minimum kinetic energy the bombarding particle must have so as to induce the reaction is called the threshold energy. At this energy, the kinetic energy of the products is a minimum, so the fraction of the incident kinetic energy that is available to create new particles is a maximum. This condition is met when all the product particles have the same velocity and the particles have no kinetic energy of motion relative to one another. (a) By using conservation of relativistic energy and momentum and the relativistic energy–momentum relation, show that the threshold kinetic energy is

$$K_{min} = \frac{[m_3^2 - (m_1 + m_2)^2]c^2}{2m_2}$$

Calculate the threshold kinetic energy for each of the following reactions: (b) $p + p \rightarrow p + p + p + \bar{p}$ (one of the initial protons is at rest, and antiprotons are produced); (c) $\pi^- + p \rightarrow K^0 + \Lambda^0$ (the proton is at rest, and strange particles are produced); (d) $p + p \rightarrow p + p + \pi^0$ (one of the initial protons is at rest, and pions are produced); and (e) $p + \bar{p} \rightarrow Z^0$ (one of the initial particles is at rest, and Z^0 particles of mass 91.2 GeV/c^2 are produced).

69. A free neutron beta decays by creating a proton, an electron, and an antineutrino according to the reaction $n \rightarrow p + e^- + \bar{\nu}$. **What If?** Imagine that a free neutron were to decay by creating a proton and electron according to the reaction $n \rightarrow p + e^-$ and assume the neutron is initially at rest in the laboratory. (a) Determine the energy released in this reaction. (b) Energy and momentum are conserved in the reaction. Determine the speeds of the proton and the electron after the reaction. (c) Is either of these particles moving at a relativistic speed? Explain.

70. The cosmic rays of highest energy are mostly protons, accelerated by unknown sources. Their spectrum shows a cutoff at an energy on the order of 10^{20} eV. Above that energy, a proton interacts with a photon of cosmic microwave background radiation to produce mesons, for example, according to $p + \gamma \rightarrow p + \pi^0$. Demonstrate this fact by taking the following steps. (a) Find the minimum photon energy required to produce this reaction in the reference frame where the total momentum of the photon–proton system is zero. The reaction was observed experimentally in the 1950s with photons of a few hundred MeV. (b) Use Wien's displacement law to find the wavelength of a photon at the peak of the blackbody spectrum of the primordial microwave background radiation, with a temperature of 2.73 K. (c) Find the energy of this photon. (d) Consider the reaction in part (a) in a moving reference frame so that the photon is the same as that in part (c). Calculate the energy of the proton in this frame, which represents the Earth reference frame.

71. Assume the average density of the Universe is equal to the critical density. (a) Prove that the age of the Universe is given by $2/(3H)$. (b) Calculate $2/(3H)$ and express it in years.

72. The most recent naked-eye supernova was Supernova Shelton 1987A (Fig. P46.55). It was 170 000 ly away in the Large Magellanic Cloud, a satellite galaxy of the Milky Way. Approximately 3 h before its optical brightening was noticed, two neutrino detection experiments simultaneously registered the first neutrinos from an identified source other than the Sun. The Irvine–Michigan–Brookhaven experiment in a salt mine in Ohio registered eight neutrinos over a 6-s period, and the Kamiokande II experiment in a zinc mine in Japan counted eleven neutrinos in 13 s. (Because the supernova is far south in the sky, these neutrinos entered the detectors from below. They passed through the Earth before they were by chance absorbed by nuclei in the detectors.) The neutrino energies were between approximately 8 MeV and

40 MeV. If neutrinos have no mass, neutrinos of all energies should travel together at the speed of light, and the data are consistent with this possibility. The arrival times could vary simply because neutrinos were created at different moments as the core of the star collapsed into a neutron star. If neutrinos have nonzero mass, lower-energy neutrinos should move comparatively slowly. The data are consistent with a 10-MeV neutrino requiring at most approximately 10 s more than a photon would require to travel from the supernova to us. Find the upper limit that this observation sets on the mass of a neutrino. (Other evidence sets an even tighter limit.)

73. A rocket engine for space travel using photon drive and matter–antimatter annihilation has been suggested. Suppose the fuel for a short-duration burn consists of N protons and N antiprotons, each with mass m. (a) Assume all the fuel is annihilated to produce photons. When the photons are ejected from the rocket, what momentum can be imparted to it? (b) **What If?** If half the protons and antiprotons annihilate each other and the energy released is used to eject the remaining particles, what momentum could be given to the rocket? (c) Which scheme results in the greater change in speed for the rocket?

Tables

Table A.1 Conversion Factors

Length

	m	cm	km	in.	ft	mi
1 meter	1	10^2	10^{-3}	39.37	3.281	6.214×10^{-4}
1 centimeter	10^{-2}	1	10^{-5}	0.393 7	3.281×10^{-2}	6.214×10^{-6}
1 kilometer	10^3	10^5	1	3.937×10^4	3.281×10^3	0.621 4
1 inch	2.540×10^{-2}	2.540	2.540×10^{-5}	1	8.333×10^{-2}	1.578×10^{-5}
1 foot	0.304 8	30.48	3.048×10^{-4}	12	1	1.894×10^{-4}
1 mile	1 609	1.609×10^5	1.609	6.336×10^4	5 280	1

Mass

	kg	g	slug	u
1 kilogram	1	10^3	6.852×10^{-2}	6.024×10^{26}
1 gram	10^{-3}	1	6.852×10^{-5}	6.024×10^{23}
1 slug	14.59	1.459×10^4	1	8.789×10^{27}
1 atomic mass unit	1.660×10^{-27}	1.660×10^{-24}	1.137×10^{-28}	1

Note: 1 metric ton = 1 000 kg.

Time

	s	min	h	day	yr
1 second	1	1.667×10^{-2}	2.778×10^{-4}	1.157×10^{-5}	3.169×10^{-8}
1 minute	60	1	1.667×10^{-2}	6.994×10^{-4}	1.901×10^{-6}
1 hour	3 600	60	1	4.167×10^{-2}	1.141×10^{-4}
1 day	8.640×10^4	1 440	24	1	2.738×10^{-5}
1 year	3.156×10^7	5.259×10^5	8.766×10^3	365.2	1

Speed

	m/s	cm/s	ft/s	mi/h
1 meter per second	1	10^2	3.281	2.237
1 centimeter per second	10^{-2}	1	3.281×10^{-2}	2.237×10^{-2}
1 foot per second	0.304 8	30.48	1	0.681 8
1 mile per hour	0.447 0	44.70	1.467	1

Note: 1 mi/min = 60 mi/h = 88 ft/s.

Force

	N	lb
1 newton	1	0.224 8
1 pound	4.448	1

(*Continued*)

Table A.1   Conversion Factors (continued)

Energy, Energy Transfer

	J	ft · lb	eV
1 joule	1	0.737 6	6.242×10^{18}
1 foot-pound	1.356	1	8.464×10^{18}
1 electron volt	1.602×10^{-19}	1.182×10^{-19}	1
1 calorie	4.186	3.087	2.613×10^{19}
1 British thermal unit	1.055×10^3	7.779×10^2	6.585×10^{21}
1 kilowatt-hour	3.600×10^6	2.655×10^6	2.247×10^{25}

	cal	Btu	kWh
1 joule	0.238 9	9.481×10^{-4}	2.778×10^{-7}
1 foot-pound	0.323 9	1.285×10^{-3}	3.766×10^{-7}
1 electron volt	3.827×10^{-20}	1.519×10^{-22}	4.450×10^{-26}
1 calorie	1	3.968×10^{-3}	1.163×10^{-6}
1 British thermal unit	2.520×10^2	1	2.930×10^{-4}
1 kilowatt-hour	8.601×10^5	3.413×10^2	1

Pressure

	Pa	atm
1 pascal	1	9.869×10^{-6}
1 atmosphere	1.013×10^5	1
1 centimeter mercury[a]	1.333×10^3	1.316×10^{-2}
1 pound per square inch	6.895×10^3	6.805×10^{-2}
1 pound per square foot	47.88	4.725×10^{-4}

	cm Hg	lb/in.^{2}	lb/ft^2
1 pascal	7.501×10^{-4}	1.450×10^{-4}	2.089×10^{-2}
1 atmosphere	76	14.70	2.116×10^3
1 centimeter mercury[a]	1	0.194 3	27.85
1 pound per square inch	5.171	1	144
1 pound per square foot	3.591×10^{-2}	6.944×10^{-3}	1

[a]At 0°C and at a location where the free-fall acceleration has its "standard" value, 9.806 65 m/s^2.

Table A.2   Symbols, Dimensions, and Units of Physical Quantities

Quantity	Common Symbol	Unit[a]	Dimensions[b]	Unit in Terms of Base SI Units
Acceleration	$\vec{a}$	m/s^2	L/T^2	m/s^2
Amount of substance	n	MOLE		mol
Angle	θ, ϕ	radian (rad)	1	
Angular acceleration	$\vec{\alpha}$	rad/s^2	T^{-2}	s^{-2}
Angular frequency	ω	rad/s	T^{-1}	s^{-1}
Angular momentum	$\vec{L}$	kg · m^2/s	ML^2/T	kg · m^2/s
Angular velocity	$\vec{\omega}$	rad/s	T^{-1}	s^{-1}
Area	A	m^2	L^2	m^2
Atomic number	Z			
Capacitance	C	farad (F)	Q^{2}T^2/ML^2	A^2 · s^4/kg · m^2
Charge	q, Q, e	coulomb (C)	Q	A · s

(Continued)

Table A.2   Symbols, Dimensions, and Units of Physical Quantities (continued)

Quantity	Common Symbol	Unit[a]	Dimensions[b]	Unit in Terms of Base SI Units
Charge density				
Line	λ	C/m	Q/L	$A \cdot s/m$
Surface	σ	C/m^2	Q/L^2	$A \cdot s/m^2$
Volume	ρ	C/m^3	Q/L^3	$A \cdot s/m^3$
Conductivity	σ	$1/\Omega \cdot m$	Q^2T/ML^3	$A^2 \cdot s^3/kg \cdot m^3$
Current	I	AMPERE	Q/T	A
Current density	J	A/m^2	Q/TL^2	A/m^2
Density	ρ	kg/m^3	M/L^3	kg/m^3
Dielectric constant	κ			
Electric dipole moment	$\vec{\mathbf{p}}$	$C \cdot m$	QL	$A \cdot s \cdot m$
Electric field	$\vec{\mathbf{E}}$	V/m	ML/QT^2	$kg \cdot m/A \cdot s^3$
Electric flux	Φ_E	$V \cdot m$	ML^3/QT^2	$kg \cdot m^3/A \cdot s^3$
Electromotive force	$\boldsymbol{\varepsilon}$	volt (V)	ML^2/QT^2	$kg \cdot m^2/A \cdot s^3$
Energy	E, U, K	joule (J)	ML^2/T^2	$kg \cdot m^2/s^2$
Entropy	S	J/K	ML^2/T^2K	$kg \cdot m^2/s^2 \cdot K$
Force	$\vec{\mathbf{F}}$	newton (N)	ML/T^2	$kg \cdot m/s^2$
Frequency	f	hertz (Hz)	T^{-1}	s^{-1}
Heat	Q	joule (J)	ML^2/T^2	$kg \cdot m^2/s^2$
Inductance	L	henry (H)	ML^2/Q^2	$kg \cdot m^2/A^2 \cdot s^2$
Length	ℓ, L	METER	L	m
Displacement	$\Delta x, \Delta\vec{\mathbf{r}}$			
Distance	d, h			
Position	$x, y, z, \vec{\mathbf{r}}$			
Magnetic dipole moment	$\vec{\boldsymbol{\mu}}$	$N \cdot m/T$	QL^2/T	$A \cdot m^2$
Magnetic field	$\vec{\mathbf{B}}$	tesla (T) $(= Wb/m^2)$	M/QT	$kg/A \cdot s^2$
Magnetic flux	Φ_B	weber (Wb)	ML^2/QT	$kg \cdot m^2/A \cdot s^2$
Mass	m, M	KILOGRAM	M	kg
Molar specific heat	C	$J/mol \cdot K$		$kg \cdot m^2/s^2 \cdot mol \cdot K$
Moment of inertia	I	$kg \cdot m^2$	ML^2	$kg \cdot m^2$
Momentum	$\vec{\mathbf{p}}$	$kg \cdot m/s$	ML/T	$kg \cdot m/s$
Period	T	s	T	s
Permeability of free space	μ_0	$N/A^2 (= H/m)$	ML/Q^2	$kg \cdot m/A^2 \cdot s^2$
Permittivity of free space	ϵ_0	$C^2/N \cdot m^2 (= F/m)$	Q^2T^2/ML^3	$A^2 \cdot s^4/kg \cdot m^3$
Potential	V	volt (V)$(= J/C)$	ML^2/QT^2	$kg \cdot m^2/A \cdot s^3$
Power	P	watt (W)$(= J/s)$	ML^2/T^3	$kg \cdot m^2/s^3$
Pressure	P	pascal (Pa)$(= N/m^2)$	M/LT^2	$kg/m \cdot s^2$
Resistance	R	ohm $(\Omega)(= V/A)$	ML^2/Q^2T	$kg \cdot m^2/A^2 \cdot s^3$
Specific heat	c	$J/kg \cdot K$	L^2/T^2K	$m^2/s^2 \cdot K$
Speed	v	m/s	L/T	m/s
Temperature	T	KELVIN	K	K
Time	t	SECOND	T	s
Torque	$\vec{\boldsymbol{\tau}}$	$N \cdot m$	ML^2/T^2	$kg \cdot m^2/s^2$
Velocity	$\vec{\mathbf{v}}$	m/s	L/T	m/s
Volume	V	m^3	L^3	m^3
Wavelength	λ	m	L	m
Work	W	joule (J)$(= N \cdot m)$	ML^2/T^2	$kg \cdot m^2/s^2$

[a]The base SI units are given in uppercase letters.

[b]The symbols M, L, T, K, and Q denote mass, length, time, temperature, and charge, respectively.

Mathematics Review

This appendix in mathematics is intended as a brief review of operations and methods. Early in this course, you should be totally familiar with basic algebraic techniques, analytic geometry, and trigonometry. The sections on differential and integral calculus are more detailed and are intended for students who have difficulty applying calculus concepts to physical situations.

B.1 Scientific Notation

Many quantities used by scientists often have very large or very small values. The speed of light, for example, is about 300 000 000 m/s, and the ink required to make the dot over an i in this textbook has a mass of about 0.000 000 001 kg. Obviously, it is very cumbersome to read, write, and keep track of such numbers. We avoid this problem by using a method incorporating powers of the number 10:

$$10^0 = 1$$
$$10^1 = 10$$
$$10^2 = 10 \times 10 = 100$$
$$10^3 = 10 \times 10 \times 10 = 1\ 000$$
$$10^4 = 10 \times 10 \times 10 \times 10 = 10\ 000$$
$$10^5 = 10 \times 10 \times 10 \times 10 \times 10 = 100\ 000$$

and so on. The number of zeros corresponds to the power to which ten is raised, called the **exponent** of ten. For example, the speed of light, 300 000 000 m/s, can be expressed as 3.00×10^8 m/s.

In this method, some representative numbers smaller than unity are the following:

$$10^{-1} = \frac{1}{10} = 0.1$$

$$10^{-2} = \frac{1}{10 \times 10} = 0.01$$

$$10^{-3} = \frac{1}{10 \times 10 \times 10} = 0.001$$

$$10^{-4} = \frac{1}{10 \times 10 \times 10 \times 10} = 0.000\ 1$$

$$10^{-5} = \frac{1}{10 \times 10 \times 10 \times 10 \times 10} = 0.000\ 01$$

In these cases, the number of places the decimal point is to the left of the digit 1 equals the value of the (negative) exponent. Numbers expressed as some power of ten multiplied by another number between one and ten are said to be in **scientific notation.** For example, the scientific notation for 5 943 000 000 is 5.943×10^9 and that for 0.000 083 2 is 8.32×10^{-5}.

When numbers expressed in scientific notation are being multiplied, the following general rule is very useful:

$$10^n \times 10^m = 10^{n+m} \tag{B.1}$$

where n and m can be *any* numbers (not necessarily integers). For example, $10^2 \times 10^5 = 10^7$. The rule also applies if one of the exponents is negative: $10^3 \times 10^{-8} = 10^{-5}$.

When dividing numbers expressed in scientific notation, note that

$$\frac{10^n}{10^m} = 10^n \times 10^{-m} = 10^{n-m} \tag{B.2}$$

Exercises

With help from the preceding rules, verify the answers to the following equations:

1. $86\ 400 = 8.64 \times 10^4$
2. $9\ 816\ 762.5 = 9.816\ 762\ 5 \times 10^6$
3. $0.000\ 000\ 039\ 8 = 3.98 \times 10^{-8}$
4. $(4.0 \times 10^8)(9.0 \times 10^9) = 3.6 \times 10^{18}$
5. $(3.0 \times 10^7)(6.0 \times 10^{-12}) = 1.8 \times 10^{-4}$
6. $\dfrac{75 \times 10^{-11}}{5.0 \times 10^{-3}} = 1.5 \times 10^{-7}$
7. $\dfrac{(3 \times 10^6)(8 \times 10^{-2})}{(2 \times 10^{17})(6 \times 10^5)} = 2 \times 10^{-18}$

B.2 Algebra

Some Basic Rules

When algebraic operations are performed, the laws of arithmetic apply. Symbols such as x, y, and z are usually used to represent unspecified quantities, called the **unknowns.**

First, consider the equation

$$8x = 32$$

If we wish to solve for x, we can divide (or multiply) each side of the equation by the same factor without destroying the equality. In this case, if we divide both sides by 8, we have

$$\frac{8x}{8} = \frac{32}{8}$$

$$x = 4$$

Next consider the equation

$$x + 2 = 8$$

In this type of expression, we can add or subtract the same quantity from each side. If we subtract 2 from each side, we have

$$x + 2 - 2 = 8 - 2$$

$$x = 6$$

In general, if $x + a = b$, then $x = b - a$.

Now consider the equation

$$\frac{x}{5} = 9$$

If we multiply each side by 5, we are left with x on the left by itself and 45 on the right:

$$\left(\frac{x}{5}\right)(5) = 9 \times 5$$

$$x = 45$$

In all cases, *whatever operation is performed on the left side of the equality must also be performed on the right side.*

The following rules for multiplying, dividing, adding, and subtracting fractions should be recalled, where a, b, c, and d are four numbers:

	Rule	Example
Multiplying	$\left(\dfrac{a}{b}\right)\left(\dfrac{c}{d}\right) = \dfrac{ac}{bd}$	$\left(\dfrac{2}{3}\right)\left(\dfrac{4}{5}\right) = \dfrac{8}{15}$
Dividing	$\dfrac{(a/b)}{(c/d)} = \dfrac{ad}{bc}$	$\dfrac{2/3}{4/5} = \dfrac{(2)(5)}{(4)(3)} = \dfrac{10}{12}$
Adding	$\dfrac{a}{b} \pm \dfrac{c}{d} = \dfrac{ad \pm bc}{bd}$	$\dfrac{2}{3} - \dfrac{4}{5} = \dfrac{(2)(5) - (4)(3)}{(3)(5)} = -\dfrac{2}{15}$

Exercises

In the following exercises, solve for x.

Answers

1. $a = \dfrac{1}{1 + x}$ $\qquad x = \dfrac{1 - a}{a}$

2. $3x - 5 = 13$ $\qquad x = 6$

3. $ax - 5 = bx + 2$ $\qquad x = \dfrac{7}{a - b}$

4. $\dfrac{5}{2x + 6} = \dfrac{3}{4x + 8}$ $\qquad x = -\dfrac{11}{7}$

Powers

When powers of a given quantity x are multiplied, the following rule applies:

$$x^n x^m = x^{n+m} \tag{B.3}$$

For example, $x^2 x^4 = x^{2+4} = x^6$.

When dividing the powers of a given quantity, the rule is

$$\frac{x^n}{x^m} = x^{n-m} \tag{B.4}$$

For example, $x^8/x^2 = x^{8-2} = x^6$.

A power that is a fraction, such as $\frac{1}{3}$, corresponds to a root as follows:

$$x^{1/n} = \sqrt[n]{x} \tag{B.5}$$

For example, $4^{1/3} = \sqrt[3]{4} = 1.587\,4$. (A scientific calculator is useful for such calculations.)

Finally, any quantity x^n raised to the mth power is

$$(x^n)^m = x^{nm} \tag{B.6}$$

Table B.1 summarizes the rules of exponents.

Table B.1	**Rules of Exponents**
	$x^0 = 1$
	$x^1 = x$
	$x^n x^m = x^{n+m}$
	$x^n/x^m = x^{n-m}$
	$x^{1/n} = \sqrt[n]{x}$
	$(x^n)^m = x^{nm}$

Exercises

Verify the following equations:

1. $3^2 \times 3^3 = 243$

2. $x^5 x^{-8} = x^{-3}$

3. $x^{10}/x^{-5} = x^{15}$

4. $5^{1/3} = 1.709\ 976$ (Use your calculator.)

5. $60^{1/4} = 2.783\ 158$ (Use your calculator.)

6. $(x^4)^3 = x^{12}$

Factoring

Some useful formulas for factoring an equation are the following:

$$ax + ay + az = a(x + y + z)\quad \text{common factor}$$

$$a^2 + 2ab + b^2 = (a + b)^2 \qquad \text{perfect square}$$

$$a^2 - b^2 = (a + b)(a - b) \qquad \text{differences of squares}$$

Quadratic Equations

The general form of a quadratic equation is

$$ax^2 + bx + c = 0 \tag{B.7}$$

where x is the unknown quantity and a, b, and c are numerical factors referred to as **coefficients** of the equation. This equation has two roots, given by

$$x = \frac{-b \pm \sqrt{b^2 - 4ac}}{2a} \tag{B.8}$$

If $b^2 \geq 4ac$, the roots are real.

Example B.1

The equation $x^2 + 5x + 4 = 0$ has the following roots corresponding to the two signs of the square-root term:

$$x = \frac{-5 \pm \sqrt{5^2 - (4)(1)(4)}}{2(1)} = \frac{-5 \pm \sqrt{9}}{2} = \frac{-5 \pm 3}{2}$$

$$x_+ = \frac{-5 + 3}{2} = -1 \qquad x_- = \frac{-5 - 3}{2} = -4$$

where x_+ refers to the root corresponding to the positive sign and x_- refers to the root corresponding to the negative sign.

Exercises

Solve the following quadratic equations:

Answers

1. $x^2 + 2x - 3 = 0$ $x_+ = 1$ $x_- = -3$

2. $2x^2 - 5x + 2 = 0$ $x_+ = 2$ $x_- = \frac{1}{2}$

3. $2x^2 - 4x - 9 = 0$ $x_+ = 1 + \sqrt{22}/2$ $x_- = 1 - \sqrt{22}/2$

Linear Equations

A linear equation has the general form

$$y = mx + b \tag{B.9}$$

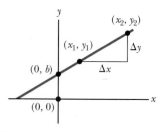

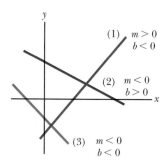

Figure B.1 A straight line graphed on an *xy* coordinate system. The slope of the line is the ratio of Δy to Δx.

Figure B.2 The brown line has a positive slope and a negative *y*-intercept. The blue line has a negative slope and a positive *y*-intercept. The green line has a negative slope and a negative *y*-intercept.

where *m* and *b* are constants. This equation is referred to as linear because the graph of *y* versus *x* is a straight line as shown in Figure B.1. The constant *b*, called the **y-intercept,** represents the value of *y* at which the straight line intersects the *y* axis. The constant *m* is equal to the **slope** of the straight line. If any two points on the straight line are specified by the coordinates (x_1, y_1) and (x_2, y_2) as in Figure B.1, the slope of the straight line can be expressed as

$$\text{Slope} = \frac{y_2 - y_1}{x_2 - x_1} = \frac{\Delta y}{\Delta x} \tag{B.10}$$

Note that *m* and *b* can have either positive or negative values. If $m > 0$, the straight line has a *positive* slope as in Figure B.1. If $m < 0$, the straight line has a *negative* slope. In Figure B.1, both *m* and *b* are positive. Three other possible situations are shown in Figure B.2.

Exercises

1. Draw graphs of the following straight lines: (a) $y = 5x + 3$ (b) $y = -2x + 4$ (c) $y = -3x - 6$
2. Find the slopes of the straight lines described in Exercise 1.

Answers   (a) 5   (b) −2   (c) −3

3. Find the slopes of the straight lines that pass through the following sets of points: (a) $(0, -4)$ and $(4, 2)$ (b) $(0, 0)$ and $(2, -5)$ (c) $(-5, 2)$ and $(4, -2)$

Answers   (a) $\frac{3}{2}$ (b) $-\frac{5}{2}$ (c) $-\frac{4}{9}$

Solving Simultaneous Linear Equations

Consider the equation $3x + 5y = 15$, which has two unknowns, *x* and *y*. Such an equation does not have a unique solution. For example, $(x = 0, y = 3)$, $(x = 5, y = 0)$, and $(x = 2, y = \frac{9}{5})$ are all solutions to this equation.

If a problem has two unknowns, a unique solution is possible only if we have *two* pieces of information. In most common cases, those two pieces of information are equations. In general, if a problem has *n* unknowns, its solution requires *n* equations. To solve two simultaneous equations involving two unknowns, *x* and *y*, we solve one of the equations for *x* in terms of *y* and substitute this expression into the other equation.

In some cases, the two pieces of information may be (1) one equation and (2) a condition on the solutions. For example, suppose we have the equation $m = 3n$ and the condition that *m* and *n* must be the smallest positive nonzero integers possible. Then, the single equation does not allow a unique solution, but the addition of the condition gives us that $n = 1$ and $m = 3$.

Example B.2

Solve the two simultaneous equations

$$(1) \quad 5x + y = -8$$

$$(2) \quad 2x - 2y = 4$$

SOLUTION

From Equation (2), $x = y + 2$. Substitution of this equation into Equation (1) gives

$$5(y + 2) + y = -8$$

$$6y = -18$$

▶ **B.2** c o n t i n u e d

$$y = \boxed{-3}$$
$$x = y + 2 = \boxed{-1}$$

Alternative Solution Multiply each term in Equation (1) by the factor 2 and add the result to Equation (2):

$$10x + 2y = -16$$
$$\underline{2x - 2y = 4}$$
$$12x \qquad = -12$$
$$x = \boxed{-1}$$
$$y = x - 2 = \boxed{-3}$$

Two linear equations containing two unknowns can also be solved by a graphical method. If the straight lines corresponding to the two equations are plotted in a conventional coordinate system, the intersection of the two lines represents the solution. For example, consider the two equations

$$x - y = 2$$
$$x - 2y = -1$$

These equations are plotted in Figure B.3. The intersection of the two lines has the coordinates $x = 5$ and $y = 3$, which represents the solution to the equations. You should check this solution by the analytical technique discussed earlier.

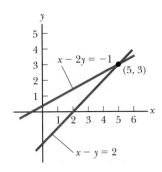

Figure B.3 A graphical solution for two linear equations.

Exercises

Solve the following pairs of simultaneous equations involving two unknowns:

	Answers
1. $x + y = 8$ $x - y = 2$	$x = 5, y = 3$
2. $98 - T = 10a$ $T - 49 = 5a$	$T = 65, a = 3.27$
3. $6x + 2y = 6$ $8x - 4y = 28$	$x = 2, y = -3$

Logarithms

Suppose a quantity x is expressed as a power of some quantity a:

$$x = a^y \qquad \text{(B.11)}$$

The number a is called the **base** number. The **logarithm** of x with respect to the base a is equal to the exponent to which the base must be raised to satisfy the expression $x = a^y$:

$$y = \log_a x \qquad \text{(B.12)}$$

Conversely, the **antilogarithm** of y is the number x:

$$x = \text{antilog}_a y \qquad \text{(B.13)}$$

In practice, the two bases most often used are base 10, called the *common* logarithm base, and base $e = 2.718\ 282$, called Euler's constant or the *natural* logarithm base. When common logarithms are used,

$$y = \log_{10} x \quad (\text{or } x = 10^y) \qquad \text{(B.14)}$$

When natural logarithms are used,

$$y = \ln x \quad (\text{or } x = e^y) \tag{B.15}$$

For example, $\log_{10} 52 = 1.716$, so antilog$_{10}$ $1.716 = 10^{1.716} = 52$. Likewise, $\ln 52 = 3.951$, so antiln $3.951 = e^{3.951} = 52$.

In general, note you can convert between base 10 and base e with the equality

$$\ln x = (2.302\,585)\,\log_{10} x \tag{B.16}$$

Finally, some useful properties of logarithms are the following:

$$\left.\begin{array}{l} \log(ab) = \log a + \log b \\[4pt] \log(a/b) = \log a - \log b \\[4pt] \log(a^n) = n \log a \end{array}\right\} \text{ any base}$$

$$\ln e = 1$$

$$\ln e^a = a$$

$$\ln\left(\frac{1}{a}\right) = -\ln a$$

B.3 Geometry

The **distance** d between two points having coordinates (x_1, y_1) and (x_2, y_2) is

$$d = \sqrt{(x_2 - x_1)^2 + (y_2 - y_1)^2} \tag{B.17}$$

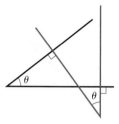

Figure B.4 The angles are equal because their sides are perpendicular.

Two angles are equal if their sides are perpendicular, right side to right side and left side to left side. For example, the two angles marked θ in Figure B.4 are the same because of the perpendicularity of the sides of the angles. To distinguish the left and right sides of an angle, imagine standing at the angle's apex and facing into the angle.

Radian measure: The arc length s of a circular arc (Fig. B.5) is proportional to the radius r for a fixed value of θ (in radians):

$$s = r\theta$$
$$\theta = \frac{s}{r} \tag{B.18}$$

Figure B.5 The angle θ in radians is the ratio of the arc length s to the radius r of the circle.

Table B.2 gives the **areas** and **volumes** for several geometric shapes used throughout this text.

The equation of a **straight line** (Fig. B.6) is

$$y = mx + b \tag{B.19}$$

where b is the y-intercept and m is the slope of the line.

The equation of a **circle** of radius R centered at the origin is

$$x^2 + y^2 = R^2 \tag{B.20}$$

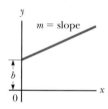

Figure B.6 A straight line with a slope of m and a y-intercept of b.

The equation of an **ellipse** having the origin at its center (Fig. B.7) is

$$\frac{x^2}{a^2} + \frac{y^2}{b^2} = 1 \tag{B.21}$$

where a is the length of the semimajor axis (the longer one) and b is the length of the semiminor axis (the shorter one).

The equation of a **parabola** the vertex of which is at $y = b$ (Fig. B.8) is

$$y = ax^2 + b \tag{B.22}$$

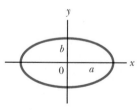

Figure B.7 An ellipse with semimajor axis a and semiminor axis b.

Table B.2　Useful Information for Geometry

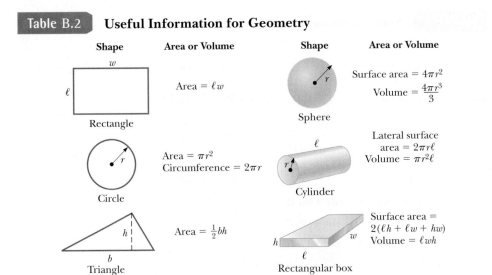

Shape	Area or Volume	Shape	Area or Volume
Rectangle	Area $= \ell w$	Sphere	Surface area $= 4\pi r^2$ Volume $= \dfrac{4\pi r^3}{3}$
Circle	Area $= \pi r^2$ Circumference $= 2\pi r$	Cylinder	Lateral surface area $= 2\pi r \ell$ Volume $= \pi r^2 \ell$
Triangle	Area $= \frac{1}{2}bh$	Rectangular box	Surface area $= 2(\ell h + \ell w + hw)$ Volume $= \ell wh$

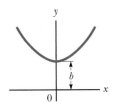

Figure B.8 A parabola with its vertex at $y = b$.

The equation of a **rectangular hyperbola** (Fig. B.9) is

$$xy = \text{constant} \tag{B.23}$$

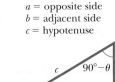

Figure B.9 A hyperbola.

B.4　Trigonometry

That portion of mathematics based on the special properties of the right triangle is called trigonometry. By definition, a right triangle is a triangle containing a $90°$ angle. Consider the right triangle shown in Figure B.10, where side a is opposite the angle θ, side b is adjacent to the angle θ, and side c is the hypotenuse of the triangle. The three basic trigonometric functions defined by such a triangle are the sine (sin), cosine (cos), and tangent (tan). In terms of the angle θ, these functions are defined as follows:

$$\sin \theta = \frac{\text{side opposite } \theta}{\text{hypotenuse}} = \frac{a}{c} \tag{B.24}$$

$$\cos \theta = \frac{\text{side adjacent to } \theta}{\text{hypotenuse}} = \frac{b}{c} \tag{B.25}$$

$$\tan \theta = \frac{\text{side opposite } \theta}{\text{side adjacent to } \theta} = \frac{a}{b} \tag{B.26}$$

The Pythagorean theorem provides the following relationship among the sides of a right triangle:

$$c^2 = a^2 + b^2 \tag{B.27}$$

From the preceding definitions and the Pythagorean theorem, it follows that

$$\sin^2 \theta + \cos^2 \theta = 1$$

$$\tan \theta = \frac{\sin \theta}{\cos \theta}$$

The cosecant, secant, and cotangent functions are defined by

$$\csc \theta = \frac{1}{\sin \theta} \qquad \sec \theta = \frac{1}{\cos \theta} \qquad \cot \theta = \frac{1}{\tan \theta}$$

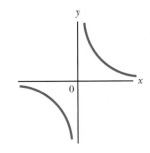

$a =$ opposite side
$b =$ adjacent side
$c =$ hypotenuse

Figure B.10 A right triangle, used to define the basic functions of trigonometry.

Table B.3	Some Trigonometric Identities

$$\sin^2 \theta + \cos^2 \theta = 1 \qquad\qquad \csc^2 \theta = 1 + \cot^2 \theta$$

$$\sec^2 \theta = 1 + \tan^2 \theta \qquad\qquad \sin^2 \frac{\theta}{2} = \tfrac{1}{2}(1 - \cos \theta)$$

$$\sin 2\theta = 2 \sin \theta \cos \theta \qquad\qquad \cos^2 \frac{\theta}{2} = \tfrac{1}{2}(1 + \cos \theta)$$

$$\cos 2\theta = \cos^2 \theta - \sin^2 \theta \qquad\qquad 1 - \cos \theta = 2 \sin^2 \frac{\theta}{2}$$

$$\tan 2\theta = \frac{2 \tan \theta}{1 - \tan^2 \theta} \qquad\qquad \tan \frac{\theta}{2} = \sqrt{\frac{1 - \cos \theta}{1 + \cos \theta}}$$

$$\sin (A \pm B) = \sin A \cos B \pm \cos A \sin B$$

$$\cos (A \pm B) = \cos A \cos B \mp \sin A \sin B$$

$$\sin A \pm \sin B = 2 \sin \left[\tfrac{1}{2}(A \pm B)\right] \cos \left[\tfrac{1}{2}(A \mp B)\right]$$

$$\cos A + \cos B = 2 \cos \left[\tfrac{1}{2}(A + B)\right] \cos \left[\tfrac{1}{2}(A - B)\right]$$

$$\cos A - \cos B = 2 \sin \left[\tfrac{1}{2}(A + B)\right] \sin \left[\tfrac{1}{2}(B - A)\right]$$

The following relationships are derived directly from the right triangle shown in Figure B.10:

$$\sin \theta = \cos (90° - \theta)$$
$$\cos \theta = \sin (90° - \theta)$$
$$\cot \theta = \tan (90° - \theta)$$

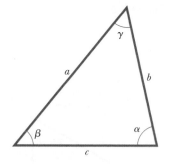

Figure B.11 An arbitrary, non-right triangle.

Some properties of trigonometric functions are the following:

$$\sin (-\theta) = -\sin \theta$$
$$\cos (-\theta) = \cos \theta$$
$$\tan (-\theta) = -\tan \theta$$

The following relationships apply to *any* triangle as shown in Figure B.11:

$$\alpha + \beta + \gamma = 180°$$

$$\text{Law of cosines} \begin{cases} a^2 = b^2 + c^2 - 2bc \cos \alpha \\ b^2 = a^2 + c^2 - 2ac \cos \beta \\ c^2 = a^2 + b^2 - 2ab \cos \gamma \end{cases}$$

$$\text{Law of sines} \qquad \frac{a}{\sin \alpha} = \frac{b}{\sin \beta} = \frac{c}{\sin \gamma}$$

Table B.3 lists a number of useful trigonometric identities.

Example B.3

Consider the right triangle in Figure B.12 in which $a = 2.00$, $b = 5.00$, and c is unknown. From the Pythagorean theorem, we have

$$c^2 = a^2 + b^2 = 2.00^2 + 5.00^2 = 4.00 + 25.0 = 29.0$$

$$c = \sqrt{29.0} = \boxed{5.39}$$

To find the angle θ, note that

$$\tan \theta = \frac{a}{b} = \frac{2.00}{5.00} = 0.400$$

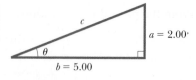

Figure B.12 (Example B.3)

▶ **B.3** continued

Using a calculator, we find that

$$\theta = \tan^{-1}(0.400) = \boxed{21.8°}$$

where $\tan^{-1}(0.400)$ is the notation for "angle whose tangent is 0.400," sometimes written as arctan (0.400).

Exercises

1. In Figure B.13, identify (a) the side opposite θ (b) the side adjacent to ϕ and then find (c) $\cos\theta$, (d) $\sin\phi$, and (e) $\tan\phi$.

Answers   (a) 3   (b) 3   (c) $\frac{4}{5}$ (d) $\frac{4}{5}$ (e) $\frac{4}{3}$

2. In a certain right triangle, the two sides that are perpendicular to each other are 5.00 m and 7.00 m long. What is the length of the third side?

Answer   8.60 m

3. A right triangle has a hypotenuse of length 3.0 m, and one of its angles is 30°. (a) What is the length of the side opposite the 30° angle? (b) What is the side adjacent to the 30° angle?

Answers   (a) 1.5 m   (b) 2.6 m

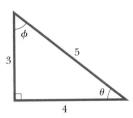

Figure B.13 (Exercise 1)

B.5 Series Expansions

$$(a+b)^n = a^n + \frac{n}{1!}a^{n-1}b + \frac{n(n-1)}{2!}a^{n-2}b^2 + \cdots$$

$$(1+x)^n = 1 + nx + \frac{n(n-1)}{2!}x^2 + \cdots$$

$$e^x = 1 + x + \frac{x^2}{2!} + \frac{x^3}{3!} + \cdots$$

$$\ln(1\pm x) = \pm x - \tfrac{1}{2}x^2 \pm \tfrac{1}{3}x^3 - \cdots$$

$$\left.\begin{array}{l}
\sin x = x - \dfrac{x^3}{3!} + \dfrac{x^5}{5!} - \cdots \\[2mm]
\cos x = 1 - \dfrac{x^2}{2!} + \dfrac{x^4}{4!} - \cdots \\[2mm]
\tan x = x + \dfrac{x^3}{3} + \dfrac{2x^5}{15} + \cdots \quad |x| < \dfrac{\pi}{2}
\end{array}\right\} x \text{ in radians}$$

For $x \ll 1$, the following approximations can be used:[1]

$$(1+x)^n \approx 1 + nx \qquad \sin x \approx x$$
$$e^x \approx 1 + x \qquad\qquad \cos x \approx 1$$
$$\ln(1\pm x) \approx \pm x \qquad \tan x \approx x$$

B.6 Differential Calculus

In various branches of science, it is sometimes necessary to use the basic tools of calculus, invented by Newton, to describe physical phenomena. The use of calculus is fundamental in the treatment of various problems in Newtonian mechanics, electricity, and magnetism. In this section, we simply state some basic properties and "rules of thumb" that should be a useful review to the student.

---

[1]The approximations for the functions $\sin x$, $\cos x$, and $\tan x$ are for $x \leq 0.1$ rad.

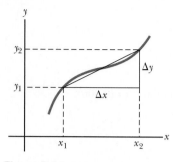

Figure B.14 The lengths Δx and Δy are used to define the derivative of this function at a point.

First, a **function** must be specified that relates one variable to another (e.g., a coordinate as a function of time). Suppose one of the variables is called y (the dependent variable), and the other x (the independent variable). We might have a function relationship such as

$$y(x) = ax^3 + bx^2 + cx + d$$

If a, b, c, and d are specified constants, y can be calculated for any value of x. We usually deal with continuous functions, that is, those for which y varies "smoothly" with x.

The **derivative** of y with respect to x is defined as the limit as Δx approaches zero of the slopes of chords drawn between two points on the y versus x curve. Mathematically, we write this definition as

$$\frac{dy}{dx} = \lim_{\Delta x \to 0} \frac{\Delta y}{\Delta x} = \lim_{\Delta x \to 0} \frac{y(x + \Delta x) - y(x)}{\Delta x} \tag{B.28}$$

where Δy and Δx are defined as $\Delta x = x_2 - x_1$ and $\Delta y = y_2 - y_1$ (Fig. B.14). Note that dy/dx does *not* mean dy divided by dx, but rather is simply a notation of the limiting process of the derivative as defined by Equation B.28.

A useful expression to remember when $y(x) = ax^n$, where a is a *constant* and n is *any* positive or negative number (integer or fraction), is

$$\frac{dy}{dx} = nax^{n-1} \tag{B.29}$$

If $y(x)$ is a polynomial or algebraic function of x, we apply Equation B.29 to *each* term in the polynomial and take $d[\text{constant}]/dx = 0$. In Examples B.4 through B.7, we evaluate the derivatives of several functions.

Special Properties of the Derivative

A. Derivative of the product of two functions If a function $f(x)$ is given by the product of two functions—say, $g(x)$ and $h(x)$—the derivative of $f(x)$ is defined as

$$\frac{d}{dx}f(x) = \frac{d}{dx}[g(x)h(x)] = g\frac{dh}{dx} + h\frac{dg}{dx} \tag{B.30}$$

B. Derivative of the sum of two functions If a function $f(x)$ is equal to the sum of two functions, the derivative of the sum is equal to the sum of the derivatives:

$$\frac{d}{dx}f(x) = \frac{d}{dx}[g(x) + h(x)] = \frac{dg}{dx} + \frac{dh}{dx} \tag{B.31}$$

C. Chain rule of differential calculus If $y = f(x)$ and $x = g(z)$, then dy/dz can be written as the product of two derivatives:

$$\frac{dy}{dz} = \frac{dy}{dx}\frac{dx}{dz} \tag{B.32}$$

D. The second derivative The second derivative of y with respect to x is defined as the derivative of the function dy/dx (the derivative of the derivative). It is usually written as

$$\frac{d^2y}{dx^2} = \frac{d}{dx}\left(\frac{dy}{dx}\right) \tag{B.33}$$

Some of the more commonly used derivatives of functions are listed in Table B.4.

Table B.4 **Derivative for Several Functions**

$$\frac{d}{dx}(a) = 0$$

$$\frac{d}{dx}(ax^n) = nax^{n-1}$$

$$\frac{d}{dx}(e^{ax}) = ae^{ax}$$

$$\frac{d}{dx}(\sin ax) = a\cos ax$$

$$\frac{d}{dx}(\cos ax) = -a\sin ax$$

$$\frac{d}{dx}(\tan ax) = a\sec^2 ax$$

$$\frac{d}{dx}(\cot ax) = -a\csc^2 ax$$

$$\frac{d}{dx}(\sec x) = \tan x\sec x$$

$$\frac{d}{dx}(\csc x) = -\cot x\csc x$$

$$\frac{d}{dx}(\ln ax) = \frac{1}{x}$$

$$\frac{d}{dx}(\sin^{-1} ax) = \frac{a}{\sqrt{1 - a^2x^2}}$$

$$\frac{d}{dx}(\cos^{-1} ax) = \frac{-a}{\sqrt{1 - a^2x^2}}$$

$$\frac{d}{dx}(\tan^{-1} ax) = \frac{a}{1 + a^2x^2}$$

Note: The symbols a and n represent constants.

Example B.4

Suppose $y(x)$ (that is, y as a function of x) is given by

$$y(x) = ax^3 + bx + c$$

where a and b are constants. It follows that

$$y(x + \Delta x) = a(x + \Delta x)^3 + b(x + \Delta x) + c$$
$$= a(x^3 + 3x^2\,\Delta x + 3x\,\Delta x^2 + \Delta x^3) + b(x + \Delta x) + c$$

so

$$\Delta y = y(x + \Delta x) - y(x) = a(3x^2\,\Delta x + 3x\,\Delta x^2 + \Delta x^3) + b\Delta x$$

Substituting this into Equation B.28 gives

$$\frac{dy}{dx} = \lim_{\Delta x \to 0} \frac{\Delta y}{\Delta x} = \lim_{\Delta x \to 0} \left[3ax^2 + 3ax\,\Delta x + a\,\Delta x^2 \right] + b$$

$$\frac{dy}{dx} = \boxed{3ax^2 + b}$$

Example B.5

Find the derivative of

$$y(x) = 8x^5 + 4x^3 + 2x + 7$$

SOLUTION

Applying Equation B.29 to each term independently and remembering that d/dx (constant) $= 0$, we have

$$\frac{dy}{dx} = 8(5)x^4 + 4(3)x^2 + 2(1)x^0 + 0$$

$$\frac{dy}{dx} = \boxed{40x^4 + 12x^2 + 2}$$

Example B.6

Find the derivative of $y(x) = x^3/(x + 1)^2$ with respect to x.

SOLUTION

We can rewrite this function as $y(x) = x^3(x + 1)^{-2}$ and apply Equation B.30:

$$\frac{dy}{dx} = (x + 1)^{-2} \frac{d}{dx}(x^3) + x^3 \frac{d}{dx}(x + 1)^{-2}$$

$$= (x + 1)^{-2}\,3x^2 + x^3\,(-2)(x + 1)^{-3}$$

$$\frac{dy}{dx} = \boxed{\frac{3x^2}{(x + 1)^2} - \frac{2x^3}{(x + 1)^3} = \frac{x^2(x + 3)}{(x + 1)^3}}$$

Example B.7

A useful formula that follows from Equation B.30 is the derivative of the quotient of two functions. Show that

$$\frac{d}{dx}\left[\frac{g(x)}{h(x)}\right] = \frac{h\dfrac{dg}{dx} - g\dfrac{dh}{dx}}{h^2}$$

SOLUTION

We can write the quotient as gh^{-1} and then apply Equations B.29 and B.30:

$$\frac{d}{dx}\left(\frac{g}{h}\right) = \frac{d}{dx}(gh^{-1}) = g\frac{d}{dx}(h^{-1}) + h^{-1}\frac{d}{dx}(g)$$

$$= -gh^{-2}\frac{dh}{dx} + h^{-1}\frac{dg}{dx}$$

$$= \frac{h\dfrac{dg}{dx} - g\dfrac{dh}{dx}}{h^2}$$

B.7 Integral Calculus

We think of integration as the inverse of differentiation. As an example, consider the expression

$$f(x) = \frac{dy}{dx} = 3ax^2 + b \tag{B.34}$$

which was the result of differentiating the function

$$y(x) = ax^3 + bx + c$$

in Example B.4. We can write Equation B.34 as $dy = f(x)\,dx = (3ax^2 + b)\,dx$ and obtain $y(x)$ by "summing" over all values of x. Mathematically, we write this inverse operation as

$$y(x) = \int f(x)\,dx$$

For the function $f(x)$ given by Equation B.34, we have

$$y(x) = \int (3ax^2 + b)\,dx = ax^3 + bx + c$$

where c is a constant of the integration. This type of integral is called an *indefinite integral* because its value depends on the choice of c.

A general **indefinite integral** $I(x)$ is defined as

$$I(x) = \int f(x)\,dx \tag{B.35}$$

where $f(x)$ is called the *integrand* and $f(x) = dI(x)/dx$.

For a *general continuous* function $f(x)$, the integral can be interpreted geometrically as the area under the curve bounded by $f(x)$ and the x axis, between two specified values of x, say, x_1 and x_2, as in Figure B.15.

The area of the blue element in Figure B.15 is approximately $f(x_i)\,\Delta x_i$. If we sum all these area elements between x_1 and x_2 and take the limit of this sum as $\Delta x_i \to 0$,

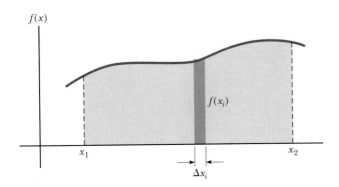

Figure B.15 The definite integral of a function is the area under the curve of the function between the limits x_1 and x_2.

we obtain the *true* area under the curve bounded by $f(x)$ and the x axis, between the limits x_1 and x_2:

$$\text{Area} = \lim_{\Delta x_i \to 0} \sum_i f(x_i)\Delta x_i = \int_{x_1}^{x_2} f(x)\, dx \qquad \text{(B.36)}$$

Integrals of the type defined by Equation B.36 are called **definite integrals.**

One common integral that arises in practical situations has the form

$$\int x^n\, dx = \frac{x^{n+1}}{n+1} + c \quad (n \ne -1) \qquad \text{(B.37)}$$

This result is obvious, being that differentiation of the right-hand side with respect to x gives $f(x) = x^n$ directly. If the limits of the integration are known, this integral becomes a *definite integral* and is written

$$\int_{x_1}^{x_2} x^n\, dx = \frac{x^{n+1}}{n+1}\Bigg|_{x_1}^{x_2} = \frac{x_2^{n+1} - x_1^{n+1}}{n+1} \quad (n \ne -1) \qquad \text{(B.38)}$$

Examples

1. $\displaystyle\int_0^a x^2\, dx = \frac{x^3}{3}\Bigg]_0^a = \frac{a^3}{3}$

2. $\displaystyle\int_0^b x^{3/2}\, dx = \frac{x^{5/2}}{5/2}\Bigg]_0^b = \tfrac{2}{5} b^{5/2}$

3. $\displaystyle\int_3^5 x\, dx = \frac{x^2}{2}\Bigg]_3^5 = \frac{5^2 - 3^2}{2} = 8$

Partial Integration

Sometimes it is useful to apply the method of *partial integration* (also called "integrating by parts") to evaluate certain integrals. This method uses the property

$$\int u\, dv = uv - \int v\, du \qquad \text{(B.39)}$$

where u and v are *carefully* chosen so as to reduce a complex integral to a simpler one. In many cases, several reductions have to be made. Consider the function

$$I(x) = \int x^2 e^x\, dx$$

which can be evaluated by integrating by parts twice. First, if we choose $u = x^2$, $v = e^x$, we obtain

$$\int x^2 e^x\, dx = \int x^2\, d(e^x) = x^2 e^x - 2\int e^x x\, dx + c_1$$

Now, in the second term, choose $u = x$, $v = e^x$, which gives

$$\int x^2 e^x \, dx = x^2 e^x - 2x e^x + 2\int e^x \, dx + c_1$$

or

$$\int x^2 e^x \, dx = x^2 e^x - 2xe^x + 2e^x + c_2$$

The Perfect Differential

Another useful method to remember is that of the *perfect differential*, in which we look for a change of variable such that the differential of the function is the differential of the independent variable appearing in the integrand. For example, consider the integral

$$I(x) = \int \cos^2 x \sin x \, dx$$

This integral becomes easy to evaluate if we rewrite the differential as $d(\cos x) = -\sin x \, dx$. The integral then becomes

$$\int \cos^2 x \sin x \, dx = -\int \cos^2 x \, d(\cos x)$$

If we now change variables, letting $y = \cos x$, we obtain

$$\int \cos^2 x \sin x \, dx = -\int y^2 \, dy = -\frac{y^3}{3} + c = -\frac{\cos^3 x}{3} + c$$

Table B.5 lists some useful indefinite integrals. Table B.6 gives Gauss's probability integral and other definite integrals. A more complete list can be found in various handbooks, such as *The Handbook of Chemistry and Physics* (Boca Raton, FL: CRC Press, published annually).

Table B.5   **Some Indefinite Integrals (An arbitrary constant should be added to each of these integrals.)**

$$\int x^n \, dx = \frac{x^{n+1}}{n+1} \text{ (provided } n \neq 1)$$

$$\int \ln ax \, dx = (x \ln ax) - x$$

$$\int \frac{dx}{x} = \int x^{-1} \, dx = \ln x$$

$$\int xe^{ax} \, dx = \frac{e^{ax}}{a^2}(ax - 1)$$

$$\int \frac{dx}{a + bx} = \frac{1}{b} \ln (a + bx)$$

$$\int \frac{dx}{a + be^{cx}} = \frac{x}{a} - \frac{1}{ac} \ln (a + be^{cx})$$

$$\int \frac{x \, dx}{a + bx} = \frac{x}{b} - \frac{a}{b^2} \ln (a + bx)$$

$$\int \sin ax \, dx = -\frac{1}{a} \cos ax$$

$$\int \frac{dx}{x(x + a)} = -\frac{1}{a} \ln \frac{x + a}{x}$$

$$\int \cos ax \, dx = \frac{1}{a} \sin ax$$

$$\int \frac{dx}{(a + bx)^2} = -\frac{1}{b(a + bx)}$$

$$\int \tan ax \, dx = -\frac{1}{a} \ln (\cos ax) = \frac{1}{a} \ln (\sec ax)$$

$$\int \frac{dx}{a^2 + x^2} = \frac{1}{a} \tan^{-1} \frac{x}{a}$$

$$\int \cot ax \, dx = \frac{1}{a} \ln (\sin ax)$$

$$\int \frac{dx}{a^2 - x^2} = \frac{1}{2a} \ln \frac{a + x}{a - x} \, (a^2 - x^2 > 0)$$

$$\int \sec ax \, dx = \frac{1}{a} \ln (\sec ax + \tan ax) = \frac{1}{a} \ln \left[\tan \left(\frac{ax}{2} + \frac{\pi}{4} \right) \right]$$

$$\int \frac{dx}{x^2 - a^2} = \frac{1}{2a} \ln \frac{x - a}{x + a} \, (x^2 - a^2 > 0)$$

$$\int \csc ax \, dx = \frac{1}{a} \ln (\csc ax - \cot ax) = \frac{1}{a} \ln \left(\tan \frac{ax}{2} \right)$$

(Continued)

Table B.5   Some Indefinite Integrals (continued)

$$\int \frac{x\,dx}{a^2 \pm x^2} = \pm\tfrac{1}{2}\ln(a^2 \pm x^2)$$

$$\int \sin^2 ax\,dx = \frac{x}{2} - \frac{\sin 2ax}{4a}$$

$$\int \frac{dx}{\sqrt{a^2 - x^2}} = \sin^{-1}\frac{x}{a} = -\cos^{-1}\frac{x}{a}\ (a^2 - x^2 > 0)$$

$$\int \cos^2 ax\,dx = \frac{x}{2} + \frac{\sin 2ax}{4a}$$

$$\int \frac{dx}{\sqrt{x^2 \pm a^2}} = \ln(x + \sqrt{x^2 \pm a^2})$$

$$\int \frac{dx}{\sin^2 ax} = -\frac{1}{a}\cot ax$$

$$\int \frac{x\,dx}{\sqrt{a^2 - x^2}} = -\sqrt{a^2 - x^2}$$

$$\int \frac{dx}{\cos^2 ax} = \frac{1}{a}\tan ax$$

$$\int \frac{x\,dx}{\sqrt{x^2 \pm a^2}} = \sqrt{x^2 \pm a^2}$$

$$\int \tan^2 ax\,dx = \frac{1}{a}(\tan ax) - x$$

$$\int \sqrt{a^2 - x^2}\,dx = \tfrac{1}{2}\left(x\sqrt{a^2 - x^2} + a^2\sin^{-1}\frac{x}{|a|}\right)$$

$$\int \cot^2 ax\,dx = -\frac{1}{a}(\cot ax) - x$$

$$\int x\sqrt{a^2 - x^2}\,dx = -\tfrac{1}{3}(a^2 - x^2)^{3/2}$$

$$\int \sin^{-1} ax\,dx = x(\sin^{-1} ax) + \frac{\sqrt{1 - a^2 x^2}}{a}$$

$$\int \sqrt{x^2 \pm a^2}\,dx = \tfrac{1}{2}x\sqrt{x^2 \pm a^2} \pm a^2\ln(x + \sqrt{x^2 \pm a^2})$$

$$\int \cos^{-1} ax\,dx = x(\cos^{-1} ax) - \frac{\sqrt{1 - a^2 x^2}}{a}$$

$$\int x(\sqrt{x^2 \pm a^2})\,dx = \tfrac{1}{3}(x^2 \pm a^2)^{3/2}$$

$$\int \frac{dx}{(x^2 + a^2)^{3/2}} = \frac{x}{a^2\sqrt{x^2 + a^2}}$$

$$\int e^{ax}\,dx = \frac{1}{a}e^{ax}$$

$$\int \frac{x\,dx}{(x^2 + a^2)^{3/2}} = -\frac{1}{\sqrt{x^2 + a^2}}$$

Table B.6   Gauss's Probability Integral and Other Definite Integrals

$$\int_0^\infty x^n e^{-ax}\,dx = \frac{n!}{a^{n+1}}$$

$$I_0 = \int_0^\infty e^{-ax^2}\,dx = \frac{1}{2}\sqrt{\frac{\pi}{a}} \quad \text{(Gauss's probability integral)}$$

$$I_1 = \int_0^\infty xe^{-ax^2}\,dx = \frac{1}{2a}$$

$$I_2 = \int_0^\infty x^2 e^{-ax^2}\,dx = -\frac{dI_0}{da} = \frac{1}{4}\sqrt{\frac{\pi}{a^3}}$$

$$I_3 = \int_0^\infty x^3 e^{-ax^2}\,dx = -\frac{dI_1}{da} = \frac{1}{2a^2}$$

$$I_4 = \int_0^\infty x^4 e^{-ax^2}\,dx = \frac{d^2 I_0}{da^2} = \frac{3}{8}\sqrt{\frac{\pi}{a^5}}$$

$$I_5 = \int_0^\infty x^5 e^{-ax^2}\,dx = \frac{d^2 I_1}{da^2} = \frac{1}{a^3}$$

$$\vdots$$

$$I_{2n} = (-1)^n \frac{d^n}{da^n} I_0$$

$$I_{2n+1} = (-1)^n \frac{d^n}{da^n} I_1$$

B.8 Propagation of Uncertainty

In laboratory experiments, a common activity is to take measurements that act as raw data. These measurements are of several types—length, time interval, temperature, voltage, and so on—and are taken by a variety of instruments. Regardless of the measurement and the quality of the instrumentation, **there is always uncertainty associated with a physical measurement.** This uncertainty is a combination of that associated with the instrument and that related to the system being measured. An example of the former is the inability to exactly determine the position of a length measurement between the lines on a meterstick. An example of uncertainty related to the system being measured is the variation of temperature within a sample of water so that a single temperature for the sample is difficult to determine.

Uncertainties can be expressed in two ways. **Absolute uncertainty** refers to an uncertainty expressed in the same units as the measurement. Therefore, the length of a computer disk label might be expressed as (5.5 ± 0.1) cm. The uncertainty of ± 0.1 cm by itself is not descriptive enough for some purposes, however. This uncertainty is large if the measurement is 1.0 cm, but it is small if the measurement is 100 m. To give a more descriptive account of the uncertainty, **fractional uncertainty** or **percent uncertainty** is used. In this type of description, the uncertainty is divided by the actual measurement. Therefore, the length of the computer disk label could be expressed as

$$\ell = 5.5 \text{ cm} \pm \frac{0.1 \text{ cm}}{5.5 \text{ cm}} = 5.5 \text{ cm} \pm 0.018 \quad \text{(fractional uncertainty)}$$

or as

$$\ell = 5.5 \text{ cm} \pm 1.8\% \quad \text{(percent uncertainty)}$$

When combining measurements in a calculation, the percent uncertainty in the final result is generally larger than the uncertainty in the individual measurements. This is called **propagation of uncertainty** and is one of the challenges of experimental physics.

Some simple rules can provide a reasonable estimate of the uncertainty in a calculated result:

Multiplication and division: When measurements with uncertainties are multiplied or divided, add the *percent uncertainties* to obtain the percent uncertainty in the result.

Example: The Area of a Rectangular Plate

$$A = \ell w = (5.5 \text{ cm} \pm 1.8\%) \times (6.4 \text{ cm} \pm 1.6\%) = 35 \text{ cm}^2 \pm 3.4\%$$
$$= (35 \pm 1) \text{ cm}^2$$

Addition and subtraction: When measurements with uncertainties are added or subtracted, add the *absolute uncertainties* to obtain the absolute uncertainty in the result.

Example: A Change in Temperature

$$\Delta T = T_2 - T_1 = (99.2 \pm 1.5)°C - (27.6 \pm 1.5)°C = (71.6 \pm 3.0)°C$$
$$= 71.6°C \pm 4.2\%$$

Powers: If a measurement is taken to a power, the percent uncertainty is multiplied by that power to obtain the percent uncertainty in the result.

Example: The Volume of a Sphere

$$V = \tfrac{4}{3}\pi r^3 = \tfrac{4}{3}\pi(6.20 \text{ cm} \pm 2.0\%)^3 = 998 \text{ cm}^3 \pm 6.0\%$$

$$= (998 \pm 60) \text{ cm}^3$$

For complicated calculations, many uncertainties are added together, which can cause the uncertainty in the final result to be undesirably large. Experiments should be designed such that calculations are as simple as possible.

Notice that uncertainties in a calculation always add. As a result, an experiment involving a subtraction should be avoided if possible, especially if the measurements being subtracted are close together. The result of such a calculation is a small difference in the measurements and uncertainties that add together. It is possible that the uncertainty in the result could be larger than the result itself!

Periodic Table of the Elements

Group I	Group II					Transition elements			

H 1 1.007 9 $1s$									
Li 3 6.941 $2s^1$	**Be** 4 9.0122 $2s^2$								
Na 11 22.990 $3s^1$	**Mg** 12 24.305 $3s^2$								
K 19 39.098 $4s^1$	**Ca** 20 40.078 $4s^2$	**Sc** 21 44.956 $3d^14s^2$	**Ti** 22 47.867 $3d^24s^2$	**V** 23 50.942 $3d^34s^2$	**Cr** 24 51.996 $3d^54s^1$	**Mn** 25 54.938 $3d^54s^2$	**Fe** 26 55.845 $3d^64s^2$	**Co** 27 58.933 $3d^74s^2$	
Rb 37 85.468 $5s^1$	**Sr** 38 87.62 $5s^2$	**Y** 39 88.906 $4d^15s^2$	**Zr** 40 91.224 $4d^25s^2$	**Nb** 41 92.906 $4d^45s^1$	**Mo** 42 95.94 $4d^55s^1$	**Tc** 43 (98) $4d^55s^2$	**Ru** 44 101.07 $4d^75s^1$	**Rh** 45 102.91 $4d^85s^1$	
Cs 55 132.91 $6s^1$	**Ba** 56 137.33 $6s^2$	57–71*	**Hf** 72 178.49 $5d^26s^2$	**Ta** 73 180.95 $5d^36s^2$	**W** 74 183.84 $5d^46s^2$	**Re** 75 186.21 $5d^56s^2$	**Os** 76 190.23 $5d^66s^2$	**Ir** 77 192.2 $5d^76s^2$	
Fr 87 (223) $7s^1$	**Ra** 88 (226) $7s^2$	89–103**	**Rf** 104 (261) $6d^27s^2$	**Db** 105 (262) $6d^37s^2$	**Sg** 106 (266)	**Bh** 107 (264)	**Hs** 108 (277)	**Mt** 109 (268)	

Symbol — **Ca** 20 — **Atomic number**
Atomic mass† — 40.078
$4s^2$ — **Electron configuration**

*Lanthanide series

La 57 138.91 $5d^16s^2$	**Ce** 58 140.12 $5d^14f^16s^2$	**Pr** 59 140.91 $4f^36s^2$	**Nd** 60 144.24 $4f^46s^2$	**Pm** 61 (145) $4f^56s^2$	**Sm** 62 150.36 $4f^66s^2$

**Actinide series

Ac 89 (227) $6d^17s^2$	**Th** 90 232.04 $6d^27s^2$	**Pa** 91 231.04 $5f^26d^17s^2$	**U** 92 238.03 $5f^36d^17s^2$	**Np** 93 (237) $5f^46d^17s^2$	**Pu** 94 (244) $5f^67s^2$

Note: Atomic mass values given are averaged over isotopes in the percentages in which they exist in nature.
†For an unstable element, mass number of the most stable known isotope is given in parentheses.

Group III	Group IV	Group V	Group VI	Group VII	Group 0
				H 1 $\quad$ 1.007 9 $\quad$ $1s^1$	**He** 2 $\quad$ 4.002 6 $\quad$ $1s^2$
B 5 $\quad$ 10.811 $\quad$ $2p^1$	**C** 6 $\quad$ 12.011 $\quad$ $2p^2$	**N** 7 $\quad$ 14.007 $\quad$ $2p^3$	**O** 8 $\quad$ 15.999 $\quad$ $2p^4$	**F** 9 $\quad$ 18.998 $\quad$ $2p^5$	**Ne** 10 $\quad$ 20.180 $\quad$ $2p^6$
Al 13 $\quad$ 26.982 $\quad$ $3p^1$	**Si** 14 $\quad$ 28.086 $\quad$ $3p^2$	**P** 15 $\quad$ 30.974 $\quad$ $3p^3$	**S** 16 $\quad$ 32.066 $\quad$ $3p^4$	**Cl** 17 $\quad$ 35.453 $\quad$ $3p^5$	**Ar** 18 $\quad$ 39.948 $\quad$ $3p^6$

Ni	Cu	Zn	Group III	Group IV	Group V	Group VI	Group VII	Group 0
Ni 28 $\quad$ 58.693 $\quad$ $3d^84s^2$	**Cu** 29 $\quad$ 63.546 $\quad$ $3d^{10}4s^1$	**Zn** 30 $\quad$ 65.41 $\quad$ $3d^{10}4s^2$	**Ga** 31 $\quad$ 69.723 $\quad$ $4p^1$	**Ge** 32 $\quad$ 72.64 $\quad$ $4p^2$	**As** 33 $\quad$ 74.922 $\quad$ $4p^3$	**Se** 34 $\quad$ 78.96 $\quad$ $4p^4$	**Br** 35 $\quad$ 79.904 $\quad$ $4p^5$	**Kr** 36 $\quad$ 83.80 $\quad$ $4p^6$
Pd 46 $\quad$ 106.42 $\quad$ $4d^{10}$	**Ag** 47 $\quad$ 107.87 $\quad$ $4d^{10}5s^1$	**Cd** 48 $\quad$ 112.41 $\quad$ $4d^{10}5s^2$	**In** 49 $\quad$ 114.82 $\quad$ $5p^1$	**Sn** 50 $\quad$ 118.71 $\quad$ $5p^2$	**Sb** 51 $\quad$ 121.76 $\quad$ $5p^3$	**Te** 52 $\quad$ 127.60 $\quad$ $5p^4$	**I** 53 $\quad$ 126.90 $\quad$ $5p^5$	**Xe** 54 $\quad$ 131.29 $\quad$ $5p^6$
Pt 78 $\quad$ 195.08 $\quad$ $5d^96s^1$	**Au** 79 $\quad$ 196.97 $\quad$ $5d^{10}6s^1$	**Hg** 80 $\quad$ 200.59 $\quad$ $5d^{10}6s^2$	**Tl** 81 $\quad$ 204.38 $\quad$ $6p^1$	**Pb** 82 $\quad$ 207.2 $\quad$ $6p^2$	**Bi** 83 $\quad$ 208.98 $\quad$ $6p^3$	**Po** 84 $\quad$ (209) $\quad$ $6p^4$	**At** 85 $\quad$ (210) $\quad$ $6p^5$	**Rn** 86 $\quad$ (222) $\quad$ $6p^6$
Ds 110 $\quad$ (271)	**Rg** 111 $\quad$ (272)	**Cn** 112 $\quad$ (285)	113†† $\quad$ (284)	**Fl** 114 $\quad$ (289)	115†† $\quad$ (288)	**Lv** 116 $\quad$ (293)	117†† $\quad$ (294)	118†† $\quad$ (294)

Eu	Gd	Tb	Dy	Ho	Er	Tm	Yb	Lu
Eu 63 $\quad$ 151.96 $\quad$ $4f^76s^2$	**Gd** 64 $\quad$ 157.25 $\quad$ $4f^75d^16s^2$	**Tb** 65 $\quad$ 158.93 $\quad$ $4f^85d^16s^2$	**Dy** 66 $\quad$ 162.50 $\quad$ $4f^{10}6s^2$	**Ho** 67 $\quad$ 164.93 $\quad$ $4f^{11}6s^2$	**Er** 68 $\quad$ 167.26 $\quad$ $4f^{12}6s^2$	**Tm** 69 $\quad$ 168.93 $\quad$ $4f^{13}6s^2$	**Yb** 70 $\quad$ 173.04 $\quad$ $4f^{14}6s^2$	**Lu** 71 $\quad$ 174.97 $\quad$ $4f^{14}5d^16s^2$
Am 95 $\quad$ (243) $\quad$ $5f^77s^2$	**Cm** 96 $\quad$ (247) $\quad$ $5f^76d^17s^2$	**Bk** 97 $\quad$ (247) $\quad$ $5f^86d^17s^2$	**Cf** 98 $\quad$ (251) $\quad$ $5f^{10}7s^2$	**Es** 99 $\quad$ (252) $\quad$ $5f^{11}7s^2$	**Fm** 100 $\quad$ (257) $\quad$ $5f^{12}7s^2$	**Md** 101 $\quad$ (258) $\quad$ $5f^{13}7s^2$	**No** 102 $\quad$ (259) $\quad$ $5f^{14}7s^2$	**Lr** 103 $\quad$ (262) $\quad$ $5f^{14}6d^17s^2$

†† Elements 113, 115, 117, and 118 have not yet been officially named. Only small numbers of atoms of these elements have been observed.

Note: For a description of the atomic data, visit *physics.nist.gov/PhysRefData/Elements/per_text.html*.

APPENDIX

D

SI Units

Table D.1 SI Units

Base Quantity	SI Base Unit	
	Name	Symbol
Length	meter	m
Mass	kilogram	kg
Time	second	s
Electric current	ampere	A
Temperature	kelvin	K
Amount of substance	mole	mol
Luminous intensity	candela	cd

Table D.2 Some Derived SI Units

Other Quantity	Name	Symbol	Expression in Terms of Base Units	Expression in Terms of SI Units
Plane angle	radian	rad	m/m	
Frequency	hertz	Hz	s^{-1}	
Force	newton	N	$kg \cdot m/s^2$	J/m
Pressure	pascal	Pa	$kg/m \cdot s^2$	N/m^2
Energy	joule	J	$kg \cdot m^2/s^2$	$N \cdot m$
Power	watt	W	$kg \cdot m^2/s^3$	J/s
Electric charge	coulomb	C	$A \cdot s$	
Electric potential	volt	V	$kg \cdot m^2/A \cdot s^3$	W/A
Capacitance	farad	F	$A^2 \cdot s^4/kg \cdot m^2$	C/V
Electric resistance	ohm	Ω	$kg \cdot m^2/A^2 \cdot s^3$	V/A
Magnetic flux	weber	Wb	$kg \cdot m^2/A \cdot s^2$	$V \cdot s$
Magnetic field	tesla	T	$kg/A \cdot s^2$	
Inductance	henry	H	$kg \cdot m^2/A^2 \cdot s^2$	$T \cdot m^2/A$

Answers to Quick Quizzes and Odd-Numbered Problems

Chapter 23

Answers to Quick Quizzes

1. (a), (c), (e)
2. (e)
3. (b)
4. (a)
5. A, B, C

Answers to Odd-Numbered Problems

1. (a) $+1.60 \times 10^{-19}$ C, 1.67×10^{-27} kg
   (b) $+1.60 \times 10^{-19}$ C, 3.82×10^{-26} kg
   (c) -1.60×10^{-19} C, 5.89×10^{-26} kg
   (d) $+3.20 \times 10^{-19}$ C, 6.65×10^{-26} kg
   (e) -4.80×10^{-19} C, 2.33×10^{-26} kg
   (f) $+6.40 \times 10^{-19}$ C, 2.33×10^{-26} kg
   (g) $+1.12 \times 10^{-18}$ C, 2.33×10^{-26} kg
   (h) -1.60×10^{-19} C, 2.99×10^{-26} kg
3. 57.5 N
5. 3.60×10^6 N downward
7. 2.25×10^{-9} N/m
9. (a) 8.74×10^{-8} N   (b) repulsive
11. (a) 1.38×10^{-5} N   (b) $77.5°$ below the negative x axis
13. (a) 0.951 m   (b) yes, if the third bead has positive charge
15. 0.872 N at $330°$
17. (a) 8.24×10^{-8} N   (b) 2.19×10^6 m/s
19. $k_e \dfrac{Q^2}{d^2} \left[\dfrac{1}{2\sqrt{2}}\,\hat{\mathbf{i}} + \left(2 - \dfrac{1}{2\sqrt{2}}\right)\hat{\mathbf{j}} \right]$
21. (a) 2.16×10^{-5} N toward the other   (b) 8.99×10^{-7} N away from the other
23. (a) $-(5.58 \times 10^{-11}\,\text{N/C})\hat{\mathbf{j}}$ (b) $(1.02 \times 10^{-7}\,\text{N/C})\hat{\mathbf{j}}$
25. (a) $\dfrac{k_e q}{a^2}(3.06\hat{\mathbf{i}} + 5.06\hat{\mathbf{j}})$ (b) $\dfrac{k_e q^2}{a^2}(3.06\hat{\mathbf{i}} + 5.06\hat{\mathbf{j}})$
27. (a) $k_e \dfrac{Q}{d^2}[(1 - \sqrt{2})\,\hat{\mathbf{i}} + \sqrt{2}\,\hat{\mathbf{j}}]$

   (b) $-k_e \dfrac{Q}{4d^2}[(1 + 4\sqrt{2})\,\hat{\mathbf{i}} + 4\sqrt{2}\,\hat{\mathbf{j}}]$
29. 1.82 m to the left of the -2.50-μC charge
31. (a) 1.80×10^4 N/C to the right   (b) 8.98×10^{-5} N to the left
33. 5.25 μC
35. (a) $(-0.599\hat{\mathbf{i}} - 2.70\hat{\mathbf{j}})$ kN/C   (b) $(-3.00\hat{\mathbf{i}} - 13.5\hat{\mathbf{j}})\,\mu$N
37. (a) 1.59×10^6 N/C   (b) toward the rod
39. (a) 6.64×10^6 N/C away from the center of the ring
   (b) 2.41×10^7 N/C away from the center of the ring

(c) 6.39×10^6 N/C away from the center of the ring
(d) 6.64×10^5 N/C away from the center of the ring
41. (a) 9.35×10^7 N/C   (b) 1.04×10^8 N/C (about 11% higher)
   (c) 5.15×10^5 N/C   (d) 5.19×10^5 N/C (about 0.7% higher)
43. (a) $k_e \dfrac{\lambda_0}{x_0}$ (b) to the left
45. (a) 2.16×10^7 N/C   (b) to the left
47.

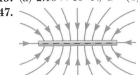

49. (a) $-\frac{1}{3}$ (b) q_1 is negative, and q_2 is positive.
51. (a) 6.13×10^{10} m/s^2 (b) 1.96×10^{-5} s   (c) 11.7 m
   (d) 1.20×10^{-15} J
53. 4.38×10^6 m/s for the electron; 2.39×10^3 m/s for the proton
55. (a) $\dfrac{K}{ed}$ (b) in the direction of the velocity of the electron
57. (a) 111 ns   (b) 5.68 mm   (c) $(450\hat{\mathbf{i}} + 102\hat{\mathbf{j}})$ km/s
59. $-\dfrac{\pi^2 k_e q}{6a^2}\,\hat{\mathbf{i}}$
61. (a) $\dfrac{mg}{|Q|}\sin\theta$ (b) 3.19×10^3 N/C down the incline
63. $-\dfrac{k_e \lambda_0}{2x_0}\,\hat{\mathbf{i}}$
65. (a) 2.18×10^{-5} m   (b) 2.43 cm
67. (a) 1.09×10^{-8} C   (b) 5.44×10^{-3} N
69. (a) $24.2\hat{\mathbf{i}}$ N/C   (b) $(-4.21\hat{\mathbf{i}} + 8.42\hat{\mathbf{j}})$ N/C
71. $-0.706\hat{\mathbf{i}}$ N
73. 25.9 cm
75. 1.67×10^{-5} C
77. 1.98 μC
79. 1.14×10^{-7} C on one sphere and 5.69×10^{-8} C on the other
81. (a) $\theta_1 = \theta_2$
83. (a) 0.307 s   (b) Yes; the downward gravitational force is not negligible in this situation, so the tension in the string depends on both the gravitational force and the electric force.
85. (a) $F_x = F_y = F_z = 1.90 k_e \dfrac{q^2}{s^2}$ (b) $3.29 k_e \dfrac{q^2}{s^2}$ (c) away from the origin
89. $(-1.36\hat{\mathbf{i}} + 1.96\hat{\mathbf{j}})$ kN/C

91. (a) $\vec{\mathbf{E}} = \dfrac{935x}{(0.062\,5 + x^2)^{3/2}}\hat{\mathbf{i}}$ where $\vec{\mathbf{E}}$ is in newtons per coulomb and x is in meters   (b) $4.00\hat{\mathbf{i}}$ kN/C (c) $x = 0.016\,8$ m and $x = 0.916$ m   (d) nowhere is the field as large as 16 000 N/C

Chapter 24

Answers to Quick Quizzes

1. (e)
2. (b) and (d)
3. (a)

Answers to Odd-Numbered Problems

1. (a) 1.98×10^6 N · m^2/C   (b) 0
3. 4.14 MN/C
5. (a) 858 N · m^2/C   (b) 0   (c) 657 N · m^2/C
7. 28.2 N · m^2/C
9. (a) -6.89 MN · m^2/C   (b) less than
11. $-Q/\epsilon_0$ for S_1; 0 for S_2; $-2Q/\epsilon_0$ for S_3; 0 for S_4
13. 1.77×10^{-12} C/m^3; positive
15. (a) 339 N · m^2/C   (b) No. The electric field is not uniform on this surface. Gauss's law is only practical to use when all portions of the surface satisfy one or more of the conditions listed in Section 24.3.
17. (a) 0   (b) $\dfrac{2\lambda}{\epsilon_0}\sqrt{R^2 - d^2}$
19. -18.8 kN · m^2/C
21. (a) $\dfrac{Q}{2\epsilon_0}$ (b) $-\dfrac{Q}{2\epsilon_0}$
23. 3.50 kN
25. $-2.48\ \mu$C/m^2
27. 508 kN/C up
29. (a) 0   (b) 7.19 MN/C away from the center
31. (a) 51.4 kN/C outward   (b) 645 N · m^2/C
33. $\vec{\mathbf{E}} = \rho r/2\epsilon_0 = 2\pi k_e \rho r$ away from the axis
35. (a) 0   (b) 3.65×10^5 N/C   (c) 1.46×10^6 N/C   (d) 6.49×10^5 N/C
37. (a) 0   (b) 5.39×10^3 N/C outward   (c) 539 N/C outward
39. $\dfrac{\sigma}{\epsilon_0}$
41. $E_{\text{glass}} = E_{\text{Al}}$
43. 2.00 N
45. (a) $-\lambda$ (b) $+3\lambda$ (c) $6k_e \dfrac{\lambda}{r}$ radially outward
47. (a) 0   (b) 7.99×10^7 N/C (outward)   (c) 0   (d) 7.34×10^6 N/C (outward)
49. 0.438 N · m^2/C
51. 8.27×10^5 N · m^2/C
53. (a) $\dfrac{\sigma}{2\epsilon_0}\hat{\mathbf{k}}$ (b) $\dfrac{3\sigma}{2\epsilon_0}\hat{\mathbf{k}}$ (c) $-\dfrac{\sigma}{2\epsilon_0}\hat{\mathbf{k}}$
55.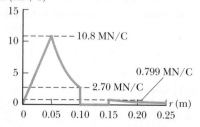

57. (a) -4.01 nC   (b) $+9.57$ nC   (c) $+4.01$ nC   (d) $+5.56$ nC
59. $\dfrac{\sigma}{2\epsilon_0}$ radially outward
61. (a) $E = \dfrac{Cd^3}{24\epsilon_0}$ to the right for $x > d/2$ and to the left for $x < -d/2$ (b) $\vec{\mathbf{E}} = \dfrac{Cx^3}{3\epsilon_0}\hat{\mathbf{i}}$
63. (a) 0.269 N · m^2/C   (b) 2.38×10^{-12} C
65. $\dfrac{a}{2\epsilon_0}$ radially outward
67. (a) $\dfrac{\rho_0 r}{2\epsilon_0}\left(a - \dfrac{2r}{3b}\right)$ (b) $\dfrac{\rho_0 R^2}{2\epsilon_0 r}\left(a - \dfrac{2R}{3b}\right)$

Chapter 25

Answers to Quick Quizzes

1. (i) (b)   (ii) (a)
2. Ⓑ to Ⓒ, Ⓒ to Ⓓ, Ⓐ to Ⓑ, Ⓓ to Ⓔ
3. (i) (c)   (ii) (a)
4. (i) (a)   (ii) (a)

Answers to Odd-Numbered Problems

1. (a) 1.13×10^5 N/C   (b) 1.80×10^{-14} N   (c) 4.37×10^{-17} J
3. (a) 1.52×10^5 m/s   (b) 6.49×10^6 m/s
5. $+260$ V
7. (a) -38.9 V   (b) the origin
9. 0.300 m/s
11. (a) 0.400 m/s   (b) It is the same. Each bit of the rod feels a force of the same size as before.
13. (a) 2.12×10^6 V   (b) 1.21×10^6 V
15. $6.93k_e\dfrac{Q}{d}$
17. (a) $-45.0\ \mu$V   (b) 34.6 km/s
19. (a) 0   (b) 0   (c) 44.9 kV
21. (a) $4\sqrt{2}k_e\dfrac{Q}{a}$ (b) $4\sqrt{2}k_e\dfrac{qQ}{a}$
23. (a) -4.83 m   (b) 0.667 m and -2.00 m
25. (a) 32.2 kV   (b) $-0.096\,5$ J
27. 8.94 J
29. $-5k_e\dfrac{q}{R}$
31. (a) 10.8 m/s and 1.55 m/s   (b) They would be greater. The conducting spheres will polarize each other, with most of the positive charge of one and the negative charge of the other on their inside faces. Immediately before the spheres collide, their centers of charge will be closer than their geometric centers, so they will have less electric potential energy and more kinetic energy.
33. $22.8k_e\dfrac{q^2}{s}$
35. 2.74×10^{-14} m $= 27.4$ fm
37. (a) 10.0 V, -11.0 V, -32.0 V   (b) 7.00 N/C in the positive x direction
39. (a) $\vec{\mathbf{E}} = (-5 + 6xy)\hat{\mathbf{i}} + (3x^2 - 2z^2)\hat{\mathbf{j}} - 4yz\hat{\mathbf{k}}$ (b) 7.07 N/C
41. (a) 0   (b) $k_e\dfrac{Q}{r^2}$
43. $-0.553k_e\dfrac{Q}{R}$

45. (a) $\dfrac{C}{m^2}$ (b) $k_e\alpha\left[L - d\,\ln\left(1 + \dfrac{L}{d}\right)\right]$

47. $k_e\lambda(\pi + 2\ln 3)$

49. 1.56×10^{12}

51. (a) 1.35×10^5 V (b) larger sphere: 2.25×10^6 V/m (away from the center); smaller sphere: 6.74×10^6 V/m (away from the center)

53. Because n is not an integer, this is not possible. Therefore, the energy given cannot be possible for an allowed state of the atom.

55. (a) $6.00\hat{\mathbf{i}}$ m/s (b) 3.64 m (c) $-9.00\hat{\mathbf{i}}$ m/s (d) $12.0\hat{\mathbf{i}}$ m/s

57. 253 MeV

59. (a) 30.0 cm (b) 6.67 nC (c) 29.1 cm or 3.44 cm (d) 6.79 nC or 804 pC (e) No; two answers exist for each part.

61. 702 J

63. 4.00 nC at $(-1.00$ m, $0)$ and 5.01 nC at $(0, 2.00$ m$)$

65. $-\dfrac{\lambda}{2\pi\epsilon_0}\ln\left(\dfrac{r_2}{r_1}\right)$

67. $k_e\lambda\ln\left[\dfrac{a + L + \sqrt{(a + L)^2 + b^2}}{a + \sqrt{a^2 + b^2}}\right]$

69. (a) 4.07 kV/m (b) 488 V (c) 7.82×10^{-17} J (d) 306 km/s (e) 3.89×10^{11} m/s^2 toward the negative plate (f) 6.51×10^{-16} N toward the negative plate (g) 4.07 kV/m (h) They are the same.

71. (b) $E_r = \dfrac{2k_e p\cos\theta}{r^3}$, $E_\theta = \dfrac{k_e p\sin\theta}{r^3}$ (c) yes (d) no

(e) $V = \dfrac{k_e py}{(x^2 + y^2)^{3/2}}$

(f) $E_x = \dfrac{3k_e pxy}{(x^2 + y^2)^{5/2}}$, $E_y = \dfrac{k_e p(2y^2 - x^2)}{(x^2 + y^2)^{5/2}}$

73. $\pi k_e C\left[R\sqrt{R^2 + x^2} + x^2\ln\left(\dfrac{x}{R + \sqrt{R^2 + x^2}}\right)\right]$

75. (a) $\dfrac{k_e Q}{h}\ln\left[\dfrac{d + h + \sqrt{(d + h)^2 + R^2}}{d + \sqrt{d^2 + R^2}}\right]$

(b) $\dfrac{k_e Q}{R^2 h}\left[(d + h)\sqrt{(d + h)^2 + R^2} - d\sqrt{d^2 + R^2} - 2dh - h^2 + R^2\ln\left(\dfrac{d + h + \sqrt{(d + h)^2}}{d + \sqrt{d^2 + R^2}}\right)\right]$

Chapter 26

Answers to Quick Quizzes

  1. (d)
  2. (a)
  3. (a)
  4. (b)
  5. (a)

Answers to Odd-Numbered Problems

  1. (a) 9.00 V (b) 12.0 V
  3. (a) $48.0\ \mu$C (b) $6.00\ \mu$C
  5. (a) 2.69 nF (b) 3.02 kV
  7. $4.43\ \mu$m
  9. (a) 11.1 kV/m toward the negative plate (b) 98.4 nC/m^2 (c) 3.74 pF (d) 74.8 pC

11. (a) $1.33\ \mu$C/m^2 (b) 13.4 pF

13. (a) $17.0\ \mu$F (b) 9.00 V (c) $45.0\ \mu$C on $5\ \mu$F, $108\ \mu$C on $12\ \mu$F

15. (a) $2.81\ \mu$F (b) $12.7\ \mu$F

17. (a) in series (b) $398\ \mu$F (c) in parallel; $2.20\ \mu$F

19. (a) $3.33\ \mu$F (b) $180\ \mu$C on the $3.00\text{-}\mu$F and $6.00\text{-}\mu$F capacitors; $120\ \mu$C on the $2.00\text{-}\mu$F and $4.00\text{-}\mu$F capacitors (c) 60.0 V across the $3.00\text{-}\mu$F and $2.00\text{-}\mu$F capacitors; 30.0 V across the $6.00\text{-}\mu$F and $4.00\text{-}\mu$F capacitors

21. ten

23. (a) $5.96\ \mu$F (b) $89.5\ \mu$C on $20\ \mu$F, $63.2\ \mu$C on $6\ \mu$F, and $26.3\ \mu$C on $15\ \mu$F and $3\ \mu$F

25. $12.9\ \mu$F

27. 6.00 pF and 3.00 pF

29. $19.8\ \mu$C

31. 3.24×10^{-4} J

33. (a) $1.50\ \mu$C (b) 1.83 kV

35. (a) 2.50×10^{-2} J (b) 66.7 V (c) 3.33×10^{-2} J (d) Positive work is done by the agent pulling the plates apart.

37. (a)

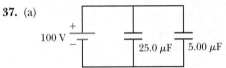

(b) 0.150 J (c) 268 V

(d)

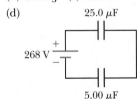

39. 9.79 kg

41. (a) $400\ \mu$C (b) 2.5 kN/m

43. (a) 13.3 nC (b) 272 nC

45. (a) 81.3 pF (b) 2.40 kV

47. (a) 369 pC (b) 1.2×10^{-10} F, 3.1 V (c) -45.5 nJ

49. (a) $40.0\ \mu$J (b) 500 V

51. $-9.43 \times 10^{-2}\,\hat{\mathbf{i}}$ N

55. (a) 11.2 pF (b) 134 pC (c) 16.7 pF (d) 67.0 pC

57. 2.51×10^{-3} m^3 = 2.51 L

59. 0.188 m^2

61. (a) volume 9.09×10^{-16} m^3, area 4.54×10^{-10} m^2 (b) 2.01×10^{-13} F (c) 2.01×10^{-14} C; 1.26×10^5 electronic charges

63. 23.3 V across the $5.00\text{-}\mu$F capacitor, 26.7 V across the $10.0\text{-}\mu$F capacitor

65. (a) $\dfrac{Q_0^2 d(\ell - x)}{2\epsilon_0\ell^3}$ (b) $\dfrac{Q_0^2 d}{2\epsilon_0\ell^3}$ to the right (c) $\dfrac{Q_0^2}{2\epsilon_0\ell^4}$

(d) $\dfrac{Q_0^2}{2\epsilon_0\ell^4}$ (e) They are precisely the same.

67. $4.29\ \mu$F

69. $750\ \mu$C on C_1, $250\ \mu$C on C_2

71. (a) One capacitor cannot be used by itself—it would burn out. The technician can use two capacitors in series, connected in parallel to another two capacitors in series. Another possibility is two capacitors in parallel, connected in series to another two capacitors in parallel. In either case, one capacitor will be left over: upper and lower (b) Each of the four capacitors will be exposed to a maximum voltage of 45 V.

73. $\dfrac{C_0}{2}(\sqrt{3}-1)$

75. $\frac{4}{3}C$

77. 3.00 μF

Chapter 27

Answers to Quick Quizzes

1. (a) > (b) = (c) > (d)

2. (b)

3. (b)

4. (a)

5. $I_a = I_b > I_c = I_d > I_e = I_f$

Answers to Odd-Numbered Problems

1. 27.0 yr

3. 0.129 mm/s

5. 1.79×10^{16} protons

7. (a) $0.632 I_0 \tau$ (b) $0.999\,95 I_0 \tau$ (c) $I_0 \tau$

9. (a) 17.0 A   (b) 85.0 kA/m^2

11. (a) 2.55 A/m^2 (b) 5.30×10^{10} m^{-3} (c) 1.21×10^{10} s

13. 3.64 h

15. silver ($\rho = 1.59 \times 10^{-8}\ \Omega \cdot$ m)

17. 8.89 Ω

19. (a) 1.82 m   (b) 280 μm

21. (a) 13.0 Ω (b) 255 m

23. $6.00 \times 10^{-15}(\Omega \cdot$ m)$^{-1}$

25. 0.18 V/m

27. 0.12

29. 6.32 Ω

31. (a) 3.0 A   (b) 2.9 A

33. (a) 31.5 n$\Omega \cdot$ m   (b) 6.35 MA/m^2 (c) 49.9 mA
(d) 658 μm/s   (e) 0.400 V

35. 227°C

37. 448 A

39. (a) 8.33 A   (b) 14.4 Ω

41. 2.1 W

43. 36.1%

45. (a) 0.660 kWh   (b) \$0.072 6

47. \$0.494/day

49. (a) 3.98 V/m   (b) 49.7 W   (c) 44.1 W

51. (a) 4.75 m   (b) 340 W

53. (a) 184 W   (b) 461°C

55. 672 s

57. 1.1 km

59. 15.0 h

61. 50.0 MW

63. (a) $\dfrac{Q}{4C}$ (b) $\dfrac{Q}{4}$ on C, $\dfrac{3Q}{4}$ on $3C$

(c) $\dfrac{Q^2}{32C}$ in C, $\dfrac{3Q^2}{32C}$ in $3C$ (d) $\dfrac{3Q^2}{8C}$

65. 0.478 kg/s

67. (a) 8.00 V/m in the positive x direction   (b) 0.637 Ω
(c) 6.28 A in the positive x direction   (d) 200 MA/m^2

69. (a) 116 V   (b) 12.8 kW   (c) 436 W

71. (a) $\dfrac{\rho}{2\pi L} \ln \left(\dfrac{r_b}{r_a} \right)$ (b) $\dfrac{2\pi L\, \Delta V}{I \ln\left(r_b / r_a \right)}$

73. 4.1×10^{-3} (°C)$^{-1}$

75. 1.418 Ω

77. (a) $\dfrac{\epsilon_0 \ell}{2d}(\ell + 2x + \kappa \ell - 2\kappa x)$

(b) $\dfrac{\epsilon_0 \ell v\, \Delta V(\kappa - 1)}{d}$ clockwise

79. 2.71 MΩ

81. (2.02×10^3)°C

Chapter 28

Answers to Quick Quizzes

1. (a)

2. (b)

3. (a)

4. (i) (b)   (ii) (a)   (iii) (a)   (iv) (b)

5. (i) (c)   (ii) (d)

Answers to Odd-Numbered Problems

1. (a) 6.73 Ω (b) 1.97 Ω

3. (a) 12.4 V   (b) 9.65 V

5. (a) 75.0 V   (b) 25.0 W, 6.25 W, and 6.25 W   (c) 37.5 W

7. $\frac{7}{3}R$

9. (a) 227 mA   (b) 5.68 V

11. (a) 1.00 kΩ (b) 2.00 kΩ (c) 3.00 kΩ

13. (a) 17.1 Ω (b) 1.99 A for 4.00 Ω and 9.00 Ω, 1.17 A for
7.00 Ω, 0.818 A for 10.0 Ω

15. 470 Ω and 220 Ω

17. (a) 11.7 Ω (b) 1.00 A in the 12.0-Ω and 8.00-Ω resistors,
2.00 A in the 6.00-Ω and 4.00-Ω resistors, 3.00 A in the
5.00-Ω resistor

19. 14.2 W to 2.00 Ω, 28.4 W to 4.00 Ω, 1.33 W to 3.00 Ω,
4.00 W to 1.00 Ω

21. (a) 4.12 V   (b) 1.38 A

23. (a) 0.846 A down in the 8.00-Ω resistor, 0.462 A down
in the middle branch, 1.31 A up in the right-hand
branch   (b) -222 J by the 4.00-V battery, 1.88 kJ by the
12.0-V battery   (c) 687 J to 8.00 Ω, 128 J to 5.00 Ω, 25.6 J
to the 1.00-Ω resistor in the center branch, 616 J to
3.00 Ω, 205 J to the 1.00-Ω resistor in the right branch
(d) Chemical energy in the 12.0-V battery is transformed
into internal energy in the resistors. The 4.00-V battery is
being charged, so its chemical potential energy is increasing at the expense of some of the chemical potential
energy in the 12.0-V battery.   (e) 1.66 kJ

25. (a) 0.395 A   (b) 1.50 V

27. 50.0 mA from a to e

29. (a) 0.714 A   (b) 1.29 A   (c) 12.6 V

31. (a) 0.385 mA, 3.08 mA, 2.69 mA   (b) 69.2 V, with c at the
higher potential

33. (a) $I_1 = 0.492$ A; $I_2 = 0.148$ A; $I_3 = 0.639$ A
(b) $P_{28.0\,\Omega} = 6.77$ W, $P_{12.0\,\Omega} = 0.261$ W, $P_{16.0\,\Omega} = 6.54$ W

35. $\Delta V_2 = 3.05$ V, $\Delta V_3 = 4.57$ V, $\Delta V_4 = 7.38$ V, $\Delta V_5 = 1.62$ V

37. (a) 2.00 ms   (b) 1.80×10^{-4} C   (c) 1.14×10^{-4} C

39. (a) -61.6 mA   (b) 0.235 μC   (c) 1.96 A

41. (a) 1.50 s   (b) 1.00 s   (c) $i = 200 + 100e^{-t}$, where i is in
microamperes and t is in seconds

43. (a) 6.00 V   (b) 8.29 μs

45. (a) 0.432 s   (b) 6.00 μF
47. (a) 6.25 A   (b) 750 W
49. (a) $\dfrac{\mathcal{E}^2}{3R}$ (b) $\dfrac{\mathcal{E}^2}{R}$ (c) parallel
51. 2.22 h
53. (a) 1.02 A down   (b) 0.364 A down   (c) 1.38 A up   (d) 0
    (e) 66.0 μC
55. (a) 2.00 kΩ (b) 15.0 V   (c) 9.00 V
57. (a) 4.00 V   (b) Point a is at the higher potential.
59. 87.3%
61. 6.00 Ω, 3.00 Ω
63. (a) 24.1 μC   (b) 16.1 μC   (c) 16.1 mA
65. (a) $q = 240(1 - e^{-t/6})$ (b) $q = 360(1 - e^{-t/6})$, where in both answers, q is in microcoulombs and t is in milliseconds
67. (a) 9.93 μC   (b) 33.7 nA   (c) 335 nW   (d) 337 nW
69. (a) 470 W   (b) 1.60 mm or more   (c) 2.93 mm or more
71. (a) 222 μC   (b) 444 μC
73. (a) 5.00 Ω (b) 2.40 A
75. (a) 0 in 3 kΩ, 333 μA in 12 kΩ and 15 kΩ (b) 50.0 μC
    (c) $i(t) = 278\, e^{-t/0.180}$, where i is in microamperes and t is in seconds   (d) 290 ms
77. (a) $R_x = R_2 - \frac{1}{4}R_1$ (b) No; $R_x = 2.75\ \Omega$, so the station is inadequately grounded.
79. (a) $\frac{2}{3}\Delta t$ (b) $3\,\Delta t$
81. (a) 3.91 s   (b) 782 μs
83. 20.0 Ω or 98.1 Ω

Chapter 29

Answers to Quick Quizzes

1. (e)
2. (i) (b)   (ii) (a)
3. (c)
4. (i) (c), (b), (a)   (ii) (a) = (b) = (c)

Answers to Odd-Numbered Problems

1. Gravitational force: 8.93×10^{-30} N down, electric force: 1.60×10^{-17} N up, and magnetic force: 4.80×10^{-17} N down.
3. (a) into the page   (b) toward the right   (c) toward the bottom of the page
5. (a) the negative z direction   (b) the positive z direction   (c) The magnetic force is zero in this case.
7. (a) 7.91×10^{-12} N   (b) zero
9. (a) 1.25×10^{-13} N   (b) 7.50×10^{13} m/s^2
11. $-20.9\hat{\mathbf{j}}$ mT
13. (a) 4.27 cm   (b) 1.79×10^{-8} s
15. (a) $\sqrt{2}\,r_p$ (b) $\sqrt{2}\,r_p$
17. 115 keV
19. (a) 5.00 cm   (b) 8.79×10^6 m/s
21. 7.88×10^{-12} T
23. 8.00
25. 0.278 m
27. (a) 7.66×10^7 s^{-1} (b) 2.68×10^7 m/s   (c) 3.75 MeV
    (d) 3.13×10^3 revolutions   (e) 2.57×10^{-4} s
29. 244 kV/m
31. 70.0 mT

33. (a) 8.00×10^{-3} T   (b) in the positive z direction
35. $-2.88\hat{\mathbf{j}}$ N
37. 1.07 m/s
39. (a) east   (b) 0.245 T
41. (a) 5.78 N   (b) toward the west (into the page)
43. 2.98 μN west
45. (a) 4.0×10^{-3} N $\cdot$ m   (b) -6.9×10^{-3} J
47. (a) north at $48.0°$ below the horizontal   (b) south at $48.0°$ above the horizontal   (c) 1.07 μJ
49. 9.05×10^{-4} N $\cdot$ m, tending to make the left-hand side of the loop move toward you and the right-hand side move away.
51. (a) 9.98 N $\cdot$ m   (b) clockwise as seen looking down from a position on the positive y axis
53. (a) 118 μN $\cdot$ m   (b) $-118\ \mu\text{J} \le U \le +118\ \mu\text{J}$
55. 43.2 μT
57. (a) 9.27×10^{-24} A $\cdot$ m^2 (b) away from observer
59. (a) $(3.52\hat{\mathbf{i}} - 1.60\hat{\mathbf{j}}) \times 10^{-18}$ N   (b) $24.4°$
61. 0.588 T
63. $3R/4$
65. 39.2 mT
67. (a) the positive z direction   (b) 0.696 m   (c) 1.09 m
    (d) 54.7 ns
69. (a) 0.713 A counterclockwise as seen from above
71. (a) mg/NIw (b) The magnetic field exerts forces of equal magnitude and opposite directions on the two sides of the coils, so the forces cancel each other and do not affect the balance of the system. Hence, the vertical dimension of the coil is not needed.   (c) 0.261 T
73. (a) 1.04×10^{-4} m   (b) 1.89×10^{-4} m
75. (a) $\Delta V_\text{H} = (1.00 \times 10^{-4})\, B$, where ΔV_H is in volts and B is in teslas

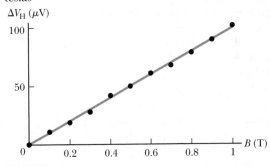

    (b) 0.125 mm
77. 3.71×10^{-24} N $\cdot$ m
79. (a) 0.128 T   (b) $78.7°$ below the horizontal

Chapter 30

Answers to Quick Quizzes

1. $B > C > A$
2. (a)
3. $c > a > d > b$
4. $a = c = d > b = 0$
5. (c)

Answers to Odd-Numbered Problems

1. (a) 21.5 mA   (b) 4.51 V   (c) 96.7 mW
3. 1.60×10^{-6} T
5. (a) 28.3 μT into the page   (b) 24.7 μT into the page

7. $5.52\ \mu$T into the page

9. (a) $2I_1$ out of the page (b) $6I_1$ into the page

11. $\dfrac{\mu_0 I}{2r}\left(\dfrac{1}{\pi}+\dfrac{1}{4}\right)$

13. 262 nT into the page

15. (a) $53.3\ \mu$T toward the bottom of the page (b) $20.0\ \mu$T toward the bottom of the page (c) zero

17. $\dfrac{\mu_0 I}{2\pi ad}\left(\sqrt{d^2+a^2}-d\right)$ into the page

19. (a) $40.0\ \mu$T into the page (b) $5.00\ \mu$T out of the page (c) $1.67\ \mu$T out of the page

21. (a) $10\ \mu$T (b) $80\ \mu$N toward the other wire (c) $16\ \mu$T (d) $80\ \mu$N toward the other wire

23. (a) 3.00×10^{-5} N/m (b) attractive

25. $-27.0\hat{\mathbf{i}}\ \mu$N

27. 0.333 m

29. (a) opposite directions (b) 67.8 A (c) It would be smaller. A smaller gravitational force would be pulling down on the wires, requiring less magnetic force to raise the wires to the same angle and therefore less current.

31. (a) $200\ \mu$T toward the top of the page (b) $133\ \mu$T toward the bottom of the page

33. 5.40 cm

35. (a) 4.00 m (b) 7.50 nT (c) 1.26 m (d) zero

37. (a) zero (b) $\dfrac{\mu_0 I}{2\pi R}$ tangent to the wall (c) $\dfrac{\mu_0 I^2}{(2\pi R)^2}$ inward

39. $20.0\ \mu$T toward the bottom of the page

41. 31.8 mA

43. (a) $226\ \mu$N away from the center of the loop (b) zero

45. (a) 920 turns (b) 12 cm

47. (a) 3.13 mWb (b) 0

49. (a) 8.63×10^{45} electrons (b) 4.01×10^{20} kg

51. 3.18 A

53. (a) $\sim 10^{-5}$ T (b) It is $\sim 10^{-1}$ as large as the Earth's magnetic field.

55. 143 pT

57. $\dfrac{\mu_0 I}{2\pi w}\ln\left(1+\dfrac{w}{b}\right)\hat{\mathbf{k}}$

59. (a) $\mu_0\sigma v$ into the page (b) zero (c) $\frac{1}{2}\mu_0\sigma^2 v^2$ up toward the top of the page (d) $\dfrac{1}{\sqrt{\mu_0\epsilon_0}}$; we will find out in Chapter 34 that this speed is the speed of light. We will also find out in Chapter 39 that this speed is not possible for the capacitor plates.

61. 1.80 mT

63. $3.89\ \mu$T parallel to the xy plane and at $59.0°$ clockwise from the positive x direction

65. (b) 3.20×10^{-13} T (c) 1.03×10^{-24} N (d) 2.31×10^{-22} N

67. $B=4.36\times10^{-4}\,I$, where B is in teslas and I is in amperes

69. (a) $\dfrac{\mu_0 IN}{2\ell}\left[\dfrac{\ell-x}{\sqrt{(\ell-x)^2+a^2}}+\dfrac{x}{\sqrt{x^2+a^2}}\right]$

71. $-0.012\,0\hat{\mathbf{k}}$ N

73. (b) $\dfrac{\mu_0 I}{4\pi}\left(1-e^{-2\pi}\right)$ out of the page

75. (a) $\dfrac{\mu_0 I(2r^2-a^2)}{\pi r(4r^2-a^2)}$ to the left (b) $\dfrac{\mu_0 I(2r^2+a^2)}{\pi r(4r^2+a^2)}$ toward the top of the page

77. (b) 5.92×10^{-8} N

Chapter 31

Answers to Quick Quizzes

1. (c)
2. (c)
3. (b)
4. (a)
5. (b)

Answers to Odd-Numbered Problems

1. 0.800 mA

3. (a) $101\ \mu$V tending to produce clockwise current as seen from above (b) It is twice as large in magnitude and in the opposite sense.

5. 33.9 mV

7. $10.2\ \mu$V

9. 61.8 mV

11. (a) 1.60 A counterclockwise (b) $20.1\ \mu$T (c) left

13. (a) $\dfrac{\mu_0 IL}{2\pi}\ln\left(1+\dfrac{w}{h}\right)$ (b) $4.80\ \mu$V (c) counterclockwise

15. (a) 1.88×10^{-7} T $\cdot$ m^2 (b) 6.28×10^{-8} V

17. 272 m

19. $\mathcal{E}=0.422\cos 120\pi t$, where $\mathcal{E}$ is in volts and t is in seconds

21. 2.83 mV

23. 13.1 mV

25. (a) $39.9\ \mu$V (b) The west end is positive.

27. (a) 3.00 N to the right (b) 6.00 W

29. (a) 0.500 A (b) 2.00 W (c) 2.00 W

31. 2.80 m/s

33. 24.1 V with the outer contact negative

35. (a) 233 Hz (b) 1.98 mV

37. $145\ \mu$A upward in the picture

39. (a) 8.01×10^{-21} N (b) clockwise (c) $t=0$ or $t=1.33$ s

41. (a) $E=9.87\cos 100\pi t$, where E is in millivolts per meter and t is in seconds (b) clockwise

43. 13.3 V

45. (a) $\mathcal{E}=19.6\sin 100\pi t$, where $\mathcal{E}$ is in volts and t is in seconds (b) 19.6 V

47. $\mathcal{E}=28.6\sin 4.00\pi t$, where $\mathcal{E}$ is in millivolts and t is in seconds

49. (a) $\Phi_B=8.00\times10^{-3}\cos 120\pi t$, where Φ_B is in T $\cdot$ m^2 and t is in seconds (b) $\mathcal{E}=3.02\sin 120\pi t$, where $\mathcal{E}$ is in volts and t is in seconds (c) $I=3.02\sin 120\pi t$, where I is in amperes and t is in seconds (d) $P=9.10\sin^2 120\pi t$, where P is in watts and t is in seconds (e) $\tau=0.024\,1\sin^2 120\pi t$, where τ is in newton meters and t is in seconds

51. (a) 113 V (b) 300 V/m

53. 8.80 A

55. 3.79 mV

57. (a) 43.8 A (b) 38.3 W

59. $\mathcal{E}=-7.22\cos 1\,046\pi t$, where $\mathcal{E}$ is in millivolts and t is in seconds

61. $283\ \mu$A upward

63. (a) 3.50 A up in 2.00 Ω and 1.40 A up in 5.00 Ω (b) 34.3 W (c) 4.29 N

65. $2.29\ \mu$C

67. (a) 0.125 V clockwise (b) 0.020 0 A clockwise

69. (a) 97.4 nV (b) clockwise

71. (a) 36.0 V (b) 0.600 Wb/s (c) 35.9 V (d) 4.32 N $\cdot$ m

73. (a) $NB\ell v$ (b) $\dfrac{NB\ell v}{R}$ (c) $\dfrac{N^2B^2\ell^2v^2}{R}$ (d) $\dfrac{N^2B^2\ell^2v}{R}$
(e) clockwise (f) directed to the left.

75. 6.00 A

77. $\mathcal{E} = -87.1\cos(200\pi t + \phi)$, where $\mathcal{E}$ is in millivolts and t is in seconds

79. 0.062 3 A in 6.00 Ω, 0.860 A in 5.00 Ω, and 0.923 A in 3.00 Ω

81. $v = \dfrac{\mathcal{E}}{Bd}\left(1 - e^{-B^2d^2t/mR}\right)$

83. $\dfrac{MgR}{B^2\ell^2}\left[1 - e^{-B^2\ell^2t/R(M+m)}\right]$

Chapter 32

Answers to Quick Quizzes

1. (c), (f)
2. (i) (b) (ii) (a)
3. (a), (d)
4. (a)
5. (i) (b) (ii) (c)

Answers to Odd-Numbered Problems

1. 19.5 mV
3. 100 V
5. 19.2 μT $\cdot$ m^2
7. 4.00 mH
9. (a) 360 mV (b) 180 mV (c) 3.00 s
11. $\dfrac{\mathcal{E}_0}{Lk^2}$
13. $\mathcal{E} = -18.8\cos 120\pi t$, where $\mathcal{E}$ is in volts and t is in seconds
15. (a) 0.469 mH (b) 0.188 ms
17. (a) 1.00 kΩ (b) 3.00 ms
19. (a) 1.29 kΩ (b) 72.0 mA
21. (a) 20.0% (b) 4.00%
23. 92.8 V
25. (a) $i_L = 0.500(1 - e^{-10.0t})$, where i_L is in amperes and t is in seconds (b) $i_S = 1.50 - 0.250e^{-10.0t}$, where i_S is in amperes and t is in seconds
27. (a) 0.800 (b) 0
29. (a) 6.67 A/s (b) 0.332 A/s
31. (a) 5.66 ms (b) 1.22 A (c) 58.1 ms
33. 2.44 μJ
35. (a) 44.3 nJ/m^3 (b) 995 μJ/m^3
37. (a) 18.0 J (b) 7.20 J
39. (a) 8.06 MJ/m^3 (b) 6.32 kJ
41. 1.00 V
43. (a) 18.0 mH (b) 34.3 mH (c) -9.00 mV
45. 781 pH
47. 281 mH
49. 400 mA
51. 20.0 V
53. (a) 503 Hz (b) 12.0 μC (c) 37.9 mA (d) 72.0 μJ
55. (a) 135 Hz (b) 119 μC (c) -114 mA
57. (a) 2.51 kHz (b) 69.9 Ω
59. (a) $0.693\left(\dfrac{2L}{R}\right)$ (b) $0.347\left(\dfrac{2L}{R}\right)$

61. (a) -20.0 mV (b) $\Delta v_C = -10.0t^2$, where Δv_C is in megavolts and t is in seconds (c) 63.2 μs

63. $\dfrac{Q}{2N}\sqrt{\dfrac{3L}{C}}$

65. (a) 4.00 H (b) 3.50 Ω
67. (a) $\frac{1}{2}\mu_0\pi N^2R$ (b) 10^{-7} H (c) 10^{-9} s
69.

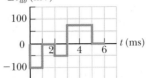

Δv_{ab} (mV)

71. 91.2 μH
73. (a) 6.25×10^{10} J (b) 2.00×10^3 N/m
75. (a) 50.0 mT (b) 20.0 mT (c) 2.29 MJ (d) 318 Pa
79. (a) $\dfrac{2\pi B_0^2R^3}{\mu_0}$ (b) 2.70×10^{18} J
81. 300 Ω
83. $\dfrac{L_1L_2 - M^2}{L_1 + L_2 - 2M}$

Chapter 33

Answers to Quick Quizzes

1. (i) (c) (ii) (b)
2. (b)
3. (a)
4. (b)
5. (a) $X_L < X_C$ (b) $X_L = X_C$ (c) $X_L > X_C$
6. (c)
7. (c)

Answers to Odd-Numbered Problems

1. (a) 96.0 V (b) 136 V (c) 11.3 A (d) 768 W
3. (a) 2.95 A (b) 70.7 V
5. 14.6 Hz
7. 3.38 W
9. 3.14 A
11. 5.60 A
13. (a) 12.6 Ω (b) 6.21 A (c) 8.78 A
15. 0.450 Wb
17. 32.0 A
19. (a) $f > 41.3$ Hz (b) $X_C < 87.5$ Ω
21. 100 mA
23. (a) 141 mA (b) 235 mA
25.

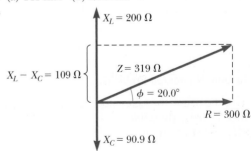

27. (a) 47.1 Ω   (b) 637 Ω   (c) 2.40 kΩ   (d) 2.33 kΩ
(e) −14.2°

29. (a) 17.4°   (b) the voltage

31. (a) 194 V   (b) The current leads by 49.9°.

33. $3R$

35. 353 W

37. 88.0 W

39. (a) 16.0 Ω   (b) −12.0 Ω

41. $\dfrac{11(\Delta V_{\text{rms}})^2}{14R}$

43. 1.82 pF

45. 242 mJ

47. (a) 0.633 pF   (b) 8.46 mm   (c) 25.1 Ω

49. 687 V

51. 87.5 Ω

53. 0.756

55. (a) 34%   (b) 5.3 W   (c) $3.9

57. (a) 1.60×10^3 turns   (b) 30.0 A   (c) 25.3 A

59. (a) 22.4 V   (b) 26.6°   (c) 0.267 A   (d) 83.9 Ω   (e) 47.2 μF
(f) 0.249 H   (g) 2.67 W

61. 2.6 cm

63. (a) X_C could be 53.8 Ω or it could be 1.35 kΩ   (b) capacitive reactance is 53.8 Ω   (c) X_C must be 1.43 kΩ

65. (b) 31.6

67. (a) 19.7 cm at 35.0°   (b) 19.7 cm at 35.0°   (c) The answers are identical.   (d) 9.36 cm at 169°

69. (a) Tension T and separation d must be related by $T = 274d^2$, where T is in newtons and d is in meters.   (b) One possibility is $T = 10.9$ N and $d = 0.200$ m.

71. (a) 0.225 A   (b) 0.450 A

73. (a) 78.5 Ω   (b) 1.59 kΩ   (c) 1.52 kΩ   (d) 138 mA
(e) −84.3°   (f) 0.098 7   (g) 1.43 W

75. 56.7 W

77. (a) 580 μH   (b) 54.6 μF   (c) 1.00   (d) 894 Hz   (e) At 200 Hz, $\phi = -60.0°$ (Δv_{out} leads Δv_{in}); at f_0, $\phi = 0$ (Δv_{out} is in phase with Δv_{in}); and at 4.00×10^3 Hz, $\phi = +60.0°$ (Δv_{out} lags Δv_{in}).   (f) At 200 Hz and at 4.00×10^3 Hz, $P = 1.56$ W; and at f_0, $P = 6.25$ W.   (g) 0.408

79. (a) 224 s^{-1} (b) 500 W   (c) 221 s^{-1} and 226 s^{-1}

81. 58.7 Hz or 35.9 Hz. The circuit can be either above or below resonance.

Chapter 34

Answers to Quick Quizzes

1. **(i)** (b)   **(ii)** (c)

2. (c)

3. (c)

4. (b)

5. (a)

6. (c)

7. (a)

Answers to Odd-Numbered Problems

1. (a) out of the page   (b) 1.85×10^{-18} T

3. (a) 11.3 GV · m/s   (b) 0.100 A

5. $(-2.87\hat{\mathbf{j}} + 5.75\hat{\mathbf{k}}) \times 10^9$ m/s^2

7. (a) 0.690 wavelengths   (b) 58.9 wavelengths

9. (a) 681 yr   (b) 8.32 min   (c) 2.56 s

11. 74.9 MHz

13. 2.25×10^8 m/s

15. (a) 6.00 MHz   (b) $-73.4\hat{\mathbf{k}}$ nT
(c) $\vec{\mathbf{B}} = -73.4 \cos(0.126x - 3.77 \times 10^7 t)\hat{\mathbf{k}}$, where B is in nT, x is in meters, and t is in seconds

17. 2.9×10^8 m/s ±5%

19. (a) 0.333 μT   (b) 0.628 μm   (c) 4.77×10^{14} Hz

21. 3.34 μJ/m^3

23. 3.33×10^3 m^2

25. (a) 1.19×10^{10} W/m^2 (b) 2.35×10^5 W

27. (a) 2.33 mT   (b) 650 MW/m^2 (c) 511 W

29. 307 μW/m^2

31. 49.5 mV

33. (a) 332 kW/m^2 radially inward   (b) 1.88 kV/m and 222 μT

35. 5.31×10^{-5} N/m^2

37. (a) 1.90 kN/C   (b) 50.0 pJ   (c) 1.67×10^{-19} kg · m/s

39. 4.09°

41. (a) $1.60 \times 10^{-10}\hat{\mathbf{i}}$ kg · m/s each second   (b) $1.60 \times 10^{-10}\hat{\mathbf{i}}$ N
(c) The answers are the same. Force is the time rate of momentum transfer.

43. (a) 5.48 N   (b) 913 μm/s^2 away from the Sun   (c) 10.6 days

45. (a) 134 m   (b) 46.8 m

47. 56.2 m

49. (a) away along the perpendicular bisector of the line segment joining the antennas   (b) along the extensions of the line segment joining the antennas

51. (a) 6.00 pm   (b) 7.49 cm

53. (a) 4.16 m to 4.54 m   (b) 3.41 m to 3.66 m   (c) 1.61 m to 1.67 m

55. (a) 3.85×10^{26} W   (b) 1.02 kV/m and 3.39 μT

57. 5.50×10^{-7} m

59. (a) 3.21×10^7 W   (b) 0.639 W/m^2 (c) 0.513% of that from the noon Sun in January

61. $\sim 10^6$ J

63. 378 nm

65. (a) 6.67×10^{-16} T   (b) 5.31×10^{-17} W/m^2
(c) 1.67×10^{-14} W   (d) 5.56×10^{-23} N

67. (a) 625 kW/m^2 (b) 21.7 kV/m   (c) 72.4 μT   (d) 17.8 min

69. (a) 388 K   (b) 363 K

71. (a) 3.92×10^8 W/m^2 (b) 308 W

73. (a) 0.161 m   (b) 0.163 m^2 (c) 76.8 W   (d) 470 W/m^2
(e) 595 V/m   (f) 1.98 μT   (g) 119 W

75. (a) The projected area is πr^2, where r is the radius of the planet.   (b) The radiating area is $4\pi r^2$.
(c) 1.61×10^{10} m

77. (a) 584 nT   (b) 419 m^{-1} (c) 1.26×10^{11} s^{-1} (d) $\vec{\mathbf{B}}$ vibrates in the xz plane.   (e) $40.6\hat{\mathbf{i}}$ W/m^2 (f) 271 nPa
(g) $407\hat{\mathbf{i}}$ nm/s^2

79. (a) 22.6 h   (b) 30.6 s

Chapter 35

Answers to Quick Quizzes

1. (d)

2. Beams ② and ④ are reflected; beams ③ and ⑤ are refracted.

3. (c)
4. (c)
5. (i) (b) (ii) (b)

Answers to Odd-Numbered Problems

1. (a) 2.07×10^3 eV (b) 4.14 eV
3. 114 rad/s
5. (a) 4.74×10^{14} Hz (b) 422 nm (c) 2.00×10^8 m/s
7. 22.5°
9. (a) 1.81×10^8 m/s (b) 2.25×10^8 m/s
   (c) 1.36×10^8 m/s
11. (a) 29.0° (b) 25.8° (c) 32.0°
13. 86.8°
15. 158 Mm/s
17. (a) $\theta_{1i} = 30°$, $\theta_{1r} = 19°$, $\theta_{2i} = 41°$, $\theta_{2r} = 77°$ (b) First sur-
    face: $\theta_{\text{reflection}} = 30°$; second surface: $\theta_{\text{reflection}} = 41°$
19. $\sim 10^{-11}$ s, $\sim 10^3$ wavelengths
21. (a) 1.94 m (b) 50.0° above the horizontal
23. 27.1 ns
25. (a) 2.0×10^8 m/s (b) 4.74×10^{14} Hz (c) 4.2×10^{-7} m
27. 3.39 m
29. (a) 41.5° (b) 18.5° (c) 27.5° (d) 42.5°
31. 23.1°
33. 1.22
35. $\tan^{-1}(n_g)$
37. 0.314°
39. 4.61°
41. 62.5°
43. 27.9°
45. 67.1°
47. 1.000 07
49. (a) $\dfrac{nd}{n-1}$ (b) $R_{\min} \to 0$. Yes; for very small d, the light
    strikes the interface at very large angles of incidence.
    (c) $R_{\min}$ decreases. Yes; as n increases, the critical angle
    becomes smaller. (d) $R_{\min} \to \infty$. Yes; as $n \to 1$, the critical
    angle becomes close to 90° and any bend will allow the
    light to escape. (e) 350 μm
51. 48.5°
53. 2.27 m
55. 25.7°
57. (a) 0.042 6 or 4.26% (b) no difference
59. (a) 334 μs (b) 0.014 6%
61. 77.5°
63. 2.00 m
65. 27.5°
67. 3.79 m
69. 7.93°
71. $\sin^{-1}\left[\dfrac{L}{R^2}(\sqrt{n^2R^2 - L^2} - \sqrt{R^2 - L^2})\right]$ or
    $\sin^{-1}\left[n\sin\left(\sin^{-1}\dfrac{L}{R} - \sin^{-1}\dfrac{L}{nR}\right)\right]$
73. (a) 38.5° (b) 1.44
75. (a) 53.1° (b) $\theta_1 \geq 38.7°$
77. (a) 1.20 (b) 3.40 ns
79. (a) 0.172 mm/s (b) 0.345 mm/s (c) and (d) northward
    and downward at 50.0° below the horizontal.
81. 62.2%

83. (a) $\left(\dfrac{4x^2 + L^2}{L}\right)\omega$ (b) 0 (c) $L\omega$ (d) $2L\omega$ (e) $\dfrac{\pi}{8\omega}$
87. 70.6%

Chapter 36

Answers to Quick Quizzes

1. false
2. (b)
3. (b)
4. (d)
5. (a)
6. (b)
7. (a)
8. (c)

Answers to Odd-Numbered Problems

1. 89.0 cm
3. (a) younger (b) $\sim 10^{-9}$ s younger
5. (a) $p_1 + h$, behind the lower mirror (b) virtual (c) upright
   (d) 1.00 (e) no
7. (a) 1.00 m behind the nearest mirror; (b) the palm;
   (c) 5.00 m behind the nearest mirror; (d) the back of
   her hand; (e) 7.00 m behind the nearest mirror; (f) the
   palm; (g) All are virtual images.
9. (i) (a) 13.3 cm (b) real (c) inverted (d) -0.333
   (ii) (a) 20.0 cm (b) real (c) inverted (d) -1.00 (iii) (a) ∞
   (b) no image formed (c) no image formed (d) no
   image formed
11. (a) -12.0 cm; 0.400 (b) -15.0 cm; 0.250 (c) both
    upright
13. (a) -7.50 cm (b) upright (c) 0.500 cm
15. 3.33 m from the deepest point in the niche
17. 0.790 cm
19. (a) 0.160 m (b) -0.400 m
21. (a) convex (b) at the 30.0-cm mark (c) -20.0 cm
23. (a) 15.0 cm (b) 60.0 cm
25. (a) concave (b) 2.08 m (c) 1.25 m from the object
27. (a) 25.6 m (b) 0.058 7 rad (c) 2.51 m (d) 0.023 9 rad
    (e) 62.8 m
29. (a) 45.1 cm (b) -89.6 cm (c) -6.00 cm
31. (a) 1.50 m (b) 1.75 m
33. 4.82 cm
35. 8.57 cm
37. 1.50 cm/s
39. (a) 6.40 cm (b) -0.250 (c) converging
41. (a) 39.0 mm (b) 39.5 mm
43. 20.0 cm
45. (a) 20.0 cm from the lens on the front side (b) 12.5 cm
    from the lens on the front side (c) 6.67 cm from the lens
    on the front side (d) 8.33 cm from the lens on the front
    side
47. 2.84 cm
49. (a) 16.4 cm (b) 16.4 cm
51. (a) 1.16 mm/s (b) toward the lens
53. 7.47 cm in front of the second lens, 1.07 cm, virtual,
    upright

55. 21.3 cm

57. 2.18 mm away from the CCD

59. (a) 42.9 cm   (b) +2.33 diopters

61. 23.2 cm

63. (a) −0.67 diopters   (b) +0.67 diopters

65. (a) Yes, if the lenses are bifocal.
(b) $f = 56.3$ cm, $P = +1.78$ diopters   (c) −1.18 diopters

67. −575

69. 3.38 min

71. (a) 267 cm   (b) 79.0 cm

73. −40.0 cm

75. (a) 1.50   (b) 1.90

77. (a) 160 cm to the left of the lens   (b) −0.800   (c) inverted

79. (a) 32.1 cm to the right of the second surface   (b) real

81. (a) 25.3 cm to the right of the mirror   (b) virtual
(c) upright   (d) +8.05

83. (a) 1.40 kW/m^2 (b) 6.91 mW/m^2 (c) 0.164 cm
(d) 58.1 W/m^2

87. 8.00 cm

89. +11.7 cm

91. (a) 1.50 m in front of the mirror   (b) 1.40 cm

93. (a) 0.334 m or larger   (b) $R_a/R = 0.025\,5$ or larger

95. (a) $n = 1.99$ (b) 10.0 cm to the left of the lens   (c) −2.50
(d) inverted

97. $d = p$ and $d = p + 2f_M$

Chapter 37

Answers to Quick Quizzes

1. (c)

2. The graph is shown here. The width of the primary maxima is slightly narrower than the $N = 5$ primary width but wider than the $N = 10$ primary width. Because $N = 6$, the secondary maxima are $\frac{1}{36}$ as intense as the primary maxima.

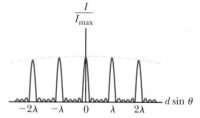

3. (a)

Answers to Odd-Numbered Problems

1. 641

3. 632 nm

5. 1.54 mm

7. 2.40 μm

9. (a) 2.62 mm   (b) 2.62 mm

11. Maxima at 0°, 29.1°, and 76.3°; minima at 14.1° and 46.8°

13. (a) 55.7 m   (b) 124 m

15. 0.318 m/s

17. 148 m

21. (a) 1.93 μm   (b) 3.00λ (c) It corresponds to a maximum. The path difference is an integer multiple of the wavelength.

23. 0.968

25. 48.0 μm

27. (a) 1.29 rad   (b) 99.6 nm

29. (a) 7.95 rad   (b) 0.453

31. 512 nm

33. 0.500 cm

35. 290 nm

37. 8.70 μm

39. 1.31

41. 1.20 mm

43. 1.001

45. 1.25 m

47. 1.62 cm

49. 78.4 μm

51. $x_1 - x_2 = (m - \frac{1}{48})650$, where x_1 and x_2 are in nanometers and $m = 0, 1, -1, 2, -2, 3, -3, \ldots$

53. $\dfrac{\lambda}{2(n-1)}$

55. 5.00×10^6 m^2 = 5.00 km^2

57. 2.50 mm

59. 113

61. (a) 72.0 m   (b) 36.0 m

63. (a) 70.6 m   (b) 136 m

65. (a) 14.7 μm   (b) 1.53 cm   (c) −16.0 m

67. 0.505 mm

69. 3.58°

71. 115 nm

73. (a) $m = \dfrac{\lambda_1}{2(\lambda_1 - \lambda_2)}$ (b) 266 nm

75. 0.498 mm

Chapter 38

Answers to Quick Quizzes

1. (a)

2. (i)

3. (b)

4. (a)

5. (c)

6. (b)

7. (c)

Answers to Odd-Numbered Problems

1. (a) 1.1 m   (b) 1.7 mm

3. (a) four   (b) $\theta = \pm 28.7°, \pm 73.6°$

5. 91.2 cm

7. 2.30×10^{-4} m

9.

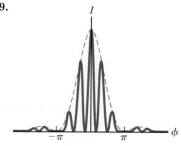

11. 1.62×10^{-2}
13. 462 nm
15. 2.10 m
17. 0.284 m
19. 30.5 m
21. 0.40 μrad
23. 16.4 m
25. 1.81 μm
27. (a) three   (b) $0°$, $+45.2°$, $-45.2°$
29. 74.2 grooves/mm
31. 2
33. 514 nm
35. (a) 3.53×10^3 rulings/cm   (b) 11
37. (a) 5.23 μm   (b) 4.58 μm
39. 0.093 4 nm
41. (a) 0.109 nm   (b) four
43. (a) $54.7°$ (b) $63.4°$ (c) $71.6°$
45. 0.375
47. (a) six   (b) $7.50°$
49. $60.5°$
51. 6.89 units
53. (a) 0.045 0   (b) 0.016 2
55. 5.51 m, 2.76 m, 1.84 m
57. 632.8 nm
59. (a) 7.26 μrad, 1.50 arc seconds   (b) 0.189 ly   (c) 50.8 μrad   (d) 1.52 mm
61. (a) $25.6°$ (b) $18.9°$
63. 545 nm
65. $13.7°$
67. 15.4
69. (b) 3.77 nm/cm
71. (a) $\phi = 4.49$ compared with the prediction from the approximation of $1.5\pi = 4.71$ (b) $\phi = 7.73$ compared with the prediction from the approximation of $2.5\pi = 7.85$
73. (b) 0.001 90 rad = $0.109°$
75. (b) 15.3 μm
77. (a) $41.8°$ (b) 0.592   (c) 0.262 m

Chapter 39

Answers to Quick Quizzes

1. (c)
2. (d)
3. (d)
4. (a)
5. (a)
6. (c)
7. (d)
8. (i) (c)   (ii) (a)
9. (a) $m_3 > m_2 = m_1$ (b) $K_3 = K_2 > K_1$ (c) $u_2 > u_3 = u_1$

Answers to Odd-Numbered Problems

1. 10.0 m/s toward the left in Figure P39.1
3. 5.70×10^{-3} degrees or 9.94×10^{-5} rad
5. $0.917c$
7. $0.866c$
9. $0.866c$
11. $0.220c$

13. 5.00 s
15. The trackside observer measures the length to be 31.2 m, so the supertrain is measured to fit in the tunnel, with 18.8 m to spare.
17. (a) 25.0 yr   (b) 15.0 yr   (c) 12.0 ly
19. $0.800c$
21. (b) $0.050\ 4c$
23. (c) 2.00 kHz   (d) 0.075 m/s $\approx$ 0.17 mi/h
25. 1.55 ns
27. (a) 2.50×10^8 m/s   (b) 4.98 m   (c) -1.33×10^{-8} s
29. (a) 17.4 m   (b) $3.30°$
31. Event B occurs first, 444 ns earlier than A
33. $0.357c$
35. $0.998c$ toward the right
37. (a) $\dfrac{2\sqrt{2}}{3} c = 0.943c = 2.83 \times 10^8$ m/s   (b) The result would be the same.
39. (a) 929 MeV/c (b) 6.58×10^3 MeV/c (c) No
41. 4.51×10^{-14}
43. $0.285c$
45. (a) 3.07 MeV   (b) $0.986c$
47. (a) 938 MeV   (b) 3.00 GeV   (c) 2.07 GeV
49. (a) 5.37×10^{-11} J = 335 MeV   (b) 1.33×10^{-9} J = 8.31 GeV
51. 1.63×10^3 MeV/c
53. (a) smaller   (b) 3.18×10^{-12} kg   (c) It is too small a fraction of 9.00 g to be measured.
55. 4.28×10^9 kg/s
57. (a) 8.63×10^{22} J   (b) 9.61×10^5 kg
59. (a) $0.979c$ (b) $0.065\ 2c$ (c) 15.0   (d) $0.999\ 999\ 97c$; $0.948c$; 1.06
61. (a) 4.08 MeV   (b) 29.6 MeV
63. 2.97×10^{-26} kg
65. (a) 2.66×10^7 m   (b) 3.87 km/s   (c) -8.35×10^{-11} (d) 5.29×10^{-10} (e) $+4.46 \times 10^{-10}$
67. 0.712%
69. (a) 13.4 m/s toward the station and 13.4 m/s away from the station.   (b) 0.056 7 rad/s
71. (a) $v/c = 1 - 1.12 \times 10^{-10}$ (b) 6.00×10^{27} J   (c) $\$2.17 \times 10^{20}$
73. (a) 21.0 yr   (b) 14.7 ly   (c) 10.5 ly   (d) 35.7 yr
75. (a) 6.67×10^4 (b) 1.97 h
77. (a) $\sim 10^2$ or 10^3 s   (b) $\sim 10^8$ km
79. (a) 0.905 MeV   (b) 0.394 MeV   (c) 0.747 MeV/c = 3.99×10^{-22} kg $\cdot$ m/s   (d) $65.4°$
81. (b) 1.48 km
83. (a) $0.946c$ (b) 0.160 ly   (c) 0.114 yr   (d) 7.49×10^{22} J
85. (a) 229 s   (b) 174 s
87. 1.83×10^{-3} eV
91. (a) $0.800c$ (b) 7.51×10^3 s   (c) 1.44×10^{12} m   (d) $0.385c$ (e) 4.88×10^3 s

Chapter 40

Answers to Quick Quizzes

1. (b)
2. Sodium light, microwaves, FM radio, AM radio.
3. (c)

4. The classical expectation (which did not match the experiment) yields a graph like the following drawing:

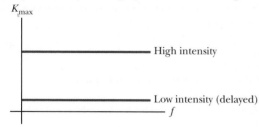

5. (d)
6. (c)
7. (b)
8. (a)

Answers to Odd-Numbered Problems

1. 6.85 μm, which is in the infrared region of the spectrum
3. (a) lightning: $\sim 10^{-7}$ m; explosion: $\sim 10^{-10}$ m   (b) lightning: ultraviolet; explosion: x-ray and gamma ray
5. 5.71×10^3 photons/s
7. (a) $\approx 2.99 \times 10^3$ K   (b) $\approx 2.00 \times 10^4$ K
9. 5.18×10^3 K
11. 1.30×10^{15} photons/s
13. (a) 0.263 kg   (b) 1.81 W   (c) $-0.015\ 3°C/s = -0.919°C/min$ (d) 9.89 μm   (e) 2.01×10^{-20} J   (f) 8.99×10^{19} photon/s
15. 1.34×10^{31}
17. (a) 295 nm, 1.02 PHz   (b) 2.69 V
19. (a) 1.89 eV   (b) 0.216 V
21. (a) 1.38 eV   (b) 3.34×10^{14} Hz
23. 8.34×10^{-12} C
25. 1.04×10^{-3} nm
27. $p_e = 22.1$ keV/c, $K_e = 478$ eV
29. 70.0°
31. (a) 43.0°   (b) $E = 0.601$ MeV; $p = 0.601$ MeV/$c = 3.21 \times 10^{-22}$ kg · m/s   (c) $E = 0.279$ MeV; $p = 0.279$ MeV/$c = 3.21 \times 10^{-22}$ kg · m/s
33. (a) 4.89×10^{-4} nm   (b) 268 keV   (c) 31.8 keV
35. (a) 0.101 nm   (b) 80.8°
37. To have photon energy 10 eV or greater, according to this definition, ionizing radiation is the ultraviolet light, x-rays, and γ rays with wavelength shorter than 124 nm; that is, with frequency higher than 2.42×10^{15} Hz.
39. (a) 1.66×10^{-27} kg · m/s   (b) 1.82 km/s
41. (a) 14.8 keV or, ignoring relativistic correction, 15.1 keV   (b) 124 keV
43. 0.218 nm
45. (a) 3.91×10^4 (b) 20.0 GeV/$c = 1.07 \times 10^{-17}$ kg · m/s   (c) 6.20×10^{-17} m   (d) The wavelength is two orders of magnitude smaller than the size of the nucleus.
47. (a) $\dfrac{\gamma}{\gamma - 1}\dfrac{u}{c}$, where $\gamma = \dfrac{1}{\sqrt{1 - u^2/c^2}}$ (b) 1.60   (c) no change   (d) 2.00×10^3 (e) 1   (f) ∞
49. (a) $v_{\text{phase}} = \dfrac{u}{2}$ (b) This is different from the speed u at which the particle transports mass, energy, and momentum.
51. (a) 989 nm   (b) 4.94 mm   (c) No; there is no way to identify the slit through which the neutron passed. Even

if one neutron at a time is incident on the pair of slits, an interference pattern still develops on the detector array. Therefore, each neutron in effect passes through both slits.
53. 105 V
55. within 1.16 mm for the electron, 5.28×10^{-32} m for the bullet
57. 3×10^{-29} J $\approx 2 \times 10^{-10}$ eV
61. 1.36 eV
63. (a) 19.8 μm   (b) 0.333 m
65. (a) 1.7 eV   (b) 4.2×10^{-15} V · s   (c) 7.3×10^2 nm
67. (a) 2.82×10^{-37} m   (b) 1.06×10^{-32} J   (c) 2.87×10^{-35} %
69. (a) $8.72 \times 10^{16}\ \dfrac{\text{electrons}}{\text{s} \cdot \text{cm}^2}$ (b) 14.0 mA/cm²   (c) The actual current will be lower than that in part (b).
71. (a) 0.143 nm   (b) This is the same order of magnitude as the spacing between atoms in a crystal   (c) Because the wavelength is about the same as the spacing, diffraction effects should occur.
73. (a) The Doppler shift increases the apparent frequency of the incident light.   (b) 3.86 eV   (c) 8.76 eV

Chapter 41

Answers to Quick Quizzes

1. (d)
2. (i) (a)   (ii) (d)
3. (c)
4. (a), (c), (f)

Answers to Odd-Numbered Problems

1. (a) 126 pm   (b) 5.27×10^{-24} kg · m/s   (c) 95.3 eV
3. (a) $A = \sqrt{3}$ (b) 0.037 0   (c) 0.750
5. (a) 0.511 MeV, 2.05 MeV, 4.60 MeV   (b) They do; the MeV is the natural unit for energy radiated by an atomic nucleus.
7. (a)

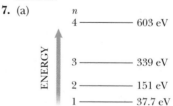

(b) 2.20 nm, 2.75 nm, 4.12 nm, 4.71 nm, 6.59 nm, 11.0 nm
9. 0.795 nm
11. (a) 6.14 MeV   (b) 202 fm   (c) gamma ray
13. (a) 0.434 nm   (b) 6.00 eV
15. (a) $(15\ h\lambda/8m_e c)^{1/2}$ (b) 1.25λ
17. (a) $A = \dfrac{2}{\sqrt{L}}$ (b) 0.409
19. (a) $\dfrac{L}{2}$ (b) 5.26×10^{-5} (c) 3.99×10^{-2} (d) In the $n = 2$ graph in the text's Figure 41.4(b), it is more probable to find the particle either near $x = L/4$ or $x = 3L/4$ than at the center, where the probability density is zero. Nevertheless, the symmetry of the distribution means that the average position is $x = L/2$.

21. (a) 0.196   (b) The classical probability is 0.333, which is significantly larger.   (c) 0.333 for both classical and quantum models

23. (a) 0.196   (b) 0.609

25. (b) $\dfrac{\hbar^2 k^2}{2m}$

27. (a) $U = \dfrac{\hbar^2}{mL^2}\left(\dfrac{2x^2}{L^2} - 3\right)$

(b)

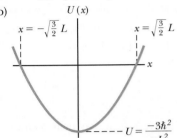

29. (a)

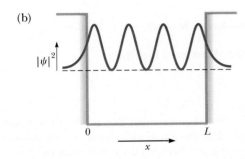

(b)

31. (a) 0.010 3   (b) 0.990

33. 85.9

35. 3.92%

37. 600 nm

39. (a) $B = \left(\dfrac{m\omega}{\pi\hbar}\right)^{1/4}$ (b) $\delta\left(\dfrac{m\omega}{\pi\hbar}\right)^{1/2}$

43. (a) 2.00×10^{-10} m   (b) 3.31×10^{-24} kg · m/s   (c) 0.171 eV

45. 0.250

47. (a) 0.903   (b) 0.359   (c) 0.417   (d) $10^{-6.59 \times 10^{32}}$

49. (a) 435 THz   (b) 689 nm   (c) 165 peV or more

51. (a) $L = \left(\dfrac{h\lambda}{m_e c}\right)^{1/2}$ (b) $\lambda' = \tfrac{8}{5}\lambda$

53. (a) $K_n = \sqrt{\left(\dfrac{nhc}{2L}\right)^2 + (mc^2)^2} - mc^2$ (b) 4.68×10^{-14} J

(c) 28.6% larger

55. (a)

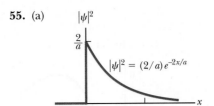

(b) 0   (d) 0.865

57. (a) 0   (b) 0   (c) $\dfrac{1}{\sqrt{2a}} = \sqrt{\dfrac{\hbar}{2m\omega}}$

59. (b) 0.092 0   (c) 0.908

61. (a) 0.200   (b) 0.351   (c) 0.376 eV   (d) 1.50 eV

63. (a) $\tfrac{3}{2}\hbar\omega$ (b) $x = 0$ (c) $x = \pm\sqrt{\dfrac{\hbar}{m\omega}}$ (d) $B = \left(\dfrac{4m^3\omega^3}{\pi\hbar^3}\right)^{1/4}$

(e) 0   (f) $8\delta e^{-4}\sqrt{\dfrac{m\omega}{\pi\hbar}}$

Chapter 42

Answers to Quick Quizzes

1. (c)

2. (a)

3. (b)

4. (a) five   (b) nine

5. (c)

6. true

Answers to Odd-Numbered Problems

1. (a) 121.5 nm, 102.5 nm, 97.20 nm   (b) ultraviolet

3. 1.94 μm

5. (a) 5   (b) no   (c) no

7. (a) 5.69×10^{-14} m   (b) 11.3 N

9. (a) 13.6 eV   (b) 1.51 eV

11. (a) 0.968 eV   (b) 1.28 μm   (c) 2.34×10^{14} Hz

13. (a) 2.19×10^6 m/s   (b) 13.6 eV   (c) -27.2 eV

15. (a) 2.89×10^{34} kg · m^2/s   (b) 2.74×10^{68} (c) 7.30×10^{-69}

17. (a) 0.476 nm   (b) 0.997 nm

19. (a) 3   (b) 520 km/s

21. (a) $E_n = -54.4$ eV/n^2 for $n = 1, 2, 3, \ldots$

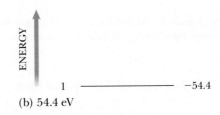

(b) 54.4 eV

23. (b) 0.179 nm

25.

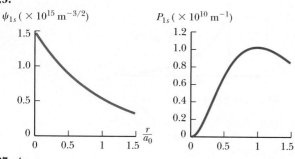

27. $4a_0$

29. 797

31. $\ell = 4$

33. (a) $\sqrt{6}\,\hbar = 2.58 \times 10^{-34}\,\text{J}\cdot\text{s}$
    (b) $2\sqrt{3}\,\hbar = 3.65 \times 10^{-34}\,\text{J}\cdot\text{s}$

35. $3\hbar$

37. $\sqrt{6}\,\hbar = 2.58 \times 10^{-34}\,\text{J}\cdot\text{s}$

39. $n = 3$; $\ell = 2$; $m_\ell = -2, -1, 0, 1,$ or 2; $s = 1$; $m_s = -1, 0,$ or 1, for a total of 15 states

41. (a) $1s^2 2s^2 2p^3$

(b)

n	ℓ	m_ℓ	m_s
1	0	0	$\frac{1}{2}$
1	0	0	$-\frac{1}{2}$
2	1	1	$\frac{1}{2}$
2	1	1	$-\frac{1}{2}$
2	1	0	$\frac{1}{2}$
2	1	0	$-\frac{1}{2}$
2	1	-1	$\frac{1}{2}$
2	1	-1	$-\frac{1}{2}$
2	0	0	$\frac{1}{2}$
2	0	0	$-\frac{1}{2}$

43. aluminum

45. (a) 30   (b) 36

47. 18.4 T

49. 17.7 kV

51. (a) 14 keV   (b) 8.8×10^{-11} m

53. (a) If $\ell = 2$, then $m_\ell = 2, 1, 0, -1, -2$; if $\ell = 1$, then $m_\ell = 1, 0, -1$; if $\ell = 0$, then $m_\ell = 0$.   (b) -6.05 eV

55. 0.068 nm

57. gallium

59. (a) 28.3 THz   (b) 10.6 μm   (c) infrared

61. 3.49×10^{16} photons

63. (a) 4.24×10^{15} W/m^2 (b) 1.20×10^{-12} J

65. (a) 3.40 eV   (b) 0.136 eV

67. (a) 1.57×10^{14} m$^{-3/2}$ (b) 2.47×10^{28} m^{-3}
    (c) 8.69×10^8 m^{-1}

69. 9.80 GHz

71. $\sim$ between 10^4 K and 10^5 K; use Equation 21.19 and set the kinetic energy equal to typical ionization energies

73. $\dfrac{1}{a_0}$, no

75. (a) 609 μeV   (b) 6.9 μeV   (c) 147 GHz   (d) 2.04 mm

77. $\dfrac{\sqrt{3}}{2}a_0 = 0.866a_0$

79. (a) 486 nm   (b) 0.815 m/s

81. (a) $\dfrac{r^2}{8a_0{}^3}\left(2 - \dfrac{r}{a_0}\right)^2 e^{-r/a_0}$

(b) $\dfrac{r}{8a_0{}^5}\left(2 - \dfrac{r}{a_0}\right)e^{-r/a_0}\left(r^2 - 6a_0 r + 4a_0{}^2\right)$

(c) $r = 0$, $r = 2a_0$, and $r = \infty$ (d) $r = \left(3 \pm \sqrt{5}\right)a_0$

(e) $r = \left(3 + \sqrt{5}\right)a_0$ where $P = 0.191/a_0$

83. (a) 4.20 mm   (b) 1.05×10^{19} photons
    (c) 8.84×10^{16} mm^{-3}

85. $\dfrac{3h^2}{4mL^2}$

87. 0.125

89. (a) $r_n = 0.106n^2$, where r_n is in nanometers and $n = 1, 2, 3, \dots$ (b) $E_n = -\dfrac{6.80}{n^2}$, where E_n is in electron volts and $n = 1, 2, 3, \dots$

91. The classical frequency is $4\pi^2 m_e k^2 e^4 / h^3 n^3$.

Chapter 43

Answers to Quick Quizzes

1. (a) van der Waals   (b) ionic   (c) hydrogen   (d) covalent

2. (c)

3. (a)

4. A: semiconductor; B: conductor; C: insulator

Answers to Odd-Numbered Problems

1. $\sim$ 10 K

3. 4.3 eV

5. (a) 74.2 pm   (b) 4.46 eV

7. (a) 1.46×10^{-46} kg $\cdot$ m^2 (b) The results are the same, suggesting that the molecule's bond length does not change measurably between the two transitions.

9. 9.77×10^{12} rad/s

11. (a) 0.014 7 eV   (b) 84.1 μm

13. (a) 12.0 pm   (b) 9.22 pm

15. (a) 2.32×10^{-26} kg   (b) 1.82×10^{-45} kg $\cdot$ m^2 (c) 1.62 cm

17. (a) 0, 3.62×10^{-4} eV, 1.09×10^{-3} eV   (b) 0.097 9 eV, 0.294 eV, 0.490 eV

19. (a) 472 μm   (b) 473 μm   (c) 0.715 μm

21. (a) 4.60×10^{-48} kg $\cdot$ m^2 (b) 1.32×10^{14} Hz   (c) 0.074 1 nm

23. 6.25×10^9

25. -7.83 eV

27. 5.28 eV

29. 2%

31. (a) 4.23 eV   (b) 3.27×10^4 K

33. (a) 2.54×10^{28} m^{-3} (b) 3.15 eV

35. 0.939

41. (a) 276 THz   (b) 1.09 μm

43. 1.91 eV

45. 227 nm

47. (a) 59.5 mV   (b) -59.5 mV

49. 4.18 mA
51. (a) 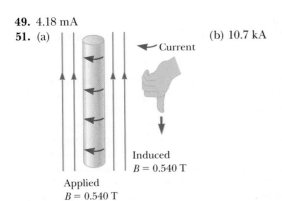 (b) 10.7 kA

53. 203 A to produce a magnetic field in the direction of the original field
55. 7
57. 5.24 J/g
61. (a) 0.350 nm (b) −7.02 eV (c) −1.20 $\hat{\mathbf{i}}$ nN
63. (a) 6.15×10^{13} Hz (b) 1.59×10^{-46} kg · m^2
(c) 4.78 μm or 4.96 μm

Chapter 44

Answers to Quick Quizzes

1. (i) (b) **(ii)** (a) **(iii)** (c)
2. (e)
3. (b)
4. (c)

Answers to Odd-Numbered Problems

1. (a) 1.5 fm (b) 4.7 fm (c) 7.0 fm (d) 7.4 fm
3. (a) 455 fm (b) 6.05×10^6 m/s
5. (a) 4.8 fm (b) 4.7×10^{-43} m^3 (c) 2.3×10^{17} kg/m^3
7. 16 km
9. 8.21 cm
11. (a) 27.6 N (b) 4.16×10^{27} m/s^2 (c) 1.73 MeV
13. 6.1×10^{15} N toward each other
15. (a) 1.11 MeV (b) 7.07 MeV (c) 8.79 MeV (d) 7.57 MeV
17. greater for $^{15}_{7}$N by 3.54 MeV
19. (a) $^{139}_{55}$Cs (b) $^{139}_{57}$La (c) $^{139}_{55}$Cs
21. 7.93 MeV
23. (a) 491 MeV (b) term 1: 179%; term 2: −53.0%; term 3: −24.6%; term 4: −1.37%
25. 86.4 h
27. 1.16×10^3 s
29. 9.47×10^9 nuclei
31. (a) 0.086 2 d^{-1} = 3.59×10^{-3} h^{-1} = 9.98×10^{-7} s^{-1}
(b) 2.37×10^{14} nuclei (c) 0.200 mCi
33. 1.41
35. (a) cannot occur (b) cannot occur (c) can occur
37. 0.156 MeV
39. 4.27 MeV
41. (a) e$^-$ + p → n + ν (b) 2.75 MeV
43. (a) 148 Bq/m^3 (b) 7.05×10^7 atoms/m^3 (c) 2.17×10^{-17}

45.
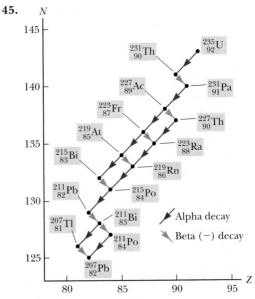

47. 1.02 MeV
49. (a) $^{21}_{10}$Ne (b) $^{144}_{54}$Xe (c) e$^+$ + ν
51. 8.005 3 u; 10.013 5 u
53. (a) 29.2 MHz (b) 42.6 MHz (c) 2.13 kHz
55. 46.5 d
57. (a) 2.7 fm (b) 1.5×10^2 N (c) 2.6 MeV
(d) r = 7.4 fm; F = 3.8×10^2 N; W = 18 MeV
59. 2.20 μeV
61. (a) smaller (b) 1.46×10^{-8} u (c) 1.45×10^{-6} % (d) no
63. (a) 2.52×10^{24} (b) 2.29×10^{12} Bq (c) 1.07×10^6 yr
65. 5.94 Gyr
67. (b) 1.95×10^{-3} eV
69. 0.401%
71. (a) $^{93}_{42}$Mo (b) electron capture: all levels; e$^+$ emission: only 2.03 MeV, 1.48 MeV, and 1.35 MeV
73. (b) 1.16 u
75. 2.66 d

Chapter 45

Answers to Quick Quizzes

1. (b)
2. (a), (b)
3. (a)
4. (d)

Answers to Odd-Numbered Problems

1. 1.1×10^{16} fissions
3. $^{144}_{54}$Xe, $^{143}_{54}$Xe, and $^{142}_{54}$Xe
5. $^1_0$n + ^{232}Th → ^{233}Th; ^{233}Th → ^{233}Pa + e$^-$ + $\bar{\nu}$; ^{233}Pa → ^{233}U + e$^-$ + $\bar{\nu}$
7. 126 MeV
9. 184 MeV
11. 5.58×10^6 m
13. 2.68×10^5
15. 26 MeV
17. (a) 3.08×10^{10} g (b) 1.31×10^8 mol (c) 7.89×10^{31} nuclei
(d) 2.53×10^{21} J (e) 5.34 yr (f) Fission is not sufficient

to supply the entire world with energy at a price of $130 or less per kilogram of uranium.

19. 1.01 g
21. (a) ^8_4Be (b) $^{12}_6\text{C}$ (c) 7.27 MeV
23. 5.49 MeV
25. (a) 31.9 g/h (b) 123 g/h
27. (a) 2.61×10^{31} J (b) 5.50×10^8 yr
29. (a) 2.23×10^6 m/s (b) $\sim 10^{-7}$ s
31. (a) 10^{14} cm^{-3} (b) 1.2×10^5 J/m^3 (c) 1.8 T
33. (a) 0.436 cm (b) 5.79 cm
35. (a) 10.0 h (b) 3.16 m
37. 2.39×10^{-3} °C, which is negligible
39. 1.66×10^3 yr
41. (a) 421 MBq (b) 153 ng
43. (a) 0.963 mm (b) It increases by 7.47%.
45. (a) $\sim 10^6$ atoms (b) $\sim 10^{-15}$ g
47. 1.01 MeV
49. (a) 1.5×10^{24} nuclei (b) 0.6 kg
51. (a) 3.12×10^7 (b) 3.12×10^{10} electrons
53. (a) 1.94 MeV, 1.20 MeV, 7.55 MeV, 7.30 MeV, 1.73 MeV, 4.97 MeV (b) 1.02 MeV (c) 26.7 MeV (d) Most of the neutrinos leave the star directly after their creation, without interacting with any other particles.
55. 69.0 W
57. 2.57×10^4 kg
59. (b) 26.7 MeV
61. (a) 5.67×10^8 K (b) 120 kJ
63. 14.0 MeV or, ignoring relativistic correction, 14.1 MeV
65. (a) 3.4×10^{-4} Ci, 16 μCi, 3.1×10^{-4} Ci (b) 50%, 2.3%, 47% (c) It is dangerous, notably if the material is inhaled as a powder. With precautions to minimize human contact, however, microcurie sources are routinely used in laboratories.
67. (a) 8×10^4 eV (b) 4.62 MeV and 13.9 MeV (c) 1.03×10^7 kWh
69. (a) 4.92×10^8 kg/h $\rightarrow$ 4.92×10^5 m^3/h (b) 0.141 kg/h
71. 4.44×10^{-8} kg/h
73. (a) 10^1 electrons (b) 10^6 (c) 10^8 eV

Chapter 46

Answers to Quick Quizzes

1. (a)
2. (i) (c), (d) (ii) (a)
3. (b), (e), (f)
4. (b), (e)
5.

6. false

Answers to Odd-Numbered Problems

1. (a) 5.57×10^{14} J (b) $\$1.70 \times 10^7$
3. (a) 4.54×10^{23} Hz (b) 6.61×10^{-16} m

5. 118 MeV
7. (b) The range is inversely proportional to the mass of the field particle. (c) $\sim 10^{-16}$ m
9. (a) 67.5 MeV (b) 67.5 MeV/c (c) 1.63×10^{22} Hz
11. (a) muon lepton number and electron lepton number (b) charge (c) angular momentum and baryon number (d) charge (e) electron lepton number
13. (a) $\bar\nu_\mu$ (b) ν_μ (c) $\bar\nu_e$ (d) ν_e (e) ν_μ (f) $\bar\nu_e + \nu_\mu$
15. (a) It cannot occur because it violates baryon number conservation. (b) It can occur. (c) It cannot occur because it violates baryon number conservation. (d) It can occur. (e) It can occur. (f) It cannot occur because it violates baryon number conservation, muon lepton number conservation, and energy conservation.
17. 0.828c
19. (a) 37.7 MeV (b) 37.7 MeV (c) 0 (d) No. The mass of the π^- meson is much less than that of the proton, so it carries much more kinetic energy. The correct analysis using relativistic energy conservation shows that the kinetic energy of the proton is 5.35 MeV, while that of the π^- meson is 32.3 MeV.
21. (a) It is not allowed because neither baryon number nor angular momentum is conserved. (b) strong interaction (c) weak interaction (d) weak interaction (e) electromagnetic interaction
23. (a) K$^+$ (scattering event) (b) Ξ^0 (c) π^0
25. (a) Strangeness is not conserved. (b) Strangeness is conserved. (c) Strangeness is conserved. (d) Strangeness is not conserved. (e) Strangeness is not conserved. (f) Strangeness is not conserved.
27. 9.25 cm
33. (a) $-e$ (b) 0 (c) antiproton; antineutron
35. The unknown particle is a neutron, udd.
39. (a) 1.06 mm (b) microwave
41. (a) $\sim 10^{13}$ K (b) $\sim 10^{10}$ K
43. 7.73×10^3 K
45. (a) 0.160c (b) 2.18×10^9 ly
47. (a) 590.09 nm (b) 599 nm (c) 684 nm
49. 6.00
51. (a) Charge is not conserved. (b) Energy, muon lepton number, and electron lepton number are not conserved. (c) Baryon number is not conserved.
53. 0.407%
55. $\sim 10^{14}$
59. 1.12 GeV/c^2
61. (a) electron–positron annihilation; e$^-$ (b) A neutrino collides with a neutron, producing a proton and a muon; W$^+$.
63. $\sim 10^3$ K
65. neutron
67. 5.35 MeV and 32.3 MeV
69. (a) 0.782 MeV (b) $v_e = 0.919c$, $v_p = 382$ km/s (c) The electron is relativistic; the proton is not.
71. (b) 9.08 Gyr
73. (a) $2Nmc$ (b) $\sqrt{3}Nmc$ (c) method (a)

Index

Standard Abbreviations and Symbols for Units

Symbol	Unit	Symbol	Unit
A	ampere	K	kelvin
u	atomic mass unit	kg	kilogram
atm	atmosphere	kmol	kilomole
Btu	British thermal unit	L	liter
C	coulomb	lb	pound
°C	degree Celsius	ly	light-year
cal	calorie	m	meter
d	day	min	minute
eV	electron volt	mol	mole
°F	degree Fahrenheit	N	newton
F	farad	Pa	pascal
ft	foot	rad	radian
G	gauss	rev	revolution
g	gram	s	second
H	henry	T	tesla
h	hour	V	volt
hp	horsepower	W	watt
Hz	hertz	Wb	weber
in.	inch	yr	year
J	joule	Ω	ohm

Mathematical Symbols Used in the Text and Their Meaning

Symbol	Meaning
$=$	is equal to
$\equiv$	is defined as
$\neq$	is not equal to
$\propto$	is proportional to
$\sim$	is on the order of
$>$	is greater than
$<$	is less than
$\gg (\ll)$	is much greater (less) than
$\approx$	is approximately equal to
Δx	the change in x
$\displaystyle\sum_{i=1}^{N} x_i$	the sum of all quantities x_i from $i = 1$ to $i = N$
$|x|$	the absolute value of x (always a nonnegative quantity)
$\Delta x \to 0$	Δx approaches zero
$\dfrac{dx}{dt}$	the derivative of x with respect to t
$\dfrac{\partial x}{\partial t}$	the partial derivative of x with respect to t
$\displaystyle\int$	integral